Informatik aktuell

Herausgeber: W. Brauer
im Auftrag der Gesellschaft für Informatik (GI)

Siegfried J. Pöppl Heinz Handels (Hrsg.)

Mustererkennung 1993

Mustererkennung im Dienste der Gesundheit

15. DAGM-Symposium
Lübeck, 27.-29. September 1993

Springer-Verlag
Berlin Heidelberg New York
London Paris Tokyo
Hong Kong Barcelona
Budapest

Herausgeber

Siegfried J. Pöppl
Heinz Handels
Medizinische Universität zu Lübeck
Institut für Medizinische Informatik
Ratzeburger Allee 160, D-23538 Lübeck

CR Subject Classification (1993): I.2-5, J.3

ISBN 978-3-540-57279-4 ISBN 978-3-642-78546-7 (eBook)
DOI 10.1007/978-3-642-78546-7

Satz: Reproduktionsfertige Vorlage vom Autor/Herausgeber

Veranstalter:

DAGM Deutsche Arbeitsgemeinschaft für Mustererkennung

Tagungsleitung:

Prof. Dr.-Ing. Dr. S. J. Pöppl
Medizinische Universität zu Lübeck
Institut für Medizinische Informatik

Programmkomitee:

H. Bunke, Bern
H. Burkhardt, Hamburg
S. Fuchs, Dresden
G. Hartmann, Paderborn
K.-H. Höhne, Hamburg
E. Hundt, München
W. A. Kropatsch, Wien
M. Oerder, Aachen
E. Paulus, Braunschweig
S. J. Pöppl, Lübeck
D.-P. Pretschner, Hildesheim
B. Radig, München
G. Sagerer, Bielefeld
B. Schleifenbaum, Wetzlar
K. Schneider, Konstanz
K. Voss, Jena

Das 15. DAGM-Symposium wird ausgerichtet vom:

Institut für Medizinische Informatik
der Medizinischen Universität zu Lübeck

DAGM Deutsche Arbeitsgemeinschaft für Mustererkennung

Die **Deutsche Arbeitsgemeinschaft für Mustererkennung** veranstaltet seit 1978 jährlich an verschiedenen Orten ein wissenschaftliches Symposium mit dem Ziel, Aufgabenstellungen, Denkweisen und Forschungsergebnisse aus den Gebieten der Mustererkennung vorzustellen, den Erfahrungs- und Ideenaustausch zwischen den Fachleuten anzuregen und den Nachwuchs zu fördern.

Die **DAGM** wird durch folgende wissenschaftliche Trägergesellschaften gebildet:

DGaO	Deutsche Gesellschaft für angewandte Optik
GMDS	Deutsche Gesellschaft für Medizinische Informatik, Biometrie und Epidemiologie
GI	Gesellschaft für Informatik
ITG	Informationstechnische Gesellschaft
DGNM	Deutsche Gesellschaft für Nuklearmedizin
IEEE	The Institute of Electrical and Electronic Engineers, Deutsche Sektion
DGPF	Deutsche Gesellschaft für Photogrammetrie und Fernerkundung

Die **DAGM** ist Mitglied der *International Association for Pattern Recognition (IAPR).*

Der mit DM 5.000,- dotierte

D A G M - Preis 1992

gestiftet von der **OLYMPUS-EUROPA-STIFTUNG** *"Wissenschaft für das Leben"*.

wurde

M. Schwarzinger, D. Noll und W. v. Seelen
Universität Bochum

für den folgenden Beitrag verliehen

Object Recognition with Deformable Models Using Constrained Elastic Nets

Die mit DM 1.000,- dotierten Anerkennungspreise für das Jahr 1992 wurden verliehen an:

F. Ade, M. Peter, **M. Rutishauser, M. Trobina** **A. Ylä-Jääski** ETH Zürich	Vision for a 3-D Objekt Manipulation System
M. Dose und G. Schöner Universität Bochum	Closed Loop Autonomous Vehicle Path Planning by Dynamical Systems
H. Kristen und O. Munkelt Technische Universität München	Markov-Feld-basierte Bildinterpretation mit automatisch generierten Datenbasen
R. Linder, E. Rinast, H.-D. Weiss, **S. J. Pöppl** Med. Universität zu Lübeck	Leistungsvergleich moderner Klassifikationsstrategien in der abdominalsonographischen Mustererkennung
I. Rothe und K. Voss Universität Jena	Orientierungsbestimmung von Objekten durch Momentinvarianten

Die Anerkennungspreise wurden gestiftet von der **OLYMPUS-EUROPA-STIFTUNG** *"Wissenschaft für das Leben"*.

Vorwort

Erstmalig findet ein DAGM-Symposium im hohen Norden, in der Hansestadt Lübeck, statt. Als Vorsitzender der DAGM ist es mir eine besondere Freude, das 15. DAGM-Symposium an der Medizinischen Universität zu Lübeck ausrichten zu können, dies auch im Hinblick auf den Beginn des neu eingerichteten Studiengangs Informatik.

Das Schwerpunktthema *Mustererkennung im Dienste der Gesundheit* fand große Beachtung, was sich anhand der überdurchschnittlich hohen Anzahl von eingereichten Beiträgen zeigte. Von den über 120 eingereichten Beiträgen konnten 44 als Vorträge und 48 als Posterbeiträge angenommen werden. Ein Drittel der Beiträge behandelt neue Methoden der Mustererkennung zur Analyse von Biosignalen und medizinischen Bilddaten. Wie auch auf den vorangegangenen Symposien werden Probleme und Problemlösungen aus allen Bereichen der Mustererkennung diskutiert und Ergebnisse der methodischen und der anwendungsorientierten Forschung vorgestellt.

In diesem Jahr wird erstmalig eine moderierte Postersession angeboten, die die inhaltliche Diskussion der in diesem Jahr in großer Anzahl vertretenen Posterbeiträgen vertiefen und intensivieren soll. Durch die Einführung dieser Postersession, die von erfahrenen Kollegen moderiert wird, soll den Posterbeiträgen ein besonderes Gewicht gegeben werden, darüberhinaus sind die Posterbeiträge gleichberechtigt zu den Vorträgen als Artikel in diesem Buch enthalten.

Ich danke allen Autoren dieses Bandes für ihre Sorgfalt und ihr Verständnis für die enge Terminsetzung. An der Vorbereitung des Symposiums haben besonderen Anteil Herr Dr. Handels als Gesamtorganisator und Mitherausgeber dieses Buches, Herr Dipl.-Phys. Hahn und Herr Dipl.-Inf. Roß bei der Betreuung der Ausstellung und des kulturellen Rahmenprogramms. Als Tagungssekretärin danke ich Frau Weber für Ihre engagierte Unterstützung. Danken möchte ich auch allen anderen Mitarbeitern meines Instituts, ohne deren Hilfe die Durchführung dieses 15. DAGM-Symposiums nicht möglich gewesen wäre. Sie alle freuen sich mit mir, wenn das 15. DAGM-Symposium 1993 in Lübeck ein Erfolg und Ihren Erwartungen gerecht wird.

Allen Teilnehmerinnnen und Teilnehmern wünsche ich einen angenehmen Aufenthalt und fruchtbaren, wissenschaftlichen Erfahrungsaustausch beim 15. DAGM-Symposium in Lübeck.

Lübeck, im Juli 1993

Siegfried J. Pöppl

Inhaltsverzeichnis

Bildfolgen

Mathematische Grundlagen

Bildanalyse in der Medizin

Spracherkennung

Mustererkennung in der Medizin

Neuronale Netze

Bildanalyse

Technische Anwendungen

POSTER

Bildanalyse und Segmentierung in der Medizin

Medizinische Anwendungen

Grundlagen der Mustererkennung

Bildanalyse/Sprach- und Texterkennung

Technische Anwendungen

Image Sequence Analysis

Goesta H. Granlund

Computer Vision Laboratory
Linköping University
581 83 Linköping, Sweden

Abstract. In this paper are certain issues regarding the analysis of image sequences discussed. We will deal with aspects of optical flow, tracking of 2-D line segments, representation of orientation in multiple dimensions and adaptive filtering in multiple dimensions. Examples will be given of representations and operations upon time sequences.

1 Introduction

In this paper we will discuss certain issues regarding the analysis of image sequences. Due to the limited format, it will by no means be possible to give any comprehensive overview of the field. Most aspects will in fact have to be omitted, and it is unfortunately not possible to give due references to all significant contributions within the area.

A typical example of an image sequence is the set of images produced by a TV camera during a certain time interval. The operations for image sequences deal with changes in images due to movements of the objects in the scene. This may be done at a low level giving movement of lines and edges, or at a high level giving movement of objects.

There are numerous applications in which tracking of moving objects is a necessary feature. Since images and consequently image sequences contain large amounts of data and are computationally demanding, the development of algorithms for image sequences has not been practically possible until quite recently, when computers with sufficient memory and computational power became available. The following is a partial list of applications which utilize image sequence processing:

- Industrial applications - Dynamic monitoring of industrial processes. Active robot vision.
- Communication - Bandwidth compression of TV conferencing and picture phone video signals.
- Medical applications - Study of cell motion by microcinematography. Study of heart motion from X-ray movies.
- Meteorology - Tracking of weather systems.
- Highway traffic monitoring.
- Vehicle systems - Giving the driver information about the surrounding traffic. Automatic systems driving the vehicle.

One of the most important issues in image sequence processing is motion estimation. In tracking of multiple objects (moving differently), motion estimation provides a powerful way of segmenting and identifying individual objects. An overview of techniques used in image sequence analysis is given by Huang [21, 22].

Over the years several algorithms have been proposed for time sequence analysis. Three different approaches can be distinguished in this field.

- **Matching** 2-D image processing techniques are used on one frame at a time to extract feature points, curves, etc. Matching of the feature extraction result is used to find the corresponding parts in the neighbor frames [21, 22, 3].
- **Derivatives** The solution to the optical flow equation is approximated by the use of gradient filters. The filters are 3-D-filters (two spatial dimensions + time) [29, 33, 24, 7].
- **Signal Processing** The 3-D data set is analyzed with tools such as the Fourier Transform to design algorithms for the estimation of the planes in the spatio-temporal data set originating from moving lines and curves originating from moving points [16, 1, 19].

The borderlines between these three methods are not always clear and it is possible to use algorithms which incorporate aspects of all of the three approaches.

Recent developments in visual system models suggest the existence of frequency and orientation channels representing the local spatial, as well as the local spatio-temporal image spectrum, [23, 2]. In the latter case the energy concentration in the local spectrum to a particular orientation and frequency channel means a velocity vector with a particular direction and coarseness range. Adelson and Bergen [1] as well as Watson and Ahumada [34] and others have proposed human visual motion sensing models based on the local spectra by means of the spatio-temporal filter responses using separable filters of Gabor type.

The main feature in matching is to locate local events such as edges, lines or other characteristic points in a frame, and then find them in the next frame. In the case of a two-dimensional translation, the latter is especially important in real time applications, because of the speed obtainable in comparison to the heavy computations needed for signal transformation and matching respectively. When the two-dimensional motion includes rotation and scaling in addition to translation, correlation methods are less attractive considered the computation power required.

Differential methods can be extended for a general tree-dimensional motion. The correlation methods could also be extended, but due to the high dimensionality of the search space, the computation requirement becomes almost impossible to fulfill, unless very powerful search algorithms are found. It seems difficult to extend Fourier methods to three dimensions.

Every algorithm for image sequences depends on a model of the scene dictating what changes may occur from one moment to the next. The type of model being used is a balance between the facts that a simple model gives a simple algorithm but is also more or less inaccurate compared to the real world. Most algorithms proposed are restricted to two-dimensional motion, especially translation. Often the algorithms ignore many important factors, such as occlusion, deformable bodies and radiometric considerations, i.e. how the objects reflect light. Usually it is assumed that the gray levels of two points in two image frames corresponding to the same physical point on an object are identical. This may be more or less valid in some problems but completely wrong in others.

Since the processing time is decreasing due to hardware development, n–D image processing with applications to time sequences of 2–D images, volume images, time sequences of volume images and segmentation of 2-D images using many feature dimensions are becoming feasible. In many applications real time image processing is already a reality. The problem of orientation detection arises in many situations in 2–D as well as in higher dimensions in image processing, notably in texture segmentation and optical flow computations. In image sequence analysis, the local spatiotemporal spectrum energy is concentrated to a tilted plane when the optical flow is possible to determine unambiguously. To compute the optical flow in this case, the tilt of the plane must be known.

In parallel with these studies, much research effort is invested in computationally economic methods which mimic the essence of the discoveries about the human visual system. The spatial frequency decomposition is shown to be achieved efficiently using a pyramid data structure in combination with separable spatial filtering [32, 8, 9] resulting in the *Laplacian pyramid.* To achieve orientation selectivity by using linear filters in combination with the Laplacian pyramids, Fleet and Jeppson [15] proposed filter design criteria.

2 Optical Flow

Depending on the chosen mathematical model, the exact meaning of optical flow has been defined differently in the computer vision literature.

One of the most common models uses variational calculus, Horn and Schunk [20]. In this approach $\frac{df}{dt}$, which depends on optical flow is minimized over a region of E_2. On the solution a smoothness constraint assuring a smooth variation of the optical flow is imposed.

$$\begin{aligned} \min_{\bar{k}\in S_{h_s}} &= \int_{E_2} (\bar{k}^t \nabla f)^2 dxdy \\ &+ \lambda \int_{E_2} (\frac{\partial k_x}{\partial x})^2 + (\frac{\partial k_x}{\partial y})^2 + (\frac{\partial k_y}{\partial x})^2 + (\frac{\partial k_y}{\partial y})^2 dxdy \qquad (1) \end{aligned}$$

where $\bar{k}$ is a point in S_{h_s} which represents a plane consisting of the points $(u, v, 1)^t$ with u and v being the x and y components of the optical flow vector. E_2 is an image region and λ is a heuristic positive coefficient regulating the importance of the smoothness in the solution. The coefficient is chosen large when the gray value measurements are noisy, small otherwise. It should be noted that there is no time integration in the expression above. The minimization is possible to perform by solving a second order elliptic partial differential equation related to this problem. A feature of the smoothness constraint is that when there is no unique optical flow in a region, then unique solutions from other parts are propagated towards this region. The propagation direction is uniform in the original approach but [30] discusses a modified version of the smoothness constraint which depends on the gray value variations and hence allows oriented smoothness resulting in improved optical flow estimates along object boundaries.

3 Tracking of 2-D Line Segments

Partly in cooperation, Crowley [10] and Faugeras and Deriche [12] have developed algorithms for tracking of line segments using Kalman filtering techniques. The applications are in robotic vison, i.e indoor scenes, to build a geometric model of the environment, e.g motion stereo, or to track features that have been established in some other way. One major demand under which the algorithms have been developed is that of real time operation. The model should be updated at least a few times per second, which results in rather simple versions of the Kalman filter.

The algorithms are similar and the steps are:

1. Extract edge elements by convolution, using for example gradient filters, at one frame in the sequence.
2. Extract the "proper" line segments of pixel width.
3. Link the line segments to polygons.
4. Predict the segments (or polygons) in the next frame using Kalman filters.
5. Match the predicted segments, from frame $k-1$, with the estimated ones, in frame k.

One crucical step where the two algorithms differ is the parameterization of the line segments. The minimal representation of a line has four parameters, e.g. the two endpoints. The endpoints of the line are not a proper representation due to uncertainties in the estimation. One of the representations is

- The midpoint coordinates,(x_m, y_m).
- The orientation, θ, of the line.
- The lenght, l, of the line.

3.1 Dynamic Monocular Machine Vision

Dickmanns and Graefe [14, 13] have by the means of a 4D model of the environment and measurements in the image plane follwed by Kalman filtering succeeded to solve a number of tasks.

- Balancing of an inverted pendelum.
- Vehicle docking to an object with known shape but unknown position and orientation.

- Autonomous vehicle driving on roads of highway type in speeds up to 100 km/h.

The main problem in these applications is not the filtering step but how to limit the data flow in order to get a real time performance. The following comments concern one of these tasks, namely autonomous vehicle guidance.

The 4D model, three spatial dimensions and time of the road and the car, is updated by a Kalman filter. All predictions are made in this 4D model while the measurements are made in the image plane. The correspondences between the model and the measurements are established by the laws of perspective projections.

The measurements in each image plane are simple correlations to establish the curbs of the road and the curvature of the curbs. From the curvature and the change of curvature it is possible to estimate how the road curves and to steer the car according to this. One further correlation in the image is used in order to avoid obstacles or other vehicles on the road. Obstacles are then modeled as a cylinder on the road and the car either stops or avoids it by changing lane.

In the total state space model of the environment the dynamics of the vehicle is necessary. A planar "bicycle model" with two degrees of freedom, the slidslip angle and the inertial yaw rate, has been applied. Together with the curvature of the road and a carefully callibrated camera, with known position on the vehicle, a fifth order state space model is used.

4 Representation of Orientation in 3-D

Motion of a point in 2-D can be viewed as a line in 3-D time-space. Correspondingly, the motion of a line in 2-D can be viewed as a plane in 3-D time-space. There are however some complications, not only due to the increased volume of data, but also from a more fundamental point of view. In two dimensions, the orientation of a line or an edge can unambiguously be represented by a vector in a "double angle" representation [17]. The mapping requirements of operations in multiple dimensions are more severe than for two dimensions [28]. With a hemisphere as the original space, an exact equvialent of the complication encountered in 2-D occurs: Surfaces that differ by a small angle can end up being represented by vectors that are very far apart, i e close to opposite sides of the rims of the hemispheres. This is of course unacceptable if the metric properties of the space are of any consequence, which will always be the case if there is a next stage where the information is to be further processed. Consider e g the case of differentiation when the vector passes in a step-like fashion from one side of the hemisphere to the other. It is necessary therefore, that a mapping is established that "closes" the space in the same manner as earlier discussed for the two-dimensional case.

It turns out that information can for this purpose be represented by *tensors* [25]. The tensor representation can be used for filtering in volumes and in time sequences, implementing spatio-temporal filters [35]. It can also be used for computation of higher level features such as curvature [5] or acceleration. The tensor mapping can be controlled by transforms to implement adaptive filtering of volume data or time sequences [27].

The tensor representation of the local orientation of a neighbourhood with one single orientation in 3-D, is given by

$$\mathbf{T} = \frac{1}{x}\begin{pmatrix} x_1^2 & x_1x_2 & x_1x_3 \\ x_1x_2 & x_2^2 & x_2x_3 \\ x_1x_3 & x_2x_3 & x_3^2 \end{pmatrix} \tag{2}$$

where $\mathbf{x} = (x_1, x_2, x_3)$ is directed as the normal vector to the plane of the neighbourhood and $x = \sqrt{x_1^2 + x_2^2 + x_3^2}$. The magnitude of $\mathbf{x}$ is determined by the local energy distribution.

The orientation estimation requires a number of precomputed quadrature filters evenly spread in one half of the Fourier space, [25, 26]. The minimum number of quadrature filters required for orientation estimation in 3-D is 6, where the filters are directed as the vertices of a hemiicosahedron, see Figure 1:

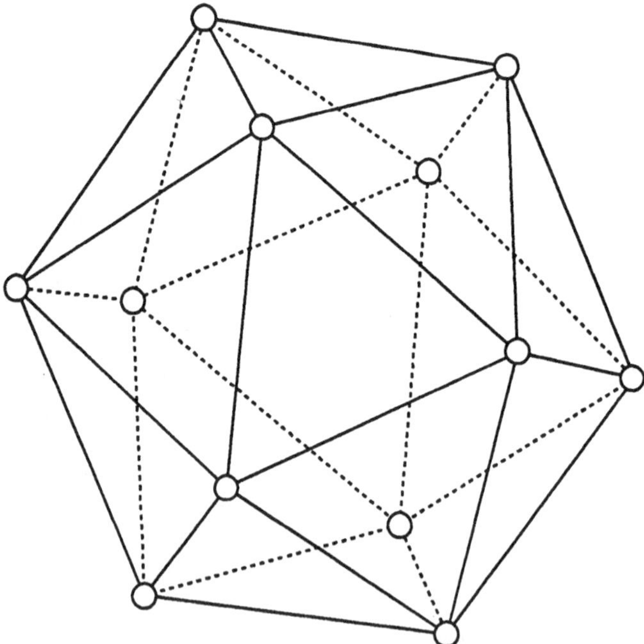

Fig. 1. An icosahedron (one of the 5 Platonic polyhedra).

$$\begin{aligned}
\hat{\mathbf{n}}_1 &= c\,(\ \ a,\ \ 0,\ \ b\ )^t \\
\hat{\mathbf{n}}_2 &= c\,(-a,\ \ 0,\ \ b\ )^t \\
\hat{\mathbf{n}}_3 &= c\,(\ \ b,\ \ a,\ \ 0\ )^t \\
\hat{\mathbf{n}}_4 &= c\,(\ \ b,-a,\ \ 0\ )^t \\
\hat{\mathbf{n}}_5 &= c\,(\ \ 0,\ \ b,\ \ a\ )^t \\
\hat{\mathbf{n}}_6 &= c\,(\ \ 0,\ \ b,-a\ )^t
\end{aligned} \tag{3}$$

with

$$\begin{aligned}
a &= 2 \\
b &= 1+\sqrt{5} \\
c &= (10+2\sqrt{5})^{-1/2}
\end{aligned} \tag{4}$$

A quadrature filter designed with a lognormal function, is given in the frequency domain by:

$$\begin{cases} F_k(\omega) \;=\; F_\omega(\omega)(\hat{\omega}\cdot\hat{\mathbf{n}}_{\mathbf{k}})^2 & if\ \ \omega\cdot\hat{\mathbf{n}}_{\mathbf{k}} > 0 \\ F_k(\omega) \;=\; 0 & otherwise \end{cases} \tag{5}$$

The spatial filter coefficients are found by a straightforward 3-D-DFT or by use of an optimization technique. The resulting spatial filter is complex-valued. This procedure is used to obtain the six quadrature filters.

It is easy to implement the orientation algorithm with these precomputed filters [25]. The tensor describing the neighbourhood is given by:

$$\mathbf{T^e} \;=\; \sum_k q_k(\mathbf{N}_k - \frac{1}{5}\mathbf{I}) \tag{6}$$

where q_k again denotes the magnitude of the output from filter k. $\mathbf{N}_k = \hat{\mathbf{n}}_k\hat{\mathbf{n}}_k^t$ denotes the direction of the filter expressed in the tensor representation and $\mathbf{I}$ is the unity tensor. A less compact description of Eq. 6 is:

1. Convolve the input data with the six complex-valued filters, i.e. perform twelve scalar convolutions.
2. Compute the magnitude of each complex-valued filter by

$$q_k = \sqrt{q_{ke}^2 + q_{ko}^2}$$

   where q_{ke} denotes the filter output of the real part of filter k and q_{ko} denotes the filter output of the imaginary part of filter k.
3. Compute the tensor $\mathbf{T^e}$ by Eq. 6, i.e.

$$\mathbf{T^e} = \begin{pmatrix} T_{11} & T_{12} & T_{13} \\ T_{12} & T_{22} & T_{23} \\ T_{13} & T_{23} & T_{33} \end{pmatrix}$$

where

$$\begin{aligned} T_{11} &= A(q_1+q_2)+B(q_3+q_4)-S \\ T_{22} &= A(q_3+q_4)+B(q_5+q_6)-S \\ T_{33} &= A(q_5+q_6)+B(q_1+q_2)-S \\ T_{12} &= C(q_3-q_4) \\ T_{13} &= C(q_1-q_2) \\ T_{23} &= C(q_5-q_6) \end{aligned}$$

for

$$\begin{aligned} S &= \frac{1}{5}\sum_{k=1}^{6} q_k \\ A &= \frac{4}{10+2\sqrt{5}} \\ B &= \frac{6+2\sqrt{5}}{10+2\sqrt{5}} \\ C &= \frac{2+2\sqrt{5}}{10+2\sqrt{5}} \end{aligned}$$

4.1 Evaluation of the Representation Tensor

It is shown in [25] that the eigenvector corresponding to the largest eigenvalue of $\mathbf{T^e}$ is the normal vector of the plane best describing the neighbourhood. This implies that an eigenvalue analysis is appropriate for evaluating the tensor. Below the eigenvalue distribution and the corresponding tensor representation are given for three particular cases of $\mathbf{T^e}$, where $\lambda_1 \geq \lambda_2 \geq \lambda_3 \geq 0$ are the eigenvalues in decreasing order, and $\hat{\mathbf{e}}_\mathbf{i}$ is the eigenvector corresponding to λ_i.

1. $\lambda_1 > 0\,;\ \lambda_2 = \lambda_3 = 0;$
   $\mathbf{T^e} = \lambda_1 \hat{\mathbf{e}}_\mathbf{1} \hat{\mathbf{e}}_\mathbf{1}^t$
   This case corresponds to a neighbourhood that is perfectly *planar*, i.e. is constant on planes in a given orientation. The orientation of the normal vectors to the planes is given by $\hat{\mathbf{e}}_1$.
2. $\lambda_1 = \lambda_2 > 0\,;\ \lambda_3 = 0;$
   $\mathbf{T^e} = \lambda_1 \left(\mathbf{I} - \hat{\mathbf{e}}_\mathbf{3} \hat{\mathbf{e}}_\mathbf{3}^t\right)$
   This case corresponds to a neighbourhood that is constant on *lines*. The orientation of the lines is given by the eigenvector corresponding to the least eigenvalue, $\hat{\mathbf{e}}_3$.
3. $\lambda_1 = \lambda_2 = \lambda_3 > 0;$
   $\mathbf{T^e} = \lambda_1 \mathbf{I}$
   This case corresponds to an *isotropic* neighbourhood, meaning that there exists energy in the neighbourhood but no orientation, e.g. in the case of noise.

The eigenvalues and eigenvectors are easily computed with standard methods such as the Jacobi method, e.g. [31]. Note that the spectral decomposition theorem states that all neighborhoods can be expressed as a linear combination of these three cases.

4.2 Velocity Estimation

If the signal to analyze is a time sequence, a plane implies a moving line and a line implies a moving point. The optical flow will be obtained by an eigenvalue analysis of the estimated representation tensor. The projection of the eigenvector corresponding to the largest eigenvalue onto the image plane will give the flow field. However, the so-called aperture problem will give rise to an unspecified velocity component, the component moving along the line. The aperture problem is a problem in all optical flow algorithms which rely on local operators. On the other hand, the aperture problem does not exist for moving points in the sequence. In this case of velocity estimation the correspondence between the energy in the spatial dimensions and the time dimension is established to get correct velocity estimation.

By examining the relations between the eigenvalues in the orientation tensor it is possible to divide the optical flow estimation into different categories, [6, 18]. Depending on the category, different strategies can be chosen, see Section 4.1. Case number two in Section 4.1, i.e. the line case, gives a correct estimation of the velocity in the image plane and is thus very important in the understanding of the motion. To give an illustration of this, a synthetic test sequence of a rotating and translating star together with a fair amount of Gaussian noise, Figure 2 is used as an example.

In Figure 3 the correct velocity field is given with arrows, white arrows correspond to the plane case and black arrows to the line case.

To do this division of different shapes of the tensor the following functions are chosen

$$p_{plane} = \frac{\lambda_1 - \lambda_2}{\lambda_1} \tag{7}$$

$$p_{line} = \frac{\lambda_2 - \lambda_3}{\lambda_1} \tag{8}$$

$$p_{iso} = \frac{\lambda_3}{\lambda_1} \tag{9}$$

These expressions can be seen as the probability for each case. In Figure 4 the division is made by selecting the case having the highest probability.

The calculation of the optical flow is done using Eq. 10 for the plane case and Eq. 11 for the line case. In neighborhoods classified as 'isotropic' no optical flow is computed. The 'true' optical flow in neighborhoods of the 'plane' type, such as moving lines, cannot be computed by optical flow algorithms using only local neighbourhood operations as mentioned earlier. The optical flow is computed by

$$\begin{aligned} \mathbf{x} &= \hat{\mathbf{e}}_1 \\ \mathbf{v}_{line} &= (-x_1 x_3 \hat{\mathbf{x}}_1 - x_2 x_3 \hat{\mathbf{x}}_2)/(x_1^2 + x_2^2) \end{aligned} \tag{10}$$

where $\hat{\mathbf{x}}_1$ and $\hat{\mathbf{x}}_2$ are the orthogonal unit vectors defining the image plane.

The aperture problem does not exist for neighborhoods of the 'line' type, such as moving points. This makes them, as mentioned, very important for motion

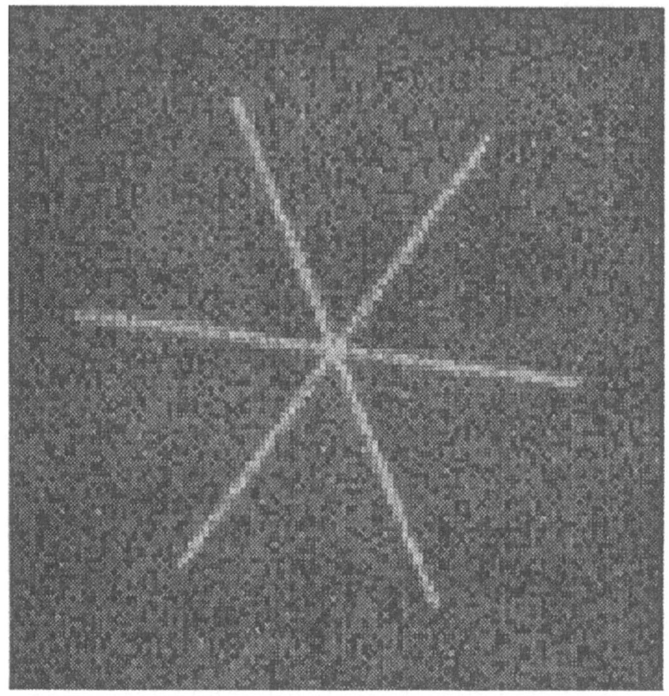

Fig. 2. One frame from the original sequence of the translating and rotating star, with white Gaussian noise added.

analysis. The optical flow is computed by

$$\begin{aligned} \mathbf{x} &= \hat{\mathbf{e}}_3 \\ \mathbf{v}_{point} &= (x_1\hat{\mathbf{x}}_1 + x_2\hat{\mathbf{x}}_2)/x_3 \end{aligned} \tag{11}$$

In Figure 5 the result from Eq. 10, white arrows, and Eq. 11, black arrows, is given.

The use of certainty measures is one of the central mechanisms in the hierarchical framework and the optical flow is not used directly. Separate estimates of the direction of movement and velocity accompanied with certainty measures computed by combining the tensor norm and the appropriate discriminant function.

5 Spatio-Temporal Channels

The human visual system has difficulties handling high spatial frequencies simultaneously with high temporal frequencies [4, 11]. This means that objects with high velocity cannot be seen sharply without tracking. One aspect of this is that the visual system performs an effective data reduction. The data reduction is made in such a way that high spatial frequencies can be handled if the temporal

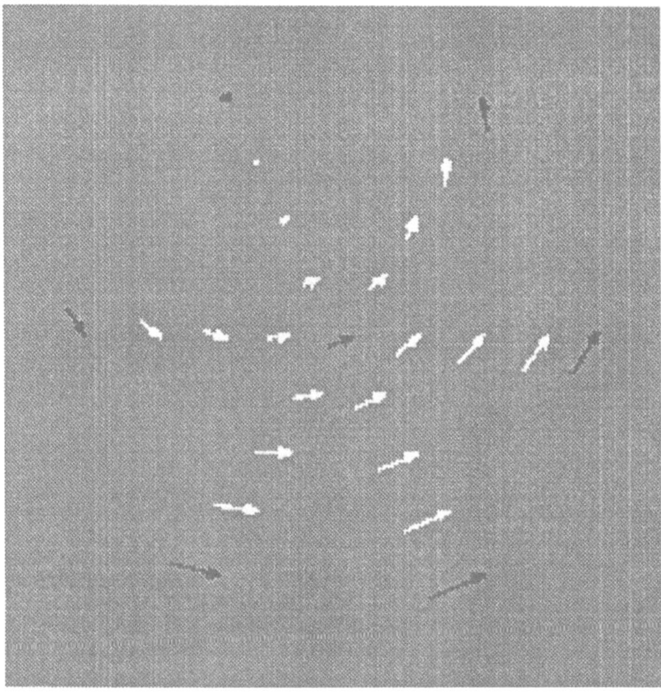

Fig. 3. The correct velocity vectors from the star sequence. Black vectors correspond to the moving point case and white ones to the moving line case.

frequency is low, and vice versa. This strategy is possible to use in a computer vision model for time sequences.

The input image sequence is subsampled both spatially and temporally into different channels. In Table 1 the data content in the different channels relatively to a reference sequence, *ch00*, is shown. The reference sequence has maximum resolution in all dimensions; typically this means a video signal of 50 Hz, height 576 and width 720 pixels. The frequency difference between adjacent channels is one octave, i.e. a subsampling factor of 2 is used.

The numbers in Table 1 indicate that a large data reduction can be made by not using the channels with high resolution in both spatial and temporal domains. For instance, the channels on the diagonal together contain approximately 1/4 of the data in the reference sequence (*ch00*).

There is a signal theoretical reason to use a pyramid representation of the image. A single filter has a particular limited pass band, both temporally and spatially, which may or may not be tuned to the different features to describe. In Figure 6a the upper cut-off frequency for a spatio-temporal quadrature filter set is indicated. The lower cut-off frequency is not plotted for the sake of clarity. Only the first quadrant in the ω_s, ω_t plane is plotted. The use of this filter set on different subsampled channels corresponds to using filters with different center

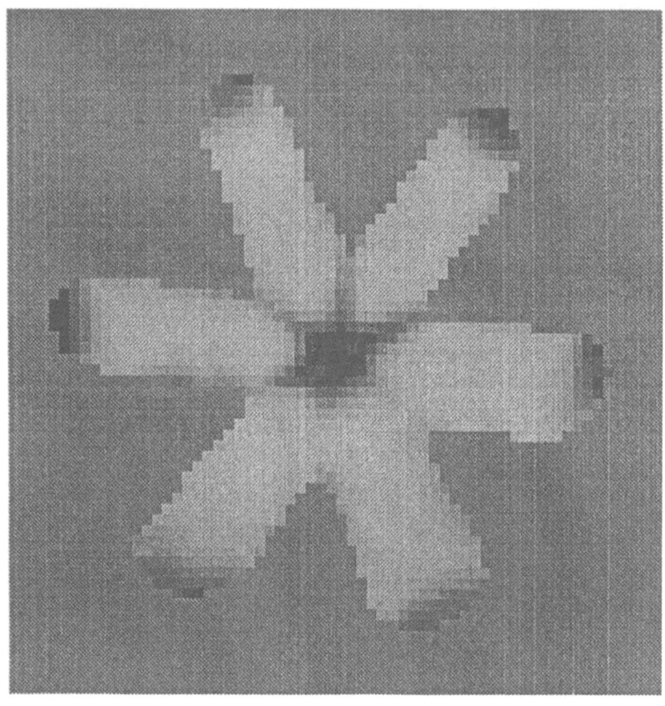

Fig. 4. The subdivision in different cases following Eq. 7 for the test sequence.

	Spatial subsampling					
	1/8	**1/4**	**1/2**	**1**		
	1/64	1/16	1/4	1	1	
Relative	1/128	1/32	1/8	1/2	**1/2**	Temporal
data content	1/256	1/64	1/16	1/4	**1/4**	subsampling
	1/512	1/128	1/32	1/8	**1/8**	

	Spatial subsampling					
	1/8	**1/4**	**1/2**	**1**		
	ch30	ch20	ch10	ch00	1	
Notation for	ch31	ch21	ch11	ch01	**1/2**	Temporal
sequence	ch32	ch22	ch12	ch02	**1/4**	subsampling
	ch33	ch23	ch13	ch03	**1/8**	

Table 1. Data content and name convention for the different spatio-temporal channels.

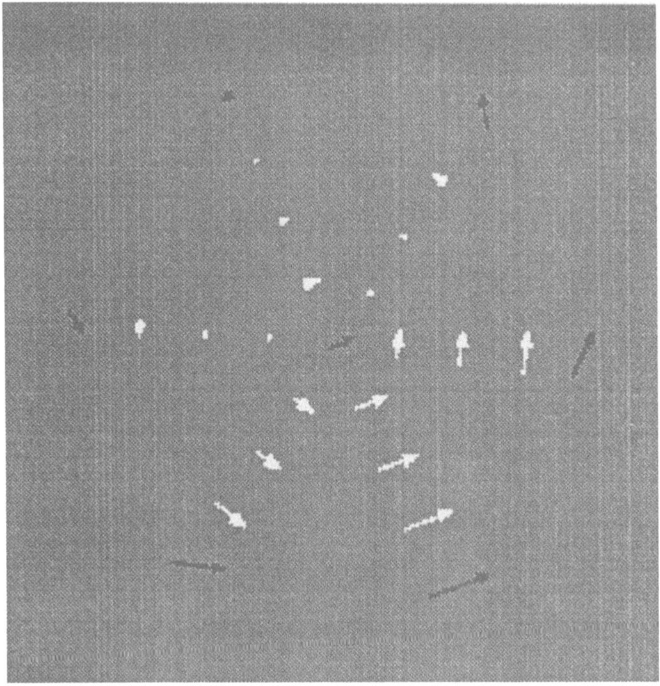

Fig. 5. The result of the optical flow algorithm. Black vectors correspond to the moving point case and white ones to the moving line case.

frequencies and constant relative bandwidth. Figure 6b indicates the upper cut-off frequency when convolving the channels on the diagonal in Table 1 with this filter set.

To avoid aliasing in the subsampling, the sequence must be prefiltered with a lowpass filter. As the resulting channel shall be further processed, the design of the lowpass filter is critical. The estimation of optical flow from Eq. 10 and Eq. 11 utilizes the relationship of energies originating from spatial variations and from temporal variations. The lowpass filter used for anti-aliasing should then *not* influence this relationship.

6 Adaptive Filtering of Image Sequences

An important property of the tensor representation described above is that averaging of such tensors is a meaningful operation [25]. It is thus possible to increase robustness by convolving the tensor field with an averaging filter. The filter used in the experiments was a truncated isotropic gaussian filter of size $7 \times 7 \times 7$ having a standard deviation of 1.3 [inter voxel distances]. Note that this filtering does not change the sensitivity to high frequency components in the original data, it only means that rapid changes in the representation of local structure

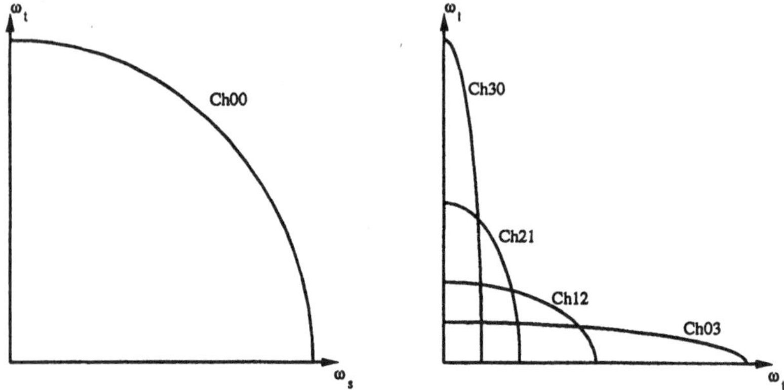

Fig. 6. Cut-off frequency for a spatio-temporal filter.

are suppressed.

The adaptivity of the filter consists in a signal controlled variation of shape and orientation. The shape of the filter has 3 degrees of freedom and can be said to span the range *plane - line - point*, corresponding to cases 1 - 3 above. The shape of the filter is rotation invariant and the filter can have any orientation.

The adaptive filter is constructed as a linear combination of 6 fixed filters having the same shape but different orientations.

$$H_k(\mathbf{u}) = H_\rho(u)(\hat{\mathbf{u}} \cdot \hat{\mathbf{n}}_k)^2 \tag{12}$$

The orientations of the filters are the same as for the quadrature filters. The angular shape is also of the same type, but the radial function can be separately chosen. A suitable radial function is:

$$H_\rho(u) = \begin{cases} \sin^2(\pi u/2) & 0<u<1 \\ 1 & 1<u<\pi-1 \\ \sin^2(\pi(\pi-u)/2) & \pi-1<u<\pi \\ 0 & \pi<u \end{cases} \tag{13}$$

The adaptive filter is given by:

$$H(\mathbf{u}) = \sum_k a_k H_k(\mathbf{u}) \tag{14}$$

The 6 weighting coefficients are given by the inner products of the tensor representing the local structure and the tensors representing the orientations of the filters. It can be shown that the filter $H(\mathbf{u})$ will correspond exactly to the local structure, as represented by $\hat{\mathbf{T}}$, if:

$$a_k = \alpha\, \hat{\mathbf{N}}_k \bullet\, {}^t\hat{\mathbf{T}} + \beta \tag{15}$$

where $\alpha = 5/4$ and $\beta = 1/6$. The $\bullet$ denotes an inner tensor product defined by:

$$\hat{\mathbf{N}}_k \bullet {}^t\mathbf{T} = \sum_{ij} n_{kij} \; {}^t t_{ij} \tag{16}$$

where n_{kij} and ${}^t t_{ij}$ are the individual tensor elements ${}^t\hat{\mathbf{T}}$ is the normalized traceless version of $\mathbf{T}$ and ${}^tT$ denotes the norm of ${}^t\mathbf{T}$¡.

To keep the low frequency components of the original data a suitable fixed low pass filter is added to the adaptive filter.

$$H_{LP}(u) = \begin{cases} \cos^2(\pi u/2) & 0 < u < 1 \\ 0 & 1 < u \end{cases} \tag{17}$$

The filters $H_{LP}(u)$ and $H_k(\mathbf{u})$ are realized as $9 \times 9 \times 9$ convolutions kernels. The final filter is then given by

$$H_{LP}(u) + H(u) \tag{18}$$

Figure 7 is one slice from an ultrasonic sequence of a beating heart. The sequence is corrupted by a great deal of measurement noise, which is significantly reduced in the enhanced image, Figure 8.

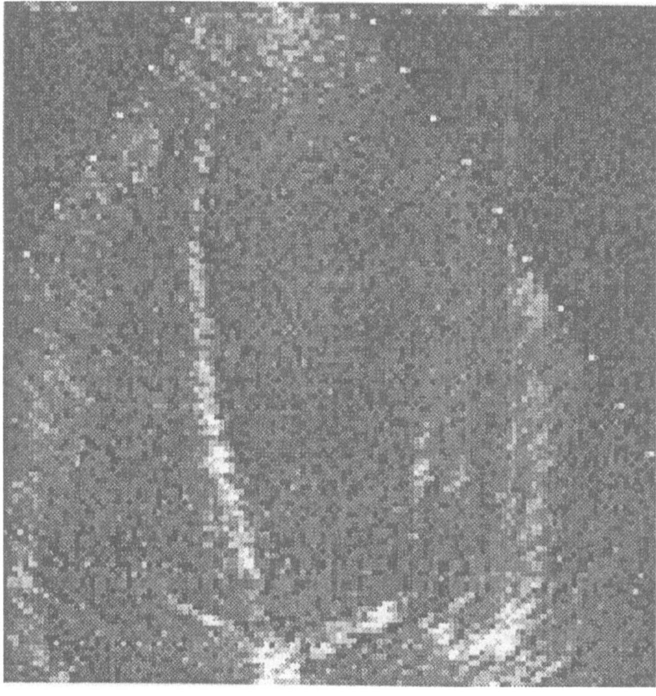

Fig. 7. Ultrasound image of a heart, original

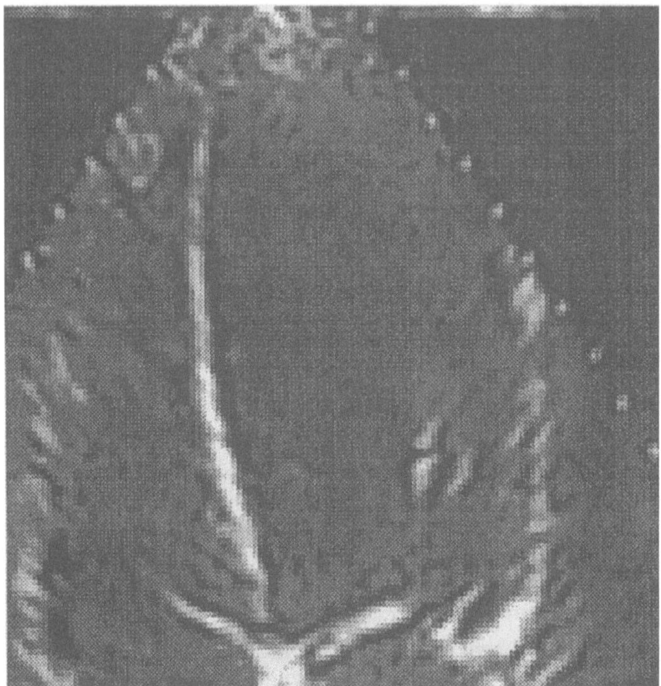

Fig. 8. Ultrasound image of a heart, filtered

7 Acknowledgements

The author wants to acknowledge the financial support of the Swedish National Board of Technical Development, which supported a great deal of the local research and documentation work mentioned in this overview. Considerable credit should be given to the staff of the Computer Vision Laboratory of Linkoeping University, for discussion of the contents as well as for text and figure contributions to different parts of the manuscript.

References

1. E. H. Adelson and J. R. Bergen. Spatiotemporal energy models for the perception of motion. *Jour. of the Opt. Soc. of America*, 2:284–299, 1985.
2. E. H. Adelson and J. A. Movshon. Phenomenal coherence of moving gratings. *Nature*, 200:523–525, 1982.
3. P. Anadan. A computational framework and an algorithm for the measurement of visual motion. *Int. J. of Computer Vision*, 2:283–310, 1989.
4. M. A. Arbib and A. Hanson, editors. *Vision, Brain and Cooperative Computation*, pages 187–207. MIT Press, 1987.

5. H. Bårman, G. H. Granlund, and H. Knutsson. Tensor field filtering and curvature estimation. In *Proceedings of the SSAB Symposium on Image Analysis*, pages 175–178, Linköping, Sweden, March 1990. SSAB. Report LiTH-ISY-I-1088, Linköping University, Sweden, 1990.
6. H. Bårman, L. Haglund, H. Knutsson, and G. H. Granlund. Estimation of velocity, acceleration and disparity in time sequences. In *Proceedings of IEEE Workshop on Visual Motion*, pages 44–51, Princeton, NJ, USA, October 1991.
7. J. Bigün, G. H. Granlund, and J. Wiklund. Multidimensional orientation: texture analysis and optical flow. In *Proceedings of the SSAB Symposium on Image Analysis*. SSAB, March 1991.
8. P. J. Burt and E. H. Adelson. The Laplacian pyramid as a compact image code. *IEEE Trans. Comm.*, 31:532–540, 1983.
9. J. Crowley and A. Parker. A representation for shape based on peaks and ridges in difference of low-pass transform. *IEEE Trans. on PAMI*, (6):156–169, 1989.
10. J. L. Crowley and P. Stelmaszyk. Measurement and integration of 3-d structures by tracking edge lines. In *Computer Vision - ECCV90*, pages 269–280. Springer-Verlag, 1990.
11. Hugh Davson, editor. *The Eye*, volume 2A. Academic Press, New York, 2nd edition, 1976.
12. R. Deriche and O. Faugeras. Tracking line segments. In *Lecture Notes in Computer Science*, pages 259–268. INRIA, Springer-Verlag, 1990.
13. E. D. Dickmanns and Volker Graefe. Applications of dynamic monocular machine vision. *Machine Vision and Applications*, 1:241–261, 1988.
14. E. D. Dickmanns and Volker Graefe. Dynamic monocular machine vision. *Machine Vision and Applications*, 1:223–240, 1988.
15. David J. Fleet. *Measurement of image velocity.* Kluwer Academic Publishers, 1992. ISBN 0-7923-9198-5.
16. David J. Fleet and Allan D. Jepson. Computation of Component Image Velocity from Local Phase Information. *Int. Journal of Computer Vision*, 5(1):77–104, 1990.
17. G. H. Granlund. In search of a general picture processing operator. *Computer Graphics and Image Processing*, 8(2):155–178, 1978.
18. O. Hansen. *On the use of Local Symmetries in Image Analysis and Computer Vision.* PhD thesis, Aalborg University, March 1992.
19. D. J. Heeger. Optical flow from spatiotemporal filters. In *First Int. Conf. on Computer Vision*, pages 181–190, London, June 1987.
20. B. K. P. Horn. *Robot vision.* The MIT Press, 1986.
21. T. S. Huang, editor. *Image Sequence Analysis.* Information Sciences. Springer Verlag, Berlin, 1981.
22. T. S. Huang, editor. *Image Sequence Processing and Dynamic Scene Analysis*, Berlin, June 1982. NATO Advanced Study Institute, Springer Verlag.
23. D. H. Hubel and T. N. Wiesel. Receptive fields of single neurones in the cat's striate cortex. *J. Physiol.*, 148:574–591, 1959.
24. Bernd Jähne. Motion determination in space-time images. In O. Faugeras, editor, *Computer Vision-ECCV90*, pages 161–173. Springer-Verlag, 1990.
25. H. Knutsson. Representing local structure using tensors. In *The 6th Scandinavian Conference on Image Analysis*, pages 244–251, Oulu, Finland, June 1989. Report LiTH-ISY-I-1019, Computer Vision Laboratory, Linköping University, Sweden, 1989.

26. H. Knutsson, H. Bårman, and L. Haglund. Robust orientation estimation in 2d, 3d and 4d using tensors. In *Proceedings of International Conference on Automation, Robotics and Computer Vision*, September 1992.
27. H. Knutsson, L. Haglund, and G. H. Granlund. Tensor field controlled image sequence enhancement. In *Proceedings of the SSAB Symposium on Image Analysis*, pages 163–167, Linköping, Sweden, March 1990. SSAB. Report LiTH–ISY–I–1087, Linköping University, Sweden, 1990.
28. Hans Knutsson. Producing a continuous and distance preserving 5-D vector representation of 3-D orientation. In *IEEE Computer Society Workshop on Computer Architecture for Pattern Analysis and Image Database Management - CAPAIDM*, pages 175–182, Miami Beach, Florida, November 1985. IEEE. Report LiTH–ISY–I–0843, Linköping University, Sweden, 1986.
29. H. H. Nagel. Constraints for the estimation of displacement vector fields from an image sequence. In *Proceedings of Int. Joint. Conf. on Artificial Intelligence in West Germany*, 1983.
30. H. H. Nagel. On the estimation of optical flow: Relations between different approaches and som new results. *Artificial Intelligence*, pages 299–324, 1987.
31. W. H. Press, B. P. Flannery, S. A. Teukolsky, and W. T. Vetterling. *Numerical Recipes*. Cambridge University Press, 1986.
32. S. L. Tanimoto and T. Pavlidis. A hierarchical data structure for picture processing. *Computer Graphics and Image Processing*, 2:104–119, June 1975.
33. S. Uras, F. Girosi, A. Verri, and V. Torre. A computational approach to motion perception. *Biological Cybernetics*, pages 79–97, 1988.
34. A. B. Watson and Jr. A. J. Ahumada. Model of human visual-motion sensing. *Jour. of the opt. soc. of America A*, 1(2):322–342, 1985.
35. J. Wiklund, L. Haglund, H. Knutsson, and G. H. Granlund. Time sequence analysis using multi-resolution spatio-temporal filters. In *The 3rd International Workshop on Time-Varying Image Processing and Moving Object Recognition*, pages 258–265, Florence, Italy, May 1989. Invited Paper. Report LiTH–ISY–I–1014, Computer Vision Laboratory, Linköping University, Sweden, 1989.

Stabile Objektverfolgung und Detektion von nicht-vorhersagbarem Verhalten in komplexen Bildfolgen

Jost Bernasch, Roman Koutny
FORWISS München, FG Kognitive Systeme
Orleonstraße 32, D-8000 München 80
bernasch@forwiss.tu-muenchen.de

Zusammenfassung

In diesem Beitrag wird ein Verfahren vorgestellt zur robusten Verfolgung von bewegten Objekten sowie zur Detektion diskontinuierlicher Bewegungen. Es arbeitet auf komplexen Bildfolgen wie Straßenverkehrsszenen bei sich bewegender Kamera.

Das Verfahren benötigt kein Vorwissen. Zur stabilen Objektverfolgung werden Objektprimitive gemäß sehr einfacher Modelle wie Quadrate oder Kreisscheiben benutzt. Durch einen Kalmanfilter werden die Objekte zeitlich stabilisiert. Darauf aufbauend wird ein Maß für die Stabilität definiert.

Mit dem Stabilitätsmaß und der Bewegungsprädiktion der zeitlich stabilisierten Objekte wird eine Aussage über die Vorhersagbarkeit *des Verhaltens dieser Objekte abgeleitet. Insbesondere werden dadurch Diskontinuitäten im Objektverhalten detektiert. Damit wird ein wichtiger Beitrag geleistet, dynamische Szenenbeschreibungen effizient zu aktualisieren.*

1 Einleitung

Für die Interpretation komplexer Bildsequenzen ist die Detektion und stabile Erkennung von Objekten notwendig [2, 15, 16]. Bisherige Ansätze verwenden oft wenig robuste Daten wie z.B. optischen Fluß [1] und Flächensegmente [8] oder sind auf aufwendige oder sehr spezifische Modelle angewiesen. Bei dem hier vorgestellten Verfahren wird dagegen kein Vorwissen benötigt und die Objektprimitive werden gemäß sehr einfacher Modelle wie Quadrat oder Kreisscheibe verfolgt. Merkmale für die Beschreibung von Objektprimitiven sind lokale Spiegel-Symmetrieachsen (LAX) [11]. Es wird gezeigt, wie durch die Kombination mit einem Differenzbildverfahren und durch die geschickte Einschränkung des Suchraumes für den Objektmittelpunkt die Objektprimitive sehr effizient in Einzelbildern gefunden werden.

Im Gegensatz zu Ansätzen von Ullman [14] oder Salari und Sethi [10, 13] sind keine Annahmen über bijektive Merkmalszuordnung oder Forderungen nach Bewegung aller Punkte notwendig. Zudem erhält man *sofort* ohne Iteration für jedes neue Bild einen Schätzwert für die aktuelle Objektposition. Von anderen Extrapolationsverfahren (z.B. von Brandt [6], Preißler [8]) unterscheidet sich das vorgestellte Verfahren darin, daß sowohl für den Objektmittelpunkt als auch für alle Randpunkte eine Varianz $\sigma^2(t)$ abgeleitet wird, die den aktuellen Schätzwert und neue Messungen gemäß ihrer Güte berücksichtigen. Dadurch kann die Optimalitätseigenschaft des Kalmanfilters genutzt werden [5].

Durch die stabile Verfolgung und die Prädiktion der Objekte ist zudem nicht-vorhersagbares Verhalten von Objekten detektierbar. Solche Objekte werden separiert und für die Aktualisierung einer Szenenbeschreibung verwendet. Diese zusätzlichen Ergebnisse werden durch die Nutzung zeitlicher Information (im wesentlichen verkleinerte Suchräume) mit weniger Aufwand erreicht als das ursprüngliche Verfahren auf Einzelbildern benötigte. Wir beschreiben zuerst das Verfahren für Einzelbilder und dessen Verbesserung, anschließend die zeitliche Stabilisierung sowie Ergebnisse auf natürlichen Straßenverkehrsszenen (Abb.1).

2 Lokale Symmetrie

Die Verwendung lokaler Spiegel-Symmetrieachsen [11] erlaubt die Erkennung von Objekten ohne eine explizite Berechnung von Kanten oder Regionensegmenten und vermeidet die damit verbundenen Probleme, siehe z.B. [9]. Stattdessen werden Merkmale nur in der *lokalen* Umgebung eines Pixels gesucht. Einfache Objektmodelle fassen lokale Merkmale zusammen.

Abbildung 1: Autobahnszene mit einfahrendem PKW: zwei Bilder im Abstand von 1.2 Sekunden.

2.1 Theorie lokaler Symmetrien

Unter einer lokalen Spiegel-Symmetrieachse versteht man eine Gerade zur x-Achse, deren Drehwinkel ϑ folgenden Ausdruck minimiert:

$$\int_0^\infty \int_0^{2\pi} v(r)(f(r,\vartheta+\varphi)-f(r,\vartheta-\varphi))^2 d\varphi dr. \tag{1}$$

Dabei ist r der Abstand zum Objektzentrum, $f(r,\varphi)$ beschreibt die Grauwerte des Bildes in Abhängigkeit von r und dem Neigungswinkel φ zur Symmetrieachse (Abb 2a). $v(r)$ ist eine mit wachsendem r abnehmende Gewichtungsfunktion.

Die Berechnung der Symmetriewinkel für diskrete Bilder erfolgt über eine $n \times n$ Umgebung mit typischerweise $n = 3$ (Abb. 2c) oder $n = 5$. Durch Unterdrückung der konstanten Terme der zu minimierenden Funktion $\sum_{j=1}^{8}(g_j - g_{i-j})^2$ wird das Maximum der sich ergebenden Autokorrelationsfunktionen G_i

$$G_i = \sum_{j=1}^{8} g_j g_{i-j} \quad \text{mit} \quad i = 1,..,8 \tag{2}$$

als Lösung gesucht. Die Genauigkeit des Symmetriewinkels ϑ wird erhöht durch die Interpolation

$$\vartheta = \frac{\pi}{8}\left(i_{max} - 2 + \frac{G_{i_{max}+1} - G_{i_{max}-1}}{2(2G_{i_{max}} - G_{i_{max}+1} - G_{i_{max}-1})}\right). \tag{3}$$

Der *Konfidenzwert* $C = G_{i_{max}} - G_{i_{min}}$ gibt an, wie stark ausgeprägt der Symmetriewinkel ist. An Kantenverläufen ist dieser Wert besonders hoch.

2.2 Suchvorgang

Um bei der Objektsuche größeninvariant zu sein, wird eine Gauß'sche Auflösungspyramide verwendet [4]. Für *jeden* Bildpunkt einer Pyramidenstufe werden in einer vorgegebenen Anzahl von Richtungen die Punkte mit den *höchsten* Konfidenzwerten gesucht. Der Suchbereich ist so begrenzt, daß keine Überlappungen in den Pyramidenstufen auftreten.

Die zu diesen Punkten gehörenden Symmetriewinkel werden mit den Symmetriewinkeln vorab erstellter Modelle verglichen. Ein Objekt gilt als erkannt, wenn ein festgelegter Prozentsatz der Symmetriewinkel mit denen des Modells innerhalb einer vorgegebenen Differenz übereinstimmt. Der Prozentsatz ist für ein Modell bei allen Bildfolgen (vier à 130 Bilder) *konstant*. Als Modelle werden Objektprimitive wie Quadrate und Kreisscheiben mit 24 Suchrichtungen verwendet.

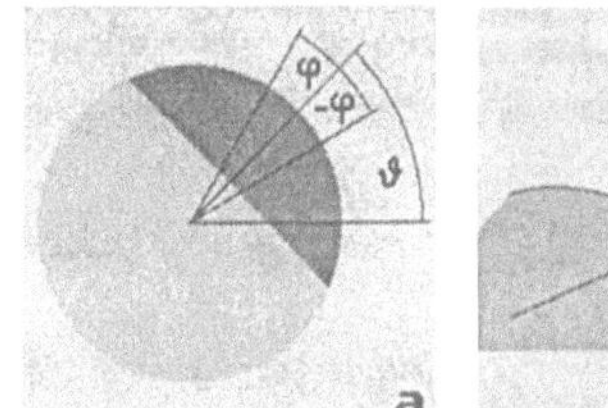
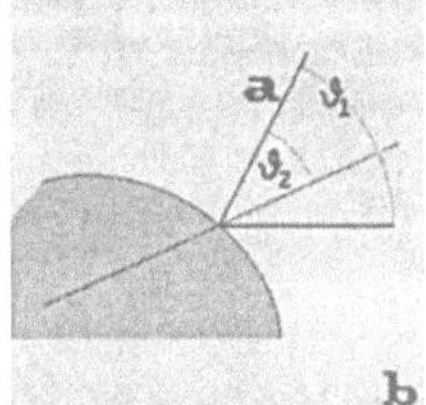
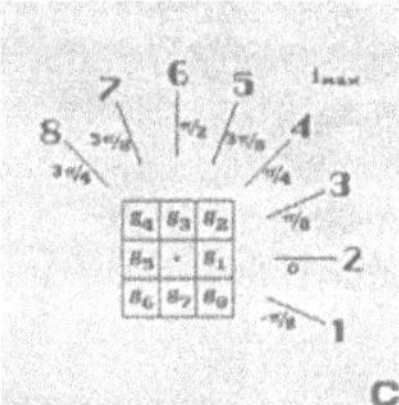
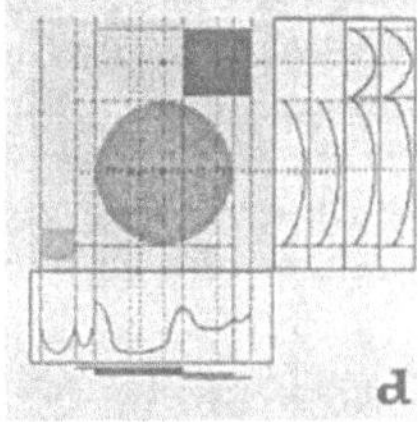

Abbildung 2: (a) Lokale Symmetrieachse unter dem Winkel ϑ. (b) Absoluter Symmetriewinkel ϑ_1 und relativer Symmetriewinkel ϑ_2 der Symmetrieachse a (rotationsinvariant). (c) Zuordnung der Symmetrieachsen zu den i_{max}-Werten einer 3×3 Umgebung. (d) Unter indirekter Nutzung der Information vertikaler und horizontaler Kantenverläufe wird die Anzahl der Objektmittelpunkte und damit der Objektkandidaten auf einer bestimmten Pyramidenstufe eingeschränkt.

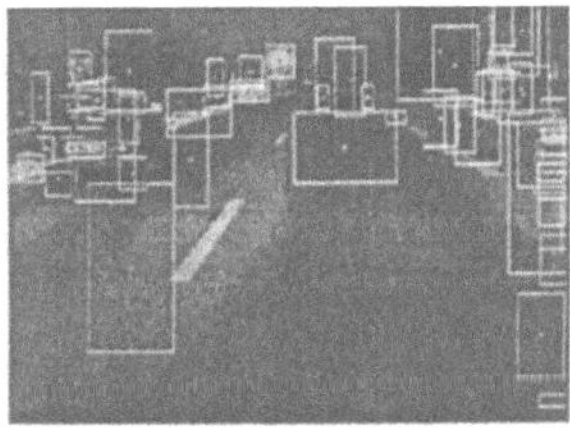

Abbildung 3: Autobahnszene: (a) LAX Merkmale allgemein (b) Differenzbild (c) Einbeziehung des Differenzbildes in LAX.

2.3 Verbesserung des Verfahrens

Das Verfahren von Seitz [11] wurde bzgl. benötigter Rechenzeit, Robustheit und Anzahl detektierter Objekte verbessert. Durch eine Verkleinerung des Wertebereichs der Symmetriewinkel auf 0 bis 180 Grad konnte die für die Suche benötigte Anzahl an Modellen reduziert werden. Durch einen Nachverarbeitungsschritt werden mehrfach erkannte Objekte ausgeschlossen und die Koordinaten der ermittelten Objektzentren rechnerisch verbessert.

Zur Reduzierung der benötigten Rechenzeit wurde eine Methode entwickelt, die mit Hilfe der Verteilung der Symmetriewinkel und der Konfidenzwerte den Suchraum der möglichen Objektkandidaten *vor* der eigentlichen Suche stark einschränkt [7]. Dabei wurde ausgenutzt, daß Symmetrieachsen senkrecht auf Kantenverläufe von Objekten stehen und die Konfidenzwerte dort am höchsten sind (siehe Abbildung 2d).

Eine weitere Reduzierung der Objektkandidatenzahl bei der Analyse von Bildfolgen brachte die Verwendung von Differenzbildern. Differenzbilder liefern Information darüber, in welchen Bildbereichen Bewegung zwischen zwei Einzelbildern stattgefunden hat. Mit diesen zusätzlichen Informationen konnte die Suche auf die Bildsegmente reduziert werden, die *potentiell* Bewegungsinformation enthalten. Eine Eigenschaft dieser Methode ist, daß Objekte, die sich (relativ zur Kamera) nicht oder nur kaum bewegen, nicht erkannt werden. Dies kann aber durch die zeitliche Stabilisierung ausgeglichen werden und ist zudem für die geplante Applikation einer Aufmerksamkeitssteuerung nicht von Nachteil [3].

Mit diesen Modifikationen benötigt eine SUN-SPARC Station 2 für die Analyse eines Bild der Größe 354 × 283 Pixel mit 256 Graustufen durchschnittlich statt 75 nur 6 CPU-Sekunden in der Bildsequenz (ohne Optimierung).

Neben der Verbesserung des Verfahrens auf Einzelbildern ist für eine robuste Objektdetektion in Bildfolgen eine zeitliche Zuordnung und damit Stabilisierung der gefundenen Objekte notwendig. Diese Stabilisierung wird durch eine Erweiterung des Verfahrens auf Bildfolgen erreicht.

3 Zeitliche Stabilisierung mittels Kalmanfilter

Werden Objektprimitive in Bildfolgen unabhängig für jedes einzelne Bild gesucht, ergibt sich der Effekt, daß aufgrund des unterschiedlichen Informationsgehalts der Differenzbilder und der Störungen in den Einzelbildern *ständig* präsente Objekte nur vereinzelt gefunden werden. Um eine lückenlose Verfolgung und Stabilisierung der Objekte zu ermöglichen, wurde das Verfahren durch einen Kalmanfilter erweitert.

Der Kalmanfilter hat die Aufgabe, die Bewegung der zu verfolgenden Objekte zeitlich zu stabilisieren. Dabei fließen alle Meßwerte in das Filter ein, ohne daß besonders gestörte Messungen vorher aussortiert oder frühere Meßwerte zwischengespeichert werden müssen. Sowohl die Objektmittelpunkte als auch jeder der n Randpunkte werden dabei verfolgt. Eine ausführliche Darstellung der Technik findet sich z.B. in Gelb [5].

3.1 Schätzung der Objektmittelpunkte

Die *Vorhersage* des neuen Schätzwertes für den Objektmittelpunkt erfolgt mit

$$\begin{pmatrix} \hat{x}(t_{k+1}^-) \\ \hat{y}(t_{k+1}^-) \end{pmatrix} = \begin{pmatrix} \hat{x}(t_k) \\ \hat{y}(t_k) \end{pmatrix} + \vec{v_k}\,[t_{k+1} - t_k], \tag{4}$$

$$\sigma_x^2(t_{k+1}^-) = \sigma_x^2(t_k) + \sigma_w^2[t_{k+1} - t_k], \tag{5}$$

wobei $\vec{v_k} = \vec{v} + \vec{w}$ die mit dem Fehler $\vec{w}$ (Varianz σ_w^2) behaftete bekannte Geschwindigkeit des Objekts ist. Einen neuen Schätzwert für die Objektposition $(\hat{x}(t_{k+1}), \hat{y}(t_{k+1}))^T$ erhält man aus

$$\begin{pmatrix} \hat{x}(t_{k+1}) \\ \hat{y}(t_{k+1}) \end{pmatrix} = \begin{pmatrix} \hat{x}(t_{k+1}^-) \\ \hat{y}(t_{k+1}^-) \end{pmatrix} + K(t_{k+1}) \begin{pmatrix} z_1(t_{k+1}) - \hat{x}(t_{k+1}^-) \\ z_2(t_{k+1}) - \hat{y}(t_{k+1}^-) \end{pmatrix}, \tag{6}$$

Dabei bezeichnet $K(t_{k+1})$ die *Kalmanverstärkung* (Kalman Gain) zum Zeitpunkt t_{k+1}, $\vec{z}(t_{k+1})$ den neuen Meßwert. Die Kalmanverstärkung gewichtet den Einfluß der Differenz des neuen Meßwertes von dem vorhergesagten Schätzwert in Abhängigkeit von der Varianz des Meßwertes und der vorhergesagten Schätzung. Sie wird berechnet durch

$$K(t_{k+1}) = \frac{\sigma_x^2(t_{k+1}^-)}{\sigma_x^2(t_{k+1}^-) + \sigma_z^2(t_{k+1})}, \tag{7}$$

Die Varianz des neuen Schätzwertes erhält man aus:

$$\sigma_x^2(t_{k+1}) = \sigma_x^2(t_{k+1}^-) - K(t_{k+1})\sigma_x^2(t_{k+1}^-). \tag{8}$$

Analog zur Objektsuche wird der Suchbereich um diese neue Position durchsucht. Wir gehen von der Annahme aus, daß jedes Objekt deutlich nach oben, unten, links und rechts zum Hintergrund begrenzt wird. Somit ermöglichen es uns die bei der Suche erhaltenen Randpunkte, die Koordinaten des eigentlichen Objektzentrums zu bestimmen. Das derart erhaltene Objektzentrum ist der Meßwert der Position in dem aktuellen Bild. Für diesen Objektmittelpunkt werden die Randpunkte und Symmetriewinkel neu berechnet.

Für jede Anwendung ist es entscheidend, eine Methode zu finden, mit der die Varianz $\sigma_z^2(t_k)$ einer Messung bestimmt werden kann. Die Varianz $\sigma_z^2(t_k)$ der gemessenen Objektposition zum Zeitpunkt t_k berechnen wir mit Hilfe der Symmetriewinkel $\vartheta_i(t_k)$ und der Modellwinkel ϑ_i^M aus

$$\sigma_z^2(t_k) = \frac{1}{n}\sum_{i=1}^{n}(\vartheta_i(t_k) - \vartheta_i^M)^2. \tag{9}$$

Diese Methode erlaubt uns, die Suche nach bereits bekannten Objekten auf jeweils einen *kleinen* Bildbereich einzuschränken. Die Berechnung der Varianz (Güte) der neu bestimmten Objektposition vermeidet die problematische Festlegung der beiden Schwellwerte maximal erlaubte Winkelabweichung und minimaler Prozentsatz übereinstimmender Symmetriewinkel. Diese bestimmen dann nur die Unterscheidung zwischen erkannten und nicht erkannten Objekten. Da die Modellobjekte von Objektprimitiven idealisiert sind, werden für die Verfolgung jedes erkannten Objekts die Modelldaten verwendet, die zunächst bei der ersten Detektion erhalten wurden (siehe Abschnitt 4). Im weiteren Verlauf der Bildsequenz werden die Modelldaten aktualisiert. Damit konnten Objekte von Fahrzeugen stabil über eine abgeschlossen Sequenz von 130 Bildern verfolgt werden.

3.2 Bewegungshypothese

Um die sich sehr unterschiedlich bewegenden Objektmittelpunkte verfolgen zu können, wurde folgendes Bewegungsmodell entwickelt. Auf der Basis der ermittelten Objektpositionen aus zurückliegenden Bildern wird die Hypothese $\vec{v}(t+\Delta t) = \vec{v}(t) + \int_t^{t+\Delta t} \vec{a}(\tau)d\tau$ für die Objektgeschwindigkeit aufgestellt. Dabei wird von annähernd linearer Bewegung aller Fahrzeuge über *kurze* Zeiträume, aber nicht notwendigerweise konstanten Beschleunigungen ausgegangen. Die Ungenauigkeiten in der Objektposition $\vec{s}_k$ zum Zeitpunkt k werden eliminiert durch die Ermittlung einer Vorhersage $\overline{v_{k+1}} = \vec{v}_k + \vec{a}_k\Delta t$ für die Geschwindigkeit über ein zeitfestes n mit

$$\vec{v}_k = \frac{1}{n}\sum_{i=k-n+1}^{k}(\vec{s}_i - \vec{s}_{i-1}) = \frac{1}{n}(\vec{s}_k - \vec{s}_{k-n}), \tag{10}$$

für die Beschleunigung gilt dann $\vec{a}_k = (\vec{v}_k - \vec{v}_{k-1})/\Delta t$.

Durch Nutzung dieser Beschreibung der Bewegung konnte gezeigt werden, daß sowohl Fahrzeuge, die sich *relativ kaum* zum eigenen Fahrzeug bewegen als auch Fahrzeuge, die sich mit hoher Geschwindigkeit auf das eigene, kameraführende Fahrzeug zubewegen, stabil verfolgt werden können (vgl. Abb. 4 und Abb. 5).

3.3 Schätzung der Randpunkte

Reflexionen auf Oberflächen und die Annäherung zweier verschiedener Objekte im Verlauf einer Bildsequenz können die Konfidenzwerte in der Umgebung eines Objekts verändern. Dadurch werden über die Randpunkte des Objekts seine Form und Größe beeinflußt. Um die Lage untereinander unabhängiger Randpunkte *relativ* zum Objektzentrum zu stabilisieren, wird für *jeden* dieser Punkte ein separates Kalmanfilter eingesetzt.

Durch die Berücksichtigung gestörter Grauwerte in Form einer Störfunktion Δf in Gleichung (1) erhält man in der diskreten Form ausgehend von $\Delta\vec{g} := (\Delta g_1, ..., \Delta g_8)^T$ über die Autokorrelationsfunktionen (2) nach Umformungen die Gleichung

$$\Delta G_k(\vec{g} + \Delta\vec{g}) := G_k(\vec{g} + \Delta\vec{g}) - G_k(\vec{g}), \tag{11}$$

die den Einfluß der gestörten Grauwerte auf die Bestimmung der Symmetriewinkel beschreiben (detaillierte Herleitung siehe [7]). Da $G_k(\vec{g})$ nicht bekannt ist, wird das bereits früher detektierte Objekt als Referenzgröße herangezogen. Die Varianz $\sigma^2 = \frac{1}{8}\sum_{k=1}^{8} \Delta G_k(\vec{g} + \Delta\vec{g})^2$ bestimmt im Kalmanfilter den Einfluß der aktuellen Messung auf die Bestimmung des neuen Schätzwertes.

Die Abschätzung der Varianz der Randpunkte ermöglicht es, durch den Kalmanfilter Objekte über mehrere Bilder hinweg zu verfolgen, *ohne* daß sich ihre Größe durch eine teilweise Verdeckung sprunghaft ändert. Aus einer stetigen Größenänderung eines Objekts kann somit qualitativ auf eine Änderung seines Abstands geschlossen werden.

3.4 Bewertungsmaße für die Stabilität

Für jedes detektierte Objekt wird ein Maß für die Stabilität eingeführt. Es drückt aus, mit welcher Sicherheit das Objekt im zeitlichen Verlauf wiedererkannt werden konnte. In die Berechnung des zeitlichen Stabilitätsmaßes fließen die Merkmale

- durchschnittliche Differenz der Symmetriewinkel und Modellwinkel,
- Grad der Änderung der Konfidenzwerte,
- Grad der Änderung der Objektform und
- Sprunghaftigkeit der Änderung der Beschleunigung

ein. Insbesondere die beiden letzten Kriterien spielen für die Beurteilung der zeitlichen Stabilität eine wichtige Rolle. Die Stabilität erhöht sich bei geringer Ausprägung der Änderungsmerkmale zwischen zwei Bildern, sie verringert sich, wenn mindestens zwei dieser Merkmale stark ausgeprägt sind, z.B. wenn eine erhöhte Winkeldifferenz und starke Änderung der Konfidenzwerte vorliegen. Der Stabilitätswert stabiler Objekte nimmt dabei nur in den Teilsequenzen zu, in denen das Objekt sehr sicher wiedererkannt werden konnte. Die exakte Berechnung findet sich in [7].

Durch dieses Stabilitätsmaß werden stabile von nicht stabilen Objekten unterschieden. Zudem werden nicht sicher wiederzuerkennende Objekte von der Verfolgung nicht sofort ausgeschlossen, sondern erst *nachdem* ihre Stabilität unter einen vorgegebenen Wert gesunken ist. Dadurch wird vermieden, daß Objekte nur deshalb aufgegeben werden, weil sie in wenigen Einzelbildern unsicher wiedererkannt wurden oder ihre Form sich z.B. durch Teilverdeckung leicht modifiziert hat.

4 Ergebnisse

In unseren Versuchen wurden die Modelldaten von zu verfolgenden Objekten dynamisch nach k Einzelbildern (typischerweise $k = 4$) durch aktuelle Werte ersetzt. Es war somit möglich, auf sich ändernde Lichtverhältnisse in der Bildfolge zu reagieren. Bei der Analyse von Bildfolgen mit Straßenverkehrsszenen konnten ständig präsente Fahrzeugdetails stabilisiert und über eine Sequenz von 130 Bilder verfolgt werden (Abb. 5). Besonderes Interesse galt der Detektion von in der Bildfolge neu erscheinenden Fahrzeugen. Abbildung 4 demonstriert den Überholvorgang eines PKWs auf der rechten Fahrspur (beschleunigte Einfahrt). An der Rückseite des Wagens konnten zwei Details detektiert und bis zum Ende der Sequenz verfolgt werden.

5 Zusammenfassung und Diskussion

In diesem Beitrag wurde gezeigt, daß auf der Basis sehr einfacher Objektprimitive, die mit lokalen Symmetrieachsen beschrieben werden, Objekte in Bildfolgen über 130 Bilder hinweg stabil verfolgt

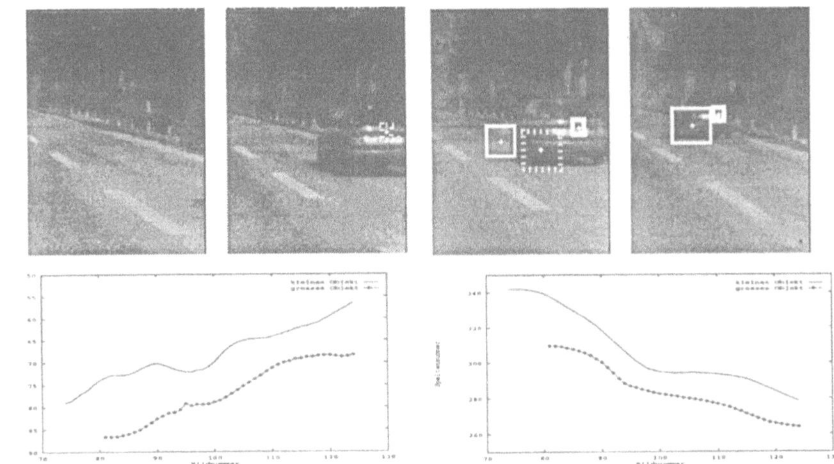

Abbildung 4: Der Koordinatenverlauf zweier Objekte an der Rückseite eines in der Bildfolge *neu* erscheinenden PKWs, der bis an das Ende der Bildfolge stetig verfolgt wurde (Ausschnittsvergrößerung aus den Bildern 50, 80, 84, 110).

werden konnten. Dazu werden keine Kanten- oder Flächensegmente benötigt. Mit Hilfe geeignet definierter Varianzen gingen die Meßwerte für jedes Bild optimal in die Fortschreibung bereits detektierter Objekte ein.

Ergebnisse wurden auf Straßenverkehrsszenen von Autobahnen und Landstraßen mit schwierigen Beleuchtungsverhältnissen und verschiedenen 3D-Objekten (Pkw, Lkw, Schilder,...) gezeigt. Die Objekte bewegten sich mit sehr unterschiedlichen und jeweils unbekannten Geschwindigkeiten. Dies konnte durch das im Kalmanfilter integrierte adaptive Bewegungsmodel erfolgreich berücksichtigt werden. Trotz erheblicher Verbesserungen konnte der Rechenaufwand gegenüber dem ursprünglichem Verfahren deutlich verringert werden. Die Berechnung der Ergebnisse in Echtzeit erscheint mit kommerziellen VLSI Chips erreichbar. Siehe dazu die Überlegungen in Seitz [12].

Als zusätzliches Ergebnis liefert das Verfahren Informationen über diskontinuierliches Verhalten stabiler Objekte. Dazu gehört das Neuerscheinen von Objekten (Abb. 4) oder auch das *nicht* vorhersagbare Verhalten stabiler Objekte. Die Information über diese diskontinuierlichen Ereignisse werden von einem Modul zur Aufmerksamkeitssteuerung in Bildfolgen verwendet [3]. Die Berücksichtigung von Diskontinuitäten ist nicht nur in der Anwendung einer Aufmerksamkeitssteuerung in Straßenverkehrsszenen sondern auch bei der Analyse anderen Bildsequenzen von besonderer Bedeutung.

In der Zukunft werden sich die Arbeiten auf folgende Punkte konzentrieren: (1) Beschreibung des Informationsgehaltes von diskontinuierlichen Bewegungen stabiler Objekte mit Methoden der Informationstheorie, (2) Bewertung der dynamischen Entwicklung des Stabilitätsmaßes und (3) Erweiterung des Ansatz auf eine 3-D Erkennung. Die bisher erreichte Leistungsfähigkeit bei der stabilen Verfolgung von Objekten ohne Nutzung spezifischen Wissens zeigt ein großes Potential für zukünftige Arbeiten. Wir erwarten, daß die Methode eine wichtigen Beitrag bei der Aktualisierung von dynamischen Szenenbeschreibungen in verschiedenen Domänen leisten wird.

Die Autoren danken der BMW AG München für die Bereitstellung des Bildmaterials im Rahmen

Abbildung 5: Verfolgung eines Lastwagens auf der Gegenfahrbahn und eines vorausfahrenden PKWs.

des Kooperationsprojektes MOVIE.

Literatur

[1] G. Adiv. Inherent ambiguities in recovering 3-D motion and structure from a noisy flow field. *IEEE Transactions on Pattern Analysis and Machine Intelligence*, 11:477–489, 1989.

[2] D. Ballard and C. M. Brown. *Computer Vision*. Prentice-Hall, Englewood Cliffs, New Jersey, 1982.

[3] Jost Bernasch. Visual attention control in image sequences using multiple knowledge sources. In *Proceedings of IS&T/SPIE Symposium on Electronic Imaging: Nonlinear Image Processing IV, San Jose CA*, Jan 31 - Feb 5 1993.

[4] P.J. Burt. The pyramid as a structure for efficient computation. In A. Rosenfeld, editor, *Multiresolution Image Processing and Analysis*, Springer Series in Information Sciences, pages 6–35. Springer-Verlag Berlin Heidelberg New York, 1984.

[5] A. Gelb. *Applied Optimal Estimation*. M.I.T Press, Cambridge, Massachusetts, 1974.

[6] K.-P. Karmann and A. von Brandt. Moving object recognition using an adaptiv background memory. In V. Cappellini, editor, *Time-Varying Image Processing and Moving Object Recognition, 2*, pages 289–296. Elsevier Science Publishers, 1990.

[7] Roman Koutny. Robuste Detektion und Verfolgung von Objektprimitiven auf der Basis lokaler Symmetrieachsen in komplexen Bildfolgen. Master's thesis, Technische Universität München, Lehrstuhl Informatik IX, 1993.

[8] M. Preißler and T. Messer. Ein Extrapolationsverfahren zur Verfolgung von Segmenten in Bildsequenzen. In B. Radig, editor, *Mustererkennung 1991, 13.DAGM-Symposium*, Informatik-Fachberichte Band 290, pages 491–498. Springer-Verlag Berlin Heidelberg New York, 1991.

[9] A. Rosenfeld and A.C. Kak. *Digital Picture Processing, Second Edition, Volume 2*. Academic Press, Orlando, FL, 1982.

[10] V. Salari and I.K. Sethi. Feature Point Correspondence in the Presence of Occlusion. *IEEE Transactions on Pattern Analysis and Machine Intelligence*, 12(1):87–91, 1990.

[11] P. Seitz. The robust recognition of object primitives using local axes of symmetry. *Signal Processing*, 18:89–108, 1989.

[12] P. Seitz, G.K. Lang, B. Gilliard, and J.C. Pandazis. The robust recognition of traffic signs from a moving car. In B. Radig, editor, *Mustererkennung 1991, 13.DAGM-Symposium*, Informatik-Fachberichte Band 290, pages 287–294. Springer-Verlag Berlin Heidelberg New York, 1991.

[13] I.K. Sethi and R. Jain. Finding Trajectories of Feature Points in a Monocular Image Sequence. *IEEE Transactions on Pattern Analysis and Machine Intelligence*, 9(1):56–73, 1987.

[14] Shimon Ullman. *The Interpretation of Visual Motion*. MIT Press, 1979.

[15] D. Wetzel, A. Zins, and H. Niemann. Edge and motion based segmentation for traffic scene analysis. In *Proceedings Third Open German-Russian Workshop on Image Interpretation and Pattern Recognition, Erlangen, Germany*, March 1993.

[16] A. Zins and H. Niemann. Filterung des Bildhintergrundes in mit bewegter Kamera aufgenommenen Bildfolgen. In S. Fuchs and R. Hoffmann, editors, *Mustererkennung 1992, 14.DAGM-Symposium*, Informatik aktuell, pages 443–448. Springer-Verlag Berlin Heidelberg New York, 1992.

Ein Mehrgitterverfahren zur Bewegungssegmentierung in Bildfolgen

Horst Haußecker[1] und Bernd Jähne[2]

[1] Institut für Umweltphysik der Universität Heidelberg
Im Neuenheimer Feld 366, 69120 Heidelberg
[2] Scripps Institution of Oceanography, PORD
La Jolla, CA 92093-0230 USA

Zusammenfassung

Es wird ein Verfahren vorgestellt, mit dem die Bewegung einzelner Bildbereiche in ausgedehnten Bildfolgen zuverlässig segmentiert werden kann. Dabei werden zuerst die ruhenden Strukturen in Bildsequenzen durch eine zeitliche Gaußpyramide ermittelt. Zum anschließenden Vergleich dieser Bilder mit den Originalbildern wird ein lokales Ähnlichkeitsmaß berechnet und dieses mit Hilfe von Fuzzy-Logic bewertet. Das Resultat ist ein fehlertolerantes Bewertungsverfahren, welches selbst mit bloßem Auge schwer zu erkennende Objekte detektiert. Der Bildinhalt wird somit, unabhängig vom Inhalt des Einzelbildes, allein aufgrund der Eigenschaft "Bewegung in der Sequenz" segmentiert.

1 Einleitung

Die exakte Bestimmung der Geschwindigkeiten von Objekten in Bildfolgen ist ein grundlegendes Problem der Digitalen Bildfolgenanalyse. Eine quantitative Bewegungsanalyse liefert jedoch selbst bei ausgedehnten Bildfolgen im allgemeinen nicht die wirkliche Bewegung der Objekte im 3D-Raum und ist bei Analyse von nur zwei aufeinanderfolgenden Bildern meist nicht möglich, [Verri, 1989]. In vielen Fällen ist es allerdings nicht notwendig, Geschwindigkeiten von Objekten in Bildern exakt zu bestimmen. Oft reicht eine mehr qualitative Analyse aus, d. h. eine *Bewegungssegmentierung*, die Bildbereiche bestimmt, in denen Bewegung stattfindet. Beispiele für solche Anwendungen sind die Überwachung von Verkehrsanlagen und Bahnsteigen [Haass, 1984] oder die Kontrolle von Fußgängerwegen [Shio, 1991]. Der hier beschriebene Anwendungsfall ist die Beobachtung der Ablösung von Sedimenten in fließenden Gewässern. Die Wasserbewegung im Untergrund kann zur Verlagerung von Bodenpartikeln und damit zu Bodendeformationen führen. Begünstigt wird diese Partikelverlagerung an Schichtgrenzen zweier Bodenarten oder der Grenze von Boden zu festen oder flexiblen Bauteilen [Köhler, 1992]. Die automatische Detektion von Partikelbewegungen in diesen Bereichen der Bodenschichten in Abhängigkeit von hydrodynamischen und mechanischen Belastungen liefert wichtige Informationen über die Stabilität von Uferbefestigungen. Typische Bilder solcher Grenzschichten zwischen einer Sandoberfläche und einem aufliegenden Geotextil sind in Abb. 6 a) und b) zu sehen.

Zur Detektion von sich bewegenden Partikeln ist es notwendig, sie von den ruhenden Strukturen im Bild zu unterscheiden. Da es sich bei beiden zum Teil um gleichartige Objekte handelt (hier Sandkörner), kann man nicht durch eine Vorsegmentierung des Bildinhaltes Teilbereiche dem ruhenden Hintergrund zuordnen, wie dies z. B. bei Straßenszenen der Fall ist. Gleichzeitig ändert sich der ruhende Hintergrund im Laufe der Zeit durch Materialverlagerungen, so daß man nicht von einem zeitlich invarianten Hintergrundbild ausgehen kann, sondern zu jedem Bild die zeitlich lokalen, ruhenden Strukturen bestimmen muß.

Das vorliegende Verfahren gliedert sich somit in zwei Schritte: Im ersten Schritt wird aus einer zeitlichen Umgebung des betrachteten Bildes der ruhende Hintergrund extrahiert. Im zweiten Schritt wird dieses Hintergrundbild mit dem Originalbild verglichen und es werden Bildbereiche bestimmt, in denen ein signifikanter Unterschied zwischen beiden Bildern besteht. Dadurch werden Gebiete detektiert, in denen mit großer Wahrscheinlichkeit Bewegung stattfindet.

2 Extraktion ruhender Strukturen mittels einer zeitlichen Gaußpyramide

Zur Entfernung von sich bewegenden Strukturen aus der Bildsequenz wird eine zeitliche Glättung der Form

$$H(\mathbf{x}, t_0) = \int_{t_0-\Delta t}^{t_0+\Delta t} w(t_0 - t) g(\mathbf{x}, t)\, dt, \tag{1}$$

durchgeführt. Als Faltungsmaske wird eine *Gaußmaske*

$$w(t) = g_\sigma(t) = \frac{1}{\sigma\sqrt{2\pi}} e^{-\frac{t^2}{2\sigma^2}} \tag{2}$$

verwendet. Sie bietet eine minimale Unschärfe der Transferfunktion im Fourierraum bei minimaler Ausdehnung der Faltungsmaske im Zeitraum. Zur diskreten Realisierung definieren wir, analog zu [Jähne, 1991], die *zeitlichen Binomialmasken*

$$\mathbf{B}_t^2 = \frac{1}{4}[1\,2\,1] \qquad \mathbf{B}_t^4 = \frac{1}{16}[1\,4\,6\,4\,1] \tag{3}$$

und den *Reduktionsoperator* $\mathbf{R}_t$, der nur jedes zweite Bild aus einer Bildfolge herausgreift.

Eine effektive und sukzessive Berechnung der Glättung mit einer Gaußmaske wird durch die in Abb. 1 skizzierte *zeitliche Gaußpyramide* mit den in (3) definierten Binomialmasken angenähert. Dazu glättet man die Originalsequenz mit der $\mathbf{B}_t^4$-Maske. Eine Reduktion der Zeitauflösung des Ergebnisses dieser Faltung durch $\mathbf{R}_t$ führt zu Aliasingeffekten, die aber dadurch kompensiert werden können, daß nach der Reduktion zusätzlich mit einer $\mathbf{B}_t^2$-Maske geglättet wird. Insgesamt wird die nächste Ebene der Pyramide durch den zusammengesetzten Operator $(\mathbf{B}_t^2\, \mathbf{R}_t\, \mathbf{B}_t^4)$ erzeugt. Die Wirkung dieser Glättung ist in Abb. 2 anhand

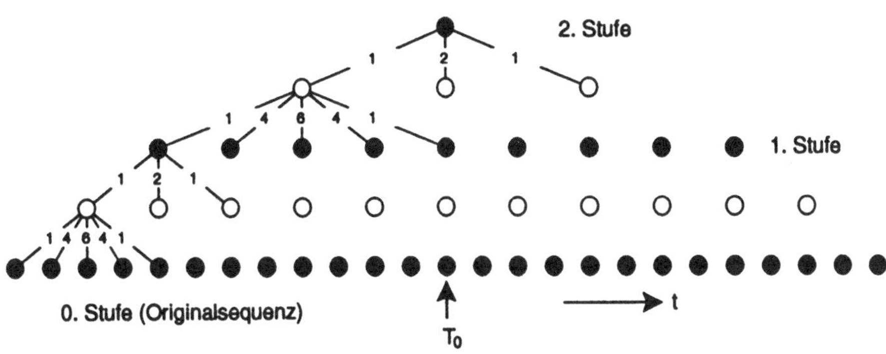

Abb. 1: *Zeitliche Gaußpyramide. Die Kreise entsprechen einzelnen Bildern.*

der resultierenden Transferfunktion im Vergleich zu einer B_t^4-Maske dargestellt. Dieses in [Jähne, 1992] beschriebene ***gezielte Unterabtasten*** ermöglicht eine sehr recheneffektive und aliasingfreie Iteration von Binomialmasken.

Zur Extraktion des Hintergrundes eines weiteren Bildes muß nicht die gesamte Pyramide neu aufgebaut werden. Verschiebt man die Pyramide um ein Bild nach rechts, so werden die meisten der bereits berechneten Glättungen erneut benötigt. Wenn man alle Zwischenstufen der Pyramide aufbewahrt, so kann man diese Zwischenergebnisse verwenden, und es genügt die am rechten Rand neu hinzukommenden Faltungen zu berechnen.

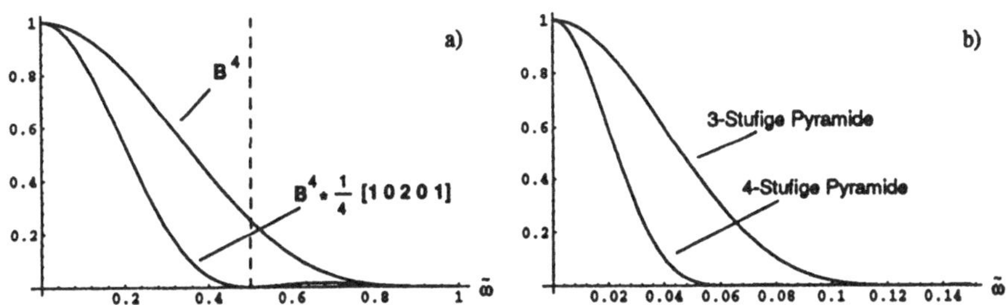

Abb. 2: a) *Transferfunktion des verwendeten Glättungsfilters im Vergleich zu* B^4. *Die gestrichelte Linie markiert die Grenzfrequenz der nächsten reduzierten Stufe.*
b) Transferfunktion der zeitlichen Gaußpyramide für verschiedene Stufen. Dabei bedeutet $\tilde{\omega} = \frac{\omega}{\omega_{Nyquist}}$.

3 Vergleich des Originalbildes mit dem Hintergrund

Nachdem mit dem in Abschnitt 2 vorgestellten Verfahren die ruhenden Strukturen in einem Bild detektiert wurden, muß nun das so erhaltene Hintergrundbild mit dem Originalbild verglichen werden, um die Bildbereiche zu segmentieren, in denen Bewegung stattfindet. Dies geschieht durch einen *qualitativen* Vergleich zwischen den *lokalen Strukturen* in beiden Bildern. Er ähnelt der menschlichen Interpretation von Bildinhalten, durch die beim Vergleich zweier Bilder Strukturen als gleich bzw. ähnlich erkannt werden, auch ohne daß die Differenz im absoluten Grauwert an isolierten Punkten quantitativ bestimmt wird. Zur Realisierung wird zuerst ein *lokales Ähnlichkeitsmaß* für jeden Punkt berechnet. Dieses wird mit einer Schwelle verglichen, die darüber entscheidet, ob es sich um einen signifikanten Unterschied zwischen beiden Bildern handelt oder um Rauschen. Es zeigt sich dabei, daß nur eine Schwelle sinnvoll ist, die sich den lokalen Bildstrukturen anpaßt.

3.1 Lokales Ähnlichkeitsmaß

Ein lokales Ähnlichkeitsmaß $S(\mathbf{x}_0)$ am Punkt $\mathbf{x}_0$ wird folgendermaßen definiert:

$$S(\mathbf{x}_0) = \int_G h(\mathbf{x}_0 - \mathbf{x})\,(H(\mathbf{x}) - O(\mathbf{x}))^2\, d^2x, \tag{4}$$

mit $H(\mathbf{x})$ = Hintergrundbild und $O(\mathbf{x})$ = Originalbild. G bezeichnet eine lokale Integrationsumgebung und $h(\mathbf{x})$ eine normierte Fensterfunktion mit Schwerpunkt bei 0, welche die punktweise Differenz der beiden Bilder in der lokalen Umgebung verschieden stark wichtet. Dadurch werden einzelne Rauschpeaks unterdrückt und es wird berücksichtigt, daß es sich bei den bewegten Objekten um ausgedehnte Strukturen handelt.

Die endgültige Entscheidung darüber, ob ein Punkt dem Hintergrund oder dem Originalbild zuzuordnen ist, ergibt sich durch einen abschließenden Vergleich des Ähnlichkeitsmaßes S mit einer lokalen Schwelle. Sie bestimmt, ob der Unterschied zwischen beiden Bildern als signifikant betrachtet wird oder nicht. [Westberg, 1992] zeigt, daß ein Likelihood-Ratio-Test zum Bewerten der Differenz zweier Bilder unter bestimmten Voraussetzungen zu einem Schwellwertverfahren führt, bei dem ein Ähnlichkeitsmaß wie in (4) mit einer Schwelle verglichen wird. Im Gegensatz zu dieser *globalen Schwelle* paßt sich die hier verwendete *lokale Schwelle* den lokalen Strukturen in beiden Bildern an. Diese *strukturadaptive Bewertung* des Ähnlichkeitsmaßes entspricht dem menschlichen visuellen System, welches abhängig vom Bildinhalt manche Unterschiede sofort erkennt, bei anderen aber "genauer hinschauen" muß. Ein Ansatz, der dieser Intention folgt, wird im nächsten Abschnitt vorgestellt.

3.2 Lokale Schwelle durch Fuzzy-Logic

[Hsu, 1984] und [Nagel, 1982] verwenden einen Likelihood-Test zum Vergleich zweier aufeinanderfolgender Bilder auf Unterschiede bezüglich modellierter Grauwertverläufe. In [Bouthemy, 1989] wird das gleiche Testverfahren angewendet, um ein 3D-Volumen im $\mathbf{x}t$-Raum auf Anwesenheit einer Kante und somit Bewegung zu testen. In der vorliegenden Arbeit werden in ähnlicher Art und Weise statistische Eigenschaften der Grauwertverläufe korrespondierender lokaler Umgebungen im Bild und im Hintergrund verglichen. In erster Näherung kommen dafür der *mittlere Grauwert* und die *Varianz* der Grauwertstichproben in Betracht. Durch Verwenden höherer Momente kann die exakte Grauwertstatistik beliebig gut angenähert werden. Gute Ergebnisse wurden auch durch Vergleich der lokalen Orientierung zwischen Originalbild und Hintergrund erzielt. Im Gegensatz zu [Nagel, 1982] werden jedoch nicht direkt Grauwertverläufe modelliert und diese statistisch verglichen, sondern die Unterschiede der lokalen Grauwertstatistiken werden zur Bewertung des zuvor berechneten Ähnlichkeitsmaßes benutzt.

Die Berechnung der lokalen Schwelle erfolgt mit Hilfe der *Fuzzy-Logic*, die der menschlichen Entscheidungsfindung nachempfunden ist, [Zadeh, 1988]. Im vorliegenden Verfahren sind die Ausgangsgrößen die *Differenz des lokalen mittleren Grauwertes* (im folgenden kurz GD) und die *Differenz der lokalen Varianz* (im folgenden kurz VD) von Hintergrund und Originalbild. Als Ergebnis der Verknüpfung dieser Größen erhält man die lokale Schwelle. Da alle möglichen Werte von GD und VD von vornherein bekannt sind, läßt sie sich als *Look-Up-Tabelle* (LUT) abspeichern. Abb. 5a zeigt die lokale Schwelle in Abhängigkeit von GD und VD. In Abb. 3 ist das gesamte Verfahren der Bewegungssegmentierung im Überblick dargestellt.

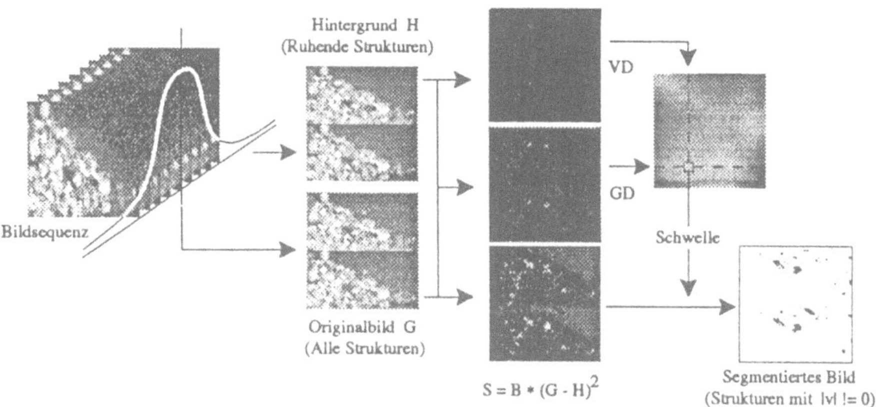

Abb. 3: *Ablaufdiagramm des Verfahrens der Bewegungssegmentierung.*

Insgesamt entspricht dies einer Verknüpfung der Struktureigenschaften der lokalen Umgebungen in beiden Bildern zu einem Binärbild. Darin bezeichnet 1 "Bewegung" und 0 "keine Bewegung". Für n verschiedene Merkmale μ_i läßt sich dies durch die Abbildung

$$f(\mu_1, \mu_2, \ldots, \mu_n) \rightarrow \{0, 1\} \tag{5}$$

des n-dimensionalen Merkmalraumes in den Binärraum der Bewegungssegmentierung beschreiben. Im unserem Fall ist $n = 3$ und $(\mu_1, \mu_2, \mu_3) = (GD, VD, S)$. Anstelle der direkten Verknüpfung von GD, VD und S nach (5) wird aus den ersten beiden Eigenschaften eine lokale Schwelle durch Fuzzy-Logic bestimmt, mit der die dritte Eigenschaft binarisiert wird. Dies hat den Vorteil, daß eine analytisch nur schwer zu bestimmende Funktion durch wenige logische Verknüpfungen aufgebaut wird und sich somit leicht optimieren läßt.

4 Ergebnisse

In Abb. 6 werden drei verschiedene Bilder mit ihren einzelnen Bearbeitungsschritten gezeigt. Man sieht, daß dabei auch Strukturen detektiert werden, deren Grauwerte fast im Rauschen untergehen, wenn es sich dabei um bewegte Teilchen handelt. Dies wird noch deutlicher, wenn man Grauwertschnitte durch die Bilder betrachtet. In Abb. 4 ist je ein Schnitt durch das Originalbild und das Hintergrundbild an der in Abb. 6 markierten Stelle dargestellt.

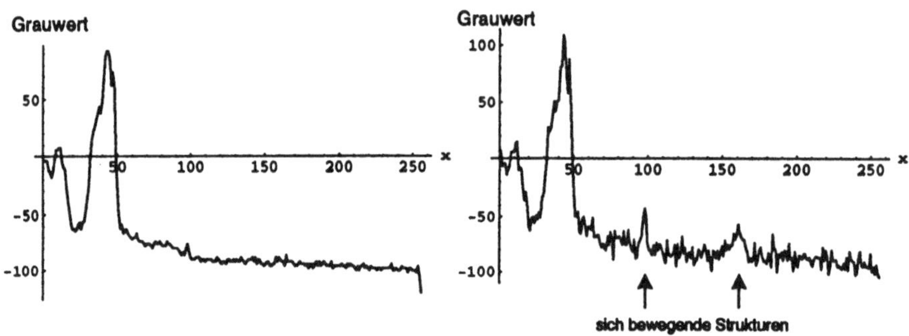

Abb. 4: *Grauwertschnitte durch das Bild in Abb. 6 a). Rechts ist der Schnitt durch das Originalbild und links der durch das zeitlich geglättete Hintergrundbild dargestellt.*

Man erkennt, daß sich die mit Pfeilen gekennzeichneten lokalen Grauwertmaxima nur knapp über den Rauschlevel erheben. Im Schnitt durch das Hintergrundbild sind diese trotz des erheblich geringeren Rauschens (das Rauschen wird bei der zeitlichen Tiefpaßfilterung unterdrückt) nicht zu erkennen. Dies deutet darauf hin, daß es sich bei den beiden Peaks um bewegte Strukturen handelt. Zusätzlich weist die lokale Schwelle an dieser Stelle ein deutliches Minimum auf (Abb. 5b). Dadurch wird die Detektion der bewegten Strukturen erleichtert. Im Ergebnisbild werden an diesen Stellen auch deutlich zwei Teilchen detektiert. Zu Testzwecken wurde der Algorithmus noch zusätzlich auf eine Verkehrsszene angewendet. Ohne eine Veränderung der Parameter der Fuzzy-Logic werden dort im rechten oberen Viertel deutlich zwei Fahrzeuge und im rechten unteren Viertel ein Fußgänger detektiert (Abb. 6c).

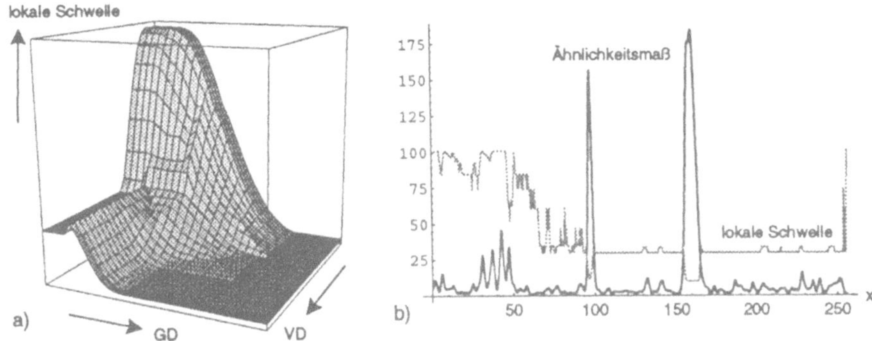

Abb. 5: *a) LUT der lokalen Schwelle. b) Ähnlichkeitsmaß und lokale Schwelle im Grauwertschnitt der Abb. 4.*

Literaturverzeichnis

Bouthemy, P.: *A Maximum Likelihood Framework for Determining Moving Edges.* IEEE Trans. Pattern Anal. Machine Intell., vol. 11, No. 5, pp. 499-511, May 1989

Haass, U.L.: *Verfolgung dynamischer Änderungen in TV-Bildszenen zur Deutung von Bewegungen.* FhG-Berichte 2-84, 12, 1984

Hsu, Y.Z., Nagel, H.-H., Rekers, G.: *New likelihood test methods for change detection in image sequences.* Comp. Vision, Graph. Image Proc. 26, 73, 1984

Jähne, B.: *Digital Image Processing - Concepts, Algorithms and Scientific Applications.* Springer 1991

Jähne, B.: *Spatio-Temporal Image Processing With Scientific Applications.* Habilitationsschrift, Technische Universität Hamburg-Harburg 1992

Köhler, H.-J.: *Materialverlagerung an Grenzschichten unter hydrodynamischer Belastung.* Projektantrag, Bundesanstalt für Wasserbau Karlsruhe 1992

Nagel, H.-H., Rekers, G.: *Moving object masks based on an improved likelihood test.* Proc. Int. Conf. Patt. Recogn., pp. 1140-1142, München 1982

Shio, A., Sklansky, J.: *Segmentation of People in Motion.* IEEE Visual Motion, Proc., pp. 325-332, 1991

Verri, A., Poggio, T.: *Motion Field and Optical Flow: Qualitative Properties.* IEEE Trans. Pattern Anal. Machine Intell., vol. 11, No. 5, pp. 545-558, May 1989

Westberg L.: *Hierarchical Contour-Based Segmentation of Dynamic Scenes.* IEEE Trans. Pattern Anal. Machine Intell., vol. 14, No. 9, pp. 946-952, September 1992

Zadeh, L.A.: *Fuzzy-Logic.* IEEE Computer April 1988

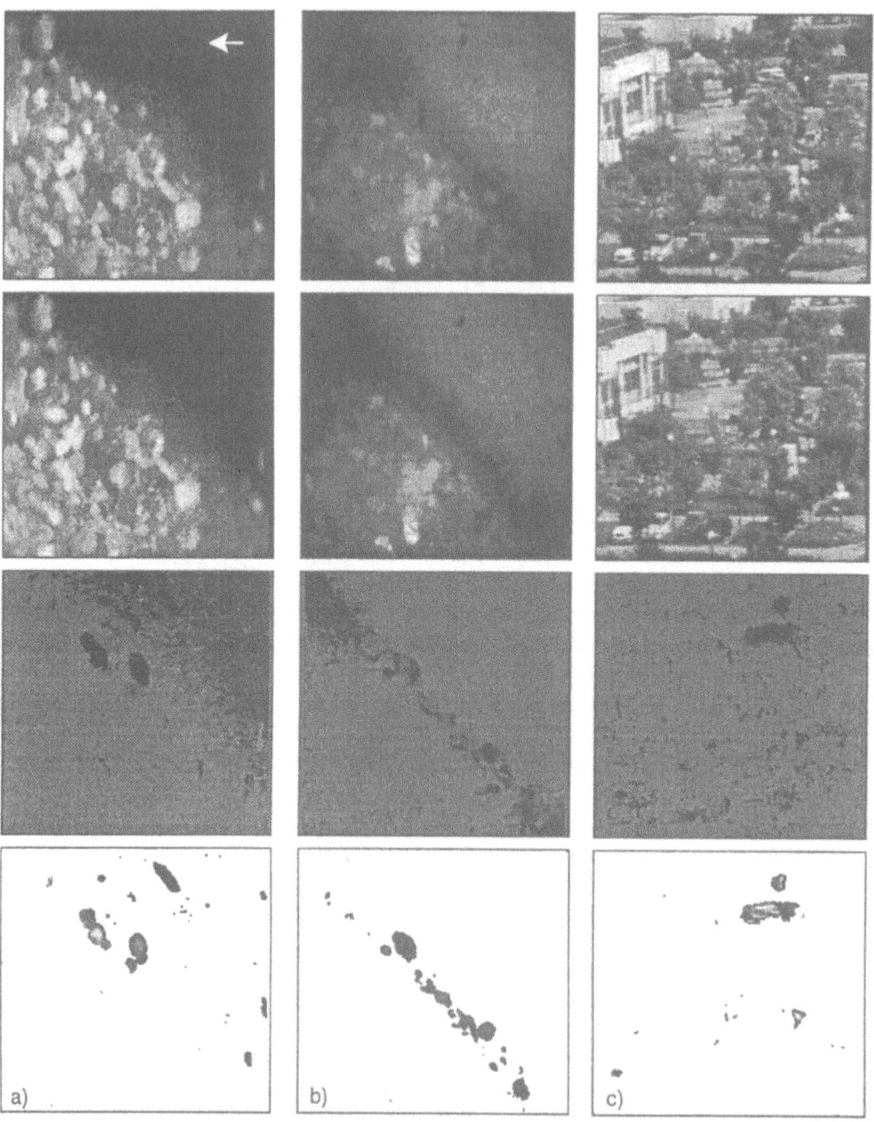

Abb. 6: *Beispiele zur Bewegungssegmentierung: Die drei Spalten zeigen verschiedene Beispielbilder. a) + b): Querschnitt einer Sandgrenzschicht. Der Bildausschnitt beträgt etwa 1 cm in horizontaler und vertikaler Richtung. c): Verkehrsszene. Von oben nach unten: Originalbild, Hintergrund, lokale Schwelle und Ergebnis der Segmentierung. Der Pfeil in Originalbild a) markiert den horizontalen Grauwertschnitt, der in Abb. 4 dargestellt ist.*

Rekursive Schätzung der relativen 3D-Bewegung einer Ebene aus längeren monokularen Bildfolgen*

Konstantinos Daniilidis
Institut für Informatik und Praktische Mathematik
Christian-Albrechts-Universität zu Kiel, Preusserstr. 1-9, 24105 Kiel
email: kd@informatik.uni-kiel.d400.de

Zusammenfassung

In diesem Beitrag wird die Fehlerempfindlichkeit der rekursiven Schätzer von 3D-Bewegung und Struktur aus längeren monokularen Bildfolgen untersucht. Ein neuer Algorithmus zur Schätzung der relativen Geschwindigkeit und der Normalen einer Ebene wird vorgeschlagen. Die Experimente aus synthetischen sowie auch Realweltbildfolgen zeigen, daß sich der Schätzfehler bei der Einführung eines Bewegungsmodells abschwächt, während die bekannte Empfindlichkeit an der Unterscheidung zwischen Translation und Rotation vorhanden bleibt.

1 Einführung

Zur Führung von autonom mobilen Fahrzeugen muß man aus bildgebenden Sensoren die Struktur der Umgebung und die relativen Bewegungen der Szenenkomponenten zur Kamera ermitteln. Die Durchführung dieser Aufgabe ist durch die Auswertung von monokularen Grauwertbildfolgen unter ganz allgemeinen Annahmen (wie Starrheit der Szenenkomponenten) im Prinzip möglich. Jedoch leidet eine Lösung dieses Problems unter Mehrdeutigkeiten. Existierende Algorithmen weisen eine hohe Empfindlichkeit gegen Meßrauschen auf, was deren Einsatz bei Realweltsituationen zur Zeit verbietet.

In [5, 4, 6] wurden analytische Nachweise für den Zusammenhang zwischen der Fehlerempfindlichkeit und der relativen Bewegung sowie der Geometrie der abgebildeten Szenenkomponenten bei der Berechnung der Bewegungsparameter aus einem Bildpaar gegeben. Die Benutzung eines Bildpaars dient als Fundament zum Verständnis der bei längeren Bildfolgen auftretenden Empfindlichkeiten und gibt direkt Leistungshinweise für die Ansätze, die eine längere Bildfolge als eine Summe von aufeinander folgenden Bildpaaren verarbeiten. In der Praxis einzusetzende Systeme müssen dreidimensionale Interpretationen während einer sich dauernd ändernden Umgebung anhand der ständig aufgenommenen Grauwertbilder ermitteln. Deshalb ist das eigentliche Ziel die Verwendung der gesamten Information, die sich während der Aufnahmezeit durch die andauernden relativen Bewegungen zwischen der Kamera und den Objekten der Umgebung erschließen läßt. Insbesondere ist der zwangsmäßig rekursive Charakter dieser Ermittlung von Bedeutung.

Die vorliegende Studie widmet sich der Ausnutzung der zeitlichen Kohärenz der Bewegung während einer längeren Bildfolge durch die Einführung von a priori Wissen über Bewegungsmodelle der Szenenkomponenten. Wir schlagen ein neues Verfahren zur rekursiven Berechnung der relativen Bewegung und der Neigung einer Ebene vor. Anhand dieses Verfahrens untersuchen wir, unter welchen Bewegungs- und Geometriekonfigurationen sich die Fehlerempfindlichkeit abschwächt oder unbeeinflußt bleibt. Die Untersuchung wird an synthetischen Experimenten sowie auch an einer Bildfolge durchgeführt, die von einer am Greifer eines Roboterarms befestigten Kamera aufgenommen wurde. Das vorgeschlagene rekursive Verfahren für Punktmerkmale, angewandt an mehreren Ebenen eines polyedrischen Objektes, in Kombination mit dem in [4]

*Diese Arbeit wurde während meiner Tätigkeit am Institut für Algorithmen und Kognitive Systeme an der Universität Karlsruhe (TH) unter finanzieller Unterstützung des Deutschen Akademischen Austauschdienstes (DAAD) durchgeführt. Mein Dank gilt meinem Betreuer Hans-Hellmut Nagel für die kritische Durchsicht der entsprechenden Teile meiner Dissertation. Ich bedanke mich bei Sami Atiya für unsere inspirationsvollen schätzungstheoretischen Diskussionen und bei Volker Gengenbach für die Hilfe bei der Durchführung der Experimente.

vorgestellten Bildpaarverfahren für geradlinige Merkmale kann die Basis für einen verallgemeinerten Algorithmus zur Schätzung der relativen Bewegung von polyedrischen Objekten bilden.

Der Leser sei für einen erschöpfenden Überblick über Verfahren zur Auswertung von längeren Grauwertbildfolgen auf [4] verwiesen. Die Ansätze unterscheiden sich nach der Art der verwendeten Bildbereichshinweise, die Annahmen über die Geometrie der Szene sowie die Bewegung und die verwendete Schätzmethode. Wir beschränken uns auf Ansätze zur Auswertung von *monokularen* Bildfolgen, die eine starre Bewegung unterstellen, und unterteilen die Algorithmen in vier Gruppen in Bezug auf die Annahmen und die verwendeten Sensoren. Die allgemeinste Problemstellung findet sich in der ersten Gruppe von Ansätzen, die eine beliebige Bewegung [8, 21, 3, 18, 10, 11, 1] oder eine Glattheit bzw. eine besondere Form der Bewegungstrajektorie [2, 15, 20, 19] unterstellen. Die zweite Gruppe besteht aus Ansätzen, die die Bewegungsinformation aus anderen Sensoren entnehmen (*Bewegungsstereo*), während die dritte Gruppe Modellinformation über die Szenenkomponenten einsetzt. Die vierte Gruppe beinhaltet Ansätze, die das Modell der orthographischen statt der perspektivischen Projektion verwenden. Aus Platzgründen wird leider auf die Literatur aus diesen Gruppen von Ansätzen, die mit dem hier vorgestellten Ansatz nicht direkt vergleichbar sind, verzichtet.

Bei der Auswertung einer längeren Bildfolge ist man gezwungen, die Schätzung der relativen Bewegung zwischen bildgebendem Sensor und einer Szenenkomponente *rekursiv* durchzuführen, d.h. zu jedem Zeitpunkt t_k den Zustandsschätzwert anhand nur der zu diesem Zeitpunkt aufgenommenen Messungen zu aktualisieren. Das Problem der Bewegungs- und Strukturschätzung leidet unter einer nichtlinearen Meßfunktion (perspektivische Abbildung) und gegebenfalls unter nichtlinearer Übertragungsfunktion. Dadurch ist man auf suboptimale Schätzer verwiesen: Die Mehrheit der obengenannten Ansätze verwenden den Erweiterten Kalman-Filter (EKF), dessen Leistung sehr stark von der Abweichung der Startwerte von den tatsächlichen Werten abhängt. Wenn die Abweichung groß ist, tritt eine Verzerrung auf, die sich nicht in der vom Filter berechneten Fehlerkovarianz erfassen läßt. Der EKF bildet nur den ersten Schritt des Iterierten Erweiterten Kalman-Filters (IEKF), der an einem bestimmten Zeitpunkt optimal im Sinne einer MAP-Schätzung ist, falls Meß- und Systemrauschen normalverteilt sind. Eine bessere nicht-iterative Approximation zum optimalen Filter ist der Modifizierte Gaußfilter zweiter Ordnung (MGSO)[12].

Das Ziel unserer Studie ist ein zweifaches: Erstens ist uns von Interesse, die Bewegungs- und Geometriekonstellationen zu ermitteln, bei denen die Einführung von a priori Wissen in Form eines Bewegungsmodells eine Abschwächung der Fehlerempfindlichkeit verursacht. Zweitens wollen wir die Eignung von jedem der obgenannten Schätzer untersuchen. Bei längeren Bildfolgen erlauben die Fülle der Bildbereichshinweise, die Anzahl der Unbekannten und die Einführung von a priori Wissen keine analytische Untersuchung wie beim Bildpaar, deshalb beschränken wir uns auf die funktionalen Zusammenhänge für den Verlauf des Schätzfehlers, die sich bei Realweltexperimenten und Simulationen beobachten lassen.

2 Relative Bewegung einer Ebene

Die Anzahl der Unbekannten, die der Struktur einer Szenenkomponente entprechen, ist proportional zur Anzahl der charakteristischen Szenenbereichshinweise (Vertizes, Kanten, markierte Punkte). Damit wir aber eine Einsicht in den Fehlerverlauf der Struktur bekommen und Rückschlüsse auf die Fehlerempfindlichkeitsergebnisse im Fall des Bildpaars [6] machen können, werden wir die grundlegende Annahme machen, daß wir nur die Menge der Szenenbereichshinweise untersuchen, die sich in einer Ebene befinden. So reduziert sich die Anzahl der Unbekannten auf zwei, den Polar- und Azimutwinkel der Normalen der Ebene. Der Abstand der Ebene vom Projektionszentrum wird wegen der Skalierungsmehrdeutigkeit im Betrag der Translation mitberücksichtigt.

Diese Annahme steht im Einklang mit aktuellen Anwendungen der Bewegungsschätzung in Realweltsituationen. Nehmen wir an, daß eine Kamera auf einem autonom geführten Fahrzeug befestigt ist, so ist die befahrbare Fläche eine Ebene. Der hier vorgestellte Algorithmus berechnet die momentane Translations- und Winkelgeschwindigkeit der Egobewegung in Bezug auf das Kamerakoordinatensystem, sowie auch die Normale der befahrbaren Ebene. Damit erspart man sich die Ermittlung der Transformation vom Straßenkoordinatensystem zum Kamerakoordinatensystem [7].

Eine zweite Anwendung ist die Auswertung von monokularen Bildfolgen aufgenommen von einer Kamera, die am Greifer eines Roboterarms befestigt ist. Eine weit verwendete Annahme ist, daß die zu manipulierenden Objekte polyedrisch sind. Der hier vorgestellte Algorithmus ermittelt die Egobewegung in Bezug auf das Kamerakoordinatensystem und die Normale der berücksichtigten Ebene. Ist die Transformation vom Kamerakoordinatensystem zum Greiferkoordinatensystem (Hand-Auge-Kalibrierung) bekannt, so läßt sich die zum beabsichtigten Zugriff auf das Objekt benötigte Orientierung bezüglich des Roboterkoordinatensystems ermitteln. Eine dritte Anwendung ist die Ermittlung der Bewegung von Objekten, die man als polyedrische Strukturen modellieren kann.

Das eingeführte a priori Wissen betrifft die Glattheit der Bewegung. Wir nehmen an, daß die Translations- sowie auch die Winkelgeschwindigkeit in Bezug auf das Kamerakoordinatensystem konstant bis auf eine Unsicherheit bleiben, die sich mittels der Kovarianz des Systemrauschens darstellen läßt. Das trifft im Fall eines autonom geführten Fahrzeugs sowie auch eines Roboterarms zu, wenn die Bewegung zwischen zwei Aufnahmen gering ist. Die Annahme ist für den Fall der Verwendung von Verschiebungsraten im Bildbereich geeignet, weil die Ermittlung von Verschiebungsraten kleine Bewegungen im Abtastintervall unterstellen. Unser Ansatz ähnelt der Methode von [17], wobei aber auf die Eignung des rekursiven Schätzers (EKF) nicht eingegangen wird. Weiterhin vereinfachen wir die Meßgleichungen durch Einführung von Hilfsparametern und modellieren korrekt die Kollisionszeit.

Wir gehen auf die Beschreibung der vorgestellten Methode im Zustandsraum ein. Wir bezeichnen mit $\boldsymbol{v}$, $\boldsymbol{\omega}$ und $\boldsymbol{N}$ jeweils die Translations-, die Winkelgeschwindigkeit und die Einheitsnormale der Ebene in Bezug auf das Kamerakoordinatensystem. Der Abstand des Kamerazentrums von der Ebene ist d, so daß die Gleichung der Ebene im Kamerakoordinatensystem lautet $\boldsymbol{N}^T\boldsymbol{X} = d$. (siehe Abb. 1). Unter der Annahme von konstanten Geschwindigkeiten erhalten wir folgende Differentialgleichungen für den zeitlichen Verlauf des Systems:

$$\begin{array}{llll} \dot{\boldsymbol{v}} & = 0 & \dot{\boldsymbol{\omega}} & = 0 \\ \dot{\boldsymbol{N}} & = -\boldsymbol{\omega} \times \boldsymbol{N} & \dot{d} & = -\boldsymbol{v}^T\boldsymbol{N} \end{array} \tag{1}$$

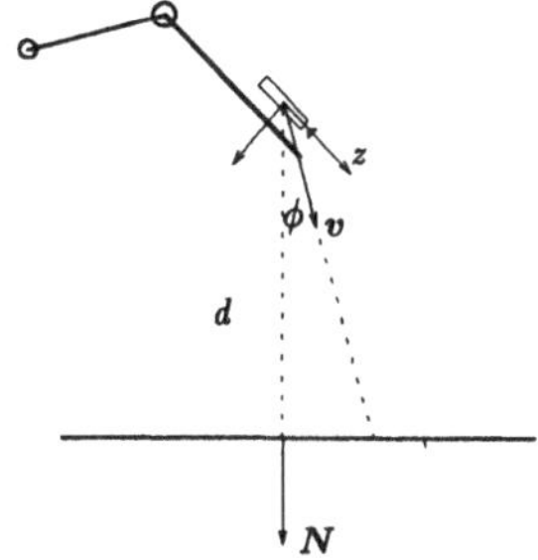

Abbildung 1: Bewegung einer an einem Greifer befestigten Kamera

Wir führen als unbekannten Vektor die durch den Abstand zur Ebene skalierte Translationsgeschwindigkeit $\boldsymbol{u} = \boldsymbol{v}/d$ ein. Der Betrag dieses Vektors wird von [17] als das Inverse der Kollisionszeit betrachtet. Wie man aber aus der Abb. 1 erkennen kann, stimmt das nicht, falls man die Kollisionszeit als die Zeit definiert, die bei konstanter Geschwindigkeit bis zum Treffen der Ebene verläuft. Nennt man den Winkel zwischen der Normalen und der Translationsgeschwindigkeit ϕ, läßt sich die Kollisionszeit als

$$\frac{1}{\rho_{ttc}} = \frac{d}{\|\boldsymbol{v}\|\cos\phi} = \frac{d}{\boldsymbol{v}^T\boldsymbol{N}} = \frac{1}{\boldsymbol{u}^T\boldsymbol{N}} \tag{2}$$

beschreiben. Wir werden das Inverse der Kollisionszeit $\rho_{ttc} = \boldsymbol{u}^T\boldsymbol{N}$ als Nebenprodukt aus unserem Algorithmus erhalten und es als Bewertungsmaß verwenden.

Wir definieren den Zustandsvektor des Systems als den neunelementigen Vektor

$$\boldsymbol{s} = \begin{pmatrix} \boldsymbol{u}^T & \boldsymbol{\omega}^T & \boldsymbol{N}^T \end{pmatrix}, \tag{3}$$

wobei $\boldsymbol{N}$ ein Einheitsvektor ist. Die entsprechende Zwangsbedingung $\|\boldsymbol{N}\| = 1$ wird in den Meßgleichungen berücksichtigt.

Zur Durchführung des Prädiktionsschrittes benötigen wir die Übergangsgleichung des Systems vom Zeitpunkt t_k auf den Zeitpunkt t_{k+1}, die wir aus den Differentialgleichungen (1) herleiten müssen. Die Übergangsgleichung für die Winkelgeschwindigkeit ist recht einfach:

$$\boldsymbol{\omega}_{k+1} = \boldsymbol{\omega}_k. \tag{4}$$

Die Integration der Differentialgleichung für die Normale ist bekannt [14]. Die Lösung ist die Rotation der Normalen um eine Achse parallel zu $\boldsymbol{\omega}$ um den Winkel $-\|\boldsymbol{\omega}\|T$, wobei $T = t_{k+1} - t_k$:

$$\boldsymbol{N}_{k+1} = \boldsymbol{N}_k - \frac{\sin(\|\boldsymbol{\omega}_k\|T)}{\|\boldsymbol{\omega}_k\|}\boldsymbol{\omega}_k \times \boldsymbol{N}_k + \frac{1-\cos(\|\boldsymbol{\omega}_k\|T)}{\|\boldsymbol{\omega}_k\|^2}\boldsymbol{\omega}_k \times (\boldsymbol{\omega}_k \times \boldsymbol{N}_k). \tag{5}$$

Die Ermittlung der Übergangsgleichung für die skalierte Geschwindigkeit $\boldsymbol{u}_{k+1} = \frac{\boldsymbol{v}_{k+1}}{d_{k+1}}$ ist aufwendiger. Für den Übergang der skalierten Geschwindigkeit läßt sich beweisen [4], daß

$$\boldsymbol{u}_{k+1} = \frac{\boldsymbol{u}_k}{1 - \boldsymbol{u}_k^T \boldsymbol{N}_k T + \frac{\boldsymbol{u}_k^T(\boldsymbol{\omega}_k \times \boldsymbol{N}_k)(1-\cos(\|\boldsymbol{\omega}_k\|T))}{\|\boldsymbol{\omega}_k\|^2} - \frac{(\boldsymbol{u}_k \times \boldsymbol{\omega}_k)^T(\boldsymbol{\omega}_k \times \boldsymbol{N}_k)(\|\boldsymbol{\omega}_k\|T - \sin(\|\boldsymbol{\omega}_k\|T))}{\|\boldsymbol{\omega}_k\|^3}}. \tag{6}$$

Zur Vervollständigung der Beschreibung des Systemübergangs benötigen wir die Jacobische Matrix der Übergangsfunktion, d.h. die Ableitung der Zustandsgrößen zum Zeitpunkt t_{k+1} nach den Zustandsgrößen zum Zeitpunkt t_k, um den Prädiktionsschritt bei einem rekursiven Schätzer durchzuführen. Die Berechnung dieser Jakobischen Matrix findet sich in [4]. Als Messung verwenden wir das Feld der Verschiebungsraten zum Zeitpunkt t_k. Aus der Annahme, daß es sich um die relative Bewegung einer Ebene handelt, lassen sich aus den Verschiebungsraten acht Hilfsparameter [16] herleiten.

$$\boldsymbol{q}_k = \begin{matrix} (-u_{kx}N_{kz} - \omega_{ky} & -u_{kx}N_{kx} + u_{kz}N_{kz} & -u_{kx}N_{ky} + \omega_{kz} & -u_{ky}N_{kz} + \omega_{kx} \\ -u_{ky}N_{kx} - \omega_{kz} & -u_{ky}N_{ky} + u_{kz}N_{kz} & -\omega_{ky} + u_{kz}N_{kx} & u_{kz}N_{ky} + \omega_{kx})^T \end{matrix} \tag{7}$$

Die acht Parameter können aus $m \geq 4$ Verschiebungsraten durch Lösung eines linearen Systems ermittelt werden, das aus folgender Minimierung folgt:

$$\sum_{i=1}^{m}(\dot{\vec{\boldsymbol{x}}}_{ki} - B_{ki}\boldsymbol{q}_k)^T C_{ki}^{-1}(\dot{\vec{\boldsymbol{x}}}_{ki} - B_{ki}\boldsymbol{q}_k) \Longrightarrow \min_{\boldsymbol{q}_k}. \tag{8}$$

Die Matrix C_{ki} ist die Kovarianz des Rauschens in den Verschiebungsraten, die abhängig vom Verfahren zur Ermittlung des optischen Flusses ist. Man kann durch die Einführung der Pseudoinversen statt der Inversen in (8) die Gewichtung nur der Projektion des optischen Flusses auf eine stabile Richtung erzwingen. Ist z.B. nur die Komponente entlang des Grauwertgradienten meßbar, so bekommen wir in (8) direkt die Minimierung der Quadrate dieser Komponenten. Die Ermittlung der Parameter $\boldsymbol{q}_k$ ist eine lineare Operation, daher bleibt der Fehler normalverteilt mit der Kovarianzmatrix $(\sum_{i=1}^{m} B_{ki}^T C_{ki}^{-1} B_{ki})^{-1}$. Zusätzlich zu den acht Parametern $\boldsymbol{q}_k$ wird als neunte Messung die Zwangsbedingung $\|\boldsymbol{N}_k\| = 1$ mit verschwindender Fehlerkovarianz berücksichtigt. Der Meßvektor zum Zeitpunkt t_k lautet dann

$$\boldsymbol{z}_k = \begin{pmatrix} \boldsymbol{q}_k \\ \|\boldsymbol{N}_k\|^2 - 1 \end{pmatrix}. \tag{9}$$

Wie man aus den Meßgleichungen erkennen kann, ist die Messung eine nichtlineare Funktion der Zustandsgrößen. Präziser gesagt sind die Messungen bilinear bezüglich der Translationsgeschwindigkeit und der Normalen und linear bezüglich der Winkelgeschwindigkeit. Zum Prädiktionsschritt der rekursiven Schätzung verwenden wir die Gleichungen (6), (4) und (5). Zum Aktualisierungsschritt werden wir drei Verfahren gegenüberstellen: den Erweiterten Kalman-Filter (EKF), den Modifizierten Gaußschen Filter zweiter Ordnung (MGSO) und den Iterierten Erweiterten Kalman-Filter (IEKF)[1]. Der rekursive Schätzer benötigt die Einstellung der Startwerte $\hat{\boldsymbol{s}}_0$ und der entsprechenden Kovarianz P_0^- sowie auch der Meß- und Prozeßrauschwerte.

[1] mit der Modifikation, daß die Minimierung nach dem Levenberg-Marquardt und nicht dem Gauß-Newton Schema abläuft (siehe auch [13]).

3 Experimentelle Untersuchung der Fehlerempfindlichkeit

Die hier zu auswertende monokulare Grauwertbildfolge besteht aus zwanzig Aufnahmen und wurde schrittweise von einer am Greifer eines Manipulators befestigten Kamera aufgenommen. Bei jedem Schritt hat der Manipulator eine Bewegung entprechend den sechs Stellgrößen unternommen, die von uns im voraus ausgerechnet wurden, so daß die erwünschte Bewegung (konstante Translations- und Winkelgeschwindigkeit) in Bezug auf das Kamerakoordinatensystem erreicht wird.

Die Stellgrößen in dem Experiment wurden so eingestellt, daß eine bezüglich des Kamerakoordinatensystems rein translatorische Bewegung mit $\boldsymbol{v} = (0, 0, 20)$ mm erfolgt. Ob der Roboter tatsächlich die erwünschte Bewegung durchgeführt hat, hängt von zwei Faktoren ab: Erstens von der Unsicherheit in der relativen Rotation zwischen dem Kamerakoordinatensystem und dem Greiferkoordinatensystem und zweitens von der Präzision, mit der der Roboter die Befehle durchführt. Wegen der reinen Translation bleibt die Normale der Ebene bezüglich des Kamerakoordinatensystems zeitlich konstant. Die ungefähren Werte $(-0.2, 0.2, 0.95)$ der Normalenkomponenten wurden mittels Kalibrierung nach [22] mit Hilfe nur einer Ebene an der ersten Aufnahme ermittelt. Die Bezeichnung „ungefähr" bezieht sich auf bekannt inhärente Ungenauigkeiten des verwendeten Verfahrens. Aus den Komponenten der Normalen erkennt man, daß es sich um eine – zumindest im Fall des Bildpaars – fehlerempfindliche Bewegungs- und Geometriekonfiguration handelt, weil der Winkel zwischen $\boldsymbol{N}$ und $\boldsymbol{v}$ niedrig ist.

Aus der aufgenommenen Bildfolge wurden Kantenelemente extrahiert, zu geradlinigen Kantensegmenten nach gruppiert und deren Schnittpunkte ermittelt. Diese Vorverarbeitungsschritte wurden automatisch nach der Aufnahme von dem Bildauswertesystem [9] durchgeführt. Das Grauwertbild und die detektierten Kantensegmente und Verschiebungsraten aus der Aufnahme zum Zeitpunkt t_{15} werden in der Abb. 2 gezeigt. Ziel unserer Untersuchung ist die Ermittlung der Fehlerverstärkung, die beim Auswertungsschritt der Bewegungs- und Strukturermittlung auftritt. Daher ist es wünschenswert, daß der Fehler in den Verschiebungsraten nur auf die Ungenauigkeit in der Ermittlung der Schnittpunkte und nicht auf falsche Zuordnungen zurückzuführen ist. Deshalb wurden die zeitlichen Zuordnungen der Schnittpunkte modellbasiert unter Verwendung der a priori Information über ihre Anordnung auf der Eichplatte ermittelt.

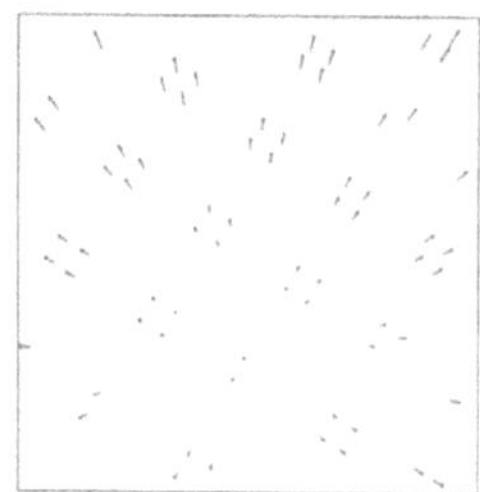

Abbildung 2: Aufnahmen zu Zeitpunkten t_0, t_5, t_{10} und t_{15} (links) und die extrahierten Geradensegmente (mitte) und Verschiebungsvektorfelder (rechts).

Der erste Eindruck aus Abb. 2 ist, daß es sich um eine rein translatorische Bewegung handelt, deren Expansionspunkt ungefähr die Bildpunktkoordinaten (260, 360) besitzt. Der Hauptpunkt hat die Koordinaten $(x_0, y_0) = (262, 267)$, daher besitzt die Translationsgeschwindigkeit eine vernachlässigbare x-Komponente. Unter Berücksichtigung des Skalierungsfaktors ($s_y = 1100$) kommt man auf einen Winkel zwischen dem $\boldsymbol{v}$-Vektor und der z-Achse von $\arctan(93/1100) \approx 5°$ Grad. Die Diskrepanz zu dem eingestellten Wert, der einer reinen Translation in z-Richtung entspricht, ist auf die Ungenauigkeit der Hand-Auge-Kalibrierung zurückzuführen. Diese Tatsache muß man in den folgenden Diagrammen berücksichtigen, wo als tatsächlicher Wert die Translationsgeschwindigkeit in z-Richtung aufgezeichnet ist.

Für den Aktualisierungsschritt werden alle drei Schätzer (EKF, MGSO und IEKF) ausprobiert. Zusätzlich wird eine Standard-Methode zur Schätzung der Bewegung einer Ebene aus einem Bildpaar [16] angewendet, die wir mit „NI" (nicht-inkrementell) bezeichnen, während die tatsächlichen Werte mit „GT" (ground-truth) bezeichnet werden.

Aus den Schätzwerten des Azimut- und Polarwinkels der Translationsgeschwindigkeit in

Abb. 3 erkennen wir eine bei allen Filtern auftretende Verzerrung, die eine Wanderung des Expansionspunktes nach links und unten verursacht. Es ist zu bemerken, daß eine Verzerrung im Polarwinkel von 2° Grad eine Verschiebung des Expansionspunkts um ca. $\tan(2°) \cdot 100 \approx 38$ Bildpunkte verursacht. Die Erklärung für diese Verzerrung findet sich in den Verläufen der Schätzwerte für die Komponenten der Winkelgeschwindigkeit. Die größte Verzerrung – der Leser sei hier auf die Skalierung der Ordinate in den Diagrammen der Abb. 3 hingewiesen – mit ca. 0.002 rad/T tritt bei ω_x auf, was aus der analytischen Fehleruntersuchung in [6] zu erwarten ist: nur die Summe $-v_y + \omega_x$ kann robust berechnet werden. Entsprechend, aber niedriger ist der Beitrag von ω_y zur Kompensierung der v_x-Komponente. Allerdings haben wir erhofft, daß sich die Anfälligkeit für eine Verwechselung zwischen Rotation und Translation entlang einer längeren Bildfolge abschwächen würde. Immerhin weist aber jeder rekursive Schätzer ein stabileres Verhalten im Vergleich zum nicht-inkrementellen Verfahren auf. Bemerkenswert ist, daß die Schätzung der Normalen von den obgenannten Effekten nicht beeinträchtigt wird. Der Fehler im geschätzten Polarwinkel liegt zwischen 3° und 5° Grad, und das ist von Bedeutung für die Praxis, weil dieser Winkel die Anfahrrichtung eines Sauggreifers auf eine zu greifende Ebene angibt.

Zur Simulation von mehreren Bewegungs- und Geometriekonfigurationen verwenden wir ein Modell der im Experiment mit realen Bilddaten benutzten Eichplatte und die gleichen Hand-Auge und Roboter-Welt Transformationen sowie auch dieselben internen Kameraparameter. Wenn nicht darauf hingewiesen wird, wird ein gleichverteiltes Meßrauschen von ± 0.5 Bildpunkten angenommen und kein Prozeßrauschen verwendet. Die letzte Annahme wurde gemacht, damit wir vollständig den Einsatz des a priori Wissens über die Glattheit der Bewegung ausnutzen. Die Startwerte sind immer Null für die Geschwindigkeiten, und der Startwert für die Normale liegt in der YZ-Ebene und hat einen Neigungwinkel von 45° Grad. Es hat sich herausgestellt, daß das Hochsetzen auf sehr große Werte der Startkovarianz, wie es bei anderen Ansätzen vorgeschlagen wird, auf Divergenz in allen drei verwendeten Filtern führt. Außerdem besitzt man a priori Wissen über den Umfang der Geschwindigkeiten: Die durch den Abstand skalierte Geschwindigkeit $\boldsymbol{u}$ und die Winkelgeschwindigkeit $\boldsymbol{\omega}$ können nicht beliebig groß werden, weil die Länge der Verschiebungsrate mitwächst. Angesichts der Tatsache, daß bei dem vorhandenen Skalierungsfaktor eine zur Bildebene parallele Winkelgeschwindigkeit von 0.01 [rad/T] eine Verschiebungsrate von 10 Bildpunkten verursacht – entsprechendes gilt für die Translation –, beschränken wir die Startkovarianz für die skalierte Translationsgeschwindigkeit $\boldsymbol{u}$ auf $10^{-4}\,[1/T]^2$ und für die Winkelgeschwindigkeit $\boldsymbol{\omega}$ auf $10^{-5}\,[\text{rad}/T]^2$.

Beim ersten Experiment variieren wir die Richtung der Translationsgeschwindigkeit. Wir stellen die Größe des effektiven Gesichtsfelds auf 6x6 Quadrate der Abb. 2. Die optische Achse bildet mit der Ebenennormalen einen Winkel von 30° Grad. Wir zeichnen dieselben Fehler wie oben für fünf verschiedene Bewegungen m_i mit entsprechenden Translationsgeschwindigkeiten $(0, i, 6)$mm (Abb. 4) auf. Der Vektor der Translationsgeschwindigkeit wandert in der YZ-Ebene von einer zur optischen Achse parallelen Lage zu einer zur Ebenennormalen parallelen Lage. Der Zusammenhang zwischen dem Fehler in Translationsrichtung und dem Fehler in ω_x ist eindeutig zu erkennen: Beide Fehler wachsen mit zunehmender Abweichung der Translationsrichtung von der optischen Achse. Weitere Experimente werden in [4] präsentiert.

4 Schlußüberlegungen

Der Verlauf der Schätzfehler bei der Realweltbildfolge zeigt, daß unter den drei rekursiven Schätzer der IEKF und der MGSO-Filter niedrigere Fehler aufweisen. Wir sollen hier darauf hinweisen, daß hier zur Erstellung eines Vergleichs ein Fall ausgewählt wurde, wobei der EKF nicht divergiert, was mehrmals der Fall ist. Aus den synthetischen Experimenten schließen wir, daß der IEKF ein stabiles Verhalten aufweist, das aber immer noch eine Verzerrung enthält, die von der Kopplung der Translation und der Rotation und von der relativen Lage der Ebenennormalen und der Translationsgeschwindigkeit abhängt. Wegen dieser Verzerrung, die dem Problem inhärent ist, müssen zusätzliche Bildinformationen eingeführt und /oder komplexere Bewegungen modelliert werden. Schon die Tatsache, daß die Dualität der Lösung bei konstanter Richtung der Translationsgeschwindigkeit erhalten bleibt, weist darauf hin, daß bei einem Bewegungsmodell mit variierender Translationsrichtung oder bei Einführung der Stellgrößen des Roboters in den Prädiktionsschritt die vom Winkel zwischen Normalen und Translation abhängige Fehlerempfindlichkeit abgeschwächt werden kann. Unsere Fehlerempfindlichkeitsuntersuchung kann

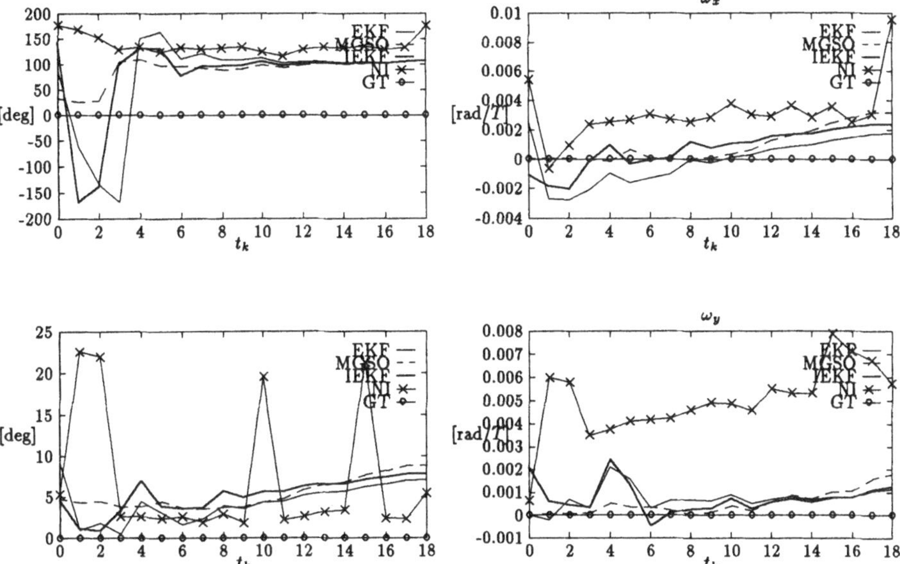

Abbildung 3: Schätzwerte für den Azimutwinkel (links oben), den Polarwinkel (links unten) der Translationsgeschwindigkeit, die x- und y-Komponente (rechts oben bzw. unten) der Winkelgeschwindigkeit.

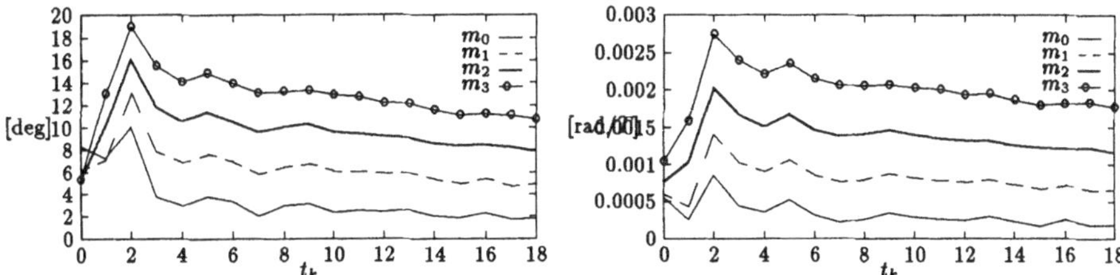

Abbildung 4: Fehlerwinkel in der vom IEKF geschätzten Translationsgeschwindigkeit (links) und absoluter Fehler in ω_x (rechts) bei variierender Translationsrichtung

bei dem Entwurf von aktiven Verfahren zur Berechnung von relativen Bewegungen verwendet werden. Es bleibt als offene Frage nachzuweisen, daß aktives Bewegungssehen (wie z.Bsp. durch Verfolgungsbewegungen des Auges) Instabilitäten und Mehrdeutigkeiten bei der Auswertung von monokularen Bildfolgen aufhebt.

Literatur

[1] H. Ando. Dynamic reconstruction of 3D structure and 3D motion. In *Proc. IEEE Workshop on Visual Motion*, pp. 101–110, Princeton, NJ, Oct. 7-9, 1991.

[2] T. Broida and R. Chellappa. Estimating the kinematics and structure of a rigid object from a sequence of monocular images. *IEEE Trans. Pattern Analysis and Machine Intelligence*, 13:497–513, 1991.

[3] N. Cui, J. Weng, and P. Cohen. Extended structure and motion analysis from monocular image sequences. In *Proc. Int. Conf. on Computer Vision*, pp. 222–229, Osaka, Japan, Dec. 4-7, 1990.

[4] K. Daniilidis. Zur Fehlerempfindlichkeit in der Ermittlung von Objektbeschreibungen und relativen Bewegungen aus monokularen Bildfolgen. Dissertation, Fakultät für Informatik, Universität Karlsruhe (TH), Juli 1992.

[5] K. Daniilidis and H.-H. Nagel. Analytical results on error sensitivity of motion estimation from two views. *Image and Vision Computing*, 8:297–303, 1990.

[6] K. Daniilidis and H.-H. Nagel. The coupling of rotation and translation in motion estimation of planar surfaces. In *IEEE Conf. on Computer Vision and Pattern Recognition 1993*, pp. 188–193, New York, NY, June 15-17, 1993.

[7] E.D. Dickmanns and V. Graefe. Applications of dynamic monocular machine vision. *Machine Vision and Applications*, 1:241–261, 1988.

[8] L. Dreschler and H.-H. Nagel. Volumetric Model and 3D Trajectory of a Moving Car Derived from Monocular TV Frame Sequences of a Street Scene. *Computer Graphics and Image Processing*, 20:199–228, 1982.

[9] V. Gengenbach. Automatischer Zugriff eines Roboters auf ungeordnete Werkstücke mit Hilfe einer 3D-Lagebestimmung durch ein Mehrkamerasystem. Diplomarbeit, Fakultät für Informatik der Universität Karlsruhe, Januar 1990.

[10] C.G. Harris and J.M. Pike. 3D positional integration from image sequences. *Image and Vision Computing*, 6:87–90, 1988.

[11] J. Heel. Dynamic motion vision. *Robotics and Autonomous Systems*, 6:297–314, 1990.

[12] A.H. Jazwinski. *Stochastic Processes and Filtering Theory.* Academic Press, New York, NY and London, UK, 1970.

[13] D. Koller, K. Daniilidis, T. Thorhallsson, and H.-H. Nagel. Model-based object tracking in traffic scenes. In *Proc. Second European Conference on Computer Vision*, pp. 437–452, Santa Margerita, Italy, May 23-26, G. Sandini (Ed.), Lecture Notes in Computer Science 588, Springer-Verlag, Berlin et al., 1992.

[14] G.A. Korn and T.M. Korn. *Mathematical Handbook for Scientists and Engineers.* McGraw-Hill, New York, 1968.

[15] R.V. Raja Kumar, A. Tirumalai, and R. C. Jain. A non-linear optimization algorithm for the estimation of structure and motion parameters. In *IEEE Conf. Computer Vision and Pattern Recognition*, pp. 136–143, San Diego, CA, June 4-8, 1989.

[16] H.C. Longuet-Higgins and K. Prazdny. The interpretation of a moving retinal image. *Proc. Royal Society of London*, B208:385–397, 1980.

[17] D.W. Murray and D.M. Pickup. Recursive updating of planar motion. In *Proc. British Machine Vision Conference*, pp. 169–177, Glasgow, UK, Sept. 24-26, 1991.

[18] J. Oliensis and J.I. Thomas. Incorporating motion error in multi-frame structure from motion. In *Proc. IEEE Workshop on Visual Motion*, pp. 8–13, Princeton, NJ, Oct. 7-9, 1991.

[19] H.S. Sawhney, J. Oliensis, and A.R. Hanson. Description and reconstruction from image trajectories of rotational motion. In *Proc. Int. Conf. on Computer Vision*, pp. 494–498, Osaka, Japan, Dec. 4-7, 1990.

[20] H. Shariat and K.E. Price. Motion estimation with more then two frames. *IEEE Trans. Pattern Analysis and Machine Intelligence*, PAMI-12:417–434, 1990.

[21] M. Spetsakis and J. Aloimonos. A multi-frame approach to visual motion perception. *International Journal of Computer Vision*, 6:245–255, 1991.

[22] R. Tsai. A versatile camera calibration technique for high accuracy 3D machine vision metrology using off-the-shelf TV cameras and lenses. *IEEE Trans. Robotics and Automation*, 3:323–344, 1987.

Ein schnelles Verfahren zur Lösung des Stereokorrespondenzproblems bei der 3D-Particle Tracking Velocimetry

T. Netzsch
BASF AG, Abt.ZI/CI
D-67056 Ludwigshafen
Tel: 0621- 6072592 Fax: 0621-60 71417

B.Jähne
Scripps Institution of Oceanography, La Jolla, CA 92093-0230, USA und
Interdisziplinäres Zentrum für Wissenschaftliches Rechnen der Universität Heidelberg
D-69120 Heidelberg

Zusammenfassung

Dreidimensionale Particle Tracking Velocimetry (3D-PTV) ist eine Meßtechnik zur Bestimmung von Geschwindigkeitsfeldern in Strömungen. Nach Einbringen eines Tracers in das strömende Medium und geeigneter Beleuchtung, werden mit Stereokameras Bildsequenzen aufgezeichnet. Um dreidimensionale Informationen zu erhalten muß das Stereokorrespondenzproblem gelöst werden. Während in der Literatur beschriebene Methoden [Ad-88, Ma-92] versuchen die Korrespondenz für Einzelbilder zu lösen, wird hier ein schneller Algorithmus vorgestellt, der auf der Bearbeitung von Bildsequenzen beruht, so daß die Stereokorrespondenz für *Spuren* nicht für einzelne Partikel zu lösen ist. Durch die Einführung von geeigneten Korrespondenzmerkmalen können Mehrdeutigkeiten weitgehend aufgelöst und die Anzahl durchzuführender Operationen, die bei n Spuren i.a. n-Fakultät beträgt, auf ~ n reduziert werden.

1 Einleitung

In vielen Bereichen, sowohl in der industriellen Meßtechnik, als auch in der Grundlagenforschung, werden Verfahren zur dreidimensionalen Strömungsmessung eingesetzt. Zur Untersuchung von Stofftransport und Mischungsvorgängen, zur Erstellung von Strömungsprofilen, aber auch zur Beschreibung des Gasaustauschs zwischen Ozeanen und Weltmeeren [Jä-85, Wk-90] werden räumlich ausgedehnte Meßdaten benötigt. Bildverarbeitungsmethoden können diese Informationen liefern. Während die Particle Imaging Velocimetry (PIV) eine zweidimensionale Bestimmung von momentanen Geschwindigkeitsfeldern ermöglicht, kann man mit der 3D-PTV durch stereoskopische Aufnahmen alle drei Geschwindigkeitskomponenten in einem Beobachtungsvolumen erfassen. Zudem ist die PIV nicht in der Lage Trajektorien zu beobachten, was mit der 3D-PTV bei deutlich kürzeren Rechenzeiten über längere Sequenzen sehr gut realisierbar ist.

2 Prinzip der 3D - Particle Tracking Velocimetry (PTV)

Bei der 3D-PTV wird die Strömung durch Tracer visualisiert und anschließend werden Bildsequenzen mit zwei Kameras aus verschiedenen Blickrichtungen aufgenommen. Dadurch ist es im Gegensatz zur PIV möglich, Trajektorien über einen längeren Zeitraum zu verfolgen. Um die Tiefeninformation zu erhalten, muß das Stereokorrespondenzproblem gelöst werden. Gängige Methoden [Ad-88, Ma-92] bearbeiten Einzelbilder um das Stereokorrespondenzproblem zu lösen. Dabei sind Mehrdeutigkeiten, insbesondere bei hohen Tracerdichten jedoch nicht auszuschließen.

Deshalb wird hier die Stereokorrespondenz durch Vergleich von Bildsequenzen bei kontinuierlicher Beleuchtung gelöst, wobei die Korrespondenz nicht für einzelne Teilchen im Einzelbild, sondern für *Spuren* in einer *Bildsequenz* hergestellt wird. Zusätzlich werden Korrespondenzmerkmale eingeführt, die die Anzahl durchzuführender Spurvergleiche deutlich reduzieren. Um die Qualität der Algorithmen zu überprüfen, werden zusätzlich Tests mit simulierten Spuren durchgeführt. Folgende Schritte sind also erforderlich:

- Visualisierung/Bildaufnahme
- Kalibrierung der Stereokamera
- 2D - Particle Tracking
- Lösung des Stereokorrespondenzproblems für *Spuren*
- Test der Algorithmen mit simulierten Spuren
- Bestimmung der Strömungsparameter

Schwerpunkt dieser Arbeit ist der Stereokorrespondenzalgorithmus. Auf die Erläuterung der Kalibrierung wird hier verzichtet. Details können [Ne-92, Po-90, Ts-86, Le-87] entnommen werden.

3 Visualisierung und Bildaufnahme

Testbildaufnahmen wurden am ringförmigen Wind-/Wasserkanal des Instituts für Umweltphysik der Universität Heidelberg durchgeführt.

Als Tracer [Wk-90] wurden Polystyrolpartikel mit einer Größe von 50 µm bis 150 µm verwendet. Die Bildaufnahme eines durch einen breiten Lichtschnitt erzeugten Beobachtungsvolumens erfolgte durch zwei extern synchronisierte CCD-Videokameras nach NTSC-Norm (30-Hz, interlaced). Die Signale beider Kameras (Kamera 1: Rot-Signal, Kamera 2: Grün-Signal) wurden überlagert und in Echtzeit analog auf einen Sony-Laserdisc-Viderecorder (YUV-Signal) aufgezeichnet, der trotz analoger Aufzeichnung Zugriff auf definierte Einzelbilder erlaubt.

Die Trennung von Rot- und Grün-Signal erfolgte bei der Digitalisierung. Die digitalisierten Bilder wurden auf WORM-Disk abgespeichert.

Es wurden Messungen mit vier Kamerapositionen durchgeführt:

- horizontale und vertikale Kameraanordnung
- jeweils parallele optische Achsen und geneigte optische Achsen

4 2D-Particle Tracking

Ziel des Particle Tracking ist es, nach erfolgreicher Segmentierung im Einzelbild, die korrespondierenden Objekte im Folgebild zuzuordnen und so aus den Objekten Spuren zu erzeugen. Zu diesem Zweck wurden zwei Algorithmen [Ne-92, Wk-92] entwikkelt. Während Algorithmus 1 [Ne-92] in Bezug auf die Rechenzeit optimiert ist und mit globalen Schwellwerten arbeitet, liefert Algorithmus 2 [Wk-92] bei hohen Tracerdichten und stark verrauschten Bildern durch die Verwendung eines regionalen Wachstumsverfahrens bei der Segmentierung gute Ergebnisse.

Bild 4.1 zeigt zwei mit Algorithmus 1 bearbeitete synchrone Sequenzen von 140 Halbbildern. Dargestellt sind die Bahnkurven von Partikeln, die länger als 10 Halbbilder verfolgt werden konnten. Wie in allen folgenden Bildern, sind in der oberen Hälfte die Spuren der linken Kamera, in der unteren Hälfte die der rechten Kamera zu sehen.

Bild 4.1 Bildsequenzen nach 2D-Particle Tracking

Das 2D-PT liefert als Ergebnis eine zeitlich sortierte, verkettete Liste von Spuren. Die zeitliche Sortierung entsteht automatisch durch die sequentielle Abarbeitung der Einzelbilder. Jede Spur enthält die Information über den Grauwertschwerpunkt jedes Spurpunktes in Rechnerkoordinaten, sowie als Zeitinformation die Bildnummer des Spuranfangs, so daß jedem Spurpunkt eine genau definierte Zeit innerhalb der Sequenz zugeordnet werden kann.

5 Stereokorrespondenz

5.1 Problemstellung

Das Stereokorrespondenzproblem besteht bei dem hier geschilderten Verfahren darin, die zu einer Spur im rechten (linken) Kamerakoordinatensystem korrespondierende Spur im linken (rechten) Kamerakoordinatensystem zu finden. Unter Berücksichtigung der Randbedingungen gilt es geeignete Kriterien zu finden, um die Spuren schnell und möglichst fehlerfrei zuordnen zu können.

5.2 Randbedingungen

- Eindeutigkeit (5.1)
- Epipolargeometrie (5.2)
- Korrespondenz muß für ganze Spur erfüllt sein (5.3)
- keine Überdeckungen über einen längeren Zeitraum (5.4)

Da jedes Partikel tatsächlich nur einmal existiert, muß die Korrespondenz eindeutig lösbar sein. Mehrdeutigkeiten sollten aufgelöst werden, oder deuten auf Fehler im Particle-Tracking bzw. zu große Teilchendichte hin.

Zu jedem einzelnen Spurpunkt liegt der korrespondierende Punkt auf der epipolaren Linie. Der Suchbereich auf der epipolaren Linie kann unter Berücksichtigung der Geometrie des Versuchaufbaus bzw. des beobachtbaren Tiefenbereichs eingeschränkt werden. Die Lage der epipolaren Linie wird durch Auswertung eines 3D-Testgitters bestimmt.

Da jede Spur von einem einzigen Partikel erzeugt wird, muß die Korrespondenz für *alle* Spurpunkte einer Spur erfüllt sein. Überdeckungen über einen längeren Zeitraum können ausgeschlossen werden.

5.3 Kriterien

Einzelne Partikel unterscheiden sich bzgl. Größe, Form und Helligkeit nur wenig oder gar nicht. Daher müssen für die Zuordnung der Spuren geometrische Kriterien gewählt werden, die sich aus der Epipolargeometrie ergeben. Berücksichtigt man die Mehrmediengeometrie [Ma-92], so muß die epipolare Linie als gebogene Linie angesehen werden, welche durch ein Polygon approximiert wird. Zu diesem Polygon wird eine Toleranz, bestimmt durch die Genauigkeit der Bildkoordinaten, addiert, so daß man ein zweidimensionales Suchfenster erhält, in dem der korrespondierende Spurpunkt liegen muß. Zu beachten ist, daß diese Bedingung für jeden einzelnen Spurpunkt korrespondierender Spuren erfüllt sein muß.

Nach Anwendung dieses Kriteriums liegt für jede Spur eine Liste mit Korrespondenzkandidaten vor, die keinen, einen oder mehrere Kandidaten enthält. Folgende Fälle können dabei auftreten:

- 1:1 Sowohl die linke als auch die rechte Spur haben jeweils einen Kandidaten, die Spuren korrespondieren

- 1:m (m:1) Eine linke (rechte) Spur hat mehrere rechte (linke) Kandidaten. Diese wiederum haben jeweils nur genau diese linke (rechte) Spur als Kandidat.
  Bsp.: die untere Spur wurde beim Particle Tracking nicht zusammenhängend erkannt

- m:m Dielinke Spur hat mehrere rechte Kandidaten und diese haben selbst mehrere linke Spuren als Kandidaten

Für den Fall (1:1) ist die Korrespondenz eindeutig gelöst. Für den Fall 1:m (m:1) ist die Korrespondenz dann eindeutig gelöst, wenn die m Kandidaten nicht zeitlich überlappen und daher tatsächlich ein Partikel repräsentieren. Für die anderen Fälle sind weitere Kriterien notwendig um Fehlkorrespondenzen zu erkennen und so Mehrdeutigkeiten auszuschließen.

Unter Berücksichtigung der Randbedingung 5.3 ergibt sich, daß Kandidaten ausgeschlossen werden können, für die die Korrespondenz nur über einen Teil der Spur bzw. für mehrere kurze Teilbereiche erfüllt ist.

Die Randbedingung 5.1 (Eindeutigkeit) erlaubt die Einführung eines weiteren Kriteriums: Sind mehrere Kombinationsmöglichkeiten vorhanden, so müssen diese so gelöst werden, daß max. viele Korrespondenzen eindeutig gelöst werden. Beispiel (Bild 5.1):

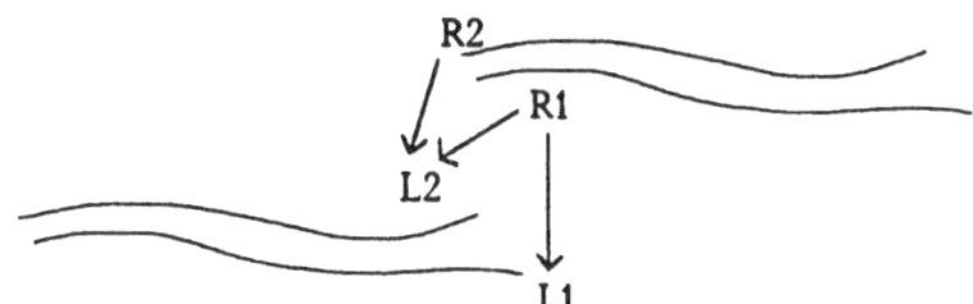

Bild 6.1 Eindeutigkeitskriterium

Spur R1 hat als Kandidat Spur L1 und L2, Spur R2 hat als Kandidat Spur L2. Es gibt zwei mögliche Lösungen:

- R1 korrespondiert mit L1, dann korrespondiert R2 mit L2
- R1 korrespondiert mit L2, dann gibt es für R2 keine Korrespondenz

Andere Möglichkeiten sind wegen der Eindeutigkeit ausgeschlossen. Da die erste Lösung die Korrespondenz für beide Spurpaare löst, ist sie die mit der größeren Wahrscheinlichkeit, unter der Voraussetzung daß beim Particle Tracking keine groben Fehler enstehen.

Falls jetzt noch Mehrdeutigkeiten vorhanden sind, kann als weiteres Kriterium Betrag und Richtung der Steigung der Spuren verwendet werden.

5.4 Algorithmus zur schnellen Lösung des Stereokorrespondenzproblems

Der Algorithmus besteht im wesentlichen aus zwei Schritten. Im ersten Schritt (Bild 5.2) werden für alle Spuren die Spurkandidaten, die innerhalb des epipolaren Suchfensters liegen in die Kandidatenlisten eingetragen. In einem zweiten Schritt (Bild 5.3) werden dann die Kandidatenlisten bearbeitet und je nach Fall (1:1, 1:m, m:m) die notwendigen Kriterien angewendet.

Die Geschwindigkeit des Algorithmus hängt von der Anzahl Spurvergleiche und der Anzahl Operationen pro Spurvergleich ab.

Um die Anzahl der Spurvergleiche möglichst gering zu halten wird als Zusatzkriterium, ein Suchfenster mit der doppelten Größe des epipolaren Suchfensters in jeder Richtung verwendet. Liegt der erste Spurpunkt der zu vergleichenden Spur außerhalb dieses Bereichs, so werden die Spuren nicht miteinander verglichen. Ein größerer Bereich muß benutzt werden, da Fehler im Particle Tracking nicht auszuschließen sind.

Um die Anzahl Operationen pro Spurvergleich zu minimieren, werden nur die Kriterien angewendet, die zur Herstellung der Eindeutigkeit notwendig sind. Im Fall 1:1, sowie im eindeutig lösbaren Fall 1:m wird daher nur das epipolare Suchfenster verwendet. In allen anderen Fällen werden der Reihe nach die anderen Kriterien eingesetzt, wobei nach Anwendung jedes Kriteriums überprüft wird, ob die Mehrdeutigkeiten beseitigt sind.

Für die Geschwindigkeit des Algorithmus sind außerdem die verwendeten Datenstrukturen entscheidend. Da die Spuren sequentiell abgearbeitet werden müssen, wurden deshalb sowohl die Spuren als auch die Kandidatenlisten in verketten Listen realisiert, die durch sequentielle Bearbeitung der Bilder automatisch zeitlich sortiert sind. Um die Zugriffszeiten auf die einzelnen Spuren und Spurpunkte zu minimieren wurden bewußt Redundanzen in Kauf genommen und wo immer möglich Pointer (teilweise mehfach) verwendet. Der Algorithmus wurde in der Sprache C realisiert.

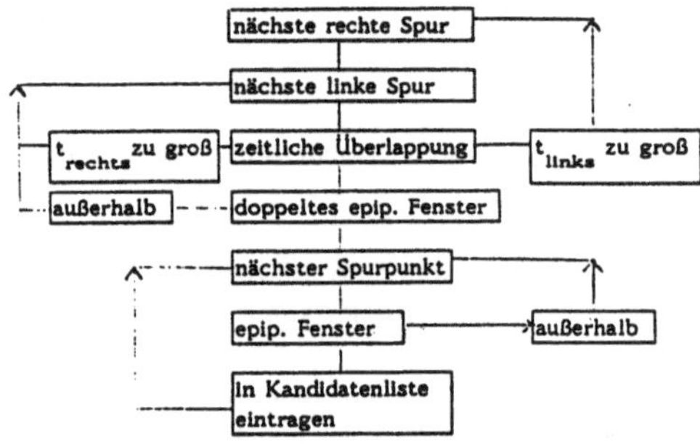

Bild 5.2 Algorithmus zum Spurvergleich

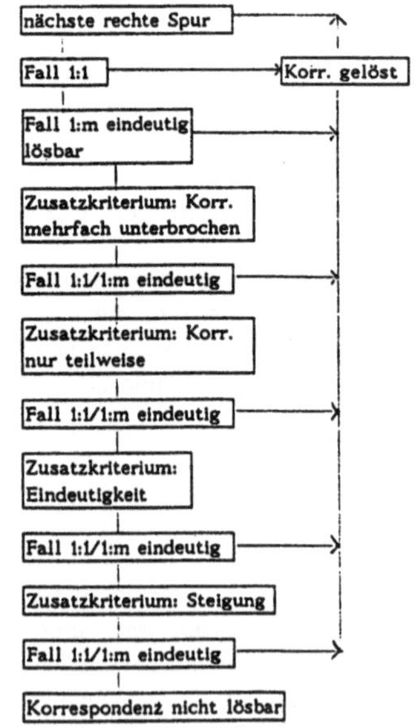

Bild 5.3 Algorithmus zur Bearbeitung der Kandidatenliste

5.5 Bilder aus realen Sequenzen

In der oberen Hälfte der folgenden Bilder sind jeweils die Spuren der linken Kamera gezeigt, in der unteren Hälfte die der rechten Kamera.

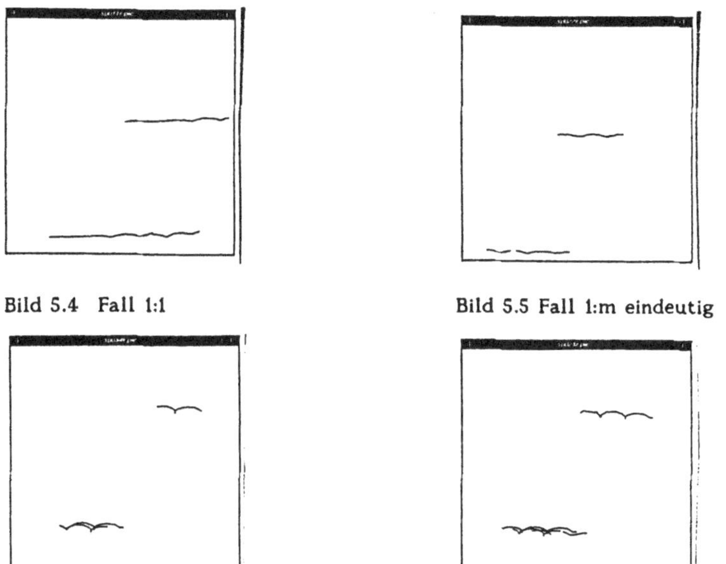

Bild 5.4 Fall 1:1

Bild 5.5 Fall 1:m eindeutig

Bild 5.6 Fall 1:m mehrdeutig

Bild 5.7 Fall m:m (Ausschnitt)

Bild 5.4 zeigt den einfach lösbaren Fall 1:1. In Bild 5.5 ist der eindeutig lösbare Fall 1:m zu sehen: Die beiden unteren Spuren wurden von einem Partikel erzeugt und korrespondieren beide mir der oberen Spur. Die Unterbrechung ist durch eine teilweise Überdeckung durch andere Partikel oder Verunreinigungen zu erklären. Die Bilder 5.6 und 5.7 zeigen mehrdeutige Fälle, die jeweils durch Anwendung von Zusatzkriterien gelöst werden konnten.

5.6 Ergebnisse

Mit obigem Algorithmus wurden mehrere Bildsequenzen ausgewertet. Es konnte durchschnittlich in 95% aller möglichen Fälle die Stereokorrespondenz gelöst werden. Für ein Texterkennungssystem wäre dies sicherlich ein sehr schlechter Wert, der einen Einsatz dieses Systems verhindern würde. Die 5 % nicht verwendbare Spuren sind hier jedoch kein Problem, da durch die Partikel selbst schon eine statistische Auswahl der Orte getroffen wird, an denen die Strömung gemessen werden kann, so daß auch eine Erfolgsquote von nur 80% ein vernünftiges Ergebnis wäre.

Die Bearbeitungszeit (ohne PT) für eine Sequenz mit 49 linken und 97 rechten Spuren (140 Halbbilder) beträgt auf einem OS-9 System mit CPU 68020, 20 MHz ca. 2 sec. Besonders bemerkenswert ist die geringe Anzahl von 1.6 Spurvergleichen pro linker Spur. Die relativ große Differenz der Anzahl rechter bzw. linker Spuren ist auf nicht identische Einstellung von Offset und Gain der beiden Kameras, sowie den kleineren Beobachtungsbereich der linken Kamera bei der Bildaufnahme zurückzuführen. Dies zeigt, daß der Stereokorrespondenzalgorithmus trotz unterschiedlicher Startbedingungen hervorragend funktioniert.

Im folgenden die Ergebnisse dieser Sequenz:

	links	rechts
Anzahl Spuren	49	97
gelöste Korrespondenzen	41	49
nicht gelöste Korrespondenzen	1	2
Spuren ohne Kandidat	7	46
davon x-Koor. zu groß	0	39
Anz. Spurvergleiche	78	-
Anz. Spurvergleiche pro Spur	1.6	-
eindeutig lösabar (1:1, 1:m eindeutig)	23	23
mehrdeutig lösabar (m:m, 1:m mehrdeutig)	18	26

6 Ausblick

Die Algorithmen wurden direkt mit realen Bildsequenzen entwickelt. Zum Test der Algorithmen wird deshalb zur Zeit ein System zur Erstellung von simulierten Bildsequenzen erarbeitet. Von besonderem Interesse ist hier die Frage, welcher Algorithmus bei steigender Teilchendichte zuerst versagt: das 2D-Particle- Tracking oder der Stereokorrespondenzalgorithmus.

Literatur

[Ad-88] A.A. Adamczyk, L. Rimai; Reconstruction of a 3-dimensional flow-field from orthogonal views of seed track video images, Experiments in Fluids, 6, S. 380 - 386

[Ho-86] B.K.P. Horn; Robot Vision, MIT Press, Cambridge 1986

[Go-85] H. Goldstein; klassische Mechanik, 8.Aufl., Wiesbaden: Aula 1985

[Jä-85] B. Jähne; Transfer processes across the air-water interface, Habilitationsschrift, Unviversität Heidelberg, 1985

[Jä-91] B. Jähne; Digitale Bildverarbeitung, Heidelberg: Springer 1991, 2.Auflage

[Le-87] R. Lenz; Linsenfehlerkorrigierte Eichung von Halbleiterkameras mit Standardobjektiven für hochgenaue 3D-Messungen in Echtzeit. Proc. 9. DAGM-Symp. Mustererkennung 1987. Informatik-Fachberichte 149, S. 212-216. Berlin: Springer 1987

[Ma-92] H.G. Maas; Digitale Photogrammetrie in der dreidimensionalen Strömungsmeßtechnik, Dissertation, ETH Zürich 1992

[Ne-92] T. Netzsch; Dreidimensionale Messung turbulenter Strömung mit Bildverarbeitung, Proc. 14. DAGM Symp. Mustererkennung 1992, Informatik aktuell, S. 150-157, Berlin: Springer 1992

[Po-90]. S. Posch; Automatische Tiefenbestimmung aus Grauwertstereobildern, Wiesbaden: Dt. Univ. Verlag 1990

[Ts-86] R. Tsai; An efficient and accurate camera calibration technique for 3D machine vision, Proc. Computer Vision and Pattern Recognition, S. 364-374, IEEE, Miami Beach 1986

[Wk-90] D. Wierzimok; Messung turbulenter Strömungen unterhalb der windwellenbewegten Wasseroberfläche mittels Bildfolgenanalyse, Dissertation, Universität Heidelberg 1990

[Wk-92] D. Wierzimok, Tracking in Strömungsbildfolgen, Proc. 14. DAGM Symp. Mustererkennung 1992, Informatik aktuell, S. 158-165, Berlin: Springer 1992

Modellunabhängige Schätzung von 3–D Attributen Während der Bildfolgensegmentierung

Włodzimierz Kasprzak

Bayerisches Forschungszentrum für Wissensbasierte Systeme (FORWISS)
Am Weichselgarten 7, D-91058 Erlangen

Zusammenfassung. Eine Methode zur Schätzung von relativen 3-D Konturenattributen wird vorgestellt, die auf dem Prinzip des länglichen Bewegungsstereo beruht. Es werden geschlossene Konturen im Bild detektiert und deren zeitliche Veränderungen quantitativ erfasst. Zwei Arten der Stabilisierung dieser Werte werden genutzt - durch gewichtete Mittelung in einer kurzen Bildfolge und durch adaptive Verfolgung in längerer Bildfolge. Unter Annahme einer projektiven Abbildung werden Tiefe und Orientierung im Raum den bewegten Konturen zugeordnet.

1 Einleitung

Im folgenden Beitrag werden Bildfolgen untersucht, die mit einer sich bewegenden Kamera aufgenommen wurden. Die Verschiebung der Bildelemente ist hier überwiegend durch die Eigenbewegung der Kamera entlang der Tiefenachse entstanden. Pixel-basierte Ansätze zur Berechnung eines dichten optischen Flußes mit nachfolgender Segmentierung ([1], [2]) sind geeignet um Bewegungen parallel zur Bildebene zu detektieren. Das Prinzip des *länglichen Bewegungsstereo* erlaubt dagegen eine passive Entfernungs- und Bewegungsmessung entlang der Tiefenachse der Kamera [3].

Gradienten- und *merkmal-basierte* Methoden zur Verschiebungsschätzung in Bildfolgen verarbeiten gewöhnlich zwei nacheinanderfolgende Bilder. Die Robustheit der Ergebnisse ist proportional zur Anzahl der beobachteten Bilder. Liu und Jain [4] bezeichnen eine gleichzeitige Segmentierung von mehreren Bildern als Bewegungsdetektion im *Bild-Zeit-Raum.* In einer langen Bildfolge können zeitlich nacheinanderfolgende Verschiebungsdaten zusätzlich stabilisiert werden, z.B. mit Hilfe von adaptiven Filtern [5] oder Neuronalen Netzen [6].

2 Konturenbasierte Verschiebungsschätzung

Eine kurze Folge von N nacheinaderfolgenden Bildern mit konstantem Zeitinterwal bezeichnen wir weiter als ein *N-Bild.* Zur Detektion von N korrespondierenden Konturen im *N-Bild* wird ein modifiziertes Verfahren zur Oberflächendetektion in tomographischen Volumenbildern verwendet [7].

Die 4 Punkte maximaler Konturenausdehnung in jeder Richtung des Bildkoordinatensystems, sowie das umfassende Rechteck werden mit Pixelgenauigkeit detektiert (Abbildung 1(a)). Dabei kann die Auswahl der Punkte nicht eindeutig sein (z.B. anstatt des Punktes P3 könnte auch P31 genommen werden). Dagegen wird der Konturenschwerpunkt C mit einer Genauigkeit lokalisiert, die unter einer Bildeinheit (BE) liegt. Damit ist eine Subpixelgenauigkeit einer elementaren Verschiebung (d.h. für zwei Bilder) nur für den Punkt C gegeben.

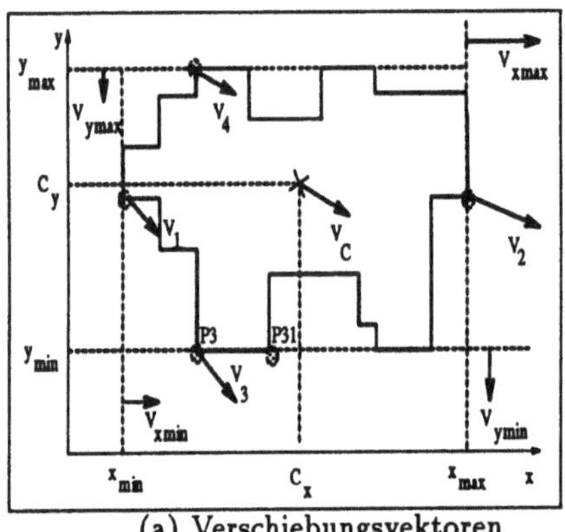

(a) Verschiebungsvektoren

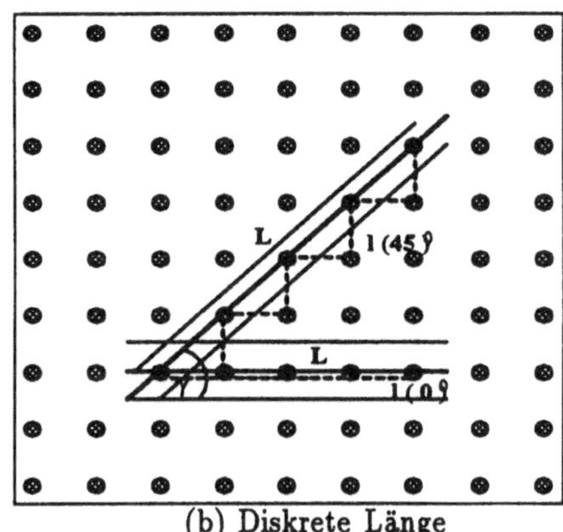

(b) Diskrete Länge

Abbildung 1. Diskrete Kontur und ihre Merkmale

Aufgrund der diskreten Bildauflösung entstehen Fehler bei der Abbildung einer kontinuierlichen Konturenlänge L auf diskrete Länge l. Für ein Geradensegment der Länge L_k ist dieser Fehler von der Richtung γ der Geraden im Bild abhängig (Abbildung 1(b)):

$$l_k - L_k \leq L_k(\sin(\gamma) + \cos(\gamma) - 1) \pm 0.5 \cos\gamma \tag{1}$$

Bei gemessenem l_k wird der Erwartungswert von L_k angenähert durch den Wert mit mittlerem Fehler: $\mathcal{E}[L_k] \simeq l_k/2(1+\sqrt{2}/2)+(1-\sqrt{2}) \simeq 0.854\, l_k - 0.414\, [BE]$.

Sei $l = l(t_j)$ die diskrete Länge einer Kontur, die im Bild zum Zeitpunkt t_j detektiert wird. Sei δt das Zeitinterval zwischen zwei nacheinanderfolgenden Bildern (z.B. $1/25s$). Der *Konturenveränderungsvektor* $v = (v_x, v_y, v_z)$ einer diskreten Kontur bildet eine Annäherung der unbekannten Veränderung $\tilde{\mathbf{v}} = (\tilde{\mathbf{v}}_x, \tilde{\mathbf{v}}_y, \tilde{\mathbf{v}}_z)$ der kontinuierlichen Kontur. Die konventionellen Komponenten $\tilde{\mathbf{v}}_x$ und $\tilde{\mathbf{v}}_y$ sind definiert als Verschiebungen des Schwerpunktes in Zeit δt ($j = 0, 1, 2..$):

$$\tilde{\mathbf{v}}_x(j) = \tilde{\mathbf{x}}_C(j+1) - \tilde{\mathbf{x}}_C(j) \qquad [BE/\delta t] \tag{2}$$

$$\tilde{\mathbf{v}}_y(j) = \tilde{\mathbf{y}}_C(j+1) - \tilde{\mathbf{y}}_C(j) \qquad [BE/\delta t] \tag{3}$$

Die dritte Komponente $\tilde{\mathbf{v}}_z$ stellt die relative Veränderung der kontinuierlichen Konturenlänge L dar:

$$\tilde{\mathbf{v}}_z(j) = \frac{L(j+1) - L(j)}{L(j)}, \; j = 0, 1, 2, ..; \qquad [1/\delta t] \tag{4}$$

3 Konturenveränderung im *N-Bild*

Um bereits während der Messungsphase relativ stabile Werte des Konturenveränderungsvektors zu ermitteln, sowie um die restlichen Verschiebungen der Kontur auch mit Subpixelgenauigkeit zu berechnen, werden die elementaren Verschiebungen zwichen dem ersten und den $N-1$ nächsten Bildern gemittelt.

3.1 Die gewichtete Mittelung

Mit $\{(x_C(i), y_C(i)) | i = 0, 1, ..., N-1\}$ seien die korrespondierenden Grawitätszentren in N Bildern bezeichnet. $\tilde{\mathbf{v}}$ wird approximiert durch v wie folgt:

$$\tilde{\mathbf{v}}_x \simeq v_x = \frac{1}{N-1} \sum_{i=1}^{N-1} \phi_i (x_C(i) - x_C(0)) \qquad [BE/\delta t] \tag{5}$$

$$\tilde{\mathbf{v}}_y \simeq v_y = \frac{1}{N-1} \sum_{i=1}^{N-1} \phi_i (y_C(i) - y_C(0)) \qquad [BE/\delta t] \tag{6}$$

$$\tilde{\mathbf{v}}_z \simeq v_z = \frac{1}{N-1} \frac{\sum_{i=1}^{N-1} \phi_i (l(i) - l(0))}{l(0)} \qquad [1/\delta t] \tag{7}$$

Im Idealfall sind die Gewichte $\{\phi_i | i = 1, ..., N-1\}$ direkt von den Tiefen $Z_C(i)$ der auf die Grawitätszentren projektierten Objektpunkte abhängig:

$$\phi_i = \frac{\tilde{\mathbf{x}}_C(1) - \tilde{\mathbf{x}}_C(0)}{\tilde{\mathbf{x}}_C(i) - \tilde{\mathbf{x}}_C(0)} = \frac{\tilde{\mathbf{y}}_C(1) - \tilde{\mathbf{y}}_C(0)}{\tilde{\mathbf{y}}_C(i) - \tilde{\mathbf{y}}_C(0)} = \frac{Z_C(i)}{i\, Z_C(1)} \tag{8}$$

Da die Tiefenwerte nicht zur Verfügung stehen, werden die Gewichte aus der z-Komponente des Veränderungsvektors abgeleitet. Eine ungewichtete Mittelung entspricht dem Fall, wenn sich das auf die Kontur abgebildete Objekt zusammen mit der Kamera bewegt ($\tilde{\mathbf{v}}_z = 0$). Für planare Flächen, die als annähernd parallel zur Bildebene klassifiziert sind, gilt:

$$\frac{L(1) - L(0)}{L(i) - L(0)} = \phi_i = \frac{Z_C(i)}{i\, Z_C(1)} \tag{9}$$

In [8] werden Diskretisierungsfehler im schlechtesten Fall des Konturenveränderungsvektors untersucht. Sie hängen u.a. von der Anzahl N der gemittelten Einzelwerte ab. Dabei zeigt sich, daß der Diskretisierungsfehler von v_z, ähnlich wie der Diskretisierungsfehler von v_x und v_y, nur von dem Lokalisierungsfehler der Vertices abhängig.

(Beispiel) Sei ein Geradensegment der Länge $L = 80$ gegeben mit $|\tilde{\mathbf{v}}_z| = 0.2$ und $|\tilde{\mathbf{v}}_x| = 2.0$. Die maximalen Diskretisierungsfehler betragen:

$$N = 2 \Rightarrow (0.50|\tilde{\mathbf{v}}_x| < |v_x| < 1.50|\tilde{\mathbf{v}}_x| \ ; \ 0.802|\tilde{\mathbf{v}}_z| < |v_z| < 1.198|\tilde{\mathbf{v}}_z|)$$
$$N = 5 \Rightarrow (0.80|\tilde{\mathbf{v}}_x| < |v_x| < 1.20|\tilde{\mathbf{v}}_x| \ ; \ 0.910|\tilde{\mathbf{v}}_z| < |v_z| < 1.090|\tilde{\mathbf{v}}_z|)$$

Dieses Beispiel verdeutlicht den Einfluß der Diskretisierung auf die Einzelverschiebungen (u.U. ein Fehler von 50%).

3.2 Bestimmung der Gewichte

Die Gewichte $\{\phi_i\}$ sind ihrerseits abhängig von v_z. Deswegen wird für ein *N-Bild* eine initiale Schätzung $v_z(Init)$, $\{\phi_i(init)|i = 1, ..., N-1\}$ benötigt:

$$v_z(Init) = \frac{1}{N-1} \sum_{i=1}^{N-1} \phi_i(Init) \frac{(l(i) - l(0))}{l(0)}; \qquad \phi_i(Init) = \frac{1}{i} \tag{10}$$

Von den initialen Werten ausgehend findet eine kurze Iterierung der Berechnung von v_z und der ϕ_i-s statt, solange sich deren Werte bedeutend ändern.

Wir beschränken uns ausschließlich auf konstante Objekteverschiebung, was insbesondere für Objekte, deren Abbildung auf die Bildebene zu relativ kleinen Konturen führt, als sinvoll erscheint. Es sei $\Delta V = (\Delta X, \Delta Y, \Delta Z)^T$ die gesuchte Verschiebung eines Objektes (in Kamerakoordinaten) im Zeitraum δt. Aus der Definition von v_z und der Projektionsgleichung kann folgendes abgeleitet werden:

$$\Delta Z \simeq Z_C \left(\frac{1}{1+v_z} - 1\right) \tag{11}$$

Aus Gleichungen (8), (9) und (11) folgt letztendlich:

$$\phi_i = \frac{Z_C(i)}{i\ Z_C(1)} \simeq \frac{1+i\beta}{i\ (1+\beta)}; \qquad \beta = \left(\frac{1}{1+v_z} - 1\right) \tag{12}$$

4 Relative 3-D Konturenattribute

4.1 v_z-basierte Schätzung von $\Delta Z/Z_C$

Im allgemeinen verlaufen die projektierten Liniensegmente nicht parallel zur Bildebene. Virtuelle Liniensegmente werden definiert, welche auf dieselben Bildkanten abgebildet werden wie die originellen Liniensegmente aber die parallel zur Bildebene verlaufen. Sei $\mathcal{L}$ die Länge eines Liniensegmentes zwischen den Punkten P_1 und P_2. Seien Z_1 und Z_2 die Tiefen der Endpunkte. Das virtuelle Segment mit Länge $\mathcal{L}_v$ verläuft zwischen den Punkten $P_{v1} = (X_{v1}, Y_{v1}, Z_C)$ und $P_{v2} = (X_{v2}, Y_{v2}, Z_C)$ mit $Z_C = (Z_1 + Z_2)/2$ (Abbildung 2). Für relativ große Tiefen $Z_C(t_j)$ verhält sich $\mathcal{L}_v(t_j)$ annähernd konstant mit der Zeit. Andernfalls muß der Einfluß der sich ändernden $\kappa_i(t_j)$-s berücksichtigt werden:

$$\mathcal{L}_v(t_{j+1}) \simeq \mathcal{L}_v(t_j) \Rightarrow \Delta Z \simeq Z_C(t_j) \left(\frac{1}{1+\bar{\mathbf{v}}_z(t_j)} - 1\right) \tag{13}$$

$$\mathcal{L}_v(t_{j+1}) = \bar{\alpha}(t_j)\mathcal{L}_v(t_j) \Rightarrow \Delta Z = Z_c(t_j) \left(\frac{\bar{\alpha}(t_j)}{1+\bar{\mathbf{v}}_z(t_j)} - 1\right) \tag{14}$$

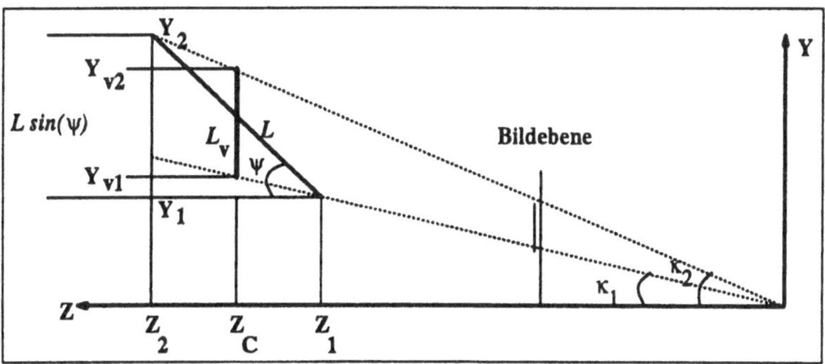

Abbildung 2. Virtuelle Objektkante

4.2 Punkte-basierte Schätzung der 3-D Attributenwerte

Gemäß dem Bewegungsstereo-Prinzip, durch Verfolgung der Lage von mehreren Punkten der Kontur, können wir bereits während der Segmentierung eine Schätzung von 3-D Konturenattributen vornehmen.

Sei ein Punkt $P = (X, Y, Z)$ (in Kamerakoordinaten) auf ein Bildpunkt $p = (x, y, 0)$ abgebildet. Sei $P' = (X', Y', Z')$ die Lage des Punktes P nach Zeitraum δt und $p' = (x', y', 0)$ die Lage seiner Projektion im nächsten Bild. Die Verschiebungen von p entlang der x und y Achsen ergeben sich aus:

$$u = x' - x = \frac{-F\Delta X - x'\Delta Z}{Z} \; ; v = y' - y = \frac{-F\Delta Y - y'\Delta Z}{Z}$$

Nach Eliminierung von Z bekommt man : $-Fv\Delta X + Fu\Delta Y = \Delta Z(xv - uy)$. Für zwei Endpunkte P_1, P_2 eines Liniensegmentes wird ein System von zwei Gleichungen der obigen Form aufgebaut. Daraus können ΔX und ΔY berechnet werden als Funktion von ΔZ (falls $u_1 v_2 - v_1 u_2 \neq 0$). Es können mehrere Konturenpunkte zu dieser Berechnung beitragen. Jetzt kann auch die Lage des Punktes P als Funktion von ΔZ bestimmt werden (falls $u \neq 0$ oder $v \neq 0$). Aus den relativen Lagewerten werden weitere 3-D Attribute der bewegten Kontur abgeleitet:

- Das Verhältnis $Z/\Delta Z$ bildet ein Kriterium für die Konturengruppierung.
- Eine Konturenklassifizierung in vertikale, horizontale bzw. schräge Konturen bezüglich der Z^w–Achse des Weltkoordinatensystems ist möglich (z.B. bezüglich der Fahrbahn).
- Eine Schätzung des Parameters α wird möglich. Für "vertikale" Konturen (aber in Kamerakoordinaten) beträgt $\alpha = 1.0$ und für "horizontale" steigt sein Wert (für $v_z > 0$) bzw. fällt (für $v_z < 0$) bis $1 - \beta$.

5 Verfolgung einer Kontur in längerer Bildfolge

In einer längeren Bildfolge gelten die $v(t_j)$ Werte für einzelne *N-Bilder* als Messungen. Zum Zeitpunkt t_j werden alle bisherigen Ergebnisse zusammengefaßt zu einem Paar von sich stabilisierenden Werten: $(v(t_j)^M, v(t_j)^P)$. Die Vorgehensweise wird erklärt am Beispiel von v_z.

Die *Vorhersage* $v_z^P(t_{j+1})$ des nächsten v_z-Wertes:

$$\frac{v_z^P(t_{j+1})}{v_z^M(t_j)} = \frac{Z_c(t_{j+1})}{Z_c(t_{j+2})} \simeq \frac{1+\gamma(t_j)}{1+2\gamma(t_j)}; \; \gamma(t_j) = \Big(\frac{\alpha(t_j)}{1+v_z^M(t_j)} - 1\Big) \quad (15)$$

Zur *Modifizierung* $v_z^U(t_j)$ des aktuell ermittelten v_z-Wertes wird ein skalares Filter verwendet - $v_z^M(t_j) \simeq (1-k(t_j))\, v_z(t_j) + k(t_j)\, v_z^P(t_j)$, wobei $k(t_j)$ die Differenz zwischen dem vorhergesagten Wert und dem gemessenen Wert wiederspiegelt.

(a) reale Szene

(b) künstliche Szene

Abbildung 3. Einzelbilder aus zwei Testfolgen

6 Ergebnisse

Ein Einzelbild einer realen Folge ist in Abbildung 3(a) dargestellt. Während der Konturendetektion wird nach Konturen gesucht, die einerseits eine homogene Region umschliessen und andererseits einen hohen Anteil an Kantenelementen des Kantenbildes besitzen. Die besten Ergebnisse wurden erzielt für eine Homogenitätstoleranz von $\pm 6 - 8$ (bei 256-Intensitätsstufen) und für einen absoluten Wert der Kantenstärke von $|40 - 70|$ (Robinson-Operator). Das segmentierte Bild (Abbildung 4(a)) enthält 268 detektierte Konturen, mit durchschnittlicher Länge von 105 BE. Davon weisen 123 Konturen eine Bewegung vor. 32 bewegte Konturen und 135 der als unbewegt geltenden Konturen (Hintergrund und weit entfernte Fahrzeuge) wurden eingeteilt zur Klasse "uberwiegend vertikal " (Abbildung 4(b)).

Eine Bewertung der Genauigkeit von berechneten Verschiebungen und Tiefe-zu-Bewegung-Merkmalen wird am Beispiel einer synthetischen Bildfolge erläutert (Abbildung 3(b)). Die Ergebnisse einer Verfolgung des Mittelstreifen auf der rechten Fahrbahn in zehn Bildern sind in Tabelle 1 zusammengefasst. Dabei wurde auch eine Störung eingebaut, als Ergebnis eines Schattenwurfes des "vorbeifahrenden" Fahrzeugs (bemerkbar für t_j von 5 bis 8). Wie erwartet war die

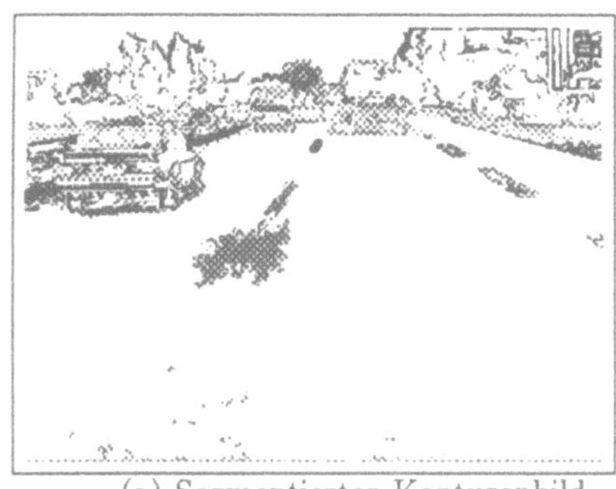
(a) Segmentiertes Konturenbild

(b) "Vertikale" Konturen

Abbildung 4. Beispiele von Ergebnisbildern

diskrete Länge l mit dem größten Fehler behaftet - bis zu 13.7% oder 12.1 BE. Aber die Qualität des Konturenveränderungsvektors oder der relativen Tiefenmessung war bedeutend besser. Bereits die gewichteten Schätzungen für einzelne *N-Bilder* lagen sehr nahe an den simulierten Daten. Die Fehler der adaptiv modifizierten v_z und v_{Cx}, v_{Cy} waren nicht grösser als $0.3BE/\delta t$ oder 7%. Bei bekannter Eigenbewegung (ca. 99 km/h) wurde die Tiefe des Streifens mit einem Fehler unter $2m$ bzw. 7.4% geschätzt. Zu diesen Ergebnissen sei noch beigefügt, daß die Auflösung der simulierten Bilder nur 256x256 Bildpunkte betragen hat.

7 Zusammenfassung

Eine Methode zur konturen-basierten Entfernungs- und Bewegungsschätzung bei bewegter Kamera wurde vorgestellt. Durch die Wahl der Merkmale - Konturen, deren Länge und Grawitätszentren - kann eine Subpixelgenauigkeit der Lokalisierung erreicht werden. Dank der Mittelung der Veränderung von Merkmalen in mehreren Bildern, können stabilere und mehr signifikante Werte detektiert werden, als es bei zwei Bildern der Fall ist. Die Stabilisierung der Verfolgung von charakteristischen Punkten einer Kontur in einer längeren Bildfolge erlaubt, unter gewissen Beschränkungen, eine modell-unabhängige Schätzung von 3-D Attributen des betrachteten Objektes.

Dank Die Arbeit wurde gefördert durch die Deutsche Forschungsgemeinschaft zu Bonn im Rahmen des Projektes: "Erkennung bewegter Objekte". Die Bildfolge ist eine freundliche Übergabe der BMW AG., München, an FORWISS.

Literatur

1. Kirchner H. : *Ein mehrstufiger Ansatz zur Bewegungserkennung in Bildfolgen*, **Dissertation**, Universität Erlangen-Nürnberg, IMMD, Erlangen, 1992.
2. Koller D.: *Detektion, Verfolgung und Klassifikation bewegter Objekte in monokularen Bildfolgen am Beispiel von Straßenverkehrsszenen*, **DISKI - Dissertationen zur Künstlichen Intelligenz**, infix, Sankt Augustin, 1992.

Konturenmerkmal	Zeitpunkt t_j =					
	0	2	4	6	8	9
Simuliertes L [BE]	43.98	51.01	60.05	72.00	88.30	98.82
Detektiertes $\mathcal{E}[L]$ [BE]	47.4	54.2	64.5	78.1	100.4	112.3
Fehler $(\mathcal{E}[L]-L)/L$ [%]	7.8	6.3	7.4	8.5	13.7	13.6
Simuliertes $\tilde{v}_z$ [$1/\delta t$]	0.0750	0.0827	0.0922	0.1040	0.1191	0.1284
- v_z(Init) [$1/\delta t$]	0.082	0.100	0.121	0.143	0.158	0.168
- ϕ-gemitteltes v_z	0.072	0.087	0.098	0.120	0.116	0.124
- adaptiertes v_z^M	0.072	0.079	0.088	0.107	0.121	0.129
Fehler $(v_z^M - \tilde{v}_z)L$ [$BE/\delta t$]	-0.132	-0.189	-0.252	0.216	0.168	0.059
$\Delta Z(sim)$ (für $Z_C = -1m$) [m/s]	0.9625	1.0418	1.1363	1.2500	1.3890	1.4705
Berechnetes ΔZ ($Z_C = -1m$) [m/s]	0.9000	0.9850	1.0975	1.3275	1.4925	1.5875
$(\Delta Z - \Delta Z(sim))/\Delta Z(sim)$ [%]	-6.5	-5.5	-3.4	6.2	7.5	8.0
Simuliertes $Z_{C(t_j)}(sim)$ [m]	-28.61	-26.41	-24.21	-22.01	-19.81	-18.71
$Z_{C(t_j)}$ für $\Delta Z = 1.1004m/s$ [m]	-30.57	-27.93	-25.07	-20.72	-18.43	-17.33
$(Z_C - Z_C(sim))/Z_C(sim)$ [%]	-6.5	-5.8	-3.6	5.9	7.0	7.4
Simuliertes $\tilde{v}_{Cx}$ [$BE/\delta t$]	1.50	1.77	2.11	2.56	3.17	3.56
- ϕ−gemitteltes $v_{Cx}[BE/\delta t]$:	1.46	1.79	2.24	2.98	3.56	3.91
- adaptiertes $v_{Cx}^M[BE/\delta t]$	1.46	1.71	2.06	2.61	3.33	3.79
Fehler $(v_{Cx}^M - \tilde{v}_{Cx})$ [$BE/\delta t$]:	-0.04	-0.06	-0.05	0.05	0.16	0.23
Fehler $(v_{Cy}^M - \tilde{v}_{Cy})$ [$BE/\delta t$]:	-0.04	-0.01	-0.02	-0.08	-0.18	-0.21

Tabelle 1. Beispiel einer Konturenverfolgung

3. Huber J., Graefe V.: *Quantitative Interpretation of Image Velocities in Real Time*, **IEEE Workshop on Visual Motion** (Princeton, New Jersey, Oct,1991), IEEE Computer Society Press, Los Alamitos CA, 211–216.
4. Liu S.-P., Jain R.: *Motion Detection in Spatio-Temporal Space.*, **Computer Vision, Graphics and Image Proc, 45**(1989), 227–250.
5. Dickmanns E.D., Graefe V., *Applications of Dynamic Monocular Machine Vision*, **Machine Vision and Applications**, 1988, No. 1, 241–261.
6. Zhou Y-T., Chellappa R.: *Artificial Neural Networks for Computer Vision*, Springer Vg., New York etc., 1992.
7. Kasprzak W.: *A Two-Step Modelling Algorithm for Tomographic Scenes*. In: [Burkhardt H., K.H.Höhne, B.Neumann (Hrsg.): **Mustererkennung 1989, 11. DAGM–Symposium**, (Hamburg, Oktober 1989), *Informatik-Fachberichte, Bd. 219*, Springer Vg., Berlin, 1989], 119–123.
8. Kasprzak W.: *Translational Motion or Depth from Segmentation in Image-Time Space*, In: Pavesic N., Niemann H., Paulus D.(eds.): **Image Processing and Stereo Analysis**, Arbeitsberichte des IMMD, Universität Erlangen-Nürnberg, Band 26 (1993), No.1, 73–97.

Bestimmung des optischen Flusses aus Bildfolgen unter Berücksichtigung von Diskontinuität und Okklusion

An Luo und Hans Burkhardt
Technische Informatik I, TU Hamburg-Harburg
Postfach 90 10 52, 21071 Hamburg, Germany

Zusammenfassung

In diesem Beitrag wird eine neue Methode zur direkten MAP-Schätzung des optischen Flusses aus einer Graubildfolge vorgestellt, wobei dessen Diskontinuität implizit berücksichtigt wird. Die resultierende nichtkonvexe Energiefunktion wird mit einem pyramidalen, zweistufigen Algorithmus global minimiert. Dabei wird die Okklusion, die normalerweise die richtige Bestimmung des optischen Flusses erschwert, durch zwei kooperative, gegeneinander gerichtete Optimierungsprozesse detektiert und zugleich zur Verbesserung der Schätzung ausgenutzt. Nach der Vorstellung der Algorithmen wird diese neue Methode anhand einiger Beispiele demonstriert.

1 Einführung

Eine zentrale Aufgabe der 3-D Vision ist die Extraktion der 3-D Bewegung und Struktur zur räumlichen Interpretation und Segmentierung bewegter Objekte in Bildfolgen. Das ebene Verschiebungsvektorfeld als Projektion der tatsächlichen 3-D Bewegung eignet sich gut zur Erfassung dieser Informationen. Aus räumlichen und zeitlichen Grauwertänderungen der Bilder kann sich eine Schätzung des Verschiebungsvektorfeldes, der sogenannte optische Fluß, ergeben, aus dem die 3-D Struktur und Bewegung bestimmt werden können. Dessen Komponente in jedem Bildpunkt wird als Verschiebungsvektor bezeichnet. Dieser Beitrag konzentriert sich auf die Berechnung des optischen Flusses.

In den letzten Jahren wurden viele Arbeiten zur Berechnung des optischen Flusses durchgeführt [AN88]. Die bekannten Verfahren unterteilen sich in vier Klassen: helligkeitsbasierte differentielle, helligkeitsbasierte Korrelations-, merkmalbasierte Matchings- und zeitliche/räumliche Filterungsverfahren. In dieser Arbeit wird nur ein helligkeitsbasiertes differentielles Verfahren verwendet, mit dem ein dicht verteilter optischer Fluß ohne Korrespondenz der aufeinanderfolgenden Bilder unmittelbar ermittelt werden kann. Es vermeidet z.B. den Nachteil merkmalbasierter Verfahren, welche i.a. nur wenige Verschiebungsvektoren ermitteln und somit nur dünn verteilte Felder bestimmen können.

Die Berechnung des optischen Flusses nur aus der Graubildfolge ist ein im mathematischen Sinne schlecht gestelltes Problem, denn außer dem Blendenproblem hängt es auch von ausreichenden Grauwertänderungen ab. Um die Mehrdeutigkeiten zu beseitigen, muß ein a-priori Modell des optischen Flusses angenommen werden. Basierend auf diesem Modell können Verfahren zur Bestimmung des Flusses entwickelt werden. Beispielsweise wurden globale Glättung [HS81], Glättung entlang Konturen [Hil84], gerichtete Glättung [NE86] und grauwertbezogene Glättung [Ais89] als a-priori Modelle verwendet.

Zur Verminderung des Rechenaufwandes und Vermeidung einer nur lokal optimalen Zuordnung können die Algorithmen zur Bestimmung des optischen Flusses als Mehrgitterverfahren implementiert werden. Pyramidale Kontrollstrategien zur Berechnung des optischen Flusses kann man z.B. in [Gla84] und in [Enk88] finden.

In dem hier vorgestellten Ansatz wird der optische Fluß mit einem Markov'schen Zufallsfeld (MRF) modelliert, wobei eine neue a-priori Energiefunktion die Eigenschaft der schwachen Kontinuität über den optischen Fluß charakteristiert. Basierend darauf werden die in [LB92] entwickelten helligkeitsbasierten, kooperativen, gegeneinander gerichteten Algorithmen mit Mehrfachauflösung auf die Berechnung des optischen Flusses erweitert, wobei keine epipolare Beschränkung mehr angenommen werden muß. Mit diesen modifizierten Algorithmen erfolgt die Bestimmung des optischen Flusses unmittelbar aus Graubildpaaren. Dabei ergeben sich Unstetigkeiten des optischen Flusses an Rändern von Objekten, wodurch verdeckte Bereiche in Bildern (Okklusionsgebiete) zusätzlich detektiert werden können. Im Vergleich zu anderen Algorithmen wird bei diesem neuen Ansatz die implizite Annähme über schwache Glattheit des optischen Flusses vorausgesetzt. Es ergibt sich eine einfache Lösungsstruktur, welche zugleich einen niedrigen algorithmischen Rechenaufwand besitzt. Außerdem wird die Okklusion, welche normalerweise die richtige Bestimmung des optischen Flusses erschwert oder sogar verhindert, gleichzeitig detektiert und sogar zur Verbesserung der Schätzung ausgenutzt.

Nach einer näheren Beschreibung der Algorithmen und ihrer Implementierung folgen einige Beispiele bei typischen Bewegungen sowie eine Diskussion ihrer Leistungsfähigkeit.

2 Ein neuer Ansatz zur Berechnung des optischen Flusses

Alle oben erwähnten herkömmlichen Ansätze können die starke Diskontinuität des optischen Flusses und die verdeckten Bereiche in Bildern nicht richtig behandlen. In diesem Fall sind die entsprechenden Punkte in einem Bild des Bildpaars unsichtbar und somit ist keine Korrespondenz vorhanden. Wenn dieser verdeckte Bereich nicht detektiert wird, wird der optische Fluß dort mit verschmierten Bewegungsgrenzen fehlerhaft bestimmt. Außerdem pflanzt sich dieser Fehler global fort. Wie in [Ais89] bestätigt wird, treten die meisten Fehler bei der Berechnung des optischen Flusses infolge der Okklusion auf, obwohl beim Entwurf der verschiedenen Glättungskriterien viel Mühe zur Erhaltung der Diskontinuitäten an den Objekträndern aufgewendet wurde.

Zur Lösung dieses Problems wird im folgenden der optische Fluß direkt durch ein MRF unter impliziter Berücksichtigung von Diskontinuität modelliert. Dabei werden die verdeckten Bereiche beider Bilder durch kooperative, gegeneinander gerichtete Prozesse detektiert. Dieser neue Ansatz, mit dem der schwach kontinuierliche optische Fluß unmittelbar aus einer Graubildfolge berechnet wird, basiert auf dem in [LB92] vorgestellten Stereoverfahren. Der neue Ansatz ist jedoch schwieriger, da keine epipolare Einschränkung angenommen wird und eine beliebige Relativbewegung möglich ist.

Diskontinuität und Okklusion Innerhalb eines einzelnen Objektes kann für den optischen Fluß eine a-priori Energiefunktion des MRF mit Hilfe des Membranmodells verwendet werden. An den Stellen, wo die Diskontinuität vorliegt, muß aber die lokale Potentialfunktion der Gibbs'schen Verteilung gelockert oder zum Verschwinden gebracht werden. Der optische Fluß des Gesamtbildes soll jedoch mit einem einzigen MRF modelliert werden, welches keine Kontrollvariablen zur Beschreibung der Diskontinuitäten enthält. Da die Medianfilterung das Rauschen in relativ glatten Bereichen ähnlich wie die lineare Glättungsfilterung effektiv unterdrückt und zugleich Ränder gut erhält, wird die folgende a-priori lokale Potentialfunktion für den optischen Fluß vorgeschlagen:

$$V_{\mathbf{x}}(\mathbf{u}_s) = \|\mathbf{u}_s(\mathbf{x}) - \mathbf{u}_{s(med)}(\mathbf{x})\|^2 \tag{1}$$

wobei mit $\mathbf{u}_1(\mathbf{x}) = (u_1(\mathbf{x}), v_1(\mathbf{x}))^T$ und $\mathbf{u}_2(\mathbf{x}) = (u_2(\mathbf{x}), v_2(\mathbf{x}))^T$ die optischen Flüsse von $g_1(\mathbf{x})$ nach $g_2(\mathbf{x})$ bzw. umgekehrt von $g_2(\mathbf{x})$ nach $g_1(\mathbf{x})$ gekennzeichnet werden.

Bei sich unabhängig bewegenden Objekten können Gegenstände hinter anderen verschwinden bzw. wieder auftauchen. Dafür werden zwei binäre Okklusionsfelder $O_1(\mathbf{x})$ und $O_2(\mathbf{x})$ in beiden Bildrahmen der Bildfolge $g_1(\mathbf{x})$ bzw. $g_2(\mathbf{x})$ definiert. $O_1(\mathbf{x})$ wird auf 0 gesetzt, wenn der Teil der Szene im betreffenden Bildpunkt $\mathbf{x} = (x, y)^T$ des Bildes $g_1(\mathbf{x})$ im Bildrahmen $\mathbf{g}_2$ verschwindet. Wenn ein Teil der Szene im Bildpunkt $\mathbf{x} = (x, y)^T$ des Bildrahmens $\mathbf{g}_2$ wieder auftaucht, wird $O_2(\mathbf{x})$ auf 0 gesetzt. Die übrigen Elemente von $O_s(\mathbf{x})$ ($s = 1$ und 2) werden auf 1 gesetzt, da dieser auf $\mathbf{x} = (x, y)^T$ projizierte Bereich nicht verdeckt ist. Dabei wird $\mathbf{u}_s(\mathbf{x})$ durch die folgende Nebenbedingung festgelegt:

$$\mathbf{u}_s(\mathbf{x}) = -\mathbf{u}_{3-s}(\mathbf{x} + \mathbf{u}_s(\mathbf{x})) \text{ wenn } O_s(\mathbf{x}) = 1 (s = 1 \text{ oder } 2). \tag{2}$$

Bei unserer Modellierung ist die Okklusion von großem Interesse. Die Information über $O_s(\mathbf{x})$ ist in dem dazugehörigen optischen Fluß $\mathbf{u}_s(\mathbf{x})$ nicht enthalten, sondern kann aus dem zweiten optischen Fluß $\mathbf{u}_{3-s}(\mathbf{x})$ ermittelt werden. Diese **Transformation** $O_s(\mathbf{x}) = \mathcal{OT}(\mathbf{u}_{3-s}(\mathbf{x}))$, die zur Bestimmungen von $O_1(\mathbf{x})$ aus $\mathbf{u}_2(\mathbf{x})$ und von $O_2(\mathbf{x})$ aus $\mathbf{u}_1(\mathbf{x})$ dient, wird durch die folgende Gleichung beschrieben:

$$\{\mathbf{x}' : O_s(\mathbf{x}') = 1, \mathbf{x}' \in B\} = \{\mathbf{x}' : \mathbf{x}' = \mathbf{x} + \mathbf{u}_{3-s}(\mathbf{x}), \mathbf{x} \in B\}. \tag{3}$$

Die Diskretisierung erfolgt auf ähnliche Weise wie in [LB92]. Sofern der optische Fluß exakt verfügbar ist, kann man die Okklussionsfelder auch genau bestimmen.

MAP-Schätzung des optischen Flusses Unter der Annahme, daß die in den Bewegungsgleichungen enthaltenen Bildstörungen gaußverteilt sind, läßt sich für $\mathbf{g}_{3-s}$ bei gegebenem $\mathbf{g}_s$, $\mathbf{u}_s$ und $\mathbf{O}_s$ eine bedingte Wahrscheinlichkeitsdichteverteilung gemäß

$$P(\mathbf{g}_{3-s}|\mathbf{g}_s, \mathbf{u}_s) = C_s exp\{-\frac{1}{2\sigma^2}\sum_{\mathbf{x}} O_s(\mathbf{x})(g_{3-s}(\mathbf{x} + \mathbf{u}_s(\mathbf{x})) - g_s(\mathbf{x}))^2\} \tag{4}$$

angeben, mit $s = 1$ oder 2. Nur außerhalb des verdeckten Bereichs gelten die Bewegungsgleichungen, daher wird eine Maskierung mit $O_s(\mathbf{x})$ vorgenommen. Aus der obigen $P(\mathbf{g}_{3-s}|\mathbf{g}_s, \mathbf{u}_s)$ und der a-priori Verteilung $P(\mathbf{u}_s)$ gemäß der Potentialfunktion in Gleichung (1) ergeben sich die a-posteriori Wahrscheinlichkeitsdichten der optischen Flüsse $\mathbf{u}_s$ (s=1 und 2) bei beiden bekannten Bildern $\mathbf{g}_1$ und $\mathbf{g}_2$ mit Hilfe der Bayes'schen Regel zu:

$$P(\mathbf{u}_s|\mathbf{g}_s, \mathbf{g}_{3-s}) = \frac{P(\mathbf{g}_{3-s}|\mathbf{g}_s, \mathbf{u}_s)P(\mathbf{u}_s)}{P(\mathbf{g}_{3-s}|\mathbf{g}_s)} = \frac{1}{Z_{ps}} exp\{-\frac{U_{ps}(\mathbf{u}_s)}{2\sigma^2}\}$$

mit den a-posteriori Energiefunktionen $U_{ps}(\mathbf{u}_s)$ unter Kenntnis der Okklusionsfelder $\mathbf{O}_s$:

$$U_{ps}(\mathbf{u}_s) = \sum_{\mathbf{x}} (O_s(\mathbf{x})(g_{3-s}(\mathbf{x} + \mathbf{u}_s(\mathbf{x})) - g_s(\mathbf{x}))^2 + \lambda \|\mathbf{u}_s(\mathbf{x}) - \mathbf{u}_{s(med)}(\mathbf{x})\|^2). \tag{5}$$

Um den optischen Fluß mit MAP-Wahrscheinlichkeit zu schätzen, sollten die a-posteriori Energiefunktionen $U_{p1}(\mathbf{u}_1)$ und $U_{p2}(\mathbf{u}_2)$ bezüglich $\mathbf{u}_1$ bzw. $\mathbf{u}_2$ unter der Bedingung minimiert werden, daß die verdeckten Bereiche in beiden Bildern erkannt sind. Man muß anmerken, daß das zur Optimierung der Energiefunktion $U_{p1}(\mathbf{u}_1)$ benötigte Feld $\mathbf{O}_1$ allein aus dem im komplementären Optimierungsprozeß zu schätzenden optischen Fluß $\mathbf{u}_2$ bestimmt werden kann und umgekehrt. Deshalb werden zwei kooperative, gegeneinander gerichtete iterative Prozesse zur gleichzeitigen Optimierung beider a-posteriori Energiefunktionen verwendet, welche die nach Gleichung (2) identischen, jedoch mit unterschiedlichem Vorzeichen, optimalen optischen Flüsse $\mathbf{u}_1$ und $\mathbf{u}_2$ ermitteln.

3 Zweistufige Algorithmen in pyramidaler Stuktur

Wegen der Medianoperation sind die Energiefunktionen (5) offensichtlich nichtkonvex, so daß die einfache Anwendung des Gradientverfahrens zu einem lokalen Minimum führen kann. Mit der statistischen Methode ist die global-optimale Schätzung nur mit sehr hoher Komplexität und langsamer Konvergenz zu erreichen. Im folgenden stellen wir eine neue deterministische Strategie vor, wobei die drei Maßnahmen dazu dienen:

1. Wenn man die aktuelle Schätzung $\mathbf{u}^*$ als Annäherung des zu erwartenden optischen Flusses betrachtet, kann das Graubild $g_{3-s}(\mathbf{x}+\mathbf{u}_s)$ mit einem planaren Modell an dieser verschobenen Stelle $(\mathbf{x}+\mathbf{u}^*)$ besser lokal approximiert werden [NE86] [Ais89]. Damit ergibt sich die approximierte a-posteriori Energiefunktion aus Gleichung (5):

$$U_{ps}(\mathbf{u}_s) = \sum_{\mathbf{x}} (O_s(\mathbf{x})(g_{3-s}(\mathbf{x}') - g_s(\mathbf{x}) + (\mathbf{u}_s(\mathbf{x}) - \mathbf{u}_s^*(\mathbf{x})) \cdot \nabla g_{3-s}(\mathbf{x}'))^2 + \lambda V_{\mathbf{x}}(\mathbf{u}_s)) \quad (6)$$

mit $\mathbf{x}' = \mathbf{x} + \mathbf{u}_s^*(\mathbf{x})$. Mit dieser Energiefunktion ist der optimale optische Fluß besser iterativ zu bestimmen.

2. Der in [LB92] entwickelte zweistufige Algorithmus wird zur Minimierung dieser nichtkonvexen Energiefunktion verwendet. Sofern ein Startpunkt in der Nähe des globalen Optimums ausgewählt wird, kann dieses Optimum durch Anwendung des Gradientenverfahrens unmittelbar erreicht werden. Beim Membranmodell, das mit einer konvexen Energiefunktion das ursprüngliche Modell annähert, gilt die Potentialfunktion $V_{\mathbf{x}}(\mathbf{u}_s) = \|\mathbf{u}_{sx}(\mathbf{x})\|^2 + \|\mathbf{u}_{sy}(\mathbf{x})\|^2$ in der a-posteriori Energiefunktion von Gleichung (6), deren Euler-Lagrange'sche Gleichung von der folgenden Form ist:

$$\lambda \Delta \mathbf{u}(\mathbf{x}) = O_s(\mathbf{x})(g_{3-s}(\mathbf{x}') - g_s(\mathbf{x}) + (\mathbf{u}_s(\mathbf{x}) - \mathbf{u}_s^*(\mathbf{x})) \cdot \nabla g_{3-s}(\mathbf{x}'))\nabla g_{3-s}(\mathbf{x}') \quad (7)$$

mit $\mathbf{x}' = \mathbf{x}+\mathbf{u}_s^*(\mathbf{x})$. Daraus ergibt sich die folgende iterative Formel bei der Approximation von $\Delta \mathbf{u} = \mathbf{u}_{(ave)} - \mathbf{u}$:

$$\mathbf{u}_s^{(n+1)}(\mathbf{x}) = \mathbf{u}_s^{*(n)}(\mathbf{x}) - O_s(\mathbf{x}) \frac{g_{3-s}(\mathbf{x} + \mathbf{u}_s^{*(n)}(\mathbf{x})) - g_s(\mathbf{x})}{\lambda + \|\nabla g_{3-s}(\mathbf{x} + \mathbf{u}_s^{*(n)}(\mathbf{x}))\|^2} \nabla g_{3-s}(\mathbf{x} + \mathbf{u}_s^{*(n)}(\mathbf{x})) \quad (8)$$

mit der Erwartung

$$\mathbf{u}_s^{*(n)}(\mathbf{x}) = \mathbf{u}_{s(ave)}^{(n)}(\mathbf{x}) \quad (9)$$

Mit Hilfe dieser iterativen Formel kann man einen glatten optischen Fluß in wenigen Iterationen ermitteln. Diese Schätzung ist im Inneren einzelner Objekte weitgehend korrekt und weicht nur in der Nähe der Okklusion oder Diskontinuität vom tatsächlichen optischen Fluß ab. Weil die Diskontinuität und Okklusion relativ selten im gesamten Bildrahmen vorkommen, wird diese Schätzung als Startpunkt für die folgende Optimierung der ursprünglichen Energiefunktion genutzt.

Beim Originalmodell des schwach kontinuierlichen optischen Flusses gilt die Potentialfunktion $V_{\mathbf{x}}(\mathbf{u}_s) = \|\mathbf{u}_s(\mathbf{x}) - \mathbf{u}_{s(med)}(\mathbf{x})\|^2$ in der a-posteriori Energiefunktion von Gleichung (6), deren Euler-Lagrange'sche Gleichung gegeben ist durch:

$$\lambda(\mathbf{u}_{s(med)}(\mathbf{x}) - \mathbf{u}_s(\mathbf{x})) = O_s(\mathbf{x})(g_{3-s}(\mathbf{x}') - g_s(\mathbf{x}) + (\mathbf{u}_s(\mathbf{x}) - \mathbf{u}_s^*(\mathbf{x})) \cdot \nabla g_{3-s}(\mathbf{x}'))\nabla g_{3-s}(\mathbf{x}') \quad (10)$$

mit $\mathbf{x}' = \mathbf{x} + \mathbf{u}_s^*(\mathbf{x})$. Nur wenn der lokale Medianwert der aktuellen Schätzung als $\mathbf{u}_s^*$ zu erwarten ist, entspricht die iterative Formel direkt Gleichung (8), wobei aber nur die Erwartung gilt:

$$\mathbf{u}_s^{*(n)}(\mathbf{x}) = \mathbf{u}_{s(med)}^{(n)}(\mathbf{x}). \quad (11)$$

Um die identischen optischen Flüsse in beiden Bildrahmen zu erhalten, werden die Gleichungen (9) und (11) gemäß Nebenbedingung (2) modifiziert:

$$\mathbf{u}_s^*(\mathbf{x}) = \frac{\mathbf{u}_{s(op)}(\mathbf{x}) - O_{3-s}(\mathbf{x} + \mathbf{u}_{s(op)}(\mathbf{x}))\mathbf{u}_{3-s(op)}(\mathbf{x} + \mathbf{u}_{s(op)}(\mathbf{x}))}{1 + O_{3-s}(\mathbf{x} + \mathbf{u}_{s(op)}(\mathbf{x}))} \quad (12)$$

wobei die Bezeichung (*op*) beim Membranmodell mit (*ave*) oder beim Originalmodell mit (*med*) zu ersetzen ist.
3. Sowohl zur globalen Optimierung als auch zur Beschleunigung der Konvergenz wird das Mehrgitterverfahren verwendet, denn je gröberer die Auflösung ist desto größer ist die Umgebung, in welche die lokalen differentiellen Operationen wirken. Weiterhin verschwinden die Fehler bestimmter Frequenzen nur bei angepaßter Bildauflösung relativ schnell [Gla84] [Enk88].

Durch die komplizierten Kontrollstrategien entstehen Wechselwirkung zwischen Auflösungsebenen, wodurch eine Oszillierung zwischen unterschiedlichen Skalierungen oder sogar eine falsche Konvergenz entstehen kann. Dagegen wird im vorgestellten Algorithmus eine einfache Kontrollstrategie verwendet, wobei die Schätzungen nacheinander von der gröbsten bis zur feinsten Auflösung durchgeführt werden und der Optimierungsprozeß jeder Auflösung die Ergebnisse der letzten gröberen Auflösung als Startpunkt nimmt.

Der folgende deterministische, zweistufige Algorithmus in pyramidaler Stuktur beinhaltet die drei obigen Maßnahmen. Er kann verwendet werden, um die kooperativen, gegeneinander gerichteten Optimierungsprozesse für die MAP-Schätzung des optischen Flusses zu implementieren.

Auf jeder Auflösungsebene k von der gröbsten l bis zur feinsten 0 sind die folgenden Operationen durchzuführen:

- *Initialisierung:*

$$
\begin{aligned}
O_{1,k}(\mathbf{x}) &\leftarrow \mathit{Interpolation}(O_{1,k+1}(\mathbf{x})) \\
O_{2,k}(\mathbf{x}) &\leftarrow \mathit{Interpolation}(O_{2,k+1}(\mathbf{x})) \\
\mathbf{u}_{1,k}(\mathbf{x}) &\leftarrow \mathit{Interpolation}(\mathbf{u}_{1,k+1}(\mathbf{x})) \\
\mathbf{u}_{2,k}(\mathbf{x}) &\leftarrow \mathit{Interpolation}(\mathbf{u}_{2,k+1}(\mathbf{x}))
\end{aligned}
$$

- *Wiederhole die folgende Iteration bis zur Konvergenz:*

$$
\begin{aligned}
O_{1,k}(\mathbf{x}) &\leftarrow \mathcal{OT}(\mathbf{u}_{2(op),k}(\mathbf{x})) \\
O_{2,k}(\mathbf{x}) &\leftarrow \mathcal{OT}(\mathbf{u}_{1(op),k}(\mathbf{x})) \\
\mathbf{u}_{1,k}(\mathbf{x}) &\leftarrow \mathbf{u}^*_{1,k}(\mathbf{x}) - O_{1,k}(\mathbf{x})\frac{g_{2,k}(\mathbf{x}+\mathbf{u}^*_{1,k}(\mathbf{x})) - g_{1,k}(\mathbf{x})}{\lambda + \|\nabla g_{2,k}(\mathbf{x}+\mathbf{u}^*_{1,k}(\mathbf{x}))\|^2}\nabla g_{2,k}(\mathbf{x}+\mathbf{u}^*_{1,k}(\mathbf{x})) \\
\mathbf{u}_{2,k}(\mathbf{x}) &\leftarrow \mathbf{u}^*_{2,k}(\mathbf{x}) - O_{2,k}(\mathbf{x})\frac{g_{1,k}(\mathbf{x}+\mathbf{u}^*_{2,k}(\mathbf{x})) - g_{2,k}(\mathbf{x})}{\lambda + \|\nabla g_{1,k}(\mathbf{x}+\mathbf{u}^*_{2,k}(\mathbf{x}))\|^2}\nabla g_{1,k}(\mathbf{x}+\mathbf{u}^*_{2,k}(\mathbf{x}))
\end{aligned}
$$

† *in den $\mathcal{OT}$-Transformationen und $\mathbf{u}^*_{s,k}(\mathbf{x})$ von Gleichung (12) wird (op) mit (ave) in ersten Iterationen bzw. danach mit (med) ersetzt.*

Im obigen Rechenschema werden die Gaußpyramiden $\mathbf{g}_{1,k}$ und $\mathbf{g}_{2,k}$ ($k = 0, 1, ..., l$) von beiden Bildern berechnet. Die Interpolationsprozesse für die optischen Flüsse und Okklusionsfelder werden ebenfalls mit Hilfe eines Gaußfilters durchgeführt [Bur84].

Dieser Algorithmus hat viele Vorteile: Erstens kann der schwach kontinuierliche optische Fluß unmittelbar aus einem Graubildpaar einer Bildfolge berechnet werden, wobei die Diskontinuität des optischen Flusses implizit berücksichtigt wird. Zweitens werden zugleich auch die verdeckten Bereiche in beiden Bildern zur Verbesserung der Schätzung des optischen Flusses detektiert. Schließlich konvergiert dieser Algorithmus aufgrund der Mehrgitterstrategie schnell, so daß nur ein moderater Rechenaufwand benötigt wird. Im folgenden werden einige Beispiele vorgestellt, um die obige Behauptung zu bestätigen.

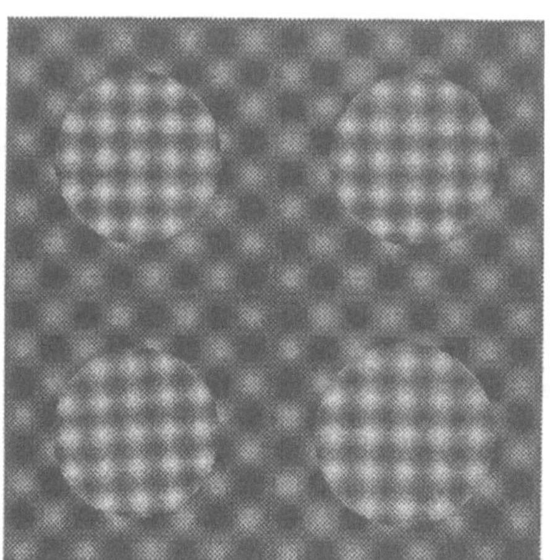

Abbildung 1: Künstlich erzeugte Bilder:
(a) das ursprüngliche Bild,
(b) Verschiebung,
(c) Drehung und
(d) Vergößerung des Objektes

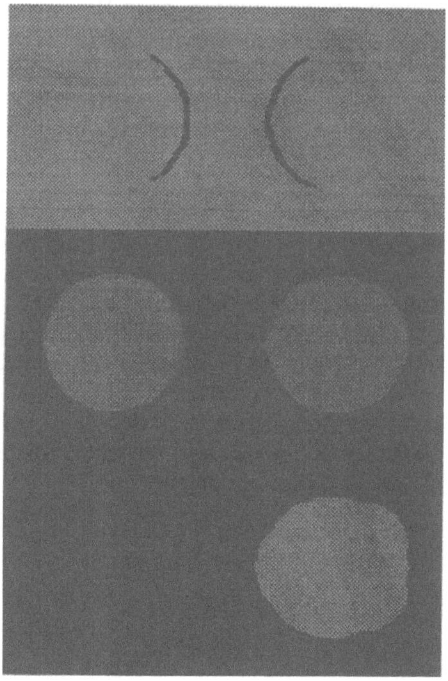

Abbildung 2: Die Ergebnisse des optischen Flusses bei der Verschiebung des Objektes:
(a) die Okklusion $\mathbf{O}_1$ und (b) $\mathbf{O}_2$,
(c) Betrag des Flusses $\|\mathbf{u_1}\|$ und (d) $\|\mathbf{u_2}\|$,
(e) Phase des Flusses $\theta(\mathbf{u_1})$ und (f) $\theta(\mathbf{u_2})$

4 Experimentelle Ergebnisse

Zunächst wenden wir den vorgestellten Algorithmus auf künstlich erzeugten Bildfolgen dreier typischer Bewegungen, und zwar Verschiebung, Drehung und Skalierung, an. Das erste Bild einer Bildfolge in Abbildung 1(a) wird so erzeugt, daß ein rundes Objekt vor einem statischen Hintergrund auftaucht. Die Abbildung 1(b), (c) bzw. (d) zeigt als zweites Bild der Folge ein verschobenes, gedrehtes bzw. skaliertes Objekt. Für diese drei Bildfolgen zeigen die Abbildungen 2, 3 bzw. 4 jeweils die optischen Flüsse und Okklusionsfelder. Es wird vereinbart, daß in (c) und (d) 30 Grauwertstufen einer Verschiebung um 1 Pixel entsprechen. In (e) und (f) zeigen die Grauwertstufen von $0 - 255$ die Orientierung des Verschiebungsvektors von $0 - 2\pi$, gemessen ab der positiven x-Achse entgegen dem Uhrzeigersinn. In (a) und (b) kennzeichnen die dunklen Teile die verdeckten Bereiche. Daran sieht man in diesen drei Fällen, daß die Okklusion genau detektiert wird, der optische Fluß mit dem tatsächlichen Feld gut übereinstimmt, und dessen Diskontinuität erhalten bleibt. Numerische Simulationen zeigen, daß auch mit überlagertem Rauschen eine gute Bestimmung des optischen Flusses mit dieser Methode möglich ist.

Diese neue Methode ist auch für reale Szenen gut geeignet. Als Beispiel wird eine Szene im Labor gezeigt, die mit einer CCD-Kamera aufgenommen wurde, wobei zwischen beiden Bildern in Abbildung 5 (a) und (b) sich nur die Micky-Mouse hinter dem Monitor bewegt. Als Ergebnisse werden beide Okklusionsfelder ((c), (d)), und die Verschiebungs-

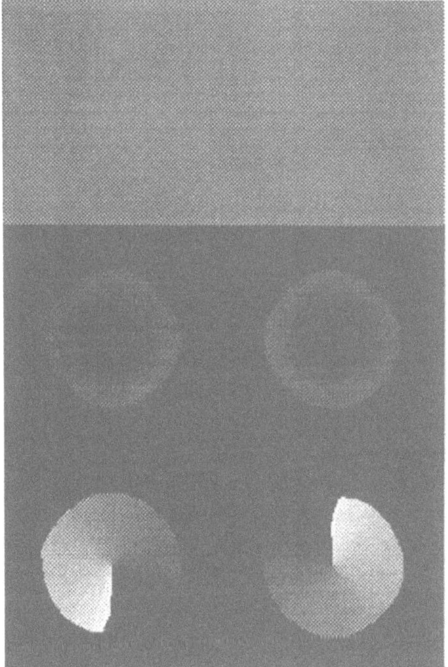

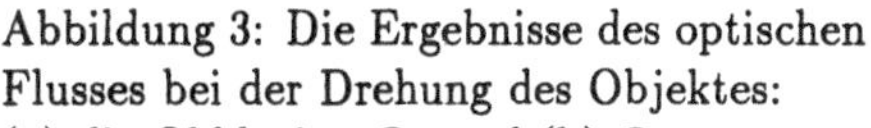

Abbildung 3: Die Ergebnisse des optischen Flusses bei der Drehung des Objektes: (a) die Okklusion $\mathbf{O}_1$ und (b) $\mathbf{O}_2$, (c) Betrag des Flusses $\|\mathbf{u_1}\|$ und (d) $\|\mathbf{u_2}\|$, (e) Phase des Flusses $\theta(\mathbf{u_1})$ und (f) $\theta(\mathbf{u_2})$

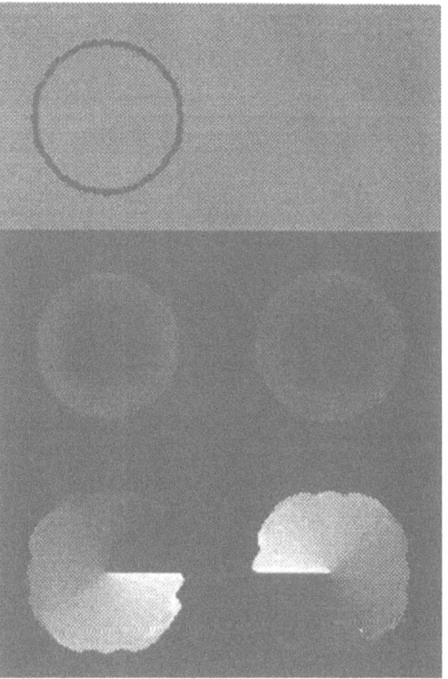

Abbildung 4: Die Ergebnisse des optischen Flusses bei der Vergrößerung des Objektes: (a) die Okklusion $\mathbf{O}_1$ und (b) $\mathbf{O}_2$, (c) Betrag des Flusses $\|\mathbf{u_1}\|$ und (d) $\|\mathbf{u_2}\|$, (e) Phase des Flusses $\theta(\mathbf{u_1})$ und (f) $\theta(\mathbf{u_2})$

vektorfelder in Betrag ((e), (f)) und Phase ((g), (h)) dargestellt. Dabei wird der optische Fluß gut rekonstruiert, und auch die verdeckten Bereiche werden in beiden Bildern detektiert. Obwohl die quantitative Bewertung in der realen Szene sehr schwer ist, kann man sehen, daß die Ergebnisse in Abbildung 5 zumindest visuell gültig sind.

5 Schlußfolgerung

Zusammenfassend ist dieser neue Ansatz gut für die Berechnung des optischen Flusses geeignet. Vorteilhaft ist, daß damit der optische Fluß unmittelbar aus Graubildfolgen unter Berücksichtigung der Diskontinuität des Feldes bestimmt werden kann, wobei zugleich die Okklusion, welche in anderen Ansätzen die Ermittlung des Flusses erschwert, detektiert und sogar ausgenutzt wird. Im Vergleich zu anderen Methoden weist diese Methode ein einfaches Konzept auf und benötigt keine Vor- oder Nachverarbeitungsstufen für Bildmerkmale oder Diskontinuitätskanten, so daß der optische Fluß ohne komplizierte Annahmen direkt und besser berechnet werden kann. Außerdem ist der zugehörige Algorithmus relativ einfach und wegen der Mehrgitterstruktur wenig rechenaufwendig. Die wesentlichen Nachteile helligkeitsbasierter Verfahren, und zwar Probleme an stark texturierten Oberflächen, Bewegungsgrenzen und Tiefendiskontinuitäten, werden bei diesem neuen Ansatz im allgemeinen gut vermieden.

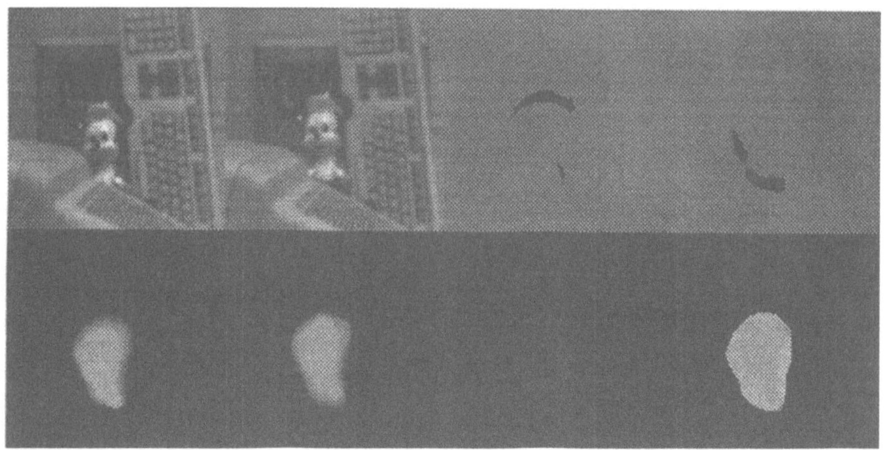

Abbildung 5: Die Ergebnisse des optischen Flusses aus einer realen Bildfolge: (a), (b) das Bildpaar aus der Bildfolge, (c) die Okklusion $\mathbf{O_1}$ und (d) die Okklusion $\mathbf{O_2}$, (e) Betrag des Flusses $\|\mathbf{u_1}\|$ und (f) $\|\mathbf{u_2}\|$, (g) Phase des Flusses $\theta(\mathbf{u_1})$ und (h) $\theta(\mathbf{u_2})$

Literatur

[Ais89] J. Aisbett, Optical flow with an intensity-weighted smoothing, IEEE Trans. Pattern Anal. Mach. Intelligence, Vol. 11, 1989, pp. 512-522

[AN88] J. K. Aggarwal and N. Nandhakumar, On the computation of motion from Sequences of images - a review, Proceedings of IEEE, Vol. 76, 1988, pp. 917-935

[Bur84] P. J. Burt, The pyramid as a structure for efficient computation, in A. Rosenfeld (eds.), Multiresolution Image Processing and Analysis, Springer-Verlag, 1984, pp. 6-35

[Enk88] W. Enkelmann, Investigations of multigrid algorithms for the estimation of optical flow fields in image sequences, Computer vision, graphics and image processing, Vol. 43, 1988, pp. 150-177

[Gla84] F. Glazer, Multilevel relaxation in low-level computer vision, in A. Rosenfeld (eds.), Multiresolution Image Processing and Analysis, Springer-Verlag, 1984, pp. 312-330

[Hil84] E .C. Hildreth, Computation underlying the measurement of visual motion, Artificial Intelligence, Vol. 23, 1984, pp. 309-354

[HS81] B.K.P.Horn and B.G.Schunck, Determining optical flow, Artificial Intelligence, Vol.17, 1981, pp. 185-203

[LB92] A. Luo and H. Burkhardt, An intensity-based cooperative bidirectional stereo matching with simultaneous detection of discontinuities and occlusions, eingereicht.

[NE86] H. Nagel and W. Enkelmann, An investigation of smoothness constraints for the estimation of displacement vector fields from image sequences, IEEE Trans. Pattern Anal. Mach. Intelligence, Vol. 8, 1986, pp. 565-593

Ein allgemeiner Zugang zur Berechnung von Invarianten

Irene Rothe und Herbert Süße
Friedrich Schiller Universität Jena
Fakultät für Mathematik und Informatik
07743 Jena

1. Einleitung

Die Berechnung von Invarianten hat für die Mustererkennung eine große Bedeutung, da bei praktischen Aufgabenstellungen Figuren, Objekte, Bilder usw. oft durch Transformationen "verfälscht" wurden. Es gilt, Größen zu berechnen, die invariant gegenüber solchen Transformationen sind. Einzelne invariante Maße sind schon sehr lange bekannt, wie z.B. das Teilverhältnis bei affinen Abbildungen oder das Doppelverhältnis bei projektiven Abbildungen. In der Mustererkennung ist es aber wichtig, vollständige Sätze von Invarianten zu berechnen, damit Objekte eineindeutig durch ihre invarianten Maße beschrieben werden können. Vollständige Sätze von Invarianten wurden z.B., wie aus Tabelle 1 ersichtlich ist, von Hu, Burkhardt, Arbter und Wang abgeleitet [Hu 62, Wa 77, Bu 79, Ar 90].

Transformation 2D-Bild	Translation	Skalierung	Rotation	affine Transformation
geordnete Punktmenge			eindim. Fourierdeskriptoren [Bu 79]	eindim. Fourierdeskriptoren [Ar 90] -projektive Invarianten
Grauwertbilder (diskrete 2D-Funktionen)	zweidim. Fourierdeskriptoren			
Grauwertbilder (stetige 2D-Funktion)		Legendre Invarianten	Hu-Momentinvarianten [Hu 62]	Wang-Invarianten [Wa 77] Tensorinvarianten [Re 93]

In der vorliegenden Arbeit soll gezeigt werden, daß alle diese Arbeiten, die unabhängig voneinander entstanden sind, auf einem gemeinsamen Grundprinzip beruhen, welches hier vorgestellt wird. Überraschenderweise lassen sich auch die Hu-Invarianten in dieses Grundprinzip einordnen. Basierend auf diesem Grundprinzip können auch andere hier nicht erwähnte Invarianten hergeleitet werden.

2. Grundlagen

In dieser Arbeit werden zweidimensionale Bilder untersucht, die entweder als analoge Funktionen $f(x,y)$ über einem Bereich $B=([0,X],[0,Y])$ oder als diskrete Funktionen über $B=([0,\ldots,M\text{-}1],[0,\ldots,N\text{-}1])$ zu interpretieren sind.
Es werden Variablentransformationen

$$x \rightarrow x', \; y \rightarrow y' \quad mit \quad x' = u(x,y) \; , \; y'=v(x,y)$$

untersucht, wobei u und v beschränkte, stetig differenzierbare Funktionen sind. Aus einem

Bild $f(x,y)$ entsteht so das transformierte Bild $f'(x',y')$. Es wird vereinbart, daß die von Null verschiedenen Funktionswerte bei einer Variablentransformation innerhalb des ursprünglichen Definitionsbereiches verbleiben. Aus praktischem Interesse untersucht man von den vielen möglichen Transformationen insbesondere die Klassen der Kongruenz-, Ähnlichkeits- und der affinen Transformationen.

3. Eine allgemeine Methode zur Berechnung von Invarianten

In dieser Arbeit soll unter einer *Invarianten* ein Funktional I der Bildfunktion $f(x,y)$ über dem Bereich B verstanden werden, das vor und nach einer Klasse von Transformationen den gleichen Wert annimmt. Der Wertebereich des Funktionals sind die komplexen Zahlen.
In dieser Arbeit haben die Funktionale die folgende Gestalt:

$$F_g[f] = \int_B f(x,y)\, g(x,y)\, dxdy,$$

wobei g ein Integralkern ist. Im diskreten Fall ersetzen wir das Integral durch eine Summe. Transformiert man die Bildfunktion $f(x,y)$ mit der Transformation T, so erhält man ein transformiertes Bild Tf mit dem Funktionalwert

$$F_g[Tf] = \int_{B'} f'(x',y')\, g'(x',y')\, dx'dy'.$$

Eine ***absolute Invariante*** bzgl. der Klasse T von Transformationen ist ein Funktional I_g, für das bezüglich beliebiger Bildfunktionen f gilt:

$$I_g[Tf] = I_g[f].$$

Die Klasse T der Transformationen werde durch n Transformationsparameter $\xi_1,\ldots,\xi_n$ eindeutig beschrieben.
Die Bildfunktion $f(x,y)$ werde durch s Bildparameter charakterisiert, z.B. durch seine $M \cdot N$ Grauwerte. Die Invariantenberechnung ist nur sinnvoll, wenn $s > n$ gilt. Eine Transformation des Bildes bedingt dann eine Transformation der das Bild charakterisierenden Parameter.
Im folgenden wird eine Methode in vier Abschnitten beschrieben, mit der man aus den Bildparametern invariante Größen ableiten kann:

1) Es wird eine eineindeutige Beschreibung des Bildes $f(x,y)$ durch Bildparameter gesucht,
2) es wird versucht, eine Funktion zu finden, die die Bildparameter nach der Transformation explizit durch die Bildparameter vor der Transformation und die Transformationsparameter ausdrückt. In dieser Arbeit werden die so erhaltenen Bildparameter *relative Invarianten* genannt,
3) die Transformationsparameter werden ebenfalls explizit durch die Bildparameter vor der Transformation ausgedrückt, indem bestimmte Bildparameter nach der Transformation normiert werden, oder man versucht direkt, durch Kombination der relativen Invarianten absolute Invarianten zu berechnen,
4) durch Einsetzen der Transformationsparameter in die explizite Darstellung aus 2) werden absolute Invarianten berechnet.

Beschreibung der einzelnen Schritte:

1) Beschreibung des Bildes durch Parameter:

Eine Möglichkeit, ein Bild durch Parameter zu beschreiben, ist die Darstellung der analogen und diskreten Funktionen nach Basisfunktionen:

$$f(x,y) = \sum_{k=-\infty}^{+\infty} \sum_{l=-\infty}^{+\infty} \gamma_{k,l}\; h_{k,l}(x,y) \; .$$

Die Koeffizienten $\gamma_{k,l}$, die die Menge Γ bilden, beschreiben dann die Funktion $f(x,y)$ als Bildparameter eineindeutig. Sie werden für die Suche nach Invarianten verwendet. Benutzt man Orthonormalbasen, dann sind die Bildparameter die verallgemeinerten Fourierkoeffizienten. Es reicht aus, nur endliche Summen zu betrachten, da alle praktisch bedeutsamen Bilder durch endliche Summen mit hinreichend vielen Summanden mit der erforderlichen Genauigkeit approximiert werden können.

2) Mit der Transformation

$$x' = u(x,y), \; y' = v(x,y)$$

werden die Parameter $\gamma_{k,l}$ in die transformierten Parameter $\gamma'_{k,l}$ umgeformt. Es mögen Parameter $\xi_1,\ldots \xi_n$ existieren, die diese Transformation beschreiben. Weiter sind Funktionen $F_{k,l}$ zu finden mit

$$\gamma'_{k,l} = F_{k,l}(\xi_1,\ldots,\xi_n,\Gamma),$$

so daß diese $\gamma'_{k,l}$ relative Invarianten darstellen.

3) Hat man solche Funktionen $F_{k,l}$ gefunden, die die neuen Koeffizienten $\gamma'_{k,l}$ durch die alten Koeffizienten $\gamma_{k,l}$ und die Transformationsparameter $\xi_1,\ldots,\xi_n$ ausdrücken, so versucht man, n von s Koeffizienten $\gamma'_{k,l}$ bzw. Funktionen $F_{k,l}$, die eine Menge Γ'_n bilden sollen, derart auszuwählen, so daß man die Transformationsparameter als Funktion von Elementen aus Γ'_n und Γ wie folgt darstellen kann

$$\xi_1 = \hat{\xi}_1(\Gamma'_n,\Gamma) \; ,$$

$$\vdots$$

$$\xi_n = \hat{\xi}_n(\Gamma'_n,\Gamma).$$

Um Invarianten, d.h. Größen unabhängig von den Transformationsparametern, zu berechnen, muß man auf so viele Koeffizienten verzichten, wie man Transformationsparameter hat. Werden alle Werte aus der Menge Γ'_n in obigen Gleichungen konstant gesetzt, wobei C die Menge der verwendeten Konstanten bilden soll, so kann man im nächsten Schritt absolute Invarianten berechnen. Das Konstantsetzen wird in dieser Arbeit als *Normierung* bezeichnet. Sollte es auf einfache Art und Weise möglich sein, aus den relativen Invarianten die Transformationsparameter direkt zu eliminieren, dann erhält man sofort absolute Invarianten.

4) In die aus dem zweiten Schritt erhaltenen Gleichungen

$$\gamma'_{k,l} = F_{k,l}(\xi_1,\dots,\xi_n,\Gamma)$$

setzt man nun die für die Transformationsparameter $\xi_1,\dots,\xi_n$ erhaltenen Gleichungen aus dem dritten Schritt ein und kann alle $\gamma'_{k,l}$ aus $\Gamma\backslash\Gamma'_n$ zu folgenden absoluten Invarianten umformen

$$\gamma''_{k,l} = F_{k,l}(\xi_1(\Gamma,C),\dots,\xi_n(\Gamma,C),\Gamma).$$

Diese Koeffizienten $\gamma''_{k,l}$ sind weder von den Koeffizienten $\gamma'_{k,l}$ noch von Transformationsparametern abhängig.
Durch die *willkürliche* Wahl der n Koeffizienten und die Art der Normierung sind die vielfältigsten Invariantenherleitungen möglich. In den Beispielen werden Möglichkeiten der Wahl der n Koeffizienten $\gamma'_{k,l}$ und der Art der Normierungen gegeben.

4. Beispiele

4.1 Fourierdeskriptoren von Burkhardt und Arbter [Bu 79, Ar 90]

In diesem Abschnitt werden Invarianten von Konturen hergeleitet, die durch die x- und y-Koordinaten von N Punkten einer geordneten, endlichen, im positiven Sinne orientierten Punktmenge beschrieben wird: $z_0=(x_0,y_0),\dots,z_{N-1}=(x_{N-1},y_{N-1})$. Sollen nun z.B. affininvariante Fourierdeskriptoren berechnet werden, dann präzisiert man die allgemeine Vorgehensweise zu:

1) Erklärt man die Ordinate zur imaginären Achse, dann erhält man aus der Punktmenge eine komplexe Funktion z, der man die folgenden Fourierkoeffizienten eineindeutig zuordnen kann:

$$\alpha(z)_k = \frac{1}{\sqrt{N}}\sum_{n=0}^{N-1}(x_n+iy_n)\,e^{\frac{-2\pi ink}{N}}\quad;\quad k=0,\dots,N-1.$$

Dies sind die Bildparameter.

2) Als Darstellung der affintransformierten Fourierkoeffizienten $\alpha(z_{affin})_k$ in expliziter Form durch die komplexen Transformationsparameter a und b und die alten Fourierkoeffizienten $\alpha(z)_k$ erhält man:

$$\alpha(z_{affin})_k = a\alpha(z)_k+b\alpha(z)^*_{-k}$$

$$\textit{mit}\quad a=\frac{1}{2}(a_{11}+a_{12})+\frac{i}{2}(a_{21}-a_{12}),\ b=\frac{1}{2}(a_{11}-a_{22})+\frac{i}{2}(a_{21}+a_{12}),$$

wobei a_{11}, a_{12}, a_{21}, a_{22} beliebige reelle Zahlen sind, die die affine Transformation beschreiben.

3) Zur Berechnung der zwei komplexen Transformationsparameter a und b werden zwei willkürlich gewählte Fourierkoeffizienten $\alpha(z_{affin})_k$ normiert. Um die Berechnung so

einfach wie möglich zu gestalten, wurde speziell der p-te und der $(-p)$-te (d.h. $(N$-$p)$-te) Fourierkoeffizient wie folgt normiert:

$$\alpha(z_{affin})_p = 1 \quad , \quad \alpha(z_{affin})_{-p} = 0.$$

Als analytische Lösung eines Gleichungssystem mit vier Gleichungen und vier Unbekannten für die beiden komplexen (also vier reellen) Transformationsparameter erhält man:

$$a = \frac{\alpha(z)_p^*}{\alpha(z)_p\alpha(z)_p^* - \alpha(z)_{-p}^*\alpha(z)_{-p}} \, , \, b = \frac{-\alpha(z)_{-p}}{\alpha(z)_p\alpha(z)_p^* - \alpha(z)_{-p}^*\alpha(z)_{-p}} \, .$$

4) Setzt man diese Parameter an die entsprechenden Stellen in $\alpha(z_{affin})_k$ ein, so erhält man die Invarianten

$$\frac{\alpha(z)_k\alpha(z)_p^* - \alpha(z)_{-k}^*\alpha(z)_{-p}}{\alpha(z)_p\alpha(z)_p^* - \alpha(z)_{-p}^*\alpha(z)_{-p}} \; ; \quad k=1,\ldots,N-1 \; , \quad \alpha(z)_p\alpha(z)_p^*-\alpha(z)_{-p}^*\alpha(z)_{-p} \neq 0,$$

welche ein vollständiges und minimales Invariantensystem nach [Ar 90] bilden. Wählt man andere Normierungskonstanten, so erhält man auch entsprechend andere affininvariante Fourierdeskriptoren.

4.2. Momentinvarianten nach Wang

1) Ein anderes Beispiel einer eineindeutigen Beschreibung einer analogen Funktion $f(x,y)$ wird durch die Menge der zugehörigen Momente m_{pq} gegeben:

$$m_{pq} = \int_{-\infty}^{+\infty} \int_{-\infty}^{+\infty} x^p y^q \, f(x,y) dx dy \; ; \quad p,q = 0,\ldots,\infty \, .$$

2) Bei der Berechnung von affininvarianten Momenten wurde die Darstellung der affinen Transformationsmatrix als Verknüpfung einer anisotropen Skalierungsmatrix, einer Scherungsmatrix, einer Rotationsmatrix und eines Verschiebungsvektors ausgenutzt [Wan 77].

3) Um sechs Parameter der affinen Transformation nur durch Momente vor der Transformation auszudrücken, wäre $m'_{10}=m'_{01}=m'_{11}=0$, $m'_{20}=m'_{02}=1$, $m'_{30}+m'_{12}=0$ und $m'_{03}+m'_{21}>0$ eine Normierung.

4) Konstruiert man aus den allgemeinen Momenten translationsinvariante Momente, bildet aus diesen dann die folgenden Momente

$$m_{pq}^I = \iint (x-\frac{m_{11}}{m_{02}}y)^p \, y^q \, f(x,y) \; dx \; dy,$$

so erhält man scherungsinvariante Momente. Leitet man nun aus obigen Momenten rotationsinvariante und später anisotropskalierungsinvariante Momente ab, so erhält man schließlich affininvariante Momente.

4.3. Momentinvarianten nach Hu

1) Die Invarianten nach Hu [Hu 62] lassen sich überraschenderweise auch in diese allgemeine Vorgehensweise einordnen. Als Bildparameter benutzt man jetzt die komplexen Momente nach [Te 88]

$$C_{pq} = \int_{-\infty}^{\infty} \int_{-\infty}^{\infty} (x+iy)^p (x-iy)^q f(x,y) dx dy$$

oder in Polarkoordinaten

$$C_{pq} = \int_{0}^{2\pi} \int_{0}^{\infty} r^{p+q+1} \, e^{i(p-q)\phi} \, f(r\cos\phi, r\sin\phi) \, dr d\phi .$$

2) Dann läßt sich eine Rotation um φ einfach durch

$$C'_{pq} = C_{pq} \, e^{i(p-q)\varphi}$$

beschreiben, womit diese gleichzeitig relative Invarianten sind.
Durch einfaches Ausrechnen zeigt man, daß die komplexen Momente mit den nichtkomplexen Momenten wie folgt m_{pq} zusammenhängen:

$$C_{pq} = \sum_{r=0}^{p} \sum_{s=0}^{q} \binom{p}{r} \binom{q}{s} \, i^{p+q-r-s} \, (-1)^{q-s} \, m_{r+s,p+q-r-s}.$$

Setzt man diese C_{pq} in die obigen relativen Invarianten ein, so kann man relative Invarianten in Abhängigkeit von den nichtkomplexen Momenten bilden, z.B.

$$C'_{00} = m_{00} \, , \; C'_{10} = (m_{10}+im_{01}) e^{i\varphi} \, , \; C'_{01} = (m_{10}-im_{01}) e^{-i\varphi} \; usw.$$

3) Durch geschickte Kombination dieser relativen Invarianten kann man den Winkel eliminieren und erhält so als absolute Invarianten gerade die Invarianten nach Hu. Da die relativen Invarianten einfach nach dem Drehwinkel auflösbar sind, können diese auch zur Drehwinkelberechnung herangezogen werden

$$\varphi + \frac{\alpha_{p-q,q}}{p-2q} = \frac{1}{p-2q} \arctan \frac{\Im\left(C'_{p-q,q}\right)}{\Re\left(C'_{p-q,q}\right)} + \frac{k\pi}{p-2q} \; ; \; 0 \leq k < p-2q \, ,$$

wie in [Ro 92] ausführlich dargestellt ist. Dabei ist $\alpha_{p-q,q}(C_{pq})$ der Drehwinkel bezüglich einer Normallage des Objektes vor der Rotation.

4.4. Legendre-,Hermite-Invarianten, ...

Je nach Wahl der Basis bezüglich der die Funktion $f(x,y)$ dargestellt wird, kann man Invarianten berechnen, z.B. Legendre-Invarianten bzgl Translation:
Die Koeffizienten der Darstellung von $f(x,y)$ mit Hilfe der Legendrepolynome werden wie folgt berechnet:

$$c_{k,l} = \sqrt{k+\frac{1}{2}}\sqrt{l+\frac{1}{2}}\frac{1}{2^k k!}\frac{1}{2^l l!}\int_{-1}^{1}\int_{-1}^{1}\frac{d^k}{dx^k}(x^2-1)^k\frac{d^l}{dy^l}(y^2-1)^l f(x,y)dxdy.$$

Zur Berechnung von Invarianten bezüglich der Translation ist es nötig, zwei Koeffizienten konstant zu setzen. Der Einfachheit halber setzt man c'_{10} und c'_{01} gleich Null, woraus folgt

$$a = -\frac{c_{10}}{\sqrt{3/2}} \quad , \quad b = -\frac{c_{01}}{\sqrt{3/2}} \ .$$

Somit sind die Merkmale

$$c^I_{k,l} = const_{k,l}\int_{-1}^{1}\int_{-1}^{1} f(x,y)\frac{d^k}{dx^k}((x-\frac{c_{10}}{\sqrt{3/2}})^2-1)^k\frac{d^l}{dy^l}((y-\frac{c_{01}}{\sqrt{3/2}})^2-1)^l\,dx\,dy$$

translationsinvariant.

4.5. Zweidimensionale translationsinvariante Fourierdeskriptoren

Auch für Grauwertbilder lassen sich mit dieser allgemeinen Herangehensweise Fourierdeskriptoren ableiten. Die folgenden Merkmale

$$\alpha(f)^I_{k,l} = |\alpha(f)_{k,l}|e^{i(\Phi(f)_{k,l}+(-kt+ltr-l)\Phi(f)_{r,s}+(k-lr)\Phi(f)_{tr-1,t})} \quad ; \quad tr-1 \geq 0$$

sind zweidimensionale verschiebungsinvariante Fourierdeskriptoren. Dabei ist die Phase entsprechend normiert worden.
Man kann nun Grauwertbilder behandeln, bei denen eine Konturdetektion nicht sinnvoll ist, man muß aber beachten, daß die Verschiebung nicht oder nur wenig über den Bildrand hinausgeht. Wie in Abbildung 1 zu sehen ist, erfüllen die Sternenbilder annähernd solche Restriktionen. Das Teilbild 2 ist gegenüber Teilbild 1 um (4,-6) Positionen verschoben, die Berechnung der Verschiebungsparameter aus den zwei speziellen relativen Invarianten

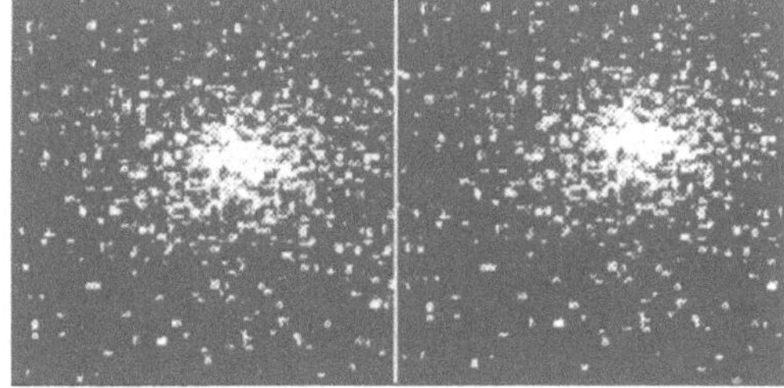
Abb.1: Verschobene Sternenhaufen

$$a = \frac{(\Phi(f)_{1,0} - \Phi(f_{transl})_{1,0})}{2\pi} M + g_a M ,$$

$$b = \frac{(\Phi(f)_{0,1} - \Phi(f_{transl})_{0,1})}{2\pi} N + g_b N \quad mit \quad g_a, g_b \epsilon Z .$$

ergab (3.85,-6.06). Die Invarianten sind aus der folgenden Tabelle zu entnehmen. Sie sind nicht exakt gleich, da die Verschiebung nicht ganz innerhalb des Bildes verlaufen ist.

Nr.der Invarianten	Betrag der I. vor der Translation	Betrag der I. nach der Translation	Phase der I. vor der Translation	Phase der I. nach der Translation
(1,-2)	150.6	150.8	177.8	178.2
(1,-1)	253.1	252.7	0.7	359.3
(1,0)	301.2	304.8	182.3	181.3
(1,1)	244.4	249.0	0.0	0.0
(1,2)	180.7	176.7	180.7	180.7

5. Literatur

Ar90 Arbter,K.:Affininvariante Fourierdeskriptoren ebener Kurven. Promotion der Techn. Uni Hamburg, 1990

Bu79 Burkhardt,K.: Transformationen zur lageinvarianten Merkmalsgewinnung. VDI-Fortschritt-Bericht, Reihe 10, Nr. 7, VDI-Verlag Düsseldorf, 1979

Hu62 Hu,M.-K.: Visual pattern recognition by moment invariants. Proc. IRE-49, (1961) pp 113-121

Mu92 Mundy,J., Zisserman,A.: Geometric Invariance in Computer Vision. MIT Press, England (1992)

Re91 Reiss,T.H.: The revised fundamental theorem of moment invariants. IEEE Trans. PAMI-13 (1991), no 8, pp 830-833

Re93 Reiss,T.H.: Recognition Planar Objects Usuing Invariant Image Features. Springer Verlag Berlin Heidelberg, 1993

Re88 Reeves,A.P.: Three dimensional shape analysis using moments and Fourier-descriptors. IEEE Trans. PAMI-10 (1988), no 6

Ro92 Rothe,I., Voss,K.: Orientierungsbestimmung von Objekten durch Moment-invarianten. Tagungsband des 14.DAGM Sympos. in Dresden, Springer Verlag 1992, S.42-49

Te88 Teh,C., Chin,R.: On image analysis by the methods of moments. IEEE Trans. PAMI-10 (1988), Nr.4

Wa77 Wang,K.: Affine-invariant moment method of three dimensional object identification. Syracuse Uni, Ph.D., 1977

Affin-invariante Erkennung von Grauwertmustern mit Fourierdeskriptoren

Axel Fenske

Technische Universität Hamburg-Harburg

Technische Informatik I, Postfach 90 10 52, 2100 Hamburg 90

Zusammenfassung

Die vorliegende Arbeit stellt eine neue Methode zur affin-invarianten Erkennung von Grauwertmustern vor. Sie beruht auf einer Verallgemeinerung der Fourierdeskriptoren auf Grauwertmuster. Bisher waren die Fourierdeskriptoren nur zur Beschreibung von Konturmustern bekannt. Durch den neuen Ansatz ist es möglich, planare zweidimensionale Grauwertobjekte von einer beliebigen räumlichen Kameraposition aufgrund ihrer unterschiedlichen Grauwertmuster zu unterscheiden, auch wenn die Objekte die gleichen Konturformen aufweisen. Dabei wird das Grauwertmuster auf die Parametrisierung der Kontur abgebildet, so daß die Fourierdeskriptoren für Grauwertmuster formal den Deskriptoren ebener Kurven gleichen und die bekannten Ergebnisse zur Bildung affin-invarianter Merkmale übernommen werden können. Außerdem wird ein neuer Konstruktionsansatz zur Bildung der Invarianten vorgestellt, der die Phasenfehler der Fourierdeskriptoren minimiert. Schließlich wird anhand zweier Experimente ein Vergleich zwischen den Grauwertfourierdeskriptoren und der Methode der affin-invarianten Momente gezogen. Spielt der Grauwertanteil gegenüber dem Formanteil bei der Unterscheidung der Objekte die größte Rolle, sind die Deskriptoren den Momenten etwas überlegen. Ist dagegen der Formanteil dominant, besitzen die Momente die größere Separationsfähigkeit.

1 Einleitung

Die lageinvariante Klassifikation planarer oder nahezu flächenhafter Objekte basiert in der Regel auf deren unterschiedlichen Formen. Besitzen die Objekte die gleiche Konturform, versagen diese Methoden. Eine Unterscheidung, zu welcher Klasse ein Testobjekt gehört, ist in diesem Fall nur über die Grauwertinformation möglich. Verändert ein planares Objekt seine Position beliebig im dreidimensionalen Raum, wird die Transformation der dazugehörigen Bildpunkte unter Annahme des parallelprojektiven Kameramodells durch die affine Gruppe $\mathcal{G}_{\mathcal{A}}$ beschrieben. Die affine Transformation approximiert die zentralprojektive Abbildung, falls der Abstand Kamera-Objekt groß ist im Vergleich zur Objektausdehnung. Die Aufgabenstellung der affin-invarianten zweidimensionalen Objekterkennung findet sich z.B. im industriellen Produktionsprozeß unter der Rahmenbedingung, daß Objekte ungeordnet unter dem Bildsensor vorbei geführt werden, oder bei der Erkennung von Schildern und Plaketten aus einer beliebigen Kameraposition. Weiterhin kann die affin-invariante zweidimensionale Mustererkennung auch ein wichtiger Vorverarbeitungsschritt der dreidimensionalen Objekterkennung sein.

Als einzige Methode, affin-invariante Merkmale *direkt* aus einem Grauwertmuster zu extrahieren, sind die Momente bekannt [1, 2]. Die Objekte müssen aber auch hier vom Hintergrund getrennt sein. Andere Verfahren zur Invariantenbildung wie z.B. geometrisches Hashing benötigen geometrische Entitäten wie Ecken oder Markierungspunkte, um affine Invarianz zu erreichen [3]. Die vorliegende Arbeit stellt einen neuen Ansatz zur affin-invarianten Erkennung vor, der auf der Methode der Fourierdeskriptoren beruht und der außer der Grauwertinformation nur die äußere Konturlinie eines Objektes benötigt. Bisher war die Methode der Fourierdeskriptoren nur zur Beschreibung ebener Kurven bekannt. Die Übertragung auf Grauwertmuster wird über die Einführung einer Grauwertparametrisierung erreicht, durch die das Grauwertmuster auf die Parameterdarstellung der Kontur abgebildet wird. Die Kombination der Grauwert- mit der Konturinformation ermöglicht es, Grauwertmuster zu separieren, die eine verschiedene Konturform oder eine verschiedene Parametrisierung haben.

In Abschnitt 2 werden die Transformationsgruppen besprochen, unter denen die Grauwertfourierdeskriptoren invariant sind. Außerdem werden kurz die Grundlagen der Fourierreihen ebener Kurven behandelt. In

Abschnitt 3 wird die Grauwertparametrisierung eingeführt. Die Anwendung der Grauwertparametrsierung führt zu den Fourierdeskriptoren für Grauwertmuster, siehe Abschnitt 4. Ein neuer Ansatz zur Bildung der invarianten Deskriptoren wird in Abschnitt 5 vorgestellt. Durch ihn verändert sich das Schema der Phasennormierung zur Bildung der Invarianten derart, daß die Normierungsphasen von dem Index der zu normierenden Größe abhängen. Durch diese Kopplung wird eine Korrelation der Phasenfehler erreicht, durch die es zu Fehlerauslöschungseffekten bei der Phasennormierung kommt. Die so entwickelten Fourierdeskriptoren streuen erheblich geringer im Merkmalsraum als die bisher bekannten. Für den Fall ebener Kurven sind einige dieser Systeme vollständig. Dies ist bemerkenswert, da sie nicht eindeutig umkehrbar sind, eine Eigenschaft, die sonst üblicherweise zum Beweis der Vollständigkeit benötigt wird.

Der Abschnitt 6 gibt die Ergebisse zweier Klassifikationsexperimente wieder. Das erste behandelt die Erkennung von Grauwertmustern, das zweite die Erkennung von Konturmustern. Es werden jeweils die neuen Invariantensysteme mit den bekannten Systemen von Fourierdeskriptoren verglichen. Zweitens wird der Einfluß verschiedener Klassifikatoren untersucht. Drittens werden die Erkennungsraten der Fourierdeskriptoren mit denen affin-invarianter Momente verglichen.

2 Grundlagen

Sei $\mathcal{O}$ ein planares zweidimensionales Objekt im dreidimensionalen Raum. Das Bild von $\mathcal{O}$ wird in der kontinuierlichen Beschreibung durch die Grauwertfunktion $f(x,y) \in \mathbb{R}$ beschrieben mit den Bildkoordinaten $x, y \in \mathbb{R}$. Es sei angenommen, daß das Objekt vom Hintergrund getrennt werden kann. Der Hintergrund wird auf $f(x,y) = 0$ gesetzt und die äußere Kontur C des Objektes sei gegeben. Verändert $\mathcal{O}$ seine Lage im dreidimensionalen Raum, werden die Bildpunkte des Grauwertmusters unter der Annahme des parallelprojektiven Kameramodells durch ein Element der affinen Transformationsgruppe $g_{\mathcal{A}} \in \mathcal{G}_{\mathcal{A}}$ transformiert. Es ist $g_{\mathcal{A}} : (x,y) \longmapsto (x',y')$, wobei gilt:

$$\begin{pmatrix} x \\ y \end{pmatrix}' = \mathbf{A}\begin{pmatrix} x \\ y \end{pmatrix} + \mathbf{b} \quad \text{mit} \quad \mathbf{A} = \begin{pmatrix} a_{11} & a_{12} \\ a_{21} & a_{22} \end{pmatrix} \quad , \quad \mathbf{b} = \begin{pmatrix} b_1 \\ b_2 \end{pmatrix} \quad , \quad \det \mathbf{A} > 0 \quad , \quad a_{kl}, b_k \in \mathbb{R} \quad . \tag{1}$$

sind. Es sind nur eigentliche affine Transformationen mit $\det \mathbf{A} > 0$ zugelassen, um Spiegelungen, die sich auf den Umlaufsinn eines Konturverfolgers auswirken, und singuläre Abbildungen auszuschließen. Die affine Gruppe läßt neben Translationen, Rotationen und Größenänderungen eines Grauwertmusters auch Scherungen zu, siehe Abb. 1.

Weiterhin sei als Transformationsgruppe noch die Gruppe der Kontraständerungen $\mathcal{G}_K$ zugelassen. Die Abbildung $g_k \in \mathcal{G}_K$ wirkt auf einem Grauwertmuster durch den Kontrastfaktor $K \in \mathbb{R}_+$ folgendermaßen:

$$g_K : f(x,y) \longmapsto f'(x,y) = K \cdot f(x,y) \quad , \quad K = \text{konst} \quad , \quad \forall x, y \in \mathbb{R} \quad . \tag{2}$$

Die Gesamttransformation eines Grauwertmusters wird beschrieben durch die Gruppe $\mathcal{G}_{\mathcal{A},K} = \mathcal{G}_{\mathcal{A}} \times \mathcal{G}_K$. Es ist

$$g_{\mathcal{A},K} : f(x,y) \longmapsto f'(x',y') = K \cdot f(x,y) \quad \forall x, x', y, y' \in \mathbb{R} \quad , \tag{3}$$

wobei Gl. (1) erfüllt ist.

Eine Äquivalenzklasse eines Objektes $\mathcal{O}$ wird erzeugt durch die Menge aller Muster, die durch Abbildungen $g_{\mathcal{A},K}$ auf einem dazugehörigem Referenzmuster entstehen, d.h. zwei Objekte $\mathcal{O}, \mathcal{O}'$ mit den Grauwertfunktionen f, f' sind äquivalent unter der Gruppe $\mathcal{G}_{\mathcal{A},K}$, falls gilt:

$$\mathcal{O}' \overset{\mathcal{G}_{\mathcal{A},K}}{\sim} \mathcal{O} \iff \exists \;\; g_{\mathcal{A},K} : f(x,y) \longmapsto f'(x',y') \quad . \tag{4}$$

Sei $\mathbf{I}$ ein reell- oder komplexwertiger Merkmalsvektor mit den Komponenten I_k. Die notwendige Bedingung für affin-invariante Merkmale ist:

$$\mathcal{O}' \overset{\mathcal{G}_{\mathcal{A},K}}{\sim} \mathcal{O} \implies \mathbf{I}' = \mathbf{I} \quad . \tag{5}$$

Falls auch die hinreichende Bedingung erfüllt ist, dann heißt die Menge der Invarianten $\{I_k\}$ vollständig. Ist der Merkmalssatz eines Mustererkennungssystems nicht vollständig, erfordert die Klassifikationsaufgabe die Prüfung der Merkmale auf Separierbarkeit.

Die Methode der Fourierdeskriptoren zur Beschreibung von ebenen Kurven beruht auf der Theorie der Fourierreihen für periodische Funktionen [4, 5]. Es sei $\mathbf{r}(t) = (x(t), y(t))^T$ mit $t \in [0, T]$ eine vektorielle Parameterdarstellung der Kontur C mit dem Parameter t, siehe Abb. 1a. Die Parametrisierung sei $t = 0$ an einem

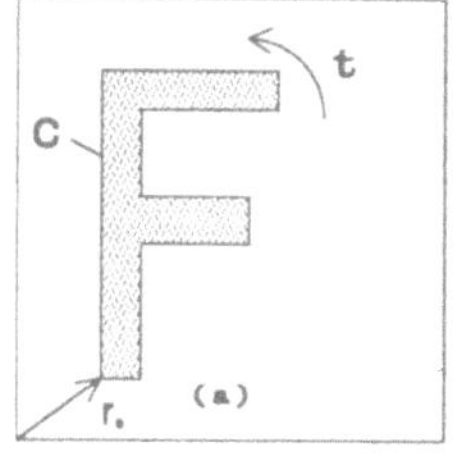

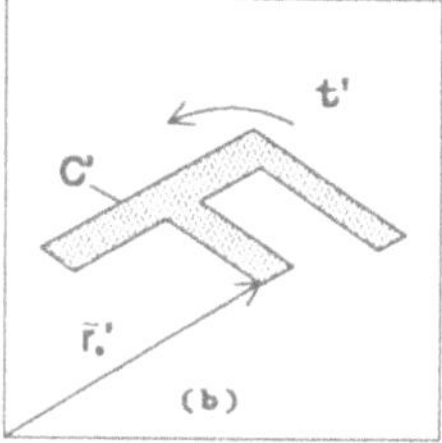

Abbildung 1: (a) Referenzobjekt $\mathcal{O}$ mit geschlossener Kontur $\mathcal{C}$; (b) Affin transformiertes Objekt $\mathcal{O}'$ mit verschiedenem Startpunkt $\tilde{\mathbf{r}}_0'$ des Konturfinders

beliebigen Startpunkt $\mathbf{r}_0 = \mathbf{r}(0) \in \mathcal{C}$ und wächst monoton auf T an, wenn die Kontur in mathematisch positiver Richtung bis zum Startpunkt durchlaufen wird. Daraus folgt, daß $\mathbf{r}(t)$ eine periodische Funktion mit der Periode T ist. Dann läßt sich $\mathbf{r}(t)$ als Fourierreihe darstellen mit den komplexwertigen Fourierkoeffizienten X_k, Y_k:

$$\mathbf{r}(t) = \begin{pmatrix} x(t) \\ y(t) \end{pmatrix} = \sum_{k=-\infty}^{\infty} \begin{pmatrix} X_k \\ Y_k \end{pmatrix} e^{j2\pi kt/T} \quad , \quad F_k = \begin{pmatrix} X_k \\ Y_k \end{pmatrix} = \frac{1}{T} \int_0^T \begin{pmatrix} x(t) \\ y(t) \end{pmatrix} e^{-j2\pi kt/T} dt \tag{6}$$

mit $k \in \mathbb{Z}$. Zusätzlich zur Transformation des Musters tritt in der Regel eine Startpunktverschiebung des Konturfolgers auf. Der Startpunkt der transformierten Kontur $\tilde{\mathbf{r}}_0'$ ist in diesem Fall nicht das affine Abbild von $\mathbf{r}_0$, d.h. $\tilde{\mathbf{r}}_0' \neq \mathbf{A}\mathbf{r}_0 + \mathbf{b}$. Die Abb. 1b verdeutlicht diesen Sachverhalt.

3 Grauwertparametrisierung

Bevor die Grauwertparametrisierung, auf der die Fourierdeskriptoren für Grauwertmuster aufbauen, definiert wird, werden allgemein die Bedingungen genannt, die eine Parametrisierung zur Bildung der Invarianten unter der affinen Gruppe erfüllen muß. Bei den meisten Anwendungen von Fourierreihen ist die Parametrisierung signalunabhängig vorgegeben, z.B. die Zeitachse. Im Unterschied dazu werden in der vorliegenden Arbeit Parametrisierungen betrachtet, die vom Signal selbst abhängen. Da die Werte der Fourierkoeffizienten gemäß Gl. (6) von der Parametrisierung abhängen, stellt diese gewissermaßen einen Freiheitsgrad bei der Beschreibung der Kurve durch die Fourierreihe dar.

Definition 1 *Sei $\mathcal{O}' = g(\mathcal{O})$ das Abbild des Objektes $\mathcal{O}$, wobei $g \in \mathcal{G}$ Element der Transformationsgruppe $\mathcal{G}$ sei. $\mathbf{r}'(t') = g(\mathbf{r}(t))$ und $\mathbf{r}(t)$ seien die Parameterdarstellungen der dazugehörigen Konturen $\mathcal{C}', \mathcal{C}$. Dann heißt die Parametrisierung* **konsistent**, *wenn sie sich unter $\mathcal{G}$ und einer Aufpunktverschiebung affin und orientierungstreu transformiert, d.h. für alle $\mathbf{r}'(t'), \mathbf{r}(t)$ gilt*

$$t' = \mu t + \nu \quad , \quad \mu \in \mathbb{R}_+ \quad , \quad \nu \in \mathbb{R} \ . \tag{7}$$

Die Bogenlänge ist z.B. konsistent unter der Ähnlichkeitsgruppe, aber nicht unter der affinen Gruppe $\mathcal{G}_\mathcal{A}$ [2]. Die Konsistenz einer Parametrisierung ist eine notwendige und hinreichende Bedingung für die Linearität der Fourierkoeffizienten ($k \neq 0$) unter der Transformationsgruppe $\mathcal{G}_{\mathcal{A},K}$. Es gilt

Satz 1 *Seien $\mathbf{r}'(t'), \mathbf{r}(t)$ die Parameterdarstellungen der geschlossenen Konturen $\mathcal{C}', \mathcal{C}$, wobei $t \in [0,T], t' \in [0,T']$ sind. F_k', F_k seien die Fourierkoeffizienten der dazugehörigen Fourierreihen. Dann gilt*

$$F_k' = \begin{cases} e^{jk\phi}\mathbf{A}F_k & : \ k \neq 0 \\ e^{jk\phi}\mathbf{A}F_k + \mathbf{b} & : \ k = 0 \end{cases} \tag{8}$$

genau dann, wenn $\mathbf{r}'(t') = g_{\mathcal{A},K}(\mathbf{r}(t))$ und die Parametrisierung konsistent ist, d.h. $t' = \mu t + \nu$. Es ist

$$\mu = \frac{T'}{T} \quad , \quad \nu = -\frac{\mu T \phi}{2\pi} \ . \tag{9}$$

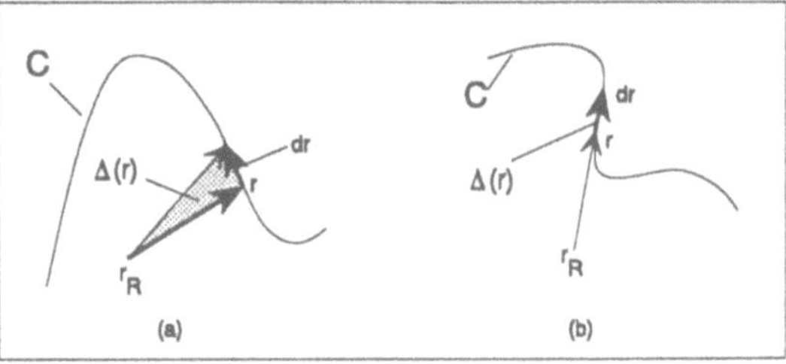

Abbildung 2: (a) Infinitesimales Dreiecksgebiet $\Delta(\mathbf{r}(\xi_i))$; (b) Kollinearer Fall

Zum Beweis siehe [6]. Die Eigenschaft der Linearität der Fourierkoeffizienten ($k \neq 0$) wird zur Bildung der invarianten Fourierdeskriptoren benötigt, siehe Abschnitt 4. Parametrisierungen für planare Kurven, die konsistent unter der affinen Gruppe sind, sind in [2, 7] hergeleitet. In [8] wird ausführlich die Flächenparametrisierung t_F untersucht, die sich durch ihre Einfachheit zur Implementation und ihre Übertragbarkeit von glatten Kurven auf Polygonzüge auszeichnet.

Sei ξ eine stetige und injektive Parametrisierung zur Parameterdarstellung $\mathbf{r}(\xi)$ einer Kurve $\mathcal{C}$. Die Bogenlänge erfüllt diese Kriterien und kann immer gewählt werden. Gegeben sei ferner ein Referenzpunkt $\mathbf{r}_R = (x_R, y_R)^T$, der sich aus der Konturlinie berechnen läßt und der sich wie die Kontur affin transformiert. Dies kann z.B. der Massenschwerpunkt der Konturlinie sein. Die Flächenparametrisierung t_F ergibt sich folgendermaßen: der Parameter t_F eines Konturpunktes $\mathbf{r}(\xi_i)$ ist

$$t_F(\xi_i) = \frac{1}{2}\int_{\xi_0}^{\xi_i} |(\mathbf{r}(\xi) - \mathbf{r}_R) \times \dot{\mathbf{r}}(\xi)|\, d\xi \;=\; \frac{1}{2}\int_{\xi_0}^{\xi_i} |(\mathbf{r}(\xi) - \mathbf{r}_R) \times d\mathbf{r}(\xi)| \quad . \tag{10}$$

Das Linienintegral wird ausgewertet von dem Startpunkt $\mathbf{r}_0 = \mathbf{r}(\xi_0)$ bis zu dem Konturpunkt $\mathbf{r}(\xi_i)$. Mit $\times$ wird das Kreuzprodukt zweier Vektoren bezeichnet. Die Begriffsbildung Flächenparametrisierung hat ihren Ursprung in der geometrischen Bedeutung des Parameters t_F. Ausgehend von dem Vektor zwischen Referenzpunkt und Startpunkt des Konturverfolgers ist der Wert von t_F an einem Konturpunkt $\mathbf{r}$ die akkumulierte absolute Fläche, die dieser Vektor überstreicht, wenn der Endpunkt die Kontur in einer Richtung bis zu $\mathbf{r}$ entlangfährt. Der Grund für die Konsistenz der Flächenparametrisierung unter der affinen Gruppe ist, daß eine Fläche eine relative Invariante unter $\mathcal{G}_A$ ist; der Wert der transformierten Fläche gleicht der ursprünglichen bis auf den multiplikativen Faktor $\det \mathbf{A}$. Zum Beweis der Konsistenz siehe [8].

Die Grauwertparametrisierung, die mit t_G bezeichnet wird, ist durch die geometrische Interpretation der Flächenparametrisierung leicht verständlich. Sie ist definiert als eine mit den überstrichenen Grauwerten gewichtete Flächenparametrisierung [9]. Sei $\Delta(\mathbf{r}(\xi_i))$ das infinitesimale Gebiet, das von den Vektoren $\mathbf{r}(\xi_i) - \mathbf{r}_R$ und $d\mathbf{r}(\xi_i)$ aufgespannt wird. Die Abb. 2a verdeutlicht die Definition, wobei der Übergang zum Differential $d\mathbf{r}(\xi_i) = \lim_{\epsilon \to 0}[\mathbf{r}(\xi_i + \epsilon) - \mathbf{r}(\xi_i)]$ gedacht werden muß. Die Grauwertparametrisierung t_G wird folgendermaßen definiert:

Definition 2 *Der* **Grauwertparameter** t_G *eines Konturpunktes* $\mathbf{r}_i = \mathbf{r}(\xi_i)$ *ist die akkummulierte Summe der Grauwerte über die Gebiete* $\Delta(\mathbf{r}(\xi))$, *wenn die Kurve von einem Startpunkt* $\mathbf{r}_0 = \mathbf{r}(\xi_0)$ *bis zu* $\mathbf{r}_i$ *durchlaufen wird, d.h.*

$$t_G(\xi_i) = \int_{\xi_0}^{\xi_i} f(\xi)\, d\xi \quad , \quad \textit{wobei mit} \quad f(\xi) = \int_{\Delta(\mathbf{r}(\xi))} f(x,y)\, dx dy \tag{11}$$

die Summe der Grauwerte über dem infinitesimalen Gebiet $\Delta(\mathbf{r}(\xi))$ *bezeichnet wird.*

Satz 2 *Die Grauwertparametrisierung* t_G *ist konsistent unter der eigentlichen affinen Gruppe und der Gruppe der Kontrasttransformationen sowie unter Aufpunktverschiebungen des Konturverfolgers.*

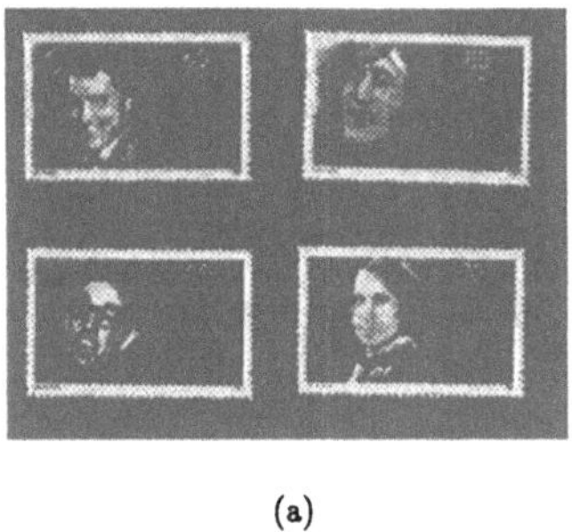

(a)

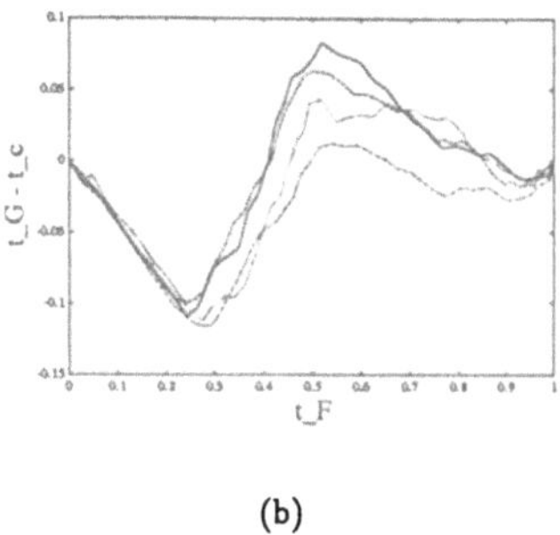

(b)

Abbildung 3: (a) Vier verschiedene Grauwertobjekte und (b) deren Grauwertparametrisierungen t_G als Differenz zu t_c : (M1) — (M2) -- (M3) $\cdots$ (M4) $-\cdot-\cdot$

Zum Beweis siehe [6]. Verläuft ein Konturstück vom Referenzpunkt aus gesehen in radialer Richtung, tragen nur die Grauwerte auf der Kontur zur Grauwertparametrisierung bei, siehe Abb. 2b. Die Grauwertparametrisierung t_G ist streng monoton, woraus die Stetigkeit der Parameterdarstellung $\mathbf{r}(t_G)$ einer Kontur C folgt [6]. Die Definition von t_G läßt sich unmittelbar auf diskrete Grauwertmuster übertragen. Da die Abbildung eines Grauwertmusters auf die Grauwertsummen $f(\xi)$ in Gl. (11) nicht injektiv ist, wird ein Grauwertmuster nicht eineindeutig auf die Parameterdarstellung $\mathbf{r}(t_G)$ abgebildet. Daraus folgt die Unvollständigkeit der Fourierdeskriptoren für Grauwertmuster [6]. Als Beispiel zeigt die Abb. 3 vier Grauwertobjekte und deren Grauwertparametrisierungen. Es handelt sich um ähnliche Briefmarken gleicher Größe und äußerer Form. Die Marken seien von links nach rechts und von oben nach unten mit M1 bis M4 bezeichnet. Als Referenzpunkt ist der Linienschwerpunkt der Kontur gewählt; der Startpunkt der Konturverfolger ist jeweils die linke obere Ecke. Die Abb. 3b zeigt die Differenz der Parameterwerte t_G zu der Grauwertparametrisierung t_c, die sich bei einem Muster mit konstanten Grauwert ergibt, als Funktion der überstrichenen Fläche. Die Parameterwerte sind dabei auf eins normiert. Trotz der gering erscheinenden Unterschiede der Parametrisierungen ist es möglich, die vier formgleichen Briefmarken nur aufgrund des unterschiedlichen Grauwertmusters zu unterscheiden.

4 Fourierdeskriptoren für Grauwertmuster

Als Fourierreihe eines Grauwertmusters wird die Fourierreihe bezeichnet, die sich durch die Anwendung der Grauwertparametrisierung t_G zur Parameterdarstellung $\mathbf{r}(t_G)$ der Kontur C eines Objektes $\mathcal{O}$ ergibt, siehe Gl. (6). Für die Fourierkoeffizienten $F_k(t_G)$ bzw. $X_k(t_G)$ und $Y_k(t_G)$ wird die Bezeichnung **Grauwertfourierkoeffizienten** verwendet. Die Grauwertfourierkoeffizienten transformieren sich gemäß Satz 1 linear ($k \neq 0$) unter der Gesamttransformation $\mathcal{G}_{A,K}$, da t_G konsistent unter $\mathcal{G}_{A,K}$ sowie unter Aufpunktverschiebungen ist, siehe Satz 2.

In [7, 8] hat Arbter ein Konstruktionsschema für die Bildung von affin-invarianten Fourierdeskriptoren ebener Kurven entwickelt, in das sich alle bekannten Invariantensätze für Fourierdeskriptoren einfügen lassen. Die Invarianzleistung umfaßt natürlich auch die Untergruppen der affinen Gruppe. Das Schema kann wegen der formalen Gleichheit für die Grauwertfourierkoeffizienten übernommen werden. Der erste Schritt zur Bildung von absoluten Invarianten ist die Definition von relativen Invarianten, den sogenannten **A-Invarianten**. Die A-Invarianten des Grauwertmusters $Q_k(t_G)$ sind definiert durch

$$Q_k\,(t_G) = \frac{X_k(t_G)Y_p^*(t_G) - Y_k(t_G)X_p^*(t_G)}{X_p(t_G)Y_p^*(t_G) - Y_p(t_G)X_p^*(t_G)} \quad , \quad k,p \in \mathbb{Z}\setminus\{0\} \quad , \quad p = \text{konst} \quad , \tag{12}$$

wobei $X_pY_p^* - Y_pX_p^* \neq 0$ Voraussetzung ist. Für p sollte ein Koeffizient mit hohem Signal-Rausch-Verhältnis gewählt werden. Dies ist üblicherweise der erste Fourierkoeffizient mit $p = 1$. Die A-Invarianten sind relative Invarianten unter der Gruppe $\mathcal{G}_{A,K}$. Es gilt

Satz 3 *Seien f, f' die zu den Objekten $\mathcal{O}, \mathcal{O}'$ gehörenden Grauwertfunktionen. $Q_k(t_G)$ und $Q'_k(t'_G)$ seien die*

jeweiligen A-Invarianten. Aus der Äquivalenz von $\mathcal{O}$ und $\mathcal{O}'$ unter der Gruppe $\mathcal{G}_{\mathcal{A},K}$ folgt für alle $k \in \mathbb{Z} \setminus \{0\}$

$$\mathcal{O}' \overset{\mathcal{G}_{\mathcal{A},K}}{\sim} \mathcal{O} \quad \Longrightarrow \quad Q'_k(t'_G) = e^{-j2\pi(k-p)\nu/\mu T}\, Q_k(t_G) \quad . \tag{13}$$

Der Beweis erfolgt durch Einsetzen mit Hilfe des Satzes 1. Aus den A-Invarianten entwickelte absolute Invarianten sind nun die **Fourierdeskriptoren**. Ein einfaches System von Fourierdeskriptoren ist $\{|Q_k|\}$, bei dem der Phasenfaktor unterdrückt wird. Soll der Phasenfaktor berücksichtigt bleiben, ist es notwendig, durch geeignete Normierungsmaßnahmen die Abhängigkeit von der Aufpunktverschiebung, die durch den Term ν ausgedrückt wird, zu eliminieren. Es sei $\varphi_k = arg(Q_k)$ die Phase von Q_k, d.h.: $Q_k = |Q_k|e^{j\varphi_k}$. Im folgenden Abschnitt wird ein Normierungsverfahren zur Bildung absoluter Invarianten gemäß der notwendigen Bedingung vorgestellt.

5 Gekoppelte Phasennormierung

Im Unterschied zu [7, 8] wird nun der folgende Konstruktionsansatz zur Phasennormierung eingeführt. Ausgehend von der Größe

$$I_k = |Q_k|e^{j\varphi_k} \prod_{i=1}^{n} e^{j\lambda_i \varphi_{k_i}} \quad , \quad \lambda_i, k_i \in \mathbb{Z} \setminus \{0\}\,, n \geq 1 \tag{14}$$

werden die Fourierdeskriptoren $I_k(t_G)$ für Grauwertmuster gebildet [10]. Der Index k_i symbolisiert feste, aber noch aus der Invarianzforderung zu bestimmende Werte.

Satz 4 *Seien $Q_k(t_G)$ die A-Invarianten eines Grauwertmusters. Dann ist die Folge*

$$I_k(t_G) = |Q_k(t_G)|e^{j\varphi_k} \prod_{i=1}^{n} e^{j\lambda_i \varphi_{k_i}} \quad , \quad k = \pm 1, \pm 2, \ldots, \pm\infty \quad , \tag{15}$$

mit $\lambda_i, k_i \in \mathbb{Z} \setminus \{0\}, n \geq 1$ ein System von Fourierdeskriptoren, wenn die λ_i, k_i die diophantische Gleichung

$$(k-p) + \sum_{i=1}^{n}(k_i - p)\lambda_i = 0 \tag{16}$$

erfüllen. Voraussetzungen seien $|Q_{k_i}| \neq 0$ und $k_i \neq p$.

Der Beweis erfolgt durch Einsetzen, siehe auch [6]. Zu beachten ist, daß die diophantische Gleichung (16) von dem Laufindex k abhängt. Es existieren verschiedene Möglichkeiten, die Gl. (16) zu lösen. Es ist möglich, die k_i oder die λ_i festzulegen. In realen Anwendungen sind die Phasen φ_k von ihren idealen Werten φ_k^0 mit einem Fehler $\delta\varphi_k$ gestört: $\varphi_k = \varphi_k^0 + \delta\varphi_k$. Wegen der Fehlerfortpflanzung sollte daher das n so klein wie möglich sein. Auch die Gewichte λ_i werden auf $\lambda_i = \pm 1$ beschränkt. Die diophantische Gleichung (16) ist in diesem Fall immer lösbar, da der größte gemeinsame Teiler der Menge $\{\lambda_1, \ldots, \lambda_n\}$ eins ist und $(k-p)$ teilt [11]. Werden nun $(n-1)$ Indizes der k_i festgelegt, ergibt sich für den verbleibenden Index k_l:

$$k_l = k + \text{konst} \quad , \tag{17}$$

d.h. es findet eine Kopplung zwischen der Normierungsphase φ_{k_l} und der zu normierenden Phase φ_k statt. Diese Art der Phasennormierung heißt deshalb **gekoppelte Phasennormierung**, und die daraus folgenden Systeme von Fourierdeskriptoren heißen **gekoppelt** [6]. Die Tabelle 1 listet als Beispiel vier gekoppelte Systeme erster oder zweiter Ordnung ($n = 1, 2$) auf.

Die bisher in der Literatur vorgestellten Systeme von Fourierdeskriptoren sind sämtlich ungekoppelt, d.h. die Normierungsphasen sind für alle Q_k gleich. Ein Beispiel dafür ist das folgende System, das mit V bezeichnet wird:

$$V \; : \quad I_k = |Q_k|e^{j(\varphi_k + (1-k)\varphi_2)} \quad , \quad k \neq -1, 0, 1 \quad . \tag{18}$$

Das System V ist innerhalb der ungekoppelten Systeme sehr robust. Es läßt sich anhand eines einfachen, linearen Fehlermodells zeigen, daß die gekoppelten Systeme den ungekoppelten bezüglich der Fehlerfortpflanzung der Phasenstreuungen überlegen sind. Da die Phasen der A-Invarianten korreliert sind, kommt es zusätzlich

System	Fourierdeskriptoren	$k \neq$
I	$I_k = \lvert Q_k \rvert e^{j(\varphi_k + \varphi_{-k+2})}$	-1,0,1,2,3
II	$I_k = \lvert Q_k \rvert e^{j(\varphi_k + \varphi_{-k+1} + \varphi_2)}$	-1,0,1,2,
III	$I_k = \lvert Q_k \rvert e^{j(\varphi_k - \varphi_{k+1} + \varphi_2)}$	-2,-1,0,1
IV	$I_k = \lvert Q_k \rvert e^{j(\varphi_k - \varphi_{k-1} - \varphi_2)}$	-1,0,1,2

Tabelle 1: Tabelle ausgewählter gekoppelter Systeme von Fourierdeskriptoren

zu Fehlerauslöschungseffekten, so daß die gekoppelten Fourierdeskriptoren teilweise geringere Phasenfehler im Merkmalsraum aufweisen als die Fourierkoeffizienten selbst [6].

Es läßt sich zeigen, daß einige der gekoppelten Systeme für den Fall ebener Kurven, also bei Verwendung der Flächenparametrisierung $I_k(t_F)$, vollständig sind, so z.B. die Systeme III und IV [6]. Dies ist bemerkenswert, da diese Systeme nicht eindeutig umkehrbar sind und die Vollständigkeit gemeinhin über die Umkehrbarkeit bewiesen wird. Für den Beweis der Vollständigkeit gekoppelter Systeme wird iterativ eine Formulierung gefunden, die auch die hinreichende Bedingung erfüllt. Damit ist auch der zweite Teil der Aussage in [8] bewiesen: „ein umkehrbares Invariantensystem ist auch vollständig, ein vollständiges nicht notwendigerweise umkehrbar".

Für die Fähigkeit der Grauwertfourierdeskriptoren zur Separation gilt:

Satz 5 *Ist $I_k(t_F)$ ein Satz von Fourierdeskriptoren, der für ebene Kurven vollständig ist, dann können mit dem entsprechenden Satz von Grauwertfourierdeskriptoren $I_k(t_G)$ alle Grauwertmuster voneinander separiert werden, die eine verschiedene Konturform oder/und eine von einer affinen Parametertransformation verschiedene Grauwertparametrisierung haben.*

Zum Beweis siehe [6].

6 Experimentelle Ergebnisse

Zur affin-invarianten Klassifikation wurden zwei Experimente durchgeführt. Einmal handelte es sich um die Erkennung der vier Briefmarken aus der Abb. 3a. Das zweite Experiment wurde mit vier Binärobjekten ausgeführt, siehe Abb. 4b. Die Binärobjekte werden von links nach rechts und von oben nach unten mit B1 bis B4 bezeichnet. Jede Objektklasse wurde insgesamt 1200 mal unter unterschiedlichen Positionen aufgenommen, jeweils 200 mal bei einem bestimmten Aufsichtswinkel Φ, der den Winkel zwischen der Objektnormalen und der optischen Achse der Kamera angibt, siehe die Abb. 4a. Der Aufsichtswinkel variierte von 0° bis 75° in Schritten von 15°. Die Briefmarken zeigen bei diesen Aufsichtswinkeln ein relativ homogenes, mattes Reflexionsverhalten und können als Lambertsche Flächen angenommen werden. Als Lichtquelle diente eine normale, großflächige Deckenbeleuchtung, die als ideal diffuse Strahlungsquelle anzusehen ist. Durch die Versuchsbedingungen sind die Modellannahmen des Abschnitts 2 gerechtfertigt. Um einen sinnvollen Vergleich mit der Methode der affin-invarianten Momente zu erlauben, wurden der Kontrast und die Helligkeit der Grauwertbilder über den Mittelwert und die Standardabweichung normiert; Details sind in [6] beschrieben.

Als Klassifikatoren wurden der gewichtete Minimum-Distanz-, der Bayes- und der Nächste-Nachbar-Klassifikator getestet [6, 12]. Der Minimum-Distanz-Klassifikators wurde über die Standardabweichungen der einzelnen Merkmale der Lernstichprobe gewichtet. Als Lernstichprobe diente pro Musterklasse eine Menge von jeweils 20 Mustern bei den Aufsichtswinkeln $\Phi = 0°, 15°, 30°, 45°, 60°$, also insgesamt 100 Muster pro Klasse.

Ausgehend von den Zentralmomenten

$$\mu_{pq} = \iint\limits_{\mathbb{R}^2} f(x,y)(x-x_s)^p(y-y_s)^q \, dx dy \quad , \tag{19}$$

die auf den Massenschwerpunkt der Grauwerte $\mathbf{r}_s = (x_s, y_s)^T$ bezogen sind, dient die Menge M_1, M_2, M_3, M_4 an affin-invarianten Momenten zum Vergleich zu der Methode der Grauwertfourierdeskriptoren:

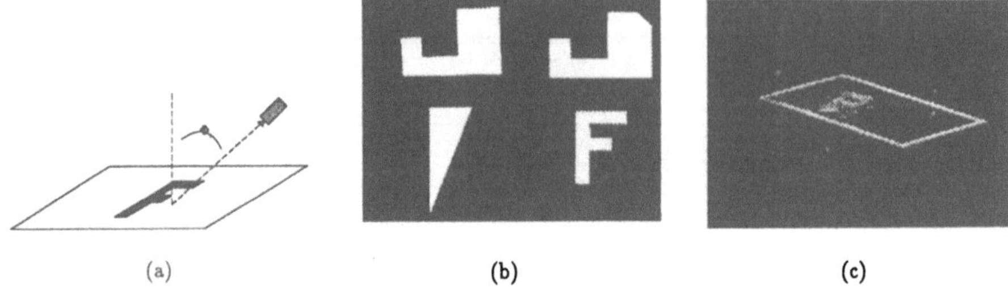

(a) (b) (c)

Abbildung 4: (a) Kameraanordnung, (b) Binärobjekte bei einem Aufsichtswinkel von $\Phi = 0°$, (c) Briefmarke M1 bei $\Phi = 75°$

Klassifikator	I	II	III	IV	V	MOM
Min.-Dist.	80,7	80,9	88,5	89,2	61,3	72,7
Bayes	88,4	91,3	93,0	92,9	89,8	87,3
N.-Nach.	89,9	89,7	88,8	90,6	90,1	62,2

Tabelle 2: Gesamtklassifikationsraten der Briefmarken

$$\begin{aligned} M_1 &= [\mu_{20}\mu_{02} - \mu_{11}^2]/\mu_{00}^4 \\ M_2 &= [\mu_{20}(\mu_{21}\mu_{03} - \mu_{12}^2) - \mu_{11}(\mu_{30}\mu_{03} - \mu_{21}\mu_{12}) + \mu_{02}(\mu_{30}\mu_{12} - \mu_{21}^2)]/\mu_{00}^7 \\ M_3 &= [\mu_{40}\mu_{04} - 4\mu_{31}\mu_{13} + 3\mu_{22}^2]/\mu_{00}^6 \\ M_4 &= [\mu_{40}\mu_{22}\mu_{04} + 2\mu_{31}\mu_{22}\mu_{13} - \mu_{40}\mu_{13}^2 - \mu_{04}\mu_{31}^2 - \mu_{22}^3]/\mu_{00}^9 \end{aligned} \tag{20}$$

Das System der Momente $M_1, \ldots, M_4$ wird mit MOM bezeichnet; es ist invariant unter der affinen Gruppe $\mathcal{G}_{\mathcal{A}}$, aber nicht unter der Gruppe der Kontraständerungen $\mathcal{G}_K$ [13]. Die Menge der Fourierdeskriptoren ist jeweils I_k ; $k \in [-20, 20]$.

Die Klassifikationsraten der beiden Versuche sind im folgenden in Prozent (%) angegeben. Sie ergeben sich aus der Anzahl der richtigen Klassenzuordnungen aus der Teststichprobe, die pro Klasse 1100 Muster umfaßt. Die Tabelle 2 zeigt die Erkennungsraten der Briefmarken. Der Bayes-Klassifikator liefert die besten Ergebnisse. Die Erkennungsraten liegen hier um 90%, wobei alle Systeme I-V der Fourierdeskriptoren bessere Klassifikationsleistungen liefern als die Momente. Die Raten der gekoppelten Systeme III und IV, die für den Fall ebener Kurven vollständig sind, sind um 6% höher als die der Momente. Für den gewichteten Minimum-Distanz- und den Nächsten-Nachbar-Klassifikator zeigt sich die Überlegenheit der Grauwertfourierdeskriptoren deutlicher. Nur das ungekoppelte System V zeigt für den Minimum-Distanz-Klassifikator schlechtere Ergebnisse als das Momenten-System. Da der Minimum-Distanz-Klassifikator ein linearer Klassifikator ist [12], werden hier die größeren Phasenstreuungen des ungekoppelten Systems gegenüber den gekoppelten Systemen I-IV besonders deutlich.

Anhand der dazugehörigen Distanzmatrizen läßt sich zeigen, daß die Momente insgesamt eine größere Dynamik zur Separation der Musterklassen besitzen. Sie sind aber empfindlicher gegenüber lokalen Grauwertstörungen als die Grauwertfourierdeskriptoren, welche durch Beleuchtungseinflüsse und Abweichungen des matt diffusen Reflexionsverhaltens hervorgerufen werden.

Bei dem Binärmuster-Experiment ist besonders die Unterscheidung der beiden „ähnlichen" Klassen B1 und B2 hervorzuheben, da die Klassifikationsraten für die Unterscheidung der drei Klassen B1∪B2, B3, B4 bei Verwendung des Bayes- und des Nächste-Nachbarn-Klassifikators für die Momente 100% und für die Fourierdeskriptoren 99-100% betragen. Die Tabelle 3 zeigt die Erkennungsraten, falls nur die beiden Klassen B1 und B2 betrachtet werden. Für die Separation der beiden Binärmuster sind die Momente bei allen drei Klassifikatoren den Konturfourierdeskriptoren im Einklang mit Ergebnissen von [13, 14] überlegen. Der Grund liegt darin,

Klassifikator	I	II	III	IV	V	MOM
Min.-Dist.	63,5	64,9	52,2	53,5	46,0	96,8
Bayes	90,0	88,9	94,3	92,1	93,2	97,8
N.-Nach.	88,7	87,2	82,0	80,4	79,1	94,8

Tabelle 3: Gesamtklassifikationsraten zur Unterscheidung des Binärmusters B1 von B2

daß hier nur der Formanteil eine Rolle spielt und nicht der Grauwertanteil, gegenüber dessen Störungen die Momente empfindlicher sind als die Grauwertfourierdeskriptoren.

Literatur

[1] M. K. Hu: Visual Pattern Recognition by Moment Invariants. IRE Trans. Inform. Theory, IT-8, S. 179–187, Feb. 1962.

[2] D. Cyganski, J. A. Orr: Object Recognition and Orientation Determination by Tensor Methods. In: T. S. Huang (Hrsg.): Advances in Computer Vision and Image Processing , Vol. 3, Jai Press, Greenwich, 1988.

[3] J. L. Mundy, A. Zisserman (Hrsg.): Geometric Invariance in Computer Vision. MIT Press, Cambridge, 1992.

[4] G. H. Granlund: Fourier Preprocessing for Hand Print Character Recognition. IEEE Trans. Computers, C-21, S. 195–201, 1972.

[5] C. T. Zahn, R. Z. Roskies: Fourier Descriptors for Plane Closed Curves. IEEE Trans. Computers, C-21, S. 269–281, 1972.

[6] A. Fenske: Affin-invariante Fourierdeskriptoren für Grauwertmuster. Dissertation (einger.), Technische Universität Hamburg-Harburg, 1993.

[7] K. Arbter, W. E. Snyder, H. Burkhardt and G. Hirzinger: Application of Affine-Invariant Fourier Descriptors to Recognition of 3-D Objects. IEEE Trans. on Pattern Analysis and Machine Intelligence, PAMI-12, Nr. 7, S. 640–647, 1990.

[8] K. Arbter: Affininvariante Fourierdeskriptoren ebener Kurven. Dissertation, Technische Universität Hamburg-Harburg, 1990.

[9] A. Fenske, H. Burkhardt: Affine Invariant Recognition of Gray Scale Objects by Fourier Descriptors. Proc. SPIE International Symposium on on Optical & Optoelectronic – Applications of Digital Image Processing XIV, San Diego, Vol. 1567, S. 53 – 65, 1991.

[10] H. Burkhardt, A. Fenske, H. Schulz-Mirbach: Invariants for the recognition of planar contour and gray-scale images. Technisches Messen, 59. Jgg., Nr. 10, S. 398–407, 1992.

[11] I. Niven, H. S. Zuckermann, H. L. Montgomery: An Introduction to the Theory of Numbers. John Wiley & Sons, New York, 1991.

[12] J. Schürmann: Polynomklassifikatoren für die Zeichenerkennung. R. Oldenbourg Verlag, München, 1977.

[13] T. H. Reiss: Recognizing Objects Using Invariant Image Features. Dissertation, University of Cambridge, 1992.

[14] C. Geiselmann: Zur automatischen Identifizierung einfacher Objekte aus einer Bildfolge. Dissertation, Universität Stuttgart, 1992.

Zur Auswahl von Merkmalen*

Olaf Munkelt

Technische Universität München
Institut für Informatik - Lehrstuhl Prof. Radig
munkelt@informatik.tu-muenchen.de

Zusammenfassung. Die Auswahl geeigneter (Regionen-) Merkmale aus einer Menge von gegebenen Merkmalen zur Unterstützung des Bildinterpretationsprozesses ist ein bisher wenig beachteter Aspekt der Bildverarbeitung. Diese Arbeit stellt ein Verfahren zur Bewertung einer beliebigen Teilmenge einer gegebenen Menge von Merkmalen vor. Diejenige Teilmenge mit der maximalen Bewertung bildet die *optimale* Kombination oder die *bestmögliche* Auswahl von Merkmalen. Es wird gezeigt, daß der Ansatz die *Erkennungsleistung* eines Bildinterpretationssystems signifikant erhöht.

1 Einleitung

Die Fragestellung der Auswahl von Merkmalen ist bisher im Zusammenhang mit der bekannten Merkmalsklassifikation [12], [7] gestellt worden. Hier nutzen überwachte und unüberwachte Verfahren zumeist stochastische Modelle, um Merkmalsvektoren zu klassifizieren. Ist die Kovarianzmatrix C bekannt, die die Korrelation der Merkmale beschreibt, so kann die Hauptachsentransformation angewendet werden, um Cluster zu trennen. Damit erhält man quasi "neue" Merkmale, die unkorreliert sind. Ein Nachteil dieser Methode ist der Effizienzverlust durch die während des Bildinterpretationsprozesses durchzuführende Hauptachsentransformation. So wird in [2] ein Merkmal dahingehend beurteilt, ob es (a) extrahierbar ist, (b) inwieweit Hypothesen von ihm gestützt werden und (c) wie groß die Änderung eines Merkmals abweichend von einem vorgegebenen Modell ist. Aus diesen drei stochastischen Maßen wird ein optimaler Algorithmus konstruiert, der jedes Merkmal bewertet. Für eine Auswahl von Merkmalen ist auch diese Methode nicht geeignet; wenn man nicht die n−besten Merkmale als solche Auswahl betrachten will. Nach diesem Prinzip stellt [8] insgesamt 5 verschiedene Verfahren vor, die ebenfalls auf stochastischen Maßen beruhen und allesamt Funktionale minimieren.

In dieser Arbeit wird unter einem Merkmal $M_j, j = 1, \ldots, m$, m Anzahl der Merkmale, eine Abbildung verstanden mit $f(M_j) : R \rightarrow \mathbb{R}$ und R eine 2-D Region, wie sie entweder als Ergebnis einer Segmentation eines Videobildes vorliegt oder als Muster in Form einer projezierten 3-D Modellfläche. Merkmale (Regionen- bzw. Formmerkmale) sind z.B. die Größe der Region, der Winkel der Hauptträgheitsachsen der umschließenden Ellipse der Region etc. (vgl. Anhang). Dieser Beitrag stellt ein Verfahren zur Auswahl der *optimalen* Kombination von Merkmalen aus diesen Merkmalen vor. Hierzu wird ein in Kap.2 erläuterter modellbasierter Ansatz verwendet. Kap.3 beschreibt das darauf aufbauende Verfahren zur Merkmalsauswahl, deren Ergebnisse in Kap.4 dokumentiert sind.

* Diese Arbeit wurde von der DFG im Rahmen des SFB 331, *Informationsverarbeitung in autonomen mobilen Handhabungssystemen*, Teilprojekt L5 gefördert.

2 Modellbildung

Die Verwendung von 3-D CAD-Modellen als Grundlage zur Erzeugung von Mustern, die zum Vergleich mit Videobildern genutzt werden, gewinnt für den Bildinterpretationsprozeß zunehmend an Bedeutung ([6], [4], [11]). Hierbei lassen sich Verfahren unterscheiden, die 3-D Modelle aus geometrischen Primitiven, wie Zylinder, Quader aufbauen [3] und Verfahren, die eine explizite Darstellung der äußeren Flächenbeschreibung von Objekten nutzen [5]. Erstere verwenden für die interne Repräsentation von Objekten die CSG, *Constructive Solid Geometry*, letztere die B-Rep, *boundary-representation*, [15]. Der Vorteil der B-Rep basierten Verfahren ist zum einen die Modellierung eines zu erkennenden Objektes als Ganzes, zum anderen die Möglichkeit, nur bestimmte für die Bildinterpretation relevante Teile des Objektes betrachten zu können. Im Gegensatz zur CSG, die sowohl den Aufbau eines Objektes, als auch die Erkennung nur über die geometrischen Primitive ermöglicht. Die Struktur des implementierten Systems zur Modellbildung ist deshalb an die B-Rep angelehnt und verwendet zur hidden-surface Berechnung ein modifiziertes Objektraumverfahren [16], mit dem sich auch nicht-konvexe Objekte modellieren lassen [13]. Weiterhin verfügt das System über eine Schnittstelle zum CAD-I Format, um Modelle aus bestehenden CAD-Systemen nutzen zu können [14].

Als Grundlage zur Erzeugung von Mustern ist die Berechnung einer endlichen Anzahl von Ansichten von Flächen/Linien erforderlich. Hierzu wird mit Hilfe einer Triangulierung der Gaußschen Sphäre (GS) [1] in 320 Dreiecke eine möglichst gleichmäßige Verteilung von Ansichten auf ein Objekt, dessen Schwerpunkt im Koordinatenursprung der GS liegt, erreicht (s. Anhang). Eine Ansicht v_i wird durch die Mittelpunkte (in Polarkoordinaten) dieser Dreiecke $(\varphi_i, \vartheta_i, r_i), i = 1, \ldots, 320$ definiert: die Blickrichtung auf ein Objekt zeigt gerade auf den Koordinatenursprung. Der Radius $r = r_i, \forall i$ wird entsprechend dem Kameramodell bestimmt [10]. Eine Ansicht v_i ist einer Ansicht $v_j, j = 1, \ldots, 320$ genau dann benachbart, wenn die zugehörigen Dreiecke eine Linie gemeinsam haben. Eine Ansicht besitzt somit höchstens drei Nachbarn. Der Abstand von je 2 Ansichten beträgt im Mittel $d = 0.17\, r$, wie man durch Nachrechnen leicht bestätigt.

3 Die Auswahl von Merkmalen

Der erste Schritt zur Auswahl der besten Merkmale aus n-Merkmalen ist die Aufstellung einer Matrix $\mathcal{W}$ für eine beliebige, aber fixe 3-D Modellfläche F_{3-D} in der folgenden Form (o.b.d.A. werden im folgenden Flächen betrachtet):

$$\mathcal{W} = ((w_{ij})) \quad w_{ij} = f_j(F_i), \quad f_j := f(M_j), \quad F_i = P(F_{3-D}, v_{k_i}) \qquad (1)$$

$$v_k = \text{Menge der Ansichten}, \quad k = 1, \ldots, 320 \quad k_i \in \{1, \ldots, 320\}$$

mit $j = 1, \ldots, m$, $i = 1, \ldots, n$ und n Anzahl der Ansichten von denen die Fläche sichtbar ist und m Anzahl der Merkmale, F_i die durch Projektion P von der Ansicht v_{k_i} berechnete 2-D Modellfläche. Die Zeilen der Matrix $\mathcal{W}$ bilden den Merkmalsvektor. Hier werden jedoch die Spalten von $\mathcal{W}$ untersucht, die gerade die einzelnen Merkmale repräsentieren. Man sieht leicht, daß zwei identische Spalten w_i, w_j redundante Information im Sinne der Verwendung dieser beiden

Merkmale für die Erkennung beinhalten; d.h. zwei verschiedene Merkmale liefern für die gewählte Fläche in den verschiedenen Ansichten gleiche Werte. Es genügt somit, nur eins von den beiden zu betrachten. Die gleiche Argumentation läßt sich auch auf den Fall $w_i = \alpha w_j$ anwenden, sofern die Erkennung invariant bezüglich der linearen Abhängigkeit von Merkmalen ist. Im mathematisch strengen Sinn sind die Spalten von $\mathcal{W}$ i.a. nicht l.a. Deshalb ist es nötig, Maße zu finden, deren Wertebereich an den Grenzen gerade wiedergibt, ob zwei Spalten l.a. bzw. l.u. sind und zwischen diesen Grenzen die Abhängigkeit stetig interpoliert. Im folgenden werden zwei entsprechende Maße konstruiert:

$$\overline{L_1}(w_i, w_j) = \frac{< w_i, w_j >}{\|w_i\|\|w_j\|}, \quad \overline{L_2}(w_i, w_j) = 1 - \frac{\|\frac{<w_i, w_j>}{\|w_i\|^2} w_i - w_j\|}{\|w_j\|} \quad i, j = 1, \ldots, m$$

$\overline{L_1}(w_i, w_j)$ leitet sich von der Definition des Winkels φ zweier Vektoren im n-dimensionalen Raum ab, $\overline{L_1}(w_i, w_j) = \cos(\varphi)$, während $\overline{L_2}(w_i, w_j)$ ein normiertes Maß für den kürzesten Abstand von w_i zu w_j ist, Ansatz: $\min_\alpha \|w_i - \alpha w_j\|$. $\overline{L_1}(w_i, w_j)$ und $\overline{L_2}(w_i, w_j)$ sind zwei linear unabhängige Maße:

$$\overline{L_2}(w_i, w_j) = 1 - \|\overline{L_1}(w_i, w_j) \frac{w_i}{\|w_i\|} - \frac{w_j}{\|w_j\|}\|$$

Durch Normierung der $\overline{L_{1,2}}(w_i, w_j)$ wird jetzt eine symmetrische Matrix $\mathcal{M}$ konstruiert, deren Elemente m_{ij} den Grad der linearen Unabhängigkeit im Intervall $[0, 1]$ des i-ten Merkmals von dem j-ten Merkmal angeben:

$$L_{1,2}(w_i, w_j) = 1 - |\overline{L_{1,2}}(w_i, w_j)| \quad \longrightarrow$$

$$L_{1,2}(w_i, w_j) = 0 \leftrightarrow w_i, w_j \quad \text{l.a.} \quad \text{und} \quad L_{1,2}(w_i, w_j) = 1 \leftrightarrow w_i, w_j \quad \text{l.u.}$$

$$\mathcal{M} = ((m_{ij})) \quad m_{ij} = \frac{m_1 L_1(w_i, w_j) + m_2 L_2(w_i, w_j)}{m_1 + m_2} \quad i, j = 1, \ldots, m \qquad (2)$$

mit $m_1, m_2 \in I\!R^+$ beliebigen Gewichtungsfaktoren. Aus dieser Matrix, die man auch als Kovarianzmatrix bezüglich dem Kriterium l.u. betrachten kann, wird jetzt diejenige Kombination von Merkmalen berechnet, die "bestmöglich" l.u. ist. Hierzu werden alle Teilmengen von Merkmalen $M_i, i = 1, \ldots, m$ (Cliquen) berechnet und mit Hilfe von $\mathcal{M}$ bewertet:

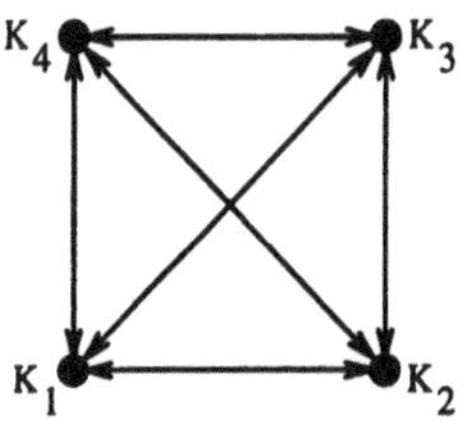

$$C = \{K_1, \ldots, K_{n_c}\} \text{ mit } K_i = M_{l_i},\ l_i \in \{1, \ldots, m\}$$

$$V_C(K_i) = \textstyle\sum_{j=1, i \neq j}^{n_c} m_{l_i l_j} \quad i = 1, \ldots, n_c \qquad (3)$$

$$V_C = \textstyle\sum_{i=1}^{n_c} \alpha_i V_C(K_i) \quad \alpha_i \in I\!R^+$$

Mit C als Clique und $V_C(K_i)$ als Wert eines Knotens (eines Merkmals) und V_C der Wert der Clique mit geeignet zu wählenden Gewichtsfaktoren $\alpha_i, i = 1, \ldots, m$ für jedes Merkmal (s.u.).

Definition 1. Eine bel. Kombination von Merkmalen $C_{opt} = \{M_{i_1}, \ldots, M_{i_l}\}$, $i_l \in \{1, \ldots, m\}$ aus einer vorgegebenen Menge von Merkmalen $\{M_1, \ldots, M_m\}$ heißt *optimal* oder *bestmögliche* Auswahl genau dann, wenn für den Wert V_C nach Gleichung (3) dieser Kombination gilt:

$$V_{C_{opt}} = \max_{C \in C_{all}} V_C \; C_{all} \text{ als Menge aller Cliquen der Größen } n_c, n_c = 1, \ldots, m$$

Bei der Wahl der Gewichtsfaktoren α_i in (3) läßt sich die durch die Modellbildung vorgegebene Nachbarschaftsbeziehung von Ansichten auf der GS nutzen. Merkmale, deren Werte sich in der Nachbarschaft einer Ansicht wenig ändern, sind im Sinne einer robusten Erkennung besser geeignet als solche, deren Werte sich stark ändern. Robust bedeutet hier, daß die Segmentierung eines Videobildes in dem Sinne nicht optimal ist, als das die resultierenden Segmente nicht eine 1:1 Entsprechung zu den Modellflächen besitzen. Merkmale, deren Werte auf kleinere Artefakte der Segmentierung auch nur mit einer dementsprechenden "kleinen" Änderung reagieren sind somit bevorzugt auszuwählen. Dieser Argumentation folgend werden zwei Änderungsmaße vorgestellt, die die Nachbarschaftsbeziehungen von Ansichten verwenden.

$$s'(x_1, x_2, \ldots, x_n) = \frac{\sqrt{\frac{1}{n-1}\sum_{i=1}^{n}(x_i - \overline{x})}}{\frac{\max_i x_i - \min_i x_i}{\sqrt{2}}}, \qquad \overline{x} = \frac{1}{n}\sum_{i=1}^{n} x_i$$

Wobei $0 \leq s'(x_1, x_2, \ldots, x_n) \leq 1$ die normierte Standardabweichung ist und $\overline{x}$ den Erwartungswert darstellt. Dann wird das erste Änderungsmaß für jedes Merkmal $M_j, j = 1, \ldots, m$ wie folgt definiert:

$$A_1(M_j) = [1 - \frac{1}{n}\sum_{i=1}^{n} s'(w_{ij}, N(v_i))] \frac{\sum_{i=1}^{n} |N(v_{k_i})|}{3n}, \quad 0 \leq A_1(M_j) \leq 1 \qquad (4)$$

$$N(v_{k_i}) = \{w_{k_l j} : v_{k_l} \text{ ist Nachbar von } v_{k_i}, k_l \in \{1, \ldots, n\}\} \quad \text{und} \quad |N(v_{k_i})| \leq 3$$

Der erste Term von $A_1(M_j), j = 1, \ldots, m$ beurteilt die Änderung des Merkmalswerts mit Hilfe der normierten Varianz, während der zweite Term (für alle Merkmale konstant) zusätzlich die überhaupt auftretende Anzahl von Nachbarschaften mitbewertet. Ansichten einer Modellfläche, die zusammenhängend auf der GS liegen, werden somit höher bewertet, als solche die verstreut liegen.

Ein zweites Änderungsmaß gewinnt man, wenn man nicht von den Ansichten ausgeht und die Merkmalswerte der Nachbaransichten betrachtet, sondern die Merkmalswerte auf die Ansichten abbildet und dabei den Abstand zwischen diesen Ansichten bewertet. Wiederum ausgehend von einer Spalte von W werden die hierin enthaltenen Merkmalswerte aufsteigend sortiert:
$w_{sort_j(1)j} \leq \cdots \leq w_{sort_j(n)j}, j = 1, \ldots, m$. Mit $i = sort_j(l), i \in \{1, \ldots, n\}$ als Sortierfunktion für das Merkmal M_j, die den Index der entsprechenden Ansicht aus v_k bestimmt. Die zu den je 3 Merkmalswerten $w_{sort_j(i+1)j}, w_{sort_j(i+2)j}, w_{sort_j(i+3)j}$ gehörigen Ansichten $v_{sort_j(i+1)}, v_{sort_j(i+2)}, v_{sort_j(i+3)}$ werden somit

als hypothetische Nachbarn der Ansicht $v_{sort_j(i)}$ betrachtet. Bewertet wird der Abstand dieser Ansichten:

$$\overline{d_M}(M_j) = \frac{1}{3(n-3)} \sum_{l=1}^{n-3} \sum_{k=l+1}^{l+3} d(v_{sort_j(l)}, v_{sort_j(k)}) \quad j = 1, \ldots, m, \quad n > 3$$

Durch Normierung auf den mittleren Abstand zweier Ansichten auf der GS ergibt sich das zweite Änderungsmaß für ein Merkmal M_j:

$$A_2(M_j) = 1 - \frac{\overline{d_M}(M_j) - 0.17r}{2r - 0.17r} \quad 0 \leq A_2(M_j) \leq 1, \quad j = 1, \ldots, m \tag{5}$$

Das zweite Änderungsmaß bewertet somit, ob bei einer "kleinen" Änderung des Wertes eines Merkmals die entsprechenden Ansichten auf der GS nahe beieinander liegen. Beide Maße werden zu einem Änderungsmaß verknüpft und damit α_j definiert:

$$\alpha_j = A(M_j) = \frac{f_1 A_1(M_j) + f_2 A_2(M_j)}{f_1 + f_2} \tag{6}$$

$$0 \leq \alpha_j \leq 1, \quad f_1, f_2 \in I\!R^+ \text{ bel. Gewichtungsfaktoren}$$

Das Änderungsmaß nach (6) bestimmt somit für jedes Merkmal ein Gütekriterium, in das sowohl die Änderungseigenschaften der einzelnen Merkmalswerte eingeht (4), als auch deren Zusammenhang mit den Ansichten auf der GS (5), indem die Nachbarschaftsbeziehung ausgenutzt wird.

4 Ergebnisse

Zunächst wird am Beispiel einer Modellfläche eines einfachen L-förmigen Objektes die Konstruktion der optimalen Merkmalsmenge gezeigt. Abb. 1 zeigt das Objekt. Die gewählte Fläche ist grau schattiert. Diese Fläche ist von den grau schattierten Ansichten auf der GS aus sichtbar. In der Darstellung sieht man diese Ansichten von hinten. Tab. 1 zeigt für die 4 dargestellten Ansichten den Ausschnitt aus der Matrix $\mathcal{W}$. Bei der Konstruktion der Matrix $\mathcal{M}$ müssen m_1, m_2 festgelegt werden (2). Wählt man die ebenfalls zu bestimmenden Gewichtsfaktoren mit $\alpha_i = 1$, $i = 1, \ldots, m$, erhält man den in Abb. 2 gezeigten Verlauf der maximalen Werte der Cliquen. Man sieht, daß die Werte von L_1 (in Abb. 2 als $m1$ bezeichnet) deutlich unterhalb von den Werten von L_2 liegen. Dies wurde

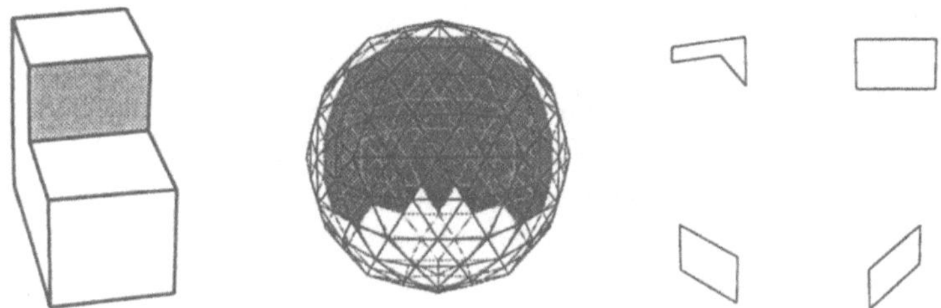

Abbildung 1. L-förmiges Objekt, Sichtbereich und 4 Projektionen

v_k	M_{16} (Circularity)	M_{20} (Structurefactor)	M_{13} (Axis_Phi)	M_9 (1.Hu-Invariant)
102	0.2558039427	3.0635237694	-0.1080196574	0.3714792105
218	0.5856804252	0.7263725400	0.0	0.1879200268
248	0.3768907487	1.0530773401	-0.7409411669	0.2059190754
275	0.2859414220	1.8005777597	0.8579329252	0.2540543720

Tabelle 1. Ausschnitt aus $\mathcal{W}$

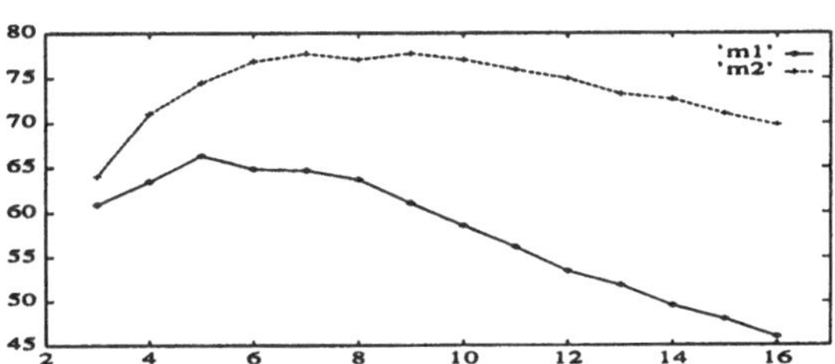

Abbildung 2. Maximale Werte ($100\ V_C$) der Cliquen mit der Größe $i = 3, \ldots, 16$

empirisch bei allen bisher betrachteten Objekten bestätigt. Im Sinnc einer Bewertung des Informationsgehalts von Merkmalen ist diejenige Kombination von Merkmalen optimal, die den größten Wert V_C liefert; somit wird $m_1 = 0, m_2 = 1$ gewählt. Hingegen beeinflussen die $\alpha_i, i = 1, \ldots, m$ die Werte V_C nicht signifikant genug, um eine eindeutige Wahl von f_1, f_2 zu treffen (6). Hier kommt es eher darauf an, ob man den Zusammenhang von Ansichten auf der GS mit sich ändernden Werten stärker gewichten will ($f_2 < f_1$) oder die Varianz von Werten in der Nachbarschaft einer Ansicht ($f_1 < f_2$). Für die betrachteten Objekte ist $f_1 = f_2 = 1$ gewählt worden. Nach Auswertung von Gleichung (3) ergab sich für die betrachtete Fläche die optimale Kombination von Merkmalen gemäß Tab. 2. Die so gefundene optimale Kombination von Merkmalen wurde genutzt, um die

M_j	V_C	α_j	$A_1(.)$	$A_1(.)$
M_{10}	0.747039	0.546897	0.536802	0.556993
M_4	0.735227	0.577278	0.490472	0.664083
M_{13}	0.680280	0.522207	0.430558	0.613856
M_5	0.623186	0.585866	0.615551	0.556182
M_6	0.620039	0.583352	0.621811	0.544893
M_1	0.619871	0.602152	0.623414	0.580889
M_3	0.553189	0.806702	0.840680	0.772723

Tabelle 2. Die bestmögliche Auswahl von Merkmalen für die Fläche aus Abb. 1.

Erkennungsleistung eines Bildinterpretationssystems [9] zu erhöhen. Dieses basiert auf Markov-Feldern und verfügt insbesondere über die Möglichkeit, die zur Interpretation notwendige Datenbasis automatisch zu generieren. Die Datenbasis besteht hierbei aus einer Menge von Beispielbildern, die die zu interpretierenden Objekte zeigen. Die hier verwendete Modellbildung (Kap.2) wurde dabei zur Generierung einer Datenbasis so genutzt, daß ein Beispielbild gerade einer Ansicht des Objektes entspricht (die Segmente eines Beispielbildes entsprechen hierbei den Flächen in einer Ansicht). Ein Videobild zu interpretieren bedeutet dann, den hieraus extrahierten Segmenten, die in der Datenbasis gespeicherten Flächen so zuzuordnen, daß jedes Segment genau einer Fläche entspricht. Der Interpretationsprozeß verwendet das simulierte Annealing. Das bedeutet, daß auf der Modellseite der einzige Freiheitsgrad bei der Erhöhung der *Erkennungsleistung* darin besteht, eine optimale Auswahl von Merkmalen zu treffen.

Definition 2. Unter der *Erkennungsleistung* eines Bildinterpretationssystems wird der Koeffizient γ verstanden, der durch

$$\gamma = \frac{\text{Anzahl korrekter Zuordnungen Segmente zu Flächen}}{\text{Anzahl Segmente}}$$

festgelegt wird.

Bei den Videobildern wurde zur Interpretation zum einen eine Datenbasis genutzt, die alle im Anhang aufgelisteten Merkmale verwendet (DB_{all}) und zum anderen eine Datenbasis generiert, die die optimale Kombination von Merkmalen gemäß Tab. 2 verwendet (DB_{opt}). Beide Datenbasen beinhalten die Ansichten, die durch $0 \leq \varphi, \vartheta \leq 90^o$ festgelegt werden. Sei V_1 die Menge von Videobildern, die das Objekt in diesem Winkelbereich zeigen und V_2 diejenige mit $90^o \leq \varphi \leq 112.5^o$ und $-22.5^o \leq \vartheta \leq 90^o$, dann ergibt sich für das Objekt aus Abb. 1 die in Tab. 3 dokumentierten Werte, wobei V_1 und V_2 jeweils 10 Videobilder umfassen.

	γ_{all}	γ_{opt}
V_1	0.8	1.0
V_2	0.7	0.95

Tabelle 3. Erkennungsleistung mit verschiedenen Merkmalen und Videobildern

5 Ausblick

Die Ergebnisse, die unter Nutzung der optimalen Auswahl von Merkmalen (1), (3) aus einer vorgegebenen Menge von Merkmalen erzielt worden sind (Tab. 3), stellen eine signifikante Verbesserung der Erkennungsleistung bei Verwendung der Markov-Feld basierten Bildinterpretation dar. Ebenfalls sind die in [2] genannten Punkte $(a) - (c)$ für die Bewertung von Merkmalen erfüllt. Trotzdem bleiben heuristische Einflußgrößen, wie m_1, m_2 (2) und f_1, f_2 (6), so daß für sie keine allgemeingültige Aussage im Sinne einer Verbesserung der Erkennungsleistung eines gewählten Bildinterpretationsystems für beliebige Objekte getroffen werden kann.

Der vorgestellte Ansatz kann dort effizient verwendet werden, wo der Bildinterpretationsprozeß invariant bezüglich der Linearität von Merkmalen ist. Diese Merkmale brauchen sich nicht wie in dieser Arbeit auf Regionenmerkmale beschränken, sondern können auch bspw. Grauwertmerkmale, wie Entropy oder mittlerer Grauwert umfassen. Abhängigkeiten $f(.)$, die nichtlinear sind, lassen sich mit dem Ansatz $||w_i - f(w_j)|| \rightarrow min$ ebenfalls in eine Matrix $\mathcal{M}$ (2) überführen. Läßt sich ein Nachbarschaftssystem auszeichnen, so lassen sich die Gewichtungsfaktoren α_i bestimmen, ansonsten gilt $\alpha_i = 1 \; \forall i$. Damit läßt sich die optimale Kombination von Merkmalen gemäß (3) einfach berechnen.

Literatur

1. Ballard, D.H., und C.M. Brown: *Computer Vision.* Prentice-Hall, Englewood Cliffs, NJ: 1982.
2. Chen, C. H., und P. G. Mulgaonkar: „CAD-Based Feature-Utility Measures for Automatic Vision Programming". In: *Directions in Automated CAD-Based Vision.* IEEE Computer Society Press: 1991: 106 - 114.
3. Dickinson, S., A. Pentland und A. Rosenfeld: „From Volumes to Views: An Approach to 3-D Object Recognition". In: *Directions in Automated CAD-Based Vision.* IEEE Computer Society Press: 1991: 85 - 96.
4. Flynn, P. J., und A. K. Jain: „CAD-Based Computer Vision: From CAD Models to Relational Graphs". *IEEE Trans. on Pattern Analysis and Machine Intelligence* **13**(2) (1991) 114–132.
5. Flynn, Patrick J., und Anil K. Jain: „BONSAI: 3-D Object Recognition Using Constrained Search". *IEEE Trans. on Pattern Analysis and Machine Intelligence* **13**(10) (Oktober 1991).
6. Hansen, C. D., und T. V. Henderson: „CAGD-Based Computer Vision". *IEEE Trans. on Pattern Analysis and Machine Intelligence* **11**(10) (1989) 1181–1193.
7. Jaehne, B.: *Digitale Bildverarbeitung.* Zweite Edition. Springer-Verlag: 1991.
8. Kittler, J.: *Feature Selection and Extraction.* Academic Press: 1986: 60–83.
9. Kristen, H., und O. Munkelt: „Markov-Feld basierte Bildinterpretation mit automatisch generierter Datenbais". In R. Hoffmann, S. Fuchs und (Ed.): *Mustererkennung.* DAGM. Springer-Verlag: 1992: 50–57.
10. Lenz, R.: „Linsenfehlerkorrigierte Eichung von Halbleiterkameras mit Standardobjektiven fuer hochgenaue 3D-Messungen in Echzeit". In Paulus, E. (Ed.): *Mustererkennung.* DAGM. Springer-Verlag: 1987: 212–216.
11. Lowe, D. G.: „Three-dimensional object recognition from single two-dimensional images". *Artificial Intell.* **31** (1987) 355–395.
12. Niemann, H.: *Pattern Analysis and Understanding.* Springer Series in Information Sciences. Second Edition. Springer-Verlag: 1989: Kap. 4.
13. Reiners, Clemens: „Ein CAD-basiertes Strukturvergleichsverfahren, Diplomarbeit,TU München". 1992.
14. Schlechtendahl, E. G. (Ed.): *Specification of a CAD*I Neutral File for CAD Geometry.* Reseach Reports ESPRIT. Springer-Verlag: 1988.
15. Shirai, Y.: *Three-Dimensional Computer Vision.* Springer Series Symbolic Computation, Computer Graphics - Systems and Application. Springer-Verlag: 1987.
16. Sutherland, I. E., R. F. Sproull und R. A. Schumaker: „A Characterization of Ten Hidden Surface Algorithms". *Computing Surveys* **6**(1) (März 1974) 2–55.

A Anhang

A.1 Triangulierung der Gaußschen Sphäre (GS)

Als Ausgangsfigur zur Triangulierung wird der Ikosaeder verwendet. Die 12 Eckpunkte des Ikosaeders berechnen sich hierbei wie folgt:

$$g = \frac{1+\sqrt{5}}{2}, \qquad a = \frac{\sqrt{g}}{\sqrt[4]{5}} r, \qquad b = \frac{1}{\sqrt[4]{5}\sqrt{g}} r$$

$$p_{1,2} = \begin{pmatrix} 0 \\ \pm a \\ b \end{pmatrix}, \; p_{3,4} = \begin{pmatrix} 0 \\ \pm a \\ -b \end{pmatrix}, \; p_{5,6} = \begin{pmatrix} \pm b \\ 0 \\ a \end{pmatrix}, \; p_{7,8} = \begin{pmatrix} \pm b \\ 0 \\ -a \end{pmatrix}, \; p_{9,10} = \begin{pmatrix} \pm a \\ b \\ 0 \end{pmatrix}, \; p_{11,12} = \begin{pmatrix} \pm a \\ -b \\ 0 \end{pmatrix}$$

mit g als goldener Schnitt und r Radius der zu approximierenden GS. Die folgenden Tripel bilden die 20 Dreiecke des Ikosaeders:

$$\{p_6, p_5, p_1\}, \{p_6, p_5, p_2\}, \{p_8, p_7, p_3\}, \{p_8, p_7, p_4\}, \{p_9, p_3, p_1\},$$
$$\{p_9, p_5, p_1\}, \{p_9, p_7, p_3\}, \{p_{10}, p_3, p_1\}, \{p_{10}, p_6, p_1\}, \{p_{10}, p_8, p_3\},$$
$$\{p_{11}, p_4, p_2\}, \{p_{11}, p_5, p_2\}, \{p_{11}, p_7, p_4\}, \{p_{11}, p_9, p_5\}, \{p_{11}, p_9, p_7\},$$
$$\{p_{12}, p_4, p_2\}, \{p_{12}, p_6, p_2\}, \{p_{12}, p_8, p_4\}, \{p_{12}, p_{10}, p_6\}, \{p_{12}, p_{10}, p_8\}$$

Jedes dieser Dreiecke wird in jeweils 16 weitere Dreiecke unterteilt, so daß die Gaußsche Sphäre in insgesamt 320 Dreiecke unterteilt ist. Am Beispiel des ersten Dreiecks wird aufgezeigt, wie die Koordinaten $d_i, i = 1, \ldots, 15$ berechnet werden (vgl. Abb. 3):

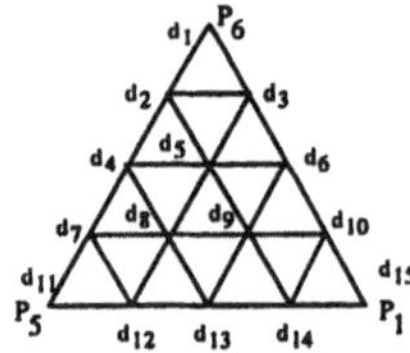

Abbildung 3. Die GS und ein ausschnittsvergrößertes Dreieck

$$l \rightarrow 0 \; \forall i = 0 \ldots 4:$$
$$\forall j = 0 \ldots i : d_l = p_6 + \tfrac{i}{4}(p_5 - p_6) - \tfrac{j}{4}(p_5 - p_1), \; l \rightarrow l+1$$

Nach Transformation der $d_i, i = 1, \ldots, 15$ in Polarkoordinaten werden nun die r-Koordinaten der Eckpunkte der Dreiecke an den Rand der GS angepaßt:

$$(d_i)_r \rightarrow \begin{cases} (d_i)_r + o_1 & : \; i = 2,3,7,10,12,14 \quad \text{und} \quad o_1 = \frac{r-\sqrt{r^2-b^2}}{2} \\ (d_i)_r + o_2 & : \; i = 4,5,6,8,9,13 \quad \text{und} \quad o_2 = 2o_1 \\ (d_i)_r & : \; i = 1,11,15 \end{cases}$$

Die Rücktransformation in kartesische Koordinaten bestimmt die Eckpunkte der 320 Dreiecke. Die Dreiecke werden beginnend von $\vartheta = -90^o$ bis $\vartheta = 90^o$ fortlaufend numeriert, indem $0^o \leq \varphi \leq 360^o$ gewählt wird.

A.2 Merkmale

Nr	Name	Formel
M_1	F	$\mathtt{F} = \sum_{(z,s)\in R} i$
M_2	$\mathtt{M}_{10}$	$\mathtt{M}_{10} = \sum_{(z,s)\in R}(z_0 - z)$
M_3	$\mathtt{M}_{01}$	$\mathtt{M}_{01} = \sum_{(z,s)\in R}(s_0 - s)$
M_4	$\overline{\mathtt{M}_{11}}$	$\overline{\mathtt{M}_{11}} = \frac{M_{11}}{\mathtt{F}^2}, \quad M_{11} = \sum_{(z,s)\in R}(z_0 - z)(s_0 - s)$
M_5	$\overline{\mathtt{M}_{20}}$	$\overline{\mathtt{M}_{20}} = \frac{M_{20}}{\mathtt{F}^2}, \quad M_{20} = \sum_{(z,s)\in R}(z_0 - z)^2$
M_6	$\overline{\mathtt{M}_{02}}$	$\overline{\mathtt{M}_{02}} = \frac{M_{02}}{\mathtt{F}^2}, \quad M_{02} = \sum_{(z,s)\in R}(s_0 - s)^2$
M_7	$\overline{\mathtt{I}_a}$	$\overline{\mathtt{I}_a} = \frac{I_a}{\mathtt{F}}, \quad I_a = h + \sqrt{h^2 - M_{20}M_{02} + M_{11}^2}, \quad h = \frac{M_{20}+M_{02}}{2}$
M_8	$\overline{\mathtt{I}_b}$	$\overline{\mathtt{I}_b} = \frac{I_b}{\mathtt{F}}, I_b = h - \sqrt{h^2 - M_{20}M_{02} + M_{11}^2}$
M_9	$\mathtt{H}_1$	$\mathtt{H}_1 = \frac{M_{20}+M_{02}}{\mathtt{F}^2}$
M_{10}	$\mathtt{H}_2$	$\mathtt{H}_2 = \frac{(M_{20}-M_{02})^2+4M_{11}^2}{\mathtt{F}^4}$
M_{11}	$\mathtt{Axis}_a$	$\mathtt{Axis}_a = \sqrt{\frac{a\mathtt{F}}{b\pi}}, \quad a = \sqrt{\frac{I_a}{\mathtt{F}}}, \quad b = \sqrt{\frac{I_b}{\mathtt{F}}}$
M_{12}	$\mathtt{Axis}_b$	$\mathtt{Axis}_b = \sqrt{\frac{b\mathtt{F}}{a\pi}}$
M_{13}	$\mathtt{Axis}_\varphi$	$\mathtt{Axis}_\varphi = \arctan(\frac{M_{20}-I_a}{M_{11}})$
M_{14}	$\mathtt{Shape}_a$	$\mathtt{Shape}_a = \sqrt{\frac{I_a}{\mathtt{F}}}$
M_{15}	$\mathtt{Shape}_b$	$\mathtt{Shape}_b = \sqrt{\frac{I_b}{\mathtt{F}}}$
M_{16}	Circ	$\mathtt{Circ} = \frac{\mathtt{F}}{[\max_{(z,s)\in R} d((z,s)(z_0,s_0))]^2\,\pi}$
M_{17}	Conv	$\mathtt{Conv} = \frac{\mathtt{F}}{F_c}, \quad F_c$ Fläche der konvexen Hülle von R
M_{18}	Anis	$\mathtt{Anis} = \frac{\mathtt{Shape}_a}{\mathtt{Shape}_b}$
M_{19}	Bulk	$\mathtt{Bulk} = \frac{4\pi \mathtt{Shape}_a \mathtt{Shape}_b}{\mathtt{F}}$
M_{20}	StFact	$\mathtt{StFact} = \frac{4\pi \mathtt{Shape}_b - \mathtt{F}}{\mathtt{F}}$
M_{21}	$\mathtt{Circ}_r$	$\mathtt{Circ}_r = [\min_r r^2 = (s - s_0)^2 + (z - z_0)^2, (z,s) \in R]$
M_{22}	$\mathtt{Rec}_\varphi$	$\mathtt{Rec}_\varphi =$ Winkel des kleinsten umschl. Rechtecks
M_{23}	$\mathtt{Rec}_{L_1}$	$\mathtt{Rec}_{L_1} =$ gr. Hauptachse des kleinsten umschl. Rechtecks
M_{24}	$\mathtt{Rec}_{L_2}$	$\mathtt{Rec}_{L_2} =$ kl. Hauptachse des kleinsten umschl. Rechtecks

Die verwendeten Merkmalen und deren formale Beschreibung.

Anwendungen von Wavelets in der Bildcodierung

Ulrich Eckhardt
Institut für angewandte Mathematik
der Universität Hamburg
Bundesstraße 55
20 146 Hamburg 13

Eckard Hundt
Siemens AG
Zentrale Forschung und Entwicklung
Otto–Hahn–Ring 6
81 739 München

Zusammenfassung

Für zahlreiche verschiedene Anwendungen wurden in der Literatur Wavelets mit unterschiedlichen Eigenschaften vorgeschlagen. Jeder dieser Vorschläge weist Vor– und Nachteile auf, so daß die Auswahl unter ihnen oft nicht leicht ist. Bei einer vorgelegten konkreten Anwendung läßt sich häufig kein Kompromiß finden, der allen Anforderungen gerecht wird. Man behilft sich daher damit, daß man ein Wavelet auswählt und dann durch geeignete Modifikationen an das Anwendungsproblem anpaßt. Im vorliegenden Beitrag sollen anhand von zwei aus der Literatur bekannten Vorschlägen, die sich auf die praktische VLSI–Implementierung von Wavelets beziehen, die Auswirkungen solcher Modifikationen quantitativ untersucht werden.

1 Einleitung

Um Bilder über Kanäle beschränkter Kapazität zu übertragen, benutzt man Codierverfahren. Diese bestehen im allgemeinen aus drei Stufen mit unterschiedlichen Eigenschaften:

- Zunächst wird das Bild *transformiert*, wobei das Ziel der Transformation darin besteht, die in dem Bild enthaltene Information so zu dekorrelieren, daß unentbehrliche (nichtredundante) und wichtige (relevante) Information im Bildraum der Transformation deutlich von entbehrlicher und unwichtiger Information getrennt erscheint.
- Durch die Transformation wird die Bildinformation, die normalerweise diskret vorliegt, auf reellwertige Größen abgebildet. Aus diesem Grunde schließt an den Transformationsschritt eine Quantisierung an. Diese beinhaltet im allgemeinen einen Verlust an Information.
- Die quantisierten Größen werden dann so codiert, daß sie sich ohne Informationsverlust mit minimalem Aufwand übertragen lassen.

Im Rahmen dieser Untersuchung befassen wir uns nur mit dem ersten Schritt, der Transformation des Ausgangsbildes.

Die Wavelet-Darstellung eines Bildsignals erscheint auf den ersten Blick als eine für die Bildcodierung besonders geeignete Transformationsmethode. Sie erlaubt nämlich auf natürliche Weise eine Subbanddarstellung des Bildsignals, und diese Darstellung ist wegen der Orthogonalität der Wavelet-Basisfunktionen nichtredundant. Weiterhin sind die Basisfunktionen lokal, so daß eine künstliche Blockbildung mit der Tendenz zu Blockartefakten nicht erforderlich ist. Außerdem ist es möglich, die Wavelet-Darstellung so zu modifizieren, daß sie sich besonders gut zur VLSI-Implementierung eignet, es lassen sich multiplikationsfreie Näherungen finden (siehe etwa [4,3]). Weiterhin weist die Wavelet-Transformation gegenüber anderen Transformationsverfahren (wie etwa der Diskreten Cosinus-Transformation) den Vorteil auf, daß sie sich besonders gut zur Vektorquantisierung eignet (siehe [1, Seite 214]). Insbesondere bei der Bildcodierung mit niedrigen Übertragungsraten erscheint die Wavelet-Transformation attraktiv (siehe [1, Seite 218]).

Bei näherer Untersuchung stellt man jedoch fest, daß es eine Anzahl von Problemen bei der Benutzung von Wavelets gibt. Zunächst einmal hat man das Problem, aus der Vielzahl der publizierten Wavelet-Vorschläge die für die vorgelegte Anwendung passende Darstellung zu finden. Zudem weist jeder der bekannten Vorschläge gravierende Nachteile auf. Nun wird man bei der praktischen Anwendung ohnehin die vorliegenden Wavelet-Ansätze geeignet modifizieren müssen, und sei es auch nur durch Rundung von nicht rationalen Filterkoeffizienten. Erwünscht wäre hier eine Aussage, welchen Einfluß eine solche Modifikation auf die Resultate hat.

Die hier vorgestellten Resultate entstanden während eines Forschungsaufenthaltes eines der Autoren am Siemens-Forschungslaboratorium in München. Für zahlreiche nützliche Hinweise und interessante Diskussionen danken die Autoren Herrn Dr. B. Hammer und Herrn Dr. F. Seytter vom Siemens-Forschungslaboratorium sowie Herrn Faber von der Universität Jena.

2 Wavelets

Im Rahmen dieser Darstellung wollen wir Wavelets als Filter verstehen. Man hat dann normalerweise ein Tiefpaß- und ein Hochpaß-Filter, die zur Transformation und zur Rekonstruktion verwendet werden. Unter den zahlreichen publizierten Wavelet-Vorschlägen betrachten wir hier nur drei Vertreter, der Einfachheit halber wird nur der eindimensionale Fall dargestellt, für die Anwendung auf zwei- und höherdimensionale Bilddaten benutzt man kartesische Produkte der entsprechenden Filter. Es sei also $\{x_n\}_{n=-\infty}^{\infty}$ eine Signalfolge.

Das Haar-Wavelet Hier bildet man die Tiefpaß-Folge $y_{2n} := \frac{1}{2}(x_{2n} + x_{2n+1})$ sowie die Hochpaß-Folge $z_{2n} := \frac{1}{2}(x_{2n} - x_{2n+1})$. Es ist sofort klar, daß man aus diesen beiden unterabgetasteten Folgen die Originalfolge wieder rekonstruieren kann. Es ist nämlich $x_{2n} := y_{2n} + z_{2n}$ und $x_{2n+1} = y_{2n} - z_{2n}$. Das Haar-Wavelet wurde übrigens schon sehr früh von G. Richter zur Datenkompression bei astronomischen Bildmaterial [7] mit Erfolg verwendet.

Das 4-Tap Daubechies-Wavelet I. Daubechies schlug 1988 eine Klasse von Wavelets vor [2], die sich insbesondere durch endlichen Träger der zugehörigen Filter auszeichnen. Das einfachste Filter aus dieser Klasse hat vier nichtverschwindende Filterkoeffizienten und erfreut sich großer Verbreitung. Die Koeffizienten des Tiefpaß-Filters sind

$$h_0 = \frac{1+\sqrt{3}}{8}, \qquad h_1 = \frac{3+\sqrt{3}}{8}, \qquad h_2 = \frac{3-\sqrt{3}}{8}, \qquad h_3 = \frac{1-\sqrt{3}}{8}$$

sowie $h_n = 0$ für $n < 0$ und für $n > 3$. Die Koeffizienten des Hochpaß–Filters sind

$$g_n = \begin{cases} (-1)^n \cdot h_{-n+1} & \text{für } n = -2, -1, 0, 1, \\ 0 & \text{sonst.} \end{cases}$$

Man bildet damit die beiden Folgen

$$\begin{aligned} y_{2n} &= h_0 x_{2n} + h_1 x_{2n+1} + h_2 x_{2n+2} + h_3 x_{2n+3} \\ z_{2n} &= g_{-2} x_{2n-2} + g_{-1} x_{2n-1} + g_0 x_{2n} + g_1 x_{2n+1}. \end{aligned}$$

Wie man leicht durch Nachrechnen verifiziert, kann man aus diesen beiden Folgen die Originalfolge exakt rekonstruieren gemäß

$$\begin{aligned} x_{2n} &= 2 \cdot (h_0 y_{2n} + h_2 y_{2n-2} + h_1 z_{2n} + h_3 z_{2n+2}), \\ x_{2n+1} &= 2 \cdot (h_1 y_{2n} - h_2 z_{2n+2} + h_3 y_{2n-2} - h_0 z_{2n}). \end{aligned}$$

Das Wavelet von Mallat Hier ist das Filter symmetrisch und hat aber keinen endlichen Träger. Wir verzichten deshalb auf Angabe der Koeffizienten und verweisen auf den Artikel von Mallat [5].

3 Wünschenswerte Eigenschaften von Wavelets

Jedes der im vorigen Abschnitt genannten Wavelets hat Vor– und Nachteile für die Anwendung zur Bildcodierung. Wir zählen einige Eigenschaften, die bei dieser Anwendung eine Rolle spielen, auf:

- Bei der Bildverarbeitung zieht man Nullphasenfilter vor, die Positionen von Kanten nicht verändern. Dies ist insbesondere wichtig bei Codierung durch Bewegungskompensation. Unter den oben genannten Wavelet–Filtern besitzt nur das symmetrische Mallat–Filter diese Eigenschaft.

- Bei Anwendungen in der Bildverarbeitung liegen die Daten üblicherweise als ganzzahlige Werte vor, beispielsweise als quantisierte Grauwertbilder. Bei Transformation mit nicht ganzzahligen Filterkoeffizienten erhält man relle Werte. Dies bedeutet, daß sich der Datenumfang zunächst einmal vergrößert, sofern man nicht rundet. Insbesondere bei VLSI–Implementierung hätte man aus Effizienzgründen gern ganzzahlige Werte oder doch einfache rationale Zahlen [4, 3]. Jedoch kann man allenfalls das Haar–Wavelet ohne einschneidende Modifikationenen so implementieren, daß Folgen ganzer Zahlen wieder auf Folgen ganzer Zahlen abgebildet werden (jedoch kann man im allgemeinen nicht erreichen, daß Folgen nichtnegativer Zahlen wieder in Folgen nichtnegativer Zahlen überführt werden).

- Der Träger des Filters, das heißt, die Menge aller der Indices, für die der entsprechende Filterkoeffizient nicht verschwindet, sollte möglichst klein sein, das heißt, die Filter sollten die Eigenschaft der Lokalität haben. Diese Eigenschaft besitzt beispielsweise das Mallat–Wavelet nicht.

- Unter Umständen ist die Eigenschaft, daß das Transformationsfilter mit dem Rekonstruktionsfilter identisch ist, nicht wichtig. Beispielsweise könnte man sich vorstellen, daß das Rekonstruktionsfilter besonders einfach sein soll, da beim Empfänger im allgemeinen einschränkende Hardwarerestriktionen vorliegen. Dagegegen könnte man beim Sender ohne Schwierigkeiten aufwendigere Transformationsalgorithmen einsetzen, sofern die exakte Rekonstruierbarkeit erhalten bleibt (siehe beispielsweise [1]) oder wenigstens näherungsweise gegeben ist.
- Ein generelles Problem bei Wavelet–Darstellung ist die im allgemeinen fehlende Translationsinvarianz [8]. Dieser Defekt macht sich insbesondere bei Bewegungdetektion störend bemerkbar.

4 Modifikation des Haar–Filters

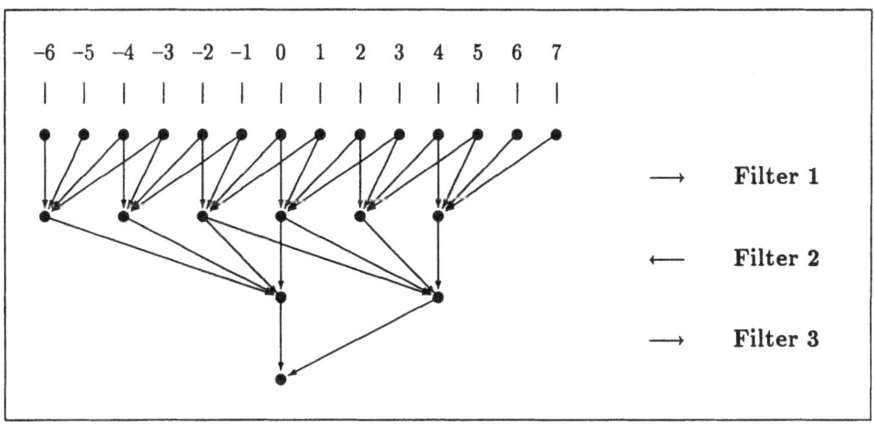

Abbildung 1: Transformation nach dem NEC–Vorschlag

Es ergibt sich aus der Diskussion im vorhergehenden Abschnitt, daß es erforderlich ist, Kompromisse einzugehen. Diese könnten so aussehen, daß man ein vorgelegtes Wavelet derart modifiziert, daß es zu einer gegebenen Anwendung paßt, ohne dabei seine vorteilhaften Eigenschaften zu verlieren.

Wir untersuchen zunächst die Verhältnisse beim Haar–Filter. Für einen Parameter λ mit $0 < \lambda < 1$ modifizieren wir dieses auf die folgende Weise:

$$y_{2n} := \lambda x_{2n} + (1-\lambda)x_{2n+1}, \qquad z_{2n} := \lambda x_{2n} - (1-\lambda)x_{2n+1}.$$

Für $\lambda = \frac{1}{2}$ erhalten wir das originale Haar–Filter. In diesem Falle können wir das Rekonstruktionsfilter auf einfache Weise berechnen, es ist nämlich

$$x_{2n} := \frac{1}{2\lambda}(y_{2n} + z_{2n}) \qquad \text{und} \qquad x_{2n+1} = \frac{1}{2(1-\lambda)}(y_{2n} - z_{2n}).$$

Wir betrachten jetzt eine spezielle Anwendung des modifizierten Haar–Wavelets.

Die Firma NEC [6] schlug einen Wavelet–Kompressionsalgorithmus vor, bei dem die Transformation in drei aufeinanderfolgenden Stufen wie folgt organisiert ist:

1. Stufe 4–Tap–Wavelet von Daubechies.

2. Stufe 4–Tap–Wavelet von Daubechies, jedoch rückwärts genommen,

3. Stufe Haar–Wavelet.

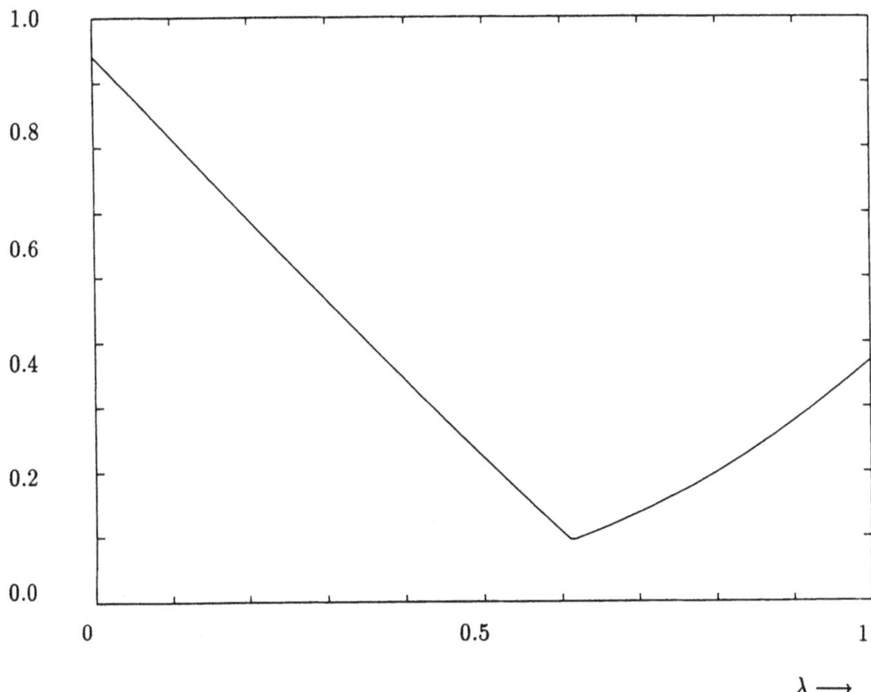

Abbildung 2: Abhängigkeit des Maximums des Phasenbetrages von dem Parameter λ

Durch Benutzung des rückwärts genommenen Filters in der 2. Stufe werden die Verschiebungen auf Grund der Asymmetrie teilweise kompensiert. Das Haar–Wavelet in der dritten Stufe soll den Einflußbereich einschränken (siehe Abbildung 1).

Man kann nun versuchen, das Maximum des Phasenbetrags durch ein modifiziertes Haar–Filter in der dritten Stufe mit geeignet gewähltem λ noch weiter zu verkleinern. In Abbildung 2 ist das Maximum des Phasenbetrages in Abhängigkeit von λ gezeigt. Man sieht, daß es ein deutliches Minimum gibt. Der Optimalwert von λ liegt bei $\lambda^* = 0.6119614298$. Die folgende Tabelle gibt für einige Werte von λ das Betragsmaximum der Phase an (vgl. auch Abbildung 3).

λ	Maximum des Phasenbetrags
0	0.941 025
0.5	0.321 548
λ^*	0.193 256
1.	0.471 251

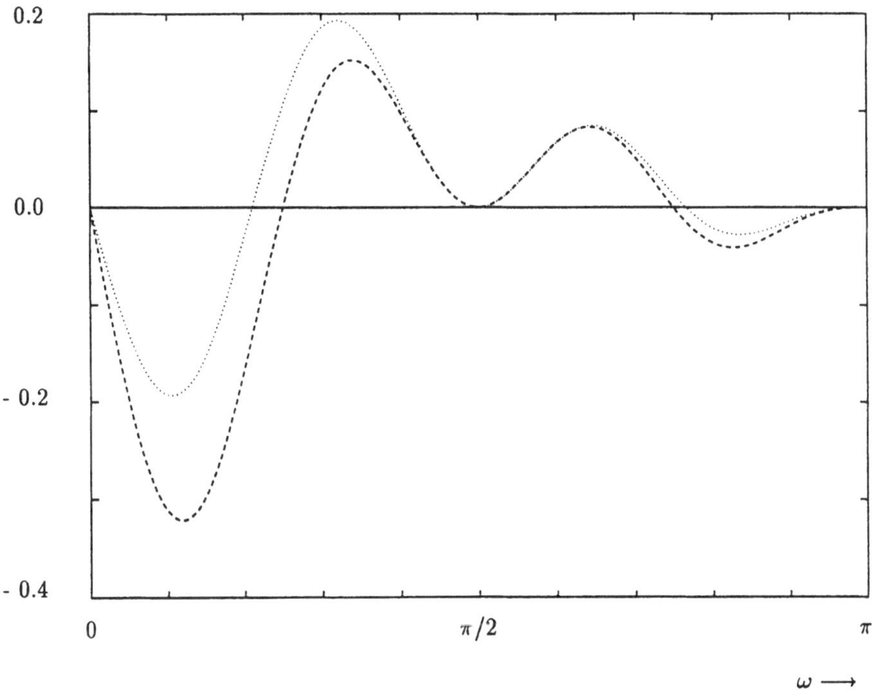

Abbildung 3: Phase des NEC–Filters für die Parameterwerte $\lambda = 0.5$ (gestrichelt) und $\lambda = \lambda^*$ (gepunktet)

Durch den einfachen Trick der Parametrisierung des Haar–Filters kann man also das Maximum des Phasenbetrags gegenüber dem ursprünglichen Ansatz ohne nennenswerten Zusatzaufwand um 40% verbessern. Man könnte die Phase noch weiter reduzieren, indem man das Daubechies–Filter in den beiden ersten Stufen geeignet abändert. Mit den damit zusammenhängenden Problemen werden wir uns im folgenden Abschnitt befassen.

5 Modifiziertes Daubechies–Filter

Lewis und Knowles [4, 3] schlugen für die VLSI–Implementierung ein auf einfache rationale Werte gerundetes Daubechies–4–Tap–Wavelet vor. Die entsprechenden Koeffizienten sind

$$h'_0 = \frac{11}{32}, \quad h'_1 = \frac{19}{32}, \quad h'_2 = \frac{5}{32}, \quad h'_3 = -\frac{3}{32}.$$

Diese Koeffizienten entstehen aus den Daubechies–Koeffizienten durch die Approximation $\sqrt{3} \approx 7/4$ Der maximale absolute Fehler bei dieser Darstellung ist kleiner als $2.2437 \cdot 10^{-3}$.

Im Unterschied zur Situation im vorhergehenden Abschnitt läßt sich hier nicht so ohne weiteres ein exaktes Rekonstruktionsfilter angeben. Zwar existiert ein exaktes Rekonstruktionsfilter, dieses läßt sich durch Matrixinversion auch explizit berechnen, jedoch ist es nicht gerade „handlich", seine Koeffizienten hängen beispielsweise von der Bildgröße ab. Die Berechnung dieses Filters dürfte sich nur dann lohnen, wenn der Rechenaufwand bei Transformation und Rücktransformation deutlich unterschiedlich sein darf (beispielsweise dann, wenn die Rücktransformation in einem mobilen Empfänger mit eingeschränkter Hardware stattfinden soll).

Man kann sich jedoch die Frage stellen, ob es möglich ist, ein Filter für die Rücktransformation zu finden, das sich durch eine leichte Modifikation aus dem Daubechies–Wavelet ergibt und das den Rekonstruktionsfehler „klein" läßt.

Setzt man $h'_i = h_i + \varepsilon_i$, dann liefert eine einfache Fehlerabschätzung, die die Vorzeichen der Approximationsfehler der $h(i)$ berücksichtigt unter der Voraussetzung $0 \leq x_n \leq 255$ für das transformierte Bild einen Maximalfehler von

$$|y'_{2n} - y_{2n}| \leq 1.1443, \qquad |z'_{2n} - z_{2n}| \leq 1.1443.$$

Das Rekonstruktionsfilter sei gegeben durch die Koeffizienten $h''_i = h_i + \delta_i$.

Wir erhalten für den Fall $\delta_i = \varepsilon_i$ (dies entspricht der Rekonstruktion mit dem Lewis–Knowles–Filter) unter der obengenannten Voraussetzung einen Rekonstruktionsfehler von 1.9922. Für $\delta_i = 0$, das heißt Transformation mit dem Lewis–Knowles Filter und Rekonstruktion mit dem Daubechies–Filter, beträgt der Fehler 1.8492.

Man kann die δ_i so bestimmen, daß der Rekonstruktionsfehler möglichst klein wird. Bezeichnet man mit $\{x'_n\}$ die aus den Näherungsfolgen $\{y_{2n}\}$ und $\{z_{2n}\}$ vermittels der h''_i rekonstruierte Folge, dann ist

$$\begin{aligned} x'_{2n} - x_{2n} = 2\Big[& \quad (h'_2\delta_0 + h'_1\delta_3 + \varepsilon_2 h_0 + \varepsilon_1 h_3) \cdot x_{2n-2} + \\ & + (h'_2\delta_1 - h'_1\delta_2 + \varepsilon_2 h_1 - \varepsilon_1 h_2) \cdot x_{2n-1} + \\ & + (h'_0\delta_0 + h'_1\delta_1 + h'_2\delta_2 + h'_3\delta_3 + \varepsilon_0 h_0 + \varepsilon_1 h_1 + \varepsilon_2 h_2 + \varepsilon_3 h_3) \cdot x_{2n} + \\ & + (h'_0\delta_1 + h'_2\delta_3 - h'_1\delta_0 - h'_3\delta_2 + \varepsilon_0 h_1 + \varepsilon_2 h_3 - \varepsilon_1 h_0 - \varepsilon_3 h_2) \cdot x_{2n+1} + \\ & + (h'_0\delta_2 + h'_3\delta_1 + \varepsilon_0 h_2 + \varepsilon_3 h_1) \cdot x_{2n+2} + \\ & + (h'_0\delta_3 - h'_3\delta_0 + \varepsilon_0 h_3 - \varepsilon_3 h_0) \cdot x_{2n+3} \quad \Big]. \end{aligned}$$

Wegen $0 \leq x_k \leq 255$ kann man den Fehler $x'_{2n} - x_{2n}$ einschließen zwischen 255×(Summe der negativen Koeffizienten der x_k) und 255×(Summe der positiven Koeffizienten der x_k). Diese Forderung führt auf ein lineares Optimierungsproblem zur Berechnung der δ_i. Durch Lösung dieses Problems erhält man als optimale Koeffizienten des Rekonstruktionsfilters

$$\begin{aligned} h_0 &= 0.341907279381863, & h_1 &= 0.588348922326149, \\ h_2 &= 0.158092720618137, & h_3 &= -0.0899755998373323. \end{aligned}$$

Unter der Annahme $0 \leq x_n \leq 255$ ergibt sich hier als Schranke für den Rekonstruktionfehler 0.9884, also eine Verbesserung von etwa 46% gegenüber der Rekonstruktion mit dem Daubechies–Filter.

6 Anwendungen

Die hier vorgeschlagene Vorgehensweise ist nicht auf die beiden angeführten Fälle beschränkt. Man kann vielmehr nach dem hier vorgestellten Muster für einen vorgelegten Anwendungsfall nach Maßgabe eines Anforderungskataloges ein „Wavelet nach Maß" konstruieren. Dabei muß der Anwendungsaufwand nicht notwendig steigen, wie sich bei der Modifikation des NEC–Vorschlages zeigte.

Anhand von Bildvorlagen aus unterschiedlichen Anwendungsgebieten wurden die hier vorgestellten Modifikationen bekannter Wavelet–Vorschläge im Hinblick auf ihre Anwendung auf Bildcodierung praktisch untersucht. Es stellte sich dabei heraus, daß die tatsächlich auftretenden Fehler in der Größenordnung der hier angegebenen Schranken liegen.

Literatur

[1] Marc Antonini, Michel Barlaud, Pierre Mathieu and Ingrid Daubechies. Image coding using wavelet transform. *IEEE Transactions on Image Processing*, 1:205–220, 1992.

[2] Ingrid Daubechies. Orthonormal bases of compactly supported wavelets. *Communications on Pure and Applied Mathematics*, XLI:909–996, 1988.

[3] A. S. Lewis and G. Knowles. Image compression using the 2–D wavelet transform. *IEEE Transactions on Image Processing*, 1:244–250, 1992.

[4] A. S. Lewis and K. Knowles. VLSI architecture for 2–D Daubechies wavelet transform without multipliers. *Electronics Letters*, 27:171–173, 1991.

[5] Stephane Mallat. Zero–crossings of a wavelet transform. *IEEE Trans.*, IT–37:1019–1033, 1991.

[6] NEC Corporation. A description of MRR transform for core–experiments, *International Organization for Standardization, ISO/IEC JTC1/SC2/WG11, coding of moving pictures and associated audio*, January 1992.

[7] G. M. Richter. Zur Auswertung astronomischer Aufnahmen mit dem automatischen Flächenphotometer. *Astron. Nachr.*, 299:283–303, 1978.

[8] Eero P. Simoncelli, William T. Freeman, Edward H. Adelson and David J. Heeger. Shiftable multiscale transforms. *IEEE Transactions*, IT–38:587–607, 1992.

Iterierte Funktionensysteme
Eine Lösung des eindimensionalen inversen Problems

Michael A. Neuhauser Walter G. Kropatsch Irene J. Leitgeb

Technische Universität Wien, Abt. für Bildverarbeitung und Mustererkennung
Institut für Automation 183/2, Treitlstr. 3, A–1040 Wien, Österreich
Telefon: +43 (1) 58801-8161, Fax: +43 (1) 569697, E-mail: man@prip.tuwien.ac.at

Diese Arbeit wurde zum Teil durch Digital Equipment Corporation unter Vertrag EERP/624/AU-027, und dem ERASMUS Austauschprogramm ICP-92-A-2007/11 unterstützt.

Zusammenfassung

Herkömmliche Codierungsmethoden erzielen unbefriedigende Resultate bei natürlichen, fraktalen Bildern. Um ein fraktales Modell zur Bildcodierung verwenden zu können, muß sowohl Codierung als auch Decodierung automatisch und effizient durchgeführt werden können. Dieser Artikel behandelt die Codierung von digitalen Bildern mittels Iterierter Funktionen Systeme (IFS). Es werden die Grundlagen von IFS erläutert und gezeigt, wie durch Diskretisierung eine effiziente Verarbeitung möglich ist. Nach einer Beschreibung des Codierungsproblems wird der eindimensionale, binäre Fall unter Ausnützung eines invarianten Merkmals gelöst. Anschließend wird ein Experiment diskutiert, dessen Resultate zeigen, daß die gestellte Aufgabe durch das beschriebene Verfahren gut gelöst wird. Die Möglichkeit zur Erweiterung auf den zweidimensionalen Fall besteht.

1 Einleitung

Bilder natürlicher Szenen sind oft selbstähnlich, das macht sie zu fraktalen Bildern. Man denke sich eine Gebirgslandschaft: ein Gebirge setzt sich aus Bergen zusammen, die sich wiederum aus von kleinen Felsen gebildeten größeren Felsen zusammensetzen ... bis hinunter zu kleinen Steinen. Auch viele Bilder von Pflanzen, speziell Bäumen, oder von Wolken sind fraktaler Natur.

Da immer mehr digitale Bilder verarbeitet und gespeichert werden, gibt es einen wachsenden Bedarf an hoch komprimierenden Codierungsmethoden. Es erscheint uns günstig Codierungsmethoden zu untersuchen, die auf fraktalen Modellen beruhen, da herkömmliche Methoden bei fraktalen Bildern versagen.

Iterierte Funktionen Systeme (IFS) sind ein fraktales Modell, das in letzter Zeit besonderes Interesse hervorgerufen hat, sowohl zur Codierung von Bildern [2] als auch zur Erzeugung realistischer Bilder [3]. IFS erzeugen strikt selbstähnliche Bilder. Das Problem der automatischen Codierung von Bildern mittels IFS ist noch nicht befriedigend gelöst [5, 6, 8]. Wir präsentieren hier eine Lösung für den eindimensionalen, binären Fall.

Der Aufbau dieses Artikel ist wie folgt: Grundlagen von IFS; Diskretisierung für effiziente Verarbeitung; Formulierung des Problems der Codierung mittels IFS; Beschreibung unserer Lösung; Resultate eines Experiments; Schluß.

2 Grundlagen Iterierter Funktionensysteme

Hier werden die wichtigsten Begriffe und Eigenschaften von IFS dargelegt. Für Beweise und eine detailiertere Behandlung siehe [1].

Sei (X, d) ein vollständiger metrischer Raum mit Distanzfunktion d. Eine Funktion $w : X \to X$ heißt *kontraktiv* auf X genau dann, wenn es ein $c \in [0,1)$ gibt, sodaß $d(w(x), w(y)) \leq cd(x,y)$ für alle $x, y \in X$. Das kleinste c, das diese Bedingung erfüllt, heißt *Kontraktionsfaktor* von w. Jede kontraktive Funktion w besitzt genau einen *Fixpunkt* $x \in X$, $w(x) = x$.

Sei $\mathcal{H}(X)$ die Klasse aller nicht leeren, kompakten Untermengen von X. Mit der *Hausdorffmetrik* d_h ist $(\mathcal{H}(X), d_h)$ ein vollständiger metrischer Raum.

Ein *IFS* in X ist eine endliche Menge von zwei oder mehr kontraktiven Funktionen auf X. Der *Hutchinsonoperator* $\mathbf{w} : \mathcal{H}(X) \to \mathcal{H}(X)$ definiert das Verhalten eines IFS $\{w_i\}_{i=1}^n$: $\mathbf{w}(A) := \bigcup_{i=1}^n w_i(A)$ für $A \in \mathcal{H}(X)$. Sind alle w_i des IFS kontraktiv auf (X, d), so ist $\mathbf{w}$ eine kontraktive Funktion auf $(\mathcal{H}(X), d_h)$. Damit existiert für $\mathbf{w}$ eines jeden IFS ein eindeutiger Fixpunkt $\mathcal{A} \in \mathcal{H}(X)$, $\mathcal{A} = \mathbf{w}(\mathcal{A})$, der *Attraktor* des IFS. Der Kontraktionsfaktor von $\mathbf{w}$ bezüglich d_h wird Kontraktionsfaktor des IFS genannt, er ist gleich dem Maximum der Kontraktionsfaktoren der w_i.

Der Fall $X = \mathbb{R}^2$ ist von besonderem Interesse, da man Elemente $A \in \mathcal{H}(\mathbb{R}^2)$ als binäre Bilder interpretieren kann: Allen Punkte der Ebene, die in A enthalten sind, wird die Farbe Schwarz zugeordnet, allen anderen Weiß. Da wir aber an digitalen Bildern interessiert sind, muß man dieses binäre Bild noch abtasten oder diskret repräsentieren, darauf wird im nächsten Abschnitt eingegangen.

Als Metrik verwendet man üblicherweise den euklidischen Abstand, der dem „natürlichen" Abstand entspricht, für die Funktionen des IFS affine Transformationen, da sie geometrische Transformationen sind, deren Wirkungsweise unmittelbar anschaulich ist. Eine *affine Transformation* w auf $\mathbb{R}$ ist eine Funktion, die jedem $x \in \mathbb{R}$ den Wert $w(x) := kx + d$ zuordnet, mit $k, d \in \mathbb{R}$. Abb. 2 zeigt den abgetasteten Attraktor des IFS $\{\frac{1}{4}x, \frac{1}{4}x + 20, \frac{1}{8}x + 35\}$ in $\mathbb{R}$ (ein Pixel hat dabei die Länge 1, das erste Pixel beginnt bei 0).

3 Diskrete IFS

Um dem Attraktor eines IFS in $\mathbb{R}^2$ auf einem Bildschirm gegebener Auflösung darstellen zu können, muß er abgetastet werden. Meist wird dies mit dem Chaosspiel ([1]) gemacht. Diese Methode hat aber einige Nachteile, da der Attraktor dabei weit genauer als nötig berechnet wird. Wir verwenden den diskreten Raum $\mathbb{P}^2 \subset \mathbb{N}^2$ der Pixel des Bildschirms. Der Einfachheit halber wird ein quadratischer Schirm der Größe $R \in \mathbb{N}$ angenommen: $\mathbb{P} := \{0, 1, \ldots, R-1\}$. Man identifiziert den Bildschirm mit $[0, R) \times [0, R) \subset \mathbb{R}^2$; jedes Pixel $p = (p_x, p_y) \in \mathbb{P}^2$ repräsentiert ein Quadrat $[p_x, p_x + 1) \times [p_y, p_y + 1) \subset \mathbb{R}^2$. Zusätzlich kann man ohne Einschränkung ([1]) annehmen, daß der Attraktor vollständig in $[0, R) \times [0, R)$ liegt.

Um den Raum der digitalen Binärbilder mathematisch zu modellieren, verwenden wir einen Isomorphismus zwischen Binärbildern und Pixelmengen. Ein Binärbild entspricht jener Pixelmengen $I \in \mathcal{H}(\mathbb{P}^2)$, die genau die schwarzen Pixel des Binärbildes enthält und umgekehrt. Die Idee ist, ein Binärbild des Attraktors nicht durch Abtastung des Attraktors zu berechnen, sondern einen Operator $\mathbf{w}' : \mathcal{H}(\mathbb{P}^2) \to \mathcal{H}(\mathbb{P}^2)$ (ein *diskretes IFS*) so zu einem gegebenen IFS zu konstruieren, daß dessen Fixpunkte Approximationen des gewünschten Bildes sind. Um dieses $\mathbf{w}'$ zu definieren, werden die affinen Transformationen $w_i : \mathbb{R}^2 \to \mathbb{R}^2$ des IFS durch diskrete Transformationen $w_i' : \mathbb{P}^2 \to \mathbb{P}^2$ approximiert. Dazu wird die Funktion $\varrho : \mathbb{P}^2 \to \mathbb{R}^2$ definiert, die ein Pixel auf seinen Mittelpunkt abbildet: $\varrho(p) := \binom{p_x+0.5}{p_y+0.5}$, $p = (p_x, p_y) \in \mathbb{P}^2$. Umgekehrt liefert die Funktion $\Delta : \mathbb{R}^2 \to \mathbb{P}^2$ das Pixel, das einen gegebenen Punkt enthält: $\Delta(\bar{x}) := (\lfloor \bar{x}_1 \rfloor, \lfloor \bar{x}_2 \rfloor)$, $\bar{x} = \binom{\bar{x}_1}{\bar{x}_2} \in \mathbb{R}^2$. Es ist

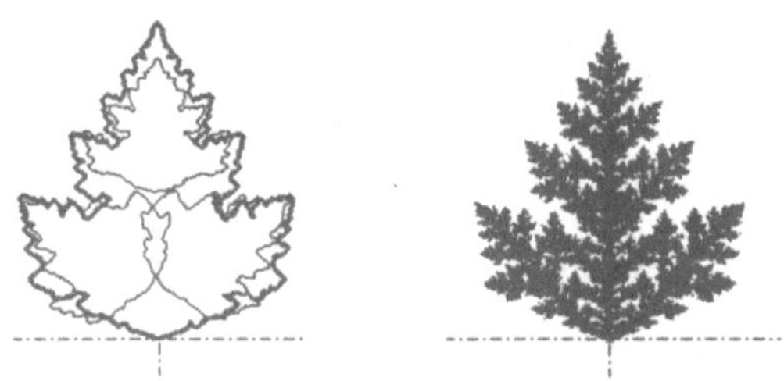

Abbildung 1: Collage: Überdeckung eines Bildes mit affinen Bildern seiner selbst.

sicher, daß das Resultat von Δ innerhalb von $\mathbb{P}^2$ liegt; alles außerhalb des Bildschirms wird verworfen. Damit ist für eine affine Transformation $w : \mathbb{R}^2 \to \mathbb{R}^2$ ihre zugehörige *diskrete Transformation* $w' : \mathbb{P}^2 \to \mathbb{P}^2$ wie folgt definiert:

$$w'(p) := (\Delta \circ w \circ \varrho)(p) \qquad p \in \mathbb{P}^2 \tag{1}$$

Der *diskrete Hutchinsonoperator* $\mathbf{w}' : \mathcal{H}(\mathbb{P}^2) \to \mathcal{H}(\mathbb{P}^2)$ eines IFS $\{w_i\}_{i=1}^n$ bildet eine Pixelmenge auf eine Pixelmenge ab und ist ähnlich wie der Hutchinsonoperator definiert:

$$\mathbf{w}'(I) := \bigcup_{i=1}^n w_i'(I) \qquad I \in \mathcal{H}(\mathbb{P}^2), \tag{2}$$

mit $w_i'(I) := \{w_i'(p) \mid p \in I\}$. Jeder Fixpunkt $\mathcal{A}_d \in \mathcal{H}(\mathbb{P}^2)$, $\mathcal{A}_d = \mathbf{w}'(\mathcal{A}_d)$, des diskreten Hutchinsonoperators wird *diskreter Attraktor* genannt. In [7] wird bewiesen, daß die Distanz zwischen jedem diskreten Attraktor $\mathcal{A}_d$ und dem Attraktor $\mathcal{A}$ begrenzt ist:

$$\mathbb{R}^2: \quad d_h(\mathcal{A}, \varrho(\mathcal{A}_d)) \leq \tfrac{1/\sqrt{2}}{1-\lambda} \qquad\qquad \mathbb{R}: \quad d_h(\mathcal{A}, \varrho(\mathcal{A}_d)) \leq \tfrac{1/2}{1-\lambda} \tag{3}$$

Dabei ist λ der Kontraktionsfaktor des IFS und $\varrho(\mathcal{A}_d) := \{\varrho(p) \mid p \in \mathcal{A}_d\}$. Durch die Diskretisierung ist $\mathbf{w}'$ im allgemeinen keine kontraktive Transformation mehr, dadurch kann es mehrere Fixpunkte besitzen, es gibt jedoch auf jeden Fall genau einen maximalen diskreten Attraktor [7]. In [7] finden sich zwei effiziente Algorithmen, die einen diskreten bzw. den maximalen diskreten Attraktor eines IFS berechnen.

4 Das inverse Problem

Es ist also möglich, ein digitales Binärbild aus einem IFS zu berechnen. Will man aber ein vorgebenes digitales Bild $I \in \mathcal{H}(\mathbb{P}^2)$ codieren, so muß man ein IFS finden, dessen Attraktor bei Umwandlung in ein digitales Bild dieses I möglichst gut approximiert. Diese Aufgabe ist als das *inverse Problem* bekannt. Formal läßt es sich als Minimierung des Wertes $d_h(\varrho(A), \varrho(I))$ beschreiben. Dabei ist $A \in \mathcal{H}(\mathbb{P}^2)$ eine bestmögliche Approximation von $\mathcal{A}$ durch eine Pixelmenge, also $d_h(\mathcal{A}, \varrho(A)) \leq d_h(\mathcal{A}, \varrho(B))$ für alle $B \in \mathcal{H}(\mathbb{P}^2)$.

Das inverse Problem ist nicht leicht zu lösen; bis jetzt ist keine automatische, vollständige Lösung publiziert worden. Die veröffentlichten Ansätze gehen fast alle vom Collage Theorem aus. Dieses besagt folgendes: Man versucht, das gegebene Bild I möglichst genau mit affinen, verkleinerten Kopien seiner selbst zu überdecken. Den affinen Kopien entsprechen affine Transformationen. Bildet man ein IFS aus diesen Transformationen, so wird dessen Attraktor dem Bild I in dem Maße ähnlich sein, wie sich I und $\mathbf{w}(I)$ ähnlich sind. Genauer: ist $d_h(I, \mathbf{w}(I)) = \epsilon$ (der *Collagefehler*), so gilt $d_h(I, \mathcal{A}) \leq \frac{\epsilon}{1-\lambda}$; wobei $\mathcal{A}$ der Attraktor und λ der Kontraktionsfaktor des IFS ist. In Abb. 1 wird dies am

Beispiel eines Blattes illustriert; links ist dort eine Collage des Blattes zu sehen, rechts der Attraktor des resultierenden IFS.

Die meisten Ansätze zur Lösung des inversen Problems versuchen den Collagefehler ϵ durch eine Suche im Parameterraum der affinen Transformationen zu minimieren. Diese Suche ist aber sehr langsam und instabil, da es viele lokale Minima in dieser Funktion gibt. Die Autoren Lévy-Véhel und Gagalowicz stellen in [5] ein Gradientenverfahren vor. Werden die Anfangswerte günstig gewählt, sind die Resultate dieser Methode gut; ansonsten bleibt der Algorithmus in einem lokalen Minimum hängen. Robuster ist der in [6] beschriebene Algorithmus, der auf der Methode des simulated annealing basiert. Der Algorithmus ist zwar langsamer, aber dafür in geringerem Ausmaß von den Startwerten abhängig. Vrscay geht in [8] einen ähnlichen Weg, er verwendet zur Minimumsuche jedoch genetische Algorithmen.

5 Der Algorithmus MatchRatio

Der hier beschriebene Algorithmus löst das inverse Problem in $\mathbb{R}$, das darin besteht, zu einer gegebenen binären Pixelzeile $P \in \mathcal{H}(\mathbb{P})$ ein IFS zu finden, dessen Attraktor die Pixelzeile möglichst gut approximiert. Das wird dadurch erreicht, daß MatchRatio ein IFS berechnet, sodaß $\mathbf{w}'(P) = P$ gilt, P also ein diskreter Attraktor des IFS ist. Dann folgt aus (3) und der Dreiecksungleichung: $d_h(\varrho(\mathcal{A}_d), \varrho(P)) \leq \frac{1}{1-\lambda}$.

Die prinzipielle Vorgehensweise von MatchRatio bei der Berechnung des IFS ist, Transformationen zu finden, die disjunkte Teilintervalle von P erzeugen. Damit ist gemeint, daß für die *Zielbereiche* $Z_i := \{p \in \mathbb{P} \mid \min w_i'(P) \leq p \leq \max w_i'(P)\}$ der Transformationen w_i, $i = 1, \ldots, n$ des zu berechnenden IFS $Z_i \cap Z_j = \emptyset$ für $i \neq j$ gilt. $w_i'(P) = P \cap Z_i$ wird *Teilüberdeckung* genannt. Die Disjunktivität der Zielbereiche erlaubt es, jeden Zielbereich unabhängig von allen anderen zu betrachten.

MatchRatio nützt die Tatsache, daß das Verhältnis von Längen invariant unter affinen Transformationen ist. Deshalb wird P zunächst als Folge von Längen schwarzer und weißer Zusammenhangskomponenten dargestellt. (Eine Zusammenhangskomponente von P ist eine Folge von gleichfarbigen Pixeln, die links und rechts von andersfarbigen Pixeln bzw. vom Anfang oder Ende von P begrenzt sind.) Schwarze Zusammenhangskomponenten heißen *Dashes*, weiße *Gaps*. $\tilde{P}$ ist die alternierende Folge der Längen von Dashes d_i und Gaps g_i von P:

$$\tilde{P} := (d_0, g_0, d_1, g_1, \ldots, g_{m-1}, d_m) \tag{4}$$

Ein *Ratio* ist ein Verhältnis der Länge eines Dashes zu der Länge eines benachbarten Gaps in $\tilde{P}$. Die Ratios von P sind unter affinen Transformationen invariant und berechnen sich wie folgt:

$$r_0 := \frac{d_0}{g_0},\ r_1 := \frac{d_1}{g_0},\ r_2 := \frac{d_1}{g_1}, \ldots,\ r_{2(m-1)} := \frac{d_m}{g_{m-1}} \tag{5}$$

Durch die Verwendung diskreter Transformationen w' ergibt sich eine Störung der Ratioinvarianz. Eventuell ist w' so stark kontraktiv, daß für manche Gaps keine weißen Pixel mehr entstehen; man spricht vom *Schließen* eines Gaps. Die Idee ist, sich dieses Verhalten nutzbar zu machen und Teilsequenzenen von Ratios zu finden, die durch diskrete Transformation von P enstanden sind. Von diesen Teilsequenzen kann man auf die erzeugenden Transformationen schließen und findet so das gewünschte IFS. Siehe Abb. 2 für eine Illustration.

Der Kontraktionsfaktor von w' ist bestimmend für das Ergebnis von $w'(P)$. Er legt die Länge fest, bis zu der Gaps offen bleiben, bzw. ab der sie geschlossen werden, allerdings

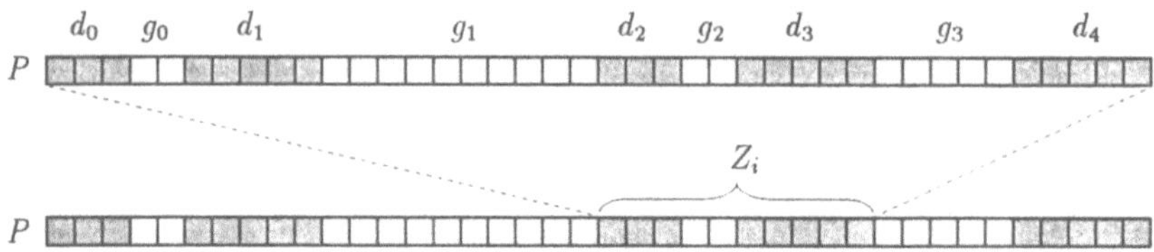

Abbildung 2: Die Pixelzeile P wird auf $P \cap Z_i$ abgebildet.

nicht vollständig, da dies auch von der Position der transformierten Pixelzeile in Bezug auf das Pixelraster abhängt. Wenn man umgekehrt zu jedem Gap entscheidet, ob er unter w' offen bleibt oder geschlossen wird (dadurch entsteht eine Folge von Dashes und Gaps, die als *Kombination* von P bezeichnet wird), so ergibt dies ein Intervall möglicher Kontraktionsfakoren von w'. Dieses Kontraktionsfaktorintervall erlaubt es aber wiederum, Abschätzungen für die Ratios der Kombination abzuleiten. Wenn eine Teilsequenz der Ratios von P in diesen Ratiointervallen enthalten ist, man spricht von einem *Match*, so besteht die Möglichkeit, daß eine diskrete Transformation existiert, die die entsprechende Teilüberdeckung erzeugt. Ob dies tatsächlich der Fall ist, und wenn ja, was die Parameter dieser Transformation sind, muß dann noch explizit festgestellt werden. In [4] ist dazu ein Verfahren beschrieben, auf daß hier jedoch aus Platzgründen nicht eingegangen werden kann.

Zur Ableitung des Ratiointervalls (detailierter in [4] zu finden): Wäre der Kontraktionsfaktor c der diskreten Transformation bekannt, dann kann ein Intervall $\mathcal{R}(d, g, c)$ für das Ratio eines Dashes der Länge d und eines Gaps der Länge g abgeleitet werden. Allerdings ist der Kontraktionsfaktor nicht bekannt. Nehmen wir zunächst an, daß wir die Kombination kennen, die untersucht werden soll, daß also für jeden Gap bekannt ist, ob er offen ist oder nicht. Es läßt sich dann ein Intervall (c_{min}, c_{max}) für den Kontraktionsfaktor einer diskreten Transformation angeben, die diese Kombination erzeugen kann. Aus $\mathcal{R}(d, g, c)$ erhält man eine Abschätzung für das Ratio $\frac{d}{g}$ unter einer solchen Transformation, kurz das *Ratiointervall*:

$$\mathcal{R}(d, g, c_{min}, c_{max}) := \left(\frac{d-1}{g+1}, \frac{\lceil c_{max}(d-1)\rceil + 1}{\max\{\lfloor c_{min}(g+1)\rfloor - 1, 1\}}\right] \tag{6}$$

Zusammengefaßt erzeugt eine Kombination ein Kontraktionsfaktorintervall, dieses erzeugt aus der Kombination eine Folge von Ratiointervallen. Findet man eine Teilsequenz der Ratios von P, die in der Folge der Ratiointervalle enthalten sind (sie *matchen*), so hat man möglicherweise den Zielbereich einer Transformation des zu berechnenden IFS gefunden.

Es wäre aber keine gute Idee, alle möglichen Kombinationen von P zu erzeugen und dann mit den Ratios zu matchen: bei m Gaps wären dies nämlich 2^m Kombinationen, was einen exponentiellen Aufwand in der Anzahl der Gaps bedeutet. Deshalb werden beide Aufgaben, das Erzeugen der Kombination und das Matchen, verzahnt ausgeführt.

Man legt zunächst nur fest, bei welchem Ratio (bzw. Dash) man startet, und daß alle Gaps offen sind. Daraus ergeben sich bereits Werte für c_{min} und c_{max}. Man kann also bereits das erste Ratio mit dem ersten Ratiointervall matchen. Ist der Vergleich positiv, geht man zum nächsten Paar weiter, ist er negativ, so schließt man den ersten Gap, bringt c_{min} und c_{max} auf den neuesten Stand, und wiederholt den Test. Wird dabei $(c_{min}, c_{max}) = \emptyset$, so gibt es kein w', das die Kombination erzeugen kann. Dann muß die letzte Entscheidung einen Gap offen zu lassen revidiert werden, also ein Backtracking eingeleitet werden. Kann auch das letzte Ratiointervall erfolgreich gematcht werden und gibt es tatsächlich

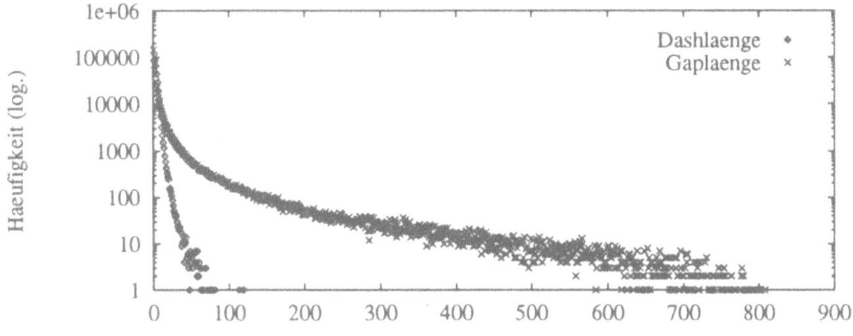

Abbildung 3: Histogramme der dash- und gap-Längen von P.

eine Transformation, die diese vermeintliche Teilüberdeckung erzeugt, so wird diese gespeichert. Es wird aber auf jeden Fall weitergesucht, damit *alle* Teilüberdeckungen ab dem Anfangsdash gefunden werden. Denn man möchte die längste Teilüberdeckung finden, weil man möglichst wenig Transformationen für die gesamte Pixelzeile aufwenden möchte.

Das zuvor beschriebene Verfahren liefert also die längste Teilüberdeckung, beginnend mit einem Anfangsdash, die durch eine nicht spiegelnde Transformation, also $k > 0$ für $w(x) = kx + t$, erzeugt wird. Sind die Richtungen, in der das „nächste" Ratio bzw. Ratiointervall liegt, gleich (bisher von links nach rechts), dann werden ungespiegelte Transformationen gefunden, sind sie gegenläufig, werden spiegelnde Transformationen gefunden.

Wir sind nun in der Lage den gesamten Algorithmus zu beschreiben:

Suche ungespiegelte und gespiegelte Transformationen: Der Anfangsdash ist der erste (ungespiegelt) bzw. der letzte (gespiegelt) Dash von links. Die längste Teilüberdeckung, beginnend mit dem Anfangsdash, wird berechnet und gespeichert; schlimmstenfalls ist diese nur einen Dash lang. Der nächste Anfangsdash ist der erste Dash, der auf den letzten überdeckten Dash folgt. Wiederhole, bis der letzte Dash von P überdeckt ist.

Nun existieren zwei Überdeckungen von P: eine Sequenz von Teilüberdeckungen ohne Spiegelung und eine mit Spiegelung. Eine Überdeckung von P mit der geringsten Anzahl von Teilüberdeckungen wird gesucht; jeder der ausgewählten Teilüberdeckungen entspricht eine Transformation des gesuchten IFS.

6 MatchRatio — Experiment

Das Verhalten des Algorithmus MatchRatio wird in diesem Abschnitt an Hand der Resultate eines Experiments genauer betrachtet. MatchRatio wurde in der Programmiersprache C implementiert. Es wurden 10000 zufällige IFS generiert, von der Art wie sie MatchRatio als Ergebnis liefert. Für jedes dieser IFS (sie hatten minimal 2, maximal 50 und durchschnittlich 15.2 Transformationen) wurde ein diskreter Attraktor in einer Pixelzeile P mit 1000 Pixeln berechnet; die Anfangs- und Endpixel von P waren dabei die minimalen und maximalen Pixel der Attraktoren. Um Informationen über die erzeugten P zu haben, wurden logarithmische Histogramme der gap- und dash-Längen erzeugt, die in Abb. 3 zu sehen sind.

MatchRatio wurde auf jede dieser Pixelzeilen P angewendet. So sind Vergleichszahlen für die Anzahl der Transformationen vorhanden, da ein erzeugendes IFS für die Ausgangspixelzeile bekannt ist. Vom Ergebnis-IFS wurde wiederum ein diskreter Attraktor berechnet. Ausgang-IFS und -pixelzeile wurden mit Ergebnis-IFS und -pixelzeile vergli-

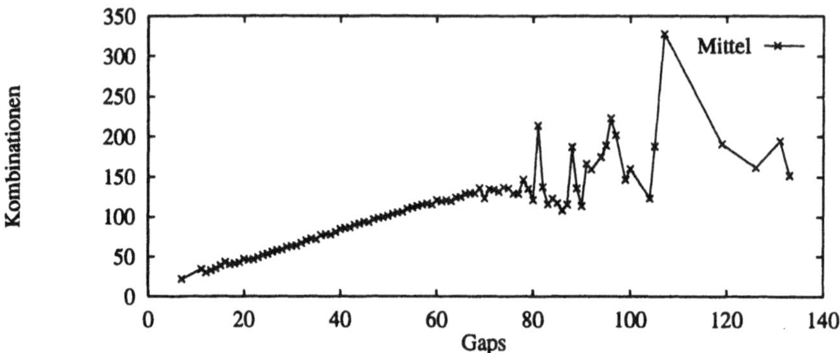

Abbildung 4: Anzahl der betrachteten Kombinationen über Anzahl von Gaps in P.

chen.

Anzahl Transformationen im Ergebnis–IFS: In der oberen Zeile der folgenden Tabelle ist die Differenz der Anzahl der Transformationen des Ausgangs–IFS und der Anzahl der Transformationen des Ergebnis–IFS eingetragen. In der unteren Zeile steht, wie oft ein Wert aufgetreten ist:

Differenz #Transf.	-4	-3	-2	-1	0	1	2	3	4	5	6	7	8	10
#Fälle	7	42	341	1778	7281	404	95	26	9	9	1	4	2	1

Es zeigt sich, daß MatchRatio meistens ein IFS findet, das gleich viele Transformationen wie das Ausgangs–IFS hat. Am zweithäufigsten wurde ein IFS mit einer Transformation weniger gefunden. Es wurden IFS mit bis zu vier Transformationen weniger als die betreffenden Ausgangs–IFS gefunden, allerdings auch solche mit bis zu zehn Transformationen mehr; das trat aber nur einmal auf.

MatchRatio findet nicht immer ein IFS mit gleich viel oder weniger Transformationen, als das erzeugende IFS, weil mitunter Teilüberdeckungen gefunden werden, die so lange sind, daß andere Teilüberdeckungen, die zu einem IFS mit weniger Transformationen führen würden, nicht gefunden werden. Eine Möglichkeit, solche Fälle zu verhindern, wäre, für *jeden* Dash von P eine möglichst lange Teilüberdeckung zu suchen und dann die Überdeckung mit möglichst wenigen Teilüberdeckungen zu ermitteln. Das bedeutet natürlich einen höheren Aufwand. Die Ergebnisse in der Tabelle haben gezeigt, daß solche Fälle nicht oft auftreten; eine Erweiterung von MatchRatio scheint also nicht sinnvoll.

Qualität der Codierung: Berechnet man einen diskreten Attraktor des von MatchRatio für P gefundenen IFS, so ist es möglich, daß dieser nicht mit P übereinstimmt. Im Experiment wurde für jedes gefundene IFS ein diskreter Attraktor berechnet und mit der Eingabe P verglichen. In 99.7% der Fälle war der diskrete Attraktor mit P ident.

Berechnungsaufwand: Würde man alle Kombinationen von P prüfen, so würde dies bei m Gaps einen Aufwand von 2^m zu prüfenden Kombinationen ergeben. In der Praxis werden jedoch viel weniger Kombinationen betrachtet. Im Experiment wurde für jede Suche nach einer Teilüberdeckung an einem bestimmten Anfangsdash gezählt, wieviele Kombinationen betrachtet wurden. In Abb. 4 sieht man das Mittel der Anzahl der betrachteten Kombinationen über der Anzahl der Gaps von P aufgetragen. Die Kurve legt den Schluß nahe, daß der Berechnungsaufwand linear von der Anzahl der Gaps abhängt, jedoch keinesfalls exponentiell. Dabei ist zu beachten, daß der mittlere Bereich der Kurve am aussagekräftigsten ist, da dort viele einzelne Werte für die Mittelwertbildung vorlagen.

Der Aufwand ist deshalb so gering, weil viele Kombinationen schon früh verworfen werden können. Wenn ein Ratio nicht mit dem Ratiointervall matcht, oder das Kontraktionsfaktorsintervall leer ist, so können alle Kombinationen verworfen werden, die aus der aktuellen Teilkombination entstehen können, weil sie durch keine diskrete Transformation erzeugt werden können.

Laufzeit: MatchRatio benötigte durchschnittlich 0.2 Sekunden Echtzeit auf einer SUN SPARC station 2 (ohne sonstige Anwenderprozesse laufen zu haben), um ein IFS für eine Pixelzeile zu finden.

Betrachtete Teilüberdeckungen pro längster Teilüberdeckung: Bei der Suche nach einer Teilüberdeckung werden als erstes Kombinationen mit möglichst wenigen geschlossenen Gaps betrachtet. Das legt nahe, daß jene Teilüberdeckung, die die meisten Dashes überdeckt, immer als erste gefunden wird. Das Experiment hat jedoch gezeigt, daß in 75% aller Fälle die Teilüberdeckung mit den meisten Dashes als erste gefunden wurde und in 25% der Fälle als zweite. Es ist also unumgänglich, alle möglichen Teilüberdeckungen zu bestimmen.

7 Schluß

Die Resultate des Experiments haben gezeigt, daß der Algorithmus MatchRatio in der Lage ist, die Parameter eines IFS aus einer diskreten Repräsentation seines Attraktors zurückzugewinnen. Er löst das eindimensionale inverse Problem und liefert zu jeder beliebigen Pixelzeile ein IFS, wobei eine untere Schranke für die Güte der Codierung angegeben wurde. Es wurde die Verwendung einer unter den Transformationen eines IFS weitgehend invarianten Eigenschaft der Pixelzeile gezeigt. Diese ist das Verhältnis der Länge von schwarzen und weißen Zusammenhangskomponenten.

MatchRatio stellt mit der Lösung des inversen Problems in $\mathbb{R}$ einen Schritt zur Lösung des inversen Problems in $\mathbb{R}^2$ dar. Das Prinzip der unter den Transformationen eines IFS invarianten Eigenschaften einer binären Pixelzeile könnte auf binäre Bilder erweitert werden. Mögliche invariante Eigenschaften in $\mathbb{R}^2$ wären Winkel zwischen Kanten, das Verhältnis der Länge von Kanten etc. Eine Fortsetzung dieser Arbeit könnte darin bestehen, einen Weg zur direkten Ermittlung von Transformationen eines IFS in $\mathbb{R}^2$ aus den obengenannten geometrischen Eigenschaften eines Bildes zu finden.

Literatur

[1] M. F. Barnsley. *Fractals Everywhere*. Academic Press, 1988.

[2] M. F. Barnsley and A. D. Sloan. A better way to compress images. *BYTE*, pages 215–223, Jan. 1988.

[3] J. C. Hart and T. A. DeFanti. Efficient antialiased rendering of 3-d linear fractals. *Computer Graphics*, 25(4):91–100, July 1991.

[4] I. J. Leitgeb. Iterierte Funktionensysteme – Das eindimensionale inverse Problem. Diplomathesis, Technical University of Vienna, Karlsplatz 13, A–1040 Vienna, Austria, Feb. 1993.

[5] J. Lévy-Véhel and A. Gagalowicz. Shape approximation by a fractal model. In *Proceedings of the EUROGRAPHICS 1987*, pages 159–179.

[6] J. Lévy-Véhel and A. Gagalowicz. Fractal approximation of 2-D object. In *Proceedings of the EUROGRAPHICS 1988*, pages 297–311.

[7] M. A. Neuhauser. Diskrete Iterierte Funktionensysteme. Diplomathesis, Technical University of Vienna, Karlsplatz 13, A–1040 Vienna, Austria, Nov. 1992.

[8] E. R. Vrscay. Moment and collage methods for the inverse problem of fractal construction with iterated function systems. In *SIGGRAPH '91 Course Notes C14*, pages 271–289.

AUDIGON – Ein medizinisches Expertensystem zur Diagnose der Kniegelenksarthrose aus kernspintomographischen Bildern

Peter Weierich Dirk Wetzel Britta Bühnemann
Heinrich Niemann

Bayerisches Forschungszentrum für Wissensbasierte Systeme (FORWISS)
Forschungsgruppe Wissensverarbeitung
Am Weichselgarten 7, D-91058 Erlangen, Tel. 0 91 31/69 1-1 94
email (*name*)@forwiss.uni-erlangen.de

Klaus Glückert
Orthopädische Universitätsklinik Erlangen–Nürnberg, Rathsbergerstr. 57, D-91054 Erlangen

Zusammenfassung

Wir stellen ein System zur Interpretation von kernspintomographischen Bildern des Kniegelenks zur objektivierten Bestimmung der Schädigungsgrade des Gelenkknorpels vor. Das deklarative medizinische Wissen wird dabei in einem semantischen Netz repräsentiert. Zur Bewertung diagnostischer Aussagen wird die Fuzzy–Mengentheorie verwendet, die um Konsistenzmaße erweitert wurde, die die Qualität der Schlußfolgerungen bewerten. Schließlich wird eine Wartungskomponente von AUDIGON vorgestellt, mit der die Fuzzy–Regelbasis an neue Anforderungen angepaßt werden kann.

1 Ziele und Aufgaben

Die Gelenksarthrose ist ein Leiden, das in höherem Alter jedermann betrifft. Durch das Auftreten dieser Volkskrankheit bei jüngeren Menschen entstehen der Volkswirtschaft jährlich Schäden in Milliardenhöhe durch Behandlungskosten, Arbeitszeitausfälle, frühzeitigen Ruhestand etc. [1] Trotzdem sind gerade die Frühstadien und Entstehungsmechanismen der Arthrose, des Verschleißes des Gelenkknorpels, bislang kaum erforscht. Mit klassischen nichtinvasiven Untersuchungsmethoden lassen sich insbesondere Frühstadien der Gonarthrose nicht unterscheiden.

Heute bietet die Kernspintomographie (MR), speziell mit schnellen 3D–Bildsequenzen (FISP, FLASH, GRASS), die Möglichkeit, den Zustand des Gelenkknorpels detailliert darzustellen, ohne einen operativen Eingriff am Patienten vornehmen zu müssen [5][9][2].

Bei der Auswertung des umfangreichen Bildmaterials, das an der Orthopädischen Universtätsklinik Erlangen in den vergangenen Jahren entstanden ist, wurden sehr schnell die Mängel deutlich, die sowohl die rein visuelle Befundung, als auch die Standard–Auswertungssoftware aufweist. Vor diesem Kontext wurden für das zu entwickelnde Expertensystem AUDIGON (AUtomatische DIagnose der GONarthrose) zwei Aufgaben definiert:

[1] In den Vereinigten Staaten von Amerika sind laut [6][4] 32.6 % der Erwachsenen von Problemen mit dem Muskoskeletalsystem betroffen. 7.14 Millionen US–Amerikaner waren schätzungsweise wegen Gelenkleiden in Behandlung, und 5 Millionen litten an der Gonarthrose.

- Extraktion der (MR–)Intensitätswerte und isolierte Darstellung des Gelenkknorpels im Kniegelenk zur weiteren Auswertung, beispielsweise im Rahmen der Grundlagenforschung auf dem Gebiet der Arthroseforschung.
- Generierung eines Befundberichts, aus dem hervorgeht, wie stark der Gelenkknorpel geschädigt ist. Der Bericht soll den klinischen Erfordernissen genügen und eine objektivierte Beobachtung des Krankheitsverlaufs erlauben.

Somit soll AUDIGON im Rahmen der medizinischen Qualitätssicherung und der klinischen Forschung einen wertvollen Dienst leisten.

2 Überblick

Der vorliegende Prototyp von AUDIGON wurde als semantisches Netz mit der Entwicklungsumgebung ERNEST (Lehrstuhl für Mustererkennung, Universität Erlangen-Nürnberg) erstellt. Das gesamte medizinische, diagnostische und technische Wissen ist auf neun Abstraktionsebenen (Abb. 1) mit insgesamt 240 Knoten und 428 Kanten repräsentiert (siehe Abb. 2).

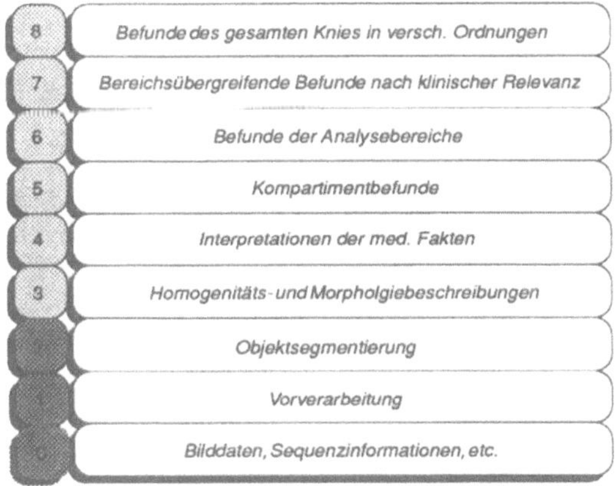

Abbildung 1: **Abstraktionsebenen in AUDIGON**

Die unteren drei Abstraktionsebenen beinhalten die Vorverarbeitung und Bildsegmentierung, die übrigen den diagnostischen Teil von AUDIGON. Die Knoten repräsentieren jeweils Bilddaten, anatomische Regionen und diagnostische Aussagen. Die Kanten verbinden als Konkretisierungskanten die Abstraktionsebenen, innerhalb der Abstraktionsebenen einzelne Konzepte als Bestandteils- und Spezialisierungskanten. Ein Graphensuchalgorithmus wird verwendet, um eine zum medizinischen Wissen konsistente Interpretation zu erzeugen, die optimal zu den Bilddaten paßt.

Bisher kann lediglich der Gelenkknorpel der Oberschenkelkondylen segmentiert und befundet werden.

3 Vorverarbeitung und Bildsegmentierung

In der Vorverarbeitung wird zunächst mit Hilfe von Schwellwertverfahren die Hintergrundluft segmentiert und ausgeblendet. Anschließend wird das sogenannte Wasserphantom — ein mit physiologischer Kochsalzlösung gefülltes Röhrchen, das in die Kniekehle

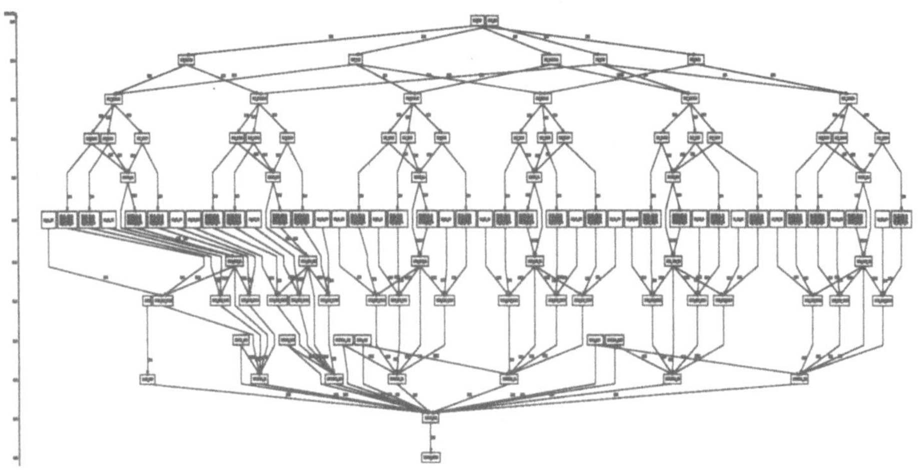

Abbildung 2: Das semantische Netz von AUDIGON

des untersuchten Gelenks gelegt wird — detektiert und segmentiert. Die darin gefunden Intensitätswerte werden zur Ermittlung des Schwellwerts zur Erkennung von Flüssigkeitsansammlungen im Kniegelenk verwendet.

Die Schritte der Segmentierung nach erfolgter Vorverarbeitung und heuristischer Bestimmung des Mittelpunkts des lateralen Kondylus sind in Abb. 3 dargestellt. Grundidee ist dabei, daß die Kondylen als (halb-)kreisförmige Körper betrachtet werden. Die linken und rechten Ränder der Kondylen werden mit Hilfe eines gerichteten Gradientenfilters ermittelt. In die Mitte zwischen den Randpunkten wird das Transformationszentrum für die anschließende Polartransformation gelegt, die die kreisförmige Kontur in eine annähernd gerade Linie transformiert. Der entscheidende Vorteil daran ist, daß man zur Konturdetektion jetzt einen gerichteten Gradientenfilter verwenden kann.

Der in der Regel durch einen hohen Kontrast gekennzeichnete Knochen–Knorpelübergang wird mit Hilfe Dynamischer Programmierung im polartransformierten Bild bestimmt. Im nächsten Schritt wird der Oberschenkelknochen ausgeblendet und analog zur Segementierung der Knochenkontur der Übergang vom Knorpel zum Gelenkbinnenraum segmentiert. Exemplarisch wird die Segmentierung in 3 dargestellt.

4 Diagnoseteil — Fuzzy Mengentheorie

4.1 Motivation

Die Anwendung der Fuzzy Mengentheorie in AUDIGON ist motiviert durch typische, durch Unschärfe und Unsicherheit geprägte Aussagen der medizinischen Experten, zum Beispiel: *Wenn das MR-Signal im Knorpel leicht inhomgen ist, die Knorpeloberfläche leicht fibrilliert ist, der Knorpel makroskopisch jedoch in Ordnung ist, dann ist er wahrscheinlich Grad-1-vorgeschädigt.*

Diese Regeln können mit Hilfe der Fuzzy Mengentheorie direkt umgesetzt werden. In Abbildung 4 sind typische Zugehörigkeitsfunktionen und Regeln dargestellt.

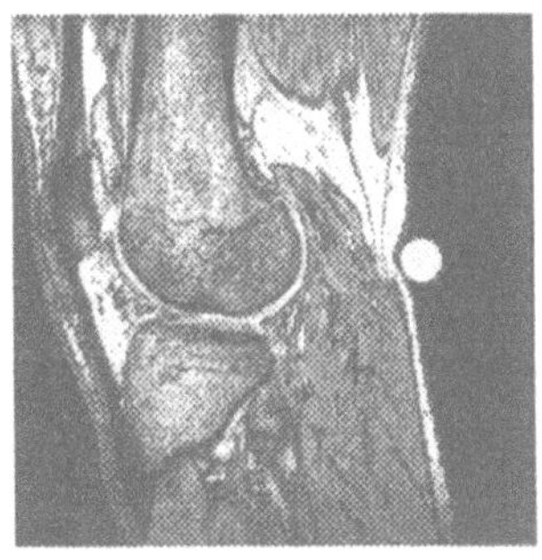
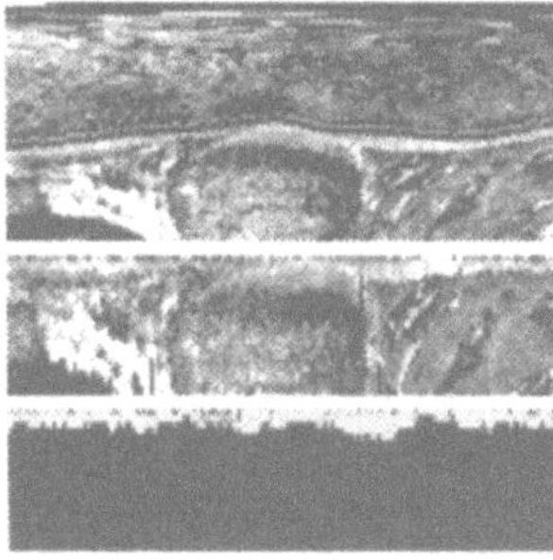
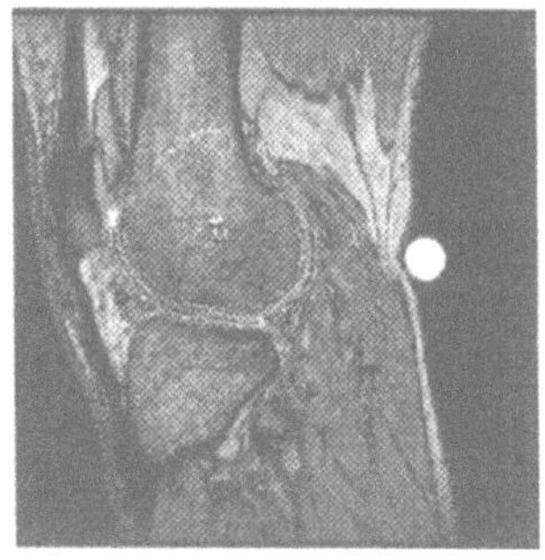

Abbildung 3: Die Verarbeitungsschritte zur Segmentierung der lateralen Kondylen des Oberschenkelknochens: Originalbild (links), polartransformiertes Bild (mitte), im oberen Teil ist der Oberschenkelknochen repräsentiert, die weiße Linie markiert die Knochenkontur in einem Pixel Abstand (mitte, oben), Polarbild mit abgeschnittenem Oberschenkelknochen (mitte, mitte), Pixel des Knorpels nach Segmentierung (mitte, unten) und das Originalbild mit retransformierter Knochen- und Knochenkontur (rechts).

4.2 Modell

Zur Generierung der Diagnose wird zunächst der segmentierte Knorpel in kleinere Kompartimente zerlegt. Für jedes Kompartiment werden die Regeln für alle sechs möglichen Schädigungsgrade ausgewertet und die Ergebnisse in dreidimensionale Schädigungsgradkarten eingetragen (Dimensionen: Kompartimentnummer, Schichtbildnummer, Schädigungsgrade). Die Schädigungsgradkarten werden im folgenden derart sortiert, daß in der vordersten Karte für jede Lokalisation der plausibelste (bestbewertete) Schädigungsgrad eingetragen wird (Abb. 5). Insgesamt werden 15 Merkmale, 90 Zugehörigkeitsfunktionen und 12 unterschiedliche Regeln ausgewertet.

Die weitere Verarbeitung beinhaltet die hierarchische Zusammenfassung der lokalen Befunde zu Diagnosen für die jeweils vom Arzt angegebene interessierende Region (z.B. Femur Lateralis).

AUDIGON generiert automatisch einen Befundbericht (Abb. 7), der neben den relevanten, aus dem Bilddatenheader gewonnenen Patientendaten, der klinischen Fragestellung und der Diagnose mit Diagnosesicherheit auch ein MR-Bild mit der maximal gefundenen Schädigung enthält.

4.3 Konsistenzmaße

Zur Bewertung der Glaubwürdigkeit der Ergebnisse und zur Bewertung der Vollständigkeit der Wissensrepräsentation wurden die folgenden drei Konsistenzmaße entwickelt:

$$CM_1 = \frac{1}{I*J}\sum_{j=1}^{J}\sum_{i=1}^{I}(CF(GS^1_{i,j}) - CF(GS^2_{i,j})) \tag{1}$$

$$CM_2 = \frac{1}{I*J}\sum_{j=1}^{J}\sum_{i=1}^{I}CF(GS^1_{i,j}) \tag{2}$$

$$CM_3 = \frac{1}{I*J}\sum_{j=1}^{J}\sum_{i=1}^{I}f(i,j), f(i,j) = \begin{cases} 0, & if \quad GS^1_{i,j} \leq 0.5 \\ 1, & else \end{cases} \tag{3}$$

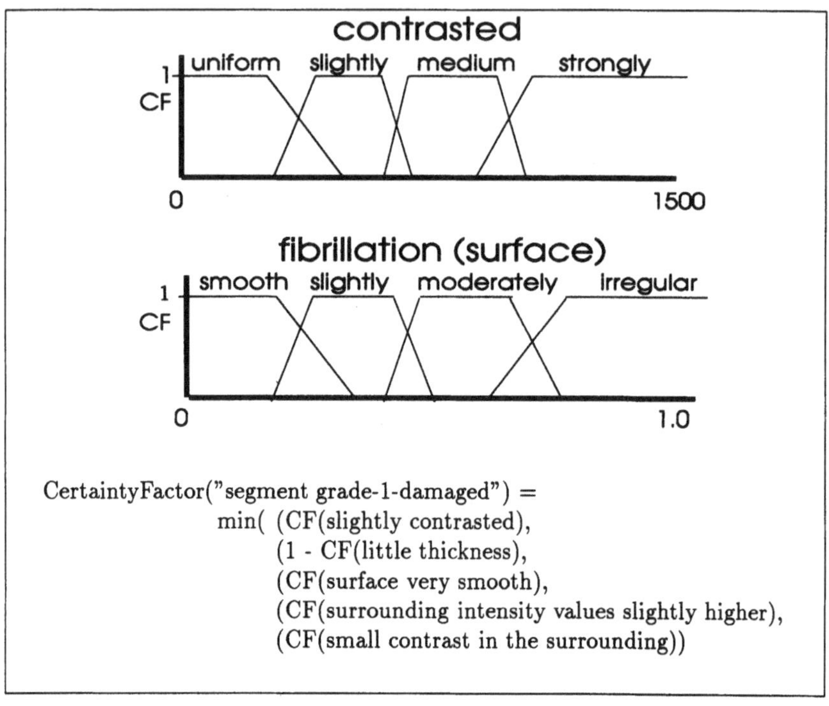

Abbildung 4: **Beispiel zweier Zugehörigkeitsfunktionen und einer Fuzzy-Regel**

$GS^k_{i,j}$ bezeichnet den Schädigungsgrad in der $k$ten Schicht der 3D–Fuzzykarte im Segment (i, j), i: Bildnummer, j: relative Position (Regionen–Achse in Abb. 5).

Gleichung 1 berechnet die Trennschärfe als mittlere Differenz zwischen der best- und zweitbestbewerteten Interpretation. In Gleichung 2 wird die Entschiedenheit berechnet: Der Mittelwert der besten Bewertungen. CM_3 schließlich wird als Überdeckung bezeichnet: Wenn CM_3 einen niedrigen Wert annimmt, dann ist das Wissen nur partiell modelliert.

4.4 Wartung der Wissensbasis

Eine graphische Benutzeroberfläche stellt die gesamte AUDIGON–Funktionalität zur Verfügung. Um jedoch die Fuzzy–Regelbasis leicht an neue Anforderungen und Erkenntnisse anpassen zu können, wurde eine Benutzeroberfläche um eine Wartungskomponente erweitert, die eine interaktive Modifikation der Fuzzy–Zugehörigkeitsfunktionen und -Regeln erlaubt. Dies ist entweder durch direkte Manipulation beispielsweise der Stützpunkte der Zugörigkeitsfunktionen oder durch textuelle Eingabe von Parametern möglich. Ein Beispiel ist in Abbildung 6 dargestellt.

5 Erste Ergebnisse und Ausblick

Erste Versuche zeigten gute Segmentierungsergebnisse wenn die Randpunkte der Knochenkontur korrekt bestimmt werden konnten. Die diagnostischen Aussagen waren für den medizinischen Experten plausibel. Eine systematische Evaluierung auf der Grundlage klinisch gesicherter Diagnosen steht noch aus.

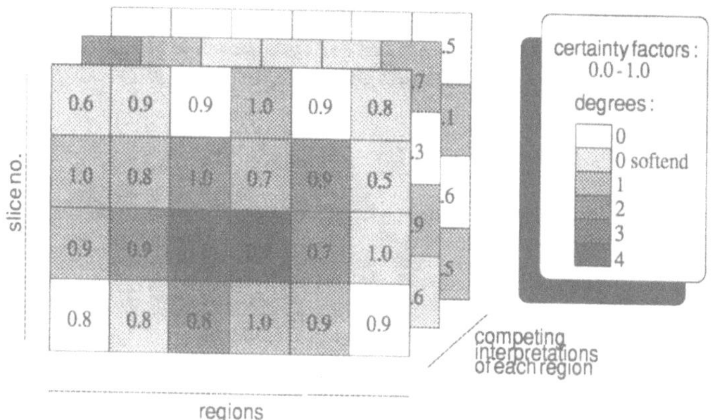

Abbildung 5: Beispiel einer 3D-Schädigungsgradkarte: Jedes Quadrat repräsentiert eine Region im Gelenkknorpel. Die Schattierungen markieren unterschiedliche Schädigungsgrade (0 – 4) in miteinander konkurrierenden Interpretationen. Auf der vordersten Karte sind jeweils die am besten bewerteten Schädigungsgrade und deren Sicherheiten eingetragen. Die teilweise verdeckten Schichten enthalten die am zweit– und drittbesten bewerteten Schädigungsgrade.

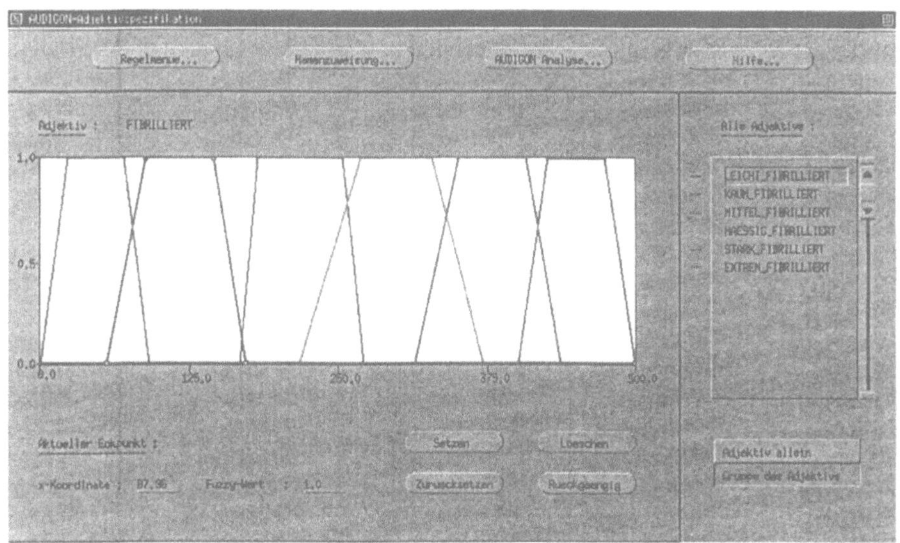

Abbildung 6: Beispiel für ein Fenster zur Manipulation von Fuzzy–Zugehörigkeitsfunktionen

Die Weiterentwicklung wird eine Segmentierung aller Gelenkknorpelanteile im Kniegelenk umfassen. Dabei wird Vorwissen über die anatomischen Strukturen im Kniegelenk ebenfalls über eine Fuzzy-Repräsentation mit eingehen.

Literatur

[1] Bühnemann, K. AUDIGON — Erstellung der Interaktionskomponente Studienarbeit, FORWISS, Universität Erlangen-Nürnberg, 1993.

[2] Glückert, K., Blank-Schäl, A., Hofmann, G., Kladny, B., Wirtz, P. "The potential of MRI to evaluate early degenerative cartilage disease". International conference on occupational muscoskeletal disorders and prevention of low back pain, 16.-18.10.1990, Milano

[3] Glückert, K., Weierich, P., Wetzel, D., Niemann, H., Kladny, B. Audigon - a knowledge based system for automated diagnosis of osteoarthrosis of the knee. Osteoarthritis and Cartilage, Vol. 1, No. 1, Bailliere Tindall, London, 1993, p. 59

[4] Maurer, K. Basic Data on Arthritis Knee, Hip, and Sacroiliac Joints in Adult Ages 25-74 Years, United States 1971 – 75 Vital Health and Statistics Series 11, No. 213 USDHEW, Hyattsville, Md., August 79

[5] Hayes, C.W., Sawyer, R.W., Conway, W.F. "Patellar Cartilage Lesions: In Vitro Detection and Staging with MR Imaging and Pathologic Correlation". *Radiology*, vol. 176, pp. 479-483, 1990.

[6] Lawrence, Reva C., Hochberg, Marc C., Kelsey, Jennifer L., McDuffie, Frederic C., Medsger, Thomas A., Felts, William R. jr., Shulman, Lawrence E. Estimates of the Prevalence of Selcted Artritic and Musculosceletal Diseases in the United States "*The Journal of Rheumatology* 16:4, pp. 427 – 441, 1989.

[7] Niemann, H., Sagerer, G., Schröder, S., Kummert, F. "ERNEST: A Semantic Network System for Pattern Analysis". *IEEE Trans. Pattern Analysis and Machine Intelligence*, vol. 12, pp. 883 - 905, 1990.

[8] Niemann, H., Wetzel, D., Weierich, P., Sagerer, G., Glückert, K. Methods of Artificial Intelligence in Medical Imaging, in: proceedings of I.CO.GRAPHICS 1992, Milan, Italy Mondadori Informatica SPS - AICOGRAPHIC, Italy, 1992, pp. 253-260.

[9] Reiser, M., et al. "Magnetic resonance in cartilaginous lesions of the knee joint with three-dimensional gradient-echo imaging". *Skeletal Radiology*, vol. 17, pp. 465 - 471, 1988.

[10] Sagerer, G. "Automatic Interpretation of Medical Image Sequences". *Pattern Recognition Letters*, vol. 8, pp. 87 - 102, 1988.

[11] Weierich P., Wetzel, D. "AUDIGON - Ein wissensbasiertes System zur Diagnose der Gonarthrose aus kernspintomographischen Schnittbildern". Diplomarbeit, Lehrstuhl für Mustererkennung, Universität Erlangen-Nürnberg. 1991.

[12] Zadeh, L.A. "Fuzzy Sets". *Information and Control*, vol.8, pp. 338 - 353, 1965.

[13] Zimmermann, H.J. "*Fuzzy Set Theory - and Its Applications*". 2nd ed., Kluwer Academic Publishers, Norwell, 1991.

Universität Erlangen-Nürnberg – Orthopädische Klinik

Leiter: Prof. Dr. Hohmann
Rathsberger Straße 57, D-W-8520 Erlangen, Tel. 0 91 31/82 2-0

Befundbericht Kernspintomographie

Patientenangaben

Name, Vorname:	G.H. M 28 Y ,
Geburtsdatum:	1961.01.27

Klinische Fragestellung

Rechtes Kniegelenk
Arthrosegrade des lateralen Femur mit Angabe der Diagnosesicherheit

Befund

Der Femur lateralis ist medial mit Arthrosegrad 2 (Sicherheit: 1.00) maximal geschaedigt.
Schaedigunguebersicht:

Sektor	Grad	Sicherheit	Schicht(en)
ventral	1	1.00	19,21 - 23,
medial	2	1.00	18 - 20,22,
dorsal	1	0.98	20,22,

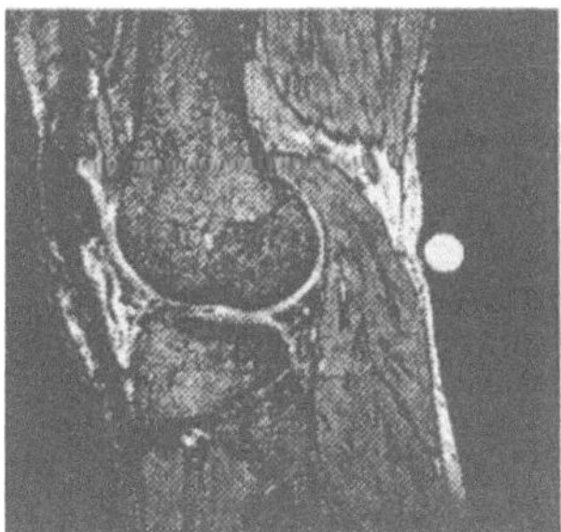

Empfohlene Maßnahmen

Sportliche Betaetigung zum Muskelaufbau, z.B. Radfahren

Untersucher: Prof. Dr. med. XYZ
Datum: 7. April 1993

Abbildung 7: Beispiel für einen automatisch generierten Befundbogen

Segmentierung und Analyse drei- und vierdimensionaler Ultraschalldatensätze

Sabine Zehetbauer[1] und Uwe Meyer-Gruhl[2]

[1]TomTec Tomographic Technologies GmbH, Breslauerstr. 1-3, 8057 Eching;
[2]Technische Universität München, Institut für Informatik, Lehrstuhl Prof. Radig

In diesem Beitrag wird ein hierarchischer Segmentierungsalgorithmus für drei- und vierdimensionale Ultraschallbilder vorgestellt, der kein a-priori-Wissen einbezieht. Eine Vorsegmentierung erfolgt durch die Berechnung der Wasserscheidentransformation im 3D- bzw. 4D-Raum. Das anschließende auf Graphenebene durchgeführte Volumenwachstumsverfahren vergleicht drei verschiedene Graphen (Regionengraph, gewichteter Regionengraph, Konturengraph) im Hinblick auf ihr Wachstumsverhalten. Durch Iteration der entwickelten Algorithmen entsteht eine Hierarchie von Segmentierungsebenen, die sich in Regionenanzahl und Regionengröße unterscheiden. Bei einer Anwendung auf Ulraschallbilder des Herzens und der Karotis erzielte der Konturengraph das beste Ergebnis hinsichtlich der Segmentierungsgenauigkeit.

1 Einleitung

Die transösophageale Echokardiographie ermöglicht die drei- bzw. vierdimensionale Darstellung des Herzens [2][4]. Ein Ultraschalltransducer, der sich in einem modifizierten Speiseröhrenspiegel befindet und jeweils um ein festes Inkrement weitergeschoben wird, zeichnet parallele zweidimensionale Schnittebenen in äquidistanten Abständen auf. Eine zusätzliche EKG-Triggerung der Aufnahme führt zu vierdimensionalen Datensätzen, die eine Auswertung der Bewegungen des Herzens erlauben.

Die Segmentierung der vierdimensionalen Ultraschallbilder in Volumenelemente, die den anatomischen Strukturen entsprechen, ist Voraussetzung für quantitative Bildanalysen, wie z.B. die Volumenberechnung von Organen. Dem Kardiologen kann das Herzschlagvolumen Informationen über Leistungsfähigkeit und Belastbarkeit des Herzens geben.

Eine Segmentierung direkt im 3D-Raum ermöglicht zum einen den vollen Gebrauch der lokalen und globalen Zusammenhänge und erzeugt zum anderen dreidimensionale Regionen, d.h. es ist kein Zusammenfügen von zweidimensionalen Segmenten erforderlich. Die in dieser Arbeit vorgestellten Segmentierungsverfahren verwenden als einziges Merkmal zur Segmentierung die Grauwerte und beziehen kein bildspezifisches Wissen in den Segmentierungsprozeß ein, was den Einsatz der Verfahren auf einem breiten Anwendungsspektrum gewährleistet.

2 Wasserscheidentransformation

Eine anfängliche Einteilung der Bilddaten in homogene Bildbereiche erfolgt mit Hilfe des morphologischen Operators der Wasserscheidentransformation, der sich auf n-dimensionale Räume erweitern läßt [7] und auf der Annahme basiert, daß jede homogene Grauwertregion in einem Bild durch einen geringen Kontrast und somit einen geringen Gradienten gekennzeichnet ist. Die höchsten Werte im Gradientenbild entsprechen den kontrastreichsten Konturen im Originalbild. Die weiteren Segmentierungsschritte werden somit auf dem Gradientenbild ausgeführt.

Begriffe

In der mathematischen Morphologie werden zweidimensionale Grauwertbilder oft als *topographische Reliefs* betrachtet [5]. Bei der topographischen Darstellung eines Bildes steht der numerische Wert (Grauwert) jedes Pixels für die Erhebung an diesem Punkt. Diese Betrachtungsweise eignet sich für eine Definition folgender Begriffe: Wenn ein Wassertropfen auf einen Ort auf der topographischen Oberfläche fällt, wird er einer Linie des steilsten Abstiegs folgen und ein *regionales Minimum* erreichen. Das mit einem regionalen Minimum verbundene *Staubecken* ist die Menge aller Punkte, von denen aus ein Wassertropfen dieses regionale Minimum erreicht (s. Abb. 1). Diese Menge wird auch als *Einflußzone* eines Minimums bezeichnet. Zwei benachbarte Staubecken haben eine gemeinsame Trennlinie *(Wasserscheide)*: ein Wassertropfen, der auf eine Wasserscheide fällt, kann in jedes der beiden Staubecken fließen (s. Abb. 1). Die *optimale Kontur* entspricht dieser Wasserscheide.

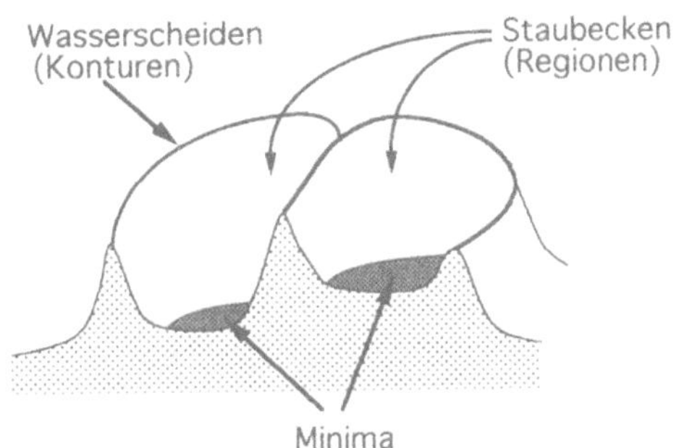

Abbildung 1: Minima, Staubecken und Wasserscheiden

Algorithmus

Der Algorithmus zur Berechnung der Wasserscheiden auf einem zweidimensionalen Bild läßt sich durch einen Prozeß des "*Untertauchens*" veranschaulichen [7]: Wir bohren in jedes regionale Minimum des Reliefs ein Loch und tauchen das Gebirge mit gleichmäßiger vertikaler Geschwindigkeit in einen See. Ausgehend vom Minimum mit der geringsten Höhe wird das Wasser, das durch die Löcher tritt, fortlaufend die verschiedenen Staubecken auffüllen. Um das Zusammenfließen des Wassers zu vermeiden, bauen wir entlang der Linien, an denen das Wasser von zwei verschiedenen Minima zusammenlaufen würde, einen "*Damm*" (s. Abb. 2). Nach dem vollständigen Untertauchen ist jedes Minimum komplett von Dämmen umgeben, die die zu den Minima gehörenden Staubecken begrenzen.

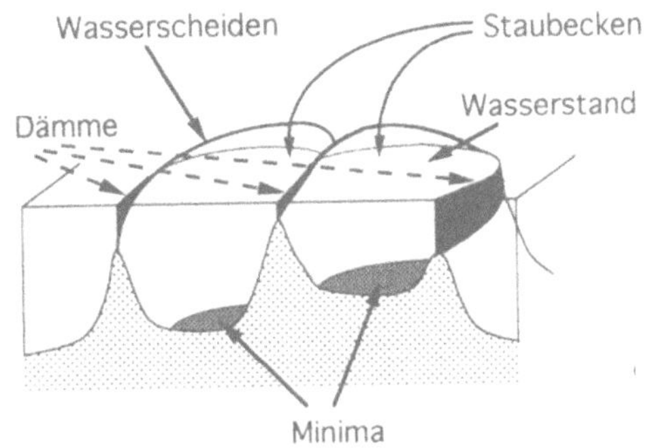

Abbildung 2: Bau von Dämmen

Die komplette Menge von Dämmen, die erbaut werden, ergibt eine Unterteilung des Bildes in seine verschiedenen Staubecken. Die Dämme entsprechen den Wasserscheiden.

Der von VINCENT & SOILLE [7] vorgeschlagene Algorithmus wurde auf den 3D- bzw. 4D-Raum erweitert [9] und liefert per constructionem volumenhafte Regionen mit geschlossenen Konturen, d.h. die Segmentierung erfolgt nicht schichtweise und erfordert somit kein "Matching" einzelner segmentierter Schnittebenen zu einem Volumen. Die als Queue-Algorithmus implementierte Wasserscheidentransformation kann in zwei Schritte zerlegt werden: einen anfänglichen *Sortierschritt* und einen *Schritt des Flutens*.

Zunächst werden die Pixel nach aufsteigenden Grauwerten sortiert, um einen direkten Zugriff auf die Pixel mit einem bestimmten Grauwert zu haben.

Diese Möglichkeit wird im zweiten Schritt, dem Fluten, genutzt. Der Prozeß des Untertauchens kann als *rekursiver Prozeß* dargestellt werden, wobei die Ausgangsmenge der Rekursion $X_{h_{min}}$ aus den Punkten besteht, die zuerst vom Wasser erreicht werden.

Im Verlauf des Flutens werden die aufeinanderfolgenden Grauwertschwellen des Bildes betrachtet und die geodesischen Einflußzonen einer Schwelle innerhalb der nächsten berechnet. Die *geodesische Einflußzone* $iz_A(B_i)$ einer verbundenen Komponente B_i von B in A[1] ist definiert als der geometrische Ort aller Punkte in A, deren geodesischer Abstand[2] zu B_i geringer ist als ihr geodesischer Abstand zu jeder anderen Komponente von B (s. Abb. 3).

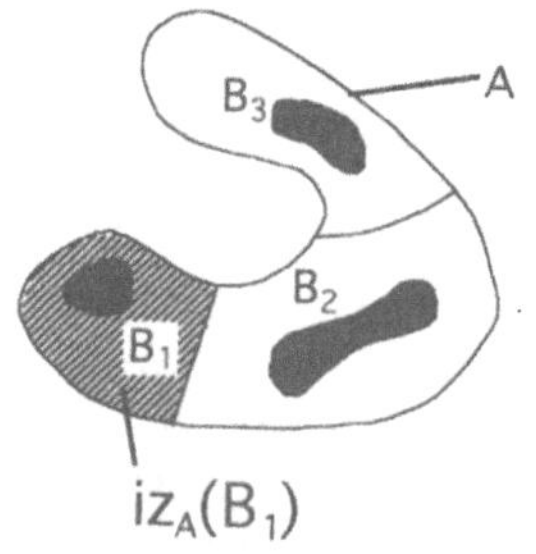

Abb. 3: Geodesische Einflußzone

In jedem Rekursionsschritt werden die bereits bis Höhe h entdeckten Staubecken in Höhe h+1 ausgebreitet, indem man die geodesischen Einflußzonen von h innerhalb von h+1 berechnet (s. Abb. 4). Zudem wird geprüft, ob neue Minima mit dem Wert h+1 auftreten. Diese sind mit keinem der bereits existierenden Staubecken verbunden (s. Abb. 4).

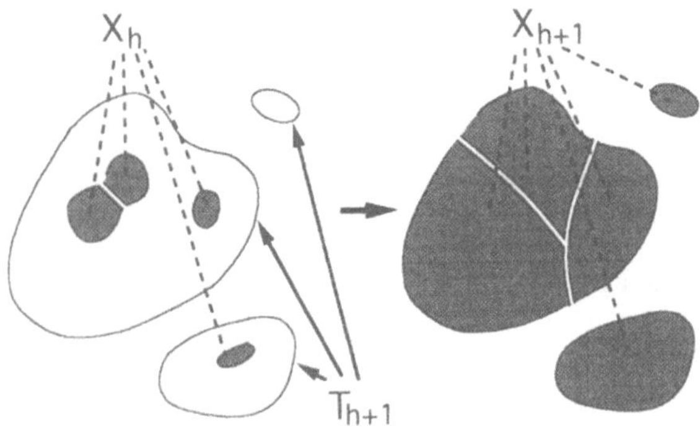

Abbildung 4: Rekursionsbeziehung zwischen X_h und X_{h+1}

[1] A sei eine zusammenhängende Menge, die eine Menge B enthält, die wiederum aus mehreren verbundenen Komponenten $B_1, B_2, \ldots, B_k$ besteht.

[2] Der *geodesische Abstand* $d_A(x,y)$ zwischen zwei Bildpunkten x und y in A ist definiert als das Infimum der Länge der Pfade, die x und y verbinden und die vollständig in A liegen.

Die Menge der Staubecken eines Grauwertbildes I entspricht der Menge $X_{h_{max}}$, die nach folgender Rekursion erreicht wird:

$$X_{h+1} = \min_{h+1} \cup IZ_{T_{h+1}(I)}(X_h), \quad \forall h \in [h_{min}, h_{max} - 1].$$

Bei der Berechnung der Wasserscheidentransformation im drei- bzw. vierdimensionalen Raum werden die regionalen Gradientenminima in alle Richtungen des Raumes ausgebreitet. Dies erfolgt mittels der Berechnung der geodesischen Einflußzonen, die auf der Bestimmung der geodesischen Abstände in drei bzw. vier Dimensionen basiert. Die Abstände werden anhand der auf der mehrdimensionalen Gitterstruktur definierten Nachbarschaft berechnet. Der Algorithmus liefert dementsprechend drei- bzw. vierdimensionale Regionen mit *geschlossenen*, gut lokalisierten Konturen, führt jedoch zu einer starken *Übersegmentierung*, d.h. die richtigen Konturen gehen in einer Masse von irrelevanten Konturen verloren.

3 Hierarchische Segmentierung auf Graphenebene

Zur Beseitigung der aus der Wasserscheidenberechnung resultierenden Übersegmentierung dient ein hierarchischer Ansatz [9], der als Volumenwachstumsprozeß interpretiert werden kann. Das Volumenwachstum wird auf einer höheren Ebene - auf Graphenstrukturen - durchgeführt und erfordert die folgenden Verfahrensschritte: Aus dem mit Hilfe der Wasserscheidentransformation vorsegmentierten Bild wird ein Mosaikbild erzeugt, wobei jedem Volumenelement ein Grauwert zugeordnet wird. Dieses Mosaikbild wird in einen Graphen überführt, auf dessen Datenstruktur das Volumenwachstum, d.h. der Zusammenschluß von Regionen, stattfindet. In jedem Graphen werden die Objekte durch Knoten und die Beziehungen zwischen Objekten durch Kanten dargestellt [6].

3.1 Mosaikbild-Erzeugung

Für die Überführung des vorsegmentierten Bildes in einen Graphen muß in einem ersten Schritt ein vereinfachtes Bild *(Mosaikbild)* erzeugt werden. Dieses Mosaikbild entsteht, indem jeder Region ein einziger Grauwert zugewiesen wird. Die gewählten Grauwerte sollten in etwa den Grauwerten im Originalbild entsprechen.

Gemäß MEYER & BEUCHER [3] ist der Grauwert, der jedem Staubecken im Mosaikbild zugeteilt wird, gleich demjenigen Grauwert im ursprünglichen Bild, der an der gleichen Position steht wie das jeweilige Minimum des Staubeckens im Gradientenbild (s. Abb. 5). Jede Mosaikfliese trägt den Grauwert des Ausgangsbildes, der im jeweiligen "homogensten" Bereich der entsprechenden Region liegt. Aus diesem Mosaikbild wurden drei verschiedene Graphen (s. Abb. 6) mit unterschiedlichen Nachbarschaftsbeziehungen generiert und verschiedenen Wachstumsprozessen unterworfen.

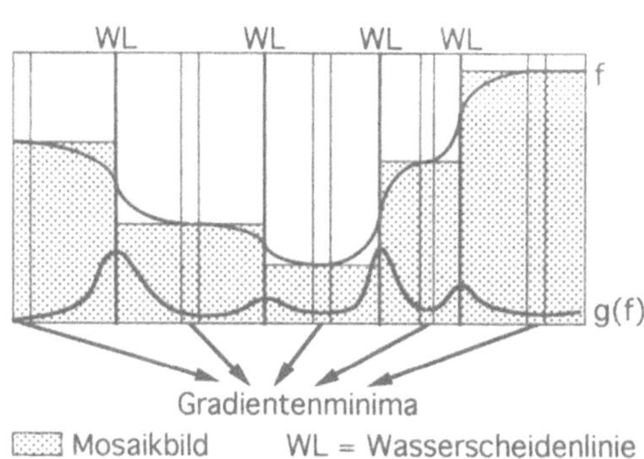

Abbildung 5: Prinzip der Berechnung des Mosaikbildes

3.2 Regionengraph

Die Idee beim Regionengraph liegt in einem Volumenwachstum, das durch die Anwendung der Wasserscheidentransformation auf Graphenebene entsteht. Dazu wird der von VINCENT & SOILLE [7] beschriebene 2D-Regionenwachstumsalgorithmus erweitert zu einem 3D-Volumenwachstum. Ein Regionengraph wird gebildet, dessen Knoten die Regionen repräsentieren und dessen Knotenwerte den Mosaikgrauwerten entsprechen. Zwischen zwei benachbarten Regionen existiert genau eine Kante (s. Abb. 6a).

Der Volumenwachstumsalgorithmus auf dem Regionengraph gliedert sich in folgende Schritte: Zunächst wird der Gradient des Graphen ermittelt und auf diesem Gradienten anschließend die Wasserscheidentransformation auf der Datenstruktur des Graphen berechnet. Die Wasserscheiden-Knoten erlauben die Ableitung eines Mosaikbildes zweiter Ordnung, das eine Zerlegung des Originalbildes auf einer höheren Ebene darstellt. Regionen, deren Grauwerte nahe beieinander liegen, werden zu größeren Regionen verbunden.

Durch eine iterative Anwendung des beschriebenen Volumenwachstumsverfahrens kann eine Hierarchie von Segmentierungsebenen erzeugt werden, die sich in Regionenanzahl und Regionengröße unterscheiden. Als Abbruchbedingung der Iterationen dient die Erreichung einer problemangepaßten Regionenzahl bzw. einer optimalen Regionengröße.

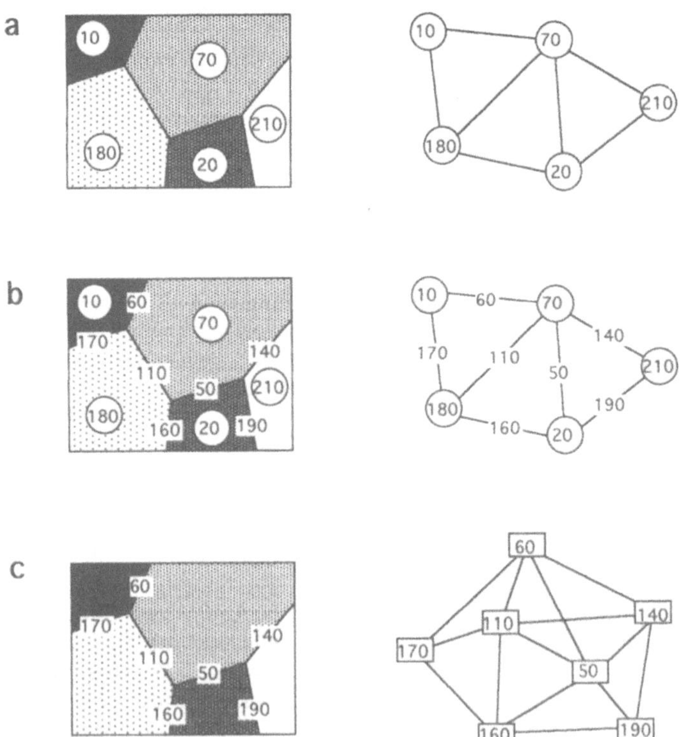

Abbildung 6: Mosaikbild und daraus abgeleitete Graphen:
(a) Regionengraph, (b) gewichteter Regionengraph, (c) Konturengraph

3.3 Gewichteter Regionengraph

Durch eine zusätzliche Bewertung der Kanten, die den Grad der existierenden Nachbarschaftsbeziehungen zwischen den Knoten repräsentiert, entsteht der gewichtete Regionengraph. Die absolute Differenz aus zwei Knotengrauwerten wird der diese beiden Knoten verbindenden Kante zugewiesen (s. Abb. 6b). Ein sequentielles Volumenwachstum, das ausgehend von den Gradientenminima (repräsentiert durch die Kantenwerte) die Regionen in Richtung aufsteigender Grauwertübergänge erweitert, dient zur Reduzierung der Regionenanzahl.

3.4 Konturengraph

Im Konturengraph hingegen werden anstelle der Regionen die Regionengrenzen durch Knoten dargestellt und alle Regionengrenzen (Knoten), die dasselbe Staubecken umschließen, durch Kanten verbunden (s. Abb. 6c). Der Wert jedes Knotens entspricht dem Gradientenwert (Grauwertübergang) der entsprechenden Regionengrenze. Auf diesem Konturengraph wird eine Wasserscheidentransformation durchgeführt, die bestimmte Konturen im Mosaikbild als Wasserscheiden identifiziert. Im Mosaikbild zweiter Ordnung bleiben nur diejenigen Regionengrenzen erhalten, deren Knoten als Wasserscheide erkannt wurden, während alle Grenzen innerhalb der Staubecken wegfallen.

4 Ergebnisse und Diskussion

Die verschiedenen Ansätze der hierarchischen Segmentierung wurden an Hand von drei- und vierdimensionalen Ultraschallbildern des menschlichen Herzens und der Karotis (Halsschlagader) getestet (s. Abb. 7).

Aufgrund des globalen Charakters der Wasserscheidentransformation wird beim Regionengraph der Zusammenschluß von Regionen nicht in Abhängigkeit von lokalen Kriterien vollzogen, sondern die gesamte Bildinformation wird dazu genutzt. Ein Problem bei diesem Verfahren bildet allerdings die Gradientenbestimmung auf der Datenstruktur des Graphen, die durch Einbeziehung der benachbarten Knoten angenähert werden muß. Eine iterative Anwendung des Volumenwachstumsalgorithmus auf dem Regionengraphen bewirkt ein schnelles Regionenwachstum, gleichzeitig aber auch den Verlust anatomisch relevanter Objektgrenzen.

Der Vorteil des gewichteten Regionengraphen gegenüber dem Regionengraphen liegt in der genaueren Behandlung der Grauwertübergänge. Das vereinfachte sequentielle Volumenwachstumsverfahren auf der Grundlage des gewichteten Regionengraphen resultiert in einer deutlichen Reduzierung der Regionenanzahl, führt jedoch ebenfalls zum Verlust wichtiger Konturen.

Der Konturengraph vereint die Vorteile des ersten und zweiten Ansatzes und erzielt somit das beste Segmentierungsergebnis hinsichtlich Segmentierungsgenauigkeit und problemangepaßter Regionenanzahl. Bei einer iterativen Anwendung dieses Wachstumsprozesses bleiben alle relevanten Objektgrenzen erhalten, das Regionenwachstum verläuft jedoch sehr langsam.

Eine Weiterentwicklung dieser Segmentierungsstrategie in Form einer Kombination dieses hierarchischen Ansatzes mit Verfahren, die der Entstehung einer Übersegmentierung entgegenwirkten, wie z.B. spezielle Glättungsfilter [1] oder Funktionen zur Änderung der Gradientenfunktion (Glättung, bzw. Reduzierung der Minima) [1][8], ist erfolgversprechend, allerdings ist eine deutliche Verbesserung der Segmentierungsergebnisse im Low-Level Bereich, d.h. ohne szenenspezifisches Wissen, bei Ultraschallbildern aufgrund der schlechten Bildqualität kaum zu erwarten.

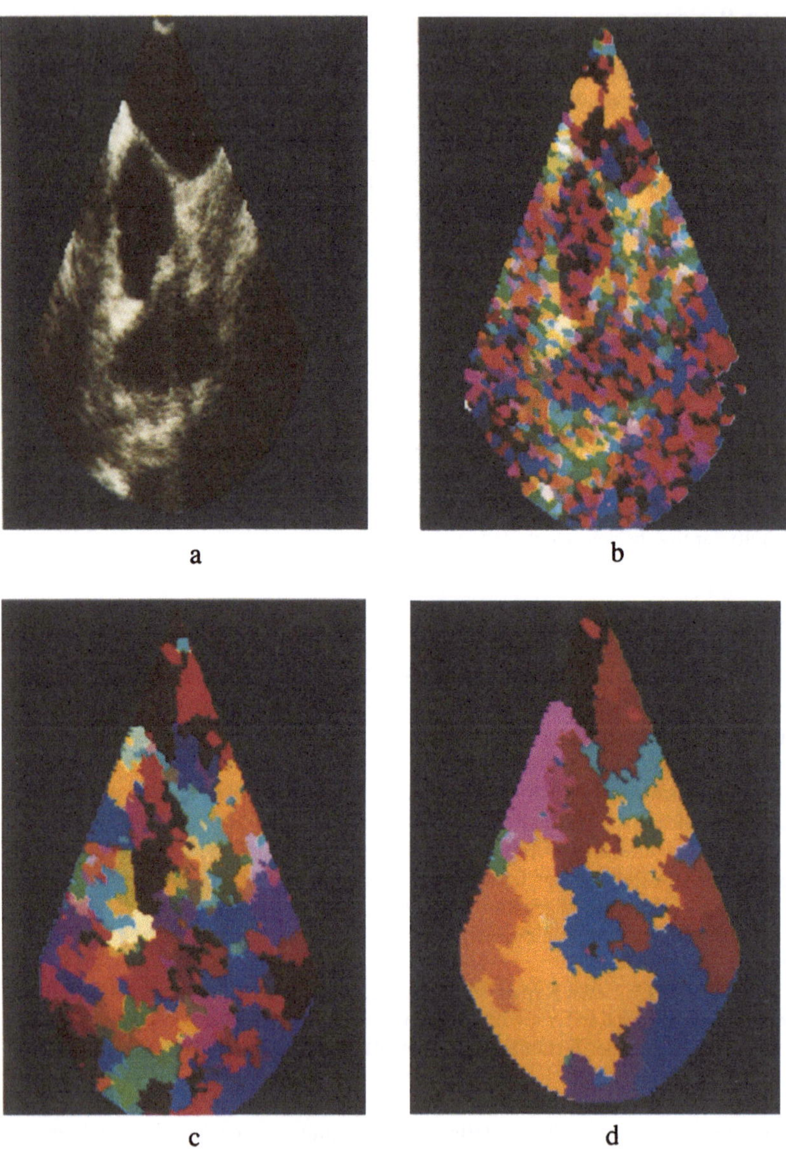

Abbildung 7:

(a) Originalschicht eines 4D-Ultraschalldatensatzes des Herzens als Grauwertbild: Ziel der Segmentierung ist vor allem die Detektion der Herzkammer, die sich als handförmiger dunkler Bereich im oberen Drittel des Kreissektors darstellt.

(b) Ergebnis der Wasserscheidenberechnung in Falschfarbendarstellung: Eine deutliche Übersegmentierung ist erkennbar.

(c) Mosaikbild nach der 2. Iteration (Konturengraph) in Falschfarbendarstellung: Die Regionenanzahl ist deutlich reduziert und die Objektgrenzen sind gut lokalisiert.

(d) Mosaikbild nach der 6. Iteration (Konturengraph) in Falschfarbendarstellung: Die Form des Ventrikels ist noch weitgehend erhalten, dieser besteht nur noch aus 2 Segmenten.

Literatur

[1] C.-S. FUH, P. MARAGOS, L. VINCENT: Region-Based Approaches to Visual Motion Correspondence; Technical Report, No. 91-18, Harvard Robotics Laboratory, Cambrigde MA, Nov. 1991.

[2] R. HAMMENTGEN: Transösophageale Echokardiographie monoplan/biplan; Springer-Verlag, Berlin, 1991.

[3] F. MEYER, S. BEUCHER: Morphological Segmentation; Journal of Visual Communication and Image Representation, Vol. 1, No. 1, pp. 21-46, Academic Press, Sept. 1990.

[4] CH. SOHN, J. GROTEPASS: Die 3-dimensionale Organdarstellung mittels Ultraschall; Ultraschall, Vol. 11, pp. 295-301, Georg Thieme Verlag, Stuttgart, 1990.

[5] S. R. STERNBERG: Grayscale Morphology; Computer Vision, Graphics, and Image Processing, Vol. 35, pp. 333-355, Academic Press, 1986.

[6] L. VINCENT: Graphs and Mathematical Morphology; Signal Processing, Vol. 16, pp. 365-388, Elsevier Science Publishers, Amsterdam, 1989.

[7] L. VINCENT, P. SOILLE: Watersheds in Digital Spaces: An Efficient Algorithm Based on Immersion Simulations; IEEE Transactions on Pattern Analysis and Machine Intelligence, Vol. 13, No. 6, pp. 583-598, June 1991.

[8] L. VINCENT: Morphological Grayscale Reconstruction in Image Analysis: Efficient Algorithms and Applications; Technical Report, No. 91-16, Harvard Robotics Laboratory, Cambridge MA, 1991.

[9] S. ZEHETBAUER: Segmentierung und Analyse drei- und vierdimensionaler Ultraschalldatensätze; Diplomarbeit, Technische Universität München, Institut für Informatik, Lehrstuhl Prof. Radig, 1992.

Segmentation of Magnetic Resonance Brain Images using Analog Constraint Satisfaction Neural Networks

Andrew J. Worth[1 2] and David N. Kennedy[2]

[1] Universität Hamburg, FB Informatik, AB KOGS, Bodenstedtstr. 16, D-22765 Hamburg, Germany

[2] Center for Morphometric Analysis, Neuroscience Center, Massachusetts General Hospital-East, Building 149, 13th St., Charlestown, MA 02129, USA***

Abstract. The Grey-White Decision Network (GWDN) is presented as an analog constraint satisfaction neural network that segments magnetic resonance brain images. Constraints on signal intensity, neighborhood interactions and edge influences are combined to assign labels of grey matter, white matter or "other" to each pixel. An improved version of this novel segmentation network that is provably stable is described. Results of the network are presented along with a comparison of these results to a collection of human segmentations.

1 Introduction

Segmentation is the first step to a quantitative analysis of brain morphometry. After labeling each pixel as grey matter, white matter, or other tissues and fluids, contiguous regions of similarly labeled pixels can be grouped together within and across brain slice images to describe anatomical structures. Subsequently the volumes, shapes and locations of these structures can be determined and these measurements can be correlated in both normal and abnormal subjects to provide insight into human brain structure and function.

Automating the image analysis process saves human time and effort and can provide accurate and reproducible results. For the purposes of this paper, we define segmentation as the low level, pixel-by-pixel labeling of grey matter, white matter, or "other" matter. The coronal MR brain slice images in this study are obtained using a T1-weighted spoiled gradient echo (FLASH or SPGR) sequence. We focus on this type of image because it is important to examine the underlying low-level pixel segmentation even though the task becomes easier with increasing scan quality and resolution and by using multiple echo images.

2 System Description

The GWDN has its origins in the parallel distributed processing mechanism characterized by Grossberg as an on-center, off-surround, recurrent competitive field

*** Correspondences should be sent to this address or to andy@cma.mgh.harvard.edu

(RCF) of units with shunting interactions [1]. As a neurally inspired dynamical system, the results of the decisions made at every pixel location by the GWDN are determined by the locations of the stable points (attractors, repellers, and saddle points) of the system. The locations of these stable points are determined by the "forces" that drive the activations of the units. These forces are derived from the following constraints or properties of the task of segmentation: 1) A pixel should only be labeled grey, white or other, 2) A grey matter pixel will tend to be found near other grey matter pixels (similarly for white and other), 3) The MR signal intensities for grey and white matter tend to be clustered around certain average values, and 4) Edges can be detected at the transitions between grey, white, and other matter pixels. These four properties are embodied as constraints or analog "rules" that the system must satisfy, and they determine the architecture of the network as explained below.

2.1 Units and Connections

The grey-white decision network uses three layers of units: "grey", "white", and "other". Every layer has a unit that corresponds to each pixel of the original image. Competition between layers ensures that only one unit becomes active at any pixel location. This is accomplished by inhibitory connections between layers. Cooperation within layers strengthens groups of similarly labeled pixels and removes isolated pixels. This is performed by excitatory connections within layers. Reciprocal (recurrent) connections from each unit to itself allow each unit to sustain its activation. Properties 1 and 2 above are realized by this configuration of units and connections.

2.2 Gaussian Intensity Similarity Function

The Gaussian "intensity similarity functions" $f_g(I)$, $f_w(I)$, and $f_o(I)$ provide input to the system that represents evidence relating the MR signal intensity with a corresponding label ("grey", "white", or "other"). For grey units, the function $f_g(I)$ is a measure of the similarity of a pixel's value to the average grey value with a Gaussian fall-off around that mean. The function $f_w(I)$ does the same for white matter units. These functions are (to a first approximation) the probability of a given grey, white or "other" label. For instance, when a pixel's intensity is near the grey function's mean, it produces a large output for the grey matter function. A large output from the grey function means that there is a lot of evidence that the pixel should be labeled "grey". Other matter units are the default and the function $f_o(I) = [2 - (f_w(I) + f_g(I))]^+$, where $[w]^+ = max(0, w)$. Property 3 above is realized by these input functions. The values for the peaks and widths of the Gaussian intensity similarity functions can be determined by examining the intensity distribution of the original image.

2.3 Edge Inputs

The Boundary Contour System (BCS) [2] is used to realize the 4th property given above. Briefly, the BCS detects and enhances edges that are supported by both short and long range evidence in the image. In the current implementation, a simplified version of the BCS is used to extract edges from the original image. More information on using the BCS to detect edges in MR brain images can be found elsewhere [3].

Edges detected by the BCS are used in the grey-white decision net to modulate the communication between units. When a strong edge is detected between two pixels, their intercommunication is reduced or shut off. Communication between units is not affected by weak or nonexistent edges. After a unit makes a decision, its activation is an influence on its neighbors in the same layer to make the same decision. In this way, a grey pixel in a region of white can be forced to change its decision to white. Edges are used to discourage such interactions between adjacent pixels; if the grey pixel is surrounded by strong edges, it may be able to ignore its dissimilar neighbors' influence and can remain grey. Edges have no effect on the communication between units at the same pixel location (i.e. the inhibition between the grey, white and "other" units at a single location is not modulated).

3 Overview of Network Operation

The grey-white decision net works as follows. After initializing all unit activations to zero, units will tend to turn on if their MR signal intensities are near the peak of their Gaussian intensity similarity functions. Consider the grey, white, and "other" units at each of two neighboring pixel locations shown in Fig. 1. At location ij, $f_g(I_{ij})$ responds strongly while at location pq, $f_w(I_{pq})$ responds strongly (see the graphs on either side of the figure). The grey unit at ij will tend to have a large activation which forces the white and other matter units off at that same location. Similarly, the white unit at location pq will shut off the grey and other matter units at its location (the inhibitory arrows in Fig. 1). At the same time, the two active units will attempt to turn on the neighboring units within their layers (the excitatory arrows in Fig. 1). A dynamic balance will result from the direct and indirect excitations and inhibitions. When equilibrium is attained, a decision has been made: the unit with the largest activation at that location wins and a pixel label is assigned accordingly. Strong edges affect the calculation by causing the two pixel's decisions to be made more independently. If there are no strong edges, neighboring units can "gang up" on isolated non-conforming pixel locations to make that location the same pixel label as the surrounding majority.

4 The Difference between RCCF and RCF

The original equations for the recurrent cooperative competitive field (RCCF) version of the GWDN are described in [4, 3]. The main equation is repeated here:

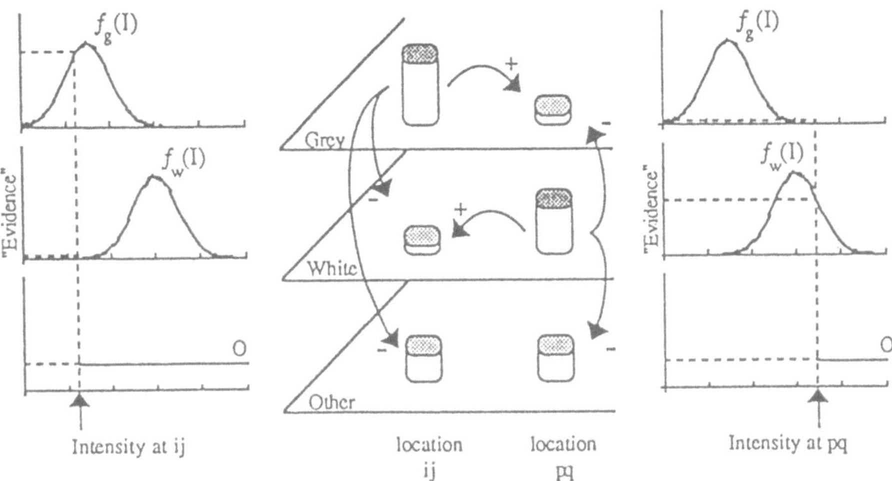

Fig. 1. Two neighboring pixel locations responding to their inputs. The arrows labeled with "+" and "-" indicate only the larger inhibitory and excitatory influences between units. The activation of each unit is represented by the cylinder height. The input functions that give rise to these activations at the two locations are shown on either side.

$$
\begin{aligned}
\frac{dx_i^g}{dt} &= Ax_i^g \\
&+ (B - x_i^g)(\sum_{q \in N} G_{qi} P_{qi} f(x_q^g) + I_i^g) \\
&- (C + x_i^g)(\sum_{q \in N} G_{qi} P_{qi} [f(x_q^w) + f(x_q^o)]),
\end{aligned}
\tag{1}
$$

where x_i^g is the activation of the ith grey matter unit, A, B, and C are constants, G_{qi} is the connection strengths between units (a decreasing function of distance), P_{qi} is a function of edges detected between two units (P_{qi} approaches zero when edges are detected), $f()$ controls the strength of communication between units, $I_i^g = f_g(I)$ is the input similarity function, and N defines a neighborhood around unit i. Similar equations control x_i^w and x_i^o. Note that the subscripts i and q stand for a *two*-dimensional index representing a pixel location in the image. The first term is the "passive decay" term and sends a unit's activation to zero in the absence of other inputs. The second, or "excitation" term tends to increase the activation of the unit and the third, or "inhibition" term tends to decrease the activation.

In the RCCF as described by (1), properties 1 and 2 are realized by competition and cooperation. However, oscillations can arise in a system that is both cooperative *and* competitive. Instability (oscillation) in such systems is a problem because it becomes difficult to determine if the network has finished its calculation. This stability problem can be avoided by the use of a recurrent competitive field (RCF) instead of a recurrent cooperative/competitive field (RCCF). In this case, the cooperative (excitatory) part of the system dictated by property 2 above can instead be used to decrease the competitive (inhibitory) influences of the system. In other words, instead of directly causing a given grey unit to become active, its active grey neighbors will prevent it from being inhibited by nearby white and other matter units: instead of excitation, cooperation is achieved by disinhibition.

This disinhibition can be incorporated into the system by two modifications to the equation given above. First, instead of adding neighbor influences to the excitatory term, they are subtracted from the inhibitory term, and second, the sum of the inhibitory terms should not be allowed (due to this new subtraction) to become negative. By using the proper substitution of variables (and by setting C equal to 0) it can be shown (see[3]) that the system is now of the same form as described by Cohen and Grossberg [5] and is stable by the Cohen-Grossberg theorem.

The new equation is

$$\begin{aligned}\frac{dx_i^g}{dt} &= -Ax_i^g \\ &+ (B - x_i^g)I_i^g \\ &- (C + x_i^g)[\sum_{q \in N} P_{qi}G_{qi}(f(x_q^w) + f(x_q^o)) - \sum_{q \in N} P_{qi}G_{qi}f(x_q^g)]^+.\end{aligned} \tag{2}$$

The rectification, $[w]^+ = max(w, 0)$, around the two sums is to ensure that the total inhibition never becomes negative.

This new equation performs the desired emergent calculation as the following argument shows. If there is evidence from neighboring units for a given pixel to be labeled "grey" and little or no evidence from neighboring units for an "other" matter or "white" labeling, then the neighboring grey units need not have an effect on this pixel: in (2), the first sum will be small and the second will be large, so the total inhibition will be rectified at zero. The white and other units at this pixel location will be inhibited by the surrounding grey units, though, so any positive evidence from the input, I_i^g, for this grey unit will cause the grey unit to have the largest activation at this pixel location, and the pixel will be labeled "grey." On the other hand, if there is evidence for a "grey" labeling at a pixel location, and the pixel's surrounding neighborhood has been labeled non-"grey", then the subtraction of the two sums in (2) will produce a positive result and this grey unit will be inhibited. This will allow the white or other units to have the largest activation.

5 Simulation Results

Simulations of both the RCCF and RCF versions of the grey-white decision net produce segmentations that are fairly consistent with segmentations by human experts with the proper choice of parameters. Figures 2 and 3 show the inputs and outputs of the system. The outputs of both instantiations of the GWDN are virtually the same: the resulting pixel-by-pixel decision is far more dependent on parameter differences than the change in architecture from RCCF to RCF. However, the RCF, as well as being provably stable, also appears to reach its decision faster than the RCCF.

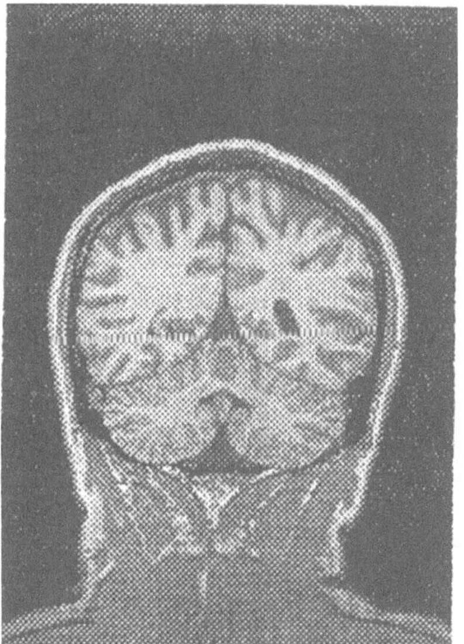

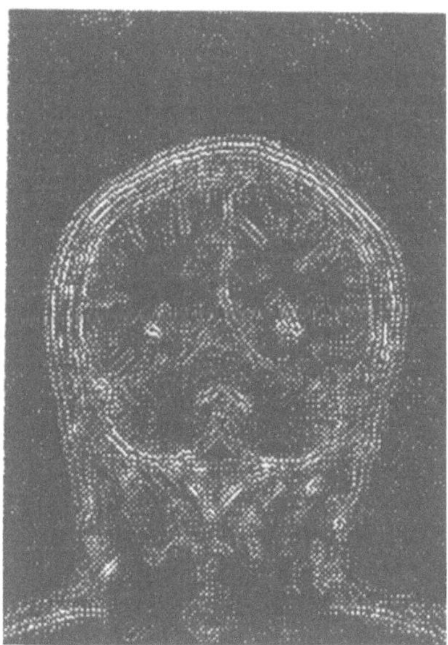

Fig. 2. (a) Original MRI brain slice image. (b) Edge image produced by the Boundary Contour System (BCS).

5.1 Problem Areas

While the results are encouraging, there are a number of problems. One of the more obvious problem areas is that the net labels the neck, shoulders, and scalp areas as grey or white matter. These are clearly not part of the brain. Another problem is the separation between the cerebrum and cerebellum. The two sweeping dark curves that separate the left and right halves of the brain and the cerebellum can easily be recognized in the original image even though the grey scale intensities do not form a continuous line. Since the grey-white decision network only takes into consideration the intensities, changes in intensities, and neighboring intensities, it has no way of compensating for these problems. Extensions to solve these and other problems are discussed in [3].

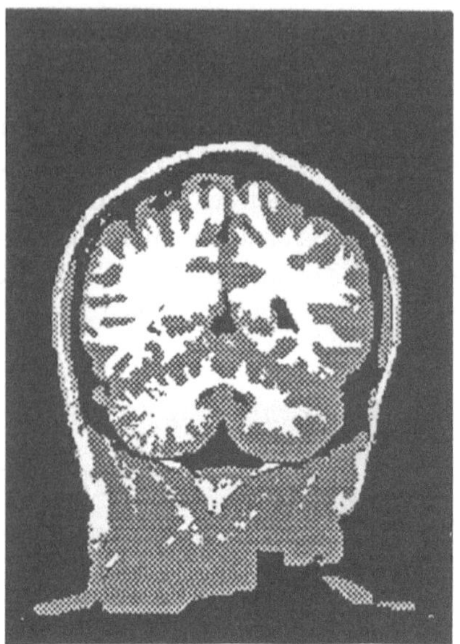

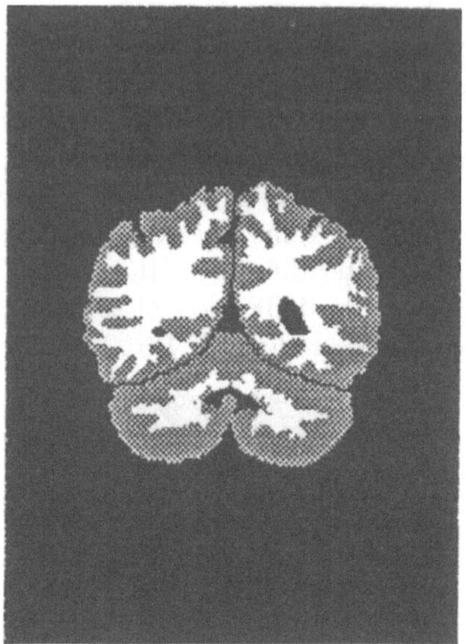

Fig. 3. (a) Output of the grey-white decision net: grey matter is colored grey, white matter is colored white and everything else is colored black. (b) The expert's segmentation of the same brain slice.

5.2 Segmentation Accuracy

In addition to a visual inspection, a comparison between segmented binary "mask" images can be made by calculating the change in individual areas or by calculating the total overlapping area of each mask. The latter is done by dividing the total number of locations where both masks are on by the total number of locations where either mask is on. Table 1 shows the averages (and standard deviations in parenthesis) of these comparison metrics for seven human segmentations compared against each other and for the GWDN output compared against these human segmentations (O_k is the overlap index for some anatomical region k and ΔA^k is the percent difference in area). Table 1 indicates that the variation between the GWDN output and the human segmentations is comparable to the variation between human segmentations. These results demonstrate that the GWDN is effective for segmenting MR brain images given fairly good parameters.

6 Conclusion

The grey-white decision network has been shown to effectively segment MR brain images. While the accuracy of the results are dependent on correct parameter settings, the architecture is appropriate for this task. The advantages of the GWDN include: its independence of the dimension of the image; its analog method for incorporating the constraints of pixel intensity, neighborhood interactions, and edge influences; its affordance of dynamical systems analysis; the

avg (sdev)	% ΔA^k		O^k	
White Matter Region	Human	GWDN	Human	GWDN
left cerebral	0.3 (8.3)	-1.8 (5.7)	0.89 (0.05)	0.86 (0.04)
right cerebral	0.7 (11.8)	-12.1 (7.5)	0.88 (0.06)	0.82 (0.05)
left cerebellar	7.3 (41.6)	6.9 (28.4)	0.73 (0.16)	0.78 (0.13)
right cerebellar	3.3 (26.8)	-27.5 (12.2)	0.82 (0.15)	0.68 (0.12)

Table 1. Comparison between human segmentations and GWDN output for white matter regions. The values shown are averages (with standard deviations in parentheses). The values listed under "Human" are the averages of the statistics comparing seven human segmentations with each other. Values listed under "GWDN" are the averages of the statistics comparing the GWDN segmentation with every human segmentation.

fact that it is stable by the Cohen-Grossberg theorem; and that it can be readily extended using additional constraints. Also, the parallel, distributed, analog nature of the GWDN makes it suitable for construction using VLSI, optical, or other specialized hardware technologies.

Acknowledgment The presentation of this work was made possible by grant 515 402 127 3 from the Deutscher Akademischer Austauschdienst. This work was supported in part by grant NS 20489 and also NS 24279 from the National Institute of Neurologic Disorders and Stroke.

References

1. S. Grossberg. The Quantized Geometry of Visual Space: The Coherent Computation of Depth, Form, and Lightness. *Behavioral and Brain Sciences*, 6:625–692, 1983.
2. S. Grossberg and E. Mingolla. Neural dynamics of surface perception: Boundary webs, illuminants, and shape-from-shading. *Computer Vision, Graphics and Image Processing*, 37:116–165, 1987.
3. A.J. Worth. Neural networks for automatic segmentation of magnetic resonance brain images. doctoral dissertation, Department of Cognitive and Neural Systems, Boston University, 1993.
4. A.J. Worth, S. Lehar, and D.N. Kennedy. A recurrent cooperative/competitive field for segmentation of magnetic resonance brain images. *IEEE Trans Knowledge Data Eng*, 4(2):156–161, 1992.
5. M.A. Cohen, and S. Grossberg. Absolute Stability of Global Pattern Formation and Parallel Memory Storage by Competitive Neural Networks. *IEEE Transactions on Systems Man and Cybernetics*, 13(5):815–826, 1983.

Nichtlineare Diffusion zur Integration visueller Daten – Anwendung auf Kernspintomogramme

Rainer Sprengel, Christoph Schnörr

Universität Hamburg, FB Informatik, AB Kognitive Systeme
Bodenstedtstraße 16, W-2000 Hamburg 50
email: sprengel@informatik.uni-hamburg.de

Zusammenfassung. Wir stellen einen nichtlinearen Diffusionsansatz zur Integration visueller Daten vor, welcher signifikante Sprünge der Daten erhält. Im Gegensatz zu verwandten, in der Literatur vorgeschlagenen Ansätzen, ergibt unser Verfahren eine eindeutige Lösung, welche kontinuierlich von den Daten abhängt und sich durch Standardverfahren der numerischen Mathematik stabil und effizient approximieren läßt. Eine erste Skizze des Ansatzes mit Anwendungen auf eindimensionale Daten ist in [16] vorgestellt worden.
In diesem Artikel wird die Anwendung des Verfahrens zur Rekonstruktion zweidimensionaler Bilder an einem exemplarischen Kernspintomogramm beschrieben. Die Ergebnisse lassen sich dann z.B. für eine Segmentation der Daten in die relevanten Bildstrukturen verwenden.

1 Einleitung

Eine wichtige Aufgabe der frühen Bildverarbeitung ist die Integration und Reduktion lokal verfügbarer Bildinformationen wie Grauwerte, Bewegungsvektoren oder Tiefenhinweise. Für eine Segmentation und Interpretation der Bilddaten ist es sinnvoll über homogene Bereiche hinweg zu glätten. Auf der anderen Seite sollen aber die Ränder dieser Bereiche erhalten bleiben und nicht geglättet werden. Dies erfordert im allgemeinen Wissen von einer höheren Ebene.

Eine Klasse von Arbeiten zur Integration von Bilddaten sind in den letzten Jahren unter dem Stichwort "Regularisierung" schlecht gestellter Probleme bekannt geworden [13, 1]. Als prototypischen Vertreter dieser Klasse betrachten wir das Problem der Minimierung des quadratischen Funktionals:

$$J : \mathcal{H} \to \mathbb{R}, \qquad J(v) = \int_\Omega \left\{(v-g)^2 + \lambda^2 \|\nabla v\|^2\right\} dx$$

Hier bezeichnet $\mathcal{H}$ den Raum geeigneter Funktionen v und $g : \Omega \subset \mathbb{R}^n \to \mathbb{R}$ die gegebenen visuellen Daten, die von irgendeinem bildgebenden Sensor oder einer Vorverarbeitungsstufe stammen. λ ist eine numerische Konstante, die den Einfluß der beiden Terme des Integranden entsprechend wichtet. Der erste Term mißt die Abweichung von den Daten, während der zweite Term die Glattheit der Lösung bewertet. Man beachte, daß diese Methode nicht nur auf die Rekonstruktion von Grauwertbildern beschränkt ist, sondern auch zur Verarbeitung unterschiedlicher lokaler Daten wie z.B. Tiefen- oder Bewegungsdaten (optischer Fluß) angewendet werden kann [17, 7].

Angenehme Eigenschaften der Minimierung quadratischer Funktionale aus der Sicht der Bildverarbeitung sind:

1. Die das Funktional J minimierende Funktion $u \in \mathcal{H}$ ist eindeutig.
2. Spezielles Vorwissen über qualitative Eigenschaften der Lösung kann über eine entsprechende Formulierung des Glattheitstermes eingebracht werden. Z.B. können der Rotations- und Divergenzanteil eines Verschiebungsvektorfeldes verschieden gewichtet werden.
3. Vom numerischen Standpunkt aus attraktiv ist die Tatsache, daß die Lösungsfunktion u durch Funktionen $u_h \in \mathcal{H}_h \subset \mathcal{H}$ so approximiert werden kann, daß

$$||u - u_h||_{\mathcal{H}} \rightarrow 0, \qquad \text{für} \qquad h \rightarrow 0.$$

   Dadurch ergeben sich viele Freiheitsgrade bei der Art der Diskretisierung des Problems (durch die Wahl geeigneter diskreter Unterräume $\mathcal{H}_h$), da die Eigenschaften jeder diskreten Repräsentation der Lösung durch diejenigen der kontinuierlichen Lösung bestimmt sind.
4. Die Berechnung der Lösung kann weitestgehend lokal und parallel erfolgen.

Weitere Details zu den Punkten 1 bis 4 finden sich in [17] und [14] und der dort zitierten Literatur.

Ein Problem bei der Rekonstruktion durch die Minimierung quadratischer Funktionale ist jedoch, daß die Integrationsgebiete, die die Daten auf die durch Signalübergänge begrenzten Gebiete einschränken, von vornherein nicht bekannt sind. Eine Integration über den gesamten Bildbereich eliminiert jedoch weitgehend signifikante Übergänge der Daten, die als Hinweise für eine Segmentation dienen könnten.

Wir stellen nachfolgend einen Ansatz vor, der die Methode der Minimierung quadratischer Funktionale so erweitert, daß signifikante Übergänge in den Daten erhalten bleiben, ohne jedoch, wie bei bisher in der Literatur vorgeschlagenen Erweiterungen (siehe nächster Abschnitt), die oben genannten attraktiven Eigenschaften dieser Klasse von Ansätzen aufgeben zu müssen.

2 Relevante Arbeiten

Es gibt verschiedene Autoren, die das Segmentationsproblem ebenfalls über die Minimierung eines Funktionals formuliert haben. Nordström [10] verändert die nichtlineare Diffusionsgleichung von Perona und Malik [12] so, daß das Ergebnis dem Minimum eines Funktionals

$$J = \int_{\Omega} f(x, v, \nabla v) dx$$

entspricht. Da aber der dabei verwendete Diffusionskoeffizient dem von Perona und Malik vorgeschlagenen entspricht, ist $f(\cdot) = f(x, v, \cdot)$ nicht konvex und die Existenz einer Lösung zumindest innerhalb des Sobolev-Raumes $\mathcal{H} = W^{1,2}(\Omega)$ fragwürdig. Aber selbst wenn eine Lösung innerhalb eines größeren Funktionenraumes existieren würde, so bliebe ungeklärt, wie diese mathematisch gesichert mit numerischen Verfahren approximiert werden könnte.

Es gibt auch Arbeiten, die zusätzliche Variablen in das Funktional einführen, welche die Orte der Kanten explizit repräsentieren sollen. So schlagen z.B. Mumford und Shah [9] vor, das Funktional

$$J(v, D) = \int_{\Omega \backslash D} \left\{ \beta(v-g)^2 + |\nabla v|^2 \right\} dx + \alpha \mathcal{L}^{n-1}(D)$$

zu minimieren, wobei v ein Element des Sobolev-Raumes $\mathcal{H}^1(\Omega \backslash D)$ ist sowie $D \subset \Omega$ eine abgeschlossene Menge in Ω ist, die die Diskontinuitäten repräsentiert. Die "Länge" der Diskontinuitäten wird durch das $(n-1)$-dimensionale Hausdorff-Maß $\mathcal{L}^{n-1}$ gemessen (die Menge D kann sehr irregulär sein). Hier werden also Unstetigkeitsstellen von v explizit modelliert und als Unbekannte in die Minimierung mit einbezogen. Das entsprechende Minimierungsproblem wird dadurch allerdings sehr schwierig. Es konnte gezeigt werden, daß Lösungen in einem sehr allgemeinen Funktionenraum existieren [6, 3]. Über die Regularitätseigenschaften der Lösung u und insbesondere D sind unseres Wissens bisher keine weitergehenden Aussagen bekannt. Auch eine gesicherte numerische Methode zur Diskretisierung und Berechnung einer Näherungslösung ist uns nicht bekannt.

Es gibt weiterhin eine Vielzahl von Ansätzen, die entsprechende Minimierungsprobleme direkt diskret formulieren [5, 2, 18]. Ausgangspunkt ist hier die Diskretisierung des quadratischen Funktionals mit Differenzen erster Ordnung. Zur Repräsentation der Positionen der Kanten werden sog. "Linienprozeße" eingeführt (es handelt sich um Binärvariable, die angeben sollen, ob eine Unstetigkeitsstelle vorliegt oder nicht). Um die Anzahl der Unstetigkeitsstellen zu begrenzen, wird ein weiterer Kostenterm eingeführt, der das Vorhandensein einer Sprungstelle "bestraft". Äquivalent zu diesen Ansätzen sind Verfahren, die den quadratischen Glattheitsterm so modifizieren, daß die Kosten ab einem bestimmten Gradienten mit zunehmendem Gradienten nicht mehr weiter ansteigen [2]. Zur Minimierung dieser hochgradig nichtkonvexen Funktionale müssen entweder sehr langsam konvergierende Techniken (simulated annealing [5]) oder aber sog. Einbettungsverfahren (z.B. der graduated non-convexity Algorithmus von Blake und Zisserman [2]) eingesetzt werden, wobei letztere nicht immer zum Ziel führen müssen und nicht geradlinig auf allgemeinere Probleme anwendbar sind.

3 Der Ansatz

In diesem Abschnitt wollen wir anhand des eingangs angegebenen prototypischen Beispiels den Ansatz der Minimierung eines quadratischen Funktionals erweitern. Wir werden zeigen, daß dabei auf die positiven Eigenschaften des quadratischen Ansatzes nicht verzichtet zu werden braucht.

Es soll ein Variationsproblem der folgenden Art gelöst werden:

$$J : \mathcal{H} \to \mathbb{R}, \quad J(v) = \int_{\Omega} \{(v-g)^2 + \lambda(\nabla v)\} dx \to \min$$

Dabei gilt:

$$\begin{aligned} \lambda : \mathbb{R}^n \to \mathbb{R}\,, &\quad \lambda(x) = \psi(|x|), \\ \psi : \mathbb{R} \to \mathbb{R}\,, &\quad \psi \text{ stetig differenzierbar.} \end{aligned}$$

Eine notwendige Bedingung für die Lösung dieses Problems ist das Verschwinden der ersten Variation dieser Gleichung:

$$\delta J(u,v) = \Phi'(0) = 0 \quad \text{mit} \quad \Phi(t) = J(u+tv) \quad \text{für alle } v \in \mathcal{H}$$

Damit gilt für unser Problem:

$$\delta J(u,v) = \int_\Omega \left\{ 2(u-g)v + \frac{\psi'(|\nabla u|)}{|\nabla u|} \nabla u \nabla v \right\} dx = 0$$

Führt man hier den sog. Diffusionskoeffizienten ρ über

$$s\rho(s) = \psi'(s), \qquad 0 \leq s \in \mathbb{R}$$

ein, so kann man die Variationsaufgabe in der folgenden Form schreiben:

$$a(u,v) = f(v), \qquad \forall v \in \mathcal{H},$$

mit

$$a(u,v) = \int_\Omega 2uv + \rho(|\nabla u|)\nabla u \nabla v dx, \qquad f(v) = \int_\Omega 2gv dx$$

Durch obige Gleichung ist ein Operator $J' : \mathcal{H} \to \mathcal{H}^*$ definiert, der eine Funktion $u \in \mathcal{H}$ auf das Element $a(u,\cdot) - f(\cdot)$ im Dualraum $\mathcal{H}^*$ abbildet. Für die Anwendung des Funktionals $J'(u) \in \mathcal{H}^*$ auf eine Funktion v schreibt man auch $\langle J'(u), v\rangle = a(u,v) - f(v)$.

Falls J' nun die folgenden Eigenschaften der Lipschitz-Stetigkeit und strengen Monotonie hat, so läßt sich beweisen, daß eine eindeutige Lösung existiert, welche durch die Methode der finiten Elemente [4] approximiert werden kann:

$$\|J'(u) - J'(v)\|_{\mathcal{H}^*} \leq L\|u-v\|_{\mathcal{H}}, \qquad L > 0, \qquad \forall u,v \in \mathcal{H},$$

$$\langle J'(u) - J'(v), u-v\rangle \geq c\|u-v\|^2_{\mathcal{H}}, \qquad c > 0, \qquad \forall u,v \in \mathcal{H}.$$

Diese Eigenschaften sind erfüllt, wenn man ρ folgendermaßen wählt [15]:

$$\rho(t) = \begin{cases} 2\lambda_h^2 & \text{falls } 0 \leq t \leq c_\rho \\ 2\lambda_l^2 + \frac{2(\lambda_h^2 - \lambda_l^2)c_\rho}{t} & \text{falls } 0 \leq c_\rho \leq t \end{cases}$$

Solange der Absolutbetrag des Gradienten der Lösung unterhalb eines Umschaltparameter c_ρ liegt, wird mit dem Parameter λ_h geglättet. Überschreitet der Gradient die Größe c_ρ, so fällt die Glättung umgekehrt proportional zur Gradientenstärke ab. Man beachte, daß jeder stärkere Abfall von ρ jedoch dazu führen würde, daß der Ansatz die geforderten Eigenschaften nicht mehr erfüllt.

Die Aufgabe wird diskretisiert, indem man einen endlichdimensionalen Teilraum $\mathcal{H}_h$ von $\mathcal{H}$ so auswählt, daß der Repräsentant $u_h \in \mathcal{H}_h$ der Lösung u des Problems diese genügend genau approximiert.

Man erhält ein Gleichungssystem zur Bestimmung von u_h, indem man v in der obigen Variationsgleichung durch geeignete Basisfunktionen ϕ_i des Raumes $\mathcal{H}_h$ ersetzt

$$\langle J'(u), \phi_i\rangle = a(u,\phi_i) - f(\phi_i) = 0.$$

Mit $u = \sum u_j \phi_j$ sowie $g = \sum g_j \phi_j$ ergibt sich

$$\langle J'(u), \phi_i \rangle = 2 \sum_j (u_j - g_j) \int_\Omega \phi_i \phi_j dx + \sum_j u_j \int_\Omega \rho(|\nabla u|) \nabla \phi_i \nabla \phi_j dx = 0.$$

Das Gleichungssystem kann mit einem nichtlinearen Gauss-Seidel bzw. SOR-Verfahren gelöst werden. Da sich die Eigenschaft der Lipschitz-Stetigkeit des Operators J' auf das Gleichungssystem überträgt, läßt sich eine Schrittweite angeben, bei der das Verfahren auf alle Fälle zur eindeutigen Lösung konvergiert [11].

4 Ergebnisse

In diesem Abschnitt werden die Ergebnisse des Verfahrens für ein Kernspintomogramm vorgestellt. Um die Ergebnisse besser beurteilen zu können werden auch die Resultate anderer (einfacher) Verfahren zur Glättung von Bilddaten gezeigt. Das Bild stammt aus der Explorationsphase des COVIRA-Projektes (Echo 2 des Testbildes 1451), in dem verschiedene Methoden zur Bildsegmentation untersucht wurden [8]. Das hier betrachtete Kernspintomogramm hat ein ziemlich schlechtes Signal-Rausch Verhältnis und läßt sich deshalb nur schwer segmentieren. Als vorbereitender Schritt ist es deshalb sinnvoll, das Bild zu glätten.

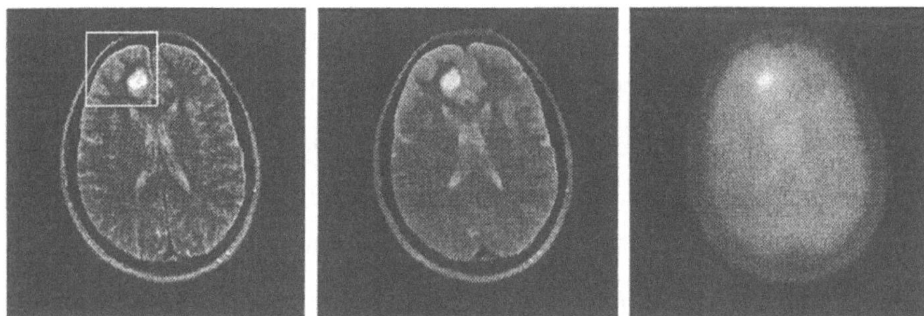

Abb. 1. Links: Kernspintomogramm eines Gehirns mit Tumor (heller Fleck), Mitte: ein diffundiertes Bild ($\lambda_h = 9.0, \lambda_l = 0.1$ und $c_\rho = 0.2$). Rechts: Bild mit quadratischem Glattheitsterm ($\lambda = 9.0$) geglättet. Im Originalbild ist der Bildausschnitt eingezeichnet, der in den Abbildungen 2 und 3 als Grauwertgebirge (von der linken oberen Ecke aus gesehen) dargestellt ist.

Bild 1 zeigt das verwendete Kernspintomogramm eines menschlichen Gehirns. Bei dem weißen Fleck oben links handelt es sich um einen Tumor. Das mittlere Bild zeigt das Ergebnis des Verfahrens für dieses Bild. Das rechte Bild der Abbildung 1 zeigt das Ergebnis der Glättung über ein quadratisches Funktional, also mit einem konstanten Diffusionskoeffizienten, der hier den gleichen Wert wie λ_h bei der nichtlinearen Glättung hat. Man sieht deutlich, daß hier über die Grauwertkanten des Bildes hinweg geglättet worden ist. Im Gegensatz dazu sind im nichtlinear geglätteten Bild die Strukturen des

Bildes ab einer über die globalen Parameter einstellbaren Größenordnung erhalten geblieben. Dies wird auch besonders in der Abbildung 2 deutlich, die einen Ausschnitt des Original- sowie des Ergebnisbildes zeigt. Wesentlich ist, daß auch die Positionen der Übergänge bei einer Variation der Parameter c_ρ und λ_l erhalten bleiben.

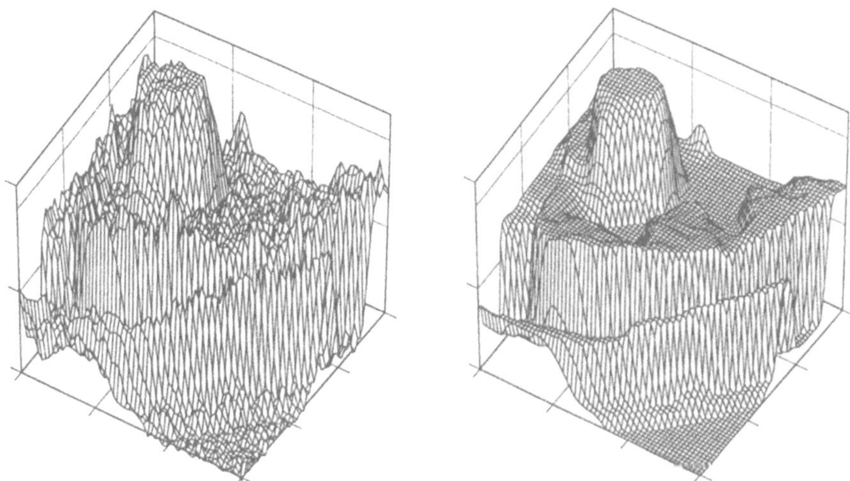

Abb. 2. Links Ausschnitt aus dem Originalbild als Grauwertgebirge. Rechts der gleiche Ausschnitt aus dem nichtlinear geglätteten Bild.

Es ist allerdings auch zu erkennen, daß der Kontrast vor allem der schmalen Strukturen wie z.B. der Haut etwas abnimmt. Rechnet man den Fall einer idealen eindimensionalen Rechteckfunktion der Höhe h und Breite a analytisch durch, so zeigt sich, daß das Produkt Höhe mal Breite den Mindestwert $c_\rho \lambda_h^2$ überschreiten muß, damit das Verfahren überhaupt auf die schwache Glättung umschaltet. Der Kontrast vermindert sich dabei um etwa

$$\frac{c_\rho \lambda_h}{\sinh \frac{a}{\lambda_h}}.$$

Dieser Effekt der Kontrastabnahme bei schmalen Strukturen ist auch bei anderen Verfahren zu beobachten. Als Beispiel haben wir das gleiche Bild auch einmal tiefpassgefiltert und mit einem 5×5-Median-Filter (beides aus der Bibliothek des Bildverarbeitungspakets "Khoros") geglättet (siehe Bild 3). Trotz eines geringeren Glättungseffektes hat der Kontrast im Bereich der Haut stark abgenommen. Interessant ist auch, daß leichte Helligkeitsunterschiede, wie z.B. innerhalb des Tumorbereiches oder auch außerhalb des Schädelbereiches, beim nichtlinear diffundierten Bild vollkommen verschwunden sind, während diese Schwankungen bei den anderen beiden Verfahren noch deutlich zu erkennen sind.

Abbildung 4 zeigt den Absolutbetrag des Gradienten (berechnet über einfache Differenzbildung direkt benachbarter Pixel) für das Originalbild, das nichtlinear diffundierte Bild und das linear diffundierte Bild aus Abbildung 1. Diese ersten Ergebnisse für ein Kernspintomogramm zeigen, daß es vielversprechend erscheint, die nichtlineare Diffusion als vorverarbeitenden Schritt z.B. für eine Segmentation zu verwenden.

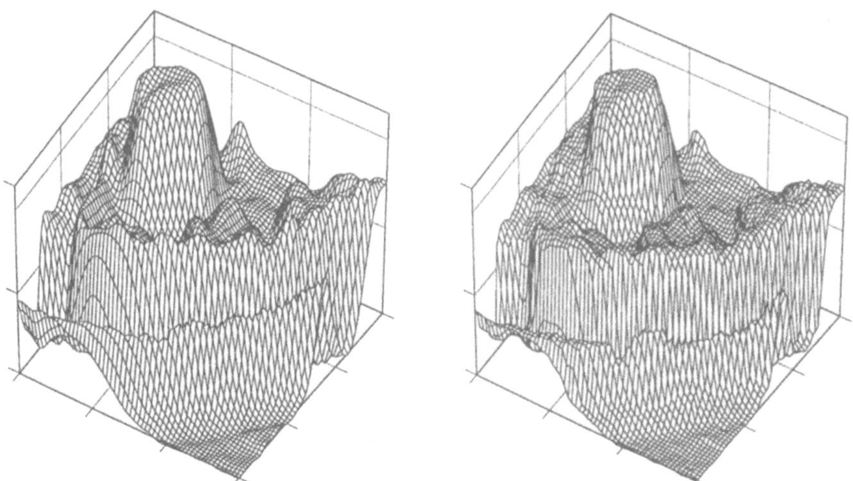

Abb. 3. Links: Tiefpassfilterung des Kernspintomogramms mit einer Grenzfrequenz des 0.3-fachen der Maximalfrequenz. Rechts: Glättung mit einem 5 × 5-Median-Operator.

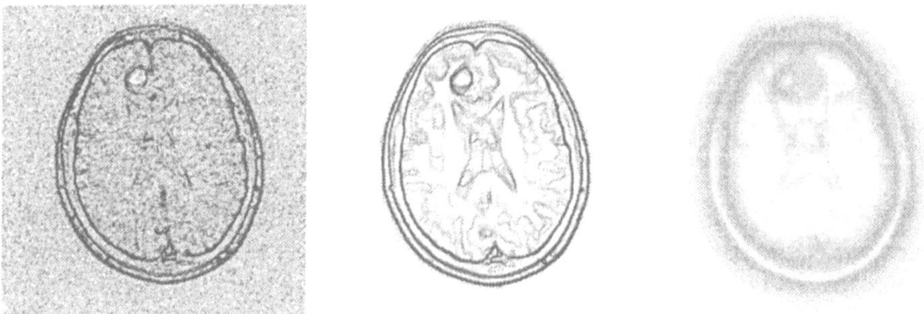

Abb. 4. Absolutbeträge der Gradienten der Bilder aus Abbildung 1.

Danksagung

Wir danken Carsten Schröder und Siegfried Stiehl für die kritische Durchsicht einer ersten Fassung dieses Papiers. Das Kernspintomogramm wurde uns vom COVIRA-Projekt[1] zur Verfügung gestellt.

Literatur

1. M. Bertero, T.A. Poggio und V. Torre. Ill-Posed Problems in Early Vision. *Proc. of the IEEE* **76** (8), August 1988.
2. A. Blake und A. Zisserman. Visual Reconstruction. The MIT Press Series in Artificial Intelligence. The MIT Press, Cambridge, Mass., 1987.

[1] Das COVIRA-Projekt (COmputer VIsion in RAdiology) wurde unter der Projektnummer A-1011 im Rahmen des AIM-Programms durch die EG gefördert

3. M. Carriero und A. Leaci. Existence Theorems for a Dirichlet Problem with Free Discontinuity Set. *Nonl. Anal. Theory Methods & Applications* **15**, 661–677, 1990.
4. C.G. und H.G. Roos. Numerik partieller Differentialgleichungen. Teubner Studienbücher. B. G. Teubner Verlag, Stuttgart, 1992.
5. S. Geman und D. Geman. Stochastic Relaxation, Gibbs Distributions, and the Bayesian Restoration of Images. *IEEE Trans. on Pattern Analysis and Machine Intelligence* **6**, 721–741, 1984.
6. E. De Giorgi, M. Carriero und A. Leaci. Existence Theorems for a Minimum Problem with Free Discontinuity Set. *Arch. Rat. Mech. Anal.* **108**, 195–218, 1989.
7. B.K.P. Horn und B.G. Schunck. Determining Optical Flow. *Artificial Intelligence* **17**, 1981.
8. M.H. Kuhn et al. COVIRA: COmputer VIsion in RAdiology. In Jaap Noothoven van Goor und Jens Philkjær Christensen (Hrsg.), Advances in Medical Informatics – Results of the AIM Exploratory Action, Band 2 der Reihe Studies in Health Technology and Informatics, Seite 88–101. IOS Press, Amsterdam – Oxford – Washington, DC – Tokyo, 1992.
9. D. Mumford und J. Shah. Optimal Approximations by Piecewise Smooth Functions and Associated Variational Problems. *Comm Pure Appl Math* **42**, 577–685, 1989.
10. N. Nordström. Biased Anisotropic Diffusion - A Unified Regularization and Diffusion Approach to Edge Detection. *Image and Vision Computing* **8** (4), 318–327, 1990.
11. J.M. Ortega und W.C. Rheinboldt. Iterative Solution of Nonlinear Equations in Several Variables. Academic Press, Orlando, 1970.
12. P. Perona und J. Malik. Scale-Space and Edge Detection Using Anisotropic Diffusion. *IEEE Trans. on Pattern Analysis and Machine Intelligence* **12** (7), 629–639, 1990.
13. T.A. Poggio, V. Torre und C. Koch. Computational Vision and Regularization Theory. *Nature* **317**, 314–319, 1985.
14. C. Schnörr. Determining Optical Flow for Irregular Domains by Minimizing Quadratic Functionals of a Certain Class. *International Journal of Computer Vision* **6** (1), 25–38, 1991.
15. C. Schnörr. Unique Reconstruction of Piecewise Smooth Images by Minimizing Strictly Convex Non-quadratic Functionals. Mitteilung FBI-HH-M-213/92, Fachbereich Informatik, Universität Hamburg, Oktober 1992. Erscheint in J. Math. Imaging and Vision.
16. C. Schnörr und B. Neumann. Ein Ansatz zur effizienten und eindeutigen Rekonstruktion stückweiser glatter Funktionen. In S. Fuchs und R. Hoffmann (Hrsg.), Proceedings 14th DAGM-Symposium, Dresden, September 1992, Informatik aktuell, Seite 411–416. Springer-Verlag, Berlin – Heidelberg – New York, 1992.
17. D. Terzopoulos. Multilevel Computational Processes for Visual Surface Reconstruction. *Computer Vision, Graphics, and Image Processing* **24**, 52–96, 1983.
18. D. Terzopoulos. The Computation of Visible-Surface Representations. *IEEE Trans. on Pattern Analysis and Machine Intelligence* **10**, 417–438, 1988.

Extraktion von Linieneigenschaften basierend auf Richtungsfeldern

Robert Kutka, Sebastian Stier
Siemens AG, Zentralabteilung Forschung und Entwicklung,
Otto-Hahn-Ring 6, 81739 München

Zusammenfassung

Es werden Algorithmen beschrieben, die auf einem Richtungsfeld aufbauend, folgende Eigenschaften von Blutgefäßen in Röntgenbildern extrahieren: Ein Linienmaß, das selbst schmale kaum sichtbare Adern in ihrer Intensität verstärkt, ein segmentiertes Binärbild der Gefäße, Breiten und Mittellinien. Da alle Module auf das gleiche Richtungsfeld zugreifen, sind sie in Summe sehr schnell. Die Linienbreite wird darüberhinaus unabhängig von Parametern bestimmt.
Die Eigenschaften sollen zur 3D-Rekonstruktion von Blutgefäßen eingesetzt werden, um Ärzte bei der Diagnose, Strahlentherapie- und Operationsplanung zu unterstützen.

1. Einleitung

In diesem Beitrag stellen wir Werkzeuge zur Gewinnung von Merkmalen und Segmentierung schmaler Blutgefäße bis in den Bereich von 2 Pixeln Durchmesser in digitalen Subtraktionsangiographie (DSA)-Aufnehmen vor.
Die dazu aus der Literatur bekannten Algorithmen variieren je nach Anwendung in ihrem Auflösungsvermögen, ihrer Rauschrobustheit, Rechenzeit und Benutzerinteraktion. Sie reichen von weitgehend automatischen [1-10] bis zu benutzergeführten Systemen [11-18]. Eines der anerkannt besten Werkzeuge ist dabei der Cannyoperator [10], auf welchem viele Algorithmen aufbauen. Wir vergleichen daher unseren nichtlinearen Ansatz mit dem Canny-Linienfinder.
Wir verwenden ein regelbasiertes Verfahren, das benutzerunabhängig, hochauflösend und relativ schnell ist.
Es besteht aus drei Hauptmodulen:
Zunächst definiert ein sternförmiges Vorfilter an jedem Punkt des Eingangsbildes Richtungen und Amplituden von möglichen Linien.
Dann werden benachbarte Bildpunkte mit ähnlichen Eigenschaften durch Vektoren verbunden.
Das nächste Modul verfolgt Linien entlang der Vektoren solange bestimmte Stetigkeitsbedingungen erfüllt sind und ordnet dabei jedem Punkt eine Linienintensität zu.
Ein nachgeschalteter Glättungsalgorithmus, der auch auf dem Richtungsfeld aufbaut, reduziert verbleibendes Rauschen innerhalb der Linienstrukturen, ohne die Ränder zu verändern. Dieses Verfahren ähnelt dem der Anisotropen Diffusion [6], ist jedoch auf andere, schnellere Weise realisiert.
Schließlich zeigen wir, wie sich mithilfe des Feldes von Richtungen auch Aderdurchmesser und Mittellinien berechnen lassen.

2. Der Verfolgungsalgorithmus

Wir suchen für jeden Bildpunkt ein Maß für die Intensität von Linienstrukturen. Es soll dort hohe Ausschläge zeigen, wo an entsprechenden Stellen des Originalbildes Linien zu finden sind, dagegen sehr niedrige Werte im Bildhintergrund, selbst wenn dieser Röntgenrauschen enthält.
Das Verfahren beruht auf der Suche nach hellen Linien auf dunklem Grund. Im umgekehrten Fall sind die Eingangsbilder zu invertieren.

2.1. Richtungen und Amplituden

Auf jeden Punkt des Eingangsbildes, außer an Randpunkten, wenden wir eine sternförmige Filtermaske, bestehend aus 8 Streifen, sogenannten Richtungsfiltern, an. Sie sind von 1 bis 8 numeriert (Abb. 1).
Entlang jedes Streifens i wird die Summe s_i der Pixelwerte bestimmt. Für jeden Bildpunkt definieren wir eine *Richtung d* als Nummer des Streifens mit der höchsten Summe, sodaß $s_d \geq s_i$, für alle $i = 1, ..., 8$. Sind keine Richtungen zugeordnet, wie an Randpunkten, werden Nullen in das Richtungsbild eingetragen.
Als nächstes berechnen wir den Mittelwert m der 8 Richtungssummen:

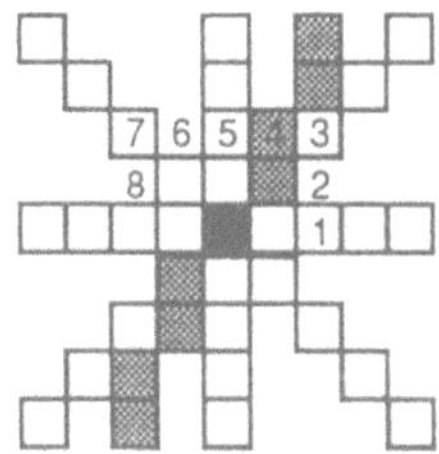

Abb. 1: Maske zur Richtungsfilterung. Zur Übersicht sind nur die ungeradzahligen und das 4. Richtungsfilter (dunkel) dargestellt.

$$m = \sum_{i=1}^{8} s_i \;/\; 8 \tag{1}$$

und die 8 Differenzen a_i der Richtungssummen zum Mittelwert:

$$a_i = s_i - m. \tag{2}$$

Diese sind ein Maß für die "Linienhöhe" der jeweiligen Richtung.
Nun addieren wir für jeden Bildpunkt jedes Richtungsstreifens i die Differenz a_i falls diese positiv ist, auf eine mit Nullen gefüllte Bildmatrix.
Das resultierende *Amplitudenbild* zeigt bereits dort hohe Ausschläge, wo an entsprechenden Stellen des Originalbildes Linien zu finden sind. In diesem Bild sind keine Überschwinger an Blutgefäßen sichtbar, wie sie bei linearen Filtern auftreten, da die Linienprofile durch Addition von Streifen in Längsrichtung angehoben, durch Streifen in Querrichtung jedoch nicht geändert werden.

2.2. Zusammenhang

Dieses und das Verfolgungsmodul sind die empfindlichsten Teile der Linienerkennung, da die Qualität entscheidend von der Wahl der Regeln und Parameter abhängt.
Grob gesagt, wird ein Bildpunkt mit einem anderen aus der Nachbarschaft durch einen Vektor verbunden, wenn ihre "Linieneigenschaften" ähnlich sind.
Für jede Richtung definieren wir zunächst einen Bereich, in dem nach geeigneten Nachbarn gesucht werden soll (Abb. 2). Zu jeder der acht Richtungen existieren zwei Suchbereiche entgegengesetzter Orientierung. Die übrigen der 16 Bereiche folgen aus den ersten drei manuell definierten aus der Symmetrie.
Wir legen nun ein *Ähnlichkeitsmaß für Richtungen* fest, das von der *Linienrichtung des aktuellen Bildpunktes* und der *Richtung* sowie dem *Ort eines Pixels aus dem zugehörigen Suchbereich* abhängt. Damit können wir zu jedem Punkt aus dem Suchbereich einen Satz von Richtungen angeben, die als ähnlich zur Richtung des aktuellen Pixels erklärt werden. Wir wählen diese Mengen heuristisch so, daß die verfolgten Pfade möglichst glatt sind. Aus Symmetriegründen genügt es wiederum, nur die Richtungsmengen für die ersten drei Suchbereiche festzulegen.
Wir untersuchten mehrere Zusammenhangskriterien. Ein geeignetes ist die minimale Amplitudendifferenz. Sie berücksichtigt die Idee, daß sich die Intensität (bzw. die Luminanz) der Blutgefäße von Punkt zu Punkt nur langsam ändert. Wir wählen daher aus den Kandidaten des betreffenden Suchbereichs, die ähnliche Richtungen aufweisen, denjenigen Bildpunkt mit der dem aktuellen Pixel ähnlichsten Amplitude und verbinden ihn durch einen Vektor.
Eine Schwelle vermeidet die Verbindung, wenn sich die Amplitude zu abrupt ändert.
Durch diese Maßnahme werden Abzweigungen des verfolgten Pfades in Strukturen mit anderer Luminanz unterbunden und "Haarartefakte" unterdrückt.

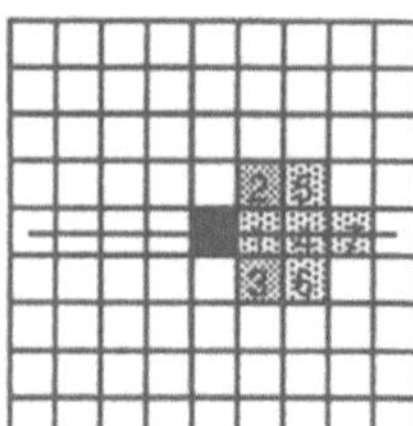

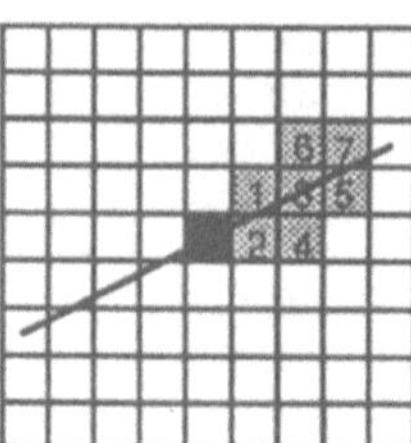

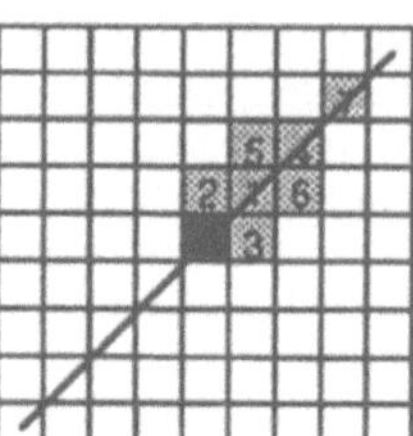

Abb. 2: Die ersten drei der 16 Suchbereiche. Die gerade Linie zeigt die Richtung, die dem Zentralpixel zugeordnet ist. Die numerierten Felder geben die Kandidaten für Verbindungsvektoren an.

2.3. Verfolgung

Der Algorithmus startet an jedem Bildpunkt und springt von Vektor zu Vektor, solange bestimmte Verfolgungskriterien erfüllt sind. Während der Verfolgung wird ein Linienmaß berechnet und dem Startpunkt zugeordnet.
Diese Kriterien sollen - im Gegensatz zu den Zusammenhangskriterien - von zwei Vektoren abhängen und können daher erst während der Verfolgung abgefragt werden. In erster Linie handelt es sich dabei um Beschränkungen des Krümmungs- und Abzweigewinkels.

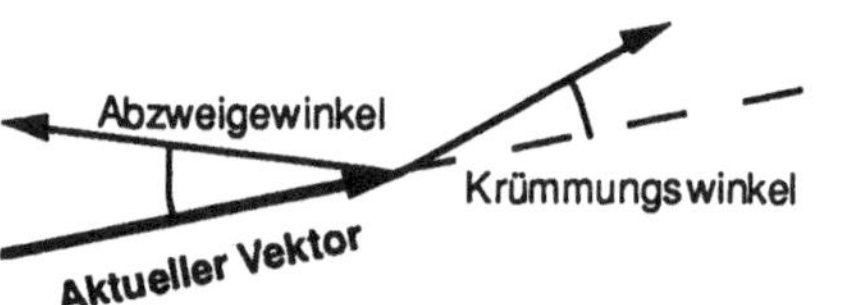

Abb. 3: Definition des Krümmungs- und Abzweigewinkels.

Zu einem während der Verfolgung durchlaufenen Vektor gibt es an seinem Endpunkt bis zu zwei weiterführende Vektoren. Der Krümmungs- und Abzweigewinkel sei wie in Abb. 3 definiert.
Eine Schwelle, angewandt auf den Krümmungswinkel, beschränkt die Krümmung des verfolgten Pfades.
Eine Beschränkung des Abzweigewinkels jedoch vermeidet, daß der verfolgte Pfad in einen anderen hineinläuft. Damit werden Rauschartefakte, wie "Haare" an Linien, unterdrückt.
Der Verfolgungsprozeß läuft so lange wie die Schwellwertbedingungen eingehalten werden oder eine gegebene maximale Länge erreicht ist.
Verschiedene Linienmaße sind getestet worden, die in erster Linie die Amplitude des aktuellen und der Nachbarpixel und die Verfolgungslänge berücksichtigen. Eine andere laufende Arbeit benutzt zusätzlich die Konstanz der Liniendurchmesser.
Wir halten die folgende Definition für geeignet:
Das *Linienmaß* $l(p)$ an einem Pixel p ist gleich der Amplitude $a(p)$, falls die Verfolgungslänge eine Schwelle überschreitet.

$$l(p) = max(länge1(p), länge2(p)) * a(p) \qquad (3)$$

Dabei bedeuten $länge1(p)$ und $länge2(p)$ die Anzahl der Verfolgungsschritte, von p aus gestartet, in den beiden entgegengesetzten Orientierungen.
Das Linienmaß ergibt dort hohe Werte, wo im Originalbild an entsprechenden Punkten Linien zu finden sind. Es unterdrückt Rauschstrukturen, da diese im allgemeinen kleine Amplituden oder kurze Verfolgungslängen aufweisen.
Es ist geeignet, um über eine Schwelle ein *binäres Bild der segmentierten Blutgefäße* zu gewinnen.

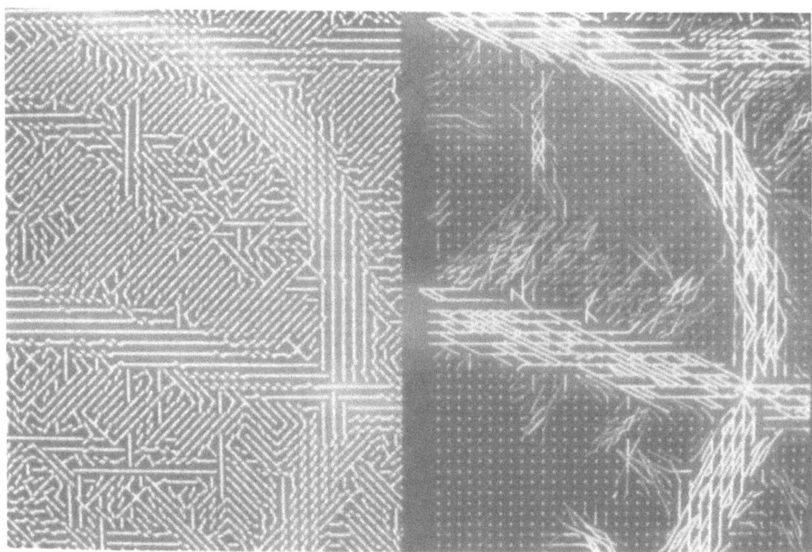

Abb. 4: Vergrößerter Ausschnitt eines cerebralen DSA-Bildes. Auf der linken Seite sind an jedem Bildpunkt die zugehörige Richtung, auf der rechten für den gleichen Bildausschnitt die gewählten Vektoren dargestellt. Der Verfolgungsprozeß überquert korrekt Aderkreuzungen.

2.4. Parallele Glättung

Um noch verbleibendes Rauschen der vorigen Verarbeitungsschritte weiter zu vermindern, entwickelten wir einen Algorithmus, der Linienstrukturen innerhalb ihrer Ränder glättet. Er baut auf dem Richtungsfeld (aus 2.1.) auf und benötigt keinen Ableitungsoperator, um Ränder zu finden.
Er benutzt die Tatsache, daß das Richtungsfeld, das durch das Vorfilter (Abb. 1) erzeugt wird, in einer schmalen Umgebung der Linien auf diesen senkrecht oder zumindest transversal steht. Dies folgt aus der Definition der Richtungen. Ein außerhalb eines Blutgefäßes gelegener Bildpunkt, dessen Richtungsmaske teilweise in die Linie hineinragt, erhält eine Richtung zugewiesen, die quer zum Blutgefäß verläuft, da die hineinragende Richtungsmaske die höchste Summe besitzt. Die Aderränder sind also Unstetigkeitsstellen des Richtungsfeldes. Abb. 4 zeigt ein Beispiel eines Richtungsfeldes.
Wir stoppen daher den Glättungsprozeß an diesen Unstetigkeitsstellen.
Wir definieren zuerst ein Maß für die Parallelität von Richtungen.
Als geeignet erweist sich, daß zwei Richtungen d_1 und d_2 "parallel" ($d_1 \parallel d_2$) heißen sollen, wenn sie um höchstens einen Richtungsschritt abweichen. Die Abweichung um eine Einheit soll erlaubt sein, um leichtes Rauschen des Richtungsbildes zu vernachlässigen.

Definition der parallelen Glättung
Wir führen an jedem Punkt eines gegebenen Bildes eine Glättung innerhalb einer Umgebung R (z. B. 3x3 oder 5x5 Pixel) durch, beziehen jedoch nur solche Bildpunkte aus der Umgebung in die Glättung ein, deren Richtung parallel zur Richtung des Zentralpixels ist.
Wir haben dazu mehrere Glättungsvarianten untersucht und finden die folgenden zwei für geeignet:

Mittelwertkriterium
Sei $l(p_0)$ der Wert des zu glättenden Bildes am Bildpunkt p_0. Der geglättete Wert $s(p_0)$ entsteht dann durch Mittelung:

$$s(p_o) = \sum_{\{p \in R \mid d(p) \parallel d(p_o)\}} l(p) \quad / \quad \left|\{p \in R \mid d(p) \parallel d(p_o)\}\right| \tag{4}$$

Die Betragstriche |...| bedeuten die Anzahl der Elemente dieser Menge, also die Anzahl der Pixel die zu p_0 "parallele" Richtungen besitzen.
Dieses Kriterium führt zwar zu einer deutlichen Hervorhebung und Glättung von Blutgefäßen gegenüber dem Linienmaß, es besitzt jedoch den Nachteil, daß das Luminanzprofil, also die Luminanzfunktion in Querrichtung zu den Adern, geebnet wurde, anstatt ein ausgeprägtes Maximum in der Linienmitte zu behalten. Es entsteht der Eindruck der Linienverbreitung.
Geeigneter ist folgender Ansatz:

Normiertes Summenkriterium

$$s(p_o) = \sum_{\{p \in R \mid d(p) \parallel d(p_o)\}} l(p) \quad / \quad const \tag{5}$$

In diesem Fall wird für alle Bildpunkte die Summe durch die gleiche Konstante geteilt. Damit wichtet dieser Operator das Resultat mit der Anzahl der "parallelen Punkte". Dies hat den Vorteil, daß Ränder von Blutgefäßen abgesenkt und "Haarartefakte" weiter unterdrückt werden.
Dieses Kriterium schlagen wir daher zur Nachverarbeitung des Linenintensitätsbildes vor.
Das Ergebnis ist ein geglättetes Intensitätsbild, das innerhalb und außerhalb von Linienstrukturen rauschreduziert wurde.

Die Methode der parallelen Glättung läßt sich auch auf andere Eigenschaften, die als Bild (Pixel-Map) vorliegen, anwenden, z. B. zum Glätten von Original-, Amplituden-, Breiten- oder segmentierten Bildern.

Das geglättete Linienmaß wurde mit dem Canny-Linienfinder [10] verglichen. Visuell beurteilt findet dieser Operator, wenn die Auflösung fein genug eingestellt ist, ebenso schmale Linien und gibt sie verstärkt aus, wie unser Verfolgungsalgorithms. Jedoch sind im Bildhintergrund mehr Artefakte ("Haare") sichtbar. Auch werden breitere Linien bereits geteilt ausgegeben, die unser Linienmaß noch als zusammenhängend ausgibt.
Was die Rechenzeit betrifft, schneidet der Cannyoperator mit 11 Sekunden besser als unser gesamter Verfolgungsalgorithmus mit 50 Sekunden ab[1].
Ein weiterer Vergleich zwischen der parallelen Glättung und der Anisotropen Diffusion [6] bietet sich an. Die Kanten werden allerdings über einen Differentialoperator gefunden, der die Glättung an Kanten dämpft, aber nicht ganz unterbricht.
Ein visueller Vergleich zeigt, daß die parallele Glättung einen schärferen Bildeindruck hinterläßt, da sie die Linienkanten scharf definiert.
Vergleicht man die Rechenzeiten, benötigt die Anisotrope Diffusion etwa 10 Sekunden zur Vorbereitung und etwa 15 Sekunden für jeden Iterationsschritt. Die parallele Glättung kommt dagegen mit 9 Sekunden aus.

3. Aderdurchmesser und Mittellinien

In diesem Kapitel sollen Liniendurchmesser und Mittellinien mithilfe des Richtungsfeldes (2.1.) berechnet werden. Wir nutzen, wie bei der parallelen Glättung, die Eigenschaft, daß das Richtungsfeld in einer Umgebung der Linien transversal zum Linienrand verläuft.

---

[1] Die Rechenzeiten wurden auf einer Sun Spark Workstation an 512 x 512 Pixel-Bildern verglichen.

Es soll an jedem Linienpunkt der *Durchmesser* eingetragen werden. Um Antworten im Bildhintergrund zu vermeiden, erfolgt die Breitenberechnung nur innerhalb des in 2.3. segmentierten Binärbildes.
Der Algorithmus startet an einem Aderpunkt und schreitet pixelweise senkrecht zur Richtung des Startpunktes, so lange, bis die Richtungen der überprüften Punkte nicht mehr parallel zu der des Startpixels sind, wobei die Schritte gezählt werden. Wir wählen dazu das in 2.4. definierte Parallelitätsmaß.
Der Vorgang wird in Gegenrichtung wiederholt und beide Richtungen werden addiert.
Den ermittelten Schrittweiten entsprechen jedoch aufgrund des Pixelrasters verschiedene Gefäßdurchmesser, je nachdem, ob in senkrechter, waagrechter, diagonaler oder schräger Richtung gemessen wurde. Deshalb muß in eine *Euklidische Metrik* umgerechnet werden. Dazu wird zu jeder der 8 Richtungen i ein Skalierungsfaktor e_i, so gewählt, daß durch Multiplikation ein Euklidischer, also unter Rotation des Eingabebildes invarianter, Durchmesser entsteht.

$$(\textit{Durchmesser in Richtung i}) = e_i * (\textit{Schrittweite in Richtung i}). \qquad (6)$$

Mit dieser Definition können extrem schmale Linien im Pixelbereich erfaßt werden, da diese durch das Richtungsfeld vorgegeben sind.
Die berechnete Breite ist unabhängig von Parametern, im Gegensatz z. B. des Cannyoperators, welcher bei Veränderung der Filterbreite und der Schwelle zu unterschiedlichen Linienbreiten führt.
Der Algorithmus arbeitet außerdem schnell (2 sec für ein 512*512 Pixelbild).
Probleme bereiten, wie bei allen maskenbasieten Operatoren, Blutgefäße, die breiter als die Filtermaske sind. Diese werden als Paar von Randlinien aufgefaßt.

Breitenschwankungen, die durch unregelmäßig erfaßte Aderränder entstehen, können durch eine parallele Glättung nach dem Mittelwertkriterium ausgeglichen werden.

Die *Mittellinien* lassen sich in einem Schritt mit den Breiten berechnen.
Wenn für einen Startpunkt beide ermittelten Schrittweiten gleich sind oder sich nur um 1 unterscheiden, wird dieser Punkt als Mittellinienpunkt markiert. Die Differenz um 1 ist notwendig, um bei geradzahligem Durchmesser die Mittellinie zu garantieren. Die auf diese Weise konstruierten Mittellinien können 1 bis 2 Pixel breit sein, je nach ungeradem oder geradem Durchmesser.

4. Schlußfolgerung

Der beschriebene Algorithmus wurde auf verschiedenen medizinischen DSA-Bildern getestet.
Es zeigt sich, daß die optimalen Parameter und Regeln weitgehend unabhängig vom Bildmaterial sind. Es wird daher dem Benutzer nur ein Parameter zur Wahl überlassen, nämlich eine Schwelle über das Amplitudenbild, um die Feinheit der Auflösung vorzugeben.
Gegenwärtig werden die extrahierten Eigenschaften schrittweise in einen Algorithmus zur 3D-Rekonstruktion von Blutgefäßen eingebaut. Sie sollen die Vieldeutigkeit bei der Zuordnung korrespondierender Punkte verringern.

5. Literatur

[1] C. Smets, G. Verbeeck, P. Suetens, and A. Oosterlinck. A knowledge-based system for the delineation of blood vessels on subtraction angiogramms. Pattern Recognition Letters 8(1988), 113-121.

[2] C. Smets. A Knowledge Based System for the Automatic Interpretation of Blood Vessels on Angiograms. Leuven University Press, Leuven (1990).

[3] R. Collorec and J.L. Coatrieux. Vectorial tracking and directed contour finder for vascular network in digital subtraction angiography. Pattern Recognition Letter 8(1988), 353-358.

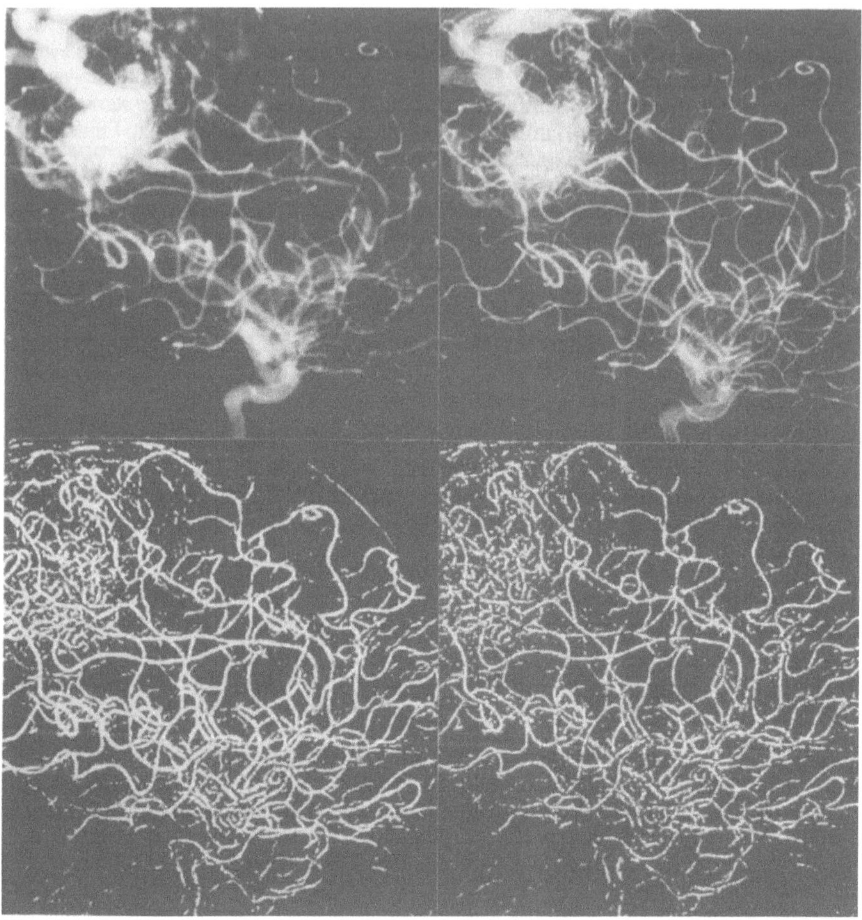

Abb. 5: Links oben: Originale DSA-Aufnahme des Kopfes. Rechts oben: Linienmaß, dem Originalbild überlagert. Links unten: Segmentierte Blutgefäße, Rechts unten: Aus dem Richtungsfeld abgeleitete Mittellinien

[4] J.Y. Catros and D. Mischler. An artificial intelligence approach for medical picture analysis. Pattern Recognition Letters 8(1988), 123-130.

[5] S.A. Stansfield. ANGY: A Rule-Based Expert System for Automatic Segmentation of Coronary Vessels From Digital Subtracted Angiograms. IEEE-PAMI 8(1986), 188-199.

[6] P. Perona and J. Malik. Scale space and edge detection using anisotropic diffusion. In Proc. of the IEEE Conf. on Computer Vision , 1987, pp. 16-22.

[7] K. Waidhas, R. Kutka. Adernextraktion durch iteratives Gradientenmatching in stark verrauschten medizinischen Bildern. Informatik Fachberichte 219, p. 201-9, 1989

[8] Suetens P., Haegemans A., Oosterlinck A., Gybels J. An attempt to reconstruct the cerebral bloodvessels from a lateral and a frontal angiogram. Pattern Recognition 16, 517-524 (1983)

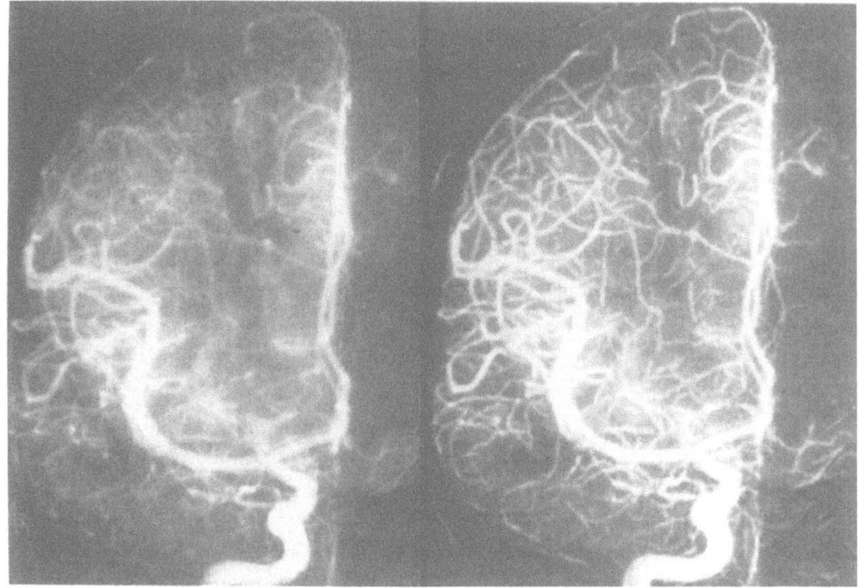

Abb. 6: Links: Originale DSA-Aufnahme des Kopfes.
Rechts: Dem Original überlagertes Linienmaß

[9] Suetens P., Oosterlinck A., Haegmans A., Gybels J. Three-dimensional reconstruction of the blood vessels of the brain. Proceedings of the ISMIII, Int. Symp. on Medical Imaging and Image Interpretation Berlin, 429-435 (1982)

[10] J.F. Canny, Finding edges and lines in images, MIT Artif. Intel. Lab., Cambridge, MA, TR-720, 1983

[11] J.H.C. Reiber, P.W. Serruys, and C.J. Slager. Structural analysis of the coronary and retinal arterial tree. In Quantitative and Coronary and Left Ventricular Cineangiography. Martinus Nijhoff, pp. 185-213, Dordrecht, 1986.

[12] D. J. Stevenson, L. D. R. Smith, G. Robinson,Working towards the automatic detection of blood vessels in X-ray angiograms, Pattern Recognition Letters 6 (1987) 107-112

[13] Rake S. T., Smith L. D. R. The interpretation of x-ray angiograms using a blackboard control architecture. Proceedings of the Int. Symposium CAR, 681-686 (1987)

[14] Y. Sun. Automated Identification of Vessel Contours in Coronary Arteriograms by an Adaptive Tracking Algorithm. IEEE Trans. Med. Imaging, vol. MI-8, pp. 78 - 88, 1989

[15] F. Mokhtarian, A. K. Mackworth. A Theory of Multiscale, Curved-Based Shape Representation for Planar Curves. IEEE PAMI, Vol. 14, pp. 789 - 805, 1992

[16] J. H. Rong, R. Collorec, J. L. Coatrieux, C. Toumoulin. Model Guided Automatic Frame-to-Frame Segmentation in Digital Subraction Angiography. SPIE, Vol. 1137, 1989

[17] D. M. Herrington, M. Siebes, G. D. Walford, R. H. Selzer. Derivative-Based Edge Detection in Quantitative Coronary Angiography is not Independent of Vessel Size. Proceedings Computers in Cardiolagy 1988

[18] D. L. Parker, D. L. Pope, R. van Bree, H. W. Marshall. Three-Dimensional Reconstruction of Moving Arterial Beds from Digital Subtraction Angiography. Compters and Biomedical Research 20 (1987) 166-185

Fraktale in der Analyse von Ultraschallbildern der Leber

Cornelia Zahlten[1], Carl J. G. Evertsz[1] und Heinz-Otto Peitgen[1]
Ivan Zuna[2], Stephan Delorme[2] und Gerhard van Kaick[2]

[1] Centrum für Complexe Systeme und Visualisierung, Universität Bremen, FB 3, Postfach 330 440, D-28334 Bremen
[2] Institut für Radiologie und Pathophysiologie, Deutsches Krebsforschungszentrum, Im Neuenheimer Feld 280, D-69120 Heidelberg

Abstract. Die fraktale Geometrie bietet eine Vielzahl von Möglichkeiten zur quantitativen numerischen Analyse medizinischer Bilddaten. In der vorgestellten Anwendung wird aus Ultraschallaufnahmen der Leber der Grad der Verfettung des Lebergewebes bestimmt, indem die Wahrscheinlichkeitsverteilung von im Bild auftretenden lokalen fraktalen Dimensionen ausgewertet wird. Zur Verifikation der Ergebnisse wird eine aus Computertomographie-Daten gewonnene Quantifizierung des Verfettungsgrades herangezogen. Im Gegensatz zu früheren Texturanalysen, die sich auf kleine Bildausschnitte begrenzen, erzielt unser Verfahren die besten Resultate, wenn das ganze Bild in die Analyse einbezogen wird.

1 Einleitung

Einer der primären Begriffe in der Theorie der Fraktale ist die *Selbstähnlichkeit*, mit der man die Eigenschaft eines Objektes, beziehungsweise eines Musters bezeichnet, dessen Aussehen als Ganzes in jedem kleinen Teil wiedererkennbar ist. Eine mathematische Beschreibung von Selbstähnlichkeit erhält man durch die *fraktale Dimension* D, die man sich in stark vereinfachter Form folgendermaßen vorstellen kann: Hat man ein Objekt, das man als Vereinigung von N kleineren Teilen auffassen kann, von denen jedes eine um den Faktor r verkleinerte Kopie des Originals ist, so bezeichnet D das Verhältnis $D = \frac{\log N}{\log 1/r}$. Wenn keine exakte Selbstähnlichkeit vorliegt, wie es bei fast allen in der Natur auftretenden Fraktalen der Fall ist, ist der Begriff der *fraktalen Massen-Dimension* D_m zutreffender. Betrachtet man einen zufällig gewählten Punkt x einer fraktalen Menge A und mißt den Anteil der Punkte von A innerhalb einer Kugel vom Radius r um x, so wächst dieser Anteil bei wachsendem r wie r^{D_m} [3].

Die hier vorgestellte Arbeit basiert wesentlich auf Ideen von R.F. Voss, der kürzlich Ideen aus der fraktalen Analyse und der Perkolationstheorie miteinander kombiniert, und den Begriff der *lokalen zusammenhängenden fraktalen Dimension* geprägt hat [8]. Sein daraus entwickeltes Verfahren konnte Chinesische Landschaftszeichnungen aus der Zeit 1000AD bis 1300AD erfolgreich in zwei unterschiedliche Epochen einteilen. Für weitere Anwendungen fraktaler Dimensionen in der Bildverarbeitung siehe zum Beispiel [4, 5, 1].

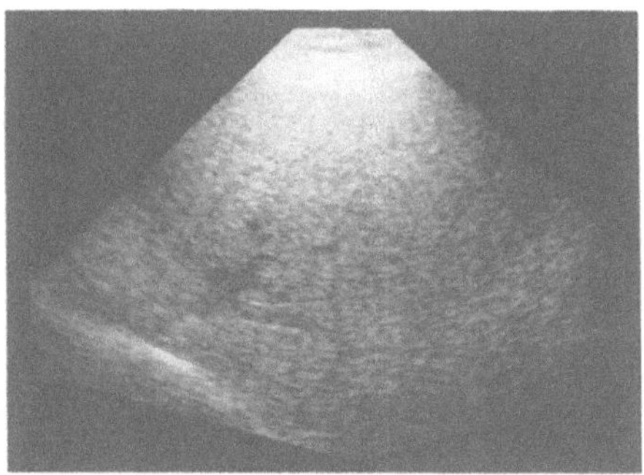
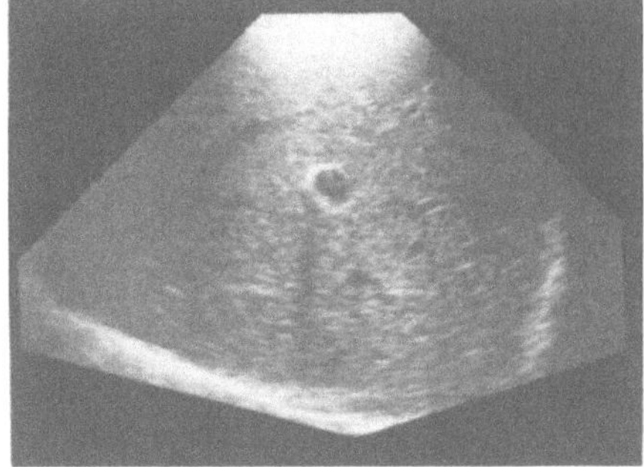

Abb. 1. Links eine Ultraschallaufnahme einer stark verfetteten Leber und rechts einer gesunden Leber.

2 Quantifizierung von Leberverfettung aus Ultraschallaufnahmen

Die Verfettung der Leber entspricht morphologisch der Einlagerung von Fetttröpfchen verschiedener Größe und Verteilung in den Zellen. Es handelt sich um eine unspezifische Reaktion der Leber auf eine Vielzahl von Reizen. Sie kann eine harmlose Veränderung sein oder aber Ausdruck einer ernsten fortschreitenden Lebererkrankung. Dies läßt sich häufig erst laborchemisch klären.

Durchgängiges sonographisches Korrelat einer stärkeren Leberverfettung ist eine homogene Zunahme der Echodichte. Hiermit verbunden ist eine verstärkte Schallabsorption in der Tiefe mit der Folge, daß die schallkopffernen Anteile der Leber dunkler erscheinen und erschwert beurteilbar sind. Der helle Reflex, der normalerweise die Pfortaderäste begleitet, hebt sich schwächer ab. Insgesamt resultiert im optischen Eindruck eine Rarefizierung der intrahepatischen Gefäße. Die Echodichte korreliert nicht stets quantitativ mit dem Verfettungsgrad. Vor allem bei niedrigen Verfettungsgraden ist das sonographische Bild sehr variabel. Dies rührt daher, daß die Echodichte nicht nur Ausdruck des Fettgehaltes sondern auch der Verteilung der Fettröpfchen ist. Im Vergleich zur gesunden Leber kann sonographisch außerdem eine Vergrößerung der Fettleber, eine veränderte Form und eine insgesamt hellere Darstellung der Fettleber festgestellt werden. Das zuverläßigste nicht-invasive Verfahren zur Abschätzung des Verfettungsgrades am Patienten ist die Dichtemessung in der Computertomographie.

Unser Ziel ist es, aus Ultraschallaufnahmen der Leber eine *Quantifizierung* der Verfettung vorzunehmen, wobei wir die Analyse lediglich auf die Gewebestruktur, die sichtbaren Gefäße und Ränder und das Echomuster beschränken. Da der Rand auf den Aufnahmen nur teilweise zu erkennen ist, kommt eine Formanalyse nicht in Betracht. Die Helligkeit soll ebenfalls kein Kriterium sein, da sie einerseits nicht ausschließlich durch den Fettgehalt beeinflußt wird und wir andererseits möglichst unabhängig von der Geräteeinstellung bleiben wol-

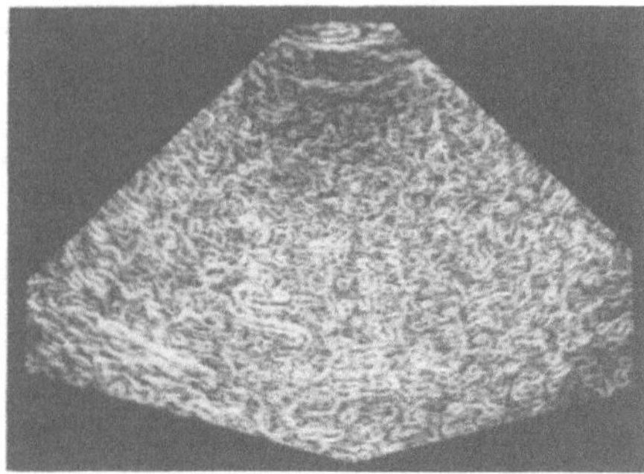
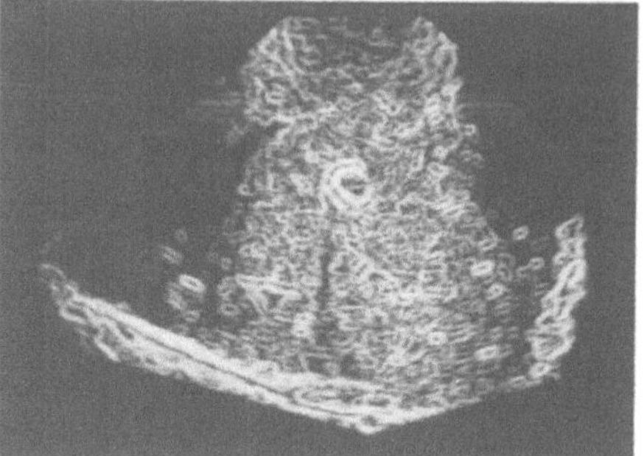

Abb. 2. Resultat nach Anwendung der Mittelwertglättung und des normierten Gradientenoperators auf die Aufnahmen aus Abb. 1.

len. Eine interessante und in der Literatur bisher nicht eindeutig geklärte Frage ist es, inwieweit die lokale Struktur des Echomusters eine klinische Bedeutung besitzt.

3 Verfahren zur Quantifizierung

3.1 Bildvorverarbeitung

Die Analyse soll der Schallabschwächung und der Echostruktur in Ultraschallaufnahmen Rechnung tragen. Als Vorverarbeitungsstufen werden die Aufnahmen deshalb zwei Transformationen unterzogen.

Um möglicherweise von der Aufnahmetechnik verursachtes Rauschen zu unterdrücken, verwenden wir eine *Mittelwertglättung zum Radius* $r = 3$. Ein großer Radius (z.B. $r = 10$) kann genutzt werden, um das lokale Echomuster zu verwischen und zu untersuchen, ob relevante Unterschiede durch Echoflecken (Speckles) oder aber durch größere Strukturen verursacht werden. Zur Kantenverstärkung von Leberrändern, Gefäßen und Speckles wird auf das geglättete Bild ein *normierter Gradientenoperator* angewandt. Zur Normierung wird die Länge des Grauwertgradienten durch den Intensitätswert des betrachteten Pixels geteilt. Dies hat den Vorteil, daß das Ergebnis unabhängig von der globalen Intensität eines Bildes ist. Das Gradientenbild ist somit unabhängig von der Einstellung des Intensitätsreglers am Ultraschallgerät, sofern man davon ausgehen kann, daß die Intensität der Ultraschallausgabe im interessierenden Bereich linear von der Einstellung abhängt. Als weiteres Ergebnis der Normierung wird die ansonsten im Vergleich zu den Gefäß/Leber- und Leberrändern schwächere Intensität der Speckle-Strukturen auf gleiches Intensitätsniveau angehoben.

Abbildung 2 zeigt die Aufnahmen aus Abb. 1 nach einer Mittelwertglättung vom Radius $r = 3$ und normierter Gradientenbildung. Es zeigt sich, daß auch Fettleberaufnahmen große Anteile hoher Gradientenwerte aufweisen können. In der gesunden Leber bilden aber die zu hohen Gradienten gehörigen Pixel größere, meist längliche, zusammenhängende Strukturen. Eine fraktale Analyse, die Aussagen zur lokalen Geometrie im Gradientenbild macht, kann dies präzisieren.

3.2 Lokale zusammenhängende fraktale Dimension

Zur Berücksichtigung zusammenhängender Strukturen bestimen wir die Zusammenhangskomponenten (Cluster) eines Bildes mithilfe des im Rahmen der Perkolations-Theorie entwickelten Hoshen-Kopelmann-Algorithmus [7]. Dabei bezeichnen wir zu gegebenem Schwellwert m_0 zwei benachbarte Pixel als *zusammenhängend*, wenn ihre Intensitäten beide größer als m_0 sind. Die *lokale zusammenhängende fraktale Dimension* D_{conn} ergibt sich dann folgendermaßen:

Man bestimmt zu jedem Punkt x des Bildes die Anzahl der Punkte des zu x gehörenden Clusters, die innerhalb eines Fensters $B_r(x)$ vom Radius r um x liegen. Bezeichnet man diese Anzahl mit $C_x(r)$, ist die Dimension D_{conn} durch das Verhältnis

$$C_x(r) \sim r^{D_{conn}}$$

definiert. Praktisch berechnet man das Verhältnis durch mehrere Schritte in r und akzeptiert einen Wert für D_{conn} nur dann, wenn eine lineare Regressions-Anpassung von $log\ C_x(r)$ versus $log\ r$ ohne große Fehler möglich ist. Für die in Kapitel 4 angegebenen Resultate wurden als Radien $r = 1, 2, 4$ verwendet. Dies entspricht Fenstern von 3x3, 5x5 und 9x9 Pixeln. Um in Nähe des Bildrandes noch Aussagen treffen zu können, berücksichtigt man den Anteil des Fensters, der noch im Bildbereich liegt.

Zur Wahl eines günstigen Schwellwertes m_0 haben wir zwischen den minimalen und maximalen Intensitätswert der Gradientenbilder 20 Schwellwerte $m_1, \ldots, m_{20}$ gleichen Abstands gelegt. Zu jedem m_i wurden die Cluster bestimmt und die Dimensionen D_{conn} berechnet. Für eine sehr niedrige Schwelle, beispielsweise m_1, wird das ganze Bild aus im wesentlichen nur einem Cluster bestehen. Als lokale zusammenhängende fraktale Dimension D_{conn} berechnet man deshalb für fast alle Punkte $D_{conn}(x) = 2$, da die lokale Geometrie des umgebenden Clusters für alle diese Punkte eine Fläche ist. Dagegen werden oberhalb des Schwellwertes m_{20} nur noch wenige vereinzelte Punkte liegen. Für die meisten dieser Punkte wird man deshalb $D_{conn}(x) = 0$ berechnen. Zu den beiden extremen Schwellen m_0 und m_{20} erhält man deshalb jeweils für alle untersuchten Leberbilder im wesentlichen gleiche Häufigkeitsverteilungen der Dimensionen D_{conn}. Es gibt aber dazwischenliegende Schwellwerte, für die deutliche Differenzen in den Verteilungen erkennbar sind. Zur Quantifizierung des Verfettungsgrades von Lebergewebe wählen wir dasjenige m_i, für das sich die Verteilungen der Dimensionen am stärksten unterscheiden.

3.3 Auswertung der Wahrscheinlichkeitsverteilung von D_{conn}

Abbildung 3 zeigt die Wahrscheinlichkeitsverteilungen der Dimension D_{conn} (bezgl. der Schwelle m_5) von 19 untersuchten Patienten, wobei pro Patient 3 Ultraschallaufnahmen in die Analyse einbezogen wurden. Die Verteilung wurde

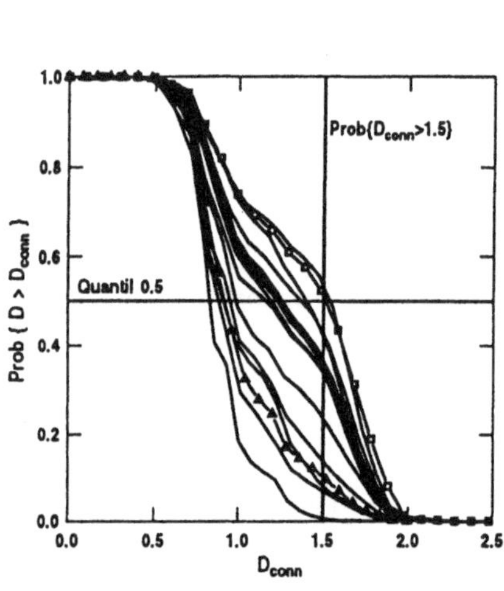

Abb. 3. Wahrscheinlichkeitsverteilungen der Dimension D_{conn} für 19 verschiedene Lebern. Die Häufigkeiten sind bei hohen Dimensionen beginnend aufaddiert und die Summe auf 1 normiert.

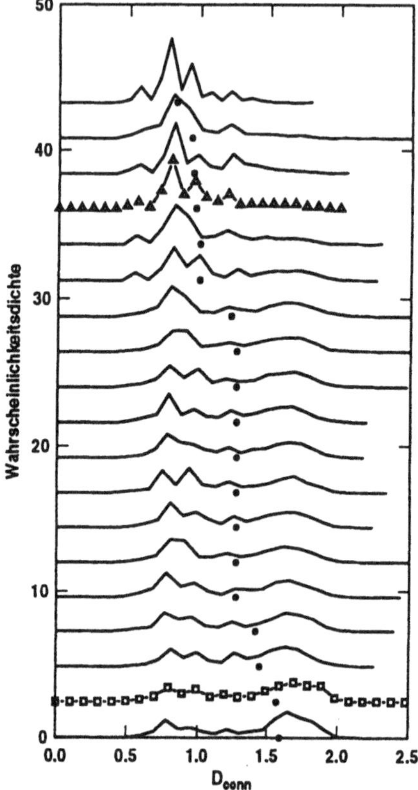

Abb. 4. Dichte der Dimensionen D_{conn} aus Abb.3, durch Addition mit einer Konstanten übereinander abgebildet. Die Kurven sind nach dem Quantil 0.5 sortiert (runde Markierung).

näherungsweise durch Addition der Häufigkeiten $h(D)$ (beginnend mit den hohen Dimensionen) und durch Normierung der Summe auf Eins bestimmt. Zur Auswertung der Verteilungen betrachten wir Quantile, wobei wir unter einem Quantil q_a diejenige Dimension D^* verstehen, für die

$$\int_{D^*}^{\infty} h(D)dD = a$$

gilt. Beispielweise läßt sich den in Abb. 3 dargestellten Verteilungen durch das Quantil $q_{0.5}$ eine eindeutige Reihenfolge zuordnen. Darüberhinaus gibt der maximale Abstand korrespondierender Quantile ein Maß für die Güte des gewählten Schwellwertes. Mit anderen Worten: Betrachtet man zu jedem Schwellwert m_i die 19 zugehörigen Verteilungs-Kurven, so zieht man zur Quantifizierung dasjenige m_i heran, für das die Kurven am weitesten auseinanderklaffen. In den betrachteten Fällen war dies stets der Schwellwert m_5.

Die Verteilungen aus Abb. 3 sind nicht-kumulativ und nach dem Quantil $q_{0.5}$ sortiert in Abb. 4 nochmals dargestellt. Die mit Quadraten markierte Kurve entspricht der gesunden Leber und die mit Dreiecken markierte Kurve entspricht der stark verfetteten Leber aus Abb. 1. Man erkennt neben der Verschiebung von D^* deutlich eine Änderung der Kurvenform und eine Verschiebung des Peaks, der für die verfettete Leber bei deutlich niedrigeren Werten von D_{conn} liegt als für die gesunde.

Anstelle von Quantilen kann man die relative Anzahl der Bildpunkte, deren Dimension oberhalb einer gewissen Schranke liegt, verwenden, um den Verteilungen eine eindeutige Reihenfolge zuzuordnen. In Abb. 3 wird eine solche Reihenfolge durch die senkrechte Linie bei $D_{conn} = 1.5$ illustriert.

4 Anwendung und Resultate

Die von uns betrachteten Ultraschalldaten wurden am Deutschen Krebsforschungszentrum mit einem 3.5 MHz Schallkopf bei konstanter Geräteeinstellung aufgenommen. Die Grauwertauflösung beträgt 8 Bit. Die Bildabmessung beträgt 640x480 Pixel. Es wurden jeweils 3 transversale Ultraschallaufnahmen der Leber von 19 Patienten ausgewertet. Alle Aufnahmen zeigen im wesentlichen nur Lebergewebe, Teile des Leberrandes, des Zwerchfells und Gefäße im Inneren der Leber. Benachbarte Organe sind nicht abgebildet.

Die Analyse wurde sowohl mit einer Mittelwert-Glättung zum Radius $r = 3$ als auch zum Radius $r = 10$ durchgeführt. Im letzteren Fall wurde die Speckle-Struktur zerstört, um zu prüfen, ob auch dann noch eine Differenzierung zwischen gesundem und fetthaltigem Lebergewebe möglich ist. Zur genaueren Einschätzung des Beitrags der Speckle-Struktur zur Klassifizierung wurden gleiche Analysen für kleine Bildregionen, die im wesentlichen nur lokale Echomuster enthalten, durchgeführt. Dazu wurden aus dem Original alles Gewebe außerhalb der Leber, die Leberränder sowie erkennbare Gefäße innerhalb der Leber ausgeblendet.

Zur Verifizierung der aus den Wahrscheinlichkeitsverteilungen der Dimensionen D_{conn} resultierenden Quantifizierung des Fettgehaltes der untersuchten Lebern wurden unsere Ergebnisse mit einer korrespondierenden Auswertung von Computertomographie-Aufnahmen (CT) der gleichen Patienten korreliert. Die Dichtemessung per CT ist das bisher zuverlässigste nicht-invasive Verfahren zur Abschätzung des Verfettungsgrades, und die gemessenen Hounsfield-Einheiten entsprechen nach [9] linear dem Grad der Leberverfettung. Zur Korrelationsmessung wurden zwei Maße verwendet. Die *Pearson-Korrelation*, die die Stärke des linearen Zusammenhangs zwischen zwei Meßreihen mißt, und die *Spearman-Korrelation*, die die Rangfolge der Meßwerte auswertet (siehe z.B. [2]). Da wir nicht annehmen können, daß die von uns getroffene Quantifizierung in linearem Zusammenhang mit dem Grad der Leberverfettung steht, ist die Spearman-Korrelation von Bedeutung.

Tabelle 1 faßt die durchgeführten Analysen und deren Korrelation mit den aus den Referenz-CT-Aufnahmen erhaltenen Hounsfield-Einheiten zusammen.

Die Anzahl der einbezogenen Fälle beträgt in allen Analysen $n = 18$. In der linken Spalte der Tabelle sind die verschiedenen, oben beschriebenen Analyseverfahren angegeben. Zum Vergleich mit bisher gebräuchlichen Verfahren sind in die Tabelle ein Helligkeitsparameter (mittlere Intensität in ausgewählten Regions of Interest) und ein Mikrostrukturparameter (Kontrast in ausgewählten Regions of Interest) aufgenommen worden. Diese und verwandte Texturparameter wurden in Kombination zu mehreren erfolgreich zur Fettleberquantifizierung verwendet [6].

Tabelle 1. Korrelation der Ergebnisse zu CT-Hounsfield-Einheiten

Analyse-Verfahren	Pearson-Korr.	Spearman-Korr.
Mittelwert-Glättung zu $r = 3$, relative Anzahl der Pixel mit $D_{conn} \geq 1.5$	0.6012 (**)	0.7415 (**)
Mittelwert-Glättung zu $r = 10$, relative Anzahl der Pixel mit $D_{conn} \geq 1.5$	0.6206 (**)	0.5698 (*)
Mittelwert-Glättung zu $r = 3$, Quantil 0.5	0.5961 (**)	0.6311 (**)
Mittelwert-Glättung zu $r = 10$, Quantil 0.9	0.4699 (*)	0.3909
Bildausschnitt ohne Rand/Gefäße, Mittelwert-Glättung zu $r = 3$, Quantil 0.1	0.6289 (**)	0.23539
mittlere Helligkeit	-0.45532	-0.33506
Kontrast	-0.41299	-0.21406

Zu jeder Analyse wird sowohl der Wert der Pearson- als auch der Spearman-Korrelation aufgeführt. Die Stärke der Korrelation ist mit (*) gekennzeichnet, falls die Ablehnwahrscheinlichkeit der Hypothese, daß eine Korrelation vorliegt, zu $p < 0.05$ berechnet wurde. Sie ist für $p < 0.01$ mit (**) markiert.

Es zeigt sich, daß beide Verfahren, die das ganze Bild einbeziehen und denen eine Mittelwertglättung vom Radius $r = 3$ zugrundeliegt, jeweils eine bessere Spearman-Korrelation erzielen als die korrespondierenden Analysen mit Glättung zum Radius 10. Obwohl wir eine Spearman-Korrelation von 0.74, wie sie als bisher bestes Ergebnis erzielt wurde, noch nicht als sehr starke Korrelation werten wollen, zeigt sich doch, daß sogar alle vier Analysen, die sich auf das ganze Bild beziehen, jeweils bessere Korrelations-Koeffizienten und ebenfalls eine höhere Güte der Korrelation erreichen als die beiden einzelnen Texturparameter Helligkeit und Kontrast. Die am ehesten mit den Texturparametern vergleichbare Analyse in der fünften Zeile der Tabelle, die sich auf vergleichbare Bildausschnitte ohne Leberrand und ohne Gefäße bezieht, zeigt ähnliche Ergebnisse wie der Texturparameter Kontrast.

Was die Bedeutung der Speckle-Struktur angeht, so erhält man eine dramatische Verschlechterung der Ergebnisse, wenn man die Analyse auf Bildteile beschränkt, die im wesentlichen nur Echomuster enthalten. Hieraus kann man schließen, daß die relevante Information nicht von den Speckles sondern von größeren Strukturen wie Rändern und Gefäßen getragen wird. Andererseits zeigen die Korrelations-Koeffizienten etwas niedrigere Werte, wenn die Speckle-Struktur durch eine Mittelwertglättung vom Radius 10 verwischt wird. Eine Erklärung hierfür ist, daß durch eine solche Glättung nicht nur das Echomuster sondern auch anderere feine Strukturen wie etwa kleine Gefäße und derern Ränder verwischt werden. Ausserdem ist anzumerken, daß sämliche Radien, die zur Berechnung der lokalen zusammenhängenden fraktalen Dimensionen herangezogen wurden, kleiner als $r = 10$ sind.

Zur Quantifizierung des Verfettungsgrades haben wir die bei weitem besten Ergebnisse erzielt, wenn das gesamte Bild in die Auswertung einbezogen wurde. Wir sind deshalb von der Untersuchung kleiner Geweberegionen zu einer globalen Betrachtung der Ultraschallbilder übergegangen.

5 Ausblick

Die vorgestellten Analyse-Methoden sind nicht auf das spezielle Problem der Fettleber-Diagnose aus Ultraschallaufnahmen begrenzt. Auch wenn die Fettleberstudie bisher noch nicht abgeschlossen ist, so zeigt sich doch schon jetzt, daß die fraktale Geometrie ein vielfältiges Werkzeug zur Bilddatenanalyse bereitstellt.

Danksagung. Diese Arbeit basiert wesentlich auf Ideen von Richard F. Voss und wurde angeregt durch seine Gastaufenthalte am Centrum für Complexe Systeme und Visualierung in Bremen, durch viele Gespräche und insbesondere durch seine Arbeit [8].

Literatur

1. Evertsz, C.J.G., Zahlten, C., Peitgen, H.-O., et al.: Distribution of fractal dimensions and the degree of liver fattiness from ultrasound. In: Fractals in Biology and Medicine. Proc. Conf. Ascona, Switzerland, 1993
2. Hays, W.L.: Statistics. Holt, Rinehart and Winston, London, 1963
3. Mandelbrot, B.B.: The fractal geometry of nature. W.H. Freeman, 1982
4. Müssigmann, U.: Homogeneous fractals and their application in texture analysis. In: H.-O. Peitgen, J.M. Henriques, L.F. Penedo (eds.) Fractals in the fundamental and applied sciences, North Holland, Amsterdam, 1991, 269-283
5. Pentland, A.P.: Fractal-based description of natural scenes. IEEE Trans. on Pattern Analysis and Machine Intelligence **6** (1984) 661-674
6. Raeth, U., Schlaps, D., Limberg, B. et al.: Diagnostic accuracy of computerized B-scan texture analysis and conventional ultrasonography in diffuse parenchymal and malignant disease. J. Clin. Ultrasound **13** (1985) 87-99
7. Stauffer, D., Aharony, A.: Introduction to Percolation Theory. Taylor&Francis, London, 1985
8. Voss, R.F.: The local connected fractal dimension: multifractal classification of early Chinese landscapes drawings. In: R.A.Earnshaw (ed.) Application of Chaos and Fractals, Springer-Verlag, London, 1992
9. Wegener, O.H.: Ganzkörpertomographie. Blackwell-Wissenschafts-Verlag, 1992

Neuronal, Statistisch, Wissensbasiert: Ein Beitrag zur Paradigmendiskussion für die Mustererkennung

Gerhard Sagerer

Universität Bielefeld, Technische Fakultät, AG Angewandte Informatik, Postfach 100131, 33501 Bielefeld

Abstract. Spätestens mit dem DAGM-Symposium 1991 wurde der Paradigmenstreit offenkundig. In dem Beitrag wird versucht, über die Zielsetzungen und die Motivation jedes Ansatzes einen Weg zur Integration zu eröffnen. Den Ausgangspunkt soll eine Selbstdefinition der Disziplin "Mustererkennung" bilden. Statt einzelne Verfahren vorzustellen oder zu vergleichen, werden die grundsätzlichen Vorgehensweisen erläutert und diskutiert. In der Konsequenz zeigt sich, daß alle drei Paradigmen in ihrer Symbiose die Entwicklung des Forschungsgebiets verstärken können.

1 Was ist Mustererkennung?

Vielleicht erscheint es zunächst überflüssig, in einem Vortrag auf der DAGM diese Frage zu stellen. Aber der Versuch, eine Diskussion über Paradigmen der Mustererkennung zu führen, zwingt dazu, von den Zielsetzungen auszugehen, die sich die Disziplin selbst gibt. Es steht mir sicherlich nicht zu, hier eine umfassende Antwort geben zu wollen, geschweige denn zu können. Ich möchte jedoch versuchen, mich der Frage in der Tradition der DAGM und der ICPR zu nähern. Auf diesem DAGM-Symposium sollen behandelt werden: "Grundlagen, Signalverarbeitung, Spracherkennen und Sprachverstehen, Bildverarbeitung, theoretisch und praktisch, Bildverstehen, Einsatz von Wissen, Neuronale Netze, Anwendungen sowie Hardware-Architekturen und Software-Werkzeuge". Die ICPR 92 war "one 'umbrella' conference consisting of a number of speciality conferences" [Gel92], im einzelnen "Computer Vision and Applications", "Pattern Recognition Methodology and Systems", "Image, Speech, and Signal Analysis", und "Architectures for Vision and Pattern Recognition". Ist die Mustererkennung also nur noch ein Dach, unter dem sich verschiedene Aktivitäten subsumieren?

Folgt man der Definition "Forschungs- und Entwicklungsaktivitäten, welche die mathematisch-technischen Aspekte der Perzeption betreffen, sind das Gebiet der Mustererkennung im weiten Sinne" [Nie83], so zeigt sich, daß sich die Disziplin als eine Zielvorstellung anstatt eines Methodenrepertoires mit potentiellen Anwendungen versteht. In die gleiche Richtung zielt die Aussage: "Die Mustererkennung gehört zu den Bemühungen der modernen Informationsverarbeitung, Wahrnehmungsleistungen zu automatisieren, wie wir sie sonst nur von natürlichen Vorbildern kennen. Damit bringen wir die Biologie und uns selbst ins Spiel" [Sch92]. Im gleichen Text wird aber auch eine Abgrenzung vorgenommen: "Die Mustererkennung hat sich auch nicht zum Ziel gesetzt, aufzuklären, wie die Biologie derartiges zustande bringt. Die Mustererkennung ist eine technische Disziplin und bemüht sich, Wahrnehmungsleistungen für bestimmte Zwecke mit technischen Mitteln zu realisieren. Sie ist eher den Ingenieurwissenschaften zuzuordnen als den Kognitionswissenschaften." Der bestimmte Zweck, die Anwendung, wird als eine zentrale Komponente einbezogen. Mit der auf eine Zielvorstellung hin ausgerichteten Definition öffnet sich die Mustererkennung für beliebige Methoden, sofern sie dem Zweck der Erkennung und Analyse sensoriell aufgenommener Daten dienen. Dementsprechend vielfältig ist die Methodenbasis: "Statistik, Kommunikationstheorie, Linguistik, Automatentheorie, Informationstheorie, Regelungs- und Systemtheorie, Untersuchung von Nervennetzen, heuristische Methoden" [Nie74], "Neuro-, Kognitionswissenschaften, Informationstechnik, Mathematik, Statistik, künstliche Intelligenz" [Sch92].

Mustererkennung zeigt sich so als die Anwendung sehr unterschiedlicher Basistechniken, deren Spezialisierung und Weiterentwicklung in Bezug auf ein vorgegebenes Ziel. Und mit diesem Ziel steht sie als Ingenieursdisziplin im Spannungsfeld zwischen biologischer Kybernetik, Kognitionspsychologie und den Anwendungen. Dabei lösen erfolgreiche Anwendungen meistens eine Methodenorientierung aus, während in erheblich geringerem Umfang Paradigmen mögliche Anwendungen motivieren.

Trotzdem oder gerade deshalb stand das DAGM-Symposium 1991 unter dem Schwerpunktthema "Vergleich der Erkennungsleistung Neuronaler Netze und Wissensbasierter Systeme". Trotzdem, weil damit eine Paradigmendiskussion initiiert wurde, und gerade deshalb, da die Erkennungsleistung das Kriterium des Vergleichs sein sollte. In der Themenstellung ausgeklammert wurde das klassische statistische Verfahrensrepertoire, das jahrelang die Symposien prägte. Beiträge, die sich gerade diesem Vergleich widmeten, stießen aber auf sehr große Resonanz.

Ich möchte im folgenden versuchen, eine Einordnung der drei Paradigmen, die aktuell die Mustererkennung prägen, in Bezug auf die skizzierte "Definition" der Disziplin vornehmen. Auch wenn ich mich bei den Zitaten beschränkt habe, so spiegeln sie doch mehr als die Meinung der beiden Autoren wieder. Die Tendenz entspricht der Herangehensweise zahlreicher Autoren seit den 60'er Jahren, wie [Seb62, Fu68, Wat69, Ros69, Dud72, Pat72, Fu76].

2 Motivation und Ziele aus der Sicht der Paradigmen

Jedes der drei Paradigmen — neuronale Netze, statistische Verfahren, wissensbasierte Systeme — basiert auf verschiedenen Annahmen, Voraussetzungen und Forderungen. Diese bestimmen teilweise implizit, teilweise explizit die Vorgehensweise für eine Problemlösung. Andererseits stehen diesen Basispostulaten Zieldefinitionen gegenüber, die wiederum Vorgehensweisen und damit letztendlich gewisse Systemarchitekturen präferieren und andere ausschließen. Da jedes Paradigma auch Anwendungen außerhalb der Mustererkennung besitzt, muß für eine Diskussion in Bezug auf die Mustererkennung zunächst eine Festlegung der Grundlagen und Ziele der Mustererkennung selbst erfolgen.

Das Ziel ist, Wahrnehmungsleistungen zu automatisieren. Für den Zweck der Perzeption ist die "Umwelt" eines Mustererkennungssystems durch physikalisch meßbare Größen beschränkt. Die Gesamtheit all dieser Größen zu betrachten ist sowohl technischen als auch biologischen Systemen unmöglich. Sinnvoll — und zumindest auf absehbare Zeit nur möglich — ist die Beschränkung auf ein begrenztes Anwendungsgebiet, den **Problemkreis** Ω. Die potentiellen Elemente eines Problemkreises werden **Muster** genannt. Als Messungen lassen sich Muster als analoge oder diskrete Funktionen $\mathbf{f}(\mathbf{x})$ charakterisieren, die sowohl in ihren Argumenten wie Ort und Zeit als auch in ihrem Wertebereich mehrdimensionale Größen erfassen. Meist wird angenommen, daß für einen Problemkreis die Zahl der Komponenten von $\mathbf{f}$ und $\mathbf{x}$ konstant ist. Diese Einschränkung ist jedoch nicht systemweit erforderlich. Auch Problemkreise wie Kombination Sehen/Hören sind nicht ausgeschlossen. Zumindest für gewisse Methodengruppen im signalnahen Bereich gilt hier die Restriktion der gleichen Dimensionalität für die beiden Subsysteme. Ein Muster ist aber an einen Problemkreis gebunden. Innerhalb des Problemkreises besitzt es eine Bedeutung. "Das tägliche Leben konfrontiert uns mit einer Unzahl von Mustern. Unsere Sinne nehmen bestimmte Signale aus der Umwelt auf: Wir sehen graphische, bildhafte Muster, hören Klangmuster und assoziieren damit Begriffe, erkennen Bedeutungen" [Sch92]. Ein Muster trägt somit inhärent eine Bedeutung, eine Beschreibung in einem Problemkreis. Zu betrachten ist es also als ein Tripel bestehend aus Messung, Problemkreis und Beschreibung

$$\mathbf{M} = (\mathbf{f}(\mathbf{x}), \Omega, \mathbf{B}) \tag{1}$$

Vorausgesetzt wird nicht, daß für einen Problemkreis alle Muster bekannt sind. Es wir "nur" postuliert, daß zur Sammlung von Informationen eine repräsentative **Stichprobe** ω zur Verfügung steht. Für die Muster dieser Stichprobe sollen die Meßdaten, die Beschreibung und natürlich ihre Zugehörigkeit zum Problemkreis bekannt sein. Die Art der Beschreibung ist aber nicht vom Problemkreis, sondern auch von der Aufgabenstellung des Systems abhängig. Die einfachste Beschreibung ist die Benennung eines Musters als Gesamtheit durch einen Begriff. Ein Muster wird einer Klasse zugeordnet. Wenn für eine Anwendung nur dieser Klassenname von Interesse ist, liegt ein **einfaches Muster** vor. Wird jedoch eine individuelle symbolische Beschreibung gefordert, dann redet man von **komplexen Mustern**. Sie bestehen aus mehreren einfachen Mustern und die Beschreibung enthält die Benennungen der enthaltenen einfachen Muster und deren Beziehungen untereinander.

Die Domäne der statistischen Mustererkennung ist die **Klassifikation**, das heißt, die Meßdaten $\mathbf{f}(\mathbf{x})$ werden als Ganzes einer Klasse Ω_k zugeordnet. Der Name Ω_k ist die Beschreibung $\mathbf{B}$. Ein Muster wird genau einer Klasse zugeordnet und der Problemkreis Ω ist durch eine festgelegte Anzahl von Klassen vollständig charakterisiert. Die Meßdaten eines Musters werden als Repräsentanten eines stochastischen Prozesses betrachtet, der eine Klasse Ω_k erzeugt. Es wird postuliert, daß die Muster Merkmale besitzen, die für die Zugehörigkeit zu einer Klasse charakteristisch sind, daß die Merkmale einer Klasse kompakt im Merkmalsraum liegen, und daß die Merkmale verschiedener Klassen getrennte Bereiche in diesem Raum einnehmen. Die Klassen und somit die Bedeutung von Mustern werden statistisch im Merkmalsraum charakterisiert. Der Klassifikator ordnet dann ein Muster einer Klasse zu, indem sein Merkmalsvektor in einen Entscheidungsraum abgebildet wird, der die Klassenzugehörigkeit festlegt. Der Zusammenhang zwischen Meßdaten und Benennungen wird durch Beispiele — die Stichprobe — beschrieben. Es wird die **Lernstichprobe** zur Optimierung der Abbildung von der **Teststichprobe** unterschieden. Mit letzterer überprüft man die Leistung des Klassifikators anhand statistischer Maße, z.B. Fehlerrate, Konfidenzen.

Die Generierung individueller symbolischer Beschreibungen für ein komplexes Muster gründet sich auf Annahmen über dessen Struktur. Ein solches Muster besitzt einfachere Bestandteile, die zueinander in bestimmten Relationen stehen. Bestandteile in diesem Sinne können vollständige Objekte aber auch Objektkomponenten sein. Die Anordnung der Komponenten als auch die Lage der Objekte zueinander ist durch den Problemkreis eingeschränkt. Hier setzt das Paradigma "Wissensbasiert" an. Die Restriktionen werden als Wissen über den Problemkreis angesehen. Durch explizite Repräsentation in geeigneten Formalismen können die strukturellen Informationen über den Problemkreis und über die zugehörigen Muster für die automatische Generierung einer individuellen symbolischen Beschreibung genutzt werden. Mit Fragestellungen zur Repräsentation und Verarbeitung von Wissen sind die "Prinzipien der künstlichen Intelligenz" angesprochen [Ric89]. Damit sind die generellen Kriterien einer adäquaten Wissensrepräsentation — hinreichende Ausdrucksstärke, Uniformität, Strukturerhaltung und Effizienz — für die Mustererkennung zu untersuchen. Die Repräsentation soll sich dabei an der Kognitionsforschung ausrichten: "Wie organisieren Menschen ihre Gedanken in rationaler Weise und formulieren sie umgangssprachlich" [Ric89].

Nach [Rit90] beschränkt sich der Einsatz künstlicher Intelligenz gerade wegen dieses Rationalitätsanspruchs "auf enge und sehr genau vorhersehbare Problembereiche und weist auf einen fundamentalen Unterschied zu biologischen Nervensystemen hin". Die Nutzung des "biologischen Know-hows" ist der Fundamentalansatz der Neuroinformatik. Die Funktionsprinzipien biologischer Gehirne sollen als Grundlage für technische Systeme dienen. Damit sind die Fragestellungen der Mustererkennung ein wichtiger Teilaspekt der Neuroinformatik. Die eingesetzten Architekturen und Algorithmen müssen hier jedoch durch biologisch experimentelle Ergebnisse begründbar sein.

3 Die Basistechniken

Jedes Paradigma stellt ein Methodenrepertoire und Architekturprinzipien für Mustererkennungsprobleme bereit. Häufig scheitert eine Bewertung der unterschiedlichen Ansätze an den verschiedenen Begrifflichkeiten, die in den Paradigmen gewählt werden. In diesem Abschnitt will ich versuchen, die grundsätzlichen

Techniken kurz zu skizzieren. Um anschließend vergleichen zu können, wird, soweit möglich, eine einheitliche Notation gewählt.

3.1 Klassische statistische Methoden

Charakterisiert sind diese Techniken dadurch, daß Ereignisse als Entitäten behandelt werden. Eine Zerlegung in Bestandteile, denen wieder Begriffe zugeordnet werden, findet nicht statt. Falls diese Verfahren auf komplexe Muster angewendet werden sollen, muß eine Segmentierung vorab erfolgen. Jedes Segmentierungsobjekt wird dann als Entität betrachtet und isoliert einer Klasse zugeordent, ohne daß andere Bestandteile des komplexen Musters berücksichtigt werden. Die Meßdaten der zu klassifizierenden Entität werden in Form eines Merkmalsvektors **c** berücksichtigt, der einen Punkt im N-dimensionalen Merkmalsraum repräsentiert. Die Architektur eines Gesamtsystems zur Klassifikation ist in Bild 1 gegeben. Im folgenden wird ausschließlich die Komponente **Klassifikation** betrachtet.

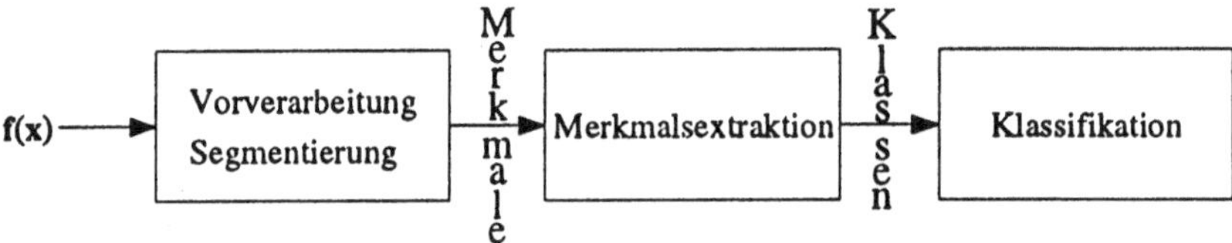

Bild 1 Systemarchitektur statistischer Klassifikatoren

Gesucht ist eine Entscheidungsfunktion, die einem Merkmalsvektor einen Klassenindex zuordnet, das heißt

$$D(\mathbf{c}) = k, \quad k = 1, \ldots, K \tag{2}$$

Zur mathematischen Optimierung dieser Funktion auf der Grundlage einer Lernstichprobe und geeigneter Kriterien erweist sich die Verwendung von **Entscheidungsvektoren** $\mathbf{d}(\mathbf{c})$ als günstigere Formulierung, wobei gilt:

$$\mathbf{d}(\mathbf{c}) = (d_1(\mathbf{c}), \ldots, d_K(\mathbf{c})) \quad \text{mit} \quad \sum_{k=1}^{K} d_k(\mathbf{c}) = 1$$
$$D(\mathbf{c}) = argmax_k\{d_k(\mathbf{c})\} \tag{3}$$

Die Wahl des Optimierungskriteriums bestimmt die Funktionen d_k. Die beiden "klassischen Ansätze" — Minimierung einer Kostenfunktion bzw. Approximation der durch die Lernstichprobe gegebenen idealen Entscheidungsfunktionen — werden im folgenden kurz skizziert.

Für die Kostenminimierung wird vorausgesetzt, daß die klassenabhängigen Dichtefunktionen $p(\mathbf{c} \mid \Omega_k)$, die a priori Wahrscheinlichkeiten der Klassen p_k

sowie Kosten für Fehlklassifikationen $r_{kl}, 0 \leq r_{kk} < r_{kl} \leq 1$ bekannt sind. Dabei sind r_{kl} die Kosten für eine Klassifikation nach k, falls die Klasse l vorlag. Die durch einen Entscheidungsvektor entstehenden mittleren Kosten sind dann bestimmt durch

$$V(\mathbf{d}) = \sum_{k=1}^{K} p_k \sum_{l=1}^{K} r_{lk} \int p(\mathbf{c}|\Omega_k) d_l(\mathbf{c}) d\mathbf{c} \tag{4}$$

Diese Funktion nimmt ihr Minimum an, wenn

$$d_k(\mathbf{c}) = \begin{cases} 1 \text{ falls} & k = argmin_l \{ \sum_{j=1}^{K} r_{lj} p_j p(\mathbf{c} \mid \Omega_j) \} \\ 0 \text{ sonst} \end{cases} \tag{5}$$

Aus dieser allgemeinen Klassifikationsvorschrift lassen sich spezielle Varianten über Restriktionen bzgl. der Kosten, der Klasse der Dichten oder der a priori Wahrscheinlichkeiten bilden. So entsteht der Bayes Klassifikator, wenn $r_{kk} = 0$ und $r_{lk} = 1, l \neq k$ gesetzt werden. Meistens werden für die $p(\mathbf{c} \mid \Omega_k)$ Gauß'sche Dichten angenommen, die sich sehr einfach mittels einer Lernstichprobe bestimmen lassen. Für die "Normalverteilungsklassifikator" genannte Variante gelten beide Einschränkungen.

Beim Polynomklassifikator werden ideale Entscheidungsfunktionen δ_k approximiert. Diese werden mittels der Lernstichprobe definiert zu

$$\delta_k(\mathbf{c}) = \begin{cases} 1 \text{ falls} \quad \mathbf{c} \quad \text{aus} \quad \Omega_k \\ 0 \text{ sonst} \end{cases} \tag{6}$$

Die Entscheidungsfunktionen d_k werden als Polynome festen aber beliebigen Grades in den Koeffizienten der Merkmalsvektoren festgelegt.

$$d_k(\mathbf{c}) = \mathbf{a}_k \mathbf{x}(\mathbf{c}) \tag{7}$$

$$\mathbf{d}(\mathbf{c}) = \mathbf{A}\mathbf{x}(\mathbf{c}) \tag{8}$$

Damit lassen sich beliebige Funktionen approximieren. Bei festliegendem $\mathbf{x}(\mathbf{c})$ ist der Klassifikator vollständig durch die Matrix $\mathbf{A}$ charakterisiert. Sie wird so bestimmt, daß der Abstand zwischen den idealen und den approximierten Entscheidungsfunktionen minimal ist, d.h. für die optimale Matrix $\mathbf{A}^*$ gilt

$$\varepsilon(\mathbf{A})^* = \min_{\mathbf{A}} E\{(\delta(\mathbf{c}) - \mathbf{A}\mathbf{x}(\mathbf{c}))^2\} \tag{9}$$

Die Schätzung ergibt sich aus

$$\mathbf{A}^* = (\frac{1}{N} \sum_{j}^{N} \mathbf{x}(\mathbf{c}_j) \mathbf{x}(\mathbf{c}_j)^T)^{-1} (\frac{1}{N} \sum_{j}^{N} \mathbf{x}(\mathbf{c}_j) \mathbf{x}(\mathbf{c}_j)^T) \tag{10}$$

wobei vorausgesetzt ist, daß die zu invertierende Matrix nicht singulär ist.

Für beide Klassifikatortypen lassen sich die Parameter der Entscheidungsfunktion aus einer Lernstichprobe direkt bestimmen. Sowohl die Parameter von

Normalverteilungen als auch die Koeffizienten der Matrix $\mathbf{A}^*$ lassen sich auch rekursiv lernen. Gängige Techniken sind dafür das **überwachte** oder das **entscheidungsüberwachte Lernen**. Im ersten Fall wird die Klasse des neuen Musters als bekannt vorausgesetzt, im zweiten werden die a posteriori Wahrscheinlichkeiten der Klassen bzw. die Werte der Klassenpolynome als Grundlage für die Zuordnung zu der aufzubessernden Klasse verwendet. Für randomisierte Zuordnungen ist die Konvergenz gezeigt, falls das Muster $\mathbf{c}$ probabilistisch gemäß den a posteriori Wahrscheinlichkeiten einer Klasse zugeordnet wird. **Unüberwachtes Lernen** im klassischen Fall ist z.B. durch Vektorquantisierung oder durch die Identifikation von Mischverteilungen gegeben.

3.2 Künstliche Neuronale Netze

Während die Systemarchitektur der "klassischen" statistischen Klassifikatoren als Sequenz von Vorverarbeitung – Merkmalsgewinnung – Klassifikation im Prinzip nicht mehr thematisiert wird, ist gerade der Architekturaspekt ein zentraler Forschungsgegenstand der Neuroinformatik. Mit Bezug auf die Biologie als Vorbild ist das Ziel, Informationen mittels Aktivitätsmuster über viele Freiheitsgrade zu verteilen. Für die einzelnen Bauelemente (**Neuronen**) wird keine hohe Genauigkeit gefordert. Berechnungen erfolgen durch Ausnutzung nichtlinearer, kollektiver Prozesse. Die geeignete Vernetzung und Synchronisation einfacher Bauelemente, von denen nur "wenige" Typen zu unterscheiden sind, manifestiert die Architektur eines Systems als enge Beziehung zwischen Struktur und Funktion. Teilnetze bilden Spezialmodule. Systeme zeichnen sich insgesamt durch hohe Vernetzungsdichte und massive Parallelität aus.

Neuronenmodelle basieren auf dem Prinzip der synaptischen Summation, die über drei Verarbeitungsschritte charakterisiert wird: die Summation eines Eingabevektors $\mathbf{x}$ mit einem Gewichtsvektor $\mathbf{w}$, eine nichtlineare Transformation f und optional den Einbezug stochastischer Prozesse. Der Ausgabewert $\mathbf{y}$ eines derartigen künstlichen Neurons ist gegeben durch

$$y(\mathbf{x}) = f(s(\mathbf{x}) - \theta) \tag{11}$$

wobei θ als Schwellwert, s z.B. als Skalarprodukt oder euklid'scher Abstand der Vektoren $\mathbf{x}$ und $\mathbf{w}$ und die Aktivierungsfunktion f z. B. als Stufen-, Vorzeichen-, Fermi-, tanh- oder Gaußfunktion definiert wird. Über das Argument $x = s(\mathbf{x}) - \theta$ der Funktion f wird der Zustand z des Neurons charakterisiert. Meist wird $z = x$ oder $z \in \{1, 0, -1\}$ für "erregt", "Ruhe" bzw. "hemmend" für $x > 0$, $x = 0$ bzw. $x < 0$ definiert.

Ein neuronales Netz ist dann als gerichteter Graph mit einer Zustandsmenge Z für jeden Knoten und einer Zustandsmenge W, mit Gewichten für jede Verbindung sowie je einer Menge von Ein- und Ausgaben gegeben. Die Zustandsmenge Σ eines Netzes mit k Knoten und v Verbindungen ist

$$\Sigma = Z^k \times W^v \tag{12}$$

wobei ein Netzzustand σ als ein Paar von Abbildungen

$$\begin{aligned} \sigma = \quad & (z, w) && \text{mit} \\ z : \{1, \ldots, k\} & \rightarrow Z && \text{Aktivitäten und} \\ w : \{1, \ldots, v\} & \rightarrow W && \text{Gewichte} \end{aligned} \tag{13}$$

von Zuordnungen aufgefaßt wird. Die Aktivitäten werden in einem Vektor $\mathbf{z} \in Z^k$ und die Gewichte in einer Matrix $\mathbf{W} \in W^k \times W^k$ zusammengefaßt. Die Grundtypen neuronaler Architekturen hängen stark von der Konnektivitätsmatrix $\mathbf{W} = [w_{ij}]$ ab, z.B.:

- voll verbundenes Netz: alle $w_{ij} \neq 0$
- k Neuronen ohne Wechselwirkung: $\mathbf{W}$ Diagonalmatrix
- dünn verbundenes: wenige $w_{ij} \neq 0$
- reine Vorwärtskopplung: $w_{ij} = 0$ falls $i < j$
- kurzreichweitige Wechselwirkung: $\mathbf{W}$ Bandmatrix

Das Verhalten neuronaler Netze ist durch die Vorschrift, wie ein neuer Zustand $\sigma^{(t+1)}$ aus dem alten Zustand $\sigma^{(t)}$ hervorgeht, charakterisiert. Wie die Zustände selbst ist auch diese Netzwerkdynamik in zwei Komponenten unterteilt. Die Änderung der Aktivitäten beschreibt die Kurzzeitdynamik. Dieser Übergang ist für ein Neuron i meist in der Form

$$z_i^{(t+1)} = f_i(s(\mathbf{z}) - \theta) \tag{14}$$

beschreibbar. Die Langzeitdynamik entspricht Änderungen der Gewichte

$$w_{ij}^{(t+1)} = w_{ij}^{(t)} + g_{ij}(w_{ij}^{(t)}, z_i^{(t)}, z_j^{(t)}) \tag{15}$$

Beide Typen von Dynamik erfüllen das Postulat der Lokalität und der Verteilung über viele Freiheitsgrade. So können im allgemeinen die Funktionen f_i individuell für jeden Knoten und g_{ij} individuell für jede Kante definiert werden. Durch die Kurzzeitdynamik ist die Arbeitsphase des Netzwerks und durch die Langzeitdynamik die Lernphase festgelegt. Die Funktionen g_{ij} sind damit von der Lernregel abhängig. Ziel des **Lernens** ist es, ein angestrebtes Antwortverhalten eines Netzwerks auf eine Eingabegröße $\mathbf{x}$ herbeizuführen. Der Weg dazu ist in der Neuroinformatik die Ermittlung geeigneter Gewichtsparameter w_{ij} aus Datenbeispielen. Wie in den Ansätzen aus Abschnitt 3.1 wird zwischen überwachtem, entscheidungsüberwachtem und unüberwachtem Lernen unterschieden.

Mit der Aufgabe, die Parameter eines Netzes zu trainieren, sind aktuell den Netzwerktopologien und den verwendeten Neuronentypen noch enge Grenzen gesetzt. Fragestellungen nach der Konvergenz und deren Geschwindigkeit sind herausfordernde Fragestellungen. Gängige Netzwerktypen wie Multilagen-Perzeptron-Netze, Autoassoziationsnetze oder Realisierungen von Rekurrenten-Netzen betrachten Netztopologien, die aus wenigen Neuronen einheitlichen Typs bestehen. SEXNET zur Gesichterkennung [Gol91] verwendet ein Autoassoziationsnetz mit ca. 2000 Neuronen und 3 Schichten, 900 - 40 - 900 zum Training, 900 - 40 - 1 für die Klassifikation. In einem System zur Navigation [Pom91] wird ein 900 - 5 - 30 Multilagen-Perzeptron-Netz eingesetzt. Dieser Netztyp ist aufgrund seiner relativ einfachen Architektur und des bekannten Backpropagation-Algorithmus zum Lernen am weitesten verbreitet. Gegenüber diesem überwachten Lernverfahren ermöglicht die Hebb-Regel unüberwachtes Lernen. Für die in diesem Kontext fast ausschließlich eingesetzen linearen Aktivitätsfunktionen wird die Karhunen-Loeve-Transformation der Eingangssignale realisiert. Der Abstand zu biologischen Systemen ist gravierend. So besteht z.B. ein Bienengehirn aus ca. 10^6 Neuronen mit ca. 10^9 Verbindungen. Das menschliche Auge besteht aus mehr als 90 Neuronentypen, das Hirn insgesamt aus 10^3 bis 10^4.

Vor allem für den Problemkreis sensomotorische Steuerungen werden Kohonen's selbstorganisierende Karten eingesetzt. Dabei werden die Neuronen in einer zweidimensionalen Struktur angeordnet. Ein Eingabesignal führt zur Auswahl eines Aktivitätszentrums und einer in der zweidimensionalen Topologie lokalen Erregungszone. In den Adaptionsgleichungen wird somit der "Abstand" zwischen Neuronen mittrainiert. In einer Erweiterung, den lokal linearen Karten, wird eine weitere Neuronenschicht für die Ausgabe hinzugefügt. Damit ist die Modellierung nichtlinearer Transformationen und so "Klassifikation" in kontinuierliche Wertebereiche möglich [Rit92], siehe auch [Kum93] in diesem Band.

3.3 Wissensbasierte Techniken

Wie im Abschnitt 2 dargestellt, ist die Generierung individueller symbolischer Beschreibungen für ein Muster eine zentrale Aufgabe der Mustererkennung, die auch als Musteranalyse, höhere Bilddeutung oder als Verstehen von Mustern bezeichnet wird. Das zentrale Problem dabei ist es, numerische Repräsentationen in symbolische zu transformieren und diese für die Generierung einer anwendungsabhängigen Beschreibung zu nutzen. Diese Transformation läßt sich nicht in geschlossener Form angeben. Sie wird in zahlreiche Teilschritte zerlegt, wobei die Reihenfolge und die Anzahl der einzelnen Transformationen zu Beginn des Prozesses nicht feststeht und von Prozeß zu Prozeß verschieden sein kann. Damit läßt sich die Aufgabe als Suchprozeß mit dem Ziel, eine Beschreibung zu generieren, und der Teilaufgabe, "Auswahl einer geeigneten Transformation in Bezug auf das Ziel", auffassen. Um die Auswahl effizient gestalten zu können und so den Suchraum zu reduzieren, müssen Informationen über Beschränkungen des Problemkreises eingesetzt werden. Dieses **Wissen** über Objekte, Ereignisse und über deren strukturelle Eigenschaften und Beziehungen muß formal repräsentiert werden, damit es für Analyseprozesse nutzbar ist.

Eine Wissensbasis für Mustererkennungssysteme muß also die Verbindung zwischen den Eingangssignalen und symbolischen Beschreibungen manifestieren. Das Spannungsfeld zwischen diesen beiden "Welten" ist in Bild 2 skizziert. Dargestellt sind dort auch die grundlegenden Voraussetzungen für die Repräsentation der beiden Welten, die erst den Einsatz einer Wissensbasis ermöglichen.

Der initiale Zustand eines Analyseprozesses ist durch das digitalisierte komplexe Muster $\mathbf{f}(\mathbf{x})$ und die Wissensbasen des Systems charakterisiert. Jede Wissensbasis besteht aus einem deklarativen Teil, den Daten oder Fakten, und einzel-

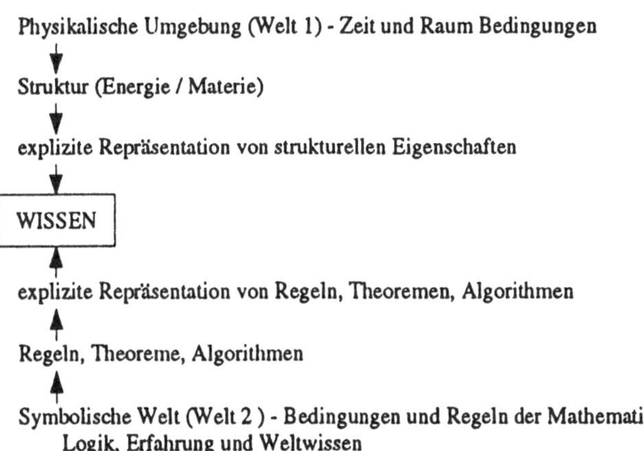

Bild 2 Wissen in Musteranalysesystemen (nach [Nie90])

nen Inferenzprozessen. Ein solcher Inferenzprozeß transformiert einen Zustand, indem neue Daten erzeugt, alte Daten gelöscht oder manipuliert werden. Bezeichnet man mit *DATEN*(i) den deklarativen Teil der Wissensbasis und die bereits erzielten Zwischenergebnisse des Zustands i und mit $\mathbf{T} = \{T_1, \ldots, T_N\}$ die Menge aller Transformationen, dann ist ein Analyseprozeß durch Bild 3 charakterisiert. Der Anfangszustand enthält das Muster, der Endzustand die Beschreibung, im allgemeinen sind in jedem Zustand mehrere Transformationen konkurrierend anwendbar.

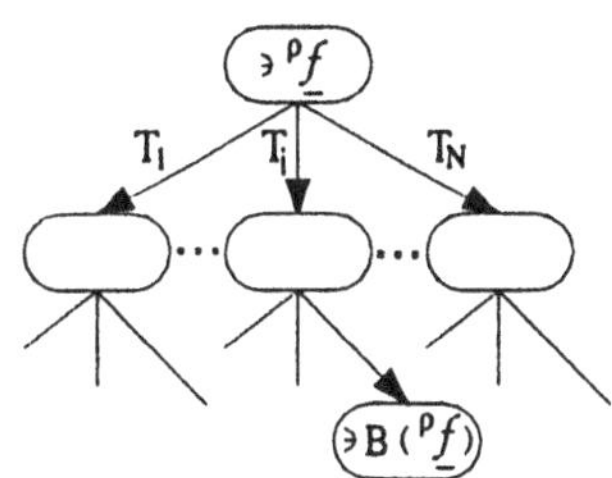

Bild 3 Suchbaum eines Analyseprozesses

Es stellt sich nun die Frage nach der Organisation der Architektur solcher Systeme. Eine Zerlegung in funktionale Komponenten wie in Bild 4 gezeigt, wird so oder ähnlich für wissensbasierte Systeme vorgeschlagen [Nie90, Gör88, Hat88]. Unter Methoden sind dabei Bildverarbeitungsalgorithmen und symbolische Inferenzprozesse subsumiert. Die zentrale Kontrolle dient der Steuerung des Analyseprozesses, d.h. in Abhängigkeit der Wissensbasis und der bereits erzielten Resultate entsprechende Transformationen anzustoßen. Die automatische Akquisition von Wissen (Lernen) und die Erklärung von Ressourcen, Handlungen und Ergebnissen des Systems bilden weitere wichtige Komponenten.

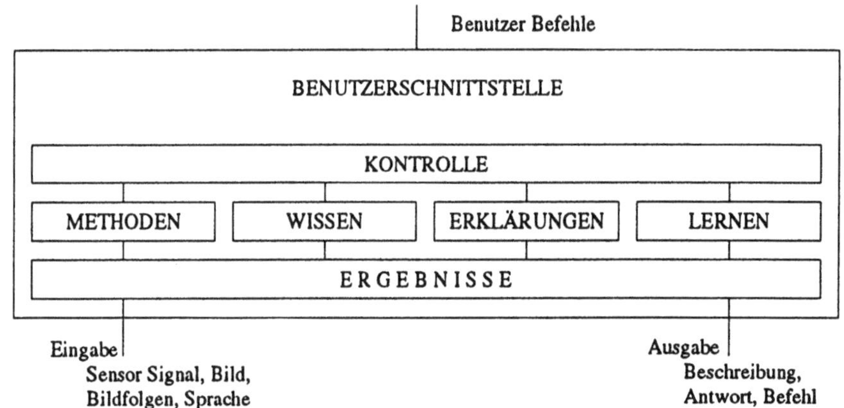

Bild 4 Funktionale Komponenten eines Musteranalysesystems (nach [Nie90])

Eine andere Organisation ergibt sich, wenn man nicht funktionale Einheiten betrachtet, sondern den Weg vom Muster zur Beschreibung in einzelne Stufen gruppiert. Ein derartiges geschichtetes Modell ist in Bild 5 skizziert. Jede Abstraktionsebene wird in drei Komponenten aufgespalten. Die Verarbeitung wird wiederum über eine zentrale Kontrolle gesteuert, die Verarbeitungsrichtung daten- oder modellgetrieben oder Mischformen aus beiden ist nicht durch das Modell fixiert.

Untersuchungen zur adäquaten Repräsentation von Wissen, siehe z.B. [Min75, Bra79, Sch86], zeigen auf, daß Wissen über einen Problemkreis nur bedingt eine Einheit bildet. Es sind mehrere **Wissensarten** zu unterscheiden. Diese sind nicht mit den Hierarchiestufen des geschichteten Modells zu verwechseln. Neben der Unterscheidung prozedural und deklarativ sind unter anderem Wissensarten wie Definitionen, Beschreibungen oder Zusicherungen zu trennen. Eine derartige Modularisierung eines Systems steht orthogonal sowohl zum funktionalen als auch zum hierarchischen Modell. Eine erste Wissensart, die insbesondere für den eigentlichen Verstehensprozeß von Bedeutung ist, bilden die Begriffe oder *Konzepte* aus der "Welt" des Anwendungsgebietes. Diese Begriffswelt muß in Form ihrer einzelnen Elemente sowie der Beziehungen zwischen diesen Elementen modelliert werden. Schemata zur expliziten Darstellung dieser Modelle sind formale Grammatiken, Prädikate, ATN's, Frames oder semantische Netze.

Schlußfolgerungen generieren neue strukturelle Informationen aus dem Signal. Zu unterscheiden sind Inferenzen, die als datengetrieben bzw. modellgetrieben oder prädiktiv charakterisierbar sind. Durch Inferenzen generierte Daten, die eine vollständige Interpretation eines Teils oder des gesamten Musters repräsentieren, werden als *Instanzen* bezeichnet. Werden aus Konzepten und dem Eingangssignal Hypothesen generiert, die zwar an einen Signalausschnitt

	WISSEN	PROZESSE	ERGEBNISSE	
Ebene n	Benutzer Anforderungen	Antwort Generierung	Ausgabe: Beschreibung auf höherer Ebene B	KONTROLLE
...	...	...	...	
Ebene j	Aufgabe	Zielorientiert	Aktion, Meldungen	
...	...	...	...	
Ebene i	Objekte	Abbildung	Symbolische Namen	
...	...	...	...	
Ebene 2	Segmentierung	Segmentierung	segmentierte Muster	
Ebene 1	Störungen	Normalisierung	verbesserte Muster	
Ebene 0	Mustergenerierung	Aufnahme	Eingabe: niedere Ebene Muster $\underline{f}(\underline{x})$	

Bild 5 Geschichtetes Modell (nach [Nie90])

gebunden sind, aber ein Konzept nicht vollständig sättigen, so können diese als datenadaptierte oder modifizierte Konzepte bezeichnet werden. Konzepte sind so den Klassen im Sinne von Abschnitt 3.1 vergleichbar. Es kann jedoch kein einheitlicher Merkmalssatz für alle Konzepte eines Problemkreises festgelegt werden. Damit ist die Auswahl der Merkmale für die Modellierung eines Begriffes deklaratives Wissen. Für jedes Merkmal (*Attribut*) muß die Struktur und die Interpretation, d.h. seine Semantik, definiert werden. Prinzipiell sind auch für Attribute modellgetriebene Inferenzen, z.B. durch Einschränkung des Wertebereichs, realisierbar. Entscheidungen während eines Analyseprozesses können, wie bei Klassifikatoren, nicht lokal als richtig oder falsch getroffen werden. Es ist erforderlich, Hypothesen zu generieren und diese zu bewerten. Deklarativ muß ein *Bewertungskalkül*, z.B. Wahrscheinlichkeiten, Prioritäten, Fuzzy-Logik, ausgewählt werden. Wiederum sind mit der Festlegung des Kalküls auch die zugehörigen Inferenzen zu definieren. Berechnungen resultieren in Werten, die nur im Zusammenhang mit dem Kalkül eine Bedeutung haben. Eine weitere Wissensart sind *Zusicherungen* oder Relationen, die Aussagen darüber machen, wie sich unterschiedliche Konzepte/Instanzen oder Attribute zueinander verhalten müssen oder es "üblicherweise" tun. Zusicherungen können auch neben dem Überprüfungsaspekt dazu genutzt werden, neue Strukturen zu erzeugen, indem modellgetrieben die Erfüllung der Zusicherung durch eine Restriktion garantiert wird. Die Berechnung der Relationen muß Werte des Bewertungskalküls liefern. Die vorgestellten Wissensarten sind mit ihren Prozessen in Bild 6 zusammengestellt. Es ist zusätzlich für eine Systemrealisierung eine Steuerungseinheit, das Kontrollmodul, erforderlich.

Auch wissensbasierte Mustererkennung ist ein Optimierungsproblem. Ziel ist

Deklaratives Wissen	Inferenzen (Prozeduales Wissen)	(Zwischen-)Ergebnisse
Konzepte (Begriffe, Objekte, Ereignisse)	datengetrieben: Instantiierung	Instanzen
	modellgetrieben: Restriktionen propagieren	modifizierte Konzepte Konzepte
Attribute (numerische, symbolische Merkmale)	datengetrieben: Extraktion, Kombination	Werte
	modellgetrieben: Restringieren	Wertebereiche für Attribute
Bewertungskalkül	Berechnungsvorschriften	Werte
Zusicherungen (strukturelle Beziehungen)	datengetrieben: Überprüfungen	erfüllt / nicht-erfüllt Bewertung
	modellgetrieben: Restringieren	Wertebereich für Attribute modifizierte Konzepte

Bild 6 Wissensarten in einem Musteranalysesystem

es, aus einer explizit oder nur implizit vorgegebenen Menge von Beschreibungen für ein Eingabemuster diejenige auszuwählen, die den Bedingungen der Wissensbasis genügt und bezüglich eines Bewertungskalküls optimal zum Eingangssignal paßt, d.h.

$$B^*(\mathbf{f}(\mathbf{x})) = argmax_B\{g(B, \mathcal{W}, \mathbf{f}(\mathbf{x}))\} \tag{16}$$

wobei g die Bewertungsfunktion, $\mathcal{W}$ die Wissensbasis und B Beschreibungen sind. Ausführliche Darstellungen von Repräsentationsformalismen im Kontext Mustererkennung sind z.B. in [Sag90, Nie90, Bal82] gegeben. Eine sehr gute Übersicht von Bewertungskalkülen gibt [Pra85].

4 Wege zur Symbiose

Alle drei Paradigmen — "klassische" Klassifikatoren, neuronale Netze, wissensbasierte Mustererkennungssysteme — versuchen auf unterschiedliche Weise Optimierungsprobleme zu lösen. Für spezielle Varianten der in Abschnitt 3 allgemein vorgestellten Ansätze wurden Vergleiche durchgeführt. In [Kre91] wird gezeigt, daß sowohl das Multilagen-Perzeptron und die Radial-Basis-Funktionen als auch der Polynomklassifikator universelle Approximatoren an ideale Entscheidungsfunktionen sind. Zur Auswahl einer der drei Ansätze für Klassifikationssysteme kommen die Autoren zu dem Ergebnis: "Ein abschließendes Urteil abzugeben, ..., fällt schwer. ... Die Auswahl eines bestimmten Verfahrens muß daher weitere Gesichtspunkte wie Geschwindigkeit der Adaption, die Durchsichtigkeit des Verfahrens, Implementierungsüberlegungen, Verfügbarkeit u.a. miteinbeziehen." Im Gegensatz zu den dort durchgeführten experimentellen Vergleichen, die eine Auswahl von Verfahren aus den Paradigmen voraussetzen, soll die Diskussion hier auf den Grundannahmen und der methodischen Basis eines jeden Paradigmas erfolgen.

4.1 Klassische Verfahren und künstliche neuronale Netze

Eine häufig gemachte Erfahrung ist, daß etliche Ansätze künstlicher neuronaler Netze mit klassischen Verfahren identisch sind oder zumindest große Ähnlichkeiten bestehen. So gibt es für die Karhunen-Loeve-Transformation, für Vektorquantisierer und gängige Klassifikatoren neuronale Darstellungen. Bild 7 zeigt einen Polynomklassifikator in der Nomenklatur neuronaler Netze. Dieses "Paradigma-wechsel-dich"-Spiel läßt sich auch in umgekehrter Richtung, z.B. für die Optimierung der Gewichtsmatrizen **W** gemäß dem Polynomklassifikator, durchführen. Sind künstliche neuronale Netze doch nur des Kaisers neue Kleider? Ich möchte diese Frage trotz der verschwindenden Grenzen mit einem Nein beantworten. Die Gründe hierfür sollen im folgenden kurz über die reichhaltigen Wechselwirkungen zwischen den Ansätzen skizziert werden.

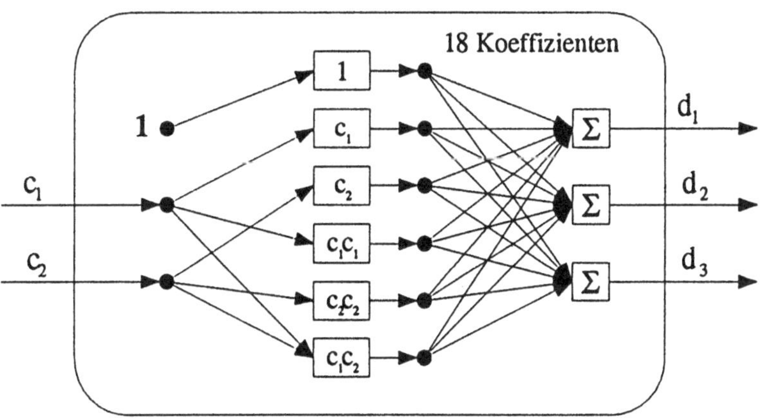

Bild 7 Vollständig quadratischer Polynomklassifikator für zwei Merkmale und drei Klassen (nach [Sch92])

Was hat die Neuroinformatik für die Mustererkennung in den letzten Jahren geleistet und was ist noch zu erwarten? In [Nie74] wird die Untersuchung von Nervennetzen als eine wichtige Grundlage der Mustererkennung genannt. Die Architektur von Klassifkationssystemen in Bild 1 resultierte aus dem Zwang der Modularisierung in einzelne Komponenten, da eine globale Optimierung des Gesamtsystems zu aufwendig ist. Über lernende Systeme wird ausgesagt [Nie74]: "Das theoretische Interesse liegt auf der Hand. Das praktische hängt von möglichen Anwendungen ab." Die Binnenoptimierung der Komponenten für technisch zuverlässige Systeme rückt in den Vordergrund. Klassifikatoren erreichen Anfang der 80'er Jahre für viele Problemfelder exzellente Leistungen. Natürlicherweise wurden die Anwendungsfelder komplexer, und das Problem der Musteranalyse rückte in den Vordergrund. Ausgehend von formalen Grammatiken [Fu82] werden in der Mustererkennung intensiv wissensbasierte Techniken bis zum symbolischen Lernen untersucht. Die Neuroinformatik dagegen setzt radikal auf das biologische Vorbild. Basierend auf empirischer Forschung und der Berechnungstheorie werden lernende informationsverarbeitende Systeme nachgebaut, die die vermuteten Prinzipien durch Synthese "beweisen" [Mal91]. Damit sind auch die Grenzen des aktuellen Forschungsstandes gekennzeichnet, nämlich durch den Stand der empirischen biokybernetischen Forschung einerseits und durch die mathematische Handhabbarkeit der Modelle andererseits. Multilagen-Perzeptron-Netze oder selbstorganisierende Karten sind "nur" Anfänge, die in weiter Ferne zur Modellierung von Gehirnen führen könnten. Doch bereits die einfachen Architekturen führten zur "Revitalisierung" des klassischen Mustererkennungsproblems. Schürmann charakterisierte in einem Vortrag [Sch93] die Leistung der Neuroinformatik für die Musterklassifkation mit der Ballade vom Schatz im Weinberg [Bür33]. Die Leistungen für die Mustererkennung reichen jedoch weiter. Die festgefahrene Sequenzialisierung in die Verfahrensgruppen Vorverarbeitung, Merkmalsgewinnung, Klassifkation wird allein durch die Forderung nach empirisch am biologischen Vorbild ausgerichteten Architekturen in Frage gestellt. Die in der "klassischen" Mustererkennung nur vereinzelt entwickelten Algorithmen zur globalen Optimierung des Verbundes von Systemkomponenten, wie z.B. in [Jäp80], werden in das Zentrum der Untersuchungen gerückt. Die Entwicklung von Klassifikationsstufen oder neuronalen Netzen, die nur aus einem oder zwei Neuronentypen aufgebaut sind, ist nur ein Zwischenschritt. Daß sich sehr viele, wenn nicht die meisten klassischen Algorithmen auch im neuronalen Paradigma formulieren lassen, eröffnet zahlreiche Chancen. Eine zunächst banale doch in ihrer Konsequenz weitreichende Perspektive ist die einheitliche Formulierung von Vorverarbeitungs-, Merkmalsextraktions- und Klassifikationsalgorithmen. Die Integration linearer und nichtlinearer Elementaroperationen in einem massiv parallelen Netzwerk sowie dessen Optimierung in Bezug auf das Ausgangsverhalten kann neue Möglichkeiten zur algorithmischen Lösung von Perzeptionsaufgaben eröffnen.

Die — zumindest intendierte — strikte Orientierung am biologischen Vorbild hat die Neuroinformatik bereits über die Betrachtung von Klassifkationsaufgaben hinausgeführt. Zur empirischen Untersuchung besser geeignet als Signal-Symbol-Transformationen sind die Verbindungen zwischen Sinnesreizen und Verhaltensmustern, wobei Rückkopplungen des Verhaltens auf die wahrnehmbaren Sinnesreize mit betrachtet werden. Damit werden Systeme realisiert, die Signale in kontinuierliche Klassensysteme wie Winkel oder Raumkoordinaten "klassifizieren", ohne eine eindeutige Lösung für die Zwischenschritte zu postulieren [vS92, Mey92]. So wird eine teilweise Integration der "computational theory" nach [Mar82], regelungstechnischen Ansätzen wie [Dic92], den Arbeiten von Brooks aus der Robotik [Bro86] sowie Ansätzen zur selektiven Perzeption [Bro92] erreicht.

4.2 Subsymbolisch und Wissensbasiert

Sowohl die Abgrenzung als auch die Symbiose zwischen Klassifikatoren bzw. neuronalen Netzen als numerisch-subsymbolische Verfahren einerseits und wissensbasierten strukturellen Verfahren andererseits ist zunächst erheblich einfacher zu charakterisieren. Die wissensbasierte Interpretation wird als viertes Modul in die Systemarchitektur gemäß Bild 1 eingeordnet [Sch92]. "Lediglich durch die in einem Spielraum freie Entwurfsentscheidung, welches denn nun die elementaren Symbole der Erkennung sein sollen, wird eine gewisse Konkurrenz zwischen den Konzepten der strukturellen Mustererkennung und der Mustererkennung im Sinne von Klassifizierung ins Spiel gebracht." Daß explizit repräsentiertes Wissen in Verbindung mit entsprechenden Inferenzprozessen und Kontrollalgorithmen für Musteranalysesysteme notwendig ist, ist weitestgehend unumstritten. "Such a high level model is often much more complex than the lower-level-representations, often has a large 'propositional' component, and is often manipulated by 'inference like' procedures. Explicit knowledge and belief structures are ... playing an ... important role." Dieses Statement aus [Bal82] hat nach wie vor Gültigkeit. Bei der Repräsentation von Wissen sind nach [Ric89] die kognitive Ebene, die Repräsentationsebene und die Implementierungsebene zu trennen. Wie bereits erwähnt, basiert die Modellierung auf der kognitiven Ebene auf dem Grundsatz, daß Gedanken in rationaler Weise organisiert sind und umgangssprachlich formuliert werden. Diesem **Rationaltitätspostulat** der KI setzt die Neuroinformatik den Aspekt der **präratioualen Intelligenz** [Cru93] entgegen.

Während für die KI die Expertenbefragung die Gesetzmäßigkeiten eines Problemkreises festlegt, basieren subsymbolische Verfahren auf dem Lernen aus Beispielen. Gerade für die Perzeptionsleistungen ist aber jeder und keiner ein "Experte" im Sinne der rationalen Intelligenz. So spielt gerade für Mustererkennungssysteme häufig der Entwickler selbst diese Rolle. Dem setzt [Mal91] gegenüber: "... können Anwendungen, die darauf angelegt sind, von der natürlichen Intelligenz zu lernen nur dann funktionieren, wenn die Arbeitsweise dieser natürlichen Intelligenz hinreichend gut verstanden ist. Es genügt sicher nicht, isolierte Ergebnisse der Neurowissenschaften oder Introspektionen in elektrischer Weise herauszugreifen und in eine Anwendung zu integrieren. Funktionierende Anwendungen setzen die empirische Erforschung der biologischen Informationsverabeitung voraus." Das Spannungsfeld zwischen der biologischen Fundierung und des symbolisch orientierten Rationalitätspostulats läßt sich aus pragmatisch technischer Sicht, aber auch aus einer mehr grundlagenorientierten Sicht teilweise auflösen.

Im Gegensatz zu symbolbasierten Algorithmen, bei denen eine explizite Programmierung der angestrebten Operationen möglich ist, ist eine solche explizite Programmierung bei neuronalen Netzen nicht möglich, da aufgrund der subsymbolischen Repräsentation von Informationen in diesen Systemen Regeln nicht auf der Begriffsebene der jeweiligen Anwendung formuliert und in das System explizit eingebracht werden können. Vielmehr erfordert die "Programmierung" neuronaler Netze das Auffinden einer meist sehr großen Anzahl von Gewichtsparametern, von denen jeder einzelne nur eine geringe Auswirkung auf die Systemeigenschaften hat, die in ihrer Gesamtheit jedoch das Transformationsverhalten des Systems "holistisch" festlegen. Daher ist die Kernfrage die Entwicklung geeigneter Verfahren, um für eine gegebene Anwendung geeignete Gewichte effektiv zu bestimmen. Aufgrund der Nichtlinearität neuronaler Systeme ist eine analytische Berechnung – abgesehen von Sonderfällen – aussichtslos. Die meisten Verfahren beruhen auf "Lernregeln", die, ausgehend von einer Anzahl von "Trainingsbeispielen", geeignete Gewichtsparameter allmählich im Verlauf eines iterativen Verfahrens bestimmen.

In der Vergangenheit konzentrierte sich der Einsatz neuronaler Netze auf die Lösung von Einzelaufgaben, wie etwa assoziative Speicherung, Klassifikation oder Transformation von Signalen. Erst in jüngerer Zeit wird zunehmend die Aufgabe betrachtet, multiple neuronale Netze zu koordinieren, bzw. die Funktionalität neuronaler Netze und "konventioneller" Verarbeitungsmodule in größeren Systemen zu integrieren [Alo89]. Dies erfordert die Entwicklung neuer Paradigmen, insbesondere auch zur Repräsentation symbolischer Informationen und zusammengesetzter Datenstrukturen [Hin90, Smo90]. Erste Ansätze bieten Arbeiten zur Organisation multipler Netze in kompetitiven oder hierarchischen Architekturen. Die meisten dieser Arbeiten sind jedoch sehr punktueller Natur und bieten lediglich Demonstrationen im Bereich sehr einfacher, künstlicher Muster. Dies gilt auch für die zitierten Ansätze, symbolische Information konnektionistisch zu repräsentieren.

Was fehlt, sind breiter angelegte Untersuchungen, die sich sowohl auf der neuronalen als auch auf der symbolbasierten Seite auf ausgereifte Systemkomponenten stützen und die die Integration solcher Komponenten für die Anwendung auf komplexe Muster in Angriff nehmen. Die stark ausgeprägten Möglichkeiten zahlreicher Wissensrepräsentationsformalismen, Begriffe in einfachere Entitäten zu zerlegen und strukturelle Beziehungen zu modellieren, bieten Ansätze zur Integration derart, daß die ganzheitlichen prärationalen Stärken neuronaler Netze mit den dekompositionellen rationalen Modellierungen expliziter Wissensrepräsentation kombiniert werden. Dafür ist es nicht erforderlich, horizontale Übergabepunkte zwischen subsymbolischen und symbolischen Repräsentationen festzulegen. Betrachtet man die in Abschnitt 3.3 diskutierten Wissensarten, so fehlt dort die analoge Repräsentation. Solche Modelle zum Erkennen eines Objekts "auf einen Blick" können gerade neuronale Netze oder statistische Klassifikatoren bereitstellen. Ob ein analoges Modell für einen Begriff erzeugt werden kann, hängt nur sekundär von seiner Komplexität ab. In erster Linie muß eine ausreichende Lernstichprobe zum Training der Parameter existieren. Damit wird auch die Trennung "low level versus high level" Verarbeitung aufgehoben. Analoge und symbolische Repräsentationen überlappen einander. Begriffe können auch in beiden Formen modelliert werden. Holistisch detektierte Objekte sind auch nachträglich in ihre Bestandteile gemäß der Wissensbasis zerlegbar.

Bleibt die Frage nach der empirischen Fundierung einer solchen Integration. Außer gemessenen Raten korrekter Ergebnisse, die zumindest für Musteranalysesysteme statistisch oft nicht sehr zuverlässig sind, bleibt nur die Überprüfung am biologischen Vorbild. Für die Neuroinformatik ist dies ja das Fundament. Die Frage nach der vollständigen Funktionsweise des Gehirns und damit nach der biologischen Realisierung von musterinterpretierenden Systemen wird aber — zumindest in absehbarer Zeit — nicht beantwortet werden. An die Stelle von Introspektionen für den Entwurf der wissensbasierten Systemteile müssen kognitionspsychologische Experimente die Grundlagen liefern. Beide Seiten, die biologischen und die psychologischen Experimente, geben Hinweise, können aber nicht das gesamte Problemspektrum abdecken. Heuristiken und mathematische Bewertungskalküle und Optimierungsverfahren müssen die fehlenden Verbindungen herstellen.

5 Ein kurzes Resümee

Das selbstgesteckte Ziel der Mustererkennung ist die Automatisierung von Perzeptionsleistungen des Menschen. Damit steht sie als ingenieurswissenschaftliche Disziplin zwischen Biokybernetik und Kognitionspsychologie, die sich als empirische Wissenschaften der Perzeptionsforschung widmen. Als weiterer Bezugspunkt ist das Anwendungsfeld eines Mustererkennungssystems von Bedeutung. Für Realisierungen ist die Performanz und die Robustheit von besonderem Interesse. Um diese Ziele zu erreichen, sind methodische Abgrenzungen wenig hilfreich. Gefragt ist vielmehr die Integration unterschiedlicher Verfahren. Der Bereich der subsymbolischen Methoden bis zur Signal-Symbol-Transformation wurde durch die zunehmende Bedeutung der Neuroinformatik gerade unter den Aspekten Robustheit und empirische Fundierung neu belebt. Während künstliche neuronale Netze wie auch die "klassischen" statistischen Verfahren holistische Methoden bereitstellen, wurden dekompositionelle, attributierte Repräsentationen als Konsequenz des Rationalitätspostulats in wissensbasierten Techniken entwickelt. Der Diskussion nach einer kanonischen Schnittstelle zwischen numerischer und symbolischer Verarbeitung im Sinne einer hierarchischen Verarbeitung sollte die Untersuchung hybrider Systeme folgen, die beide Techniken bis in die Extreme Signal bzw. Beschreibung verzahnen. Nur so können die anstehenden Aufgaben gelöst werden. Ein Schritt wären Systeme, die sich sowohl auf subsymbolischer als auch auf symbolischer Seite auf ausgereifte Komponenten stützen und die Integration solcher Komponenten für die Analyse komplexer Muster in Angriff nehmen. Ein solches notwendigerweise auch technisch realisiertes System sollte, soweit möglich, durch die Ergebnisse biologischer und psychologischer Kognitionsexperimente fundiert sein. Auf diesem Weg kann die Mustererkennung sehr wichtige Beiträge bei der Entwicklung von Maschinen leisten, die an den Menschen angepaßt sind und nicht verlangen, daß sich der Mensch gedanklich der Maschine unterzuordnen hat.

Danksagung

Dieser Beitrag basiert auf zahlreichen Gesprächen insbesondere mit H. Niemann, H. Ritter und J. Schürmann. Ohne ihre Beiträge wäre diese Diskussion so nicht möglich gewesen. Viele Impulse kamen auch von meinen Kollegen im Sonderforschungsbereich 360 "Situierte Künstliche Kommunikatoren" an der Universität Bielefeld. Bei der Ausarbeitung wurde ich von F. Kummert und S. Posch unterstützt.

Literatur

[Alo89] J. Aloimonos, D. Shulman: *Integration of visual Modules*, Academic Press, 1989.

[Bal82] D. Ballard, C. Brown: *Computer Vision*, Prentice Hall, New York, 1982.

[Bra79] R. J. Brachman: *On the Epistemological Status of Semantic Networks*, in N. V. Findler (Hrsg.): *Associative Networks*, Academic Press, New York, 1979, S. 3–50.

[Bro86] R. A. Brooks: *Achieving AI through Building Robots*, MIT AI Memo, 1986.

[Bro92] C. Brown: *Issues in Selective Perception*, in *Proceedings of the International Conference on Pattern Recognition*, Bd. 1, 1992, S. 21–30.

[Bür33] G. A. Bürger: *Gesammelte Schriften*, 1829 – 1833.

[Cru93] H. Cruse, H. Ritter: *Prärationale Intelligenz, Programm für das Forschungsjahr 1994 des ZIF*, Zentrum für Interdisziplinäre Forschung der Universität Bielefeld, 1993.

[Dic92] E. D. Dickmanns: *A general dynamic vision arcitecture for UGV and UAV*, *Journal of Applied Intelligence*, Bd. 2, 1992, S. 251–270.

[Dud72] R. Duda, P. Hart: *Pattern Classification and Scene Analysis*, J. Wiley, New York, 1972.

[Fu68] K. Fu: *Sequential Methods in Pattern Recognition and Machine Learning*, Academic Press, New York, 1968.

[Fu76] K. S. Fu: *Digital Pattern Recognition*, Springer Verlag, Berlin, Heidelberg, New York, 1976.

[Fu82] K. S. Fu (Hrsg.): *Pattern Recognition Theory and Application*, Prentice Hall, Englewood Cliffs, N.J., 1982.

[Gel92] E. S. Gelsema, E. Backer, I. T. Young: *Message of the Conference Chairman*, in *11th International Conference on Pattern Recognition*, Bd. I, The Hague, 1992, S. V.

[Gol91] B. Golomb, D. Lawrence, T. Sejnowski: *SEXNET: A Neural Network Identifies Sex from Human Faces*, in R. Lippmann, J. Moody, D. Touretzky (Hrsg.): *Advances in Neural Information Processing Systems 3*, Morgan Kaufman Publishers, San Mateo, CA, 1991, S. 572–577.

[Gör88] G. Görz: *Strukturanalyse natürlicher Sprache*, Addison-Wesley, Bonn, 1988.

[Hat88] J. P. Haton: *Knowledge-Based Approaches in Acoustic-Phonetic Decoding of Speech*, in H. Niemann, M. Lang, G. Sagerer (Hrsg.): *Recent Advances in Speech Understanding and Dialog Systems*, NATO ASI Series F, Vol. 46, Springer-Verlag, Berlin, 1988, S. 51–70.

[Hin90] G. E. Hinton: *Mapping part-whole hierarchies into connectionist networks*, *Artificial Intelligence*, Bd. 46, 1990, S. 47–76.

[Jäp80] D. Jäpel: *Klassifikatorbezogene Merkmalsauswahl*, Bd. 13 Nr. 4 von *Arbeitsberichte des IMMD*, Universität Erlangen-Nürnberg, 1980.

[Kre91] U. Kreßel, J. Schürmann, J. Franke: *Neuronale Netze für die Musterklassifikation*, in B. Radig (Hrsg.): *Mustererkennung 91, 13. DAGM-Symposium München, Informatik-Fachberichte*, Springer-Verlag, Berlin, 1991, S. 1–18.

[Kum93] F. Kummert, E. Littmann, A. Meyering, S. Posch, H. Ritter, G. Sagerer: *A Hybrid Approach to Signal Interpretation Using Neural and Semantic Networks*, 1993, in diesem Band.

[Mal91] H. A. Mallot: *Frühe Bildverarbeitung in neuronaler Architektur*, in B. Radig (Hrsg.): *Mustererkennung 91, 13. DAGM-Symposium München, Informatik-Fachberichte*, Springer-Verlag, Berlin, 1991, S. 19–34.

[Mar82] D. Marr: *Vision*, Freeman, San Francisco, 1982.

[Mey92] A. Meyering, H. Ritter: *Learning 3D-Shape Perception with Local Linear Maps*, in *Proceedings of the International Joint Conference on Neural Networks*, Baltimore, 1992, S. 432–436.

[Min75] M. Minsky: *A Framework for Representation Knowledge*, in P. H. Winston (Hrsg.): *The Psychology of Computer Vision*, McGraw-Hill, New York, 1975, S. 211–277.

[Nie74] H. Niemann: *Methoden der Mustererkennung*, Akademische Verlagsgesellschaft, Frankfurt, 1974.

[Nie83] H. Niemann: *Klassifikation von Mustern*, Springer-Verlag, Berlin, 1983.

[Nie90] H. Niemann: *Pattern Analysis and Understanding*, Springer-Verlag, Berlin, zweite. Ausg., 1990.

[Pat72] E. A. Patrick: *Fundamentals on Pattern Recognition*, Prentice Hall, Englewood Cliffs N. J., 1972.

[Pom91] D. Pomerleau: *Rapidly Adapting Artificial Neural Networks for Autonomous Navigation*, in R. Lippmann, J. Moody, D. Touretzky (Hrsg.): *Advances in Neural Information Processing Systems 3*, Morgan Kaufman Publishers, San Mateo, CA, 1991, S. 429–435.

[Pra85] H. Prade: *A Computational Approach to Approximate and Plausible Reasoning with Applications to Expert Systems*, *IEEE Transactions on Pattern Analysis and Machine Intelligence*, Bd. 7, 1985, S. 260–283.

[Ric89] M. M. Richter: *Prinzipien der Künstlichen Intelligenz*, Teubner Verlag, Stuttgart, 1989.

[Rit90] H. Ritter, T. Martinetz, K. Schulten: *Neuronale Netze*, Addison-Wesley, Reading, MA, 1990.

[Rit92] H. Ritter, T. Martinetz, K. Schulten: *Neural Computation and Self-organizing Maps*, Addison-Wesley, Reading, MA, 1992.

[Ros69] A. Rosenfeld (Hrsg.): *Picture processing by computer*, Academic Press, New York, 1969.

[Sag90] G. Sagerer: *Automatisches Verstehen gesprochener Sprache*, Bd. 74 von *Reihe Informatik*, Bibliographisches Institut, Mannheim, 1990.

[Sch86] P. Schefe: *Künstliche Intelligenz – Überblick und Grundlagen*, Bibliographisches Institut, Mannheim, 1986.

[Sch92] J. Schürmann, U. Kreßel: *Mustererkennung mit statistischen Methoden*, Vorlesungsskript sommersemester 1992 und 1993, Daimler-Benz AG, Forschungszentrum Ulm, Institut für Informationstechnik, 1992.

[Sch93] J. Schürmann: *Konzepte der Mustererkennung*, Vortrag im Rahmen des Seminars Sprach- und Bildverarbeitung, Universität Bielefeld, 1993.

[Seb62] G. Sebestyen: *Decision making processes in pattern recognition*, Macmillan, New York, 1962.

[Smo90] P. Smolensky: *Tensor Product variable binding and the representation of symbolic structures in connectionist systems*, *Artificial Intelligence*, Bd. 46, 1990, S. 159–216.

[vS92] W. von Seelen, H. Janßen: *Structural principles in visually guided autonomous vehicles*, in *11th International Conference on Pattern Recognition*, Bd. I, The Hague, 1992, S. 302–311.

[Wat69] S. Watanabe (Hrsg.): *Methodologies of Pattern Recognition*, Academic Press, New York, 1969.

Spracherkennung in der medizinischen Dokumentation

J.-P. Pietsch
Geschäftsfeld Neue Medien
SerCon GmbH - Region Mitte
Kreuzberger Ring 56
65205 Wiesbaden

1 Einleitung

Maschinelle Spracherkennung bietet mehr als nur automatische Diktiergeräte. Sie verbindet das Medium der gesprochenen Sprache mit der Funktionalität der Computer. In diesem Vortrag stelle ich ein bereits existierendes und angewendetes Spracherkennungssystem vor. In der Medizin kann die automatische Spracherkennung überall da am sinnvollsten eingesetzt werden, wo große Mengen ähnlicher Texte zu diktieren sind, also z. B. für Befundungen, Arztbriefe und OP-Berichte in sämtlichen medizinischen Bereichen (Radiologie, Chirurgie, Endoskopie, Pathologie usw.). Der größte Vorteil ergibt sich dabei aus der Integration der Spracherkennung in existierende medizinische Informations- und Dokumentationssysteme. So können die Befunde nach dem Diktieren direkt in eine Dokumentationsdatenbank übernommen werden und ohne Zeitverzug an anderer Stelle abgerufen werden. Aber es kann nicht nur diktiert werden, denkbar ist auch die Steuerung medizinischer Geräte (z. B. Operationsmikroskop) oder natürlichsprachliche Datenbankabfragen.

2 Ein System zum Diktieren von Texten mit großem Vokabular

Die IBM Speech Server Serie ist ein sprecherabhängiges System zur Erkennung isoliert gesprochener Worte, das aus dem Forschungsprototyp TANGORA hervorging. Ein Vokabular von etwa 20.000 Wortformen kann in Echtzeit erkannt werden. Das System arbeitet mit statistischen Methoden ohne Verwendung linguistischen Wissens. Die Erkennungsrate liegt bei über 95 %.

2.1 Der Erkennungsprozeß

Der Erkennungsprozeß gliedert sich in vier Stufen (siehe hierzu auch die Literaturangabe):

1. **Signalverarbeitung und Vektorquantisierung**. Das Sprachsignal wird mit 12 Bit Abtastgenauigkeit bei 20.000 Abtastungen pro Sekunde digitalisiert. Aus den so

entstandenen Daten wird in einer speziellen Hardware (Spracherkennungskarte) durch Fourier-Analyse alle 10 ms ein 20-elementiger Merkmalsvektor gewonnen, der die Energien in 20 Frequenzbändern zwischen 200 und 8000 Hz beschreibt. Die Frequenzbänder wurden anhand eines "Ohr-Modells" festgelegt. Dieser Vektor wird mit 200 sprecherspezifischen Prototypvektoren verglichen. Diese Prototypvektoren werden in einer ca. 30-minütigen Trainingsphase gewonnen, die jeder Sprecher und jede Sprecherin einmal durchführen muß (siehe unten). Somit werden die Eingangsdaten in dieser Stufe auf 100 Byte/s komprimiert.

2. **Schnelles akustisches Modell**. Jedes Wort des Vokabulars ist als eine Folge von phonetischen Symbolen beschrieben (ähnlich einer Lautschrift). Es wird nun im Prinzip für jedes Wort des Vokabulars die Wahrscheinlichkeit ermittelt, mit der die Aussprache des Wortes die beobachtete Symbolfolge produziert. Die 100 bis 300 Worte mit der größten Wahrscheinlichkeit sind dann die Kandidaten für die weitere Verarbeitung. Für die Vergleichsoperationen wird ein sogenanntes *Hidden Markov Model* verwendet, auf das hier nicht näher eingegangen werden soll.

3. **Sprachmodell**. Hier werden nun die Wahrscheinlichkeiten berücksichtigt, mit denen verschiedene Worte aufeinander folgen. Ein Sprachmodell wird aus der Analyse großer Textsammlungen gewonnen. Es werden die Wahrscheinlichkeiten für Worte (*Unigramm*), Folgen von zwei Worten (*Bigramm*) und Folgen von drei Worten (*Trigramm*) berücksichtigt.

4. **Detailliertes akustisches Modell**. Dieses Modell ähnelt dem schnellen akustischen Modell. Es werden aber anstelle der Phoneme künstliche Lauteinheiten von 10 ms Dauer, sogenannte Pheneme verwendet. Die so gewonnenen Wahrscheinlichkeiten werden mit denjenigen aus der bisherigen Auswertung kombiniert und somit die Erkennungsrate weiter erhöht.

2.2 Die Client-Server-Architektur

Das Spracherkennungssystem ist als Client-Server-Architektur realisiert. Der Server stellt die grundlegenden Spracherkennungsfunktionen bereit. Sie beruhen auf einer Basiskomponente (Sprachdekoder), einem sprachspezifischen Zusatz (deutsch, englisch, italienisch oder französisch) und einem anwendungsspezifischen Sprachmodell (z. B. Radiologie).

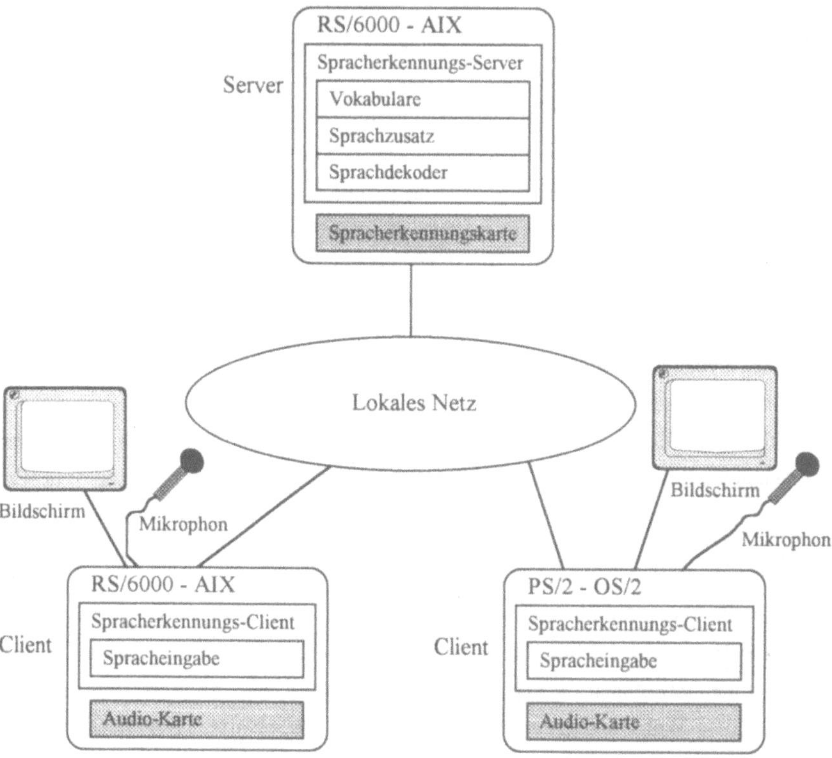

Der Client stellt die graphische Benutzeroberfläche und eine Audiokarte zur Verfügung, an die das Mikrophon zur Spracheingabe angeschlossen wird. Die Oberfläche ist unten näher beschrieben.

2.3 Technische Restriktionen

Das System ist sprecherabhängig, d. h. der Sprecher oder die Sprecherin muß vor dem ersten Diktat dem System die Charakteristika seiner Stimme beibringen. Dies geschieht durch das Sprechen von 84 Trainingssätzen. Nach Abschluß des Trainings steht gleich die volle Erkennungsrate von über 95 % zur Verfügung. Es besteht an dieser Stelle auch die Möglichkeit des Anhörens einer Musteraussprache der Trainingssätze.

Die einzelnen Worte müssen isoliert ausgesprochen werden, d. h. es müssen zwischen den Worten kurze Pausen gemacht werden. Dies ist notwendig, um die Erkennung in Echtzeit durchführen zu können. Die Erfahrung zeigt, daß eine Gewöhnung an die isolierte Sprechweise sehr schnell eintritt und kaum Probleme bereitet. Insbesondere wird die isolierte Sprechweise bereits beim Diktieren der Trainingssätze gelernt.

3 Anbindung der Spracherkennung an existierende Software

Zur Texteingabe mittels natürlicher Sprache dient ein Spracheingabefenster auf einer graphischen Benutzeroberfläche, in dem der über das Mikrophon eingegebene Text in Echtzeit angezeigt wird. Es bietet auch Möglichkeiten, um eventuelle Wortkorrekturen durchzuführen. Zum nochmaligen Anhören der diktierten Worte zu Kontrollzwecken kann ein Lautsprecher an das System angeschlossen werden. In einem weiteren Anwendungsfenster läuft eine herkömmliche medizinische Datenbank oder eine beliebige andere Anwendung. Sogenannte Anwendungsprofile beschreiben die Interaktion der Spracheingabe mit der Anwendung. Sie erlauben es, Texte, Kommandos und Formatierungsanweisungen an das Anwendungsprogramm zu übergeben. Rückmeldungen von der Anwendung an das Spracheingabefenster sind nicht möglich.

Bei der Neuentwicklung oder Erweiterung von Anwendungsprogrammen kann die Spracherkennung über eine Anwendungsprogrammierschnittstelle (*Application Programming Interface*, *API*) auch direkt in die Anwendung integriert werden.

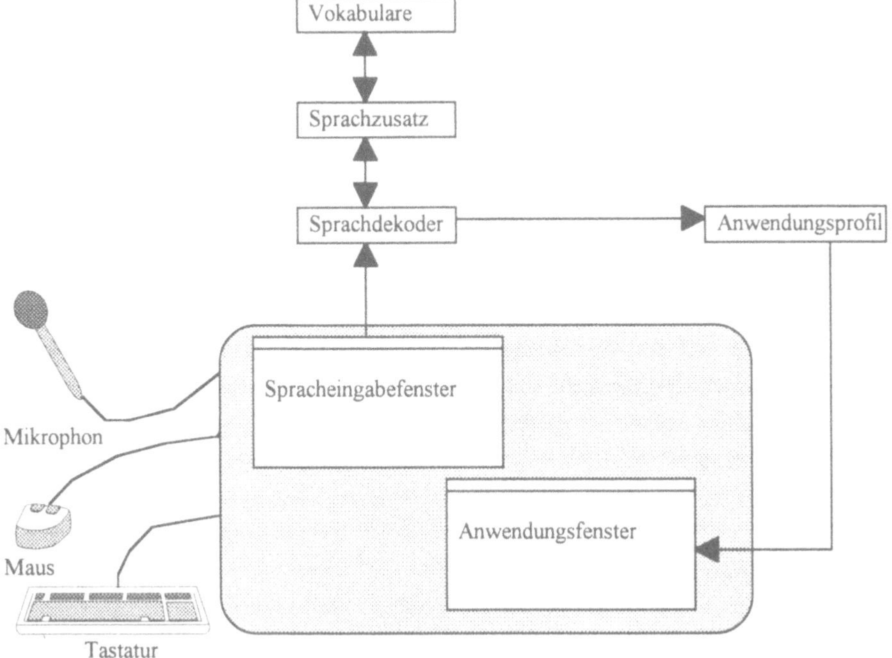

3.1 Ein Anwendungsbeispiel

Nehmen wir an, im Krankenhaus existiere bereits eine Datenbank zur Befunddokumentation. Dann stellt sich ein möglicher Arbeitsablauf nach der Integration der Spracherkennung wie folgt dar:

Ein Arzt sitzt vor einem Arbeitsplatzrechner mit graphischer Oberfläche und trägt ein Kopfmikrophon. Somit hat er die Hände frei, um beispielsweise Röntgenaufnahmen beim Betrachten in die Hand zu nehmen oder ein Mikroskop zu bedienen. Er kann nun mit einem Kommandowort einen neuen Befund eröffnen, z. B. spricht er ins Mikrophon "Neuer Befund".

Nun kann er die Patientendaten aus der Datenbank abrufen, indem er entweder den Namen des Patienten buchstabiert, eine andere Identifikation spricht oder eine Tastatur- oder Mauseingabe vornimmt. Das Spracherkennungsprogramm erwartet jetzt den eigentlichen Befundtext. Der Arzt diktiert den Befund, während er die Röntgenaufnahme oder andere Untersuchungsergebnisse betrachtet. Die dafür aufzuwendende Zeit liegt etwas über derjenigen, die er sonst zum Diktieren in ein Diktiergerät benötigt. Die durchschnittliche Diktiergeschwindigkeit bei isolierter Sprechweise liegt bei 40 bis 60 Worten pro Minute, je nach Sprecher oder Sprecherin.

Ist der Befund fertig diktiert, liest er ihn im Spracherkennungsfenster zur Kontrolle. Der Arzt kann dann mittels einfachen Anklickens falsch verstandener Wörter ein Korrekturfenster aufrufen und das korrekte Wort aus einer Alternativenliste auswählen oder über die Tastatur eingeben. Hier hat er auch die Möglichkeit, sich die akustische Aufnahme des diktierten Wortes noch einmal anzuhören.

Ist der Befund überprüft und gegebenenfalls korrigiert, kann er den Befund in die Datenbank übernehmen (z. B. mit den Worten "Befund fertig"). Mehr ist nicht zu tun. Der Befund befindet sich nun in der Datenbank und kann von dort aus weiterverarbeitet werden. Ein sofortiges Ausdrucken ist natürlich ebenfalls über ein Sprachkommando möglich, z. B. mit den Worten "Befund drucken".

3.2 Fachspezifische Vokabulare

Für einzelne Anwendungen können spezifische Vokabulare erstellt werden. Diese bestehen aus den Wortformen und den statistischen Informationen (Worthäufigkeiten, Bigramme, Trigramme), sind also mehr als bloße Wortsammlungen. Daher werden sie auch als Sprachmodelle bezeichnet. Die Sprachmodelle werden aus großen Textmengen im Umfang von mindestens 500.000 Worten extrahiert. Solche Textkorpora können Sammlungen von früher erstellten Befunden sein. Die Verwendung fachspezifischer Sprachmodelle bringt den wesentlichen Vorteil, daß die verfügbaren 20.000 Wortformen den im Fachgebiet am häufigsten genutzten Worten entsprechen. Somit bemerkt der Anwender kaum etwas von einer Beschränkung der Anzahl der Worte. Außerdem berücksichtigt die Wortfolgestatistik typische Formulierungen (Wortfolgen) in einzelnen Fachgebieten und erhöht somit die Erkennungsrate.

Bis zu 2000 zusätzliche Worte lassen sich während des Diktierens über das Korrekturfenster hinzufügen. Das System lernt also neue Worte beim Diktieren. Ebenso werden beim Diktieren die statistischen Informationen ergänzt und modifiziert. Dadurch ist das System adaptiv und die Erkennungsrate verbessert sich im Laufe der Zeit noch über die anfänglichen 95 % hinaus.

3.3 Kommandos und Textbausteine

Ebenso wie die oben beispielhaft genannten Kommandos "Neuer Befund", "Befund fertig" und "Befund drucken" lassen sich beliebige sprachliche Kommandos zur weiteren Automatisierung der Arbeitsabläufe über Anwendungsprofile definieren. Alle Kommandos, die jetzt noch über die Tastatur eingegeben werden, lassen sich dann über natürliche Sprache eingeben. Beliebige Anwendungsprogramme können so mit natürlicher Sprache gesteuert werden. Das System unterscheidet zwischen gewöhnlichem Text und Kommandoworten. Kommandoworte müssen so gewählt werden, daß sie nicht im Text auftauchen. Kann dies nicht gewährleistet werden, können Schlüsselworte definiert werden, die zwischen einem Diktier- und einem Kommandomodus umschalten (z. B. "Tangora" nach dem Namen des vorangegangenen Forschungsprototypen).

Eine weitere Möglichkeit zur Vereinfachung von Routinevorgängen bietet die Integration von Textbausteinen für häufig auftauchende Textabschnitte. So kann zum Beispiel mit dem Kommando "Normalbefund" eine Auswahlliste von Normalbefunden angezeigt werden, aus denen dann einer mittels Sprache ausgewählt und in den Befund eingefügt werden kann. Da die Textbausteine ebenso wie die Anwendungsprofile gewöhnliche Textdateien sind, läßt sich das System auch vom Endanwender bzw. der Endanwenderin problemlos erweitern. Alle Möglichkeiten existierender Textverarbeitungsprogramme stehen voll zur Verfügung.

Literatur

Walch, G., K. Mohr, U. Bandara, J. Kempf, E. Keppel, K. Wothke. "Der IBM Spracherkennungsprototyp TANGORA: Anpassung an die deutsche Sprache". *Informatik Fachberichte* 219, Proc. 11. DAGM-Symposium Mustererkennung (1989): 543-550.

Spracherkennung mit TANGORA: Medizinische Sprachmodelle und ihre Erweiterung auf Komposita in der deutschen Sprache

Marcus Spies
Abt. Spracherkennung;
Wissenschaftliches Zentrum Heidelberg
IBM Deutschland GmbH
Vangerowstr. 18
6900 Heidelberg

Abstract:

Zu Beginn wird eine kurze Einführung in die wesentlichen Aspekte der Spracherkennungsmethode gegeben, die dem IBM System TANGORA zugrundeliegt. Wesentlich ist hier die Aufteilung des Erkennungsprozesses in einen auf akustischen Daten basierenden Teil (Decodierung) und einen auf Korpora aus dem jeweiligen Anwendungsbereich zurückgreifenden sprachstatistischen Teil (Sprachmodell). Es werden dann die Prinzipien der Sprachmodellierung in TANGORA erläutert. In einem abschließenden Teil werden neuere Ansätze angesprochen, um die beschriebene Methodik auf deutsche Komposita anzuwenden. Dies würde erlauben, nur die Wortbestandteile der Komposita in das anwendungsspezifische Vokabular zu übernehmen, und so die Flexibilität des Sprachmodells noch einmal erweitern.

Allgemeines

TANGORA ist ein Echtzeit-Spracherkennungssystem für große Vokabulare (mehr als 20000 Wortformen), das mit geringem Aufwand vom Benutzer sprecherspezifisch trainiert werden kann. Dieses System hat mittlerweile Eingang in ein Produkt der IBM gefunden: ISSS, IBM Speech Server Series.

Das TANGORA-System basiert auf Forschungen am IBM Research Laboratory in Yorktown (s. Jelinek und Mercer, 1980, Jelinek, Mercer und Bahl, 1982) zur statistischen Spracherkennung (s. Rabiner und Juang, 1986). Seit 1988 wurden durch Zusammenarbeit mit den wissenschaftlichen Zentren der IBM in Paris, Rom und Heidelberg die Grundlagen für eine simultane Produktentwicklung des Spracherkenners in vier Sprachen gelegt (s. Walch et al., 1989; Cerf-Danon et al., 1991).

Anwendungen der Spracherkennung werden derzeit nach Feldversuchen mit Kunden in den Bereichen Medizin, Juristik, Electronic Banking aufgebaut.

Für medizinische Anwendungen liegen, nach Kooperationen mit den Universitätskliniken in Aachen und Mainz, Sprachmodelle in den Bereichen Radiologie und Orthopädie vor. In der Medizin sollen durch die Unterstützung mit ISSS vor allem die Befundung und das Diktieren von Berichten den Arbeitsablauf in Kliniken und größeren Praxen erleichtert werden, da der diktierende Arzt sein Dokument sofort nach dem Diktat schriftlich vor sich sieht, keine Rechtschreibfehler korrigieren muß, und etwaige inhaltliche Abänderungen in einem normalen Editierprozeß sofort vornehmen kann.

Grundlagen der Spracherkennung im TANGORA-System

TANGORA ist ein Echtzeit-Erkennungssystem für diskret gesprochene Sprache; d.h. für Spracheingabe mit kurzen Pausen zwischen den Worten.

Grundlage des Tangora-Systems ist ein Bayesscher Klassifikator, der auf digital vorverarbeitete akustische Signale angewandt wird (s. etwa Walch et al., 1989). Bezeichnet *C* eine Folge derart vorverarbeiteter akustischer Signale zwischen zwei Wortpausen, und *W* einen Wortkandidaten, so wird gerade der Wortkandidat mit der höchsten a-posteriori-Likelihood ausgewählt. Diese ergibt sich nach dem Bayesschen Theorem für einen speziellen Kandidaten aus der Beziehung

$$\Pr(W|C) = \frac{\Pr(C|W)\Pr(W)}{\Pr(C)}$$

indem man den für alle Kandidaten gleichen Nenner als Proportionalitätsfaktor gleich 1 setzt (zu Eigenschaften eines solchen Klassifikators s. etwa Anderson, 1971, pp. 195 ff.).
Den beiden übrig bleibenden Termini in der Formel entsprechen grundlegende Komponenten des Spracherkenners:

- $\Pr(C|W)$ bezeichnet die klassenbedingte Wahrscheinlichkeit, die sich unter Annahme des Wortes *W* für die Signalfolge *C* ergibt. Diese Wahrscheinlichkeit wird in TANGORA durch den sogenannten *Decoder* mithilfe des Viterbi-Algorithmus berechnet (s. Rabiner und Juang, 1986).

- $\Pr(W)$ bezeichnet die Basisrate des Wortkandidaten *W* in der betreffenden Spracherkennungsanwendung. Diese Basisraten werden kontextabhängig in *Sprachmodellen* berechnet, die das statische Wissen für eine Anwendung zusammenfassen.

Man kann also kurz sagen, daß sich die Entscheidung über Wortkandidaten in TANGORA jeweils aus Decoder- und Sprachmodell-Wahrscheinlichkeit ergibt.

Für den Anwender ist bei dieser Architektur die Möglichkeit der Anpassung des vom Spracherkenner verarbeiteten Wortschatzes an branchenspezifische oder sogar individuelle Anforderungen von besonderer Bedeutung. Die Grundlage dieser Anpassungsmöglichkeit sind die im TANGORA-System verwendeten statistischen Sprachmodelle.

Sprachmodelle für Anwendungslösungen

Das Spracherkennungssystem der IBM benutzt, wie erläutert, zur Entscheidung über konkurrierende Worthypothesen aus der akustischen Dekodierung Sprachmodelle. Dies sind aus anwendungsspezifischen Korpora gewonnene Schätzungen von Wortfolgenhäufigkeiten.

Sprachmodelle basieren auf einer Sammlung von Textproben aus dem gewünschten Anwendungsbereich. Aus diesen Textproben werden die häufigsten Wortformen und Wortfolgestatistiken generiert. Die Wortfolgestatistiken werden mithilfe eines ursprünglich aus der Biometrie stammenden Verfahrens geschätzt. Die so gewonnenen Parameter werden schließlich in effizienter Weise dem Laufzeitsystem bei der Spracherkennung verfügbar gemacht.

Das hier angewandte Verfahren zur Häufigkeitssschätzung basiert auf Good, 1953, siehe Jelinek & Mercer, 1980, Nadas, 1948 und 1985, Maltese, 1990.
Geschätzt werden sollen die Häufigkeiten von Wortformtrigrammen in einem gegebenen Korpus. Bei einem Wortschatz von 20000 Wortformen, wie er derzeit in dem Spracherkennungssystem TANGORA genutzt wird, wären ca. 8 Billionen Trigramme möglich; die in Heidelberg gesammelten Korpora umfassen derzeit etwa 1 Milliarde Wörter, sind also immer noch um drei Zehnerpotenzen zu klein, um überhaupt alle Trigramme auch nur beobachten zu können. Dieses Problem der knappen Daten hat für die Biometrie Good (a.a.O.) gelöst, indem er Klassen von Objekten gebildet hat, die in der Stichprobe mit gleicher Häufigkeit vorkommen. Akzeptiert man die sog. Symmetrievoraussetzung, so werden für alle Objekte einer solchen Häufigkeitsklasse auch gleiche Populationshäufigkeiten geschätzt. Die Schätzung basiert auf der Annahme einer Binomialverteilung der Zufallsvariablen, welche die Ziehung eines Objektes aus der Häufigkeitsklasse beschreibt.

Der von Jelinek und Mercer (a.a.O.), sowie von Maltese (a.a.O.) eingeschlagene Weg geht von einem Hidden-Markov-Modell für die zu schätzenden Wahrscheinlichkeiten aus. Dabei werden mehrere im Korpus beobachtete Häufigkeiten benutzt. Für jedes Trigramm *uvw* sind dies

- ein Nullgramm-Term f_0
- ein Unigramm-Term $f(w)$
- ein Bigramm-Term $f(w|v)$
- ein Trigramm-Term $f(w|uv)$.

Diese Terme entsprechen den im Korpus beobachteten relativen Häufigkeiten (der Nullgramm-Term hat nur korrektive Bedeutung).
Faßt man nun diese Terme als Wahrscheinlichkeiten des Wortes *w* unter verschiedenen Bedingungen (= Knoten eines Markovnetzes) auf, so kann man eine latente Variable (oder einen verborgenen Knoten im Markovnetzwerk) hinzufügen, von der aus durch Zustandsübergänge eine der vier Bedingungen erreicht wird, die das Wort *w* erzeugen. Bezeichnet man die Übergangswahrscheinlichkeiten für die betreffenden Terme mit $\lambda_0\ \lambda_1\ \lambda_2\ \lambda_3$, so ergibt sich folgender Ansatz zur Darstellung der gesuchten Trigrammwahrscheinlichkeit:

$$\Pr(w|uv) = \lambda_0 f_0 + \lambda_1 f(w) + \lambda_2 f(w|v) + \lambda_3 f(w|uv)$$

Hierbei gilt, daß

$\lambda_i \geq 0 \quad (i = 0,\ldots,3) \quad$ und

$$\sum_{i=0}^{3} \lambda_i = 1$$

Wir schätzen also die gesuchte Trigrammwahrscheinlichkeit von *w* nach den Worten *u* und *v* durch eine Konvexkombination der im Korpus beobachteten relativen Häufigkeiten der Uni- bis Trigrammterme. Dabei werden die beobachteten relativen Häufigkeiten als Marginalwahrscheinlichkeiten entsprechend vieler Senken eines Netzwerkes angesehen, zu denen Übergänge von einem verborgenen Knoten als Quelle führen. Die korrespondierenden Übergangswahrscheinlichkeiten erlauben eine Gewichtung des Beitrags der beobachteten Häufigkeiten zu der zu schätzenden Wahrscheinlichkeit. Theoretisch müßte man nun für jedes einzelne Trigramm ein solches Modell aufstellen und aus mehreren Korpora die Parameter schätzen. Dies ist wiederum unrealistisch. Darum bildet man (s. Maltese, 1990) analog zu dem Verfahren Goods aufgrund der Beobachtungen in einem Korpus eine Partition der Wortformen nach Häufigkeitsgruppen beobachteter Bigrammterme. Die Annahme ist dann, daß innerhalb solcher hinsichtlich der Häufigkeit beobachteter Bigramme ähnlicher Gruppen die Gewichtung der einzelnen Beiträge der Häufigkeiten gleich ist.

Die eigentliche Schätzung der Übergangswahrscheinlichkeiten (fortan Koeffizienten genannt) von dem verborgenen Knoten erfolgt dann mittels der "deleted estimation" (s. Jelinek und Mercer, 1980). Bei diesem Verfahren werden durch Weglassung von Korpusteilmengen mehrere kleinere Stichproben erzeugt (analog zu Tukey´s jackknifing, s. Jelinek und Mercer, 1980, p. 397); ein Anfangswert der Koeffizienten ergibt für jede Stichprobe eine Likelihood, die maximiert wird. Nadas (1985) zeigt, daß dieses Verfahren dem ursprünglichen nach Good in der Praxis sehr nahe kommt. Alternativ kann man mit dem EM-Algorithmus nach Dempster, Laird und Rubin (1977) abwechselnd erwartete und beobachtete Häufigkeiten generieren, die Differenz der betreffenden Häufigkeiten weist dann gerade in die Richtung des steilsten Anstiegs der Likelihood-Funktion. Es ist allerdings zu beachten, daß bei beiden Vorgehensweisen lokale Optima auftreten können. Es sei angemerkt, daß diese Art der Parameteranpassung äquivalent zum Lernalgorithmus in einem bestimmten Typ neuronaler Netzwerke ist (den Boltzmann-Maschinen, s. Aarts und Korst, 1989).

Als Gütemaß für die Qualität der geschätzten Häufigkeiten läßt sich die von Ferretti, Maltese, und Scarci (1990) diskutierte Perplexität verwenden, die eine verallgemeinerte Entropieberechnung darstellt.

Komposita im Sprachmodell

Ein besonderes Problem bei der Spracherkennung im Deutschen sind die Komposita. Es ist im Deutschen weitaus leichter als in anderen Sprachen möglich, Komposita zu bilden. Gerade im medizinischen Bereich fällt auf, daß Fachtermini, die in anderen Sprachen durch Genitivattribute o.ä. ausgedrückt werden, häufig als deutsche Komposita auftreten. Komposita lassen sich im derzeitigen TANGORA-System problemlos eingeben; sie werden wegen geringer Verwechslungswahrscheinlichkeit leicht erkannt. Allerdings tritt jedes Kompositum als eine Wortform im Wortschatz des Systems auf und belastet ihn dadurch. So wären etwa

Rippenfraktur

und

Femurfraktur

zwei Komposita, die nach derzeitigem Ansatz im Sprachmodell getrennt aufgeführt werden müßten. Unser Ziel ist nun, zunächst im Sprachmodell Komposita nur nach ihren Wortbestandteilen zu speichern. Wir hätten also den Eintrag "Fraktur" und die Einträge "Femur", sowie "Rippen", und eine Schätzung, die uns sagt, wie wahrscheinlich die entsprechenden Komposita sind, aufzustellen.

Wie läßt sich nun aber die Methode der Trigrammstatistiken auf Komposita anwenden? Es scheint nicht sinnvoll, für jeden Kompositabestandteil genauso wie für jedes einzeln stehende Wort Trigrammhäufigkeiten aufstellen zu wollen.U.a. würde dann das Vorkommen beider Kompositabestandteile als einzelnen Worten mit den Kompositahäufigkeiten konfundiert (Bsp.: "Am re. Femur Fraktur mit Splitterbildung"). Außerdem würde sich der tatsächlich berücksichtigte Kontext vor einem Kompositum verkleinern (da man nicht über TRI-gramme hinausgehen will). Daher entsteht die Frage, ob es nicht möglich ist, eine sinnvolle Zerlegung der Wahrscheinlichkeiten vorzunehmen, so daß Kontext und Bestandteile eines Kompositums getrennt berücksichtigt werden können.

Einen Schlüssel zur Lösung dieses Problems liefert die von der Linguistik her bekannte Tatsache, daß im Deutschen die grammatisch bestimmende Teile eines Kompositums immer hinten stehen. Der letzte Wortbestandteil des Kompositums gibt Auskunft über Genus, Kasus, Numerus, wenn das Kompositum eine Substantiv ist. Analoges gilt über Verbkomposita. Wenn man diese Tatsache verallgemeinert, so kann man annehmen, daß der vorausgehende Kontext, in dem ein Kompositum auftritt, die Wahrscheinlichkeit des letzten Bestandteils stark beeinflußt und daß umgekehrt, kennt man diesen letzten Bestandteil, der vorausgehende Kontext wenig über die übrigen Kompositabestandteile aussagt.

In Termini der Wahrscheinlichkeitstheorie läßt sich eine derartige Annahme durch bedingt unabhängige Ereignisse formulieren. Man würde dann sagen, daß, gegeben den letzten Kompositumbestandteil, die vorausgehenden Bestandteile und der vorausgehende Kontext bedingt unabhängig sind. Bezeichnet man mit W den letzten Kompositumbestandteil, mit A die vorausgehenden Bestandteile, und mit C den vorausgehenden Kontext, so kann man diese Annahme wie folgt formulieren:

$$\Pr(AC|W) = \Pr(A|W)\Pr(C|W)$$

beziehungsweise

$$\Pr(A|CW) = \Pr(A|W)$$

Derartige bedingte Unabhängigkeiten sind ein wichtiges Werkzeug in der Aufstellung probabilistischer wissensbasierter Systeme, vgl. Pearl (1988), Spies (1993). Akzeptiert man diese Unabhängigkeitsannahme, so würde sich die gesuchte Trigrammwahrsscheinlichkeit des Wortes W als Kompositumbestandteil hinter A , im Kontext C, folgendermaßen ausdrücken lassen:

$$\Pr(W|CA) = \frac{\Pr(A|CW)\Pr(W|C)}{\Pr(A|C)} = \frac{\Pr(A|W)\Pr(W|C)}{\Pr(A|C)}$$

Hierin treten zwei bekannte Trigrammwahrscheinlichkeiten auf: $\Pr(A|W)$ und $\Pr(W|C)$, also die Wahrscheinlichkeit des letzten Bestandteils *W*, gegeben den Kontext *C*, und die des Kompositumsanfanges *A*, gegeben den Kontext *C*. Neu zu erheben ist dann lediglich die Wahrscheinlichkeit "*innerhalb des Kompositums*" $\Pr(A|W)$. Insgesamt läßt sich also die gesuchte Bestimmung der Trigrammwahrscheinlichkeiten auf zwei bereits aus dem bisherigen Sprachmodell bekannte Terme und einen kompositumssspezifischen Term reduzieren. Damit ist die Möglichkeit gegeben, das Sprachmodell in relativ einfacher Weise für deutsche Komposita zu erweitern.

Diese Möglichkeit basiert allerdings auf der zuvor gemachten Unabhängigkeitsannahme. Derzeit versuchen wir in der Heidelberger Spracherkennungsgruppe, diese Annahme anhand unserer Korpora zu überprüfen. Dabei stellt sich allerdings wieder das Problem geringer Häufigkeiten, sodaß wir nur mit Schätzungen analog zum Verfahren Good's operieren können. Ergebnisse dieser Prüfung werden in Kürze vorliegen.

Abschließend sei gesagt, daß die Behandlung von Komposita im Sprachmodell nur einen der zwei notwendigen Schritte darstellt, um unser Spracherkennungssystem zu erweitern. Der andere Schritt ist eine Modifizierung des TANGORA-Decoders. Hierfür finden sich Vorschläge in Bandara et al., 1991.

Danksagung

Dem Leiter der Spracherkennungsgruppe der IBM in Heidelberg, Herrn Dr. E. Keppel, danke ich für seine Unterstützung. Ferner danke ich meinen Kollegen Dr. U. Bandara, , G. Möse, und Dr. K. Wothke für Gespräche und Informationen, die mir bei der Formulierung des Kompositumproblems geholfen haben. Schließlich danke ich Herrn Mathijs Kadijk, Frau Barbara Kaup und Herrn Michael Mende für die Vorbereitung von Programmroutinen zur Häufigkeitserfassung von Komposita und ihren Bestandteilen.

Literatur

Aarts, E., Korst, J. (1989): Simulated Annealing and Boltzmann Machines. New York, Wiley.

Anderson, T. (1971): An Introduction to Multivariate Statistical Analysis. New York, Wiley.

Bandara, U., Mohr, K., Hitzenberger, L. (1991): Handling German compound Words in an isolated-word speech recognizer. IEEE Workshop Speech Recognition, Arden House, Harriman.

Cerf-Danon, H., DeGennaro, S., Ferretti, M., Gonzalez, J., Keppel, E. (1991): TANGORA - A large vocabulary Recognition System for five Languages. EUROSPEECH '91, pp. 183 - 192.

Dempster, A., Laird, N., Rubin, D. (1977): Maximum Likelihood from Incomplete Data via the EM Algorithm. J. Roy. Stat. Soc., B, 39, pp. 1 -38.

Ferretti, M., Maltese, G., Scarci, S. (1989): Language Model and Acoustic Model Information in Probabilistic Speech Recognition. Proc. IEEE, pp. 707 - 710.

Good, I. (1953): The Population Frequencies of Species and the Estimation of Population Parameters. Biometrika, 40, 3-4, pp. 237 - 264.

Jelinek, F., Mercer, R. (1980): Interpolated Estimation of Markov Source Parameters from sparse Data. In: E. Gelsema, L. Kanal (Hrsgg.): Pattern Recorgnition in Practice. Amsterdam, North Holland, pp. 381 - 397.

Jelinek, F., Mercer, R., Bahl, L. (1982): Continuous Speech Recognition: Statistical Methods. In: P. Krishnaiah, L. Kanal (Hrsgg.): Handbook of Statistics, Vol. 2, Amsterdam, North Holland, p.. 549 - 573.

Maltese, G., (1990): Statistical Models for Italian Language. Res. Rep., IBM Rome Scientific Center. SCR - 009.

Nadas, A. (1984): Estimation of Probabilities in the Language Model of the IBM Speech Recognition System. IEEE Proc. ASSP, 32, 4, pp. 859 - 861.

Nadas, A. (1985): On Turing's Formula for Word Probabilities. IEEE Proc. ASSP, 33, 6, pp. 1414 - 1416.

Pearl, J. (1988): Probabilistic Reasoning in Intelligent Systems. Networks of Plausible Inference. San Mateo, CA, Morgan Kaufman.

Rabiner, L., Juang, B. (1986): An Introduction to Hidden Markov Models. IEEE ASSP Magazine, pp. 4 - 16.

Spies, M. (1993): Unsicheres Wissen. Berlin, Heidelberg, Spektrum Akademischer Verlag.

Walch, G., Mohr, K., Bandara, U., Kempf, J., Keppel, E., Wothke, K. (1989): Der IBM Spracherkennungsprototyp TANGORA. - Anpassung an die deutsche Sprache. 11. DAGM Symposium Mustererkennung. Hamburg, 2. -4. 10. 1989.

Hidden Markov Models – A Unified Approach to Recognition of Spoken and Written Language

Alfred Kaltenmeier, Fritz Class, Peter Regel-Brietzmann, Torsten Caesar, Joachim Gloger, Eberhard Mandler

Abstract

Hidden Markov models (HMM) are currently the most widely used and most successful approach to automatic recognition of *spoken* utterances. Hence it seems straightforward to use HMMs for automatic recognition of *unrestricted handwritten* words. This paper describes a unified system well suited for both recognition tasks. Except for feature extraction which transforms either a binary image or speech samples into a sequence of feature vectors, all system modules are the same for both tasks. Classification consists of a transformation based on linear discriminant analysis and soft-decision vector quantizers which transform feature vectors into sets of symbols and associated likelihoods. Symbols and likelihoods form the input to HMM training and recognition. Word models are concatenated HMMs of subword units which are either letters or context-dependent phones. Results will be presented for both handwriting and speech recognition.

1 Einleitung

Natürlich gesprochene Sprache und kursiv geschriebene Handschrift gehören zu den wichtigsten Formen menschlicher Kommunikation. Beide Formen weisen eine Reihe von Gemeinsamkeiten auf: es handelt sich um sprachliche, sequentielle Äußerungen von Menschen, die starken individuellen Variationen unterliegen.

Gesprochene Sprache wird oft stark vereinfachend angesehen als Folge von Sätzen → Wörtern → Lauten. Die physikalische Form ist ein eindimensionales Signal – der Sprachschall – auf dem die automatische Spracherkennung aufsetzt. Wesentliche Probleme sind hierbei: es gibt keine formalen Regeln, wie Laute zu bilden sind; im Sprachsignal können Laute oder ganze Wörter fehlen, obwohl diese bei der Produktion vorgesehen waren und bei der Perzeption vom Menschen auch nicht vermißt werden; bei spontaner Sprache ist die Sprachkompetenz praktisch nicht zu begrenzen; Satzabbrüche, Wiederholungen und Hesitationen können an beliebiger Stelle vorkommen. Heute ist die Technik noch weit davon entfernt, spontane Sprache automatisch zu erkennen und zu verstehen. Unter gewissen Einschränkungen, vor allem hinsichtlich Anwendungsbereich und Wortschatz, scheint diese Aufgabe jedoch bereits heute lösbar.

Handschrift kann als die verbleibende Spur einer sequentiellen, weitgehend kontinuierlichen Handbewegung eines Schreibers aufgefaßt werden, die vom Menschen als zweidimensionales Muster wahrgenommen und verstanden wird. Dieser Erkennungsprozeß ist nur in Ausnahmefällen buchstabenweise möglich; in der Regel erfordert er ein ganzheitliches Vorgehen, das ganze Wörter, den individuellen, aktuellen Schreibstil, Wissen über den Entstehungsprozeß der Schrift sowie ein Verstehen des Textes erfordert. Beim gegenwärtigen Stand der Technik ist die Handschrift-Erkennung daher auch nur begrenzt einsetzbar; beispielsweise beim Erfassen von postalischen Namen in Adressen. Das Lesen handgeschriebener Adreßbestandteile hat den Vorteil, daß wichtige Teile der Information in der formalen Schreibweise und Anordnung der Wörter enthalten sind.

Obwohl die physikalischen Erscheinungsformen von Handschrift und Sprache völlig unterschiedlich sind, bietet es sich an, die prinzipiellen Gemeinsamkeiten des sequentiellen, sprachlichen Kommunkationsprozesses auf ein einheitliches Klassifikationskonzept abzubilden. Unser Konzept basiert auf semi-kontinuierlichen Hidden Markov Modellen (SCHMM) [HUA89] in Verbindung mit Linearer Diskriminanzanalyse (LDA) [FUK90] und soft-decision Vektorquantisierung.

2 Systemübersicht

Bild 1 zeigt unsere Gesamtsystem; oben sind die Eingangssignale – Handschrift bzw. Sprachsignal – dargestellt, die mit unterschiedlichen Modulen zur Merkmalsgewinnung und Normierung vorverarbeitet werden, Abschnitt 3. Die folgenden Stufen sind gemeinsam für Schrift und Sprache. Der erste Block ist die *Merkmalstransformation und Klassifikation*, Abschnitt 4. Dieser Modul liefert pro Zeit- bzw. Bildfenster mehrere Vektorsymbole zusammen mit Wahrscheinlichkeiten. Im Prinzip erfolgt hier eine Vektorquantisierung für SCHMMs. Merkmalstransformation und -klassifikation sind auf die Modelle abgestimmt und werden bei der *Systemadaption*, Abschnitt 5, optimiert. Die Systemadaption erfolgt off-line anhand von repräsentativen Stichproben. Die *Erkennung*, Abschnitt 6, hypothetisiert mögliche Wörter in der einkommenden Sequenz von bewerteten Vektorsymbolen. Wortmodelle werden dabei durch Verkettung von HMMs entsprechender Wortuntereinheiten gebildet. Bei isolierten Wörtern wird dabei eine *Liste*, bei kontinuierlicher Sprache ein *Netz* (gerichteter Graph) mit Alternativen ausgegeben.

In der Notation von [RAB89] wird ein HMM mit N Zuständen beschrieben durch zwei Wahrscheinlichkeits-Matrizen $[\boldsymbol{A}, \boldsymbol{B}]$ und einen Wahrscheinlichkeits-Vektor $\boldsymbol{\pi}$:

- $\boldsymbol{A}= [a_{ij}]$, $i,j = 1,2,\ldots,N$ ist die Matrix der Übergangswahrscheinlichkeiten, $[a_{ij}]$ die Wahrscheinlichkeit des Übergangs von einem Zustand i zu einem Zustand j;
- $\boldsymbol{B} = [\boldsymbol{b}_j(O_k)], j = 1,2,\ldots,N$ ist die Matrix der diskreten Ausgabewahrscheinlichkeiten;
- $\boldsymbol{\pi}$ ist der Vektor für die Anfangsbelegung der Zustände (hier gilt $\boldsymbol{\pi}^T=[1,0,\ldots,0]$).

Die Struktur der Modelle kann vom Entwickler – in gewissen Grenzen – frei und aufgabenspezifisch definiert werden, Bild 2. Bei nicht-restriktiver Handschrift-Erkennung ist beispielsweise zu beachten, daß jeder Buchstabe in mindestens vier verschiedenen Gestaltklassen auftreten kann: Schreibschrift oder Blockschrift, jeweils Groß- oder Kleinbuchstaben. Diese unterschiedlichen Schreibstile müssen im Modell berücksichtigt werden. Der Modelltopologie von Bild 2, die für alle 26 Buchstaben-HMMs gleich ist, liegt folgende

Idee zugrunde: die *vier* Zeilen eines Modells sollen die vier verschiedenen Schreibstile, die drei Spalten den Anfangs-, Mittel- bzw. Endteil eines Zeichens modellieren. Der angehängte Einzelzustand modelliert den Zwischenraum, der zwischen aufeinanderfolgenden Zeichen auftreten kann.

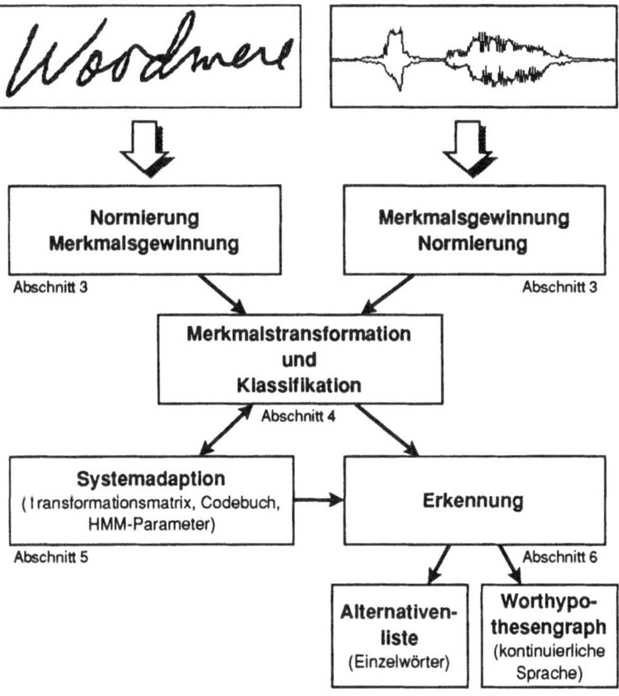

Abbildung 1: Blockschaltbild des Erkennungssystems mit Merkmalsgewinnung, Merkmalstransformation und -klassifikation, HMM-Parametertraining und Viterbi-Erkennung.

Bei der Spracherkennung sind die Modelle einfacher aufgebaut, sie enthalten nämlich nur eine Zeile mit drei Zuständen. Diese sollen – wie bei der Handschrift-Erkennung – den Anfangs-, Mittel- bzw. Endteil eines Lauts modellieren. Da sich benachbarte Sprachlaute gegenseitig stark beeinflussen können (Elisionen, Assimilationen) verwenden wir hier eine größere Anzahl (246) sog. *kontextabhängiger* Modelle. Im Prinzip ließe sich die Kontextabhängigkeit auch in einer Parallelstruktur wie bei der Schrifterkennung modellieren; die 3-Zustands-Modelle sind jedoch bei der Erkennung algorithmisch einfacher zu handhaben.

3 Merkmalsgewinnung

Spracherkennung: Im akustischen Modul des Erkennungssystems wird das Sprachsignal in eine geeignete parametrische Darstellung umgesetzt. Allgemein verwendet man hierzu spektrale Parameter, die in irgendeiner Form aus dem Leistungsdichtespektrum des Sprachsignals abgeleitet werden. Wir verwenden beispielsweise das Mel-Cepstrum

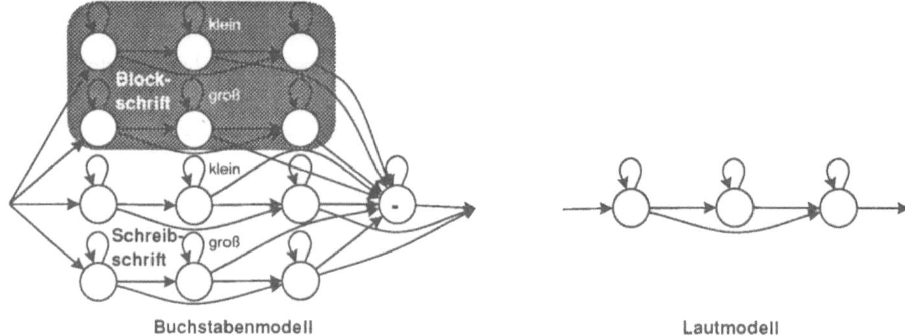

Abbildung 2: HMM-Modelle für Sprach- und Schrifterkennung.

[CLA90] mit 12 Cepstralkoeffizienten **c** und einem logarithmischen Energiekoeffizienten *e*. Ein Sprachsignal liegt damit als zeitliche Folge von Merkmalsvektoren $\{\boldsymbol{v}^T = [\boldsymbol{c}^T, e]\}$ mit 13 Komponenten vor.

Diese Merkmalsvektoren enthalten noch den störenden Einfluß unterschiedlicher Mikrofon- und Sprechercharakteristiken, die möglichst frühzeitig – noch vor der Klassifikation – eliminiert werden sollten. Aus Benutzersicht am angenehmsten ist hierbei eine *nicht-überwachte*, d.h. unbemerkt mitlaufende Adaption. Wir verwenden hierzu ein Verfahren, das im Prinzip auf einer Normierung der Merkmalsvektoren auf das Langzeitspektrum beruht [CLA93]. Hierbei wird durch fortlaufende Mittelwertschätzung und laufenden Mittelwertabzug ein gutes Adaptionsverhalten erreicht, bei dem über einstellbare Zeitkonstanten eine ständige Anpassung an Sprecher-, Mikrofon- und Übertragungscharakteristiken möglich ist.

Schrifterkennung: Im ikonischen Modul des Erkennungsssystems wird ein vorliegendes Handschrift-Muster abgetastet und binär quantisiert. Das Muster ist danach nicht nur mit schreiber-typischen Variationen behaftet, sondern auch von Bildstörungen überlagert. Lokale Bildstörungen können mit ikonischen Bildverarbeitungsmethoden größtenteils ausgeglichen werden; eine weitere Filterung des Bildes ist nach einer Zusammenhangsanalyse möglich [MO89].

Aufgrund der zweidimensionalen Ausprägung von Handschrift ist es keineswegs trivial, den zugrundeliegenden sequentiellen Schreibprozeß, d.h. die Bewegung des Schreibstifts nachzuempfinden. Einfacher, wenn auch sicher nicht so leistungsfähig, ist eine Zerlegung entlang der Hauptschreibrichtung von links nach rechts. Hier kann man mit einer geeigneten Fensterfunktion Untereinheiten bestimmen, die kleiner als ein Buchstabe sind. Jede dieser Untereinheiten wird vertikal in fünf verschiedene Bereiche – den geschätzten typografischen Kennwerten an dieser Stelle des Wortes entsprechend – unterteilt. Diese Bereiche werden auf vordefinierte strukturelle Merkmale hin untersucht, für die im Intervall [0,1] liegende Bewertungsmaße zurückgeliefert werden. Insgesamt wird jede Untereinheit durch einen Vektor mit 19 Komponenten beschrieben; pro Buchstabe werden bei normaler Schreibweise etwa vier Vektoren ausgegeben. Ein Handschrift-Muster liegt damit als quasi-zeitliche Folge von Merkmalsvektoren $\{\boldsymbol{v}\}$ mit 19 Komponenten vor.

Globale Zusammenhänge (Regularitäten) von Handschrift werden statistisch mit Hilfe von Klassifikatoren und HMM-Modellen erfaßt. Lokale Aspekte, d.h. schreiber-typische Variationen des Schriftbildes sollen ebenfalls mit statistischen Methoden erfaßt werden, wobei die entsprechenden Meßdaten jeweils aus einem einzelnen Adreßwort ermittelt werden müssen. Wichtige normierende Maße sind hierbei die Lage der Schreiblinie, die Höhe der Kleinbuchstaben, der Scherwinkel und die Strichdicke; für Details siehe [CGM93]. Mit Hilfe dieser Parameter kann eine Bildnormierung vorgenommen werden. Bild 3 zeigt ein Adreßwort nach Vorverarbeitung und Normierung, vgl. Bild 1.

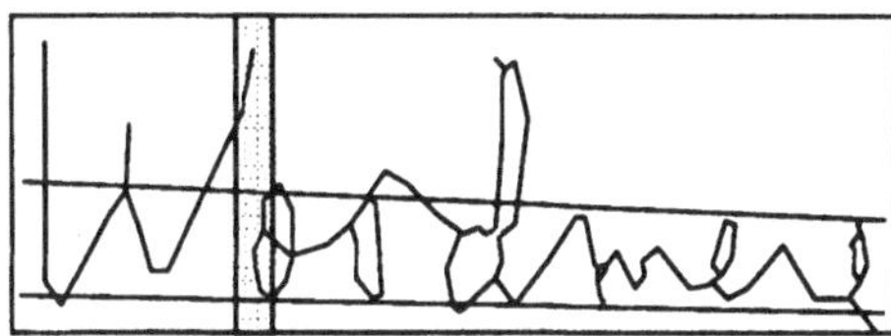

Abbildung 3: Adreßwort nach Vorverarbeitung und Normierung.

Zusätzliche Merkmale: Aus den primären Merkmalen $\boldsymbol{v}$ werden anschließend mit Hilfe einer segmentweisen Regression die dynamischen Merkmale $\triangle\boldsymbol{v}$ und aus diesen wiederum die Merkmale $\triangle\triangle\boldsymbol{v}$ erzeugt. Ein Merkmalsvektor $\boldsymbol{x}^T$=[$\boldsymbol{v}^T$,$\triangle\boldsymbol{v}^T$, $\triangle\triangle\boldsymbol{v}^T$] hat somit $3 \cdot N$ Koeffizienten ($N = 39$ bei Sprach- bzw. $N = 57$ bei Schrifterkennung). Die Länge des Regressionsfensters ist unterschiedlich bei Sprache und Schrift (5 bzw. 3 Rahmen/Segment); ein einem Rahmen t zugeordneter Merkmalsvektor $\boldsymbol{x}_t$ enthält demzufolge Information aus der Vektorfolge $[\boldsymbol{v}_{t-4}, \ldots, \boldsymbol{v}_t, \ldots, \boldsymbol{v}_{t+4}]$ bzw. $[\boldsymbol{v}_{t-2}, \ldots, \boldsymbol{v}_t, \ldots, \boldsymbol{v}_{t+2}]$. Die Gesamt-Segmentlänge stimmt also recht gut mit der mittleren Laut- bzw. Buchstabenlänge (9 bzw. 4 Rahmen) überein.

4 Merkmalstransformation und -klassifikation

Während die Merkmalsgewinnung für Sprache und Schrift aufgrund ihrer physikalisch unterschiedlichen Erscheinungsformen naturgemäß verschieden ist, sind die beiden Module *Merkmalstransformation* und *-klassifikation* identisch, da ja Sprachsignal und Handschriftbild in die gleiche Form – Folge von Merkmalsvektoren – umgesetzt wurden.

Merkmalstransformation: Bei der Sprach- oder Schriftzeichenerkennung handelt es sich um Mehrklassenprobleme, bei denen die einzelnen Klassen durch Merkmalsvektoren repräsentiert werden, die einen hochdimensionalen Merkmalsraum aufspannen. Die Repräsentation eines Musters durch Merkmale ist dabei grundsätzlich unabhängig von Kriterien, die die Trennbarkeit der Klassen im Merkmalsraum festlegen. Wir versuchen hier die Klassenunterscheidbarkeit mit Hilfe einer linearen Transformation zu verbessern, die die einzelnen Klassengebiete im Merkmalsraum möglichst weit voneinander trennt. Diese Transformation soll gleichzeitig die Dimension des Merkmalsraums reduzieren, ohne die Erkennungssicherheit allzu sehr zu beeinträchtigen. Dies leistet die *Lineare Diskriminanzanalyse* (LDA).

Für die LDA muß man *klassenspezifische* Mittelwertvektoren und Kovarianzmatrizen schätzen; man benötigt also eine gekennzeichnete Lernstichprobe. Da wir jedoch keine in Laute und Buchstaben (oder noch kleinere Einheiten) segmentierte Stichproben zur Verfügung haben, setzen wir ein automatisches Verfahren zur Kennzeichnung ein. Die Klassen sind hierbei direkt HMM-Modellen oder einzelnen Modellzuständen zugeordnet.

Üblicherweise wird die LDA-Transformation auf die einzelnen Merkmalssätze $\boldsymbol{v}$, $\Delta\boldsymbol{v}$ und $\Delta\Delta\boldsymbol{v}$ jeweils getrennt angewendet. Wir beschreiten jedoch den einfacheren Weg, die Δ- und $\Delta\Delta$-Merkmale implizit durch ein größeres Zeitfenster über die Primärmerkmale $\boldsymbol{v}$ zu berücksichtigen. Dazu fassen wir 9 bzw. 5 primäre Merkmalsvektoren (bei Sprache bzw. Schrift) zu einem einzigen Vektor mit $9 \cdot 13 = 117$ bzw. $5 \cdot 19 = 95$ Koeffizienten zusammen und transformieren den erweiterten Merkmalsvektor dann mit der LDA-Transformationsmatrix auf 32 Koeffizienten.

Merkmalsklassifikation: Die Worterkennung basiert auf HMMs mit semikontinuierlicher Modellierung der Emissionswahrscheinlichkeiten (SCHMM). Bei diesem Ansatz wird die Verteilung der Merkmalsvektoren mit Hilfe N-dimensionaler, klassenspezifischer Normalverteilungen modelliert. Die Gesamtheit dieser Modellverteilungen bezeichnen wir als Codebuch $\widehat{\boldsymbol{X}} = [\hat{\boldsymbol{x}}_1, \hat{\boldsymbol{x}}_2, \ldots, \hat{\boldsymbol{x}}_L; \boldsymbol{K}_1, \boldsymbol{K}_2, \ldots, \boldsymbol{K}_L]$. Jede Codebuch-Klasse ist dabei bestimmt durch die *Kovarianzmatrix* $\boldsymbol{K}_l$ und den *Mittelwertvektor* $\hat{\boldsymbol{x}}_l$ der zur Klasse l zugehörigen Merkmalsvektoren.

Klassenzahl, Kovarianzmatrizen und Mittelwertvektoren von LDA-Transformation und Codebuch müssen *nicht identisch* sein. Weiterhin sind bei der LDA-Transformation keinerlei Annahmen über klassenspezifische Verteilungen notwendig, während hier Normalverteilungen der einzelnen Codebuch-Klassen angenommen sind.

5 Systemadaption

Unser gesamtes HMM-Erkennungssystem basiert auf statistischen Ansätzen und Modellen, deren Parameter aus einer Lernstichprobe geschätzt werden müssen. Das Vorgehen ist beim Handschrift-Erkenner nur unwesentlich anders als beim Spracherkenner; die Adaption erfolgt im wesentlichen in vier Schritten mit Hilfe einer (bei Handschrift z.T. nach Schreibstil gekennzeichneten) Lernstichprobe:

1. Erzeugung getrennter SCHMM-Codebücher für primäre, Δ- und $\Delta\Delta$-Merkmale und Klassifikation der gesamten Lernstichprobe. Die Codebücher werden mit dem LBG-Verfahren [LIN80] erzeugt, das auf die soft-decision Adaption von Normalverteilungsklassifikatoren erweitert wurde [CLA93].
2. Initialisierung von 37 HMM bei Sprache und 26 bei Schrift.
   - Modelltraining mit dem iterativen Forward/Backward-Algorithmus [BAU67]. Bei Schrift wird hier nur der gekennzeichnete Teil der Lernstichprobe verwendet.
   - Initialisierung von 246 kontextabhängigen Modellen (nur bei Sprache).
   - Weitertraining mit der gesamten Lernstichprobe bei Sprache und Schrift.
   - Ausgabe der klassenspezifischen Mittelwerte und Kovarianzmatrizen am Ende der letzten Trainingsiteration (Merkmalsvektoren mit 117 bzw. 95 Koeffizienten).
4. - Clusterung von Mittelwerten und Kovarianzmatrizen (Normalverteilungen) auf eine vorgebene neue Klassenzahl. Dieses reduzierte Codebuch entspringt einem Kompromiß zwischen Umfang der Lernstichprobe, erreichbarer Erkennungsleistung sowie Rechenzeit- und Speicherplatzbedarf des Erkenners.

- Berechnung der LDA-Transformationsmatrix und Codebuch-Transformation.
- LDA-Transformation und Klassifikation der gesamten Lernstichprobe mit dem *LDA-transformierten* Codebuch. Im Gegensatz zu 1. wird hier *ein* Codebuch verwendet.

5. Wiederholung von Schritt 2 ⇒ endgültige HMM-Parameter.

6 Erkennung

Die Erkennung basiert auf HMMs von Wortuntereinheiten. Hierzu werden Wortmodelle – nach den Vorschriften des Wortlexikons – aus HMMs ihrer entsprechenden Untereinheiten zusammengesetzt und als ganzes verarbeitet. Ein Wortmodell ist demnach ein Graph, bestehend aus einer Menge von Knoten (HMM-Zuständen) und Kanten (Übergängen), vgl. Bild 2. Mit Hilfe des Viterbi-Algorithmus wird der beste Pfad durch ein solches Wortmodell berechnet, gegeben die Observationsfolge von bewerteten Vektorsymbolen. Das Erkennungsergebnis ergibt sich aus der besten Pfadbewertung *aller* Wortmodelle.

Bei geschriebenen oder gesprochenen Einzelwörtern sind Wortgrenzen relativ sicher auffindbar, was bei kontinuierlich gesprochener Sprache jedoch nicht mehr möglich ist. Im Prinzip kann ja jedes Lexikonwort an jeder Stelle des Signals beginnen und an jeder nachfolgenden Stelle enden; dies ergäbe eine riesige Menge von Hypothesen. Man muß den Erkennungsalgorithmus daher modifizieren. Wir verwenden beispielsweise den aus der Literatur bekannten 'wortabhängigen N-best Viterbi-Algorithmus'. Die Ausgabe besteht hier nicht mehr aus einer Liste mit Wortalternativen wie bei der Einzelwort-Erkennung, sondern aus einem Wortnetz – einem Graphen mit Zeitmarken als Knoten und bewerteten Worthypothesen als Kanten [CLA93]. Dieser Graph bildet die Schnittstelle zur linguistischen Verarbeitung unseres 'sprachverstehenden' Gesamtsystems.

Die folgenden Ergebnisse gelten für die Einzelwort-Erkennung; damit ist ein direkter Vergleich zwischen Sprach- und Schrifterkennung möglich. Ergebnisse für kontinuierlich gesprochene Sprache finden sich in [CLA93] und [CLA93a].

7 Ergebnisse

Das Spracherkennungssystem wurde trainiert an einem 1000-Wörter Vokabular, jeweils gesprochen von 10 Sprechern. Das Testvokabular bestand aus 400 *anderen* Wörtern, gesprochen von 4 *anderen* Sprechern. Tabelle 1 vergleicht die 3-Codebücher-Version und die 1-Codebuch-LDA-Version und zeigt den Einfluß kontextabhängiger Modelle (128 bzw. 246). Dargestellt sind die mittleren Fehlerraten bei verschiedenen Testbedingungen, die Kennzeichnungen N_0, N_{100} und N_{140} entsprechen verschiedenen Geräuschsituationen.

Das Handschrift-Erkennungssystem wurde mit 4340 handgeschriebenen US-Postortsnamen trainiert und mit 460 anderen Adreßwörtern getestet. Lern- und Teststichproben sind – genau wie bei den Experimenten zur Spracherkennung – vollkommen getrennt. Tabelle 2 zeigt die Fehlerraten in Abhängigkeit von einer vorgegebenen Lexikongröße. Die Ergebnisse sind bei der Handschrift-Erkennung zwar merklich schlechter als bei der Spracherkennung (ungefähr 20% bei einer Lexikongröße von 350 bzw. 400 Wörtern), die Eignung

Testbedingung	N_0	N_{100}	N_{140}
128 Modelle, 3 CB	9.1%	8.3%	15.5%
246 Modelle, 3 CB	7.3%	7.9%	13.7%
37 Modelle, LDA	4.2%	6.1%	12.0%
128 Modelle, LDA	3.2%	4.1%	7.0%
246 Modelle, LDA	2.5%	2.9%	6.0%

Tabelle 1: **Spracherkennung**: Fehlerraten für verschiedene Testbedingungen (Lexikongröße: 400 Wörter).

des HMM-Ansatzes ist jedoch eindeutig bewiesen. Der Unterschied zwischen Handschrift-Erkennungsraten bei Lern- und Teststichproben, auf den wir bisher noch nicht eingegangen sind, ist ein Indiz dafür, daß dieses Defizit größtenteils aus einer nicht ausreichenden und repräsentativen Lernstichprobe resultiert. Daher ist schon bald mit weiteren Verbesserungen zu rechnen.

Lexikongröße	100	200	350	1000
Fehlerrate	16.0%	19.0%	23.0%	31.2%

Tabelle 2: **Handschrift-Erkennung**: Fehlerraten für verschiedene Lexikongrößen

Literaturverzeichnis:

[*BAU*67] L.E. Baum, J.A. Eagon: *An Inequality with Applications to Statistical Prediction for Functions of Markov Processes and to a Model for Ecology*, Bull. Amer. Math. Soc., Vol. 73, 1967, pp. 360–363.

[*CGM*93] T. Caesar, J.M. Gloger, E. Mandler: *Design of a System for Off-line Recognition of Handwritten Word Images*, Proc. 1st European Conference Dedicated to Postal Technologies, Vol. 1, June 14-16, 1993, Nantes, France.

[*CLA*90] F. Class: Standardisierung von Sprachmustern durch vokabular-invariante Abbildungen zur Anpassung an Spracherkennungssysteme, Fortschrittber. VDI, Reihe 10, Nr. 131, VDI-Verlag Düsseldorf, 1990.

[*CLA*93] F. Class, A. Kaltenmeier, P. Regel-Brietzmann: *Optimization of an HMM-based Continuous Speech Recognizer*, subm. to EUROSPEECH'93, Berlin.

[*CLA*93*a*] F. Class, A. Kaltenmeier, P. Regel-Brietzmann: Evaluation of an HMM Speech Recognizer with Various Continuous Speech Databases, subm. to EUROSPEECH'93, Berlin.

[*FUK*90] K. Fukunaga: Introduction to Statistical Pattern Recognition, Academic Press, New York, Second Edition, 1990.

[*HUA*89] X. D. Huang, M. A. Jack: *Semi-continuous Hidden Markov Models For Speech Signals*, Computer Speech and Language, Vol. 3, 1989, pp. 239–251.

[*LIN*80] Y. Linde, A. Buzo, R.M. Gray: *An Algorithm for Vector Quantizer Design*, IEEE Trans. COM, Vol. 28, No. 1, Jan. 1980, pp. 84–95

[*MO*89] E. Mandler, M.F. Oberländer: *Ein single-pass Algorithmus für die schnelle Konturkodierung von Binärbildern*, Proc. 12th DAGM-Symposium Mustererkennung, ed. by R.E. Grosskopf, Aalen, 1990.

[*RAB*89] L.R. Rabiner: *A tutorial on Hidden Markov Models and Selected Applications in Speech Recognition*, Proc. IEEE, Vol. 77, No. 2, Febr. 1989, pp. 257–285.

Detektion und Klassifikation der P-Wellen im EKG durch Gabor-Wavelets

M. Michaelis[1], S. Perz[1], C. Black[1] und G. Sommer[2]

[1] GSF - MEDIS-Institut, D-8042 Neuherberg, Email: michaeli@gsf.de
[2] Institut für Informatik, Christian-Albrechts-Universität, D-2300 Kiel

Abstract. Durch eine Wavelettransformation mit einem komplexen Gabor-Filter kann ein zeitabhängiges Signal in eine Darstellung transformiert werden, die durch Zeit, Skala und Energie bzw. Zeit, Skala und Phase parametrisiert ist. Die Energie ist dabei ein Maß für das Auftreten von Struktur einer bestimmten Größe im Signal. Dies wird in der vorliegenden Arbeit für die Detektion von P-Wellen in EKG-Signalen genutzt. Durch eine Interpretation des Phasenbildes werden die gefundenen P-Wellen klassifiziert.

1 Einleitung

Die Untersuchung des Elektrokardiogramms (EKG) ist ein in der klinischen Routine und ärztlichen Praxis weitverbreitetes Verfahren zur Diagnostik kardialer Erkrankungen. Der erfolgreiche Einsatz automatisierter Verfahren der EKG-Analyse erfordert die Bereitstellung geeigneter Algorithmen der Signalverarbeitung zur Merkmalsgewinnung für die Klassifikation der komplexen, medizinisch relevanten Muster.

Zu den Domänen der EKG-Diagnostik zählen die Erfassung des Rhythmus, der Herzfrequenz, der physiologischen und pathologischen Reizbildungszentren, die zeitliche Erfassung der Vorhofsystole und der AV-Überleitungsstörungen. Die Voraussetzung hierfür ist die zuverlässige Erkennung und Klassifikation der Vorhoferregungspotentiale, die im EKG als P-Wellen bezeichnet werden (siehe Abb.1). Die Erkennung und Klassifikation von P-Wellen ist wegen der niedrigen Amplituden und des damit verbundenen ungünstigen Signal-zu-Rauschverhältnisses in vielen Fällen eine schwierige Aufgabe.

Im Folgenden werden die Möglichkeiten der Wavelettransformation zur Detektion und Klassifikation der P-Wellen diskutiert. Während insgesamt bei der Verarbeitung nicht stationärer Signale Wavelettransformationen eine zunehmende Bedeutung gewonnen haben [1], gibt es im Bereich der automatischen EKG-Verarbeitung diesbezüglich bislang nur vereinzelte Untersuchungen [6].

2 Methodik der Wavelettransformation

2.1 Quadraturfilter

Verwendet man lineare Filter zur Detektion von Struktur in Signalen, tritt im Antwortsignal immer eine Vermischung der beiden Informationen 'Struktur ist

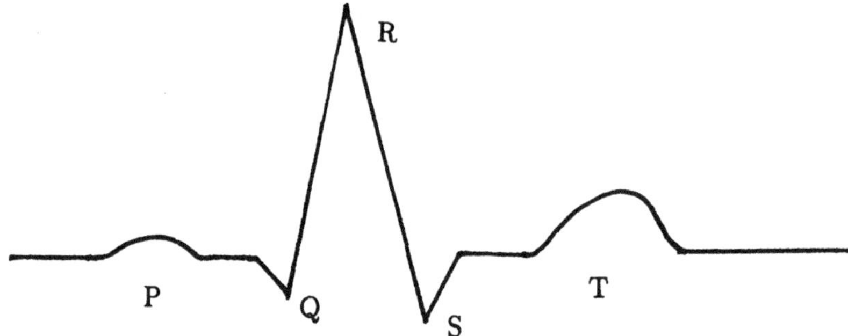

Fig. 1. (a) Schematische Darstellung der elektrischen Herzaktion (EKG). Die P-Welle charakterisiert die Erregung der Vorhöfe.

vorhanden' (Energie) und 'Abstand von Filter und Strukur' (Phase) auf. Dies ist in Abb.2 veranschaulicht. Eine verschwindende Filterantwort kann sowohl bedeuten, daß keine Struktur im Signal ist, als auch, daß Filter und Struktur gegeneinander um eine bestimmte Strecke verschoben sind. Diese Vermischung kann mit linearen Filtern nicht vermieden werden. Um trotzdem Energie und Phase trennen zu können, wird ein komplexer Filter verwendet, dessen Realteil eine gerade Funktion und dessen Imaginärteil eine ungerade Funktion ist. Im Idealfall bilden beide Funktionen ein Quadraturpaar, d.h. sie gehen durch eine Hilberttransformation [7] auseinander hervor. Real- und Imaginärteil des Antwortsignals des komplexen Filters können zu Energie und Phase kombiniert werden [3] (vgl. Abb.2):

$$\text{Energie} = |\text{RE}|^2 + |\text{IM}|^2 \qquad \text{Phase} = \arg\left(\frac{\text{IM}}{\text{RE}}\right). \tag{1}$$

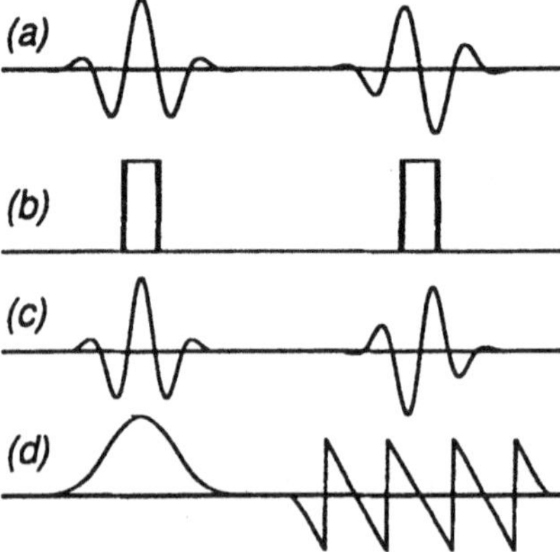

Fig. 2. Beispiel zur Anwendung von Quadraturfiltern: (a) gerader und ungerader Filter, (b) Signal, (c) Filterantwort und (d) Energie und Phase.

2.2 Gabor-Wavelets

Die Detektion von Mustern in nicht stationären Signalen ist eng verbunden mit einer **lokalen** Frequenzanalyse. Die verwendeten Filter sollten daher sowohl im Zeit- als auch im Frequenzbereich gut lokalisiert sein. Diese gleichzeitige Lokalisierung wird durch das aus der Quantenmechanik bekannte und 1946 von D. Gabor [2] in die Informationstheorie eingeführte Unschärfeprinzip begrenzt. Falls σ_t und σ_ω die Varianzen einer Funktionen im Zeit- und Frequenzbereich bezeichnen gilt: $\sigma_t \sigma_\omega \geq 1/2$. In jüngster Zeit wurden eine Reihe von Zugängen entwickelt, um den durch die Unschärferelation limitierten Aufgaben Lokalisation und Detektion von Signalstrukturen auf geeigneten Repräsentationen zu begegnen [8]. Hierzu gehören auch die von J. Morlet eingeführten Gabor-Wavelets [5]:

$$g_{\sigma,t_0}(t) = e^{ic\frac{(t-t_0)}{\sigma}} e^{-\frac{1}{2}\frac{(t-t_0)^2}{\sigma^2}} \quad . \tag{2}$$

Dabei ist 'c' eine fest gewählte Konstante, welche die Zahl der Schwingungen der Modulation unter der Gaußglocke festlegt (Abb.4). Die Parameter σ und t_0 legen die Skala der Funktion und deren Verschiebung im Zeitbereich fest. Die Projektion auf ein so parametrisiertes Funktionensystem bezeichnet man als Wavelettransformation.

3 Spezifikation der EKG-Daten

Für die nachfolgenden Analysen wurden exemplarisch EKG-Daten mit normal konfigurierten (monophasischen), doppelgipfligen und biphasischen P-Wellen ausgewählt. Entsprechend der Morphologie in den einzelnen Ableitungen charakterisieren diese P-Ausprägungen verschiedene Vorhoferkrankungen (siehe Abb.3).

Die verwendeten Ruhe-Elektrokardiogramme bestehen aus den 12 konventionellen Ableitungen, mit einer Registrierdauer von 5 Sekunden und einer Abtastrate von 4ms. Die Amplitudenauflösung beträgt 12 Bit, bezogen auf den Bereich von ±5mV des an der Körperoberfläche abgeleiteten Signals. Die im folgenden angegebenen Werte für σ sind auf die Zeiteinheit 4ms bezogen.

4 Anwendung der Gabor-Wavelets zur P-Wellen-Analyse

4.1 Detektion der P-Wellen

Die Wahl des Parameters 'c' in (2), der die Anzahl der Oszillationen der Gabor-Funktionen festlegt, ist von elementarer Bedeutung. Die Abbildung 4 zeigt zwei Gaborfunktionen mit dem von uns gewählten Wert $c = 1.9$ und dem in anderen Arbeiten meist verwendeten Wert $c \approx 5$. Die Wahl eines größeren c verkleinert die Varianz der Funktion im Frequenzbereich, allerdings, wie es das Unschärfeprinzip vorschreibt, auf Kosten der Breite im Zeitbereich. Dies wird in Abb.5 veranschaulicht, die die Energieverteilung über den Parametern t_0 und σ für

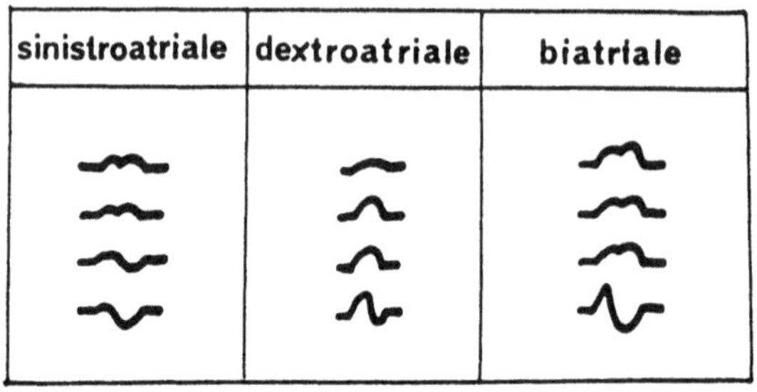

sinistroatriale	dextroatriale	biatriale

Fig. 3. (b) Schematische Darstellung verschiedener Ausprägungen der P-Welle als Folge der Überlastung der Vorhöfe in den Ableitungen I, II, III und V1. (nach Wagner, J.: Praktische Kardiologie, de Gruyter, 1985, S.49)

eine P-Welle zeigt. Auf der Skala des Energiemaximums entspricht die halbe Wellenlänge der modulierenden harmonischen Funktion in (2) in etwa der Breite der P-Welle. Das heißt, daß die Seitenmaxima der Funktion in Abb.4b ($c = 5.3$) weit über die P-Welle hinaus in den energetisch wesentlich bedeutenderen QRS-Komplex hineinreichen.

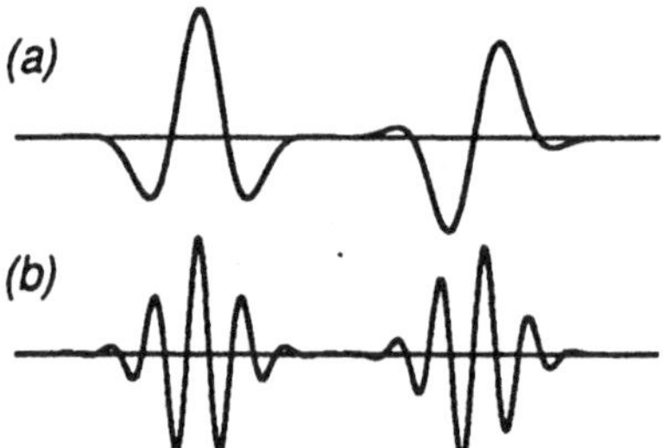

Fig. 4. Gaborwavelets (a) $c = 1.9$, (b) $c = 5.3$ (siehe Text).

Da die interessierenden Strukturen Steigungen, Halbwellen (normale P-Wellen) und Wellen mit einer Schwingung (biphasische P-Wellen) sind, ist eine Wahl von $c = 1.9$ angebracht, wodurch die Gaborfunktion im wesentlichen auf eine Oszillation beschränkt bleibt. Die Wavelettransformation ist hierdurch weniger als lokale Frequenzanalyse, sondern vielmehr als ein 'template matching' aufzufassen.

Diese Wahl wird von zwei Nachteilen begleitet. Zum einen hat der Realteil von (2) einen nicht mehr zu vernachlässigenden Gleichanteil, was durch folgende Modifikation behoben wird (N ist eine Normierung auf $L^1(g) = 1 + i$):

$$g_{\sigma,t_0}(t) = N(e^{ic\frac{(t-t_0)}{\sigma}} - e^{-\frac{1}{2}c^2})e^{-\frac{1}{2}\frac{(t-t_0)^2}{\sigma^2}} \quad . \tag{3}$$

Zum anderen sind Real- und Imaginärteil keine exakten Quadraturpartner. Dies wird jedoch in Kauf genommen, da die echte Hilberttransformierte des Imaginärteils im Zeitbereich nur sehr langsam abfällt. Vor der Transformation werden die EKG-Signale mit einer Gaußfunktion ($\sigma = 4 \ldots 8$ms) geglättet. Dies

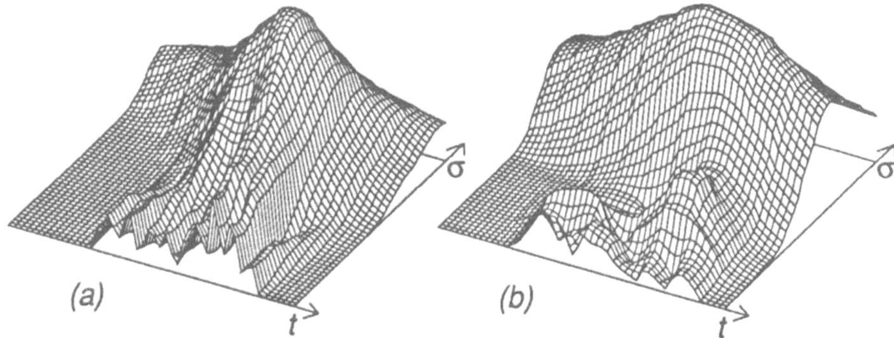

Fig. 5. Energieverteilungen einer P-Welle zu den Gaborwavelets mit (a) $c = 1.9$ und (b) $c = 5.3$.

zerstört keine der interessierenden Strukturen, vermindert jedoch den hochfrequenten Rauschanteil.

Die Abbildung 6a,b zeigt ein mit den Funktionen (3) gewonnenes Energie- und Phasenbild. Zur Detektion der P-Welle wird das Energiesignal auf einigen, empirisch als günstig ermittelten Skalen ausgewertet. Abb.6e-g zeigen die Energie auf den Skalen $\sigma = 4.8, 9.6$ und 22ms. Zuerst wird auf einer großen Skala (σ = 22ms) das absolute Maximum auf einer Länge von ± 0.4s gesucht, welches zum QRS-Komplex gehört. Die Zeit von 0.4s wurde so gewählt, daß ein nicht zum QRS-Komplex gehörendes Maximum in diesem Intervall sicher kein absolutes Maximum ist.

Die P-Welle findet man durch ein Maximum der Energie auf einer großen Skala (σ = 31.2ms) vor dem QRS-Komplex. Auf dieser Skala werden die Feinheiten der P-Welle und das Rauschen nicht mehr aufgelöst, wodurch man einen Überblick über den interessierenden Bereich bekommt. Die nächst kleineren Skalen werden jetzt sukzessive benutzt, um den Bereich der P-Welle weiter einzuschränken. Das Energiesignal zerfällt dann in mehrere Maxima deren Zusammenhang sich aus der groben Skala ergibt. Das Verfahren bricht ab, wenn das Signal nicht mehr vom Rauschen getrennt werden kann. Es soll, auf Kosten der genauen Lokalisierung von Anfang und Ende, sichergestellt sein, daß der ausgeschnittene Bereich die P-Wellen vollständig enthält aber auch keine andere energetisch bedeutende Struktur mit angeschnitten ist.

Abb.7 zeigt ein Beispiel eines gestörten EKG-Signals mit schwach ausgeprägter P-Welle. Die P-Welle konnte trotzdem richtig detektiert werden, die Lokalisierung ist jedoch beeinträchtigt. Das zur Detektion verwendete Energiesignal auf einer großen Skala (σ = 31.2ms) wird von den Störsignalen nicht beeinflußt.

4.2 Klassifikation der P-Wellen

Die in Abb.3 dargestellten Ausprägungen der P-Welle werden morphologisch in drei Klassen unterschieden: monophasisch, doppelgipflig und biphasisch. Die

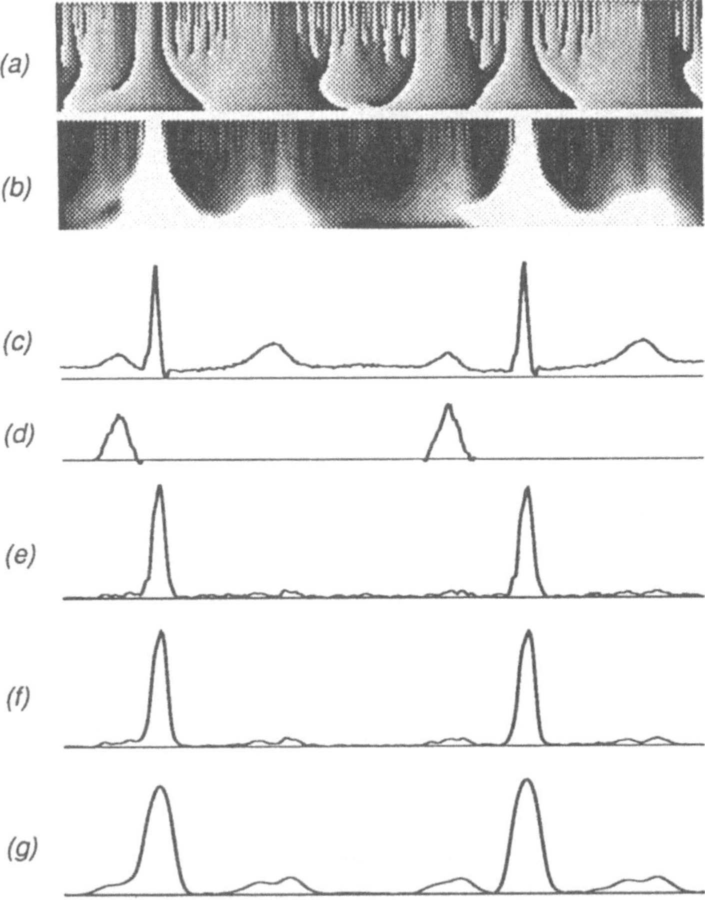

Fig. 6. Waveletanalyse eines EKG-Signals: (a) Phase und (b) Energie. Der Skalenbereich reicht von σ = 3ms (oben) bis σ = 96ms (unten). Im Energiebild sind P- und T-Welle gegenüber dem QRS-Komplex in der Helligkeit angehoben. (c) EKG-Signal mit kurzer PQ Strecke im ersten Zyklus, (d) detektierte P-Wellen, (e) - (g) Energiesignal zu den Skalen $\sigma = 4.8, 9.6, 22$ms.

Abbildung 9a zeigt exemplarisch jeweils ein Beispiel dieser Klassen. In Ergänzung zur Energie des Signals, die zwar die Existenz aber nicht die Art der P-Welle anzeigt, enthält die Phase die notwendige Information zur Klassifikation.

Die analysierende Funktion (3) besteht aus einem geraden Realteil und einem ungeraden Imaginärteil. Der relative Anteil dieser beiden Funktionen an der Energie kommt in der Phase zum Ausdruck. Ein ungerades Signal wird einen großen imaginären Anteil an der Energie haben, d.h. eine Phase um $+\pi/2$ oder $-\pi/2$, während bei einem geraden Signal der Realteil dominiert (Phase $0, +\pi$ oder $-\pi$). Der Zeitpunkt bezüglich dessen das Signal als gerade oder ungerade klassifiziert wird ist der Mittelpunkt t_0 des Filters. Die Phase wird also bestimmt

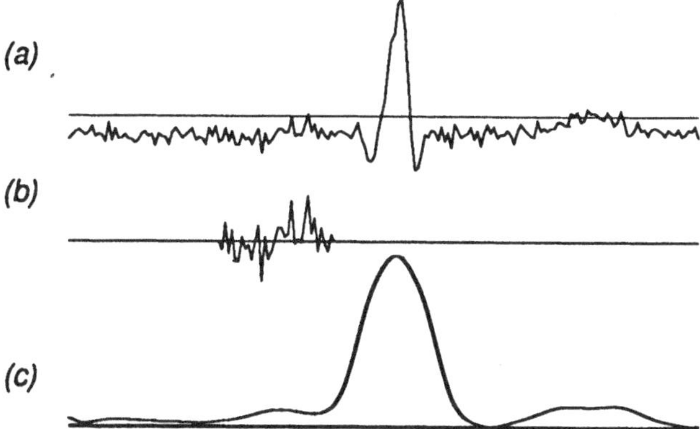

Fig. 7. Beispiel einer verrauschten Ableitung mit schwach ausgeprägter P-Welle. (a) Signal, (b) detektierter P-Wellenbereich, (c) Energiesignal zur Skala $\sigma = 31.2$ms.

von der Struktur selbst, dem Abstand zu ihr und der Skala auf der man sie betrachtet. Zur Klassifikation der P-Welle wurden zwei Verfahren untersucht:

A: Interpretation der Phase am absoluten Energiemaximum

Das Ziel dieser Methode ist es, auf schnelle und einfache Weise P-Wellen in die beiden Klassen 'biphasisch' und 'nicht biphasisch' einzuteilen. Sucht man das absolute Maximum der Energie, gibt die Phase an dieser Stelle Auskunft über die Symmetrie der P-Welle. Das Maximum der Energiefunktion sollte dazu möglichst ausgeprägt sein, wofür sich unsere Wahl des Parameters $c = 1.9$ aus (3) als geeignet erweist (vgl. Abb.5). Das Ergebnis der Klassifikation von ca. 1100 P-Wellen, die nach dem in 4.1 beschriebenen Verfahren automatisch detektiert und ausgeschnitten wurden ist in Abb.8 dargestellt. Es zeigt sich, daß die 'geraden' P-Wellen (monophasisch und doppelgipflig) durch die Phase nicht sauber einzugrenzen sind. Dies hat mehrere Ursachen: (1) schwach ausgeprägte Energiemaxima, (2) Anfang und Ende der P-Wellen sind nicht genau genug bestimmt, und (3) ein asymmetrischer Aufbau von monophasischen und doppelgipfligen P-Wellen.

Da jedoch biphasische P-Wellen weniger häufig sind, als nicht biphasische, lassen sich ca. 75% der P-Wellen sicher als nicht biphasisch klassifizieren.

B: Bestimmung der lokalen Extrema anhand des Phasenbildes

Um eine robuste Klassifikation aller drei Ausprägungen von P-Wellen zu erhalten, wird das Phasenbild als ganzes untersucht. Abb.9b zeigt hierzu ein Phasenbild, bei dem alle Bereiche negativer Phase dunkel und die positiver Phase hell dargestellt sind. Von Interesse sind die Bereiche, in denen die Phase ihr Vorzeichen wechselt. Dies sind die Bereiche, in denen die Energie vom symmetrischen Realteil dominiert wird. Im Signal sind zu diesem Zeitpunkt (und auf der entsprechenden Skala) lokale Maxima (Phase 0) oder Minima (Phase π) zu erkennen. Für dieses Verfahren interessieren nur Bereiche mit einer hohen Energie, da die Phase in den anderen Bereichen von unbedeutenden Strukturen bestimmt wird. Auf kleiner Skala ist dies die Breite der P-Welle, für größere

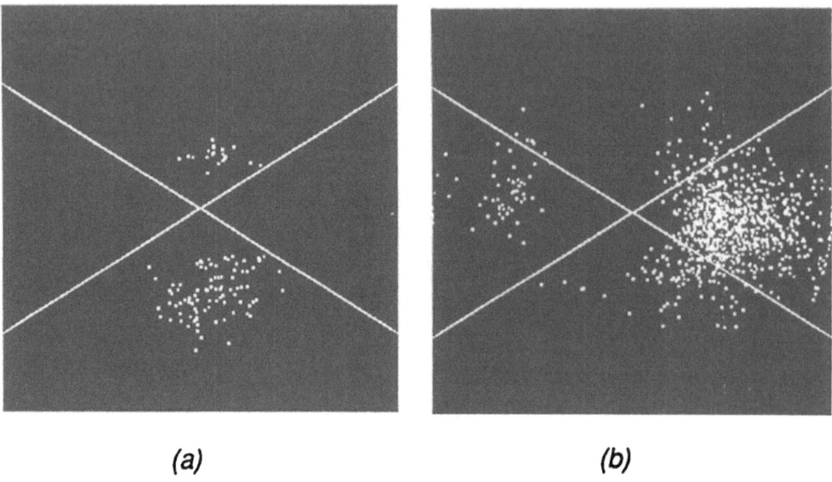

(a) (b)

Fig. 8. Phase (Winkel) und Energie (Abstand vom Zentrum) der nach dem Verfahren A ausgewerteten 1100 P-Wellen. Die Phasenbereiche der oberen und unteren Quadranten sind den unsymmetrischen Signalen zugeordnet (biphasische P-Wellen), der linke und rechte Quadrant den symmetrischen Signalen. (a) biphasische P-Wellen und (b) nicht biphasische P-Wellen.

Skalen wächst die Breite mit den analysierenden Gaborfunktionen an.

Biphasische P-Wellen zeichnen sich dadurch aus, daß sie ein Maximum und Minimum auch auf sehr großen Skalen besitzen. Entsprechend sieht man im Phasenbild zwei Linien eines Vorzeichenwechsels im Bereich der P-Welle, die sich über alle Skalen fortsetzen. Verfolgt man diese Linien in Richtung kleine Skalen findet man den genauen Ort der Extrema. Die Anwendung einer Grob-Fein-Strategie garantiert hierbei Robustheit gegenüber Rauschen.

Monophasische und doppelgipflige P-Wellen haben dementsprechend nur eine Linie eines Vorzeichenwechsels der Phase, die sich über alle Skalen erstreckt. Auf kleineren Skalen tauchen bei den doppelgipfligen P-Wellen dann weitere Maxima oder Minima auf. Die Länge dieser Linien im Skalenraum gibt Auskunft über die Ausgeprägtheit der Extrema. Verfolgt man, ausgehend von den großen Skalen, die Vorzeichenwechsel zu den kleinen Skalen, findet man auch hier die genauen Zeitpunkte der Maxima oder Minima. Der Abstand der Gipfel doppelgipfliger P-Wellen ist in Verbindung mit der P-Dauer eine diagnostisch relevante Größe.

5 Ausblick

Die durchgeführten Untersuchungen am Beispiel der P-Wellenanalyse zeigen, daß im Bereich der automatischen Klassifikation von Elektrokardiogrammen die speziellen Eigenschaften der Wavelettransformation interessante Möglichkeiten in Zusammenhang mit der Lokalisation und Detektion signifikanter Ereignisse bieten. Die geforderte lokale Adaptivität der Analyse wird mit Hilfe einer Wavelettransformation in einer Skalenhierarchie erreicht. Hierbei gestattet die Anwen-

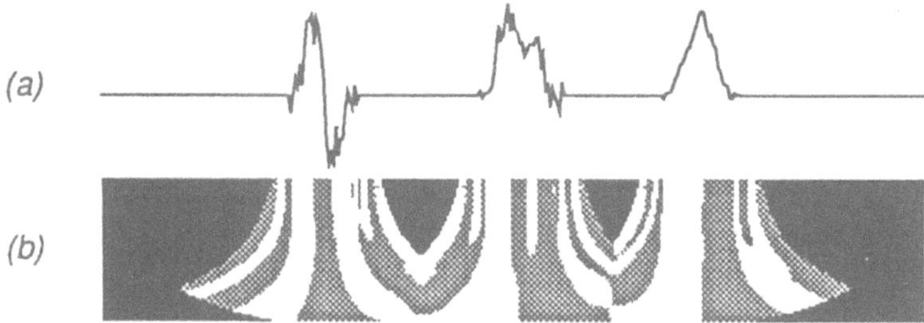

Fig. 9. (a) biphasische, doppelgipflige und monophasische P-Welle. (b) Das im Klassifikationsverfahren B verwendete Phasenbild. Positive Phasen sind hell dargestellt, negative Phasen dunkel.

dung von Quadraturpaaren Amplituden- und Phaseninformation der Signale zu nutzen. Eine abschließende Bewertung dieser Methode bzw. deren Einbindung in Routineanwendungen setzt allerdings noch weitere Untersuchungen voraus, insbesondere die Erprobung an großen Fallzahlen.

Refcrences

1. Combes, J.M., Grossmann, A. und Tchamitchian, Ph. (Eds): *Wavelets: Time-Frequency Methods and Phase Space.* Proc. of the International Conference, Marseille 1987, Springer (1989).
2. Gabor D.: *Theory of Communication.* IEE Part III **93** (1946) 429-457
3. Hauske, G. und Zetzsche, C.: *Die Bedeutung des Analytischen Signals in Bildanalyse und Bildcodierung.* FREQUENZ **44** (1990) 68-73
4. Khadra, L., Matalgah, M., El-Asir, B. und Mawagdeh, S.: *The Wavelet Transform and its Applications to Phonocardiogram Signal Analysis.* Medical Informatics **16** (1991) 271-277
5. Kronland-Martinet, R., Morlet, J. und Grossmann, A.: *Analysis of sound patterns through Wavelet Transforms.* Int. Journal of Pattern Recognition and Artificial Intelligence **1** (1987) 97-126
6. Morlet, D., Peyrin, F., Desseigne, P., Touboul, P. und Rubel, P.: *Time-Scale Analysis of High-Resolution Signal-Averaged Surface ECG Using Wavelet Transformation.* IEEE Computers in Cardiology (1991) 393-396
7. Oppenheim, A. und Schafer, R.: *Digital Signal Processing.* Prentice-Hall International (1975).
8. Sommer, G.: *Bivariate Local Signal Representations*, in Klette, R. und Kropatsch, W. (Eds.): *Theoretical Foundations of Computer Vision.* Akademie Verlag, Berlin (1992) 119-130

KLASSIFIKATION VON BIOSIGNALEN AM BEISPIEL VISUELL EVOZIERTER POTENTIALE MIT HILFE VON WAVELET-NETZEN

H. HEINRICH, H. DICKHAUS, U. KLAUCK *
STUDIENGANG MEDIZINISCHE INFORMATIK
UNIVERSITÄT HEIDELBERG / FACHHOCHSCHULE HEILBRONN
MAX-PLANCK-STRAßE 39, 7100 HEILBRONN
EMAIL: HACKI@INDIGO.BIO.FH-HEILBRONN.DE

Mit Wavelet-Netzen, einer Kombination aus Neuronalem Netz und der Wavelet-Decomposition, können beliebige zeitveränderliche Signale approximiert und gleichzeitig Parameter aus dem Signal extrahiert werden. Die Eignung dieser Methode speziell für den Bereich der Biosignalverarbeitung wird in diesem Beitrag an einem klinischen Beispiel mit visuell evozierten Potentialen untersucht. Die Klassifikationsergebnisse, die im Vergleich zu den in der klinischen Praxis üblichen Parametern erzielt werden, sind beachtenswert und lassen daher den Einsatz der Wavelet-Netze für andere Fragestellungen in diesem Bereich ebenfalls sinnvoll erscheinen.

1. Einleitung

Eine der Hauptaufgaben im Bereich der Biosignalverarbeitung ist das Bestimmen einiger weniger, aber dafür trennscharfer Merkmale für die klinische Diagnostik. Diese Parameter sollen zu einem besseren Verständnis der zu untersuchenden Fragestellung führen und nach Möglichkeit den Klassifikationsprozeß transparent gestalten. Außerdem ist ein automatisches Extrahieren dieser Maßzahlen aus dem Zeitsignal wünschenswert, um den Zeitaufwand und die Fehleranfälligkeit einer manuellen Auswahl zu umgehen.

Methoden, die derartige Parameter liefern, sollten universell einsetzbar sein, damit eine empirische und zeitaufwendige Suche nach Merkmalen bei unterschiedlichen Problematiken vermieden werden kann. Für Biosignale müssen diese Verfahren den zeitabhängigen Verlauf der Signale berücksichtigen, wie er beispielsweise bei evozierten Potentialen im Elektroencephalogramm oder bei Spätpotentialen im Elektrokardiogramm zu finden ist. Hierfür eignen sich prinzipiell Methoden, mit denen Funktionen approximiert werden können.

In jüngster Zeit werden für diese Approximation Neuronale Netze oder auch die Wavelet-Decomposition verwendet. Beide Verfahren liefern aber nicht die gewünschte Merkmalsreduktion [1]. Zhang/Benveniste [1] bzw. Szu et al. [2] schlagen für die Signalrepräsentation eine Kombination dieser beiden Verfahren vor, nämlich **Wavelet-Netze**. Diese Netze liefern neben der Approximation des Signals automatisch eine überschaubare Zahl aussagekräftiger Merkmale, die direkt als Eingabegrößen für einen Klassifikator verwendet werden können.

Neben einer theoretischen Vorstellung dieser Methode mit eigenen Modifikationen bzw. Erweiterungen sollen in diesem Beitrag an einem klinischen Beispiel die Parameter der Wavelet-Netze mit konventionellen Maßzahlen verglichen werden: Für die Diskriminierung von nierenkranken und gesunden Kindern werden gemittelte visuell evozierte Potentiale (VEP's) verwendet, also transiente Signale. Als Vergleichsparameter werden mit den Amplituden- und Latenzwerten einzelner Wellen (N80- und P100-Welle) rein phänomenologische Größen betrachtet, die in der klinischen Diagnostik gemittelter evozierter Potentiale als Standard zu sehen sind [3].

---

* bei Fa. Bruker Analytische Meßtechnik GmbH, Rheinstetten

2. Wavelet-Netze

Ein Signal s(t) wird bei einem Wavelet-Netz durch die lineare Überlagerung von K Wavelons (Wavelet-Knoten) repräsentiert, die sich durch Modifikation eines Basis-Wavelets h(t) ergeben (vgl. Abb. 1):

$$\hat{s}(t) = \sum_{k=1}^{K} w_k h\left(\frac{t-b_k}{a_k}\right) = \sum_{k=1}^{K} w_k \sin\left(\omega_k \frac{t-b_k}{a_k}\right) \exp\left[-0.5\left(\frac{t-b_k}{a_k}\right)^2\right]$$

wobei es für jeden Knoten 4 Parameter gibt, nämlich

- einen Verschiebungsfaktor b_i, der Aufschluß über das Auftreten der Schwingung gibt,
- einen Scale-Parameter a_i, der das Abklingverhalten der Schwingung beschreibt,
- einen Frequenzparameter ω_i und
- einen Gewichtungsfaktor w_i.

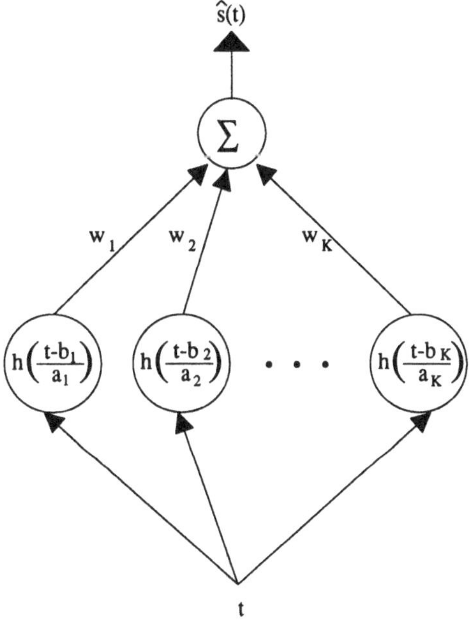

Abb. 1 Struktur eines Wavelet-Netzes: Die lineare Überlagerung von K Wavelons ergibt für jeden Zeitpunkt t (Netzinput) einen approximierten Signalwert für s(t).

Das in dieser Studie verwendete Basis-Wavelet orientiert sich an der von Morlet et al. vorgeschlagenen modulierten Gaußkurve [4]. Zusätzlich wurde die Frequenz der Schwingung als Parameter eingeführt, da damit bei einer geringeren Anzahl von Knoten eine bessere Approximation von Funktionen erzielt werden konnte. Wilson et al. [5] schlagen bei der Wavelet-Transformation ebenfalls vor, den Scale-Parameter und die Frequenz getrennt zu betrachten, was als Analogie betrachtet werden kann.

Die Schätzung der Parameter kann z.B. über einen Ansatz der kleinsten Fehlerquadrate erfolgen :

$$E = \frac{1}{2}\sum_{i=1}^{N}\left[s(t_i)-\hat{s}(t_i)\right]^2 = \min.$$

Die Lösung dieses nichtlinearen Problems kann - wie bei den NN's - über ein Gradientenabstiegsverfahren erfolgen, wofür die partiellen Ableitungen von E benötigt werden:

$$\frac{\partial E}{\partial w_k} = -\sum_{i=1}^{N}\left[s(t_i)-\hat{s}(t_i)\right] \sin\left(\omega_k \frac{t_i - b_k}{a_k}\right) \exp\left[-0.5\left(\frac{t_i - b_k}{a_k}\right)^2\right]$$

$$\frac{\partial E}{\partial \omega_k} = -\sum_{i=1}^{N}\left[s(t_i)-\hat{s}(t_i)\right] w_k \cos\left(\omega_k \frac{t_i - b_k}{a_k}\right) \exp\left[-0.5\left(\frac{t_i - b_k}{a_k}\right)^2\right]\left(\frac{t_i - b_k}{a_k}\right)$$

$$\frac{\partial E}{\partial b_k} = -\sum_{i=1}^{N}\left[s(t_i)-\hat{s}(t_i)\right] w_k \exp\left[-0.5\left(\frac{t_i - b_k}{a_k}\right)^2\right]\left[\sin\left(\omega_k \frac{t_i - b_k}{a_k}\right)\left(\frac{t_i - b_k}{a_k^2}\right) - \cos\left(\omega_k \frac{t_i - b_k}{a_k}\right)\frac{\omega_k}{a_k}\right]$$

$$\frac{\partial E}{\partial (a_k^{-1})} = -\sum_{i=1}^{N}\left[s(t_i)-\hat{s}(t_i)\right] w_k \exp\left[-0.5\left(\frac{t_i - b_k}{a_k}\right)^2\right](t_i - b_k) \, * \\ \left[-\sin\left(\omega_k \frac{t_j - b_k}{a_k}\right)\left(\frac{t_i - b_k}{a_k}\right) + \cos\left(\omega_k \frac{t_i - b_k}{a_k}\right)\omega_k\right]$$

Um das Konvergenzverhalten bei diesem iterativen Verfahren zu verbessern, wird hierbei nicht a_i direkt, sondern a_i^{-1} optimiert. Als Optimierungsverfahren wird der von Fahlman entwickelte Quickpropalgorithmus [6] verwendet, der sich durch sehr rasches Konvergenzverhalten auszeichnet. Die relativ schlechte Generalisierungsfähigkeit dieses Verfahrens spielt hierbei keine Rolle, da nur ein vorgegebener Kurvenverlauf gelernt werden muß. Es empfiehlt sich außerdem, einen Pool von Kandidaten (z.B. N=16) zu trainieren, um möglichst die optimale Lösung zu finden.

Für die Approximation von gemittelten VEP's (250 ms Poststimuluszeit, f_a = 256 Hz) hat es sich gezeigt, daß 2 Knoten und damit 8 Parameter genügen, um das Signal hinreichend genau zu beschreiben. Für die Initialisierung der Parameter werden für die VEP's folgende Verteilungen verwendet:

- b_i gleichverteilt über [25 ms, 175 ms]
- a_i^{-1} gleichverteilt über [20, 60]
- ω_i gleichverteilt über [$30 \frac{1}{s}$, $120 \frac{1}{s}$]
- $w_i = 0$.

Abb. 2a zeigt einen typischen Vertreter eines über 50 Aufnahmen gemittelten VEP's vom Auftreten des Reizes bis 250 ms Poststimuluszeit. Approximierte Kurve (fette Linie) und reales VEP (dünne Linie) sind nahezu identisch. In Abb. 2b sind die Kurvenverläufe der beiden Knoten dargestellt. Es ergeben sich eine höherfrequente Schwingung (11 Hz) und eine mit niedrigerer Frequenz (5 Hz), die zu unterschiedlichen Zeitpunkten auftreten. Die Frequenzwerte stimmen mit den von Leo et al. [7] ermittelten Frequenzwerten überein. Außerdem berücksichtigen die unterschiedlichen Verschiebungsparameter b_i den zeitvariablen Charakter des Signals, so daß die Wavelet-Netze zumindest ein geeignetes Modell für die VEP's bilden.

Um die 16 Kandidaten zu trainieren, benötigt ein PC (486-33) für das Finden der optimalen Lösung im übrigen 45 s bei einer maximalen Iterationszahl von 200.

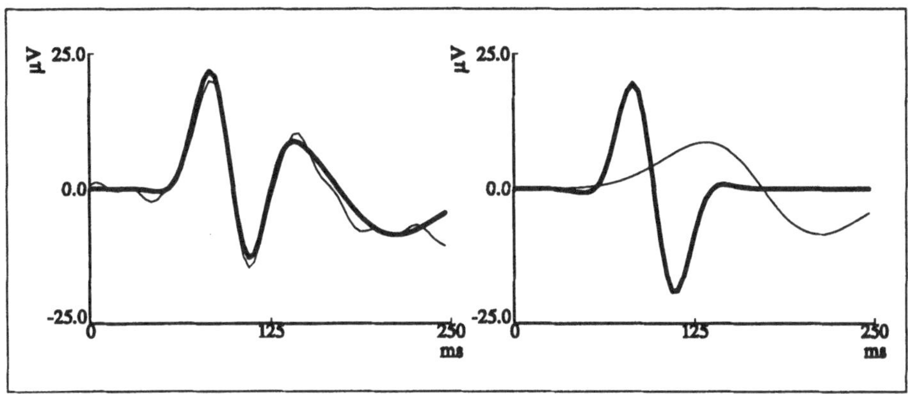

Abb. 2: (a) Reales VEP (dünne Linie) und Approximation durch Wavelet-Netz (fette Linie) sind nahezu identisch.
(b) Kurvenverlauf der beiden Wavelons: Das eine Wavelon (dünne Linie) besitzt im Vergleich zum anderen Knoten (fette Linie) eine niedrigere Frequenz.

Anhand eines klinischen Beispiels mit VEP's sollen die Parameter der Wavelet-Netze hinsichtlich ihrer Trennschärfe untersucht werden.

3. Klinische Studie

Um die Entwicklungsstörungen von Kindern mit Niereninsuffizienz durch objektive neurophysiologische Parameter zu dokumentieren und verfolgen zu können, werden in der Literatur [8] u.a. Untersuchungen mit VEP's oder P300-Experimenten vorgeschlagen.

Die in unserer Studie verwendeten VEP's sind durch invertierende Schachbrettmuster evoziert und mit 256 Hz abgetastet. Durch Filterung mit einem digitalen Tiefpaß (f_g = 35 Hz) sind höherfrequente Störungen abgeschwächt. Das gemittelte VEP wird aus 50 Einzelantworten gebildet. Das Patientenkollektiv setzt sich aus 16 nierenkranken Kindern zusammen, die sich entweder einer kontinuierlichen Peritonealdialyse (CAPD) oder einer Hämodialyse (HD) unterziehen müssen. Mit den 16 gesunden Kindern der Kontrollgruppe ergibt sich also ein Stichprobenumfang von 32.

Als Vergleichsparameter für die Merkmale der Wavelet-Netze werden die Latenzen und Amplituden der N80- und der P100-Welle verwendet, die manuell bestimmt werden müssen. Die entsprechenden Werte für die N140-Welle sind des öfteren nicht bestimmbar, so daß sie für eine Klassifikation nicht herangezogen werden können. Bei den Wavelet-Netzen stehen für jedes Individuum 8 Parameter (4 Parameter für jedes der beiden Wavelons) zur Verfügung.

Abb.3 zeigt die Verteilung für den Gewichtungsfaktor und den Shift-Faktor des höherfrequenten Wavelons, das als zweiter Knoten bezeichnet wird. Das Cluster, das die Gruppe der nierenkranken Kinder (schwarze Vierecke) in der Shifts-Gewichts-Ebene bildet, steht nahezu senkrecht zu dem Cluster der Kontrollgruppe (gepunktete Kreise). Dies läßt auf eine unterschiedliche Struktur der Kovarianzmatrix schließen. Daher liegt es nahe, die quadratische Diskriminanzanalyse als Klassifikationsverfahren einzusetzen.

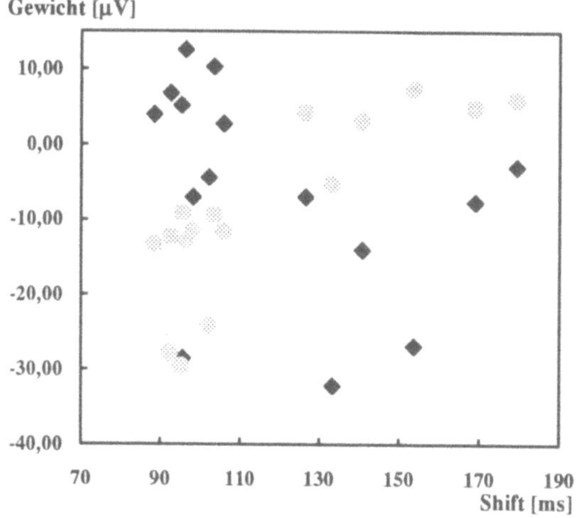

Abb.3: Scatterdiagramm der Verteilung des Gewichts- und des Shift-Parameters des höherfrequenten Knotens (Kontrollen: gepunktete Kreise, CAPD/HD: schwarze Vierecke). Das Cluster, das die Gruppe der nierenkranken Kinder (schwarze Vierecke) in der Shifts-Gewichts-Ebene bildet, steht nahezu senkrecht zu dem Cluster der Kontrollgruppe (gepunktete Kreise).

Merkmalsarten	Merkmalskombinationen	korrekte Klassifikationen		
		Kontrolle	**CAPD/HD**	**gesamt**
Latenz- und Amplitudenwerte	Latenz N80 u. P100 Amplitude N80 u. P100	14 (87.5%)	13 (81.2%)	27 (84.4%)
	Latenz u. Amplitude N80	12 (75%)	14 (87.5%)	26 (81.2%)
Parameter der Wavelet-Netze	Node 1: alle 4 Parameter Node 2: alle 4 Parameter	14 (87.5%)	15 (93.75%)	29 (90.6%)
	Node 1: Gewicht u. Shift Node 2: Gewicht u. Shift	13 (81.2%)	16 (100%)	29 (90.6)
	Node 1: Frequenz u. Shift	14 (87.5%)	13 (81.2%)	27 (84.4%)
	Node 2: Gewicht u. Shift	13 (81.2%)	14 (87.5%)	27 (84.4%)

Tab. 1: Klassifikationsergebnisse (quadratische Diskriminanzanalyse) für verschiedene Merkmalskombinationen (Stichprobenumfang 32): Node 1 bezeichnet das Wavelon mit der niedrigeren Frequenz, Node 2 den höherfrequenten Knoten. Bei gleicher Parameterzahl gibt es für die Wavelet-Netze Konstellationen, bei denen die Anzahl der korrekten Klassifikationen über denen der Latenz- und Amplitudenwerte liegt.

In Tabelle 1 sind die Klassifikationsergebnisse zusammmengefaßt, die eben mit der quadratischen Diskriminanzanalyse bei unterschiedlichen Merkmalskombinationen erzielt wurden. Sowohl bei der Verwendung von zwei als auch bei vier Merkmalen klassifizieren die Parameter der Wavelet-Netze besser als die Latenz- und Amplitudenwerte. Zu beachten ist hierbei, daß eine Merkmalskombination der Gewichts- und Shiftparameter beider Wavelons alle CAPD/HD-Patienten erkennt. Die Verwendung der insgesamt verfügbaren acht Parameter liefert keine weitereVerbesserung der Klassifikationsergebnisse.

4. Diskussion

Im vorliegenden Beitrag wurde mit den Wavelet-Netzen eine Methode vorgestellt, mit der durch die lineare Überlagerung einzelner Wavelons visuell evozierte Potentiale, also zeitabhängige Biosignale, gut approximiert werden können und die anschauliche Parameter (Frequenz, Scale, Shift und Gewicht) liefert, die - wie an einem klinischen Beispiel gezeigt - für eine Klassifikation geeignet sind.

Bei der Diskriminierung von nierenkranken und gesunden Kindern anhand von VEP's liegen die Klassifikationsergebnisse mit den Wavelet-Parametern bei einer quadratischen Diskriminanzanalyse leicht über den Ergebnissen, die mit konventionellen Parametern (Latenzen und Amplituden der N80- und P100-Welle) erzielt werden können. Mit vier Parametern wird hierbei eine Klassifikationsrate von 90,6% (29 von 32) erzielt. Wenn der geringe Stichprobenumfang (16 Individuen pro Gruppe) auch keine statistisch fundierte Aussage zuläßt, so können die Parameter der Wavelet-Netze doch zumindest als Alternative zu den bisher verwendeten Amplituden und Latenzen gesehen werden.

Die Wavelet-Netze selbst wurden über einen Ansatz der kleinsten Fehlerquadrate geschätzt, wobei die Anzahl der Wavelons für die VEP's mit 2 Knoten vorgegeben werden konnte. Für andere Anwendungen, bei der sich die Anzahl der Knoten nicht so einfach bestimmen läßt, empfiehlt sich eine Implementierung, bei der die Wavelons dynamisch angelegt werden. Zudem erscheint es lohnenswert, anstatt des Fehlerkriteriums ein Korrelationskriterium zu testen, da man hiermit bei gestörten Signalen u.U. robustere Schätzungen erzielen kann.

Die Anwendung der Wavelet-Netze muß nicht auf eindimensionale Biosignale beschränkt bleiben. Zhang/Benveniste [1] führen die Wavelet-Netze in vektorieller Schreibweise ein, so daß damit der Übergang z.B. zur Bildverarbeitung schon gegeben ist, wo eine Anwendung dieser Methodik ebenfalls sinnvoll erscheint.

Literatur :

[1] Q. Zhang, A. Benveniste, "Wavelet Networks", IEEE Trans. NN, vol. 3, no. 6, pp. 889-898, November 1992

[2] H.H. Szu, B. Telfer, S. Kadambe, "Neural network adaptive wavelets for signal representation and classification", Optical Engineering, vol. 31, no. 9, pp. 1907-1916, September 1992

[3] K. Maurer, K. Lowitzsch, M. Stöhr, "Evozierte Potentiale: AEP-VEP-SEP; Atlas mit Einführungen", Stuttgart: Enke, 1990

[4] J. Morlet, G. Arens, I. Fourgeau, D. Giard, "Wave propagation and sampling theory", Geophys., vol. 47, pp. 203-236, 1982

[5] R. Wilson, A.D. Calway, E.R.S. Pearson, "A generalized Wavelet transform for Fourier analysis: The multiresolution Fourier transform and its application to image and audio signal analysis", IEEE Trans. Inform. Theory, vol. 38, no. 2, pp. 674-690, March 1992

[6] S.E. Fahlman, "An empirical study of learning speed in back-propagation networks", Technical Report CMU-CS-88-162, School of Computer Science, Carnegie Mellon University, Pittsburgh, PA, September 1988

[7] T. Leo, E. Pisani, A. Uncini, "PC-based classification of cerebral activity", in: H. Dickhaus, T. Leo: Proceedings of the second joint seminar on medical informatics and bioengineering, pp. 148-160, 1988

[8] J.T. Marsh, W.S. Brown, D. Wolcott, J. Landsverk, A.R. Nissenson, "Electrophysiological indices of CNS function in hemodialysis and CAPD", Kidney Int., vol. 30, pp. 957-963, 1986

Beatmungsoptimierung mit neuronalen Netzen bei unscharfer Patientenzustandsbeschreibung

K. Gärtner, B. Baumung, F. Reisner
Technische Universität Dresden
Lehrstuhl Erkennende Systeme und Bildverarbeitung
Mommsenstr. 13; 01062 Dresden

1 Einführung

Mit der Entwicklung von Expertensystemen für den medizinischen Bereich wird der Versuch unternommen, die äußerst komplizierten Entscheidungsprozesse von Ärzten bei Diagnose- und Therapieaufgaben zu unterstützen. Ein spezielles Anwendungsgebiet stellt dabei die maschinelle Beatmung von Patienten auf Intensivstationen dar. Zur Unterstützung des Arztes bei der Bestimmung der optimalen Steuerparameter wird das medizinische Beratungssystem IBEUS (**I**ntensivtherapeutisches **B**eatmungs- **E**ntscheidungs- **U**nterstützungs**S**ystem) entwikkelt. Dieses System soll eine ausreichende Sauerstoffversorgung des Patienten ohne schädigende Nebenwirkungen auf andere Organe gewährleisten. Dazu werden für die mit dem Analysator ermittelten Patientendaten (Merkmale) neue Steuerparametervorschläge durch das Beratungssystem ermittelt, die eine gute Prognose (Zustandsverbesserung) bei der Beatmung erwarten lassen.

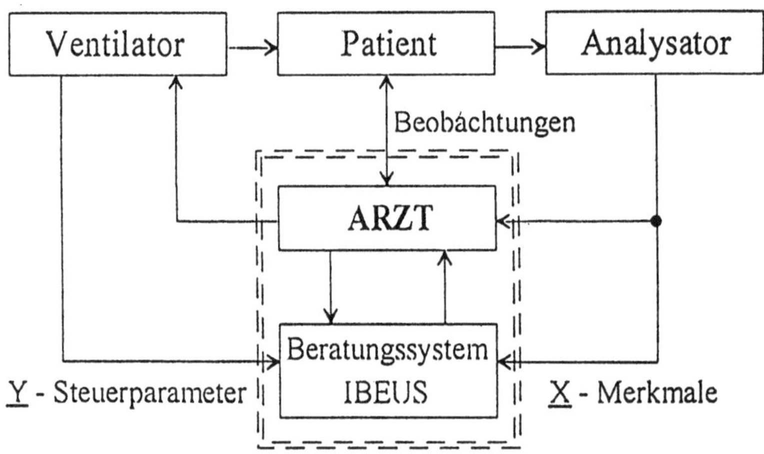

Abb. 1: Blockdiagramm mit IBEUS als interaktives System

Den Erkennungsprozeß charakterisieren dabei folgende Merkmale:

1. unsicherer aktueller Patientenzustand
2. keine klaren Richtlinien für die Therapieauswahl
3. unsichere Folgen der Therapie
4. keine volle Befriedigung geplanter Therapieziele möglich

Um das Beatmungstherapiemuster (bestehend aus 6 Steuerparametern) bei diesem sehr komplizierten und komplexen Prozeß richtig zu erkennen, werden folgende Aufgaben der ärztlichen Entscheidungsfindung für eine Inferenzstrategie des Beratungssystems umgesetzt:

1. Beschreibung des aktuellen Patientenzustandes mit Gewißheitswerten für die Angabe des pathophysiologischen Befundes
2. Erleichterung der Therapieauswahl durch Berechnung von Funktionsmerkmalen, die den inneren Gasaustausch bzw. die Beatmungseffektivität näher bestimmen
3. Vergleich mit Therapiebeispielen zur Bestimmung der Therapiefolgen bzw. Prognoseschätzung bei der Therapiefestlegung

2 Systemkonzeption

Aus den genannten Aufgaben entstand eine hybride Inferenzkomponente zur Problemlösung, die aus drei Teilkomponenten besteht (Abb. 2):

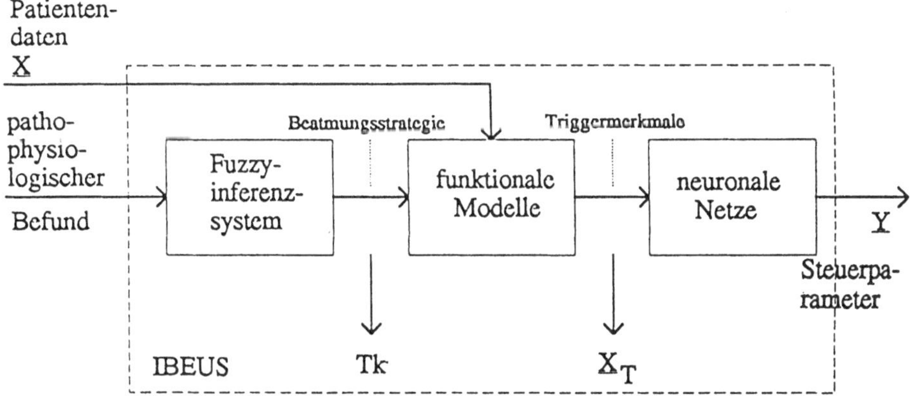

Abb. 2: IBEUS - Inferenzteilkomponenten

- Aus der Fülle der zur Verfügung stehenden Patientendaten sind diejenigen zu selektieren, die aufgrund des pathophysiologischen Befundes für den jeweiligen Patienten relevant sind. Dabei soll ein Fuzzyinferenzsystem die Angabe und Verarbeitung einer Auftretensgewißheit (7 Therme zwischen sicher und ausgeschlossen) für jede Indikation und Komplikation ermöglichen und über eine Regelbasis brauchbare Schlußfolgerungen zur Beatmungsstrategie (Therapieklasse Tk) entwickeln.
- Aus den vorliegenden medizinischen Datenmaterial zur Beatmung des Patienten werden anschließend die nach der Beatmungsstrategie vorgegebenen Triggermerkmale über funktionale Modellbeziehungen berechnet. Diese Modelle, die unter Fachexperten anerkannt werden, bestimmen verschiedene Parameter zur Beurteilung des Gasaustausches sowie zur Beurteilung der Hämodynamik.
- Ein Vergleich mit Therapiebeispielen und die Festlegung des Beatmungsmusters erfolgt über neuronale Netze, die mit Stichproben von Patientendaten belehrt werden.
  Nach dem Training schlagen sie einen geeigneten Beatmungsparametervektor für den aktuellen Fall vor.

2.1 Fuzzyinferenzsystem (FIS)

Da es zur Gewinnung der Beatmungsstrategie aus einem unsicheren pathophysiologischen Befund kein exaktes mathematisches Modell gibt, eignet sich neben neuronalen Netzen die Theorie unscharfer Mengen besonders zur Modellierung solcher Fälle. In unscharfen Regelbasen kann das Wissen des Arztes fast unverändert abgespeichert werden, da durch Verwendung linguistischer Variablen eine recht wirklichkeitsgetreue Abbildung der intuitiven, vom Arzt angegebenen (und damit sprachlich formulierten) Regeln in eine vom Rechner verarbeitbare Form möglich ist.
Eine realisierte, rein regelbasierte alternative Lösung dieser Aufgabe mit Certainty Factors erwies sich hier als inkonsistent.

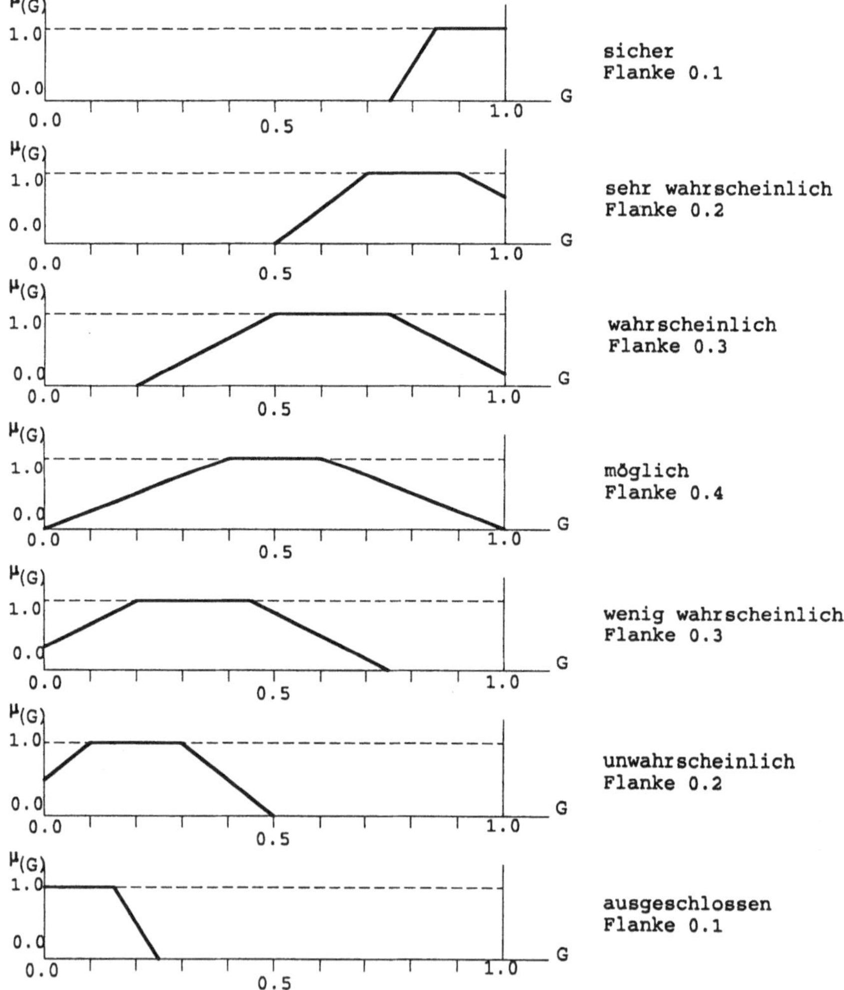

Abb. 3: Linguistischer Variablentyp GEWISSHEIT

Mit Hilfe von IBEUS ist es möglich, für 14 Grunderkrankungen, die zu einer künstlichen Beatmung führen können, spezifische Beatmungsstrategien (z.B. Lungenödem, Intoxikation)

zu verfolgen. Der Arzt gibt den pathophysiologischen Befund mit Hilfe von Pull-Down-Menüs ein. Dabei kann er jeder der 14 möglichen Erkrankungen einen Gewißheitswert zuordnen, von denen das Fuzzy-System die wichtigsten auswählt (Abb. 3).
Neben der vom Arzt angegebenen Gewißheit (G) ist auch die Priorität der angegebenen Indikation von Bedeutung. Es wird immer die Indikation beatmet, die für den Patienten am gefährlichsten ist.
Aufgabe des FIS ist es, aus mehreren vorliegenden Fakten diejenigen herauszufinden, die im vorliegenden Beatmungsfall berücksichtigt werden müssen. Mit Hilfe der Regeln soll herausgefunden werden, welche der vorliegenden Erkrankungen die Beatmungsstrategie bestimmt. In den Prämissen werden deshalb die Gewißheiten der Erkrankungen aufgeführt, in der Konklusion wird die Wichtigkeit der entsprechenden Beatmungsindikation (z. B. pul_ards, pul_pneu) bestimmt. Die Regeln des FIS wurden in Zusammenarbeit mit Ärzten aufgestellt (Abb. 4).

Prämisse - WENN ..	Konklusion - DANN ...
ARDS ≥ wenig_wahrscheinlich	pul_ards = absolut_zwingend
ARDS = unwahrscheinlich	pul_ards = wichtig
ARDS = sicher_nicht	pul_ards = unwichtig
Pneumonie ≥ sehr_wahrscheinlich	pul_pneu = absolut_zwingend
Pneumonie = wahrscheinlich	pul_pneu = zwingend
Pneumonie = ungewiß	pul_pneu = sehr_wichtig
Pneumonie = wenig_wahrscheinlich	pul_pneu = wichtig
Pneumonie = unwahrscheinlich	pul_pneu = weniger_wichtig
Pneumonie = sicher_nicht	pul_pneu = sehr_unwichtig

Abb. 4: Ausschnitt aus dem Entwurf der Regelbasis (ARDS - akutes Atemnotsyndrom des Erwachsenen, Pneumonie - Lungenentzündung)

Im System soll nun das plausible, aber auch das einfache Schließen nach der Fuzzy-Set-Theorie untersucht werden, um den für dieses System geeigneten Inferenzmechanismus herauszufinden.
Beim plausiblen Schließen wird beim Matching mit Näherungsformeln nach Tilli /2/ jede vorliegende Symptomkonstellation des Patienten mit den Regelprämissen verglichen, die daraus berechneten Ergebnisse gehen in die Inferenz (Schließen mit jeder Regel) ein, die aus Zeitgründen ebenfalls mit Näherungsformeln arbeitet. Danach werden die Ergebnisse mit einem geeigneten UND-Operator akkumuliert und mittels Schwerpunktverfahren defuzzyfiziert, um als scharfe Parameter an den Arbeitsspeicher weitergegeben zu werden.
Beim einfachen Schließen wird das Matching mit einem Kombinationsoperator (z.B. Min-Max) durchgeführt; das Schließen mit jeder Regel erfolgt mit dem Produktoperator. Nach der Akkumulation des Gesamtergebnisses mit einem geeigneten UND- oder ODER-Operator wird das scharfe Ergebnis wie beim plausiblen Schließen durch das Schwerpunktverfahren ermittelt und dem Arbeitsspeicher übergeben.
Erst dann werden die konkreten Parameterwerte vom Patienten in den Arbeitsspeicher zur Gewinnung der Triggermerkmale mittels funktionalen Modellen übertragen.

2.2 Funktionale Modelle

Je nach Wahl der Beatmungsstrategie werden verschiedene Modelle zur Berechnung noch notwendiger Funktionsmerkmale aktiviert. Für IBEUS sind 16 verschiedene Modelle (physikalische Beziehungen, Abschätzungen) zur Berechnung der jeweils erforderlichen beatmungsrelevanten Funktionsparameter (sogenannte Triggermerkmale) generiert worden. Die Auswahl der Berechnungsmodelle erfolgt über die ermittelte Beatmungsstrategie des konkreten Patienten. Für die Berechnung der Triggermerkmale werden folgende Eingangsparameter benötigt: gemessene Funktionsmerkmale, Ventilatorparameter und Umgebungsbedingungen. Die Triggermerkmale bilden dann die Grundlage für die Bestimmung der Ventilatorparameter.
Als Beispiel sei die Bestimmung des alveolären Sauerstoffpartialdrucks PAO_2 mit aktueller inspiratorischer Sauerstoffkonzentration FIO_2 genannt:

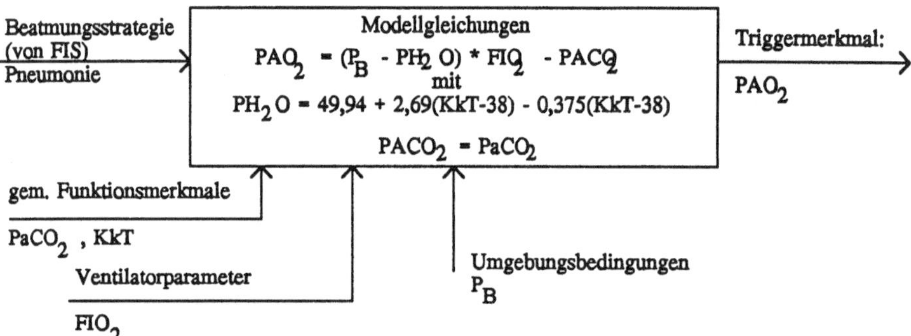

Abb. 5: Bestimmung des alveolären Sauerstoffpartialdrucks PAO_2 (in kPa) ($PaCO_2$ - arterieller Kohlendioxidpartialdruck (in Torr), KkT - Körperkerntemperatur (in °C), PH_2O - Wasserdampfdruck (in Torr); P_B - Luftdruck (in Torr))

2.3 Neuronale Netze

Für die Therapievorschlagsermittlung wurden unabhängig voneinander drei Ansätze verfolgt. Der erste, rein regelbasierte Ansatz ermöglichte nur die Angabe von ungefähren Richtwerten zur Beatmungstherapie und erlaubte keine konkreten Werteangaben. Der zweite Ansatz war ein Verfahren aus der statistischen Entscheidungstheorie, das die Prognose des Patientenzustandes optimiert. Die Ergebnisse konnten aufgrund eines zu geringen Stichprobenumfangs den Anforderungen noch nicht entsprechen. Deshalb soll hier ein dritter Ansatz mit neuronalen Netzen vorgestellt werden.
Da das System mehrere Vorschläge für die Beatmung eines Patienten entsprechend seines pathophysiologischen Befundes liefern soll, muß für jede vom System erkennbare Beatmungsstrategie (Therapieklasse) ein separates neuronales Netz bereitstehen.
Die einzelnen Netze werden zu einer neuronalen Netzwerkkomponente (NNK) zusammengefaßt.
Je nach Therapieklasse enthält der Triggermerkmalvektor zwischen 7 und 12 Parameter. Als weitere Eingangsinformation für die neuronalen Netze dienen die zuvor eingestellten Werte der Ventilatorparameter. Die Triggermerkmalwerte bilden zusammen mit den zuvor eingestellten Beatmungsparametern das Eingangsmuster für die neuronalen Netze.
Da diese Werte einen relativ großen Wertebereich belegen, erfolgt vor der Verarbeitung durch ein neuronales Netz eine Normierung aller Musterdaten.

Für den Aufbau der Netzwerke ist das Backpropagation-Netzwerkmodell mit jeweils 4 Neuronenschichten ausgewählt worden. Die Neuronenzahl der Eingangsschicht jedes Netzwerkes richtet sich nach der Anzahl der Eingangsmusterwerte und liegt zwischen 11 und 16 Neuronen. Die Ausgabeschicht jedes Netzwerkes besitzt 6 Neuronen, je eines für jeden Beatmungsparameter. Von der Eingangsschicht zur Ausgangsschicht erfolgt eine stetige Reduktion der Neuronenzahl.

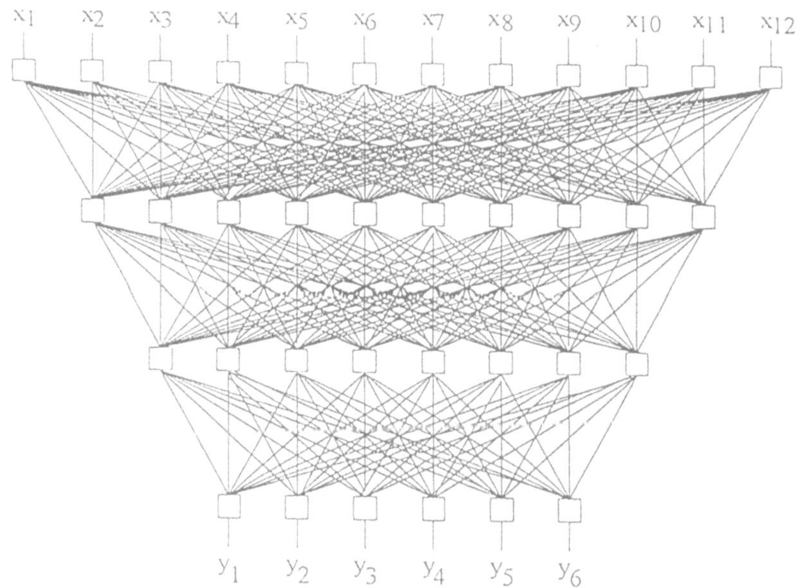

Abb. 6: Struktur eines Netzwerkes zur Problemlösung

Für die Berechnung innerhalb eines Neurons werden die folgenden Funktionen eingesetzt:

1. Inputfunktion : Summationsfunktion
2. Aktivierungsfunktion : Fermifunktion (sigmoid)
3. Outputfunktion : Identitätsfunktion

Das Trainieren der Netzwerke erfolgt durch epochenweises Lernen. Dabei werden die Verbindungsgewichte zwischen den Neuronenschichten erst nach einem kompletten Duchlauf aller Lernmuster angepaßt. Das hat den Vorteil, daß die Reihenfolge der Lernmuster in der Lernstichprobe keinen Einfluß auf das Lernverhalten ausübt.

Die zum Training der neuronalen Netzwerke benötigten Lernmuster werden aus einer medizinischen Stichprobe selektiert. Diese Stichprobe ist im Routinebetrieb in mehreren medizinischen Einrichtungen erfaßt worden. Ein Datensatz der Stichprobe besitzt folgenden Aufbau:

1. Nr laufende Nummer des Datensatzes
2. KBNR Krankenblattnummer
3. Indikation Grunderkrankung und Lungenschädigung
4. $\underline{X}$ meß- und berechenbare Funktionsparameter
5. $\underline{Y}_{old}$ aktuell eingestellte Beatmungsparameter
6. $\underline{Y}_{new}$ neu einzustellende Beatmungsparameter
7. PW durch den Arzt eingeschätzter Prognosewert nach einer Stunde Beatmung mit $\underline{Y}_{new}$ (PW=(1 ... 5))
8. TK festgelegte Therapieklasse TK=(1 ... 14)

Der Prognosewert PW stellt eine Einschätzung des Patientenzustandes nach erfolgter einstündiger Beatmung dar. Für die Lernproben können nur Datensätze Verwendung finden, für die eine gute Prognose (PW=4,5) nach dem Beatmungsvorgang eingeschätzt wurde.
Als Ergebnis liefert ein neuronales Netzwerk einen konkreten Therapievorschlag (Ausgangsmuster), mit welchen Ventilatorparametereinstellungen der Patient während der nächsten Stunde beatmet werden soll. Die von den Ausgangsneuronen des Netzwerkes ausgegebenen Werte sind vor ihrer Ausgabe auf ihren normalen Wertebereich zurückzukonvertieren.

3 Rechentechnische Umsetzung

Prolog 2 wird als Implementierungssprache für IBEUS (Steuerung, Dialog, Regelbasis) verwendet. Da das unscharfe, insbesondere das plausible Schließen mit einem hohen Rechenaufwand verbunden ist, der in der KI-Sprache Prolog nicht mehr effizient implementiert werden kann, bietet sich in dieser Situation die Nutzung externer Codemodule an, die in imperativen Programmiersprachen geschrieben werden. Prolog 2 stellt u.a. eine Schnittstelle zu Microsoft-C bereit, die von FIS und NNK genutzt wird.
Der in der Programmiersprache C geschriebene Teil des Fuzzyinferenzsystems wurde nach softwaretechnologischen Gesichtspunkten in mehrere Module zergliedert. Die Grafik veranschaulicht Zusammenhänge zwischen den Moduln:

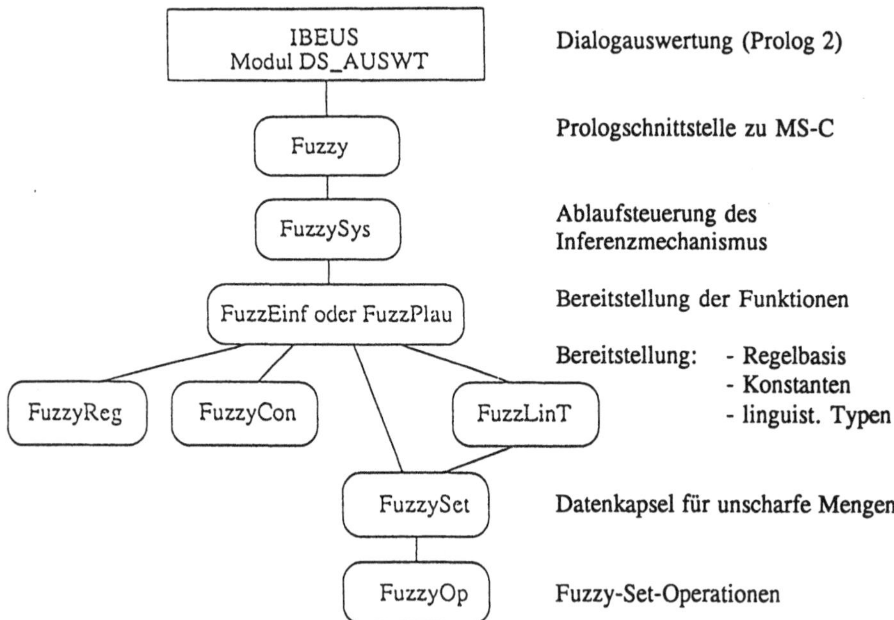

Abb. 7: Zusammenhänge zwischen den C-Modulen im FIS

Die entsprechenden Programme zur Selektion, Normierung und Belehrung der neuronalen Netze sind in Turbo-Pascal geschrieben.
Das kopplungsfähige Programm zur Bereitstellung der nach der Beatmungsstrategie erforderlichen neuronalen Netze liegt in MS-C vor (Abb. 8).

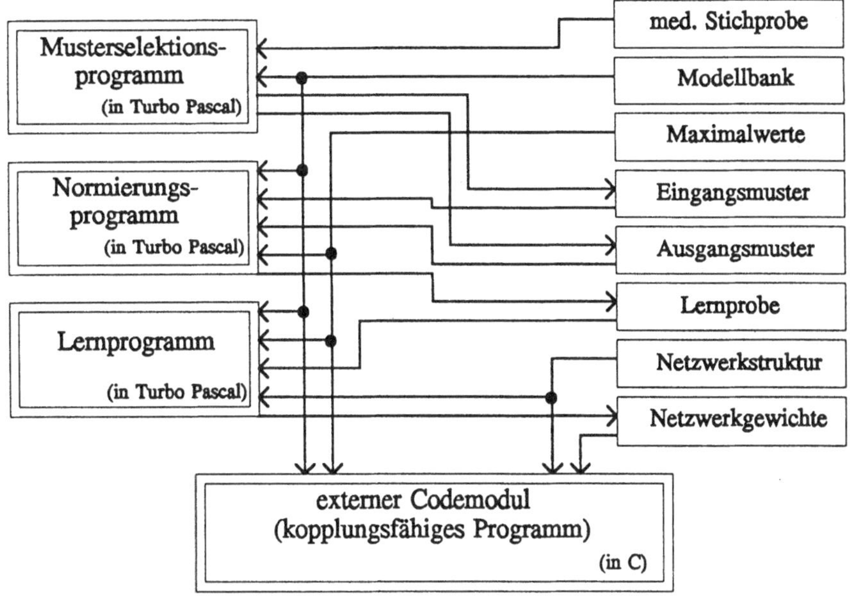

Abb. 8: Überblick über Komponenten der NNK

4 Literatur

1. Kruse, R.; Schwecke, E.; Heinsohn, J.:
   Uncertainty and Vagueness in Knowledge-Based Systems.
   Springer Verlag 1991
2. Tilli, Th.:
   Fuzzy-Logik. Grundlagen, Anwendungen, Hard- und Software.
   Franzis-Verlag München 1992
3. Gärtner, K.:
   Interaktive Beatmungsoptimierung mit dem Beratungssystem IBEUS.
   In: Biometrie und Informatik in Medizin und Biologie 23, Heft 4,
   Gustav Fischer Verlag Stuttgart 1992, S. 236-241
4. Gärtner, K; Fuchs, S.; Jauch H.:
   Hybrid Inference Components for monitoring of artificial respiration.
   In: Data fusion. Springer Verlag 1993
5. Baumung, B.:
   Verarbeitung unscharfer Informationen am Beispiel der Regelung der maschinellen Beatmung.
   Diplomarbeit, TU Dresden, 1993
6. Kruse, H.; Mangold, R.; Mechler, B.; Penger, O.:
   Programmierung Neuronaler Netze, Eine Turbo Pascal Toolbox.
   Addison Wesley, 1. Auflage, 1991
7. Reisner, F.:
   Realisierung einer hybriden Inferenzkomponente für die Lösung des Problems der Beatmungsoptimierung unter Nutzung neuronaler Netze.
   Diplomarbeit, TU Dresden, 1993

Quantitative Beschreibung der 3-dimensionalen Struktur des Zytoskeletts

S. Karnaoukhova[+*], D. Komitowski[+], J. Bille[*]

[+]Deutsches Krebsforschungszentrum, DKFZ, INF 280, 69120 Heidelberg
[*]Institut für Angewandte Physik der Universität Heidelberg,
Albert-Überle-Straße 3-5, 69120 Heidelberg

Schönheit rettet die Welt
Dostojewski

Zusammenfassung

Die Struktur des Zytoskeletts von in vitro kultivierten Hepatomzellen wird mit Methoden der Digitalen Bildverarbeitung beschrieben. Die Intermediär-Keratinfilamente werden mittels indirekter Immunfluoreszenz sichtbar gemacht. Dreidimensionale Bilder des Zytoskeletts werden mit einem konfokalen Laser-Scanning-Mikroskop (CLSM) aufgenommen. Die Bilder mit 256 Grautonabstufungen der Lichtintensität werden anschließend binärisiert und skelettiert. Für die verschiedenen Zellen werden folgende Parameter berechnet: Größe der Zellen, Dichte der Zytoskelettfilamente, Länge der Filamente in der xy-Ebene und in z-Richtung, sowie die Anzahl der Verknüpfungen mit 3 bzw. 4 Linienabzweigungen. Aufgrund dieser Parameter besteht die Möglichkeit, die Zellstruktur und den Differenzierungsgrad zu analysieren.

1 Einleitung

Das in allen Zellen vorhandene Netz von Filamenten – das Zytoskelett – ist für verschiedene Zellfunktionen wie Formerhaltung, Bindung von RNS und Proteinen, intrazytoplasmatische Anordnung von Organellen und Zellbewegung verantwortlich [1]. Das Zytoskelett zeigt biochemische und auch strukturelle Merkmale, die für die Prozesse der Zelldifferenzierung und Tumorentstehung charakteristisch sind. Intermediärfilamente (IF) sind kräftige und dauerhafte Proteinfasern des Zytoskeletts, welche einen "Korb" um den Kern bilden und sich in sanft geschwungener Anordnung bis zum Zellrand erstrecken. Zu den stabilsten und langlebigsten IF gehören jene auf Keratinbasis. Diese stellen einen charakteristischen "Fingerabdruck" dar, welcher es ermöglicht, die Ursprünge von Tumoren bis in bestimmte Typen von Epithelzellen zurückzuverfolgen [2].

Die Struktur des Zytoskeletts erweist sich als außerordentlich komplex und ist durch die dreidimensionale Ausdehnung der Analyse mittels konventioneller Lichtmikroskopie unzugänglich. Um die strukturellen Merkmale des Zytoskeletts zu analysieren, bieten sich Methoden der computergestützen Bildverarbeitung an. Bildverarbeitung bedeutet, die Wirklichkeit in Komponenten zu zerlegen und mathematisch zu beschreiben. Die mathematischen Methoden werden dabei so gewählt, daß ihre Komplexität auf das Notwendige beschränkt bleibt [3], [4].

Durch die mikroskopische Aufnahme entsteht eine enorme Datenfülle. Um die wesentlichen Strukturen herausarbeiten zu können, muß eine Datenkompression durchgeführt werden. In unserem Fall wird diese durch Binärisierungs- und Skelettierungsalgorithmen durchgeführt. Die erhaltenen Bilder werden anhand verschiedener Parameter analysiert. Dafür kommen zunächst Dichte und Länge der Filamente und Anzahl der Verknüpfungen in Frage. Ziel ist es, eine Klassifizierung nach Zelltyp und Differenzierungsgrad zu erreichen. Welche Parameter letztendlich die entscheidenden Kriterien für diese Klassifizierung werden, muß an dieser Stelle noch offen bleiben.

2 Material und Methoden

2.1 Färbung

Das Zytoskelett wird durch die indirekte Immunfluoreszenz dargestellt. PLC (menschliche Hepatomzellen) werden auf Deckgläsern kultiviert und in Methanol bei −20°C fixiert und aufbewahrt. Die Zellen werden in PBS (phosphate buffered saline) rehydriert; unspezifische Proteinbindung durch Inkubation in 1% BSA (bovine serum albumin) in PBS blockiert, und es wird für eine Stunde bei Raumtemperatur mit dem monoklonalen Pancytokeratin-Antikörper der Maus Lu5 (Dianova, Hamburg) inkubiert. Anschließend werden die Zellen gespült, und der gebundene Lu5 wird durch Ziege-anti-Maus-FITC (Dianova, Hamburg) nachgewiesen. Diese Zellen werden in Glycerin/PBS (50:50) mit 0,2% Propylgallat als Anti-Bleichmittel eingedeckt und mit einem CLSM (Leica Lasertechnik, Heidelberg) untersucht.

2.2 Die Aufnahme des Zytoskeletts mittels eines CLSM

Bilder des Zytoskeletts werden durch ein konfokales Laser-Scanning Mikroskop aufgenommen. Einzelne Bildebenen werden Bildpunkt für Bildpunkt (512 × 512 Abtastpunkte) durch einen Laserstrahl (Ar^+-Laser mit $488nm$ Wellenlänge) beleuchtet, und das emittierte Fluoreszenzlicht wird mit einem Photomultiplier detektiert.

Die Zelle wird in 47 aufeinanderfolgenden Ebenen fokussiert; die Fluoreszenzintensität von acht aufeinanderfolgenden Beleuchtungen wird gemittelt, und die Bilder der detektierten Fluoreszenzverteilung werden in einer dreidimensionalen Matrix gespeichert. Die durch die Ebenen 1 bis 4 in Abb. 1 bestimmten Schnitte sind in Abb. 2 dargestellt.

3 Bildverarbeitung

3.1 Binärisierung

Die Grauwertbilder werden mit Hilfe des statistischen Ansatzes der Klassentrennbarkeit binärisiert [5]. Jede Ebene des Bildes wird durch die Statistik erster Ordnung, d.h. durch ein normiertes Histogramm $p(0), p(1), \ldots p(i), \ldots p(255)$, mit $0 \leq p(i) \leq 1$, beschrieben, wobei $p(i)$ die Auftrittswahrscheinlichkeit des Grauwertes i darstellt. Jeder mögliche Schwellenwert S, mit $0 \leq S \leq 255$, trennt die Grauwerte in 2 Klassen, K_0 und K_1, die Klassen der Grauwerte, die nach der Binärisierung dem Objekt bzw. dem Hintergrund zugewiesen werden. Dieser Schwellenwert S wird für jede Ebene unabhängig

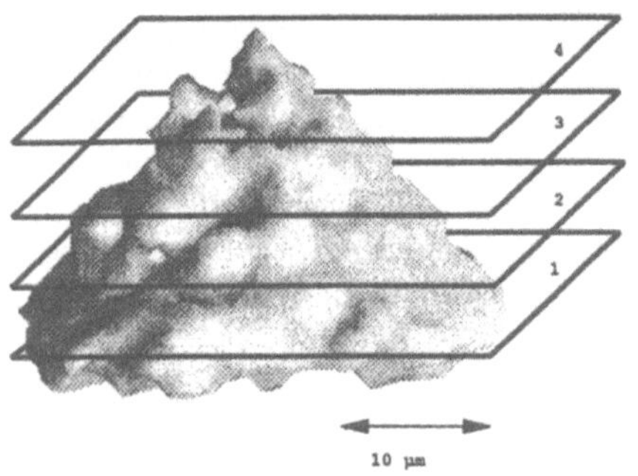

Abb. 1. 3-D Repräsentation einer Zelle

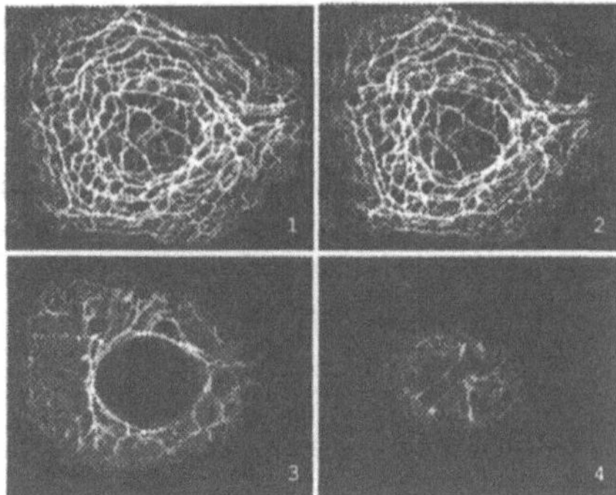

Abb. 2. Scanebenen einer Zelle

berechnet, da die entsprechenden Schnitte unabhängig voneinander entstanden sind. Die Auftrittswahrscheinlichkeiten der Elemente dieser Klassen sind

$$P_0 = \sum_{i=0}^{S} p(i) \qquad \text{und} \qquad P_1 = \sum_{i=S+1}^{255} p(i) = 1 - P_0. \tag{1}$$

Weiterhin werden die folgenden Größen berechnet: der mittlere Grauwert m_k der Klasse K_k, $k = 0, 1$, somit der Grauwert des gesamten Bildes,

$$m = m_0 P_0 + m_1 P_1, \tag{2}$$

und die entsprechenden Varianzen

$$s_0^2 = \sum_{i=0}^{S} (i - m_0)^2 p(i) \qquad s_1^2 = \sum_{i=S+1}^{255} (i - m_1)^2 p(i). \tag{3}$$

Alle diese Größen hängen vom Parameter S ab. Das Kriterium für die Festlegung von S ist die Maximierung der Varianz s_B^2 zwischen K_0 und K_1 und die Minimierung der Varianz s_W^2 innerhalb von K_0 und K_1. Dafür werden die Größen

$$s_B^2 = P_0 (m_0 - m)^2 \qquad \text{und} \qquad s_W^2 = P_0 s_0^2 + P_1 s_1^2. \tag{4}$$

berechnet und das von S abhängende Verhältnis $\frac{s_B^2}{s_W^2}$ maximiert.

3.2 Skelettierung

Binärisierte Bilder werden durch das Verfahren nach Arcelli et.al. [6] skelettiert. Jedes Mal wird der aktuelle Bildpunkt P=1 getilgt, wenn ein Muster lokaler 3 × 3-Konfigurationen aus Abb. 3 im entsprechenden Nord-,Süd-,West- oder Ost-Durchlauf auftritt.

Die Abb. 4 zeigt die Anwendung dieses Algorithmus auf ein Bild des Zytoskeletts. Skelettierte Bilder ermöglichen die weitere Analyse der auftretenden komplexen Strukturen.

Nord			Süd			West			Ost		
0	0	0	1	1	X	0	X	1	X	X	0
X	P	X	X	P	X	0	P	1	1	P	0
X	1	1	0	0	0	0	X	X	1	X	0
X	0	0	X	1	X	0	0	X	X	1	X
1	P	0	0	P	1	0	P	1	1	P	0
X	1	X	0	0	X	X	1	X	X	0	0
X=0 oder 1											

Abb. 3. Für alle diese Muster ist P=1 zu tilgen

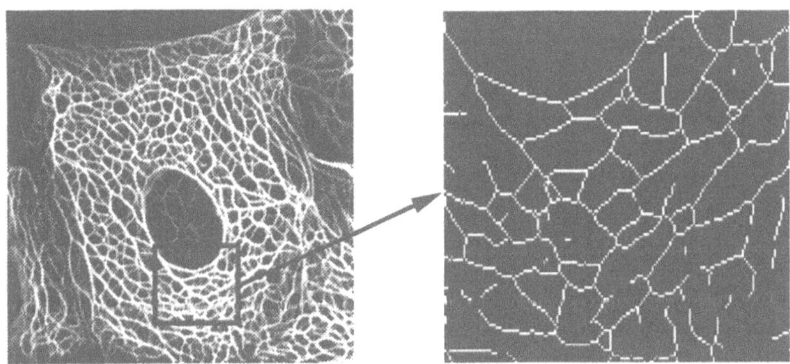

Abb. 4. Skelettierung einer Ebene. Ein Grauwertbild (links) und sein Skelett (rechts)

4 Beschreibung der Netzstruktur

4.1 Größe und Dichte

Von 20 Zellen wurden Bilder aufgenommen, und diese wurden entsprechend der Zellgröße in vier Gruppen eingeteilt (Abb. 5, ein Balken entspricht einer Zelle). Die Größe der Zelle wird aus den Endpunkten der Filamente berechnet. Abb. 6 zeigt die entsprechende Dichte.

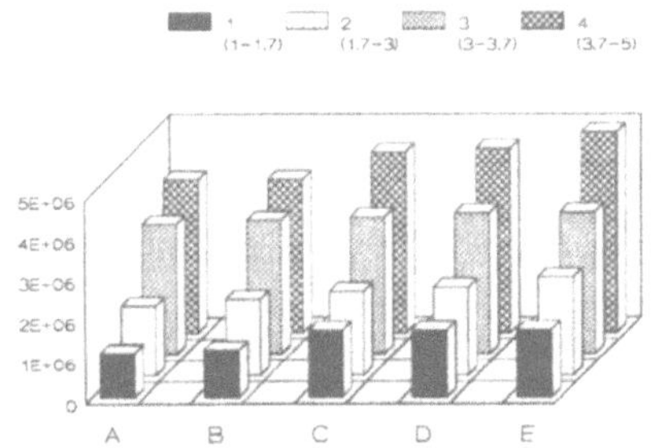

Abb. 5. Größe der Zellen in Voxel

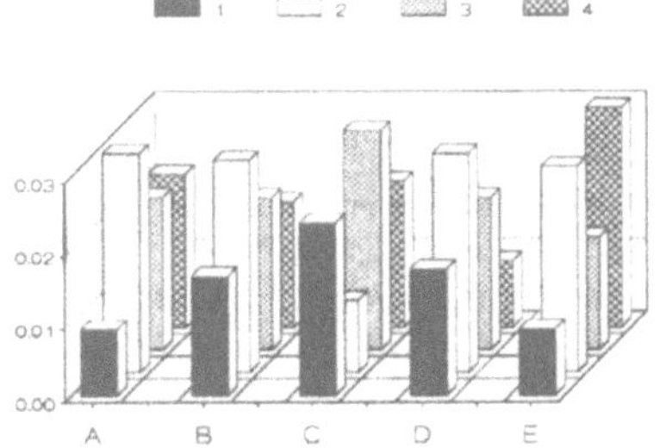

Abb. 6. Dichte der Linien in Voxel

4.2 Anzahl der Verknüpfungen

In Abb. 7 ist die Anzahl der Verknüpfungen mit 3 Abzweigungen für dieselben Zellen gezeigt, in Abb. 8 die Anzahl der Verknüpfungen mit 4 Abzweigungen.

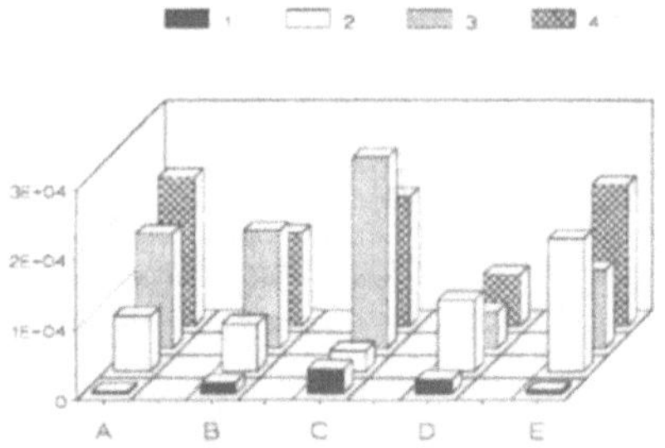

Abb. 7. Anzahl der 3er-Verknüpfungen

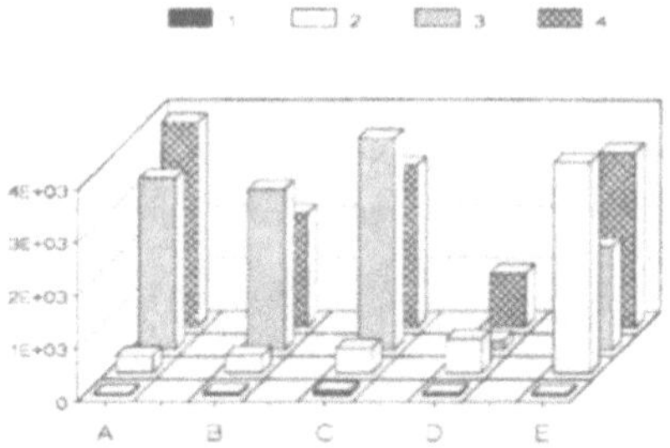

Abb. 8. Anzahl der 4er-Verknüpfungen

4.3 Länge der Linien

Die Länge der Linien in der x-y-Ebene ist in Abb. 9 dargestellt, Abb. 10 zeigt die Länge der Linien parallel zur z-Achse.

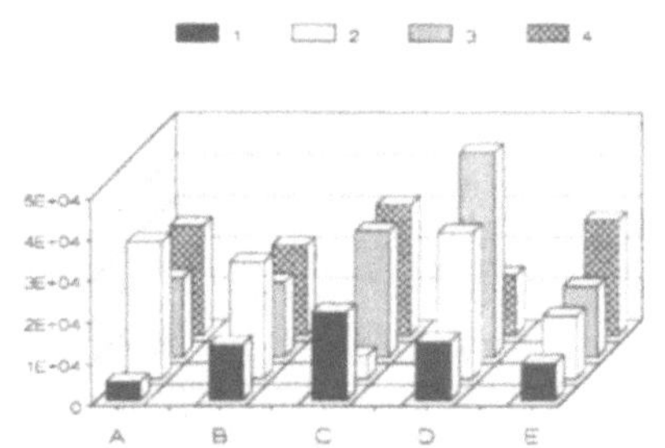

Abb. 9. Länge der Linien in der x-y-Ebene

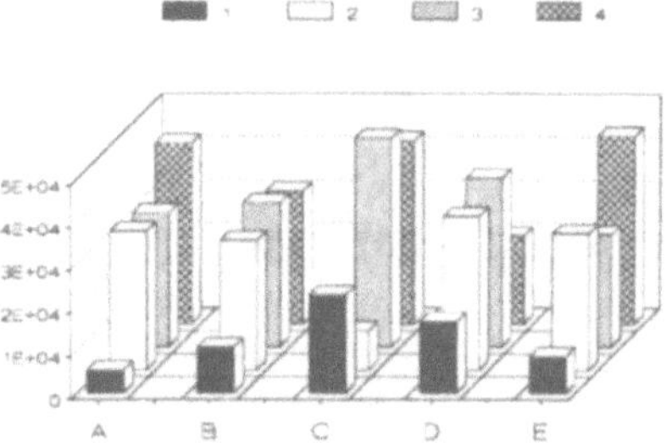

Abb. 10. Länge der Linien in z-Richtung

5 Diskussion

Unsere Ergebnisse zeigen am Beispiel von in vitro kultivierten menschlichen Hepatomzellen, daß das Zellwachstum mit einem Ausbau des Zytoskeletts einhergeht. Dabei bleibt die Filamentdichte in etwa konstant, wärend die Zahl der Filamentverknüpfungen erheblich zunimmt. Möglicherweise kann der Wachstumsprozeß durch eine sich verändernde Zytoskelettstruktur beschrieben und klassifiziert werden. Inwiefern die bisher nachgewiesenen Veränderungen weitergehende Schlußfolgerungen erlauben, wird in weiteren Analysen untersucht.

Danksagung

Die Autoren danken Herrn Dr. G. König für die Anfertigung der Präparate sowie die konstruktive Zusammenarbeit.

Literatur

[1] W. M. Bement, G. I. Gallicano, D. G. Capco, *Role of the Cytoskeleton During Early Development*, Microscopy research and technique, 22:23-48, 1992.

[2] B. Albert, D. Bray, J. Lewis u.a., *Molekularbiologie der Zelle*, Weinheim, Cambrige, NY, 1990.

[3] T. Marr, *Vision*, San-Francisco, Freeman, 1982.

[4] D. Ruelle, *Zufall und Chaos*, Springer-Verlag, 1992.

[5] N. Otsu, *A threshold selection method from gray-level histograms*, IEEE Trans., Vol. Systems, Man and Cybernetics-9, S. 62-66, 1979.

[6] C. Arcelli, L. Cordella, S. Levialdi, *More about a thinning algorithm*, Electronic Letters, Vol. 16, N. 2, S. 51-53, 1980.

On Ramp-Edge Preserving Weak Membrane Reconstruction of Smooth Intensity Functions from 2-D Computed Tomograms*

Martin Fick, Heiko Neumann[1], Karsten Ottenberg[2], H.Siegfried Stiehl[1]

[1] Univ. Hamburg, FB Informatik, Bodenstedtstr. 16, D-22765 Hamburg
[2] Philips GmbH Forschungslaboratorien, Postfach 630565, D-22315 Hamburg

Abstract. Differently scaled intensity discontinuities along the gradient direction of contrast edges define significant X-ray CT and MR image structure. Such contrast edges are intrinsically related to the partial-volume phenomenon in a local neighborhood of anisotropic voxels. Reliable detection and quantitative description of such scaled intensity discontinuities (with a priori unknown loci, relative contrasts, local widths, and orientations) can be achieved with a scale-space approach. The result is an explicit token representation which can be understood as a compact symbolic description of significant image domain structure, viz a set of quantitative attributes per discontinuity (in contrast to e.g. the wavelet representation). Upon such a token representation subsequent computational processes for e.g. either reconstruction of smooth regions or smooth contour grouping can be carried out. Based upon our earlier research, we present a multi-resolution approach to reconstruct the image intensity function from the token representation alone by using an extended membrane formalism as prior image model. The reconstructed smooth intensity function then may serve as input to a region definition process, e.g. based upon supervised or unsupervised local classification. We also report on current experimental results from the entire processing cascade, which combines a quantitative step and ramp discontinuity description with a novel membrane-based intensity reconstruction.

1. Introduction

Accurate detection and rich description of local CT and MR image structure, such as scaled discontinuities (step and ramp contrast edges) and homogeneous regions, is a prerequisite to accurate and reliable extraction of local evidence from which e.g. hypotheses about organs and lesions in the human body can be inferred. With respect to our medical application domain, although the methodologies are independent of the image class, one primary goal of our research was

* Part of this research has been carried out in the COVIRA (Computer Vision in Radiology) consortium funded within the AIM Main Phase by the Commission of the European Communities under contract A 2003.

to develop a method to measure image structure in 2-D intensity functions acquired from different imaging modalities, e.g. MR, CT, or PET scanners. Here, image structure (i.e. intensity discontinuities of varying contrast and local slope, a variety of junctions of different order as well as homogeneous regions) can directly be related to the underlying physical cause, e.g. spatial tissue distributions in the presence of the partial-volume phenomenon. Thus, for instance, the transition width of locally scaled 2-D intensity discontinuities provides information about the local orientation of 3-D organ surfaces with respect to the transmission plane of the image acquisition process.

To assure accurate and reliable results from an edge detection/localization scheme, an either implicitely assumed or explicitly formulated edge model as well as a theoretical investigation into the derivation and performance of an edge detector are required. For ramp-type intensity variations in anisotropic computed tomograms of different specifity, the locus, width, orientation, and relative contrast determine the set of parameters to be estimated from the discrete intensity function. Optimum operator response of edge detection operators depends on both the presence of image function discontinuities matching the characteristics of the edge model and the existence of an operator support (viz, kernel size) adapted to the local width of the discontinuity.

In previously reported research work (e.g. [1], [9], [12]), we have developed a quantitative multi-scale edge detection scheme complemented by an analysis of the intrinsic localization accurracy of regularized linear 1-D and 2-D first and second order derivative operators ([10]). Based upon this (see below for a brief review and [2] for a recent synopsis), we here report on further research extending our approach towards computational segmentation of locally smooth regions in 2-D computed tomograms of different specifity.

2. Scale-space-based extraction of a token representation

The 1-D model and the scale integration scheme. We proposed the generalized error integral curve $\Phi_g(x,{}^1\sigma) = (a + c \cdot \mathcal{H}(x)) \otimes G(x,{}^1\sigma)$ (with scaled Gaussian $G(x)$ of space constant ${}^1\sigma$; $\mathcal{H}(x)$: Heaviside function) as an explicit 1-D discontinuity model since *i)* the complete range of differently scaled discontinuities (with arbitrary local contrast c) can be approximated and *ii)* the intrinsic local parameters of a prior unknown 1-D image function discontinuity, e.g. ${}^1\sigma$ and c, can be efficiently and accurately estimated via scale-space processing (see [1], [9], [12]).

Using the normalized first order partial derivative of a Gaussian with space constant ${}^2\sigma$, $\hat{G}_x(x, {}^2\sigma) = k({}^2\sigma) \cdot \partial G(x,{}^2\sigma)/\partial x$, we derive the operator response $M_g(x,{}^1\sigma,{}^2\sigma) = \Phi_g(x,{}^1\sigma) \otimes \hat{G}_x(x,{}^2\sigma)$. After application of some Fourier theorems (see e.g. [4]) and — without loss of generality — defining $x = x_{max} \equiv 0$, one yields

$$M_g(x_{max} \equiv 0,{}^1\sigma,{}^2\sigma_i) = \frac{c \cdot {}^2\sigma_i}{\sqrt{{}^1\sigma^2 + {}^2\sigma_i^2}} \qquad (1)$$

for operator responses at points of local maxima in the first derivative. It has been shown in our previous papers that a fine-to-coarse scale-space search approach allows for the selection of an optimum operator scale such that both $M_g(x_{max},{}^1\sigma,{}^2\sigma)$ is optimal *and* the operator support fits the local width of the discontinuity, e.g. ${}^1\sigma = {}^2\sigma = \sigma_\Phi$ in our case. As a consequence of this approach, the intrinsic local parameters of a prior unknown 1-D image intensity discontinuity (viz the locus, the space constant ${}^1\sigma$, and the contrast c) can be estimated via scale-space processing.

Elements of the 2-D computational edge process. The theoretically derived approach to modelling of 1-D scaled intensity discontinuities is canonically extendable to 2-D, since the scaled non-linear operator $||\nabla\hat{G}(x,y,{}^2\sigma_i)||$ is an invariant of the 1-jet (implying both a local gauge (u,v)-coordinate system oriented along the gradient direction and existence of a non-vanishing gradient; see [7] for a thorough treatise and [2] for further implications). Hence the first order directional derivative G_v is an invariant and satisfies $G_v = ||\nabla G||$. The extension to 2-D images can also be realized by starting out from the concept of steerable filters on the basis of a scaled circularly symmetric Gaussian [8]. This concept allows the directional derivative of the Gaussian to be steered in gradient direction which again yields $G_v = ||\nabla G||$. Thus the different notations are identical and, w.r.t. the favourable computational gain, oriented discrete filter kernels can simply be avoided for the 2-D case (see [2] for further details).

In order to find the edge point candidates on each individual discrete scale, a local non-maximum suppression operation consequently must be applied at each location along the gradient direction using bilinear interpolation of the gradient magnitudes in a 2×2 neighborhood (assuming also isotropic Gaussian noise). Since corresponding edge point candidates in two adjacent discrete scales can be shifted with respect to their location due to Gaussian diffusion blurring, we restricted a possible shift to be in a 3×3 neighborhood and to be orthogonal to the tangent of the contour (see also [9] for further details).

3. Weak membrane reconstruction constrained by a token representation

Our attempt to reconstruct the intensity surface was on one hand motivated by the need to assess the fidelity of the token representation, on the other hand by the need for a smooth as well as ramp-type edge maintaining reconstruction prior to a computational process for region definition (see also [2] for further details as well as relationships to anisotropic diffusion). Furthermore, our investigation offers new insights into solving the problem of combining advanced ramp-type edge with region-oriented segmentation schemes. In this paper, we *i)* briefly sketch the initial motivation and notation, *ii)* discuss implicit deficiencies of our initial weak membrane approach [11] to ramp-maintaining reconstruction, and *iii)* give an improved membrane definition with favourite properties w.r.t., for instance, both reconstruction accuracy and numerical behaviour. We also present some results from an extensive comparative/competitive evaluation of the initial as well as the novel membrane approach.

Membrane regularization. As we have described above, the token representation can be considered a rich but sparse description of the image structure. The estimated discontinuity attributes of the token representation can now be used as local constraints to the solution of the inverse problem of reconstructing the smooth intensity function. Since this inverse problem is ill-posed, it has to be regularized by incorporating a membrane as the a priori image model into the computational solution (see [3]). Current state-of-the-art research results on weak membrane reconstruction inherently assume step-edge discontinuities (as is also the case in anisotropic diffusion; see [2] for a survey), which however violates basic assumptions about image structure in anisotropic computed tomograms from clinical routine examinations. Hence we extended the by now classical membrane notation towards constraints incorporating ramp-edge attributes into the data-related term. Consequently the novel data similarity term is evaluated at edge locations only, meaning that just approx. $\frac{1}{8}$ of the total pixel number, N^2, enters into the solution process of the reconstruction problem which in turn is of computational advantage. The constraints on the membrane are defined at locations in the $< xy >$-plane utilizing the recovered local relative contrast, the local gradient direction, and the local transition width of the discontinuities. As additional constraints, we use both the fact that the image function takes only positive values and the simple assumption that the ground level at the borders of the image is of zero value (the mean of either the image or the brain intensities could also be a valid choice).

We have defined a quadratic functional to be minimized which incorporates a data dependent contrast similarity term and a model dependent smoothness term, where the influence of both is controlled by a classical regularization parameter α. Ergo

$$F(I) = \frac{1}{2}\sum_{xy}[\sum_{ij\in\lambda_{xy}} ((I_{ij}-I_{xy})-c_{ij,xy})^2+\alpha\cdot((I_{x+1,y}-I_{xy})^2+(I_{x,y+1}-I_{xy})^2)] \longmapsto \min \tag{2}$$

with $\lambda_{xy} = \{[(ij), c_{ij,xy}]\}$ as the set of image locations that terminate a contrast transition emanating from the discrete location $< xy >$, and where $c_{ij,xy}$ defines the local contrast between the two points $< xy >$ and $< ij >$. Naturally, each contrast transition is represented twice using $< xy >$ and $< ij >$ as reference points, respectively, ensuring that the equality $c_{ij,xy} = -c_{xy,ij}$ holds.

The application of this initial ramp-preserving membrane regularization approach to routine MR images (see [11]) experimentally revealed deficiencies due to intrinsic membrane properties. At image locations with locally abrupt high-contrast locations (e.g. inner and outer skull; see Figures below), the membrane notation from above with data-dependent contrast constraint alone yields a reconstruction which is satisfying both the contrast and smoothness constraint, *but* does not yield space-invariant similarity with the intensity data in the neighbourhood of ramp loci. Hence the global solution, which is in contrast to conventional anisotropic diffusion, preserved ramp discontinuities to a maximum extent as anticipated due to the contrast constraint, but it lacked local intensity-related spatially isotropic simililarity. Thus to allow for an appropriate tailoring of the membrane regularization approach to the image structure at hand, the quadratic functional from above has been extended by an additional data term which embodies a "spring force" binding the contrast constraint for a ramp to its locally neighboured intensity plateaus. The local mean intensity around the foot and shoulder point of a ramp can simply be estimated via e.g. a Parzen estimator with Gaussian window. The new intensity-related data term D_{new} is given by $D_{new} = (I_{pq} - I_{mean,pq})^2$ and, together with the old data term, we arrived at the extended novel data term $D = (1-\beta)D_{old} + \beta D_{new}$, where $D_{old} = ((I_{ij} - I_{xy}) - c_{ij,xy})^2$ from above, and β is an additional weighting parameter (see below).

In order to find the optimal solution of this minimization problem, an inhomogeneous linear system of equations, defined by the equations $\frac{\partial F}{\partial I_{pq}}(I) = 0$, has to be solved simultaneously by gradient descent. To do so, we now solve the above equation for I_{pq} and apply a successive over-relaxation algorithm (SOR) as an iterative scheme, which for every pixel at an arbitrary location $< pq >$ corrects the current value of I at a time t:

$$I_{pq}^{t+1} = I_{pq}^{t} - \frac{\omega}{T_{pq}} \frac{\partial F}{\partial I_{pq}}(I)^t \tag{3}$$

The method is convergent for $0 < \omega < 2$, where $\omega = 1$ corresponds to Gauss-Seidel iteration. We derive a local estimate for T_{pq} based on the second derivative of the functional F (see e.g. [3]), i.e.

$$T_{pq} \geq \frac{\partial^2 F}{\partial I_{pq}^2}(I) = 2N_{pq} + 4\alpha, \qquad \text{with} \qquad N_{pq} = card(\lambda_{pq}) \tag{4}$$

In the discrete formulation of the problem as a linear algebraic equation system $A\mathbf{x} = \mathbf{b}$, the new membrane data term both results in a quantitatively better vector $\mathbf{b}$ and, more significantly, by entering into matrix A improves the convergence properties of our iterative method.

Multi-resolution approach. Considering the behaviour of the numerical solution scheme to the reconstruction problem, we observe that the SOR approach has unacceptable slow convergence rate. Since, for instance, entire local plateaus of the intensity surface must be lifted to a certain level, the initial error is dominated by low frequency components of the intensity surface to be reconstructed. To overcome these types of problems in order to get an efficient solution to the optimization task, the solution has been reformulated as a multi-resolution approach (based upon e.g. [5], [14]). Here we define a *set* of functionals ${}^{(\sigma)}F, \sigma = 0, 1, .., M$, each corresponding to an image function ${}^{(\sigma)}I$ on the level of resolution σ, where ${}^{(\sigma)}I$ results from ${}^{(\sigma-1)}I$ by local averaging of pixel blocks of $2^\sigma \times 2^\sigma$ size ($\sigma = 0$ corresponds to the original image resolution). Beginning at the level of coarsest resolution, the adapted SOR algorithm is successively applied for solving the optimization task ${}^{(M)}F$ such that this solution ${}^{(\sigma=M)}I$ is used as an initialization for ${}^{(M-1)}F$ at the next finer level of resolution, etc., until the minimization problem ${}^{(0)}F$ on the finest level of resolution has been solved. Investigations into the natural problem of integrating the multi-scale edge process and the multi-resolution membrane process are under way (also we are investigating into the applicability of the thin plate model).

4. Experiments, results, and brief assessment

In previously published articles ([1], [9]) we have reported on results from processing cranial CT and MR images for the purpose of recovery of a token representation as well as for our initial weak membrane reconstruction [11]. Here we give results from applications of the entire processing cascade to routine MR images such as to demonstrate the clearly visible superiority of the novel extended membrane. Fig. 1(top, left) shows the original 256×256 routine MR brain image with a frontal lesion (T_1-weighted, low contrast). The results of *i)* the initial membrane with D_{old} (for $\alpha = 1.0$, $\beta = 0.0$, and $\omega = 1.0$) are given in Fig. 1 (top, right), *ii)* the extended membrane (for $\alpha = 0.1$, $\beta = 0.5$, and $\omega = 1.0$) are given in Figs. 1 (bottom, left) and (for $\alpha = 0.1$, $\beta = 0.9$, and $\omega = 1.0$) in (bottom, right), respectively. Figs. 2a and 2b display comparisons of the profile of row 128 in Fig. 1 (top, left) (black dots) with corresponding profiles (grey dots) in Figs. 1 (top, right) and 1 (bottom, left). It can be easily grasped from both the pictorial and graphical results that the extended membrane approach is superior w.r.t. the initial membrane definition. However, the selection of the classical regularization parameter α is, as usual, critical although the value range can be on empirical grounds limited to 0.1 up to maximally 2.0 (thus making the choice of *alpha* for the extended membrane more noncritical than for the initial membrane). Further empirical evidence from our experimental studies showed that the new parameter β is amenable to a robust and constant parameter setting, e.g. $\beta = 0.5$ (see also Fig. 3). Moreover we observed for $\beta = 0.5$ a minimum global error reconstruction (where the error ϵ is given as a fidelity measure in terms of the mean absolute intensity deviation of the original image and its reconstruction ; see Fig. 3) for constant $\alpha = 1.0$ but ω varying over the admissible range for the SOR method (abscissa in Fig. 3). For further details we refer the reader to [6].

5. Discussion

We have reported on ongoing research towards as-automated-as-possible, inherently parallel, and theoretically derived low level image segmentation processes drawing upon a novel combination of multi-scale edge extraction and multi-resolution image function reconstruction. The primary goal of the multi-scale approach was to derive a symbolic (token set) description of the image function (see also [9] for clinical relevance). In order to allow for subsequent computational region definition, we investigated into the inverse problem, viz the synthetization or reconstruction of a smooth 2-D intensity function from a token set alone. The reconstruction result then in turn will serve as a starting point for subsequent computational segmentation processes. For instance, the token representation can trigger both contour grouping and filling-in of small edge gaps by utilizing local geometric constraints derived from the differential geometry of the reconstructed intensity surface ([13]). Also the reconstruction of a smooth intensity function by preserving discontinuous image structure is expected to validly constrain the solution space of local intensity-based classification schemes. Apart from the promising initial results, however, a number of problems still remain to be tackled, e.g. incorporation of physical knowledge sources encoding e.g. domain specific parameters derived from MR and CT image formation knowledge, hysteresis approach based upon knowledge about the respective imaging modality, optimal scale-space sampling, reduction of the number of iterations per resolution level by multigrid methods such as coarse grid correction, and suitable microelectronic devices as well as parallel architectures for the different interconnected layers of the processing cascade such as to gain computational speed.

References

1. S. Back, H. Neumann, and H.S. Stiehl. On scale-space edge detection in computed tomograms. In H. Burckhardt, K.-H. Höhne, and B. Neumann, editors, *Mustererkennung 1989*, Informatik Fachberichte 219, pages 216 – 223, Berlin, 1989. Springer.
2. W. Beil, H. Neumann, K. Ottenberg, and H.S. Stiehl. On the recovery of image structure in 2-D discrete X-ray and magnetic resonance computed tomograms. Technical Report, Universität Hamburg, Fachbereich Informatik, 1993. (forthcoming).
3. A. Blake and A. Zisserman. *Visual Reconstruction*. The MIT Press, Cambridge (MA/USA), 1987.
4. R.N. Bracewell. *The Fourier Transform and its Applications*. McGraw-Hill, New York, 1978.
5. A. Brandt. Multi-level adaptive solutions to boundary-value problems. *Mathematics of Computation*, 31(138):333 – 390, 1977.
6. M. Fick. Membranrekonstruktion der Bildfunktion aus attributierten Kantenlisten. Diplomarbeit, Universität Hamburg, FB Informatik, 1993.
7. L.M.J. Florack, B.M. ter Haar Romeny, J.J. Koenderinck, and M.A. Viergever. Scale and the differential structure of images. *Image and Vision Computing*, 10(6):376 – 388, 1992.
8. W.T. Freeman and E.H. Adelson. The design and use of steerable filters. *IEEE Transactions on Pattern Analysis and Machine Intelligence*, 13(9):891 – 906, 1991.
9. G. Gabrielides, H. Neumann, and H.S. Stiehl. Estimating partial volume induced local blurring of organ contours in computed tomograms. In H.U. Lemke, M.L. Rhodes, C.C. Jaffee, and R. Felix, editors, *Computer Assisted Radiology, Proc. CAR '91*, pages 556 – 562, Berlin, 1991. Springer.
10. H. Neumann, K. Ottenberg, and H.S. Stiehl. Accuracy of regularized differential operators for discontinuity detection in 1-D and 2-D intensity functions. In R.M. Haralick and W. Förstner, editors, *IEEE International Workshop on Robust Computer Vision*, pages 214 – 260, Seattle (WA/USA), Oct. 1-3 1990.
11. H. Neumann, K. Ottenberg, and H.S. Stiehl. Finding and describing local structure in discrete two-dimensional computed tomograms. In *11th IAPR Int. Conf.*

on Pattern Recognition (ICPR-92), Vol. I-IV, pages (III) 408 – 412, Le Hague, NL, Aug. 30-Sept. 3 1992. IEEE Computer Society Press.
12. H. Neumann, K. Ottenberg, and H.S. Stiehl. Quantitative description and reconstruction of intensity functions using scale-space and multiresolution processing. In J. Vandevalle, R. Boite, M. Moonen, and A. Oosterlinck, editors, *6th European Signal Processing Conference (EUSIPCO-92), Aug. 24-27, Brussels (Belgium)*, Signal Processing VI: Theories and Applications, Vol. I - III, pages (III) 1425 – 1429, Amsterdam, 1992. Elsevier.
13. K. Ottenberg. Contour grouping via brightness-contrast diffusion. (submitted).
14. K. Stüben and U. Trottenberg. Multigrid methods: Fundamental algorithms, model problem analysis and analysis. In W. Hackbusch and U. Trottenberg, editors, *Multigrid Methods*, Lecture Notes in Mathematics 960, pages 1 – 176, Berlin, 1982. Springer.

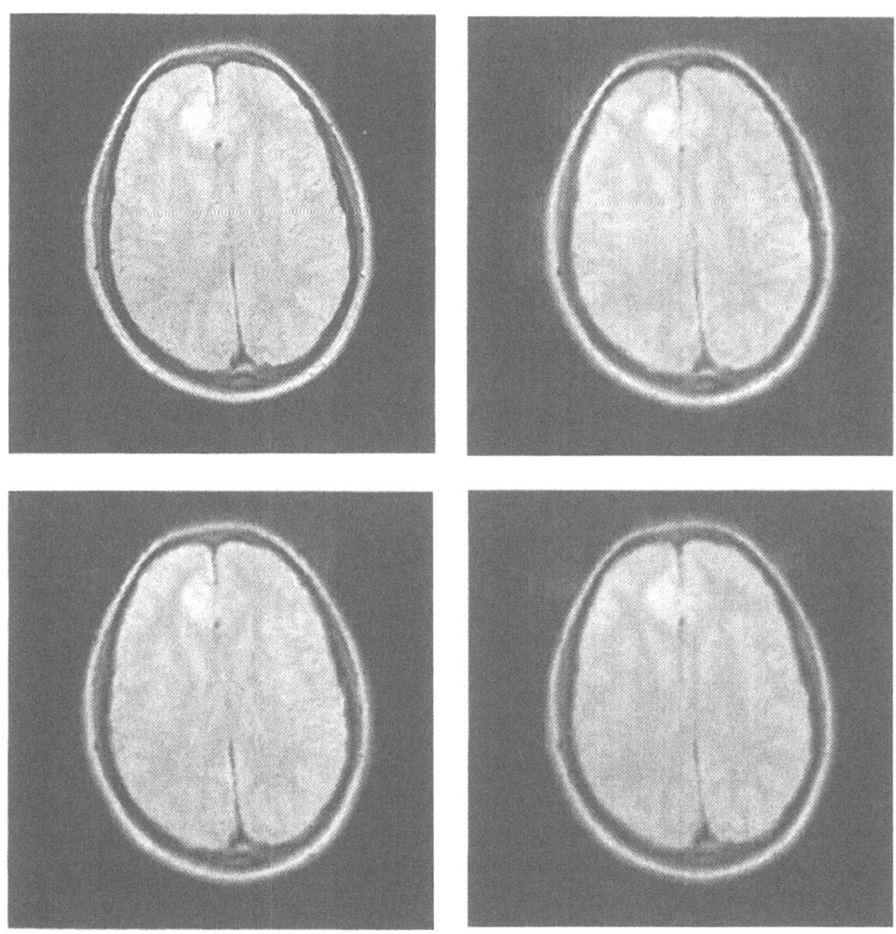

Figure 1: Original image (top,left); membrane reconstruction (top, right): $\beta = 0$, $\alpha = 1.0$; (bottom,left): $\beta = 0.5$, $\alpha = 0.1$ and (bottom,right): $\beta = 1.0$, $\alpha = 0.1$

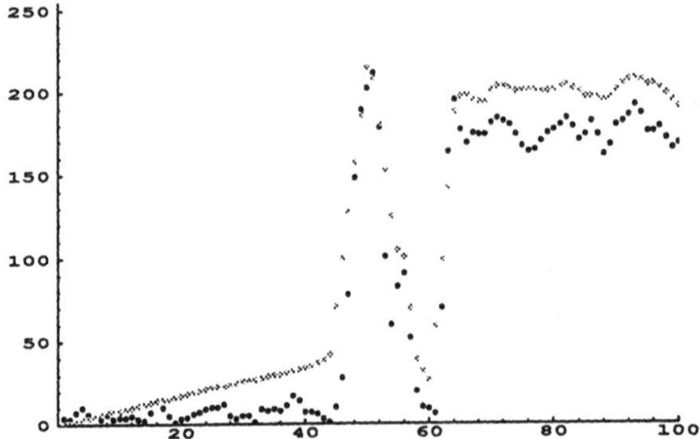

Figure 2: Profiles of the original image (black) and reconstruction image (gray) in row 128

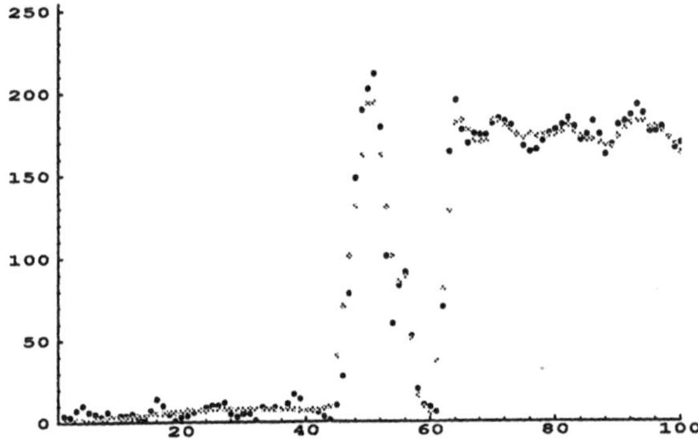

Figure 3: Same profiles with new data term reconstruction (gray, $\beta = 0.5$)

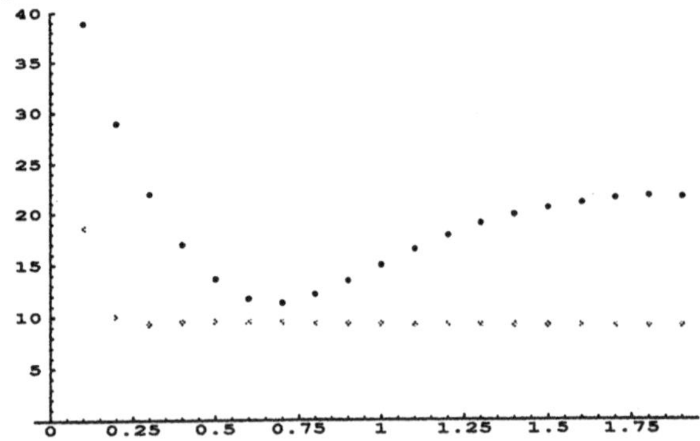

Figure 4: Comparision of the global error in relation to the relaxation parameter ω for $\beta = 0$ (black) and $\beta = 0.5$ (gray) (membrane, $\alpha = 1.0$)

Automatische Schielwinkelmessung[1] durch Hough–Transformation und Kovarianzfilterung

T. Lehmann[2], A. Kaupp, D. Meyer–Ebrecht, R. Effert[3]

Lehrstuhl für Meßtechnik (LfM)
Rheinisch–Westfälische Technische Hochschule Aachen
Templergraben 55, D - 5100 Aachen

Zusammenfassung

Zur Messung des Schielwinkels eines Patienten werden seine Augen mit zwei CCD–Kameras fotografiert. Bei der Aufnahme werden drei Blitzlichtgeräte ausgelöst. Durch die Vermessung der Reflexpositionen im digitalen Bild kann der Schielwinkel bestimmt werden.

In diesem Beitrag wird ein schwellwertfreies Segmentierungsverfahren zur automatischen Detektion der Reflexe vorgestellt. Mit einem auf der Hough–Transformation basierendem Algorithmus wird der Mittelpunkt der Iris lokalisiert. In der daraus bestimmten *region of interest* werden die Reflexpositionen mit einer Kreuz–Kovarianz–Filterung lokalisiert. Danach erfolgt eine Klassifikation und Plausibilitätskontrolle zum Ausschluß von Fehlmessungen. Der Algorithmus wurde an 191 willkürlich ausgewählten klinischen Bildern mit fehlerfreien Resultaten erprobt und wird in der Praxis eingesetzt.

1 Einleitung

Schielen (Strabismus) nennt man die beständige oder immer wieder auftretende Fehlstellung eines Auges. Bei einem schielenden Menschen kann das Gehirn die von den Augen erzeugten Einzelbilder nicht richtig zur Deckung bringen und unterdrückt das vom schielenden Auge gelieferte Bild. Dies führt bei kleinen Kindern in kürzester Zeit zu einer irreversiblen Herabsetzung der Sehkraft des schielenden Auges, die nur durch eine frühzeitige Behandlung vermieden werden kann.

Daher wurde an der RWTH Aachen ein Schielwinkel–Meßgerät entwickelt [1,2,3,4]. Während der Patient eine Markierung anvisiert, werden seine Augen mit zwei CCD–Kameras aufgenommen. Dabei werden drei über den Kameras montierte Blitzlichtgeräte ausgelöst (Abb. 1). Aus den Positionen der entstehenden Reflexe an der Hornhautvorderfläche (Purkinje Bilder 1. Ordnung, kurz PK1) und der Rückseite der Augenlinse (PK4) kann der Schielwinkel berechnet werden [5].

Da für das Schielwinkelmeßgerät eine hohe Betriebssicherheit gefordert ist, darf die automatische Positionsbestimmung keine falschen Werte liefern. Also müssen 'nicht auswertbare' Bilder eigenständig als solche erkannt werden und alle als 'auswertbar' klassifizierten Bilder richtig ausgewertet werden. Der Detektionsalgorithmus soll möglichst viele Aufnahmen auswerten können. Er muß daher unanfällig gegen Störreflexe in anderen Bildbereichen sein und auf veränderte Aufnahmebedingungen flexibel reagieren. Insbesondere muß er von der Belichtung der Aufnahme unabhängig und an geänderte Aufnahmeapparaturen anpaßbar sein.

---

[1]Das Projekt *Schielwinkelmessung* wird gefördert durch die Deutsche Forschungsgemeinschaft DFG

[2]jetzt am Institut für Medizinische Informatik und Biometrie (Komm. Direktor: Prof. Dr. R. Repges)

[3]Augenklinik der Medizinischen Einrichtungen der RWTH Aachen (Direktor: Prof. Dr. M. Reim)

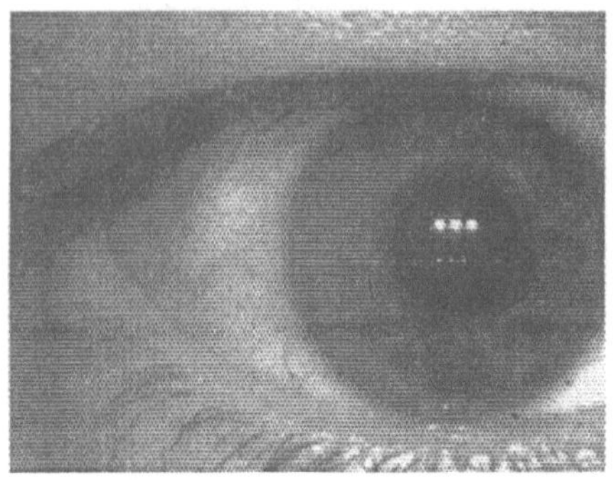

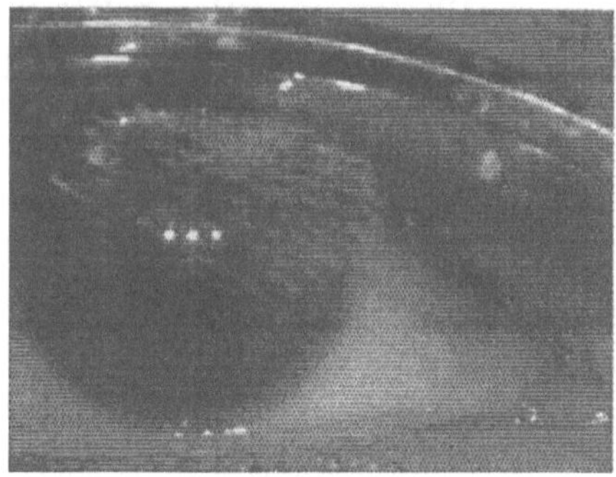

Abbildung 1: Originalbilder

Die Abbildung zeigt zwei Bilder, wie sie bei der Schielwinkelmessung aufgenommen werden. Die Teilbereiche Augapfel, Iris und Pupille und im rechten Bild eine Brille mit Fingerabdruck sind deutlich zu erkennen. Innerhalb der Pupille sieht man jeweils drei große Reflexe und ein Stück tiefer, etwas versetzt, drei kleine. Dies sind die zu detektierenden PK1 und PK4. Die in ihrer Intensität ca. 50 fach schwächeren PK4 sind im linken Bild hervorgehoben.

2 Modellierung der Bildentstehung

Beschreibt man das Auge durch seine lichtbrechenden Schichten: Hornhaut, Kammerwasser, Augenlinse und Glaskörper [6], so läßt sich der optische Strahlengang berechnen [7,4]. Die PK1 liegen als virtuelle und die PK4 als reelle Bilder der Blitzblende ca. 5 mm hinter der Hornhaut und damit im Schärfentiefebereich der Objektive. Die Blitzblende wird bei der Spiegelung um den Faktor 150 verkleinert, so daß die PK's als Punktlichtquellen angesehen werden können. Die Intensität der PK4 ist dabei 50 mal schwächer als die der PK1. In der Sensorebene lassen sich die Intensitätsverteilungen $I(x, y)$ der PK als Fraunhofer Beugungsfiguren [8] der Objektivblende beschreiben. Bei einer kreisförmigen Blende mit Radius q im Abstand D vor dem Sensor ergibt sich für weißes Licht mit der Wellenzahl $k = \frac{2\pi}{\lambda}$ und der Wellenlänge λ:

$$I(x,y) = I_0 \int\limits_{\lambda=380nm}^{\lambda=780nm} \left(\frac{J_1\left(\frac{kqr}{D}\right)}{\left(\frac{kqr}{D}\right)} \right)^2 d\lambda, \quad \text{mit} \quad \begin{aligned} r &= \sqrt{x^2+y^2} \\ I_0 &= \text{konstant} \\ J_1 &= \text{Besselfunktion 1. Ordnung.} \end{aligned} \tag{1}$$

3 Konzeption des Algorithmus

Entscheidend für die Konzeption des Verfahrens ist, welches *a priori* Wissen zur Detektion benutzt und welches zur Plausibilitätskontrolle eingesetzt wird. Das Wissen, das zur Detektion herangezogen wird, darf die Flexibilität des Verfahrens nicht einschränken. Hierzu kann also nur das Wissen über den Signalverlauf der PK's, nicht jedoch das Wissen über die möglichen Reflexpositionen im Bild verwendet werden.

Den PK ähnliche Reflexe entstehen durch die Tränenflüssigkeit oder den Fettfilm auf der Gesichtshaut, also in anderen Bildbereichen. Dies macht die Bestimmung einer *region of interest* zur Eingrenzung des Suchbereiches für die PK notwendig.

In Abbildung 2 ist die Verwendung von *a priori* Wissen dem Strukturgramm des Algorithmus gegenübergestellt. Bei der Bildanalyse zur automatischen Schielwinkelmessung wird zuerst die Iris als *region of interest* lokalisiert. Die Iris hat gegenüber der Pupille den Vorteil einer größeren Randkurve mit stärkerem Kontrast und konstantem Radius[4]. Innerhalb der Iris werden dann die PK1 und PK4 unabhängig voneinander detektiert. Dabei soll sowohl bei den PK1 als auch bei den PK4 jeweils nach einzelnen Reflexen, nicht nach der Reflexgruppe gesucht werden. Aufgrund der ermittelten Reflexpositionen und des *a priori* Wissen über die Aufnahmegeometrie wird entschieden, ob die Aufnahme als 'auswertbar' oder 'nicht auswertbar' eingestuft wird. Nur die als 'auswertbar' klassifizierten Aufnahmen werden dem letzten Verarbeitungsschritt, der Vermessung, zugeführt, in dem die Schielwinkelberechnung aufgrund der Reflexpositionen erfolgt.

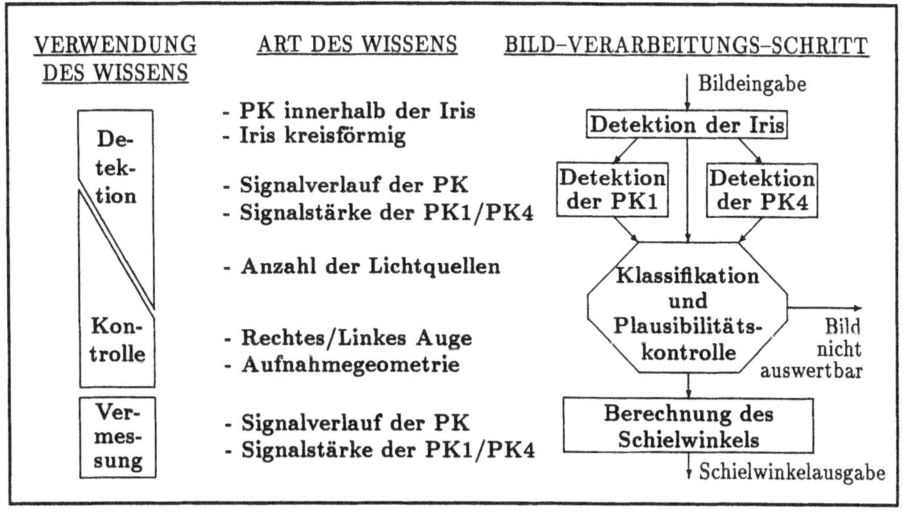

Abbildung 2: Einsatz von *a priori* Wissen und Struktur des automatischen Algorithmus
Die drei Spalten in der Abbildung sind inhaltlich horizontal zueinander ausgerichtet.

4 Irisdetektion durch Hough–Transformation

Das Ergebnis der Irisdetektion soll ihr Mittelpunkt im Ausgangsbild sein. Die Iris läßt sich dabei als dunkle Kreisfläche mit konstantem Radius auf hellem Grund (Augapfel) beschreiben. Es kann nicht davon ausgegangen werden, daß die Randkurve der Iris geschlossen und störungsfrei auf dem Bild vorhanden ist. In der Regel überdecken das obere und untere Lid jeweils einen Teil der Iris oder die Iris ragt über die Bildgrenzen hinaus (Abb. 1). Wegen ihrer Unempfindlichkeit gegen Unterbrechungen der Randkurve eines Objektes ist die Hough–Transformation zur Irisdetektion besonders geeignet [9].

Die Hough–Transformation für Kreise: Die von Hough 1962 entwickelte Methode zur Erkennung komplexer Strukturen [10] basiert auf der Idee, alle zur Struktur gehörenden Randpunkte in einen Punkt des Abbildungsraumes, dem sogenannten *accumulator array*, abzubilden. Nach der Transformation wird dort das globale Maximum bestimmt und wieder in den Bildraum zurückgerechnet, womit die gesuchte Struktur lokalisiert ist. Eine Hough–Transformation läßt sich für jede parameterisierbare Randkurve angeben,

[4]Bei einem einjährigen Kind ist das Auge bereits ausgewachsen und damit der Radius der Iris konstant.

wobei die Parameter der Kurve in die Achsen des *accumulator array* übergehen. Für Kreise führt dies auf ein dreidimensionales *accumulator array*, denn die Kreisgleichung

$$(x - X_m)^2 + (y - Y_m)^2 = R^2 \qquad \text{mit} \qquad X_m, Y_m, R = const. \tag{2}$$

enthält die drei Parameter: Mittelpunkt (X_m, Y_m) und Radius R. Nach Bereichsbegrenzung und Quantisierung dieser Parameter werden die Voxel des *accumulator array* mit Null initialisiert. Für einen festen Radius $R = R_1$, also einem Schnitt durch den Abbildungskubus, kann die Umkehrung von (2) angegeben werden:

$$(X_m - x_k)^2 + (Y_m - y_k)^2 = R_1^2 \qquad \text{mit} \qquad x_k, y_k, R_1 = const. \tag{3}$$

Bei der Transformation des Pixels $P_k = (x_k, y_k)$ werden alle Zellen des *accumulator array* $A = (X_m, Y_m, R)$, die (3) erfüllen, einheitlich inkrementiert.

Berücksichtigung der Gradientenrichtung: Das Ausgangsbild zur Berechnung der klassischen Hough–Transformation ist ein binäres Kantenbild, welches durch Gradientenberechnung im Originalbild entsteht. Die Berücksichtigung der Richtung des Gradienten erhöht die Genauigkeit der Hough–Transformation [9]. Bei Verwendung des Sobel–Operators[5] zur Gradientenberechnung lassen sich acht Richtungen unterscheiden, die Inkrementierung nach (3) erfolgt dann nur noch für ein achtel Kreissegment.

Die Berücksichtigung der Gradientenrichtung macht die Irisdetektion auch unempfindlich gegen Störungen, wie sie durch Brillengläser entstehen können. Brillenreflexe sind helle Flecken auf dunklem Grund und können deshalb, selbst wenn sie einen der Irisgröße entsprechenden Radius haben sollten, nicht zu Fehldetektionen führen.

Berücksichtigung der Gradientenamplitude: Die Unabhängigkeit vom Kontrast im Originalbild, und damit von der Aufnahmebelichtung, kann durch Berücksichtigung der Amplitude des Gradienten erreicht werden. Die Punkte im *accumulator array* werden nicht mehr einheitlich inkrementiert, sondern die Erhöhung wird mit der Amplitude des Gradienten gewichtet. Damit kann das vorrausgesetzte binäre Kantenbild entfallen und die Hough–Transformation kann völlig ohne Schwellwerte für jedes Pixel im Originalbild (Abb. 1) durchgeführt werden.

Quantisierung des accumulator array: Die Bereichsbegrenzung und Quantisierung der Mittelpunktskoordinaten im *accumulator array* entspricht der Auflösung des verkleinerten Bildes[6]. Damit lassen sich die Mittelpunkte mit einer ausreichenden Genauigkeit in das Originalbild übertragen. Aufgrund der konstanten Größe der Iris reichen drei Stufen zur Quantisierung des relevanten Radiusbereiches aus.

Definition eines Gütekriteriums: In jeder Quantisierungsstufe des Radius werden im *accumulator array* unterschiedlich viele Punkte (gleiche Raumwinkel der Kreissegmente) mit unterschiedlichen Gewichten (Berücksichtigung der Gradientenamplitude) inkrementiert. Die absolute Höhe des globalen Maximums allein läßt daher keine Aussage mehr darüber zu, ob im Originalbild die Iris gefunden werden konnte[7]. Bei vorhandener Iris

[5] Der Sobel–Operator berechnet den Gradienten bei gleichzeitiger Binomial–Tiefpaßfilterung 1. Ordnung in der zum Gradienten orthogonalen Richtung und liefert somit geglättete Kanten.

[6] Zur drastischen Reduktion der Rechenzeit, wird vor der Hough–Transformation eine Verkleinerung des Originalbildes durch Binomial–Tiefpaßfilterung und Unterabtastung durchgeführt.

[7] Das Maximum im Schnittbild R_1 des *accumulator array* bei (X_{m1}, Y_{m1}) ist notwendig für die Existenz der Iris mit Mittelpunkt an dieser Stelle, jedoch nicht hinreichend.

ist in dem Schnittbild mit passendem Radius ein deutliches Maximum zu erwarten, bei nicht vorhandener Iris oder nicht passendem Radius ist dagegen mit einem Schnittbild mit mehr oder weniger gleichverteilten Werten zu rechnen (Abb. 3). Dies legt die folgende Definition einer Güte $\mathcal{G}_{\text{HOUGH}}(R)$ der Hough–Transformierten nahe:

$$\mathcal{G}_{\text{HOUGH}}(R) \stackrel{\text{def}}{=} \frac{\text{Absolutes Maximum im Schnittbild}}{\text{Mittlerer Grauwert des Schnittbildes}} \tag{4}$$

Mit diesem Gütekriterium können nicht nur die einzelnen Schnitte mit $R = const.$ des *accumulator array* miteinander verglichen werden, es ist auch eine Aussage über die Existenz der Iris im Originalbild möglich.

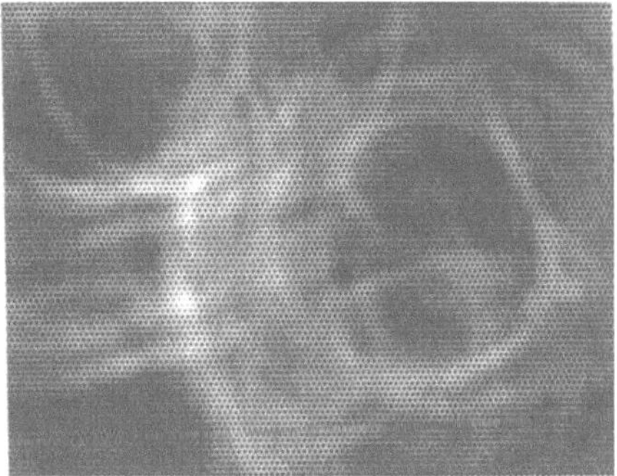

Abbildung 3: Hough–Transformierte mit den Güten 7,79 und 2,86

Der linke Teil der Abbildung zeigt die Hough–Transformierte der Schielwinkel–Meßaufnahme aus dem linken Teil der Abbildung 1, bei der die Randkurve der Iris fast ungestört ist. Der rechte Teil zeigt die Hough–Transformierte einer Porträtsaufnahme, welche das ganze Gesicht zeigt. Da in der Porträtaufnahme eine der Iris entsprechende Struktur fehlt, ist bei der Transformation kein globales Maximum entstanden. (Die Bilder wurden für die Darstellung histogrammoptimiert.)

Zusammenfassung des Algorithmus zur Bestimmung der region of interest: Das verkleinerte Bild wird mit dem Sobel–Operator gefaltet. Unter Berücksichtigung des Vorzeichens lassen sich nach der Filterung mit den vier Sobel–Masken acht Richtungen unterscheiden. Es wird also für jeden Bildpunkt im verkleinerten Ausgangsbild vier mal ein, mit der Amplitude des entsprechenden Sobels gewichteter, Achtelkreis in den drei Radiusebenen des *accumulator array* inkrementiert.

Nach der Transformation wird die Radiusebene mit der maximalen Güte nach (4) bestimmt und als Ergebnis der Irisdetektion werden die Werte dieser Ebene für Mittelpunkt, Radius und Güte an die nachfolgenden Verarbeitungsschritte weitergegeben (Abb. 2).

5 PK–Detektion durch Kreuz–Kovarianz–Filterung

Um die Koordinaten des Irismittelpunktes wird die *region of interest* $u(x,y)$ als kreisförmiger Ausschnitt des Originalbildes definiert. In diesem Ausschnitt müssen die PK nur mit *a priori* Wissen über den Signalverlauf unabhängig von einander detektiert werden. Mit (1) kann dazu eine rotationssymetrische Musterfunktion $v(x,y)$ generiert werden, wobei die Maskengröße von v durch

$$x'_{max} = y'_{max} = 2 \cdot D_{PK} - B_{PK} - 1 \tag{5}$$

bestimmt wird. D_{PK} steht dabei für den Mittelpunktsabstand zweier benachbarter PK-Impulse der Breite B_{PK}.

Das einfache *matched filtern*[8] von u mit v reicht wegen der geringen Signalauflösung bei der Digitalisierung von (1) mit nur wenigen Pixeln zur Detektion der PK nicht aus [4]. In [12] wird die Berechnung der normierten Kreuz-Kovarianz-Funktion

$$KKV(x,y) = \frac{\sum_{k=0}^{x'_{max}} \sum_{l=0}^{y'_{max}} \{[u(x+k,y+l)-\overline{u}]\,[v(k,l)-\overline{v}]\}}{\sqrt{\left(\sum_{k,l}[u(x+k,y+l)-\overline{u}]^2\right)\left(\sum_{k,l}[v(k,l)-\overline{v}]^2\right)}} \tag{6}$$

vorgeschlagen. $\overline{u}$ ist dabei der jeweilige Mittelwert des Ausschnitts von u unter v und $\overline{v}$ der Mittelwert von v. Das Ergebnisbild wird nach mindestens L Bereichen größter Amplitude durchsucht, wobei L die Anzahl der verwendeten Lichtquellen ist. Die Schwerpunkte dieser Segmente

$$S_x = \frac{\sum_{x=0}^{x_{max}} x \sum_{y=0}^{y_{max}} [u(x,y)-u_{min}]}{\sum_{x,y}[u(x,y)-u_{min}]} \quad \text{und} \quad S_y = \frac{\sum_{y=0}^{y_{max}} y \sum_{x=0}^{x_{max}} [u(x,y)-u_{min}]}{\sum_{x,y}[u(x,y)-u_{min}]} \tag{7}$$

werden als Koordinaten der PK der im nächsten Abschnitt beschriebenen Plausibilitätskontrolle unterzogen (Abb. 2).

Für die PK1 kann das angegebene Verfahren vereinfacht werden. Aufgrund der hohen Intensität liegen ihre Maxima bei 'auswertbaren' Aufnahmen immer in der Sättigung des CCD-Sensors. Die Kreuz-Kovarianz-Filterung kann daher durch eine einfache Binarisierung ersetzt werden. Mögliche Störreflexe auf Brillengläsern können durch ihre unterschiedliche Größe als solche erkannt werden.

6 Klassifikation und Plausibilitätskontrolle

In diesem Verarbeitungsschritt soll entschieden werden, ob eine Aufnahme 'auswertbar' oder 'nicht auswertbar' ist. Die Gesamtkonzeption des Algorithmus stellt dafür eine Reihe unabhängiger Kontrollkriterien zur Verfügung, die noch nicht zur Detektion der Reflexe benutzt wurden. Die einzelnen Tests wurden hierarchisch gegliedert. Zuerst werden die Bedingungen getestet, die sich nur auf eine Reflexart beziehen. Danach wird das *a priori* Wissen aus der Aufnahmegeometrie über die möglichen Positionen der Reflexgruppen zueinander überprüft:

1. Stimmt die Anzahl der gefundenen PK1 mit der Zahl L der verwendeten Blitzlichtgeräte überein?
2. Sind die PK1 in X-Koordinatenrichtung äquidistant?
3. Liegen die PK1 in Y-Koordinatenrichtung auf einer Horizontalen?
4. Tests 1 bis 3 für die PK4.
5. Liegen die PK4 entsprechend der Aufnahmegeometrie ober- bzw. unterhalb der PK1?
6. Liegen die PK4 zu den PK1 entsprechend einer Aufnahme des linken oder rechten Auges rechts oder links versetzt?

[8] Das *matched filtern* entspricht der Berechnung der Kreuz-Korrelations-Funktion [11].

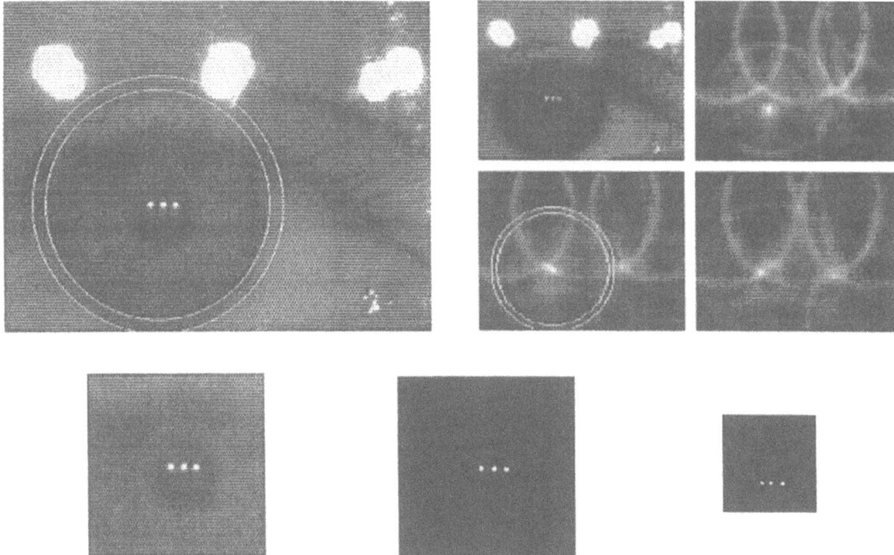

Abbildung 4: Zwischenergebnisse des vollständigen Algorithmus

Das Ausgangsbild (oben links) wird verkleinert und Hough–transformiert (oben rechts). Die Schnittebene des Hough–Raumes mit maximaler Güte nach (4) ist um die detektierten Mittelpunktskoordinaten mit einem Doppelkreis entsprechend ihres Radiusbereiches markiert. Der gefundene Radiusbereich ist in das Originalbild (oben links) zurück transformiert worden. Die so bestimmte *region of interest* ist unten links abgebildet, das Ergebnis der PK1–Detektion unten in der Mitte und das der PK4–Detektion unten rechts. Die drei hellen weißen Flecken sind Blitzlichtreflexe auf dem Brillenglas. Sie können zu einer, wenn auch geringen, Verfälschung des Detektionsergebnisses führen. In diesem Beispiel wäre das deutliche Maximum im ersten Schnittbild des Hough–Kubus (ganz rechts oben) das richtige Ergebnis gewesen.

Ist ein Kriterium nicht erfüllt, so wird die Aufnahme als 'nicht auswertbar' eingestuft. Nur wenn alle Bedingungen erfüllt sind, wird die Aufnahme als 'auswertbar' klassifiziert und aus den $2 \cdot L$ Koordinaten der PK wird der Schielwinkel berechnet.

7 Experimentelle Ergebnisse und Diskussion

Abbildung 4 verdeutlicht die Teilschritte des vollständigen Detektionsalgorithmus. Das Originalbild (oben links) wird verkleinert und in drei Radienbereiche Hough–transformiert (oben rechts). Der Mittelpunkt der Iris läßt sich eindeutig detektieren. Im unteren Teil der Abbildung ist die ermittelte *region of interest* sowie das Ergebnis der PK1–Detektion durch Binarisierung bei Sensorsättigung und das der PK4–Detektion durch normierte Kreuz–Kovarianz–Filterung dargestellt. Die sechs Reflexe wurden trotz der zahlreichen Störungen in der Meßaufnahme korrekt lokalisiert. Abbildung 5 zeigt die verkleinerten Versionen der Originalbilder aus Abbildung 1 sowie deren Hough–Transformierten.

Der Algorithmus zur automatischen Bildauswertung der Schielwinkelmeßaufnahmen wurde auf einem 486er AT unter MS-DOS mit TURBO C 2.0 implementiert. Die Berechnung des gesamten Algorithmus dauert ca. 2 Minuten.

Der Algorithmus wurde mit insgesamt 191 willkürlich ausgewählten klinischen Bildern getestet. Darunter waren nicht nur Bilder, in denen Iris, PK1 und PK4 gut erkennbar sind, sondern auch schlecht belichtete und unscharfe, bzw. Aufnahmen ohne erkennbaren Augapfel. Alle 34 manuell nicht auswertbaren Bilder wurden auch automatisch als solche

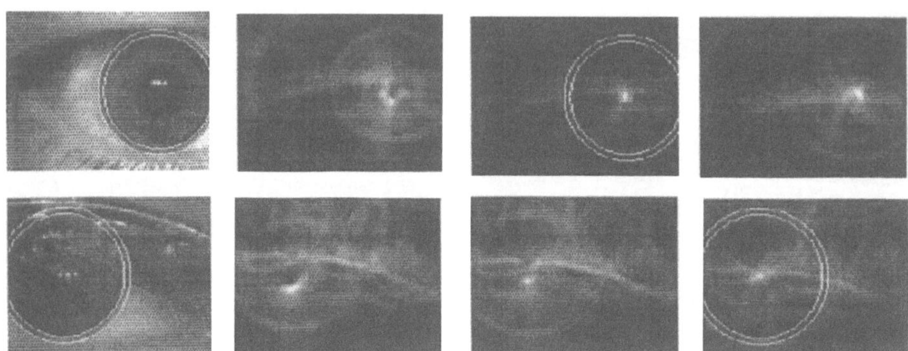

Abbildung 5: Ergebnisse der Hough-Transformation für Kreise
Die zwei Reihen der Abbildung zeigen die verkleinerten Versionen der Bilder aus Abbildung 1 sowie die drei Schnitte durch den Abbildungskubus der Hough-Transformation. Die detektierten Parameter der Irisrandkurve sind durch Doppelkreise entsprechend der Radiusquantisierung der Hough-Transformation markiert.

klassifiziert. Zwanzig manuell auswertbare Aufnahmem (10%) wurden als 'nicht auswertbar' eingestuft, die verbleibenden 137 Bilder wurden richtig ausgewertet. Bei keiner Aufnahme wurde ein falscher Schielwinkel ausgegeben.

Probleme entstehen durch Reflexe an Brillengestellen, die wegen der Berücksichtigung der Gradientenamplitude die Hough-Transformation derart verfälschen können, daß die Detektion der Iris mißlingt (bei 4 der 20 'nicht auswertbaren'). Bei sehr unscharfen Aufnahmen können die PK1 in einander verlaufen und die Musterfunktion v 'paßt' nicht mehr auf die ebenfalls verbreiterten PK4.

8 Literatur

[1] Effert, R., Barry, J. C., Kaupp, A., und Dahm, M.: Eine neue photographische Methode zur Messung von Schielwinkeln bei Säuglingen und Kleinkindern. *Klinische Monatsblätter für Augenheilkunde*, 198:284–289, 1991.

[2] Barry, J. C., Effert, R., und Kaupp, A.: Objective measurement of strabismus in infants and children through photographic reflection pattern evaluation. *Ophthalmology*, 1992.

[3] M. Kleine: *Konzeption und Aufbau einer Einrichtung zur Schielwinkelmessung unter Verwendung digitaler Bildverarbeitung.* Diplomarbeit, RWTH Aachen, 1991.

[4] T. Lehmann: *Automatisierung der Schielwinkelmessung durch digitale Bildverarbeitung.* Diplomarbeit, RWTH Aachen, 1992.

[5] J. C. Barry: *Ein quantitatives Meßverfahren zur Bestimmung von kleinen Schielwinkeln bei Säuglingen und Kleinkindern.* Med. Diss., RWTH Aachen, 1991.

[6] Bennett, A. G. und Rabbetts, R. B.: *Clinical Visual Optics.* Butterworths, London, 1988.

[7] G. Schröder: *Technische Optik.* Vogel-Buchverlag, Würzburg, 1987.

[8] Hecht, E. und Zajac, A.: *Optics.* Addison-Wesley Publ. Comp., Reading Ma., 1974.

[9] Ballard, D. H. und Brown, C. M.: *Computer Vision.* Prentice-Hall, Inc., Englewood Cliffs, New Jersey, 1982.

[10] P. V. C. Hough: *A method and means for recognizing complex pattern.* US Patent 3069654, 1962.

[11] H. D. Lüke: *Signalübertragung.* Springer-Verlag, Berlin, 3. Auflage, 1988.

[12] W. K. Pratt: *Digital Image Processing.* Wiley-Interscience Publication, New York, 1978.

A Hybrid Approach to Signal Interpretation Using Neural and Semantic Networks

F. Kummert, E. Littmann, A. Meyering, S. Posch, H. Ritter, G. Sagerer

Technische Fakultät, Universität Bielefeld
Postfach 100131, D-33501 Bielefeld, FR Germany

We propose a hybrid approach to signal interpretation combining semantic and neural network techniques. Neural networks are attached to concepts of the semantic network. In this way the advantage of holistic and robust recognition of objects by neural networks and the decompositional power of semantic networks are combined. The approach is applied to the visual recognition of a hand and the characterization of its orientation in complex images. Results from 200 real images are presented.

1 Introduction

Recognition and description of objects is one of the main problems in computer vision. Furthermore, this task is one of the key capabilities required for the successful operation of both biological organisms and artificial robots. Two different methods to solve this problem are often discussed as competing approaches: On the one hand artificial neural networks (ANNs) based on trained parameters, and on the other hand semantic networks with knowledge based techniques. With this work we do not intend to argue which one of these approaches is better suited to the object recognition task. Rather we aim at utilizing the advantages of both techniques by combining them in a hybrid approach to knowledge representation and utilization. As an application we chose the visual recognition of a hand and the characterization of its orientation. This poses a challenging vision problem the solution of which is of great practical interest, for instance to facilitate the control of multifingered anthropomorphic manipulators. However, the techniques developed here can also be applied to other problems in computer vision and tasks of signal interpretation in general like speech understanding.

In the next section the advantages and disadvantages of both techniques and our approach for their combination in a hybrid system are discussed. Section 3 gives a brief overview of the neural and semantic networks used as the starting point for this work. Then the application of the hybrid system to recognize 3D-hand orientations is presented in detail and in section 5 results are discussed. We conclude with a discussion and plans for future work.

2 A Hybrid Approach

Common to all semantic network approaches is the explicit structuring of the domain knowledge using a decomposition and specialization hierarchy of concepts. Likewise, knowledge about attributes and relations between parts of the concepts is modeled in an explicit way. The analysis process is strongly influenced by this decompositional view of the world. Based on intensity information at pixel level the sensor data are transformed into increasingly abstract representational levels. First, simple elements like lines or regions are extracted. These are combined to more complex parts such as surface patches, which should correspond to elementary parts of an object and form the basis for object identification and description. This process may proceed in both a data or model driven way, as well as in a mixed strategy. While the decomposition of objects and the explicit description of their attributes is one of the main advantages of semantic networks, it also may cause drawbacks. The inherent ambiguity of signal interpretation, especially for primitive parts of the segmentation hierarchy may lead to many competing interpretations for more complex objects of the knowledge base. Additionally, the acquisition and adaptation of the knowledge base imposes a considerable effort on semantic network approaches.

In contrast to semantic networks the neural net approach does not attempt a decomposition into symbolic object parts. Instead, the properties of objects are modeled "holistically" in the weight parameters of an artificial neural network (ANN). Thus, the use of heuristics and world knowledge in the first approach is replaced by learning from examples. The trained network represents implicit knowledge about the attributes required for the identification of the object. This allows for a fast recognition of the learned objects that is also robust with regard to noise and variations in the signal. Another advantage is the inherent potential for parallelism as the representation of knowledge and computation is distributed over many simple processing units. However, being a "holistic" system it is not feasable to build a single ANN that can cope with all possible configurations of many simultaneous objects in a complex scene. Rather, multiple ANNs have to be applied to certain regions of interest, but their coordination is not (yet) well understood in the neural paradigm. Another possible problem is the necessary size of the training set for ANNs. For complex images, the number of required training samples may be too large for realistic applications.

To overcome the disadvantages of both approaches we propose a hybrid system combining neural and semantic network techniques. The main idea is to associate or attach ANNs as holistic models to concepts of the semantic net, with both components modeling the same object [1]. That is, the interface between the different network types is not defined at one fixed level of the segmentation hierarchy, rather it is determined as appropriate for the given task, knowledge base, or the current state of the analysis process. Given such a hybrid knowledge base, different options are available to recognize a modeled object in a model-driven strategy. If a concept node is to be instantiated the associated ANN can be activated and the object is recognized in a fast and robust way without the necessity to detect the parts of the object as modeled by the semantic network. If no ANN has been attached to a concept node the analysis works in the usual manner pursuing the decomposition hierarchy. In this mode of operation the semantic network is mainly utilized to control the analysis process and focus the various ANNs attached to the semantic network on different image regions. If in a later phase of the analysis process information about parts and attributes of parts is required the knowledge about the structure of objects modeled in the semantic network can still be exploited. An example for such a situation is the detection of gripping positions to guide a robot hand after the object has been detected holistically by a neural network. In a data driven analysis strategy the interaction works in a similar way. After an object has been recognized by an ANN the corresponding concept can be instantiated even if its parts are not (yet) detected. In a mixed strategy the instantiated objects recognized by ANNs can be used to select appropriate goal concepts from more abstract levels of the semantic network. In this way the number of competing interpretations is drastically reduced and the analysis process can be restricted propagating the contraints from the estimated goal concepts and the instantiated objects.

As indicated above, it is not necessary to attach a ANN to each concept of the semantic network. Rather, one might choose to first train and associate ANNs for objects that occur frequently or that are difficult to recognize by a semantic network. In cases when sufficient training data are not available for a successful training of an ANN, no ANN is bound to the corresponding concept. On the other hand, the hybrid approach gives the option not to fully decompose some of the objects alleviating the effort to acquire and adapt the knowledge base of the semantic network.

Further extensions of the hybrid approach include the utilization of neural networks to compute attributes and judgments during analysis as well as to learn control information to guide the analysis.

---

[1] The same applies to other concepts modeled in the semantic network, like events or abstract conceptions. For simplicity, however, we only refer to objects in the following.

This gives more possibilities to exploit the learning capabilities and robustness of neural networks for semantic nets. Another option is to explore additional ways to adapt ANNs: As indicated above it is usually not feasible to train an ANN for each object to be expected in a complex scene. However, analysing an image sequence the results of the analysis can be used to adapt ANNs to objects occuring frequently in the sequence.

3 Formalisms for Neural and Semantic Networks

The neural network used in the following is the *Local-Linear-Map network (LLM)* [5, 2, 3]. This sort of network consists of units that are significantly more complex than the usually employed sigmoid neurons. Therefore, a moderate number of units is sufficient for many tasks.

Each of these units processes the same input vector $\mathbf{x}$ of dimensionality L and computes a node response $\mathbf{y}$ of dimensionality M. Each LLM-unit is characterized by three components: an input weight vector $\mathbf{w}_r^{(in)} \in \mathbb{R}^L$, an output weight vector $\mathbf{w}_r^{(out)} \in \mathbb{R}^M$ and a MxL-matrix $\mathbf{A}_r$. The matrix $\mathbf{A}_r$ implements a locally valid linear mapping. The node response $\mathbf{y}$ of a unit r is determined by

$$\mathbf{y}_r = \mathbf{w}_r^{(out)} + \mathbf{A}_r(\mathbf{x} - \mathbf{w}_r^{(in)}).$$

For computing the final net output two different variants can be distinguished. If the LLM network acts like a "winner-takes-all" network the node response of one single unit is used as final output. Otherwise, a weighted superposition of several node responses is used. The contribution of each node to the superposition can depend e.g. on the distance between the input vector and the input weight vector of the node and can also be influenced by the previous node response. In this way, a short-term memory-effect might be produced.

In our approach the "winner-takes-all" variant is used. The "winner" unit s is determined in this case by comparing the distances $d_r = \|\mathbf{x} - \mathbf{w}_r^{(in)}\|$ between the input vector $\mathbf{x}$ and the input weight vectors $\mathbf{w}_r^{(in)}$, $r = 1, 2, 3 \ldots$, and selecting the node with the minimal distance.

To perform the required transformation between input and output space the necessary values of the network parameters are learned during a training phase. For this purpose correct input-output pairs $(\mathbf{x}^{(\alpha)}, \mathbf{y}^{(\alpha)})$, $\alpha = 1, 2, \ldots T$, of a set of T training samples are presented repeatedly and in a random sequence. The input and output weight vectors as well as the matrix coefficients are adapted according to the following simple error-correction rules:

$$\begin{aligned}
\Delta \mathbf{w}_s^{(in)} &= \epsilon_1(\mathbf{x}^{(\alpha)} - \mathbf{w}_s^{(in)}), \\
\Delta \mathbf{w}_s^{(out)} &= \epsilon_2(\mathbf{y}^{(\alpha)} - \mathbf{w}_s^{(out)}) - \mathbf{A}_s \Delta \mathbf{w}_s^{(in)}, \\
\Delta \mathbf{A}_s &= \epsilon_3(d_s^2)^{-1}(\mathbf{y}^{(\alpha)} - \mathbf{y}^{(net)})(\mathbf{x}^{(\alpha)} - \mathbf{w}_s^{(in)})^T.
\end{aligned}$$

During this adaptation process each training step parameter ϵ_i decays exponentially from a large initial value (typically $\epsilon_i^{initial} = 0.9$) to a small final value (typically $\epsilon_i^{final} = 0.01$).

In the following the semantic network is described. In contrast to other approaches like KL-ONE or PSN in the ERNEST semantic network language only three different types of nodes and three different types of links exist. They have well defined semantics and we believe that these structures are adequate to represent the knowledge for different pattern understanding tasks. *Concepts* represent classes of objects, events, or abstract conceptions having some common properties. In the context of image understanding an important step is the interpretation of the sensor signal in terms modeled in the knowledge base. The second node type, called *instance*, represents these extensions of a concept. It associates certain areas of the image with concepts of the knowledge base. It is a copy of the related

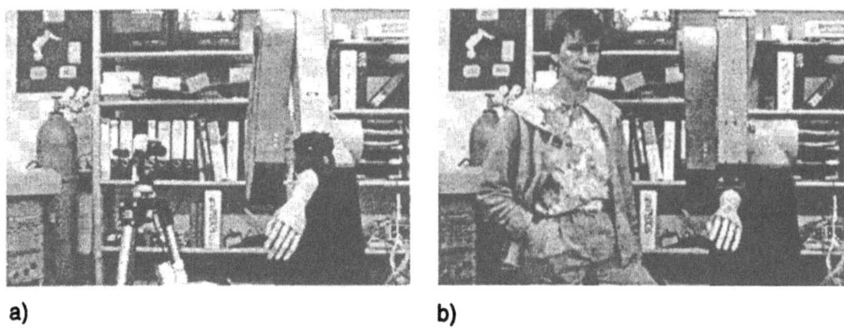

a) b)

Figure 1: Complex scenes with a roboter hand

concept where common property descriptions of a class are substituted by values derived from the signal. In an intermediate state of processing instances of some concepts may not be computable because certain prerequisites are missing. Nevertheless, the available information can be used to constrain an uninstantiated concept. This is done via the node type *modified concept.* As in all approaches to semantic networks the *part* link decomposes a concept into its natural components. Another well-known link type is the *specialization* with a related inheritance mechanism by which a special concept inherits all properties of the general one. For a clear distinction of knowledge of different levels of abstraction the link type *concrete* is introduced. In addition to its links, a concept is described by attributes representing mainly numerical features and restrictions on these values according to the modeled term. Furthermore, relations defining constraints for the attributes can be specified and must be satisfied for valid instances.

The creation of modified concepts and instances constitutes the knowledge utilization in the semantic network. For the creation of instances, this process is based on the fact that the recognition of a complex object needs the detection of all its parts as a prerequisite. For concepts which model terms only defined within a certain context the instantiation process must proceed in the opposite direction. In this case the context must exist before an instance of the context-dependent concept can be created. In the network language, these ideas are expressed by six problem-independent inference rules. Context-independent parts, contexts, and concretes are the prerequisites for the creation of instances and modified concepts in a data-driven strategy. The opposite link directions are used for model driven inferences. Since the results of an initial segmentation are not perfect, the definition of a concept is completed by a judgment function estimating the degree of correspondence of an image area to the term defined by the related concept. On the basis of these estimates and the inference rules an A*-like control algorithm is applied. For a detailed description of the network language see [4, 1].

4 Application

The explicit description of terms relevant in the task-domain via attributes and via links to other terms is the main advantage of semantic networks. However, the complexity and variety of real images makes it necessary to constrain the scene description to the relevant conceptions for which we need an interpretation. To obtain a feasible method for the processing of complex images as shown in Fig. 1 we choose an incremental design of the system. For the visual control of robots with multifingered manipulators the hand position and orientation are essential descriptions. Therefore,

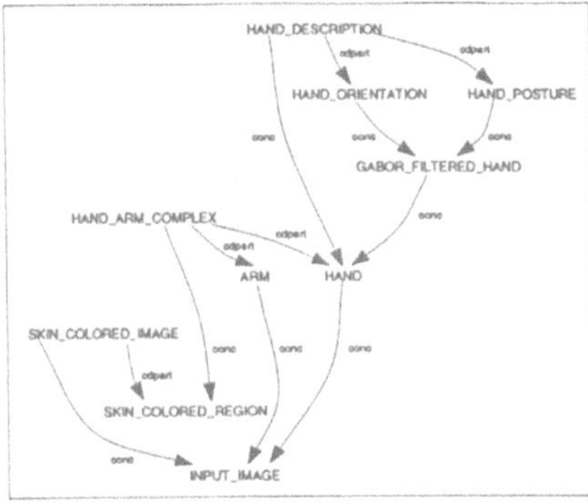

Figure 2: A hybrid network for the calculation of hand orientation

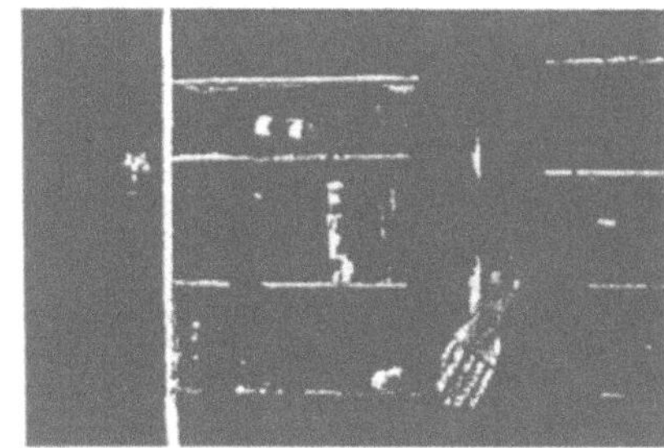

Figure 3: Skin-color activity map for the image of Fig. 1-a

the knowledge base as outlined in Fig. 2 follows the paradigm of selective perception and models only conceptions relevant to this subtask.

Some of these concepts, like *HAND_ORIENTATION*, represent rather complex object classes. The description of these concepts via parts and attributes poses a difficult problem, as there is no simple mapping between image pixels and roll, pitch, or yaw of the hand[2]. Following the decompositional paradigm, the fingers of a hand have to be modeled explicitly as parts. In turn, the fingers are composed e.g. of line segments which can directly be extracted from the image. Additionally, the mapping between finger positions in the image and the related angles have to be modeled as well.

Adaptive neural networks are an intriguing method to satisfy these requirements. Given a set of images containing a hand with known orientation they can be trained to extract roll, pitch, and yaw of the hand directly from simple features of the pixel image, thus providing the necessary attributes for the concept *HAND_ORIENTATION* in one single step. A similar problem poses the separation of hand pixels and background pixels into two classes. As represented by the concept *INPUT_IMAGE* the pixels show a large overlap in the intensity space as well as in the RGB space. Therefore, the use of simple band-pass filter techniques leads to unsatisfactory segmentation results. Again, a neural network can be trained to identify skin-colored regions and to represent the two classes "skin-colored" and "background" separately in the concept *SKIN_COLORED_IMAGE*. The functioning of the network may be described as an adaptive multiple band-pass filter technique with some interpolation extensions. The neural network shows better classification results and is more robust to variances in lighting conditions.

Fig. 2 gives an overview of the resulting hybrid knowledge base for the description of a hand including the position and the orientation. The lowest level builds the concept *INPUT_IMAGE* which represents the interface to the input data. Two typical examples of such images are shown in Fig. 1. The concept *SKIN_COLORED_IMAGE* is realized by a LLM-network. The network was trained to map the local color information in the surround of each pixel onto a real-valued "skin-color activity" value (see Fig. 3 for an example). This training was based on samples from subregions of two of the training images. The 9-dimensional input for each pixel contains the RGB values of the pixel itself as well as the average RGB values within a 3×3 and a 5×5 neighborhood, thus introducing a

[2]We describe the angles of the hand orientation according to the common notation in robotics.

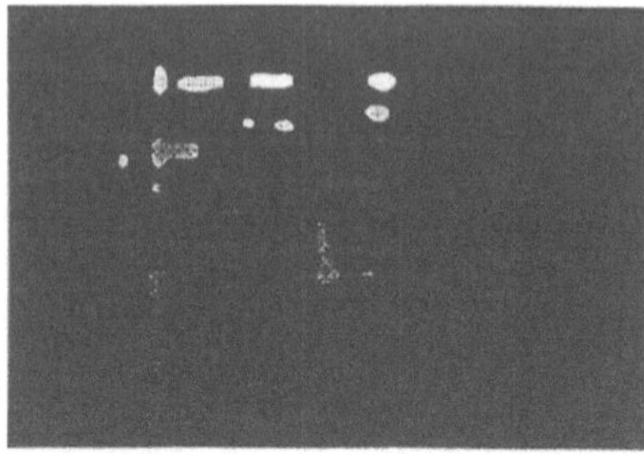

Figure 4: Segmented image with instances of hand-arm-complex (every instance is marked by one grey value)

Figure 5: Optimal hand instance

moderate amount of spatial context information into the network. The resulting skin-color activity values of all image values form the "skin-color activity map" that is the basis for the detection of the *HAND_ARM_COMPLEX* and its concrete *SKIN_COLORED_REGION*. The corresponding concept employs first a median filtering to eliminate small regions of activity. On the basis of the histogram of the filtered skin-color activity map a threshold is calculated. For the resulting binary image a region labeling is performed (see Fig. 4). Each labeled region gives rise to a competing instance for the *SKIN_COLORED_REGION* and hence for a *HAND_ARM_COMPLEX*. The judgment of these instances is based on heuristic restrictions on boundaries of hand-arm regions. Due to the A*-driven control, the best-judged instance is selected for further processing in the next step.

The concept *HAND* is modeled as a context-dependent part of the *HAND_ARM_COMPLEX*. To separate the arm from the hand the hand-arm-complex is transformed into the normalized orientation by a Karhunen-Loeve transformation. Based on general knowledge about the shape and proportions of hands and arms the hand is extracted from the complex. The resulting region is remapped and represents the area of interest for evaluating the orientation of the hand (see Fig. 5). This is done by the concept *HAND_ORIENTATION* which is modeled again by a LLM-network. For extracting the hand orientation from the pixel information, the required mapping of a suitable feature space into the orientation space is learned by a LLM network. Therefore, the region of interest is preprocessed first. After thresholding this region, a laplace filter operates on the image and negative filter values are clipped. The resulting image mainly contains the edge information of the region and forms the basis for a subsequent operation with gabor filters. These filters are orientation sensitive and well suited to exploit the edge information. We use a 3×3 "filter grid" centered at the area of interest with four different orientations at each grid location. The resulting 36-dimensional feature vector is represented in the concept *GABOR_FILTERED_HAND* and forms the input for the LLM network (see also [3]). At the present state this network calculates two of the three relevant angles—roll and yaw—, whereas the third (pitch) is a constant through all images.

5 Results

For the validation of our approach a data set consisting of 200 real images was used. For the training of the orientation network the end effector of a robot was positioned in each image in a random orientation. Therefore, the orientation angles varied in the range of $[-35°, 35°]$ for roll and $[-55°, 0°]$ for yaw. Pitch was held constant at 0°. Figure 6 demonstrates the range of possible hand poses.

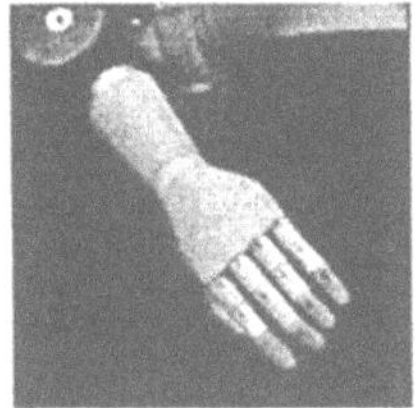
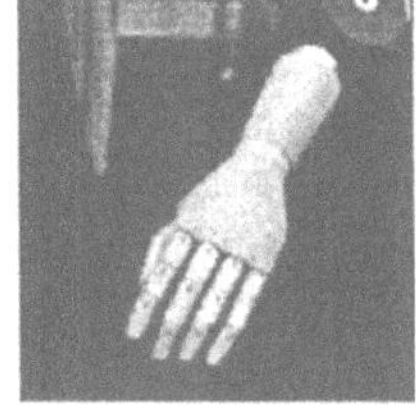
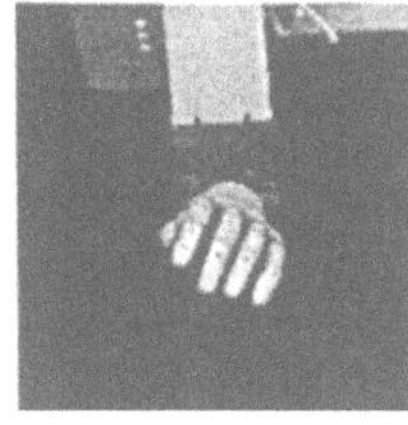
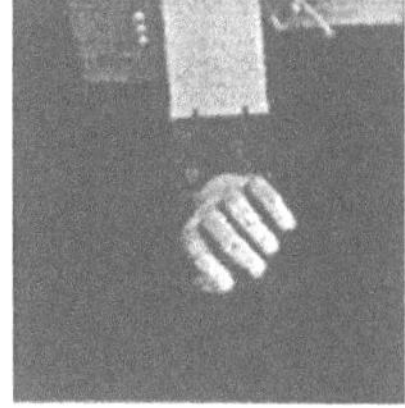

Figure 6: Possible hand orientations with extreme values for roll and yaw whereas pitch was fixed at 0°. (roll,yaw) have the values (a) (35°, −55°), (b)(−35°, −55°), (c)(−35°, 0°), (d)(35°, 0°).

These 600 × 400 RGB images show a laboratory. After each taking of 10 camera images we rearranged the scene. Thus, 20 different types of images are generated. Figure 1 shows two typical examples.

These images are used for the training of LLM networks for the detection of skin-colored pixel as well as for the recognition of orientation. The current LLM network for the former task consists of one layer containing 10 nodes. It is trained on two 300 × 200 pixel subregions taken from two of the training images to provide an output value of 1 for pixels belonging to the hand-arm-complex and −1 otherwise. 200.000 adaptation steps were carried out by randomly drawing pixels from the training images. For the evaluation of the network performance pixels with a positive activity value were classified as skin-colored. Table 1 shows the total recognition rate and the percentage of recognized pixels belonging to the hand region (≈ 8% of all image pixels), resp. the background (≈ 92%). The results were obtained for both training regions and for an independent test image. An example of an activity map is shown in Figure 3.

	Total	Background	Hand
training image1	96.0 %	98.9 %	62.5 %
training image2	96.9 %	99.0 %	70.4 %
test image	96.7 %	97.9 %	81.3 %

Table 1: Detection of skin-colored pixels: Performance of the LLM network. Total recognition rate and percentage of recognized pixels belonging to the hand region (≈ 8%) resp. the background (≈ 92%). Results are given for both training images and an independent test image.

	MSRE	NMSRE
training	2.32°	0.091
test	4.34°	0.172

Table 2: Performance of the LLM network for the recognition of orientation on the training/test set presenting the mean square root error as an absolute value as well as normalized by the standard deviation of the data sets.

In the next step the semantic net separates the hand from the background on the base of the activity image. In 190 of the 200 images the correct regions are detected, i.e. in 95% of the data set the correct region of attention is found. Only one type of images gives rise to errors. The corresponding 10 images show a box with a face in front of a light background. This object has a similar color and form as the hand and yields the best-judged instance for a hand.

The subsequent orientation detection works on these areas of interest. The data set is split into two subsets: 130 images are used for training, the remaining 60 images form the test set. The task is managed by a LLM-network consisting of five units. For the 36-dimensional input space the training set contains very few examples. Therefore, the network is trained with the relatively small number of 10.000 adaptation steps only since more training iterations would lead to overfitting.

The performance of the orientation LLM network is represented in Table 2. Evaluated on the independent test set, the mean square root error (MSRE) of 3.55° for roll and 2.48° for yaw shows

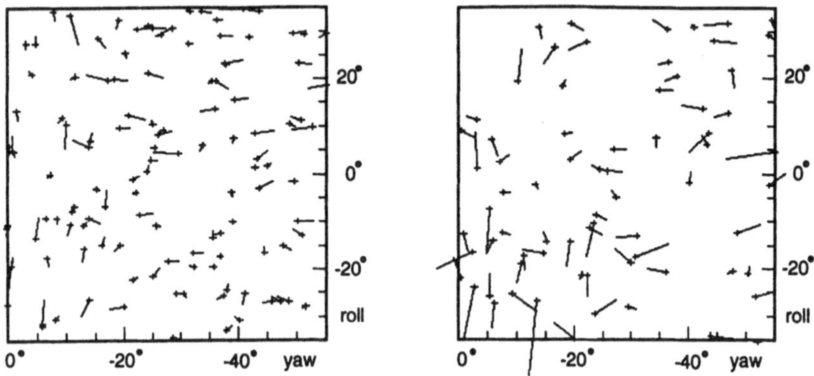

Figure 7: Representation of the NMSRE on a) the training set respectively b) the test set. The center of a cross is the target point whereas the length of the tail represents the size of the total error for roll and yaw in degree.

the deviation between target and net result as absolute values. This leads to a total MSRE of 4.34° for the euclidean distance in both dimensions. Normalizing the MSRE with the corresponding standard deviations of the data sets we obtain values of 0.184 for roll and 0.152 for yaw. This yields a total NMSRE of 0.172.

6 Conclusion

We presented a hybrid system integrating semantic and neural networks for scene analysis. This approach was successfully applied to the recognition of hands and the determination of their 3D-orientation in complex real world scenes. Further research will include a more sophisticated representation scheme for the shape of the hand, and a verification of the detected hand using eigenhands. Investigations of additional ways for the cooperation of neural and semantic networks as well as other applications are under way.

References

[1] F. Kummert, G. Sagerer, and H. Niemann. A Problem-Independent Control Algorithm for Image Understanding. In *11th International Conference on Pattern Recognition*, volume I, pages 297–301, The Hague, 1992.

[2] E. Littmann, A. Meyering, H. Ritter. Cascaded and Parallel Neural Network Architectures for Machine Vision — A Case Study. In *Proc. 14. DAGM-Symposium, Dresden*, eds. S. Fuchs, R. Hoffmann, pages 81-87, Springer, Heidelberg, 1992.

[3] A. Meyering, H. Ritter. Learning 3D-Hand Postures from Perspective Pixel Images. In *Artificial Neural Networks II*, eds. I. Aleksander, J. Taylor, pages 821-824, Elsevier Science Publishers (North Holland), 1992.

[4] H. Niemann, G. Sagerer, S. Schröder, and F. Kummert. ERNEST: A Semantic Network System for Pattern Understanding. *IEEE Transactions on Pattern Analysis and Machine Intelligence*, 12:883–905, 1990.

[5] H. Ritter. Learning with the Self-Organizing Map. In *Artificial Neural Networks*, eds. T. Kohonen, K. Mäkisara, O. Simula, J. Kangas, pages 379-384, Elsevier Science Publishers (North Holland), 1991.

Lernen mit Hilfe des Maximum-Entropie-Prinzips

A. Grauel
Universität Paderborn, Abt.Soest, FB 16
Elektrische Energietechnik/Automatisierungstechnik
Fachgebiet Mathematische Methoden und Systemtheorie
Steingraben 21, 4770 Soest

Abstract: We propose a new approach to neural networks which can be important to e.g. pattern recognition. To that we introduce the maximum entropy principle into the theory. We give an outline of the approach with a formulation of the constraints and indicate the solution area which depends on some network parameters. For our study of the dynamical behaviour of a neural system we identify the conditional probability with the stochastic distribution of the Boltzmann machine where for simplicity the Hebbian rule is taken into account. The requirements that for e.g. the states in the next time step of the system can be predicted if the mapping properties are fixed at the beginning of the calculation whereby these can be seen as a new learning rule. By choosing proper constraints we can calculate distributions for network states, coupling parameters, threshold values etc. In this contribution the method is tested for an auto-associative memory.

1 Einleitung

Die thermodynamische Formulierung für die freie Energie setzt die Existenz einer Energiefunktion voraus. Existiert keine Energiefunktion um z.B. die Netzwerkzustände, die Speicherfähigkeit etc. zu berechnen, dann müssen andere Prinzipien benutzt werden. Eine Möglichkeit besteht darin wahrscheinlichkeitstheoretische Methoden zu verwenden, wie z.B. das Entropieprinzip. Dieses Prinzip der Maximierung der Entropie liefert ein Berechnungsverfahren basierend auf derjenigen Verteilung, die das "Nichtwissen" über das System (die Entropie) maximiert. Wir bezeichnen dieses Berechnungsverfahren als Maximum-Entropie-Formalismus. Dieses Verfahren beruht auf der Verwendung der Shannon-Entropie in Kombination mit der Methode der Lagrange-Multiplikatoren. Dabei werden die zu berücksichtigenden Nebenbedingungen mit Lagrange-Multiplikatoren multipliziert und in der Entropiegleichung berücksichtigt. Man spricht deshalb auch von einem erweiterten Entropie-Prinzip. Nebenbedingungen können Parameterverteilungen Schwellwert, Kopplungskonstanten etc., Verteilung der Outputzustände etc. sein.

Die Existenz einer Energiefunktion setzt symmetrische Kopplungsstärken zwischen den Neuronen voraus. Biologisch sind symmetrische Kopplungsstärken aufgrund der Wirkungsweise einer chemischen Synapse nicht motivierbar. Netzwerke die asymmetrische Kopplungen berücksichtigen sind bekannt[1-3] ebenso asymmetrische verdünnte Netzwerke[4] und Netzwerke die musterinduzierte Übergänge liefern[5]. In einer vorangegangenen Arbeit[6] haben wir musterinduzierte Übergänge von einfachen geometrischen Mustern und von Sprachmustern in einem time-delayed Netzwerk mit asymmetrischen Kopplungen untersucht.

In diesem Beitrag benutzen wir die Hebbsche Regel, die dargestellte Methode ist aber unabhängig von der Hebbschen Lernregel.

2 Entropieformalismus

Annahmen für den Entropie-Formalismus mit Nebenbedingungen sind:

i) Es sei ein System von Zustandsvektoren $\{x^i\}$ im diskreten Raum gegeben.

ii) Das System nimmt den Zustand x^i mit der Wahrscheinlichkeit p_i an.

Die Entropie H, die das Nichtwissen bezüglich eines Systemzustandes beschreibt, ist definiert durch

$$H = -\alpha \sum_{i=1}^{N} p_i \cdot \ln p_i \quad .$$

Hierbei ist N die Zahl der Systemzustände und α eine geeignet zu wählende Konstante. Für ein thermodynamisches System ist $\alpha = k$ die Boltzmann Konstante.

Mit Hilfe von Nebenbedingungen können wir weitere Informationen über das System berücksichtigen, indem wir die Nebenbedingungen multipliziert mit Lagrange-Multiplikatoren in der Darstellung für die Entropie H berücksichtigen. Diese Methode ist wohlbekannt aus anderen Bereichen, beispielsweise in der Mechanik[7,8], Statistik[9], Thermodynamik[10], Oberflächen- und Grenzflächenthermodynamik[11,12] etc. Ist der Mittelwert $<0>$ einer Mikroobservablen $0(x^i)$ als Meßwert M bekannt, dann ist eine Nebenbedingung

$$M = \sum_{i=1}^{N} 0(x^i) \cdot p_i \quad .$$

Sind m Nebenbedingungen zu erfüllen, dann gilt

$$M_k = \sum_{i=1}^{N} O_{ki} \cdot p_i \quad , \quad k=1,\ldots,m,$$

wobei $O_{ki} := O_k(x^i)$. Falls rang $(O) < N$, kann das Gleichungssystem nicht eindeutig nach den Verteilungen p_i aufgelöst werden. Für jede Wahrscheinlichkeitsverteilung p_i gilt stets:

$$\sum_{i=1}^{N} p_i = 1 \quad .$$

Insgesamt haben wir $m+1$ Nebenbedingungen. Die Entropiemaximierung bezüglich der Wahrscheinlichkeitsverteilung p besagt, daß ein p so auszuwählen ist, daß die Entropie H maximal wird. Die erweiterte Entropiegleichung mit den Nebenbedingungen besitzt die Form

$$F(\underline{p}) = H(\underline{p}) + \sum_{k=1}^{m} \Lambda_k \left(\sum_{i=1}^{N} O_{ki}\, p_i - M_k \right) .$$

Mit der notwendigen Bedingung für ein Maximum

$$\left. \frac{\partial F(p_i)}{\partial p_i} \right|_{\underline{p}=\underline{p}_0} = 0 \quad ,$$

folgt:
$$\left.\frac{\partial F(p_i)}{\partial p_i}\right|_{p=p_0} = \alpha\cdot(\ln p_i+1)+\Lambda_{m+1} + \sum_{k=1}^{m} \Lambda_k O_{ki} = 0 .$$

Für die Wahrscheinlichkeitsverteilung folgt
$$p_i = \exp(-\lambda - \sum_k \lambda_k O_{ki}) \quad , \quad i=1,\ldots,n .$$

Als Abkürzungen wurden eingeführt
$$\lambda := \frac{1}{\alpha}\Lambda_{m+1} + 1 \text{ und } \lambda_k := -\frac{1}{\alpha}\Lambda_k .$$

Damit ergibt sich für den Meßwert M_k
$$M_k = \sum_i O_{ki} \exp(-\lambda - \sum_{r=1}^{m} \lambda_r O_{ri}) \quad , \quad k=1,\ldots,m, \quad \text{und}$$
$$1 = \sum_i \exp(-\lambda - \sum_{r=1}^{m} \lambda_r O_{ri}) .$$

Aus der letzten Gleichung läßt sich der Lagrange-Multiplikator λ eliminieren. Explizit gilt:
$$\exp(\lambda) = \sum_i \exp(- \sum_{r=1}^{m} \lambda_r O_{ri}) .$$

Diese aufgelöste Form ermöglicht es λ in der Gleichung für den Meßwert M_k zu eliminieren. Es entsteht das Gleichungssystem
$$\sum_{i=1}^{N} (O_{ki}-M_k) \exp(- \sum_{r=1}^{m} \lambda_r O_{ri})=0 \quad , \quad k=1,\ldots,m,$$

bestehend aus m Gleichungen für die m unbekannten Lagrange-Multiplikatoren. Diese Gleichung repräsentiert ein nichtlineares Gleichungssystem für die Lagrange-Multiplikatoren. Sind die λ-Werte gefunden, dann ist die Wahrscheinlichkeitsfunktion p_i bestimmt und unsere Forderungen erfüllt.

3 Anwendung des Entropieformalismus

Es sei ein neuronales Netz als Input-/Output-System betrachtet, wobei der Output durch den Netzwerkzustand S(t+1) gegeben sei und dieser als funktionell abhängig von den Netzwerkparametern ($\{w_{ij}\}$, ϑ_i, T) und dem Netzwerkzustand S(t) angenommen wird. Unter der Annahme einer stochastischen Output-Funktion ergibt sich für die Wahrscheinlichkeit eine Komponente s des Outputs zu finden zu
$$P(O_k) = \sum_i P(O_k|I^i)\cdot P(I^i) \quad , \quad s=1,\ldots,m .$$

$P(O_k \mid I^i)$ ist die bedingte Wahrscheinlichkeit, O_k als Output zu erhalten, wenn der Input I^i vorliegt. $P(I^i)$ ist die Wahrscheinlichkeit, mit der der Input bezüglich i auftritt. Die Berechnung der Mittelwerte erfolgt mit vorstehender Gleichung:
$$\langle O_k\rangle = \sum_{O_k} O_k\cdot P(O_k) .$$

Mit $O_k \in \{0,1\}$ folgt:
$$\langle O_k\rangle = P(O_k=1) \quad \text{bzw.} \quad \langle O_k\rangle = \sum_i P(O_k=1|I^i)\cdot P(I^i)$$

Ist $M_k := \langle O_k \rangle$ und $O_{ki} := P(O_k = 1 \mid I^i)$, so erhalten wir:

$$M_k = \sum_i O_{ki} \, P(I^i) \qquad \text{mit } k=1,\dots,m.$$

Hieraus folgt, daß unter gewissen Voraussetzungen bei einer Vorgabe von M_k und O_{ki}, die gesuchte Wahrscheinlichkeit $P(I^i)$ berechnet werden kann. Mit Hilfe dieser Wahrscheinlichkeit können wir dann beispielsweise den Erwartungswert des Inputs $\langle I \rangle$ berechnen.

Reale Lösungen verlangen rang $(O_{ki}\text{-}M_k) = m$ und

$$\min_i O_{ki} < M_k < \max_i O_{ki} \quad .$$

Für das Studium des dynamischen Verhaltens eines neuronalen Systems, kann man die Wahrscheinlichkeit O_{ki} mit der Akzeptanzwahrscheinlichkeit einer Boltzmann-Maschine

$$P(S_i(t+1)=1) = \frac{1}{1+\exp(-h_i/T)} \quad .$$

identifizieren. Dabei ist h_i das postsynaptische Potential

$$h_i = \sum_{k=1}^{N} w_{ij} \, S_k - \vartheta_i \quad .$$

Für die Kopplungsstärke kann man die Hebbsche Regel in der folgenden Form benutzen

$$w_{ij} = \sum_{z=1}^{n} S_i^z \, S_j^z \quad ,$$

wobei n die Zahl der zu memorisierenden Zustände ist. Identifizieren wir die Wahrscheinlichkeit O_{ki} mit der Akzeptanzwahrscheinlichkeit der Boltzmann-Maschine $O_{ki} = P(S_i(t+1)=1)$, so folgt mit dem postsynaptischen Potential h_i und der Kopplungsstärke:

$$O_{ki} = \left(1+\exp\left(-\left(\sum_z S_k^z \, \underline{S}^z \cdot \underline{S}^i - \vartheta_k\right)/T\right)\right)^{-1} \quad .$$

Die Größe

$$\sum_z S_k^z \, \underline{S}^z \cdot \underline{S}^i$$

kann mit Hilfe der Speichervektoren berechnet werden.

Wählen wir zunächst drei orthogonale sechskomponentige Speichervektoren, so erhalten wir

$$\max_i \sum_z S_k^z \, \underline{S}^z \cdot \underline{S}^i = 2 \qquad \text{und} \qquad \min_i \sum_z S_k^z \, \underline{S}^z \cdot \underline{S}^i = 0 \quad .$$

Bei Wahl einer Schwelle ϑ_k und in Abhängigkeit von der Temperatur ergeben sich verschiedene Lösungsflächen im Parameterraum. Geometrisch ergibt sich für den maximalen Wert $\max_i \sum_z S^z \, \underline{S}^z \cdot \underline{S}^i$ eine untere Schranke (untere Grenzfläche) und bezüglich des minimalen Wertes $\min_i \sum_z S^z \, \underline{S}^z \cdot \underline{S}^i$ erhalten wir eine obere Grenzfläche. Wählen wir kleine Schwellwerte, z.B. $\vartheta_k = -0{,}1$, dann erhalten wir für das postsynaptische Potential positive Werte und der Erwartungswert O_{ki} für niedrige Temperaturen liegt bei 1 (siehe Fig.1a). Neuronenzustände mit der Wahrscheinlichkeit Null sind nicht möglich. Die obere und die untere Grenzkurve begrenzt die möglichen Werte, nur Werte zwischen beiden Grenzflächen gehören zu realen Lösungen des Systems. Analoge Überlegungen sind auch für andere Schwellwerte gültig. Wählen wir Schwellwerte zwischen max O_{ki} und min O_{ki} ($\vartheta_k = 1{,}5$), dann ist bei niederer Temperatur jeder Wert zwischen Null und 1 möglich (Fig.1b). Wählen wir solche Schwellwerte, daß das postsynaptische Potential sein Vorzeichen ändert, dann können keine Zustände mit Wahrscheinlichkeiten verschieden von Null für $T \rightarrow 0$ existieren (Fig.1c).

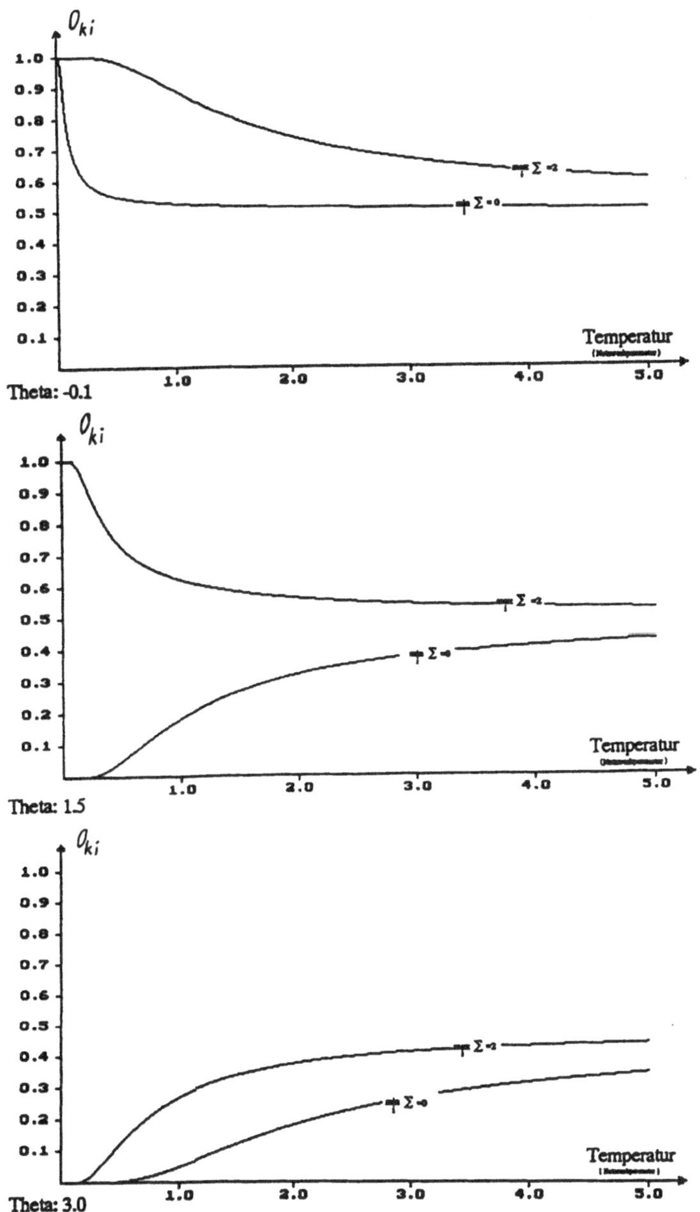

Fig.1 Abhängigkeit der Wahrscheinlichkeit O_{ki} von der Temperatur für verschiedene Schwellwerte, die jeweils obere Kurve gehört zu min O_{ki} und die untere zu max O_{ki}. Als Schwellwerte ϑ_k wurden gewählt: a) $\vartheta_k = -0,1$, b) $\vartheta_k = 1,5$ und c) $\vartheta_k = 3,0$.

Fig. 2 zeigt, daß für die Abspeicherung nichtorthogonale Speichervektoren, bei der gleichen Zahl von Komponenten wie im orthogonalen Fall, sich der Bereich der Lösungsfläche vergrößert.

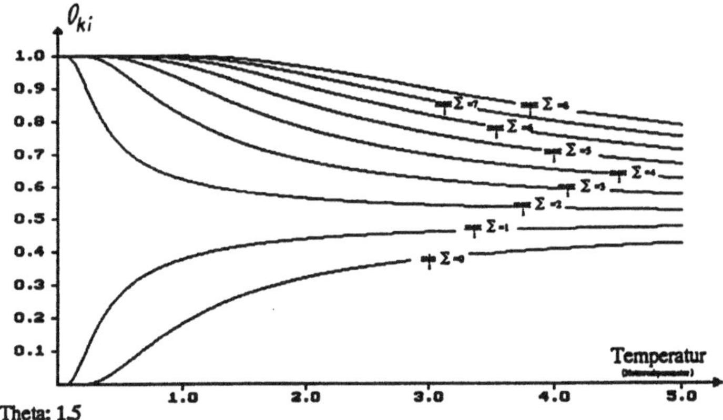

Fig.2 Bei nichtorthogonalen Speichervektoren ergeben sich Überlappungen, die bei gleicher Komponentenzahl der Vektoren, größere Werte $\max_i O_{ki}$ ergeben. Dieses hat zur Folge, daß der Bereich zwischen den Lösungsflächen sich vergrößert. Die untere Grenzfläche ($\min_i O_{ki} = 0$) repräsentiert als Referenzfläche im Parameterraum den überlappungsfreien Fall.

Mit Hilfe des dargestellten Formalismus läßt sich z.B. die Verteilung der Netzwerkzustände für einen Autoassoziativspeicher berechnen. Es zeigt sich, daß den berechneten Inputverteilung maximaler Entropie Inputzustände entsprechen, die im Einzugsgebiet des Speichervektors liegen, wenn der Speichervektor mit hoher Wahrscheinlichkeit als Outputzustand vorlag (siehe Fig.3 und Fig.4). Dieses gilt für orthogonale als auch nichtorthogonale Zustände.

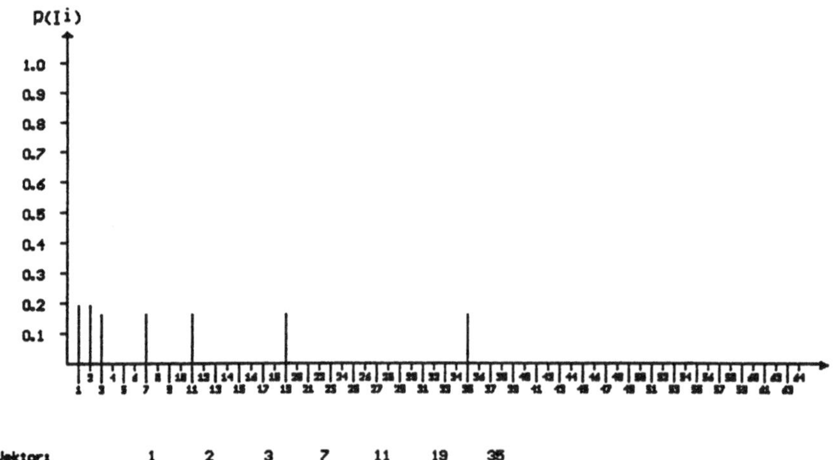

Fig.3 Numerisch berechnete Inputverteilung maximaler Entropie für einen orthogonalen Speichervektor. Gefordert wurde eine Abbildungswahrscheinlichkeit von 95%. Als Parameter wurden gewählt: T = 1 und $\vartheta_k = -0{,}1$ sowie als Hamming-Distanz eins. Auf

der Abzisse sind die möglichen Inputvektoren aufgetragen, jeder Zahl entspricht einem ganz bestimmten Inputvektor.

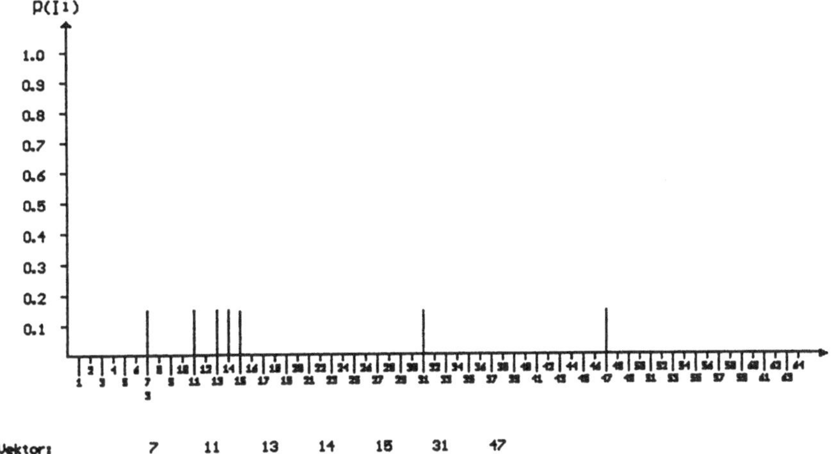

Fig.4 Berechnete Inputverteilung für einen nichtorthogonalen Speichervektor. Wahl einer Abbildungswahrscheinlichkeit von 95 %, T=1 und $\vartheta_k = -0{,}1$ sowie Hamming-Distanz eins. Es ist gegenüber Fig.3 ersichtlich, daß die Verteilung der Inputvektoren andere Einzugsgebiete besitzen mit der sie auf die Speichervektoren abgebildet werden.

Zusammenfassung: Da keine Einschränkungen bezüglich der Architektur in den Formalismus eingehen, besitzt dieser ein konstruktives Element in dem Sinne, daß bestimmte Parameter über ein gefordertes Systemverhalten berechnet werden können. Dieses hat zur Folge, daß bestimmte Muster über ein gefordertes Systemverhalten erzeugt werden können. Ferner besteht die Möglichkeit, z.B. auch für geschichtete Netzwerke (hardwaremäßige Hierarchien) über ein gefordertes Systemverhalten (z.B. geforderte Assoziationen) die Kopplungsstärken zu berechnen.

4 Literaturverzeichnis

1. G.Parisi: J.Phys. A19 (1986) L675.
2. I.Kanter, H.Sompolinsky: Phys. Rev. Lett. 57(1986)2861.
3. L.Personnaz, I.Guyon, G.Dreyfus, G.Toulouse: J.Stat.Phys. 43(1986)411.
4. R.Kree, A.Zippelius: Phys. Rev. A36(1987)4421.
5. R.Kühn, J.L.van Hemmen: Temporal Association in Models of Neural Networks (E. Domany, J.L.van Hemmen, K.Schulten, Eds.), Seite 213, Springer-Verlag, Heidelberg (1991).
6. A.Grauel, H.-G.Grundmann, R. Pels: Informatik aktuell (S. Fuchs, R. Hoffmann eds.), Springer-Verlag, Heidelberg, 1992 , to be published.

7. H. Goldstein: Classical Mechanics (Second Edition), Addison-Wesley Publ. Company, London (1980).

8. R.Abraham, J.E.Marsden: Foundations of Mechanics, The Benjamin/ Cummings Publ. Company, London (1978).

9. R.Peters: Einführung in die allgemeine Informationstheorie Springer-Verlag, Heidelberg (1967).

10. I.Müller: Thermodynamik (Grundlagen der Materialtheorie), Bertelsmann Universitätsverlag, Düsseldorf (1973).

11. A.Grauel: Physica, 103A, 468(1980) International Journal of Theoretical Physics,27, 861(1988).

12. A.Grauel: Feldtheoretische Beschreibung der Thermodynamik für Grenzflächen, Lecture Notes in Physics, No. 326, Springer-Verlag, Heidelberg (1989).

13. W.Wecker: Lernalgorithmus für neuronale Netzwerke unter Zwangsbedingungen (Diplomarbeit), Universität Paderborn, Abt. Soest (1992).

14. A.Grauel: Neuronale Netze, Grundlagen und mathematische Modellierung, B.I. Wissenschaftsverlag, Mannheim (1992).

A Neural Architecture for Emergent Visual Contrast, Brightness, and Space Perception*

Ian R. Billups[2], Heiko Neumann[1]**, H. Siegfried Stiehl[1], Lothar Wieske[1]

[1] Univ. Hamburg, FB Informatik, Bodenstedtstr. 16, D-22765 Hamburg
[2] Univ. of Birmingham, School of Computer Science, GB-Birmingham B15 2TT

Abstract. We describe ongoing research towards a unified theory for the coherent segmentation of contrast and brightness in static achromatic monocular images. A computational architecture is proposed that serves as a framework for a theory of emergent contrast, contour, luminance, and space perception. At the first stage of processing, the computational mechanism uses a centre-surround antagonism based on shunting interactions. This architecture is shown to multiplex local contrast and luminance data, which are subsequently demultiplexed into segregated data streams that signal both local contrast information of each polarity and a scaled low-pass filtered version of the luminance distribution. In correspondence with recent findings about the major processing channels in the primate visual system, the ON and OFF contrast channels feed into a subsystem for recurrent contrast processing, perceptual organization, and grouping. This subsystem is in general able to segment figural shape outlines, to close fragmented contours, and to integrate contour segments into globally coherent arrangements. We suggest that activity generated in this subsystem both feeds into an ART-based long-term memory (for invariant object recognition) and acts as a modulation mechanism that controls the diffusion coefficients for lateral activity spreading within the segregated brightness/darkness channel.

1. Introduction

Recent neurophysiological and neuroanatomical findings in mammalian vision research suggest the existence of two parallel afferent processing channels. These so-called ON and OFF channels transmit visual information along projection fibres that are segregated at least up to the striate cortex (V1, area 17). The retinal processing can be adequately explained on a functional level by the ganglion cell layer, whose cells comprise a centre-surround antagonistic lateral coupling. The segregated ON and OFF channels receive separate input from activations generated by lateral interaction described in terms of on-centre/off-surround (on/off) and off-centre/on-surround (off/on) receptive field (RF) profiles. The major axonal projections make up a processing path that enters the LGN to reach area 17 in the visual cortex (e.g. [5]). We hypothesize that there are lateral interactions along this path in order to generate segregated processing channels for

** To whom correspondence should be sent.

* Work has been partially supported by the Federal German Ministry of Research and Technology (BMFT), grant 413-5839-01 IN 101 C/1 and the DAAD/BRC research cooperation ARC fund, grant AZ 313-ARC-VI-92/107/scu.

signalling both local contrast information of each polarity (the ON and OFF channels) and luminance information (the brightness and darkness (B&D) channel). The contrast-related channels are assumed to feed into one of the major subsystems dedicated to achromatic shape and form perception, the luminance channel is assumed to feed into the B&D (or luminance) system ([17], [15]).

The basic architecture of the subsystem for the segmentation of form is modelled on the basis of the Boundary Contour System originally proposed by Grossberg & Mingolla (e.g. [12]) a functional extension to which has been described by Neumann & Stiehl ([20], [21]). The computational mechanism accounts for a response to straight stimulus configurations, corners, and junctions: features which, along with others, define invariants in the optic array. According to Gibson and Koenderink & vanDoorn ([7], [16]), classes of invariants are causally related to illumination, viewpoint position, flow fields of an actively moving perceiver, and local discontinuities in the visual structure. Gibson ascribed ecological significance to contrast edges and their local aggregation into junctions, since termination, junction or crossing points of lines and contrast edges, which most commonly result from non-accidental projections of discontinuous properties in the sensed scene ([25]), remain stable in their configuration over a great variety of perspective projections of surfaces ([7]). Furthermore, corner and junction configurations have been found to play an important role in optical flow computation such as to solve the aperture problem; to serve as salient features for attention focussing, fovealization and visually driven guidance; and for visual object recognition (see e.g. [1]).

In this paper, an architecture is proposed for coherent monocular contour and brightness perception and grouping, which contributes to the FACADE theory defined by Grossberg and colleagues (e.g. [11]) by extending their Boundary Contour System and unifying their monocular preprocessing stage and the Feature Contour System (FCS). It provides a novel hypothesis of the functional role of co-existing parallel ON and OFF channels (see [5] for a recent discussion), suggests a three-stage process in brightness reconstruction, provides a simplified computational explanation for the generation of orientation-sensitive simple cells in area 17, and extends the implementation described in [19], [22] to the 2D case. Also, the capability of the embedded layered feedback architecture to perform local contrast and line processing is demonstrated.

2. Centre-surround Antagonism and the Segregation of the ON, OFF, and B&D Channels

Based on an investigation into the response properties of retinal ganglion cells in cats ([4]), a computational model of a centre-surround interaction has been derived that accounts for various phenomena. Modelling the responses of cells by means of shunting membrane equations of the general form $y_i = -Ay_i + (B - y_i)\text{net}_i^+ - (C + y_i)\text{net}_i^-$ (where A, B, C are scalar constants, and net_i^+ and net_i^- denote excitatory and inhibitory input activation derived from the input to the cell layer under consideration), accounts for some observed functional properties such as their (minimal) response to homogeneous illumination, the saturation of their activity levels, their suppression of featural noise and their reflectance processing ([9]). Physiological data suggest that the spatial extents of on/off and off/on RF profiles are symmetric in that the weighted inputs that form centre and surround enter into both the ON and the OFF channels ([5]). Using these weightings to describe ON and OFF activities (net_i^c, net_i^s), using the above mentioned membrane equation, and solving for steady state, we get:

$$\begin{aligned} {}^{(ON)}x_i &= \gamma(\text{net}) \cdot (B \cdot (\text{net}_i^c - \text{net}_i^s) + (B - C) \cdot \text{net}_i^s) \quad \text{and} \\ {}^{(OFF)}x_i &= \gamma(\text{net}) \cdot (B \cdot (\text{net}_i^s - \text{net}_i^c) + (B - C) \cdot \text{net}_i^c) \end{aligned} \tag{1}$$

where $\gamma(\text{net}) = (A + \text{net}_i^c + \text{net}_i^s)^{-1}$. Net activations net^c and net^s are generated by simple convolution of the stimulus luminance distribution I with scalable weighting functions λ^c and λ^s, respectively. The constants A, B, and C describe the passive decay and the excitatory and inhibitory saturation levels, respectively. Activations described by ${}^{(ON)}x_i$ and ${}^{(OFF)}x_i$ are subsequently rectified.

As one can see, each channel encodes both a scaled difference-of-low-pass (DoLP) filtered version of I and a scaled low-pass filtered version of I. This pair of channels multiplexes three data streams, which represent *i)* local ON and *ii)* local OFF contrast information, and *iii)* (low spatial frequency) B&D information.

Starting from the steady state equations for the ON and the OFF channels, a simple set of mechanisms realize the segregation into these three streams, which are then used for contrast detection and brightness reconstruction. ON-OFF competition eliminates the DC component from each channel to generate representations of DoLP filtered I for contrast detection, i.e. ${}^{(ON)}y_i$ and ${}^{(OFF)}y_i$ respectively. Ground-level activity s_i, a low-pass filtered version of the input luminance distribution, is generated in the B&D system by simply adding together the ON and OFF channel responses. Since the activity distribution s_i is only a low-pass filtered version of the input, the activations must be contrast-enhanced to generate sharply bounded region outlines in the visual perception of object surfaces. We postulate that such a compensatory mechanism for modulation of B&D system activity is realized by excitatory and inhibitory interactions between the ON and OFF contrast channels and the smoothed activity distributions ${}^{(ON)}y_i^s$ and ${}^{(OFF)}y_i^s$. See Fig. 1(a) for an overview of the architecture of this preprocessing stage.

It is postulated here that divergence (blurring) of the y^s activities occurs in the separate channels, which can be formalized as a convolution with a spatial low-pass filter (e.g. a Gaussian). Due to the segregation of the ON and OFF channels, selected activities in the response distribution show sensitivity to contrast polarity ascribed to cortical simple cells. The properties of cortical complex cells can now be synthesized very easily by again blurring and then summing (spatially pooling) the simple cell reponses in the ON and OFF channels. These activations – which in coincidence with the empirical electrophysiological evidence show no specificity to contrast polarity or selectivity to local phase (i.e. to odd/even symmetry) – form the basis for contrast detection and localization (e.g. [14]). The responses of these simple and complex cells can be described by the following equations:

$${}^{(ON)}y_{i\varepsilon}^s = \sum_j {}^{(ON)}y_j^s \cdot \lambda_{ij\varepsilon} \tag{2}$$

$${}^{(OFF)}y_{i\varepsilon}^s = \sum_j {}^{(OFF)}y_j^s \cdot \lambda_{ij\varepsilon} \quad \text{and} \tag{3}$$

$$y_{i\varepsilon}^c = \sum_j ({}^{(ON)}y_{j\varepsilon}^s + {}^{(OFF)}y_{j\varepsilon}^s) \cdot \lambda_{ij\varepsilon}, \tag{4}$$

where $\lambda_{ij\varepsilon}$ is a blurring function with elongation in direction ε to account for the orientation selectivity of simple cells. The y^c activations form an orientation field that feeds into a nonlinear feedback network described in the following chapter.

The contrast-enhancement of the B&D channel using the ON and OFF contrast channels again utilizes a shunting interaction which equation is solved for equilibrium:

$$u_i = \frac{s_i + E\,{}^{(ON)}y_i^s}{D + F\,{}^{(OFF)}y_i^s}, \quad \text{where} \quad {}^{(ON/OFF)}y_i^s = \sum_\theta {}^{(ON/OFF)}y_{i\theta}^s. \tag{5}$$

3. Recurrent Processing for Emergent Contour-Junction Segmentation and Brightness Filling-In

Based both on the results of an analysis of competence ([18], [23]) as described in [21] and on the general architecture proposed for the FACADE theory ([10]), a computational mechanism has been defined that performs stable extraction of orientation fields representing either straight and curvilinear shape outlines or junctions.

The inputs to this competitive/cooperative (CC) feedback network are the activations $y_{i\varepsilon}^c$ from the preprocessing stage described in the previous chapter. The orientation fields defined by the activations of the oriented contrast-sensitive cells can be thought of as an abstract representation of the hypercolumnar structure of the visual cortical area 17 ([14], see [21] for a discussion). The basic processing stages for the local competitive interactions in our computational mechanisms can be summarized as follows: *(1)* local intra-orientational positional competition between activations of contrast-sensitive cells, and *(2)* intra-orientational directional competition between activations from stage *(1)* (with two separate substages for *i)* mutual inhibition of activations in nearly-orthogonal directions at a spatial location, and *ii)* directional competition between activations within a local hypercolumnar neighbourhood; see [20], [21]). The sequential order of these competitive processes is required for the emergence of end-cuttings as described in e.g. [12]. The inhibitory interactions altogether account for a reduction of local uncertainty based on the initial coarse estimates of contrast orientation and locus. The equations used to realize these competitive processes are (with the rectification $[x]^+ = \max[x, 0]$):

$$\dot{w}_{i\varepsilon} = -w_{i\varepsilon} + \sum_j h(f(y_{j\varepsilon}^c)) \cdot \lambda_{ij}^+ + v_{i\varepsilon} - w_{i\varepsilon} \cdot \sum_j h(f(y_{j\varepsilon}^c)) \cdot \lambda_{ij}^- \tag{6}$$

$$x_{i\varepsilon} = \frac{[\sum_\theta w_{i\varepsilon} \cdot \gamma_{\varepsilon\theta}^+ - \sum_\theta w_{i\bar{\varepsilon}} \cdot \gamma_{\varepsilon\theta}^-]^+}{1 + \sum_\theta \sum_j ([\sum_\theta w_{j\varepsilon} \cdot \gamma_{\varepsilon\theta}^+ - \sum_\theta w_{j\bar{\varepsilon}} \cdot \gamma_{\varepsilon\theta}^-]^+)}, \tag{7}$$

where $v_{i\varepsilon}$ are feedback signals (specified below), $\lambda_{ij}^\pm$, $\gamma_{\varepsilon\theta}^\pm$ are excitatory and inhibitory weighting functions (for the spatial and orientation dimensions respectively), ε, θ are orientations in the range $[0 \,..\, \pi]$, $f(.)$ is the sigmoidal signal function $f(x) = 1/e^{-(x-m)s}$ and $h(.)$ is a threshold function $h(x) = r \cdot [x - T]^+$. The second equation acts to normalize the local activation with respect to the total activation for all orientations within a local neighbourhood. This normalizing effect over a local circular neighbourhood is an extension of the equation proposed in e.g. [12], where the stationary solution of the directional competition is taken at each location separately *without* the contextual neighbourhood (the formulation in [12] has the drawback that the characteristics of the local distribution of activations may get lost, see [21]). The emergence of a segmentation can only be guaranteed by lateral excitatory cell interactions that mediate local cooperation to further support local evidence of significant visual structures. The definition of a mechanism for 2D lateral excitation requires specific components that have been discussed in [21]. Each two-point cooperative link used in the current model is defined between pairs of spatial positions based on an anisotropic, purely excitatory weighting function, which can be understood as one half of a bipolar cell as defined in [12]. The anisotropic weighting functions can be separated to define *i)* a partition of the circular neighbourhood into equal elongated sectors, and *ii)* weightings of the spatial locations in the neighbourhood according to their distance from the reference point. The function that specifies the feature-linking capacity simply takes into account the deviation of a cell's preferred direction from the direction of the virtual line to the reference center point. The stimulus-dependent activations that feed into the cooperation at each location are defined by the dipole $(x_{i\theta}, \bar{x}_{i\theta})$, where $\bar{x}_{i\theta} = x_{i\bar{\theta}}$ $(\bar{\theta} = (\theta + \frac{\pi}{2}\text{mod}\pi)$. This realizes the spatial impenetrability postulate originally described in [12]. In order to activate a cooperation activity, at least two distinct support functions must contribute excitatory input (see [20], [21]). In this way, the cooperation mechanism adaptively acts as both an Nth-order junction 'detector' (for e.g. T-, X-, Y- or L-junctions (corners)) and – in the presence of artifacts, gaps, and subjective contours – a mechanism for enhancement or synthesis of activations induced by straight or curvilinear contours ([20], [21]). Finally, the resulting cooperation activations enter a further competitive stage to generate excitatory v-potentials, which *feed back* to the *first* competitive stage. The cooperative mechanism (including the positional competition for feedback) is described by:

$$z_{i\varepsilon} = [\sum_j \sum_\theta (x_{j\theta} - \bar{x}_{j\theta}) \cdot F_{ij}^{\varepsilon\theta}]^+ + [\sum_j \sum_\theta (x_{j\theta} - \bar{x}_{j\theta}) \cdot F_{ij}^{(\varepsilon+\pi)\theta}]^+ \quad (8)$$

$$v_{i\varepsilon} = \frac{h(g(z_{i\varepsilon}))}{1 + \sum_j h(g(z_{j\varepsilon})) \cdot \lambda_{ij}^-}, \quad (9)$$

where $F_{ij}^{\varepsilon\theta} = \mathrm{w}_{ij}^{\varepsilon} \cdot |\cos(\phi - \theta)|^p$ ($\mathrm{w}_{ij}^{\varepsilon}$ is an anisotropic weighting function, $p \in IN$ the tuning parameter, $p \geq 2$) and ϕ denotes the direction of the virtual line between i and j. In the equation that specifies v, the following functions are defined: $g(.)$ is a sigmoidal signal function, $g(x) = [x]^+/(s + [x]^+)$ and $h(x) = r \cdot [x - T]^+$ as in eq. 6. A sketch of the CC feedback architecture is shown in Fig. 1(b). As outlined in the previous chapter, the overall architecture proposed in this paper is concerned with *both* contrast *and* brightness processing. Within the preprocessing stage, two processes have been defined for the reconstruction and subsequent contrast-enhancement of a diffused and sclaed brightness distribution. Based on empirical findings, a permanent diffusive activity within the B&D system has been postulated, in which local diffusion coefficients (conductivities) are controlled by activations in the CC network (e.g. [3], [6] for psychophysical evidence, and [2], [13] for computational models). This diffusion of local activity within this final brightness representation stage is modelled by a simple local averaging process. The generation of the local activations in the B&D system is described by an anisotropic diffusion process which again can be solved at equilibrium

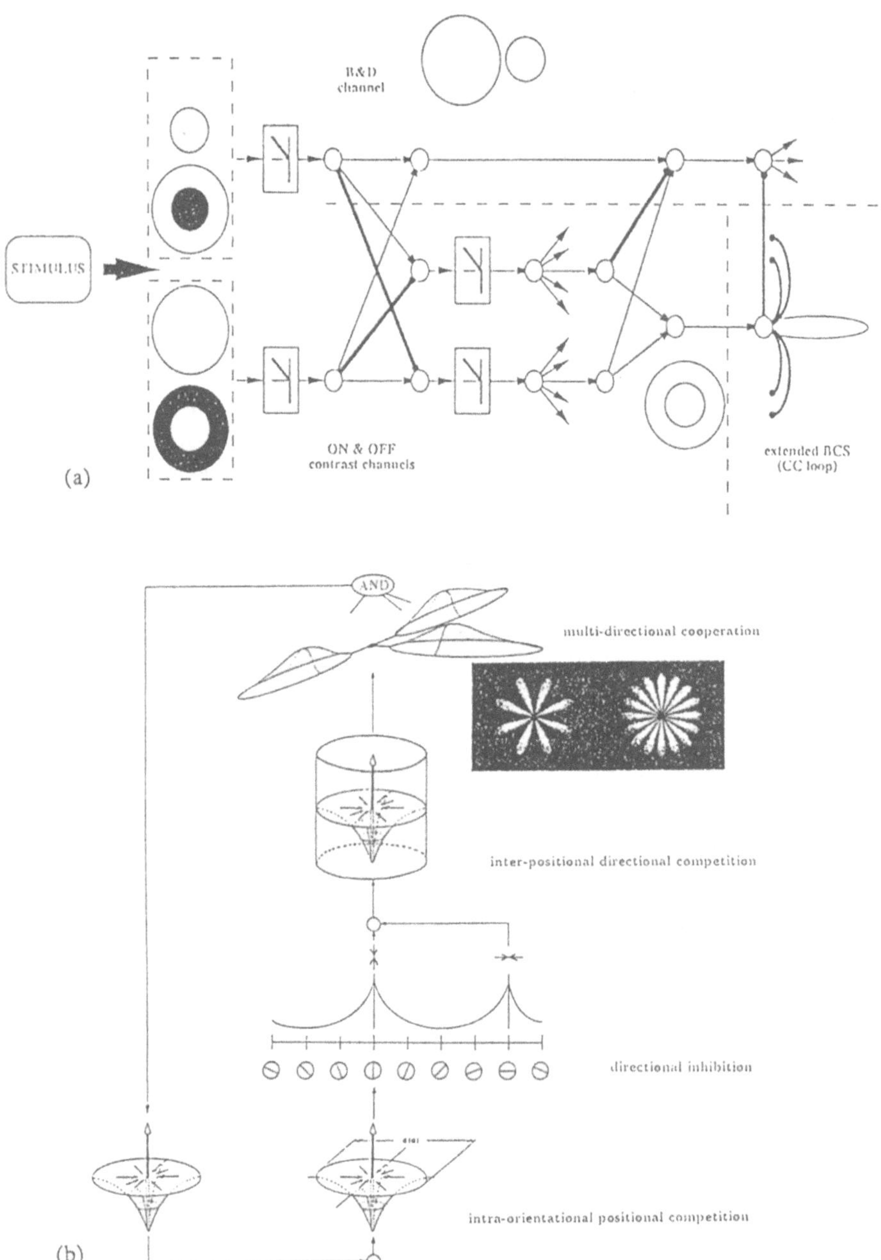

Fig. 1. Overview of the computational architecture: (a) preprocessing stage for contrast and brightness perception (thin lines with arrows denote exitatory connections, broad lines with knobs denote inhibitory connections), (b) CC feedback stage of the extended Boundary Contour System

(see also [13], [22], [19]):

$$\bar{u}_i = \frac{\sum_{j \in \mathcal{N}_i} (\bar{u}_j \cdot P_{ij}) + u_i}{G + \sum_{j \in \mathcal{N}_i} P_{ij}} \quad \text{with} \quad P_{ij} = \frac{\rho}{1 + \varepsilon(w_i + w_j)}, \tag{10}$$

where w_i, w_j represent the total activations at particular spatial positions of the orientation field.

4. Computational Experiments

Some computational experiments have been performed to demonstrate the intended functional behavior of the proposed architecture. For all numerical experiments, Gaussian weighting functions have been utilized for low pass filtering of the input stimuli and for subsequent smoothing of activities (see Fig. 1). Figure 2 depicts an example that shows the orientation field and the related brightness pattern after initial enhancement that result from the input of a bright square on a dark background (fig. 2(a,b)). Figure 2(c) shows the corresponding activation distribution in the CC feedback network, notably the emergence of the corners of the square. The corners – as special case of junction configurations of arbitrary order – furthermore serve as key features for three-space inference and 3D object recognition (see e.g. [21], [1]). The differential equation of the CC-loop has been solved by fixpoint iteration over the steady state solution. It should be noted, that numerical stability and robustness by now depends on fine tuning of the parameters for sigmoidal signal function and threshold function within the feedback path of the CC-loop (see [24]). Figure 3 shows a 1D profile through the centre of a Hering pattern that demonstrates the simultaneous contrast phenomenon.

5. Further Work

In the present paper, a framework for a computational architecture for contrast and brightness processing has been proposed that accounts for various constraints derived from psychophysical data and electrophysiological as well as neuroanatomical findings. Research is underway to further develop the cooperation stage within the CC loop to take into account long-range as well as short-range phenomena (as reported in e.g. [8]). To study the principles involved in the spatial processing and grouping of contrast patterns, psychophysical experiments were performed (in collaboration with the Department of Psychology at the University of Birmingham) to investigate the role of spatial frequency decomposition, segregation of processing channels and cooperation mechanisms for feature linking. Furthermore, the proposed architecture of the preprocessing stage will being used to investigate psychophysical phenomena of brightness perception more rigorously. In addition, an analysis of choice of parameters, as well as stability of the overall system needs to be also investigated in the near future.

References

1. I. Biederman. Human image understanding: Recent research and a theory. *Computer Vision, Graphics, and Image Processing*, 32(1):29 – 73, 1985.

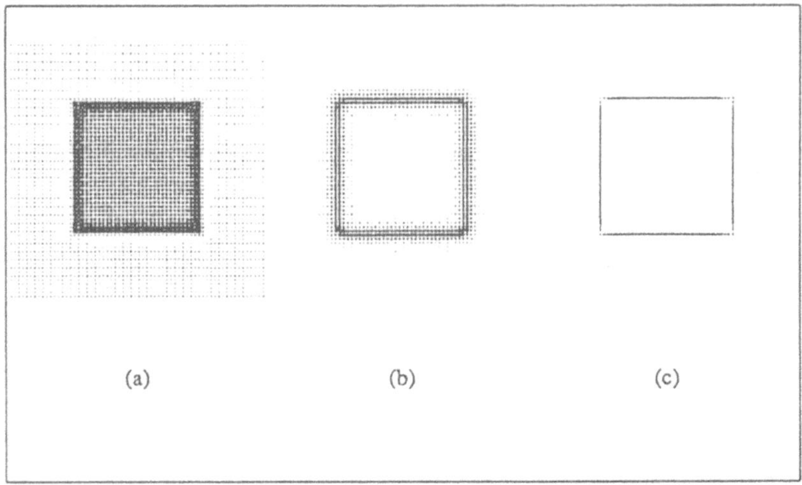

Fig. 2. Processing of a grey level image with a bright square on dark background: (a) B&D channel activation after initial enhancement (diameters of circles denote brightness), (b) initial orientation field (lengths of lines denote activation in each orientation, though individual lines may not be visible due to the photographic process), (c) emergent segmentation of the square pattern in the CC loop

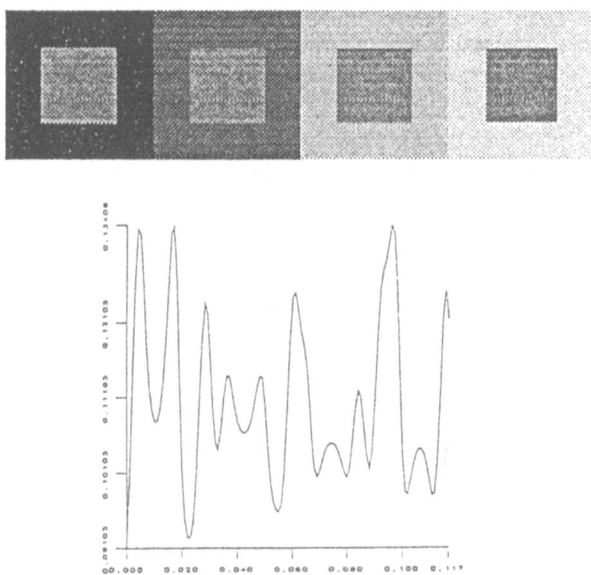

Fig. 3. Brightness reconstruction that accounts for the simultaneous contrast phenomenon with original input pattern (after Hering) (top) and profile of activation distribution at the second processing stage in the B&D system (bottom)

2. M.A. Cohen and S. Grossberg. Neural dynamics of brightness perception: Features, boundaries, diffusion, and resonance. *Perception and Psychophysics*, 36(5):428 – 456, 1984.
3. H.D. Crane and T.P. Piantanida. On seeing reddish green and yellowish blue. *Science*, 221:1078 – 1080, 1983.
4. C. Enroth-Cugell and J.G. Robson. Functional characteristics and diversity of cat retinal ganglion cells. *Investigative Ophtalmology and Visual Science*, 25:250 – 267, 1984.
5. A. Fiorentini, G. Baumgartner, S. Magnussen, P.H. Schiller, and J.P. Thomas. The perception of brightness and darkness — relations to neuronal receptive fields. In L. Spillmann and J.S. Werner, editors, *Visual Perception — The Neurophysiological Foundations*, chapter 7, pages 129 – 161. Academic Press, San Diego (FL/USA), 1990.
6. H.J.M. Gerrits and G.J.M.E.N. Timmerman. The filling-in process in patients with retinal scotoma. *Vision Research*, 9:439 – 442, 1969.
7. J.J. Gibson. *Wahrnehmung und Umwelt*. Urban und Schwarzenberg – Psychologie, München, 1982. (Original: "The Ecological Approach to Visual Perception", Houghton Mifflin Company, Boston, 1979).
8. C.M. Gray, P. König, A.K. Engel, and W. Singer. Oscillatory response in cat visual cortex exhibit inter-columnar synchronization which reflects global stimulus properties. *Nature*, 338:334 – 337, 1989.
9. S. Grossberg. How does a brain build a cognitive code? *Psychological Review*, 87:1 – 51, 1980. (reprinted in: S. Grossberg. "Studies of Mind and Brain". R.S. Cohen and M.W. Wartofsky (eds.). Boston Studies in the Philosophy of Science, Vol. 70. D. Reidel Publishing Co., Boston (USA), 1982.).
10. S. Grossberg. A model cortical architecture for the preattentive perception of 3-d form. In E.L. Schwartz, editor, *Computational Neuroscience*, System Development Foundation Benchmark, pages 117 – 138. MIT Press (Bradford Book), Cambridge (MA/USA), 1990.
11. S. Grossberg. Neural FACADES: Visual representations of static and moving form-and-color-and-depth. *Mind and Language*, 5(4):411 – 456, 1990. (Special Issue on "Understanding Vision: An Interdisciplinary Perspective").
12. S. Grossberg and E. Mingolla. Neural dynamics of perceptual grouping: Textures, boundaries, and emergent segmentation. *Perception and Psychophysics*, 38(2):141 – 171, 1985.
13. S. Grossberg and D. Todorovic. Neural dynamics of 1-d and 2-d brightness perception: A unified model of classical and recent phenomena. *Perception and Psychophysics*, 43:241 – 277, 1988.
14. D. Hubel and T.N. Wiesel. Functional architecture of macaque monkey visual cortex. *Proceedings Royal Society of London*, 198:1 – 59, 1977.
15. R. Jung. Visual perception and neurophysiology. In R. Jung, editor, *Handbook of Sensory Physiology*, volume VII/3: Central Processing of Visual Information, Part A, pages 3 – 152. Springer, Berlin, 1973.
16. J.J. Koenderink and A.J. van Doorn. The singularities of the visual mapping. *Biological Cybernetics*, 24:51 – 59, 1976.
17. M. Livingstone and D. Hubel. Segregation of form, color, movement, and depth: Anatomy, physiology, and perception. *Science*, 240:740 – 749, 6. May 1988.
18. D. Marr. *Vision – A Computational Investigation into the Human Representation and Processing of Visual Information*. W.H. Freeman and Co., San Francisco, 1982.
19. H. Neumann. Toward a computational architecture for unified visual contrast and brightness perception: I. Theory and model. In *Proc. World Conference on Neural Networks (WCNN-93)*, 1993. (in print).
20. H. Neumann and H.S. Stiehl. A competitive/cooperative (artificial neural) network approach to the extraction of n-th order junctions. In H. Burckhardt, K.-H. Höhne, and B. Neumann,

editors, *Mustererkennung 1989*, Informatik Fachberichte 219, pages 256 – 263, Berlin, 1989. Springer.

21. H. Neumann and H.S. Stiehl. Emergent segmentation of monocular visual invariants for space perception. In *Proc. Int. Joint Conf. on Neural Networks (IJCNN-92), Vol. I-IV*, pages (III) 266 – 271, Baltimore (Md/USA), June 7 - 11 1992. IEEE.
22. H. Neumann and H.S. Stiehl. Towards a neural architecture for unified visual contrast and brightness perception. In *Proc. Int. Conf. on Artificial Neural Networks (ICANN-93)*, Amsterdam (NL), Sept. 1993. (in print).
23. W. Richards. The approach. In W. Richards, editor, *Natural Computation*, pages 3 – 13. MIT Press (Bradford Book), Cambridge (MA/USA), 1988.
24. L. Wieske. Numerische Simulation von BC-Systemen. Tr, Universität Hamburg, Fachbereich Informatik, 1993. (in print).
25. A. Witkin and J.M. Tenenbaum. On the role of structure in vision. In J. Beck, B. Hope, and A. Rosenfeld, editors, *Human and Machine Vision*, pages 481 – 543. Academic Press, New York, 1983.

Optimization of adaptive preprocessing units during the learning process of Neural Networks. Application in EEG pattern recognition

H. Witte, M. Galicki, J. Dörschel, G. Grießbach, M. Eiselt*, A. Doering and D. Hoyer*
Institute of Medical Statistics, Computer Sciences and Documentation; *Institute of Pathophysiology; Medical Faculty; Friedrich Schiller University Jena; D-07740 Jena

1 Introduction

The final aim of the strategy given in this study is to train the Neural Network (NN) classifier and the adaptive preprocessing units (APU) simultaneously, i.e. that properties of preprocessing will be chosen automatically during the training phase. The strategy was developed to make efficient EEG monitoring in neonates possible. Therefore, the adaptive algorithms and the signal simulations used were adapted to realize this medical application.
The quantification of the **discontinuous EEG** may be a first step in the direction of automatic EEG monitoring in neonates [1]. The discontinuous EEG is characterized by the occurrence of periods with high amplitudes of EEG activity, named burst, and separated by periods of low amplitude, named interburst. Additionally, recognition of spike activity must be taken into consideration to prevent false positive detections of burst activity.

2 Stepwise concept for the introduction of adaptive preprocessing [2]

Before the adaptive preprocessing algorithms were introduced the whole strategy of coupling preprocessing and NN was tested with other algorithms and the results were given by [2].
The following steps can be separated:
a) Training of NN: The mean power values within the frequency bands **0.8-2.8 Hz, 2.8-4.8 Hz, 4.8-6.8 Hz and 6.8-8.8 Hz** were

used as input data (4 input neurons/channel) for the training of NN. These parameters were calculated from the specific time interval of the occurrence of the pattern. Both spectral parameters derived from the power spectrum, as well as those computed via Hilbert transformation [3], could have been used.

b) Classification: By using the trained NN, burst and interburst patterns were recognized.

Our aim was to realize the segmentation (designation of the onset and end of pattern) and the classification (burst, interburst) of the discontinuous EEG. For this aim, the time courses of the spectral parameters given above were used as input signals of the NN. Such time courses (= momentary power within frequency bands) could be calculated via the Hilbert transformation (DHT). They are dynamic equivalents of the spectral parameters derived from the power spectrum.

For our first tests, the calculation of the momentary power was performed within intervals by means of DHT on the basis of FFT.

c) On-line classification: For continuous recognition of the patterns, this interval-related concept is useless. For the purpose of on-line pattern recognition, the DHT, on the basis of FFT, can be replaced by a method using FIR filters. On this basis, an on-line calculation of the momentary power and frequency is possible. The momentary frequency can be used for automatic detection of artefacts and spikes [4].

d) Use of adaptive algorithms: The courses of momentary power as well as momentary frequency, can be approximated by adaptive recursive estimation algorithms. The properties of these estimations can be controlled by adaptation factors. This is an important fact in realizing the optimization strategy of the preprocessing units in connection with and during the training (learning phase) of the NN.

3 Neural Networks

The Neural Networks (NN) used were multi-layer networks [5] (error backpropagation; 2(1) hidden layers; logistic activation function; learning rate = 0.2; momentum term = 0.2). The network structure was created empirically.

Additionally, the common learning algorithm (APU and NN) uses a

modification of the backpropagation algorithm for simultaneous optimization of adaptive preprocessing. The modifications differ between the applications 6.1 and 6.2.
As an alternative, a global learning algorithm based on a random search technique [6] was used.

4 Description of the adaptive algorithms

The adaptive basic algorithms used are:

* **mean value**
  $M_{n+1} = M_n + c*(x_{n+1} - M_n) \Rightarrow$ (operator $\mathbf{M^c(X)}$), (1)
* **second statistical moment**
  $E_{n+1} = E_n + c*(x^2_{n+1} - E_n) \Rightarrow$ (operator $\mathbf{E^c(X)}$), (2)
* **alpha-quantile**

$$Q_{n+1} = \begin{cases} Q_n + c\,\alpha & \text{if } x_{n+1} \geq Q_n \\ Q_n - c\,(1-\alpha) & \text{if } x_{n+1} < Q_n \end{cases} \qquad (3)$$

Adaptive estimations for the parameters calculated by the Hilbert transformation can be given using these basic algorithms (1) and (2).
For the momentary power within a frequency band, the formula

$$E^{c1}\{M^{c2}[X - M^{c3}(X)]\} \qquad (4)$$

($\mathbf{c_1}$ determines the time resolution, $\mathbf{c_2}$ the upper, and $\mathbf{c_3}$ the lower cut-off frequency of the pass band)
and for momentary frequency

$$M^{c2}\{W[K(L(X), M^{c1}(L(X)), K(X, M^{c1}(X))]\} \qquad (5)$$

can be given, where $L(X)$ is a lag(delay)-operator and
$K(X,Y) = \{K(x_i, y_i)\}_{i \in N}$,
$W(X,Y) = \{W(x_i, y_i)\}_{i \in N}$
are comparators (operators for comparison).
In this way, other adaptive parameter estimation procedures can be constructed (e.g. the parameters of an ARMA-model, the probability density function, and the correlation coefficient).

Such adaptive estimations can be viewed as processing units with one signal input (sequence X) and one or more control input(s) to vary the adaptation factor(s).

Through the variation of the adaptation factors, essential properties of the estimation quality can be influenced. **Such properties include** estimation accuracy and time resolution, the quantity of signal power above an adaptive threshold (Fig. 1) as well as filter properties of the adaptive estimations.

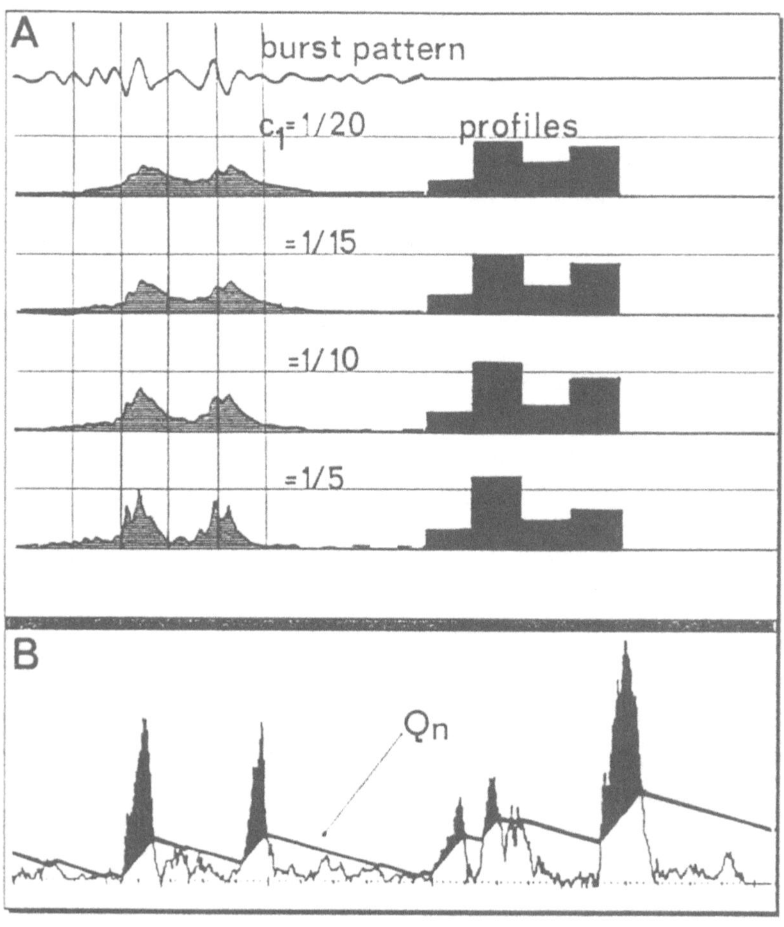

Fig. 1: Examples of adaptive preprocessing [2]
A: Calculation of the momentary spectral band power (E_n) from a burst pattern with different adaptation parameters (c_1, equ. 4). The adaptation parameter determines the time resolution.
B: The use of the adaptive quantile estimation Q_n [7] as a threshold of detection for EEG transients. Momentary power was computed and the overcrossings were used as one parameter for spike detection [4].

5 Instantaneous optimization of the adaptation parameter

An error function e(**c**) was introduced which depends on the scalar c and it presents a value obtained after the one-step backpropagation algorithm. Variations of the parameter c result in different values of this error function e(**c**), this is the consequence of different learning examples.

The high complexity and nonlinearity of the error function e(**c**) prevents an analytical solution. Simple numerical methods will be introduced which find a c-value for the minimization of this modified error function.

Firstly, a gradient technique was used if the error function e(**c**) was differentiable by c. Secondly, a gradient-free method can be used if a differentiation of e(**c**) by c is not possible. These numerical optimization methods were embedded into the whole optimization of the error function which can be described as follows (Fig. 2):

A) Starting stage: The correction term for the weight matrix $\Delta\mathbf{W}'(\mathbf{W}_0,\mathbf{Y}(\mathbf{X},\mathbf{c}_0))$ was obtained after performing one backpropagation step, where $\mathbf{W}_0$ is the starting matrix of the weights (stochastical choice) and c_0 is the starting value of c. The momentum term $\Delta\mathbf{W}$ equals zero.

B) Modification stage for the adaptation parameter $c'=c+\Delta c$ where c' is the actual value, c the previous value and Δc a correction term.

C) Learning stage of modification: In this stage the usual backpropagation learning step is performed, but using the modified value of the adaptation parameter $c'= c+\Delta c$.

To avoid instability of the algorithm, a combination of the usual backpropagation steps (**C**) with the modification steps (**B**) was used. Following quantities are used: Δ^{back}-number of backpropagation steps; Δ^{mod}-number of modification steps according to stage **B)**.

Furthermore, the adaptive estimation (adaptation parameter k) of the mean value of usual backpropagation error **e** in each iteration step

$$M_{n+1}= M_n + k*(\mathbf{e}_{n+1}-M_n) \qquad (6)$$

and a performance index

$$I(m,\Delta^{back}) = \frac{M_m - M_{m+\Delta^{back}}}{M_m} , \qquad (7)$$

which is a measure of the relative decrease of the error function **e**, were introduced (m is the current modifying iteration from which Δ^{back}-steps of the backpropagation (stage C) starts). The whole optimization procedure can be described as follows (gradient method). Firstly, the learning is performed using the standard backpropagation algorithm with Δ^{back}-steps and the current modifying iteration m=0 (starting stage A). Secondly, the value of $I(m=0,\Delta^{back})$ will be computed (assuming that $M_m \geq \varepsilon$ for m=0; ε is a small positive number). If $I(m,\Delta^{back}) \geq \delta$ (δ is a small positive number) then no modification (modification stage B) is done and the algorithm proceeds again with Δ^{back}-steps of the backpropagation algorithm (stage C; $m+\Delta^{back}$ is set to the current modifying iteration m). Otherwise, Δ^{mod}-steps of the modification algorithm will be carried out which update each time the parameter c based on the gradient of modified error function *e*(**c**) with respect to c. During each updating of the parameter c, the conditions

$$c - c_l \geq \Delta c \geq c - c_u$$

should be checked. If $c - c_u < \Delta c$ or $c - c_l < \Delta c$, then it can be assumed that $c=c_u$ or $c=c_l$. Otherwise the updated parameter c will be computed according to the gradient of modified error function *e*(**c**) with respect to c.

After the modification (stage B), the Δ^{back}-steps (stage C) will be carried out ($m+\Delta^{back}+\Delta^{mod}$ is set to the current modification iteration m) and the whole process can be repeated (C->B).

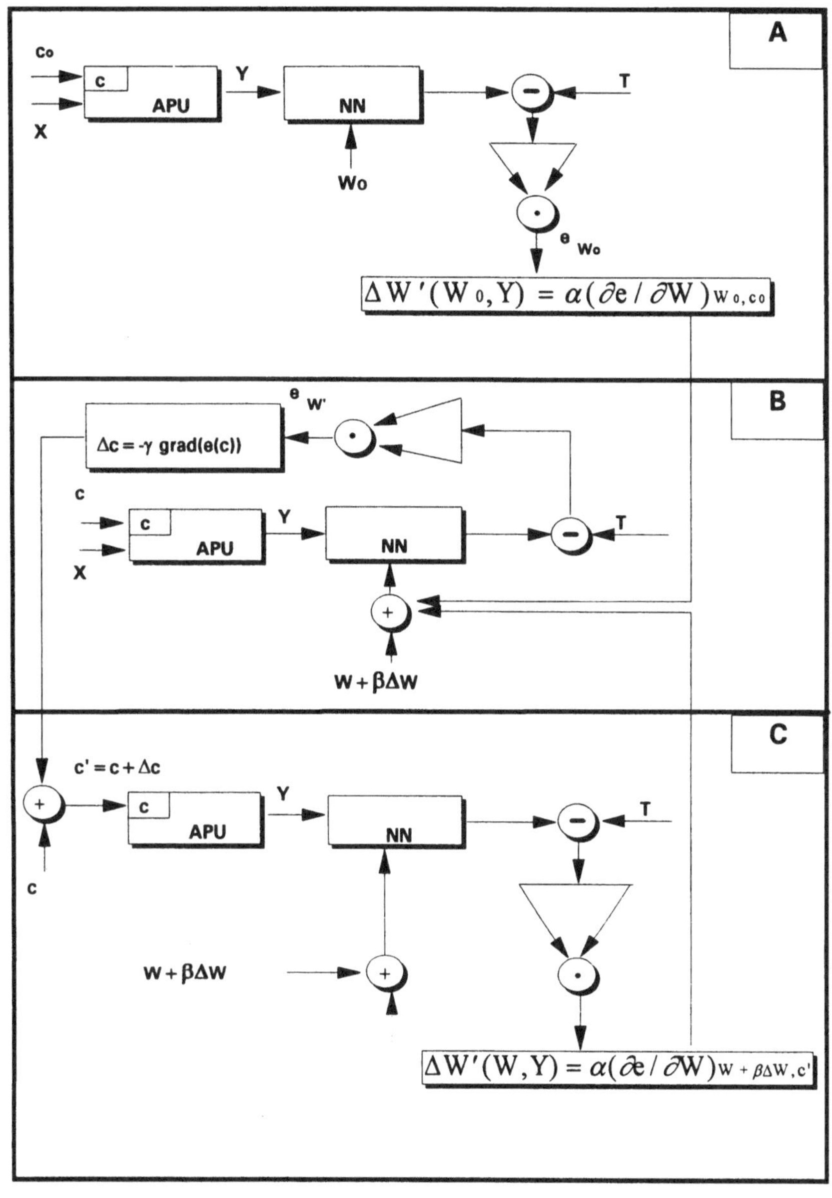

Fig. 2: Scheme of the processing stages of the learning algorithm (gradient technique) **W'** is the actual (current stage) and **W** the previous (from previous stage) weight matrix.

6 Results

6.1 EEG pattern recognition

Patterns as simple time-limited and amplitude-modulated waveforms with different based frequencies (1.6 Hz and 3.2 Hz) were applied. This simulation is related to the burst and interburst pattern of the discontinuous neonatal EEG [2]. To obtain a variable set of patterns (each class consists of 21 examples), the amplitude modulation (envelope) was changed randomly within defined limits (10% of the amplitude maximum). The momentary power within a frequency band was calculated according to equation (4).

The single preprocessed pattern was divided into three subsequent intervals. Within these intervals, the average of the momentary power was calculated. These resulting subsequent parameters (power profile) were given simultaneously at input neurons of NN.

The first investigations were carried out by using simple narrow band pass characteristics ($c_2=c_3$).In this case the learning algorithm changes the location of the mid-frequency of the band pass. Two minima (local) were found by the common learning algorithm (around $c_2=c_3=0.15$=> mid-frequency=1.6 Hz and $c_2=c_3=0.35$=> mid-frequency=3.2 Hz) depending on the starting point of the value of c (0 or 1).

This finding can be confirmed by EEG data derived from animal experiments during different experimental supply situations of the brain (normal and ischemia). The method given above was tested by these real data. The frequency band with the distinguishing differences between both classes is characterized by the mid-frequency around 5 Hz. The result of simultaneous learning of APU (band pass with the characteristics given in Fig. 3) and NN for this application is a value of about $c_2=c_3=0.48$=> mid-frequency = 6-7 Hz, i.e. that the algorithm has found the best discriminating frequency band.

6.2 Learning of an optimal adaptive threshold

Both momentary power and frequency are used in a spike recognition scheme [4]. During the occurrence of spikes, higher amplitude values (power values) and low variance of momentary frequency estimation can be observed, in comparison to the surrounding background activity. The joining of the characteristics of these parameters during spike and surrounding EEG activity can be done by an adaptive threshold calculation. The adaptive estimation of the α-quantile (Q_n) can be used as such a threshold. An α-quantile (Q_n) estimation was used to detect high power values (course of momentary power), in comparison to the surrounding activity (Fig. 1).
It can be demonstrated by simulated and real EEG data that an optimization of the adaptive threshold can be carried out to find an optimum between the rejection of false positive detections (e.g. by EOG interferences) and the prevention of false negative detections. For the optimization the parameters **c** and $\boldsymbol{\alpha}$ were controlled (equ. 3). Additionally, both band pass (6.1) and threshold can be integrated simultaneously into the training of the NN for the optimization of the spike detection.

7 Discussion

From the point of view of diagnostic value, the automatic recognition of the burst and interburst patterns of the discontinuous EEG in neonates seems to be an important task in neonatal monitoring [1]. The task to distinguish only between two well-defined (visual analysis) patterns forms the starting point for the development and testing of new processing technologies.
In this context, the Neural Network classifier (supervised learning) was used, because

* experiences exist in using dynamic spectral parameters as input parameters of NN [2]
* expert knowledge can be utilized via the preparation of the trainings set by a physician, and
* the learning algorithm of NN can be modified to optimize characteristics of the preprocessing simultaneously during the training phase.

Therefore, the processing strategy given in this study was connected with the strategy of classification by artificial Neural Networks. The results show that an efficient classification is possible by using dynamic spectral parameters as input for the NN-classifier.

8 References

[1] Rother,M., Teubner,B., Winkler,P., Eiselt,M., Clausner,B.: Predictive value of pattern selective spectral analysis of neonatal EEG. Brain Dysfunction **4**, 1(1991).

[2] Witte,H., Dörschel,J., Grießbach,G., Eiselt,M., Galicki,M., Arnold,M.: The combination of on-line preprocessing with Neural Network classification for EEG monitoring in neonates. Beitr.Anaesth.Intensivmed., 1993 (in press).

[3] Witte,H., Stallknecht,K., Ansorg,J., Griessbach,G., Petranek,S., Rother,M.: Using Discrete Hilbert Transformation to realize a general methodical basis for dynamic EEG mapping .A methodical investigation. Automedica **13**, 1(1990).

[4] Witte,H., Eiselt,M., Patakova,I., Petranek,S., Griessbach,G., Krajca,V., Rother,M.: Use of Discrete Hilbert Transformation for automatic spike mapping. A methodical investigation. Med.&Biol.Eng.&Comput. **29**, 242(1991).

[5] Rumelhart,D.E., Winton,G.E., Williams,R.J.: Learning internal representation by error backpropagation. In: Rumelhart,D.E., McClelland,J.L. (eds.): Parallel Distributed Processing. Explorations in the Microstructure of Cognition. The MIT Press, Cambridge (Mass.),1986.

[6] Solis,F.J. and Wets, J.B.: Minimization by random search techniques. Mathematics of Operations Research 6(1981),19-30.

[7] Grießbach,G., Schack,B.: Adaptive quantile estimation and its application in analysis of biological signals. Biom.J. 35(1993),165-179.

Acknowledgements: The study was supported by the Federal Ministry of Research and Technology (project 01ZZ9104) and the special Neural Network application by the CEC (ESPRIT, NEUFODI II-CPL, 7534).

Ein verteiltes modulares System zur eigenvektorbasierten Analyse natürlicher Szenen mit selbstorganisierten rezeptiven Feldern*

T. Pomierski, H. M. Gross, D. Wendt

Technische Universität Ilmenau
Fachgebiet Neuroinformatik
98684 Ilmenau, Postfach 327
e-mail: pomi@informatik.tu-ilmenau.de

Zusammenfassung

Es wird ein verteiltes modulares System vorgestellt, das in Anlehnung an die corticale Informationsverarbeitung in columnaren Strukturen eine intracolumnare Hauptkomponentenanalyse realisiert. Dieses neuartige Architekturprinzip erlaubt durch intercolumnar verteilte, parallele Analyse eine ähnlichkeitserhaltende, lokal translationsinvariante Beschreibung lokaler Aufmerksamkeitsregionen (Bildregionen) mit minimaler Parameteranzahl.

1. Einleitung

Kognitive Leistungen, wie sie für die Analyse oder Interpretation natürlicher visueller Szenen erforderlich sind, können durch rein paralleles Patternmatching mit Sicherheit nicht erklärt werden. Naheliegender ist vielmehr eine gesteuerte Dekomposition der hochparallelen visuellen Szene in eine Sequenz niedrigdimensionaler Komponenten (bedeutungsvolle Bruchstücke), wobei datengetriebene und auf internem Systemwissen basierende Verarbeitungsprinzipien (preattentive, attentive vision) von entscheidender Bedeutung für die Steuerung des Dekompositionsprozesses sind [1]. Diese Regionen, die entweder von syntaktischer Bedeutung (datengetriebene Bildanalyse) oder aufgrund aktivierbarer interner Hypothesen über die Szene oder einzelne Inputsegmente (attentive oder wissensbasierte Komponente) von hohem Interesse für ein aktives Szenenanalyseregime (active vision system) sind, werden im folgenden als *Aufmerksamkeitsregionen* (regions of interest) bezeichnet. Das Hauptaugenmerk dieser Arbeit richtet sich auf ein neuartiges Architekturprinzip zur verteilten parallelen Analyse der internen *Aufmerksamkeitsregionen*, die durch oben genannte spezielle Aufmerksamkeitsmechanismen innerhalb der visuellen Szene selektiert wurden.
In dem hier präsentierten Modellkonzept wird diese Analyse durch ein Array corticaler Verarbeitungscolumnen, deren Aufbau sich strukturell und funktionell an den Minicolumnen des primären visuellen Cortex [7] orientiert, realisiert. Dieses columnare Array analysiert parallel die durch präattentiv oder attentiv gesteuerte Aufmerksamkeitsmechanismen nacheinander selektierten *Aufmerksamkeitsgebiete* (internal foci) innerhalb der visuellen Szene. Jede Verarbeitungscolumne extrahiert elementare Bildmerkmale aus ihrem lokalen *Analysegebiet* innerhalb der *Aufmerksamkeitsregion* mit Hilfe komlexer rezeptiver Felder. Diese werden jedoch nicht algorithmisch vorgegeben, sondern in einem adaptiven Selbstorganisationsprozeß auf der Basis einer neuronal motivierten Hauptkomponentenanalyse strukturiert. Oja zeigte in [3], daß ein spezieller Hebb-ähnlicher Selbstorganisationsprozeß ein rezeptives Feld mit synaptischen Wichtungen herausbildet, das gerade die Hauptkomponente des gesamten dargebotenen Eingangsdatensatzes darstellt. In Erweiterung nach [5] führt die Selbstorganisation der eigenwertgrößten Hauptkomponenten für zufällig gewählte lokale Bildausschnitte auch für sehr unterschiedliche natürliche Grauwertszenen zu qualitativ ähnlichen Sätzen rezeptiver Felder (siehe Abb. 1). Diese Ergebnisse führten zu der Vermutung, daß ein universeller Satz von komplexen rezeptiven Feldern für jede corticale Verarbeitungscolumne unabhängig von ihrer Lokalisation

*gefördert durch das Bundesministerium für Forschung und Technologie (BMFT), Förderkennz. 413-5839-01 IN 101D - NAMOS-Projekt

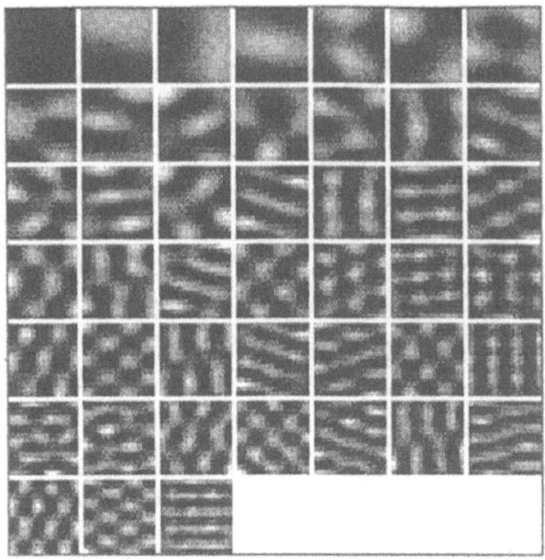

Abbildung 1: *Die genutzten rezeptiven Felder (Hauptkomponentenarrays)* $\mathbf{w}^{(1)}, ..., \mathbf{w}^{(i)}, ..., \mathbf{w}^{(n)}$ *(links), die durch zufällig gewählte Eingangsmuster aus der (rechts) dargestellten natürlichen Grauwertszene innerhalb eines neuronalen Selbstorganisationsprozesses herausgebildet wurden und für nachfolgende Analyseprozesse innerhalb einer jeden Verarbeitungscolumne des columnaren Arrays zur Verfügung stehen (nähere Erläuterungen im Text).*

innerhalb der Modellarchitektur für die Bildanalyse optimal geeignet wäre. Anliegen dieser Arbeit ist die Vorstellung der neuronalen Mechanismen, die erstmalig eine ähnlichkeitserhaltende Hauptkomponentenanalyse innerhalb einer jeden Verarbeitungscolumne garantieren und so eine lokale translationsinvariante Beschreibung des zugehörigen lokalen *Analysegebietes* mit minimaler Anzahl beschreibender Parameter zuläßt.

2. Theoretische Betrachtung der Hauptkomponentenanalyse

In natürlichen Bildern sind benachbarte Bildpixel meistens stark miteinander korreliert. Somit ist eine wesentlich kompaktere Kodierung der Bildinformation z.B. mittels einer Hauptkomponentenanalyse möglich. Im allgemeinen Fall liegt eine Anzahl von Bildausschnitten $\mathbf{x}^{(1)}, \mathbf{x}^{(2)}, ...$ vor, wobei $\mathbf{x} = (x_1, ..., x_j, ..., x_m)^T$ die Folge von Grauwertpixeln eines Bildausschnittes darstellt. Die einzelnen Grauwertpixel x_j variieren nicht völlig unabhängig voneinander, sondern sind mehr oder weniger korreliert, wobei die genaue Art der Abhängigkeit oft unbekannt ist. Von Interesse ist es nun festzustellen, inwieweit man die beobachtete Variation von Grauwertpixeln auf eine Abhängigkeit der x_j von einer kleineren Anzahl unbekannter Parameter $y_1, ..., y_i, ..., y_n$, $n < m$ zurückführen kann. Ist das möglich, so kann man m Funktionen $f_1, ..., f_j, ..., f_m$ der unbekannten Parameter finden, für die

$$x_j = f_j(y_1, ..., y_i, ..., y_n), \; j = 1, 2, ..., m, \tag{1}$$

gilt. Gegenüber den direkt zugänglichen Pixelgrauwerten x_j ermöglichen die Variablen y_i eine sparsamere Beschreibung der beobachteten Grauwertverteilungen. Insbesondere entsprechen sie eher den wirklich beteiligten Freiheitsgraden, deren Anzahl in vielen Fällen geringer als die der beobachteten Meßparameter (Grauwertpixel) x_j ist. Mathematisch kann man diese Parametrisierungen als Koordinatensystem auf einer abstrakten Mannigfaltigkeit deuten. Dabei wird die einschränkende Annahme eines linearen Zusammenhanges zwischen y_i und x_j gemacht. Diese Annahme läßt sich geometrisch als Einführung einer n-dimensionalen Hyperebene [4] deuten, die im m-dimensionalen Raum der Daten liegt und deren Lage und Orientierung so gewählt werden, daß sich jeder Bildausschnitt möglichst gut durch einen Punkt der Hyperebene approximieren läßt. Das entspricht einer

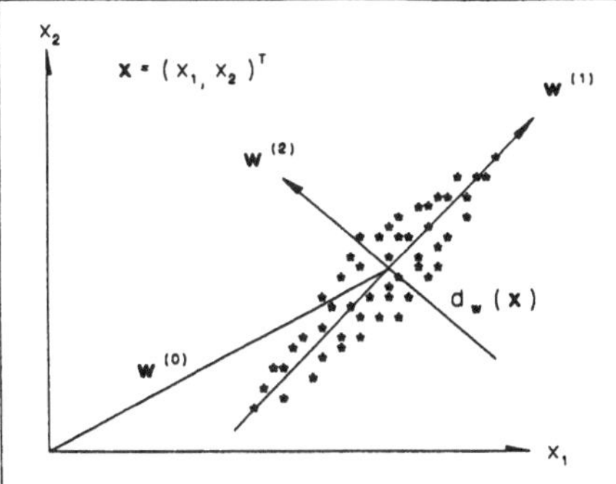

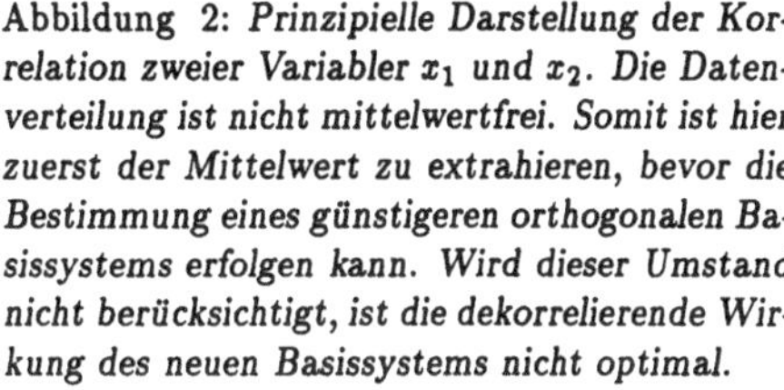
Abbildung 2: *Prinzipielle Darstellung der Korrelation zweier Variabler x_1 und x_2. Die Datenverteilung ist nicht mittelwertfrei. Somit ist hier zuerst der Mittelwert zu extrahieren, bevor die Bestimmung eines günstigeren orthogonalen Basissystems erfolgen kann. Wird dieser Umstand nicht berücksichtigt, ist die dekorrelierende Wirkung des neuen Basissystems nicht optimal.*

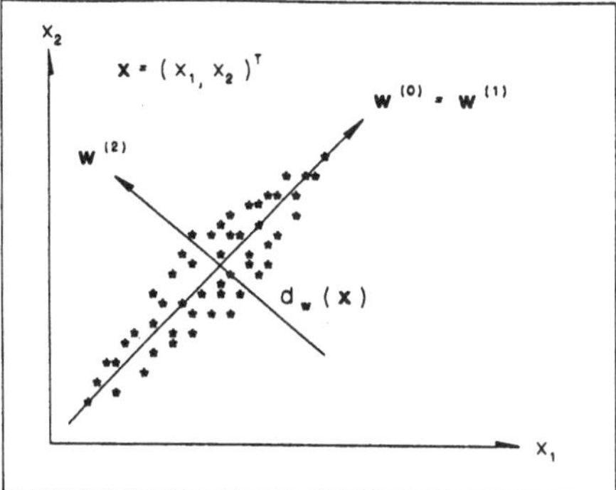

Abbildung 3: *Prinzipielle Darstellung der Korrelation zufällig gewählter Paare nebeneinanderliegender Grauwertpixel x_1 und x_2 der visuellen Szene aus Abb. 1. Hier sind Schwerpunktsvektor der Datenverteilung $\mathbf{w}^{(0)}$ und eigenwertgrößter Eigenvektor $\mathbf{w}^{(1)}$ identisch (nähere Erläuterungen im Text).*

Darstellung jedes Bildausschnittes in der Form

$$\mathbf{x} = \mathbf{w}^{(0)} + \sum_{i=1}^{n} \mathbf{w}^{(i)} y_i(\mathbf{x}) + d_{\mathbf{w}}(\mathbf{x}), \tag{2}$$

wobei $\mathbf{w}^{(0)} ... \mathbf{w}^{(n)}$ Vektoren sind, die die Hyperebene spezifizieren und $y_1(\mathbf{x}) ... y_n(\mathbf{x})$ die zum Bildausschnitt $\mathbf{x}$ gehörenden neuen Parameter sind. In der Regel fallen nicht alle Bildausschnitte in die Hyperebene, so daß sich für jede Hyperebene zu den meisten Bildausschnitten ein nichtverschwindender senkrechter Abstand ergibt, der durch $d_{\mathbf{w}}(\mathbf{x})$ gegeben ist. Die optimale Wahl der Hyperebene ergibt sich dann, wenn die Vektoren $\mathbf{w}^{(i)}$ so bestimmt werden, daß der mit der Wahrscheinlichkeitsdichte $P(\mathbf{x})$ der Daten gewichtete mittlere quadratische Restfehler $\langle d_{\mathbf{w}}(\mathbf{x})^2 \rangle$ seinen kleinstmöglichen Wert annimmt, d.h.

$$\int \Big\| \mathbf{x} - \mathbf{w}^{(0)} - \sum_{i=0}^{n} \mathbf{w}^{(i)} y_i(\mathbf{x}) \Big\|^2 P(\mathbf{x})\, d^m\mathbf{x} = Min! \tag{3}$$

Man kann zeigen, daß die Lösung dieser Minimierungsaufgabe für $\mathbf{w}^{(0)}$ den Schwerpunkt der Datenverteilung ergibt.

$$\mathbf{w}^{(0)} = \int \mathbf{x} P(\mathbf{x})\, d^m\mathbf{x} \tag{4}$$

Die übrigen Vektoren $\mathbf{w}^{(i)}$, $i = 1, 2, ..., n$ bilden eine Basis des von den n eigenwertgrößten Eigenvektoren der Korrelationsmatrix

$$\mathbf{C} = \int (\mathbf{x} - \mathbf{w}^{(0)})(\mathbf{x} - \mathbf{w}^{(0)})^T P(\mathbf{x})\, d^m\mathbf{x} \tag{5}$$

aufgespannten Eigenraumes [4]. Eine möglichst spezielle Wahl für die $\mathbf{w}^{(i)}$ $(i > 0)$ sind die n normierten, eigenwertgrößten Eigenvektoren von $\mathbf{C}$. Es ergeben sich die neuen Parameter y_i als die Projektionen der Datenvektoren entlang der n Hauptachsen ihrer Verteilung.

$$y_i = \mathbf{w}^{(i)} \mathbf{x}, \; i = 1, 2, ..., n \tag{6}$$

Geometrisch bedeutet dies, daß die Hyperebene durch den Schwerpunkt $\mathbf{w}^{(0)}$ der Datenverteilung verläuft und von den n Hauptachsen der Verteilung, die zu den größten Eigenwerten gehören, aufgespannt wird. Man kann zeigen, daß die dadurch festgelegte Orientierung der Hyperebene die

Varianz der senkrechten Projektion der Datenpunkte maximiert. Die Variablen y_i lassen sich also dadurch charakterisieren, daß sie den mit n Parametern größtmöglichen Anteil der Datenvariation beschreiben. Die Güte einer derartigen Beschreibung wird jedoch entscheidend vom Grad der Gültigkeit der zugrunde gelegten Linearitätsannahme bestimmt. Je mehr die tatsächliche Verteilung der Bildausschnitte von einer Hyperebene abweicht, desto größer wird der mit der Projektion auf die Hauptachsen der Verteilung erzwungene Kompromiß [4].

3. Selbstorganisation rezeptiver Felder zur neuronalen Hauptkomponentenanalyse

Während der Trainingsphase wurde eine Vielzahl von Bildern mit 512 × 512 Pixeln und jeweils 8 bit Auflösung untersucht. Dabei wurde ganz bewußt kein Versuch unternommen, optische Unregelmäßigkeiten zu korrigieren. Es wurden Bildausschnitte der Dimension 16 × 16 zufällig aus der Inputszene ausgewählt. Die Vorverarbeitung bestand lediglich in einer Normierung der Grauwerte der Bildausschnitte auf das Intervall <0, 1>. Das wurde möglich, da der Schwerpunktsvektor $\mathbf{w}^{(0)}$ bei Grauwertbildern gerade dem ersten Eigenvektor $\mathbf{w}^{(1)}$ entspricht (siehe Abb. 3). Da somit eine Bestimmung des Schwerpunktsvektors nicht mehr erforderlich ist, ist es auch nicht notwendig, die gesamte Lernstichprobe a-priori kennen zu müssen. Das ist die Voraussetzung für einen adaptiven Lernprozeß zur gleichzeitigen Bestimmung aller Hauptkomponenten eines günstigeren Basissystems. Für die Simulationen wurde eine von Oja [3] 1982 für ein einzelnes Neuron vorgestellte Lernregel verwendet und im weiteren Verlauf der Arbeiten modifiziert.

$$w_j(t+1) = w_j(t) + \gamma(t)\, y(t)\, [x_j(t) - y(t)\, w_j(t)],\ j = 1,2,...,m \tag{7}$$

Hierbei bezeichnet $\mathbf{x} = (x_1, ..., x_j, ..., x_m)^T$ den Eingangsvektor der auf das Intervall <0, 1> normierten Grauwertpixel eines zufällig gewählten Bildausschnittes, y das Ausgangssignal, w_j die Wichtung des Eingangsknotens x_j und γ die Lernrate. Man kann nun zeigen, daß diese Lernregel einen Wichtungsvektor $\mathbf{w}$ erzeugt, der gerade dem Eigenvektor der Korrelationsmatrix aller dargebotenen Bildausschnitte $\mathbf{x}$ mit dem größten Eigenwert entspricht. Sie extrahiert somit die dominanteste Hauptkomponente der Eingangsdaten. Gleichzeitig gewährleistet die Lernregel eine Normierung des Wichtungsvektors $\mathbf{w}$ auf die Einheitslänge [3]. Die Lernregel wurde auf n Ausgangsneuronen erweitert [5], um so nacheinander die stärksten Hauptkomponenten bestimmen zu können. Dabei handelt es sich für jedes der n Neuronen um einen zweiphasigen Prozeß, in dem zunächst vom Eingangsvektor $\mathbf{x}^{(i-1)}$ die Anteile abgezogen werden, die durch den Wichtungsvektor $\mathbf{w}^{(i)}$ bereits dargestellt werden können.

$$\mathbf{x}^{(i)}(t) = \mathbf{x}^{(i-1)}(t) - y_i(t)\mathbf{w}^{(i)}(t),\ \mathbf{x}^{(0)} = \mathbf{x},\ i = 1,2,...,n \tag{8}$$

Erst danach erfolgt die eigentliche Aktualisierung der Komponenten $w_j^{(i)}$ des Wichtungsvektors $\mathbf{w}^{(i)}$.

$$w_j^{(i)}(t+1) = w_j^{(i)}(t) + \gamma_i(t)\, y_i(t)\, x_j^{(i)}(t),\ i = 1,2,...,n,\ j = 1,2,...,m \tag{9}$$

Die Anzahl der Neuronen wurde experimentell bestimmt. Dabei zeigte sich, daß 45 Hauptkomponenten genügen, um nahezu die gesamte Information aus beliebigen 16 × 16 Pixel großen Bildausschnitten natürlicher Grauwertbilder zu extrahieren. Das Netz wurde durch Setzen aller Wichtungen $w_j^{(i)}$ auf einen zufällig kleinen Wert initialisiert, so daß die Quadratsumme der Komponenten eines jeden Wichtungsvektors $\mathbf{w}^{(i)}$ gleich eins war. Das Training bestand in der Analyse der zufällig gewählten Bildausschnitte und einer sofortigen Aktualisierung aller Wichtungen. Durch die Rückkopplung des Outputs eines Neurons auf den Input des folgenden:

$$\mathbf{x}^{(i)} = \mathbf{x}^{(i-1)} - y_i\mathbf{w}^{(i)},\ i = 1,2,...,n \tag{10}$$

wirkt sich jede Änderung der Wichtungen des Neurons i auf das Neuron $i+1$ aus.

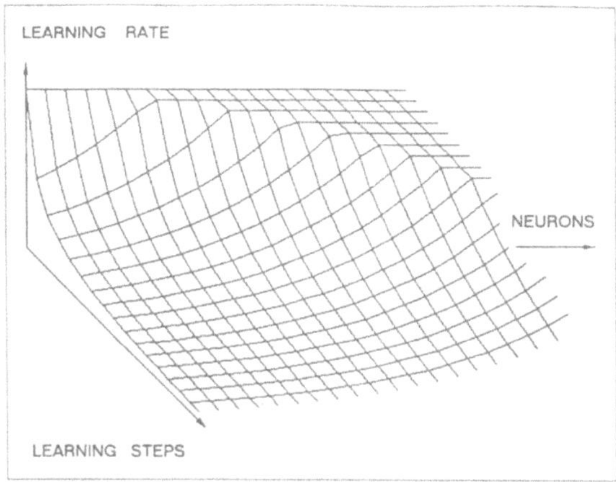

Abbildung 4: *Der kontinuierlich abfallende Lernratenverlauf für die Neuronen i.*

Daher muß sich der Wichtungsvektor $\mathbf{w}^{(i)}$ des Neurons i erst stabilisieren, bevor der Wichtungsvektor $\mathbf{w}^{(i+1)}$ des Neurons $i+1$ mit der gleichen Effektivität belehrt werden kann (siehe Abb. 4). So wird die Konvergenz der Wichtungen durch eine sich kontinuierlich verringernde Lernrate bestimmt:

$$\gamma_0(t) = \frac{a}{t+1} \qquad (11)$$

$$\gamma_i(t) = \gamma_{i-1}(t)\ v,\ i = 1,2,...,n \qquad (12)$$

Hierbei ist $\gamma = (\gamma_1, ..., \gamma_i, ..., \gamma_n)^T$ die Lernrate für die Neuronen $\mathbf{n} = (n_1, ..., n_i, ..., n_n)^T$, v der Abklingparameter, t der aktuelle Lernschritt und a der Ausgangswert. Nach zirka 50.000 Lernschritten kam es zu einer Stabilisierung der Wichtungen auch des 45. Neurons (siehe Abb. 1). Die Wichtungsänderungen unterschritten jetzt einen festgelegten Schwellwert, so daß hier der Lernvorgang abgebrochen werden konnte. Dieser Selbstorganisationsprozeß, hervorgerufen durch visuelle Stimulationen der Umwelt und resultierend in der Herausbildung komplexer rezeptiver Felder könnte funktionell mit der Bildung orientierungsselektiver rezeptiver Felder im visuellen Cortex junger Kätzchen verglichen werden [6]. Die Fähigkeit der gelernten rezeptiven Felder zur nahezu vollständigen Informationsextraktion zeigen die folgenden Abbildungen.

Abbildung 5: *Beide Grauwertszenen wurden wie angedeutet segmentartig unter Nutzung der 45 für die in Abb. 1 dargestellte Grauwertszene gelernten Hauptkomponentenarrays (rezeptive Felder) analysiert. Die Projektion der 16 × 16 Grauwertpixel eines Bildsegments entlang der 45 Hauptkomponentenarrays ergab pro Segment einen 45-dimensionalen Fitwertvektor, der nahezu die gesamte Information des analysierten Grauwertsegments beinhaltete. Die Ergebnisse der Rücktransformation der 45 Fitwerte eines Segments entlang der Hauptkomponentenarrays sind bildlich dargestellt.*

4. Intracolumnare Verarbeitung mit lokalen rezeptiven Feldern

In unserer Modellarchitektur (siehe Abb. 6) analysiert jede Verarbeitungscolumne einen definierten Bereich der *Aufmerksamkeitsregion* mit den innerhalb ihres *Analysebereiches* lokalisierten und durch den oben beschriebenen Selbstorganisationsprozeß entstandenen rezeptiven Feldern. Ein *Analysegebiet* umfaßt in unserem Modell einen Bereich von 32 × 32 Quadratpixeln. 16 × 16 Modellcolumnen analysieren parallel bei 75 %iger Überlagerung ihrer *Analysegebiete* eine *Aufmerksamkeitsregion* von 152 × 152 Grauwertpixeln innerhalb einer natürlichen Grauwertszene. Setzt man die Grauwertverläufe in natürlichen Grauwertbildern als additive Überlagerung periodischer Teilverläufe voraus, entspricht somit die Faltung der 16 × 16 Quadratpixel umfassenden rezeptiven Felder mit diesem Bereich, der Analyse einer durch das rezeptive Feld erster Ordnung $\mathbf{w}^{(1)}$ bestimmten Periode. Diese Faltung war erforderlich, da die zwei Fitwertvektoren für ein um lediglich einen Pixel verschobenes *Analysegebiet* keine Zusammenhänge erkennen ließen. Die Faltung gleicht lokale Translationen aus, indem sie für ähnliche Grauwertverläufe auch ähnliche Fitwertvektoren liefert, was jedoch mit einem Präzisionsverlust sowie einer Unumkehrbarkeit dieses Analyseschrittes einhergeht. Somit bestimmt jede der 16 × 16 Verarbeitungscolumnen die mittleren Aktivierungen aller in ihrem *Analysegebiet* innerhalb der *Aufmerksamkeitsregion* befindlichen Neuronen mit gleichen rezeptiven Feldern.

$$f_i = \frac{1}{N^2}\sum_{k=0}^{N-1}\sum_{l=0}^{N-1} \| \mathbf{w}^{(i)}\mathbf{x}_{kl} \|, \; i = 1,2,...,45 \tag{13}$$

Ein auf diese Weise für jede Verarbeitungscolumne bestimmter 45-dimensionaler Fitwertvektor $\mathbf{f}$ beschreibt die mittlere Übereinstimmung der gelernten rezeptiven Felder mit dem korrespondierenden *Analysegebiet*. Hierbei beschreibt $\mathbf{f} = (f_1, ..., f_i, ..., f_{45})^T$ den mittleren Fitwertvektor, $\mathbf{w}^{(1)}, ..., \mathbf{w}^{(i)}, ..., \mathbf{w}^{(45)}$ die zweidimensionalen Hauptkomponentenarrays oder rezeptiven Felder, $\mathbf{x}$ die Grauwertpixel des *Analysegebietes*. Die Information der 1024 Grauwertpixel eines *Analysegebietes* wird somit auf 45 Fitwerte reduziert. Die erste, den Gleichanteil beschreibende Komponente f_1 des Fitwertvektors wird für die weitere Verarbeitung unterdrückt, um eine vom lokalen mittleren Grauwert unabhängige Analyse zu ermöglichen.

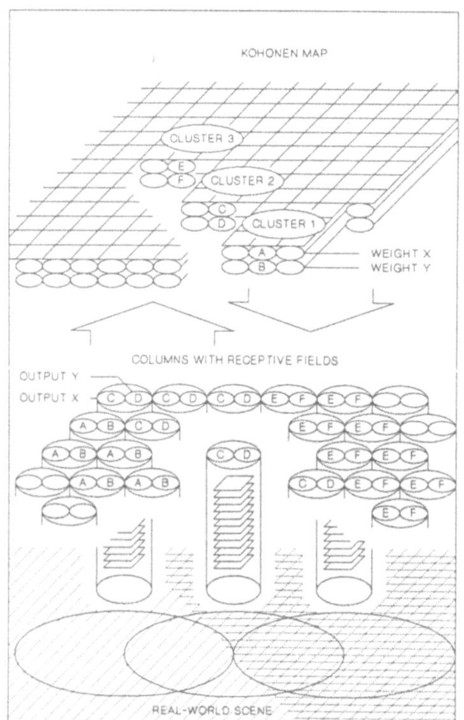

Abbildung 6: *Das verteilte multicolumnare System zur primären Analyse natürlicher Szenen, bestehend aus einem Teilsystem für die zweistufige Hauptkomponentenanalyse (siehe Text) sowie einem Teilsystem für die sequentielle Auswahl topologisch benachbarter columnarer Analyseergebnisse. AB und EF sind columnare Aktivierungen (Analyseergebnisse), die verschiedene Grauwertverteilungen (z.B. Texturen) in natürlichen visuellen Szenen beschreiben. Die Aktivierung CD beschreibt eine Grenzregion sich überlappender visueller Strukturen.*

Eine einfache Clusterung der so bestimmten 16 × 16 Fitwertvektoren innerhalb verschiedener *Aufmerksamkeitsgebiete* einer visuellen Szene zeigte, daß dies mit dem visuellen Empfinden des Menschen übereinstimmt. Weitere Untersuchungen zeigten recht bald einen prinzipiell exponentiell fallenden Verlauf für alle ermittelten Fitwertvektoren. Somit war eine zweite Hauptkomponententransformation möglich. Die jetzt 44 Fitwerte eines Vektors konnten so sogar auf 2 (x, y) pro Verarbeitungscolumne reduziert werden. Eine weitere Clusteranalyse der jetzt zweikomponentigen Fitwertvektoren zeigte, daß die Beschreibung der *Analysegebiete* immer noch repräsentativ war, um eine dem visuellen Empfinden des Menschen äquivalente Gruppierung zu erzeugen. Anschließend überträgt jede Columne ihre Fitwerte x und y auf eine zweidimensionale verteilte Kohonenkarte [2]. Alle Aktivierungen innerhalb eines *Aufmerksamkeitsgebietes* (d.h. 16 $\times$ 16 Aktivierungen (x, y)) führen innerhalb der zweidimensionalen Kohonenkarte durch Superposition zu lokalen Aktivierungshügeln. Die vorab mit (x, y)-Datenpaaren trainierte Kohonenkarte beinhaltet ein Alphabet aller möglichen natürlichen Grauwertverläufe in Form von Wichtungspaaren (x, y). Im Ergebnis interner Konkurrenz- und Selektionsmechanismen innerhalb der Kohonenkarte kann sich nur ein Cluster für eine bestimmte Zeit an einem durch die columnaren Aktivierungen definierten Ort durchsetzen, während die anderen bis zur Deaktivierung dieses Clusters unterdrückt werden. Eine topografisch korrekte Rückprojektion von der Kohonenkarte zum multicolumnaren System sorgt für eine sequentielle Aktivierung der zum jeweiligen Aktivierungshügel und damit zum jeweiligen Cluster gehörenden Verarbeitungscolumnen (siehe Abb. 7). Durch diesen rückgekoppelten Kohonen-basierten Clustermechanismus können nacheinander zusammengehörende Bildsegmente hervorgehoben und zu nachfolgenden Verarbeitungsebenen weitergeleitet werden.

5. Ergebnisse und Schlußfolgerungen

Die Nutzung einer zweifachen intracolumnaren Hauptkomponentenanalyse ermöglichte eine ähnlichkeitserhaltende Beschreibung natürlicher Grauwertverteilungen (32 $\times$ 32 Quadratpixel) mit lediglich zwei Fitwerten (x, y). Damit wurde eine topologieerhaltende Abbildung innerhalb einer zweidimen-

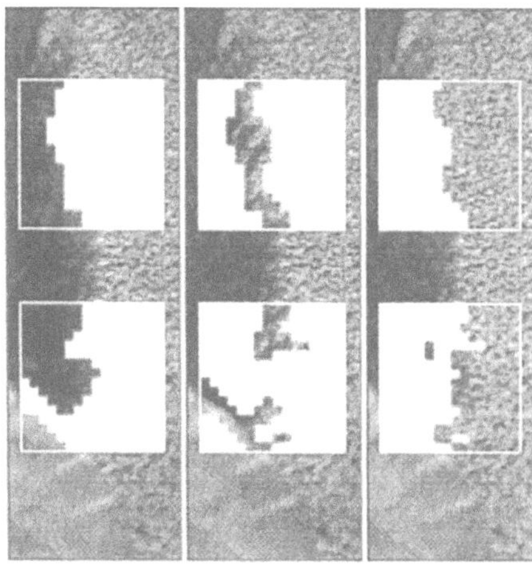

Abbildung 7: *Zwei ausgewählte Aufmerksamkeitsregionen innerhalb einer natürlichen Grauwertszene und die drei durch columnare Analyse nacheinander aktivierten Hypothesen über ähnliche Grauwertverteilungen innerhalb dieser Aufmerksamkeitsregionen.*

sionalen verteilten Kohonenkarte sinnvoll. Ein Überlapp der columnaren *Analysegebiete* von 75% führt bei ähnlichen Grauwertverläufen innerhalb der zu analysierenden *Aufmerksamkeitsregion* zu sehr kompakten Aktivierungshügeln innerhalb der Karte. Die hier vorgestellte multicolumnare Architektur wurde anhand einer umfangreichen Stichprobe natürlicher Grauwertszenen getestet. Ein typisches Klassifikationsergebnis ist in Abb. 7 dargestellt.

Zur Beurteilung der Leistungsfähigkeit des vorgestellten Verfahrens, ist es zweckmäßig festzustellen, welche Information über die ursprüngliche Grauwertszene mittels der gewonnenen Fitwerte (x, y) extrahiert werden kann. Einen ersten Eindruck vermittelt die Darstellung der Fitwerte aller über einer Grauwertszene operierenden Verarbeitungscolumnen in Form eines Grauwertbildes. Die mögliche Aktivierung einer Verarbeitungscolumne im Bereich <0.0, ..., x, ..., 1.0> wird dabei auf das Grauwertintervall <0, 255> abgebildet. Dies wird in Abb. 8 anhand zweier Grauwertszenen für

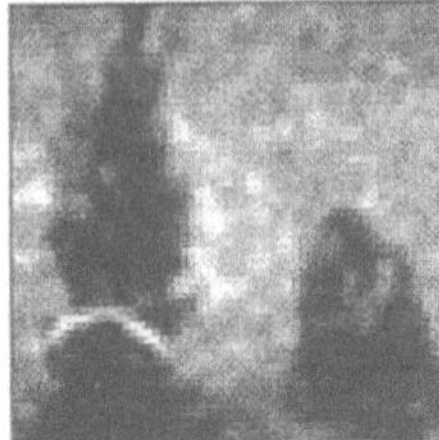

Abbildung 8: *Zwei zu analysierende Grauwertszenen und die als Grauwert dargestellten resultierenden columnaren Aktivierungen (Fitwertkomponente x) aller über der Grauwertszene operierenden Columnen. Bei einer Grauwertszene von 512 × 512 Pixeln sind dies 60 × 60 Verarbeitungscolumnen (nähere Erläuterungen im Text).*

die Aktivierungskomponente x gezeigt. Abschließend sollte erwähnt werden, daß diese Modellarchitektur auch zur Analyse von Farbszenen in speziellen Farbräumen (HSI oder $WM_{rg}M_{yb}$) geeignet erscheint, was derzeit Inhalt weiterführender Untersuchungen ist.

Literatur

[1] **Gross, H. M., Koerner, E., Boehme, H. J., Pomierski, T.** A Neural Network Hierarchy for Data and Knowledge Controlled Selective Visual Attention. In Proceedings of ICANN92, vol. 1, pp. 825 - 828, Brighton, 1992.

[2] **Kohonen, T.** Self-Organization and Associative Memory, 3rd Edition, Springer Verlag, Berlin, 1989.

[3] **Oja, E.** A Simplified Neuron Model as Principal Component Analyzer. In J. Math. Biol., vol. 15, pp. 267 - 273, 1982.

[4] **Ritter, H., Schulten, K., Martinetz, T.** Neuronale Netze, 2. erweiterte Auflage, Addison-Wesley, Bonn, 1991.

[5] **Sanger, T. D.** Optimal unsupervised learning in a single-layer linear feedforward neural network. In Neural Networks., vol. 2, pp. 459 - 473, 1981.

[6] **Singer, W., Rauschecker, J. P.** The Effects of Early Visual Experience on the Cat's Visual Cortex and their Possible Explanation by Hebb Synapses. In J. Physiol., vol. 310, pp. 215 - 239, 1981.

[7] **Szenthagothai, J.** Theorien zur Organisation und Funktion des Gehirns. In Naturwissenschaften, pp. 303 - 309, 1985.

Line-finding in 2-D and 3-D by multi-valued non-linear diffusion of feature maps

G. Gerig, G. Székely and Thomas Koller
Communication Technology Laboratory, Image Science,
ETH-Zentrum, CH-8092 Zurich, Switzerland
email: gerig@vision.ethz.ch

Abstract

The major problem of line filtering using directional second derivative operators is their multiple response, creating a relative extremum and two side-lobes of opposite sign. Furthermore, line-filters produce an asymmetric response at step-like discontinuities. Thus, the common technique of line-filtering followed by non-maximum suppression does not suffice to detect lines of different polarity and to separate lines from other types of discontinuity features. This paper presents a new approach for line detection based on the second directional derivative of the Gaussian and an inhibition of the side-lobes near lines and edges.

We propose a new hybrid technique that applies multi-valued variable conductance diffusion to multiple feature maps by coupling image intensity and matched filter output. The locally adaptive diffusion of the intensity image simultaneously guide the diffusion of the line-filter image, resulting in a suppression of the side-lobes in homogeneous regions near edges and lines. The novel technique allows the detection of lines of opposite polarity and separates. line-like from step-like discontinuity features. The technique must be considered as an experimental system and needs further exploration.

The algorithm is implemented in 2-D and 3-D. Applications to 2-D test test patterns degraded by noise illustrate the principal mechanism. The 3-D version is demonstrated with the detection of blood vessels from medical volume data.

1 Introduction

A central issue in image segmentation is the detection of local discontinuities in the image brightness function. Of particular interest are discontinuities related to boundaries between adjacent regions and to line-like features. Techniques developed to characterize the two types of features are most often referred to as "edge-detection" or "line-detection", although the variety of discontinuity profiles like steps, ramps, smooth steps, ridges, valleys, bars and mixtures thereof make a simple assignment of a particular location to be an edge– or line–feature questionable [1].

Specifically, we try to solve the problem of detecting thin, curvilinear features in 2-D and 3-D image data. The discontinuities are characterized by roof- or ridge-like profiles across and linear extensions along the features. An "optimal" operator for line and ridge detection is symmetric and can be approximated, e.g., by the second directional derivative of a Gaussian [2].

The key problem in applying second derivative operators for line detection is the interpretation of the filter response. A symmetric operator, for example a second directional derivative of a Gaussian, gives a maximum response at the location of the spine of a ridge-like structure, but additionally two side-lobes of opposite sign to its main peak located outside the structure; its response to step-like discontinuities presents a zero crossing at the location of maximum gradient and positive and negative side lobes nearby. A non-maximum suppression technique as proposed by Canny would create multiple negative and positive signals which do not allow a distinction between line features, side-lobes and edge responses.

Solutions were proposed by applying filter pairs, for example cosine- and sine-Gabor filters [3] [4] or a filter and its Hilbert transform [5]. The squared sum of components represents the local energy measure, whereas the phase between the dual filter responses at the locations of energy maxima can be used to decide about the type of features. Own tests showed that the use of the energy representation is less sensitive to small structures than the use of the individual odd and even filters, and that the phase information is prone to errors introduced by interferences of neighboring structures and by noise.

A linear filter like the second directional derivative of a Gaussian can be easily extended to 3-D. If one could find a solution to suppress the response of side-lobes and edges while preserving the line-response, Canny line-filtering could represent a useful and efficient technique to detect curvilinear structures in 3-D images.

The study of non-linear diffusion processes, introduced by Grossberg [6] and Perona [7], forms the basis of our new image-enhancing and feature extracting strategies. Recent developments focus on the interaction between geometry and conductivity by controlling the conductivity with higher order geometric properties [8, 9].

We propose a new technique based on a multi-valued nonlinear diffusion. The intensity image and the image filtered with a second directional derivative of a Gaussian represent two layers in a coupled system. The diffusion process, adaptively controlled by the intensity image, inhibits the undesired responses in the line filtered image.

2 Feature detection by matched filtering

Common solutions for optimal filtering to enhance structural features can be found in digital signal processing. The most appropriate filters often match the characteristics of the structures themselves. A typical analysis by Canny [10] [2] presented an "optimal" operator for finding ridge and roof profiles. The symmetric operator can be represented as the second derivative of a function of finite extent. Canny proposed to use a *directional second derivative of a Gaussian* to approximate the optimal operator. Extended to three dimensions, a filter for the detection of line features can be designed as an isotropic 3-D Gaussian filter summing second derivatives from two orthogonal directions (Laplacian of Gaussian). Figure 1 bottom right illustrates the 3-D iso-density surface of a line-filter. The ellipsoidal central part is weighted negatively as opposed to the positive torus volume, calculating the difference between a lineal structure and its background. Along the filter direction, voxels are integrated with a Gaussian mask, whereas second derivatives become active orthogonal to the filter direction.

3 Multi-valued nonlinear diffusion of feature maps

Perona and Malik [11, 7] developed a nonlinear smoothing and edge detection scheme which became an important new concept for multi-scale image processing. Their *anisotropic*

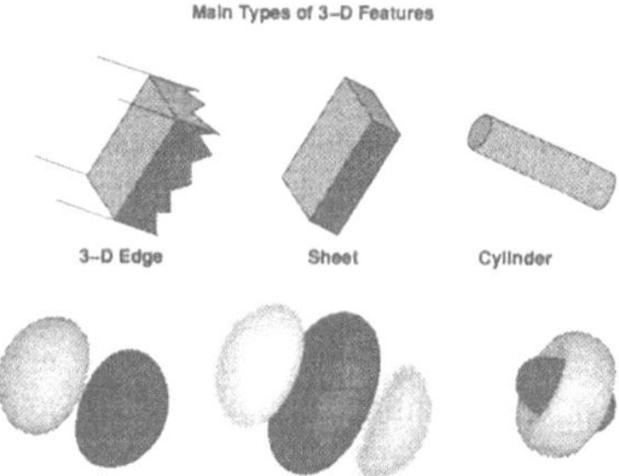

Figure 1: Detection of 3-D features by matched filters: The 3-D features edge, sheet and cylinder can be detected by applying directional derivatives of a 3-D Gaussian. A first directional derivative matches a 3-D edge (lower left), a second directional derivative a sheet (middle) and the sum of two orthogonal second derivatives a tube (right).

diffusion filtering method is mathematically formulated as a diffusion process and encourages intra-region smoothing in preference to smoothing across the boundaries: $I_t(\bar{x}, t) = div\,(c(\nabla I)\nabla I(\bar{x}, t))$. The non-linear diffusion process can be generalized to be applied to multi-valued functions. Vector-valued functions are processed using a system of coupled diffusion equations. In an earlier project, we have developed a diffusion scheme that enhances the multiple channels of vector-valued image data by a simultaneous diffusion [12]. The diffusion scheme shares a common conductance function $c_{com} = c(f_1 \ldots f_n)$ which considers discontinuities in multiple channels, thus leading to a better discrimination of low-contrast borders.

It becomes interesting to apply a multi-valued diffusion scheme not only to intensity images but to a set of different geometric features. A nonlinear scale-space which evolves by including information about the geometry could reflect meaningful structures on different scales [9]. Geometry can either be characterized by the development of separate feature maps out of a single one (as shown by Shah [13], for example) or by applying multi-valued diffusion to the output of filtered images. The following discussion describes a new experimental diffusion system that integrates intensity features and second derivative filtering.

Our new method was mainly inspired by the work of Whitaker [9] on geometry-driven diffusion. Multi-valued functions can be processed by a system of coupled diffusion equations that share common conductance coefficients.

Figure 2 explains the basic idea of suppressing side-lobes. The left figure shows an idealized 1-D line-edge profile (solid line) together with the filter response (dotted line) of the appropriately scaled second derivative of Gaussian kernel. The response shows a local maximum at the middle of the line-edge, while the two local minima could be interpreted as the response of the filter to (locally step-like) edges at each side of the line. The right figure illustrates a situation where the high intensity region is too wide to be detected as a line by the selected filter scale. At this scale, the correct interpretation is the presence of two step edges with opposite polarity. Each step-edge is represented by two local extrema (one maximum and one minimum). Both of the extrema are shifted away from the step-edges locations into homogeneous image regions.

Our approach uses the observation that line-responses occur *within* the line feature whereas edge-responses are always found within homogeneous regions *apart from* the

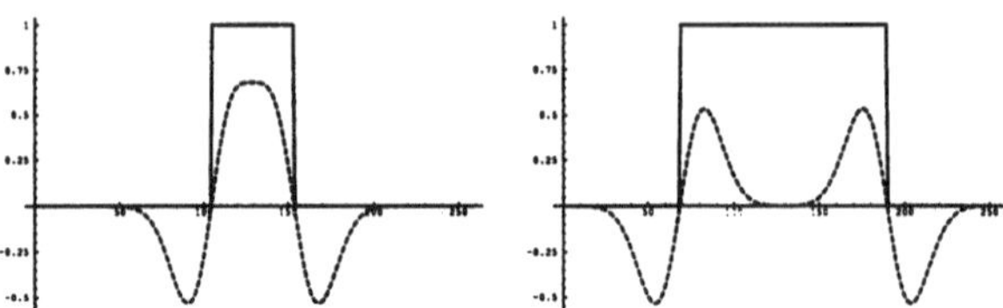

Figure 2: Filter response of an appropriately scaled second derivative of Gaussian kernel to idealized edges. The left hand side shows a line edge (solid line) with overlay of the filter response (dotted line). The right hand side shows two step-edges of opposite polarity with the response of the same filter kernel.

feature. Variable conductance diffusion of a function f does stop diffusion at locations of large discontinuities of f and let diffusion run within regions of relatively small ∇f. If f is chosen to be the image intensity function, the diffusion performs intra-region smoothing while enhancing borders between regions. Figure 3c illustrate this behavior leading to a restoration of the original image signal. In homogeneous regions near edge- and line-features where the second derivative presents edge-responses, diffusion of the original signal takes place. This observation is the key for our new experimental system. If we *apply diffusion to the second-derivative signal but control the conductances by the intensity image*, the undesired edge-responses (side-lobes) would be diffused.

The processing scheme can be described as a double-layer nonlinear diffusion sharing common conductance coefficients, as expressed in equation 1.

$$\begin{aligned} \nabla \cdot c(\|\nabla f_1\|)\nabla f_1 &= \frac{\partial f_1}{\partial t} \quad | \; f_1 : \text{intensity image} \\ &\vdots \\ \nabla \cdot c(\|\nabla f_1\|)\nabla f_2 &= \frac{\partial f_2}{\partial t} \quad | \; f_2 : \text{second derivative of the Gaussian} \end{aligned} \qquad (1)$$

As Figure 3 **b** shows, the second derivative presents side-lobes of opposite polarity to the main peak in a region outside the line discontinuities, at locations where diffusion of image intensity occurs. Iterative application of the nonlinear diffusion restores the piece-wise homogeneous original function while smoothing the side-lobes of line- and edge-response of the second derivative simultaneously (compare Figures 3 **b** and **d**). The edge response is not completely suppressed, since diffusion is blocked at locations of maximum gradient. However, the relative maxima and minima near the edges are changed into a zero-crossing of significantly lower signal but higher slopes. Lines of both polarities are preserved.

4 Implementation

The line detection scheme presented herein is based on a simultaneous filtering with a set of directional filters and a local integration of the multiple outputs. The test implementation is based on filtering in the spatial domain. We applied differentiation to the Gaussian smoothed image and used the separability of the Gaussian to obtain an efficient implementation. The 3-D line-filter is designed by the sum of two-orthogonal second derivatives of the 3-D Gaussian (see Figure 1 bottom right).

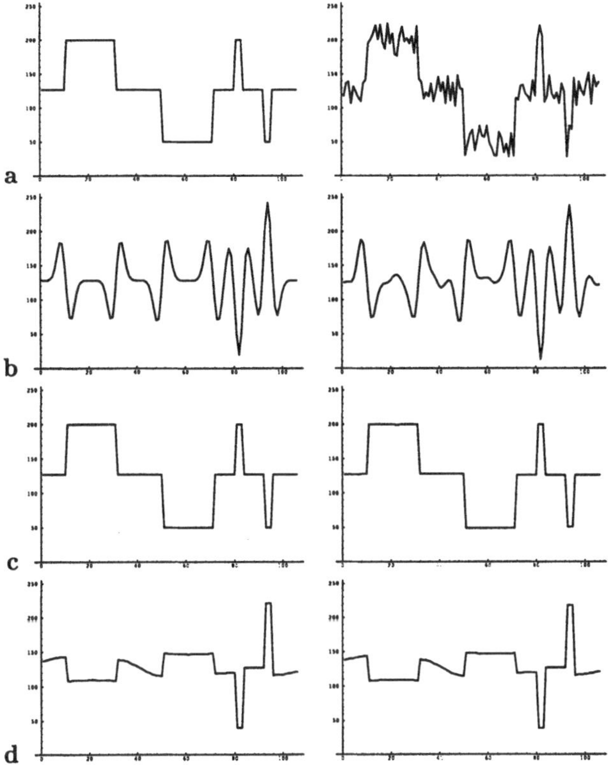

Figure 3: Multi-valued diffusion applied to one-dimensional test functions. The left and right column represent profiles of the original and the noisy test patterns, respectively. Figures **a** to **d** top to bottom: Original intensity function (**a**), second derivative of Gaussian filtering (**b**), intensity function after nonlinear diffusion (**c**), second derivative of Gaussian after multi-valued nonlinear diffusion (**d**).

The filtering scheme consists of the following steps:

1. Convolution with isotropic Gaussian filter of width σ
2. Calculation of second directional derivatives
3. [1]Half-wave rectification (discrimination between dark and bright features)
4. Integration of multiple filter channels by taking the maximum filter response at each pixel location

Figure 4 left column illustrates the results of the 3-D line filtering (c) applied to medical volume data (a). The strong filter response not only at locations of lines but also near step-like discontinuities makes clear that simple thresholding or detection of relative maxima does not suffice for finding line structures.

[1]Half-wave rectification is only applied if lines of one type of polarity have to be extracted. However, it is not a necessary component of the multi-valued diffusion scheme.

Variable conductance diffusion requires estimates of local gradients and the calculation of the conductances. The time required for calculation at each pixel or voxel, respectively, can be reduced by considering precalculated flow-tables as look-up-tables (see [12] for a detailed discussion).

5 Applications

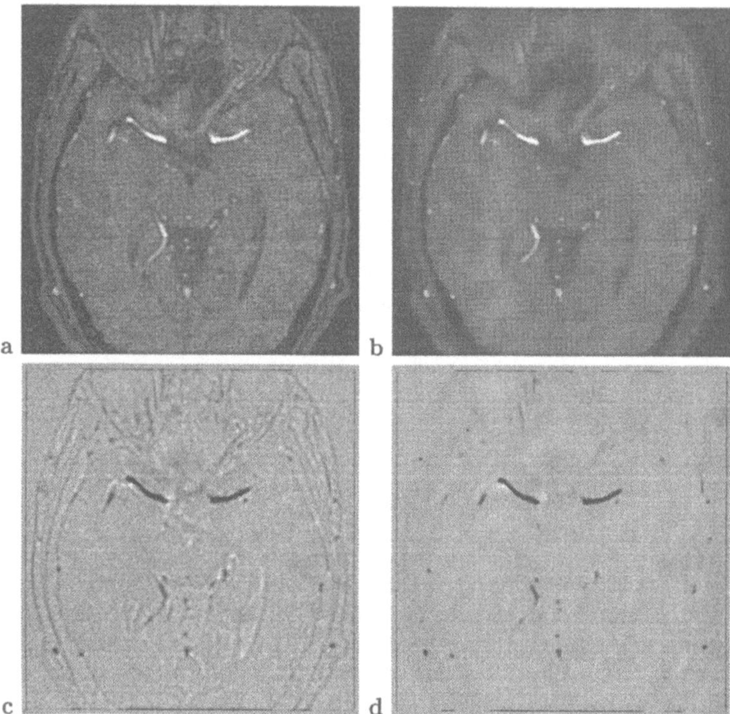

Figure 4: Coupled diffusion simultaneously applied to 3-D volume (a) and filtered (c) data. a) Axial slice through magnetic resonance angiography (MRA) volume data enhancing blood vessels as bright line-like features. b) Variable conductance diffusion applied to image a (10 iterations, parameter Kappa=6.5). c) Filtered volume data (line-filter based on directional second derivatives of Gaussian). d) Diffused filtered data where the conductivity is controlled by the intensity image. Important line-response of matched filtering is preserved while edge-response is suppressed.

The multi-valued diffusion scheme is implemented in 2-D and 3-D and applied to very different types of scenes taken from robot vision and medicine. Herein, a medical application will be presented. Magnetic resonance angiography (MRA) is a volume acquisition technique that suppresses static tissue while enhancing flowing substance. Blood vessels appear as bright lines within low-contrast soft tissue (Figure 4a). A typical data set consists of 256x256x100 voxels with a voxel dimension of approximately $1mm^3$. 3-D line-filtering with directional masks as described previously is applied to the volume data ($\sigma = 1.2$). We use the a priori knowledge that vessels always appear as bright line-features on a darker background. This allows us to apply a half-wave rectification of only the negative filter responses. An axial cut through the filtered volume data is shown in Figure 4c). The edge-response of the 3-D line filter is clearly visible as dark lines along region boundaries. Multi-valued diffusion of the intensity image performs intra-region smoothing while preserving high-contrast line structures (Fig. 4b). The second derivative filter output gets a significant change due to the smoothing of the edge-responses (Fig. 4d). Only the significant line-structures remain. Surface rendering of the segmented cerebral blood vessels in comparison to maximum intensity projection is shown in Figure 5.

6 Discussion and Conclusions

The new technique combining multi-valued diffusion and matched filtering is designed to preserve the filter signal at locations of the sought features but suppresses its multiple responses. The algorithm must be considered as an experimental system which still needs a careful evaluation of the robustness and the creation of possible artifacts. Our experience with different kind of scenes from robot vision and from medicine demonstrated the interesting properties of multi-valued diffusion.

The fact that the edge-response doesn't vanish completely but is transformed into a zero-crossing with increased slope must be studied more carefully (Fig. 3 **c**). Further, the implementation of 3-D line filters as described here still needs to be improved. An new implementation will be based on a set of orthogonal second derivative filters that would allow interpolation of response in arbitrary directions [14].

Variable conductance diffusion is applied for the purpose of feature extraction rather than for image restoration. Because diffusion of the matched filter output is driven by the boundary features of the intensity images, one can characterize the proposed system as a kind of *geometry-driven diffusion.* Recent work by Whitaker [9] and groups at other places clearly demonstrate that geometry-driven, nonlinear diffusion could represent a powerful new concept for multi-scale low-level image processing.

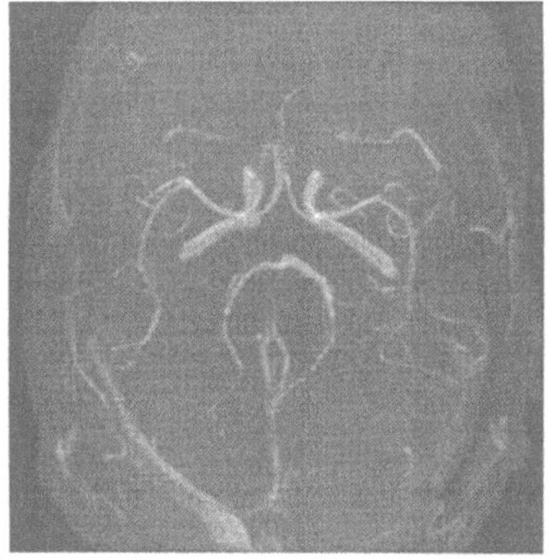
a

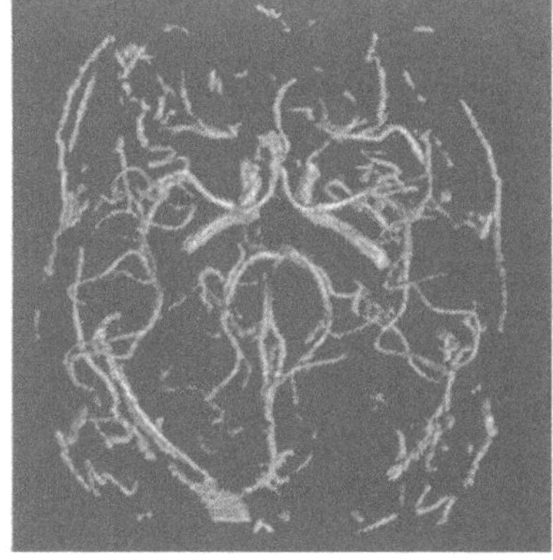
b

Figure 5: Maximum intensity projection (MIP) in comparison to 3-D rendering of the segmented cerebral vascular system (segmented by 3-D line filtering and multi-valued diffusion). Image **a** illustrates an axial MIP of the original MRA volume data. Figure **b** represents a 3-D surface rendering of the complete set of segmented structures.

A multi-scale analysis has not been considered yet. Scale comes in by variable conduction diffusion which can be considered as an image-induced representation at a continuum of scales. Lines of different width, however, must be segmented by line filters at different scales. One could consider a coupling of a multi-scale nonlinear diffusion (proposed by Whitaker [8] and by Alvarez et al. [15]) with a multi-scale second derivative of Gaussian filtering.

References

[1] P. Perona and J. Malik. Detecting and localizing edges composed of steps, peaks and roofs. In *Proc. Third Int. Conf. of Computer Vision ICCV'90*, pages 52–57. IEEE Computer Society, 1990.

[2] J.F. Canny. A computational approach to edge detection. *IEEE Transactions on Pattern Analysis and Machine Intelligence*, 8(6):679–698, 1986.

[3] E.H. Adelson and J.R. Bergen. Spatio-temporal energy models for the perception of motion. *Journal of the Optical Society of America*, A(2):284–299, 1985.

[4] M.C. Morrone and D.C. Burr. Feature detection in human vision: A phase-dependent energy model. In *Proceedings of the Royal Society of London*, number 235 in B, pages 221–245, 1988.

[5] P. Perona. Deformable kernels for early vision. Technical Report MIT-LIDS-P-2039, Caltec, Pasadena, October 1991.

[6] S. Grossberg. Neural dynamics of brightness perception: Features, boundaries, diffusion and resonance. *Perception and Psychophysics*, 36(5):428–456, 1984.

[7] P. Perona and J. Malik. Scale-space and edge detection using anisotropic diffusion. *IEEE Transactions on Pattern Analysis and Machine Intelligence*, 12(7):629–639, July 1990.

[8] R.T. Whitaker and S.M. Pizer. A multi-scale approach to nonuniform diffusion. *Computer Vision, Graphics, and Image Processing: Image Understanding*, 57(1):99–110, January 1992.

[9] R.T. Whitaker. Geometry-limited diffusion in the characterization of geometric patches in images. *Computer Vision, Graphics, and Image Processing: Image Understanding*, 1(1):111–120, January 1992.

[10] J.F. Canny. Finding edges and lines in images. Technical Report 720, MIT Artificial Intelligence Laboratory, Dept. of Electrical Engineering and Computer Science, Cambridge, MA, 1983.

[11] P. Perona and J. Malik. Scale space and edge detection using anisotropic diffusion. *Proceedings of IEEE Workshop on Computer Vision, Miami*, pages 16–22, November 1987.

[12] R. Kikinis G. Gerig, O. Kübler and F.A. Jolesz. Nonlinear anisotropic filtering of MRI data. *IEEE Transactions on Medical Imaging*, 11(2):221–232, June 1992.

[13] J. Shah. Segmentation by nonlinear diffusion. In *Proc. Conf. on Computer Vision and Pattern Recognition CVPR'91*, pages 202–207, 1991.

[14] J.J. Koenderink and A.J. van Doorn. Receptive field families (generic neighborhood operators). *Biol. Cybern.*, 63:291–297, 1990.

[15] L. Alvarez, P-L. Lions, and J-M Morel. Image selective smoothing and edge detection by nonlinear diffusion. ii. *SIAM J. Numer. Anal.*, 29(3):845–866, June 1992.

A Fast Hybrid Color Segmentation Method

Lutz Priese and Volker Rehrmann

Institut für Informatik
Universität Koblenz-Landau
Rheinau 1
D-56075 Koblenz

Abstract. We introduce a very general method of achieving stable and fast color segmentation. This method works on different hierarchical data structures and combines local bottom-up region growing segmentation with top-down separation techniques. As our method supports exploitation of inherent parallelism, a real-time object detection has been designed and is in the implementation phase.

1 Introduction

Image segmentation is an important step towards an object detection in image analysis. In the literature several major methods for segmentation are distinguished [4]. Common are *edge-detection, region-growing* and *clustering* techniques. Whilst clustering uses mainly statistical methods, syntactical methods are more popular for edge-detection and region-growing techniques. Region-growing methods are usually classified as *local, global* or *splitting-and-merging* techniques. Local techniques are simple and fast, but have the problem of chaining: two very dissimilar pixels may be connected by a chain of similar pixels and thus are mismatched into one region. Global techniques solve this problem by using $k \times k$-neighborhoods with sufficiently large parameters k. However, one has to pay the price of an increase in required computation time compared to local techniques. As a consequence, global techniques play no role in real-time or only close-to-real-time applications for color vision. The clustering methods [1] also demand high computational costs that lead (at the moment) to computation times far away from real time [3].
As our aim was the development of a real-time object detection system utilizing color information we had to develop a new method combining the advantages of local (simplicity and quickness) and global (robustness and accuracy) techniques. Therefore we introduce some hierarchical structure. At the lowest level we follow local region-growing and obtain connected regions of a small size. These connected regions are treated as pixels at higher levels. We thus result in some *hybrid* approach: by bottom-up local region growing and repeating distance measurements at higher levels we stay simple and fast but nevertheless detect chaining mismatches at some level, n. These mismatches are now solved using the global aggregated information of the connected regions up to this level n. Furthermore this method supports an implementation with some parallel programming

language with distributed data on a MIMD architecture. This will help to implement very fast versions. To result in a fast system we nevertheless have to take care that only few chaining mismatches occur at higher levels. Therefore we also spend some effort on a good preprocessing, involving adequate edge-preserving and single pixel error repairing filters. To get good color distance measurements we investigated several color spaces. Most reasonable results were found in the HSV model.

The new hybrid method was applied successfully in implementations of two image analysing systems, namely the CSC and the dpe. Experiments with images of natural scenes led to surprisingly good results. At the moment we are developing a real-time version for traffic-sign recognition in a vehicle on a transputer network (TIP, transputer image processing system)[1].

2 The Method: Hierarchical Segmentation

2.1 Hierarchical structure of islands

We now explain the principles of this hybrid, hierarchical segmentation method in some detail.

A basic concept in this approach is that of an **island**. An *island of level* 0 simply denotes a single pixel. An *island of level* 1 is a set of pixels, usually within some simple geometrical shape. An *island of level* $n+1$ is a set of islands of level n following the same geometrical shape. Figure 1 presents some examples. Note that the squared and triangular islands of example 1 and 2 overlay an orthogonal grid of pixels (the standard camera topology), whilst example 3 needs a different grid of pixels, where the euclidian distance between any two neighboring pixels is constant, the so-called hexagonal topology. Islands of level n may overlap to form an island of level $n+1$ (Figure 1,c) or not (Figure 1,a,b).

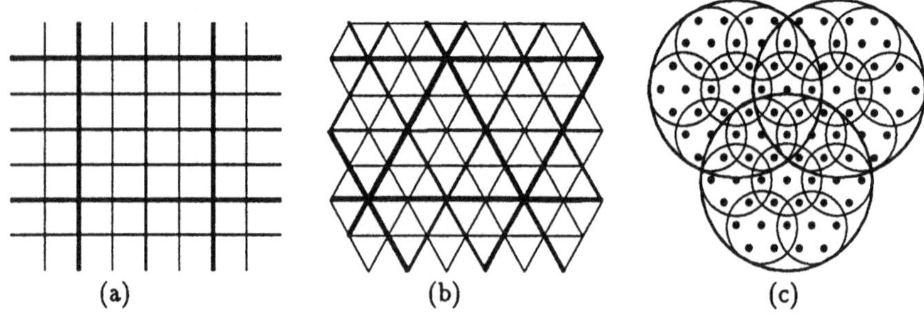

Fig. 1. Islands of three different levels in hierarchical structures based on an orthogonal, triangular and hexagonal topology, respectively.

[1] This traffic sign recognition is supported by Daimler-Benz AG and the BMFT within the European Prometheus project.

The islands, $I_1, \cdots, I_k$, of level n that form an island, I, of level $n+1$ are called the **sub-islands** of I. We usually work with homogeneous structures, where any island of the same level is equally shaped and consists of the same number of sub-islands, and the same construction for islands by sub-islands holds at any level. The number of sub-islands forming one island in such a homogeneous structure is called its *degree*. Note, that this concept of islands and levels is a pure topological concept independent of the content of the image under analysis. To handle this image information we introduce the concept of **regions of level** n:
A *region of level* 0 simply denotes a single pixel.
A *region of level* 1 is a set of connected pixels within one island of level 1.
A *region of level* $n+1$ is a set of connected regions of level n within one island of level $n+1$. All regions, $r_1, \cdots, r_s$ of level n that are connected to a region, r, of level $n+1$ are called the **sub-regions** of r. Note that s may exceed the degree of a structure, see figure 2.

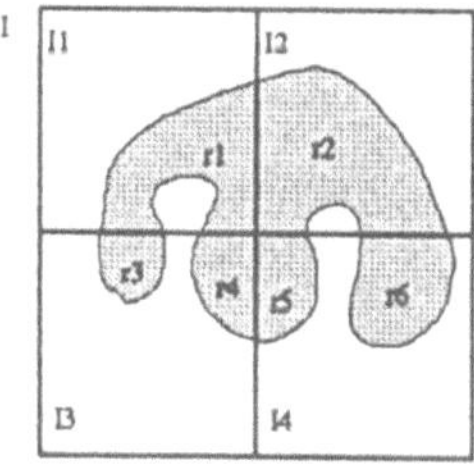

Fig. 2. A region r with 6 sub-regions in a degree 4 structure.

Regions that never find a partner region to be connected to a region of some higher level are called *complete*. A pointer structure in both directions (upward and downward) is a canonical data structure for regions. The complete regions form the segmentation.

2.2 Segmentation in island structures

One phase in a segmentation within island structures thus becomes the process to determine which regions of level n will be connected into a region of level $n+1$. Note that any region, r, of level $n+1$ must be a set of sub-regions within a single island, I, of level $n+1$. Thus only regions within the sub-islands, $I_1, \cdots, I_k$, of I are candidates to be connected. Of course, local region growing methods as described in the introduction are within this scope: the decision as to whether two neighboring regions r_1, r_2 will be connected depends solely on the presence of two pixels $p_1 \in r_1$, $p_2 \in r_2$ that have been connected on lowest level, where only the distance of the color or gray-level contents of p_1 and p_2 are measured. But this may lead to the unwanted chaining mismatches. Our hierarchical structures obviously indicate a way of avoiding these mismatches. In order to decide whether two neighbored regions r_1, r_2 become connected, we measure the distance of the *color content* $c(r_1)$ and $c(r_2)$ of r_1 and r_2 and decide according to this measurement.

There are various ways of defining a color content $c(r)$ of a region r, of measuring the distance of such color contents, and of evaluating this measurement. Let $c(p)$ denote the color content of a pixel p. As any $c(p)$-value of any pixel p of a region r has been used somewhere in the process of building r one may simply define $c(r)$ to be the *mean* of the color contents $c(p)$ of all pixels within r. However, further $c(r)$ definitions may be used, representing some color texture, e. g. Let C denote the set of all possible color contents of regions and pixels. Usually one regards metrices, d, on C and defines the distance of two color content values c_1, $c_2 \in C$ by $d(c_1, c_2)$. The decision whether r_1 and r_2 will be connected now may depend on whether $d(c(r_1), c(r_2)) < t$ for some threshold t. As $c(r)$ contains more information than $c(r')$ for a region r' of lower level, one may use different thresholds, t, for different levels.

There is a second approach disregarding metric spaces. Simply regard the decision, whether r_1 and r_2 will be connected, as a predicate $D(c(r_1), c(r_2))$. As all color models for practical purposes (like the RGB,YIQ, HSV, HSI spaces, etc.) are 3-dimensional, D thus becomes a 6-dimensional predicate. Instead of using metrices we prefer a navigation in these six-dimensional spaces, to compute whether D holds, as a general principle. This approach is also homogeneous as the location of r_1, r_2 within the image, i.e. their $x - y$-coordinates, play no role for D, only their C-values. As the level n may influence D, D becomes a 7-dimensional predicate $D(c(r_1), c(r_2), n)$.

One point is still missing in these simple considerations. We avoid chaining mismatches by refusing to connect two regions r_1 in I_1 and r_2 in I_2 into some region r of I (with sub-islands I_1, I_2) of level $n+1$, according to $D(c(r_1), c(r_2), n)$. Thus the borderline of I_1 and I_2 will define the border between r_1 and r_2. As the island structure is an artificial one, independent of the contents of an image, this may lead to possibly large artificial boundaries. As such results are a main criticism of split-and-merge methods, we have to add more intelligence. Note, that one should only compute $D(c(r_1), c(r_2), n)$ for two regions r_1, r_2 that are **candidates** for becoming connected. One simple criterion for being candidates is of course that r_1 and r_2 have a common boundary (of at least one pixel). As a second criterion one should use r_1 and r_2 linked by a chain $p_1, \cdots, p_l$ of pixels that are pairwise connected due to $D(c(p_i), c(p_{i+1}), 0)$ for $1 \leq i < l$, $p_1 \in r_1$, $p_l \in r_2$. If two candidates r_1 and r_2 are not similar according to $D(c(r_1), c(r_2), n)$, a mismatch of their aggregated color values $c(r_1)$ and $c(r_2)$ is detected despite a linking chain between r_1 and r_2. Of course, to be candidates implies that r_1 and r_2 belong to different neighboring sub-islands I_1, I_2 of level n of some island I of level $n + 1$. To avoid the artificial boundary between I_1 and I_2 as a part of the borderline b between r_1 and r_2 we try to rearrange r_1

and r_2 close to b. Up to now we have a lot of information about the color values $c(r_1), c(r_2)$ and $c(r_1^i), c(r_2^i)$, where $r_1^i(r_2^i)$ are sub-regions of $r_1(r_2, respectively)$ close to b. This information may be used to design fast methods to rearrange r_1, r_2 in a neighborhood of their common borderline b.

3 CSC

This rather general method has been implemented in our *CSC* (Color Structure Code) that is briefly described now. The *CSC* is related to Hartmann's *HSC*. Hartmann and his group have developed the HSC during the last decade. HSC stands for a hierarchical structure code. It operates on the hierarchical hexagonal island structure as shown in Figure 1,c. Based on grey-level images the HSC detects and analyses mainly three kinds of subjects: edges, lines and parts of areas near edges. We distinguish four phases within an HSC system: preprocessing, detection, linking and evaluation. For details see [5].
In our *CSC* we also use the hexagonal hierarchical island structure and the ideas of a preprocessing, detection, linking and evaluation phase. Instead of edges, lines and parts of areas we now operate with color regions.
In the preprocessing phase we use symmetric-nearest-neighbor (s-n-n) filters with good edge-preserving properties, see [8]. As s-n-n filters are non-linear and a special purpose hardware for these filters is not available yet, we had to introduce a simplified version that is much faster but results in comparably good filtered images. This simplified s-n-n filter operates as follows. Group all neighbors $p_1, \cdots, p_6$, of a pixel, p, into three pairs $(p_1', p_2'), \cdots, (p_5', p_6')$, of opposite pixels and mark the pixel p_{2i-1}' or p_{2i}', of any group i that is "nearer" to p with respect to its color value. Now substitute the color value of p by the mean of the color values of all marked pixels. To compute the nearest pixel of p_{2i-1}', p_{2i}' to p a histogram method is applied in [8]. Here we use simplified methods to define "nearer" in different color spaces.
In the detection phase only the 7 pixels within one island I of level 1 determine the color connected regions within this island. At least two pixels are required to form a region. Thus, at most 3 regions may be found within one island I of level 1. Note that the overlapping structure (see Figure 1,c) allows us to ignore pixels outside I, as any border pixel of I is also a border pixel within a neighboring island I' of I.
In the linking phase those regions of level n are aggregated to regions of level $n+1$ as described before. However, we also apply the region splitting and border smoothing algorithms mentioned. Thus a *linking-and-splitting* phase might be a better name. Although it might not be obvious, the hexagonal topology is helpful in this phase. Let us explain this fact by a simple example, see Figure 3.
Let I^1, I^2 be two neighboring islands of some level $n+1$, let I be the common sub-island of I^1 and I^2. r_1 denotes a region within I^1, r_2 a region within I^2. Let us start with r_1 trying to find all its partner regions of level n to be linked into one region, r, of level $n+1$. r_1 and r_2 are coded up to now by a double-linked pointer structure. Thus one simply descends r_1 into an outer sub-island of I^1,

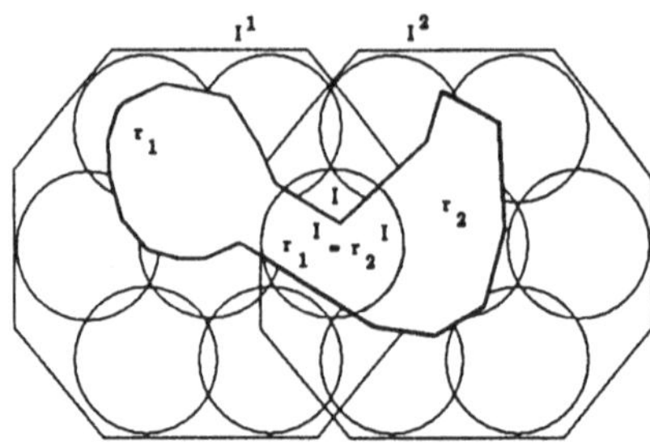

Fig. 3. Two neighbored islands I^1, I^2 with a common sub-island I and the common sub-region r_1^I of r_1 and r_2 in I.

as I, e.g.. Let r_1^I denote such a son of r_1 in I. Note that exactly this son is also a sub-region r_2^I of r_2. As we store into any son also a pointer to its fathers, we simply have to ascend from r_1^I to its second father, r_2. Thus r_2 is found from r_1 by a simple navigation in the double pointer structure. This resembles a bootstrapping procedure and works very well in practice. Only from level 1 to 2 a search is required as a kind of "induction beginning" to set the required pointers for a first time. But at this level only very few regions (1.5 on average) are candidates so that this search is trivial.

In an analogous manner the overlapping is very helpful in descending the border regions in the border smoothing phase of a region splitting algorithm (if two found candidates r_1, r_2 mismatch, i.e. $D(c(r_1), c(r_2), n) =$ false, in order to compute a smooth boundary between r_1 and r_2). The linking of r_1 and r_2 in the overlapping hexagonal island structure means that both have a common sub-region s (s $= r_1^I = r_2^I$, see Fig. 3). Hence for the splitting of r_1 and r_2 s has to be divided between them. Therefore the pointer to s at the region r_i (i = 1,2) with the less fitting color will be deleted. As s was a sub-region of r_i, s is still linked with other sub-regions of r_i. So s has to be splitted from those recursively. With this method a new well fitting borderline between r_1 and r_2 is found, as a lot of experiments show.

Altogether, the CSC provides a forest of objects. An object, i.e. a color connected region, becomes a tree. Its root describes roughly its size, location and mean color plus pointers to its sub-regions on one level less. At each node of this tree the information about the mean color, size and location of the underlying object plus all pointers to its sub-regions is found.

Of special importance is the color distance function, which determines the linking or splitting of regions. A lot of experiments in RGB and HSV [2] color spaces proved that a non-linear locus dependent color distance function in the HSV space leads to good segmentations. This function follows the theory of human color perception and color metric (e.g. the darker a color the greater is the allowed difference in hue).

The quality of the linked regions determines strategies for an evaluation of the

CSC. Thus, the better the segmentation, the easier the evaluation becomes. Here one may use standard techniques from AI. In the following chapter we will describe this evaluation with an example.

4 Results

Obviously, the techniques in the evaluation phase depend on the given task and cannot be described abstractly. However, the better a segmentation one achieves, the easier this task becomes. We will present one example.

In co-operation with Daimler-Benz AG an experiment has been started to develop a real-time traffic sign recognition system (TSR). We will provide a traffic sign classificator based on the CSC. Our main idea is very simple. We apply a CSC to aggregate all color connected regions into objects represented by trees. From the CSC data forest we compute the shape of all red, blue and white objects and decide whether their forms become (more or less) circles, segments, triangles, rectangles, arrows etc. Using information on included and/or neighboring objects provides a surprisingly fast and robust detection, location and pre-classification of all traffic signs in an image. A pictogramm recognizer (for the few remaining pictogramms of the preclassified traffic signs) now leads to an identification.

However, everybody who has ever worked in image recognition will note at once a lot of difficult problems. A sign may be in full sunlight or in a shady place, or a shadow may run across the sign. Parts of signs may be covered. Solving this problem costs a lot of time and no real time traffic sign recognition has been possible up to now. Because of the good quality of our segmentation we get most regions of a common color (e.g. no matter whether they are in a shadow or in bright sun, at any time of the day, even with a shadow running across them) as single objects with always the same algorithm (no parameters have been changed for the special circumstances). To evaluate a 512 $\times$ 512 pixel color image the detection and linking-and-splitting phase costs at the moment 4 sec on a SUN Sparc 10. In this time a complete segmentation is computed, i.e. all regions of any color are represented as single objects (a tree). This is done independently of the intended task (TSR, here). The task is defined solely in the evaluation phase. The time for the evaluation phase to find e. g. **all** objects of a predefined color and a predefined form within some set of allowed forms (as circles, segments, triangles, squares, etc.) is within a few milliseconds. The preprocessing phase also requires 2 sec, mainly to compute an s-n-n filtered image. As our algorithms have been designed to run on a parallel machine with an excellent speed-up (as has already been checked for the HSC, see[6]), we will implement a TSR on a TIP-system (**T**ransputer **I**mage **P**rocessing system) with 8-16 T9000 transputers. As a single T9000 allows for 150 Mips we thus expect a software solution for a TSR to recognize all traffic signs in natural scenes in approximately 200 ms.

5 Conclusion

We have introduced a new hierarchical color segmentation method that combines the advantages of local (simplicity and quickness) and global region growing methods (robustness and accuracy). This method has been implemented in the image recognition system CSC. We are not aware of any color segmentation that yields results of such a good stability in comparable time. With further methods (of shape analysis of our detected objects) we have applied the CSC to detect traffic signs in natural scenes. This traffic sign recognition (TSR) is very stable and fast and becomes implemented in a parallel version on a transputer image processing machine. Thus, for the middle of 1994 we expect a software solution for TSR on existing machines to detect and identify traffic signs in less than 200 ms. However, this is only a first application of our method.

6 References

1. M. Celenk, A Color Clustering Technique for Image Segmentation, *Computer Vision, Graphics and Image Processing*, **52**, 1990, 145-170.
2. J. D. Foley, A. van Dam, S. K. Feiner, J. F. Hughes, Computer Graphics, *Addison-Wesley*, 1990.
3. R. Gershon, Aspects of Perception and Computation in Color Vision, *Computer Vision, Graphics and Image Processing*, **32**, 1985, 244-277.
4. R.M. Haralick and L.G. Shapiro, Survey: Image Segmentation Techniques, *Computer Vision, Graphics and Image Processing*, **29**, 1985, 100-132.
5. G. Hartmann, Recognition of Hierarchically Encoded Images by Technical and Biological Systems, *Biological Cybernetics*, **57**, 73-84, Springer, 1987.
6. L. Priese, V. Rehrmann and U. Schwolle, An Asynchronous Parallel Algorithm For Picture Recognition Based On A Transputer Net, *Journal of Microcomputer Applications*, **13**, 57-67, Academic Press, 1990.
7. L. Priese and V. Rehrmann, A Fast Hybrid Color Segmentation Method, *Fachberichte Informatik*, Universität Koblenz-Landau, 1992.
8. T. Westman, D. Harwood, T. Laitinen and M. Pietikäinen, Color Segmentation By Hierarchical Connected Components Analysis With Image Enhancement By Symmetric Neighborhood Filters, *Proc. 10th International Conference on Pattern Recognition*, Atlantic City, NJ, 1990, 796-802.

Modellbasierte Beschreibung von Farbhistogrammen und Segmentation von Farbbildern

R. Schuster * †, S. Ahmad †

* Technische Universität München, Lehrstuhl Prof. B. Radig
Orleansstraße 34, 81667 München

† Siemens AG, Zentralabteilung Forschung und Entwicklung[1]
Otto-Hahn-Ring 6, 81730 München

Im vorliegenden Artikel werden drei Modelle vorgestellt, die die Repräsentation von Farbobjekten im Farbhistogrammraum beschreiben. Die drei Modelle (Ellipsiod-, Zylinder- und Mischungsdichte-Modell) sind in ihrer Struktur und Komplexität so gewählt, daß sie die typischen Histogramme von realen Farbobjekten gut approximieren. Im Gegensatz zu früheren Arbeiten, in denen lediglich die Hauptkomponenten der Objekthistogramme untersucht wurden, geht es hier um eine vollständige und exakte Beschreibung der Objekthistogramme einschließlich sämtlicher Störeinflüsse, wie Kamerarauschen, Glanzlichter, Übersteuerung etc.

Anhand von realen Bilddaten wird gezeigt, wie die Modellparameter bestimmt werden können und wie geeignet die Approximationen jeweils sind. Diese exakte Histogramm-Modellierung kann für viele Bereiche der Farbbildverarbeitung von Bedeutung sein. In einem Anwendungsbeispiel wird gezeigt, daß sich die Modelle für die Objekt-Hintergrund-Segmentierung in Farbbildern gut eignen.

1 Einleitung

Das Interesse an Farbbildverarbeitung ist in den letzten Jahren deutlich gestiegen. Das resultiert aus dem Wunsch, Farbe als zusätzliche Informationsquelle nutzbar zu machen und wird unterstützt durch deutlich gewachsene Rechnerleistung und neue Erkenntnisse auf dem Gebiet der Reflexionsmodelle [1,2]. Ein wichtiges und in vielen Bereichen der Farbbildverarbeitung eingesetztes Werkzeug ist das Farbhistogramm (z.B. Farbbild-Segmentierung [3], Farbobjekterkennung [4], dreidimensionale Rekonstruktion [5]). In der hier vorliegenden Arbeit werden drei Modelle entwickelt, die die Repräsentation von realen Farbobjekten im Farbhistogramm beschreiben. Die Modelle sind so aufgebaut, daß sie der Komplexität der Farbrepräsentation angepaßt werden können und eine möglichst genaue und vollständige Beschreibung der Histogramme liefern. Damit erweitern die hier vorgestellten Modelle andere Arbeiten [6,7,8], die im wesentlichen nur die Hauptkomponenten der Histogramme untersuchen.

Der Aufsatz gliedert sich wie folgt: In Kapitel 2 wird der grundsätzliche Ansatz erläutert und drei Modelle von unterschiedlicher Komplexität vorgestellt. Das darauffolgende Kapitel beschreibt die mathematische Formulierung der Modelle und die verwendeten Approximationsverfahren. In Kapitel 4 wird die Anwendung der Modelle in einem

---

[1] In Zusammenarbeit mit Abteilung ZFE ST SN 61, Projekt GestikComputer [17]

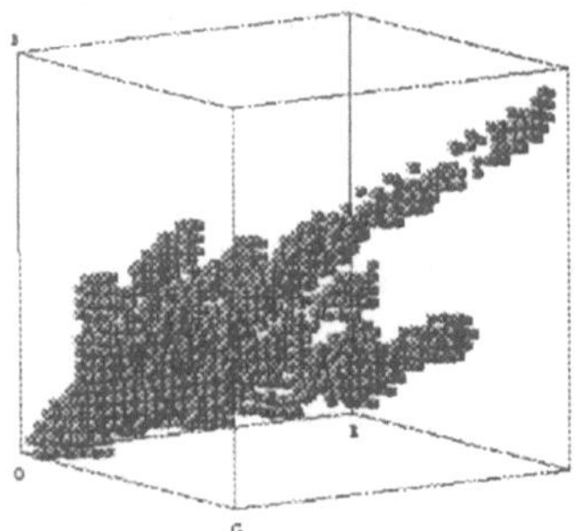

Bild 1: RGB-Histogramm von Farbwiedergabetafel (MacBeth ColorChecker, Bild 9)

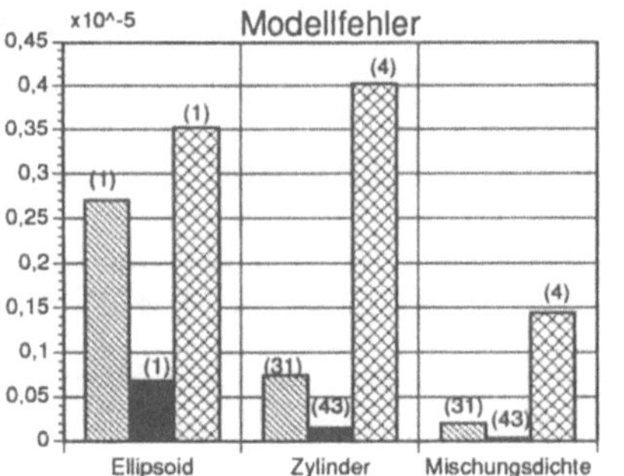

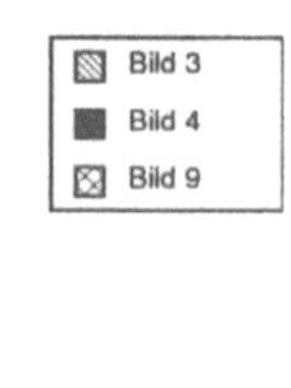

Bild 2: Modellfehler nach Approximation an reale Objekthistogramme (Anzahl der verwendeten Funktionen jeweils in Klammer angegeben)

Farbsegmentationsverfahren beschrieben. Kapitel 5 enthält nähere Informationen zu den verwendeten Bilddaten und ausführliche experimentelle Ergebnisse.

2 Modellbeschreibung im Histogrammraum

Die Farbe eines Objektes, die mit einer Farbkamera aufgenommen wird, ist abhängig von zahlreichen Einflußgrößen (z.B. Spektrum der Lichtquelle, Geometrie der Anordnung, Reflexionseigenschaften der Objektoberfläche, spektrale Eigenschaften der Kamera). Trotzdem gibt es bei natürlichen zwei- und dreidimensionalen Objekten typische Farbverteilungen und damit auch definierte Strukturen im Farbhistogramm (Bild 1). 1984 zeigte L. A. Shafer [1], daß dichromatische Materialien mit Glanzlichtern im Farbhistogramm ein Parallelogramm bilden (bestimmt von Körper- und Oberflächen-Reflexionsvektoren). Drei Jahre später beobachteten Klinker und Gershon unabhängig voneinander, daß das Parallelogramm nicht gleichmäßig ausgefüllt ist [6,7], sondern durch eine T-Form ("dog-leg") beschrieben werden kann. Novak formulierte im Jahre 1990 den Zusammenhang zwischen Oberflächenrauheit und der T-förmigen Histogrammstruktur und Nayar stellte 1991 ein Reflexionsmodell vor, das drei Komponenten umfaßt [2]. Im folgenden wird ein zusammenhängender, mit signifikanten Werten belegter Bereich im Farbhistogramm als \`Cluster´ bezeichnet.

Ausgehend von den genannten Histogrammstrukturen werden hier drei Modelle definiert, die in ihrer unterschiedlichen Komplexität den unterschiedlichen Reflexionseigenschaften von Objekten Rechnung tragen. Im Unterschied zu den oben genannten Arbeiten wird eine exakte mathematische Beschreibung der Verteilung innerhalb der Cluster entwickelt, die sowohl die Richtungskomponenten als auch die Varianzen der Cluster beinhaltet.

Ellipsoid-Modell: Cluster mit einfacher ellipsoider Ausdehnung und einer internen Verteilung, die durch eine einfache oder eine multivariate, dreidimensionale Gaußfunktion beschrieben werden kann. Dieses Modell ist für einfache Farbobjekte geeignet, die relativ gleichmäßig beleuchtet werden.

Zylinder-Modell: Lange, annähernd achsensymmetrische Cluster-Form, die abschnittsweise durch Zylinderscheiben beschrieben werden kann. Dieses Modell eignet sich besonders für dreidimensionale Farbobjekte ohne ausgeprägte Glanzlichter.

Mischungsdichte-Modell: Dieses Modell beschreibt den Cluster als Mischungsdichte von Dichtefunktionen. Das Modell kann beliebig geartete Cluster beschreiben und eignet sich daher für komplexe Farbobjekte mit Glanzlichtern, Textur etc. Auch das oben erwähnte T-förmige Reflexionsmodell kann mit dem Mischungsdichte-Modell beschrieben werden.

Man beachte, daß für die vollständige Repräsentation eines Farbobjektes auch mehrere Modelle gleicher oder unterschiedlicher Art kombiniert werden können. Ein zweifarbiges dreidimensionales Objekt mit einem Glanzlicht kann sich z.B. aus zwei zylinderförmigen und einem ellipsoiden Cluster zusammensetzen.

3 Mathematische Formulierung

Aus einem digitalen Farbbild kann ein Objekthistogramm $H(\vec{v})$ für alle $\vec{v}=(x,y,z)^T$ erzeugt werden. Die Achsen x, y und z bezeichnen einen beliebigen linearen, dreidimensionalen Farbraum. Das Objekthistogramm wird als diskrete, dreidimensionale Wahrscheinlichkeitsverteilung betrachtet, deshalb gelte die Normierung

$$H(\vec{v})=\{h(\vec{v}_1),\ldots,h(\vec{v}_n)\}, \qquad \sum_{i=1}^{n} h(\vec{v}_i)=1. \tag{1}$$

Im folgenden wird die mathematische Beschreibung der Histogrammwerte für das Ellipsoid-, Zylinder- und Mischungsdichte-Modell dargestellt.

3.1 Ellipsiod-Modell

Zur Beschreibung relativ einfacher, kompakter Cluster eignet sich die dreidimensionale Gaußfunktion der Form

$$f_1(x,y,z)=\frac{1}{(2\pi)^{3/2}\,\sigma_x\sigma_y\sigma_z}\exp\left(-\frac{1}{2}\left(\frac{(x-\mu_x)^2}{\sigma_x^2}+\frac{(y-\mu_y)^2}{\sigma_y^2}+\frac{(z-\mu_z)^2}{\sigma_z^2}\right)\right). \tag{2}$$

Einerseits hat diese Funktion durch ihre achsensymmetrischen Eigenschaften nur begrenzte Anpassungsfähigkeit. Andererseits müssen nur 6 freie Parameter bestimmt werden. Erweitert man die Funktion um die Kovarianzen, so erhält man die multivariate Normalverteilung,

$$f_2(\vec{v})=\frac{1}{(2\pi)^{3/2}\sqrt{|\Sigma|}}\exp\left(-\frac{1}{2}(\vec{v}-\vec{\mu})^T\,\Sigma^{-1}\,(\vec{v}-\vec{\mu})\right), \tag{3}$$

$$\text{mit}\quad \vec{v}=\begin{pmatrix}\mathrm{x}\\ \mathrm{y}\\ \mathrm{z}\end{pmatrix},\quad \vec{\mu}=\begin{pmatrix}\mu_x\\ \mu_y\\ \mu_z\end{pmatrix},\quad \Sigma=\begin{bmatrix}\sigma_x^2 & \mathrm{Cov(x,y)} & \mathrm{Cov(x,z)}\\ \mathrm{Cov}(\mathrm{y},x) & \sigma_y^2 & \mathrm{Cov(y,z)}\\ \mathrm{Cov}(\mathrm{z},x) & \mathrm{Cov(z},y) & \sigma_z^2\end{bmatrix}.$$

Mit dieser Funktion können auch Cluster beschrieben werden, die z.B. achsensymmetrisch zur Diagonalen der xy-Ebene sind. Beachtet man, daß $Cov(a,b)=Cov(b,a)$, dann sind bei dieser Funktion 9 freie Parameter zu bestimmen. Prinzipiell lassen sich die Parameter mit den Standardgleichungen der Wahrscheinlichkeitsrechnung direkt bestimmen. Zur Verbesserung der Genauigkeit benutzen wir zusätzlich ein iteratives Verfahren, das die Fehlerquadrate

$$E(\vec{p}) = \sum_{i=1}^{n} \left[f(\vec{v}_i, \vec{p}) - h(\vec{v}_i) \right]^2 \tag{4}$$

minimiert. Dabei enthält $\vec{p}$ die unbekannten Funktionsparameter. Als Startwerte für das iterative Verfahren werden die Ergebnisse der direkten Parameterbestimmung benutzt. Zur Minimierung von $E(\vec{p})$ kann z.B. ein statisches Optimierungsverfahren für mehrere Variablen ohne Nebenbedingungen eingesetzt werden (für Einzelheiten hierzu siehe [9]).

3.2 Zylinder-Modell

Eine weitere typische Cluster-Form hat längliche Ausdehnung und ist abschnittsweise zylinderförmig. Für die mathematische Beschreibung dieser Cluster-Form wird der Cluster-Schwerpunkt $\vec{b}$ und die Richtung der Hauptachse $\vec{a}$ bestimmt. Danach wird eine Koordinatentransformation angewendet, so daß die x-Achse in Richtung der Cluster-Hauptachse zeigt. Der gewichtete Cluster-Schwerpunkt berechnet sich aus

$$\vec{b} = \frac{1}{n} \sum_{i=1}^{n} \vec{v}_i h(\vec{v}_i) \, . \tag{5}$$

Durch die Berechnung der Eigenvektoren,

$$A = Q S R^T \quad \text{mit} \quad A = \left[h(\vec{v}_1)(\vec{v}_1 - \vec{b}) \ldots h(\vec{v}_n)(\vec{v}_n - \vec{b}) \right], \tag{6}$$

$$\text{und} \quad Q = \left[\vec{q}_1 \; \vec{q}_2 \; \vec{q}_3 \right], \; S = \begin{bmatrix} \lambda_1 & 0 & 0 & 0 \\ 0 & \lambda_2 & 0 \cdots 0 \\ 0 & 0 & \lambda_3 & 0 \end{bmatrix}, \; R = \left[\vec{r}_1 \ldots \vec{r}_n \right],$$

wird die Hauptachse $\vec{a}$ bestimmt

$$\vec{a} = \vec{q}_1 \, . \tag{7}$$

Im Unterschied zur üblichen Eigenwertanalyse im dreidimensionalen Vektorraum werden hier die Histogrammwerte als Gewichte der einzelnen Vektoren mit einbezogen. Die Gleichung für die Koordinatentransformation lautet dann

$$\vec{v}_i' = (\vec{v}_i - \vec{b}) \, R_y \, R_z \quad \text{für } i = 1 \ldots n \, , \tag{8}$$

dabei sind R_y bzw. R_z die Rotationsmatrizen für die Rotation um die y- bzw. z-Achse. Die Rotationswinkel ergeben sich direkt aus dem Richtungsvektor $\vec{a}$. Jetzt wird auf der x'-Achse eine neue Skalierung mit Δl-Schritten und eine Diskretisierung mit $l \in Z$ vorgenommen. Damit wird der Cluster in dünne Scheiben eingeteilt, deren Histogrammwerte $h_l(x', y')$ z.B. mit einer zweidimensionalen Gaußfunktion approximiert werden können

$$f_l(y', z') = \frac{1}{2 \pi \sigma_{y'} \sigma_{z'}} \exp\left(-\frac{1}{2} \left(\frac{(y' - \mu_{y'})^2}{\sigma_{y'}^2} + \frac{(z' - \mu_{z'})^2}{\sigma_{z'}^2} \right) \right). \tag{9}$$

Wie beim Ellipsoid-Modell kann diese Beschreibung durch die Kovarianzen erweitert werden. Für erhöhte Genauigkeit wird auch hier ein iteratives Verfahren eingesetzt. Für kleine Δl kann das Zylinder-Modell achsensymmetrische Cluster-Formen sehr genau beschreiben.

3.3 Mischungsdichte-Modell

Das Mischungsdichte-Modell wird durch die Summation von i gewichteten Dichtefunktionen beschrieben,

$$p(\vec{v}_k|\Theta) = \sum_i \alpha_i\, n(\vec{v}_k, \theta_i) \quad mit \ \sum_i \alpha_i = 1\,,\ \ \alpha_i \geq 0\,, \tag{10}$$

wobei $n(\vec{v}_k, \theta_i)$ eine beliebige Dichtefunktion sein kann. θ_i enthält die Parameter einer Dichtefunktion, Θ enthält die Parameter θ_i und die Gewichte α_i von allen i Dichtefunktionen. Die Parameter werden mit dem EM-Algorithmus ("Expected Maximum") bestimmt. Bei diesem iterativen Verfahren sind die Aktualisierungsgleichungen für die nächste Iteration gegeben durch

$$\alpha_i^+ = \frac{1}{\sum_{k=1}^{n} H(\vec{v}_k)} \sum_{k=1}^{n} H(\vec{v}_k)\, p^c(i|\vec{v}_k), \quad \vec{\mu}_i^+ = \frac{\sum_{k=1}^{n} \vec{v}_k\, H(\vec{v}_k)\, p^c(i|\vec{v}_k)}{\sum_{k=1}^{n} H(\vec{v}_k)\, p^c(i|\vec{v}_k)} \quad \text{und} \tag{11}$$

$$\Sigma_i^+ = \frac{\sum_{k=1}^{n} (\vec{v}_k - \vec{\mu}_i^c)(\vec{v}_k - \vec{\mu}_i^c)^T H(\vec{v}_k)\, p^c(i|\vec{v}_k)}{\sum_{k=1}^{n} H(\vec{v}_k)\, p^c(i|\vec{v}_k)}, \quad \text{wobei } p^c(i|\vec{v}_k) = \frac{\alpha_i^c\, n(\vec{v}_k|\theta_i^c)}{p(\vec{v}|\Theta^c)}. \tag{12}$$

Im Unterschied zum Standard-EM-Algorithmus berücksichtigt diese Formulierung auch die Gewichtung der Histogrammelemente $h(\vec{v})$. Aus Platzgründen wird nicht näher auf Einzelheiten des Mischungsdichteproblems bzw. des EM-Algorithmus eingegangen. Ein guter Überblick über den Problembereich und den EM-Algorithmus findet man in [10], ein Anwendungsbeispiel mit etwas vereinfachter Formulierung in [11].

4 Modellbasierte Segmentierung

Bei vielen Verfahren zur histogrammbasierten Farbbild-Segmentierung wird der Intensitätsanteil der Farbe durch die Berechnung von Farbwertanteilen eliminiert [12] und dann die Klassifikation in der zweidimensionalen Ebene (Farbtafel) durchgeführt [13]. Diese Normierung bedingt einerseits gute Farbkonstanz [14, 15], man reduziert aber andererseits den Farbraum um eine Dimension und verliert dadurch u. U. wichtige Klassifikationsmerkmale. In diesem Zusammenhang eignen sich die oben beschriebenen Modelle für die Farbild-Segmentierung im dreidimensionalen Farbraum mit den beiden Klassen Objekt und Hintergrund. Dazu muß der kontinuierliche Wertebereich $f(\vec{v})$ der Modellfunktionen in einen binären Wertebereich $f_b(\vec{v})$ umgewandelt werden. Bei allen drei Modellen wird die Binarisierung durch einfache Schwellwertbildung erreicht

$$f_b(\vec{v}) = \begin{cases} 1 & \text{für } f_{1,2}(\vec{v}) \geq \varepsilon,\ f_l(y', z') \geq \varepsilon \text{ bzw. } p(\vec{v}_k|\Theta) \geq \varepsilon \\ 0 & \text{sonst} \end{cases} \tag{13}$$

Der Schwellwert ε ist abhängig von der angestrebten Genauigkeit der Farbobjekt-Segmentierung und der Farbverteilung im Objekthintergrund. Ist der Objekthintergrund bekannt, so kann die Schwellwertbestimmung automatisiert werden. Die binarisierten, dreidimensionalen Modellfunktionen lassen sich direkt dazu verwenden, in einem Farbbild die Klassifikation Objekt-Hintergrund durchzuführen. Diese Lösung ist aber weder speicher- noch zeiteffizient.

In unserer echtzeitfähigen Implementierung übersetzten wir die Modellfunktionen in Tabellen mit effizientem Zugriffsmechanismus (siehe Kapitel 5).

5 Ergebnisse

Alle drei Modelle wurden auf reale Daten angewendet und die jeweiligen Approximationsfehler miteinander verglichen. Die verwendeten RGB-Farbbilder wurden mit einer CCD-Kamera (Sony XC-711P) aufgenommen. Als Lichtquellen wurden Neonitrit-Lampen mit 3200K Farbtemperatur verwendet. Die Histogramme haben jeweils eine Auflösung von 64x64x64 Farbwerte. In Bild 9 wurden 24 Farbfelder mit definierten spektralen Eigenschaften aufgenommen (MacBeth ColorChecker [16]). Bild 7 zeigt die vergrößerte Darstellung des Histogramms vom Farbfeld "Gelb". Für dieses Histogramm wurden die Parameter des Ellipsoid-Modells bestimmt. Dabei wurden die Startwerte der Lage- bzw. Streuungsparameter direkt berechnet und anschließend durch die iterative Minimierung der Fehlerfunktion in Gleichung (4) bestimmt (Quasi-Newton Verfahren mit BFGS-Formel und gemischter quadratischer und kubischer Linienoptimierung). Bild 8 zeigt die binarisierte Modellfunktion des approximierten Ellipsoid-Modells (multivariate Gaußfunktion, Gleichung (3)).

Das Zylinder-Modell wurde an einem einfachen dreidimensionalen Objekt (grüner Ball, Bild 3) getestet. Bild 5 zeigt das Objekthistogramm mit Hauptachse (unten) und das approximierte Zylinder-Modell (oben). Man erkennt deutlich die gute Approximation der realen Daten im oberen Teil des Zylinder-Modells. Die Krümmung im unteren Bereich des Objektshistogramms werden im Zylinder-Modell durch eine symmetrische Verdickung beschrieben. Das Mischungsdichte-Modell wurde auf ein komplexes Objekthistogramm (grüner Ball, Bild 4) angewendet. Bild 6 zeigt das Objekthistogramm (unten) und das approximierte Mischungsdichte-Modell (oben), das aus 30 überlagerten Gaußfunktionen besteht. Sowohl die Krümmung des Objekthistogramms (Kameraübersteuerung) als auch das Glanzlicht (Ausläufer nach unten) werden gut approximiert.

Die Resultate der Modellapproximation wurden für die Farbbild-Segmentierung angewandt. Bild 10 zeigt die Segmentierung des Farbfeldes "Gelb" durch Anwendung des approximierten Ellipsoid-Modells (Bild 8). Aufgrund der zahlreichen Fehlereinflüsse der Kamera und der begrenzten Bandbreite der RGB-Übertragungskanäle erfaßt das Modell nicht alle Farbwerte des Farbfeldes "Gelb". Durch Vergrößerung der Streuungsparameter bzw. Verkleinerung des Binarisierungs-Schwellwerts kann die Segmentierung verbessert werden. In unserer Implementierung des Verfahrens für Farbbildfolgen werden die Histogrammkoordinaten der binarisierten Modellfunktionen in einer Tabelle mit effizientem Zugriffsmechanismus abgelegt. Mit diesen Tabellen und Bildunterabtastung (128x128 Pixel) wird Farbbild-Segmentierung im Echtzeitbereich erreicht (auf Sun Sparc2).

Die "Qualität" der drei Modelle wurde anhand von realen Bildern verglichen (Bild 2). Als Maß für die Qualität wurde die Summe der quadratischen Fehler verwendet (Gleichung (4)). Wie erwartet erreicht das Mischungsdichte-Modell für alle Bilder die jeweils geringsten Abweichungen und ist damit am besten geeignet Farbobjekthistogramme zu approximieren.

Danksagung

Wir möchten uns bei D. Goryn, S. Hein und C. Maggioni für wertvolle Hinweise und Unterstützung bedanken.

Literatur

1. Shafer, S.A. Optical phenomena in computer vision. In *Proceedings CSCSI-84*, Canadian Society for Computational Studies of Intelligence, London, Ontario, Kanada, Mai 1984.
2. Nayar, S.K., Ikeuchi, K., und Kanade, T. Surface Reflection: Physical and Geometrical Perspectives. *PAMI 13*, 7 (July 1991), 611-634.
3. Celenk, M. A Color Clustering Technique for Image Segmentation. *Computer Vision, Graphics, and Image Processing* 52 (1990), 145 - 170.
4. Swain, M.J. und Ballard, D.H. Color Indexing. *International Journal of Computer Vision* 7 (1991), 11-32.
5. Sato, Y. und Ikeuchi, K., "Temporal-Color Space Analysis," , Technical Report, Nr.. CMU-CS-92-207, November 1992.
6. Klinker, G.J., Shafer, S.A., und Kanade, T. Using a color reflection model to separate highlights from object color. In *Image Understanding Workshop Proceedings*, Los Angeles, February 1987.
7. Gershon, R. *The Use of Color in Computational Vision*, Ph.D. Dissertation, University of Toronto, 1987.
8. Novak, C.L. und Shafer, S.A., "Anatomy of a Histogram," , Technical Report, Nr.. CMU-CS-91-203, November 1991.
9. Papageorgiou, M. *Optimierung: statische, dynamische, stochastische Verfahren für die Anwendung*, R. Oldenbourg Verlag (1991).
10. Redner, R.A. und Walker, H.F. Mixture Densities, Maximum Likelihood and the EM Algorithm. *SIAM Review* 26, 2 (April 1984), 195-239.
11. Nowlan, S.J. Maximum Likelihood Competitive Learning. In *Advances in Neural Information Processing Systems 4.*, 574-582, 1990.
12. Richter, M. *Einführung in die Farbmetrik*, Walter de Gruyter, 2. Aufl. (1980).
13. Batchelor, B.G. Color Recognition in Prolog. In *SPIE Machine Vision Applications, Architectures and Systems Integration*, SPIE, 1992.
14. Petrov, A.P. Color and Grassman-Cayley coordinates of shape. In *Human Vision, Visual Processing and Digital Display II*. SPIE, 342-352, 1991.
15. D'Zamura, M. und Lennie, P. Mechanisms of color constancy. *Journal of the Optical Society of America* 3, 10 (Oktober 1986), 1662 - 1672.
16. McCamy, C.S., Marcus, H., und Davidson, J.G. A Color-Rendition Chart. *Journal of Applied Photographic Engineering* 2, 3 (1976), 95-99.
17. Maggioni, C. A Novel Device for Using the Hand as a Human-Computer Interface. In *HCI-93, Human Computer Interaction*, Loughborough, Großbritannien, 1993.

Farbseite

Bild 3	Bild 4
Bild 5	Bild 6
Bild 7	Bild 8
Bild 9	Bild 10

Bild 3 und 4: Reale RGB-Farbbilder mit grünem Ball (Objekt) vor blauem Hintergrund. Links normale Aufnahme, rechts Aufnahme mit Glanzlicht und Übersteuerung.

Bild 5: Unten RGB-Histogramm (mit Hauptachse) von Objekt in Bild 3 und darüber approximiertes Zylinder-Modell (in B-Richtung verschoben). **Bild 6:** Unten RGB-Histogramm von Objekt in Bild 4 und darüber approximiertes Mischungsdichte-Modell (in B-Richtung verschoben).

Bild 7: Vergrößertes RGB-Histogramm von Farbfeld "Gelb", ausgeschnitten aus Bild 9. **Bild 8:** Ellipsoid-Modell nach Approximation an RGB-Histogramm von Farbfeld "Gelb".

Bild 9: Reales RGB-Farbbild von Farbwiedergabetafel (Macbeth Color-Checker [16]). **Bild 10:** Segmentation des Farbfeldes "Gelb" aus Bild 9 mit Ellipsoid-Modell in Bild 8.

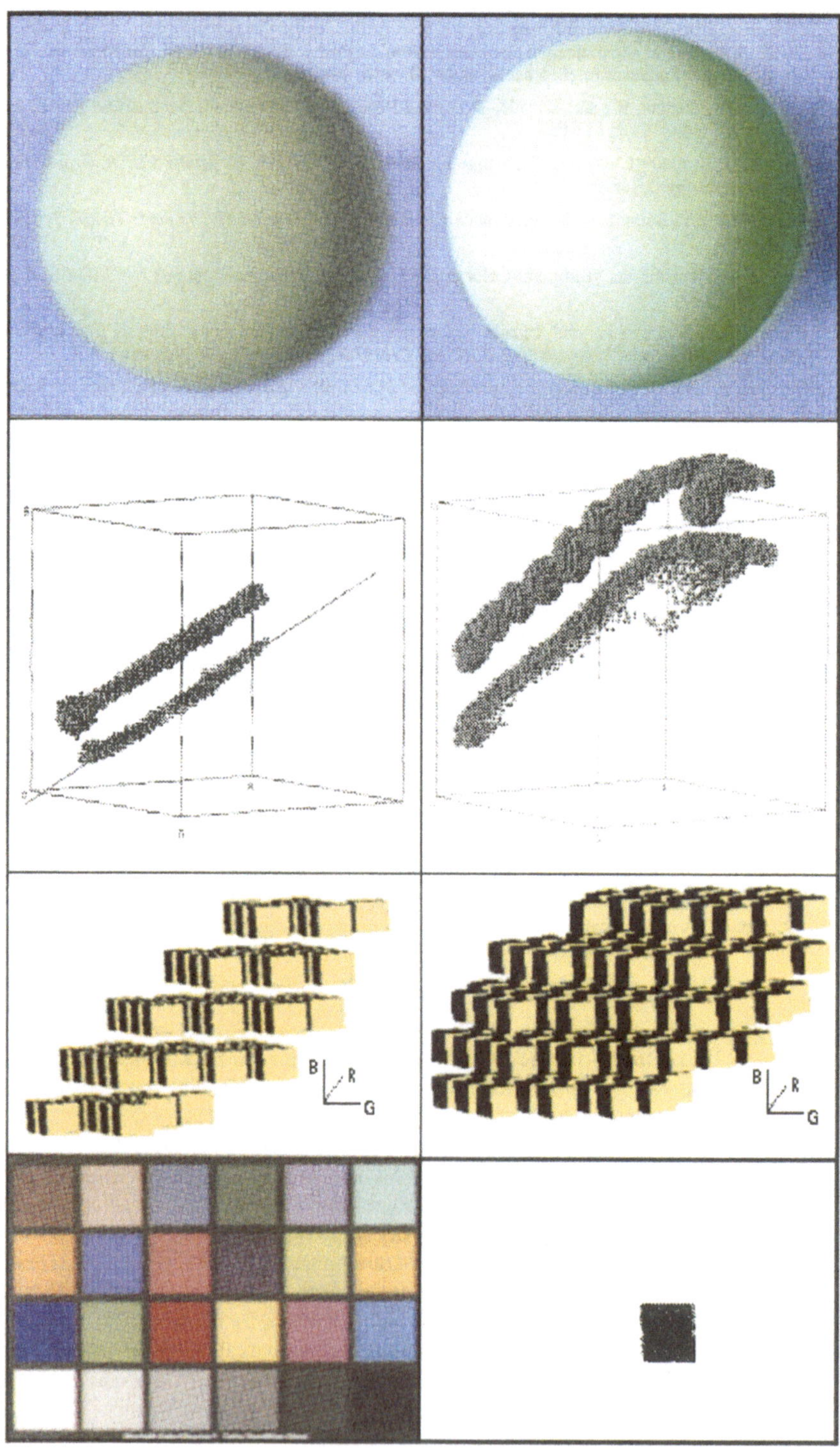
B
R
G
B
R
G

A Framework for Information Fusion and an Application to Remote Sensing Data

Renate Bartl * Axel Pinz † Werner Schneider *

Abstract

Information Fusion integrates several inputs at different levels of abstraction. The aim is to improve the quality of image analysis concerning reliability, robustness, completeness, etc. Several fusion methods exist for very specific problems (e.g. [10], [9], [8]). A framework therefore is needed in which the essential characteristics of a fusion process are described and which supports the selection of a proper fusion method. In this paper we propose such a framework and present an application to remote sensing. On a Landsat TM image and on a MKF6 photographic satellite image of the AUSTROMIR mission forest regions and pond areas are identified and then fused using morphological and spatial features to find out how the regions (and consequently the images) correspond.

1 Introduction

In recent years, the trend in image processing goes away from analysing single images as proposed by Marr [11] or Ballard and Brown [3]. On the one hand this is motivated by the fact that the output of a single image analysis task normally is not sufficient in all aspects. A combination of results originating from several image processing tasks may therefore be useful (see e.g. [12], [14]). On the other hand a single image usually does not contain enough information to reconstruct its contents. Consequently, the use of multiple images has increasingly obtained interest. Images acquired under slightly varying conditions by an active sensor (see e.g. [1], [2]) as well as images from different sensors (see e.g. [21], [7]) are combined in several ways. A basic theory is missing how to perform this combination, i.e. how to decide when to apply which method. In [4] and [15] we have already presented a framework for information fusion. An updated and extended version of this is given in the following section.

2 Fusion Framework

The framework for information fusion we present is designed for general use in image understanding. It combines knowledge about visual sources (physical models, sensor models), several levels of abstraction (image, symbolic description of image contents) and several classes of image understanding processes. In this framework the process of fusion is able to *actively select* the sources to analyse and the processes to be perfomed on these data, so that we term it *active fusion.* Figure 1 gives an overview of our framework.

The *real world* is infinite in 4 dimensions (3-D space and time), therefore only small portions of it can be observed (scene selection). Processing of images of these portions leads to a world description which, due to the infinity of the real world, is incomplete, and varies with the aims of the analysis and the methods applied.

The process of *scene selection* is related to the two questions: 'What to look at?' and 'At which moment to observe?' (when). In many cases, portions of the selected scenes are

*Inst. for Surveying and Remote Sensing, BOKU Wien, Austria, email:renate@sunivf1.boku.ac.at

†Dept.f. Pattern Recognition and Image Processing, TU Wien, Austria, email:api@prip.tuwien.ac.at

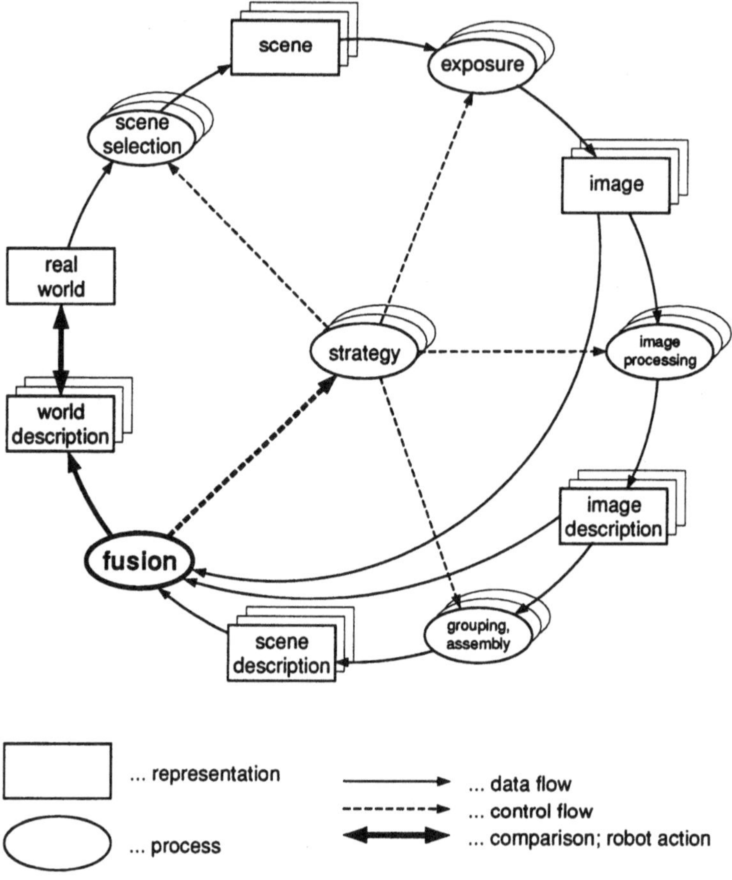

Figure 1: Our framework of Active Fusion

overlapping in 3-D space, or an identical object is present in several scenes. For example multiple scenes can be taken at different seasons or from different viewpoints (satellite, ground, etc.). If an active sensor (e.g. a moving camera) is available, it can be controlled to actively select and look at the portion of the world that is of interest in the current stage of processing.

In this framework, we define a *scene* as a 3-D portion of the world at a certain moment. A scene has important physical properties which have to be modeled and represented explicitly for the subsequent interpretation processes. The scene parameters can contain e.g. exact time and location.

The scene is exposed to sensing devices. Properties and parameters of the sensors have to be modeled (sensor models). Here, the question 'How to look at a scene?' is answered, so that the process of *exposure* can also be viewed as a sensor selection process. Besides the control of an active sensor, there are several other possibilities of 'quasi active' sensor selection. For example, an interpretation system can select the most appropriate out of several sources of information (satellite image, aerial photograph, radar image, scanner image, certain spectral band).

An *image* as the result of the exposure process is a 2-D representation of the 3-D scene. Thus the inverse problem of the recovery of the scene often is underdetermined or ill-posed. In our framework we try to find a unique solution by combining (fusing) information from several different sources. Scene and exposure parameters that lead to an image have to be represented explicitly together with the image itself to enable proper access by the subsequent fusion process.

At the stage of *image processing*, many different processes and sequences of processes (e.g. image enhancement, classification) are applied to the input images. This leads to a segmentation of the images into tokens (e.g. regions of a distinct land use class). Features are extracted and assigned to the tokens (e.g. location in the image, intensity mean), resulting in a kind of 'intermediate symbolic representation' called image description. To control the complete image understanding environment, these processes and sequences of processes have to be grouped into categories (e.g classification by maximum likelihood, nearest neighbourhood, and neural nets) and their properties have to be represented explicitly.

In our fusion based image understanding approach, the control strategies allow multiple processes at all levels resulting in multiple outputs. Even in the most simple case of only one input image, we can extract many different *image descriptions* by applying different sequences of processes to it. In general, we need a representation that enables us to compare these descriptions, i.e. to match certain image objects (e.g. rivers, forest regions) from different image descriptions, by comparing features of objects (shape, color, texture, size, location,etc.) and relations between objects (spatial, temporal). With image descriptions, all features and relations are 2-D.

Grouping and assembly processes work at a higher level of abstraction than image processing. They use 2-D objects and their features to derive 3-D scene descriptions (e.g computation of the height of an object from the length of its shadow). As for image processing, these grouping and assembly processes must be categorized and their properties must be represented.

Everything that has been said about image descriptions applies for *scene descriptions* as well. The major difference is that 3-D object models as well as 3-D features and relations have to be represented.

So far, we have described four different levels of processing and of representation. Diversity is possible and even desired at all of these levels, so that multiple processing and multiple representations occur at all levels of abstraction. The purpose of the process of *fusion* in our framework is twofold: Strategy selection, and combination of information from several representations. The strategy selection is required to limit the number of parallel paths, thereby controlling the image understanding system and avoiding well known problems of computational complexity and combinatorial explosion [Tsotsos, 1990]. The core problem of fusion, the combination of information from several representations, can be approached in many different ways and has not yet been solved in a general way. In our framework, the process of fusion produces as an output a world description (e.g in form of a thematic map). To be able to select the most valuable portions of information from the many sources of input, quality measures on the basis of probabilities for all elements of all representations are required.

A *strategy* consists of at least one sequence of processes (scene selection, exposure, image processing, grouping, assembly). It is selected and controlled by the fusion process in a top down manner.

As a final result of processing and representation at all levels, the process of fusion comes up with a current description of the world as it was sensed by the system. From the point of view of the fusion process, this *world description* is better than any single source used as fusion input, and the best new output that can be found with respect to certain quality measures. The connection between this world description and the real world situation can be established, either (indirectly) by a human interpreter performing a comparison or (directly) by having a robot interacting with the real world. If there is a robot action changing the real world, the whole circle of processing may start again.

Different aspects of this general framework of active fusion are subject of our ongoing research. In the remainder of this paper we report on an application of the framework at the image description level. We give two examples for fusion of objects from different image descriptions by matching and combination of information.

3 Input Data and Preprocessing

Our test data set consists of two images: a Landsat TM image and a digitized MKF6 photo (more details in [6]). These two images differ in scale (approx. 1.6 : 1) as well as in orientation (rotation approx. 95°), but they are geometrically similar (apart from small distortions). Fig. 2 shows the near infrared channels of both images.

(a) Landsat TM (channel 4, 0.76-0.9 μm)

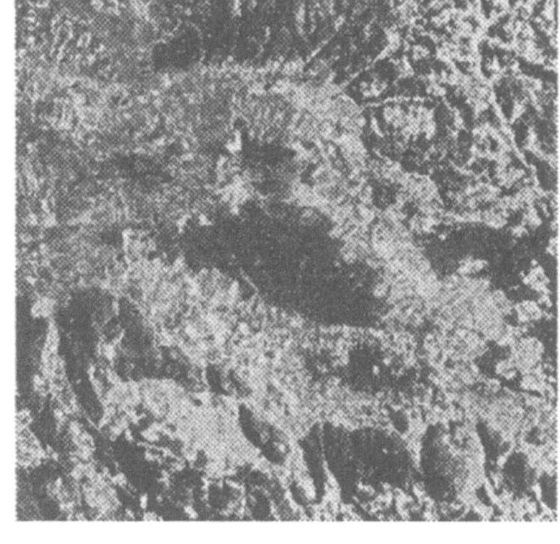

(b) MKF6 (channel 6, 0.76-0.91 μm)

Figure 2: Near-Infrared Channels of Input Data

A land use classification was performed on these images [16, 5], and forest areas and pond areas were extracted. They are represented symbolically as regions with *region features* describing morphological properties (see section 4.1) and *cluster features* describing spatial relations (see section 4.2).

4 Fusion of Regions

Fusion of regions includes the search for corresponding regions, i.e. for regions representing the same phenomenon in nature, the explicit representation of these correspondences (see e.g. [13]) so that they can be easily accessed for further use, and the integration of information from corresponding regions (e.g. improvement of thematic maps by fusion).

The first step towards fusion of regions is to find significant region features which are required to be (almost) invariant to scale, rotation, and translation. This is important since we assume to have only sparse *a priori* information about the relation between the images. The features are computed independently for regions of both images to be compared in the subsequent fusion process which outputs corresponding regions, i.e. regions representing the same phenomenon in nature.

4.1 Fusion by Region Features

Region features are features describing the shape (morphological structure) of individual regions. Examples are *bounding-rectangle-fill* which is computed as the number of pixels in the region, divided by the number of pixels of its minimum bounding rectangle, and *elongation* which is defined as difference between the lengths of major and minor axes of the ellipse with the best fit, divided by the sum of the lengths. Fig. 3 gives an illustration of morphological region features.

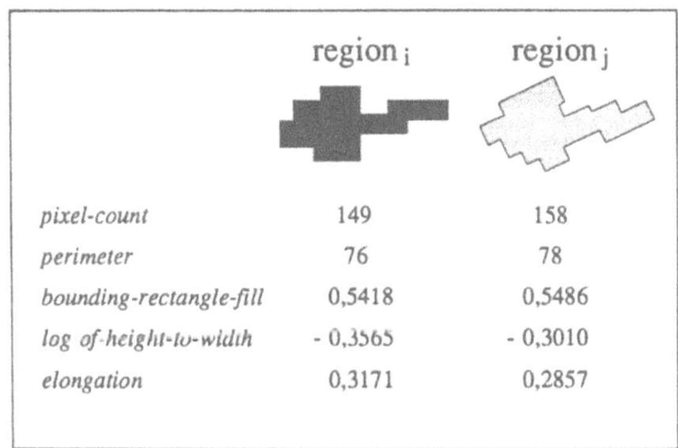

Figure 3: Morphological Region Features

These features are computed independently for all forest regions in both images. Then the features of each region of one image are compared with the features of each region of the other image, yielding a number of corresponding regions in both images.

Additionally, region features depending on scale, rotation and translation help to clarify correspondence ambiguities. Estimates for the scaling factor, the rotation angle, and the translation vector between the images can be computed. We calculate these values for each pair selected before. Assuming that the correct values are the most frequent ones, pairs with strongly diverging values can be removed.

Fig. 4 shows the fusion result where the grey-shaded areas belong to forest regions. Corresponding region pairs are marked in darker grey. We only use large regions (with more than 100 pixels) to demonstrate the method.

Further examinations are performed, both for verifying the correctness of the fusion result, and for ascertaining the correspondence of regions in more detail. Based on the estimated values of the differences between the images (scaling, rotation, translation), regions of one image are transformed into the other image, where they are expected to fit (coarsely) to the corresponding regions. The number of overlapping pixels is a simple measure for the quality of fit.

Apart from remote sensing images considered here, this type of fusion is of special interest in cases of scenes with changing content and topology. If, for instance, the objects under consideration are moving, only their intrinsic structure remains constant and is therefore appropriate for use in the fusion process. The correspondence established by morphologic region features can also be used for subsequently computing the motion of objects.

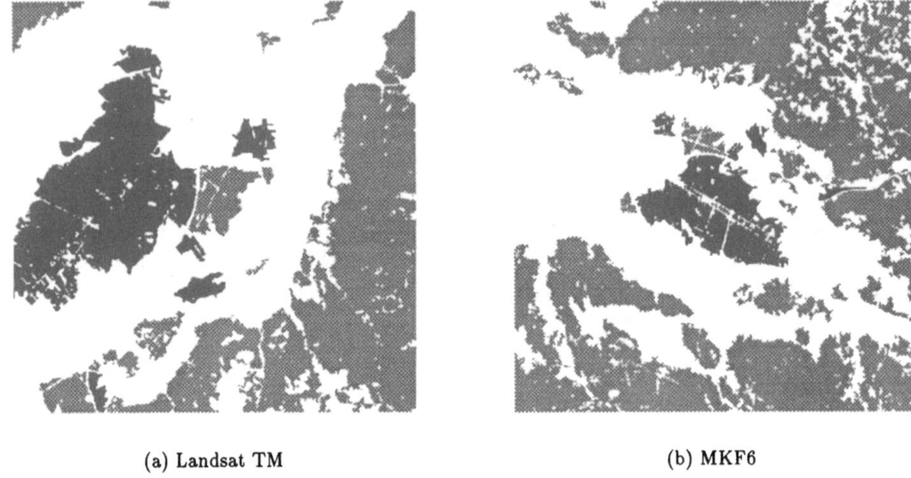

(a) Landsat TM (b) MKF6

Figure 4: Fusion by Region Features: Corresponding Forest Regions

4.2 Fusion by Cluster Features

We represent ponds symbolically as regions, which are then grouped into *clusters* [22]. We assume a region to belong to a cluster if its relative distance to at least one other region in the cluster is below a certain limit, depending on the compactness of the clusters (e.g. to be described in terms of the average region diameter).

In a first step it is reasonable to find correspondences between the clusters by comparing their size, the number of regions inside and their relative position. Since in our test example there are only two clusters (and one is very small), we did not concentrate on this point, simply taking the larger cluster out of each image for further examinations.

Inside the cluster, the geometrical structure needs to be represented in geometrical cluster features invariant to translation, rotation and scale. Based on the length and direction of the junction vector between (the centers of) two pond regions, for any three regions, relative distance values and angles of the formed triangle can be computed. Comparing these values creates hypotheses for corresponding triangles. Estimates for the scaling factor, the rotation angle and the translation vector between the images can be computed for every pair of (hypothetically) corresponding triangles. These values must be in reasonable agreement with each other and with the values derived from region features.

Various strategies for selecting triangles can be followed, e.g. forming all triangles (if the number of regions is small) or forming triangles with one corner region next to the center of gravity of the cluster. From the correspondence of triangles, finally, the correspondence of regions can be deduced.

In Fig. 5 the area around the ponds is shown in enlarged form. The extracted pond regions are highlighted in white. Because the original pond pixels are very dark, the bright pond regions appear surrounded by dark grey. In the Landsat image (Fig. 5.a) 10 pond regions can be found, whereas in the MKF6 image (Fig. 5.b) there are only 7 pond regions. In our present tests we have found 6 corresponding regions using the strategy of forming all possible combinations of triangles. Since the computational effort for this was high, we are currently trying to optimize the search strategy in hypotheses testing.

With the fusion result, an improvement of the classification can be achieved by com-

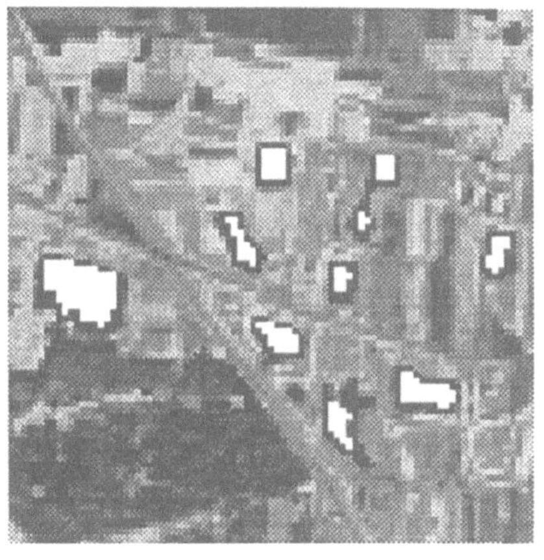

(a) Landsat TM (zoomed)

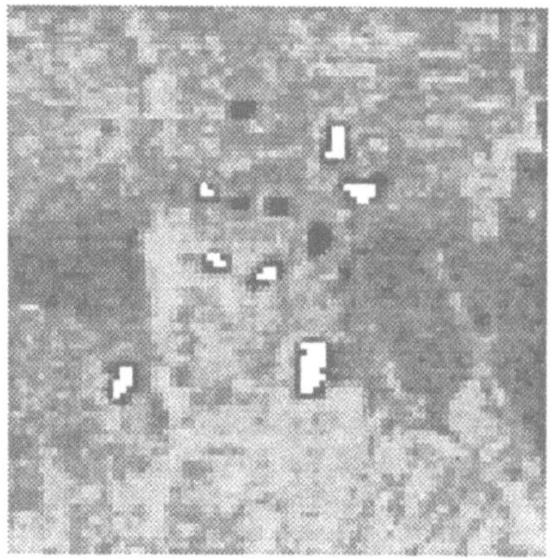

(b) MKF6 (zoomed)

Figure 5: Fusion by Cluster Features: Corresponding Pond Regions

bination of information. For the regions remaining unfused in the center of the cluster of Landsat TM, corresponding regions in MKF6 are searched explicitly. The classification is repeated and influenced by the knowledge where ponds are found in the other image. This additional information increases significantly the probability for classification as water in this *focus of attention.* If the (radiometric) pixel values contradict a water classification strongly, the Landsat TM classification may be reconsidered.

5 Conclusion and Outlook

In this paper we have presented a framework for information fusion as well as an application to remote sensing. Two approaches for finding correspondences between two images by fusing symbolic information have been described. The advantages of these methods are that the amount of data to be processed is reduced and that differences due to diverse sensing devices are diminished. Depending on the type of objects, morphological region features as well as geometrical cluster features have been shown to be useful for comparison in the fusion process.

In our next experiments we want to refine the output of fusion. Having found corresponding regions and estimates for their relative scaling, translation and rotation, specific points of correspondence should be located with high accuracy. Candidates are contour edges or center of gravity points. Also larger line features, extracted from the region contours, may be useful, especially if their accuracy can be improved by subpixel analysis [20]. These correspondences can for example be utilized for automated image registration [18] and radiometric calibration [19]. The methods developed can also be utilized for another field of application concerning the automated registration of scanning laser ophthalmoscope images of the fundus of the human eye [17].

Acknowledgement: This work was partly supported by the *Austrian National Fonds zur Förderung der wissenschaftlichen Forschung* under grant P9785.

References

[1] J. Aloimonos and D. Shulman. *Integration of Visual Modules. An Extension of the Marr Paradigm.* Academic Press, 1989.

[2] R. Bajcsy. Active perception. In *Proc. DARPA Image Understandung Workshop*, pages 279–288. Cambridge, 1988.

[3] D. Ballard and C. Brown. *Computer Vision.* Prentice Hall, 1982.

[4] R. Bartl and A. Pinz. An extension of the fusion concept. In P. Mandl, editor, *Modelling and New Methods in Image Processing and Geographic Information Systems*, volume 61 of *OCG-Schriftenreihe*, pages 111–122. Oldenbourg, 1992.

[5] R. Bartl and A. Pinz. Information fusion in remote sensing: Land use classification. In *Multisource Data Integration in Remote Sensing for Land Inventory Applications, Proc Int.IAPR TC7 Workshop*, pages 9–17, 1992.

[6] R. Bartl, W. Schneider, and A. Pinz. Information fusion in remote sensing: Combining Austromir data and Landsat data. In *Proc. of the 25th ERIM Symposium*, 1993. in press.

[7] J. J. Clark and A. L. Yuille. *Data Fusion for Sensory Information Processing Systems.* Kluwer Academic Publishers, 1990.

[8] S. Ford and D. McKeown. Information fusion of multispectral imagery for cartographic feature extraction. Presented at the 17th ISPRS Congress, Washington, DC, 1992.

[9] R. Krishnapuram and J. Lee. Fuzzy-set-dased hierarchical networks for information fusion in computer vision. *Neural Networks*, 5:335–350, 1992.

[10] I. Kweon and D. Kanade. High-resolution terrain map from multiple sensor data. *IEEE Transactions on Pattern Recognition and Machine Intelligence*, 14(2):278–292, 1992.

[11] D. Marr. *Vision: A Computational Investigation into the Human Representation and Processing of Visual Information.* W.H. Freeman and Company, New York, 1982.

[12] P. Meer, D. Mintz, A. Montanvert, and A. Rosenfeld. Consensus vision. In *Proceedings of the AAAI-90 Workshop on Qualitative Vision*, pages 111–115, 1990.

[13] L. Pau. Knowledge representation approaches in sensor fusion. *Automatica*, 25(2):207–214, 1989.

[14] F. Perlant and D. McKeown. Improved disparity map analysis through the fusion of monocular image segmentations. In J. Tilton, editor, *Multisource Data Integration in Remote Sensing, Proc. IAPR TC7 Workshop*, pages 83–97, 1991.

[15] A. Pinz and R. Bartl. Information fusion in image understanding. In *Proceedings of the 11.ICPR*, volume I, pages 366–370. IEEE Computer Society, 1992.

[16] A. Pinz and R. Bartl. Information fusion in image understanding: Landsat classification and ocular fundus images. In *SPIE Sensor Fusion V, Boston 92*, volume 1828, pages 276–287. SPIE, 1992.

[17] A. Pinz and P. Datlinger. Digital image analysis and scanning laser ophthalmoscope I: Superimposition of essential retinal features. In H. Bischof and W. G. Kropatsch, editors, *Pattern Recognition 1992*, volume 62 of *OCG-Schriftenreihe*, pages 45–55. Oldenbourg, 1992.

[18] E. Rignot, R. Kwok, and J. Curlander. Towards operational mulisensor registration. In J. Tilton, editor, *Multisource Data Integration in Remote Sensing, Proc. IAPR TC7 Workshop*, pages 39–58, 1991.

[19] W. Schneider. Problems of radiometric calibration of multilayer photographic remote sensing images. In M. Buchroithner, editor, *Proc. 11th EARSeL Symposium*, pages 5–13, 1991.

[20] W. Schneider. Landuse mapping with subpixel accuracy from Landsat TM image data. In *Proc. of the 25th ERIM Symposium*, 1993. in press.

[21] S. C. Thomopoulos. Sensor integration and data fusion. *Journal of Robotic Systems*, 7(3):337–372, 1990.

[22] J. T. Tou and R. C. Gonzalez. *Pattern Recognition Principles.* Addison-Wesley, 1974.

Detection of Prominent Boundary Points Based on Structural Characteristics of the Hierarchic Medial Axis Transform

R. L. Ogniewicz, G. Székely, and O. Kübler

Communication Technology Laboratory, Image Science
ETH Zurich, 8092 Zurich, Switzerland
Tel: +41 1 256 5375, Fax: +41 1 261 3429, E-mail: rlo@vision.ethz.ch

Abstract. The analysis of curvature characteristics of an object outline is an important issue in many Computer Vision applications. The goal is to either approximate the outline by curves which are mathematically simpler or which can be further treated by the tools and methods of differential geometry or to perform the decomposition of the object into simpler subparts. Such operations can be performed much more reliably and in agreement with human perception if additional information about the region circumscribed by the outline is available. We show how region-oriented shape analysis can profit from information gained from the Hierarchic Medial Axis Transform (HMAT) or Hierarchic Voronoi Skeleton of the shape. The skeleton's structure, in particular its branching points, allow to indicate prominent points of the boundary. Approximation techniques or smoothing operations can make explicit use of this knowledge by assuring the preservation of such prominent loci. Moreover, the hierarchic organization of the skeleton constituents facilitates the combination of different scales of form representation and a simultaneous decomposition of the shape.

1 Introduction

One of the central issues of most object recognition systems is the encoding of a shape by means of a shape descriptor which (a) leads to an efficient and robust (invariant) representation and (b) which captures the salient features of the shape. In the sequel, this shape representation should allow an efficient analysis of the shape, for example, by applying the instruments of differential geometry. Moreover, we require that such a description should be hierarchical, i.e., it should allow to first investigate the shape's gross features and then to set the focus on details.

Many shape descriptions rely on outline characteristics, in particular on curvature, such as polygonal approximations [1], spline approximations, Fourier descriptors [2], or *codons* [3]. A successful transformation into one of the above representations requires that the boundary should be more or less free of noise. This is usually achieved by first preprocessing the shape, typically by blurring or smoothing operations [4, 5]. Yet smoothing bears the potential of not only removing noise but also of destroying salient shape features which otherwise would lead to a more robust and intuitively appealing segmentation of the outline. Scale-space filtering can be employed to obtain a hierarchically organized sequence of shape representations, each corresponding to a different level of detail. Unfortunately, the resulting scale-space images [6] are often hard to interpret and the task to combine the information from different levels of the scale-space (correspondence problem) turned out to be a difficult algorithmic and conceptual problem.

It has to be pointed out that even an outline which is free of noisy artifacts does not guarantee a reasonable segmentation of the boundary. The computation of codons or polygonal approximations is basically controlled by the occurrence of specific curvature conditions, typically extrema (negative and positive) and zeroes of curvature. However, it is not assured that the resulting segmentation of the boundary agrees with human perception which rather puts the emphasis on the regions circumscribed by an outline, nor that it lays the foundation for a subsequent decomposition of the shape. Let us consider, e.g., the lizard shape in Fig. 3. The most basic and evident decomposition of this shape is to separate the lizard's legs from its body. Considering solely curvature properties, we state that (negative) curvature extrema occur at locations which are clearly no candidates for separation (indicated by arrows in Fig. 3(b)), namely near the animal's head and along its tail, and, even more troublesome, within the outline describing the legs. Obviously, some more information about the regions constituting the shape of the lizard is needed. Fig. 1 depicts another example of a biological form, the silhouette of a maple leaf. Due to its jagged boundary, local curvature extrema do not contribute much to the understanding of the shape, since they are distributed rather arbitrarily over the outline.

The presentation of a technique which can resolve this type of ambiguities by combining curvature characteristics with region information derived from the *Hierarchic Medial Axis Transform* (HMAT) is at the focus of this paper. This technique comprises shape decomposition and form-adaptive smoothing where knowledge about the regions circumscribed by the shape's outline controls the local degree of smoothing. Our method consists of the following steps:

(1) *Evaluation of the HMAT* upon the object's discrete set of boundary points (Section 2).

(2) *Mapping of regularization information, level of hierarchy and region information* onto the boundary by analyzing the structural properties of the skeleton graph. Basically, each boundary point is assigned a tuple of values describing the prominence of the boundary point with respect to the shape's global characteristics (Section 3).

(3) *Nonmaximum suppression* removes clusters of adjacent prominent boundary points (Section 4). The preserved point within a cluster is the one which corresponds to the most salient portion of the medial axis with respect to the underlying shape.

(4) *Subpart extraction and resampling* (Section 5). The attributes of the boundary points allow to extract the basic form of the object during a single scan of the outline and then gradually incorporate features at lower hierarchical levels. A resampling operation is necessary in order to obtain again a path length parameterized curve.

(5) *1D distance transform*: for each boundary point the distance to the nearest prominent boundary point is computed (Section 6).

(6) *Adaptive Gaussian smoothing* of the outline (Section 7). The width σ of the Gaussian kernel is locally adapted taking into account the distance to the

nearest prominent boundary point and a sensitivity parameter. User-defined minimal and maximal values of σ are used to keep σ within a defined interval.

2 The Hierarchic Skeleton

Skeletonization in the plane denotes a process which transforms a two-dimensional shape into a one-dimensional graph-like representation, while preserving the original connectivity of the shape. There exist several substantially different skeletonization algorithms [7, 8, 9, 10, 11]). Our approach [12, 7, 8] is based on the close relationship between the medial axis transform (MAT) according to Blum's famous grassfire concept [13], distance transforms [14] (DT) and the Voronoi diagram (VD) [15] of a shape's boundary points. In [7, 8], we have proposed so-called residual functions to attribute each edge of the VD with a measure of prominence, which expresses the structural saliency of this edge with respect to distortions of the shape introduced by geometric transformations and by noise. Analysis of typical artifacts along the boundary and their effects upon the medial axis allows us to predict the threshold necessary to separate spurious from significant branches. The resulting skeletons appear largely invariant with respect to typical noise conditions in the image and geometric transformations.

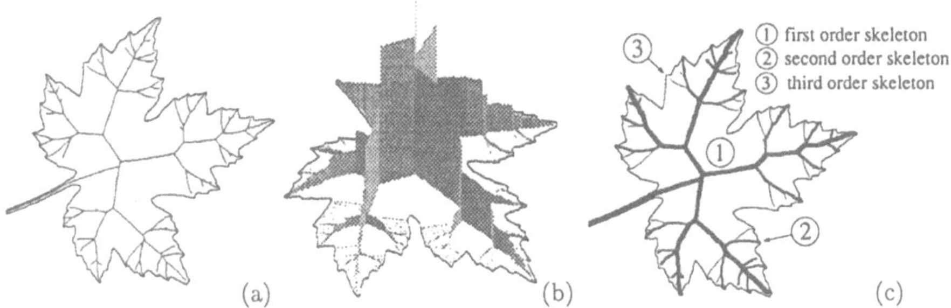

Fig. 1. The Hierarchic Skeleton. (a) The Voronoi skeleton computed for the interior of the maple leaf shape. The skeleton has been pruned by applying one of our regularization functions to the complete Voronoi diagram, such that the remaining skeleton allows to reconstruct the original shape with a boundary deviation of less than 3 pixels. (b) The values of the prominence measure are rendered as the height of each Voronoi edge. The high ridges running along the object indicate skeleton components which belong to higher levels of the skeleton pyramid (first order and second order skeleton). (c) The skeleton pyramid extracted from (b), consisting of three levels, as denoted by the numbers (the hierarchy starts at level one) and the thickness of the skeleton branches. The thickest branches belong to the first order skeleton and indicate the most significant features such as the five 'fingers' of the leaf. The second order skeleton is obtained by adding all skeleton branches which are directly connected to the first order skeleton. Branching points on the skeleton occurring both within the same level and between consecutive levels of the hierarchy flag candidate locations for separation of the shape into subparts. The number of skeleton branches at higher levels of detail reflects the fractal dimension of the shape.

We have extended the basic regularization scheme by establishing a multiresolution representation of the skeleton, henceforth termed *Skeleton Pyramid*, *Hierar-*

Fig. 2. First Order Skeletons of Distorted Shapes. It has to be stressed that (1) a continuous MAT of the above shapes according to Blum would be clobbered by numerous spurious branches induced by the jagged boundary, and (2) conventional pruning techniques (i.e., which do not establish a hierarchic representation) would not be able to remove such spurious components without accidental deletion of globally salient portions. The first order skeletons, on the contrary, appear almost identical to the continuous MAT of ideal rectangles and triangles.

chic Medial Axis Transform (HMAT) or *Hierarchic Voronoi Skeleton*: After each skeleton component has been attributed with its significance residual subsequent depth-first traversal of the skeleton graph allows to establish a local hierarchy of the skeleton branches by ordering the (Voronoi) edges, for every vertex, according to decreasing structural importance as reflected by the residua.

Fig. 1(a) depicts an example of the Voronoi skeleton, computed for a maple leaf shape; Figs. 1(b)-(c) illustrate the concept of the HMAT.

An illustrative understanding of the skeleton pyramid can be afforded if the prominence measure is interpreted as the 'height' of each skeleton component.. The attributed skeleton represents a 'residual mountain' as depicted in Fig. 1(b), where the highest ridges correspond to the most salient local axes of symmetry. The generation of the hierarchy amounts to a downhill ridge-riding according to the strategy of the least steep descent. The details of this procedure can be found in [7, 8].

The benefit of the skeleton pyramid is the fact that its structure permits to access and analyse the medial axis in a systematic way, proceeding from lower to higher levels of detail. Since each segment of the medial axis is associated with a particular section of the outline, the skeleton pyramid also defines a hierarchic organization of boundary segments.

Fig. 2 exemplifies that the organization of the HMAT indeed captures the global characteristics of a shape: The first order skeletons computed for considerably distorted rectangular and triangular shapes appear almost identical to the continuous MAT of ideal rectangles and triangles.

Here, we shall exploit the wealth of information contained in the HMAT in order to achieve the following goals:

(1) Indicate globally salient points on the boundary of a shape which should be preserved during smoothing.
(2) Flag candidate loci for the separation of the shape into subparts.

3 Mapping Region Information onto the Boundary

Our goal is to assign each boundary point a n-tuple of values which express the point's prominence with respect to the global representation of the shape.

According to the explanations of Section 2, each component of the HMAT (i.e. each Voronoi edge e_i) supplies the following attributes:

(i) A residual $r(e_i)$ which expresses the saliency of e_i, i.e., whether e_i constitutes a rather global symmetry axis ($r(e_i)$ is large) or has been generated by protrusions on the boundary ($r(e_i)$ is small).
(ii) The level $h(e_i)$ of the skeleton pyramid e_i belongs to. Note that the top level of the skeleton pyramid (i.e., with coarse resolution) is ranked 1, increasing numbers denote *higher levels of detail.*
(iii) It is an advantage of the HMAT that both the skeleton of the foreground (*endoskeleton*), and the skeleton of the background (*exoskeleton*) can be computed simultaneously without computational overhead. A boolean operator $\mathsf{foreground}(e_i) \in \{\mathsf{true}, \mathsf{false}\}$ indicates whether edge e_i belongs to the endoskeleton or the exoskeleton.
(iv) $\mathsf{left\_point}(e_i)$ and $\mathsf{right\_point}(e_i)$ supply the boundary points which have generated e_i: according to the definition of the VD, e_i is a subset of the perpendicular bisector of $\mathsf{left\_point}(e_i)$ $\mathsf{right\_point}(e_i)$.
(v) A boolean value $\mathsf{is\_node}(e_i)$, which indicates whether one of the endpoints of e_i belongs to a branching point (or node) of the HMAT.

This information can be mapped onto the set of boundary points during a single scan of all e_i. Each boundary point p_i is assigned one or two 4-tuples $\mathcal{T}^{F,B}(p_i)$, where the superscript denotes whether $\mathcal{T}$ has been derived from an endoskeleton ($\mathcal{T}^F$) or exoskeleton ($\mathcal{T}^B$):

$$\mathcal{T}^{F,B}(p_i) = \{r_{\max}(p_i), h_{\max}(p_i), h_{\min}(p_i), \mathsf{is\_node}(p_i)\}\,, \tag{1}$$

where $r_{\max} \in \mathbb{R}$, $h_{\min}, h_{\max} \in \mathbb{N}$, $\mathsf{is\_node} \in \{\mathsf{true}, \mathsf{false}\}$.

Initially, $r_{\max}$ is assigned a sufficiently small value indicating 'negative infinity', while $h_{\max}$ and $h_{\min}$ are set to 0 and $\mathsf{is\_node}$ receives the value of false. Thereafter, the final values $r_{\max}, h_{\max}, h_{\min}$, and $\mathsf{is\_node}$ are computed as follows: Let us assume that we want to compute $\mathcal{T}^F$. For each Voronoi edge e_i of the (pruned) HMAT we first check if $\mathsf{foreground}(e_i)$ is true. If e_i belongs to the foreground, we extract the values of $r(e_i)$ and $h(e_i)$. Thereafter, the contents of $r_{\max}(\mathsf{left\_point}(e_i))$, $r_{\max}(\mathsf{right\_point}(e_i))$, $h_{\min}(\mathsf{left\_point}(e_i))$, $h_{\min}(\mathsf{right\_point}(e_i))$, $h_{\max}(\mathsf{left\_point}(e_i))$, and $h_{\max}(\mathsf{right\_point}(e_i))$ are updated. E.g., in order to update $r_{\max}(\mathsf{left\_point}(e_i))$ we evaluate $r_{\max}(\mathsf{left\_point}(e_i)) = \max(r_{\max}(\mathsf{left\_point}(e_i)), r(e_i))$. Consequently,

$$r_{\max}(p_i) = \max_{\substack{e_i \in \mathrm{HMAT},\\ \mathsf{left\_point}(e_i)=p_i \,\vee\\ \mathsf{right\_point}(e_i)=p_i}} r(e_i), \quad h_{\max}(p_i) = \max_{\substack{e_i \in \mathrm{HMAT},\\ \mathsf{left\_point}(e_i)=p_i \,\vee\\ \mathsf{right\_point}(e_i)=p_i}} h(e_i), \quad h_{\min}(p_i) = \min_{\substack{e_i \in \mathrm{HMAT},\\ \mathsf{left\_point}(e_i)=p_i \,\vee\\ \mathsf{right\_point}(e_i)=p_i}} h(e_i). \tag{2}$$

The value of $\mathsf{is\_node}(p_i)$ is set true if (1) there exists an edge $e_i \in$ HMAT, such that $\mathsf{left\_point}(e_i) = p_i \;\vee\; \mathsf{right\_point}(e_i) = p_i$, and (2) $\mathsf{is\_node}(e_i) = \mathsf{true}$.

In this paper, we shall basically investigate the information provided by endoskeletons (with one exception in Section 8). The contents of the $\mathcal{T}^F$ are used to define the set of prominent boundary points $\mathcal{Q}$ of a shape S as

$$\mathcal{Q}(S, H_{\mathrm{lo}}, H_{\mathrm{hi}}, T_{\mathrm{residual}}) \stackrel{\mathrm{def}}{=} \{p_i | h_{\min}(p_i) \in [H_{\mathrm{lo}}, H_{\mathrm{hi}}] \;\wedge\; r_{\max}(p_i) > T_{\mathrm{residual}}\}, \tag{3}$$

where H_{lo} (lower level of detail, e.g. 1) and H_{hi} (higher level of detail, e.g. 3) select the levels of the skeleton pyramid, and $T_{residual}$ allows to remove less salient points which have not been excluded from the mapping process by the preceding pruning of the HMAT. Typical values of $T_{residual}$ are of magnitude 2.0...4.0; see [7, 8] for details. A set $\mathcal{N}$ of prominent n-furcations-generating boundary points (also termed $\mathcal{N}$-points) can be then defined as

$$\begin{aligned} &\mathcal{N}(S, H_{lo}, H_{hi}, T_{residual}) \stackrel{\text{def}}{=} \\ &\stackrel{\text{def}}{=} \{p_i | h_{\min}(p_i) \in [H_{lo}, H_{hi}] \ \wedge \ r_{\max}(p_i) > T_{residual} \ \wedge \ \text{is\_node}(p_i) = \text{true}\}. \end{aligned} \tag{4}$$

4 Nonmaximum Suppression

We observe that very often two or more prominent $\mathcal{N}$-points lie very close to each other, i.e., a few pixels apart. This effect is typically due to noise influence upon the sections of the boundary which generate the n-furcation. Since cluster of prominent points may obstruct postprocessing steps such as locally-adaptive smoothing and shape decomposition, $\mathcal{N}$ is submitted to a nonmaximum suppression operation: If two or more members of $\mathcal{N}$ are found to be located within a certain interval along the boundary, only the one prominent point with the highest $r_{\max}$ value is preserved.

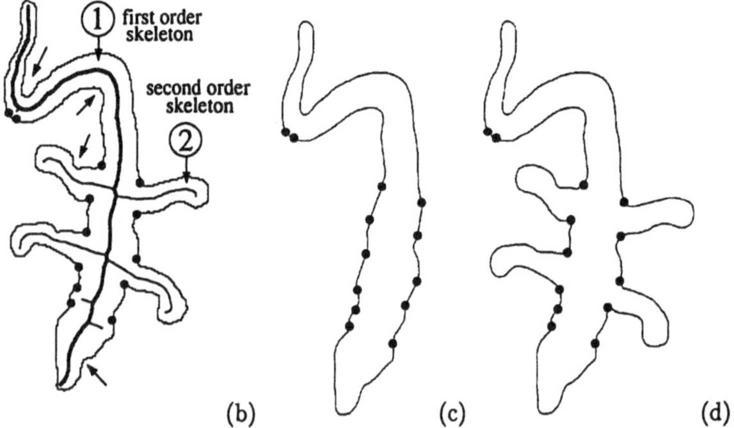

Fig. 3. HMAT-Based Shape Decomposition. (a) Branching points between the first order skeleton describing the 'spine' of the lizard and the second order skeleton of its legs indicate where a syntactically correct splitting between body and legs should take place (tiny black dots = members of $\mathcal{N}(S, H_{lo} = 1, H_{hi} = 1, T_{residual} = 3.0)$). Note that curvature analysis does not necessarily provide correct information, since strong curvature extrema (indicated by arrow signs) may occur at positions which does not lead to an intuitively appealing decomposition. (b) First order outline S_1 of the animal shape. Locally adaptive Gaussian smoothing (Section 7) has been applied in order to preserve the prominent boundary points while suppressing noisy artefacts within the remaining parts of the outline (Smoothing parameters: $\sigma_{\min} = 2.0, \sigma_{\max} = 6.0, \eta = 0.5$, see Section 7 for details). (d) Second order outline S_2, same degree of smoothing as in (c).

5 Subpart Extraction and Resampling

The decomposition of the shape into subparts requires a single traversal of the attributed boundary. In order to extract, for example, the most basic form of object S, we first compute $\mathcal{N}(S, H_{\text{lo}} = 1, H_{\text{hi}} = 1, T_{\text{residual}} = 3.0)$. Then, the first order outline S_1 comprises all boundary segments which (1) connect two elements of $\mathcal{N}$ and (2) which do not contain any other prominent boundary points p_i with $h_{\min}(p_i) > 1$. In our current implementation, gaps between adjacent members of $\mathcal{N}$ are simply closed by straight line connections. Analogously, we may extract increasingly finer descriptions of the shape, e.g., S_2 and S_3 by using $\mathcal{N}(S, H_{\text{lo}} = 1, H_{\text{hi}} = 2, T_{\text{residual}} = 3.0)$ and $\mathcal{N}(S, H_{\text{lo}} = 1, H_{\text{hi}} = 3, T_{\text{residual}} = 3.0)$, respectively. Fig. 3 contains the S_1 and S_2 representations of the 'lizard' shape.

During the extraction of the S_k, resampling assures that the distance between successive points of S_k is equal to 1, such that S_k represents a path length parameterized curve (necessary for Gaussian smoothing).

6 1D Distance Transform

The extracted contours are smoothed with a locally adaptive Gaussian filter, i.c., with a kernel that preserves the course of the outline in the vicinity of $\mathcal{N}$-points and provides a higher degree of smoothing when operating upon boundary sections far away from $\mathcal{N}$-points. Therefore, we have to know for each boundary point p_i, the distance $\mathcal{D}(p_i)$ along the boundary to the nearest $\mathcal{N}$-point: $\mathcal{D}(p_i) = \min_{a \in \mathcal{N}} d(a, p_i)$. This information can be obtained by computing a 1D distance transform along the boundary.

7 Form-Adaptive Smoothing

Let us assume that the curve $\mathrm{s}(t)$ to be smoothed is described by two coordinate functions of a path parameter t: $\mathrm{s}(t) = (x(t), y(t))$. Commonly, curve smoothing is done by convolving $\mathrm{s}(t)$ with a one-dimensional Gaussian $G_\sigma(t)$ of standard deviation σ.

A locally adaptive version of the above filter is obtained by requiring that σ becomes a function of path parameter t: $\sigma(t) = \sigma(\mathrm{s}(t))$. Therefore, the expression for the smoothed representation $\tilde{\mathrm{s}}(t)$ of $\mathrm{s}(t)$ is given by

$$\tilde{\mathrm{s}}(t) = \int_{-\infty}^{\infty} \mathrm{s}(\tau) \frac{1}{\sigma(\mathrm{s}(t))\sqrt{2\pi}} e^{-(t-\tau)^2/2[\sigma(\mathrm{s}(t))]^2} d\tau. \tag{5}$$

The value of σ is set according to the expression below,

$$\sigma(\mathrm{s}(t)) = \max(\min(\eta \mathcal{D}[\mathrm{s}(t)], \sigma_{\max}), \sigma_{\min}), \tag{6}$$

where $\mathcal{D}[\mathrm{s}(t)]$ represents the distance from $\mathrm{s}(t)$ to the nearest $\mathcal{N}$-point, η defines the linear increase of σ when moving away from a $\mathcal{N}$-point, and $\sigma_{\max}, \sigma_{\min}$ are employed to keep σ within the interval $[\sigma_{\min}, \sigma_{\max}]$. We have experimented with parameter settings of $\sigma_{\min} = 1.5 \ldots 2.0$, $\sigma_{\max} = 6.0 \ldots 8.0$, $\eta = 0.2 \ldots 0.5$. In future, the selection of smoothing parameters could be based, e.g., on a priori knowledge about the investigated shape or on scale-space analysis. Moreover, Gaussian filtering could be replaced by a smoothing operator which does not introduce the shrinkage problem [5].

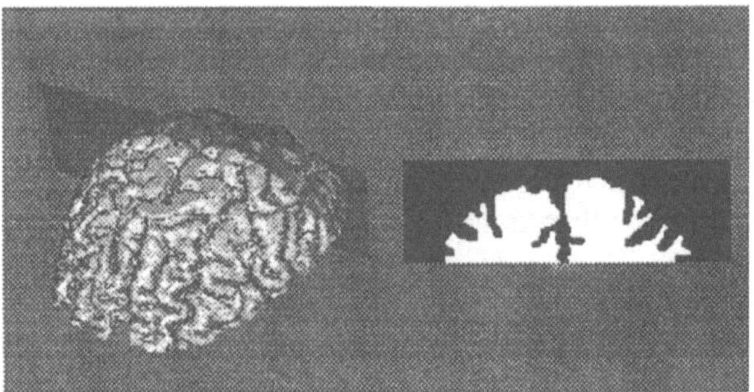

Fig. 4. 3D Coronal Slice through the Investigated Brain Dataset.

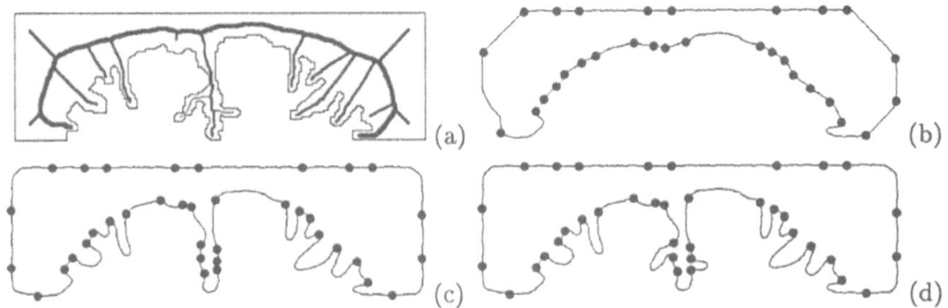

Fig. 5. Structural Decomposition of a Slice through a Human Brain Dataset. (a) Hierarchic Voronoi skeleton of the slice from Fig. 4. The arrangement of sulcal structures is mirrored by the structure and attributes of skeleton branches at hierarchy levels 2 and 3 above the base line which constitutes the first order skeleton (drawn as thick line). (b) First order approximation of the shape. The tiny black dots in the lower section are $\mathcal{N}$-points and indicate the separation between the outer hull of the brain and its inner structures. Moreover, they facilitate the transition from coarse to finer resolution by acting as reference points. Adaptive smoothing was applied to compensate for remaining noise and discretization effects upon the boundary. (c) Second order approximation of the shape. (d) Third order approximation of the shape. The sequence of approximations in (b), (c), and (d) appears to simulate the folding process during brain development (Smoothing parameters: $\sigma_{\min} = 1.5, \sigma_{\max} = 8.0, \eta = 0.5$).

8 Example: Structural Decomposition of a Slice through a Human Brain Dataset

Fig. 5 depicts the shape of a coronal slice through a human brain dataset as shown in Fig. 4. The analysis of brain structures is motivated by several hypotheses which express a possible relationship between the form and distribution of the *sulci*

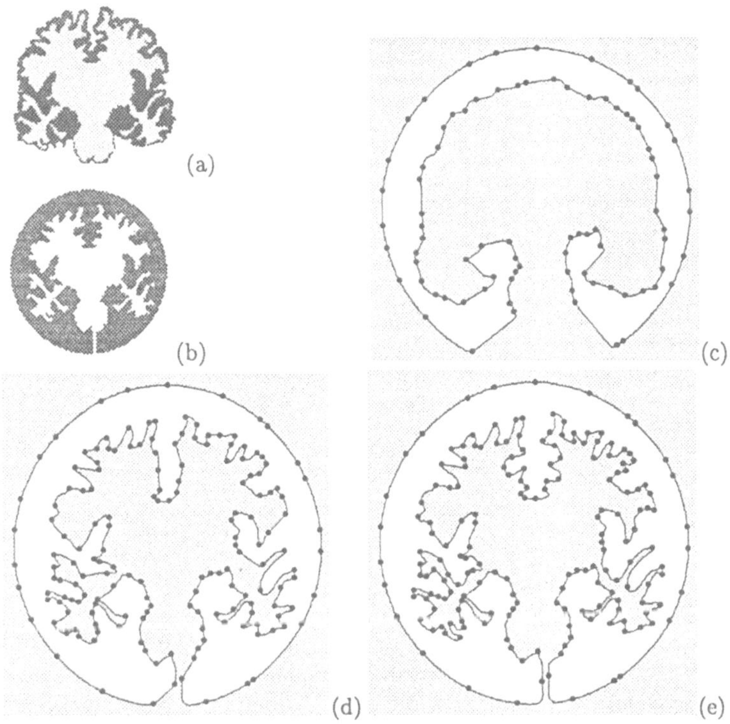

Fig. 6. HMAT-Based Decomposition of Brain Structures. (a) Cross-section from a manually segmented MRI dataset of the human brain. White matter is drawn light gray, gray matter dark gray. (b) In order to analyze the grooves of the white matter, it is flagged as background, while the gray matter is embedded within a simple shape (disk) to form the foreground object. Thus the cavities of the white matter correspond to the protrusions of the (gray) foreground object, which is then passed to the skeleton-pyramid-generating algorithm. Note that the enclosing shape has been dissected at the bottom in order to obtain one single closed outline. This is a feature of our current implementation and not a general prerequisite. (c) First order outline of the brain. (d) Second order outline. (e) Third order outline.

and *gyri* and degenerative symptoms of the brain. First, only region information from the endoskeleton is used. The HMAT of the slice is presented in Fig. 5(a). The structure of the HMAT allows to extract a sequence of approximations of the original shape S_1, S_2, S_3, while proceeding from coarse to finer descriptions. Adaptive smoothing was applied after decomposition and resampling in order to provide a smooth outline. In the sequel, the analysis of sulcal structures can set its focus on different properties of the shape: (1) applying the instruments of differential geometry to the preprocessed outline, (2) analysis of the distribution of $\mathcal{N}$-points along the boundary, or (3) structural analysis of the HMAT, in particular the length of the branches describing the sulcal fissures (second and third order skeleton) or the position of branching points between first order skeleton (base line) and second order branches.

Fig. 6 shows the corresponding results for a complete cross-section extracted

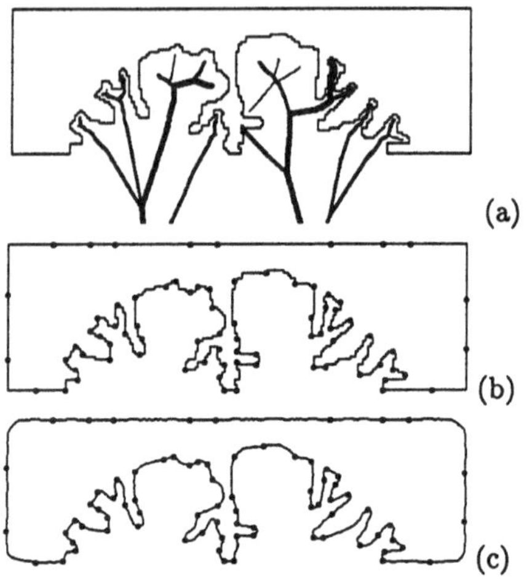

Additional information from exoskeletons allows to mark further salient loci along the boundary. (a) Hierarchic Voronoi exoskeleton. (b) The mapping procedure introduces additional $\mathcal{N}$-points (tiny black dots) at convex (with respect to the foreground) locations. (c) Outline of (b) after adaptive Gaussian smoothing. (Smoothing parameters: $\sigma_{\min} = 1.5, \sigma_{\max} = 8.0, \eta = 0.5$)

Fig. 7. Integrating Region Information from Exoskeletons.

from a different MRI dataset. This time, the structure of the white matter of the brain is analyzed.

Further information can be introduced by repeating the above process with the exoskeleton of the slice, as shown in Fig. 7(a). The complementary information of the exoskeleton introduces additional $\mathcal{N}$-points at convex locations along the boundary (Fig. 7(b)). Again, adaptive Gaussian filtering provides a smooth boundary representation; this time, the newly introduced $\mathcal{N}$-points will preserve more of the initial curvature characteristics (Fig. 7(c)).

9 Conclusions

Region information derived from hierarchic skeletons can be employed to support the analysis of an object outline. Structural properties of the HMAT can be mapped onto the boundary and serve for multiple purposes: The knowledge of prominent boundary points can help to set a focus of interest for subsequent processes such as curvature analysis by means of differential geometry. It increases the evidence of a globally salient curvature extremum, since such curvature extrema are likely to occur in the vicinity of prominent boundary points. Therefore, the establishment of prominent points could corroborate the computation of robust approximations of outlines by geometric primitives or to stabilize energy-minimization-driven processes (energy-minimizing splines or *snakes*). Moreover, it is possible to design locally adaptive filtering operations which preserve the shape's characteristics around prominent points (simultaneously reducing the need for shrinkage correction in the case of Gaussian smoothing) while only preserving gross features within globally less significant boundary sections. After the information derived from the HMAT has been assigned to the boundary points, it is possible to define a hierarchically ordered sequence of shape outlines S_k, thus performing a shape decomposition. With k increasing, each S_k incorporates more detail features of the original shape. The problem of finding a correspondence between consecutive approximations S_k, S_{k+1} of the original shape is eased by using the prominent boundary points as reference loci. Since HMAT-assisted shape analysis incorporates region information, it is much less dependent on curvature properties and therefore especially suited for the examination of complex shapes such as biological forms with a significantly jagged boundary. Its performance was demonstrated by decomposing the complex surface structures extracted from cross-sections of the human brain.

References

1. D. H. Ballard and C. M. Brown, *Computer Vision.* Prentice Hall, 1982.
2. F. P. Kuhl and C. R. Giardina, "Elliptic fourier features of a closed contour," *Computer Graphics and Image Processing*, vol. 18, pp. 236–258, 1982.
3. W. Richards and D. D. Hoffman, "Codon constraints on closed 2d shapes," in *Readings in Computer Vision: Issues, Problems, Principles, and Paradigms* (M. A. Fischler and O. Firschein, eds.), pp. 700–708, Los Altos, CA: Morgan Kaufmann Publishers, Inc., 1987.
4. D. G. Lowe, "Organization of smooth image curves at multiple scales," *International Journal of Computer Vision*, vol. 3, pp. 119–130, 1989.
5. J. Oliensis, "Local reproducible smoothing without shrinkage," in *Proc. Int. Conf. on Computer Vision and Pattern Recognition, Champaign, Illinois*, pp. 277–282, 1992.
6. F. Mokhtarian and A. Mackworth, "Scale-based description and recognition of planar curves and two-dimensional shapes," *IEEE Transactions on Pattern Recognition and Machine Intelligence*, vol. 8, no. 1, pp. 34–43, 1986.
7. R. Ogniewicz and M. Ilg, "Voronoi skeletons: Theory and applications," in *Proc. Conf. on Computer Vision and Pattern Recognition, Champaign, Illinois*, pp. 63–69, June 1992.
8. R. L. Ogniewicz, *Discrete Voronoi Skeletons.* PhD thesis, ETH Zurich, Switzerland, 1992.
9. W. Niblack, P. B. Gibbons, and D. Capson, "Generating connected skeletons for exact and approximate reconstruction," in *Proc. Int. Conf. on Computer Vision and Pattern Recognition, Champaign, Illinois*, pp. 826–828, 1992.
10. F. Leymarie and M. D. Levine, "Simulating the grassfire transform using an active contour model," *IEEE Transactions on Pattern Recognition and Machine Intelligence*, vol. 14, no. 1, pp. 56–75, 1992.
11. J. W. Brandt and V. R. Algazi, "Continuous skeleton computation by voronoi diagram," *Computer Vision, Graphics, and Image Processing*, vol. 55, no. 3, pp. 329–338, 1992.
12. O. Kübler, F. Klein, R. Ogniewicz, and U. Kienholz, "Isolation and identification of abutting and overlapping objects in binary images," in *Progress in Image Analysis and Processing*, Singapore: World Scientific Publishing Co., 1990.
13. H. Blum, "A transformation for extracting new descriptors of shape," in *Models for the Perception of Speech and Visual Form* (W. Wathen-Dunn, ed.), Cambridge MA: MIT Press, 1967.
14. L. Vincent, "Exact euclidean distance function by chain propagation," in *Proc. Computer Vision and Pattern Recognition, Maui, Hawaii*, pp. 521–525, 1991.
15. F. P. Preparata and M. I. Shamos, *Computational Geometry.* Texts and Monographs in Computer Science, New York: Springer-Verlag, second ed., 1990.

Benutzerfreundliche Bildanalyse mit *HORUS*: Architektur und Konzepte

W. Eckstein, G. Lohmann, U. Meyer-Gruhl, R. Riemer,
L. Altamirano Robles, J. Wunderwald

Technische Universität München
Institut für Informatik - Lehrstuhl Prof. Radig
{eckstein,lohmanng,meyergru,riemer,robles,wunderwa}@informatik.tu-muenchen.de

Zusammenfassung. Vorgestellt werden Architektur, Datenstrukturen und Hauptmodule des Bildanalysesystems *HORUS*. *HORUS* ist ein Werkzeug mit dem alle Stufen des Analyseprozesses von der Vorverarbeitung bis zur Interpretation problemorientiert durchgeführt werden können. Basierend auf hierarchischen Datentypen (von der Bildmatrix zu relationalen Strukturen) und funktionalen Operatoren ermöglichen verschiedene Module (Benutzerschnittstelle, Operatordatenbasis, automatische Konfiguration etc.) eine schnelle und sichere Programmentwicklung.

1 Einleitung

Das *HORUS*-System geht auf Arbeiten bei Prof. Pöppl [Eck86] zurück, die bei Prof. Radig in den letzten sechs Jahren weiterentwickelt wurden [Eck88]. Nach einer mehrfachen Reimplementierung stellt sich *HORUS* heute als ein ausgereiftes Werkzeug mit einem breit gestreuten Anwenderkreis dar, das eine große Palette von Operationen (ca. 460) aus den Bereichen Vorverarbeitung, Segmentation, Morphologie, Merkmalsgewinnung, Klassifikation, Visualisierung etc. umfaßt. Über 60 Diplomarbeiten, eine Reihe von Dissertationen ([Mes92] [Pau93] [Eck93] [Klo93]) und Industrieprojekten wurden für bzw. mit *HORUS* realisiert. Drei der für *HORUS* entwickelten Verfahren sind bei der DAGM prämiert worden [Hae86] [Lan91] [Kri92].

In der vorliegenden Arbeit soll der Aufbau von *HORUS*, wie er sich gegenwärtig darstellt, einschließlich der Erweiterungen für das nächste Jahr, vorgestellt werden.

2 Datenstrukturen

Aufbauend auf primitiven Datentypen (Basistypen) werden in *HORUS* hierarchische Strukturen erzeugt. Die Basistypen sind in numerische und ikonische Daten unterteilt. Als numerische Daten werden (variante) Tupel von ganzen Zahlen, Gleitpunktzahlen und Zeichenreihen unterstützt. Diese dienen als Steuerparameter für Operatoren und zur Konstruktion komplexer Strukturen wie Merkmalsräume oder Neuronale Netze. Auf die ikonischen Daten soll genauer eingegangen werden, da sie für das System von zentraler Bedeutung sind. Als Basistypen stehen *Bilder* und *Regionen* zur Verfügung.

Bilder sind als eine (Teil-)Menge von Pixeln eines rechteckigen Indexbereiches definiert. Als Pixeltypen gibt es Byte, Ganz- und Gleitpunktzahlen. Man kann ein Bild als eine Matrix mit einer Binärmaske interpretieren, wobei ein gesetztes Pixel in der Maske angibt, daß der Grauwert an dieser Position definiert ist. Die Maskierung der Bilder wird zur Laufzeitverbesserung von Filter- und Segmentationsoperationen verwendet.

Regionen sind beliebige Teilmengen von $\mathbb{Z} \times \mathbb{Z}$. Sie sind also nicht auf den Indexbereich der Bilder beschränkt. Dies hat entscheidenden Einfluß auf die Wirkung von Operationen (z.B. Morphologie), da per Definition keine Artefakte an Bildrändern auftreten (siehe [Eck93]). Weiterhin können sich Regionen überlappen, was einen flexiblen Umgang z.B. mit Segmentationsergebnissen oder ikonischen Modellen ermöglicht. Für die Darstellung unendlicher Regionen (die durch Komplementbildung entstehen) wurde folgende Lösung gefunden: Da sich die Komplementoperation durch Vereinigung, Durchschnitt und Differenz darstellen läßt (z.B. $\overline{A} \cup \overline{B} = \overline{A \cap B}$ oder $A \cap \overline{B} = A \backslash B$) reicht es, die Komplementbildung virtuell auszuführen (durch Setzen eines Bits) und immer mit der endlichen Version zu rechnen. Hierdurch erhält man eine formal saubere Implementierung von Regionen[1] und somit eine *korrekte* Implementierung der Morphologie.

Basierend auf den Typen Bild und Region und den darauf anwendbaren Operationen werden Objekthierarchien aufgebaut. So werden z.B. Graubilder zu Farbbildern, Einzelbilder zu Bildfolgen und zweidimensionale Bilder zu Voxelstrukturen aggregiert. Die Konstruktion erfolgt dabei in einer Objekthierarchie (Smalltalk), bei der Operationen (für Verarbeitung und Darstellung) und Eigenschaften vererbt werden. Für die ikonischen Objekte hat sich der Name *Hodget* ($\mathcal{HORUS}$-widget) eingebürgert.

Zur Interpretation von Bildern werden semantische Netze verwendet. Die Knoten der Netze sind Hodgets (z.B. Kanten, Bögen, Flächen etc.), die mit Attributen versehen sind. Die Hodgets können mit beliebigen numerischen Relationen verknüpft werden.

3 Modulstruktur

Grundlage des $\mathcal{HORUS}$-Systems ist die *Operatordatenbasis*. Auf sie greifen alle übrigen Module zu, um dynamisch Informationen über Konfiguration und Status des Systems zu erhalten. Aussagen über Operationen (z.B. bei automatischer Konfiguration von Operatorfolgen, dynamische Struktur der Benutzerschnittstelle oder Hypertext-Manual) werden aus ihr abgeleitet. Die Abwicklung des Aufrufs von Operationen (Typisierung, Wertebereiche, Fehlermeldungen etc.) wird ebenfalls mit Hilfe der Operatordatenbasis realisiert. Das nächste Modul ist das *Kernsystem*. Es beinhaltet alle Operationen, die Ein-/Ausgabe und die

[1] Abgesehen von den Regionen R, für die gilt: $|R| = |\overline{R}| = \infty$.

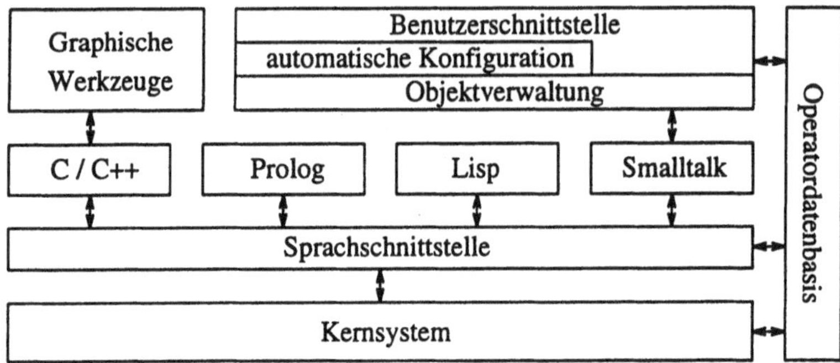

Abbildung 1. Schichtenarchitektur von *HORUS*

Verwaltung der komplexen Datenstrukturen. Angesprochen werden die Operationen durch die *Sprachschnittstelle*, die mit Hilfe eines Compilers aus der Operatordatenbasis generiert wird. Sie ermöglicht den Zugriff auf unterschiedliche, für die Bildanalyse gängige Wirtssprachen. Basierend auf *HORUS*/C werden „einfache" graphische Werkzeuge zur Entwicklung (Programmierung, Visualisierung und Online-Manual) zur Verfügung gestellt. *HORUS*/Prolog und *HORUS*/Lisp eignen sich zur Entwicklung komplexer Bildinterpretationssysteme ([Mes92] [Pau93]). Die Schnittstelle zu Smalltalk ist die Basis zur Entwicklung objektorientierter Tools.

4 Operatordatenbasis

In der Operatordatenbasis liegen bisher der Name und die Stelligkeit der Operatoren, sowie die Typisierung, Zusicherungen und Vorschlagswerte für die Parameter formalisiert vor. Zu den einzelnen Operatoren existiert eine umfangreiche natürlichsprachliche Beschreibung, die auf die Wirkung des Operators eingeht, seine Komplexität, die Operatorklasse sowie Querverweise und Alternativen angibt.

Die geplante Erweiterung des *HORUS*-Systems zu einem Werkzeug für Computer Aided Vision Engineering (CAVE) [Rad92] macht es erforderlich, auch die natürlichsprachliche Information zu formalisieren. Besondere Aufmerksamkeit verdient in diesem Zusammenhang die Wirkung der Parameteränderungen. Ein Beispiel hierfür ist etwa der Einfluß der Filtergröße eines Glättungsfilters auf die minimale Größe von Regionen. Bei einer Segmentation interessieren z.B. die Anzahl und Größe der Objekte sowie der Anteil der Rauschobjekte. Die Ergebnisse von statistischen und analytischen Untersuchungen über den Einfluß der Eingabeparameter auf das Operatorverhalten, sowie Abhängigkeiten und Wechselwirkungen zwischen Operatoren werden in der Datenbasis bereitgestellt.

Besondere Anforderungen werden an die Struktur der Datenbasis gestellt, da unterschiedliche Schichten des Systems Informationen von ihr benötigen. So erwartet beispielsweise eine Sprachschnittstelle den Namen, die Stelligkeit und die Typisierung einer Operation, während die Benutzerschnittstelle Anfragen über die Wirkung einer Parameteränderung stellt.

Eine Strukturierung der Daten ergibt sich mit der Einführung von *Molekülen*, die aus einfachen Operationen zusammengesetzte Operatoren sind (siehe Abschnitt Graphische Oberfläche). Informationen über benutzerdefinierte Moleküle z.B. über die Semantik der Parameter (Wirkungsweise, Restriktionen, Voreinstellungen) müssen in die Datenbasis eingefügt werden können. Das Wissen über die erzeugten Moleküle wird dabei, soweit möglich, aus dem Wissen über die verwendeten $\mathcal{HORUS}$-Operationen (*Atome*) propagiert, und wo dies nicht möglich ist, vom Programmierer spezifiziert. Informationen sind von unterschiedlicher Dauer:

- fest für $\mathcal{HORUS}$-Atome und für Moleküle, die zum Standardsatz gehören
- variabel für von einer Arbeitsgruppe bzw. einem Benutzer definierte Moleküle
- temporär für Moleküle, die nur in einer Sitzung Verwendung finden.

Es werden auch verschiedene Sichtweisen auf die Datenbasis angeboten, einerseits um bei der Anpassung an eine bestimmte Zielhardware die Menge der zur Verfügung stehenden Operatoren einzuschränken und andererseits um Spezialwissen von unterschiedlichen Anwendungsdomänen aufzunehmen.

Um den genannten Anforderungen gerecht zu werden und zur Anpassung an die Benutzeroberfläche ist die Realisierung der erweiterten Datenbasis in objektorientierter Technologie geplant. Die zuletzt genannten unterschiedlichen Sichten werden mit Hilfe von Vererbung realisiert.

Die Sammlung von Bildinterpretations- und Bildanalysewissen wird ergänzt (bisher 1.5 MByte) und auch unabhängig von $\mathcal{HORUS}$ einsetzbar sein. Die ausführliche Dokumentation der einzelnen Operatoren unterstützt die Wiederverwendbarkeit der Software. Auch bietet die Operatordatenbasis die Grundlage für ein verbessertes Hilfesystem (Hypertext-Manual) und ein Tutorsystem [Klo93].

5 Automatische Konfigurierung

Das künftige $\mathcal{HORUS}$-System wird die Produktivität bei der Erstellung von Bildanalyseanwendungen weiter steigern. Standardmäßige, einfach spezifizierbare Bildanalyseaufgaben muß der Entwickler nicht mehr en detail ausführen, sondern er kann dies einem System zur automatischen Konfiguration überlassen.

[Mes92] stellt einen prototypischen Konfigurierer für $\mathcal{HORUS}$-Operatoren (FIGURE) vor, der beispielsweise bei der Analyse von Straßenverkehrsszenen mit morphologischen Operatoren gute Ergebnisse erzielt.

FIGURE folgt dem hierarchischen Planungsparadigma ([McD82]): Der Entwickler spezifiziert seine Bildanalyseaufgabe in symbolischer Weise oder indem er das zu erkennende Objekt im Eingabebild interaktiv markiert. FIGURE plant nun zunächst eine Folge abstrakter Bildverarbeitungsoperatoren, verfeinert diese zu *HORUS*-Operatoren und wählt für alle in Betracht kommenden Operatorsequenzen geeigenete Parameter. Schließlich entscheidet die Anwendung der synthetisierten Programme auf das Eingabebild über das Ergebnis des Konfigurierungsprozesses, wobei die Übereinstimmung der eingezeichneten Kontur mit der entdeckten Region den Ausschlag gibt.

Die obige Vorgehensweise funktioniert nur für eine eingeschränkte Menge von Operatoren und Bildern. Deshalb verzahnt der neue Konfigurierer die Planung mit der Ausführung des Bildanalyseprogramms. Für die schrittweise Planung entwickeln wir Beurteilungskriterien für Zwischenergebnisse der Bildanalyse. Die Fähigkeit, einen nächsten Operator im Hinblick auf ein Analyseziel zu wählen, erweist sich als nützlich beim graphisch arbeitenden CAVE zur kontextsensitiven Vorschlagsgenerierung.

Im Gegensatz zum bisherigen Prototypen arbeitet der künftige Konfigurierer mit dem vollen Umfang der *HORUS*-Operatoren, wie er in der weiterentwickelten Operatordatenbasis repräsentiert ist. Besonderes Augenmerk legen wir auf die Gewinnung von Wissen über günstige Operatorenwahl und –reihenfolgen, sowie Parametrisierung. Dieses Wissen hängt stark von der Aufnahmetechnik und den Bildinhalten ab. Wir untersuchen, inwieweit sich hier Klassen bilden lassen um zu einer sinnvollen Taxonomie zu gelangen.

FIGURE erlaubt nur die Erkennung einzelner, unstrukturierter Objekte. Meistens interessieren jedoch strukturierte Objekte, wobei erst die Lagebeziehungen zwischen Teilobjekten eine robuste Objekterkennung ermöglichen. Darum verwendet der neue Konfigurierer die neuerdings in *HORUS* vorhandene Möglichkeit, relationale Strukturen darstellen zu können.

Zu den Fähigkeiten des geplanten Synthesesystems soll auch die Erfüllung gewisser Restriktionen bei der Planung gehören. Beispielsweise erfordern viele Anwendungsbereiche die Einhaltung von Zeitschranken. Ebenso wollen wir für spezielle Bildverarbeitungsrechner mit einer eingeschränkten Zahl von Operatoren konfigurieren können. Für diese Zwecke soll der eigentliche Planungsprozeß unverändert bleiben, aber eine modifizierte Sicht auf die Wissensbasis haben.

6 Automatische Generierung von Objektmodellen

Ein weiteres Modul in *HORUS* dient zur automatische Generierung von Modellen für zweidimensionale, strukturierte Objekte. Darüberhinaus werden die Datenstrukturen für die Repräsentation der relationen Strukturen aufgebaut und verwaltet. Im einzelnen werden folgende Funktionen realisiert:

Automatische Verbesserung: Ausgangspunkt ist ein Einzelbild, welches das zu modellierende Objekt ohne Fremdobjekte beinhaltet. Es wird vorausgesetzt, daß das Objekt zweidimensional ist. Durch eine Analyse des Hindergrundes wird das Rauschverhalten geschätzt, um gegebenenfalls geeignete Operatoren und deren Parametrisierung auszuwählen und die Bildqualität zu verbessern ([Kun90]).

Automatische Kantenberechnung: In dem verbesserten Bild werden Kanten für die Objektsegmentation extrahiert. Zur Bewertung des Kantenbildes werden Gütemaße wie Kantenlänge, Kantenanzahl, usw. abgeleitet um eine automatische Adaption der Filterparameter (Filtergröße, Trennfrequenz etc.) zu erhalten.

Perceptual Grouping: Zur Reduktion irrelevanter Kanten wird das Verfahren des *Perceptual Grouping* eingesetzt. Als Ergebnis erhält man eine Menge von zusammenhängenden Kanten, d.h. Kanten, die die Gruppierungsbedingungen von Kolinearität, Parallelität und Adjazenz nicht erfüllen, werden entfernt. Wichtig dabei ist, daß die Gruppierung ohne Vorwissen über Gestalt der Objekte durchgeführt wird.

Merkmalsgewinnung: Der erste Schritt zum Aufbau des Modells ist die Generierung der Knoten in der zugrundeliegenden relationalen Struktur. Dies sind Merkmale, die aus den Kanten und dazwischenliegenden Regionen abgeleitet werden. Beispiele hierfür sind Geradenstücke, Bögen, Kreise, Polygone, sowie Fläche, Form, Intensität, Farbe etc.

Berechnung der Schlüsselmerkmale: Um einen effizienten Zugriff auf die Knoten (Merkmale) für das Matching zu ermöglichen, wird deren Signifikanz nach dem Ansatz von [Kno86] bewertet und entsprechend sortiert. Verwendet wird hierfür beispielweise die Häufigkeit oder die Komplexität eines Merkmals.

Relationen zwischen Merkmalen: Im nächsten Schritt werden Relationen zwischen Merkmalen, d.h. Kanten zwischen den Knoten in der relationalen Struktur, aufgebaut. Ein Beispiel für eine solche Relation ist die Lagebeziehung „`Merkmal A ist oberhalb von Merkmal B`“ oder „`Merkmal A berührt Merkmal B`“. Hierdurch wird die räumliche Struktur des Modells aufgebaut. Anfragen an das System, wie etwa: „`Was liegt in der Nähe des Merkmales A`“ können hiermit beantwortet werden. Der gleiche Ansatz kann verwendet werden, um Relationen zwischen Modellen zu erzeugen.

Modellbewertung: Abschließend wird das Modell bewertet. Hierzu wird ein Matching des Modells mit Objekten (d.h. mit deren Modell) in anderen Bildern durchgeführt. In diesen Bildern können auch Objekte auftreten, die nicht trainiert wurden. Das Matching wird mit Hilfe einer heuristischen Graphsuche durchgeführt ([Gri90]). Durch eine Bewertung des Vergleichs erhält man einen Faktor, der angibt, wieviel Prozent der gesamten Objekte erkannt wurden. Dieser Faktor wird verwendet, um die Qualität eines Modells zu bewerten. Aus der Graphsuche werden weiterhin Informationen abgeleitet, die eine Aussage über den Grund eines teilweisen oder vollständigen Mismatch erlauben. Hiermit kann bei Bedarf die Modellgenerierung iterativ adaptiert werden.

Automatischer Aufbau der Modellbasis: Das Modell wird nach der Generierung in der Datenbank abgelegt, die eine effiziente Verwaltung vieler Modelle erlaubt. Günstig hierfür erweist sich eine Binärkodierung der Schlüsselmerkmale und Relationen, wodurch eine speicherplatzsparende Ablage und gleichzeitig ein schneller Zugriff ermöglicht wird ([Bre90]).

7 Graphische Oberfläche

Von besonderer Wichtigkeit für die Akzeptanz eines Bildanalysesystems ist die Möglichkeit zur einfachen, intuitiven Bedienung. Ohne eine geeignete graphische Oberfläche ist es heute nicht mehr praktikabel, ein komplexes System (*HORUS* umfaßt über 6 MByte Quellcode für die Bildanalyse) zu beherrschen.

Dies gilt für Anwender, die sich z.B. für Schulungszwecke mit dem Thema Bildanalyse befassen und die deshalb ein Werkzeug zum einfachen Einarbeiten mit größtmöglicher Unterstützung benötigen, aber genauso für professionelle Benutzer, die in sehr kurzer Zeit prototypische Lösungen erstellen wollen, ohne dabei auf die Mächtigkeit von einer Programmiersprache verzichten zu wollen.

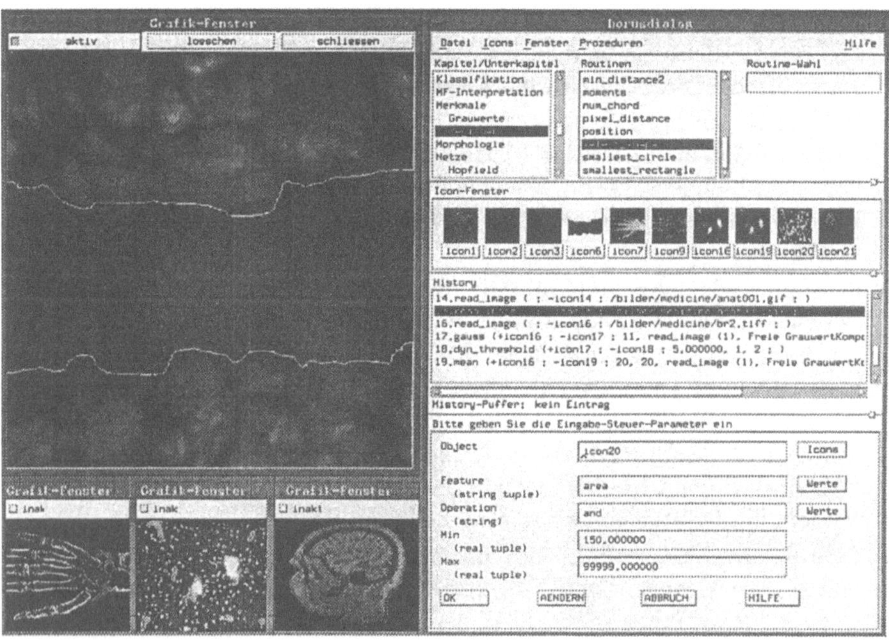

Abbildung 2. *HORUS*dialog

Für *HORUS* gibt es seit einiger Zeit eine solche Oberfläche namens *HORUS*dialog (siehe Abb. 2), die Rapid Prototyping im Bereich der Bildanalyse erlaubt. Der Benutzer kann auf die nach taxonomischen Gesichtspunkten gegliederten *HORUS*-Operatoren zugreifen und bekommt Parameter vorgegeben, die er abändern kann. Die bei den einzelnen Schritten entstehenden Hodgets werden als Icons dargestellt und das jeweils aktuelle wird im aktiven Fenster ausgegeben. Diverse Parameter, wie z.B. Darstellungsfarben oder Darstellungsform (konvexe Hülle, Regionenrand usw.) sind bequem einstellbar. Zur Zeit wird *HORUS*dialog u.a. im Bildanalyse-Praktikum der TU-München eingesetzt.

In Zukunft wird in Smalltalk eine Programmierumgebung erstellt werden, die noch konsequenter graphisch orientiert ist als $\mathcal{HORUS}$dialog. Mit diesem Werkzeug wird es möglich sein, auch komplexe Programme zu kreieren. Soweit sinnvoll, bleibt dabei die textuelle Ebene der Programmierung aus dem Spiel.

Die $\mathcal{HORUS}$-Operatoren werden für Smalltalk *funktional*, d.h. möglichst ohne Seiteneffekte, aufbereitet. Die $\mathcal{HORUS}$-Bibliothek kann zur Zeit 2D-Bilder mit beliebig vielen Grauwertkanälen und Regionendaten verarbeiten. Ein Nebeneffekt der Vererbungsmechanismen besteht in der Möglichkeit zur Überdefinition der meisten Methoden, so daß z.B. Bildfolgen oder Farbbilder mit demselben Operator gefiltert werden können wie Einzelbilder, indem Objekte (Hodgets) aus 2D-Bildern zusammengesetzt werden.

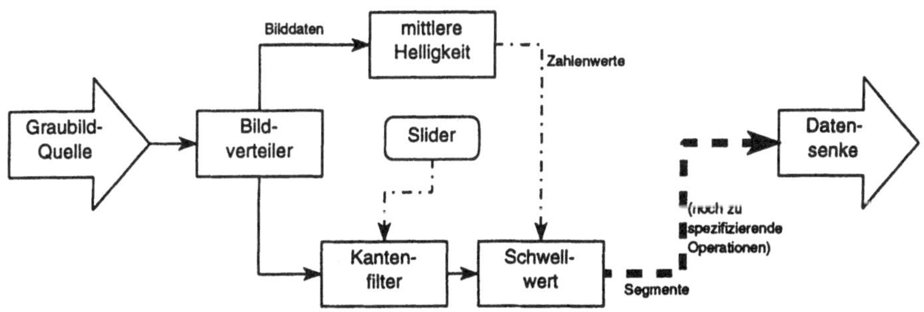

Abbildung 3. Graphisches Editieren einer Kantensegmentation

Die $\mathcal{HORUS}$-Operatoren (Methoden) werden in dem neuen Werkzeug als Icons dargestellt, während die Kanten zwischen den Icons den Datenfluß repräsentieren. Zudem werden Datenquellen für die verschiedenen Eingabeobjekte (z.B. Ganzzahlen: Slider, Listen: Popups, Bilder: Fileselector) und Datensenken zur Visualisierung (z.B. Graphikfenster) verwendet (siehe Abb. 3).

Die Visualisierer für die verschiedenen Datentypen können an jede gewünschte Kante „angehängt“ werden. Während der Programmerstellung werden Veränderungen an Parametern sofort sichtbar gemacht, um eine interaktive Rückkopplung zu gewährleisten.

Der Benutzer wird auf verschiedenen Ebenen bei der Auswahl und Parametrisierung unterstützt. Einige Kriterien lassen sich bereits lokal, andere aber nur global entscheiden. Wenn z.B. in einem längeren Segmentierungsprozeß Grauwerte auf Regionendaten abgebildet werden sollen, ist a priori nicht klar, an welcher Stelle konkret dieser Übergang stattfinden soll. Solche Probleme können nur mittels einer Planungskomponente gelöst werden, die sich modular in die Oberfläche einfügt (siehe: Automatische Konfigurierung). Lokale Restriktionen können statisch, wie z.B. Anzahl und Typ der Parameter oder dynamisch, wie z.B. Maximum und Minimum von Schwellwerten in Abhängigkeit vom Pixeltyp (long oder byte) eines Bildobjektes sein. Aus der Operatordatenbasis werden auch Informationen über die Wirkungsweise von Parametern abgeleitet, z.B. ob der Wirkungszusammenhang eher linear oder logarithmisch ist, um zu „zielorientierten“ Parametern zu kommen. Für die höheren Ebenen der Benutzerunterstützung, z.B. Hilfe bei der Auswahl des nächsten Schrittes im Bildanalyseprozeß, ist die Planungskomponente zuständig.

Diese Programmierumgebung schafft die Möglichkeit, ein erklärtes Ziel unserer Arbeit zu realisieren, nämlich einen Satz von Operatoren zu kreieren, die der menschlichen Perzeption näherkommen, als dies die „Low-Level-Operatoren“ heute tun. Dazu sollen mit dem graphischen Editor erstellte Programm-Module nicht nur als ablauffähige Programme, sondern als Moleküle wieder in die Operatordatenbasis abgelegt werden, so daß Operationen auf höherer Abstraktionsebene erzeugt und später auch wiederverwendet werden können.

Literatur

[Bre90] T. M. Breuel. *Indexing for visual recognition from a large model base.* AI Lab. Memo 1108, 1990.

[Eck86] W. Eckstein, S.J. Pöppl. *PSIWAG - A Language for Logic Programming in Image Analysis.* Proc. 8th ICPR, Paris, 1986

[Eck88] W. Eckstein. *Das ganzheitliche Bildverarbeitungssystem HORUS.* Proc. 10. DAGM Symposium, Zürich, 1988

[Eck93] W. Eckstein. *Die Bildanalysesprache TRIAS.* Dissertation. Infix-Verlag, St. Augustin, 1993.

[Gri90] W. E. L. Grimson. *Object Recognition by Computer: The Role of Geometric Constraints.* MIT Press, Cambridge, Massachusetts, London, England, 1990.

[Hae86] S. Haenel, W. Eckstein. *Ein Arbeitplatz zur halbautomatischen Luftbildauswertung.* Proc. 8. DAGM Symposium, Paderborn, 1986

[Klo93] K. Klotz. *Eine mehrschichtige Architektur zur Fehlerdiagnose und Fehlerbehebung bei der Entwicklung von logischen Programmen.* Dissertation, Infix-Verlag, St. Augustin, 1993.

[Kno86] T. F. Knoll, R. C. Jain. *Recognizing partial visible objects using features indexed hypotheses.* Trans. Robotics Automation RA-2(1), 3-14, 1986.

[Kri92] H. Kristen, O. Munkelt. *Markov-Feld-basierte Bildinterpretation mit automatisch generierter Datenbasis* Proc. 14. DAGM Symposium, Dresden, 1992.

[Kun90] M. K. Kundu, S. K. Pal. *Automatic selection of object enhancement operator with quantitative justification based on fuzzy set theoretic measures.* Pattern Recognition Letters, 11:811-829, Dezember 1990.

[Lan91] S. Lanser, W. Eckstein. *Eine Modifikation des Deriche-Verfahrens zur Kantendetektion.* Proc. 13. DAGM Symposium, München, 1991

[McD82] J. McDermott *R1: A Rule-Based Configurer of Computer Systems.* Artif.Intell. 19, S.39–88, 1982

[Mes92] T. Messer. *Wissensbasierte Synthese von Bildanalyseprogrammen.* Dissertation, Infix-Verlag, St. Augustin, 1992.

[Pau93] J. Pauli. *Erklärungsbasiertes Computer-Sehen von Bildfolgen.* Dissertation, Infix-Verlag, St. Augustin, 1993.

[Rad92] B. Radig, W. Eckstein, K. Klotz, T. Messer, J. Pauli. *Automatization in the Design of Image Understanding Systems.* Proc. 5th International Conference, IEA/AIE-92, Paderborn, Springer-Verlag 1992, LNAI 604, S. 35-45

Anwendungen von Mustererkennungstechniken bei der Interpretation digitalisierter Fertigungszeichnungen

Thomas Bartsch
SerCon GmbH, Pappelallee 33, 22089 Hamburg
Dierk G. Feldmann, Volker Junge
AB Konstruktionstechnik I, TU Hamburg-Harburg, Denickestr. 17, 21071 Hamburg

Zusammenfassung
Die rechnerunterstützte Erfassung konventionell erstellter technischer Zeichnungen zur Weiterbearbeitung in CAD-Systemen konnte bisher noch nicht befriedigend gelöst werden. Dieser Beitrag stellt ein wissensbasiertes Bildverarbeitungssystem vor, das mit objektorientierten Techniken prototypisch in C++ realisiert werden konnte. Ausführlicher werden dabei Techniken zur rotations- und skalierungsinvarianten Zeichenerkennung und zur Rekonstruktion *feature*-basierter 3D-Volumenmodelle diskutiert.

Abstract
The computer-aided capture of manually produced technical drawings into CAD databases as full-functional models has yet not been solved sufficiently. This paper presents an new concept of a knownledge-based image processing system, which has been implemented as a prototype by using object-oriented techniques in C++. Two topics are discussed in extend: rotational- and scale-invariant character recognition and feature-based reconstruction of 3D volume models.

1 Einführung

Um auch in Zukunft bei steigenden Ansprüchen der Abnehmer und kürzer werdenden Produktzyklen wettbewerbsfähig zu sein, müssen Industrieunternehmen die Qualität der hergestellten Produkte verbessern und gleichzeitig die Herstellungskosten und Durchlaufzeiten minimieren. Betrachtet man einmal genauer, wo die Herstellkosten eines Produktes festgelegt und verursacht werden, so stellt man fest, daß ein Großteil der Kosten durch den Entwurf und die Konstruktion festgelegt werden. Diese Kosten können durch ein strategisches Informationsmanagement für Entwicklung, Konstruktion und Fertigung reduziert werden, das jedem Bearbeiter von seinem Arbeitsplatz aus einen direkten Zugriff auf alle für seine Tätigkeiten relevanten Informationen ermöglicht.

Betrachtet man die Ist-Situation allein in den Bereichen Entwicklung und Konstruktion (Bild 1), so zeigt sich, daß es sich bei den heute eingesetzten DV-Werkzeugen überwiegend

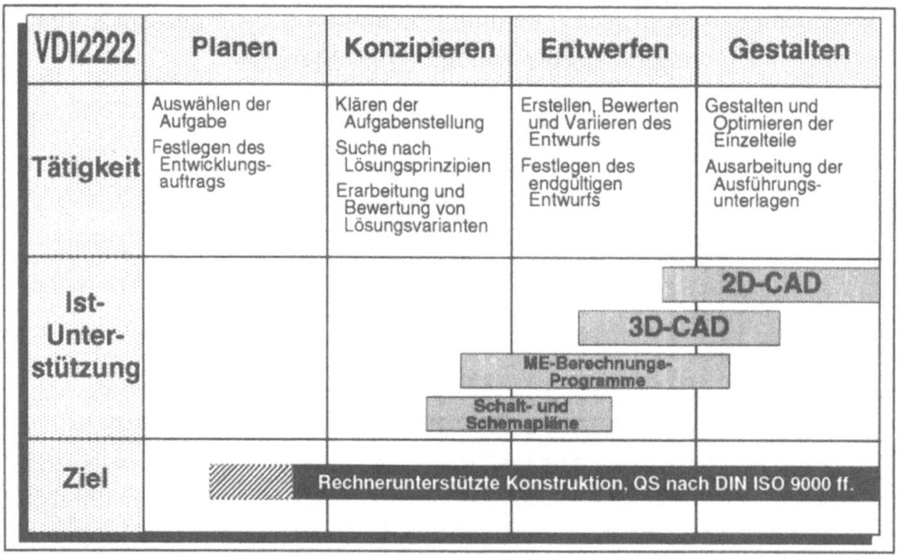

VDI2222	Planen	Konzipieren	Entwerfen	Gestalten
Tätigkeit	Auswählen der Aufgabe Festlegen des Entwicklungsauftrags	Klären der Aufgabenstellung Suche nach Lösungsprinzipien Erarbeitung und Bewertung von Lösungsvarianten	Erstellen, Bewerten und Variieren des Entwurfs Festlegen des endgültigen Entwurfs	Gestalten und Optimieren der Einzelteile Ausarbeitung der Ausführungsunterlagen
Ist-Unterstützung		Schalt- und Schemapläne ME-Berechnungs-Programme	Schalt- und Schemapläne ME-Berechnungs-Programme 3D-CAD 2D-CAD	ME-Berechnungs-Programme 3D-CAD 2D-CAD
Ziel	Rechnerunterstützte Konstruktion, QS nach DIN ISO 9000 ff.			

Bild 1 Ist-Situation der DV in Entwicklung und Konstruktion.

um lose gekoppelte Insel-Lösungen handelt, in denen jede Applikation ihre eigenen Datenbasen verwaltet (Bild 2, links).

Dabei auftretende Inkonsistenzen zwischen den Datenbeständen können vermieden werden, wenn alle Applikationen auf einem gemeinsamen Produktmodell arbeiten (Bild 2, rechts), das als Menge aller Daten definiert ist, die zur Herstellung und Anwendung eines Produktes benötigt werden und während der Zeit seiner Nutzung anfallen. Ein Produktmodell wird aber in vielen Fällen nur dann vollständig sein, wenn auch vorhandene, konventionell erstellte Dokumente aufgenommen werden können. In diesem Beitrag wird eine Lösung vorgestellt, mit der technische Zeichnungen mit voller Funktionalität in den Datenbestand von CAD-Systemen übernommen und dort wie andere Dokumente weiterbearbeitet werden können.

2 Produktmodelle unter dem Aspekt der Qualitätssicherung

Bei der Einführung von Qualitätssicherungssystemen nach DIN ISO 9000 ff. und der Erarbeitung von Prozeßbeschreibungen wird ein besonderes Augenmerk auf die lückenlose Dokumentation gelegt, um Ursachen für Qualitätsmängel und Potentiale für Qualitätsverbesserungen einfacher erkennen zu können. Diese Anforderungen können durch die Verwendung von Produktmodellen unterstützt werden:

- Der Zeitaufwand für Suchen im Archiv und Kosten für Vervielfältigungen werden reduziert.

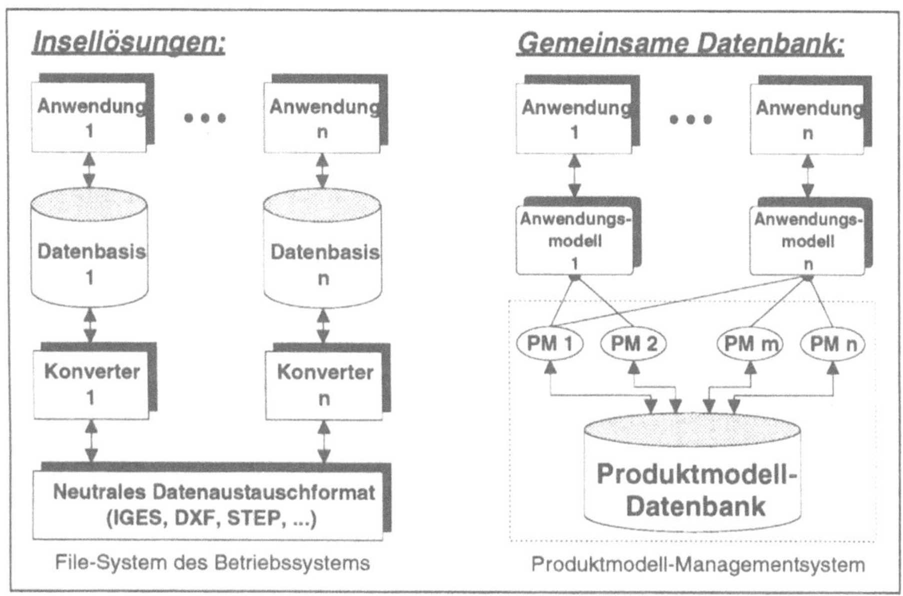

Bild 2 Integration von DV-Werkzeugen.

- Der Zugriff auf Dokumente kann über Benutzerprofile gesteuert werden.
- Benutzer müssen nicht den physikalischen Aufbewahrungsort eines Dokumentes kennen.
- Kataloge, Resultate von Berechnungen und Werkstoffdaten können ebenso wie Normen und Richtlinien in der Produktmodell-Datenbank abgelegt und allen Mitarbeitern zur Verfügung gestellt werden.

Die heute übliche Vorgehensweise bei der Erfassung vorhandener Papierdokumente ist das Einscannen und die Archivierung in Form von Rasterdaten. Insbesondere für technische Anwendungen wäre es aber wünschenswert, wenn die Dokumente mit voller Funktionalität in die vorhandene DV-Umgebung übernommen und gemeinsam z.B. mit CAD-Zeichnungen weiterverarbeitet werden könnten. Für technische Zeichnungen bedeutet 'volle Funktionalität', daß die in der Zeichnung vorhandenen Linien- und Texttypen richtig erkannt werden und die Geometrie maßstäblich erfaßt wird. Dazu müssen technische Zeichnungen nach dem Einscannen auf der Grundlage der für diese Dokumente gültigen Normen analysiert und interpretiert und die Geometrie anhand der Bemaßung ausgerichtet werden.

3 Konzept einer rechnerunterstützten Zeichnungserfassung

An der TU Hamburg-Harburg wird im Rahmen des Projektes MERLIN (Maschinelle Erfassung und Interpretation technischer Linienzeichnungen) ein System zur Erfassung, Analyse und Interpretation digitalisierter Linienzeichnungen entwickelt, das in Bild 3 dargestellt ist *[1,2]*. Die Bearbeitung ist in Vektorisierung, Bildanalyse, Zeichnungsanalyse und 3D-Rekonstruktion untergliedert.

In der Vektorisierung wird die digitalisierte Zeichnung mit einem schnellen, parallelen Verfahren skelettiert, segmentiert und polygonalisiert. Kreise und Kreisbögen werden mit einem Relaxationsverfahren erkannt. Die Zeichenerkennung wird über das unten beschriebene *Structure-Matching*-Verfahren durchgeführt. Bei der Bildanalyse werden die erzeugten Elemente zu CAD-Objekten wie Konturlinien, Mittellinien, Maßen, Symbolen etc. zusammengefaßt. Die dabei entstehende CAD-Zeichnung wird durch die Zeichnungsanalyse auf Vollständigkeit, Eindeutigkeit und Einhaltung der Normen überprüft: die dargestellte Geometrie wird nach der vorhandenen Bemaßung - auch ansichtenübergreifend - ausgerichtet, so daß als Resultat eine vollfunktionale 2D-CAD-Zeichnung entsteht. Im vierten Schritt wird aus der CAD-Zeichnung ein *feature*-basiertes Volumenmodell *[5]* erzeugt.

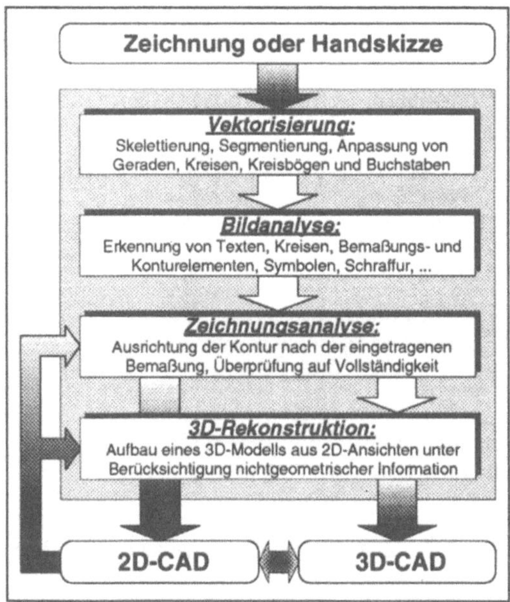

Bild 3 Schematischer Aufbau von MERLIN.

4 Rotations- und skalierungsinvariante Zeichenerkennung

Die Zeichenerkennung ist bei technischen Zeichnungen im Vergleich zur Zeichenerkennung bei zeilenorientierten Textdokumenten, für die bereits seit einiger Zeit leistungsfähige OCR-Software verfügbar ist, in mehrfacher Hinsicht erschwert:

- Zeichen können unter verschiedenen Winkeln ausgerichtet sein.
- Zeichen können unterschiedliche Schrifthöhen besitzen. Die Schrifthöhe kann dabei einen Hinweis auf die Verwendung des Zeichens in der Zeichnung geben, wie z.B. bei der Unterscheidung von Maßzahlen und Toleranzangaben.
- Zeichen können von Linien geschnitten werden.

- Neben Buchstaben und Ziffern enthalten technische Zeichnungen auch Sonderzeichen und Symbole wie z.B. Symbole für Form- und Lagetoleranzen oder Bearbeitungsanforderungen.

Aus den ersten beiden Punkten ergibt sich, daß die für die Zeichenerkennung zu entwickelnden Verfahren rotations- und skalierungsinvariant sein müssen. Darüber hinaus muß die Zeichenerkennung wegen der großen Zahl der zu erkennenden Zeichen verschiedener Schriftarten (Norm-, Handschriften) so strukturiert werden, daß eine leichte Erweiterbarkeit gewährleistet ist.

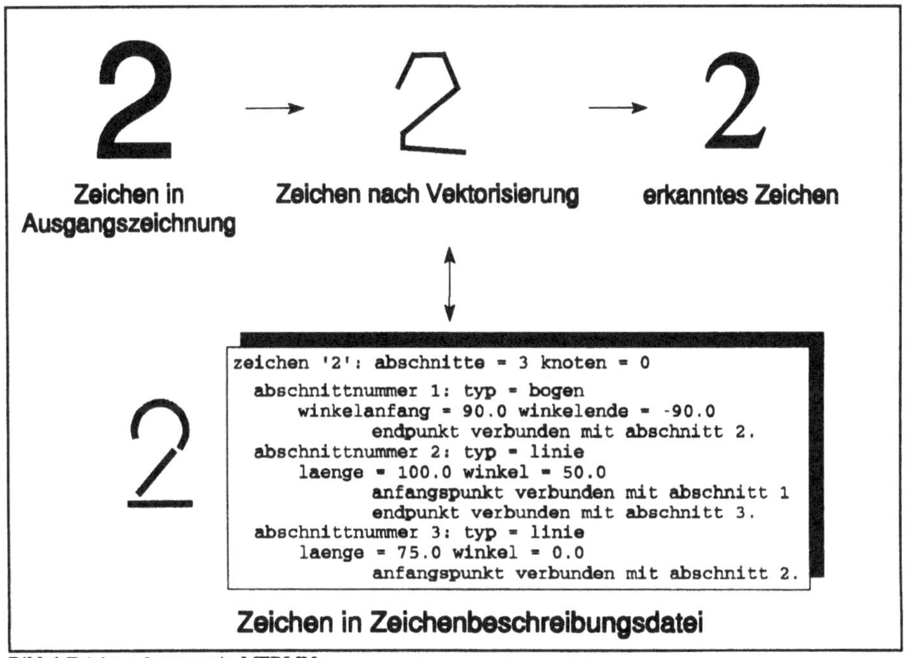

Bild 4 Zeichenerkennung in MERLIN.

Bei der Vektorisierung einer nach dem Scannen zunächst als Rastergrafik vorliegenden technischen Zeichnung werden aus den Pixeln Basiselemente wie Linien, Kreise und Kreisbögen rekonstruiert. Die in MERLIN verwendete Zeichenerkennung vergleicht die durch die Vektorisierung erhaltene Anordnung der Basiselemente mit Mustern, die in einer Zeichenbeschreibungsdatei abgelegt sind. Zeichen werden über charakteristische Abschnitte wie Linien oder Kreisbögen und deren Topologie beschrieben. Da die Form und die Anordnung der bei der Vektorisierung erhaltenen Abschnitte weitgehend unabhängig von der Schrifthöhe und -richtung ist, erfüllt dieser Ansatz die Anforderungen an Rotations- und Skalierungsinvarianz. Bild 4 zeigt am Beispiel des Zeichens '2' die Aufteilung in Basiselemente bei der Vektorisierung und die Beschreibung des Zeichens in der Zeichenbeschreibungsdatei. Die charakteristischen Abschnitte des Zeichens '2' sind ein Kreisbogen und zwei gerade Linien.

Die bei MERLIN verwendete Zeichenerkennung steht vor dem Problem, daß sich die Ergebnisse der vorangehenden Vektorisierung für das gleiche Zeichen oft in Einzelheiten unterscheiden. So ist in Bild 4 (oben mitte) zu erkennen, daß sich zwischen den beiden langen Linien der '2' noch ein kurzes Linienstück befindet. Aufgrund des bei der Vektorisierung verwendeten Verfahrens liegt außerdem der Kreisbogen der '2' nur als Polygonzug vor, der nicht interpretiert und in einen Kreisbogen umgewandelt wurde. Da der Polygonzug nur aus 3 Linien besteht, wäre eine solche Umwandlung in einen Kreisbogen auch sehr unsicher.

Um die Zeichenerkennung nicht durch solche Probleme zu erschweren, werden Linienstücke erst ab einer vorgegebenen Mindestlänge für die Zeichenerkennung ausgewertet. Ferner wird geprüft, ob sich Polygonzüge an Kreise oder Kreisbögen anpassen lassen. Diese Überprüfung wird gezielt für die Zeichen vorgenommen, für die in der Zeichenbeschreibungsdatei Kreise oder Kreisbögen eingetragen sind. Durch die gezielte Überprüfung kann auch für einen wie in Bild 4 nur aus 3 Linien bestehenden Polygonzug eine relativ zuverlässige Zuordnung zu einem Kreisbogen durchgeführt werden. Schließlich können durch den gezielten Vergleich mit einem vorgegebenen Muster auch Linien herausgefiltert werden, die durch ein Zeichen hindurchlaufen. Aufgrund dieser Vorverarbeitung vor der eigentlichen Zeichenerkennung muß für jedes Zeichen nur die charakteristische Grundform beschrieben werden, ohne daß Details der Vektorisierung beachtet werden müssen.

Die zu erkennenden Zeichen werden in der von den Autoren entworfenen Sprache *CDL* (*character description language*) beschrieben. Die Beschreibung eines Zeichens enthält Angaben über die Anzahl der Abschnitte und deren Verbindungen miteinander, die relativen Längen und Winkel der Abschnitte sowie die Typen (gerade Linie, Kreis, Kreisbogen). Mehrere Zeichenbeschreibungen werden in einer Zeichenbeschreibungsdatei zusammengefaßt, die eine normale Textdatei ist.

CDL ist eine Sprache mit einer LR(1)-Grammatik, Scanner und Parser werden mit Hilfe der Tools *bison (yacc)* und *flex (lex)* erzeugt. Damit kann die Sprache bei Bedarf einfach um neue Sprachelemente erweitert werden. Die Reihenfolge der Auswertung wird über die Vergabe von Prioritäten für die Spezifikationen der einzelnen Zeichen gesteuert und kann auf zwei Weisen erfolgen.

- Bei der interpretierenden Version wird der Parser selbst in das Bildverarbeitungssystem integriert, die *CDL*-Datei wird zur Laufzeit interpretiert. Die Lösung bietet eine hohe Flexibilität, da eine Anpassung an verschiedene Zeichensätze durch Austausch der Beschreibungsdatei erfolgen kann. Nach einer Änderung der Zeichenbeschreibungen muß lediglich die *CDL*-Datei neu geladen werden.
- Bei der kompilierenden Version wird die *CDL*-Datei mit einem Compiler in C++-Code übersetzt, der fest an das Bildverarbeitungssystem gebunden wird. Diese Lösung

bietet das beste Laufzeitverhalten, nach einer Änderung der *CDL*-Datei muß das Bildverarbeitungssystem aber teilweise kompiliert und neu gebunden werden.

Die Zeichenerkennung findet in mehreren Schritten statt. Zuerst wird eine kontextfreie Zeichenerkennung durchgeführt. Gruppen von Basiselementen werden ohne Berücksichtigung ihrer Umgebung daraufhin überprüft, ob sich eine Übereinstimmung mit einem der in der Zeichenbeschreibungsdatei abgelegten Muster erkennen läßt. Bei einem Zeichen wie z.B. der '1', das oft nur in eine einzelne, kurze Linie vektorisiert wird und mit einem Linienabschnitt einer Strich- oder Strichpunktlinie verwechselt werden könnte, ist dagegen eine kontextabhängige Betrachtung erforderlich. Dabei wird geprüft, ob sich eine Gruppe von Basiselementen, bei der es sich um ein Zeichen handeln könnte, in der Nähe von bereits erkannten Zeichen befindet und sich mit diesen zu einer Zeichenkette zusammenfassen läßt. In diesem Fall wird vermutet, daß es sich um ein Zeichen handelt, anderenfalls wird zunächst kein Zeichen erkannt. Mit diesem Verfahren können eine '1' oder ein Dezimalpunkt, die sich neben anderen Zeichen befinden, identifiziert werden.

Während der Bildanalyse wird eine vollständige Klassifizierung aller Basiselemente durchgeführt, also eine Interpretation der Basiselemente nach ihrer Bedeutung in Konturlinien, Mittellinien, Schraffurlinien, Bemaßung etc. Die bei den ersten beiden Schritten der Zeichenerkennung noch nicht erkannten Zeichen führen bei dieser Klassifizierung zu Unstimmigkeiten, da sie nicht zugeordnet werden können. Bei Zeichen, die aufgrund von Unsauberkeiten in der Ausgangszeichnung oder Fehlinterpretationen bei der Vektorisierung nicht eindeutig erkannt werden können, ist ein Eingriff des Benutzers erforderlich.

Eine Möglichkeit, Inkorrektheiten bei der Zeichenerkennung zu beheben, bildet die Zeichnungsanalyse, bei der die Geometrie anhand der Bemaßung ausgerichtet wird. Bei der Zeichnungsanalyse können fehlerhaft erkannte Maßtexte dadurch gefunden werden, daß eine deutliche Abweichung zwischen Maß und Geometrie vorhanden ist, die über die sich durch das Scannen und die Vektorisierung ergebende Unmaßstäblichkeit hinausgeht. Da in einem solchen Fall nicht eindeutig zu entscheiden ist, ob der Maßtext wirklich falsch erkannt wurde oder ob die Geometrie in der Ausgangszeichnung bereits falsch, d.h. abweichend von der Bemaßung eingetragen wurde, ist wiederum ein Eingriff des Benutzers vorgesehen.

5 3D-Rekonstruktion

Der letzte Funktionsblock in MERLIN ist die 3D-Rekonstruktion. Bisherige Techniken zur Generierung von Volumenmodellen aus zweidimensionalen Darstellungen setzen zwei oder mehr orthogonale Projektionen voraus, genutzt werden nur die geometrischen Inhalte einer Zeichnung *[4]*. Die Erzeugung dreidimensionaler Darstellungen von Maschinenteilen aus zweidimensionalen Darstellungen setzt aber viel mehr voraus: Wissen um den Inhalt und den

Aufbau technischer Zeichnungen und Wissen über die zeichnerische Darstellung von Konstruktionselementen *[2,3]*. Durch Ausnutzung beider Quellen können Zuverlässigkeit des Verfahrens und die Funktionalität des erzeugten 3D-Modells verbessert werden.

Mit dem hier skizzierten Verfahren zur Rekonstruktion von Volumenmodellen lassen sich insbesondere sog. 2½D-Teile bearbeiten. Die Menge der erkennbaren Formelemente (*"features"*) umfaßt in der aktuellen Implementation Längsbohrungen, Querbohrungen in der xy-Ebene, zylindrische Querbohrungen in z-Richtung, Innen- und Außengewinde, Innen- und Außen-Paßfedernuten einschließlich Befestigungsbohrungen, Nuten für Sicherungsbleche, Planflächen auf Zylindern und Kegeln, prismatische und konische Innen- und Außenvierkante, Innen- und Außensechskante, Stirnverzahnungen, einfache Hinterschneidungen, Fasen, Radien, Freistiche, exzentrische Wellenabschnitte und Zentrierbohrungen.

Bei den Darstellungsarten können Seitenansichten ebenso wie Vollschnitte, Halbschnitte, Halbdarstellungen und Ausbrüche behandelt werden. Die Ausgabe des 3D-Modells erfolgt wahlweise als Liste von Formelementen mit Normbezeichnung, Position und Dimensionierung oder als Generierungsvorschrift für einen 3D-CSG-Modellierer auf der Basis der Grundelemente Quader, Zylinder, Kegel, Torus und Sechskantprisma.

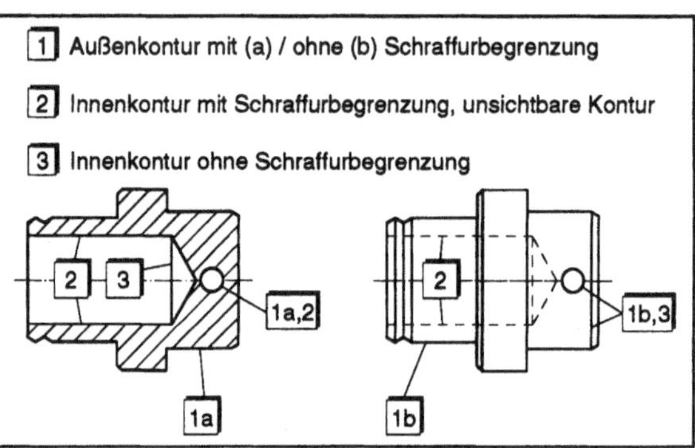

Bild 5 Linienattribute im CAD-System PROREN1.

Ausgangspunkt der Erzeugung von 3D-Modellen sind 2D-CAD-Fertigungszeichnungen, die mit dem CAD-System PROREN1 erstellt wurden. PROREN1 trennt strikt zwischen Geometrie (Konturen, Mittellinien) und sonstigen Inhalten (Bemaßung, Toleranzen, Symbole). Für jedes Geometrieelement werden vier Attribute zur Verfügung gestellt, die während der Rekonstruktion ausgewertet werden (Bild 5):

- Linienart: voll, gestrichelt, strichpunktiert, unsichtbar, Freihand.
- Linienstärke: dünn, mittel, dick.
- Konturtyp: Außen-, Innenkontur.
- Schraffurbegrenzung: mit, ohne.

Das neu entwickelte Rekonstruktionsverfahren ist ein reines *bottom-up*-Verfahren. In jeder Ansicht der Zeichnung werden zuerst die (strichpunktierten) Symmetrielinien bearbeitet. Sie legen die Haupt- und Nebenachsen symmetrischer Formelemente fest. Die übrigen Linien können dann den Symmetrielinien aufgrund ihrer relativen Lage - auch zu anderen Linien - zugeordnet werden. Mustererkennungsverfahren, die auf einem strukturellen Vergleich von vorhandenen Linien mit vorgegebenen Darstellungen von Konstruktionselementen basieren, werden zur Erkennung dieser Elemente verwendet. Ziele sind sowohl eine zuverlässige Erkennung aller definierten Konstruktionselemente als auch die Erzeugung eines korrekten, technisch sinnvollen Körpers. Dabei müssen drei Aspekte berücksichtigt werden:

- Bei der Erstellung der technischer Zeichnungen können vereinfachte Darstellungen verwendet werden, um den zeichnerischen Aufwand zu reduzieren, insbesondere bei Schnittkonturen an Flächen 2. Ordnung.
- Die exakten Abmessungen der zeichnerischen Ausführung können von den Werten der Normen abweichen.
- Es können Inkorrektheiten in der Darstellung vorhanden sein, insbesondere nichtgeschlossene Konturen, überlappende Linien, Überschneidungen und Abweichungen von Parallelität und Rechtwinkligkeit auftreten.

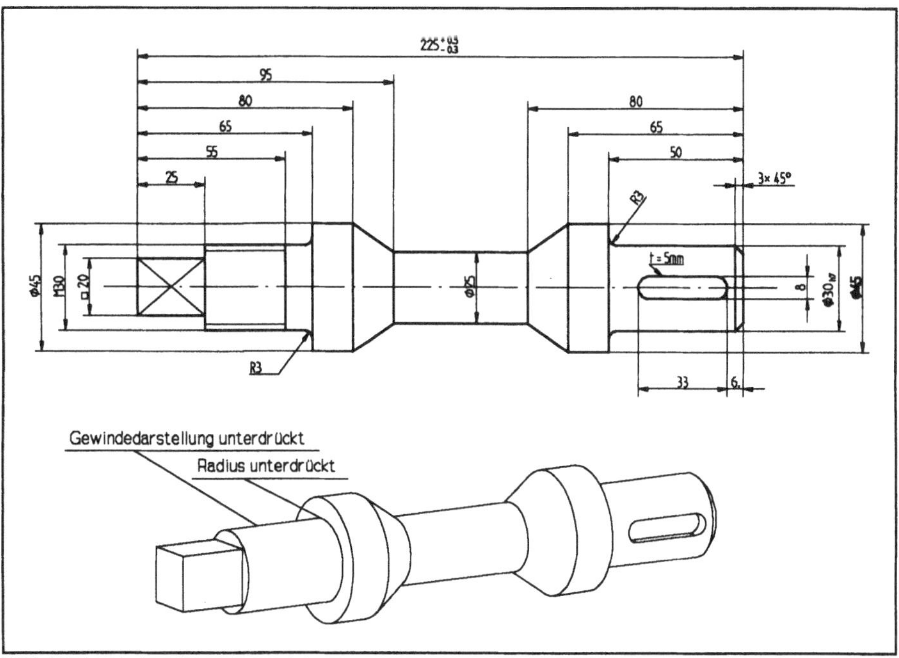

Bild 6 Beispiel einer 3D-Rekonstruktion.

Eine wesentliche Anforderung an die Erkennungsverfahren ist daher die Toleranz gegenüber solchen Abweichungen. Die Größe der Toleranzfelder kann über Parameter eingestellt werden. Im Gegensatz zur detailreichen 2D-Zeichnung, in der die Geometrie möglichst exakt dargestellt werden sollte, darf ein 3D-Modell nicht mit Details überladen werden, da darunter die Anschaulichkeit des Modells leidet (Bild 6). Fertigungs- oder montagebedingte Einzelheiten wie kleine Radien und Fasen, Freistiche und Zentrierbohrungen werden nicht als reale Geometrie erzeugt, sondern nur als Attribute an Flächen des 3D-Modells angeheftet. Auf dieselbe Art und Weise werden auch Gewinde behandelt, für die es keine adäquate 3D-Darstellung gibt. Einschränkend ist zu sagen, daß die Interpretation aus technischen Gründen zur Zeit noch auf die attributierten Geometrieelemente beschränkt ist, da die Bemaßung nicht direkt aus PROREN1 ausgelesen werden kann. Als Erweiterung ist aber eine IGES-Schnittstelle geplant. Das Rekonstruktionsverfahren ist dann auch für andere CAD-Systeme nutzbar. Eine Erweiterung auf die - zumindest teilweise - Rekonstruktion beliebiger Körper ist ebenfalls geplant. Als Erweiterung eines allgemeinen Rekonstruktionsverfahrens, z.B. *[4]*, bieten sich Möglichkeiten zur technisch sinnvollen Interpretation beliebiger zweidimensionaler Darstellungen.

6 Ausblick

Mit den durchgeführten Arbeiten konnte gezeigt werden, daß es prinzipiell möglich ist, technische Zeichnungen in einem überwiegend automatisierten Prozeß mit voller Funktionalität in den Datenbestand von CAD-Systemen zu übernehmen. Ein Prototyp wurde mit objektorientierten Techniken in C++ realisiert und ist unter DOS, OS/2 und verschiedenen UNIX-Derivaten verfügbar. Die Arbeiten werden in den kommenden drei Jahren in Form eines Kooperationsprojektes fortgeführt, an dem die TU Hamburg-Harburg, das Labor für Künstliche Intelligenz der Uni Hamburg und die Firmen SerCon GmbH, Deutsche Aerospace Airbus GmbH, Körber AG und ASCAD GmbH beteiligt sind.

Literaturverzeichnis

[1] Thomas Bartsch: *Automatisierte Erfassung technischer Zeichnungen mit einem wissensbasierten Bildverarbeitungssystem.* Fachtagung "Datenverarbeitung in der Konstruktion '92", VDI-Berichte 993.3, VDI-Verlag: Düsseldorf, 1992, 113-128.

[2] Thomas Bartsch: *Ein wissensbasiertes Bildverarbeitungssystem zur rechnerunterstützten Erfassung und Interpretation technischer Zeichnungen.* Dissertation, Technische Universität Hamburg-Harburg, 1992.

[3] Boris Pasternak, Gabriel Gabrielides, Rainer Sprengel: *WIZ - Design and Development of a Prototype for Knowledge-Based Interpretation of Technical Drawings.* LKI-M-92/1, Labor für Künstliche Intelligenz, Fachbereich Informatik, Universität Hamburg, 1992.

[4] Jörg Schultze, Mathias Feindt: *Entwurf dreidimensionaler Geometriemodelle aus technischen Zeichnungen.* Diplomarbeit, Fachbereich Informatik, Universität Hamburg, 1991.

[5] F.-L. Krause, S. Kramer, E. Rieger: *Featurebasierte Produktentwicklung.* Zeitschrift für wirtschaftliche Fertigung ZwF-CIM **87**, 247-251 (1992).

Codierung stereoskopischer Fernsehbilder unter Berücksichtigung der visuellen Eigenschaften des Menschen[1]

Johannes Steurer
Institut für Rundfunktechnik
Floriansmühlstr. 60
80939 München

Zusammenfassung

Zur Übertragung von digitalisierten stereoskopischen Bildern kann die große Ähnlichkeit der beiden Bilder ausgenützt werden. Zunächst werden relevante Eigenschaften der binokularen Wahrnehmung beim Menschen untersucht. Darauf aufbauend wird ein prädiktiver Stereobildcoder entwickelt und anhand von natürlichen Bildern getestet. Eines der beiden Bilder bleibt unverändert (kompatible Erweiterung des 2D-Fernsehens). Für das zweite Bild wird eine Kompression gegenüber dem unkomprimierten Bild um den Faktor 32 erzielt.

1 Einleitung

Übertragung und Speicherung stereoskopischer Bilder erfordern den doppelten Aufwand von einfachen 2D-Bildern. Im Anwendungsfall des stereoskopischen Fernsehens sollte beachtet werden, daß das stereoskopische Bild kompatibel zum konventionellen zweidimensionalen Fernsehen (2DTV) übertragen wird. Das Ziel dieser Arbeit besteht darin, den zusätzlichen Aufwand für das stereoskopische Bild zu minimieren, ohne daß der Mensch eine Veränderung des wahrgenommenen Raumbilds empfindet. Nach dem Wissensstand des Autors handelt es sich hierbei um das erste vollständige Codierungskonzept für kompatibles Stereofernsehen, das auf den Wahrnehmungseigenschaften beim binokularen Sehen beruht.

2 Die binokulare Wahrnehmung beim Menschen

2.1 Modell der binokularen Wahrnehmung

Der Mensch nimmt seine visuelle Umwelt nicht als zwei ähnliche Einzelbilder wahr, wie sie auf den beiden Netzhäuten abgebildet sind. Die Netzhautbilder werden in einem mentalen Prozeß kombiniert. Die sich ergebende Repräsentation kann in drei Komponenten untergliedert werden, wenngleich der Abstraktionsgrad der neuronalen Verarbeitung hierfür unbekannt ist:

- Zunächst werden die Netzhautbilder zu einem *einfachen Bild* fusioniert, das so wirkt, als würde es von einem zentralen Auge gesehen (Zyklopenauge, Julesz 1971).
- Zusätzlich zu diesem 2D-Bild ist die stereoskopische Tiefeninformation - etwa in Form einer *Tiefenkarte (Disparitätsfeld)*[2] - wahrnehmbar.

Diese beiden Komponenten sind jedoch nicht ausreichend, um alle binokularen Unterschiede - wie sie z. B. bei Verdeckungen oder glänzenden Oberflächen auftreten - zu erklären.

- Die dritte Komponente enthält die nicht fusionierbaren binokularen Unterschiede und kann als Bild des *binokularen Wettstreits* bezeichnet werden.

Tiefenkarte und *Bild des Wettstreits* beinhalten vollständig die stereoskopische Zusatzinformation. In Abb. 1 wird das geschilderte Wahrnehmungsmodell verdeutlicht.

---

[1] Entstanden während der Tätigkeit am Lehrstuhl für Nachrichtentechnik der TU München.

[2] Die Begriffe *Tiefe* und *Disparität* werden synonym verwendet. Disparität bezeichnet die in Winkelgraden meßbare Entsprechung der Tiefe.

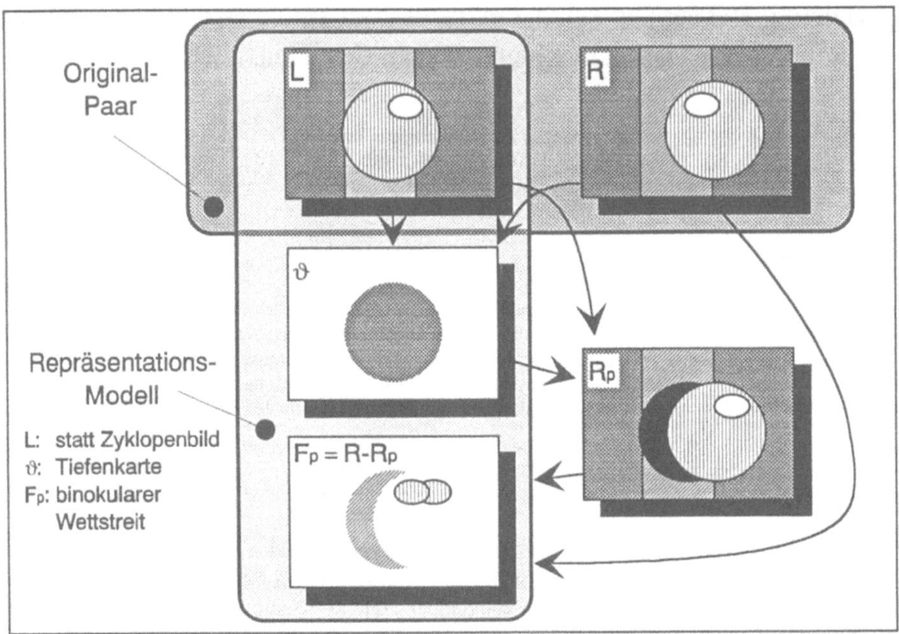

Abb. 1 Vereinfachtes, asymmetrisches Repräsentationsmodell für das binokulare Bild in der Vorstellung des Menschen: Linksbild L (vereinfachend statt Zyklopenbild), Tiefenkarte ϑ, nicht fusionierbare Unterschiede (beim Coder: Prädiktionsfehler F_P). Pfeile veranschaulichen den Signalfluß. Erklärung von R_P im Abschnitt 4.

Der realisierte Coder benützt die Komponenten des Modells (Linksbild L, Disparitätsfeld ϑ, Prädiktionsfehler F_P) zur Übertragung. Auf den ersten Blick wird der Übertragungsaufwand vergrößert (3 statt 2 Bilder). Eine wichtige Frage zur Stereobildcodierung lautet daher:

> Welche Informationsmenge ist in den zusätzlichen Komponenten *Disparitätsfeld* und *binokularer Wettstreit* wahrnehmbar?

Der maximale Informationsgehalt eines 2D-Signals wird bestimmt aus den Merkmalen *Ortsfrequenzgehalt*, der die Abtastrate festlegt, und benötigte *Quantisierung* der Abtastwerte. Da diese Parameter von den visuellen Eigenschaften des Menschen abhängen, muß die Frage - getrennt nach den beiden Komponenten *binokulare Tiefenwahrnehmung* und *binokularer Wettstreit* - durch psychophysische Experimente beantwortet werden.

2.2 Experiment 1: Signaleigenschaften bei der binokularen Tiefenwahrnehmung

Die zu untersuchenden Parameter sind die höchste laterale Ortsfrequenz f_y und die kleinste Tiefenvariation (Tiefenauflösung) ϑ_{ss} einer 3D-Struktur, die noch zur Wahrnehmung eines fusionierten 3D-Bildes führen. Abb. 2 veranschaulicht die experimentellen Bedingungen, wie diese Wahrnehmungsschwellen ermittelt wurden.

Methoden

Es wurden Stereobilder mit zufälliger Helligkeitstextur (modified random dot stereogram) und sinusförmigem Disparitätssignal variabler Ortsfrequenz f_y und Amplitude ϑ_{ss} dargeboten. Der Betrachter erhält den Eindruck einer gewellten, zufällig texturierten Oberfläche. Die Ermittlung

der Wahrnehmungsschwellen erfolgt als 2AFC-Experiment (2 alternative forced choice) mit einem anerkannten adaptiven Schätzverfahren (maximum likelihood estimation) (Treutwein 1991). Ein früheres Experiment von Tyler (1974) bleibt unbeachtet, da es nicht unter den Betrachtungsbedingungen des Fernsehens durchgeführt wurde und eine relativ ungenaue Testmethode verwendete.
Bei allen Experimente sind freie Augenbewegungen zugelassen.

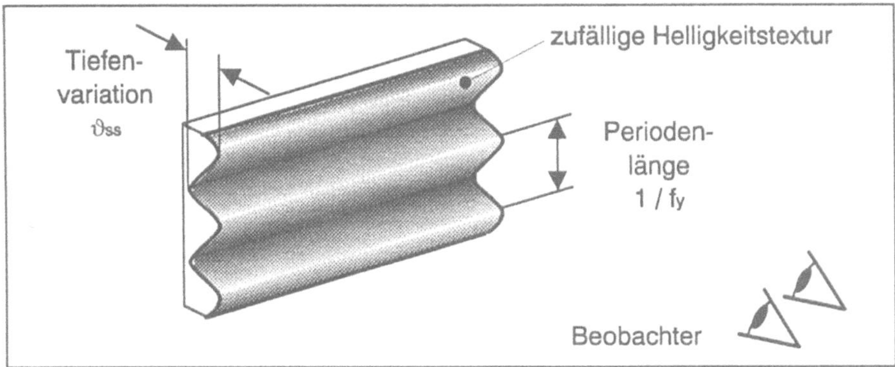

Abb. 2 Betrachtung eines Stereogramms, das eine gewellte Oberfläche mit zufälligem Helligkeitsmuster zeigt.

Ergebnisse
15 Versuchspersonen mit normaler Sehschärfe und Tiefensehschärfe nahmen am Experiment teil. In Abhängigkeit der untersuchten Parameter wurden drei Arten der Tiefenwahrnehmung registriert:

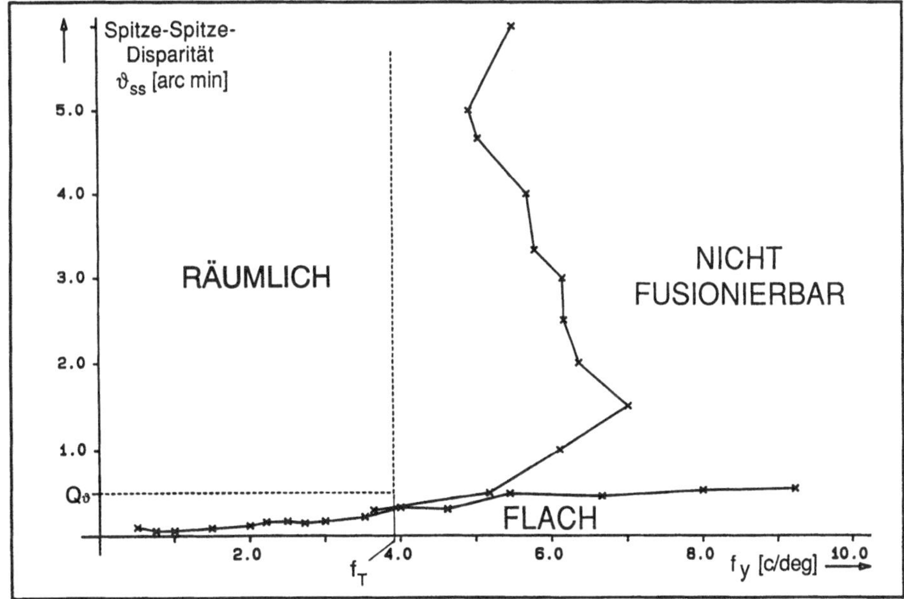

Abb. 3 Wahrnehmungsschwellen der binokularen Fusion (gestrichelt: Coderparameter).

- *flach*: bei niedrigen Werten der Disparitätsamplitude ϑ_{ss},
- *fusionierbar, eindeutige Tiefe*: bei niedriger Ortsfrequenz f_y und höheren Amplitudenwerten,
- *nicht fusionierbar, mehrdeutige Tiefe*: bei höheren Werten von Ortsfrequenz und Amplitude.

In Abb. 3 wird dargestellt, wie die Wahrnehmungsschwellen der binokularen Fusion verlaufen.

Verallgemeinernd ist die stereoskopische Tiefenwahrnehmung gekennzeichnet durch eine deutliche laterale Bandbegrenzung verbunden mit einer hohen Tiefenauflösung. Ein Disparitätssignal mit einer lateralen Ortsfrequenz f_y über 6 c/deg ist nicht als eindeutige 3D-Oberfläche wahrnehmbar. Diese Grenze entspricht einem Fünftel der lateralen Auflösung des Auges (30 c/deg). Die gemessene Tiefenauflösung (Disparitätsamplitude ϑ_{ss}) entspricht mit 0,1 arc min der zehnfachen lateralen Auflösung des Auges und ist damit vergleichbar der Nonius-Sehschärfe. (Steurer 1993).
Anschaulich gesprochen werden Areale von mindestens 5 x 5 Rezeptoren auf jeder Netzhaut ausgewertet, um einen einzigen Disparitätswert zu ermitteln. Die erzielte Auflösung erreicht den hohen Wert von bis zu 1/10 Rezeptorabstand.

2.3 Experiment 2: Signaleigenschaften bei der Wahrnehmung des binokularen Wettstreit

Die Wahrnehmungsschwelle für nicht fusionierbare binokulare Unterschiede wurde mit natürlichem Bildmaterial (8 Grauwert-Bilder) erforscht. Binokulare Helligkeitsunterschiede zwischen rechtem und linkem Bild werden dadurch erzeugt, daß einem stereoskopisch dargebotenen 2D-Bild einseitig (nur rechts) ein stationäres Zufallssignal (Rauschbild) mit uniformer Amplitudenverteilung überlagert wird. Die Störung ist ab einer Rauschamplitude von ±6 Graustufen (von 256) wahrnehmbar, entsprechend einer Standardabweichung von ± 3,5 Graustufen.
Es sei angemerkt, daß dieses Ergebnis nur die Größenordnung der zulässigen Störung des Prädiktionsfehlers F_P bei der Codierung angibt. Die Korrelation zwischen Signal und Störung ist für die Bildqualitätsbeurteilung von wesentlicher Bedeutung und kann nur für den konkreten Coder angemessen berücksichtigt werden (siehe Abschnitt 4).
Jüngere psychophysische Versuche (Kissner 1992) enttäuschen die Hoffnung, daß eine stärkere Bandbegrenzung (Verunschärfung) *eines* Bildes im Stereopaar unter der Wahrnehmungsschwelle bleibt. Daher muß das Bild des binokularen Wettstreits (Prädiktionsfehler) in der vollen lateralen Auflösung repräsentiert werden.

3 Ermittlung der Stereodisparitäten

Verwendet man einen hochgenauen Kamera-Aufbau, so entstehen nur diejenigen stereoskopischen Bildunterschiede, die auf den unterschiedlichen Kamerapositionen beruhen, aber nicht solche, die von verschiedenen Eigenschaften der abbildenden Systeme herrühren (z.B. Unterschiede der Fokussierung, Vignettierung, Vertikaldisparitäten).
Ein neues Verfahren zur Ermittlung der Stereodisparitäten wurde entwickelt. Seine wesentlichen Merkmale sind: mehrstufiger Ansatz unter Verwendung von Bildpyramiden, eindimensionale Suche (Epipolarbedingung) mit einem korrelativen Vergleichsmaß (least square fit) und sub-Pixel-Genauigkeit, Interpolation zu einem dichten Disparitätsfeld. Die Originalbilder werden dazu in Laplace-Pyramiden (Bandpaßauszüge) transformiert. In jeder Auflösungsstufe werden die Disparitätswerte für ausgewählte Punkte durch eine eindimensionale Suche ermittelt. Die Werte eines zuvor bestimmten Disparitätsfelds (gröbere Auflösungsstufe) dienen dabei als Startwerte der Suche. Die gemessenen Disparitätswerte werden zu einem dichten Feld interpoliert und zum bisherigen Disparitätsfeld addiert. Dieses Vorgehen erfolgt rekursiv für fortlaufend verfeinerte Auflösungsstufen der Bildpyramiden (coarse-to-fine, Abb. 4).

Die Disparitätsmessung ist auf die Codierung hin optimiert: Ortsfrequenz und Quantisierung des Disparitätsfelds entsprechen den gefundenen psychophysischen Parametern der binokularen Fusion.

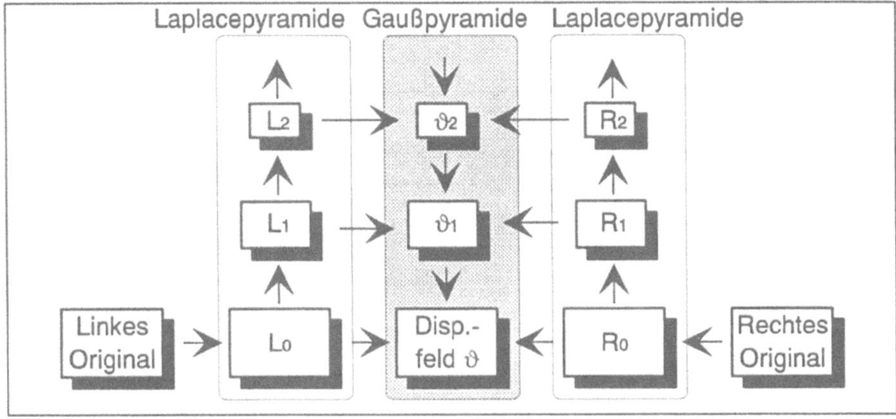

Abb. 4 Bestimmung der Stereodisparität, Blockschaltbild.

4 Codierung stereoskopischer Bilder

4.1. Codierungskonzept

Das vorgestellte Codierungskonzept basiert auf den Eigenschaften der binokularen Wahrnehmung und benützt daher die Ähnlichkeiten im stereoskopischen Bild auf eine vergleichbare Weise. Der Sender transformiert das Original-Bildpaar[3] zunächst in die drei Komponenten, die dem Empfinden des Menschen entsprechen (siehe auch Abb. 5):

- *Linksbild* L: uncodiert, kompatibel zum 2DTV,
- *Disparitätsfeld* ϑ: Information, die zur binokularen Fusion benötigt wird und eine Stereoprädiktion R_P erlaubt: $R_P(x,y) = L(x + \vartheta(x,y), y)$,
- *Prädiktionsfehler* F_P: binokulare Unterschiede, die nicht fusionierbar sind, entsprechend dem binokularen Wettstreit: $F_P(x,y) = R(x,y) - R_P(x,y)$.

Anschließend erfolgt die Datenkompression der einzelnen Komponenten.

4.2. Codierung des Disparitätsfelds

Unter Beachtung der psychophysischen Ergebnisse und mit dem Ziel einer einfachen Realisierung wird das Disparitätsfeld mit einer Ortsfrequenz $f_T = 0{,}125$ [pixel^{-1}][4] abgetastet. Die Quantisierung der Disparitätsamplituden beträgt $Q_\vartheta = 0{,}5$ [pixel]. Siehe dazu auch die gestrichelten Linien in Abb. 3.

Berücksichtigt man durch eine Entropie-Codierung die ungleichmäßige Amplitudenverteilung der Disparitätswerte, so ist eine Bitrate von 0,06 [bit/pixel] zu erzielen (Mittelwert von 8 Grauwert-Stereobildern).

---

[3] Aufgrund der genormten Betrachtungsbedingungen beim Fernsehen werden die Originalbilder mit 1 Pixel pro Winkelminute abgetastet.

[4] Pixelbezogene Angaben sind immer auf die Pixelgröße der Originalbilder skaliert.

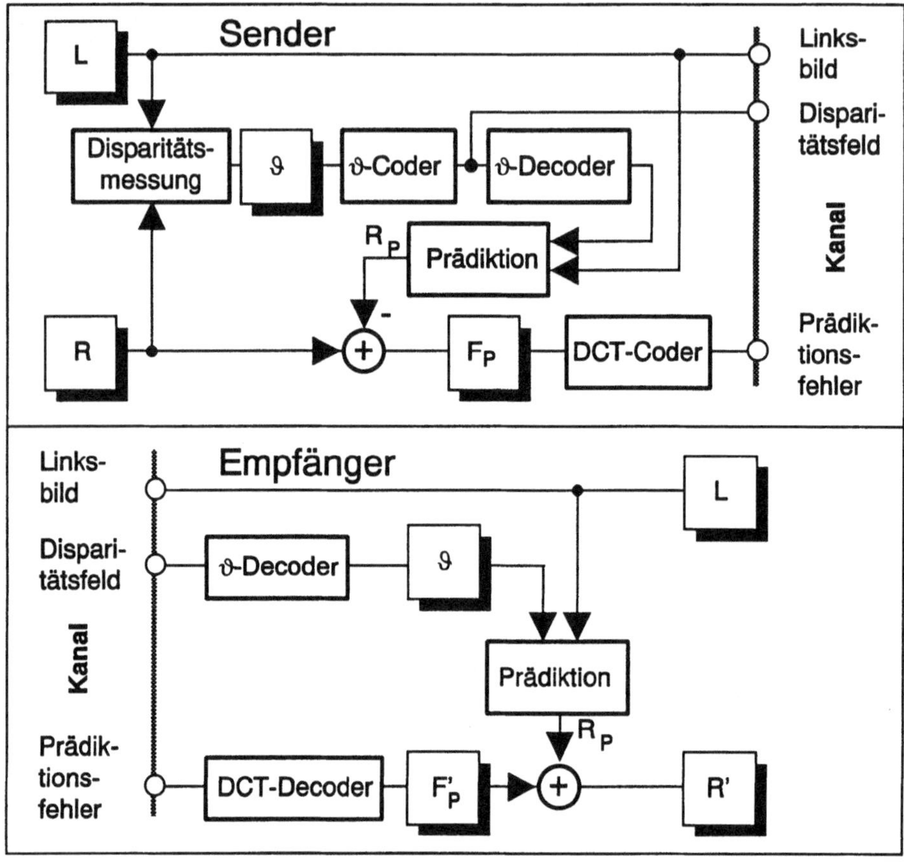

Abb. 5 Blockschaltbild der Stereobildcodierung.

4.3. Codierung des Prädiktionsfehlers

Der Prädiktionsfehler wird einer konventionellen DCT-Codierung (Blockgröße 16 x 16) mit blockadaptiver Tiefpaßfilterung unterzogen. Die Parameter des Coders werden wie folgt festgelegt: Alle DCT-Koeffizienten werden gleichmäßig quantisiert mit einer Stufenhöhe, die beim DC-Koeffizienten der Grauwertquantisierung im Originalsignal entspricht. Die Wahrnehmungsschwelle der Verzerrungsenergie, die den adaptiven Tiefpaß charakterisiert, wurde durch psychophysische Tests mit 14 Versuchspersonen ermittelt. Eine neue Testmethode wurde dazu entwickelt, die im Vergleich zur bekannten EBU-Methode wesentlich empfindlicher und einfacher zu handhaben ist (Abb. 6).

Original und codiertes Stereobild werden bei einer Wechselfrequenz von 12 Hz alternierend präsentiert. Auftretende Coderstörungen werden dabei als Flimmern wahrgenommen. Die Wahrnehmung von Flimmern ist einfacher als der qualitative Vergleich von Bildern, die zeitlich oder örtlich getrennt präsentiert werden. Ein 2AFC-Experiment dient wiederum dazu, die Wahrnehmungsschwelle der Coderstörungen zu ermitteln.

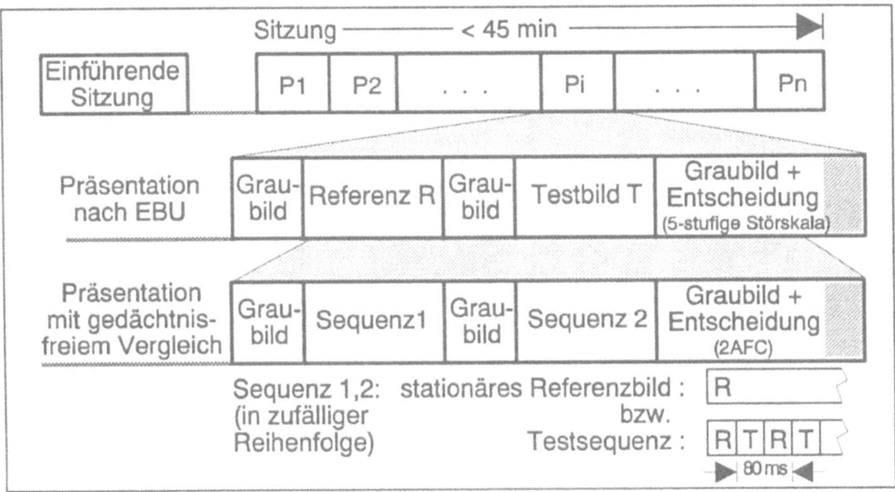

Abb. 6 EBU-Methode und gedächtnisfreies Vergleichsverfahren zur subjektiven Bildqualitätsbeurteilung.

Experimentelle Ergebnisse
In der Abb. 7 werden die Codierungsergebnisse für 8 natürliche Grauwert-Stereobilder gezeigt. Im Mittel kann der Prädiktionsfehler mit einer Bitrate von 0,19 [bit/pixel] übertragen werden.

5 Schlußfolgerung

Die Zusatzinformation, die für ein Stereobild gegenüber einem 2D-Bild nötig ist, kann mit einer Bitrate von 0,25 [bit/pixel] übertragen werden und entspricht damit 3% eines gewöhnlichen digitalen PCM-Bildes. Der erzielbare Reduktionsfaktor beträgt etwa 32 (Abb. 7).
Unter Verwendung des entwickelten Codierungskonzepts könnte das stereoskopische Fernsehen als ein voll kompatibler Zusatzdienst zum gewöhnlichen 2D-Fernsehen angeboten werden.

6 Literatur

Julesz, Bela (1971) *Foundations of Cyclopean Perception.* Chicago: The University of Chicago Press.

Kissner, Alfred (1992) *Subjektive Beurteilung von bandbreite-begrenzten Stereobildern.* Lehrstuhl für Nachrichtentechnik. Technische Universität München. Diplomarbeit.

Steurer, Johannes (1992) *Stereobildcodierung für den Menschen als Beobachter.* Technische Universität München. Dissertation. (Auch erschienen in: *Nachrichtentechnische Berichte.* Hrsg: Marko, H. und G. Binkert, Schriftenreihe des Lehrstuhl für Nachrichtentechnik, Technische Universität München. Band 23.)

- (1993) "Stereoscopic depth acuity subject to 3D structure." *Investigative Ophthalmology and Visual Science.* Vol.34. No.4. March 15. 1189.

Treutwein, Bernhard (1991) "Adaptive psychophysical methods." *Frontiers in knowledge based computing.* Hrsg. Bhatkar, Vijay P. und K. M. Rege. New Delhi: Nasosa Publishing. 101-111.

Tyler, William C. (1974) "Depth perception in disparity gratings." *Nature.* Vol.251. Sept. 13. 140-142.

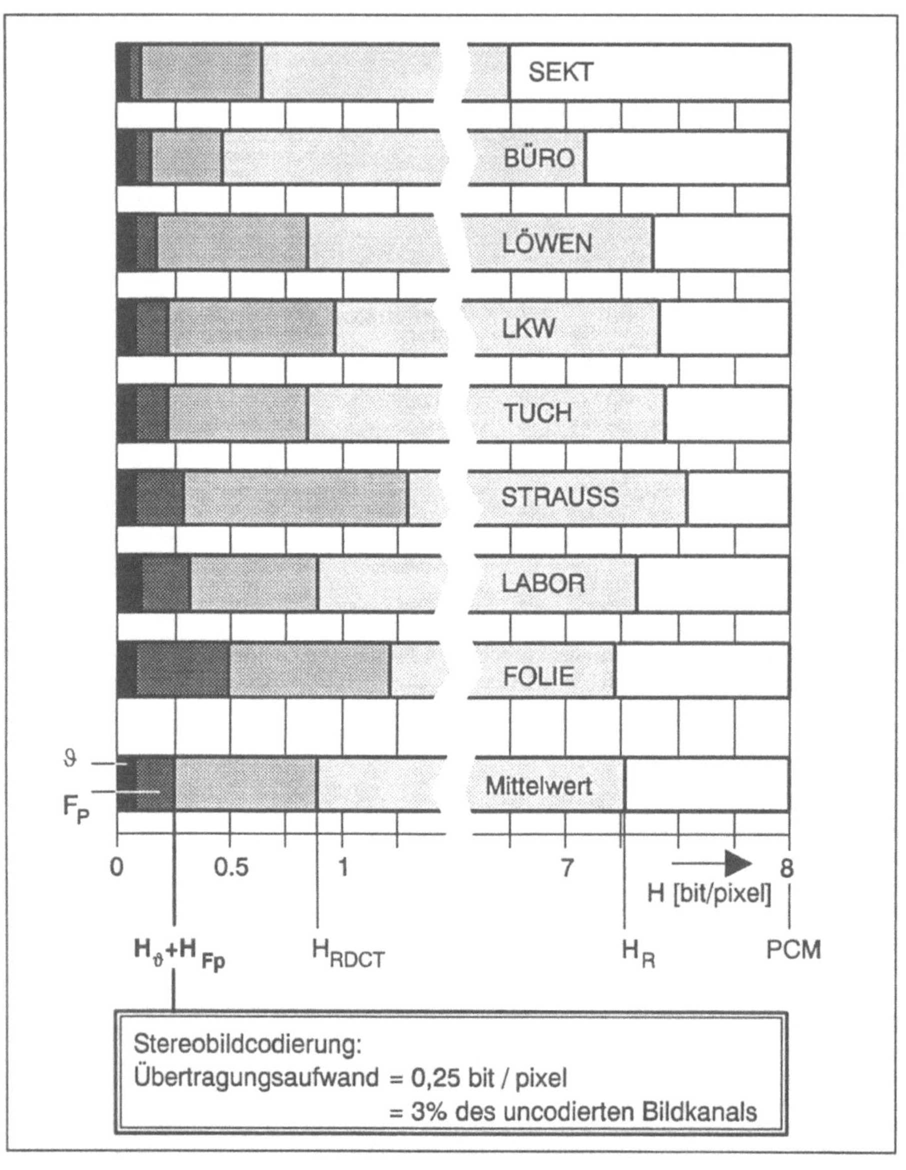

Abb. 7 Ergebnisse der Stereobildcodierung für 8 Szenen einzeln und als Mittelwert.
Es bedeuten:

$H_\vartheta + H_{FP}$: Entropie von Disparitätsfeld plus Prädiktionsfehler,

zum Vergleich:

H_{RDCT}: Entropie des DCT-codierten Rechtsbilds,

H_R: Entropie des originalen Rechtsbilds,

PCM: Bitbedarf bei reiner PCM-Codierung.

Bestimmung dreidimensionaler Geschwindigkeitsfelder aus Strömungstomographiesequenzen

Hans-Gerd Maas
Institut für Geodäsie und Photogrammetrie
ETH - Hönggerberg, CH - 8093 Zürich
Tel. +41-1-3773058, Fax +41-1-3720438, e-mail gerd@p.igp.ethz.ch

Kurzfassung:

Zur Untersuchung von chemischen Reaktionen in turbulenten Strömungen wurde ein auf einer Hochgeschwindigkeits-Halbleiterkamera basierendes System zur quasi-simultanen Gewinnung von Tomographiedaten-Sequenzen in durch Fluorescine visualisierten Mischprozessen in Strömungen implementiert. In diesen Daten werden durch dreidimensionale Kleinste-Quadrate-Korrelation Informationen über das Geschwindigkeitsfeld und über das Deformationsverhalten von Fluidelementen gewonnen.

Diese Publikation stellt die Hardwarekonfiguration für die Tomographiesequenzgewinnung, die dreidimensionale Implementation der Kleinste-Quadrate-Korrelation sowie Ergebnisse aus Beobachtungen von Mischvorgängen in turbulenten Strömungen in einem hydromechanischen Versuchstank vor.

1. Einleitung

Die Strömungsmeßtechnik hat im Verlauf der vergangenen Jahrzehnte eine ganze Reihe verschiedenster Meßverfahren zur Bestimmung von Strömungsgeschwindigkeiten entwickelt. Grundsätzlich kann man hier unterscheiden zwischen Einpunktmeßverfahren, welche den Geschwindigkeitsvektor (oder einzelne Vektorkomponenten) lokal mit hoher zeitlicher Auflösung bestimmen, und Mehrpunktmeßverfahren, welche die Bestimmung simultaner Geschwindigkeitsfelder erlauben. Letztere bedingen i.d.R. die Visualisierung von Strömungen, welche entweder diskret durch Zugabe von Partikeln oder kontinuierlich durch Zugabe von Farbstoffen erfolgen kann. Die Visualisierung durch Partikel (Abb. 1a) ist vor allem bei der Bestimmung von Geschwindigkeitsfeldern in turbulenten Strömungen üblich; dabei können durch Particle-Tracking-Velocimetry (PTV) zweidimensionale und - bei stereoskopischer Aufnahme - auch dreidimensionale Geschwindigkeitsvektoren und Partikeltrajektorien bestimmt werden (Maas, 1991, 1992a). Dagegen wird kontinuierliche Visualisierung (Abb. 1b) vor allem bei der Untersuchung von Durchmischungsprozessen eingesetzt. Eine oft benutzte Methode der kontinuierlichen Markierung ist die Zugabe von Fluorescin. Fluorescin hat die Eigenschaft, Licht einer bestimmten Wellenlänge zu absorbieren und dafür Licht einer anderen (i.d.R. höheren) Wellenlänge zu emittieren (LIF - laser induced fluorescence). Wird wie in Abb. 1b bei Mischvorgängen in Flüssigkeiten das eine der beiden Medien durch Fluoreszin markiert, so erhält man bei durch eine radiometrische Kalibrierung gegebenem funktionalem Zusammenhang

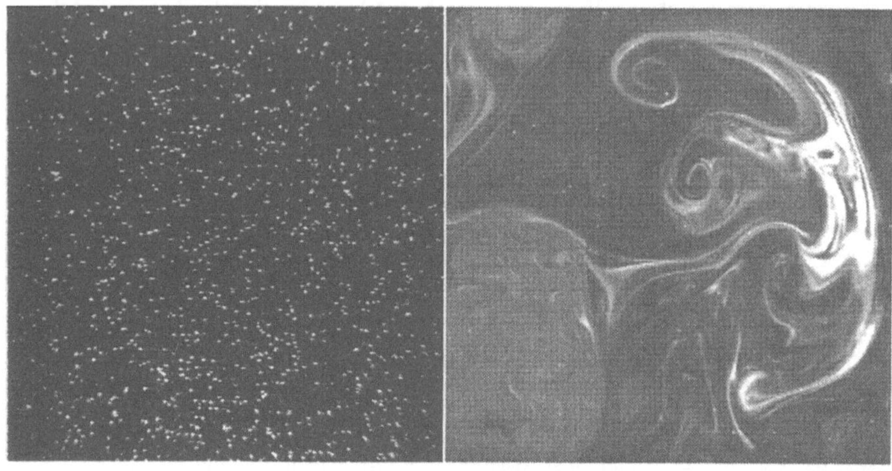

a. b.

Abb. 1: Strömungsvisualisierung durch Partikel und durch Fluorescin

zwischen den Grauwerten im digitalen Bild und Fluoreszinkonzentrationen in der Flüssigkeit Konzentrationsbilder der Mischvorgänge. Wenn dabei ein Beobachtungsvolumen in Tiefenrichtung mit einer Laserlichtschicht abgetastet und dabei mit einer Hochgeschwindigkeitskamera aufgenommen wiird, können quasi-simultan Volumendaten der Strömung gewonnen werden. Da das Fluorescin nur in geringen Konzentrationen zwischen 10^{-6} und 10^{-5} mol zugegeben wird, ist gewährleistet, daß Fluorescin außerhalb der beleuchteten Schicht keinen Einfluß auf das Bild hat, während das animierte Fluorescin - bei entsprechender Kameraempfindlichkeit und Laserleistung - den gesamten Dynamikbereich des Bildes abdecken kann.

Werden nun durch mehrmaliges aufeinanderfolgendes Scannen des Beobachtungsvolumens Sequenzen von Volumendatensätzen gewonnen, so lassen sich aus diesen Daten neben Konzentrationsgradienten (Dahm, 1990) und Trennflächendimensionen durch Kleinste-Quadrate-Korrelation auch Informationen über das Geschwindigkeitsfeld und über das Deformationsverhalten von Fluidelementen gewinnen.

2. Hardware

Die hardwaretechnische Realisierung der Gewinnung von Strömungstomographiesequenzen ist aus Abb. 2 ersichtlich, welche den Versuchsaufbau für die Untersuchung turbulenter Durchmischungsvorgänge zeigt: Der Strahl eines 26W Argon-Ionen Lasers wird über eine Zylinderlinse zu einer Lichtschicht aufgeweitet und durch einen mit einem Piezoelement verstellbaren Spiegel in Tiefenrichtung verschoben (Merkel et al, 1993); ein Fokussiersystem sorgt dafür, daß die Laserlichtschicht im Bereich des Beobachtungsvolumens eine möglichst geringe Dicke hat. Das so mit der Laserlichtschicht in Tiefenrichtung abgetastete Beobachtungsvolumen wird durch eine mit dem Piezospiegel synchronisierte Hochgeschwindigkeitskamera aufgenommen. Dabei wird, um Refexionen, z.B. an Luftblasen, zu vermeiden, ein optischer Filter mit einer Kante bei ca. 525 nm eingesetzt, der die Wellenlänge des Lasers (514 nm) komplett abschneidet und nur die Emission des Fluorescins (bei 540 nm) durchläßt.

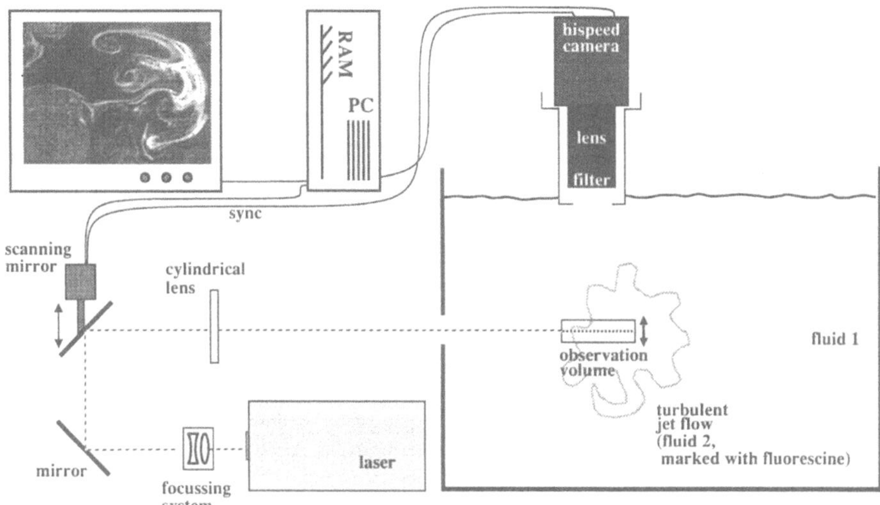

Abb. 2: Experimentaufbau

Das Beobachtungsvolumen wurde, um auch die kleinsten Skalen der Turbulenz (Kolmogorov-Scales) sichtbar zu machen, mit 15 x 15 x 4 mm^3 sehr klein gewählt. Als Kamera wurde eine EG&G Reticon MC4256 ausgewählt (Maas, 1992b, 1993), deren Bilddaten auf einem 64 MB großen Sequenzspeicher in RAM abgelegt werden. Die maximale Bildrate der Kamera beträgt 500 Bilder/Sekunde, die Auflösung 256 x 256 Pixel. Auf das o.g. Beobachtungsvolumen bezogen ergibt sich so eine Voxelgröße von 60 x 60 x 60 µm^3. Durch die Fokussierung des Laserstrahls konnte dort auch eine Schichtdicke von etwa 60 µm erreicht werden. Die Größe des Sequenzspeichers limitiert die Länge eines Experimentes auf 1024 Bilder, was beispielsweise die Gewinnung von 16 aufeinanderfolgenden Volumendatensätzen mit je 256 x 256 x 64 Voxeln zuläßt.

Abb. 3 zeigt ein Beispiel einer Strömungstomographiesequenz in einem turbulenten Mischvorgang. Die Sequenz besteht aus 5 Volumina zu je 50 Schichten à 256 x 256 Pixeln in einem Beobachtungsvolumen von 15x15x3 mm^3, die zeitliche Auflösung beträgt 10 Volumina je Sekunde.

3. Kleinste-Quadrate-Korrelation mit Volumendaten

Kleinste-Quadrate-Korrelation (Least Squares Matching) ist ein Korrelationsverfahren, welches seit etwa 10 Jahren häufig von Photogrammetern zur Bestimmung der Bildkoordinaten signalisierter Punkte mit Subpixelgenauigkeit und zur Herstellung stereoskopischer Korrespondenzen angewandt wird. Der zweidimensionale Algorithmus ist beispielsweise bei Grün/Baltsavias (1988) ausführlich beschrieben. Least Squares Matching ist jedoch ebenso für Tracking-Anwendungen geeignet. Eine erste Anwendung auf zweidimensionale LIF-Daten findet sich in einer Untersuchung über Dichteströmungen bei Papantoniou et al. (1990). Zur Anwendung auf Volumendaten soll Least Squares Matching im folgenden zum dreidimensionalen Algorithmus erweitert werden.

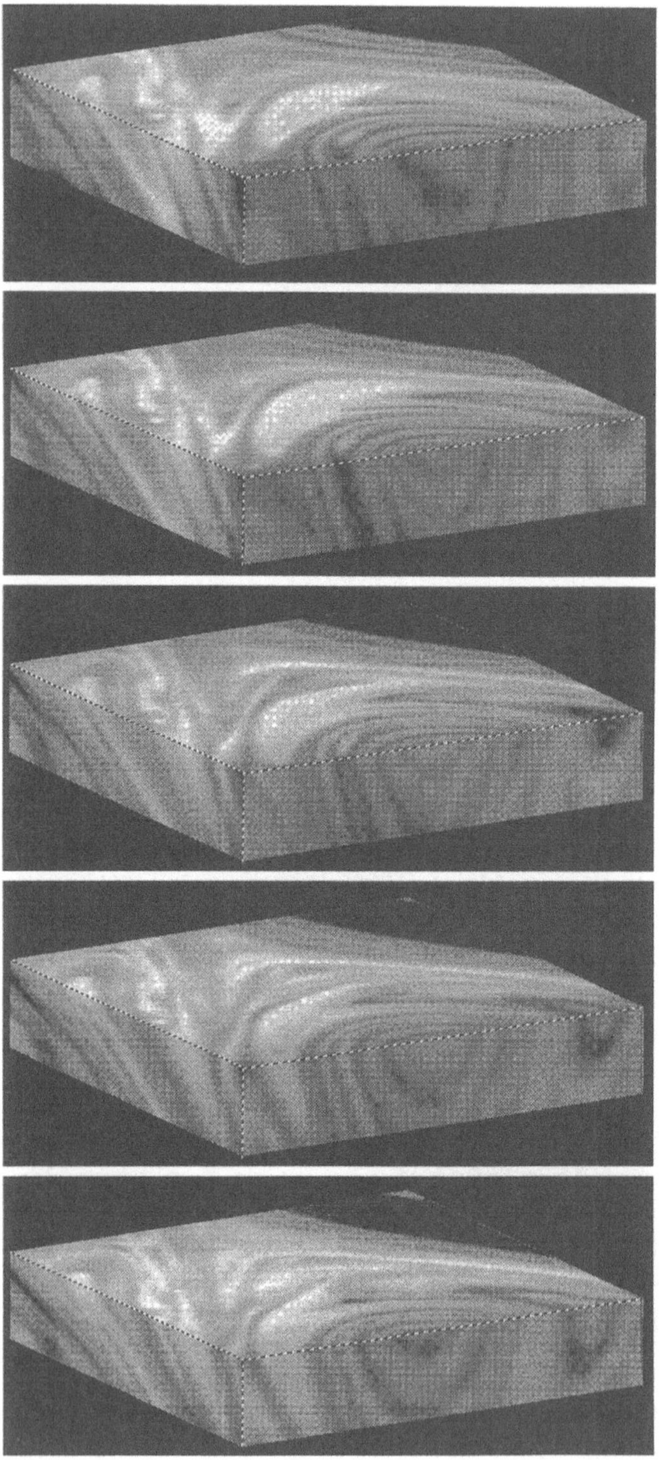

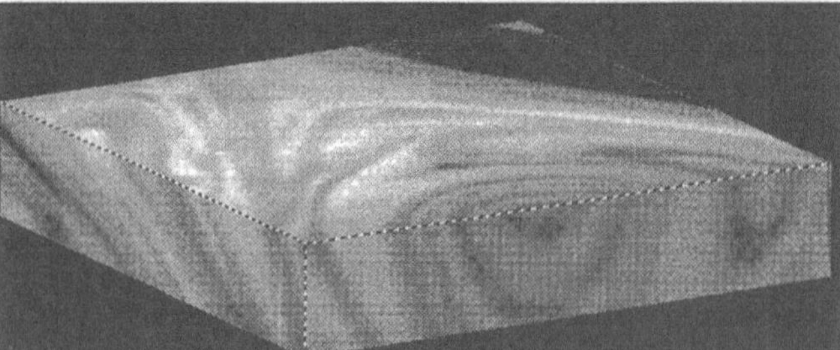

Abb. 3: Volumendatensätze:

5 aufeinanderfolgende Volumendatensätze mit je 50 Schichten zu 256 x 256 Pixel (0.5 sec. Strömungsdaten = 16 MB)

Das Grundprinzip des Least Squares Matching kann folgendermaßen erklärt werden: Zwei 'Patches' von Voxeldaten $g_1(x,y,z)$, $g_2(x,y,z)$ wählbarer Größe werden aus zwei zeitlich aufeinanderfolgenden Volumendatensätzen $\mathcal{V}_1$, $\mathcal{V}_2$ geschnitten und so aufeinander transformiert, daß für jedes Voxel gilt:

$$g_1(x, y, z) = g_2(x, y, z). \qquad \text{(Eq 1)}$$

Aufgrund der zeitlichen Entwicklung des Strömungsfeldes sowie Rauscheffekten ist dies in der Realität nicht exakt möglich, sodaß ein Fehlervektor addiert werden muß:

$$g_1(x, y, z) - e(x, y, z) = g_2(x, y, z). \qquad \text{(Eq 2)}$$

Für die geometrische Transformation zwischen g_1 und g_2 wird beim zweidimensionalen Least Squares Matching üblicherweise eine Affintransformation angesetzt; dem soll auch hier gefolgt werden:

$$\begin{aligned} x_2 &= a_0 + a_1x_1 + a_2y_1 + a_3z_1 \\ y_2 &= b_0 + b_1x_1 + b_2y_1 + b_3z_1 \\ y_2 &= c_0 + c_1x_1 + c_2y_1 + c_3z_1 \end{aligned} \qquad \text{(Eq 3)}$$

mit (a_0,b_0,c_0): Verschiebungsparameter
(a_1,b_2,c_3): Maßstabsparameter
$(a_2,a_3,b_1,b_3,c_1,c_2)$: Rotationen, Scherungen.

Die 12 Transformationsparameter der dreidimensionalen Affintransformation werden durch die Minimierung der Summe der Quadrate der Residuen $e(x, y, z)$ in (Eq 1) im Gauss-Markov Modell (z.B. Koch, 1980) bestimmt:

$$\begin{aligned} &-e(x, y, z) = Ax - l \quad ; P \\ &E(e) = 0 \ , \ E(e'e) = \sigma_0^2 P^{-1} \\ &\Rightarrow \hat{x} = (A'PA)^{-1}(A'Pl) \ , \ \hat{\sigma}^2 = \frac{e'Pe}{n-u} \\ &\text{with } x' = (da_0, da_1, da_2, da_3, db_0, db_1, db_2, db_3, dc_0, dc_1, dc_2, dc_3) \end{aligned} \qquad \text{(Eq 4)}$$

Die Tatsache, daß $g_2(x, y, z)$ tatsächlich stochastische Werte enthält, wird hier üblicherweise vernachlässigt, um den Gebrauch des Gauss-Markov Modells zu ermöglichen (Grün/Baltsavias, 1988).

In diesem Modell können die Beobachtungsgleichungen wie folgt formuliert werden:

$$\begin{aligned} g_1(x, y, z) - e(x, y, z) &= g_2^{(0)}(x, y, z) \\ &+ \frac{\partial(g_2^{(0)}(x, y, z))}{\partial x} \cdot dx + \frac{\partial(g_2^{(0)}(x, y, z))}{\partial y} \cdot dy + \frac{\partial(g_2^{(0)}(x, y, z))}{\partial z} \cdot dz \end{aligned} \qquad \text{(Eq 5)}$$

Mit

$$dx = \frac{\partial x}{\partial p_i} dp_i \qquad dy = \frac{\partial y}{\partial p_i} dp_i \qquad dz = \frac{\partial z}{\partial p_i} dp_i$$

$$p_i \in \{da_0, da_1, da_2, da_3, db_0, db_1, db_2, db_3, dc_0, dc_1, dc_2, dc_3\}$$

und den Notationen

$$\frac{\partial(g_2^{(0)}(x,y,z))}{\partial x} = g_x \qquad \frac{\partial(g_2^{(0)}(x,y,z))}{\partial y} = g_y \qquad \frac{\partial(g_2^{(0)}(x,y,z))}{\partial z} = g_z$$

erhält man für die Beobachtungsgleichungen

$$\begin{aligned} g_1(x,y,z) - e(x,y,z) = g_2^{(0)}(x,y,z) &+ g_x da_0 + g_x x_2^{(0)} da_1 + g_x y_2^{(0)} da_2 + g_x z_2^{(0)} da_3 \\ &+ g_y db_0 + g_y x_2^{(0)} db_1 + g_y y_2^{(0)} db_2 + g_y z_2^{(0)} db_3 \\ &+ g_z dc_0 + g_z x_2^{(0)} dc_1 + g_z y_2^{(0)} dc_2 + g_z z_2^{(0)} dc_3 \end{aligned}$$

(Eq 6)

und als Eingabe in das Gauss-Markov Modell

$$A = \left[g_x\ g_x x_2^{(0)}\ g_x y_2^{(0)}\ g_x z_2^{(0)}\ g_y\ g_y x_2^{(0)}\ g_y y_2^{(0)}\ g_y z_2^{(0)}\ g_z\ g_z x_2^{(0)}\ g_z y_2^{(0)}\ g_z z_2^{(0)}\right]_i \qquad \text{(Eq 7)}$$

$$l = g_1(x,y,z) - g_2^{(0)}(x,y,z). \qquad \text{(Eq 8)}$$

$$x' = (da_0, da_1, da_2, da_3, db_0, db_1, db_2, db_3, dc_0, dc_1, dc_2, dc_3),$$

Die Werte g_x, g_y, g_z sind numerische Grauwertdifferenzen, die in g_1 oder in g_2 oder gemittelt über beide berechnet werden können. Im Gegensatz zu photogrammetrischen Anwendungen verbietet sich hier eine radiometrische Anpassung der Patches, da sie die radiometrische Kalibrierung des Systems ungültig machen würde.

Der Lösungsvektor x' enthält die Information über das Geschwindigkeitsfeld mit Subpixelgenauigkeit in den Verschiebungsparametern:

$$v = \left(\frac{a_0}{\Delta t}, \frac{b_0}{\Delta t}, \frac{c_0}{\Delta t}\right) \qquad \text{(Eq 9)}$$

sowie Informationen über das Deformationsverhalten von Fluidelementen in den restlichen 9 Transformationsparametern $(a_1, a_2, a_3, b_1, b_2, b_3, c_1, c_2, c_3)$.

4. Ergebnisse

Wegen des teilweise geringen lokalen Kontrastes der in Abb. 3 gezeigten Strömungstomographiesequenz und der Konvektionsgeschwindigkeit mußten die Patches für die Kleinste-Quadrate-Korrelation mit 21x21x21 Voxeln relativ groß gewählt werden. Die Kontrastverhältnisse erlaubten nur an relativ wenigen Stellen die Bestimmung aller 12 Parameter der 3-D Affintransformation, häufig mußten einer oder mehrere Parameter wegen Nicht-Bestimmbarkeit (Grün/Stallmann, 1991) ausgeschlossen werden, was durch Signifikanztests entschieden werden kann.

Abb. 4 zeigt die Transformationen eines Volumenelementes, welches die Bestimmung des kompletten Parametersatzes erlaubte; die große Anzahl von Iterationen ist durch die Wahl des Iterationskriteriums begründet.

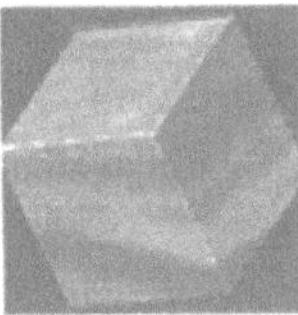

σ_0 = 4.443 nach 10 Iterationen
3-D Affintransformationsparameter und Standardabweichung:

a0: 4.641 +- 0.020,	a1: 0.987 +- 0.004,	a2: 0.036 +- 0.004,	a3: -0.270 +- 0.004
b0: 2.392 +- 0.019,	b1: 0.143 +- 0.003,	b2: 0.900 +- 0.003,	b3: -0.510 +- 0.003
c0: 0.034 +- 0.011,	c1: -0.084 +- 0.002,	c2: 0.055 +- 0.002,	c3: 0.702 +- 0.002

Abb. 4: Beispiel für 3-D Affintransformation (für Darstellungszwecke kontrastoptimiert)

Abb. 5 zeigt einen kleinen Teil der Verschiebungsvektoren zwischen den ersten beiden Volumina aus Abb. 3. Aus Gründen der Übersichtlichkeit wurde nur eine Schicht des Datensatzes und die dazugehörigen Geschwindigkeitsvektoren dargestellt. Da vor dem Least-Squares-

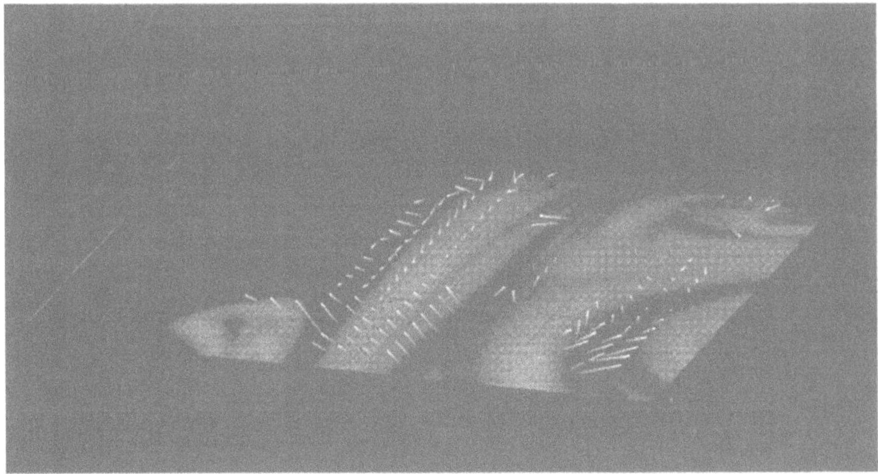

Abb. 5: Teilergebnis

Matching eine Kontrastevaluation durchgeführt wurde, wurden große Teilbereiche (oben links sowie in der Mitte in Abb. 5) ausgeschlossen; im linken unteren und rechten oberen Bereich konvergierte das Matching wegen der dort zu großen Konvektion nicht.

5. Ausblick

Es konnte gezeigt werden, daß dreidimensionale Kleinste-Quadrate-Korrelation ein geeignetes Instrument zur Ableitung von Informationen über Geschwindigkeitsfelder in turbulenten Mischvorgängen ist. Das Konvergenzverhalten des Algorithmus und die Genauigkeit der Ergebnisse sind an Stellen mit hohem lokalen Kontrast gut. Da sich in den Experimenten zeigte, daß bei sehr kleinen Beobachtungsvolumina der Kontrast in manchen Regionen unzureichend ist, wird die zukünftige Arbeit darin bestehen, zur Verdichtung der Resultate radiometrische und geometrische Zusatzbedingungen zur dreidimensionalen Kleinste-Quadrate-Korrelation analog den geometrischen Zusatzbedingungen bei der Mehrbildzuordnung durch zweidimensionale Kleinste-Quadrate-Korrelation (Grün, 1985; Rosenholm, 1986; Grün/Baltsavias, 1988) zu formulieren.

Literatur:

1. Dahm, J.A., Southerland, K.B., Buch, K.A., 1990: Four-Dimensional Laser Induced Fluorescence Measurements of Conserved Scalar Mixing in Turbulent Flows. 5th Int. Symposium on the Application of Laser Techniques to Fluid Mechanics, Lissabon, 9.-12.7.1990
2. Grün, A., 1985: Adaptive kleinste Quadrate Korrelation und geometrische Zusatzinformation. Vermessung, Photogrammetrie und Kulturtechnik 9/85, pp 309-311
3. Grün, A., Baltsavias, E., to1988: Geometrically Constrained Multiphoto Matching. Photogrammetric Engineering, Vol. 54, No. 5, pp. 633 - 641
4. Grün, A., Stallmann, D., 1991: High-accuracy edge-matching with an extension of the MPGC matching algorithm. Proceedings Industrial Vision Metrology, SPIE Volume 1526, pp.42-55
5. Koch, K.R., 1980: Parameterschätzung und Hypothesentests in linearen Modellen. Dümmler Verlag, Bonn.
6. Maas, H.-G., 1991: Digital Photogrammetry for Determination of Tracer Particle Coordinates in Turbulent Flow Research. Photogrammetric Engineering & Remote Sensing Vol. 57, No. 12, S. 1593-1597
7. Maas, H.-G., 1992a: Digitale Photogrammetrie in der dreidimensionalen Strömungsmeßtechnik. ETH Zürich - Dissertation Nr. 9665
8. Maas, H.-G., 1992b: High-Speed Solid State Camera Systems for Digital Photogrammetry. International Archives of Photogrammetry and Remote Sensing, Vol. XXIX, Part B5, pp. 709-713
9. Maas, H.-G., 1993: Techniques for Destriping of Digital Images. ISPRS Com. I Workshop on Digital Sensors and Systems, Trento, 21-25. 6. 1993
10. Merkel, G., Dracos, T., Rys, P., Rys, F., 1993: Flow Tomography by Laser Induced Fluorescence. XXV IAHR Congress, Tokyo, 30. 8. - 3. 9. 1993
11. Papantoniou, D., Bühler, H. und Dracos, T., 1990: On the Internal Structure of Thermals and Momentum Puffs. International Conference on Physical Modelling of Transport and Dispersion, MIT, 7.-10. August
12. Rosenholm, D., 1986: Accuracy improvement of digital matching for evaluation of digital terrain models. IAPRS Vol. 26, Part 3

Wissensbasierte automatische Extraktion von semantischer Information aus Katasterkarten

H. Mayer[1], C. Heipke[1], G.Maderlechner[2]

[1]TU München, Lehrstuhl für Photogrammetrie und Fernerkundung, Arcisstr. 21, 8000 München 2

[2]Siemens AG, ZFE ST SN 6, Otto-Hahn-Ring 6, 8000 München 83

Zusammenfassung

Es wird ein System vorgestellt, mit dem semantische Information wie Flurstücke oder Straßen automatisch aus gescannten Katasterkarten extrahiert werden können. Das System beruht auf einem aus vier Ebenen aufgebauten Modell, das mit Hilfe eines semantischen Netzes repräsentiert wird. Das Modell und die verwendete Vorgehensweise werden vorgestellt. Ergebnisse für die Extraktion von Flurstücken, Straßen, Gehwegen und Grünstreifen aus zwei Katasterkarten unterschiedlichen Maßstabs werden präsentiert.

1 Einleitung

Geo-Informationssyteme (GIS) dienen u.a. zur Verarbeitung, Analyse und Präsentation von geographischer Information. Eines der Haupthindernisse für einen verstärkten Einsatz von GIS im großmaßstäblichen Bereich ist ein Mangel an entsprechender digitaler geographischer Information.

Vorhandene Karten können hier einen Ausweg bieten. Ihre Information ist nicht so genau und aktuell, wie die mit geodätischen Verfahren oder photogrammetrisch erzeugte, aber die Erfassungskosten für eine manuelle Digitalisierung sind sehr viel niedriger.

Da auch die manuelle Digitalisierung aufwendig ist, entstanden praktisch einsetzbare automatische Verfahren zur Umsetzung von Rasterdaten in attributierte Vektordaten mit manueller Nachbearbeitung. Damit können eine Reihe von GIS Anwendern zufriedengestellt werden. Von Seiten der Verkehrsleitsysteme, dem Umweltschutz etc. wird aber über die attributierten Vektordaten hinausgehende Information wie Straßennetze oder Flächennutzung benötigt.

Mit einer Verbesserung der automatischen Interpretation von Karten setzt sich [Mulder 1988] auseinander. Dort werden topographische Objekte und deren Relationen beschrieben. Allerdings wird dieses Wissen nur auf idealisierte Strichzeichnungen angewendet. In [Antoine 1991] werden "reale Objekte" (z.B. Haus, Straße), ihre graphische Repräsentation und die Relationen zwischen ihnen dargestellt, das Modell für die "realen Objekte" ist allerdings stark vereinfacht. Weitere Arbeiten beschäftigen sich mit der Interpretation von Topographischen Karten [Ebi et al. 1992], von Straßenkarten basierend auf dem Exoskelett [Ilg 1990] oder mit der Verbesserung der Interpretation der Graphik von Flurkarten ([Illert 1990] und [Klauer 1993]).

In diesem Papier stellen wir ein System vor, mit dem semantische Information wie Flurstücke oder Straßen automatisch aus gescannten Katasterkarten extrahiert werden können. Hierbei gehen wir davon aus, daß eine möglichst umfassende Modellierung aller semantischen und graphischen Objekte zu einer Verbesserung der Interpretation führt. Damit fällt Information wie Straßennetze quasi nebenbei an.

Das Modell und die vier Ebenen werden vorgestellt. Nach einer Darstellung der verwendeten Vorgehensweise und der Repräsentation des Modells werden Ergebnisse für zwei verschiedene Arten von Karten präsentiert.

2 Modell einer großmaßstäblichen Karte

2.1 Ebenen des Modells

Hier wird ein Modell für Karten großen Maßstabs vorgestellt, das in vier Ebenen (siehe Abb. 1) unterteilt ist, die aus Objekten mit deren Eigenschaften und Relationen bestehen. Eine detailliertere Darstellung des Modells wird in [Mayer et al. 1992] gegeben.

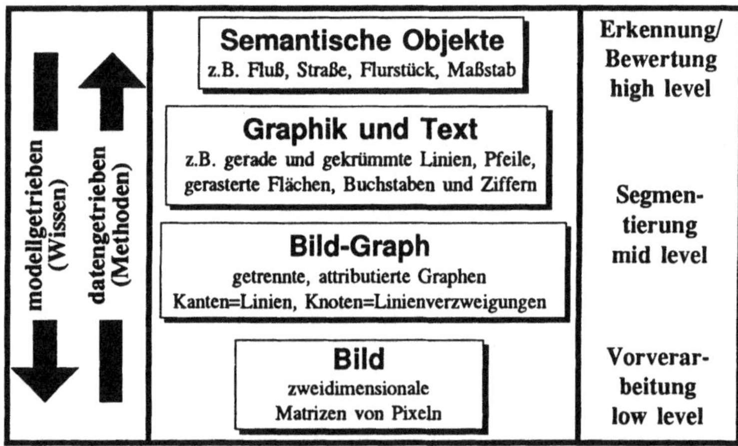

Abb. 1: Ebenen und Vorgehensweisen des Modells

Die Ebene "Semantische Objekte" korrespondiert mit der Ebene Erkennung/Bewertung (high-level) der Bildanalyse [Marr 1982]. In ihr werden alle Objekte mit ihren Eigenschaften und Relationen beschrieben, die in der Legende der Karte vorkommen (z.B. Grenzstein) oder aber, die aus der Karte ersichtlich sind und ihre Entsprechung in der Wirklichkeit besitzen (z.B. Straße, Grünstreifen).
Die Objekte können z.B. wie folgt beschrieben werden: Flurstücke bestehen aus Flurstücksnummer und Flurstücksfläche. Letztere ist aus Grenzlinien und Grenzsteinen aufgebaut. Straßen, Grünstreifen am Straßenrand und Gehwege sind oft langgestreckte Flächen und besitzen jeweils eine für sie typische, konstante Breite. Straßen bilden Netze. Gehwege und Grünstreifen sind zu den Straßen parallele Strukturen.

Die Ebenen "Graphik und Text" sowie "Bild-Graph" entsprechen zusammen der Ebene Segmentierung (mid-level) der Bildanalyse. Die Ebene "Graphik und Text" stellt Zeichnungsinformation dar. Dazu gehören sowohl Graphik- (z.B. gerade und gekrümmte Linien, Kreise, schraffierte und punktgerasterte Flächen) als auch Textobjekte (Buchstaben und Ziffern) und Symbole (z.B. Pfeile) mit ihren Eigenschaften und Relationen.
Die Ebene "Bild-Graph" besteht aus attributierten Graphen, die aus Linien und Linienverzweigungen aufgebaut sind. Linien bestehen aus mit Dickewerten versehenen Geradenstücken. Sie

enden in Linienverzweigungen. Diese besitzen eine Dicke sowie eine richtungsmäßig geordnete Liste der von ihnen abgehenden Linien.

Die Ebene "Bild" korrespondiert mit der Ebene Vorverarbeitung (low-level) der Bildanalyse. Sie besteht aus zweidimensionalen Matrizen von Pixelwerten (Rasterbilder). Diese Ebene wird durch das Scannen der Karte und nachfolgende Binarisierung erzeugt.

2.2 Repräsentation des Modells und Vorgehensweise

Das Modell mit seinen Ebenen und den vorliegenden Objekten und Relationen wird zuerst mittels eines semantischen Netzes der konzeptuellen Ebene [Brachman 1979] dargestellt. Objekte und Relationen dieses Netzes werden dann in Konzepte eines semantischen Netzes der epistemologischen Ebene abgebildet, das nur die Relationen Teil / Teil-von und Spezialisierung / Abstraktion enthält.

Zuerst werden dann mittels Bildverarbeitungsoperationen und Linienverfolgung Linien und Linienverzweigungen der Ebene "Bild-Graph" datengetrieben generiert. Nach der Vorgabe eines Konzepts des semantischen Netzes wird der Ausschnitt des semantischen Netzes bestimmt, der für die Instanziierung des Konzepts notwendig ist. Dieser Ausschnitt wird dann instanziiert.

Da sich gezeigt hat, daß sowohl eine rein modellgetriebene als auch eine rein datengetriebene Instanziierung nicht, oder nur mit großem Rechenaufwand zum gewünschten Ergebnis führen, wird eine gemischte Vorgehensweise (siehe Abb. 1) verwendet.

3 Implementierung und Ergebnisse

Das semantische Netz und die die Vorgehensweise bestimmenden Kontrollstrukturen sowie alle Funktionen für die Interpretation wurden in Common-LISP und CLOS implementiert.

Für die Interpretation der Flurkarte 1 : 1000 wurde ein Ausschnitt aus der Beispielkarte der Bayerischen Zeichenanweisung [Bayer. Staatsm. 1982] gewählt. Dieser wurde mit 400 dpi gescannt und binarisiert. Das Ergebnis ist ein Rasterbild der Ebene "Bild" (1024 x 1024 Pixel; siehe Abb. 2). In diesem Ausschnitt sollen einerseits die Flurstücke und andererseits die Straßen, Gehwege und Grünstreifen gefunden werden.

In dem Rasterbild werden mittels der Analyse der Größe der Zusammenhangskomponenten die Störungen und das die Gebäude symbolisierende Druckraster entfernt. Das Druckraster kann später zum Aufsuchen der Gebäude verwendet werden.
Auf das verbleibende Rasterbild wird eine Verdünnungsoperation angewandt. Jedem Pixel des entstehenden Bildes wird mittels einer Distanztransformation ein Dickewert zugeordnet. Die Linien werden verfolgt und es erfolgt eine Polygonapproximation (Vektorisierung). Das Ergebnis sind die Linien und Linienverzweigungen (siehe Abb. 3) der Ebene "Bild-Graph", die die Eingangsdaten für das semantische Netz bilden.
Nach der modellgetriebenen Bestimmung des Ausschnitts des semantischen Netzes für das Konzept Flurstück werden aus den Linien und Linienverzweigungen datengetrieben Graphen instanziiert. Diese werden nach der Fläche des umschreibenden Rechtecks in Kleine (Hypothesen für Symbole und Text) und Große Objekte (Hypothesen für Liniengraphik) unterschieden. Hierdurch wird die Grundlage zur Instanziierung der Objekte der Ebene "Graphik und Text" gelegt.

In den Großen Objekten werden die Linien entsprechend der Zeichenanweisung in zwei Klassen (dicke und dünne Linien; 0.35mm und 0.18mm) unterteilt. Dann wird versucht aus den dicken Linien zusammenhängende Geraden zu bilden (siehe Abb. 4). Diese sind Hypothesen für Grenzlinien (Grenzliniensymbole). An einigen Stellen (markiert in Abb. 4) entstehen dicke Linien (falsche Grenzliniensymbole) auf Grund von graphischen Überlagerungen.

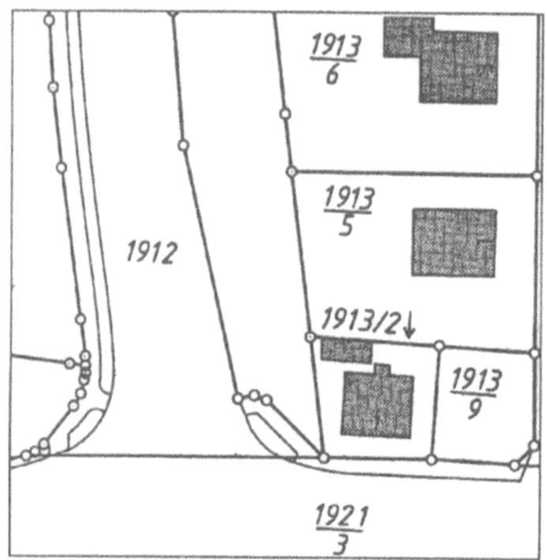

Abb. 2: Gescannter und binarisierter Ausschnitt einer Bayerischen Flurkarte im Maßstab 1 : 1000 [Bayerisches Staatsm. 1982]

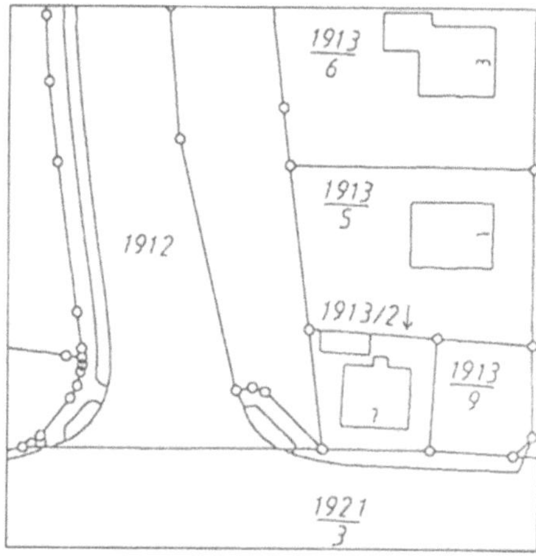

Abb. 3: Ergebnis der Vektorisierung des bereinigten Rasterbildes: Linien und Linienverzweigungen

Die Grenzliniensymbole werden zusammen mit dem Regel, daß Grenzlinien in Grenzsteinen enden, verwendet, um an ihren Enden modellgetrieben nach Grenzsteinsymbolen (Kreisen) zu suchen. Dazu wird jeweils in einer kleinen Umgebung der Linienendpunkte der Grenzliniensymbole ein Template-Matching im Rasterbild durchgeführt. Beim Template-Matching werden hierbei nur die schwarzen Pixel des Templates und des Ausschnitts des Rasterbildes auf Übereinstimmung geprüft. Es wird verwendet, da es Symbole mit gleicher Größe und Orientierung auch in der Liniengraphik mit hoher Sicherheit findet (siehe z.B. [Illert 1990]). Durch die heuristische Einschränkung auf einen kleinen Bereich ist das Matching nicht rechenintensiv. Ist das Template-Matching erfolgreich, so wird die Kreislinie in die Graphik zurückgematcht. Dort werden alle zu dem Kreis gehörigen Linien und Linienverzweigungen gefunden und dem Grenzsteinsymbol zugeordnet. Damit wird die Lage der Grenzsteine auf deren Mittelpunkte bezogen, und falsche Grenzliniensymbole werden beseitigt (siehe Abb. 5). Außer an den markierten Stellen in Abbildung 5 wurden alle weiteren Grenzsteinsymbole erkannt. Zwischen den Grenzsteinsymbolen wurden nicht alle Verbindungen hergestellt (siehe Abb. 5; die Mittelpunkte sind nicht verbunden). Diese beiden Probleme resultieren daraus, daß sich die Grenzsteinsymbole auch berühren oder überschneiden können.

Verwendet man zusätzliche Regeln für die Überschneidungen, so können die fehlenden Grenzsteine gefunden und die Geraden zwischen ihnen erzeugt werden (siehe Abb. 6).

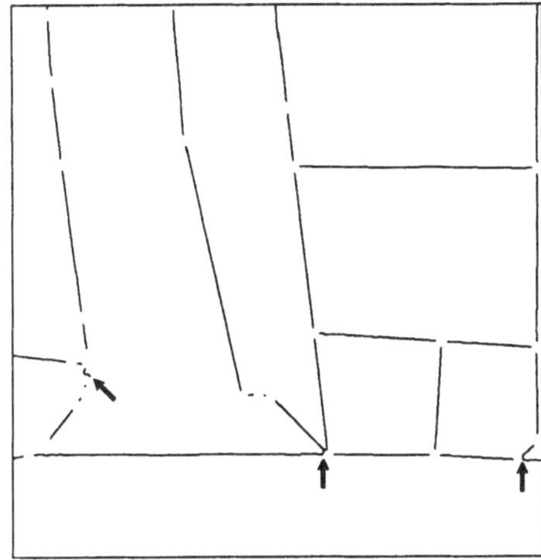

Abb. 4: Dicke gerade Linien (Grenzliniensymbole)

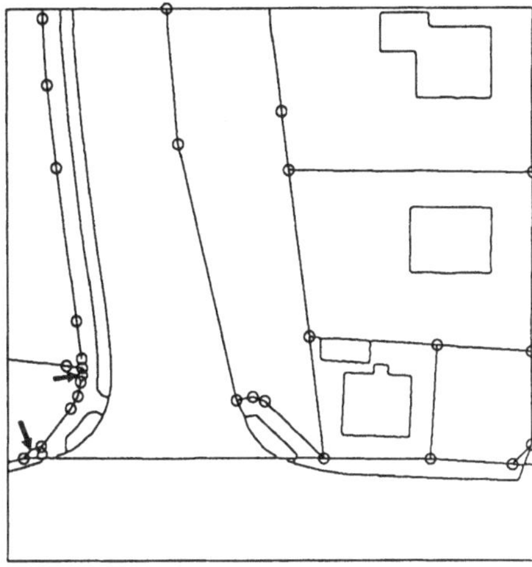

Abb. 5: Grenzsteinsymbol erkannt an den Enden von Grenzliniensymbolen

Die Kleinen Objekte sind in diesem Ausschnitt Hypothesen für Teile der Flurstücksnummern und Hausnummern. Die Bruchstriche und der Pfeil (dieser deutet auf das einer Flurstücksnummer zugeordnete Flurstück; siehe Abb. 1, rechts neben der Nummer 1913/2) werden auf Grund der Struktur ihrer Linien erkannt. Für die Erkennung der restlichen Kleinen Objekte wird ein Zeichenerkennungssystem (OCR) verwendet. Dazu werden die Mittelpunkte der Kleinen Objekte auf geometrische Nachbarschaft untersucht und Ketten von Kleinen Objekten gebildet. Die den umschreibenden Rechtecken der Ketten von Kleinen Objekten entsprechenden Ausschnitte des Rasterbildes werden dann der OCR übergeben.

Aus den Grenzstein- und den Grenzliniensymbolen werden Flurstücksflächen gebildet. Diesen werden die Flurstücksnummern zugeordnet, die aus Zeichenketten, Bruchstrichen und Pfeilen aufgebaut sind, und so Flurstücke der Ebene "Semantische Objekte" gebildet (siehe Abb. 6).

Für die Instanziierung der Straßen und Gehwege der Ebene "Semantische Objekte" wird ein Ausschnitt aus dem semantischen Netz bestimmt. Die Instanziierung der Graphikobjekte entspricht der bei den Flurstücken. Dann werden alle aus dicken und dünnen Linien (letztere entsprechen topographischen Abgrenzungen) gebildeten Flächen bestimmt (siehe Abb. 7).
Da Straßen bzw. Gehwege und Grünstreifen in der Regel langgestreckt sind, werden alle Flächen auf Langgestrecktheit untersucht (Verhältnis der Fläche zum Quadrat der Umringlänge), und für die langgestreckten Flächen die Mittelachsen berechnet (siehe Abb. 8).
Die Mittelachsen werden nach Dickewerten und Geradheitsbedingungen zu Hypothesen für Gehwege und Grünstreifen (siehe Abb. 9) bzw. für Straßen (siehe Abb. 10) klassifiziert.
Aus den Hypothesen für die Straßen werden Straßennetze gebildet. Mit dem Vorwissen, daß es sich um einen Siedlungsbereich handelt und unter der Annahme, daß zwei Straßennetze nicht parallel laufen, kann das kleinere Straßennetz gelöscht werden (siehe Abb. 10, 11).

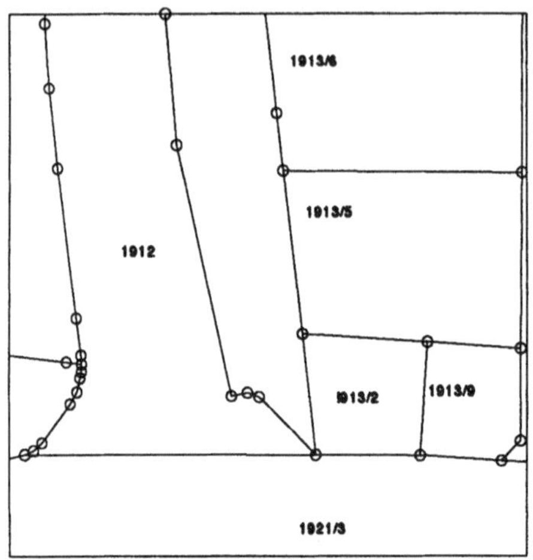

Abb. 6: Flurstücke mit Flurstücksnummern nach der Erkennung sich berührender und überschneidender Grenzsteinsymbole

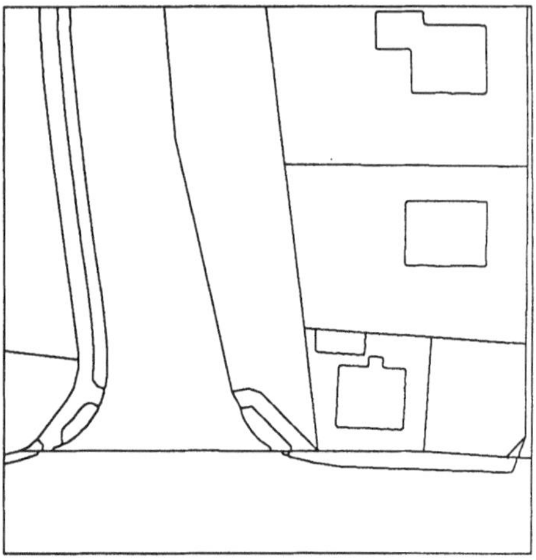

Abb. 7: Allgemeine Flächen

Abb. 8: Langgestreckte Flächen und deren Mittelachsen

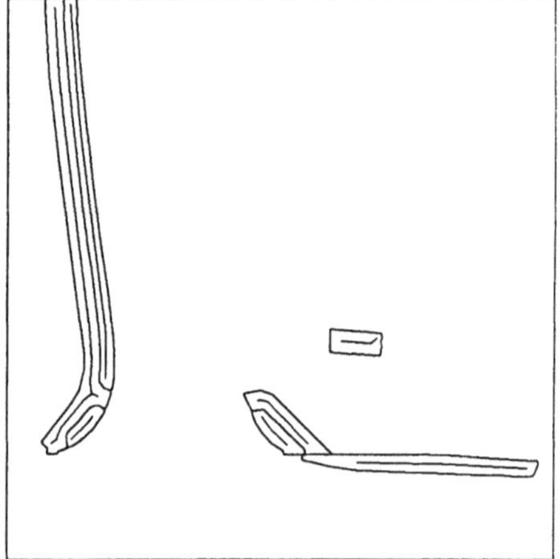

Abb. 9: Hypothesen für Geh-, Radweg und Grünstreifen

Die Hypothesen für Gehwege und Grünstreifen wurden auf ihre Parallelität zu den Straßen untersucht und gemäß der Hierarchie der Parallelität und deren Funktion (Gehwege bilden bevorzugt Netze) in Grünstreifen (siehe Abb. 12) und Gehwege (siehe Abb. 13) unterteilt.

Die Generierung von Straßennetzen wurde auch für einen Ausschnitt aus der Flurkarte 1 : 5000 der gleichen Zeichenanweisung durchgeführt (siehe Abb. 14 und 15). In diesem Ausschnitt sind keine Flurstücksnummern vorhanden.

Neben den durch die Maßstabsänderung hervorgerufenen Schwellwertänderungen mußten im semantischen Netz nur Änderungen für die Erkennung der Graphik (z.B. werden hier Grenzlinien nicht durch die Liniendicke hervorgehoben) durchgeführt werden.

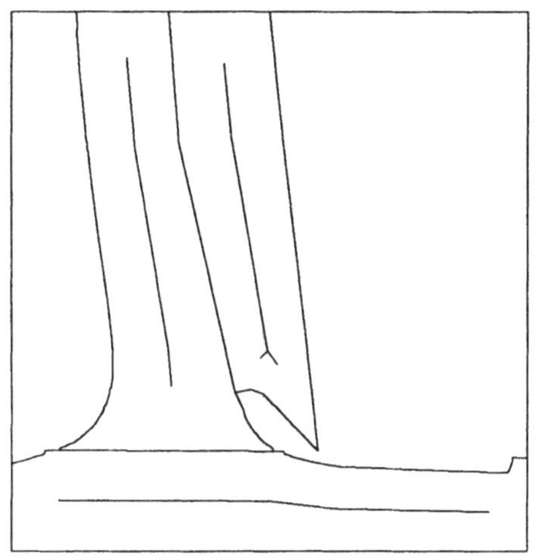

Abb. 10: Hypothesen für Straßen

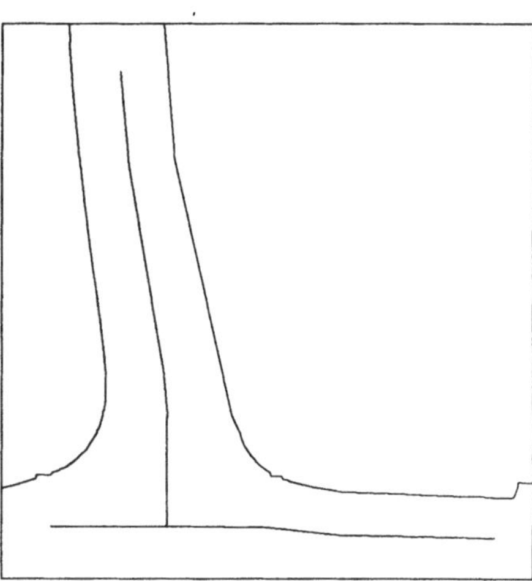

Abb. 11: Straßennetz

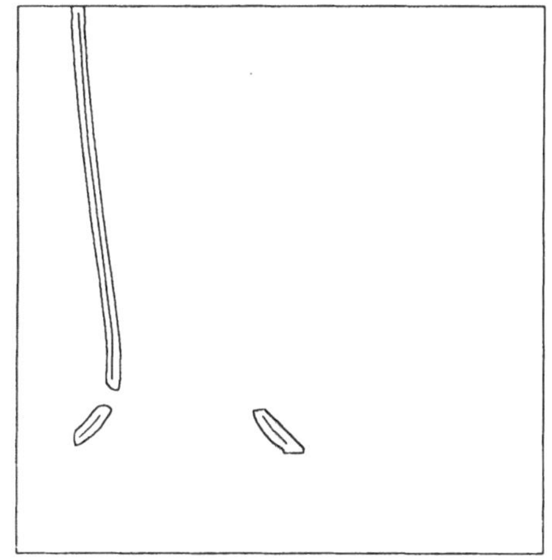

Abb. 12: Grünstreifen

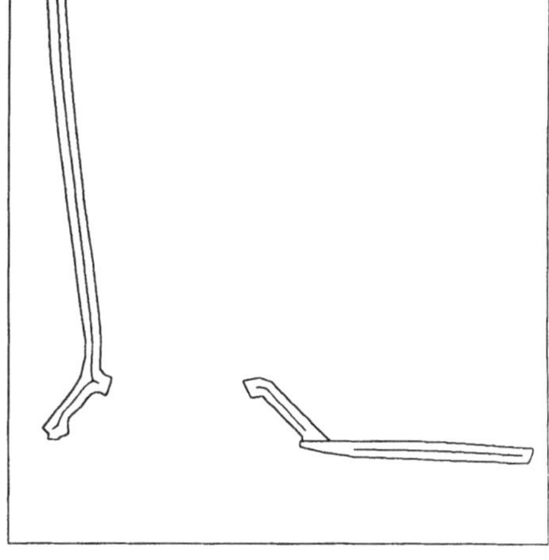

Abb. 13: Gehwege

4 Wertung und Dank

Das vorgestellte System geht mit der Beschreibung und Extraktion von semantischer Information aus Flurkarten 1 : 1000 und 1 : 5000 über die aus der Literatur bekannten Systeme hinaus. Durch eine detaillierte Modellierung erhält man für den bearbeiteten Ausschnitt eine sehr hohe Erkennungsqualität. Überdies ergibt sich durch die Modellbildung für eine Reihe von Anwendern von GIS interessante zusätzliche Information wie Straßennetze.

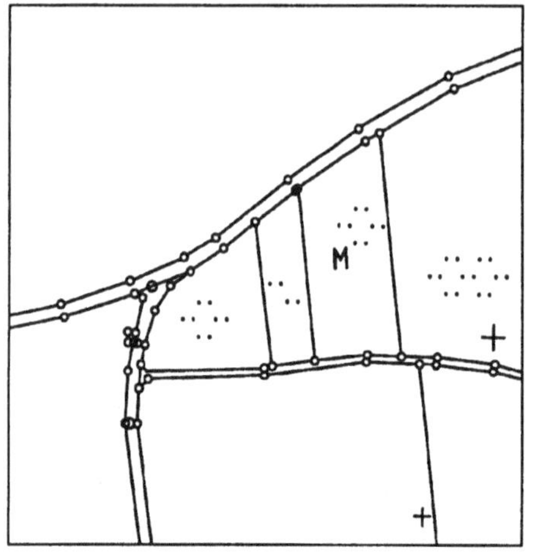

Abb. 14: Gescannter und binarisierter Ausschnitt einer Bayerischen Flurkarte im Maßstab 1 : 5000 [Bayerisches Staatsm. 1982]

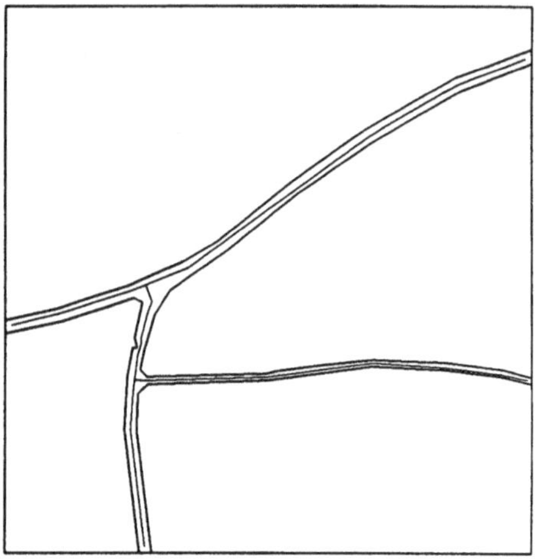

Abb. 15: Straßennetz

Ein Nachteil des hier vorgestellten Ansatzes ist allerdings der hohe Modellierungsaufwand. Eine Erweiterung des Modells für beliebige Kartenausschnitte ist aufwendig.
Insgesamt hat sich aber, wie die geringen Änderungen des Modells für die Interpretation der Flurkarte 1 : 5000 belegen, gezeigt, daß dieser hohe Modellierungsaufwand für die gestellte Aufgabe gerechtfertigt ist.

Der Siemens Nixdorf Informationssysteme AG sei für die Unterstützung von Helmut Mayer mit einem Promotionsstipendium gedankt.

Literatur

Antoine, D. (1991): CIPLAN: A model-based system with original features for understanding french plats. Proc., ICDAR '91, First Intern. Conf. on Document Analysis and Recognition, S. 647-655.

Bayerisches Staatsministerium der Finanzen (1982): Anweisung für das Zeichnen von Vermessungsrissen und Katasterkarten in Bayern (Bayer. Zeichenanweisung 82 - ZeichA82).

Brachman, R.J. (1979): On the epistemological status of semantic networks. In: Findler N.V. (Hrsg.), Associative Networks, Academic Press, New York.

Ebi, N., Lauterbach, B., Besslich, Ph. (1992): Automatic data acquisition from topographic maps using a knowledge-based image analysis system. Int. Arch. Photogr RS, Vol. 29, Part B4/IV, S. 655-663.

Ilg, M. (1990): Knowledge-based interpretation of road maps. Proc. 4th Int'l Symposium on Spatial Data Handling, Zürich, S. 25-34.

Illert, A. (1990): Automatische Erfassung von Kartenschrift, Symbolen und Grundrißobjekten aus der Deutschen Grundkarte 1:5000. Wissensch. Arb. der Fachr. Vermessungsw. der Uni Hannover, Nr. 166.

Klauer, R. (1993): Untersuchungen zur Optimierung von Verfahren bei der automatisierten Digitalisierung von Flurkarten. Wissensch. Arb. der Fachr. Vermessungsw. der Uni Hannover, Nr. 183.

Mayer, H., Heipke, C., Maderlechner, G. (1992): Knowledge-based interpretation of scanned large-scale maps using multi-level modelling. Int. Arch. Photogr RS, Vol. 29, Part B3/III, S. 578-585.

Marr D. (1982): Vision. W.H. Freeman and Co., San Francisco.

Mulder, J.A. (1988): Discrimination Vision. CVGIP 43, S. 313-336.

Ein distanz- und orientierungsinvariantes lernfähiges Erkennungssystem für Robotikanwendungen[1]

G. Hartmann, K.O. Kräuter, H. Wiemers, E. Seidenberg, S. Drüe
Fachbereich 14 Elektrotechnik, Universität-GH Paderborn
Pohlweg 47-49, D 33098 Paderborn

Abstract

Es wird die Architektur eines Robot-Vision-Systems vorgestellt, das in der Lage ist, beliebig orientierte und positionierte Objekte aus unterschiedlichen Entfernungen mit einer neben dem Greifer montierten Kamera zu erkennen. Objektname, Koordinaten und Orientierung bekannter Werkstücke werden bestimmt, so daß diese vom Roboter gegriffen werden können. Unbekannte Objekte werden ebenfalls vom Erkennungssystem erfaßt und gelernt und gelten nach Eingabe einer Bezeichnung als bekannt. Die Distanz- und Orientierungsinvarianz wird in diesem lernfähigen System durch gesteuerte Abbildungen bewirkt, die eine geeignete Repräsentation der Objekte erfordern. Eine für dieses System entwickelte Retina erlaubt es, diese Repräsentation zu erzeugen.

1. Einführung

Ausgangspunkt für die hier vorgestellte Architektur ist die Beobachtung, daß beim Sehen Objekte bis zu einer recht hohen Komplexität, etwa Gesichter, Werkzeuge, Verkehrszeichen usw. "mit einem Blick" erkannt werden. Das Ziel, Objekte ganzheitlich zu lernen und mit großer Robustheit wiederzuerkennen, läßt sich durch Verwendung eines Assoziativspeichers realisieren ([1]). Die gewünschte Lageinvarianz ist in einem aktiven Erkennungssystem mit einer beweglichen, neben dem Greifer des Roboters montierten Kamera dadurch erreichbar, daß diese immer auf einen gleichbleibenden Punkt des Objekts (z.B. den Flächenschwerpunkt) ausgerichtet werden kann. Völlig ausgeschlossen erscheint es hingegen, Objekte für alle Drehlagen und alle entfernungsbedingten Bildgrößen in einem assoziativen Netzwerk zu lernen. Mehrschichtige Perceptrons ([2]), die derartige Fähigkeiten ansatzweise besitzen, erfordern unzumutbar lange Trainingssequenzen und zerstören mit ihrer Invarianzleistung das Wissen über Position, Distanz bzw. Orientierung des Objekts.

Wir haben deshalb den Weg beschritten, über gesteuerte Abbildungen normierte Repräsentationen der Objekte zu erzeugen und zu lernen, die dann unabhängig von Distanz und Drehlage wiedererkannt werden können. Der Steuerparameter "Distanz" kann auf unterschiedliche Weise bestimmt werden und ist in einem Robotik-System ohnehin bekannt. Die präzise Bestimmung des Steuerparameters "Orientierung" für unterschiedliche Objekte ist Gegenstand dieses Beitrags. Eine Erkennung perspektivisch verzerrter Objekte, bei der die normierenden Abbildungen durch die Blickrichtungsparameter gesteuert werden, scheint aufgrund erster Ergebnisse ebenfalls möglich.

Die skizzierte Vorgehensweise ist im Prinzip nicht neu ([3], [4]), führt aber nur dann zu zuverlässigen Ergebnissen, wenn die unvermeidlichen Unterschiede zwischen der gelernten und der zu erkennenden Repräsentation eines Objekts toleriert werden. Die Auswertung eines flächenbezogenen Ähnlichkeitsmaßes (das später auf Farbe erweitert werden soll) ist in dieser Hinsicht "gutartig" und toleriert leichte Verschiebungen oder Normierungsfehler. Ein

[1] Gefördert durch den Bundesminister für Forschung und Technologie (BMFT), ITN 910 506

flächenbezogenes Ähnlichkeitsmaß kann Objekte mit deutlich unterschiedlicher Größe sicher trennen und ist deshalb in unserem System unverzichtbar notwendig. Es ist jedoch nicht hinreichend, weil die Formunterscheidung relativ schlecht ist. Bei der erforderlichen Toleranz sind z.B. ein Quadrat und eine Kreisscheibe gleicher Größe nicht mehr sicher zu unterscheiden.
Deshalb wird in unserem System zusätzlich ein konturbezogenes Ähnlichkeitsmaß ausgewertet, das die Orientierungen von Kantenelementen berücksichtigt. Die Repräsentation von Objekten durch orientierte Kantenelemente bietet zwar eine ausgezeichnete Formbeschreibung, aber praktisch keine Toleranz gegenüber kleinsten Verschiebungen und Normierungsfehlern. Deshalb wird diese Repräsentation in eine "ortstolerante" Form gebracht, die der durch komplexe Neuronen im visuellen Cortex erzeugten ähnlich ist und im Kapitel 3 näher beschrieben wird.
Schließlich muß gewährleistet sein, daß die Zahl der Komponenten der gelernten Repräsentation mit der zu erkennenden normierten Repräsentation übereinstimmt. Da die gesteuerten Abbildungen selbst weder Komponenten erzeugen noch weglassen dürfen, muß durch eine geeignete Retina und eine darauf abgestimmte Verarbeitung gewährleistet werden, daß für alle entfernungsbedingten Bildgrößen Repräsentationen mit gleicher Komponentenzahl erzeugt werden.
Die Beschreibung dieser Retina und der damit verbundenen Vorverarbeitung, die Erzeugung ortstoleranter Konturrepräsentationen und die präzise Ermittlung des Steuerparameters "Objektorientierung" sind Gegenstand des folgenden Beitrags.

2. Die künstliche Retina

Wie bereits Reitböck ([3]) und Schwarz ([4]) gezeigt haben, wird durch die besonderen Eigenschaften einer logarithmisch-polaren Abtastung eine orientierungs- und distanzinvariante Erkennung erleichtert. So wird z.B. eine Drehung des Objekts um das Abtastzentrum durch eine Translation auf dem polaren Abtastgitter in tangentialer Richtung kompensiert. Ebenso wird eine Vergrößerung bzw. Verkleinerung des Bildes aufgrund einer sich ändernden Entfernung zwischen Objekt und Kamera durch eine Translation in radialer Richtung kompensiert.
Die gesteuerten Abbildungen reduzieren sich in diesem Fall auf ein Umadressieren der Bildpunkte, und die Komponentenzahl der daraus gewonnenen Repräsentation bleibt erhalten. Obwohl unsere Retina von dieser einfachen Form abweicht, wird das Grundprinzip übernommen, und deshalb soll zunächst die Konstruktion eines entsprechenden Abtastrasters erläutert werden.
Für die nachfolgenden Operationen im Bildbereich hat es sich als vorteilhaft erwiesen, in einem hexagonalen Punktraster zu arbeiten. Im Gegensatz zu einem kartesischen Punktraster, hat man im hexagonalen Raster je sechs Punkte mit gleicher Distanz zum Zentrum, außerdem standen für das hexagonale Raster bewährte Detektorsätze zur Verfügung, die inzwischen auch in Hardware realisiert sind. Um im Bildbereich der logarithmischen Abbildung ein gleichmäßiges hexagonales Raster zu erhalten, muß das Abtastraster im Originalbereich in radialer Richtung exponentiell gedehnt sein. Das verwendete Abtastraster ist aus konzentrischen Kreisen aufgebaut. Jeder dieser Abtastkreise soll eine bestimmte Anzahl (q) von Abtastpunkten besitzen (Fig. 1, links). Die Tatsache, daß jeder zweite Abtastkreis um einen Winkel von π/q um das Abtastzentrum gedreht ist, garantiert die hexagonale Struktur des Punktrasters im Bildbereich. Die exponentielle Dehnung des Abtastrasters in radialer Richtung bedeutet, daß der Quotient der Radien zweier aufeinanderfolgender Abtastkreise (ρ_{n+1}/ρ_n) einer Konstanten (p) entspricht. Nur wenn die beiden Parameter p und q auf eine bestimmte Weise miteinander verknüpft werden, erhält man das gleichmäßige hexagonale Raster (Fig. 1, rechts). Dieser Zusammenhang soll im folgenden hergeleitet werden.

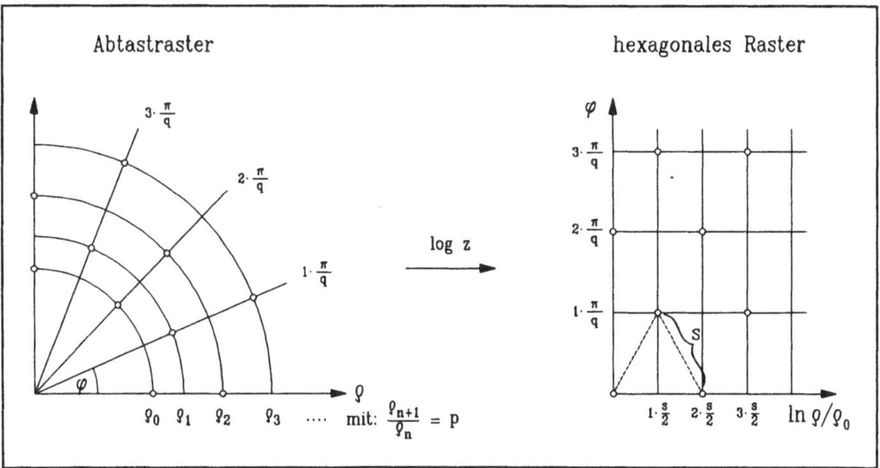

Fig. 1: Konstruktion eines Teilabtastrasters mit dem radialen Parameter p und dem tangentialen Parameter q

Drei benachbarte Punkte im hexagonalen Raster bilden die Eckpunkte eines gleichseitigen Dreiecks mit der Seitenlänge s (Fig. 1, rechts). Die Höhe dieses Dreiecks entspricht dem Wert $\pi/q = s/2 \cdot \sqrt{3}$. Der Parameter s im Bildbereich läßt sich im Originalbereich folgendermaßen ausdrücken:

$$\frac{s}{2} = \ln\frac{\rho_{n+1}}{\rho_n} \quad \text{oder} \quad \frac{\rho_{n+1}}{\rho_n} = e^{\frac{s}{2}} \equiv p \tag{1}$$

Bei vorgegebenem q ist p auf folgende Art festgelegt:

$$p = e^{\frac{s}{2}} = e^{\frac{\pi}{q\sqrt{3}}} \tag{2}$$

Um eine gewünschte Auflösung zu garantieren, muß eine bestimmte Dichte der Abtastpunkte überall im Bild gewährleistet sein. Durch das exponentielle Wachstum der Radien der Abtastkreise von innen nach außen, erhält man einen exponentiellen Abfall der Abtastpunktdichte proportional zu p^{-2n}, wobei n die Abtastkreise von innen nach außen nummeriert (Fig. 2). Eine Möglichkeit, auch in der Peripherie eine hinreichende Dichte der Abtastpunkte zu garantieren, bestünde darin, mit einer sehr hohen Punktdichte im Zentrum zu arbeiten, so daß die gewünschte Auflösung in der Peripherie trotz des exponentiellen Abfalls mit der Ringzahl vorliegt. Dies führt aber zu einer starken Überabtastung im zentralen Bereich und damit zu einem sehr hohen Rechenaufwand. In unserer künstlichen Retina verwenden wir deshalb eine Überlagerung von verschiedenen Abtastrastern mit unterschiedlicher Punktdichte. Beginnend mit einer für den zentralen Bereich ausreichenden Dichte, wird immer dann ein neues dichteres Raster überlagert, wenn die Dichte des alten unter einen bestimmten Wert fällt (Fig. 2).

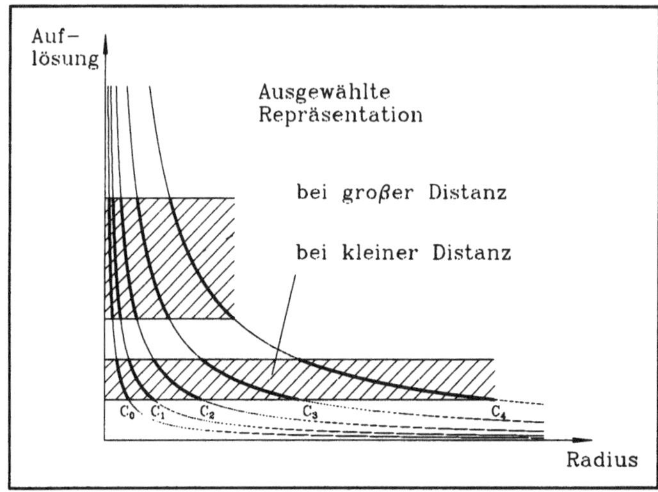

Fig. 2: Für eine verschiedene Anzahl von Punkten pro Abtastkreis ist die Dichte des resultierenden Abtastrasters von innen nach außen aufgetragen. Die einzelnen Kurven entsprechen den verschiedenen Ortsfrequenzkanälen C_k. Abhängig von der Entfernung zwischen Objekt und Kamera findet man die normierte Repräsentation an verschiedenen Stellen (dicke Linien).

In unserem Fall besitzt der Ring mit der geringsten Auflösung (R_0) $q_0 = 16$ Punkte auf einem Abtastkreis. Hieraus ergibt sich nach Gleichung (2) der Wert $p_0 = 1.120$, was fast genau dem Wert von $\sqrt[6]{2}$ entspricht. Mit dieser Näherung verdoppelt sich der Radius des innersten Abtastkreises (ρ_0) nach sechs Schritten ($\rho_6 = 2 \cdot \rho_0$), und die Dichte der Abtastpunkte verringert sich auf ein Viertel. Ab diesem Radius wird ein vierfach dichteres Abtastraster überlagert (R_1), was sich durch Verdoppeln der tangentialen und radialen Punktdichte ergibt ($q_1 = 2 \cdot q_0$, $p_1 = \sqrt{p_0}$). Nachdem sich auch bei R_1 mit wachsendem Abstand zum Zentrum die Punktdichte auf den vierten Teil reduziert hat, wird ein weiterer Ring (R_2) hinzugenommen, der wiederum die vierfache Punktzahl von R_1 aufweist. Auf diese Weise werden die Ringe R_0 bis R_4 erzeugt ($q_{n+1} = 2 \cdot q_n$, $p_{n+1} = \sqrt{p_n}$, für $n = 0,..,3$) (Fig. 3a).

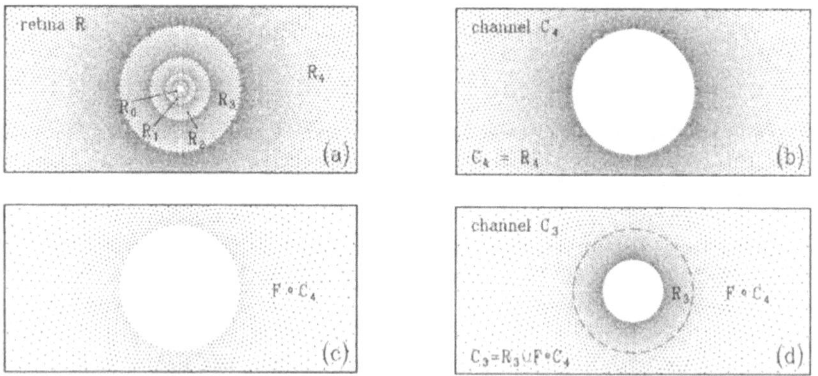

Fig: 3: Die künstliche Retina (a). Der Kanal C_3 (d) wird aus der Überlagerung des gefilterten C_4 (b) mit R_3 (c) erzeugt.

Aus den Ringen R_0 - R_4 läßt sich nun durch Tiefpaßfilterung und Unterabtastung ein Satz von retinalen Repräsentationen mit über der Retina wachsendem Pixelabstand erzeugen (Fig. 3b, 3d). Von diesen sich teilweise überlappenden Repräsentationen werden gleiche Bildpunkte in unterschiedlicher Auflösung dargestellt. Daher werden diese Repräsentationen mit dem Namen "Ortsfrequenzkanal" C_k bezeichnet. Der Kanal (C_4) ist identisch mit dem äußerstem Ring (R_4) (Fig. 3b). Die nachfolgenden Kanäle (C_k) ergeben sich aus der Überlagerung des entsprechenden Ringes (R_k) mit dem tiefpaßgefilterten und in der Auflösung entsprechend reduzierten höheren Kanal (Fig. 3c, 3d). Beschreibt man die Kombination aus Tiefpaßfilterung und Verringerung der Abtastdichte formal mit dem Operator F, ergeben sich die Kanäle C_k durch den Ausdruck $R_k \cup F \circ C_{k+1}$ (Fig. 3). Eine schematische Darstellung dieser Verarbeitungsstufe ist in Fig. 4 links gezeigt.

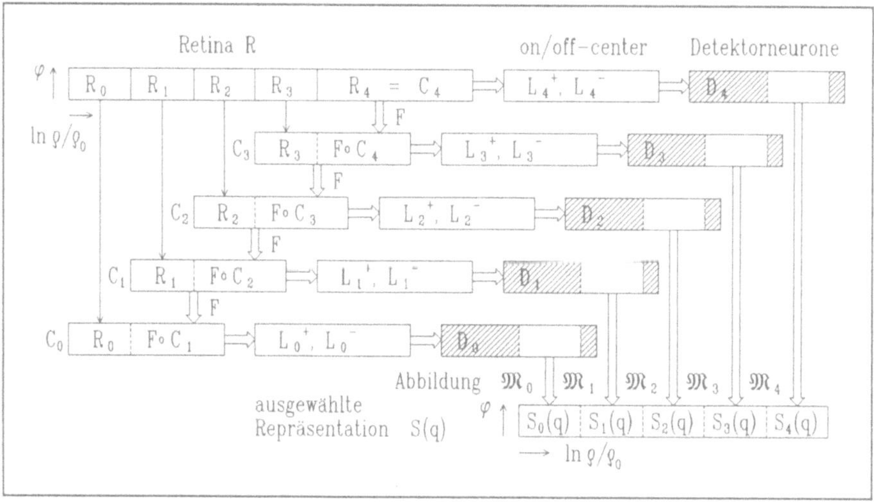

Fig. 4: Blockschaltbild des Gesamtsystems. Aus dem Bild des Objekts auf der künstlichen Retina (oben links) werden durch Filtern und Unterabtasten die Ortsfrequenzkanäle (C_0 - C_4) erzeugt. Hieraus wird das dreiwertige Laplacebild (L_k^+, L_k^-) gebildet, wodurch Bereiche mit starken Helligkeitsunterschieden markiert sind. Aus dem dreiwertigen Laplacebild ergibt sich dann die Information über lokale Kantenverläufe (D_i). Durch mehrere gesteuerte Abbildungen (rechts) wird schließlich die distanz- und orientierungsinvariante Repräsentation des Objekts erzeugt (links unten).

Der Satz von retinalen Repräsentationen C_k erlaubt es nun, auf einfache Weise distanzunabhängige Repräsentationen mit konstanter Komponentenzahl durch Auswahl zu erzeugen. Dies läßt sich leicht an Fig. 2 verdeutlichen, in der die mit dem Radius abfallende Auflösung der Ortsfrequenzkanäle C_k gezeigt ist, aus denen die Ringe der ausgewählten Repräsentationen für bestimmte Objektentfernungen (dick ausgezogen) entnommen sind. Wird nun ein Objekt gelernt, dessen Bild den oberen dick markierten Teil der Retina füllt und soll ein durch kleinere Entfernung größeres Bild erkannt werden, so muß aus jeder Repräsentation C_k ein Ring entsprechend größeren Durchmessers und größerer Breite ausgewählt werden (unterer dick markierter Bereich). Dabei verkleinert sich der mittlere Pixelabstand und die Pixelzahl bleibt konstant. Die gesteuerte Abbildung zur Abstandsnormierung ist auch in unserer zusammengesetzten Retina durch reines Umadressieren zu ereichen.

3. Ortstolerante Kantenrepräsentationen

Flächenbasierte Ähnlichkeitsmaße können nun direkt aus den retinalen Repräsentationen gewonnen werden. Die Erzeugung ortstoleranter Konturrepräsentationen erfordert hingegen noch weitere Vorbereitungsschritte, die auch in Fig. 4 gezeigt sind. Die retinalen Grauwertbilder C_0 - C_4 werden durch Zentrum-Peripherie Operationen und Schwellwertbildung in dreiwertige Laplacebilder umgesetzt. Aus den positiven und negativen Krümmungswerten L_k^+ und L_k^- extrahiert ein vollständiger Satz von Kantendetektoren durch logische Verknüpfung orientierte Kantenelemente. Die Kantendetektoren decken einen Bereich von 19 Pixeln des hexagonalen Rasters ab und haben eine Winkelauflösung von $\pm$ 15°. Jedem Bildpunkt ist ein Satz von solchen Detektoren zugeordnet, deren Aktivität orientierte Kantenelemente codiert. Interpretiert man diese Detektoren als Neuronen mit orientierten rezeptiven Feldern, so entsprechen diese recht gut den simplen Neuronen des visuellen Cortex ([5]).

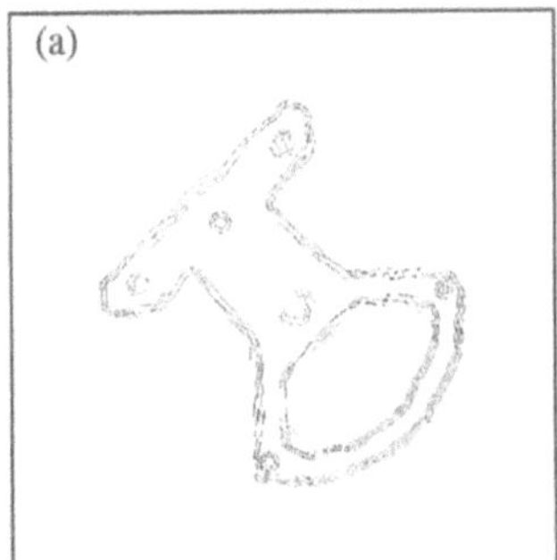

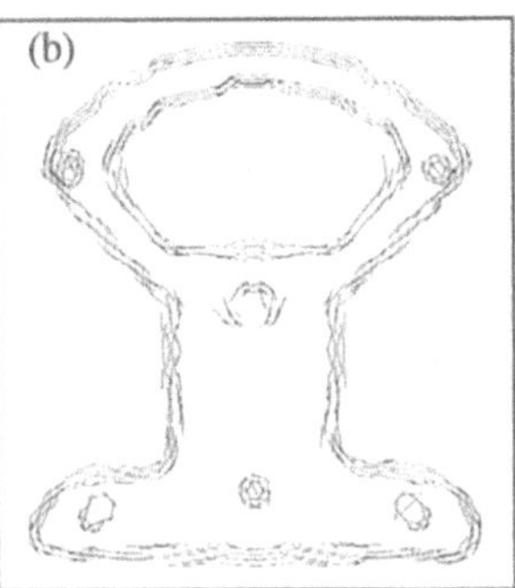

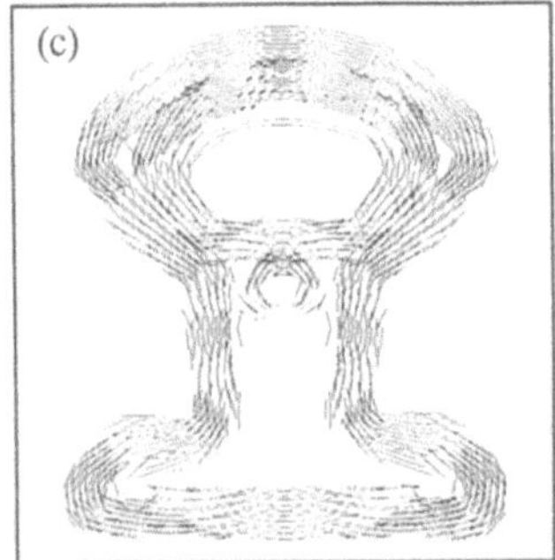

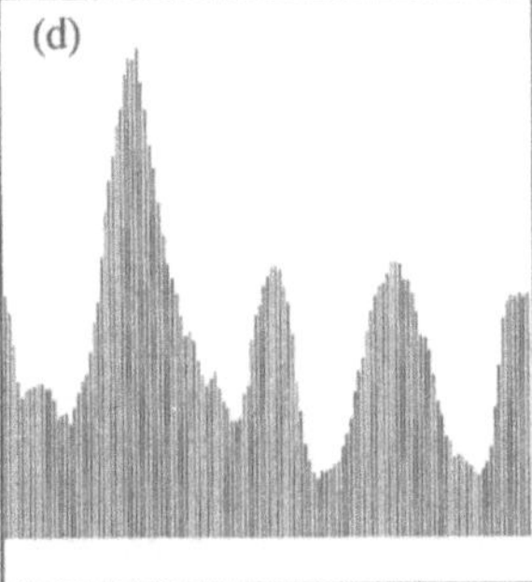

Fig. 5: Darstellung von orientierten Kantenelementen erster Stufe (simple Zellen im visuellen Cortex) (a, b) und höherer Stufe (komplexe Zellen im visuellen Cortex) (c). Orientierung und Entfernung eines Objekts werden durch gesteuerte Abbildungen normiert (a $\mapsto$ b bzw. a $\mapsto$ c). Die Orientierung eines Objekts läßt sich mit einer Genauigkeit von $\pm 1{,}4°$ durch ein Gruppe von 128 Neuronen bestimmen, die die Häufigkeit des Auftretens lokaler Kantenorientierungen auswerten (d).

Würde man das Aktivitätsmuster dieser Neuronen als Eingangsrepräsentation eines Assoziativspeichers wählen, so würde bereits die diagonale Verschiebung eines Quadranten um einen Bildpunkt jede Übereinstimmung des gezeigten und gelernten Bildes zunichte machen. Wegen dieser hohen Anforderung an Positionsgenauigkeit, die im System nicht gewährleistet werden kann, muß eine Kantenrepräsentation mit höherer Ortstoleranz erzeugt werden. Die von uns gewählte Repräsentation lehnt sich an die des visuellen Cortex an, der Konturinformationen zusätzlich durch sog. komplexe Neuronen codiert. Komplexe haben, wie simple Neuronen, orientierte rezeptive Felder mit ca. $\pm 15°$ Winkeltoleranz. Komplexe Neuronen zeigen jedoch - bezogen auf die Breite einer Linie oder eines Kantenprofils - wesentlich höhere Ortstoleranz bzw. größere rezeptive Felder. Ordnet man nun wieder jedem Bildpunkt einen Satz komplexer Neuronen zu, so wird ein und dasselbe Kantenstück auch von Neuronen codiert, die seitlich der Kontur liegenden Pixeln zugeordnet sind. Konturverläufe werden so durch örtlich verteilte "Orientierungswolken" codiert, die auch mit einer leicht verschobenen gelernten "Orientierungswolke" überlappen (Fig. 5c).

Unabhängig von der - bis heute unbekannten - Verschaltung des visuellen Cortex erzeugen wir komplexe "Neuronen" durch Kontinuitätsprüfung zwischen Kantendetektoren in Bereichen von z.B. 7x15 Bildpunkten. Somit spricht ein komplexes Neuron - in Übereinstimmung mit dem biologischen - immer dann an, wenn eine kontinuierliche Kontur den vorgegebenen Bereich durchläuft. Beträgt die Differenz zwischen Eingangs- und Ausgangsorientierung an einer Stelle mehr als z.B. 90°, so wird ein "Eckenneuron aktiviert. Aus dieser ortstoleranten Repräsentation, die in Fig. 4 mit D_k bezeichnet ist, wird durch die distanznormierende Abbildung ein Satz von Ringen S(q) entsprechend der Größe ausgewählt.

4. Orientierungsmessung

Voraussetzung für eine sichere Erkennung ist eine interne Repräsentation eines Objektes unabhängig von seiner tatsächlichen Orientierung. Um eine solche Repräsentation zu erhalten, wird eine gesteuerte Abbildung verwendet, die als Eingangsparameter den Winkel benutzt, um den das angebotene Objekt von einer objekteigenen Vorzugsorientierung abweicht. Ein wichtiger Schritt besteht somit in der Bestimmung dieses Verdrehungswinkels, der vom gewählten Fovealisierungspunkt nicht aber von der Bildgröße abhängt. Die Vorzugsorientierung entspricht der lokalen Kantenorientierung, die für das entsprechende Objekt am häufigsten vertreten ist und läßt sich daher leicht aus einem Häufigkeitshistogramm der vorhandenen lokalen Orientierungen finden (Fig. 5d). Bei symmetrischen Objekten, wo sich mehrere gleich hohe Maxima ergeben, wird für jede Orientierung eine Repräsentation gelernt.
Obwohl die lokalen Kantendetektoren nur über eine Winkelauflösung von ± 15° verfügen, kann die Orientierung eines angebotenen Objekts auf ± 1,4° genau bestimmt werden. Diese Eigenschaft ergibt sich direkt aus der polaren Abtastung, bei der die Differenz der lokalen Orientierungen zweier benachbarter Punkte auf einem peripheren Abtastkreis (in R_4) 1,4° (= 360°/q_4) beträgt.

5. Musterklassifizierung

Unser Assoziativspeicher, das CLAN (Closed Loop Antagonistic Network), lernt ohne Lehrer. Es teilt Eingangsmuster nach einmaliger Präsentation in verschiedene Musterklassen ein, wobei falls nötig neue Klassen gebildet werden ([1]). Das CLAN hat für jede Position im Raster der internen Repräsentation soviele binäre Eingänge, wie es mögliche Merkmale gibt. Ein CLAN-Eingang ist genau dann aktiv, wenn das entsprechende Merkmal an dieser Position vorhanden ist und kann als Komponente eines binären Eingangsvektors aufgefaßt werden. Neben diesem Vektor wird dem CLAN zusätzlich der inverse Vektor angeboten (antagonistische Repräsentation der Information). Das Netzwerk ermittelt mit Hilfe des angebotenen und des zusätzlich erzeugten antagonistischen Merkmalsvektors das Ähnlichkeitsmaß ("inneres Produkt" minus "Hamming-Distanz"). Das angebotene Muster wird aufgrund dieses Ähnlichkeitsmaßes einer Musterklasse zugeordnet oder eröffnet eine neue Musterklasse, falls eine hinreichende Übereinstimmung mit bereits vorhandenen Klassen nicht gegeben ist.

6. Implementierung

Für die Bildaufnahme wird eine CCD-Farbkamera mit 440.000 Pixeln verwendet. Damit auch bei kontinuierlicher Bewegung des Roboters Bilder in den Bildspeicher eingelesen werden können, muß während der Digitalisierung mit Hilfe eines elektronischen Shutters das

Bild "eingefroren" werden. Das aus der Kameraeinheit kommende Videosignal wird mit einer Auflösung von 8 Bit für jeden der drei Farbkanäle (**R**ot, **G**rün, **B**lau) digitalisiert. Es entsteht somit ein digitales Bild von 512x512 Pixeln, das über eine Look-Up-Table am Eingang des Bildspeichers abgelegt wird. In der Bildvorverarbeitung wird das digitalisierte Bild zunächst mit einem Tiefpaß gefiltert, was mit Hilfe einer 3x3 Gaußfiltermaske im Ortsbereich durchgeführt wird. Für die Bestimmung des Zentrums der nachfolgenden Transformation (Fovealisierungspunkt) wird der Flächenschwerpunkt des Objekts bestimmt. Hier wirken sich Störungen durch ungleichmäßig ausgeleuchtete Bildbereiche, Schatten und Refexionen ungünstig auf die korrekte Bestimmung des Fovealisierungspunktes aus. Aufgrund unserer ortstoleranten Repräsentation läßt sich jedoch der Fovealisierungspunkt mit hinreichender Genauigkeit bestimmen. Mit Hilfe von lokal adaptierten Schwellen, die durch eine vorherige Grauwerthistogrammbildung ermittelt werden, können die zu einem Objekt gehörenden Pixel von Hintergrund bzw. von Fehlstellen getrennt werden ("region-growing" Verfahren).

Das nachgeschaltete Erkennungssystem ist auf einem Parallelrechnersystem auf Transputerbasis implementiert. Für eine schnelle Verteilung der Bildinformation auf die einzelnen Prozessorknoten sorgt ein echtzeitfähiger Videobus (Fig. 6).

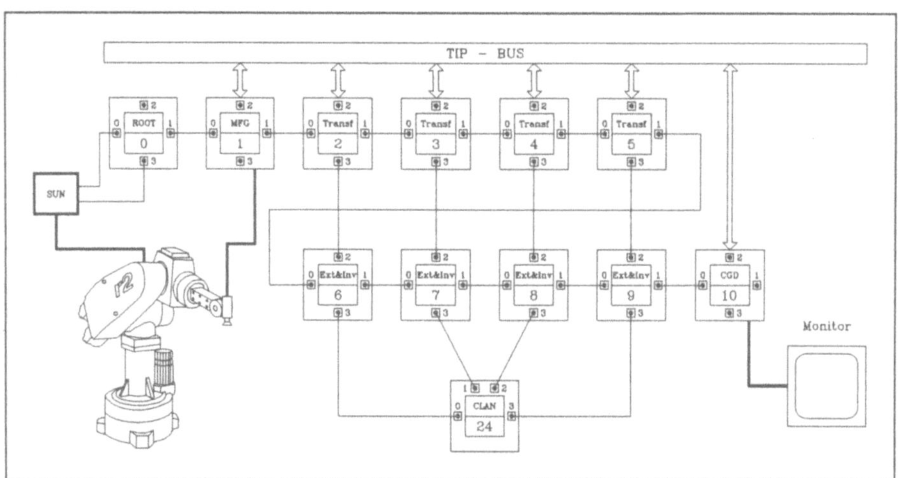

Fig. 6: Das Kamerabild wird über den Videobus auf vier Transputer, die die logarithmische Transformation durchführen (Transf) verteilt. Das transformierte Bild wird an vier weitere Transputer verschickt, in denen die Merkmalsextraktion durchgeführt und die Invarianzen erzeugt werden (Ext&Inv). Die extrahierten Merkmale werden in der invarianten Darstellung zur Klassifikation an den Assoziativspeicher (CLAN) weitergeleitet.

7. Literatur

[1] Hartmann, G. *Learning in a Closed Loop Antagonistic Network*. In: Kohonen, T. (ed.): Artificial Neural Networks, Elsevier Publishers 1991, 239-244.

[2] Fukushima, K. et al. *Neocognitron: A Neural Network Model for a Mechanism of Visual Pattern Recognition*. IEEE-SMC, 13, 1983, 826-834.

[3] Reitböck, H.J. Altmann, J. *A Model for Size- and Rotation-Invariant Pattern Processing in the Visual System*. Biol. Cybernetics, 51, 1984, 113-121.

[4] Schwarz, E.L. *Cortical Anatomy, Size Invariance and Spatial Frequency Analysis*. Perception, 10, 1981, 831-835.

[5] Hubel, D.H. Wiesel T.N. *Receptive Fields of Single Neurons in the Cat's Striate Cortex*. Journal of Physiology, 148, 1959, 574-579.

Quantitative Kurserfassung einer Fahrspur durch Befahren mit einem sehenden Fahrzeug

Reinhold Behringer

Institut für Systemdynamik und Flugmechanik
Fakultät für Luft- und Raumfahrttechnik
Universität der Bundeswehr München; D-85577 Neubiberg
email: ld2brbeh@rz.unibw-muenchen.de

Zusammenfassung Ein System zur autonomen Fahrzeugführung mittels Rechnersehen wurde verwendet, um eine Teststrecke, die zuvor manuell und elektronisch vermessen worden war, zu vermessen. Das Fahrzeug, in dem das System integriert ist, wurde dabei manuell über den Testkurs gefahren. Als Bildaufnehmer dient eine hinter der Fahrzeugfrontscheibe auf einer drehbaren Plattform befestigte CCD-Kamera, welche die Videobilder in den Framegrabber eines Transputerclusters einliest. Dort werden durch gesteuerte Bildauswertung alle 80 ms die Positionen der Fahrspurmarkierungen im Bild ermittelt, und mittels rekursiver Schätzverfahren durch ein Kalmanfilter die Zustandsvektoren der dynamischen Modelle von Fahrzeug und Fahrspurverlauf aktualisiert und auf Festplatte aufgezeichnet. Die anschließende Auswertung im Labor zeigt eine gute Übereinstimmung der geschätzten Daten des Kursverlaufes mit den tatsächlichen Werten.

1 Einleitung

Bereits seit 1977 wird an der UniBw München an der Erforschung und Entwicklung des Rechnersehens gearbeitet. Gegenstand der Arbeiten ist u.a. die Implementation eines Systems zum autonomen Führen eines Straßenfahrzeuges durch Rechnersehen im Testfahrzeug VaMoRs (Versuchsfahrzeug für autonome Mobilität und Rechnersehen) [4]. Seit Anfang 1992 werden mit VaMoRs autonome Fahrten auf öffentlichen Autobahnen in normalem Verkehr durchgeführt [2]. Grundlage für die hierbei angewandten rekursiven Schätzverfahren ist eine Modellierung der Objekte (Fahrspur, autonomes Fahrzeug und Hindernisse) in Raum und Zeit (4D-Ansatz).

Vor der Integration in VaMoRs wurde das System in einem Closed-Loop-Simulationskreis getestet. Dabei bestand eine sehr gute Möglichkeit des Vergleiches zwischen den simulierten und den geschätzten Daten.

Bei autonomen Fahrten auf öffentlichen Straßen besteht diese Möglichkeit der quantitativen Validierung nicht, man kann lediglich die qualitative Plausibilität der geschätzten Werte beurteilen.

Ab Mitte 1991 ergab sich jedoch die Möglichkeit, auf einem Platz des ehemaligen Militärflughafens Neubiberg einen durch weiße Fahrbahnmarkierungen normgerecht markierten Fahrkursus aufzumalen und damit eine quantitative Validierung der Schätzergebnisse durchzuführen.

2 Aufbau des Systems zum Rechnersehen

2.1 Übersicht

Das Testfahrzeug VaMoRs ist ein Mercedes Kastenwagen D-508 mit einer an Bord befindlichen 220 V-Spannungsversorgung durch einen 8.5-kW-Generator.

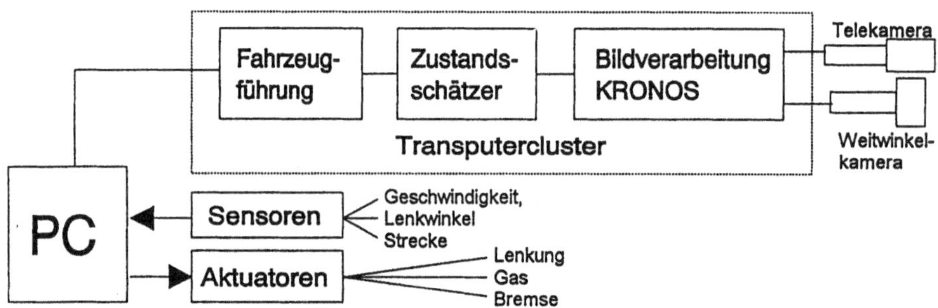

Abb. 1. Blockschaltbild des Systems für autonome Fahrzeugführung mittels Rechnersehen, wie es im autonomen Fahrzeug VaMoRs integriert ist

Abb. 1 zeigt den prinzipiellen Aufbau des Rechnersystems für autonome Spurhaltung. Die zentrale Rechnereinheit ist dabei der Transputercluster (bis zu 30 T-222 und T-800), auf dem die gesamte Bildverarbeitungs- und Zustandsschätzungssoftware läuft. Ein PC dient als Nutzer- und Fahrzeugschnittstelle und zur Datensicherung.

2.2 Kamera und Plattform

Hinter der Frontscheibe ist in etwa 2 m Höhe eine Kameraplattform aufgehängt, welche die Blickrichtungssteuerung und inertiale Blickrichtungsstabilisierung von zwei Kameras für bifokales Sehen ermöglicht.

Die verwendeten CCD-Kameras sind handelsübliche 2/3-Zoll-Chip-Kameras. Für die sichere Erfassung der Fahrspur im Bereich von 5 bis 25 m wird ein Weitwinkelobjektiv (8 mm) verwendet. Bei Fahrten auf Autobahnen wird zusätzlich eine Telekamera verwendet, um die Vorausschau zu erhöhen, wofür bisher ohne inertiale Stabilisierung Brennweiten bis zu 25 mm verwendet wurden. Damit erhöht sich der Vorausschaubereich auf etwa 70 m, was bei höherer Geschwindigkeit zu einer stabileren Schätzung führt.

2.3 Bildverarbeitung KRONOS

Ein Videohalbbild wird in einen Transputer-Framegrabber eingelesen und mit einer Auflösung von 320×256 Bildpunkten und einer Grauwerttiefe von 8 Bit digitalisiert. An den Framegrabber ist das Bildverarbeitungsmodul KRONOS [1] angeschlossen, das aus einem Array von zwei oder mehr T-222-Transputern besteht. Durch geeignete Maskenoperatoren werden an ausgewählten Stellen des Digitalbildes Grauwertkanten gesucht, deren Position, Richtung und Intensität über einen Link an das übergeordnete Zustandsschätzungsmodul geschickt werden. Die derzeitige Leistungsfähigkeit ermöglicht eine Zykluszeit von 80 ms.

3 Das Schätzverfahren

Aus der möglicherweisen großen Menge solcher Kanten werden durch geeignete Kriterien diejenigen herausgefiltert, welche der Fahrspurmarkierung entsprechen. Solche Kriterien sind u.a. die Breite der weißen Linie und die Entfernung von der vorhergesagten Position im Bild.

Die Bildpositionen der selektierten Kanten gehen als Meßwerte in das Kalmanfilter ein, welches rekursiv aus der Abweichung von prädiziertem Fahrspurrand im Bild und diesen Meßwerten die Zustandsvektoren vom Modell des Fahrzeugs und des Fahrspurverlaufs aktualisiert.

3.1 Die dynamischen Modelle

Das Fahrspurmodell geht von einer Klothoidengestalt des Fahrbahnabschnittes innerhalb des Kameravorausschaubereiches aus [5]. Dies impliziert die Annahme, daß die Krümmung sich linear über der Lauflänge ändert (1). Eine solche Gestalt der Fahrbahn ist in der Bundesrepublik Planungsgrundlage für den Bau von Fernstraßen höherer Ordnung [7].

$$c(l) = c_0 + c_1 \cdot l \quad \text{mit} \quad c_1 = \frac{dc}{dl} \tag{1}$$

Der sich daraus ergebende Zustandsvektor besteht aus den c_0 und c_1-Werten für das horizontale und vertikale Krümmungsmodell der Fahrspur, jeweils bezogen auf die maximale Vorausschau der Kamera.

Das Modell der Eigenbewegung des Fahrzeuges wird durch ein ebenes lineares Einspurmodell beschrieben [6], wodurch sich ein Zustandsvektor aus den Elementen **Ablage** zur Fahrspurmitte, **Gierwinkel** und **Schwimmwinkel** ergibt.

3.2 Kalman-Filter

Das implementierte rekursive Schätzverfahren basiert auf einem vierstufigem Zyklus, der die direkte Inversion der Abbildung vermeidet. Zunächst wird aus dem aktuellen Zustandsvektor eine Prädiktion des im nächsten Zyklus zu erwartenden Zustandsvektors durchgeführt. Daraus wird mittels einer vorwärtsgerichteten Perspektivabbildung das erwartete Erscheinungsbild der Straße, resp. die Position der erwarteten Fahrspurränder errechnet. Aus der Menge der gemessenen Grauwertkanten werden jetzt durch Anwendung geeigneter Kriterien diejenigen selektiert, die der Fahrspurmarkierung entsprechen [3]. Die Abweichung der Positionen dieser Kanten von den prädizierten Kantenpositionen bewirkt nun eine Aktualisierung des prädizierten Zustandsvektors.

Durch Tuning der entsprechenden Kovarianzen kann das Kalmanfilter auf die unterschiedlichen Situationen wie Autobahnfahrt oder Feldweg eingestellt werden.

4 Die Aufnahme der Referenzmessungen

Als Teststrecke existiert seit Sommer 1991 auf einem Teil des stillgelegten Militärflughafens Neubiberg ein aufgemalter Rundkurs. Die Markierungen sind auf einer Betonfläche mit weißer Farbe als 15 cm breite Linien aufgebracht. Der Fahrkurs liegt dabei in einer Ebene, d.h. es treten keine vertikalen Krümmungen auf. An einigen Stellen jedoch, besonders im nördlichen Teil der Strecke, sind lokale Vertiefungen durch Absenkungen von Betonplatten entstanden.

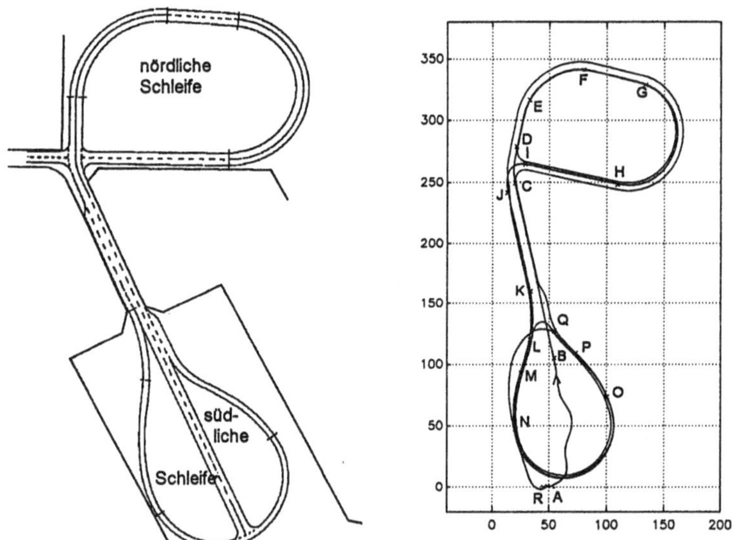

Abb. 2. Plan des Testkurses. Links: Manuell gezeichneter Plan. Rechts: Elektronisch vermessener Umrißplan der Teststrecke (Einheiten in m) und gefahrene Route, beginnend am Punkt A bis zum Punkt R

4.1 Manuelle Vermessung der Teststrecke

Zunächst wurden die Radien der aufgemalten Strecke manuell vermessen. Der Fehler bei den Kreisbögen mit konstantem Radius (C-D, E-F, G-H, I-J) beträgt für die Krümmung etwa 5 %. In den Klothoidenbögen dagegen ist eine Bestimmung der Krümmung nicht möglich, da das Aufzeichnen der Kurvenbögen nicht exakt nach der Klothoidenformel (1) vorgenommen wurde. Lediglich durch Interpolation zwischen Punkten mit konstanten Radien kann ein ungefährer Verlauf der Strecke aufgezeichnet werden.

4.2 Elektronische Referenzvermessung: Teldix-System

Im Herbst 1992 bot sich die Gelegenheit, mit einem Navigationssystem der Firma Teldix, welches eine genaue Lagebestimmung eines Fahrzeuges (in einem inertialen Koordinatensystem) ermöglicht, durch Abfahren der Strecke deren Koordinaten aufzuzeichnen (Abb. 2, rechts). Das System wurde in VaMoRs installiert und ermittelte etwa alle 1.2 m die Position, die Geschwindigkeit und (durch spezielle Sensoren) den Schwimmwinkel. Diese Daten wurden per Funk an einen zentralen Rechner gesandt und dort aufgezeichnet.

Die geglätteten Positionskoordinaten der Teldix-Messung wurden in einen Krümmungswert umgerechnet und über der gefahrenen Strecke aufgetragen, so daß sie mit der gleichzeitig laufenden Krümmungsschätzung verglichen werden können (Abb. 3).

Die absolute Genauigkeit der Positionskoordinaten ist ± 0.1 m. Durch Schwankungen des Systems Fahrzeug-Antenne kommen Störungen in der Größenordnung $+0.1$ m dazu, die jedoch durch Glättung herausgefiltert werden können. Nicht filterbar sind dagegen die tieffrequenten Schwankungen, die durch das Bestreben des Fahrers, das Fahrzeug in der Spurmitte zu halten, hervorgerufen werden. Dadurch erklären sich die großen Schwankungen der Krümmungswerte nach der Teldix-Messung in den Abschnitten E-F, G-H und N-O.

5 Ergebnisse

5.1 Validierung der Krümmungsschätzung

Vergleich mit Referenzdaten. Der Testkurs wurde, wie in Abb. 2 eingezeichnet, von A bis R abgefahren und die geschätzten Krümmungswerte in Abb. 3 aufgetragen.

Die Kurven in der nördlichen Schleife beginnen abrupt ohne Klothoidenübergang. Da das dem Schätzprozeß zugrundeliegende Modell mit einer über den gesamten Vorausschaubereich (hier 25 m) gemittelten Krümmung arbeitet [3], ergibt sich für die geschätzte Krümmung ein stetiger Anstieg bis auf den Maximalwert. Bei den langgezogenen Kurven der südlichen Schleife tritt dies nicht auf, da diese klothoidenähnlich konstruiert sind und sich die Krümmung nicht sprunghaft ändert. Nach dem Ansteigen auf den konstanten Krümmungswert der Kurven E-F, G-H, K-L und N-O wird die tatsächliche Krümmung sehr gut durch den Schätzer reproduziert. Der Schätzwert liegt hier größtenteils im Bereich $\pm 0.001\,\mathrm{m}^{-1}$ um den tatsächlichen Krümmungswert, was der unteren Grenze für die Genauigkeit der Krümmungsschätzung infolge Pixelrauschen und Meßungenauigkeit entspricht [8].

In dem klothoidenähnlichen Stück L-M-N der südlichen Schleife ist die Übereinstimmung zwischen Referenz und Schätzwert gut, da hier der Übergang der

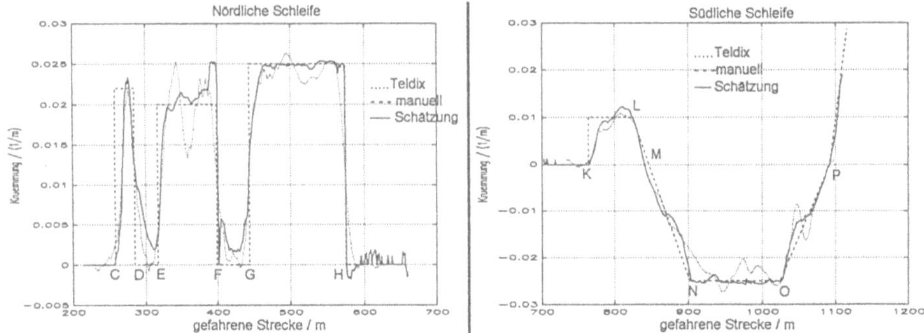

Abb. 3. Die geschätzten Fahrspur-Krümmungsdaten im Vergleich zu den Referenz-Messungen nach dem Teldix-Verfahren und der manuellen Messung. Die Buchstaben entsprechen dabei dem gefahrenen Kurs nach Abb. 2.

Kurvenradien „sanft" erfolgt. Somit bewirkt der im Zustandsvektor mitgeschätzte c_1-Wert eine modellkonsistente Schätzung.

Reproduzierbarkeit der Krümmungsschätzung. Um zu sehen, inwieweit sich die Schätzergebnisse bei wiederholter Meßfahrt reproduzieren lassen, wurden mehrere Fahrten auf dem Testkurs durchgeführt.

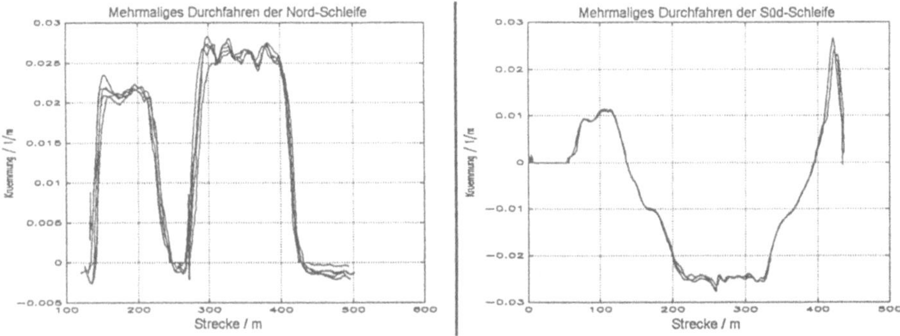

Abb. 4. Ergebnisse der Krümmungsschätzung bei mehreren Fahrten hintereinander. Links der nördliche Teil der Teststrecke, rechts die südliche Schleife

Die in Abb. 4 aufgezeichneten Fahrten wurden mit verschiedenen Kamera-Nickwinkeln und -Vorausschauentfernungen aufgenommen und sind deshalb repräsentativ für die Vielzahl der Versuche.

Reproduzierbar sind in der nördlichen Schleife die Unexaktheiten der Kurvenradien. Dabei spielen die Bodenunebenheiten eine Rolle, die in diesem Streckenteil ausgeprägt sind und die eigentlich konstante Krümmung leicht verzerren. Die Schwankungsbreite der Krümmungswerte von $\pm 0.002\,\mathrm{m}^{-1}$ bei verschiedenen Fahrten hängt auch mit eben diesen Unebenheiten zusammen, da das Fahrzeug bei mehreren Umläufen nie exakt dieselbe Position durchfährt und die Krümmungsschätzsfehler deshalb nicht reproduzierbar sind.

Die südliche Schleife weist viel weniger Unebenheiten auf, daher ist der Schätzwert für die Krümmung wesentlich reproduzierbarer als im nördlichen Teil. Dis Schwankungsbreite für den Schätzwert liegt bei $\pm 0.001\,\mathrm{m}^{-1}$.

Problematisch erweist sich die Messung der gefahrenen Wegstrecke, die über die Umdrehungen des linken Vorderrades aufgenommen wird. Die so gemessene Wegstrecke und die Geschwindigkeit müssen in Kurven auf die Werte für die Bewegung des Fahrzeug-Schwerpunkts korrigiert werden. Zusätzlich jedoch hängen diese Werte (wegen des Reifenlatsches) vom Reifenluftdruck und vom Beladungszustand des Fahrzeugs ab. Bei der Teldix-Referenzmessung zeigte sich, daß dadurch eine um 6.5 % zu große Wegstrecke gemessen wurde. Deshalb mußte die Weglänge um diesen Faktor korrigiert werden.

5.2 Integration der Krümmungsdaten zu einer Karte des Kursverlaufes

Aus der Aufzeichnung der Krümmungswerte über die zurückgelegte Wegstrecke läßt sich prinzipiell der Verlauf des Kurses, beschrieben durch den Ortsvektor $\mathbf{r_i} = (x_i, y_i)$ mittels Integration ermitteln. Der Kurswinkel χ in einem inertialen Koordinatensystem ändert sich innerhalb eines Wegstückes Δl folgendermaßen:

$$\Delta\chi = \int c(l)dl \qquad \approx c \cdot \Delta l \quad \text{(für kleine } \Delta l\text{)} \tag{2}$$

Unter der Annahme einer kleinen Winkeländerung $\Delta\chi$ ist dabei die Positionsänderung in Längsrichtung gleich der Länge des gefahrenen Bogenstückes Δl, in Querrichtung

$$\Delta q = \frac{1}{2} \cdot c \cdot (\Delta l)^2 \tag{3}$$

Dabei ergeben sich aus dem Vektor $\Delta \mathbf{r_i} = (\Delta l_i, \Delta q_i)$ durch Koordinatentransformation die Koordinaten $(\Delta x_i, \Delta y_i)$, woraus der Ortvektor $\mathbf{r_i}$ ermittelt werden kann.

Ein Kriterium für die Güte der Schätzergebnisse ist der in Kurven integrierte Winkel nach (2). Für Kurve E-F sollte sich dabei 90°, für G-H 180° ergeben. Ohne Korrektur der Weglängenmessung ergeben sich durchweg um 5 bis 10 % zu große Winkel. Erst die jeweilige Korrektur der Weglänge verkleinert den Fehler.

Bei der Auswertung zeigte sich, daß der integrierte Kurs sehr empfindlich auf Fehler in der Krümmung reagiert, da ein einmal falsch integrierter Kurswinkel χ bei der weiteren Integration erhalten bleibt. Um diese Fehler korrigieren zu können, ist es von Vorteil, wenn der gesamte durchfahrene Kurs eine geschlossene Kurve bildet. Da die Position des Schnittpunktes der Ortskurve bekannt ist, können daraus die Fehler der Schätzung eliminiert werden.

Bei der Integration der nördlichen Kurve zeigte sich, daß durch Störungen bedingte Fehlmessungen der Krümmung auf gerader Strecke eliminiert werden müssen. Krümmungswerte unterhalb eines Schwellwertes werden zu 0 gesetzt.

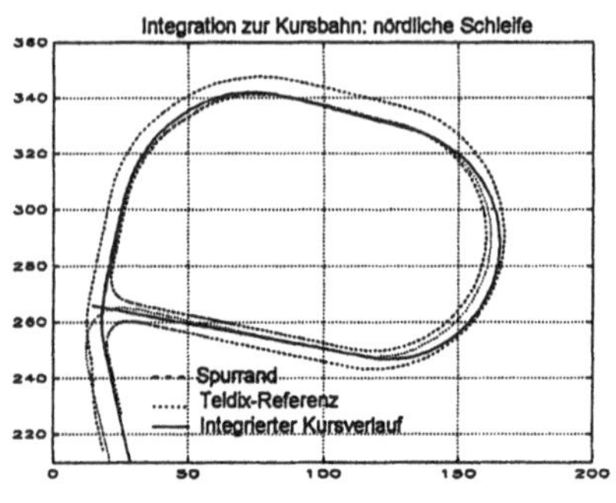

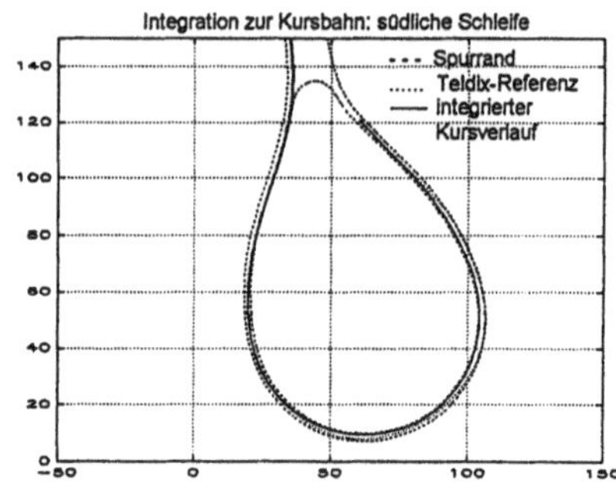

Abb. 5. Kursverlauf nach Integration der Krümmungsschätzwerte in der nördlichen und südlichen Schleife. Die Koordinaten sind in m aufgetragen.

6 Zusammenfassung

Es wurde gezeigt, daß es grundsätzlich möglich ist, mit einem bildverarbeitenden System, das in einem Straßenfahrzeug installiert ist, den Kursverlauf einer Straße durch manuelles Nachfahren aufzuzeichnen. Durch Integration der in jedem Zyklus geschätzten Werte für die Fahrspurkrümmung kann der Verlauf der Fahrspur rekonstruiert werden. Problematisch für die Integration ist hierbei die ungenaue Erfassung von Weglänge und Geschwindigkeit über eines der Fahrzeugräder, da dessen Umlaufradius von Reifenluftdruck und Fahrzeugbeladung abhängt. Durch Integration von geschlossenen Kurven kann jedoch eine Korrektur der Weglängenmessung errechnet werden.

Literatur

1. D. DICKMANNS. „KRONOS – Kantenvermessung in Videobildfolgen (Benutzerhandbuch)". Universität der Bundeswehr München, INF-2 (1992).
2. E. D. DICKMANNS, R. BEHRINGER, C. BRÜDIGAM, D. DICKMANNS, F. THOMANEK UND V. V.HOLT. An all-transputer visual autobahn-autopilot/copilot. In „Proc. ICCV", Berlin, Germany (Mai 1993).
3. E. D. DICKMANNS UND B. MYSLIWETZ. Recursive 3-D state and relative ego-state recognition. *Pattern Analysis and Machine Intelligence* **14**(2), 199–213 (Feb. 1992).
4. E. D. DICKMANNS UND A. ZAPP. Guiding land vehicles along roadways by computer vision. In „Proc. Congres Automatique", S. 233–244 (1985).
5. E. D. DICKMANNS UND A. ZAPP. A curvature-based scheme for improving road vehicle guidance by computer vision. In „Mobile Robots", Bd. 727, S. 161–168. SPIEE (1986).
6. E. DONGES. Der Fahrsimulator des Forschungsinstitutes für Anthropotechnik. Forschungsbericht 41, Forschungsinstitut für Anthropotechnik, Wachtberg-Werthhoven (Juli 1978).
7. FORSCHUNGSGESELLSCHAFT FÜR STRASSEN- UND VERKEHRSWESEN. Richtlinien für die Anlage von Straßen RAS-L-1 (1984).
8. B. MYSLIWETZ. „Parallelrechner-basierte Bildfolgen-Interpretation zur autonomen Fahrzeugsteuerung". Dissertation, Universität der Bundeswehr München, Fakultät für Luft- und Raumfahrttechnik (1990).

Analyse mehrkanaliger Meßreihen im Fahrzeugbau mit „Hidden Markovmodellen“

G. Schmid[1], E.G. Schukat-Talamazzini[2], H. Niemann[1,2]

[1] Bayerisches Forschungszentrum für Wissensbasierte Systeme (FORWISS)
Forschungsgruppe Wissensverarbeitung
Am Weichselgarten 7, 91058 Erlangen-Tennenlohe

[2] Lehrstuhl für Mustererkennung (Informatik 5), Universität Erlangen–Nürnberg
Martensstr. 3, 91058 Erlangen

Zusammenfassung

Die Festigkeitsanalyse im Fahrzeugbau beschäftigt sich mit der kundennahen Erfassung und Bewertung der an Fahrzeugteilen hervorgerufenen Beanspruchungen. Die Basisdaten hierfür sind mehrkanalige Meßreihen, die auf Testfahrten gewonnen werden. Zentrale Aufgaben bei der Analyse der Meßreihen sind die Erzeugung einer symbolischen Beschreibung auf einem adäquaten Abstraktionsniveau und/oder die Gewährleistung der Datenkonsistenz.

Wir stellen einen auf Hidden Markovmodellen basierenden Ansatz für eine Signalschnittstelle vor, mit dem wir die eine Messung charakterisierenden Beanspruchungsprofile automatisch trainieren können und der zur Identifikation *verdächtiger* Signalabschnitte geeignet ist. Die Ergebnisse einer ersten Realisierung des Verfahrens werden vorgestellt und diskutiert.

1 Einführung

Die betriebssichere Auslegung von Bauteilen, Baugruppen und Konstruktionen ist im Fahrzeugbau von größter technischer und wirtschaftlicher Bedeutung. Diese müssen den strengen gesetzlichen Vorschriften zur Betriebssicherheit genügen, sollten aber aus Kosten- und Gewichtsgründen auch nicht überdimensioniert sein. Ihre Auslegung muß sich an den Beanspruchungen orientieren, die im täglichen Einsatz beim Kunden auf das Fahrzeug ausgeübt werden.

Neue Baugruppen im Fahrzeug erfordern aufwendige Verfahren zum Nachweis der vom Markt geforderten Gebrauchseigenschaften über lange Laufstrecken und Nutzungszeiten [PT91] [STT91]. Eine der Hauptaufgaben der Festigkeitsanalyse ist das Erfassen kundennaher Beanspruchungen von Fahrzeugbauteilen und das Bewerten dieser Beanspruchungen. Zur Bereitstellung der für eine Bewertung notwendigen empirischen Daten werden Meßfahrten durchgeführt, wobei unterschiedlichste Meßsignale (z.B. Längskräfte, Seitenkräfte, Geschwindigkeit, Bremsdruck, ...) mit verschiedensten Meßaufnehmern im Fahrzeug erfaßt werden (mobile Meßtechnik). Das Ergebnis einer Meßfahrt sind große Datenmengen, die ausgewertet werden müssen. Bei einer 20-minütigen Meßfahrt und 32 Meßkanälen fallen beispielsweise ungefähr 25 MB Rohdaten an.

Im Rahmen des Projektes **Messpert** beschäftigen wir uns mit Ansätzen zur symbolischen Beschreibung von Messungen in anwendernahen Begriffen wie „Fahrt über Pflaster“ und Ansätzen zur Identifikation *verdächtiger* Signalabschnitte (zur Illustration siehe Bild 1). Eine symbolische Beschreibung ist besonders dann von Bedeutung, wenn Meßfahrten auf neuen Meßstrecken durchgeführt werden. Um diese mit bekannten Meßstrecken vergleichen und einordnen zu können, ist man daran interessiert, welche beanspruchungsrelevanten Ereignisse mit welcher Häufigkeit aufgetreten sind. Eine Fahrt über eine gute Landstraße führt zu anderen Beanspruchungen als eine Fahrt über einen Schotterweg oder etwa ein Anfahrmanöver. Die Identifikation *verdächtiger* Signalabschnitte tritt in den Vordergrund, wenn wir gewährleisten wollen, daß die aufgezeichneten Meßsignale plausibel sind. Signalabschnitte werden als verdächtig bezeichnet, wenn sie nicht den Erwartungen entsprechen, die durch den Kontext der Messung vorgegeben sind. Ein verdächtiges Signal kann bespielsweise *Spikes*, *Drifts* oder *Shifts* enthalten, deren Ursache darin liegen kann, daß Meßaufnehmer zeitweise ausgefallen, Anschlüsse vertauscht oder ganz allgemein unerwartete Ereignisse eingetreten sind. Der Begriff „Verdächtigkeit" ist unscharf formuliert, da a priori nicht klar ist, welche Ausprägungen diese annehmen kann.

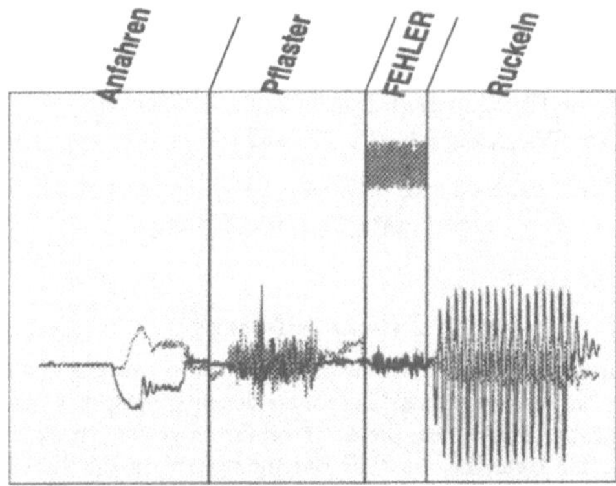

Bild 1: Symbolische Beschreibung mehrkanaliger Meßreihen

Wir stellen im folgenden einen HMM-basierten Ansatz vor, der eine symbolische Beschreibung mehrkanaliger Meßreihen in anwendernahen Begriffen erlaubt und zur Identifikation verdächtiger Signalabschnitte geeignet ist. Das hier vorgestellte System soll den für die Auswertung von Messungen verantwortlichen Ingenieur durch geeignete Aufbereitung der gemessenen Daten unterstützen, indem es ihn von einer zeitaufwendigen symbolischen Beschreibung und Plausibilitätsprüfung entlastet. Die Festigkeitsanalyse selbst bleibt weiterhin dem Meßingenieur vorbehalten, dem dafür geeignete Programme zur Verfügung stehen.

2 Die HMM-basierte Signalschnittstelle

Aus der Problembeschreibung ist offensichtlich, daß wir daran interessiert sind, mehrkanaligen Meßreihen individuelle symbolische Beschreibungen auf einem problemadäquaten Abstraktionsniveau zuzuordnen. Wir haben es also mit einem klassischen Musteranalyseproblem zu tun [Nie90].

Aus den Randbedingungen, die durch eine Vielzahl von zum Teil stark korrelierten Meßkanälen, ähnliche, aber nicht identische Testumgebungen und ganz wesentlich dadurch charakterisiert sind, daß es sich bei den Rohdaten um große Mengen an Sensordaten handelt, ergeben sich gewisse Anforderungen an die zu konzipierende Schnittstelle zu den Meßsignalen. Sie sollte Möglichkeiten zur Verarbeitung unsicherer und ungenauer Daten beinhalten, die Integration von Signalvariabilitäten erlauben und sich weitgehend automatisch an die gemessenen Signale adaptieren können.

Diese Signalschnittstelle wurde auf der Basis der vor allem in der Sprachverarbeitung sehr erfolgreich eingesetzten Hidden Markovmodelle (HMMs) [Rab88] realisiert. Dabei handelt es sich um einen statistischen Ansatz, der die Möglichkeit zur Repräsentation zeitlicher Ordnung bietet, die Verarbeitung unsicherer und ungenauer Daten zuläßt und für den zudem effiziente Verfahren zum Lernen der statistischen Parameter zur Verfügung stehen.

2.1 Hidden Markovmodelle

Ein HMM ist ein stochastischer Automat, der aus einer Menge von Zuständen und Übergängen zwischen diesen Zuständen besteht. Ein einfaches Beispiel eines HMMs zeigt Bild 2. HMMs werden als *hidden* bezeichnet, weil die Modellzustände s nicht beobachtbar sind, sondern nur die

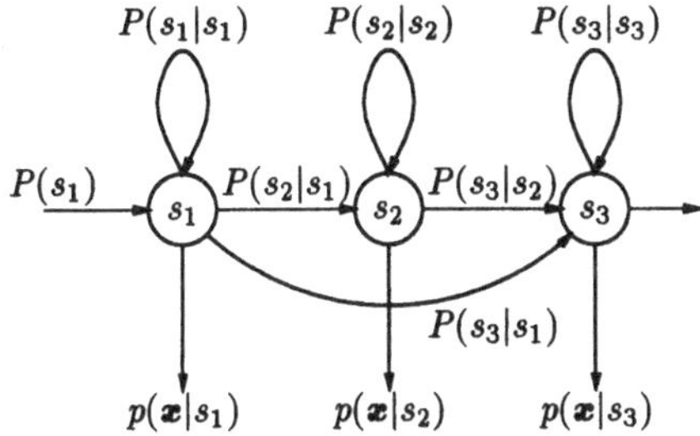

Bild 2: Beispiel für ein Links-Rechts-Markovmodell

Ausgabe $\boldsymbol{x}$ von den Zuständen zugeordneten stochastischen Ausgabeprozessen, die im kontinuierlichen Fall durch Verteilungsdichten $p(\boldsymbol{x}|s_j)$ beschrieben werden. Das Verhalten des zugrundeliegenden Markovprozesses wird beschrieben durch die Übergangswahrscheinlichkeiten $P(s_i|s_j)$ und die Einsprungswahrscheinlichkeiten $P(s_j)$. Wir haben damit zwei nebenläufige Prozesse: einen Markovprozeß zur Modellierung der zeitlichen Signalstruktur und eine Menge von Ausgabeprozessen zur Modellierung des augenblicklichen Signalcharakters. Eine ausführliche Darstellung gibt z.B. [Rab88].

2.2 Systemarchitektur von ISADORA

Unsere Untersuchungen wurden mit dem ISADORA-System [STN91], einem automatisch lernenden, HMM-basierten System zur Analyse komplexer eindimensionaler Muster, durchgeführt, das im folgenden allerdings nur in bezug auf die Analyse von Fahrzeugmeßdaten behandelt wird (siehe Bild 3). Das System gliedert sich in einen anwendungsabhängigen deklarativen Teil, bestehend aus einem hierarchischen Konstituentennetzwerk und den statistischen Parametern der HMMs, und einen prozeduralen Teil, der Module zur Vorverarbeitung von Merkmalen, zur (harten und weichen) Vektorquantisierung und zum Trainieren und Erkennen enthält. Die Ausgaben für die Markovmodelle können durch *kontinuierliche* Dichten, durch *diskrete* Verteilungen oder *semikontinuierlich* [HAJ90] modelliert werden. Im diskreten und im semikontinuierlichen Falle ist ein Vektorquantisierer erforderlich, der die kontinuierlichen Merkmalvektoren auf diskrete Klassen abbildet.

Die Repräsentation problemabhängiger Konzepte und deren struktureller Zusammenhänge erfolgt in einem *Konstituentennetzwerk*, dessen Knoten den zu modellierenden Einheiten entsprechen. Die zentrale Modellierungstechnik hierbei ist die systematische Verknüpfung kleinerer Einheiten zu komplexeren. Die textuelle Darstellung des Inhalts eines Netzwerks geschieht durch eine zyklenfreie Menge von *Regeln*. Jeder Knoten des Netzwerks wird charakterisiert durch seinen Namen, seinen Typ und eine Liste von Nachfolgerknoten (Konstituenten). In einem Netzwerk gibt es atomare Einheiten (A-Knoten), die ein dediziertes Markovmodell zur Modellierung der gewünschten Einheit haben. Die komplexeren Einheiten erhalten ein Markovmodell durch geeignete Verknüpfung der Konstituentenmodelle. Beispiele sind die Hintereinanderschaltung, die Parallelschaltung, die Rückkopplung oder allgemein die Einbettung in Zustandsnetzwerke. S-Knoten dienen der zeitlichen Dekomposition einer Einheit in seine Bestandteile. Die Modellverknüpfung erfolgt durch sequentielle Konkatenation (Hintereinanderschaltung) der Konstituentenmodelle. P-Knoten bieten eine Auswahl zwischen unterschiedlichen Konzepten. Die Realisierung erfolgt durch Parallelschaltung der Konstituentenmodelle. R-Knoten erlauben die mindestens einmalige Wiederholung (Repetition) seines einzigen Bestandteils. Die Realisierung erfolgt durch Rückkopplung des Konstituentenmodells. F-Knoten betten die Modelle seiner Konstituenten in ein endliches Zustandsnetzwerk ein. Die Realisierung erfolgt durch Verknüpfung der Konstituentenmodelle entsprechend einer Adjazenzmatrix.

Die Parameterschätzung der Markovmodelle (*Lernen*) geschieht mit einem *Maximum-Likelihood*-Schätzer, der diese schrittweise an eine vorgelegte *Lernstichprobe* anpaßt. Die Lernstichpro-

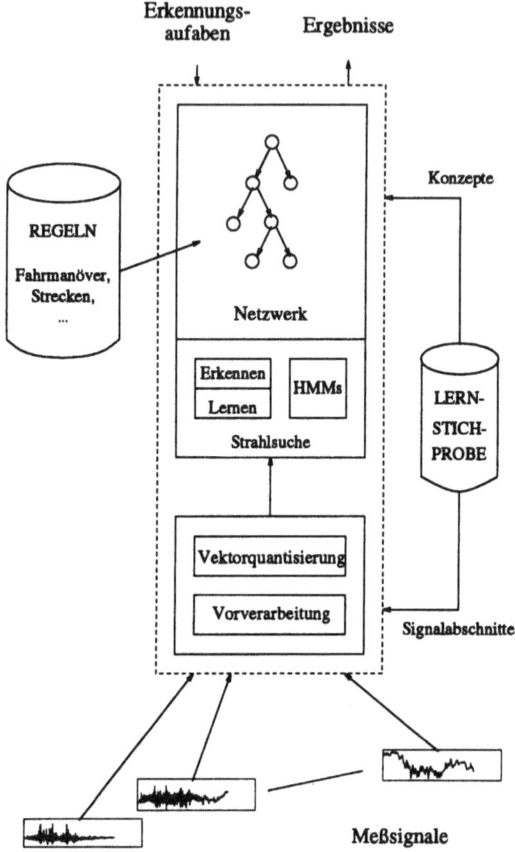

Bild 3: Systemarchitektur von ISADORA

be ist eine Sequenz von Signalen, deren jedes mit den Ausdrucksmitteln des ISADORA-Systems beschrieben ist. Da das einen Signalabschnitt bezeichnende Konzept beliebige Detaillierungsgrade von den atomaren bis zu den komplexesten Einheiten annehmen kann, schöpfen wir das ganze Kontinuum zwischen überwachtem und unüberwachtem Lernen aus.

Die Erzeugung einer symbolischen Beschreibung (*Erkennen*) und das Lernen erfolgen mit dem *Viterbi*-Algorithmus, dem zur Reduktion des Trainings- und Erkennungsaufwandes eine *Strahlsuche* überlagert ist. Als Modelltypus zur Repräsentation der atomaren Knoten werden Links-Rechts-Modelle verwendet, in denen nur Übergänge der Art $s_i \to s_i$ und $s_i \to s_{i+1}$ gestattet sind.

2.3 Symbolische Beschreibung einer Messung

Grundlage unseres Erkennungssytems ist eine statistische Modellierung des (mehrkanaligen) Meßsignals, die im folgenden kurz beschrieben wird. Wir gehen dabei von der Beobachtung aus, daß sich charakteristische Beanspruchungsphänomene in typischen Signalverläufen widerspiegeln. Unser Ansatz zur Beschreibung von Messungen orientiert sich daran, daß die am Fahrzeug verursachten Beanspruchungen im wesentlichen durch die Topologie der Meßstrecke (leichte LINKSKURVE, scharfe LINKSKURVE, ...), die Fahrbahnbeschaffenheit (Pflasterstrecke, Schotterstrecke, Autobahn, ...) und die vom Fahrer durchgeführten Fahrmanöver (BESCHLEUNIGEN, BREMSEN, LENKEN, ...) charakterisiert sind. Den auf diese Weise identifizierten elementaren Beanspruchungseinheiten haben wir dedizierte Markovmodelle zugeordnet. Diese kleinsten Einheiten haben wir über die in ISADORA gegebenen Beschreibungsmöglichkeiten zu komplexeren Einheiten zusammengefaßt, die dann im allgemeinen unsere eigentlichen Erkennungseinheiten bilden. Im folgenden geben wir ein kleines Beispiel zur Illustration unseres Modellierungansatzes, wobei wir davon ausgehen, daß für nicht weiter spezifizierte Konzepte detaillierte Beschreibungen vorliegen:

```
P:  P-Pflaster    A-Pflaster B-Pflaster G-Pflaster L-Pflaster R-Pflaster ;
R:  Pflaster      P-Pflaster ;
A:  G-Pflaster ;
S:  Messung1      A-Pflaster G-Pflaster B-Pflaster ;
```

Wir drücken über die Parallelschaltung (P) der ersten Regel aus, daß wir für eine Fahrt über `Pflaster` die Konzepte `Anfahren`, `Bremsen`, `Geradeausfahrt`, `Linkskurve` und `Rechtskurve` unterscheiden wollen. Die Repetition (R) der zweiten Regel beschreibt, daß eine Fahrt über Pflaster aus einer mindestens einmaligen Wiederholung dieser Konzepte besteht. Die dritte Regel beschreibt, daß eine Geradeausfahrt auf Pflaster eine atomare Einheit darstellt, die wir nicht mehr weiter zerlegen. Die vierte Regel besagt, daß `Messung1` aus der Hintereinanderschaltung der angegebenen Bestandteile besteht.

Die zeitliche Zuordnung auf die Modellzustände und die Schätzung der Modellwahrscheinlichkeiten erfolgt in der Trainingsphase anhand der Beschreibung unserer Stichprobe durch unsere so definierten Konzepte. Die Erkennung (symbolische Beschreibung) besteht nun darin, zu einer gegebenen Folge von Merkmalvektoren eine optimale Folge von Einheiten zu finden. Das Spektrum unserer Beschreibungsmöglichkeiten reicht von einer sehr detaillierten Beschreibung mit Hilfe unserer kleinsten Entscheidungseinheiten bis hin zu sehr allgemeinen, wo wir uns beispielsweise nur dafür interessieren, um welche Streckentypen es sich gehandelt hat, aber nicht mehr, welche Fahrmanöver im einzelnen ausgeführt wurden.

2.4 Identifikation verdächtiger Signalabschnitte

Eine zentrale Randbedingung bezüglich der Identifikation verdächtiger Signalabschnitte ist, daß keine negative Stichprobe zur Verfügung steht, die ein automatisches Training erlauben würde. Die Bereitstellung einer solchen ist aufgrund des hohen Aufwandes, der mit Messungen verbunden ist, ausgeschlossen. Eine direkte Modellierung verdächtiger Signalabschnitte ist somit nicht ohne weiteres möglich. Diese wäre ohnehin nur bedingt geeignet, da Verdächtigkeit nur ungenau definiert ist.

Bevor wir detaillierter auf den von uns gewählten Ansatz zur Modellierung von Verdächtigkeit eingehen, zunächst einige allgemeine Bemerkungen. Bei der Aufgabe der Identifikation verdächtiger Signalabschnitte können wir davon ausgehen, daß die Meßstrecke bekannt ist. Messungen werden im allgemeinen auf definierten Strecken durchgeführt, da nur dieses Vorgehen einen Vergleich mit bekannten Ergebnissen zuläßt. Dies erlaubt uns, Vorerwartungen für die Ergebnisse zu generieren und diese zu nutzen. Eine längere Schotterstrecke auf einem autobahnähnlichen Abschnitt einer bekannten Meßstrecke wäre im Sinne der Aufgabenstellung durchaus ungewöhnlich genug, diesen Abschnitt als „verdächtig" einzustufen. Eine eingeschränkte Möglichkeit zur Identifikation verdächtiger Abschnitte besteht somit darin, in einem einer Beschreibung nachgeordneten Schritt Ergebnisse zurückzuweisen, die nicht den Erwartungen entsprechen.

Im folgenden stellen wir einen Ansatz zur Integration von Verdächtigkeit vor. Wir setzen hierzu einen anhand einer repräsentativen Stichprobe durch unüberwachtes Lernen trainierten Vektorquantisierer, dessen Codebuchklassen i wir durch multivariate Normalverteilungen $\mathcal{N}(\boldsymbol{x}, \boldsymbol{\mu}_i, \boldsymbol{K}_i)$ repräsentieren, und eine semikontinuierliche Modellierung der Ausgabedichten $p(\boldsymbol{x}|s_j)$ gemäß

$$b_j(\boldsymbol{x}) = \sum_i c_{ji} \mathcal{N}(\boldsymbol{x}, \boldsymbol{\mu}_i, \boldsymbol{K}_i)$$

voraus. Anstatt nun aber zu versuchen, das Problem der Rückweisung auf der Ebene des Vektorquantisierers zu lösen, haben wir nach einer Möglichkeit gesucht, diese Entscheidung erst in der

Erkennungsphase zu treffen. Die Einführung einer oder mehrerer Rückweisungsklassen in den Vektorquantisierer ist, wie unsere Erfahrungen gezeigt haben, wenig geeignet, da sie zum einen in der Regel zu sehr auf die Unterscheidung von Klassen abzielen und zum anderen zu schwellwertabhängig sind. Abhängig von der Wahl eines Schwellwertes für die Rückweisung werden entweder zu viele Merkmalvektoren zurückgewiesen oder aber zu viele unzulässige nicht erfaßt.

Der von uns gewählte Ansatz zur Integration von Verdächtigkeit geht von der Beobachtung aus, daß für verdächtige Signalabschnitte die jeweiligen Merkmalvektoren dazu tendieren, großen Abstand von den trainierten Klassenzentren zu haben. Zum anderen ist festzustellen, daß diese gehäuft auftreten. Wir führen hierzu zunächst eine — durch die in [Bro87] vorgestellten *Richtermodelle* zur Modellierung von Ausreißern motivierte — Erweiterung unseres Vektorquantisierers derart durch, daß wir für jede Klasse i des Codebuches eine zusätzliche, sogenannte *Kontraklasse* i' einführen gemäß folgender Vorschrift:

$$\mathcal{R}(\boldsymbol{x}, \boldsymbol{\mu}_{i'}, \boldsymbol{K}_{i'}) = \mathcal{N}(\boldsymbol{x}, \boldsymbol{\mu}_i, r\boldsymbol{K}_i), \quad r > 1$$

Eine Kontraklasse i' hat also den gleichen Mittelwert wie die Klasse i, die Streuung ist aber um so größer, je größer wir r wählen. Durch die Zusammenfassung der Klassen und Kontraklassen erhalten wir einen Vektorquantisierer, der die doppelte Anzahl an Codebuchklassen besitzt. Dieser neue Vektorquantisierer bildet einen gewissen Prozentsatz der Merkmalvektoren unserer Stichprobe auf Kontraklassen ab. Bei Vorgabe dieses Prozentsatzes läßt sich r automatisch bestimmen. Damit können wir unsere semikontinuierliche Dichtemodellierung für die Ausgaben wie folgt schreiben:

$$b_j(\boldsymbol{x}) = \sum_{i'} c_{ji'} \mathcal{R}(\boldsymbol{x}, \boldsymbol{\mu}_{i'}, \boldsymbol{K}_{i'}) \; + \; \sum_i c_{ji} \mathcal{N}(\boldsymbol{x}, \boldsymbol{\mu}_i, \boldsymbol{K}_i)$$

Verwenden wir beim Training nun den um die Kontraklassen erweiterten Vektorquantisierer, so erreichen wir dadurch, daß im aktuellen Kontext zulässige Ausreißer in den jeweiligen Markovmodellen mit erfaßt werden.

Nachdem wir nun gezeigt haben, wie wir Verdächtigkeit auf der Ebene des Vektorquantisierers repräsentieren, müssen wir noch klären, wie diese in die Welt der Markovmodelle zu integrieren ist. Dies erreichen wir durch die Einführung *kontraklassensensitiver* Markovmodelle. Das sind Markovmodelle, die hinreichend sensitiv sind für Folgen von Merkmalvektoren, die vom Vektorquantisierer auf Kontraklassen abgebildet werden. Hierzu führen wir für jede Kontraklasse i' ein synthetisches, nicht trainierbares Markovmodell $\lambda_{i'}$ mit n Zuständen $s_j^{(\lambda_{i'})}$ und festen Einsprungs- und Übergangswahrscheinlichkeiten ein, dessen Ausgabedichte hinreichend sensitiv für die Kontraklasse i' ist. Eine einfache Form hierfür erhalten wir dadurch, daß wir den klassenspezifischen Summanden durch Nullsetzen der jeweiligen Koeffizienten eliminieren und nur den kontraklassenspezifischen behalten, d.h. wir setzen

$$b_j^{(\lambda_{i'})}(\boldsymbol{x}) = \sum_{k'} c_{jk'}^{(\lambda_{i'})} \mathcal{R}(\boldsymbol{x}, \boldsymbol{\mu}_{k'}, \boldsymbol{K}_{k'})$$

Ein Markovmodell mit Ausgabedichten dieser Art ist um so sensitiver für die Kontraklasse i', je größer der Koeffizient $c_{ji'}^{(\lambda_{i'})}$ im Vergleich zu den Koeffizienten $c_{jk'}^{(\lambda_{i'})}, k' \neq i'$ für die anderen Kontraklassen ist. Unsere momentane Implementierung besteht darin, den Koeffizienten $c_{ji'}^{(\lambda_{i'})}$ auf einen festen Wert θ, $0 < \theta < 1$, zu setzen und alle $c_{jk'}^{(\lambda_{i'})}, k' \neq i'$ gleich δ zu wählen, das automatisch so bestimmt wird, daß die Stochastizitätsbedingung

$$\sum_{k'} c_{jk'}^{(\lambda_{i'})} = 1$$

erfüllt ist. Die Anzahl der Zustände n dient der Dauermodellierung. Diese zusätzlichen Modelle werden in der Erkennungsphase unseren Erkennungseinheiten parallelgeschaltet. Signalabschnitte, die auf diese Modelle abgebildet werden, werden als verdächtig eingestuft und zurückgewiesen.

r	VQ		$\theta = 0.05$		$\theta = 0.1$		$\theta = 0.5$	
	FPA [%]	R [%]	FPA [%]	R [%]	FPA [%]	R [%]	FPA [%]	R [%]
5	17.1	93.2	0	45.1	0	67.2	3.5	94.7
10	12.7	92.0	0	51.1	0.2	76.9	3.8	91.0
20	8.0	90.8	0	76.5	0	81.5	2.5	87.0
30	6.1	90.1	0	75.7	0	79.3	1.8	85.7

Tabelle 1: Falsch positive Alarme (FPA) bei ungestörten im Vergleich zur Rückweisung (R) bei gestörten Messungen

3 Ergebnisse

Unsere Stichprobe für die im folgenden vorgestellten Ergebnisse bestand aus 21 relativ kurzen Referenzmessungen je ca. 10 sec Dauer (Kurztest) und einer längeren Messung mit ca. 1400 sec, die auf repräsentativen Meßstrecken durchgeführt wurden. Von den insgesamt 32 Meßkanälen (Abtastrate 320 Hz) wurden 8 repräsentative ausgewählt. Im Vorverarbeitungsschritt wurde lediglich eine Tiefpaßfilterung der einzelnen Kanäle durchgeführt. Anschließend wurden die gefilterten Signale zu einem Merkmalvektor zusammengefaßt und eine diskrete Karhunen-Loève-Transformation [Nie83] durchgeführt. Auf den so gewonnenen Merkmalen haben wir durch ein unüberwachtes Clusterverfahren einen Vektorquantisierer (VQ) mit 128 Klassen trainiert, dessen Codebuchklassen durch multivariate Normalverteilungen repräsentiert werden. Die Signalmodellierung erfolgte durch diskrete HMMs mit *weicher* VQ-Eingabe, d.h. jeder Eingabevektor wird auf eine Menge von probabilistisch bewerteten Kandidatenklassen abgebildet, was im wesentlichen der oben erwähnten semikontinuierlichen Modellierung entspricht.

Die Auswertung auf unserer Stichprobe hat gezeigt, daß wir die eine Messung charakterisierenden Beanspruchungseinheiten zuverlässig trainieren können. Eine detaillierte Untersuchung der 21 Abschnitte des Kurztests hat ergeben, daß in allen Messungen die wesentlichen Beanspruchungseinheiten (im Mittel fünf pro Abschnitt) korrekt identifiziert wurden und keine gravierenden fehlerhaften Ergebnisse aufgetreten sind.

Zur Untersuchung unseres Ansatzes zur Erkennung verdächtiger Signale haben wir für unterschiedliche Einstellungen von r und θ je 8x21 Experimente durchgeführt, indem wir jeweils einen Kanal einer Kurztestmessung zufällig gestört und die anderen unverändert gelassen haben. Dem prozentualen Anteil zurückgewiesener Merkmalvektoren (Rückweisungsrate R) auf der gestörten Stichprobe stellen wir den prozentualen Anteil falsch positiver Alarme FPA auf der ungestörten Stichprobe gegenüber. Die Ergebnisse bei unterschiedlichen Einstellungen zeigt Tabelle 1. Die Werte für VQ geben den Prozentsatz der Merkmalvektoren an, die auf Kontraklassen abgebildet werden und somit bei einer Entscheidung auf der Basis des Vektorquantisierers zurückgewiesen werden müßten. In allen Experimenten wurden für die kontraklassenspezifischen Modelle 3 Zustände angenommen.

Unsere Ergebnisse belegen, daß sich durch die Einführung kontraklassensensitiver Markovmodelle die Rückweisung verdächtiger Signalabschnitte bei gleichzeitig geringem Prozentsatz für falsch positive Alarme erreichen läßt. Der Anteil falsch positiver Alarme wird ganz wesentlich dadurch reduziert, daß wir unsere Entscheidung nicht auf der Basis des VQ treffen. Die Rückweisungsrate läßt sich weiter erhöhen, wenn wir (strecken)modellgetrieben analysieren. Bei realistischeren Störungen sind tendenziell sogar noch bessere Ergebnisse zu erwarten, da unser Störmechanismus nicht nur unzulässige Signale generiert hat. Die Auswertung simulierter Fehler wie Offsets, Drifts usw. hat die obigen Ergebnisse bestätigt. Nur hat sich erwartungsgemäß gezeigt, daß gewisse Fehler aufgrund der Nichtberücksichtigung von Frequenzinformation in den Merkmalen vom Vektorquantisierer überhaupt nicht erfaßt werden, so daß zu erwarten ist, daß sich durch eine aufwendigere Vorverarbeitung noch Verbesserungen ergeben.

Für einen praktischen Einsatz ist für FPA nur ein Wert annähernd Null akzeptabel, da eine hohe Anzahl falsch positiver Alarme zum einen die Akzeptanz schmälert und zum anderen die Gefahr in sich birgt, daß echte Fehler übersehen werden. Aus unseren Experimenten ergibt sich, daß unter den gegebenen Randbedingungen $r \approx 20$ und $\theta \approx 0.1$ günstige Einstellungen darstellen.

Weitere Untersuchungen werden zu klären haben, ob durch eine echt semikontinuierliche Modellierung, in der die Codebuchparameter des Vektorquantisierers trainierbarer Bestandteil des HMM werden, noch Verbesserungen zu erzielen sind und welchen sytematischen Einfluß die Anzahl der Zustände und die Einsprungs- und Übergangswahrscheinlichkeiten der kontraklassensensitiven Markovmodelle haben.

4 Schlußbemerkungen und Ausblick

Wir haben einen HMM-basierten Ansatz vorgestellt, der eine anwendernahe Beschreibung mehrkanaliger Meßreihen auf verschiedenen Abstraktionsebenen erlaubt und zur Identifikation verdächtiger Signalabschnitte geeignet ist. Durch die nicht an subjektiven Beurteilungsmaßstäben orientierte Beschreibung von Messungen ergibt sich die im Hinblick auf standardisierte Testumgebungen wünschenswerte Möglichkeit zur objektiven Erstellung von Meßstreckenprofilen. Der vorgestellte Ansatz beinhaltet aufgrund seiner Fähigkeit zur Beschreibung von Strecken bei geeigneter Abbildung einer Straßenkarte in ein Konstituentennetzwerk das Potential, auf der Basis weniger Sensoren eine automatische Standortbestimmung des Fahrzeuges durchzuführen.

Alternativ zu den hier vorgestellten HMMs werden im Rahmen des Projektes **Messpert** Ansätze auf der Grundlage künstlicher neuronaler Netze untersucht [Wei93].

Die Autoren danken der Audi AG, Ingolstadt, und der BMW AG, München, für die Kooperation im Rahmen des FORWISS-Projektes **Messpert**.

Literatur

[Bro87] P. Brown: *The Acoustic-Modeling Problem in Automatic Speech Recognition*, Carnegie Mellon University, Pittsburgh, 1987.

[HAJ90] X. Huang, Y. Ariki, M. Jack: *Hidden Markov Models for Speech Recognition*, Nr. 7 in Information Technology Series, Edinburgh University Press, Edinburgh, 1990.

[Nie83] H. Niemann: *Klassifikation von Mustern*, Springer, Berlin, 1983.

[Nie90] H. Niemann: *Pattern Analysis and Understanding*, Bd. 4 von *Series in Information Sciences*, Springer, Berlin Heidelberg, 2. Ausg., 1990.

[PT91] J. Petersen, H. Tunker: *Vom Meßaufnehmer zur Datenbank, 1. Teil: Systemanalyse, Abläufe, Anforderungen*, *Informatik Spectrum*, Bd. 14, Nr. 2, 1991, S. 69–73.

[Rab88] L. Rabiner: *Mathematical Foundations of Hidden Markov Models*, in H. Niemann, M. Lang, G. Sagerer (Hrsg.): *Recent Advances in Speech Understanding and Dialog Systems*, Bd. 46 von *NATO ASI Series F*, Springer, Berlin, 1988, S. 183–205.

[STN91] E. Schukat-Talamazzini, H. Niemann: *Das ISADORA-System — ein akustisch-phonetisches Netzwerk zur automatischen Spracherkennung*, in *Proc. 13. DAGM-Symposium*, Springer, Berlin, 1991, S. 251–258.

[STT91] N. Sentürk, U. Tabbert, H. Tunker: *Vom Meßaufnehmer zur Datenbank: 2. Teil: Hard- und Software für die mobile Meßtechnik des AUDI-Festigkeitsversuches*, *Informatik Spectrum*, Bd. 14, Nr. 2, 1991, S. 74–80.

[Wei93] P. Weierich: *Fault Detection in Multivariate Time Series with a Coding Approach*, in *Proc. of the International Conference on Artificial Neural Networks*, Springer, London, September 1993.

Ein bildanalytisches Verfahren zur Bestimmung dreidimensionaler morphometrischer Gewebestrukturmerkmale im Lichtmikroskop

R. Albert[1], Th. Schindewolf[1], J. Müller[2], H. Harms[1]

[1] Universität Würzburg, Institut für Virologie, Abteilung Bildverarbeitung, Versbacher Straße 7, D-97078 Würzburg

[2] Universität Würzburg, Institut für Pathologie

Kurzfassung

Mit dem menschlichen Auge ist es schwierig, die räumliche Struktur eines Zellverbandes nur mithilfe eines Durchlichtmikroskopbildes (2D) reproduzierbar zu analysieren. Die Darstellung in 3D-Bildern erfordert jedoch einen hohen zeitlichen und computertechnischen Aufwand. Dieser Beitrag stellt ein Verfahren vor, mit dem es möglich ist, dreidimensionale morphometrische Gewebestrukturmerkmale mit einem Bildanalysesystem und einem Lichtmikroskop zu bestimmen. Gewebeschnitte des Gebärmutterplattenepithels mit einer Dicke von 20-40 μm werden in 6-15 Fokusebenen aufgenommen und digitalisiert. Nach Bestimmung der Zellkernzentren und deren Koordinaten werden mit graphentheoretischen Methoden morphometrische Merkmale berechnet und als Datensatz an ein Klassifikationsprogramm übergeben. Mithilfe dieser Merkmale trennt ein binärer Entscheidungsbaum die verschiedenen histologischen Diagnosen der Präparate.

1. Einführung

In den letzten Jahren wurden in der medizinischen Bildverarbeitung, besonders im Bereich der Zellklassifikation von Ausstrichpräparaten, enorme Fortschritte erzielt.
Bei vielen Krankheitsbildern spielt jedoch nicht nur die Morphologie der einzelnen Zellen eine Rolle, sondern auch die Morphometrie des Gewebes. Zusammen mit der Zellmorphologie führen Zelldichte, Anordnung der Zellen, Verteilung der verschiedenen Zelltypen und andere Merkmale zu einem Gesamteindruck.
Die Morphometrie eines Zellverbandes kann nur an Gewebeblöcken analysiert werden. Voraussetzung für deren Beurteilung ist die Erkennung der Zellkerne. Die Problematik bei der Messung morphologischer Eigenschaften des Gewebes liegt darin, daß der Betrachter am Mikroskop nur mit einem 2D-Bild arbeiten kann und

somit keinen 3D-Eindruck bekommt. Dies ist nur mit zusätzlichem zeitlichen und hohem computertechnischen Aufwand möglich (Computertomographie, konfokaler Laserscanner).
Diese Arbeit zeigt, daß es mithilfe der Bildverarbeitung möglich ist, auch mit dem Lichtmikroskop quasi dreidimensional zu arbeiten, wenn die Schnittdicke der Präparate von üblicherweise 1-4 μm auf 20-40 μm erhöht wird. Hierdurch wird erreicht, daß der größte Teil der Zellkerne komplett im Gewebeblock enthalten ist und Fragmente stark angeschnittener Kerne in die Bearbeitung nicht einbezogen werden müssen.
Die Bestimmung von relevanten morphometrischen Merkmalen wird an Gewebeschnitten des Gebärmutterplattenepithels durchgeführt, da hier die Veränderungen - seien es Entzündungen, Dysplasien oder Karzinome - morphologisch gut definiert sind und deren Klassifikation international standardisiert ist.

2. Medizinische Problemstellung

Obwohl die verschiedenen histologischen Diagnosen eindeutig definiert sind, ist die reproduzierbare Klassifikation bei Grenzfällen ein Problem (10, 14).
Von besonderem Interesse ist die Unterscheidung von Entzündungen und leichten Dysplasien und die Trennung schwerer Dysplasien und Karzinoma in situ von Karzinomen. Die aus der Diagnose resultierenden Therapieformen sind für den Patienten lebensentscheidend.
In dieser Arbeit werden 25 Präparate untersucht, die in sechs verschiedene Gruppen unterteilt sind: gesundes Epithelgewebe (GE), reaktive Veränderungen (RV), leichte bis mäßige Dysplasien (LD), schwere Dysplasien (SD), Karzinoma in situ (CS) und Karzinome (CA).
Wir definieren für diese Studie drei Gruppen MSG1-MSG3, um die medizinisch relevanten Untergruppen zu trennen:

MSG1: GE + RV
MSG2: LD + SD + CS
MSG3: CA

3. Aufnahmetechnik

Für diese Untersuchung werden 20-40 μm dicke feulgengefärbte Schnittpräparate mit einer 100W Halogenlampe beleuchtet und mit einem Axiomat Mikroskop (Zeiss, Oberkochen) aufgenommen. Der Kondensor hat eine numerische Apertur NA_{cond} = 1.4 . Für die Messungen wird ein 40× Planapochromat Öl Objektiv (NA_{obj} = 1.0) verwendet. Die Feldblende muß zur Erhöhung des Streulichts in der Objektebene voll geöffnet sein (4).

Ein Color Frame Grabber (DT2871, Data Translation, Marlboro, MA, USA) digitalisiert die RGB-Ausgangssignale der 3-Chip-CCD-Kamera (DXC-750P Kamera und Kontrolleinheit, Sony, Japan). Die Bildgröße ist 512×512 Pixel mit einer Auflösung von 8 Bit je Farbkanal. Die absolute Abtastdichte beträgt 5.2 Pixel/µm. Mit einem Positionsdecoder (Spezialanfertigung, Zeiss, Oberkochen) ist es möglich, die Schnitte mit einer Genauigkeit von 0.1 µm zu fokusieren. Von jedem Präparat werden drei voneinander unabhängige Szenen in jeweils 6-15 Fokusebenen digitalisiert. Weitere Einzelheiten über die Auflösung des Systems in x-, y- und z-Richtung sind bereits an anderer Stelle veröffentlicht (7, 8).
Alle Bilder werden auf wiederbeschreibbaren magneto-optischen Platten gespeichert. Die für die Bildverarbeitung notwendigen Algorithmen sind in FORTRAN auf einer Workstation (DECstation 5000/200, Digital Equipment Corporation, Maynard, MA, USA) implementiert.

4. Segmentierung

Um in jedem Bild automatisch die Zellkerne vom Hintergrund zu trennen und Artefakte zu erkennen, wurde ein Segmentierungsalgorithmus entwickelt.
Der erste Schritt ist eine CIE-DIN Farbtransformation (12) der RGB-Werte in das xyz-Normvalenzsystem. In Abhängigkeit von der Größe der Zellkerne in der jeweiligen Serie wird eine n×n-Umgebung definiert mit $n \in \{30, 40, 50\}$, um für jedes Pixel den Wert x/y, den Erwartungswert E_{xy} aller x/y in dieser Umgebung und die zugehörige Standardabweichung S_{xy} zu berechnen. Liegen einige der Pixel in der n×n-Umgebung außerhalb des Bildes, so beschränkt sich die Berechnung auf die innenliegenden Punkte.
Für alle Bildpunkte wird die folgende Bedingung getestet:

$$x(P)/y(P) > E_{xy}(P) + \delta \cdot S_{xy}(P)$$

Ein Bildpunkt wird als einem Kern zugehörig angesehen, wenn die Werte dieses Punktes die Bedingung erfüllen. Mit $\delta = 0.2$ können 70% der Masken korrekt erstellt werden. Bei den übrigen Bildern genügt eine Korrektur mit $\delta \in \{0.0, 0.4, 0.6\}$. Innerhalb einer Bildserie für eine Szene bleibt δ in allen Fällen konstant. Somit kann δ in einem Testlauf mit einem Bild der Serie bestimmt werden, um dann die komplette Serie mit diesem δ zu segmentieren.
Nach Bearbeitung der auf diese Weise bestimmten Areale mit einem digitalen Tiefpaßfilter werden verbleibende kleine Löcher geschlossen. Durch das Löschen zu kleiner Flächen und die Trennung von verbundenen segmentierten Kernen ergeben sich die endgültigen Kernmasken. Die detaillierte Beschreibung dieser Schritte zur automatischen Maskengenerierung findet man in (1).

5. Fokuskriterium

Der Abstand zwischen den einzelnen Ebenen beträgt in unserer Untersuchung 2.0 µm. Da der durchschnittliche Zellkerndurchmesser wesentlich größer ist, ist jeder Zellkern in mehr als einer Ebene sichtbar (Abb. 1).

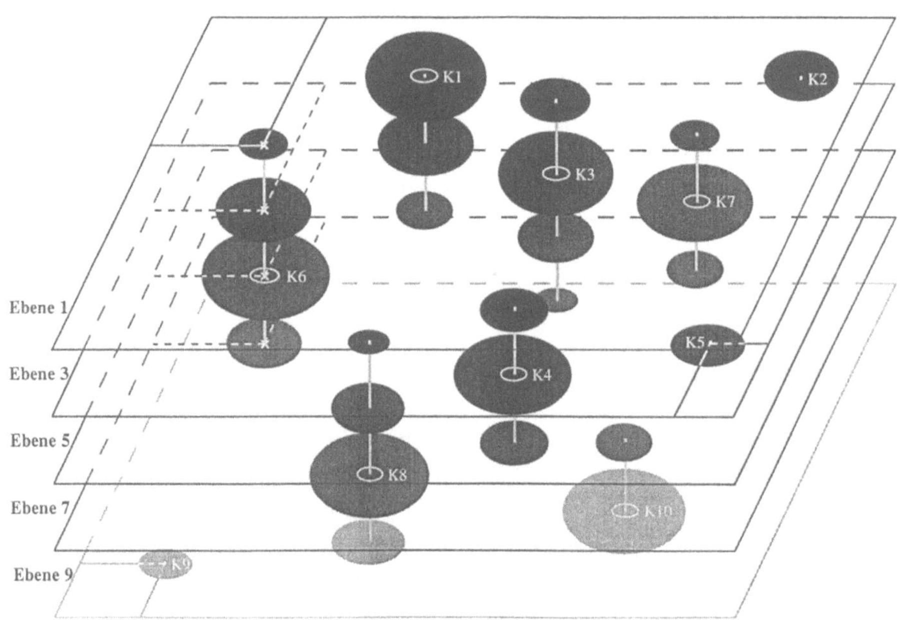

Abb. 1: Schematische Darstellung eines in 10 Fokuseinstellungen digitalisierten Szenenausschnittes mit idealisierten Kernen. Um die Übersichtlichkeit der Skizze zu gewährleisten ist nur jede zweite Ebene abgebildet. Die Hilfslinien der Kerne K5, K6 und K9 verdeutlichen die zu den verschiedenen Ebenen gehörenden Graustufen. Bei den Kernen K1, K3, K4, K6-K8 und K10 kennzeichnen die weißen Ellipsen die Ebene, in der das reale geometrische Zentrum vermutet und berechnet wird.

Einige der detektierten Flächen werden in den weiteren Ablauf nicht mit einbezogen. Es handelt sich dabei um solche Kerne, die nur in den beiden ersten oder den beiden letzten Ebenen detektiert werden können und deren exakter Mittelpunkt deshalb außerhalb des Schnittes liegen muß (Abb. 1: K2 und K9). Andere Flächen werden als Artefakte klassifiziert, da sie entweder in den benachbarten Ebenen nicht mehr zu sehen (Abb. 1: K5) oder für einen Zellkern zu klein sind.

Um den Mittelpunkt der verbleibenden Kerne so exakt wie möglich zu bestimmen, muß die dem Mittelpunkt nächstgelegene Ebene gefunden werden (Abb. 1: K1, K3-K4, K6-K8, K10). Mithilfe des Kriteriums von J. F. Brenner (3) läßt sich eine Bestimmung durchführen:

Die Differenzen der gegenüberliegenden Nachbarn der einzelnen Pixel sind umso größer, je näher das Bild in seiner optimalen Fokusposition abgetastet wird.

Unter verschiedenen automatischen Fokusierungsmethoden zeichnete sich dieses relativ einfach zu berechnende Kriterium durch seine hohe Präzision aus (5, 6).
Der Grünkanal des Farbbildes zeigt bei feulgengefärbten Zellen den höchsten Kontrast und ist damit der beste Farbkanal für die Auswahl der optimalen Fokusebene. Für eine Serie S berechnet sich in jeder Fokusebene f für jeden Kern k der Wert S_{fk} nach folgender Formel:

$$S_{fk} = \frac{1}{2 \cdot |A_{fk}|} \sum_i \sum_j \left[\left(G_{i-1,j} - G_{i+1,j} \right)^2 + \left(G_{i,j-1} - G_{i,j+1} \right)^2 \right] \quad \text{für alle } (i,j) \in A_{fk}$$

G_{ij} = Wert des Pixels (i,j) im Grünkanal
$|A_{fk}|$ = Anzahl der Pixel vom Kern k in der Fokusebene f
mit $1 \leq f \leq F$, F = Anzahl der Fokusebenen
und $1 \leq k \leq K_f$, K_f = Anzahl der Kerne in der Fokusebene f

Die Berechnung der Koordinaten eines Kernzentrums erfolgt durch die Bestimmung des Mittelpunktes des größten eingeschriebenen Kreises in A_{fk} in der Fokusebene mit dem höchsten Wert S_{fk}.

6. Graphentheoretische Methodik

Für die Anwendung graphentheoretischer Algorithmen bilden die Zentren der Zellkerne die Knoten. Mit den Koordinaten aller Zellzentren einer Serie läßt sich für diese Zellformation der minimale Spannbaum (1, 9, 15) und der O´Callaghan Baum (1, 13) konstruieren.

Der minimale Spannbaum (MSB) ist definiert durch zwei Regeln:
1. Es existiert genau eine Verbindung über eine oder mehrere Kanten von jedem Knoten zu jedem anderen Knoten.
2. Die Summe über die Länge aller Kanten in diesem Baum ist minimal.

Der O´Callaghan Baum (OCB) ist definiert durch zwei Bedingungen:
1. Ein Knoten wird mit allen anderen Knoten verbunden, deren Entfernung kleiner ist als das d-fache des Abstands zum nähesten Nachbarn.
2. Für jeden Knoten K müssen die Winkel $\alpha_{ij} := KN_iN_j$ zwischen einem Nachbarn N_i und einem Nachbarn N_j von K kleiner sein als ein frei wählbares α.

In unserem Fall erwies sich $d = 1.75$ und $\alpha = 60°$ als optimal zur Trennung der verschiedenen Entitäten.

Nach Konstruktion dieser beiden Bäume wurden die folgenden 15 morphometrischen Merkmale berechnet:
1. Zellkerndichte pro Volumen.
2. Durchschnittliche Zellgröße und

3. zugehörige Standardabweichung.
4. Durchschnittliche Kantenlänge im MSB und
5. zugehörige Standardabweichung.
6. Standardabweichung der durchschnittlichen Anzahl von Nachbarn im MSB.
7. Anzahl der Kanten je 10 µm Schnittdicke im OCB.
8. Durchschnittliche Kantenlänge im OCB und
9. zugehörige Standardabweichung.
10. Durchschnittliche Anzahl von Nachbarn im OCB und
11. zugehörige Standardabweichung.
12. Durchschnittliche Kantenlänge zum nähesten Nachbarn im OCB und
13. zugehörige Standardabweichung.
14. Durchschnittlicher Wert des kleinsten Winkels zwischen zwei Nachbarn eines Knotens im OCB und
15. zugehörige Standardabweichung.

Diese Werte zusammen mit der histologischen Diagnose stellen die Eingabe für das Klassifikationsprogramm CART dar.

7. CART (Classification And Regression Trees)

CART (2) ist ein kommerzielles Klassifikationsprogramm, das verschiedene Datensätze in Bezug auf eine Trennvariable - in unserem Fall die Gruppen MSG1-MSG3 - in verschieden Untergruppen aufteilt. Hierzu wird ein binärer Entscheidungsbaum konstruiert. Jedem Knoten wird der in ihm am häufigsten vorkommende Wert der Trennvariable zugeordnet. An jedem Knoten wählt das Programm eines der Merkmale aus, um die Datensätze dieses Knotens in zwei Untergruppen zu trennen. Die Untergruppen sind bezüglich der Trennvariable reiner als die Datensätze im Knoten selbst. Um die Stabilität der Auswahlkriterien kontrollieren zu können, wird automatisch eine Crossvalidierungs-Matrix ausgegeben - eine bekannte Methode zur Bewertung eines Klassifikators (11).

8. Ergebnisse und Ausblick

Es ist möglich mit einem Lichtmikroskop 20-40 µm dicke Schnitte zu analysieren. Die Gruppen MSG1-MSG3 können im Lerndatensatz mit einer Wahrscheinlichkeit von 96% (MSG1), 85% (MSG2) und 100% (MSG3) getrennt werden (Tabelle LS). Es zeigt sich, daß für diese Differenzierung nur die Zellkerndichte (Merkmal 1 zur Trennung von MSG3) und OCB-Nachbaranzahl (Merkmal 10 zur Unterscheidung MSG1-MSG2) benötigt werden. Die Crossvalidierungs-Matrix bestätigt mit 93%, 83% und 100% die Aussagekraft der morphometrischen Merkmale (Tabelle CV).

LS	MSG1	MSG2	MSG3
MSG1	**96**	9	0
MSG2	4	**85**	0
MSG3	0	6	**100**

CV	MSG1	MSG2	MSG3
MSG1	**93**	11	0
MSG2	7	**83**	0
MSG3	0	6	**100**

Klassifikationstabellen für Lerndatensatz (LS) und Crossvalidierung (CV). In den Spalten stehen die tatsächlichen histologischen und in den Zeilen die mit CART bestimmten Diagnosegruppen. Alle Angaben sind in Prozent.

Probleme treten auf, wenn die segmentierten Kernflächen einer Ebene mehr als 50% des Bildes einnehmen. Da die Kerne aufgrund ihrer Färbung Licht absorbieren, erhöht sich die Fokustiefe. Die Kernstrukturen sind dadurch nicht nur in einer, sondern in mehreren Ebenen sichtbar (Abb. 2).

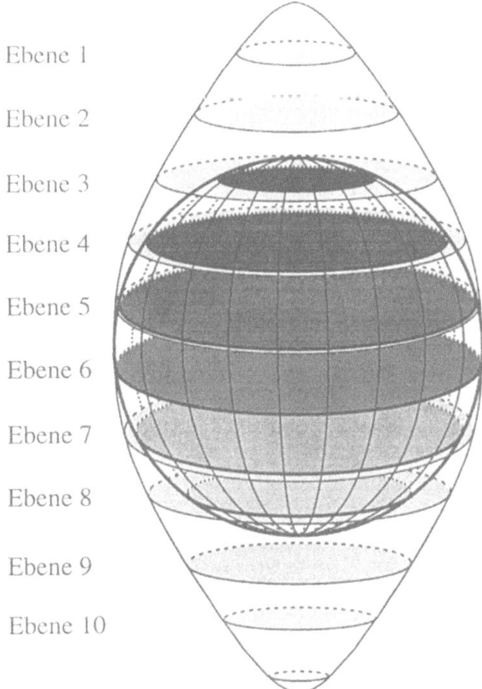

Abb. 2:
Darstellung eines idealisierten kugelförmigen Zellkerns in den Ebenen 3-8, der jedoch in den Fokuseinstellungen 1-10 detektiert wird. In jeder Fokuseinstellung sind zusätzlich zu den scharf abgebildeten Kernstrukturen der aktuellen Ebene auch die überlagerten Strukturen aus tieferen und/oder höheren Ebenen sichtbar. Durch diesen Effekt werden außer den tatsächlichen Kernflächen zusätzliche Areale segmentiert, was dazu führt, daß sich die segmentierten Kernflächen überschneiden können. In den Kernebenen 3-8 vergrößern sich die in der Zeichnung dunkelgrau dargestellten Kernflächen um die hellgrau abgebildeten Ringe. In den über bzw. unter dem Kern liegenden Ebenen 1-2 und 9-10 sind nur durchscheinende Strukturen sichtbar (hellgraue Ellipsen). Die in der imaginären Ebene 11 eingezeichnete weiße Ellipse würde in jedem Fall aufgrund ihrer zu geringen Größe eliminiert.

Die detektierten Kernflächen sehr eng zusammenliegender Zellen können sich infolgedessen berühren. Bei sehr dichtem Gewebe ist es dann kaum möglich, in den zusammenhängenden Flächen die einzelnen Kerne zu trennen. Durch diese Einschränkung konnten anfangs zehn Karzinompräparate mit sehr hoher Kerndichte nicht in der beschriebenen Form bearbeitet werden. Trotzdem ist die Diagnose

eindeutig, da diese Schnitte eine sicherlich noch höhere Kerndichte haben als die bearbeiteten Karzinome und diese als Gruppe MSG3 anhand der Zellkerndichte vollständig richtig klassifiziert werden.
Es wird somit gezeigt, daß sich sowohl bei den verschiedenen Krankheitsbildern RV und LD als auch in den unterschiedlichen Krankheitsstadien SD, CS und CA nicht nur die Morphologie der einzelnen Zellen verändert, sondern auch die des gesamten Gewebes. Im 2D-Bild des Lichtmikroskops ist die Messung relevanter Werte kaum möglich (1). Die hier dargestellte Methode bietet die Möglichkeit die räumliche Struktur eines Gewebeschnittes mithilfe von reproduzierbaren Daten quantitativ zu beschreiben.
Es wird versucht dieses Verfahren auch bei anderen medizinischen Fragestellungen zu nutzen, wie zum Beispiel in retrospektiven und prospektiven bildanalytischen Untersuchungen von Mammakarzinomen.

9. Literaturverzeichnis

1. Albert R, Schindewolf T, Baumann I, Harms H: Three-Dimensional Image Processing for Morphometric Analysis of Epithelium Sections. Cytometry 13:759-765, 1992.
2. Breiman L, Friedman JH, Olshen RA, Stone CJ: Classification And Regression Trees. Wadsworth, Inc., Belmont, California, 1984.
3. Brenner JF, Dew BS, Horton JB, al. e: An automated microscope for cytologic research. Journal of Histochemistry and Cytochemistry 24:100-111, 1976.
4. Dyer DL: Optical limits in TV microscopy. Research/Development Sept.:40-44, 1973.
5. Firestone L, Cook K, Culp K, Talsania N, Preston K: Comparison of Autofocus Methods for Automated Microscopy. Cytometry 12:195-206, 1991.
6. Groen CA, Young IT, Ligthart G: A Comparison of Different Focus Functions for Use in Autofocus Algorithms. Cytometry 6:81-91, 1985.
7. Harms H, Aus HM: Comparison of Digital Focus Criteria for a TV microscope System. Cytometry 5:236-243, 1984.
8. Harms H, Aus HM: Estimation of Sampling Errors in a Highresolution TV Microscope Image Processing System. Cytometry 5:228-235, 1984.
9. Jungnickel D: Graphen, Netzwerke und Algorithmen. B.I. Wissenschaftsverlag, Mannheim, 1987.
10. Kurman RJ: Blaustein´s Pathology of the Female Genital Tract. Springer Verlag, Berlin, 1987.
11. Lachenbruch PA: Discriminant Analysis. Hafner Press, New York, 1975.
12. Lang H: Farbmetrik und Farbfernsehen. Oldenburg Verlag GmbH, München, 1978.
13. O´Callaghan JF: An Alternative Definition for "Neighbourhood of a Point". IEEE Transactions on Computers November:1121-1125, 1975.
14. Remmele W: Pathologie 3. Springer Verlag, Berlin, 1984.
15. Rodenacker K, Chaudhuri BB, Bischoff P, Gais P, Jütting U, Oberholzer M, Gössner W, Burger G: Strukturbeschreibung und Merkmalsgewinnung in der Histometrie am Beispiel von Plattenepithelien. In: Morphometrie in der Klinischen Pathologie, Burger G, and Oberholzer M (eds). Springer-Verlag, Heidelberg, 1989, pp. 179-199.

Früherkennung von Hautkrebs durch Farbbildverarbeitung

K.-H. Franke, F. Gaßmann, R. Gergs
TU Ilmenau, Fakultät EI, Fachgruppe Bildverarbeitung
O-6300 Ilmenau

Der folgende Beitrag beschreibt und diskutiert die Anwendung der Farbbildverarbeitung zur Unterstützung der ärztlichen Diagnose bei der Früherkennung von Hautkrebs. Es wird ein Überblick über die zur Themenbearbeitung eingesetzten Algorithmen gegeben. Die Gewinnung quantifizierter Merkmale maligner Pigmentmale wird exemplarisch am Beispiel des retikulären Musters beschrieben.

1 Einleitung

Eine in ihrer Bedeutung wachsende Problemstellung der Medizin ist die Früherkennung maligner Melanome. Die Prognose dieser auch als "Schwarzer Hautkrebs" bekannten pathologischen Hauterscheinung weist aus, daß durch verschiedene Umwelteinflüsse - unter anderem durch die Zerstörung der Ozonschicht in der höheren Atmosphäre - eine Verdopplung dieser Art des Hautkrebses in allen Teilen der Welt mit weißer Bevölkerung alle 10 Jahre zu verzeichnen ist [Rass88]. Aufgrund der zunehmenden Inzidenz sowie der wegen fehlender wirksamer Behandlungsmethoden hohen Mortalitätsrate von Tumoren im fortgeschrittenen Stadium gewinnen Verfahren zur sicheren Erkennung auch initialer Melanome verstärkt an Bedeutung.

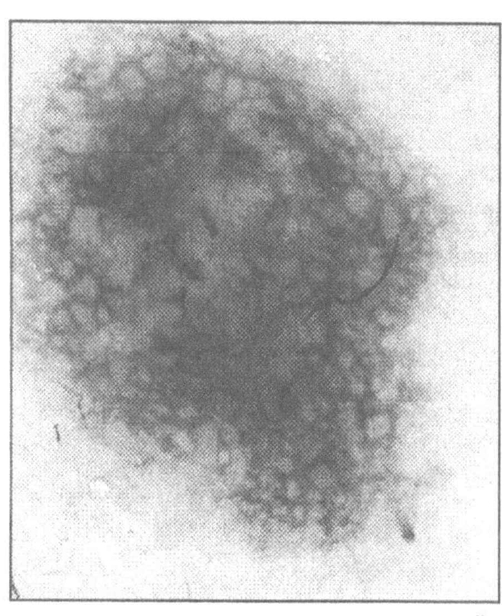

Abb. 1 Aufnahme einer Hautläsion unter Ölimmersion und Deckglas

Von den für diesen Zweck vorgeschlagenen Verfahren hat sich in der Praxis die intravitale Auflichtmikroskopie als nichtinvasives, schnelles und leistungsfähiges Verfahren durchgesetzt. Insbesondere die stark vergrößerte Betrachtung von Hautpartien, durch Ölimmersion und Deckglas transparent und planar gemacht, ist für den Dermatologen von Interesse. Die heute günstigere Prognose melanozytärer Läsionen ist nicht zuletzt auf die durch den Einsatz von Auflichtmikroskopen verbesserte Frühdiagnose, verbunden mit der Exzision bei Melanomverdacht und einer subtilen histologischen Untersuchung, zurückzuführen. Wegen der enormen morphologischen Variabilität der Pigmentmale erfordert die Untersuchung dennoch beträchtliche Erfahrung. Trotz der zur Verfügung stehenden Hilfsmittel werden vom praktizierenden Dermatologen nur etwa 70 % der initialen Melanome als solche erkannt. Andererseits wird in vielen Fällen unzutreffend Melanomverdacht geäußert [Kreu90]. Erschwerend macht sich hier der fehlende standardisierte und systematische Untersuchungsgang bemerkbar. Zudem stützt sich die Diagnose des Hautarztes allein auf seine individuellen Erfahrungen. Die Entscheidungen sind geprägt durch den Zwang, auch bei nicht eindeutiger Identifikation eines Melanoms einen operativen Eingriff vorzunehmen, zumal es aus Archivierungs- und Vergleichbarkeitsgründen schwierig ist, aussagekräftige Langzeitbeobachtungen melanozytärer Läsionen vorzunehmen.

Zur Unterscheidung benigner und maligner melanozytischer Veränderungen stehen dem Arzt eine Anzahl von Merkmalen zur Verfügung, die zum großen Teil aus der Häufigkeitsverteilung der mikroskopisch erkennbaren Strukturen auf die verschiedenen Läsionstypen resultieren. Die Merkmalsanalyse melanozytischer Pigmentmale führt unter dem Aspekt der Gewinnung von Bildeigenschaften zur folgenden Systematik nach Merkmalskategorien:

- globale Geometrie (Größe, Form, Berandung),
- Hautoberfläche (Relief, Felderung),
- Pigmentmuster / -textur (retikuläres Muster, globuläres Muster, scholliges Muster, diffuse Pigmentierung, Melanophagen, Pseudopodien, regressiver Umbau),
- Farbe (Brauntöne, schwarz, grau, blau, rot),
- Blutgefäße (Anzahl, Größe, Architektur).

Mit Hilfe der rechnergestützten Analyse soll die klinische Diagnostik maligner Melanome objektiver, reproduzierbarer und sicherer gemacht werden. Es gilt, die rein subjektiven, visuellen Merkmale durch objektiv meßbare zu verifizieren oder sogar neu zu definieren. Bei der Unterstützung der ärztlichen Diagnostik bietet sich die Farbbildanalyse in besonderer Weise an. Einerseits wird die Farbmessung an sich erforderlich, andererseits sind weitere Merkmale an die Differenzierung von Farbvalenzen gebunden oder lassen sich mittels Farbe einfacher und sicherer extrahieren als auf der Basis skalarer Größen wie Helligkeit oder Intensität.

2 Werkzeuge zur Themenbearbeitung

Eine Forderung bei dermatologischen Inspektionen besteht darin, daß aufgenommene Körperfarben möglichst identisch zu ihren Originalfarben sind. Obwohl der Mensch in der Lage ist, Farbabweichungen beim Betrachten einzelner Bilder weitestgehend zu

tolerieren, machen sich Farbverstimmungen und Helligkeitsunterschiede bei einer vergleichenden Bildanalyse störend bemerkbar. Besonders der Wunsch nach einer objektivierten Verlaufskontrolle, auch über längere Zeiträume, setzt daher die colorimetrische Korrektur der gewonnenen Aufnahmen voraus. Die Kalibrierung des Bildaufnahmekanals kann dabei in eine Helligkeitskalibrierung (Schwarz- und Weißabgleich) und eine Farbkalibrierung aufgeteilt werden. Der Schwarzabgleich hat die Aufgabe den Offset-Fehler des AD-Wandlers zu minimieren. Er beeinflußt maßgeblich die Güte der linearen Matrizierung bei der Farbkorrektur. Der Weißabgleich dagegen soll für eine optimale Aussteuerung der Kamera und des AD-Wandlers sorgen. Die Abweichungen der aufgenommenen Farbwerte können durch eine lineare Transformation korrigiert werden. Die Aufgabe des Kalibrieralgorithmus besteht nun darin, die Elemente der Korrekturmatrix so zu bestimmen, daß die Differenzen aller korrigierten Farbwerte zu den jeweils zugehörigen Sollfarbwerten möglichst gering werden. Die direkte Testfarbenmethode geht dabei vom Vorhandensein eines Testfarbensatzes (z.B. Testfarben nach DIN 6169) aus, dessen Farbwerte bekannt sind. Mittels Kameraaufnahme werden die Ist-Farben unter den jeweils herrschenden Bedingungen ermittelt. Ein Optimierungskriterium ist die Minimierung der Fehlerquadratsumme im RGB-Farbraum über alle Testfarben.
Da sich das Farbabstandsempfinden der menschlichen Sinneswahrnehmung deutlich vom euklidischen Abstand im Farbraum RGB unterscheidet, wurde zur Bewertung des eigentlichen Farbabstandes für Körperfarben der Farbraum $L^*a^*b^*$ genutzt. Die nichtlineare Transformation zwischen beiden Farbräumen führt allerdings dazu, daß die Methode der kleinsten Fehlerquadrate nicht mehr explizit anwendbar ist. Realisiert wurde daher ein Gradientenabstiegsverfahren auf der Basis eines Gauß-Algorithmus zur Lösung nichtlinearer Gleichungssysteme.
Einen wesentlichen Aspekt bei der Bewertung und Einordnung medizinischer Bilder stellt die visuelle Analyse durch den Arzt dar. Oft sind jedoch Details trotz sorgfältiger Gestaltung des Bildaufnahmekanals nicht sichtbar. Besonders bei der Begutachtung dermatologischer Bildinhalte ist zu beobachten, daß sich die betrachteten Szenen in stark eingegrenzte Farbbereiche (Hautfarben) abbilden und damit die visuelle Auswertung erschweren.
Ein Weg zur Verbesserung der visuellen Auswertbarkeit stellt die gezielte Manipulation der Farbinformation in den Farbräumen RGB und HSI dar. Da keiner der beiden Farbräume empfindungsmäßig gleichabständig ist, muß allerdings eine Manipulation der Bildstatistik nicht notwendigerweise eine Bildverbesserung im Sinne der visuellen Auswertbarkeit zur Folge haben.
Ein weiteres Hilfsmittel, besonders zur besseren Beurteilbarkeit kleiner Bilddetails, ist die Ausschnittsvergrößerung (Zoom-Funktion). Dabei wurde aus Rechenzeitgründen auf die Methode der bilinearen Interpolation zurückgegriffen. Vergleichende Untersuchungen zu anderen Methoden wie der Ebenenapproximation und der Spline-Interpolation haben zudem gezeigt, daß die Qualität der bilinear interpolierten Bildausschnitte als völlig ausreichend angesehen werden kann.

Die weitergehende strukturelle Analyse pigmentierter Hautbereiche setzt die Detektion der im Bild enthaltenen Pigmentstrukturen voraus. Dabei steht zumeist das Ziel, die wesentlichen Bildelemente in eine symbolische Beschreibung zu überführen. Der erste Schritt ist die Trennung der zur Läsion gehörenden Bereiche von der umgebenden Haut in Form eines Binärbildes unter Nutzung einer der folgenden Methoden:

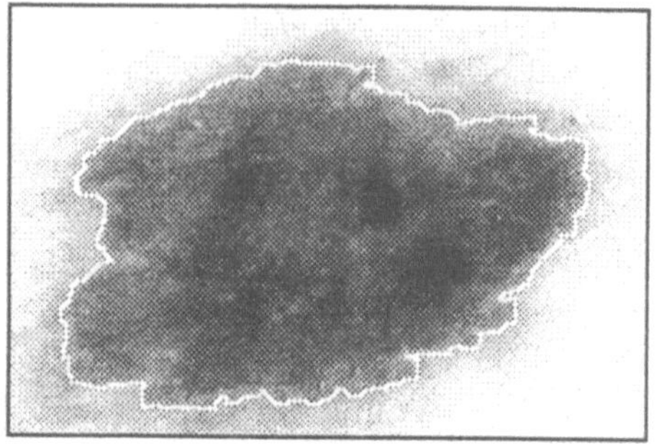

Abb. 2 Läsion mit detektierter Kontur

a) automatische Bestimmung von Triggerschwellen aus Histogrammen mit anschließendem Bereichsclipping,
b) überwachte Farbklassifikation (interaktive Clusterung anhand von Trainingsgebieten),
c) unüberwachte Farbklassifikation (automatische Clusterung),
d) Segmentierung durch Bereichswachstum.

Die vereinfachende Annahme, daß im Bild tatsächlich nur zwei zusammenhängende Farbbereiche auftreten, gilt für eine große Zahl der betrachteten Läsionen und führt unter Verwendung der Methode a) direkt zum gewünschten Binärbild. Die Segmentierung von Pigmentmalen mit inhomogenem Gesamtaufbau bzw. einer im Vergleich zur umgebenden Haut schwachen Pigmentierung erfordert jedoch die Anwendung einer der Methoden b), c) oder d). Da hierbei in der Regel mehr als zwei zusammenhängende Gebiete entstehen, ist eine nachfolgende Bereichszuordnung notwendig. Das gewonnene Binärbild bzw. die als Graph zur Verfügung stehende Kontur stellen den Ausgangspunkt für die morphometrische Analyse der Gesamtläsion dar und definieren zugleich den interessierenden Bereich für weitere strukturelle und farbliche Untersuchungen innerhalb der Läsion. Aus der Kontur werden neben rein geometrischen Größen wie Umfang, Fläche u.ä. auch Aussagen über die Art der Berandung abgeleitet. Auf der Basis der Konturanalyse durch Fourierdeskriptoren kann der Grad der Übereinstimmung mit kreisförmigen oder elliptischen Grundformen bestimmt werden. Diese regelmäßigen Formen weisen im Gegensatz zu stark zerklüfteteten Berandungen auf gutartige Hautgebilde hin. Ein weiterer Anhaltspunkt für den Charakter einer Läsion wird aus der Analyse der auftretenden Farbarten und deren Häufigkeitsverteilung gewonnen. Die Zuordnung der Farben zu den dermatologisch bedeutsamen Farbtönen erfolgt dabei über den Weg der pixelbezogenen Farbklassifikation. Aus der Anzahl der auftretenden Farben und deren örtlicher Verteilung wird außerdem ein Maß für die farbliche Homogenität der Läsion als Zeichen gutartiger Pigmentmale berechnet.

Die strukturelle Analyse einer Läsion kann an dieser Stelle nur exemplarisch diskutiert werden. Aus der Vielfalt möglicher Pigmentstrukturen wurde das retikuläre Muster ausgewählt, dessen Untersuchung im nächsten Kapitel erläutert wird.

3 Untersuchung der Netzstruktur

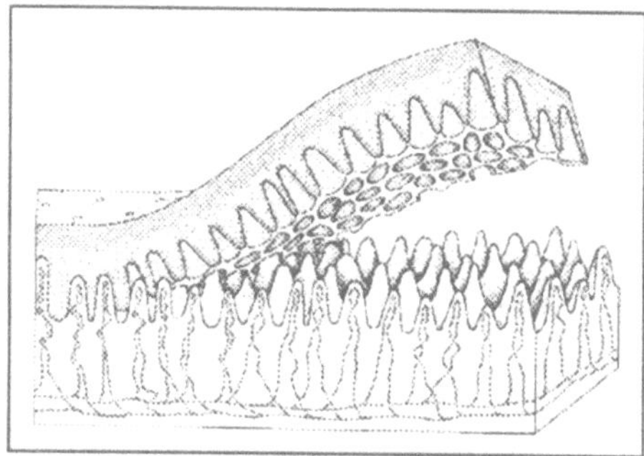

Abb. 3 Schematische Darstellung der Mikroanatomie der Haut (nach [Kreu91])

Ein wesentliches Merkmal zur Unterscheidung benigner und maligner Pigmentmale stellt das bei senkrechtem Blick auf die Hautoberfläche sichtbare, je nach Bräunungsgrad des Individuums mehr oder weniger stark ausgeprägte Netzmuster (retikuläres Muster) dar. Seine Erklärung findet diese Struktur durch die Mikroanatomie der obersten Hautschichten. Beim Blick auf die Papillen wird dem Beobachter durch Überlagerungserscheinungen an den Steilkanten eine höhere Pigmentkonzentration vorgetäuscht. Da in den Talsohlen der Kurvenradius kleiner ist als auf den Papillenkuppen, erscheint der Raum zwischen den Papillen insgesamt intensiver pigmentiert als die Spitzen, so daß diese helle Maschenöffnungen in einem bräunlichen Netz bilden [Kreu91].

Das Netzmuster ist für viele benigne und maligne Pigmentmale sehr charakteristisch. Die Breite der Netzstege (Reteleisten) und die Maschengröße sind einerseits von der Körperregion und vom Individuum selbst abhängig. Andererseits weisen eine Vergröberung des Netzes, eine Verbreiterung der Reteleisten, Störungen der Regelmäßigkeit des Netzmusters, eine starke Variation der Maschengrößen und weitere strukturelle Auffälligkeiten auf das Vorliegen eines malignen Melanoms hin [Kreu91]. Zur Unterstützung der Diagnosestellung bezüglich des Merkmals retikuläres Muster durch Bildverarbeitungsmethoden sind also Maßzahlen zur statistischen Beschreibung der Breite der Reteleisten und der Maschengröße abzuleiten.
Aufgrund der morphologischen Vielfalt möglicher Pigmentmale erfordert die Extraktion des Netzmusters grundsätzlich sehr robuste Verfahren. Im Verlauf der Forschungsarbeit wurden deshalb verschiedene Kontur- und Liniendetektionsverfahren implementiert und getestet. Zwei geeignete Verfahren werden nachfolgend beschrieben.

3.1 Verfahren der Non-Maxima-Unterdrückung

Das Verfahren der Non-Maxima-Unterdrückung basiert auf einem Vergleich der Beträge der Konturpunkte quer zur Richtung der Kante. Konturpunkte, die nicht den Maximalwert aufweisen, werden unterdrückt. Die Auswahl der Vergleichspunkte erfolgt in Abhängigkeit von der Gradientenrichtung durch eine Maskierung. Im Falle der Achternachbarschaft ergeben sich vier verschiedene Masken.

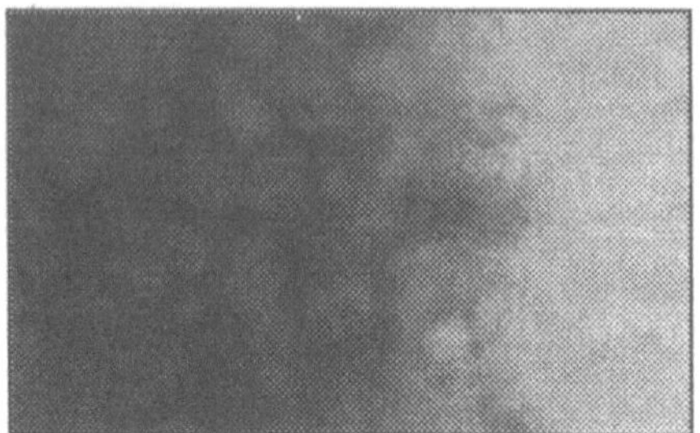

Abb. 4 Ausschnitt mit Netzstruktur

Abb. 5 mittels Non-Maxima-Unterdrückung gefundene Maschenkonturen

Anschließend wird eine Verdünnung der Gradientenbeträge, gesteuert über zwei Parameter, vorgenommen. Eine globale Gradientenbetragsschwelle dient der Unterdrückung flacher Kanten. Durch einen Toleranzbereich der Gradientenrichtung werden Kantenpunkte, deren Orientierung zu stark von der Lage der Kante abweichen, abgewiesen. Gradientenbetrag und -richtung eines jeden Bildpunktes werden in bekannter Art und Weise aus den partiellen Ableitungen des Quellbildes berechnet. Als Parameter erwiesen sich eine Gradientenbetragsschwelle von 8 LSB und eine Gradientenrichtungstoleranz von ±45 Grad als optimal.

3.2 Binarisierung mit gleitender Schwelle

Bei der Binarisierung mit gleitender Binarisierungsschwelle wird das Quellbild von einem stark tiefpaßgefilterten Zwischenbild abgezogen. Das Ergebnisbild wird anschließend mit einer niedrigen Schwelle binarisiert. Bezüglich des Quellbildes liegt die Binarisierungsschwelle also abzüglich eines Offsets auf dem Niveau des Zwischenbildes. Das Zielbild ist damit unabhängig von Bildsignalkomponenten mit niedrigen Ortsfrequenzen, die dem inhomogenen Hintergrund zuzuschreiben sind. Als Tiefpaßfilter wurde ein Gaußtiefpaß gewählt. Im Gegensatz zum Spalttiefpaß besitzt dieses ein begrenztes Ortsfrequenzspektrum und ruft daher keine Aliasing-Störungen hervor.

3.3 Ergebnisse

Abb. 4 zeigt einen Bildausschnitt mit ausgeprägter Netzstruktur aus Abb. 1. Das Non-Maxima-Verfahren erzeugt erwartungsgemäß ein System von Maschenrändern, jedem Netzsteg ist hier eine Doppellinie zugeordnet (Abb. 5). Da Unterbrechungen von Kanten mit diesem Verfahren jedoch nicht gefüllt werden können, ist eine Nachbearbeitung zwecks Konturpunktverkettung notwendig.

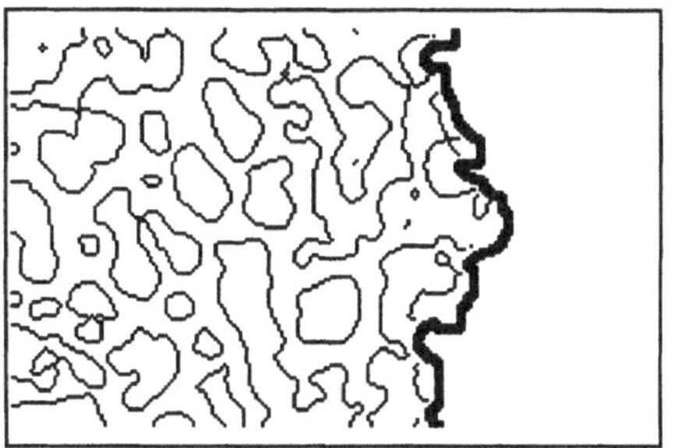

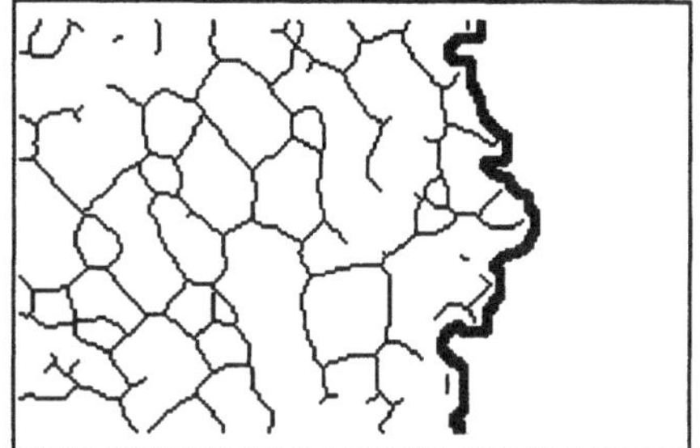

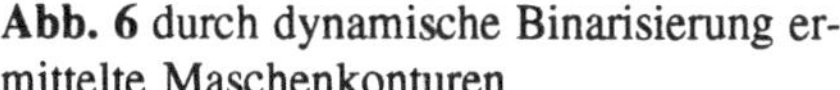

Abb. 6 durch dynamische Binarisierung ermittelte Maschenkonturen

Abb. 7 zugehörige Netzstruktur

Das Verfahren der Binarisierung mit gleitender Schwelle liefert ein Zweipegelbild bei welchem die Netzsteg- und Maschenflächen jeweils einem Pegel zugeordnet sind. Aus dem Binärbild lassen sich in einfacher Weise die Maschenkonturen (Abb. 6) durch Konturisierung der Pegelgrenzen und die Netzstruktur (Abb. 7) durch Skelettierung der Netzstege gewinnen. Wegen seiner Detektionssicherheit, aber auch aus Gründen der Zeiteffizienz (schnelles Verfahren, keine Nachbearbeitung notwendig) wurde letztlich das Verfahren der dynamischen Binarisierung priorisiert.
Die Ermittlung der diagnoserelevanten Merkmale Breite der Netzstege und Maschenfläche erfolgt innerhalb eines durch den Arzt interaktiv festzulegenden Bildbereichs. Mit Hilfe des Zeilenkoinzidenzverfahrens werden alle vollständig im gewählten Fenster liegenden Netzmaschen ausgezählt und deren Fläche bestimmt. Die durchschnittliche Breite der Reteleisten ergibt sich aus dem Verhältnis der Differenz von Fenstergröße und Summe der Maschenflächen zur Gesamtlänge des Pigmentnetzes (Abb. 7) im Fenster.
Die Quantifizierung von Merkmalen anderer bereichsorientierter Pigmentstrukturen, wie globuläres und scholliges Muster kann analog der des retikulären Musters vorgenommen werden.

4 Ausblick

Die grundlegende Bedeutung einer effektiven Langzeitkontrolle wurde durch die medizinischen Projektpartner besonders hervorgehoben. Erste Überlegungen führten hier auf die Notwendigkeit einer geometrischen Transformation der einzelnen Aufnahmen als Voraussetzung für visuelle und automatische Vergleiche, da Verdrehungen und Verschiebungen des Objektes im Bildfeld gegenüber den zeitlich zurückliegenden Aufnahmen sehr wahrscheinlich sind. Die automatische Bestimmung der notwendigen Transformationsparameter gestaltet sich hierbei jedoch wegen der Spezifik der Bilder - strukturlose Bereiche, Fehlen markanter Objekte - und der möglichen Größen-,

Struktur- und Farbänderungen schwierig. Neben die interaktive Ermittlung dieser Parameter müssen zukünftig leistungsfähige Algorithmen zur automatischen Paßpunktdetektion gestellt werden.
Auf der Basis der geometrisch korrigierten Bilder sind Algorithmen denkbar, die Änderungen der Läsion erfassen und visualisieren. Ansatzpunkte werden hier im Erkennen von Wachstumsvorgängen sowie bei Merkmalen gesehen, deren Bewertung dem menschlichen Auge-Hirn-System schwerfällt oder unmöglich ist. Dazu zählen langfristige Änderungen der Farbstatistik und strukturelle Änderungen im statistischen Sinne.

Eine Forderung, die sich unmittelbar aus Überlegungen zur Langzeitkontrolle ergibt, ist die systematische und redundanzarme Verwaltung aller im Verlauf einer Untersuchung anfallenden Daten. Eine effiziente Verwaltung dieses umfangreichen und in seinen Beziehungen sehr komplexen Datenmaterials, das neben Bildern und Verarbeitungsergebnissen auch Patientendaten, dermatologische Befunde und die Ergebnisse histologischer Analysen enthält, erfordert den Einsatz eines kommerziellen Datenbank-Managment-Systems. Erste, noch zu erweiternde Betrachtungen haben gezeigt, daß dabei das technische und rechtliche Umfeld im medizinischen Bereich berücksichtigt werden muß. Darüber hinaus sollte auch der Zugang zu Datenbanken anderer medizinischer Einrichtungen bis hin zum Zentralregister Malignes Melanom der Deutschen Dermatologischen Gesellschaft möglich sein.

5 Literatur

[Kreu90] Kreusch, J.; Rassner, G.; Henke, D.; Pietsch-Breitfeld, B.; Selbmann, H.K.: Bedeutung der Auflichtmikroskopie zur Früherkennung des malignen Melanoms. In: Orfanos, C.E.; Garbe, C.: Das maligne Melanom der Haut. München: Zuckschwerdt, 1990, S. 68-79.

[Kreu91] Kreusch, J.; Rassner, G.: Auflichtmikroskopie pigmentierter Hauttumoren, ein Bildatlas. Stuttgart; New York: G. Thieme Verlag, 1991.

[Rass88] Rassner, G.: Früherkennung des malignen Melanoms der Haut. In: Der Hautarzt 39(1988). München; New York: Springer-Verlag, S.396-401.

Segmentierung von MR-Volumendaten mit Bereichswachstumsverfahren

Gangolf Mittelhäußer, Gerhard Zimmermann
Fachbereich Informatik
Universität Kaiserslautern
Erwin-Schrödinger-Straße
67663 Kaiserslautern

Zusammenfassung: Die Segmentierung von kernspinresonanztomographischen (MR-) Volumendaten ist ein noch nicht befriedigend gelöstes Problem. In diesem Artikel werden Bereichswachstumsverfahren als Lösung im Hinblick auf eine nachfolgende Visualisierung der segmentierten Daten untersucht. Es wird definiert, was unter Segmentierung verstanden wird, dann werden kurz die Überlegungen, die zur Wahl von Bereichswachstumsverfahren geführt haben, dargestellt, die Algorithmen erklärt, die Ergebnisse vorgestellt und deren Qualität diskutiert.

1.0 Einleitung

Die Segmentierung von Volumendaten hat sich als notwendiger Verarbeitungsschritt vor der Visualisierung von Volumendaten durch Direct Volume Rendering oder Ray Casting Techniken [Todd Elvins 92] erwiesen. Segmentierung ermöglicht dem Anwender eines Visualisierungssystems, die gewaltige Informationsmenge, die auf den Bildschirm projiziert wird, durch Ausblenden von Segmenten zu reduzieren und durch Färben von Segmenten zu strukturieren.

Die Segmentierung von MR-Daten ist deshalb schwierig, weil bei diesen Daten keine eindeutige Zuordung Gewebetyp → Grauwert existiert, sondern verschiedene Gewebetypen auf denselben Grauwert abgebildet werden. Eine Segmentierung mittels Schwellwertverfahren, wie z.B. von Knochen in Computertomographien, ist daher nicht möglich. Analoges gilt für Clusteranalyseverfahren. Wegen genannter Nichteindeutigkeit ist eine eindeutige Segmentierung der MR-Daten nur durch zusätzliche Information (Vorwissen) oder zusätzliche Annahmen über die Segmentierung möglich. Da der hier beschriebene Segmentierungsalgorithmus möglichst vielseitig einsetzbar sein soll, wird kein Vorwissen, z.B. in Form von anatomischen Daten, vorausgesetzt, sondern lediglich Annahmen über die Eigenschaften der Segmente in Form eines Segmentmodells gemacht. Segmentierungen gemäß dieses Modells lassen sich mit entsprechend modifizierten Bereichswachstumsverfahren erzielen. Zusammen mit der Steuerbarkeit durch Parameter ergibt sich eine große Flexibilität, die es dem Benutzer erlaubt, die Segmentierungen an verschiedene Datensätze anzupassen und gemäß seinen Bedürfnissen (z.B. Segmentanzahl, -größe) zu verändern.

Das Ergebnis einer Segmentierung ist das segmentierte Volumen, das sogenannte Segmentvolumen, das jedem Voxel eine Segmentnummer zuordnet. Dieses läßt sich, indem jeder Segmentnummer über eine Farbtabelle eine Farbe zugeordnet wird, ebenso wie das originale MR-Volumen, sliceweise betrachten. Anschließend wählt der Benutzer in diesen Segmentslices interaktiv die Segmente aus, die visualisiert bzw. nicht visualisiert werden sollen, und führt die Visualisierung durch.
Dies soll mit den vier folgenden Abbildungen verdeutlicht werden. Abbildung 1 zeigt ein Slice des MR-Datenvolumens, Abbildung 2 das dazugehörige Slice des Segmentvolumens. In

Abbildung 3 sind die für die Visualisierung ausgewählten Segmente gezeigt und in Abbildung 4 die Visualisierung. Bei dem Visualisierungsalgorithmus handelt es sich um eine modifizierte Version [Lichtermann 91] des in [Levoy 88] beschriebenen Algorithmus. Vor der Visualisierung wurde den selektierten Segmenten eine Farbe und Durchsichtigkeit zugeordnet. Zusätzlich kann bei diesem Algorithmus noch ein quaderförmiger Bereich gewählt werden, in dem die Segmentselektion wirksam wird, außerhalb dieses Quaders wird das originale Datenvolumen visualisiert.

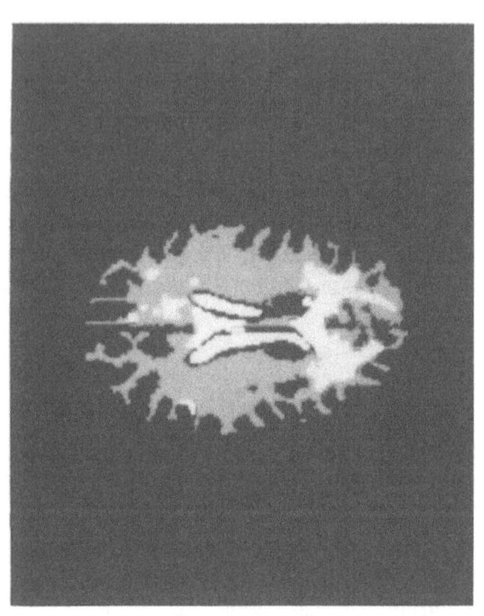

Abbildung 1

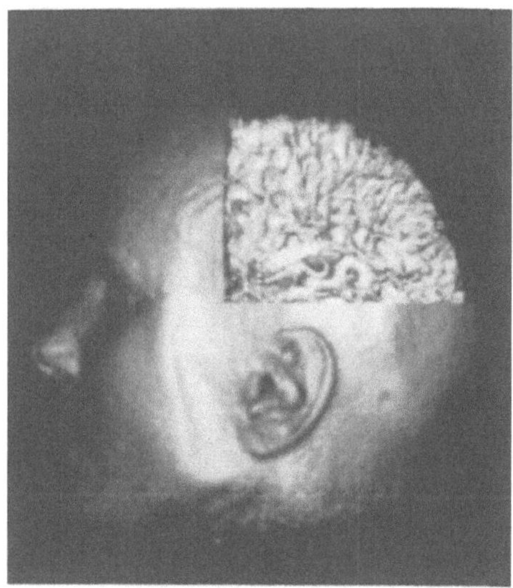

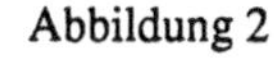

Abbildung 2

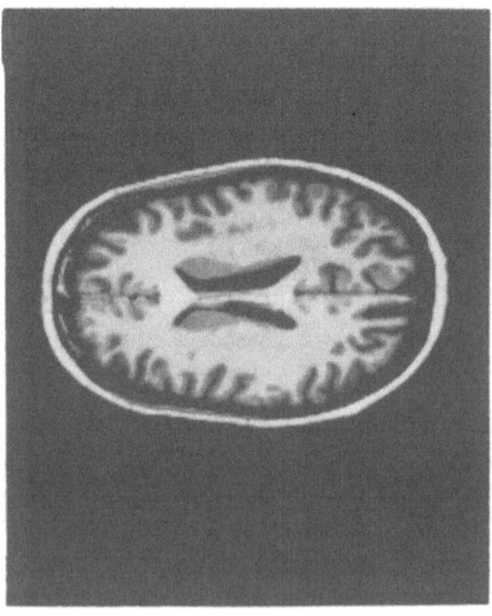

Abbildung 3

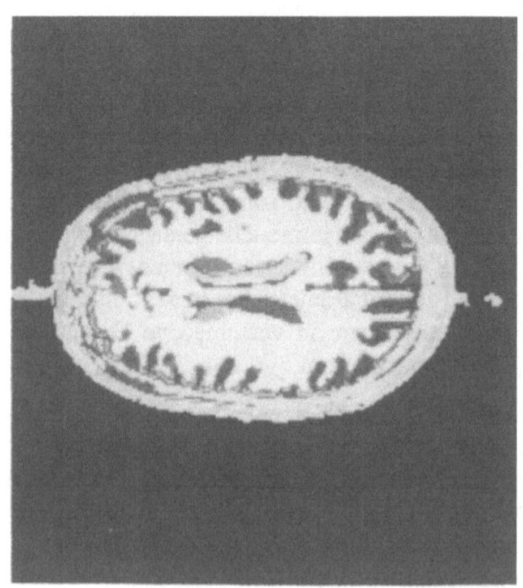

Abbildung 4

2.0 Problemdefinition

Ein Datenvolumen D wird definiert als dreidimensionales Array der Kantenlänge n. Die Arrayelemente heißen Voxel. Ein Voxel eines Datenvolumens wird also durch einen dreifachen Index angesprochen: D[i, j, k]. Im folgenden wird die Abkürzung D[x] mit x als dreidimensionalem Indexvektor x = (i, j, k) verwendet.
Schnitte (zweidimensionale Arrays) durch das Datenvolumen, bei denen eine Koordinate konstant ist, heißen Slices.

Eine Segmentierung wird definiert als Unterteilung des Datenvolumens D in Teilvolumen (Segmente S_i, i = 1, ..., n), so daß gilt:

1. Jedes Segment ist zusammenhängend (nach üblicher Definition, z.B. [Pavlidis 90]).
2. Jedes Segment S_i erfüllt ein Homogenitätsprädikat H, d.h. $H(S_i)$ = true.
3. Die Segmente sind disjunkt, d.h. für alle Segmente S_i, S_j gilt $S_i \cap S_j = \emptyset$ für $i \neq j$.
4. Die Segmentierung ist vollständig, d.h. für alle Voxel D[x] gibt es ein Segment S_i mit $D[x] \in S_i$.

Offensichtlich ist diese Definition nur eine notwendige Voraussetzung für Segmentierungen, jedoch nicht hinreichend, um Eindeutigkeit zu erzielen. Es gibt sehr viele verschiedene Segmentierungen, die dieser Definition genügen. Um zu einer eindeutigen Segmentierung zu gelangen, ist eine genauere Definition nötig. Dies ist möglich über globale Eigenachaften, die die Segmentierung haben soll.

5. Die Segmente S_i sollen möglichst homogen sein, d.h. es gibt ein (kontinuierliches) Homogenitätsmaß $M(S_i)$ und es soll gelten $\sum_{i=1}^{n} M(S_i)$ ist minimal, wobei $0 \leq M(S_i)$ und 0 Homogenität, große Werte von $M(S_i)$ große Inhomogenität bedeuten.

Da Minimalität erreicht wird, wenn alle Segmente nur noch ein Voxel enthalten, ist außerdem zu fordern:

6. Die Anzahl der Segmente soll minimal sein, d.h. für i = 1, ... , n gilt $H(S_i)$ = true mit n minimal.

Damit ist das Segmentierungsproblem auf ein Optimierungsproblem zurückgeführt.
Ein effizienter Algorithmus zur Lösung dieses Optimierungsproblems existiert allerdings nicht. Deswegen wird eine Näherungslösung durch eine Heuristik gesucht.

Da der Algorithmus als Vorverarbeitungsschritt für die Visualisierung eingesetzt werden soll, ist noch eine sehr wichtige Forderung zu erfüllen:

7. Zusätzlich zur Homogenitätsbedingung soll für die Segmente gelten, daß Kanten Segmentgrenzen bilden.

Diese Forderung ist die Voraussetzung dafür, daß die bei der Segmentierung gebildeten Grenzen mit den im Bild vorliegenden Kanten, soweit diese existieren, übereinstimmen. Erst dadurch wird für den Benutzer die Visualisierung des Volumens mit der durchgeführten Segmentierung nachvollziehbar, da die bei den verwendeten Rendering Algorithmen sichtbaren Strukturgrenzen den Kanten im Volumen entsprechen.

Diese Definitionen werden zu einem Segmentmodell zusammengefaßt, welches ein Segment definiert.

Definition: Ein Segment ist ein zusammenhängender homogener Bereich, der von Kanten, soweit vorhanden, begrenzt wird. Dabei soll die Segmentanzahl minimal sein, d.h. die Segmente maximal groß sein.

Diesem Modell entsprechende Segmente lassen sich mit kantenbegrenzten Bereichswachstumsverfahren finden. Diese Verfahren sind besonders geeignet, da sie a priori zusammenhängende Bereiche finden. Die Homogenität wird während des Wachstums ständig als wachstumsbegrenzende Größe kontrolliert, ebenso das Vorhandensein von Kanten. Maximale Größe der Segmente kann durch die Steuerung der Richtung des Wachstums und geeignete Startpunkte annähernd erreicht werden.
Punkt 5. der Definition wird nicht in das Segmentmodell übernommen, da das globale Optimum nicht in realistischer Rechenzeit gefunden werden kann. Die Eindeutigkeit geht damit verloren, stabiles Verhalten wird jedoch durch geeignete Wahl der Startpunkte erreicht (siehe unten).

3.0 Kantenbegrenztes Bereichswachtumsverfahren

3.1 Beschreibung des Algorithmus

In dem nachfolgenden Algorithmus wird vor der Segmentierung das Kantenvolumen `K` berechnet, mit dem zusätzlich zu dem Homogenitätsprädikat Segmentgrenzen gebildet werden. Ebenfalls vorher berechnet wird die Voxelmenge der Saatzellen `Saat`. Das Segmentvolumen `S`, welches am Ende jedem segmentierten Voxel `x` eine Segmentnummer > 0 zuordnet, wird anfangs mit 0 initialisiert. `S` und `K` sind also von der gleichen Größe wie das MR-Datenvolumen `D`.

Im Prinzip läuft der Algorithmus 2fach sequentiell ab. Es wird aus `Saat` nacheinander mit `get_next` der nächste Startpunkt gesucht und von dort beginnend das Segment durch Bereichswachstum gebildet.
Dazu wird nacheinander für alle Nachbarvoxel in `Nachbarn` des aktuellen Segments geprüft, ob das Segment nach Hinzunahme dieses Nachbarvoxels noch das Homogenitätsprädikat erfüllt. Wenn ja, wird das Voxel dem Segment zugeordnet und dessen noch nicht segmentierte Nachbarn `y` kleiner Kantenstärke zur Menge der Nachbarn des Segments mit `store` hinzugefügt. Wenn nein, wird mit dem nächsten Nachbarvoxel des Segments fortgefahren.
Dabei sucht die Funktion `find_best` das am besten zum Segment passende Voxel aus der Menge aller Nachbarvoxel in `Nachbarn`.

Algorithmus in C-Notation (vereinfachte Version):

```
Voxel_Menge Saat;
Datenvolumen D;
Kantenvolumen K;
Segmentvolumen S;
Homogenitätsprädikat H;
Voxel_Menge Nachbarn;
int segmentnr = 0, Kmax = 400, Hmax = 300;
Voxel x, y;
```

// die äußere Schleife startet nacheinander Wachstum der einzelnen Segmente:

```
Segmentation (Hmax, Kmax) {
  Voxel Saatpunkt;
  Initialisiere (K, S, Saat);
  segmentnr = 1;
  while (not_empty(Saat)) {
     Saatpunkt = get_next(Saat);
     start_growing (Saatpunkt, segmentnr, Hmax, Kmax);
     segmentnr = segmentnr + 1; } }
```

// die innere Schleife bildet ein Segment durch voxelweises Wachstum:

```
start_growing (Saatpunkt, segmentnr, Hmax, Kmax) {
  if ( S[Saatpunkt] == 0 && K[Saatpunkt] < Kmax ) {
     store (Saatpunkt, Nachbarn);
     while (not_empty(Nachbarn)){ // Falls vorhanden
        x = find_best (Nachbarn); // hole besten Nachbarn.
        if (H(segmentnr, D[x], Hmax) == true) {
           S[x] = segmentnr;        // Voxel zu Seg. dazufügen.
           while (finde_Nachbar(x, y) == true)
              if (S[y] == 0 && K[y] < Kmax)
               store (y, Nachbarn); } } } }
```

3.2 Homogenitätsprädikate und Segmentwachstum

Das Homogenitätsprädikat H berechnet auf Basis eines Merkmals oder mehrerer Merkmale, ob ein Segment homogen ist. Dieser Test wird vor jedem Wachstumsschritt durchgeführt, um zu entscheiden, ob ein Nachbarvoxel dem Segment zugeordnet werden kann. Das einfachste Homogenitätsprädikat testet, ob der maximale und minimale Grauwert der Voxel eines Segments S_i um nicht mehr als einen vorgegebenen Wert k voneinander abweichen:

$$\mathrm{MaxGrau}(S_i) - \mathrm{MinGrau}(S_i) < k.$$

Statt des Grauwertes des Voxels kann ein beliebiges anderes Merkmal verwendet werden, ebenso sind Linearkombinationen mehrerer Merkmale O_1, O_2 möglich:

$$\mathrm{Max}(\, a\,O_1(S_i) + b\,O_2(S_i)\,) - \mathrm{Min}(\, a\,O_1(S_i) + b\,O_2(S_i)\,) < k, \qquad a,b \in R.$$

Die Wahl des Homogenitätsprädikats beeinflußt stark das Ergebnis der Segmentierung. Der Benutzer muß entscheiden, mit welchem Merkmal oder welchen Merkmalen Homogenität gemessen werden soll und was für ein Segmentmodell damit implizit gewählt wird. Die Wahl der Parameter (Schwellwert k, Koeffizienten a, b) verändert insbesondere die Größe und Anzahl der Segmente.

Die Segmentwachtumsrichtung und damit die Segmentform sind abhängig von den Kriterien, nach denen das nächste Voxel aus der Menge der aktuellen Nachbarvoxel gesucht wird. Testet man die Voxel in der Reihenfolge, in der sie der Menge der Nachbarvoxel zugefügt wurden (first in first out), entstehen bevorzugt kugelförmige Segmente.
Eine implizite Formvorgabe läßt sich nur vermeiden, wenn man aus der Menge der Nachbar-

voxel das im Sinne der Homogenität bestpassende heraussucht. Dadurch wird das Segment zunächst beliebig geformte homogene Bereiche des MR-Volumens ausfüllen und dadurch eine maximale Größe erreichen, bevor eine Begrenzung durch Inhomogenitäten auftritt. Im Algorithmus ist dieser Suchvorgang durch die Funktion `find_best` realisiert.

3.3 Kantendetektion

Für das zur Segmentgrenzenbildung verwendete Kantenvolumen K kann im Prinzip jeder Kantendetektor verwendet werden. Wichtig ist, daß von den im Volumen vorhandenen Kanten jeweils nur die Kantenmitte detektiert wird und somit möglichst wenige Voxel bei Bildung der Segmentgrenzen durch die Schwellwertoperation K[x] < Kmax keinem Segment zugeordnet werden.

Ein geeigneter Kantendetektor ist z.B. die Maximumsnorm des Gradienten an den durch den Marr-Hildreth Operator gefundenen Konturpunkten [Bomans 90].

3.4 Startpunkte

Die Wahl der Voxel, von denen aus das Wachstum der einzelnen Segmente gestartet wird (Saatzellen), beeinflußt ebenfalls stark das Segmentierungsergebnis. Um ein stabiles Verhalten zu erreichen, sollten die Saatzellen in möglichst (nach dem Segmentmodell) homogenen Bereichen des MR-Datenvolumens liegen. Das bedeutet (für das als einfachstes Beispiel angegebene Homogenitätsprädikat) geringe Kantenstärke (1.Ableitung) und möglichst kleine Krümmung (2.Ableitung). Voxel dieser Art liegen anschaulich betrachtet auf Grauwertplateaus und somit in potentiell homogenen Bereichen. Diese können zusätzlich geordnet werden, so daß das Wachstum auf den flachsten Plateaus beginnt.

4.0 Ergebnisse und Diskussion

Eine der berechneten Segmentierungen und ihre Visualisierung wird am Ende exemplarisch gezeigt. Andere MR-Datensätze wurden ebenfalls segmentiert.

Die Rechnungen wurden auf einer SUN SPARCstation 2 durchgeführt. Die Laufzeit beträgt ca. 40 min, abhängig von der Parametereinstellung und insbesondere vom Algorithmus `find_best`, der je nach Homogenitätsprädikat unterschiedliche Komplexität hat.

Prinzipiell lassen sich für sehr verschiedenartige Datensätze und Benutzeranforderungen mit dem vorgestellten Algorithmus Segmentierungen berechnen. Eine vollautomatische Durchführung ist allerdings schwierig, dazu müßten nach Angabe eines Segmentmodells durch den Benutzer automatisch der passende Kantendetektor und das passende Homogenitätsprädikat sowie die Parameter gewählt werden.
Bei ausreichend kleiner Rechenzeit ist eine halbautomatische (interaktive) Vorgehensweise jedoch möglich, dazu müssen auch Kantendekektor und Homogenitätsprädikat nicht für jede Segmentierung neu gewählt werden, sondern können bei einem Typ von Daten (hier MR-Volumendaten) beibehalten werden.
Die Anpassung an ein bestimmtes MR-Datenvolumen erfolgt dann ausschließlich durch die Parameterwahl.

Um die für Interaktivität notwendige kurze Antwortzeit (ca. 1 sec) zu erreichen, wird derzeit an einer Parallelisierung des Algorithmus gearbeitet. Diese ist nicht unproblematisch, da die Ergebnisse von der Reihenfolge der Segmentbildung abhängen.

4.1 Qualität der Ergebnisse

Generell wird die Beurteilung der Ergebnisse dem Benutzer überlassen. Es läßt sich folgendes feststellen:

1. Dort, wo Segmentgrenzen wegen der Kantenstärke gefunden werden, liegt eine gute Übereinstimmung mit den Grenzen einer Handsegmentierung vor.
Schlechter nachvollziehbar sind Grenzen, die wegen Verletzung des Homogenitätsprädikats gezogen wurden. Diese teilen möglicherweise einen Bereich mit einem sanften Grauwertübergang großer Amplitude in mehrere Segmente, hier ist dann die Lage der Grenze nur noch von der Wahl der Startpunkte bzw. der Reihenfolge der Segmentbildung abhängig.
Durch ein komplexeres Homogenitätsprädikat läßt sich aber eine Verbesserung erzielen, z.B. durch zusätzliche Gewichtung der im Homogenitätsprädikat verwendeten Grauwerte mit der Kantenstärke, um so auch bei sanften Übergängen die Wahrscheinlichkeit der Grenzenbildung an kleinen Kanten zu erhöhen.

2. Die gefundenen Segmentierungen stimmen i.a. nicht mit im MR-Datenvolumen vorhandenen anatomischen Organen überein, da diese oft (nach dem Segmentmodell) inhomogen sind. Es läßt sich aber durch geeignete Parameterwahl erreichen, daß die Grenze eines Organs auch als Segmentgrenze gefunden wird. Damit ist dann sichergestellt, daß ein Segment sich nicht über mehrere Organe erstreckt bzw. daß jedes Organ aus einem Segment oder mehreren Segmenten besteht.
Diese müssen dann vom Benutzer entsprechend gewählt werden (dies ist allerdings mühselig, wenn das Organ sehr viele Segmente umfasst). An dieser Stelle könnte ein wissensbasiertes System aufsetzen.

4.2 Stabilität der Ergebnisse

Eine interaktive Wahl der Segmentierungsparameter ist nur dann sinnvoll, wenn das System sich bzgl. dieser Parameter stabil verhält, d.h. kleine Änderungen der Parameter auch nur kleine Änderungen des Ergebnisses bewirken. Die Stabilität wurde in [Weber 92] genauer untersucht. Dazu wurden an einem Testdatensatz die Parameter variiert und Abweichungen der gewonnenen Segmentierungen bzgl. der Anzahl der verschieden segmentierten Voxel verglichen.

In Kurve 1 ist die prozentuale Abweichung gegen die maximal erlaubte Grauwertdifferenz k des verwendeten Homogenitätsprädikats aufgetragen. Größere Abweichungen in den Segmentierungen treten immer dann auf, wenn ein Segment aufgespalten wird bzw. zwei Segmente zu einem fusioniert werden. Die Segmentierungen wurden mit der Segmentierung mit Parameterwert 300 verglichen.
In Kurve 2 wurde die maximal erlaubte Kantenstärke Kmax variiert. Auch hier ist das Verhalten ausreichend stabil. Für große erlaubte Kantenstärken findet überhaupt keine nennenswerte Änderung mehr statt, da dann die Kanten die Segmentgrenzenbildung nicht mehr wesentlich beeinflußen.

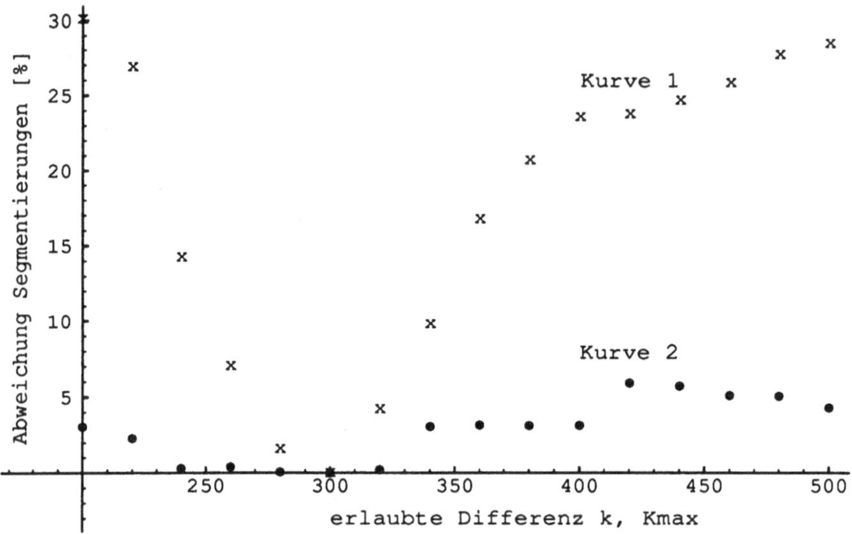

5.0 Folgerungen

Die Segmentierung von Volumendaten mit Bereichswachstumsverfahren ist möglich, Hauptschwierigkeit des Verfahrens ist die Wahl der Parameter, die starken Einfluß auf das Ergebnis hat. Da hier der Benutzer entscheiden muß, ob das Ergebnis seinen Vorstellungen entspricht, ist ein interaktives System notwendig. Dazu wiederum muß der Algorithmus auf paralleler Hardware ausgeführt werden, eine Parallelisierung des Verfahrens wird deshalb gerade durchgeführt.

6.0 Literatur

Bomans M., Höhne K.-H., Tiede U., Riemer M.: 3-D Segmentation of MR Images of the Head for 3-D Display, IEEE Transactions on Medical Imaging Vol. 9, No. 2, June 1990, pp. 177.

Levoy M.: Display of Surfaces from Volume Data, IEEE Computer Graphics & Applications, May 1988, pp. 29.

Lichtermann J., Mittelhäußer G.: Eine Hardwarearchitektur zur Echtzeitvisualisierung von Volumendaten durch "Direct Volume Rendering", GI-Workshop Visualisierungstechniken, Rechenzentrum Universität Stuttgart, Juni 1991.

Pavlidis T.: Algorithmen zur Grafik und Bildverarbeitung, Heise Verlag, Hannover 1990.

Todd Elvins T.: A Survey of Algorithms for Volume Visualization, Computer Graphics, Vol. 26, No. 3, August 1992, pp. 194.

Weber G.: Implementierung und Analyse verschiedener Verfahren zur Segmentierung von Volumendaten, Diplomarbeit, Universität Kaiserslautern, Fachbereich Informatik, Juli 1992.

Zwei Verfahren zur automatischen Konturfindung eines Herzventrikels aus angiographischen Bildern

B. Nigbur, J.H. Builtjes, H. Oswald, E. Fleck

Deutsches Herzzentrum Berlin
Augustenburgerplatz 1
13353 Berlin

Einleitung

Die digitale Bildverarbeitung in der Kardiologie gewinnt immer mehr an Bedeutung zur Diagnoseunterstützung von Herzerkrankungen. Eine automatische Konturerkennung eines Herzventrikels kann hier die Bestimmung von morphologischen und funktionellen Parametern des Herzens (wie z. B. Volumenbestimmung, Herzwanddicke, Kontraktion u.a.) unterstützen.
Die bisher im Deutschen Herzzentrum Berlin benutzte Methode zur Konturbestimmung eines Herzventrikels basiert auf der interaktiven Eingabe von Stützpunkten durch den Arzt (Herzklappe und -spitze (Apex)). Diese Punkte werden anschließend mittels einer Spline-Funktion interpoliert. Entlang dieser geschätzten Kontur wird mit Hilfe einer Kostenmatrixanalyse die Ventrikelkontur bestimmt [Bei93].
Für eine Auswertung von Bildsequenzen (z.B. zur Beschreibung der Wandbewegung eines Herzventrikels) in der klinischen Praxis ist diese Methode aufgrund des interaktiven Aufwandes nicht mehr einsetzbar. Eine vollständig automatisierte Konturfindung ist dann notwendig.
Bild 1 zeigt ein typisches Beispiel eines zu bearbeitenden Angiokardiogramms eines Herzventrikels. Das verwendete Kontrastmittel zeigt den Ventrikel als helles Gebiet gegenüber dem Hintergrund. Links oben befindet sich die Herzklappe, rechts unten der Apex. Das Ziel ist es, einen geeigneten Algorithmus zu entwickeln, der mit Hilfe von Bildverbesserungs- und Konturfindungsverfahren zum gewünschten Ergebnis führt. Um eine eindeutige

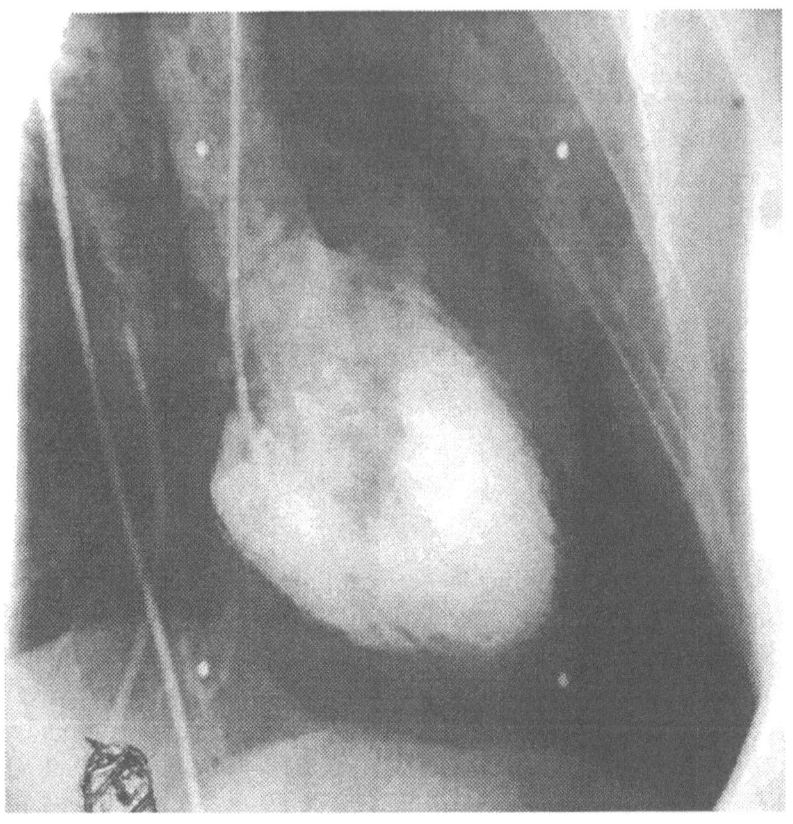

Bild 1 : Angiokardiogramm eines Herzventrikels

Konturerkennung durchführen zu können, sind folgende Arbeitsschritte notwendig: Nach einem Filter zur Rauschunterdrückung folgt eine Kantenverschärfung, danach können Konturverfolgungsverfahren angewendet werden. In der vorliegenden Arbeit werden zwei mögliche Methoden präsentiert, die eine automatische Konturfindung eines Herzventrikels ohne Interaktion des Arztes erlauben. Die erste Methode basiert auf eine Konturlinienverdünnung und -verkettung, während bei der zweiten Methode ein sogenanntes Pyramidenverfahren benutzt wird.

Rauschunterdrückung und Kantenverschärfung

Zur Rauschunterdrückung wird ein Medianfilter in Kombination mit einer Abfrage auf den Gradientenbetrag benutzt. Hierdurch werden große, rauschbedingte Grauwertunterschiede unterdrückt, während die Kanten erhalten bleiben.
In realen Bildern sind Kanten oft über mehrere Pixel verschmiert. Zur Kantenverschärfung kann ein Verfahren angewendet werden, das von Moran [Mor89] vorgeschlagen wird. Pixel mit geringen Grauwertsprüngen zu ihren direkten Nachbarn (d.h. Pixel mit geringem Grauwertgradienten) werden an diese Nachbarn angeglichen. Dagegen werden Pixel mit großen Grauwertsprüngen nicht an ihre Nachbarn angeglichen. Resultat ist ein Bild, in dem die lokalen Kontraste verschärft sind.

1. Methode: Konturfindung über Konturverdünnung und -verkettung

Eine Grauwertkante, die das Objekt (hier den Ventrikel) vom Hintergrund trennt, kann durch Bildung des Grauwertgradienten detektiert werden. Eine Grauwertkante befindet sich dort, wo der Grauwertgradient maximal ist. Die gesuchte Kontur verläuft senkrecht zur Richtung des maximalen Grauwertgradienten. Die Verarbeitung besteht aus drei Hauptschritten: Detektion, Verdünnung und Verkettung [Bäs89].

Zur Detektion von Konturen werden die Gradientenbeträge und die Gradientenrichtung bestimmt. Sei $f(x,y)$ die Grauwertfunktion eines Bildes, $\delta f / \delta x$ und $\delta f / \delta y$ die partielle Ableitung in x - bzw. in y - Richtung, dann ist der Betrag des Gradienten $\nabla f(x,y)$:

$$|\nabla f(x,y)| = \sqrt{\left(\frac{\delta f}{\delta x}\right)^2 + \left(\frac{\delta f}{\delta y}\right)^2}$$

und die Gradientenrichtung:

$$\Theta(f(x,y)) = \arctan\left(\frac{\delta f}{\delta y} / \frac{\delta f}{\delta x}\right)$$

Die Ergebnisse der Gradientenoperation werden im zweiten Schritt, der Konturverdünnung, weiterverarbeitet.

Da nur Pixel mit maximalem lokalen Grauwertgradienten den Verlauf der Kontur repräsentieren, muß im nächsten Verarbeitungsschritt eine Verdünnung der Konturlinien stattfinden. In einer 8ter-Nachbarschaft werden nur die Pixel übernommen, die in gleicher Gradientenrichtung den größten Gradientenbetrag aufweisen. Das Resultat sind Konturlinien mit einer Breite von nur einem Pixel.

Nach der Verdünnung folgt der letzte Bearbeitungsschritt, die Verkettung von Konturpunkten. Entlang der im Verdünnungsbild übriggebliebenen Linien findet eine Konturverfolgung statt. Die Suchrichtung steht senkrecht auf dem lokalen Grauwertgradienten. Damit werden nur die Pixel zu einer Kontur zusammengefügt, die auch tatsächlich auf einer Grauwertkante im Originalbild liegen.

2. Methode: Konturfindung mit einem Pyramidenverfahren

Eine weitere Möglichkeit der Konturdetektion ist die Anwendung von hierarchischen Datenstrukturen, die durch Pyramiden realisiert werden. Im Gegensatz zu dem oben genannten sequentiellen Verfahren durch Linienverfolgung stellen die Pyramidenstrukturen ein paralleles Verfahren dar. Diese Methode erlaubt es, die Kanten eines Bildes gleichzeitig zu suchen, ohne den Rechenaufwand zu erhöhen.

Eine Bildpyramide besteht aus Bildebenen von immer gröber werdender Auflösung des Originalbildes. Dazu wird eine 2x2/2-Pyramidenstruktur benutzt. Sie bewirkt, daß vier Zellen eines 2x2-Blockes in einer Pyramideebene zu einer Zelle der nächsthöheren Ebene zusammengefaßt wird. Die Größe der neu entstandenen Ebene wird dabei um den Faktor 2 reduziert, da sich je zwei diagonal benachbarte 2x2-Blöcke eine Zelle teilen. In den Elementen der

Pyramidenstruktur sind zur Konturfindung insbesondere Kurvenrelationen abgelegt. Diese Kurvenrelationen werden mit Hilfe von Kantenoperatoren aus den Grauwerten des Originalbildes abgeleitet. Ein Beispiel wäre ein 2x2-Operator, der die Grauwertdifferenzen der von ihm überdeckten Punkte des Originalbildes berechnet. Bild 2 zeigt die 2x2-Operatormatrix. Diese Differenzen werden der Größe nach sortiert, und die beiden größten als Kurvenrelation definiert.

Bild 2 : Differenzbildung zwischen (A,B), (B,C), (C,D) und (D,A).

Beim Aufbau der Pyramide werden aus den Kurvenrelationen eines 2x2-Fensterblockes die Kurvenrelationen des darüberliegenden Fensters der nächsthöheren Ebene bestimmt. Dazu wird jedes Fenster durch eine Diagonale geteilt und der Kurveninhalt auf die zwei dabei entstehenden rechtwinkligen Dreiecke aufgeteilt. Dann werden die vier äußeren Dreiecke zu einem neuen um 45 Grad gedrehten Quadrat vereinigt (siehe Bild 3). Dieses Verfahren wird bis zur Spitze der Pyramide fortgeführt.

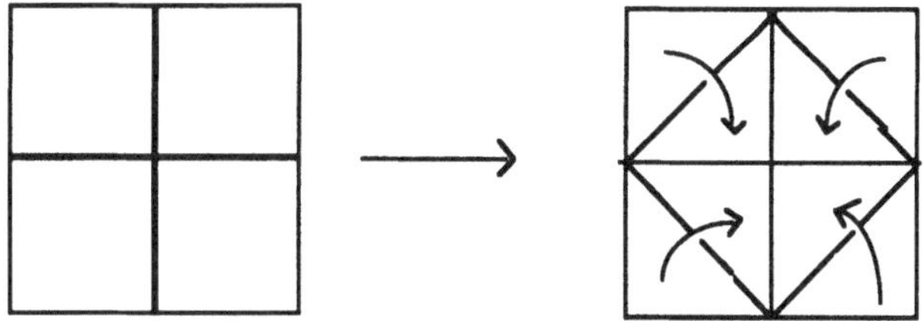

Bild 3 : Reduktion eines 2x2 Fensterblocks

Anschließend folgt eine Umkehrung der oben beschriebene Reduktion, ein sogenannter Verfeinerungsprozeß, durch den der genaue Verlauf der Kurven

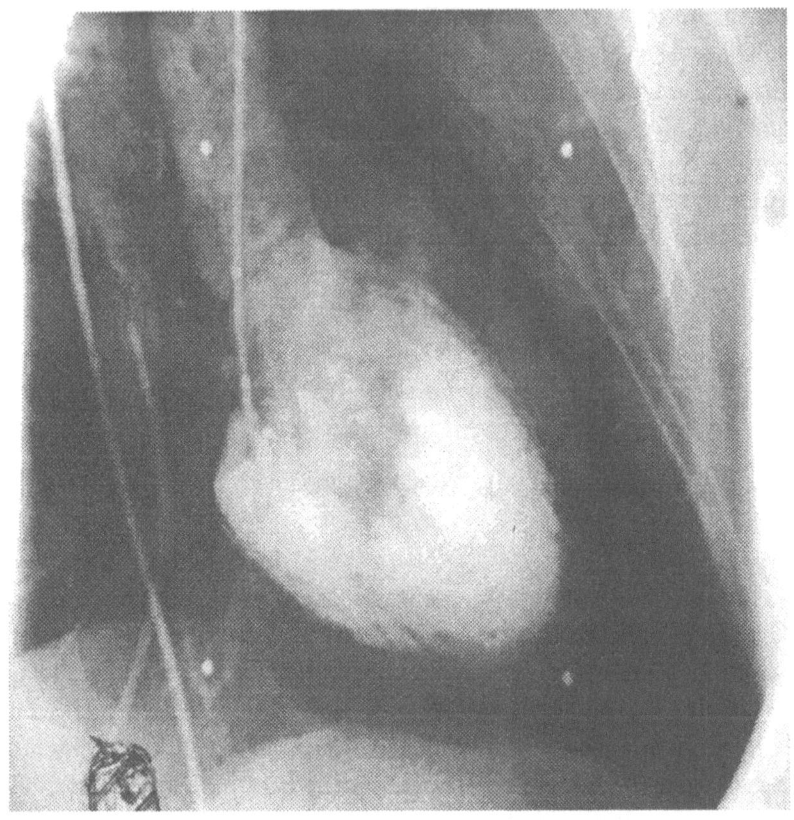

Bild 4 : Ergebnis der Konturdetektion über Konturverdünnung und -verkettung

im Bild ermittelt wird. Dabei wird Schritt für Schritt die Pyramide von oben nach unten abgearbeitet und die Kurven mit jeder Hierarchieebene weiter verfeinert, bis schließlich alle Konturteilstücke ermittelt werden.[Kro86]

Resultate

Das Ergebnisbild beider Verfahren enthält alle im Originalbild vorhandenen Konturteilstücke. Die Ergebnisse der Konturverdünnung und -verkettung und

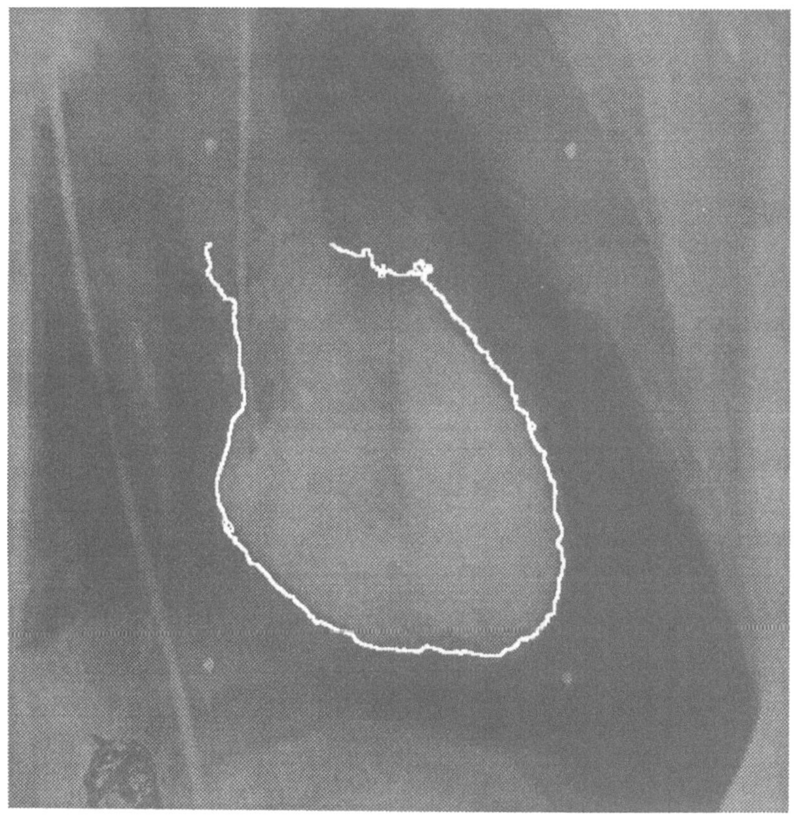

Bild 5 : Ergebnis einer automatisch gefundenen Herzventrikelkontur.

des Pyramideverfahrens unterscheiden sich nur geringfügig voneinander. Diese Unterschiede sind auf eine unterschiedliche Gradientenbildung zurückzuführen. Bild 4 zeigt das Resultat einer Konturfindung über Konturverdünnung und -verkettung (zur optische Unterstützung sind Kontrast und Helligkeit verändert worden). Aufgrund von Bildrauschen und Gradientenschwankungen werden die Konturen meistens nur als Teilstücke detektiert. Benachbarte Konturteilstücke können durch einen Schließungsalgorithmus miteinander verbunden werden. Dieser Algorithmus berücksichtigt sowohl die Verlaufsrichtung der Teilstücke als auch den Grauwertgradientenverlauf zwischen den zu verbindenen Teilstücken. Aufgrund der Lage und Form der

Konturen kann die Suche nach möglichen Kandidaten der Ventrikelkontur eingeschränkt werden. Beispiele hierfür sind sehr kurze Konturen mit nur wenigen Pixeln, die durch Bildrauschen entstanden sind, Konturen am Bildrand oder die geraden Konturen des Herzkatheters (siehe Bild 4). Da die ungefähre Form und Lage der Ventrikelkontur bekannt ist, ist eine richtige Zuordnung der gefundenen Konturen an Hand eines einfachen parametrisierbaren Modells (z.B. aufgrund der Form) ohne weiteres möglich (siehe Bild 5). Weitere Untersuchungen werden zeigen, welches der beiden hier vorgestellten Verfahren für die klinische Praxis am besten geeignet ist.

Literatur

[Bäs89]: H. Bässmann, Ph.W. Besslich, Konturorientierte Verfahren in der Digitale Bildverarbeitung, Springer-Verlag, 1989

[Bei93]: J. Beier, B. Nigbur, S. Lempert, H. Oswald, E. Fleck, Automatic detection of ventricle contours in angiograms, CAR'93, International Symposium on Computer Assisted Radiology, Berlin, June 1993

[Kro86]: W.G. Kropatsch, Kurvenrepräsentation in Pyramiden, Mustererkennung ´86 Schriftenreihe der OCG, Band 36, R. Oldenburg, 1986.

[Mor89]: C.J. Moran , A morphological transformation for sharpening edges of features before segmentation, Computer Vision, Graphics and Image Processing, Vol. 49 (85-94), 1990

Segmentierung der Substantia corticalis von Lendenwirbelkörpern aus CT-Aufnahmen

R. Petzold, M. Werner, W. Luth*
Universität Rostock - Institut für Biomedizinische Technik und Medizinische Informatik
Abt. Biomedizinische Technik, E.-Heydemann-Str. 6, 18057 Rostock
* Fraunhofer - Institut für Graphische Datenverarbeitung - Außenstelle Rostock
J.-Jungius-Str. 9, 18059 Rostock

1 Einleitung

Zur nichtinvasiven Bestimmung der Festigkeit von Knochenstrukturen, hier Lendenwirbelkörper (LWK), wird die Kopplung von bildgebenden diagnostischen Verfahren der Medizin zur Bestimmung ihrer Morphologie und der Methode der Finiten Elemente (FEM) zur Durchführung von rechnergestützten mechanischen Simulationen verwendet. Im Bereich der Geometrieverarbeitung des Prozesses der FEM-Modellierung sind die Bildverarbeitungsverfahren zur Bestimmung der Bildkomponenten aus den Ergebnissen der bildgebenden Diagnostik ein Teil, der die Güte der Simulationsergebnisse wesentlich beeinflußt. Daraus ergibt sich die Forderung nach Verfahren, die die Bildkomponenten, hier die Substantia corticalis (Corticalis) in LWK-Röntgen-CT-Aufnahmen, automatisch detektieren.
Meist wird diese Aufgabe über Schwellwertbinarisierung der CT-Zahlen gelöst. Im folgenden wurde die Einstellung des Schwellwertes nach verschiedenen Verfahren untersucht. Ein in der digitalen Bildverarbeitung oft benutzter Ansatz ist die Bestimmung des Schwellwertes aus der statistischen Analyse des Bildmaterials. Dabei ist nur bei vorliegender multimodaler CT-Zahlverteilung mit einem befriedigendem Ergebnis zu rechnen. Wie aus Abb. (1) ersichtlich, ist hier die Festlegung eines Schwellwertes mit Hilfe des Histogramms der CT-Zahlen nur schwer möglich.

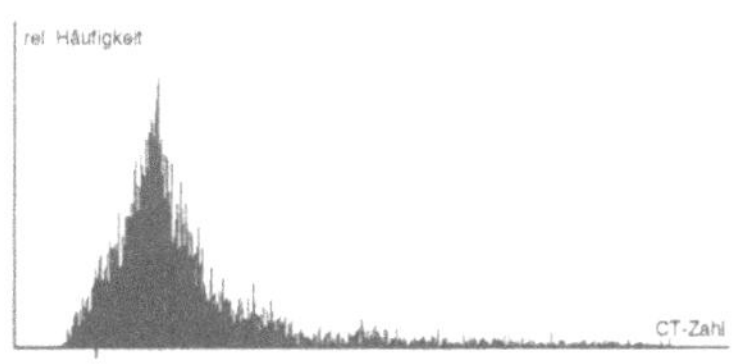

Abb. (1)
Histogramm der CT-Zahlen einer typischen CT-Aufnahme eines LWK

In der Radiologie wird als ein weiterer Ansatz die Binarisierung mittels einem aus a-priori-Wissen (CT-Richtwerte) abgeleiteten Schwellwert gewählt. Dieser Ansatz wurde in vielen

Anwendungen im CT-Umfeld umgesetzt. Durch individuelle Schwankungen und pathologische Veränderungen des Wirbelkörpers ist eine Angabe eines allgemeingültigen Schwellwertes jedoch nicht möglich, wobei nicht auszuschließen ist, daß im Einzelfall gute Ergebnisse bei Verwendung eines Richtwertes als Schwellwert erzielt werden können.

Abb. (2)
Schwellwertbinarisierung
nach CT-Richtwert

Abb. (3)
Schwellwertbinarisierung
mit individuell eingestelltem Schwellwert

Ausgehend von dem in Abb. (2) dargestellten Ergebnis wurde der Schwellwert individuell so angepaßt, daß eine zufriedenstellende Segmentierung der Substantia corticalis Abb. (3) erreicht wurde.
Die Technik der individuellen Einstellung eines Schwellwertes stellt aber einen hohen subjektiv variierenden Unsicherheitsfaktor dar, welcher einen objektiven Vergleich von Segmentationsergebnissen erschwert. Ein Nachteil der Benutzung von global definierten Schwellwerten ist also eine nicht der morphologischen Struktur des Wirbelkörpers entsprechende Segmentierung im corticalen und subcorticalen Bereich.
Zur Vermeidung dieser Nachteile ist es notwendig, die Bildpunkte nicht nur über die CT-Zahl zu klassifizieren, sondern auch Merkmale zu berücksichtigen, die die Geometrie der Corticalis als dünnwandige Ringstruktur beschreiben.

2 Bestimmung der Substantia corticalis durch Bildpunktklassifikation

Im nachfolgenden wird ein Verfahren zur Segmentation der Bildkomponente Corticalis dargestellt und auf wesentliche Ergebnisse eingegangen. Abb. (4) stellt typische CT-Bilder dar, auf die sich die nachfolgenden Abbildungen der Klassifikationsergebnisse beziehen.
Die Grundlage des Verfahrens ist die Erarbeitung eines Bildpunktklassifikators anhand des Ergebnisses einer Clusteranalyse in einem definierten Merkmalsraum. Zur Stabilisierung des Verfahrens werden nichtrelevante Bildpunkte vor der Merkmalsextraktion entfernt. Die Clusteranalyse wurde mit dem k-means-Verfahren, einem unüberwachten Minimal-Distanz-Verfahren zur partionierenden Clusteranalyse durchgeführt. [Mucha 92]

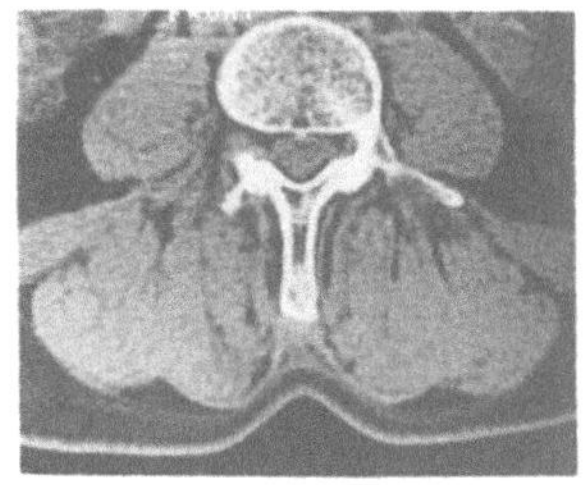
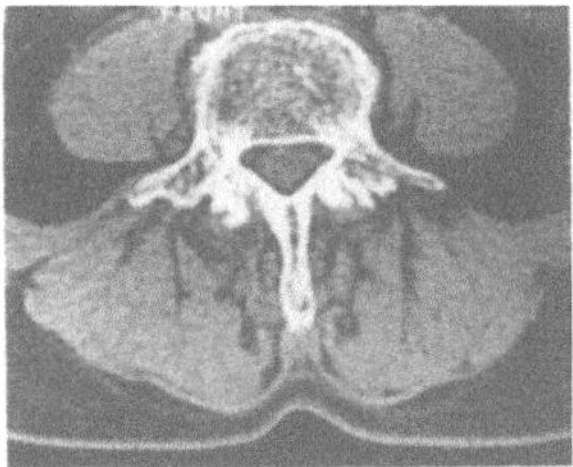
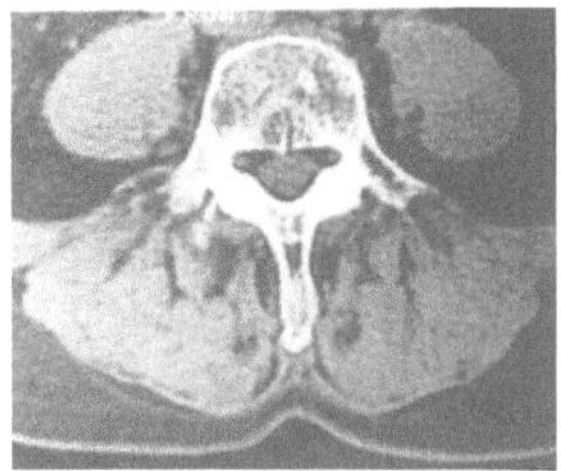

Abb. (4)
CT-Aufnahmen eines Wirbelkörpers

Bei der statistischen Auswertung des Bildmaterials wurde festgestellt, daß sich aufgrund der Materialeigenschaften der Corticalis die CT-Zahlen ihrer Bildpunkte in einem bestimmten Intervall bewegen, wobei aber die Intervallbreite individuell verschieden sein kann. Daher ist davon auszugehen, daß die CT-Zahl einen wesentlichen Beitrag zur Clusterbildung leistet.
Um die Geometrieeigenschaften der Corticalis bei der Klassifizierung der Bildpunkte zu berücksichtigen, ist es notwendig, strukturbeschreibende Komponenten zur Definition des Merkmalsraums zu verwenden. Eine umfassende Darstellung dafür in Frage kommender Merkmale für medizinisches Bildmaterial ist in [Schlaps 86] dargestellt. Von strukturbeschreibenden Merkmalen wird grundsätzlich gefordert, daß die Bildpunktumgebung in die Merkmalsgewinnung einbezogen wird. Es wurden die Kantendetektoren Gradienten-, Sobel- und Laplaceoperator untersucht, wobei die Klassifikationsergebnisse qualitativ ähnlich waren, aber als Nachteil eine Abhängigkeit der Qualität von einer Gewichtung bestand.
Als ein weiteres Merkmal zur Klassifikation der Bildpunkte wurde die lokale Varianz untersucht. Die lokale Varianz [Jähne 91] wird durch (2) bestimmt, wobei auf die Bionominalwichtung der Maskenelemente hingewiesen sei.

$$\mu^{*}_{\mathrm{LOKAL}_{3x3}}(s(x,y)) = \sum_{i=-1}^{1}\sum_{j=-1}^{1} w(i+1,j+1)\, s(x+i,y+j) \tag{1}$$

$$\sigma^{*2}_{\mathrm{LOKAL}_{3x3}}(s(x,y)) = \sum_{i=-1}^{1}\sum_{j=-1}^{1} w(i+1,j+1)(s(x+i,y+j) - \mu^{*}_{\mathrm{LOKAL}_{3x3}}(s(x,y)))^2 \tag{2}$$

$$[w(i,j)] = \tfrac{1}{16}\begin{bmatrix} 1 & 2 & 1 \\ 2 & 4 & 2 \\ 1 & 2 & 1 \end{bmatrix} \tag{3}$$

Dieser Operator zur Bestimmung der lokalen Varianz arbeitet rotationsinvariant. Die lokale Varianz beschreibt eine Schätzung der lokalen Grauwertverteilung und stellt somit ein Maß für die Homogenität eines lokalen Bildbereichs dar. Dadurch ergeben sich für die lokale Varianz innerhalb der Corticalis kleine Werte (große Homogenität). Betrachtet man die lokale Varianz an einem CT-Zahl-Übergang, so liefert der Operator hohe Werte, welche ähnlich eines Kantenoperators eine Abgrenzung der Bildpunktklasse Corticalis ermöglichen. Allgemein ist die lokale Varianz zur Klassifikation von texturierten Objekten geeignet, bei denen das Texel kleiner gleich und die Objektgröße größer gleich der Maskengröße ist. In mehreren Versuchsserien konnte die gute Eignung der lokalen Varianz als Merkmal (in Kombination mit der CT-Zahl) nachgewiesen werden.

Abb. (5)
Klassifikation nach Clusteranalyse im Merkmalsraum **M** = CT-Zahl × lokale Varianz

Um kleine Fehler der Klassifikation zu eliminieren, erfolgte eine Nachbearbeitung durch morphologische Operatoren.

Abb. (6)
Bearbeitung der Klassifikationsergebnisse durch morphologische Operatoren

Die so klassifizierten Bildpunkte werden an die nachfolgenden Bearbeitungsschritte übergeben. Hier erfolgt die Auswertung der detektierten Bildkomponente zur FEM-Modellierung des Wirbelkörpers, in dem aus den geometrischen Informationen und dem CT-Wert die Modellelemente ausgewählt und miteinander verknüpft werden.
Das dargestellte Verfahren wurde im Prototyping auf einer SUN SPARC Workstation unter der Entwicklungsumgebung KHOROS erstellt.
Der Test des Verfahrens wurde mit mehreren CT-Serien durchgeführt. Dabei konnte durch dieses Verfahren eine zufriedenstellende Segmentation der Corticalis erreicht werden. Der Vorteil, den dieser Ansatz gegenüber den auf einfacher Schwellwertsegmentierung basierenden Ansätzen hat, liegt in der Vermeidung von subjektiven Einflüssen bei der Bestimmung der

Bildkomponente und der sich daraus ergebenden besseren Vergleichbarkeit. Eine Verbesserung der Ergebnisse ist bei einer Erhöhung der räumlichen Auflösung (Schnittbreite, Tischvorschub, Schnittbildauflösung) der verwendeten CT-Technik zu erwarten, wobei im klinischen Einsatz Nutzen des Verfahrens und daraus resultierende Erhöhung der Röntgenstrahlenbelastung verantwortungsvoll abzuwägen sind. Unter diesen Voraussetzungen wäre ein Ansatz denkbar, bei dem die Clusteranalyse für benachbarte Schnitte gemeinsam erfolgt.
Zur Vervollkommnung der Wirbelkörpermodellierung mit der FEM erscheint eine Untersuchung des Spongiosabereiches des Wirbelkörpers unerläßlich. Dazu sind leistungsfähige Lösungsansätze zur Texturanalyse zu verfolgen, die jedoch von einer hohen Schnittbildauflösung abhängig sind. Beachtenswert erscheinen die auf diesem Gebiet veröffentlichten Arbeiten von Preteux et al. [Preteux 85], [Preteux 87].

3 Literatur

[Jähne 91] Jähne, B.: Digitale Bildverarbeitung. Springer-Verlag, Berlin, 1991.

[Mucha 92] Mucha, H. J.: Clusteranalyse mit Mikrocomputern. Akademie-Verlag, Berlin, 1992.

[Preteux 85] Preteux, F.; Bergot, C.; Laval-Jeantet, A.M.: Automatic quantification of vertebral cancellous bone remodeling during aging. Anat. Clin. 1985, 7: 203-208.

[Preteux 87] Preteux, F.; Schmitt, M.; Laval-Jeantet, M.; Laval-Jeantet, A.M.:The Random Boolean Function: A Probabilistic Model for Quantifying Vertebral Spongious Texture. In Computer Assisted Radiology, Proc. of the Int. Symposium, Springer-Verlag, Berlin, 1987, 739-743.

[Schlaps 86] Schlaps, D. et al.: Ultrasonic Tissue Characterization using a Diagnostic Expert System. In Information Processing in Medical Imaging, Martinus Nijhoff Publ., The Hague, 1986, 343-363.

Klassifikation melanozytärer Hautveränderungen anhand makroskopischer Farbaufnahmen

Th. Schindewolf [a], R. Albert [a], W. Stolz [b], W. Abmayr [c], H. Harms [a]

[a] Universität Würzburg, Institut für Virologie,
Abteilung Bildverarbeitung,
Versbacher Straße 7, D-97078 Würzburg

[b] Universität Regensburg, Dermatologische Klinik
[c] Fachhochschule München, Fachbereich Informatik

Kurzfassung

Die richtige und frühzeitige klinische Diagnose ist eine wichtige Voraussetzung für eine erfolgreiche Therapie von Hautkrebs. In diesem Beitrag wird ein Bildanalysesystem vorgestellt, das die Früherkennung von malignen Melanomen erleichtern soll. Melanozytäre Läsionen werden direkt am Patienten mit einer CCD-Farbkamera aufgenommen und digitalisiert. Die Segmentierung der Farbbilder nutzt das mathematische Verfahren der Hauptkomponentenanalyse und eine histogrammbasierende Methode. Bildanalytische Größen, die dermatologische Kriterien nachbilden, werden extrahiert. Als Ergebnis der statistischen Auswertung der Daten entstehen binäre Entscheidungsbäume, die eine Klassifikation in benigne und maligne Läsionen ermöglichen. Die erzielten Ergebnisse zeigen, daß sich durch den Einsatz der digitalen Bildverarbeitung die präoperative Diagnostik von pigmentierten Hautveränderungen verbessern läßt.

1. Medizinische Problemstellung

Das maligne Melanom ist die bösartigste Form aller Hauterkrankungen beim Menschen. In den USA werden jährlich etwa 22000 Fälle diagnostiziert, die jährliche Sterberate beträgt ca. 5500 .[4] Das Lebenszeitrisiko, an einem malignen Melanom zu erkranken, wird auf 1%-2% geschätzt.[7] Da bei malignen Melanomen in metastasierten Stadien nahezu keine Therapiemöglichkeiten mehr bestehen und weil das Metastasierungsrisiko eindeutig von der Tumordicke abhängt, stellt die Früherkennung die einzige Maßnahme zu Senkung der Melanommortalität dar.
Durch die visuelle Beurteilung der pigmentierten Hautveränderungen versucht der Dermatologe, die malignen Melanome zu erkennen. Einige Kriterien dazu wurden in

Dieses Projekt wurde gefördert vom Bundesministerium für Forschung und Technologie (BMFT).

der ABCD-Regel der Dermatologie zusammengefaßt: Das Vorliegen von Asymmetrie, eine unregelmäßige Begrenzung, unterschiedliche Farbschattierungen (Color) und ein Durchmesser von mehr als 6 mm gelten als Hinweise auf ein malignes Melanom.[5] Selbst mit dieser Regel liegt die diagnostische Treffsicherheit auch bei Dermatologen mit großer klinischer Erfahrung nur bei knapp über 70% und ist bei Ärzten anderer Fachrichtungen noch niedriger.[12]

2. Bildanalytisches Vorgehen

Durch Einsatz der digitalen Bildverarbeitung wird versucht, die präoperative Diagnostik von malignen Melanomen zu verbessern. Abbildung 1 zeigt den schematischen Ablaufplan für die Analyse der Hautläsionen in dieser Studie.

Die Patienten werden an der Dermatologischen Klinik untersucht, beraten und behandelt. Von den Läsionen werden CCD-Aufnahmen gemacht. Besteht der Verdacht auf ein malignes Melanom, wird die Läsion exzidiert und an dem entnommenen Gewebe eine histologische Untersuchung vorgenommen, um die exakte Diagnose zu erhalten.

Für die Auswertung der digitalisierten Läsionen wird eine DECstation 5000/200 eingesetzt; als Programmiersprache wird FORTRAN verwendet. Nach der Segmentierung der Bilder findet eine Merkmalsextraktion statt, die sowohl Kriterien der Dermatologen mit speziellen Algorithmen nachbildet, als auch gängige Merkmale der Bildverarbeitung berechnet: Die Farbe, die Symmetrie, die Textur und die Begrenzung jeder Läsion werden quantitativ bewertet. Mit der histologischen Diagnose und den Merkmalen wird ein Klassifikator trainiert, der eine computerunterstützte Diagnose ermöglicht.

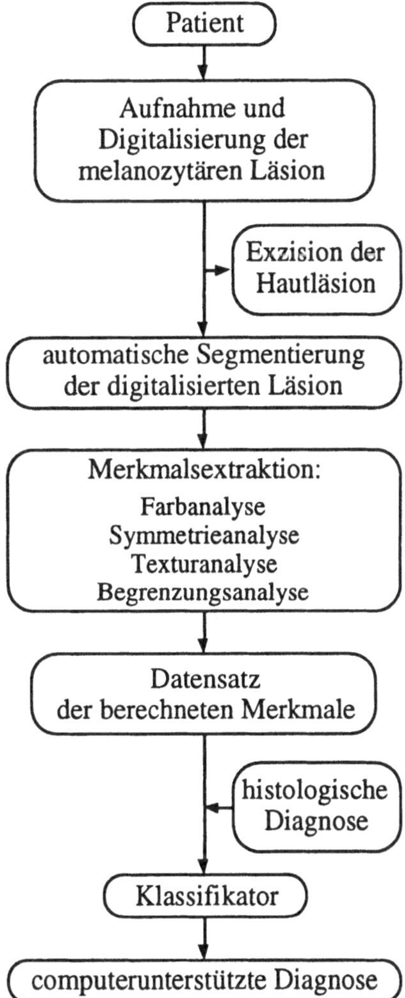

Abbildung 1: Schematischer Ablaufplan für die Analyse der melanozytären Hautläsionen in dieser Studie.

3. Bildaufnahme und Bildverarbeitung

3.1. Aufnahme, Digitalisierung

Für die direkte Digitalisierung der Hautläsionen wurde ein Aufnahmesystem entwickelt. Es beruht im wesentlichen auf einem IMCO-10-System der Firma Kontron, welches mit einer 3-Chip-CCD-Farbkamera der Firma Sony verbunden ist. Die Hautoberfläche wird mit flexiblen Lichtleitern von zwei Seiten beleuchtet. Bei einer Bildgröße von 512×512 Pixeln kann als Auflösung 25 Pixel/mm oder 44 Pixel/mm gewählt werden.
Jeder der drei Farbkanäle rot, grün und blau (RGB) wird in 8 Bit/Pixel kodiert. Zur Archivierung und zum Datenaustausch werden wiederbeschreibbare magnetooptische Platten verwendet.

3.2. Segmentierung

Die Segmentierung einer Bildszene in Läsionbereich und umgebende gesunde Haut ist eine Grundvoraussetzung für die Bildanalyse. Die Automatisierung dieses Prozesses erfordert aufgrund der in Farbe und Struktur sehr unterschiedlichen Hauttypen ein flexibles Verfahren. Ein rein manuelles Umfahren der Läsion scheidet wegen des hohen Zeitaufwandes und der subjektiven Kriterien des Beobachters aus. Experimente haben gezeigt, daß der Bereich eines Bildes mit geringer Helligkeit eine sehr gute Näherung für die Bildmaske ist. Die Berechnung des Schwellwertes zur Trennung von Läsion und umgebender Haut wird für jedes Bild wie folgt durchgeführt. Um ein optimales Grauwertbild zu erhalten, wird auf die drei Eingangsfarbkanäle die Hauptkomponentenanalyse angewendet.[1, 11] Der erste Schritt hierfür ist die Bestimmung der Kovarianzmatrix:

$$\text{col}, \text{col1}, \text{col2} \in (\text{red}, \text{green}, \text{blue}) \qquad \text{(Farben)}$$

$$m_{col} = \frac{1}{\text{columns} \cdot \text{rows}} \cdot \sum_{x=1}^{\text{columns}} \sum_{y=1}^{\text{rows}} \text{image}_{col}(x, y) \qquad \text{(Mittelwerte)}$$

$$C_{col1\,col2} = \frac{1}{\text{columns} \cdot \text{rows}} \cdot \sum_{x=1}^{\text{columns}} \sum_{y=1}^{\text{rows}} \left(\text{image}_{col1}(x, y) \cdot \text{image}_{col2}(x, y)\right) - m_{col1} \cdot m_{col2}$$

$$\text{(Kovarianzen)}$$

$$\text{Cov} = \begin{bmatrix} C_{red\,red} & C_{green\,red} & C_{blue\,red} \\ C_{red\,green} & C_{green\,green} & C_{blue\,green} \\ C_{red\,blue} & C_{green\,blue} & C_{blue\,blue} \end{bmatrix} \qquad \text{(Kovarianzmatrix)}\ .$$

Mit dem Jacobi-Verfahren können für Matrizen Näherungen für die Eigenwerte und Eigenvektoren parallel berechnet werden.[10]

Da die Eigenwerte einer quadratischen Matrix die Nullstellen ihres charakteristischen Polynoms sind, werden sie hier - im speziellen Fall einer 3×3 Matrix - mit der Cardanischen Lösungsformel exakt bestimmt.

Benutzt man die Eigenvektoren der Kovarianzmatrix als eine Transformationsmatrix für das Originalfarbbild, so lassen sich drei neue Bilder berechnen, deren Pixel weitgehend unkorreliert sind. Die berechneten Eigenwerte lassen Schlüsse auf den Informationsgehalt dieser Bilder zu: Je größer der Eigenwert, desto größer die entsprechende Bildinformation. Die Abbildung 2 zeigt die drei transformierten Bilder einer Läsion als normierte Grauwertbilder.

Die wichtigste Bildkomponente ist anschließend das Ausgangsbild für den Algorithmus zur Schwellwertbestimmung. Für alle Pixel, die im Bild unterhalb einer Testschwelle liegen, wird die Koordinatenstreuung berechnet. Diese Testschwelle wird durch den gesamten Grauwertbereich verschoben. Somit erhält man eine Funktion der Pixelstreuung in Abhängigkeit des Grauwertes. Die Koordinatenstreuung steigt stark an, wenn die Grauwertschwelle zwischen Läsion und umgebender Haut überschritten wird. Das Maximum der ersten Ableitung der ermittelten Streuungsfunktion liefert dann den Wert für die beste Schwelle zur Segmentierung. Zusätzlich müssen noch durch digitale Tiefpaßfilter Löcher in der Maske geschlossen werden. Störende Haare und Artefakte werden erkannt und gelöscht. Als letzter Schritt wird die Maskengrenzlinie geglättet.

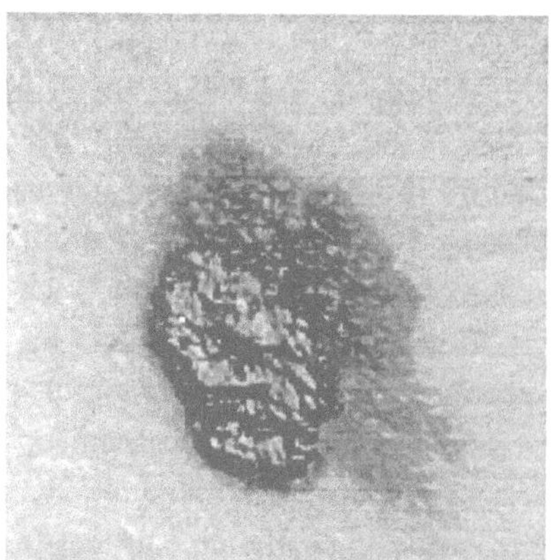

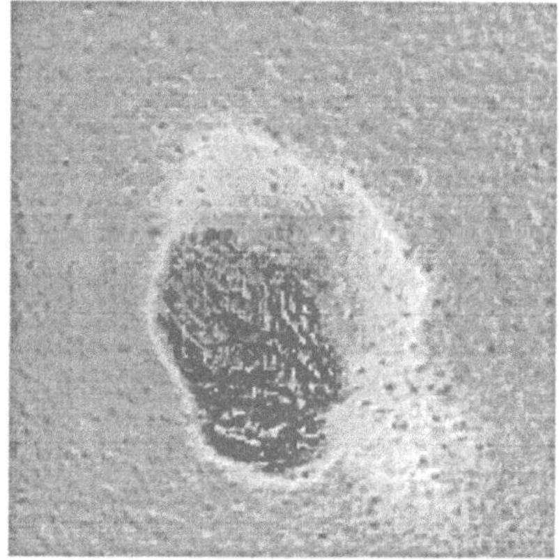

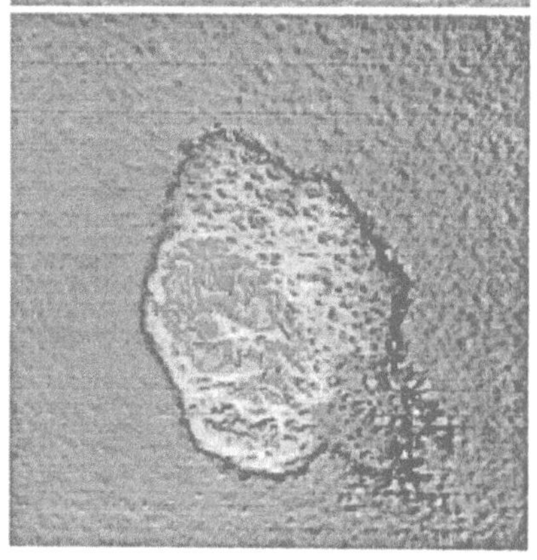

Abbildung 2: Die Bilder veranschaulichen das Ergebnis nach der Hauptkomponententransformation. Der Eigenwert für die wichtigste Hauptkomponente ganz oben ist 5260. Die zweite und die dritte Komponente darunter haben die Eigenwerte 77.55 und $-7.276 \cdot 10^{-12}$.

3.3. Merkmalsberechnung

Die Regeln der Dermatologie zur Erkennung von malignen Melanomen sind die Basis für die Auswahl und die Konstruktion der bildanalytischen Größen. Da die Signifikanz der einzelnen Merkmale im voraus nicht abschätzbar ist, werden alle berechnet, die im weitesten Sinn mit den dermatologischen Kriterien in Verbindung stehen. Die Auswahl der wirklich notwendigen Merkmalsvariablen wird dem Klassifikationsprogramm überlassen.

Neben rein geometrischen Maßzahlen (z.B. Größe, Form, Durchmesser), die sich aus der Maske der Läsion bestimmen lassen, werden weitere Merkmale berechnet, welche die Farbe, die Textur, die Symmetrie und die Begrenzung beschreiben.

Die Farbanalyse beruht auf der Auswertung statistischer Größen (Mittelwert, Standardabweichung, Maximum-Minimum, usw.) sowohl im originalen RGB-Farbraum als auch unter Verwendung des CIE-DIN-Farbsystems.[6] Die Texturmerkmale werden in einem feinen überlagerten Gitternetz ermittelt, das mit vier Gradientenfiltern und einem anschließenden Linienverfolgungsalgorithmus konstruiert wird.[8]

Um die Symmetrie zu bewerten, wird das Bild der Läsion in 4 Ringe und 8 Sektoren eingeteilt, die 32 Regionen bilden. Durch Vergleiche von Textur und Farbe innerhalb dieser Regionen lassen sich Merkmale berechnen, die eine Aussage über die Symmetrieeigenschaften geben. Um Merkmale zu erhalten, die die Begrenzung quantifizieren, werden u.a. eine Fouriertransformation des Maskenrandes durchgeführt und das Skelett der Bildmaske bestimmt.

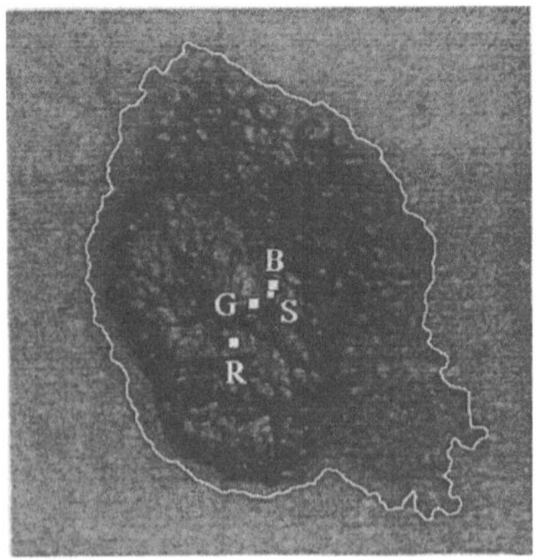

Abbildung 3: Läsion mit Maskenrand und eingezeichnetem Schwerpunkt (S), sowie den drei Farbzentren für die Bildkanäle rot, grün und blau (R, G, B). Der größte euklidische Abstand zwischen den drei Farbzentren in einer Läsion ist ein wichtiges Merkmal bei der Klassifikation. Er beträgt in diesem Bild 52 Pixel.

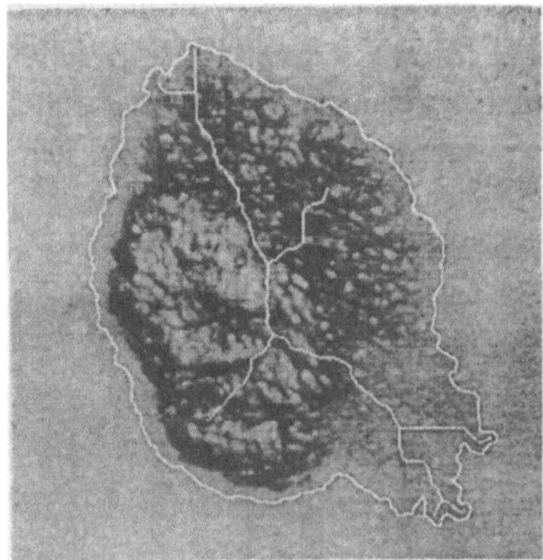

Abbildung 4: Der Maskenrand und das Skelett der Bildmaske sind dargestellt. Das Skelett wurde nach den Verfahren von Eckhardt und Maderlechner berechnet.[3] Als Merkmal dient z.B. die Anzahl der Endpunkte. In diesem Fall sind es sieben.

Als Beispiel für die Merkmalsextraktion zeigt Abbildung 3 eine Läsion mit eingezeichneten Farbzentren (Formeln im Anhang). Der größte euklidische Abstand zwischen den drei Zentren hat sich bisher als bestes einzelnes Merkmal zur Klassifikation von melanozytären Läsionen herausgestellt.
In Abbildung 4 ist das resultierende Skelett der Maskenfläche eingezeichnet.

4. Klassifikation und Ergebnisse

Zur statistischen Auswertung der berechneten Merkmale wird zur Zeit das kommerzielle Klassifikationsprogramm CART (Classification and Regression Trees) eingesetzt.[2] Dieses Programm konstruiert binäre Entscheidungsbäume. An jedem Knoten des Baumes wird der Datensatz in zwei bezüglich der Diagnose homogenere Untergruppen aufgeteilt. Zur Trennung an einem Knoten kann ein einzelnes Merkmal oder eine Linearkombination mehrerer Merkmale bestimmt werden. Zur Klassifikationsfehlerbestimmung wird das Verfahren der Cross-Validierung eingesetzt.
Bis Anfang 1993 konnten 309 direkt digitalisierte Läsionen ausgewertet werden. Davon waren 229 benigne melanozytäre Nävi und 80 maligne Melanome. Eine einfache Trennung anhand einer Linearkombination von Merkmalen ergab eine Sensitivität von 91%, eine Spezifität von 83% und eine Übereinstimmung von 85%. Die mit 10-facher Cross-Validierung berechnete Übereinstimmung war 80%. Diese Ergebnisse decken sich mit denen vorheriger Untersuchungen, die anhand von digitalisierten Farbdias durchgeführt wurden.[8, 9]

5. Ausblick

Die bereits erzielten Ergebnisse zeigen, daß es möglich ist, mit Hilfe der digitalen Farbbildanalyse die diagnostische Treffsicherheit von Dermatologen bei der Erkennung von malignen Melanomen zu erreichen und sogar zu übertreffen. Aus der statistischen Analyse der berechneten Daten lassen sich zu den bereits existierenden Regeln weitere Kriterien ableiten, die dem Dermatologen auch ohne Bildanalysesystem bei einer genaueren Auswertung der Läsionen helfen können.
Mittelfristig wird es möglich sein ein System zu etablieren, daß in der dermatologischen Routine eingesetzt werden kann, um sowohl zu dokumentieren, als auch diagnostisch zu unterstützen. Durch den Einsatz des Systems soll die Diagnostik maligner Melanome objektiver, reproduzierbarer und sicherer gemacht werden. Dies wird zu einer verbesserten Beratung und Behandlung der Patienten führen und damit zur Senkung der Melanommortalität beitragen.

6. Anhang

Die folgenden Formeln beschreiben mathematisch die Berechnung der Farbzentren $center^{col}$. Wir definieren einige Parameter und Variablen für die drei Farbkanäle rot, grün und blau:

$$
\begin{aligned}
&columns = 512 \\
&rows = 512 \\
&x \in \{1,..,columns\} \\
&y \in \{1,..,rows\} \\
&coord \in \{x,y\} \\
&col \in \{red, green, blue\} \\
&image_{col}(x,y) \in \{0,..,255\} \quad .
\end{aligned}
$$

Die Funktion sgn_{mask} beschreibt die Maske der segmentierten Läsion. Der Wert von sgn_{mask} ist 1 für alle Bildpixel, die innerhalb des Maskengebietes liegen und 0 sonst:

$$sgn_{mask}(x,y) = \begin{cases} 1 & \text{, wenn Pixel} \in \text{Läsionfläche} \\ 0 & \text{, wenn Pixel} \in \text{umgebende Haut} \end{cases} \quad .$$

Als Zwischenwerte werden für jeden Farbkanal die Variablen a_x^{col}, a_y^{col} und b^{col} eingeführt:

$$
\begin{aligned}
a_{coord}^{col} &= \sum_{x=1}^{columns} \sum_{y=1}^{rows} coord \cdot \frac{1}{image_{col}(x,y)+1} \cdot sgn_{mask}(x,y) \\
b^{col} &= \sum_{x=1}^{columns} \sum_{y=1}^{rows} \frac{1}{image_{col}(x,y)+1} \cdot sgn_{mask}(x,y) \quad .
\end{aligned}
$$

Diese Formeln sind denen für die Schwerpunktberechnung einer geometrischen Fläche ähnlich. Durch den zusätzlich eingeführten Bruch werden die dunklen Bildpixel innerhalb der Bildmaske stärker gewichtet. Der Summand (+1) im Nenner verhindert eine Division durch 0. Mit den Variablen a und b lassen sich die Koordinaten

$$center_{coord}^{col} = \left[\frac{1}{b^{col}} \cdot a_{coord}^{col} \right]$$

leichter berechnen. Die drei Farbzentren für die Farbkanäle rot, grün und blau schreiben sich schließlich als:

$$center^{col} = \left(center_x^{col}, center_y^{col} \right) \quad .$$

7. Literatur

1. Jahresbericht 1976. Forschungsinstitut für Informationsverarbeitung und Mustererkennung, Karlsruhe, 1977.
2. Breiman L, Friedman JH, Olshen RA, Stone CJ: Classification And Regression Trees. Wadsworth, Inc., Belmont, California, 1984.
3. Eckhardt U, Maderlechner G: Invariant Thinning. Hamburger Beiträge zur Angewandten Mathematik, Reihe A, Preprint 65. Institut für Angewandte Mathematik der Universität Hamburg, Hamburg, 1993.
4. Fitzpatrick TB: Hautkrebs. In: Das Krebsbuch der American Cancer Society, Holleb AI (Ed). Rowohlt, Reinbek, 1990, pp. 653-673.
5. Friedman RJ, Rigel DS, Kopf AW: Early Detection of Malignant Melanoma: The Role of Physician Examination and Self-Examination of the Skin. CA 35:130-151, 1985.
6. Lang H: Farbmetrik und Farbfernsehen. Einführung in die Nachrichtentechnik, R. Oldenbourg Verlag, München, 1978.
7. Rassner G: Früherkennung des malignen Melanoms der Haut. Hautarzt 39:396-401, 1988.
8. Schindewolf T, Stolz W, Albert R, Abmayr W, Harms H: Classification of Melanocytic Lesions with Color and Texture Analysis Using Digital Image Processing. Anal Quant Cytol Histol 15:1-11, 1993.
9. Schindewolf T, Stolz W, Albert R, Abmayr W, Harms H: Comparison of classification rates for conventional and dermatoscopic images of malignant and benign melanocytic lesions using computerized colour image analysis. EJD 3:299-303, 1993.
10. Stoehr J, Bulirsch R: Einführung in die Numerische Mathematik II. Springer Verlag, Berlin, 1978.
11. Umbaugh SE, Moss RH, Stoecker WV: An automatic color segmentation algorithm with application to identification of skin tumor borders. Comput Med Imaging Graph 16:227-235, 1992.
12. Weidner F, Hornstein OP, Bischof GM: Zur Treffsicherheit der klinischen Diagnose bei malignen Hautmelanomen. Dermatol. Monatsschr. 169:706-710, 1983.

Mustererkennung bei auflichtmikroskopischer Untersuchung pigmentierter Hauttumoren

A. Stockfisch, J. Kreusch, H.-J. Friedrich
Med. Univ. Lübeck, Klinik für Dermatologie und Institut für Medizinische Statistik und Dokumentation

Einleitung

Bei auflichtmikroskopischer Untersuchung von Pigmenttumoren der Haut sind eine Reihe von Strukturen erkennbar, die dem unbewaffneten Auge verborgen bleiben. Unter Ölimmersion lassen sich Blutgefäße und Pigmentstrukturen in der Epidermis und im oberen Korium viel genauer erkennen. Diese Befunde scheinen jedoch zu einer sehr viel besseren diagnostischen Genauigkeit zu führen, als dies mit der üblichen klinischen Diagnosestellung möglich ist. Es ist für die noch recht junge Untersuchungstechnik noch nie untersucht worden, inwieweit diese Befunde allgemein erkennbar sind, und ob sie auch von wenig erfahrenen Untersuchern konstant erhoben werden können. Auch eine universell akzeptierte Nomenklatur hat sich noch nicht durchsetzen können.

Fragestellung

Ziel der vorliegenden Arbeit war es, für die auflichtmikroskopisch beobachteten Phänomene eine auch Anfängern verständliche Nomenklatur zu entwickeln. Weiterhin sollte die Konstanz der Erkennung der verschiedenen Befunde durch ungeübte Probanden überprüft und mit Expertenmeinung verglichen werden.

Material und Methoden

Zunächst wurden aus der Literatur die für gleichartige Befunde verwendeten unterschiedlichen Bezeichnungen zusammengestellt (1, 2, 3, 4, 5, 6, 7). Um hieraus die am besten passenden Begriffe auszuwählen, wurden Probeaufnahmen von Pigmenttumoren verschiedenen Testpersonen mit unterschiedlichem dermatologischem Wissensstand vorgelegt und diese gebeten, den am besten zutreffenden Begriff für insgesamt 21 Befunde auszuwählen. Zu den Testpersonen gehörten 12 dermatologisch unerfahrene Mediziner (Internisten, Studenten im letzten Abschnitt der medizinischen Ausbildung) sowie die gleiche Anzahl an Ärzten mit fast abgeschlossener dermatologischer Ausbildung.
Aus den gewählten Begriffen (wahlweise mit einem häufig genannten Alternativbegriff) wurde ein Fragebogen mit insgesamt 21 Fragen erstellt. Dieser Fragebogen sollte dazu dienen, die Erkennung bestimmter Merkmale durch 13 Testpersonen zu ermitteln. Die Probanden

(Medizinstudenten im 2 klinischen Semester) wurden unter Verwendung der ermittelten Präferenzbegriffe in einem audiovisuellen Kurs in die auflichtmikroskopische Untersuchungsmethodik eingeführt. Die wesentlichen Befunde an Pigmenttumoren wurden dabei demonstriert. Nach einem Zeitintervall von einigen Wochen erhielten die Probanden auflichtmikroskopische Aufnahmen von insgesamt 33 Pigmenttumoren und jeweils einen Fragebogen pro Fall vorgelegt. Dabei handelte es sich bei fast jedem Tumor um Übersichtsaufnahmen ohne und mit Ölimmersion, sowie um eine Detailaufnahme; zusammen also um möglichst drei Aufnahmen (Hochglanz-Farbabzüge von Dias, Format 13 x 18 cm) je Fall. Die Reihenfolge der Abbildungen wurde nach Zufallskriterien von Proband zu Proband variiert. Sie wurden gebeten, zur Anwesenheit von insgesamt 21 Befunden ja-nein Entscheidungen zu treffen. Die Probanden sollten außerdem entscheiden, ob ihrer Meinung nach das jeweils gegebene Pigmentmal aus Melanozyten bestehe. Die Kriterien hierfür sind lediglich die Anwesenheit eines melanozytentypischen Pigmentmusters und eines Farbtones (s.Kreusch 1990). Eine abschließende Diagnose wurde nicht verlangt.
Die Ergebnisse wurden verglichen mit der Meinung zweier auflichtmikroskopisch erfahrener Beurteiler.
Den Testpersonen wurden die gleichen Farbbilder nach einem Zeitabstand von zwei bis vier Wochen erneut zur Beurteilung vorgelegt.

Die abgefragten Befunde waren folgenden Kategorien zuzuordnen:

- Geometrie des Tumors (4 Befunde)
- Tumoroberfläche (3 Befunde)
- Pigmentmuster (7 Befunde)
- Farbtöne und -sättigung (2 Befunde)
- Einheitlichkeit des Tumoraufbaus (2 Befunde)
- Struktur der tumoralen Blutgefäße (3 Befunde)
- Frage nach melanozytentypischer Hautveränderung

Die Auswertung der Antworten erfolgte in Zusammenarbeit mit dem Institut für medizinische Informatik der Med. Universität Lübeck.

Ergebnisse

Die Häufigkeit der Antworten sind folgenden Tabellen zu entnehmen:

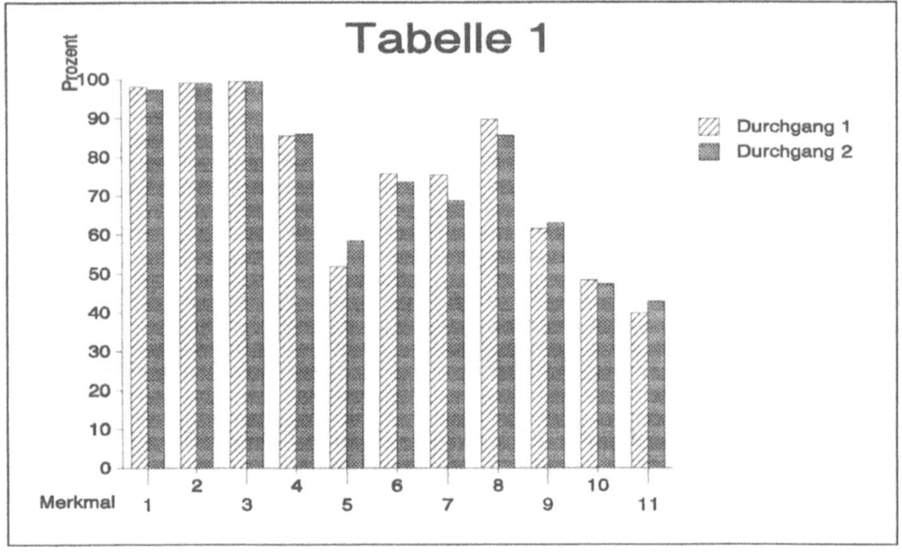

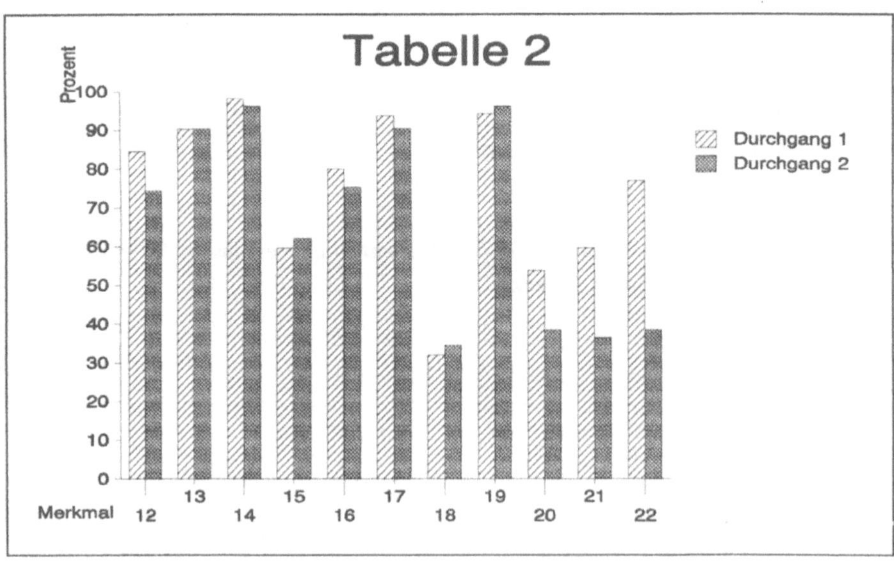

Merkmale:

1) Durchmesser > 5mm (n=28)
2) Unregelmäßige Begrenzung (n=27)
3) Asymmetrie (n=28)
4) Hautfelderung verändert / Reliefstörung (n=27)
5) Schuppen, Erosion, Ulzeration (n=15)
6) Pigmentverdichtungen in der oberen Epidermis (n=21)
7) Pigmentmal über Hautniveau (n=20)
8) Retikuläres Muster (n=17)
9) Globuläres Muster (n=5)
10) Scholliges Muster (n=7)
11) Homogene Pigmentierung / einheitliche Pigmentierung (n=21)
12) Pseudopodien (n=3)
13) Melanophagen, Melanophagenagglomerate (n=4)
14) Melanozytentypische Hautveränderung (n=25)
15) Grauer Farbton (n=12)
16) Pigmentgehalt erhöht (n=24)
17) Inhomogener Aufbau (n=21)
18) Polyklonaler Aufbau, verschiedene, klar voneinander abgrenzbare Wachstumsmuster (n=6)
19) Regressive Veränderungen (n=4)
20) Punktförmige Kapillaren (n=3)
21) Kommaförmige Kapillaren (n=4)
22) Breite, verzweigte, aufliegende Gefäße (n=1)

n: Anzahl der Fälle, in denen das Merkmal erkennbar war

Die Tabellen 1 und 2 zeigen die prozentuale Häufigkeit, mit der die Probanden die Merkmale erkannt haben, bei tatsächlichem Vorhandensein der Befunde laut Urteil erfahrener Untersucher.
Das Merkmal "Unregelmäßige Begrenzung" war beispielsweise in 27 Fällen feststellbar. Von 13 Testpersonen hätte es also höchstens 351mal (27x13) identifiziert werden können. Da es schließlich 348mal in jeweils beiden Durchgängen erkannt wurde, ergibt sich eine Häufigkeit von 99,1 %.
Die Probanden konnten bis auf ganz wenige Ausnahmen zu sämtlichen Punkten des Fragebogens Stellung nehmen. Es zeigte sich, daß Fragen zur Geometrie des Tumors gut und einheitlich beantwortet werden konnten. Hingegen war die Beurteilung der Tumoroberfläche schwieriger, insbesondere die Frage nach Epitheldefekten, sowie nach Pigmentanreicherungen in den oberen Hautschichten wurde unterschiedlich beantwortet. Die Fragen zur Einheitlichkeit des Tumoraufbaus zeigten Probleme bei dem Merkmal "Polyklonaler Aufbau", welches nur in 32,1 % (1.Durchgang) bzw. 34.6% (2.Durchgang) erkannt wurde. Dagegen ergab die Auswertung ausgesprochen gute Ergebnisse bei der Frage nach dem "Inhomogenen

Aufbau"(93,8% bzw. 90,5%). Erwartungsgemäß traten Probleme bei der Bewertung von Farbtönen und dem Sättigungsgrad der Farben auf. Unerwartet war die Tatsache, daß es Schwierigkeiten gab, die Pigmentmuster stets einheitlich zu beurteilen. Dies betraf allerdings überwiegend Befunde, die häufiger fließende Übergänge zwischen verschiedenen, verwandten Strukturen zeigten, aber welche alle melanozytischen Ursprungs waren, so daß geringe Abweichungen die Einschätzung "melanozytisches Pigmentmal" nicht beeinflußten. Auch die Vaskularisierungsmuster waren nicht leicht und immer eindeutig zu beurteilen. Hierbei muß allerdings die geringe Häufigkeit in den gezeigten Fällen sowie die problematische Wiedergabe sehr feiner Strukturdetails im roten Spektralbereich berücksichtigt werden, die in Diapositiven stets klarer als auf Papierbildern zu erkennen waren.

Auf die Frage, ob ein Tumor aus Melanozyten bestehe (Frage nach Merkmal 14), wurde zu häufig auch bei nichtmelanozytischen Tumoren (v.a. Basaliomen) mit "ja" geantwortet. Äußerst selten (1,8% bzw. 3,7%) wurde dagegen ein melanozytischer Tumor als nichtmelanozytisch bewertet. Dies ist entscheidende Voraussetzung dafür, maligne Melanome nicht als harmlosere nichtmelanozytische Tumoren falsch einzustufen.

Wesentlich erscheint das Ergebnis, daß die aus Voruntersuchungen bekannten melanomspezifischen Merkmale (Pseudopodien, Melanophagen, Regressive Veränderungen) mit hoher Konstanz erkannt wurden.

Der Schwierigkeitsindex (relativer Anteil der richtigen Antworten gemessen am Gesamtkontingent der beantworteten Fragen) lag im Bereich 0.400 - 0,599 (geringe Übereinstimmung der Probanden) bei folgenden Merkmalen: Schuppen, Erosion, Ulzeration, Pigmentverdichtungen in der oberen Epidermis, Pigmentmal über Hautniveau, Globuläres Muster, Grauer Farbton.

Ein niedriger Schwierigkeitsindex von 0,000 - 0.199 ergab sich bei der Frage nach dem Befund "Homogene Pigmentierung". Somit war bei diesem Merkmal die Definition unzureichend und der Befund schwer erkennbar.

Die intraindividuelle Konstanz der ungeübten Testpersonen war sehr hoch. Pro Fragebogen erzielten die Probanden Ergenisse von durchschnittlich 75% - 100% gemessen an den Referenzwerten der erfahrenen Beurteiler.

Diskussion

Die Untersuchung ist ein erster Schritt zur Erstellung einer einheitlichen Nomenklatur für die noch junge Untersuchungstechnik "Auflichtmikroskopie" dar. Die bestehende uneinheitliche Begriffsverwendung stellt ein nicht zu unterschätzendes Hindernis für die Verbreitung der Methode dar.

Die semantische Problematik ist von nicht unerheblicher Bedeutung, da in der bisher vorliegenden Literatur die Vielfalt an Termini für gleiche Phänomene der Auflichtmikroskopie zu Schwierigkeiten, insbesondere bei der Vermittlung der Merkmalskriterien führt.

Es kann jedoch gezeigt werden, daß viele der gewählten Befunde gut zu erklären und eher leicht zu erlernen sind. Probleme und interobservatorische Streuung ergaben sich vor allem bei stark subjektiver Einschätzung unterliegenden Merkmalen, vor allem Farbtöne und -sättigung betreffend. Eine einheitliche Nomenklatur scheint besonders für Tumoroberflächenmerkmale wünschenswert. Die Schwierigkeiten, Pigmentmuster korrekt einzuordnen, schränkt die Verwertbarkeit der Methode solange nicht ein, wie die Erkennung melanozytentypischer Pigmentmale nicht beeinträchtigt wird. Dies ist unabdingbare Anforderung an eine hohe diagnostische Treffsicherheit bei der Erkennung maligner Melanome. Die hierzu wichtige Identifikation melanomspezifischer Merkmale war dagegen durchaus befriedigend.

Die Ergebnisse lassen den Schluß zu, daß sich bei guter und interindividuell konstanter Beurteilung wesentlicher Merkmale in pigmentierten Hauttumoren die Befunde standardisiert und automatisch auswerten lassen, so daß dem Untersucher nicht unbedingt eine Diagnose abverlangt werden muß. Somit ließe sich mit Hilfe eines einfachen Rechenverfahrens eine diagnostische Zuordnung treffen. Eine objektive Befunderstellung für solche Tumoren scheint sich abzuzeichnen.

Literatur

1. Kreusch, J., Rassner, G., Henke, D., Pietsch-Breitfeld, B., Selbmann, H.K., (1990) Bedeutung der Auflichtmikroskopie zur Früherkennung des malignen Melanoms. In: Das maligne Melanom der Haut. Hrsg. C.E. Orfanos, C. Garbe, W. Zuckschwerdt Verlag, München, Bern, Wien, San Francisco,pp 68-79

2. Kreusch, J., Rassner, G., (1991) Standardisierte auflichtmikroskopische Unterscheidung melanozytischer und nichtmelanozytischer Pigmentmale. HAUTARZT 42: 77-83

3. Kreusch, J., (1992) Merkmalsanalyse melanozytischer Hautveränderungen mittels Auflichtmikroskopie. Inaugural- Dissertation

4. Bahmer, F.A., Fritsch, P., Kreusch, J., Pehamberger, H., Rohrer, C., Schindera, I., Smolle, J., Soyer, H.-P., Stolz, W., (1990) Diagnostische Kriterien in der Auflichtmikroskopie. Konsensus-Treffen in Hamburg. HAUTARZT 41: 513-514

5. Steiner, A., Pehamberger, H., Wolff, K., (1987) In vivo epiluminescence microscopy of pigmented skin lesions. II. J AM ACAD DERMATOL 17: 584-91

6. Braun-Falco, O., Stolz, W., Bilek, P., Merkle, T., Landthaler, M., (1990) Das Dermatoskop. HAUTARZT 41: 131-136

7. MacKie, R.M., (1972) Cutaneous microscopy in vivo as an aid to preoperative assesment of pigmented lesions of the skin. BR J PLAST SURG 25: 123-129

Segmentierung von Röntgenbildern fokaler Knochenläsionen durch neuronale Netzwerke - Optimierung durch Quality Metrics und modifizierte Contribution Analysis *

Vogelsang[1], F., Pelikan[1], E., Egmont-Petersen[2], M., Tolxdorff[3], T., Bohndorf[4], K.

[1]Institut für Medizinische Informatik und Biometrie, RWTH Aachen (Prof. Dr. med. Dipl. math. R. Repges)
[2]Institut of Computer and System Science DASY, Kopenhagen
[3]Institut für Med. Statistik und Informationsverarb. / Bereich Med. Informatik, FU Berlin (Prof. T. Tolxdorff)
[4]Institut für Röntgendiagnostik, Zentralklinikum Augsburg (Prof. Dr. med K. Bohndorf)

1 Einleitung

Mit dem vorgestellten Verfahren sollen radiologisch dokumentierte fokale Knochenläsionen aufgrund verschiedener im Filmröntgenbild unterscheidbarer morphologischer Strukturen segmentiert werden. Dies geschieht unter Verwendung einer Kombination von Bildverarbeitungsalgorithmen und Klassifikatoren (Neuronale Netzwerke vom Typ Backpropagation) zur Erkennung von die Textur implizit beschreibenden Vektoren. Existenz, Klassenzugehörigkeit und die räumliche Relation dieser morphologischen Strukturen zueinander können dann zur medizinische Interpretation herangezogen werden.

Zunächst wird das Eingabebild durch eine Bildvorverarbeitung um (n-1) abgeleitete Bilder ergänzt, wobei die Bildverarbeitungsoperationen so ausgewählt wurden, daß sie bestimmte medizinisch relevante Merkmale im Bild überhöhen. Somit wird jedem Bildpunkt des ursprünglichen Grauwertbildes G ein n-dimensionaler Grauwertvektor $m_{xy} = (m_{xy1}, ..., m_{xyn})$ zugeordnet. Die erhaltenen Merkmalvektoren m_{xy} spannen einen n-dimensionalen Merkmalraum auf. Durch die Vektorisierung werden kontext- und strukturbasierte Informationen aus größeren Bildregionen zusammengeführt, über relevante Bereiche zu Merkmalen der morphologischen Struktur zusammengefaßt, und dem Klassifikator zur Verfügung gestellt. Die aus einem Grauwertbild G extrahierten Merkmalvektoren m_{xy} bilden die Eingabe des zur Segmentierung verwendeten Klassifikators, der zu jedem Vektor durch eine nichtlineare Transformation einen Klassenindex berechnet und damit die Zugehörigkeit zu einer von m bestimmten morphologischen Strukturen kennzeichnet.

Für das Training des Netzwerks werden vom Radiologen erstellte exemplarische Befundungs- oder Lernmasken verwendet. Sie wurden auf Basis einer durch Lodwick [Lod65, Lod66, Lod71, Lod80] und Bohndorf [Boh93, Pel93] entwickelten Konvention zur systematischen Deskription von Knochenläsionen erstellt. Bild 3.1 zeigt eine verwendete Lernmaske.

Durch anschließende Qualitätsbetrachtungen soll die Qualität der erzielten Segmentierungen verbessert werden. Die Optimierung des Merkmalraums erfolgt mittels Qualitätsmaßen [Egm93] und einer modifizierten Contribution Analysis. Die Contribution Analysis mißt den Einfluß einzelner Komponenten der Merkmalvektoren respektive Bildverarbeitungsoperationen auf die Zuordnung der Vektoren zu den Strukturklassen. In der kumulativen Darstellung können dann die Eingabebilder hinsichtlich ihrer Relevanz bzw. ihres Informationsgehalts für die Klassifikation bewertet werden (Bild 1.1).

* DFG-Projekt TO 108/3-2: "Wissensbasierte Bildanalyse in der Diagnostik von Knochenprozessen"

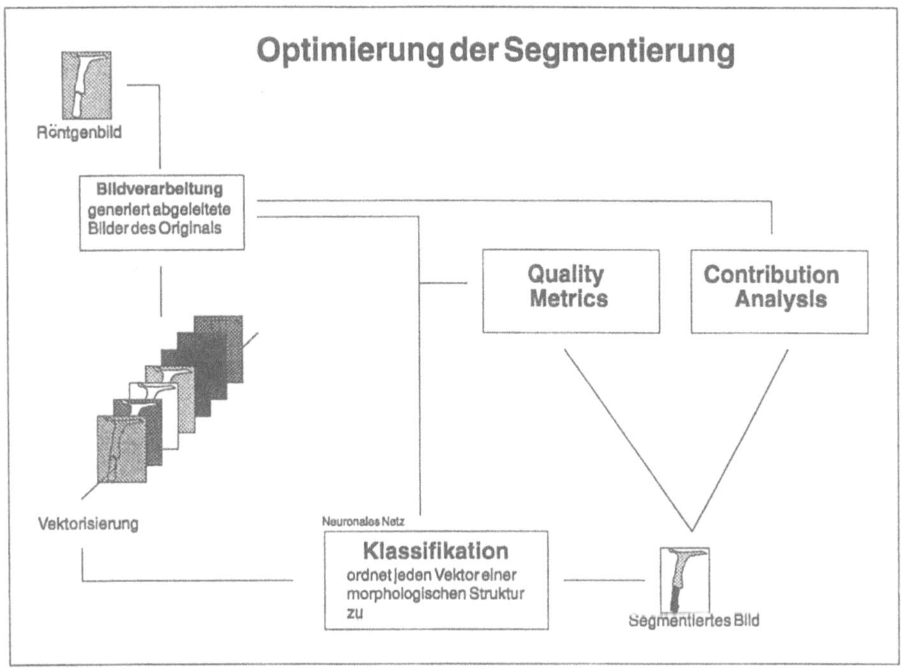

Bild 1.1: Vektorisierung und Optimierung des Verfahrens

2 Optimierung der Segmentierung

2.1 Qualitity Metrics

Es wurden durch Egmont-Petersen [Egm93] anhand von Kontingenztafeln [Har92] definierte Qualitätsmaße verwendet. Sie bieten gegenüber klassischen Maßen wie dem quadratischen Gesamtfehler [Moo92] oder dem Informationskriterium von Akaike [Aka74, Fog91] eine problemadäquate Qualitätsbeobachtung. In [Egm93] werden auch die Streuung und Verschiebung zur Untersuchung von Diskreminationsproblemen als Maße aus der parametrischen Statistik definiert.

2.1.1 Kontingenztafel

Die Eigenschaften eines Klassifikators werden in einer (m+1) x (m+1) - dimensionalen Kontingenztafel erfaßt (Bild 2.1). In ihr werden tatsächlicher Klassenindex und gefundener Klassenindex sowie die marginalen Verteilungen gegeneinander aufgetragen. Die Menge W enthält die Merkmalvektoren m_{xy}, zu denen ein der zugehörigen Maske zu entnehmender Zielvektor z_{xy} existiert. Der Eintrag ct_{ij} gibt die Anzahl der Merkmalvektoren aus W an, die durch den Klassifkator als zur Klasse K_i gehörend erkannt wurden, aber zur Klasse K_j gehören.

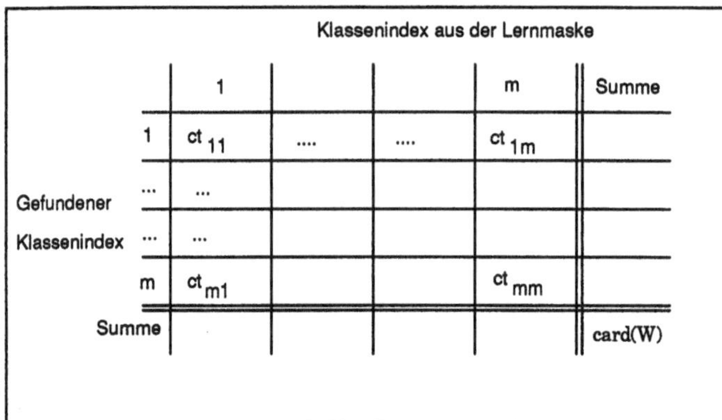

Klassenindex aus der Lernmaske

		1			m	Summe
	1	ct_{11}	....	....	ct_{1m}	
Gefundener	...	...				
Klassenindex	...	...				
	m	ct_{m1}			ct_{mm}	
	Summe					card(W)

Bild 2.1: Aufbau einer Kontingenztafel

Auf der Diagonalen der Matrix CT befinden sich die Anzahlen der korrekt klassifizierten Vektoren. Die (m+1)-te respektive letzte Spalte der Matrix gibt die Verteilung der zugewiesenen Klassenindizes an, die (m+1)-te Zeile die Verteilung der realen, der Maske entnommenen, Klassenindizes.

2.1.2 Korrektheit ρ

Die *Korrektheit* ρ gibt den Anteil der gemäß ihres Maskeneintrags klassifizierten Merkmalvektoren an der Summe der Merkmalvektoren an ($\rho \in [0,1]$; 1 : maximale Korrektheit). Die Definition lautet entsprechend :

$$\rho = \frac{r}{r+w}\ ,wobei\ r = \sum_{i=1}^{m} ct_{ij}\ \ und\ \ w = \sum_{i=1}^{m}\ \sum_{j=1,j\neq i}^{m} ct_{ij} \tag{2.1}$$

Hierbei ist r die Summe aller vom Klassifikator richtig klassifizierten Vektoren (Elemente der Hauptdiagonalen), und w die Summe aller nicht richtig klassifizierten Vektoren. Durch die Korrektheit wird die Übereinstimmung zwischen den beiden in der Kontingenztafel vorhandenen marginalen Verteilungen gemessen. Um die Klassifikationsleistung einzelner Klassen genauer zu untersuchen, wurde die bedingte Korrektheit ρ_i wie folgt definiert :

$$\rho_i = \frac{r_i}{r_i + w_i}\ ,wobei\ \ r_i = ct_{ii}\ \ und\ \ w_i = \sum_{j=1,j\neq i}^{m} ct_{ji} \tag{2.2}$$

Hierbei ist r_i die Zahl der korrekt in Klasse i klassifizierten Merkmalvektoren und w_i die Anzahl der Vektoren aus K_i, die einer anderen Klasse zugewiesen wurden. Für die bedingte Korrektheit ρ_i gilt die gleiche Metrik, wie für die Korrektheit ρ.

2.1.3 Genauigkeit κ

Die *Genauigkeit* κ ist ein ähnliches Maß wie die Korrektheit ρ [Coh60]. Sie erlaubt eine Aussage darüber, ob ein Klassifikator Vektoren zufällig richtig klassifiziert, indem sie die Randverteilung der Vektoren der (m+1)-ten Spalte der Kontingenztafel CT berücksichtigt. Die erwartete Anzahl der Vektoren in der Diagonale der Matrix wird hierbei als Referenz verwendet. κ mißt also den Grad der Übereinstimmung zwischen der Klassenzugehörigkeitsverteilung in W und der Klassifikation, die durch den Klassifikator vorgenommen wurde. Hierzu wird zunächst die erwartete Häufigkeit e in den diagonalen Positionen der Kontingenztafel CT, also die erwartete Größe für r (2.1) bestimmt. Sie berechnet sich nach:

$$e = \sum_{k=1}^{m} \frac{\sum_{i=1}^{m} ct_{ik} \cdot \sum_{j=1}^{m} ct_{kj}}{N^2} \tag{2.3}$$

wobei N = w + r = card(W). Die Größe e wird durch die erwartete Übereinstimmung zwischen den Verteilungen in der (m+1)-ten Spalte respektive Zeile von CT bestimmt, während die Korrektheit ρ die wirkliche Übereinstimmung mißt. Die Genauigkeit κ ist definiert durch

$$\kappa = \frac{\rho - e}{1 - e} \tag{2.4}$$

Wenn die Genauigkeit nahe ihrem Maximum ist, bedeutet dies eine gute Leistung des Klassifikators, Werte für κ nahe bei 0 verweisen auf eine zufällige Klassifikation ($\kappa \le 1$, für m=2 gilt $\kappa \in [-1;1]$. Für negative Werte der Genauigkeit läßt sich der Klassifikator hinsichtlich seiner Klassifikationsleistung als ungenügend einstufen. Nach [Fle81] sind Werte von $0.4 \le \kappa \le 0.75$ als gut einzustufen, Werte oberhalb von 0.75 als sehr gut.

Ebenso wie eine bedingte Korrektheit kann auch eine bedingte Genauigkeit κ_i angegeben werden, um die Klassifikationsleistung bezüglich einzelner Klassen zu untersuchen. Sie berechnet sich nach

$$\kappa_i = \frac{\rho_i - e_i}{1 - e_i} \quad mit \quad e_i = \sum_{j=1}^{m} \frac{ct_{ji}}{N} \tag{2.5}$$

wobei e_i die bedingte marginale Wahrscheinlichkeit für die Zugehörigkeit eines Merkmalvektors zur Klassenmenge K_i ist. Die Metrik der bedingten Genauigkeit entspricht der der Genauigkeit κ. Damit ergänzen sich die Maße Korrektheit und Genauigkeit zu einer guten Möglichkeit, Klassifikatoren zu bewerten.

2.2 Contribution Analysis

Die Contribution Analysis mißt die Änderung einer Ausgabeeinheit in Abhängigkeit von dem dem Klassifikator präsentierten Merkmalvektor m_{xy}. Untersuchungen zur Relevanz einzelner Einheiten der Hidden-Layer eines neuronalen Netzwerkes wurden bereits in [Moz89] und [San89] durchgeführt. In [Har91] wird ein Verfahren zur Bestimmung des Einflusses (Contribution) einzelner Eingabeeinheiten auf die Ausgabeeinheiten beschrieben. Dieses wurde modifiziert um eine Beurteilung und Auswahl von Merkmalvektorkomponenten zu erlauben.

Es wird zunächst der Einfluß einer Komponente m_{xyi} eines Merkmalvektors m_{xy} auf die Ausgabe einer Ausgabeeinheit j, insbesondere auf die Klassifizierung des Merkmalvektors m_{xy} als zur Klasse j gehörend, bestimmt. Der Einfluß C_{xyji} einer Eingabeeinheit i auf eine Ausgabeeinheit j für einen Merkmalvektor m_{xy} aus dem Merkmalraum M^n läßt sich als Änderung der Aktivität der Ausgabeeinheit bei Anlegen des Vektors m_{xy} messen [Har91].

$$C_{xyij} = \frac{do_{xyj}}{di_{xyi}} = \frac{do_{xyj}}{dneto_j} \cdot \frac{dneto_j}{di_{xyi}} \tag{2.6}$$

wobei o_{xyj} die Ausgabe der j-ten Ausgabeeinheit, i_{xyi} die Eingabe der i-ten Eingabeeinheit, d.h. die i-te Komponente des Merkmalvektors m_{xy} und $neto_j$ die Eingabe der j-ten Ausgabeeinheit ist. Es folgt damit insgesamt für den Einfluß C_{xyij} mit der verwendeten Aktivierungsfunktion tanh:

$$C_{xyij} = \frac{1}{\cosh^2(neto_j)} \cdot \sum_{k=1}^{h} (wo_{kj} \cdot \frac{1}{\cosh^2(neth_k)} \cdot wh_{ik}) \tag{2.7}$$

Es folgt

$$C_{xyij} = (1 - o_{xyj}{}^2) \cdot \sum_{k=1}^{h} (wo_{kj} \cdot (1 - h_{xyk}{}^2) \cdot wh_{ik}) \quad mit \quad \cosh^2(x) = \frac{1}{1 - \tanh^2(x)} \tag{2.8}$$

Für Klassifikationsaufgaben mit einer kleinen Menge von Merkmalvektoren ist dieses Maß ausreichend, da die C_{xyij} aufgrund ihrer relativ kleinen Anzahl in ihrer Gesamtheit interpretiert werden können. Es soll hier jedoch sukzessive der Einfluß jedes der abgeleiteten Bilder insgesamt, d.h. der Einfluß über alle $m_{xy} \in M_G \subset M^n$, wobei M_G die zu einem Röntgenbild G abgeleiteten Merkmalvektoren seien, auf die Klassifikation bestimmt werden.

Sei $M_G{}^i$ die Menge der Merkmalvektoren aus M_G, die als zur Strukturklasse K_j gehörig klassifiziert wurden. Dann ist C_{ij} definiert als Einfluß der Komponente i der Merkmalvektoren, d.h. aller an der Eingabeeinheit i vorkommenden Werte m_{xyi} respektive des abgeleiteten Bildes G_i auf die Klassifikation als Klasse j wie folgt:

$$C_{ij} = \frac{1}{card\,(M_G^i)} \cdot \sum_{m_{xy} \in M_G^i} C_{xyij} \tag{2.9}$$

Für die Merkmalvektoren eines Röntgenbildes, zusammengefaßt in der Menge M_G, läßt sich nun eine Contribution Matrix C_G definieren, an deren Stelle (j,i) die Contribution C_{ij} des i-ten Eingabebildes auf die Klassifikation als zur Klasse j gehörig steht ($C_{ij} < 0 \Rightarrow$ Eingabebild i hat hemmenden Einfluß auf die Klassifikation von m_{xy} zur Klasse j; $C_{ij} > 0 \Rightarrow$ Eingabebild i hat fördernde Wirkung). Kleine Werte für C_{ij} deuten auf einen geringen Einfluß der Vektorkomponente i zur Klassifikation als zur Klasse j gehörend hin ([San89]). Die Matrix C_G ermöglicht eine differenzierte Bestimmung der Einflüsse der assoziierten Bilder auf die Klassifikation bestimmter morphologischer Strukturklassen und damit eine gezielte Korrektur von Fehlklassifikationen durch einen geeigneten Austausch der Bildverarbeitungsoperationen.

3 Experimentelle Ergebnisse

In einem Marvin III genannten Versuch wurde mit der in folgender Tabelle aufgeführten Klassendefinition eine in Bild 3.1 dargestellte Segmentierung erreicht. Die in der Tabelle aufgeführten Strukturklassen 1 (Multiple Läsion) und 4 (Sklerotischer Randsaum) sind in den dargestellten Segmentierungen und Röntgenbildern nicht enthalten. Für sie können daher bei den anschließenden Betrachtungen keine Quality Metrics und Contributions berechnet werden.

Klassendefinition Marvin III		
Klassenindex	**Beschreibung**	**Farbe**
0	Hintergrund	schwarz
1	Multiple Läsion	magenta
2	Knochenveränderung vorhanden	weiß
3	Tumor / Knochenübergang	rot
4	Sklerotischer Randsaum	dunkelgrün
5	Reaktion der Kortikalis	gelb
6	Periostreaktion solide	blau
7	Periostreaktion lamellär	grün
8	Expansion des Knochens	orange
9	Weichteilinfiltration vorhanden	cyan
10	Knochen / Knorpelneubildung	braun
11	Hintergrund	dunkelgrau
12	Gesunder Knochen	grau

3.1 Quality Metrics und Contribution Analysis

Die durchgeführten Qualitätstests für die Segmentierung des Röntgenbilds (Bild 3.1 rechts) ergab für die Strukturklassen 5, 6, 7 und 8 nur unbefriedigende Werte für die bedingten Korrektheiten ρ_i und Genauigkeiten κ_i (Bild 3.6, net_0_B). Offensichtlich konnte dem Klassifikator zur Klassifikation dieser Klassen keine ausreichende Information zur Verfügung gestellt werden.

Mit 8 Klassifikatoren, die laut Qualitätsmessungen gute Segmentierungen des Röntgenbildes lieferten, wurden nun die Matrizen C_{Gk} (k sei der Index über die 8 Klassifikatoren) der Contribution Analysis aufgestellt. Es wurde eine neue Matrix C_{gesamt} erstellt, die an der Stelle (i,j) mit

$$C_{Gesamtij} = \sum_{k=1}^{8} \frac{C_{ijk}}{8} \qquad (3.1)$$

besetzt ist (Bild 3.4, Balkengrafik).

Aus der Contribution Matrix respektive Grafik sind Aussagen über die Einflüsse der Merkmalvektorkomponenten auf die Klassifikation zu gewinnen. Es ist leicht zu erkennen, daß die Eingabebilder 1, 2, 3 und 6 (Leerbild mit nur einem Grauwert) nur einen sehr geringen Einfluß auf die Klassifikation hatten.

Die Contribution Matrix bietet desweiteren eine Erklärung für die schlechte Bewertung der Klassen 5 und 6 durch die Maße der bedingten Korrektheit und Genauigkeit bei der Segmentierung des Bildes (Bild 3.6). Kein Eingabebild respektive keine Merkmalvektorkomponente erreicht einen relativ hohen positiven oder negativen Wert für eine der beiden Klassen. Es ist also nicht gelungen, dem Klassifikator eine hinreichende Information zur Detektion dieser Klassen zur Verfügung zu stellen.

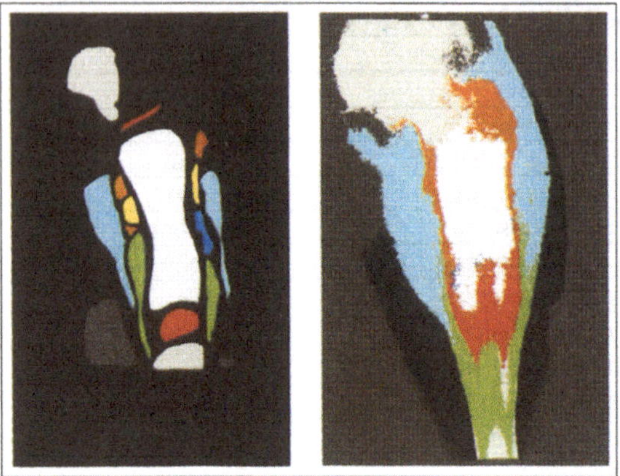

Bild 3.1: Maske (links) und Segmentierung (rechts)

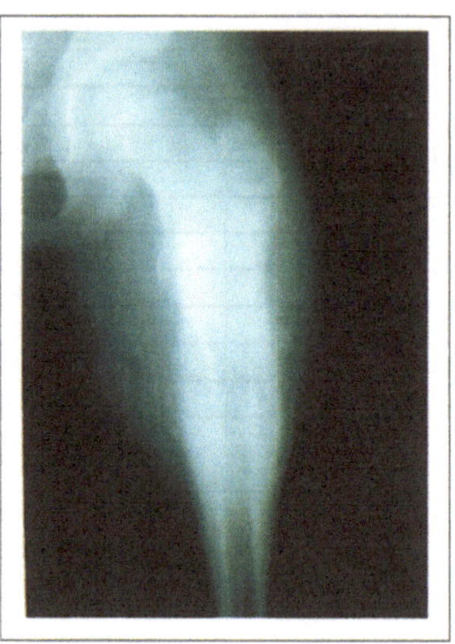

Bild 3.2 : Röntgenbild Ewing-Sarkom

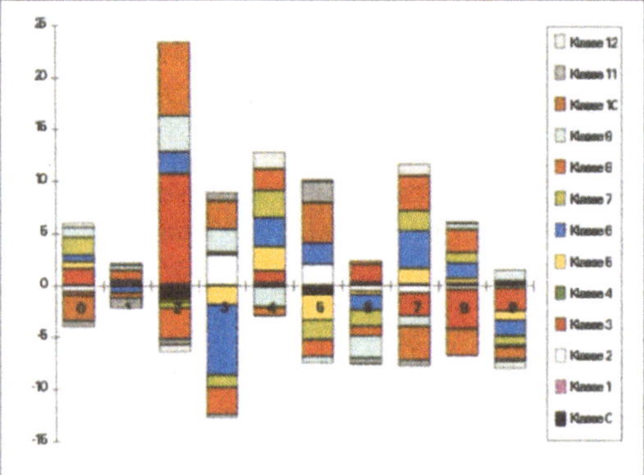

Bild 3.3: Contribution Analysis Marvin III Expanded

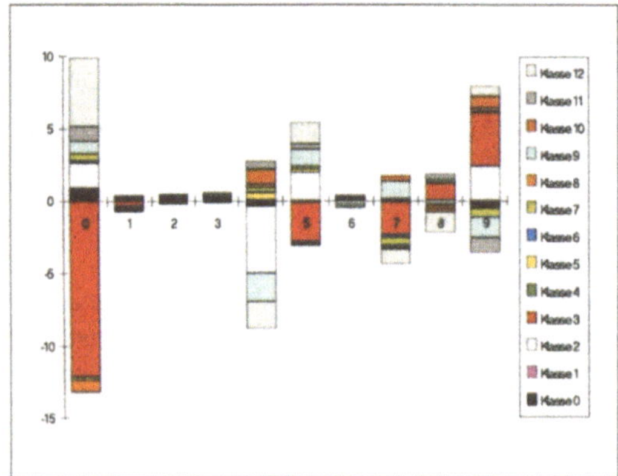

Bild 3.4: Contribution Analysis Marvin III normal

Bild 3.5: Segmentierung M III Expanded

3.2 Neue Klassifikatoren mit modifizierten Merkmalvektoren

Die in 3.1 gewonnenen Informationen über die Eingabebilder wurden genutzt, um Klassifikatoren mit besseren Qualitätsmerkmalen für die gleiche Aufgabenstellung zu parametrieren. Dies geschieht durch den Austausch der als unrelevant erkannten Eingabekomponenten. Ziel eines solchen Austausches ist insbesondere eine Verbesserung der Qualitätswerte für die in Marvin III schlecht segmentierten Klassen 5, 6, 7 und 8. In einem mit "Marvin III Expanded" bezeichneten Versuch wurden die Eingabekomponenten 1, 2, 3, und 6 durch vier in [Wei93] vorgestellten und auf expliziter Texturbeschreibung basierende Merkmale ausgetauscht. Die Klassendefinition blieb unverändert. Bild 3.5 zeigt eine der erhaltenen Segmentierungen.

3.3 Vergleich der Klassifkationen mit den Qualitätsmaßen

Im Vergleich zu den zuvor in Marvin III vorgenommenen Segmentierungen sind die zusammenhängenden Flächen kleiner geworden, es wurden größere Bereiche in die Klasse 0 (Hintergrund) klassifiziert. Ein Vergleich der beiden Segmentierungen erfolgte mit den Qualitätstests Korrektheit und Genauigkeit.

Es ist Bild 3.6 zu entnehmen, daß durch den Austausch der Merkmalvektorkomponenten eine bessere Segmentierung vorgenommen wurde (M III Expanded). Diese Beurteilung fällt aufgrund des visuellen Vergleichs schwer. Die zuvor sehr schlecht segmentierten Klassen 5, 6, und 8 wurden von dem neuen Klassifikator wesentlich besser erkannt, die anderen Klassen vergleichbar gut.

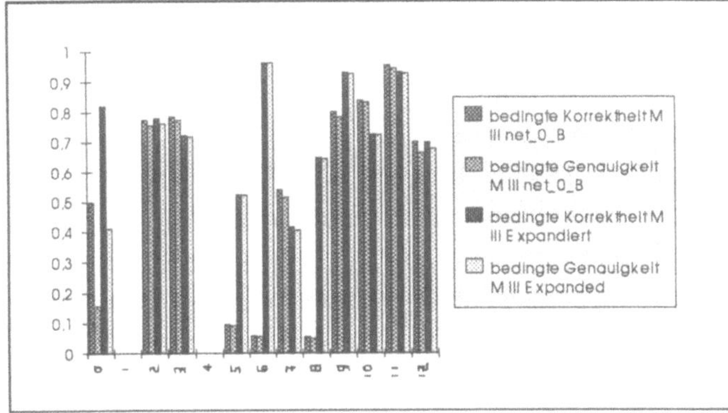

Bild 3.6: Qualitätsmaße Marvin III

Aufgrund der Contribution Analysis zu "Marvin III Expanded" (Bild 3.3) ist zu beobachten, daß bis auf die Eingabekomponente 1 alle Komponenten einen mehr oder weniger wichtigen Anteil an der Klassifikation haben. Insbesondere die Klassen, die nach der Verwendung der neuen Merkmalvektorkomponenten einen Qualitätsgewinn zu verzeichnen haben, werden stark von den neu hinzugekommenen Eingabebildern beeinflußt.

4 Zusammenfassung und Diskussion

Die Verknüpfung der beiden Komponenten Contribution Analysis und Quality Metrics bildet ein Werkzeug zur Lösung des Klassifikationsproblems unter Verwendung eines neuronalen Netzwerks als adaptivem Klassifikator (Bild 1.1). Mit den vorgestellten Qualitätstests ist es möglich, die Klassifikatoren auf ihre Eigenschaften hin eingehend zu untersuchen und Klassifikatoren für bestimmte Aufgabenstellungen auszuwählen. Sie können miteinander

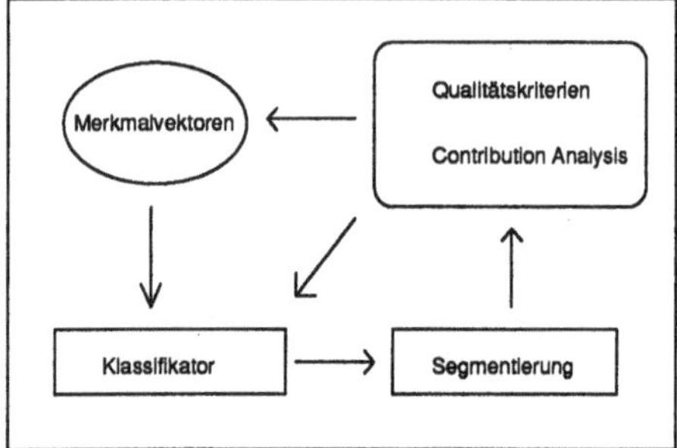

Bild 4.1: Zyklus der Segmentierungsoptimierung

hinsichtlich ihrer Klassifikationsleistung verglichen werden. Die vorgestellten Maße können auch auf andere, klassische Arten von Klassifikatoren angewendet werden, da sich auch für diese leicht die zugrunde liegenden Kontingenztafeln aufstellen lassen und daher auch dem Vergleich von auf verschiedenen Prinzipien beruhenden Klassifikatoren dienen.

Es wird eine Selektion von redundanten Merkmalvektorkomponenten ermöglicht, um entweder durch eine kleinere konnektionistische Struktur beschleunigte Klassifikatoren verwenden zu können, oder die Steigerung der Genauigkeit und Korrektheit der Segmentierung durch den Austausch dieser Komponenten zu ermöglichen. Im Experiment ist es durch einen Austausch der Merkmalvektorkomponenten auf der Grundlage der durchgeführten Contribution Analysis gelungen, die Klassifikationsleistung der Klassifikatoren insbesondere für die zuvor schlecht klassifizierten Klassen entscheidend zu verbessern. Dies konnte auch in anderen Versuchen bestätigt werden.

Es ergibt sich ein Zyklus, durch den das Verfahren bis zu einem gewissen Grad hin optimiert werden kann. Dieser ist in Bild 4.1 dargestellt.

Literaturverzeichnis

[Aka74] Akaike, H.: A New Look at the Statistical Model Identification. IEEE Transactions on Automatic Control Vol 19 No. 6, 716, 1974

[Boh93] Bohndorf, K., Pelikan, E., Tolxdorff, T., Zarrinnam, D., Wein, B., Günther, R.W.: Computerassistierte Diagnose von Knochentumoren: neue Entwicklungen. 76. Dt. Röntgenkongress, Wiesbaden, 1993

[Coh60] Cohen J.A.: A Coefficient of Agreement for Nominal Scales. Educational and Psychological Measurement Vol. 20, 37, 1960

[Egm93]Egmont-Petersen, M.: Quality Assessment of Neural Networks. Dissertation, Institute of Computer and Systems Science, Copenhagen Business School, 1993

[Fle81] Fleiss, J.L.: Statistical Methods for Rates and Proportions. John Wiley & Sons, New York, 1981

[Fog91] Fogel, D.B.: An Optimal Criterion for Optimal Neural Network Selection. IEEE Transactions on Neural Networks Vol.2 No.5, 490, 1991

[Har91] Harrison, R.F., Marshall, S.J., Kennedy, R.L.: A Connectionist Aid to the Early Diagnosis of Myocardial Infarction. Proceedings of the Third Conference on Artificial Intelligence in Medcine (Stefanelli, M., Hasman, A., Fieschi, M., Talmon, J. ,eds.), 119, Maastricht, 1991

[Har92] Hartung, J., Elpelt, B.: Multivariate Statistik. Oldenbourg Verlag, 4. Auflage, München, Berlin, 1992

[Lod65] Lodwick, G.S.: A Systematic Approach to the Roentgen Diagnosis of Bone Tumors. In: Tumors of Bone and Soft Tissue. Papers, M.D. Anderson Hospital, Year Book Med Pub, Chicago, 1965

[Lod66] Lodwick, G. S., Turner, A. H., Lusted, L. B.;,Templeton, A. W.: Computer-Aided Analysis of Radiographic Images. J.chron.Dis. Vol19, 485, Pergamon Press Ltd. Great Britain, 1966

[Lod71] Lodwick, G.S.: The Bones and Joints: An Atlas of Tumor Radiology. Year Book Medical Publishers, Chicago, 1971

[Lod80] Lodwick, G.S., Wilson, A.J., Farrel, C.: Determining Groth Rates of Focal Lesions from Radiographs. Estimating Rate of Groth in Bone Lesions : Observer Performance and Error. Rad. 134, 577, 1980

[Moo92]Moody, J. E.: The Effectice Number of Parameters: An Analysis of Generalization and Regularization in Nonlinear Learning Systems. In: Advances in Neural Information Processing Systems 4, (Moody, J.E., Hanson, S.J., Lippmann, R.P., eds), Morgan Kaufmann Publishers, San Mateo, 1992

[Moz89]Mozer, M. C., Smolensky, P.: Using Relevance to Reduce Network Size Automatically. Connection Science Vol.1 No.1, 3, 1989

[Pel93] Pelikan, E., Bohndorf, K., Tolxdorff, T. Zarrinam, D.Wein, B.: Computer-Assisted Diagnosis of Bone Tumors. Proc. CAR 93, Berlin, 630, 1993

[Rie89] Riede, U.N., Schaefer, H.E., Wehner, H.: Allgemeine und spezielle Pathologie. Thieme, 2. Auflage, Stuttgart, New York, 1989

[San89] Sanger, D.: Contribution Analysis: A Technique for Assigning Responsibilities to Hidden Units in Connectionists Networks. Connection Science, Vol.1, No. 2, 115, 1989

[Wei93] Weiler, F.: Merkmalsextraktion aus Filmröntgenbildern zur Abschätzung der Dignität fokaler Knochenläsionen. Diplomarbeit, Fakultät für Informatik, RWTH Aachen, 1993

Texturbasierte Extraktion medizinischer Merkmale aus Filmröntgenbildern*

F. Weiler[1], E. Pelikan[1], T. Nobis[1], T. Tolxdorff[2], K. Bohndorf[3]

[1]Institut für Medizinische Informatik und Biometrie, RWTH Aachen (R. Repges)
[2]Institut für Medizinische Statistik und Informationsverarbeitung, Abt. Medizinische Informatik, FU Berlin (T. Tolxdorff)
[3]Institut für Röntgendiagnostik, Zentralklinikum Augsburg (K. Bohndorf)

1 Einleitung

Die Diagnostik fokaler Knochenläsionen, welche bezogen auf die Gesamtheit der Krebserkrankungen sehr selten auftreten, stützt sich vornehmlich auf Filmröntgenbilder. Ihre relative Seltenheit und die Vielzahl der möglichen Erscheinungsformen (z.Zt. 45 Knochentumore bekannt) führt dazu, daß von einem auf diesem Gebiet unerfahrenen Radiologen oft keine präzise Differentialdiagnose gestellt werden kann. Eine Fehlbeurteilung der untersuchten Erkrankung kann im Falle der Knochentumore aber mit schwerwiegenden Folgen für den Patienten verbunden sein.

Die Knochentumore führen zu einer Veränderung des Knochengewebes, die sich im Röntgenbild in sogenannten röntgenologischen Merkmalen manifestiert und als Abweichung vom regelrechten Erscheinungsbild des Knochengewebes zu beobachten ist. Zur Erstellung einer Differentialdiagnose aus dem Röntgenbild gilt es nun, diese radiologischen Merkmale der Knochenveränderungen zu extrahieren und den Zusammenhang zwischen den beobachteten Merkmalen herzustellen.

Es soll ein Ansatz vorgestellt werden, der eine Extraktion der medizinisch relevanten Merkmale auf Basis expliziter Texturanalyse leistet.

2 Modellbildung

Folgende grundlegende Eigenschaften des Bildmaterials können formuliert werden:

- Durch die Dichteunterschiede der Gewebearten ist Knochengewebe im allgemeinen heller als das umgebende Weichteilgewebe und der Hintergrund. Durch Beleuchtungs- und Überdeckungseffekte kann aber auch in einem Teil des Bildes Knochengewebe dunkler abgebildet sein als Weichteilgewebe in einem anderem Teil des Bildes.

- Aufhellungen im Knochengewebe, die ein medizinisch relevantes Merkmal sind, sind dunkler als das umliegende Knochengewebe - analog sind Verdichtungen heller.

- Die in einem Röntgenbild sichtbaren Konturen und Strukturen sind organisch, d.h. sieht man von Überlagerungen einzelner Knochen im Bild oder Frakturen ab, kommen keine spitzen Winkel vor.

- Knochengewebe ist im Röntgenbild immer eine größere, zusammenhängende Region. Es können keine einzelnen, isolierten Pixel als Knochengewebe gelten. Dies gilt auch für die anderen medizinischen Merkmale.

*Das Projekt wird von der DFG unterstützt TO108/3-2

- Kortikalis und Kompakta bilden bei unauffälligem Befund eine klar abgrenzbare Kontur.
- An den Stellen, wo sich verschiedene Knochen im Bild überlagern entstehen Pseudokanten, die keine Grenze zwischen verschiedenen Gewebearten darstellen.
- Knochengewebe, insbesondere im Bereich der Spongiosa, hat eine eigene Struktur, die anders ist als die Struktur des Hintergrundes. Die Unterscheidung stützt sich also nicht nur auf die Helligkeit eines Bildbereiches, sondern auch auf die räumliche Anordnung der Bildpunkte. Dies gilt auch für medizinische Merkmale wie die lamelläre Periostreaktion, spikuläre Periostreaktion, Mottenfraß etc. .

Knochengewebe und krankhafte Veränderungen unterscheiden sich zum einen durch Helligkeitsunterschiede und zum anderen durch verschiedene räumliche Konzepte voneinander bzw. vom umgebenden Hintergrund. Für einen zu implementierenden Algorithmus ist zu beachten, daß ein Radiologe a-priori Wissen über das regelrechte Erscheinungsbild in den diagnostischen Prozeß einbringt, das dem Algorithmus nicht zur Verfügung gestellt werden kann.

Die zugrundeliegenden maximal 2048 mal 1684 Pixel großen, gescannten Röntgenbilder haben eine Tiefe von 12 Bit, wobei wegen des hohen Rauschanteiles nur Bilder mit 8 Bit Tiefe verwendet werden. Dazu werden die vier niedrigstwertigen Bits eines Bildpunktes entfernt, nachdem man das 12 Bit tiefe Bild auf den vollen Wertebereich von 0 bis 4095 normiert hat (histogram stretching). Zu diesen Bildern stehen vom Radiologen erstellte Befundungen zur Verfügung. Die Befundung umfaßt einen von unserer Arbeitsgruppe entwickelten Fragebogen zur Deskription fokaler Knochenläsionen sowie eine auf Folie gezeichnete Maske, die die Positionen der im Deskriptionsbogen aufgeführten Merkmale markiert [Boh92, Boh93, Pel93].

Das Ziel des Verfahrens ist eine Segmentierung eines Röntgenbildes unter medizinischen Gesichtspunkten, d.h. es sollen zusammenhängende Bildpunktmengen gefunden und der korrespondierenden medizinischen Merkmalsklasse zugeordnet werden. Der erste Schritt kann in einer Segmentierung Knochen-Hintergrund bestehen. Hier bieten sich auf Grund der Modellbildung Thresholding Verfahren an [Har85, Sah88]. Weder globale noch die die Beleuchtungseffekte berücksichtigenden lokalen Thresholding Algorithmen lieferten wegen der im Modell erwähnten Bildcharakteristika eine zufriedenstellende Segmentierung. Ein Ansatz mittels k-means Clusteranalyse unter Verwendung klassischer bildverarbeitungstechnischer Merkmale (Median, Gradient, Varianz, etc.) erwies sich ebenfalls für die vorliegende Aufgabenstellung als nur bedingt geeignet.

Der sich daraus ergebende Ansatz zur besseren Modellierung der medizinischen Merkmale basiert auf der expliziten Texturanalyse von Bildbereichen.

3 Texturmerkmale

In der Literatur werden strukturelle und statistische Texturmodelle verwendet [Van85], wovon im weiteren nur das statistische Modell verfolgt wird. Hierbei wird ein gegebenes Textursignal als Musterfunktion eines zweidimensionalen stochastischen Prozeßes aufgefaßt. Die Beschreibung der charakteristischen Eigenschaften dieser Textur kann durch die Bestimmung statistischer Maßzahlen erfolgen. Eine solche statistische Beschreibung eines Textursignals kann immer nur unvollständig sein, erwies sich aber für die vorliegende Augabenstellung der Unterscheidung verschiedener morphologischer Strukturen als ausreichend.

Die einfachste Bestimmung solcher Maßzahlen benutzt die Statistik 1. Ordnung der Grau- oder Merkmalswerte, welche aber keine räumlichen Konzepte berücksichtigt. Merkmale, die auch die räumlichen Aspekte von Texturen beschreiben, nutzen die Statistik zweiter Ordnung. Diese betrachtet die Wahrscheinlichkeit $p_{\vec{d}}(i,j)$, daß zwei Bildpunkte mit Abstand $\vec{d}$ die Grauwerte i und j haben (für alle Werte i, j und $\vec{d}$). Der Vektor $\vec{d}$ wird als Displacement bezeichnet. Für die $p_{\vec{d}}(i,j)$ gilt bei festem $\vec{d}$ und $(N+1)$ möglichen Grauwerten:

$$0 \leq p_{\vec{d}}(i,j) \leq 1 \quad \forall \quad i,j \qquad \text{und} \qquad \sum_{i=0}^{N}\sum_{j=0}^{N} p_{\vec{d}}(i,j) = 1 \tag{1}$$

Für jedes Displacement $\vec{d}$ können diese Auftrittswahrscheinlichkeiten in einer sogenannten Co-Occurence Matrix $M_{\vec{d}}(i,j)$ zusammengefaßt werden:

$$M_{\vec{d}}(i,j) = \begin{pmatrix} p_{\vec{d}}(0,0) & p_{\vec{d}}(1,0) & \cdots & p_{\vec{d}}(N,0) \\ p_{\vec{d}}(0,1) & p_{\vec{d}}(1,1) & \cdots & p_{\vec{d}}(N,1) \\ \vdots & \vdots & \ddots & \vdots \\ p_{\vec{d}}(0,N) & p_{\vec{d}}(1,N) & \cdots & p_{\vec{d}}(N,N) \end{pmatrix} \tag{2}$$

Da diese Texturbeschreibung über die Statistik zweiter Ordnung nicht rotationsinvariant ist, benutzen *Gotlieb et. al.* als Co-Occurence Matrix die Summe der Matrizen für die vier möglichen Richtungen $\{0, \frac{\pi}{4}, \frac{\pi}{2}, \frac{3\pi}{4}\}$ der Achter-Nachbarschaft eines Bildpunktes [Got90]. Das Displacement $\vec{d}$ kann auch über den Winkel θ und die Distanz δ angegeben werden, d.h. es gilt $M_{\vec{d}}(i,j) = M_{\delta,\theta}(i,j)$. Die richtungsunabhängige Co-Occurence $M_\delta(i,j)$ ist dann definiert als:

$$M_\delta(i,j) = \sum_{\theta=0,\frac{\pi}{4},\frac{\pi}{2},\frac{3\pi}{4}} M_{\delta,\theta}(i,j) \tag{3}$$

In einer $n \times n$ großen Umgebung um jeden Bildpunkt des Bildes können nun die Co-Occurence Matrizen für verschiedene Distanzen δ berechnet werden. Damit hätte man eine sehr gute Beschreibung der Textur für die Umgebung, in der der zu klassifizierende Bildpunkt liegt. Als Merkmal sind solche Matrizen aber ungeeignet, da einem Bildpunkt statt eines Merkmalswertes alle N^2 Werte der Matrix (für eine Distanz δ) zugeordnet wären. Aus diesem Grunde wurden von *Haralick* vierzehn Maße eingeführt, die die Eigenschaften einer Co-Occurence Matrix beschreiben [Har78, Got90]. Die Beschreibung der Matrix über solche Maße kann natürlich nur einen Teil der Information wiedergeben, die in der Matrix enthalten ist. Von den vierzehn definierten Maßen finden die folgenden vier Merkmale Verwendung:

Second Angular Moment:

$$\text{SAM}(\delta) = \sum_{i=0}^{N}\sum_{j=0}^{N} p_\delta(i,j)^2 \tag{4}$$

Inverse Difference Moment:

$$\text{IDM}(\delta) = \sum_{i=0}^{N}\sum_{j=0}^{N} \frac{p_\delta(i,j)}{1+(i-j)^2} \tag{5}$$

Contrast:

$$\text{CON}(\delta) = \sum_{i=0}^{N}\sum_{j=0}^{N} (i-j)^2 p_\delta(i,j) \tag{6}$$

Entropy:

$$\text{ENT}(\delta) = \sum_{i=0}^{N}\sum_{j=0}^{N} p_\delta(i,j) \log p_\delta(i,j) \tag{7}$$

Die sinnvolle Gewinnung von texturbeschreibenden Merkmalen aus einer Co-Occurence Matrix setzt ein Mindestmaß an Stationarität der Textur in der betrachteten Umgebung voraus. Dies ist jedoch nur durch große Fenster um einen Bildpunkt zu ereichen. Je größer das Operatorfenster gewählt wird, desto geringer wird die räumliche Auflösung. Es hat sich herausgestellt, daß Fenstergrößen von 15x15 bis 25x25 den besten Kompromiß darstellen und für die vorliegenden Filmröntgenbilder gute Ergebnisse liefern.

Mittels der in [Egm93] entwickelten Qualitätsmaße konnte nachgewiesen werden, daß die Merkmale SAM, IDM, CON und ENT gegenüber den klassischen Merkmalen bezüglich der Diskriminanzleistung eine Verbesserung darstellen.

4 Prototypen von Texturklassen

Als Alternative zur Texturbeschreibung über eine Co-Occurence Matrix bieten sich die Summen- und Differenzhistogramme an. Der Aufwand ihrer Bestimmung ist im Vergleich zur Co-Occurence Matrix geringer, sie lassen sich aber sehr einfach aus dieser herleiten. Das Summenhistogramm $p_S(0), \cdots, p_S(i), \cdots, p_S(510)$ ist aus einer gegebenen Co-Occurence Matrix $M_{\vec{d}}(i,j)$ gegeben durch:

$$p_S(i) = \sum_{k=0}^{N} \sum_{l=0, k+l=i}^{N} p_{\vec{d}}(k,l) \tag{8}$$

Das Differenzhistogramm $p_D(-255), \cdots, p_D(i), \cdots, p_D(255)$ ist mit gleicher Notation wie folgt gegeben:

$$p_D(i) = \sum_{k=0}^{N} \sum_{l=0, |k-l|=i}^{N} p_{\vec{d}}(k,l) \tag{9}$$

Im Summenhistogramm sind also die Wahrscheinlichkeiten $p_{S_{\vec{d}}}(i)$ aufgetragen, daß die Summe der Grauwerte zweier Bildpunkte mit Abstand $\vec{d}$ gleich i ist. Analog bezeichnet das Differenzhistogramm mit den Einträgen $p_{D_{\vec{d}}}(i)$ die Wahrscheinlichkeit, daß die Differenz der Grauwerte zweier Bildpunkte mit Abstand $\vec{d}$ gleich i ist.

Der Zusammenhang beider Histogramme zur Co-Occurence Matrix wird deutlich, wenn die Matrix in Richtung ihrer Hauptdiagonalen transformiert wird. Die so erhaltene Matrix hat als Marginalverteilungen gerade das Summen- und Differenzhistogramm.

Für eine Klassifikation mittels der Summen- und Differenzhistogramme wird ein Satz Prototypen, ähnlich dem Trainingsdatensatz bei der Diskriminanzanalyse, für die unterschiedlichen Texturen der Gewebeklassen erstellt. Jeder Textur-Prototyp ist durch sein Summen- und Differenzhistogramm beschrieben.

Es wird nun ein Ähnlichkeitsmaß für zwei verschiedene Summen- und Differenzhistogramme definiert. Der verwendete Ansatz besteht darin, daß zu jedem Bildpunkt des zu klassifizierenden Bildes das Summen- und Differenzhistogramm in einem Fenster um dieses Pixel berechnet wird. Dann werden diese beiden Histogramme auf Übereinstimmung mit den Summen- und Differenzhistogrammen der Textur-Prototypen untersucht. Der Bildpunkt wird mit dem Label der Klasse versehen, mit der die größte Ähnlichkeit bezüglich der Summen- und Differenzhistogramme besteht.

Die Summen- und Differenzhistogramme für die Distanz $\delta = 1$ und die vier möglichen Winkel $\theta \in \{0, \frac{\pi}{4}, \frac{\pi}{2}, \frac{3\pi}{4}\}$ sind für einen Textur-Prototyp gegeben durch die Auftrittswahrscheinlichkeiten:

$$p_{S_\theta}^{pt}(i) \quad \text{mit} \quad 0 \leq i \leq 510 \tag{10}$$

$$p_{D_\theta}^{pt}(i) \quad \text{mit} \quad -255 \leq i \leq 255 \tag{11}$$

Dabei ist keine der Auftrittswahrscheinlichkeiten $p^{pt}_{S_\theta}(i)$ oder $p^{pt}_{D_\theta}(i) = 0$, sondern mindestens $\frac{1}{m \times m}$ bei einem $m \times m$ großen Textur-Prototyp. Die Notwendigkeit dieser Modifikation wird bei der Definition des Ähnlichkeitsmaßes deutlich.

Für das $n \times n$ große Operatorfenster um den zu klassifizierenden Bildpunkt sind die Summen- und Differenzhistogramme entsprechend für die Distanz $\delta = 1$ und die vier möglichen Winkel $\theta \in \{0, \frac{\pi}{4}, \frac{\pi}{2}, \frac{3\pi}{4}\}$ gegeben durch:

$$p^{op}_{S_\theta}(i) \quad \text{mit} \quad 0 \leq i \leq 510 \tag{12}$$

$$p^{op}_{D_\theta}(i) \quad \text{mit} \quad -255 \leq i \leq 255 \tag{13}$$

Das Ähnlichkeitsmaß $\tau^{pt,op}$, das die Übereinstimmung der Summen- und Differenzhistogramme von Prototyp und Operatorfenster angibt, ist definiert als:

$$\tau^{pt,op} = \sum_{\theta \in \{0, \frac{\pi}{4}, \frac{\pi}{2}, \frac{3\pi}{4}\}} \tau^{pt,op}_\theta \tag{14}$$

mit den richtungsabhängigen Ähnlichkeitsmaßen:

$$\tau^{pt,op}_\theta = \sum_{i=0}^{510} -[p^{op}_{S_\theta}(i) \log p^{pt}_{S_\theta}(i)] + \sum_{i=-255}^{255} -[p^{op}_{D_\theta}(i) \log p^{pt}_{D_\theta}(i)] \tag{15}$$

Je größer die Übereinstimmung zweier Bildausschnitte (Texturprototyp und Operatorfenster) bezüglich ihrer Summen- und Differenzhistogramme ist, desto kleiner ist der Wert von $\tau^{pt,op}$. Die Ergebnisse dieses Algorithmus sind in Abb. 1 dargestellt. Alle verwen-

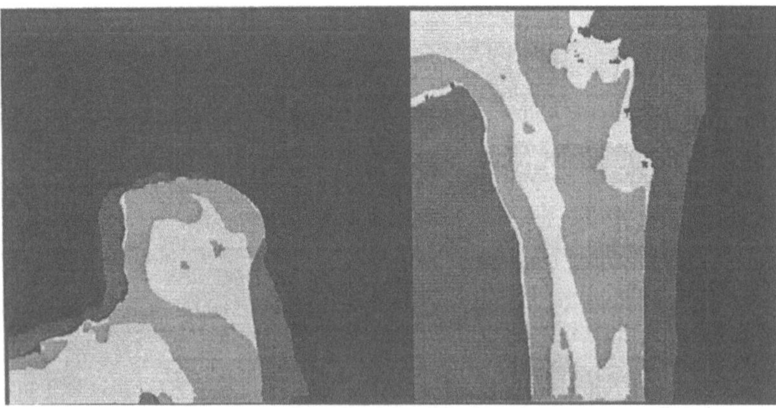

Abb. 1: Klassifikation mit Textur-Prototypen (links: Fußausschnitt, rechts: Femur). Die drei dunkelsten Grauwerte bezeichnen Prototypen der Klasse Hintergrund, die beiden hellsten Grauwerte Prototypen der Klasse Knochengewebe. (Windowgröße 16x16)

deten Textur-Prototypen stammen aus einem Fremdbild, wodurch die Robustheit des Verfahrens deutlich wird (Generalisierungsfähigkeit).

5 Agglomeratives Region Growing

In einem ersten Schritt wird das Röntgenbild bezüglich eines Homogenitätskriteriums in Regionen unterteilt. Die Bildpunkte der einzelnen Regionen sollen dem gleichen Texturprozeß entstammen, da Pixel des gleichen Texturprozeßes auch derselben medizinischen

Merkmalsklasse angehören. Dazu soll ein agglomeratives Region Growing Verfahren benutzt werden [Mes89]. Die für das Verfahren entscheidende Frage, ob zwei benachbarte Regionen verschmolzen werden dürfen, wird auf ein statistisches Entscheidungsproblem zurückgeführt. Das daraus resultierende Ähnlichkeitsmaß ist explizit mit der Fehlerrate für dieses Entscheidungsproblem verknüpft. Unter der Voraussetzung normalverteilter Texturprozeße und der statistischen Unabhängigkeit der einzelnen Bildpunkte ergibt sich folgendes Distanzmaß (verallgemeinertes Likelihood Verhältnis) für das Region Growing:

$$\lambda = \frac{(\hat{\sigma}_C)^{N_C}}{(\hat{\sigma}_A)^{N_A} \cdot (\hat{\sigma}_B)^{N_B}} \tag{16}$$

Die Größen σ_A, σ_B und σ_C sind die Maximum-Likelihood Schätzungen der Standardabweichungen für die Bildpunktwerte der Regionen A, B und der Zusammenfassung beider Regionen zu einer Region C. N_x bezeichnet die Anzahl der Bildpunkte in der Region x.

Es ist offensichtlich, daß der Wert des Distanzmaßes λ groß ist, wenn die Varianz der Zusammenfassung beider Regionen deutlich größer ist als die Varianz der einzelnen Regionen. Zusätzlich bewirkt der exponentielle Charakter der Funktion, daß eine solche Erhöhung der Gesamtvarianz $\hat{\sigma}_C^2$ gegenüber den Einzelvarianzen $\hat{\sigma}_A^2$ und $\hat{\sigma}_B^2$ umso mehr gewichtet wird, je größer der Stichprobenumfang ist. Dies ist sinnvoll, da bei kleinen Stichproben eine solche Abweichung weniger signifikant ist.

Für die Überprüfung der Ähnlichkeit zweier Regionen soll nicht nur ein Merkmal (Grauwert) herangezogen werden, sondern es sollen die vorgestellten Texturmerkmale, die in einem Merkmalsvektor zusammengefasst sind, berücksichtigt werden. Zu jedem Bildpunkt existiert somit ein Merkmalsvektor. Unter der Annahme, daß die Komponenten eines jeden Merkmalsvektors voneinander statistisch unabhängig sind, ergibt sich die Verteilungsdichte eines solchen Merkmalsvektor $\vec{y} = (y_1, \cdots, y_n)$ als Produkt der jeweiligen Verteilungsdichten der einzelnen Vektorkomponenten (Merkmale). Damit ist das verallgemeinerte Likelihood-Verhältnis λ aus dem Produkt der einzelnen Likelihood-Verhältnisse λ_i, die jeweils innerhalb der Merkmalsvektorkomponente y_i nach der Gleichung 16 gebildet werden, gegeben als:

$$\lambda(R_A, R_B) = \prod_{i=1}^{n} \lambda_i(R_A, R_B) \tag{17}$$

Das Distanzmaß λ ist damit auch für vektorielle Merkmale geeignet. Allerdings wird für dieses Distanzmaß vorausgesetzt, daß die Werte der einzelnen Regionen normalverteilt und voneinander statistisch unabhängig sind. Ob solch ein 'vereinfachendes' Modell zulässig ist, wird auch in der Literatur diskutiert [Mes89, Ros82] und bei natürlichen Bildern näherungsweise als gültig angesehen.

5.1 Kontrollstruktur und Regionenwachstum

Für das agglomerative Region Growing wird eine Kontrollstruktur verwendet, die die Abfolge der Verschmelzungsoperationen nicht durch eine vorgegebene Reihenfolge von Bildpunkten oder Regionen bestimmt, sondern allein die Merkmalsdistanzen der im Bild vorhandenen Regionen auswertet. Dazu wird für alle Paare von benachbarten Regionen das Distanzmaß λ berechnet. Es werden dann die beiden Regionen R_A und R_B miteinander verschmolzen, die die geringste Distanz $\lambda(R_A, R_B)$ zueinander haben (falls $\lambda(R_A, R_B)$ kleiner als eine vorgegebene Schwelle T ist). Danach muß die Liste der Distanzen aktualisiert werden. Darauf folgend werden wieder die beiden Regionen mit der kleinsten Distanz zueinander verschmolzen. Dieser Vorgang wird solange wiederholt, bis es keine Distanz $\lambda(R_A, R_B)$ mehr gibt, die kleiner als die vorgebene Schwelle T ist.

Wird die Initialpartitionierung so festgelegt, daß jeder Bildpunkt zuerst eine eigene Region bildet, ist das Verfahren vollständig beschrieben. Der Algorithmus hat als einzigen

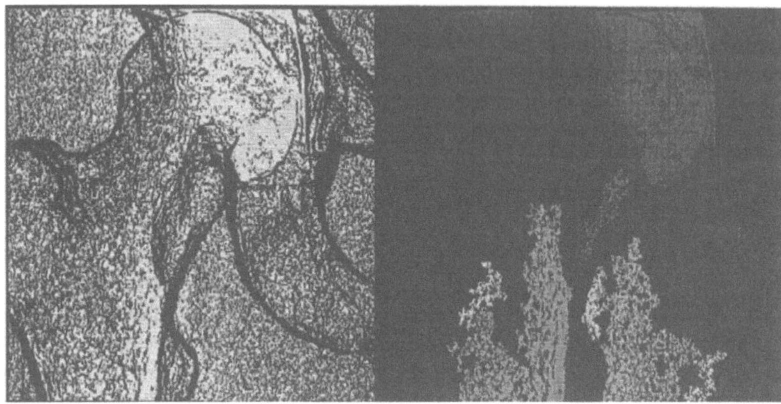

Abb. 2: Binarisierung eines mit dem Kirsch-Operator bearbeiteten Röntgenbildes (Hüftgelenk mit Femurkopf) und Anwendung des Region Growing/Labeling mit einer minimalen Regionengröße von 1000 Pixeln.

Parameter die Schwelle T. Ist dieser klein, wird das Bild übersegmentiert. Das heißt die Bildregionen, die mit medizinischen Merkmalen korrespondieren, werden nicht als ein einziges Segment gefunden, sondern bestehen aus mehreren kleinen Segmenten. Wird der Parameter T zu groß gewählt, hat dies eine Untersegmentierung zur Folge. Dies bedeutet, daß zwei oder mehr medizinisch relevante Bildregionen zu einem Segment zusammengefaßt werden. Der Fall der Untersegmentierung darf nicht auftreten, aus diesem Grunde sollte T eher zu klein gewählt werden.

Die Initialpartitionierung mit jedem Bildpunkt als eigenständige Region führt im ersten Schritt bei einem 512x512 Pixel großen Bild zu etwa einer Million Distanzmaßen λ, die zwischen benachbarten Regionen zu berechnen und nach Größe zu sortieren sind. Zur Beschleunigung des Verfahrens sollen größere Regionen als Initialpartitionierung vorgegeben werden. Um dies zu erreichen wird ein durch einen Kirsch-Operator gewonnenes Kantenbild über einen Schwellwert binarisiert, um die Bildpunkte von zusammenhängenden Bildpunktmengen (Regionen), die eine vorgegebene minimale Größe haben, mit einem Label zu versehen. Wird der Schwellwert sukzessive erhöht, wachsen die Regionen in Richtung des Gradienten. Für eine gute Initialpartitionierung wird der Schwellwert für die Binarisierung so gewählt, daß die Regionen möglichst groß sind, aber trotzdem jeder Bildpunkt einer einzelnen Region zu der gleichen medizinischen Merkmalsklasse gehört. Der Schwellwert für die Binarisierung ist so klein zu wählen, daß keine Übersegmentierung eintritt, jedoch auch so groß zu wählen, daß die Regionen möglichst umfangreich sind. Eine Abschätzung eines geeigneten Schwellwertes liefert das globale Maximum des zum Kantenbild gehörenden unimodalen Histogrammes. Eine solche Initialpartitionierung zeigt Abb. 2.

6 Klassifikation

Wurde das Bild durch den Segmentierungsalgorithmus in Regionen unterteilt, werden die einzelnen Regionen mit den entsprechenden medizinischen Merkmalsklassen assoziiert.

Statt der Klassifikation eines Bildpunktes und des zugehörigen Merkmalsvektors, muß bei dieser Vorgehensweise eine Menge von Bildpunkten, die alle zur gleichen medizinischen Merkmalsklasse gehören, klassifiziert werden. Liegt ein Satz von Textur-Prototypen für die medizinischen Merkmalsklassen vor, wird für eine Region das Summen- und Differenzhistogramm berechnet und diese Histogramme auf Ähnlichkeit mit den Summen- und Differenzhistogrammen der Prototypen untersucht. Die Klassifikation erfolgt dann zu dem Texturprototyp, für den das Ähnlichkeitsmaß $\tau^{pt,reg}$ am kleinsten ist. Der Vorteil dieses Verfahrens gegenüber dem oben beschriebenen ist, daß die Histogramme nicht über einem quadratischen Operatorfenster berechnet werden, in dem mehrere Texturen vorhanden sein können, sondern die Pixel der Region alle dem gleichen Texturprozeß entstammen. Für diese Art der Klassifikation ist es vorteilhaft, die vorgestellten auf der Co-Occurence Matrix basierenden Texturmerkmale als Merkmale für den agglomerativen regionenorientierten Segmentierungsalgorithmus zu verwenden, da für gleich texturierte Bildausschnitte die Werte dieser Merkmale eine geringe Varianz aufweisen. Für die Grauwerte des Originalröntgenbildes muß dies nicht gelten.

Literatur

[Boh92] Bohndorf, K., Tolxdorff, T., Pelikan, E., Zarrinnam, D., Günther, R.W. : Neue Ansätze zur computerassistierten Diagnose von fokalen Knochenläsionen. Radiologe 32, 416, 1992

[Boh93] Bohndorf, K., Pelikan, E., Tolxdorff, T., Zarrinnam, D., Wein, B., Günther, R.: Computerassistierte Diagnose von Knochentumoren: neue Entwicklungen. 76. Deutschen Röntgenkongress, Wiesbaden, 987, 1993

[Egm93] Egmont-Petersen, M. : Quality Assessment of Neural Networks. Dissertation, Institute of Computer and Systems Science, Copenhagen Business School, 1993

[Got90] Gotlieb, C.C., Kreyszig, H.E. : Texture Descriptors Based on Co-occurence Matrices. CVGIP 51, 70, 1990

[Har78] Haralick, R.M. : Statistical and Structural Approaches to Texture. Proc. 4th ICPR, Kyoto, 45, 1978

[Har85] Haralick, R.M., Shapiro, L.G. : Image Segmentation Techniques. CVGIP 29, 100, 1985

[Mes89] Mester, R. : Regionenorientierte Bildsegmentierung unter Verwendung stochastischer Bildmodelle. VDI Reihe 10 Nr. 106, VDI, Düsseldorf, 1989

[Pel93] Pelikan, E., Bohndorf, K., Tolxdorff, T., Zarrinnam, D., Wein, B.: Computer-Assisted Diagnosis of Bone Tumors. Proc. CAR 93, Berlin, 630, 1993

[Ros82] Rosenfeld, A., Kak, A.C. : Digital Picture Processing Volume 2. Academic Press, New York, 1982

[Sah88] Sahoo, P.K., Soltani, S., Wong, A.K.C., Chen, Y.C. : A Survey of Thresholding Techniques. CVGIP 41, 233, 1988

[Van85] Van Gool, L., Dewaele, P., Oosterlinck, A. : Texture Analysis Anno 1983. CVGIP 29, 336, 1985

[Wei93] Weiler, F. : Merkmalsextraktion aus Filmröntgenbildern zur Abschätzung der Dignität fokaler Knochenläsionen. Diplomarbeit, Math.-Naturwiss. Fakultät, RWTH Aachen, 1993

3D-Rekonstruktion mikroskopischer Bilder unter Berücksichtigung der Lichtabsorption

Xiangchen Wu, Peter Schwarzmann

Universität Stuttgart, Institut für Physikalische Elektronik
Pfaffenwaldring 47, 70569 Stuttgart

1 Einleitung

In der konventionellen Mikroskopie bedingt eine Erhöhung der lateralen Auflösungsgrenze zugleich eine Abnahme des Schärfentiefebereiches. Ein herkömmliches Mikroskopiebild besteht deshalb fast immer aus der scharf erscheinenden Objektinformation innerhalb des Schärfentiefebereiches und Überlagerten, unscharf abgebildeten Objektanteilen außerhalb dieses Bereiches.

Ein reales Objekt, wie eine Zelle, hat eine dreidimensionale Ausdehnung im Raum. Normalerweise wird dieses Objekt durch ein optisches System, dessen Schärfebereich klein im Vergleich zur Ausbreitung des Objekts der Tiefe nach ist, auf einen Flächendetektor abgebildet. In diesem Fall kann das dreidimensionale Objekt in Objektschichten, die senkrecht zur optischen Achse sind, aufgeteilt werden. Mit dem Mikroskop kann man aber nur eine bestimmte Ebene fokussieren, und die defokussierten Ebenen überlagern sich im Bild. Dieses Bild enthält nur einen Bruchteil von der gesamten Information, die das Objekt beschreibt. Um die gesamte Information zu gewinnen, die für die Rekonstruktion notwendig ist, wird der Abbildungsvorgang mit variierter Fokussierebene des Objektivs mehrmals wiederholt. So bekommt man eine Fokusserie. Aus dieser Fokusserie werden Schnittbilder berechnet, die von den aus darüber und darunter liegenden defokusierten Bildanteilen befreit sind.

Der vorliegende Beitrag erläutet eine Näherung mit einen Ansatz aus der linearen Systemtheorie[1,2,3,7,11,12] und geht dann auf 2 gegenüber dem linearen Ansatz effektivere, nichtlineare Ansätze über. Dabei wird das Problem der Lichtabsorption in dem 3D-Objekt behandelt. Die so gewonnenen Lösungen werden anschließend zur Rekonstruktion eines Bildes des Objektes mit unendlicher Schärfentiefe herangezogen. In einem Ausblick wird noch die Kombination der Methodik mit tomographischen Verfahren angesprochen.

2 Methoden

2.1 Modellierung

In Abb.1 ist das Prinzip der optischen Abbildung mit einem Linsensystem (Objektiv) skizziert.

Kennzeichend darin ist, daß jede Ebene des Objektes sich in eine Ebene des Bildraumes abbildet und jede Bildebene durch die davor und dahinter liegenden Ebenen gestört ist.

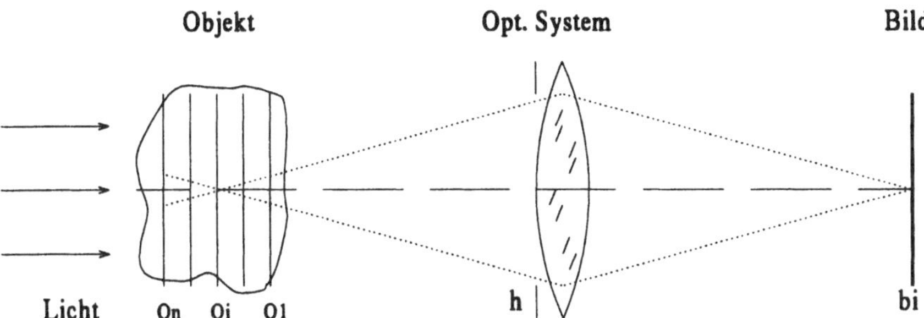

Abb. 1: Modellierung des Abbildungsvorgangs

Die beiden wesentlichen zur Abbbildung beitragenden physikalischen Vorgänge sind:

- Die Fortpflanzung des Lichtes durch das Objekt
- Die Abbildungseigenschaften der Optik.

Es ist offensichtlich, daß bei bekannten optischen Eigenschaften der Beleuchtung, des Objektes und der Optik die Bildebenen im Abbildungsraum zumindest im Prinzip eindeutig und direkt berechenbar sind.

Mit Hilfe der Impulsantwort (oder der Übertragungsfunktion im Ortsfrequenzraum) des optischen Abbildungssystems kann man im linearen Modell den Einfluß der defokussierten Ebenen, bzw. der Defokussierungen, quantitativ beschreiben. Die Übertragungsfunktion kann sowohl berechnet werden mit der Formel von Stokseth[10] oder gemessen werden.

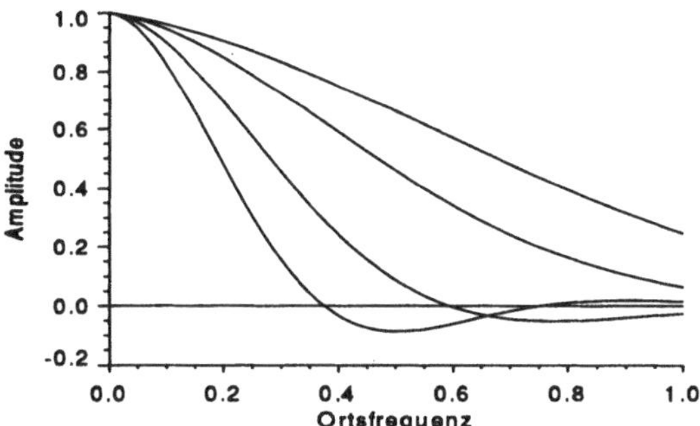

Abb. 2: Verlauf der Übertragungsfunktionen bei verschiedenen Defokussierungen

2.2 Lineares Modell: Intensität als Objektinformation

Wenn es sich um die Abbildung von selbstleuchtenden Objekten ohne Eigenabsorption oder undurchsichtigen Objekten, die auftreffendes Licht reflektieren, handelt, wird das lineare Modell verwendet[2,3,7].

$$
\begin{aligned}
b_1 &= h_{11} * o_1 + h_{12} * o_2 + \ldots + h_{1n} * o_n \\
b_2 &= h_{21} * o_1 + h_{22} * o_2 + \ldots + h_{2n} * o_n \\
&\ldots \\
b_n &= h_{n1} * o_1 + h_{n2} * o_2 + \ldots + h_{nn} * o_n
\end{aligned} \tag{1}
$$

darin bedeutet b_i die Bildintensität in der Bildebene i , o_j die Objektintensität in der Objektebene j und h_{ij} die Impulsantwort der Objektebene j in der Bildebene i. Das Objekt ist als Intensitätsverteilung charakterisiert.

In diesem Falle dürfen sich Intensitätsverteilungen der Objektebenen gegenseitig nicht beeinflußen. Natürlich kann diese Voraussetzung in realer Durchlichtmikroskopie kaum erfüllt werden. Die Lichtabsorption ist ein nichtlinearer Vorgang. Wenn man trotz starker Absorption das lineare Modell für die Durchlichtmikroskopie einsetzt, kann man keine zufriedenstellende Lösung erwarten.

2.3 Nichtlineares Modell: Transmission als Objektinformation

Unter Berücksichtigung der Lichtabsorption im Objekt wurde ein nichtlineares Modell von uns entwickelt[6], das eine bessere Annäherung an die physikalischen Verhältnisse darstellt. Dabei werden die selben Übertragungsfunktionen benutzt.

$$
\begin{aligned}
b_1 &= \{h_{11} * o_1\} \cdot \{h_{12} * o_2\} \cdot \ldots \cdot \{h_{1n} * o_n\} \\
b_2 &= \{h_{21} * o_1\} \cdot \{h_{22} * o_2\} \cdot \ldots \cdot \{h_{2n} * o_n\} \\
&\ldots \\
b_n &= \{h_{n1} * o_1\} \cdot \{h_{n2} * o_2\} \cdot \ldots \cdot \{h_{nn} * o_n\}
\end{aligned} \tag{2}
$$

In diesem nichtlinearen Ansatz wird die Objektinformation durch die Verteilung von Transmissionen beschrieben. Darum überlagern sich die einzelnen Objektebenen, gefaltet mit den entsprechenden Impulsantworten, multiplikativ. Dies kann wie folgt interpretiert werden:

Wir betrachten die i te Aufnahme b_i. Bevor das Licht die Objektebene o_i erreicht, wird es schon mit den davor liegenden Transmissionen der Objektebenen o_k mit $k > i$ multipliziert. Das Licht kommt dann durch die dahinter liegenden o_k mit $k < i$ Objektebenen hindurch bis in die Bildebene i. Das dedeutet wiederum eine Lichtabschwächung.

Um alle Transmissionen der Objektebenen zu kennen, ist das Gleichungssystem (2) zu lösen. Wegen der Nichtlinearität der Gleichungen ist es ein großes Problem, eine analytische Lösung zu finden. Zusätzlich sind auch denkbare Probleme wie Sigularitäten und Rauscheinflüsse zu überwinden. Wahrscheinlich kann das Problem nur durch iterative Methoden gelöst werden.

2.4 Nichtlineares Modell: Extinktion als Objektinformation

In diesem Modell wird die räumlich verteilte Extinktion der einzelnen Objektebenen als Objektinformation betrachtet. Unter Berücksichtigung des exponetiellen Zusammenhangs zwischen Transmission und Extinktion können wir das Gleichungssystem(2) abwandeln in

$$\begin{aligned} b_1 &= exp[-h_{11} * o_1] \cdot exp[-h_{12} * o_2] \cdot \ldots \cdot exp[-h_{1n} * o_n] \\ b_2 &= exp[-h_{21} * o_1] \cdot exp[-h_{22} * o_2] \cdot \ldots \cdot exp[-h_{2n} * o_n] \\ &\ldots \\ b_n &= exp[-h_{n1} * o_1] \cdot exp[-h_{n2} * o_2] \cdot \ldots \cdot exp[-h_{nn} * o_n] \end{aligned} \tag{3}$$

oder

$$b_i = exp[-\sum_{j=1}^{n} h_{ij} * o_j] \qquad i = 1, ..., n \tag{4}$$

wobei $o_j =$ Ebene j des Objekts
$b_i =$ Intensitätsverteilung in Bildebene i
$h_{ij} =$ Impulsantwort für die Abbildung der Objektebene o_j in die Bildebene b_i

Durch Logarithmieren der beiden Seiten von Gl.(4) ergibt sich

$$a_i = -ln\, b_i = \sum_{j=1}^{n} h_{ij} * o_j \qquad i = 1, ..., n \tag{5}$$

Durch Übergang in den Ortfrequenzbereich erhält man die zu Gl.(5) äquivalenten linearen Gleichungen

$$A_i = \sum_{j=1}^{n} H_{ij} \cdot O_j \qquad i = 1, ..., n \tag{6}$$

Gl.(6) kann vereinfacht in Matrixform dargestellt werden:

$$\mathbf{A} = \mathbf{H} \cdot \mathbf{O} \tag{7}$$

Zur Restaurierung der " in-focus " Information $o_j (j = 1, ..., n)$ werden Inverse-Filter, Wiener-Filter (Unter Berücksichtigung des Rauschens) und iterative Methoden angewandt[1,2,3,7,11,12].

Gl.(7) hat eine Lösung, wenn die Matrix **H** nicht singulär ist, d.h. die Determinante von **H** nicht Null ist. Es gibt offensichtlich keine Lösung für die Gleichanteile wegen der Singularität der Matrix **H** wie aus Abb.2 ersichtlich ist. Alle Übertragungsfunktionen für niedrige Frequenzen unterscheiden sich nicht sehr. Darum ist die Determinante $|\mathbf{H}|$ dafür sehr klein und damit ist das Gleichungsystem schlecht konditioniert. Aus diesem Grunde sind die rekonstruierten Objektebenen nicht frei von niederfrequenten Störungen anderer Objektebenen(Abb.4c, Abb.5b).

2.5 Bilder mit erhöhter Schärfentiefe

Analog wie in [7] können die rekonstruierten Objektebenen einfach anhand

$$b = exp[-\sum_{j=1}^{N} o_j] \tag{8}$$

überlagert werden. Damit haben wir ein Bild mit allen " in-focus " Informationen. Die niederfrequenten Störungen in Objektebenen kompensieren sich komplett. Nun können wir sagen, durch eine Rekonstruktion der einzelnen Objektebenen und eine anschließende Überlagerung der rekonstruierten Objektebenen können wir ein Bild mit unendlicher Schärfentiefe in der Mikroskopie gewinnen.

3 Computersimulation und Experimentelle Ergebnisse

Wir haben eine Computersimulation für den nichtlinearen Ansatz durchgeführt. Abb.4a zeigt die originalen Objektebenen o_1 bis o_4 von einem synthetischen Objekt, Abb.4b die Fokusserie des Objektes, Abb.4c die rekonstruierten Objektebenen nach dem nichtlinearen Ansatz von Gl.(5). Für die graphische Darstellung sind die Objektebenen (Extinktion) in Intensitätsbilder umgewandelt. Zum Schluß ist in Abb.4d das Bild mit unendlicher Schärfentiefe nach einer Überlagerung der rekonstruierten Objektebenen anhand der Gl.(8) dargestellt.

Als Objekte wurden 4 Ebenen (128×128 Pixel) mit je einem Kreis gewählt. Der Abstand zwischen zwei benachbarten Objektebenen ist $2\mu m$. Die Wellenlänge des inkohärenten Lichtes ist $0.53\mu m$. Der Vergroßungsfaktor des Mikroskops beträgt $100\times$ und die numerische Apertur ist 1.3.

In einem weiteren Experiment wurde eine Fokusserie von einem realen mikroskopischen Präparat mit einer TV-Kamera aufgenommen. Es handelt sich um eine Algenart (Wassernetz). Abb.5a zeigt diese Fokusserie von 4 Ebenen mit je 128×128 Pixel. Der Abstand zwischen zwei benachbarten Objektebenen ist $10\mu m$. Bei der Aufnahme benutzten wir ein Objetiv mit Vergrößung $10\times$ und $N.A. = 0.22$. Die Systemübertragungsfunktionen können sowohl berechnet[9] als auch gemessen werden. Für diese Szene wurde die berechneten benutzt. In Abb.5b sind die rekonstruierten Objektebenen, wiederum umgewandelt in Intensitätsform, dargestellt. Obwohl die rekonstruierten Objektebenen mit niederfrequenten Störungen behaftet sind, enthält das zusammengesetzte Bild (Abb.5c) nur die scharfe Information aller Objektebenen.

4 Schlußfolgerung

In diesem Beitrag haben wir ein nichtlineares Modell für die Abbildung eines Semitransparenten Objektes durch ein Mikroskop erstellt, und zwar unter Berücksichtigung der Lichtabsorption. Das anhand dieses nichtlinearen Modells rekonstruierte Bild liefert gute Ergebnisse.

Wenn das zu betrachtende Objekt hoch transparent ist, dann ist die Gesamtlichtabschwächung gering. In diesem Falle kann man Gl.(5) in Gl.(1) umwandeln mit der Überlegung $-ln(x) \approx 1 - x$ für $x \approx 1$.

Wegen der Bandbegrezung der 3D-Übertragungsfunktion[4,10] ist es nicht möglich die komplette 3D-Information aus nur einer Fokusserie zu gewinnen. Außdem kann bei unserem Verfahren durch das Zusammensetzen der rekonstruierten Objektebenen die Tiefeninformation nicht ausgenutzt werden.

Zur Abhilfe wird das Vorgehen der Computertomographie(CT) eingeführt. Gl.(9) ist dieselbe wie die der Transmissionstomographie und beschreibt die parallele Projektion mit

Strahlungsabschwächung. Wenn wir das Objekt um seine eigene Achse drehen, können wir eine Projektionsbildserie bekommen. Mit bekannten CT-Algorithmen angewandt auf diese Projektionsbildserie können wir die echte 3D-Information gewinnen. Gegenüber den bekannten CT-Verfahren bestehen in unserem Falle noch Informationsanteile entlang der optischen Achse, die wir für die 3D-Rekonstruktion sicherlich ausnutzen können, um die Rekonstruktion mit weniger Projektionen zu erhalten, obwohl diese Information nicht vollständig ist.

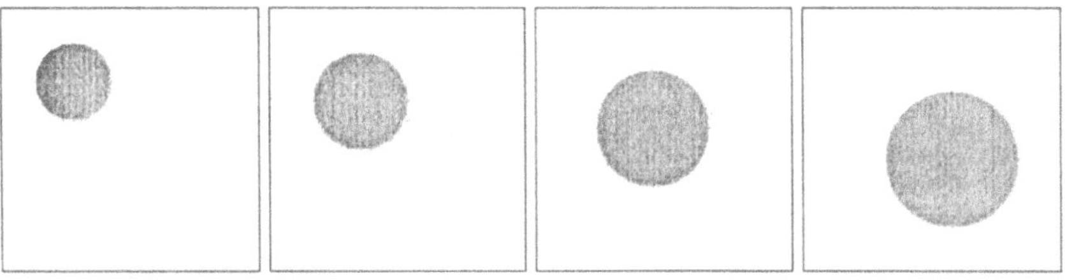

Abb.4a: Synthetische Objektebenen

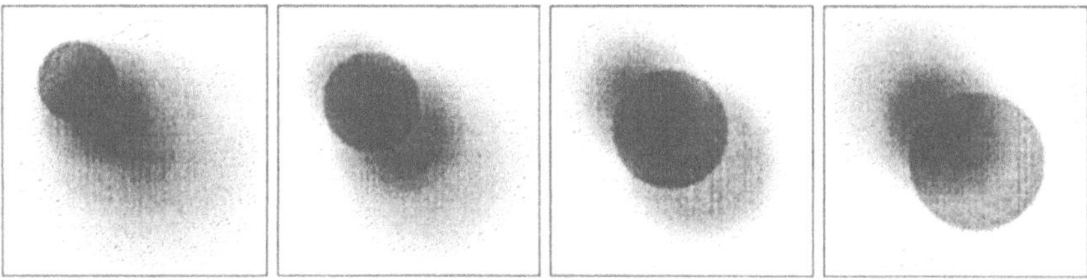

Abb.4b: Fokusserie von Abb.4a

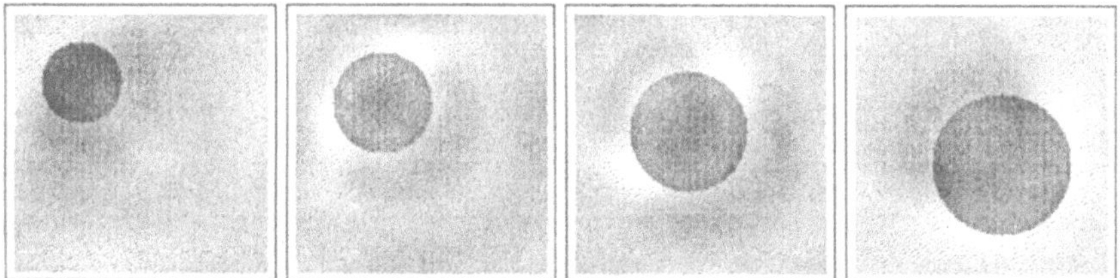

Abb.4c: rekonstruierte Objektebenen

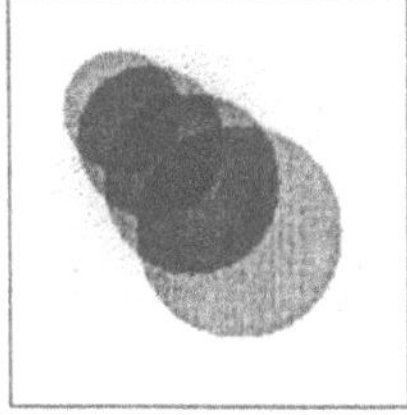

Abb.4d: Bild mit erhöhter Schärfentiefe von Abb.4c

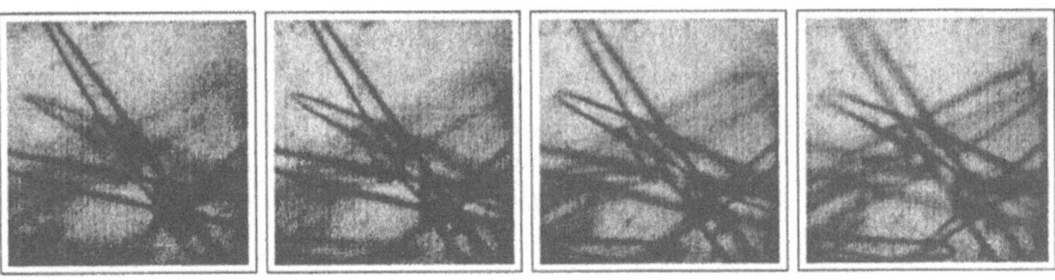

Abb.5a: Fokusserie von einem realen Wassernetz

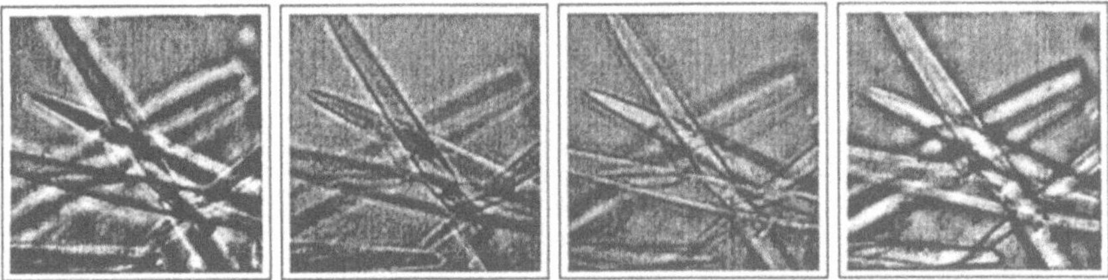

Abb.5b: rekonstruierte Objektebenen vom Wassernetz

Abb.5c: Bild mit erhöhter Schärfentiefe von Abb.5b

Literatur

[1] D.A. Agard, "Optical sectioning microscopy: Cellular architecture in three dimensions", Ann.Rev.Biophys.Bioeng.,Vol.13,s.191-219,1984

[2] K.R. Castleman, *Digital Image Processing.* Englewood Cliffs, NJ: Prentice-Hall, 1979

[3] G.A. Laub, R. Lenz, E.R. Reinhardt, "Three-dimensional object representation in microscopic imaging system", Opt. Eng., Vol.24, no., s.901-905, 1985

[4] A. Erhardt, G. Zinser, D. Komitowski, J. Bille, "Reconstructing 3-D light-microscopic images by digital image processing", Applied Optics, Vol.24, No.2, s.194-200, Jan. 1985

[5] F. Macias-Garza, A.C. Bovik, K.R. Diller, S.J. Aggarwal, J.K. Aggarwal, "Digital reconstruction of three-dimensional serially sectioned optical images", IEEE Transactions on ASSP, Vol.36, No.7, s.1067-1075, July 1988

[6] P. Schwarzmann, X.C. Wu, "Nonlinear approach to the reconstruction of microscopic objects", Acta Stereologica, Vol.8, No.2/2, s.563-568,1989

[7] X.C. Wu, D. Jördens, "Erhöhung der Schärfentiefe eines optischen Systems", DAGM 1990, Springer Verlag, s. 370-376

[8] X.C. Wu, P. Schwarzmann, "Nonlinear approach to the 3d-reconstruction of microscopic objects", SPIE, Conference on Biomedical Image Processing II, 1991

[9] P.A. Stokseth, "Properties of a defocused optical system", J. Opt. Soc. Amer. 1969; Vol. 59, No.10, s.1314-1321, 1969

[10] N. Streibl, "Three-dimensional imaging by a microscope", J. Opt. Soc.Amer., vol.2, No.4, s.121-127, 1985

[11] A. Diaspro, M. Sartore, C. Nicolini, "3D representation of biostructures imaged with an optical microscope", Image and Vision computing, Vol.8, No.2, s.130-141, 1990

[12] C. Preza, M.I. Miller, L.J. Thomas Jr., J.G. McNally, "Regularized linear methode for reconstruction of three-dimensional microscopic objects from optical sections", J. Opt. Soc. Am. A, Vol.9, No.2, 1992

Quantenrauschen und visuelle Detailerkennbarkeit in medizinischen Röntgenaufnahmen

C. Herrmann, E. Buhr, D. Hoeschen
Physikalisch-Technische Bundesanstalt, Bundesallee 100, 38116 Braunschweig

Zusammenfassung: In der konventionellen Radiologie wird die visuelle Erkennbarkeit von diagnostisch wichtigen Bilddetails wesentlich durch die physikalisch meßbare Bildgüte von Röntgenfilm-Folien-Systemen bestimmt. Ein wichtiger Bildgüteparameter ist das Wienerspektrum, das quantitativ das Bildrauschen dieser Systeme beschreibt. Die Quantennatur der Röntgenstrahlung trägt dabei am meisten zum Bildrauschen bei. Das Wienerspektrum ist bei einem empfindlichen System, das für eine bestimmte Filmschwärzung nur eine geringe Röntgendosis braucht, in der Regel höher als bei einem unempfindlichen System. In der vorliegenden Arbeit wird der Einfluß untersucht, den zwei verschiedene Wienerspektren auf die visuelle Erkennbarkeit von tumorartigen Rundherden in Lungenaufnahmen haben. Dazu wurden Originalröntgenaufnahmen des Thorax mit hoher Ortsauflösung digitalisiert. Anschliessend wurden die Wienerspektren der Aufnahmen mit Hilfe eines digitalen Bildverarbeitungssystems modifiziert. Nach dem Einbringen von simulierten Rundherden wurden die modifizierten Bilder mit einem Laser-Scanner in Originalgröße und mit Originalkontrast wieder auf photographischen Film ausgegeben. Die Erkennbarkeit der Krankheitsherde wurde in einer ausgedehnten Studie unter Mithilfe von Radiologen nach der Receiver-Operating-Characteristic-Methode (ROC) ermittelt. Dabei hat sich gezeigt, daß ein um ca. 50% höheres Wienerspektrum die Erkennbarkeit der Rundherde nicht beeinträchtigt. Daher wäre es möglich, durch die Verwendung von entsprechend empfindlicheren Röntgenfilm-Folien-Systemen bei Thoraxaufnahmen die Patientendosis um ca. 50% zu reduzieren, ohne Information für die Diagnose von Rundherden zu verlieren.

1. Einleitung

In der Bundesrepublik Deutschland gibt es jährlich ca. 80 Millionen röntgendiagnostische Anwendungen mit einer mittleren Röntgendosisbelastung von jeweils 1-2 mSv /1/. Aus Gründen des Strahlenschutzes schreibt die Röntgenverordnung /2/ vor, daß „jede unnötige Strahlenexposition von Menschen vermieden wird" (§15(1)1), und daß jede Strahlenexposition auch unterhalb der gesetzlich festgelegten Grenzwerte so gering wie möglich zu halten ist (§15(1)2). Außerdem ist die durch eine Röntgenuntersuchung bedingte Strahlenexposition „soweit einzuschränken, wie dies mit den Erfordernissen der medizinischen Wissenschaft zu vereinbaren ist" (§25(1)). Dieser Forderung nach minimaler Dosisbelastung steht entgegen, daß mit geringerer Röntgendosis die Güte der Röntgenaufnahme im allgemeinen abnimmt. Diese Güte bestimmt ganz wesentlich die Sicherheit der Diagnose, die der Radiologe aus der Röntgenaufnahme ableitet. Die Abbildung 1 zeigt den qualitativen Zusammenhang zwischen der Röntgendosisbelastung und der Diagnose-Sicherheit: Je höher die Bildgüte (z.B. hoher Kontrast, hohe Ortsauflösung, niedriges Bildrauschen), desto genauer die Diagnose. Allerdings läßt sich die Diagnose-Sicherheit nicht beliebig steigern, da z. B. die Leistungsfähigkeit des menschlichen Sehapparates begrenzt ist. Daher ist eine solche Steigerung der Bild-

bereits vorhandene Röntgenaufnahmen zu digitalisieren, hinsichtlich ihrer Bildgüte mit Methoden der digitalen Bildverarbeitung nachträglich zu bearbeiten und anschließend wieder auf photographischen Film auszugeben /7/. Durch diese Verfahrensweise entsteht keine zusätzliche Strahlenbelastung für den Patienten, die abgebildete Szenerie entspricht der radiologischen Praxis, und die Bildgüte der Aufnahmen kann beliebig variiert werden. Allerdings muß sicher gestellt sein, daß das zur Digitalisierung eingesetzte System die Vorlagen möglichst originalgetreu reproduziert.

Bei den hier vorgestellten Arbeiten wurde der letztgenannte Weg gewählt: Originalaufnahmen des Thorax (Film: T-Mat L; Folie: Kodak Lanex Medium), die ohne Befund waren, wurden mit einem Trommelscanner (HELL C 299L) mit hoher Ortsauflösung digitalisiert. Das Abtastintervall betrug 45 μm, die resultierende Bildmatrix umfaßte 7680*7680 Bildpunkte. Bei der Quantisierung wurde der Bereich von 0 bis 2,5 optische Dichten (über Filmschleier) linear in 250 Graustufen umgesetzt. Im nächsten Schritt wurden in jedem Bild mehrere Bereiche als mögliche Lokalisierungen für Testobjekte markiert. In ungefähr jedem zweiten, zufällig ausgewählten Bereich wurde je ein Testobjekt zur Bestimmung der Diagnose-Sicherheit eingebracht. Als Testobjekte dienten im Rechner simulierte Krebsmetastasen. Bei der Simulation dieser sogenannten Rundherde wurde als Modell eine Kugel mit konstantem Absorptionskoeffizienten zugrunde gelegt. Zusätzlich wurden die Rundherde mit einer dem Film-Folien-System entsprechenden Modulationsübertragungsfunktion gefiltert. Da Objekte unter 5-8 mm Größe nicht mehr von normalen anatomischen Strukturen unterschieden werden können /8/, wurde für die Rundherde ein Durchmesser von einheitlich 10 mm gewählt. Der Kontrast der Rundherde war zwischen 0,015 und 0,11 optischen Dichten und wurde in Abhängigkeit vom Bildhintergrund so gewählt, daß die Rundherde an der Erkennbarkeitsgrenze lagen. Radiologen konnten die auf diese Weise simulierten Rundherde nicht von echten Rundherden unterscheiden.

Auf der Basis der digitalen Bilder mit den eingebauten Rundherden wurden zwei Bildersätze erstellt: Ein Bildersatz bestand aus Aufnahmen, die bis auf die markierten Felder und die Rundherde unverändert blieben. Der zweite Bildersatz bestand aus den gleichen Thoraxaufnahmen mit den gleichen markierten Bereichen und Rundherden, jedoch wurde in den Bildern ein höheres Quantenrauschen simuliert. Beide Sätze mit digitalisierten Thoraxaufnahmen wurden anschließend in Originalgröße und mit Originalkontrast wieder auf photographischen Film übertragen.

Die Simulation des höheren Quantenrauschens erfolgte in drei Schritten: Im ersten Schritt wurde ein zweidimensionales Grauwertbild im Format der digitalisierten Röntgenbilder berechnet. Das Grauwerthistogramm dieses Bildes entsprach einer Normalverteilung, das Wienerspektrum des Bildes war über den gesamten Frequenzbereich konstant. Im zweiten Schritt wurde dieses Grauwertbild fouriertransformiert und im Fourierraum gefiltert, so daß nach der Rücktransformation in den Ortsraum ein Grauwertbild mit einem bestimmten spektralen Verlauf des Wienerspektrums vorlag. Im dritten Schritt wurde das gefilterte Grauwertbild zur digitalisierten Röntgenaufnahme addiert. Die Abbildung 4 zeigt drei verschiedene Wienerspektren in Abhängigkeit von der Ortsfrequenz: Das mit „Original-System" bezeich-

güte, die nicht mit einer Steigerung der Diagnose-Sicherheit verbunden ist, nicht nur sinnlos, sondern wegen der gleichzeitig erhöhten Dosisbelastung für den Patienten unerwünscht. Stattdessen wäre es in einem solchen Fall möglich, die Patientendosis ohne Informationsverlust für den Radiologen zu vermindern.

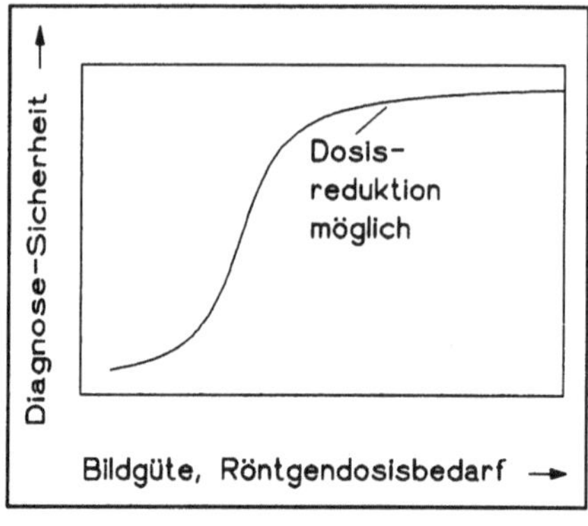

Abb. 1: Qualitativer Zusammenhang zwischen Bildgüte bzw. Röntgendosis und Diagnose-Sicherheit

In der vorliegenden Arbeit wird konkret untersucht, wie eine Änderung des Bild- bzw. Quantenrauschens in konventionellen Röntgenaufnahmen die Erkennbarkeit von Krankheitsherden beeinflußt. Dazu werden konventionelle Röntgenaufnahmen des Thorax mit hoher Auflösung digitalisiert. Das Bildrauschen wird mit Methoden der digitalen Bildverarbeitung so erhöht, als ob die Aufnahmen mit einem empfindlicheren Röntgenfilm-Folien-System aufgenommen worden wären. Zusätzlich werden in die digitalisierten Aufnahmen im Rechner simulierte Krebsmetastasen als Krankheitsherde eingebracht. Anschließend werden die Aufnahmen in Originalgröße und mit Originalkontrast wieder auf photographischen Film ausgegeben. Außerdem werden in gleicher Weise Aufnahmen ohne zusätzliches Bildrauschen erzeugt. Beide Aufnahmeserien werden Radiologen mit der Aufgabe vorgelegt, die simulierten Krebsmetastasen zu diagnostizieren. Die Diagnose-Ergebnisse werden nach dem Receiver-Operating-Characteristic-Verfahren (ROC) /3/ ausgewertet. Aus dem Vergleich der Ergebnisse für beide Bildmodifikationen wird abgeleitet, in welchem Maße eine Dosisreduktion ohne Informationsverlust in der Radiologie möglich ist.

2. Röntgenfilm-Folien-Systeme

Die Abbildung 2 zeigt einen schematischen Querschnitt durch ein heute in der Radiologie übliches Röntgenfilm-Folien-System: Der Röntgenfilm bildet zusammen mit je einer Folie an seiner Vorder- und Rückseite ein „Sandwich". Ein Teil der einfallenden Röntgenquanten wird in den Folien absorbiert, wobei durch die Folie an der Rückseite die Gesamtabsorption gegenüber einem System mit nur einer Folie erhöht wird. Die Lichtquanten belichten die Filmemulsion und verursachen die Filmschwärzung nach der Filmentwicklung. Die Ab-

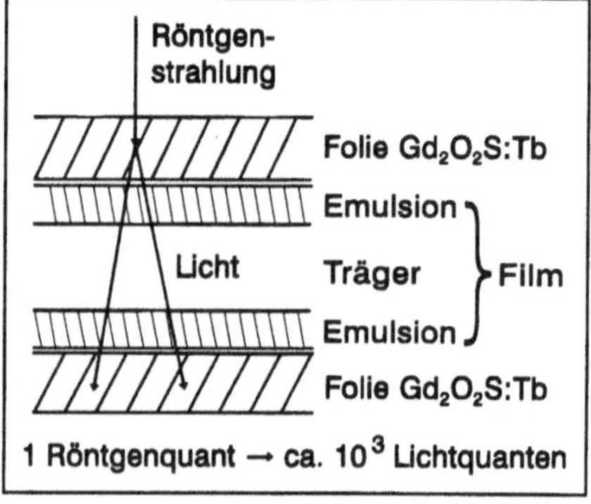

Abb.2: Aufbau eines Röntgenfilm-Folien-Systems

bildung 3 zeigt qualitativ die spektrale Emission einer Folie sowie die spektrale Empfindlichkeit eines darauf abgestimmten Films.
Die lokale statistische Verteilung der in der Folie absorbierten Röntgenquanten führt zu statistischen Schwärzungsschwankungen auf dem Film. Im Vergleich zu diesem „Quantenrauschen" ist der vom Film selbst herrührende Anteil zum gesamten Bildrauschen zunächst vernachlässigbar. Quantitativ wird das auf dem Röntgenfilm sichtbare Bildrauschen durch das Rauschleistungs- bzw. Wienerspektrum beschrieben. Das Wienerspektrum gibt die Abhängigkeit der Rauschleistung von der Ortsfrequenz wieder. Der Zusammenhang zwischen dem Wienerspektrum und den Parametern des Film-Folien-Systems kann durch die folgende Gleichung näherungsweise beschrieben werden /4/:

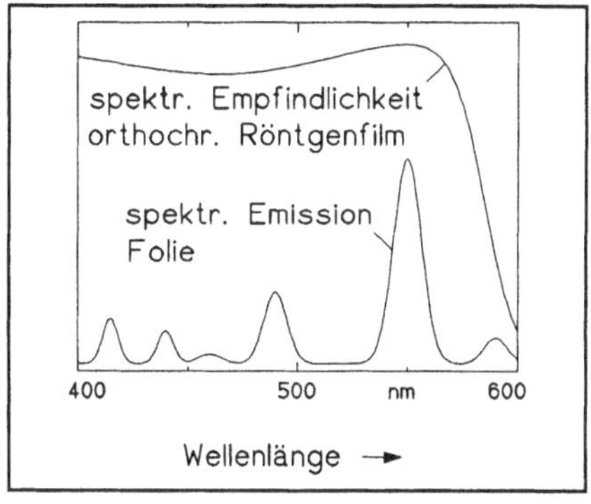

Abb. 3: Spektrale Emission der Folie und spektrale Filmempfindlichkeit

$$W(R) = K * \frac{G^2}{N} * [MTF(R)]^2 \propto \frac{1}{N} \propto \frac{1}{\alpha * Dosisbedarf}$$

Das Wienerspektrum W(R) ist also proportional zum Quadrat des Filmgradienten G und zum Quadrat der Modulationsübertragungsfunktion MTF(R) des Film-Folien-Systems. Außerdem ist es umgekehrt proportional zur Anzahl N der pro Fläche absorbierten Röntgenquanten und damit umgekehrt proportional zum Produkt aus Absorptionskoeffizient α und Dosisbedarf.

3. Erstellen von Aufnahmen unterschiedlicher Bildgüte

Um den in Abbildung 1 gezeigten Zusammenhang zu untersuchen, müssen Aufnahmen eines Objektes mit verschiedenen Bildgüten gemacht werden. Die Aufnahmen müssen zur Bestimmung der Diagnose-Sicherheit Krankheitsherde enthalten, deren Erkennbarkeit möglichst nahe an der Erkennbarkeitsgrenze liegt. Zur Herstellung solcher Aufnahmen gibt es verschiedene Möglichkeiten: Z. B. können Röntgenaufnahmen mit verschiedenen Röntgenfilm-Folien-Systemen von Patienten oder Phantomen angefertigt werden /5, 6/. Dieses Verfahren hat einige Nachteile: Man muß sich auf kommerziell verfügbare Aufnahmesysteme beschränken, Zwischenstufen z. B. zu den bekannten Empfindlichkeitsklassen sind nicht möglich; bei der Verwendung von Phantomen ist die abgebildete Szenerie in der Regel relativ weit entfernt von der Realität der Röntgenpraxis, so daß sich die gewonnenen Aussagen nur schwer auf die Praxis übertragen lassen; mehrmaliges Röntgen ohne medizinische Indikation ist aus Strahlenschutzgründen nicht möglich.
Eine weitere Möglichkeit zur Herstellung von Aufnahmen unterschiedlicher Bildgüte ist,

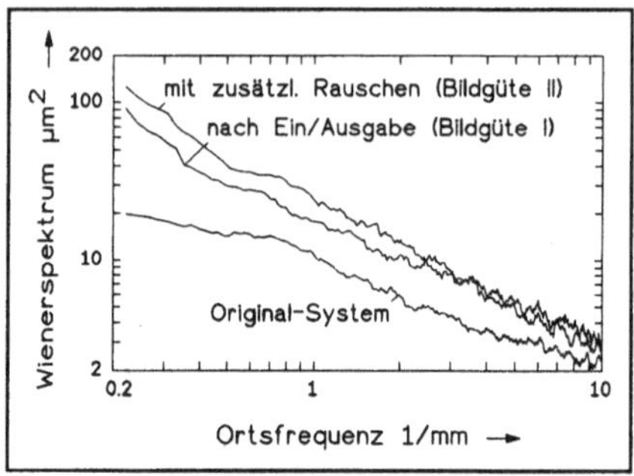

Abb.4: Wienerspektren vor bzw. nach der Bildbearbeitung

nete Spektrum gehört zu einer Aufnahme, die mit dem System Lanex Medium-Folie und T-Mat L-Film gemacht wurden, welches bei der Erstellung der Original-Thoraxaufnahmen eingesetzt worden war. Diese Aufnahme wurde dann mit dem Trommelscanner digitalisiert und anschließend unverändert wieder auf Film geschrieben. Das zugehörige Wienerspektrum ist in Abbildung 4 mit „Bildgüte I" bezeichnet. Zusätzlich wurde zur digitalisierten Aufnahme das oben beschriebene Bildrauschen addiert und das resultierende Bild ebenfalls wieder auf Film ausgegeben. Das zugehörige Wienerspektrum ist in Abbildung 4 mit „Bildgüte II" bezeichnet. Das Wienerspektrum der Aufnahmen mit zusätzlichem Rauschen ist etwa 50% höher als das Spektrum der entsprechenden Bilder ohne Zusatzrauschen und entspricht dem Wienerspektrum eines ungefähr um den Faktor 2 empfindlicheren Film-Folien-Systems.

Vergleicht man das Wienerspektrum des Originalsystems mit dem zur Bildgüte I gehörenden Spektrum, d.h. mit dem Spektrum nach dem Digitalisieren und anschließenden erneuten Ausgeben auf Film, so zeigt sich eine deutliche, durch das digitale Abtast- und Wiedergabesystem verursachte Erhöhung des Rauschens insbesondere bei kleinen Ortsfrequenzen. Dieses Zusatzrauschen ist unerwünscht, es muß jedoch als dem digitalen System inhärent hingenommen werden. Da in der vorliegenden Arbeit jedoch zwei Bildmodifikationen verglichen werden, die beide das durch den Trommelscanner verursachte Zusatzrauschen enthalten, wird die Aussagekraft der gewonnenen Ergebnisse nicht beeinträchtigt. Außerdem war das gemessene, unerwünschte Zusatzrauschen in den Reproduktionen der Thoraxaufnahmen praktisch nicht visuell wahrnehmbar: Die Reproduktionen waren so gut gelungen, daß sie nur bei detaillierter Betrachtung im direkten Vergleich mit den Originalvorlagen von diesen unterschieden werden konnten.

4. Beurteilen der Bildgüte durch Radiologen

In einer Studie mit 16 Radiologen wurde untersucht, inwieweit sich die Diagnosen der simulierten Rundherde in den Bildern mit bzw. ohne Zusatzrauschen unterscheiden. Für jede der beiden Bildgüten wurden zwölf verschiedene Thoraxaufnahmen mit insgesamt 324 markierten Suchfeldern und 159 Rundherden präpariert. Die 24 Aufnahmen wurden den Radiologen in zufälliger Reihenfolge vorgelegt. Bei den Betrachtungsbedingungen gab es keinerlei Einschränkungen bezüglich der Betrachtungszeit, der Beleuchtung oder eventuell benutzter Hilfsmittel: Jeder Radiologe konnte die Bilder in seiner eigenen Praxis unter den von ihm gewohnten Bedingungen beurteilen. Bei der Beurteilung der Bilder mußte jedes der markierten Felder in eine von fünf Kategorien eingeordnet werden: 1 = sicher kein Rundherd vorhanden; 2 = wahrscheinlich kein Rundherd vorhanden; 3 = weiß nicht; 4 = Rundherd wahrscheinlich vorhanden; 5 = Rundherd sicher vorhanden. Die Bewertungen der einzelnen Felder wurden für jeden der Radiologen nach dem „Receiver-Operating-Characteristic-Verfahren (ROC)" /3, 9/ ausgewertet. Dabei wurden für die einzelnen Bewertungsstufen die Anteile der falsch-positiven Entscheidungen, FPF (*false-positive fraction*), und die Anteile der wahr-positiven Entscheidungen, TPF (*true-positive fraction*), berechnet. Die Auftragung von TPF gegen FPF wie in Abbildung 5 wird als ROC-Kurve bezeichnet /3/. Anhand des Verlaufs der ROC-Kurve und der Fläche unter der ROC-Kurve wird die Erkennbarkeit der Rundherde beurteilt. Beispielsweise würde die ROC-Kurve eines Beobachters, der alle Rundherde richtig erkennt, mit den Geraden FPF=0 und TPF=1 zusammenfallen, die Fläche unter dieser Kurve ist dann gleich eins. Eine Bewertung der Bilder durch reines Raten führt dagegen zu gleichvielen wahr-positiven wie falsch-positiven Entscheidungen. Die zugehörige ROC-Kurve ist die Hauptdiagonale mit der Fläche 0,5.

5. Ergebnisse und Diskussion

Für die Beurteilung der 24 Thoraxaufnahmen mit insgesamt 648 Suchfeldern und 318 Rundherden benötigte jeder Radiologe etwa zwei Stunden. Dieser erhebliche Aufwand ist erforderlich, da nur durch die große Anzahl von Testobjekten die statistische Unsicherheit jeder einzelnen ROC-Kurve möglichst klein gehalten werden kann: Nur bei kleinen statistischen Unsicherheiten der Einzelkurven ist es möglich, relativ kleine Unterschiede zwischen zwei ROC-Kurven als statistisch signifikant nachzuweisen.
Die Abbildung 5 zeigt die ROC-Kurven für die Erkennbarkeit der Rundherde am Beispiel von zwei Beobachtern. Die ROC-Kurven der anderen Beobachter sind den gezeigten Beispielen ähnlich. Die Symbole „x" und „+" gehören zu den experimentell ermittelten (FPF,TPF)-Wertepaaren für die Bilder ohne und mit Zusatzrauschen. Die durchgezogenen Kurven repräsentieren die Anpassung eines mathematischen Modells /3, 9/ an die experimentellen Werte. Die gestrichelten Kurven markieren den $\pm 1\sigma$-Fehlerbereich der angepaßten ROC-Kurve für die Bilder ohne Zusatzrauschen /9/. Versuchswiederholungen mit einem Teil der Beobachter haben gezeigt, daß die einzelnen ROC-Kurven im Rahmen der angegebenen Fehlerbereiche reproduzierbar sind.
Die ROC-Kurven zeigen, daß die Kontraste der Rundherde sinnvoll gewählt waren: Die ROC-Kurven sind genügend weit von den beiden Extremen „alle Rundherde richtig erkannt" bzw. „nur Zufallstreffer" entfernt. Die Rundherde waren also weder zu leicht noch zu schwer zu erkennen. Damit kann erwartet werden, daß sich bereits relativ kleine Unterschiede in der Erkennbarkeit der Rundherde in unterschiedlichen ROC-Kurven bemerkbar machen.
Wie man anhand der Beispiele bereits sehen kann, unterscheidet sich die Erkennbarkeit der Rundherde in den Bildern mit dem zusätzlichen Rauschen nur wenig von der Erkennbarkeit der Rundherde in den Bildern ohne dieses Rauschen. Die zu den beiden Bildgüten gehörenden ROC-Kurven wurden mittels statistischer Signifikanztests miteinander verglichen:

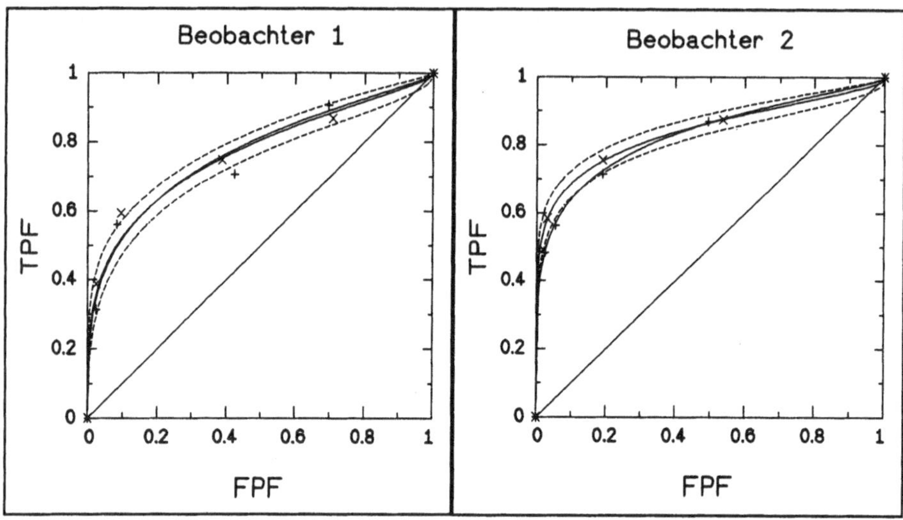

Abb.5: ROC-Kurven der Beobachter 1 und 2: x = Bildgüte I; + = Bildgüte II; durchgezogene Linien = Modellrechnung, -- = ±1σ-Fehlerbereich; Diagonale = ROC-Kurve für den Fall rein zufälliger Entscheidungen

Bei jedem einzelnen der 16 Radiologen wurde in einem χ^2-Test /9/ die Nullhypothese „Die zu den beiden Bildgüten gehörenden ROC-Kurven sind statistisch verschieden" überprüft. Dabei wurde bei keinem der Radiologen ein signifikanter Unterschied zwischen den beiden ROC-Kurven gefunden (Signifikanzniveau α=0,05).
Bei der hier vorgestellten Studie ist festzustellen, daß sich die ROC-Kurven von unterschiedlichen Beobachtern deutlich voneinander unterscheiden können, obwohl allen Beobachtern dasselbe Bildmaterial mit derselben Aufgabenstellung präsentiert wurde. Aus diesem Grunde wurden beim Vergleich der Erkennbarkeit der Rundherde bei den beiden Bildgüten nicht die über alle Beobachter gemittelten ROC-Kurven verwendet, sondern die beiden ROC-Kurven eines jeden einzelnen Beobachters miteinander verglichen /11/.

Bei den bisherigen Auswertungen des vorliegenden Datenmaterials wurde der Thorax als einheitliches „System" betrachtet, d.h., es wurde nicht nach unterschiedlichen Bereichen des Thorax differenziert. Bei der Festlegung der Rundherdkontraste (siehe Kap. 3) hat sich gezeigt, daß Rundherde in stark strukturierten Bildhintergründen, z. B. die Thoraxbereiche nahe der Wirbelsäule, bei gleichen Kontrasten schlechter zu erkennen sind als Rundherde in homogenen Hintergründen. Die Erkennbarkeit der Rundherde wird also wesentlich durch die anatomische Struktur des jeweiligen Bildhintergrundes beeinflußt. Gegenüber diesem Strukturhintergrund oder „Strukturrauschen" /12/ ist vermutlich der Einfluß gering, den Unterschiede im eigentlichen Quantenrauschen auf die Erkennbarkeit der Testobjekte haben.
Für eine detaillierte Analyse der Ergebnisse ist eine weitergehende Auswertung des Datenmaterials erforderlich, bei der die Rundherde nach der Komplexität des Bildhintergrundes klassifiziert werden.

6. Folgerungen

Die vorliegende Untersuchung hat insgesamt gezeigt, daß ein um etwa 50% höheres Wienerspektrum die Erkennbarkeit von 1 cm großen Rundherden in Thoraxaufnahmen nicht beeinträchtigt. Mit anderen Worten, die Diagnose-Sicherheit ist trotz eines Unterschiedes von etwa 50% in den Wienerspektren in beiden Fällen gleich hoch. Da die Bilder mit dem höheren Wienerspektrum Röntgenaufnahmen mit einem um den Faktor 2 empfindlicheren Röntgenfilm-Folien-System simulieren, kann aus dem Ergebnis gefolgert werden, daß sich in der radiologischen Praxis durch die Verwendung von entsprechend empfindlicheren Systemen die Patientendosis um ca. 50% reduzieren läßt. Dabei ist zumindest bei der Diagnose von Rundherden kein Informationsverlust zu befürchten.

Die Untersuchungen zeigen, daß die Erkennbarkeit der Rundherde in den Röntgenaufnahmen offensichtlich nicht entscheidend von dem Bildrauschen auf den Aufnahmen abhängt. Die Ergebnisse deuten vielmehr an, daß die anatomische Hintergrundstrukur auf den Aufnahmen die Erkennbarkeit der Rundherde begrenzt. Dies soll durch weitere Arbeiten überprüft werden.

7. Literaturverzeichnis

/1/ C.Reiners: „Strahlenexposition und Risiko in der nuklearmedizinischen und Röntgendiagnostik", Röntgenpraxis 46, 19-21 (1993)

/2/ „Die neue Röntgenverordnung", H. Hoffmann GmbH, Berlin (1987)

/3/ J.A.Swets, R.M.Pickett: „Evaluation of Diagnostic Systems", Academic Press, New York (1982)

/4/ E.Krestel: „Bildgebende Systeme für die medizinische Diagnostik", 2. Auflage, Siemens-Aktiengesellschaft (Abteilung Verlag), Berlin und München (1988)

/5/ J.E.Gray, K.W.Taylor, B.B.Hobbs: „Detection accuracy in chest radiographs", Am. J. Roentgenol. 131, 247-253 (1978)

/6/ D.J.Goodenough, K.Rossmann, L.B.Lusted, „Factors affecting the detectability of a simulated radiographic signal", Invest. Radiol. vol. 8, 339-344 (1973)

/7/ H.P.Chan, K.Doi, C.E.Metz: „Digital image processing: Effects of Metz filters and matched filters on detection of simple radiographic objects", SPIE 454 „Application of optical instrumentation in medicine XII", 420-432 (1984)

/8/ C.A.Kelsey, R.D.Moseley, B.G.Brogdon, D.G.Bhave, J.Hallberg: „Effect of size and position on chest lesion detection", A. J. Roentgenol. 129, 205-208 (1977)

/9/ C.E.Metz, P.L.Wang, H.P.Kronman: „A new approach for testing the significance of differences between ROC curves measured from correlated data", in F.Deconinck: „Information processing in medical imaging", Nijhoff, Kluwer Academie, 432-445 (1984)

/10/ I.N.Bronstein, K.A.Semendjajew: „Taschenbuch der Mathematik", Nachdruck der 20. Auflage, Verlag Harri Deutsch, Thun und Frankfurt/Main, 725-726 (1979)

/11/ C.Herrmann, E.Buhr, D.Hoeschen, S.-Y.Fan: „Comparison of ROC and AFC methods in a visual detection task", Med. Phys. 20 (1993), in Druck

/12/ G.Revesz, H.L.Kundel, M.A.Graber: „The influence of structured noise on the detection of radiologic abnormalities", Invest. Radiol. Vol.9, 479-486 (1974)

Fraktale Dimension der Kontur endoskopisch ermittelter Farbbilder von Geschwüren des Magens

D. W. R. Paulus, H. Niemann, C. Lenz
Universität Erlangen–Nürnberg
Lehrstuhl für Mustererkennung (Informatik 5)
Martensstraße 3, D–91058 Erlangen

L. Demling[1], C. Ell
Universität Erlangen–Nürnberg
Medizinische Klinik
D–91054 Erlangen

paulus@informatik.uni-erlangen.de

Zusammenfassung. In diesem Beitrag wird dargestellt, wie die fraktale Dimension von Konturen in endoskopischen Bildern ermittelt werden kann. Dieser Wert kann dazu verwendet werden, die Entscheidung des Arztes zu unterstützen, ob ein Geschwür gutartig oder bösartig ist. Die Lokalisierung eines Geschwürs in endoskopischen Bildern erfordert speziell angepaßte Verfahren, die beschrieben werden. Ziel der Untersuchungen ist es, durch rechnergestützte Bildverarbeitung dem Arzt einen objektiven Parameter zur Dignitätsbeurteilung endoskopischer Befunde zur Verfügung zu stellen.

1 Einleitung

Magensgeschwüre (Ulcera ventriculi) stellen eine häufige und wichtige Erkrankung des Menschens dar. Circa 1–5% der Magengeschwüre sind bösartig (maligne). Bei einer geschätzen Inzidenz von 1–2 Millionen Ulcera pro Jahr ist es Aufgabe des Arztes, die malignen Ulcerationen treffsicher zu erkennen. Nur die frühzeitige Erkennung bedeutet eine Chance auf Heilung.

Methode der Wahl zur Untersuchung ist die Endoskopie des Magens über flexible Glasfaserinstrumente (Endoskop). Bisher hängt die Treffsicherheit in erster Linie von der subjektiven Erfahrung des untersuchenden Arztes ab. Lediglich die durch den Untersucher veranlaßte und durchgeführte Gewebeentnahme (Biopsie) kann dann zur Objektivierung des Befunds führen. Ein wesentlicher Fortschritt bestünde darin, bereits während der Untersuchung einen objektiven Parameter zu erhalten, der über die Dignität des Ulkus eine zuverläßliche, Untersucher–unabhängige Aussage machen kann.

Bösartige Geschwüre sind in der Regel in ihrer Randkontur zerklüfteter und unregelmäßiger als gutartige. Ziel dieser Arbeit ist es, einen Zahlenwert für den Grad der Komplexität der Konturbegrenzung eines Geschwürs zu finden. Ein möglicher Lösungsweg ist dabei die fraktale Geometrie, die mit der fraktalen Dimension eine Maßzahl für die Zerklüftung von Küstenlinien, Inseln oder allgemein Linien liefert.

Dieser Beitrag beschreibt eine erste technische Realisierung der in [Dem92] dargestellten Idee, fraktale Methoden auf *endoskopische* Bilder im Gastrointestinaltrakt anzuwenden. Aufgrund der niedrigen Schärfe des Bildmaterials und der Unmöglichkeit, die Grenze eines Geschwürs exakt zu definieren, müssen hierzu spezielle Verfahren entwickelt werden.

---

[1] em. Direktor der med. Klinik mit Poliklinik, D–96132 Schlüsselfeld, Weidweg 5

2 Fraktale Geometrie diskreter Bilder

Im Gegensatz zur herkömmlichen Geometrie ist die fraktale Geometrie in der Lage, natürliche Formen wie Wolken, Küstenlininen, Wellen usw. zu beschreiben. Ein Fraktal ist eine geometrische Figur, in der sich das gleiche Motiv in stets kleinerem Maßstab wiederholt [Lau89]. Für selbstähnliche Liniengebilde kann der Begriff der *fraktalen Dimension* eingeführt werden [Man83], die im Gegensatz zur gebräuchlichen topologischen Dimension nicht unbedingt ganzzahlige Werte hat.

Formal kann man sich die Berechnung der fraktalen Dimension einer beliebigen Linie folgendermaßen vorstellen [Pei88]: Zu einer Menge A, die man sich der Einfachheit halber als Teilmenge einer Ebene vorstellen kann, und einer kleinen Zahl $a > 0$ bildet man die Überdeckungszahl $N(A, a)$; dazu wird A mit endlich vielen Kugeln $K(x_i, a)$, deren Radius maximal a ist, überdeckt. Es gilt:

$$A \subset \bigcup_{x_i \in A} K(x_i, a) \qquad (1)$$

Dann sei $N(A, a)$ die minimale Zahl von Kugeln, die für eine solche Überdeckung benötigt werden. Zwischen $N(A, a)$ und a gibt es eine Beziehung, die man als Potenzgesetz [Pei88] ansetzt:

$$N(A, a) = a^{-D} \qquad D = lim_{a \to 0} \frac{\log N(A, a)}{-\log a} \qquad (2)$$

Falls der Grenzwert in Gleichung 2 existiert, wird D die *fraktale Dimension* von A genannt [Pei88]. Abbildung 1 (a) zeigt eine experimentelle Interpretation des formalen Verfahrens. Man legt über die Menge A — hier die Linie — Gitter mit verschiedenen Maschenweiten und zählt, wieviele der Maschen die Linie treffen. Trägt man alle Werte in ein doppelt logarithmisches Koordinatensystem (Abbildung 1 (b)) ein, kann man die sogenannte Box–Dimension D als Steigung einer Geraden, welche die Messungen interpoliert, ablesen [Pei91]. Das Verfahren läßt sich auch bei zweidimensionalen Bildern realer Objekte wie Flußsystemen oder den Zotten der Darmwände anwenden.

Ein erweitertes Verfahren zur Bestimmung der fraktalen Dimension von dreidimensionalen Oberflächen wird in [Mog89] vorgestellt. Mathematische Grundlage ist ein homogenes Kraftgesetz, das die Masse einer Menge als eine Funktion angibt, die einen expandierenden Kreis um den interessierenden Punkt beschreibt.

Dazu wird um den Punkt (x, y) mit Intensität $f(x, y)$ ein Würfel der Seitenlänge L gelegt und die Anzahl N der Nachbarpunkte innerhalb des Würfels ausgezählt. Das Kräftegesetz für die Masseverteilung ist nun durch

$$M(L) = kL^D \qquad (3)$$

gegeben. Der expandierende Radius wird durch die sich vergrößernde Seitenlänge des Würfels ausgedrückt, der um den Punkt $(x, y, f(x, y))$ zentriert ist. (Abbildung 1 (c) zeigt das Verfahren für den eindimensionalen Fall). Die Anzahl der

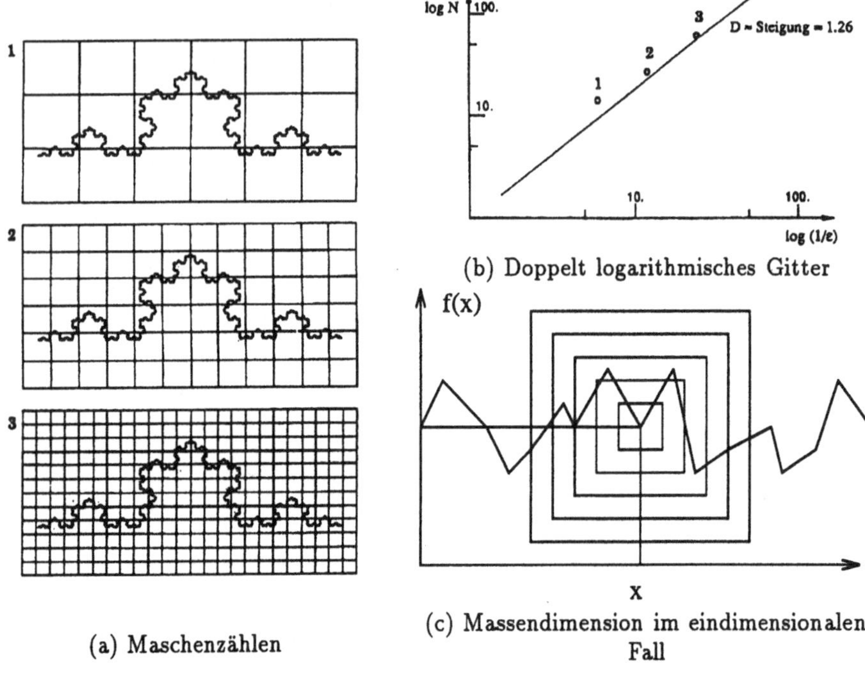

(a) Maschenzählen

(b) Doppelt logarithmisches Gitter

(c) Massendimension im eindimensionalen Fall

Abbildung 1. Fraktale Dimension ((a), (b) aus [Pei88] (c) aus [Mog89])

Punkte im Würfel wird ebenfalls in ein doppelt logarithmisches Gitter eingetragen. Der Exponent D ergibt sich wiederum als Steigung der Ausgleichsgeraden durch diese Punkte.

Weitere Probleme der Behandlung von Chaos und Fraktalen in diskreten Bildern werden in [Nie93] angesprochen. Es existieren zudem weitere Algorithmen zur Berechnung der fraktalen Dimension (vgl. z.B. [Hoe92, Sch92]).

3 Kantendetektion in Farbbildern

Ausgangspunkt der Untersuchungen waren 18 Farbbilder von 6 verschiedenen Krankheitsfällen. Abhängig von Weißabgleich, Empfindlichkeit und sonstigen Einstellungen der Digitalisiereinrichtung fallen die einzelnen Farbkanäle unterschiedlich aus (vgl. Abbildung 2).

Mit der bekannten Definition des Sobel–Operators können Approximationen der partiellen Ableitungen der Bildfunktion gebildet werden (Gleichung 4,5). Dazu werden Differenzen S von je zwei Bildpunkten $f(x_i, y_i)$ einer Nachbarschaft

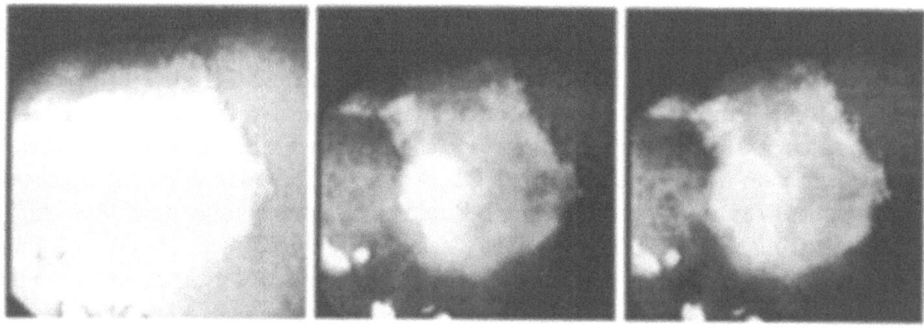

Abbildung 2. Farbbild zerlegt in seine drei Farbkanäle.

errechnet und gewichtet.

$$d_{x,y}^{(h)} = S(f_{x-1,y-1}, f_{x+1,y-1}) + 2S(f_{x-1,y}, f_{x+1,y}) + S(f_{x-1,y+1}, f_{x+1,y+1}) \quad (4)$$

$$d_{x,y}^{(v)} = S(f_{x-1,y-1}, f_{x-1,y+1}) + 2S(f_{x,y-1}, f_{x,y+1}) + S(f_{x+1,y-1}, f_{x+1,y+1}) \quad (5)$$

Kantenstärke und Kantenrichtung der Pixel im Kantenbild ergeben sich aus $d^{(h)}$ und $d^{(v)}$. Für Farbpixel $f_1 = (r_1, g_1, b_1)$ und $f_2 = (r_2, g_2, b_2)$ muß dazu ein Abstandsmaß definiert werden. Wegen der unterschiedlichen Informationsdichte in den einzelnen Farbkanälen wurde eine Gewichtung $(\beta_R, \beta_G, \beta_B) = (1, 1, 4)$ der einzelnen Kanäle durchgeführt und die folgende Differenz verwendet:

$$S(f_1, f_2) = (\beta_R * (r_1 - r_2)^2 + \beta_G * (g_1 - g_2)^2 + \beta_B * (b_1 - b_2)^2)^{\frac{1}{2}} \quad (6)$$

Abbildung 3 (a) zeigt das Ergebnis des Farbsobeloperators. Alle untersuchten Verfahren zur Kantendetektion erkennen neben den Geschwürkanten noch sehr viele weitere Kanten, die nur aus einer Oberflächenstruktur stammen [Len92]. Zur Berechnung der fraktalen Dimension muß daher die Berechnungsfläche auf die wesentlichen Kanten begrenzt werden.

4 Ellipsen und Geschwüre

Die praktische Erfahrung zeigt, daß sich Geschwüre durch eine mehr oder weniger große Ellipse annähern lassen. Zeichnet man in das Bild zwei Ellipsen mit einem größeren und einem kleineren Radius ein, dann kann man die meisten Kanten des Geschwürs mit einem *Ellipsenschlauch* abdecken. Ein oft benutztes Verfahren, um in verrauschten Bildern Gebilde wie Kreise, Ellipsen oder Geraden zu erkennen ist die *Hough-Transformation* [Nie90]. Mit der *inversen Hough-Transformation* [Hor93] ist es möglich, einzelne Ellipsen im Bild ohne aufwendige Suche im Parameterraum zu finden:

In einem Kantenbild des verwendeten objektorientierten Bildverarbeitungssystems ἵππος ([Pau92]) sind in jedem Punkt die Kantenstärke und Kantenrichtung bekannt. Mit Kenntnis der Kantenrichtung kann man Geraden $y = a_i x + b_i$ senkrecht zur Kantenrichtung assoziieren (parallel zum Gradienten). Betrachtet man nun ein ellipsenförmiges Objekt, so schneiden sich diese Geraden in einem Bereich, in dem der Schwerpunkt des Objekts liegt. Diese Geraden entsprechen im Parameterraum (a, b) jeweils einem Punkt. Der Schwerpunkt der Ellipse läßt sich approximativ mit Hilfe einer Ausgleichsgerade durch diese Punkte im Parameterraum bestimmen. Steigung $-x$ und Abschnitt y der Ausgleichsgerade berechnen sich nach Gleichung 7.

$$x = \frac{\sum a_i b_i * n - \sum a_i * \sum b_i}{\sum a_i^2 * n - \sum a_i * \sum a_i} \qquad y = \frac{\sum a_i^2 * \sum b_i - \sum a_i * \sum a_i b_i}{\sum a_i^2 * n - \sum a_i * \sum a_i} \tag{7}$$

Durch Rückkehr in den Bildraum bekommt man mit den gefundenen Werten x und y die Koordinaten des Schwerpunkt eines Objekts, und damit — innerhalb der hier betrachteten Anwendung — den Schwerpunkt eines Geschwürs.

Brennpunkte und Radius der Ellipse werden unter Verwendung einiger Heuristiken errechnet. Dazu wird der Koordinatenursprung in den berechneten Schwerpunkt der Ellipse verschoben. Für alle Kanten in den Quadranten eins und drei wird die Ausgleichsgerade im Parameterraum nach obigem Verfahren berechnet; im Bildraum werden so zwei Punkte ermittelt. Der mit der Kantenanzahl gewichtete Mittelwert dieser zwei Punkte ergibt einen Brennpunkt. Der zweite Brennpunkt entsteht durch Spiegelung am Schwerpunkt. Die Halbradien ergeben sich durch Mittelung der Abstände aller Kantenpunkte zu den Brennpunkten. Weitere Heuristiken finden sich in [Hor93].

5 Fraktale Dimension der Kontur im Kettencode

Nach dem obigen Verfahren werden nun der Mittelpunkt und eine Näherungsellipse für das Geschwür berechnet. Daraus werden zwei Ellipsen ermittelt, deren Radien gegenüber der errechneten Ellipse vergrößert, bzw. verkleinert werden. Die Kontur des Geschwürs wird nun innerhalb des Raumes zwischen diesen beiden Ellipsen angenommen. Dazu wird die Bildinformation zwischen den beiden Ellipsen aus dem Kantenbild herausgeschnitten (Abbildung 3 (b)).

Im maskierten Kantenbild erfolgt eine Kantenverfolgung mit Standardverfahren; als Ergebnis entsteht ein *Segmentierungsobjekt* (Abbildung 3 (a), [Pau92]) bestehend aus Linien im Kettencode. Von jedem der kleinen Segmentstücke im Segmentierungsobjekt ist die fraktale Dimension zu bestimmen und festzuhalten. Die gesamte fraktale Dimension ist dann der Mittelwert über die Dimensionen aller Kettenstücke.

Auf den sehr kurzen Ketten lieferte diese Verfahren Werte unter Eins (Abbildung 4 (a)). Dieses Verfahren verlangt jedoch sehr viel längere Ketten. So sind die kurzen Kettenstücke aufgrund ihrer Dimension eher als Staub (*Cantor-Staub*) als als Linien anzusehen. Die Schwierigkeiten liegen allerdings auch in

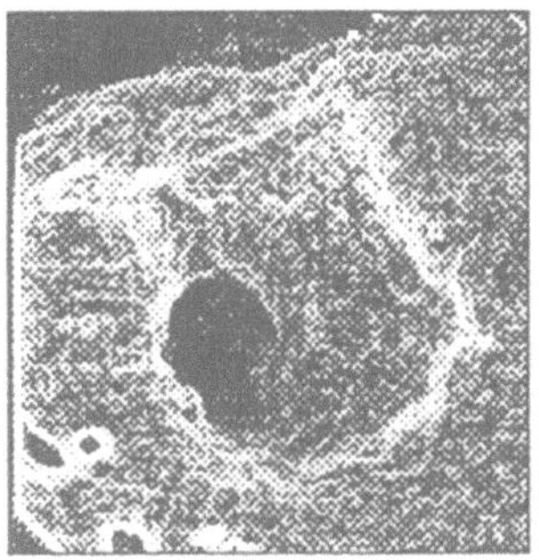

(a) Kantenstärkebild

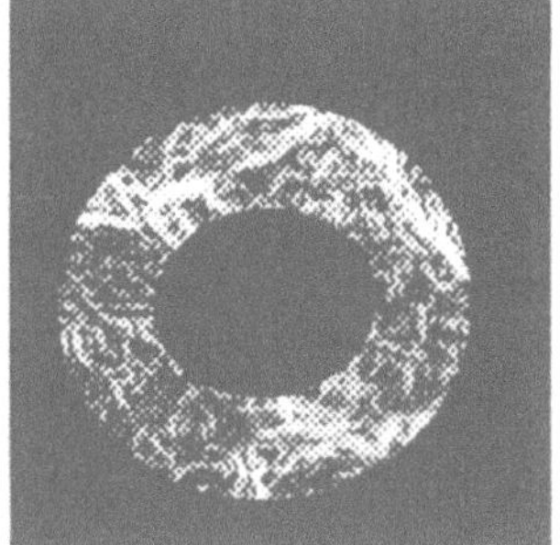

(b) Maskiertes Kantenbild

(c) Segmentierungsobjekt

Abbildung 3. Liniensuche im maskierten Kantenbild (Dimension 0.6725)

der Art des Bildes selbst. Die feineren, tatsächlich vorhandenen Fransen kommen auf den Bildern nicht deutlich genug heraus. Dies liegt beispielsweise an den feuchten, glänzenden Magen- und Darmwänden, die die Kontur der Geschwüre verwischen. Daher wurde das im folgenden vorgestellte Verfahren zur Bestimmung der fraktalen Dimension untersucht.

6 Fraktale Massendimension des Geschwürs

Die Berechnungsfläche für die fraktale Dimension ebenfalls nach dem Verfahren der Massendimension kann nach dem oben beschriebenen Verfahren auf einen Ellipsenschlauch beschränkt werden. Diese Dimension kann dann alternativ auch auf dem gesamten Bild berechnet werden. Zunächst wird aus dem Farbbild ein Grauwertbild errechnet. Dann wird die fraktale Massendimension für jedes Pixel in der Berechnungsfläche ermittelt. Um jedes Pixel wird dazu ein Würfel mit expandierender Größe gelegt, wobei als dritte Dimension die Intensität genommen wird. Innerhalb des Würfels wird dann die Anzahl der benachbarten Pixel ausgezählt. Dieser Vorgang wird für verschiedene Kantenlängen des Würfels wiederholt. Die Werte werden wiederum in ein doppelt logarithmisches Koordinatensystem aufgetragen. Die Steigung der Ausgleichsgeraden ist die gesuchte fraktale Dimension in dem aktuellen Bildpunkt. Die gesamte fraktale Dimension ist der Mittelwert über die fraktalen Dimensionen der einzelnen Pixel. Das Ergebnis ist dann ein Zahlwert, der zwischen 1.6 und 2.3 liegt. Dieser Wertebereich wurde auch in [Mog89] auf Testoberflächen gemessen.

7 Ergebnisse und Diskussion

Aufgrund des vorliegenden Bildmaterials läßt sich noch keine endgültige Aussage über die fraktale Dimension der Kontur endoskopische ermittelter Farbbil-

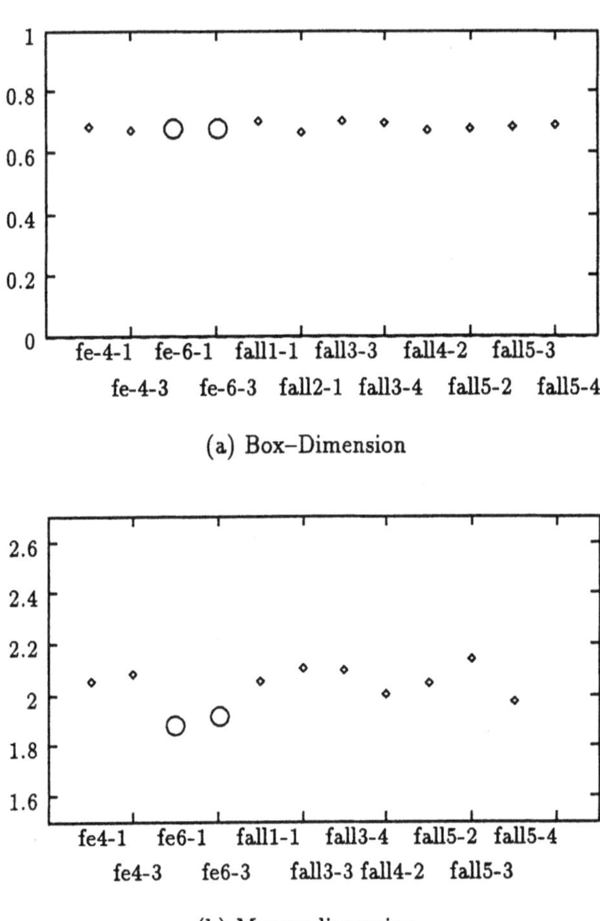

Abbildung 4. Messungen der fraktalen Dimension. ○ bösartiger, ◇ gutartiger Fall

der von Geschwüren treffen. Die Ergebnisse ermutigen jedoch zu weiteren Untersuchungen.

Die Berechnung der fraktalen Dimension aus den Kettencodes des Ellipsenbildes hat nicht die erwarteten Werte gebracht. Die Werte lagen nicht wie erwartet zwischen Eins und Zwei sondern waren durchweg kleiner als Eins (Abbildung 4 (a)). Auch die Interpretation der Werte als Stäube, die eine fraktale Dimension zwischen Null und Eins haben, zeigte keinen signifikanten Unterschied zwischen den einzelnen Bildern. Der einzige bösartige Fall, die Bilder mit der Bezeichnung "fe–6", lieferte ähnliche Werte, wie die Fälle, in denen es sich bei dem Geschwür um ein harmlosen Ulcus handelt. In Abbildung 4 (a) ist die Gleichmäßigkeit der Werte deutlich zu sehen. Eine Verlängerung der Kettencodesegmente durch Aneinanderhängen der kurzen Lininensegmente hat ebenfalls keine verbesserten Werte bewirkt.

Diese Ergebnisse haben die Berechnung der fraktalen Dimension nach dem Verfahren der Massendimension motiviert. Die Kontur wurde nur noch insoweit berücksichtigt, als in einem Vorschritt die Geschwürkanten durch eine elliptische Approximation herausgeschnitten wurden. Im Gegensatz zu der Gitter–Zählmethode, wurden bei der Ermittlung der Massendimension die Erwartungen an die Größe der Werte erfüllt. Wie aus Abbildung 4 (b) ersichtlich ist, fallen lediglich die Werte des bösartigen Falls *fe–6–1* beziehungsweise *fe–6–3* geringfügig aus der Reihe. Dies werten wir als Indiz dafür, daß eine Unterscheidung in gut– und bösartige Geschwüre möglich ist. Allerdings wurden für das maligne Karzinom *höhere* Werte erwartet. Anhand einer größeren Stichprobe sollen nun statistisch aussagekräftige Werte ermittelt werden.

Literatur

[Dem92] L. Demling: *Fractals in Endoscopy*, *Endoscopy*, Bd. 24, Nr. 6, 1992, S. 590–591.

[Hoe92] S. Hoefer, R. Kumaresan, W. Ohley: *Fractal dimension in the analysis of medical images*, *IEEE Engineering in medicine and biology*, Bd. 11, Nr. 2, June 1992.

[Hor93] J. Hornegger, D. W. R. Paulus: *Detecting Elliptic Objects Using Inverse Hough–Transform*, in G. Vernazza, A. N. Venetsanopoulos, C. Braccini (Hrsg.): *Image Processing: Theory and Applications*, Elsevier, 1993, S. 155–158, Proceedings of the IPTA '93.

[Lau89] H. Lauwerier: *Fraktale verstehen und selbst programmieren*, Wittig-Fachbuchverlag, Hückelhoven, 1989.

[Len92] C. Lenz: *Fraktale Dimension der Kontur endoskopisch ermittelter Farbbilder von Geschwüren*, Diplomarbeit, Lehrstuhl für Informatik 5 (Mustererkennung), Universität Erlangen–Nürnberg, Erlangen, 1992.

[Man83] B. B. Mandelbrot: *The fractal geometry of nature*, Freeman, New York, 1983.

[Mog89] B. Moghaddam: *Local fractal dimension operators and relaxation techniques for image segmentation*, Georg Mason University, Fairfax, Virginia, 1989, Master Thesis.

[Nie90] H. Niemann: *Pattern Analysis and Understanding*, Springer-Verlag, Berlin, Heidelberg, 1990.

[Nie93] J. Nievergelt, P. Schorn: *Numerik des Chaos oder Chaos der Numerik?*, *Informatik Spektrum*, Bd. 16, Nr. 5, 1993, S. 39–41.

[Pau92] D. W. R. Paulus: *Objektorientierte und wissensbasierte Bildverarbeitung*, Vieweg, Braunschweig, 1992.

[Pei88] H. O. Peitgen, H. Jürgens: *Fraktale: Gezähmtes Chaos*, Siemens Stiftung, München, 1988.

[Pei91] H. O. Peitgen, H. Jürgens, D. Saupe: *Fractals For The Classroom*, Springer-Verlag, 1991.

[Sch92] H. E. Schepers, J. H. G. M. van Beek: *Four Methods to estimate the fractal Dimension from self-affine signals*, *IEEE Engineering in Medicine and Biology*, Bd. 11, Nr. 2, 1992, S. 57–64.

Unterdrückung von Reflexionsartefakten in Ultraschall-Bildern

W. Pomrehn und M. Joswig

Lehrstuhl für Meßtechnik, RWTH Aachen, D-52056 Aachen

Einleitung

Die Modalität Ultraschall erzeugt im B-Scan Modus eine Art von Bildern, die mit denen anderer sog. bildgebender Verfahren der medizinischen Diagnostik nicht vergleichbar sind. Klassische Röntgenaufnahmen bilden die durchleuchteten Strukturen im Schattenriß ab; tomographische Rekonstruktionen jedweder Modalität (CT, MR, PET) sind zurückgerechnete Schichtbilder der erzeugenden Strukturdaten. Beide stellen eine begrenzte Struktur als ebenso begrenztes Bildareal dar. Demgegenüber zeigt der B-Scan eine zweidimensionale Grauwertverteilung der Echoimpulse. Das Ultraschallbild stellt nur eine indirekte Abbildung der zu untersuchenden Strukturen dar; eine Vielzahl von Artefakten verbietet die einfache Zuordnung gemessener Echoamplituden zu einer an derselben Stelle zu lokalisierenden Ursache. Bevor also eine weitergehende Bildverarbeitung auf Ultraschalldaten stattfindet wie z. B. Konturverfolgung oder Texturanalyse (Thomas et al., 1991), führt man vorteilhaft verschiedene artefakt-kompensierende Verarbeitungsschritte durch.

Bildartefakte bei Ultraschall

Ultraschall-spezifische Artefakte besonderer diagnostischer Relevanz sollen im folgenden aufgeführt und ihr Einfluß auf die Bildqualität abgeschätzt werden. Dabei unterscheiden wir spezifische Effekte des Meßaufbaus, die eine Kompensation in der Signalaufbereitung erfordern von Störungen, die mittels Bildnachbearbeitung kompensiert werden können.

Ein Effekt der ersten Art basiert auf Phasenaberration. Die Bilddarstellung im B-Scan impliziert die Annahme einer konstanten Geschwindigkeit der Impulsausbreitung im Gewebe. Diese Annahme ist aber nur bedingt erfüllt. In Weichteilen variiert die Schallgeschwindigkeit von 1470 m/s (Fett) bis zu 1570 m/s (Muskel). Die Abweichungen sind noch größer in Luft (331 m/s) und kompakten Knochen (3600 m/s). Dies führt zu Phasenverzerrungen, die sich in einer Verschlechterung des Bildkontrastes sowie einer ortsabhängigen Auflösung niederschlagen. Der Effekt ist besonders stark bei den heute fast ausschließlich eingesetzten großflächigen Schallköpfen mit "Phased Array" und elektronischer Fokussierung. Eine Kompensation ist möglich durch Korrelationsfilter im Signalpfad des Impulsempfängers sowie statistische Signalauswahlverfahren (Trahey et al., 1990; Zhao et al., 1992).

Auch Streuquellen, deren Ausdehnung kleiner ist als die minimale Auflösung des Abbildungssystems, erzeugen aufgrund konstruktiver und destruktiver Interferenzen lokale Intensitätsmaxima und -minima, die quasistochastisch verteilt sind. Sie können als dem Bild überlagertes Rauschen aufgefaßt werden und man bezeichnet diesen Effekt als "Speckle-Rauschen". Gegenmaßnahmen wie Tiefpaßfilterung oder Scharmittelung über phasenverschobene Impulsantworten führen ihrerseits zur Verschlechterung der Ortsauflösung (Bamber und Daft, 1986; Bamber und Tristan, 1988; Hall et al., 1992; Trahey et al., 1986).

Nun zu den Effekten der zweiten Art: Die Eindringtiefe des Ultraschalls wird von der Dämpfung im Gewebe bestimmt. Sie variiert mit der Schallfrequenz und beträgt bei 10 MHz in Wasser typisch 0.2 dB/cm, in Muskelgewebe typisch 100 dB/cm. Die Dämpfung ist also auch stark gewebeabhängig. Um trotz Dämpfung eine homogene Verteilung des mittleren Grauwertes im gesamten Bild zu erhalten, wird eine Amplitudenkorrektur für eine mittlere Gewebedämpfung mittels "Time Dependend Gain Control" eingesetzt. Diese führt ihrerseits zu dorsalen Schallschatten hinter Regionen mit starker Dämpfung (Konkremente) bzw. Überkompensation hinter geringer Dämpfung, z. B. den stark wasserhaltigen Zysten (Schuy und Leitgeb, 1992; Krestel, 1988).

Ultraschall-Reflexionen entstehen an Impedanzsprüngen unterschiedlicher Gewebeschichten; sie bilden die Grundlage aller Ultraschallbilder. In Weichteilen sind diese Impedanzsprünge vergleichsweise klein, so daß an Grenzschichten das 10-bis 50-fache der reflektierten Intensität transmittiert wird. Reflexionen höherer Ordnung treten praktisch nicht auf, denn bereits ab der ersten Mehrfachreflexion ist die Intensität um typisch 60 dB gefallen. Ganz andere Verhältnisse herrschen an Grenzflächen zu Luft. Hier wird nur rund 1/100 der eingestrahlten Intensität transmittiert und der Rest zurückgeworfen. Deshalb haben auch Reflexionen höherer Ordnung noch eine signifikante Amplitude. Dies zeigt sich besonders deutlich bei der Spiegelung der Leberstruktur am Diaphragma, die als Artefakt die Lungenstrukturen überlagert. Ein weiteres Beispiel sind die Phantomechos in der Harnblase durch den Impedanzsprung Luft-Wasser (Schuy und Leitgeb, 1992; Bamber und Tristan, 1988).

Alle hier aufgeführten Artefakte haben gemeinsam, daß ihr Auftreten stark von der gegebenen Anatomie beeinflußt ist und daher je nach medizinischer Applikation bestimmte Artefakte hervor- bzw. zurücktreten.

Modellierung der Mehrfachreflexionen

Die Existenz von Mehrfachreflexionen ist ein unvermeidbarer Effekt in jeder geschichteten Struktur. Die Rückrechnung auf die ursprünglichen Schichtmatrizen wird "Inversion" genannt und stellt in vielen Disziplinen der Fernerkundung ein Standardverfahren dar. Besonders ausführliche Behandlung erfährt das Thema in der Geophysik bei der Exploration nach Kohlenwasserstoffen (vgl. Robinson und Treitel, 1980). Die angegebenen Lösungen sind in vielerlei Hinsicht hilfreich, da es sich bei den Phänomenen der Seismik ebenso wie beim Ultraschall um Effekte der Ausbreitung mechanischer Wellen im Festkörper handelt.

In unserem Fall sind wir jedoch zunächst an einer Vorwärtsmodellierung interessiert, um eine quantitative Abschätzung der zu erwartenden Bildartefakte vornehmen zu können. Diese Modellierung bedeutet die Berechnung der kompletten Impulsantworten einschließlich der Mehrfachechos aus einer gegebenen Struktur. Hierfür existieren bereits Methoden wie das "Wellendigitalfilter". Seine Anwendung auf akustische Wellen in der Sprachsignalverarbeitung beschreibt Rabiner und Schafer (1978). Folgende Annahmen werden gemacht: Die Schallausbreitung erfolgt in stückweise homogenen, dispersionsfreien Medien, die Ausbreitungsgeschwindigkeit ist konstant, Streuung wird nicht berücksichtigt. Dann kann das in Abb. 1 dargestellte Wellendigitalfilter zur Beschreibung der Reflexionen benutzt werden. Man erkennt in der Struktur des Filters die axiale Näherung des Gewebes durch eine Folge äquidistanter Impedanzsprünge, repräsentiert durch die r_i, wobei n Schichten angenommen werden. In den idealen Laufzeitgliedern spiegelt sich die Annahme der verlustlosen Ausbreitung wider, in der Eindimensionalität die Vernachlässigung der Streuung. Durch Einführung eines komplexen Exponenten können die Laufzeitglieder in der gegebenen Struktur des WDF auch gewebespezifische Dämpfungsverluste nachbilden.

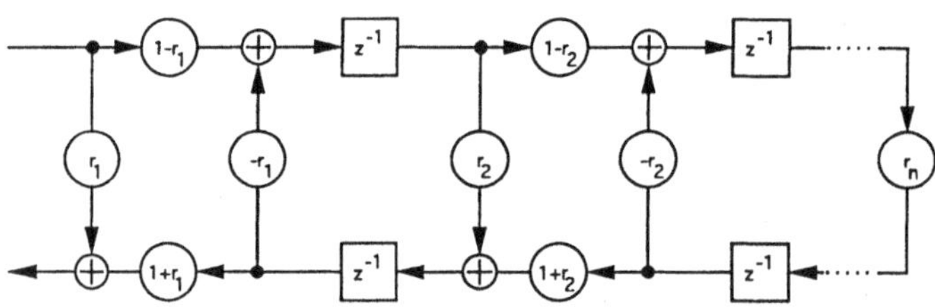

Abb. 1 Struktur des Wellendigitalfilters

Zur Berechnung der Mehrfachreflexionen wurde das Wellendigitalfilter in einem Simulationsprogramm implementiert. Dabei wurden für die Reflexionskoeffizienten typische Werte für menschliche Gewebe eingesetzt. Um das Prinzip einer Leberspiegelung am Diaphragma nachzubilden, wurde in Bild 1 eine große, elliptisch gekrümmte Grenzschicht über einer homogenen, grau dargestellten Verteilung von Luft (Lungengewebe) angenommen. Darüber befinden sich zwei Flächenobjekte, deren Impedanzen und Dämpfung typische Werte von Fett haben. Zusätzlich liegt unterhalb des simulierten Diaphragmas noch eine kleine runde Fettstruktur. In der Simulation wird weiter angenommen, daß das gesamte übrige, schwarz markierte Gebiet mit Wasser gefüllt ist. Bild 2 zeigt dann das mit dem Wellendigitalfilter erzeugte Ultraschallbild, wobei der Schall von oben in die Struktur eintritt. Deutlich zu sehen sind die Spiegelungen des Rechteckes sowie der Ellipse in die Luftblase unterhalb des simulierten Diaphragmas und die durch dessen Krümmung erzeugten geometrischen Verzerrungen. Auch der simulierte Rundherd im Lungengewebe zeigt Reflexionen höherer Ordnung, die zusätzliche Strukturen vortäuschen.

Unterdrückung der Mehrfachreflexionen

Die Unterdrückung der Mehrfachreflexionen, d.h. die Zurückrechnung auf die ursächlichen Impedanzsprünge ist wie oben erwähnt eine Inversionsaufgabe. Die wohl eleganteste Lösung ist als modifizierter Randwertansatz einer Fredholm'schen Integralgleichung formuliert (Sondhi und Gopinath, 1971); eine Anwendung auf simulierte Impulsantworten sowie ein Rechenalgorithmus finden sich in Joswig (1993). Der besondere Vorteil liegt in der großen numerischen Stabilität. Bild 3 zeigt das Ergebnis der Berechnung, man erkennt klar die Grenzlinien der Impedanzverteilung, wie sie durch die Originalstruktur in Bild 1 vorgegeben sind.

Abschließende Einschätzung

Die Unterdrückung der Mehrfachreflexionen bearbeitet die Impulsechos des Systems, wie sie im B-Scan des Ultraschallgerätes dargestellt werden. Der Rechenaufwand ist nicht von der Komplexität der Strukturen in den Bilddaten, sondern von der gewünschten Auflösung des nachbearbeiteten Bildes abhängt, also der Anzahl n zu bestimmender Reflexionskoeffizienten pro Bildspalte. Zusätzliche Meßdaten sind zur Inversion nicht erforderlich. Das hier vorgestellte Inversionsverfahren zeichnet sich durch eine numerische Stabilität aus, die es unempfindlich gegen Rauschen macht. Die Streuung wurde in der Simulation bewußt nicht berücksichtigt, da sie auch mit einem partiellen Modell nicht praxisgerecht zu beschreiben ist. Physikalische Größen wie effektive Rauhtiefe von Grenzflächen und mittlerer Krümmungsradius von Streukörpern sind praktisch nicht quantifizierbar, wenn diagnostisch relevante Impedanzverteilungen nachgebildet werden sollen. Gerade die diagnostische Relevanz der hier vorgestellten Artefaktkompensation steht aber im Vordergrund.

Praktische Bedeutung kann das Verfahren dabei in der Auswertung der aus der Urologie bekannten Transrektalbilder gewinnen (Khoury, 1985). Hier ist es für den Arzt von großem Interesse, exakte Konturen der Harnblase und der Prostata zu finden. Häufig verhindern Mehrfachreflexionen jedoch eine genaue Konturfindung verursacht durch die großen Impedanzsprünge einer zum Teil luftgefüllten Blase.

Literatur

J. C. Bamber and C. Daft, "Adaptive filtering for reduction of speckle in ultrasonic pulse-echo images", *Ultrasonics,* **24,** 41-44, 1986.

J. C. Bamber and M. Tristan, "Diagnostic ultrasound", in *The physics of medical imaging,* S. Webb ed., IOP Publ., Bristol, 1988.

T. J. Hall, S. J. Rosenthal, M. F. Insana and A. W. Templeton, "Computers in ultrasonic imaging", *J. Digit. Imag.*, **5,** 1-6, 1992.

M. Joswig, "Impulse response measurement of individualear canals and impedances at the eardrum in man", *Acustica,* **77,** 270-282, 1993.

S. Khoury, *Ultrasound in Urology*, Europ. Soc. Urology, Paris, 1985.

E. Krestel, *Bildgebende Systeme für die medizinische Diagnostik,* Siemens, Berlin, 1988.

L. R. Rabiner and R. W. Schafer, *Digital processing of speech signals,* Prentice-Hall, Englewood Cliffs, NJ, 1978.

E. A. Robinson and S. Treitel, *Geophysical signal analysis*, Prentice-Hall, Englewood Cliffs, NJ, 1980.

S. Schuy and N. Leitgeb, "Ultraschall zur Bilderzeugung", in *Biomedizinsche Technik,* Bd 1, H. Hutten ed., Springer, Berlin, 1992.

M. M. Sondhi and B. Gopinath, "Determination of vocal-tract shape from impulse at the lips", *J. Acoust. Soc. Amer.*, **49,** 1867-1873, 1971.

J. G. Thomas, R. A. Peters II and P. Jeanty, "Automatic sgmentation of ultrasound images using morphological operators", *IEEE trans. Med. Imag.*, **10,** 180-186, 1991.

G. E. Trahey, D. Zhao, J. A. Miglin et al., "Experimental results with a real-time adaptive ultrasonic imaging system for viewing through distorting media", *IEEE trans. Ultrasonics Ferroelec. Freq. Contr.*, **37,** 418-427, 1990.

G. E. Trahey, S. W. Smith and O. T. von Ramm, "Speckle pattern correlation with lateral aperture translation: experimantal results and implications for spatial compounding", *IEEE trans. Ultrasonics Ferroelec. Freq. Contr.*, **33,** 257-264, 1986.

D. Zhao and G. E. Trahey," A statistical analysis of phase aberration correction using image quality factors in coherent imaging systems", *IEEE trans. Med. Imag.*, **11,** 446-452, 1992.

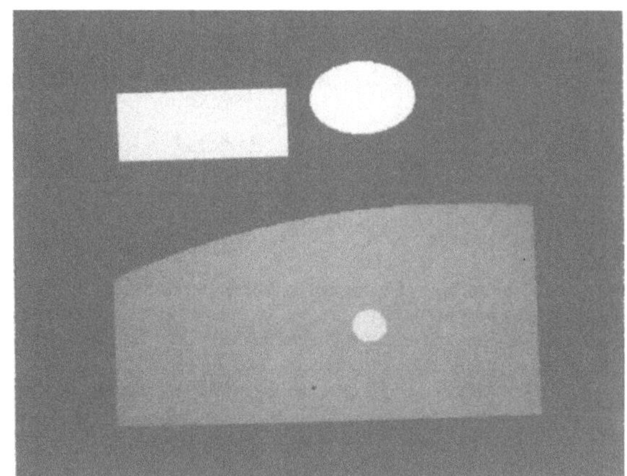

Bild 1

Modell einer Wasser / Luft - Schichtung mit eingelagerten Fettkörpern

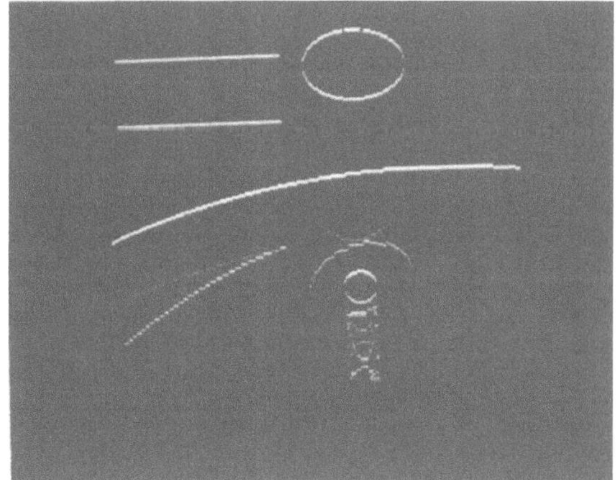

Bild 2

Echobild mit TGC-Korrektur

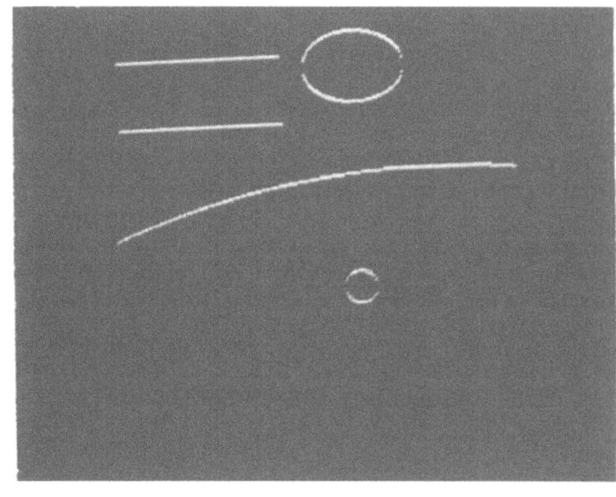

Bild 3

Rückgerechnetes Echobild ohne Mehrfachreflexionen

Recognizing and Simulating Neuronal Spike Patterns with Hidden Markov Models

G. Radons* J. D. Becker†

Abstract

The problem of recognizing different visual stimuli from the multielectrode measurement of the elicited neuronal spike patterns is treated with Hidden Markov Models (HMMs). Beyond being a pattern recognition tool, HMMs provide us with a dynamical probabilistic generator of the spatio-temporal discharge patterns.

We compare HMMs obtained from vector quantized data with models where a multivariate Poisson distribution is assumed for the output process. In the latter case the visual stimuli are recognized at very high rates from the spike patterns. An analysis of the obtained models reveals important aspects of the coding of information in the brain.

1 Introduction

The analysis of multielectrode data and the extraction of information about the coding principles in the brain is a difficult task. This is mainly due to the following characteristics of the measured data: The observed processes are in general non-stationary and they exhibit large variations, which usually cannot be explained by simple noise but may show systematic fluctuations. In addition the data are high-dimensional, e.g. we have 30 degrees of freedom corresponding to the measurement with 30 electrodes.

The experimental data treated in this work consist of simultanous recordings of spike trains with 30 microelectrodes from the visual cortex of an anesthetisized monkey [10]. We are interested in the neuronal responses recorded during the application of various visual stimuli consisting of a bar moving in 16 different directions. Our data consist of 21 repetitions of each stimulus. This gives rise to 16 classes of neural activity at the electrode array which are in some sense characteristic for the corresponding stimuli. It turned out [11] that the thirty mean firing rates, respectively the spike counts at each electrode during the movement of the bars contain no or only little information about the currently applied stimulus. It was rather found that this information is contained in the recorded spatio-temporal discharge patterns. This is in accordance with findings in other experiments [14]. Our goal consists in analyzing and characterizing such spatio-temporal excitation patterns.

The first question is, if one can recognize the visual stimulus from the elicited neuronal discharges? This is a pattern recognition task which can be tackled with classical methods. For our data one can successfully apply linear classifiers [11] or nonlinear classifiers such as artificial neural networks. These methods, however, do not take into account that the patterns are generated dynamically. Further it is difficult to infer which properties of the patterns led to the discrimination, to which extend they are of statistical nature, and what are the characteristics of the various stochastic components of the patterns.

Here we present results based on the use of Hidden Markov Models (HMMs) and corresponding parameter estimation techniques [1]. These models are currently very sucessfully applied in various speech recognition tasks (for reviews see e.g. [12, 9]), but also in biophysical problems like the analysis of ion currents through channels of cell membranes [4, 7, 2] and oscillatory neuronal responses in cat´s visual cortex [5].

The advantage of using Hidden Markov Models for multielectrode data analysis is threefold: Beyond being a pattern recognition and classification tool, it provides us with probabilistic dynamical models of the pattern generating process. This implies the possibility to reproduce the data with a reduced set of parameters, and thus serves on the one hand as a data compression method, but on the other hand it preserves the possibility to extract e.g. various correlation functions or correlograms. Finally, the extracted models, although of abstract nature, can be analysed in terms of subprocesses which contribute to the pattern generating process as a whole.

*Institut für Theoretische Physik, Universität Kiel, Olshausenstr. 40, D-2300 Kiel

†Fakultät für Physik, Universität Freiburg, Hermann-Herder-Str. 3, D-7800 Freiburg

2 Principles of Hidden Markov Models and their Application to Spike Trains

2.1 HMMs as Probabilistic Dynamical Models

A Hidden Markov Model is an abstract object consisting of a given number N of states and transitions between these states. Transitions occur with probabilities a_{ij}, i.e. a_{ij} is the conditional probability $p(j \mid i)$ for making a transition to state j, if the system is in state i. The a_{ij} have the property $\sum_{j=1}^{N} a_{ij} = 1$ for all i, which means that some transition occurs with probability one. Therefore they can be considered as elements of a stochastic matrix A, the transition matrix.

So far this defines a simple Markov process because the probability for the next state j depends only on the current state i. Hidden Markov Models are characterized by the additional ingredient that on every state i one defines a probability distribution $b_i(S)$ for emitting a symbol S of some alphabet $\{S\}$ of length $|S|$. The alphabet may also consist of infinitely many symbols $\mid S \mid= \infty$. Some symbol is generated with certainty on every state i which means that $\sum_{\{S\}} b_i(S) = 1$ for every i. The $b_i(S)$ are in general different functions depending on i. The meaning of the symbols is application dependent and is introduced for our problem in section 2.3 below.

In order to generate symbol sequences with such models one also has to specify an initial probability distribution $\vec{\pi} = (\pi_1, \ldots, \pi_N)$ over the states i. A symbol sequence of length T, $S_1 S_2 \ldots S_T$ with $S_t \in \{S\}, t = 1, \ldots, T$ is generated as follows: One randomly selects an initial state according to the distribution $\vec{\pi}$, e.g. state i with probability π_i, and one emits a symbol S_1 with probability $b_i(S_1)$, then one jumps to another state, say j with probability a_{ij} and emits symbol S_2 with probability $b_j(S_2)$, and so on. In this way a sequence $S_1 \ldots S_T$ is generated with a certain probability $p(S_1 \ldots S_T)$, which can be calculated as

$$p(S_1 S_2 \ldots S_T) = \vec{\pi} B(S_1)\ AB(S_2)\ A \ldots AB(S_{T-1})\ AB(S_T)\vec{\eta} \tag{1}$$

where we introduced the diagonal matrices $B(S), S \in \{S\}$, with elements $b_{ij}(S) = b_j(S) \cdot \delta_{ij}$ and the vector $\vec{\eta} = (1, 1, \ldots, 1)^T$, which reflects the fact that the system is allowed to be in any of the N states j while it emits the last symbol S_T. In the speech recognition literature the evaluation of expression (1) is called forward-backward algorithm.

To summarize, a Hidden Markov Model is defined by a set of parameters, which includes the initial probability distribution $\vec{\pi}$, the transition matrix A, and the symbol generating matrices $B(S)$ with S from the alphabet $\{S\}$. These parameters are conveniently collected in one vector denoted $\vec{\lambda}$, that is

$$\vec{\lambda} = (\vec{\pi}, A, \{B(S)\}) \tag{2}$$

In view of eqs. (1) and (2) one can identify a given HMM also with a parametrized probability distribution over all symbol sequences $S_1 S_2 \ldots S_T$ of arbitrary length T. We denote this probability distribution in the following by $P(S_1 S_2 \ldots S_T \mid \vec{\lambda})$.

2.2 HMMs as Classifiers

In a classification task one has a given number of classes of patterns, and one wants to decide for some pattern to which class it belongs. This is often done by calculating the distance (within some metric) of the given pattern to some representative of each class. Then one decides that it belongs to the class where the distance is minimal. HMMs can also be used for pattern classification. If one has K classes of patterns one has to design K Hidden Markov Models, each of which is a representative of a different class of patterns. In our application $K = 16$ corresponding to the 16 different visual stimuli, which produce 16 classes of neuronal responses. In order to use HMMs for pattern classification, one has to translate the pattern into a symbol string, which may also be considered as a possible output of the given HMMs. We call this symbol string, which corresponds to some input pattern to be classified, the observation sequence $O = S_1 S_2 \ldots S_T$. Now one simply decides that O belongs to that class for which the corresponding representative HMM produces O with the highest probability. Thus one has to calculate for each HMM $\vec{\lambda}_k$, $k = 1, ..., K$ the probability $P(O \mid \vec{\lambda}_k)$ that model $\vec{\lambda}_k$ produces the observation sequence O. This is done with the forward-backward algorithm (1). The model $\vec{\lambda}^*$ for which $P(O \mid \vec{\lambda})$ is maximal is regarded as the representative of the correct class, formally

$$\vec{\lambda}^* = \text{argmax}_{\lambda_i}\ P(O \mid \vec{\lambda}_i). \tag{3}$$

This comparison can in principle be done with an arbitrary ensemble of K HMMs, and one would always get an answer, i.e. a classification of the input pattern. In order to make sense the classification process has to be carried out with an ensemble where the HMMs really represent in some sense the corresponding pattern classes. This is achieved by a training or learning procedure where the parameters of each HMM are optimized in the following sense: If for each pattern class one had given only one representative symbol sequence $O^{(k)}, k = 1, \ldots, K$, one would train each HMM such that model $\vec{\lambda}_k$ produces $O^{(k)}$ with maximal probability. This is usually done iteratively.

In general, and also in our application, a class is not represented by only one string $O^{(k)}$ but by an ensemble of samples $\{O^{(k)}\}$ which are known to belong to class k. Since the members of $\{O^{(k)}\}$ may occur with different frequencies the class k is actually characterized by a probability distribution $P(O^{(k)})$ over the samples. In this case one wants the probability distribution $P(O^{(k)} \mid \vec{\lambda}_k)$ over sequences produced by model $\vec{\lambda}_k$ to be as similar as possible to $P(O^{(k)})$. This can be achieved by iteratively optimizing $\vec{\lambda}_k$, such that the Kullback-Leibler distance D,

$$D(P(O)^{(k)} \parallel P(O^{(k)} \mid \vec{\lambda}_k)) \equiv \sum_{\{O^{(k)}\}} P(O^{(k)}) \log \frac{P(O^{(k)})}{P(O^{(k)} \mid \vec{\lambda}_k)} \tag{4}$$

is minimized during the optimization procedure. One has $D \geq 0$ with equality only if $P(O^{(k)})$ and $P(O^{(k)} \mid \vec{\lambda}_k)$ are identical.

In principle any of the known optimization algorithms could be used for finding the optimal parameter vector $\vec{\lambda}_k$ which minimizes the Kullback-Leibler distance $D(\vec{\lambda}_k)$ eq. (4) or maximizes the likelihood $P(O^{(k)} \mid \vec{\lambda}_k)$ for a given class k. The most widely used method for HMMs is the so called Baum-Welch reestimation algorithm [1]. It has the general form

$$\vec{\lambda}_k(t+1) = \vec{f}[\vec{\lambda}_k(t), \{O^{(k)}\}] \tag{5}$$

where $\vec{f}$ is a relatively complicated function of the old parameters $\vec{\lambda}_k(t)$ and the observation samples. The explicit form of $\vec{f}$ can be found e.g. in [12, 9]. One should remark that with any optimization algorithm there exist the possibility to get stuck in local minima of $D(\vec{\lambda}^{(k)})$ respectively maxima of $P(O \mid \vec{\lambda}^{(k)})$. To avoid this one has to run the optimization algorithm with several different initial conditions $\vec{\lambda}^{(k)}(t=0)$.

2.3 Application to Spike Data

The first problem one encounters in applications of Hidden Markov Models to continous time problems such as speech recognition and also spike data analysis is the segmentation problem i. e. the dividing of the signal into time bins or segments of some length in order to obtain a discrete "time" process described by HMMs as a jump process between the available states. The various possibilities we explored are described in 3.1. The second problem in using discrete HMMs consist in obtaining a symbolic or discrete representation of the data for each time segment. For spike data there exists a very natural alphabet: One has to assign to every time segment simply the number of spikes contained in it. For one electrode the data are then represented as a sequence of numbers, which vary between zero and the maximum number of spikes encountered in the segments. The problem lies in the fact that in our experiment we have 30 electrodes. If the number of spikes per segment and electrode varies for instance from 0 to 9, this would lead to an alphabet with 10^{30} symbols if every possible combination of spike counts on the electrodes is taken into account by a distinct symbol. As explained in section 2.1 and 2.2 one has to optimize a probabilty distribution $b_j(S)$ on every state j of a given model. For $\mid S \mid = 10^{30}$ this is clearly impossible. There are basically two approaches to circumvent this problem. The first consists in a suitable coarse-graining of the alphabet to obtain a smaller set of symbols. This method is basically the same as vector quantizing continuous variables [8]: One has to design a code book which tells which spatial firing pattern, i. e. the vector of spike counts at each electrode in some time segment, including some neighborhood is coded by which symbol. Inevitably there is some information loss, and also the individuality of the single electrodes is given up and replaced by an overall pattern. The methods we used for the design of the code books and the results for the quality of the vector quantization (VQ) are presented in 3.1. The HMMs based on this method are called VQ-HMMs in the following and the performance of these models is reported in 3.2 and compared with a second alternative. This alternative in reducing the parameters, which characterize the output probability distributions, lies in making assumptions about the functional form of the $b_j(S)$. This amounts to parametrizing these functions and optimizing the corresponding parameters. It turned out that assuming a multivariate Poisson process on the nodes of the HMMs accounts well for the actually observed spike statistics in individual time segments. We can even assume independency of the spike distributions for each electrode on each node j. This means that we assume the following form of the output probabilities

$$b_j(\vec{k}) = \Pi_{l=1}^{30} \; b_j^{(l)}(k_l) = \Pi_{l=1}^{30} \; P(k_l, \mu_j^{(l)}) \tag{6}$$

where $P(k, \mu)$ is a Poisson distribution with mean value μ

$$P(k, \mu) = \frac{1}{k!} \, \mu^k \, e^{-\mu}. \tag{7}$$

Thus $P(k_l, \mu_j^{(l)})$ is the probability of observing k_l spikes on electrode l in some time segment of length Δ, if the system is in state j. The corresponding mean number of spikes is $\mu_j^{(l)}$. Thus we have to optimize only 30 parameters for every state j instead of 10^{30} as in the original formulation. We emphasize that although we assume independency of the single electrode processes in every state j, this does not mean that the electrodes are treated as independent. Correlations between electrodes are taken into account by the transitions between subsequent states. A HMM with output probabilities (6) is still a discrete Hidden Markov Model with an infinite alphabet since the symbol vector $\vec{k} = (k_1, ..., k_{30})$ consists of interger values $k_l = 0, 1, 2, \ldots, \infty$. The parametrization of the output function and the corresponding optimization procedure is rather in the spirit of continuous density Hidden Markov Models (CD-HMMs) [12, 9]. The Baum-Welch reestimation formulas for Poisson densities can be found in [3, 13]. We call HMMs based on this second approach PD-HMMs, where PD stands for parametrized (output) density.

3 Results

3.1 Preparation of the Data

The onset of the neuronal responses is not sharply connected with the onset of the bar's motion. This onset depends on the direction of the bar due to different distances from the starting point to the receptive field of the neurons. Therefore we decided for each stimulus to define the relevant response interval symmetrically around the maximal spike rate averaged over all 21 trials and 30 electrodes. For all 16 stimuli this point was sharply peaked with an accuracy of 10 ms, and it was no problem to find the maximum point. We analyzed responses in intervals of 300, 500, and 800 ms length.

It is an open problem on which time scale relevant neuronal information is processed. Therefore we investigated the data on different (coarse-grained) timescales. We divided the response intervals into a varying number of time bins and counted the spikes in each bin. We used non-overlapping segments of 25 and 50 ms length in all cases, and in addition we used 100 ms bins for the responses of length 500 ms and 800 ms. As a result of the segmentation of the response intervals a stimulus is now represented as a sequence of 30-dimensional spike-count vectors, where each component is a integer value.

As argued in section 2.1 one has for VQ-HMMs to solve the problem of finding an alphabet of reasonable size $|S|$. This is achieved by vector quantization (VQ), where one divides all spike-count vectors into $|S|$ groups and identifies all vectors in the group with the same symbol. An alphabet of 10 to 30 symbols seems appropriate, for we deal with a maximum of $16 \times 21 \times 32 (= 10752)$ spike-count vectors, which should be partitioned into groups. We tested several methods for finding reasonable groups [13]. The most sucessful approach was a standard parallel cluster analysis [6].

With this method one tries to find a minimal distance partition. Each group is represented by a 30-dimensional vector. One starts with arbitrary initial vectors. All spike-count vectors are attached to the group vector which has the shortest Euklidean distance to the spike-count vector. The average of all vectors of each group defines the new group vector. This is iterated until there are no more changes. It is indeed possible to get groups of more or less equal size. The partitions that arise depend on the initial values, which is due to a quite unstructured distribution of the spike-count vectors, and implies that the allocation of the symbols is arbitrary. This is of course not a disadvantage for our purpose, though one should keep in mind that the symbols have no data intrinsic meaning. We repeated this method for $|S|$ variing from 8 to 22 in steps of 2. For all numbers of groups the sum of the variances of the groups is more or less the same. Therefore we have no indication that one should use a special number of symbols, which is in accordance with the uniform distribution of the spike-count vectors.

We also tested, if the chain of symbols still decribes the stimulus. Taking the responses of length 500 ms segmented into 25 ms bins, and using standard classification methods we get a rate of 92.3% correct recognitions [11]. If we use the chain of symbols ($|S| = 20$) and replace each symbol by its group vector, we can also apply standard classification methods to these data. As a result we get 65.3% correct recognition, which is far less than the result with the original data. So we loose information about the stimuli in the clustering process.

3.2 Comparison of the Models

With the segmented and vector quantized data we are now in the position to obtain and compare various Hidden Markov Models. In all cases we used 11 repetitions of each stimulus for training the corresponding HMM and tried to classify the remaining 10 repetitions. For every tested response interval and segmentation we obtained for each of the 16 different stimuli 2 HMMs – a PD-HMM which according to section 2.3 describes the spike-count vectors directly, and a VQ-HMM which uses the vector quantized data. We found that using the PD-HMMs one always gets higher recognition rates than with VQ-HMMs. The best results were obtained for a left-right PD-HMM with 4 states trained on 300 ms response intervals which were segmented into 50 ms bins. The corresponding recognition rate was 73.8%, whereas comparable VQ-HMMs yielded only rates of about 40-50%. This appearently means that the information loss in vector quantizing the data is more serious than the assumptions made for the output probability densities in the PD-HMMs. To obtain a better understanding of this result we investigated the segregation quality of the obtained models. For that we checked, whether the 16 HMMs for given length and segmention parameters are able to distinguish self-generated data. This means that we simulated or generated 20 new data sets with each HMM and classified these data in the same manner as the original data. The resulting self-classification rate is always 100% for the PD-HMMs. For VQ-HMMs these rates vary between approximately 60% for long segments and almost 100% for short time segments. This implies that the PD-HMMs are in general better adapted to the data than the VQ-HMMs although they contain a comparable number of free parameters. We also realized that the recognition rates decrease, if we use longer response intervals. By a closer look at the data one finds that there are only small regions (about 300 ms) that are not comparable with spantoneous activity . Therefore longer response intervals contain more noise and are therefore more difficult to classify. The fact that the recognition rates increase with a finer segmentation of the data is in accordance with earlier results [11]. A detailed account of the results for all HMMs we tested can be found in [13].

To get an impression of the quality of the obtained HMMs, we present in Fig. 1 examples of original and simulated responses to one of the stimuli. We used a PD-HMM with 12 states for the generation of the data on the right of this figure. The spike patterns within the time bins were generated by randomly distributing spikes in each segment according to the Poisson law (7) with the mean values prescribed by the HMM. For each repetition a path through the HMM is chosen randomly with probabilities determined by the transition matrix of the HMM. The recognition rates for the corresponding HMM ensemble are 55.6%, which is not the highest value reached. Nevertheless, we see that the overall patterns are rather similar. In the simulated data the transitions between low and high activity regions are not as sharp as in the original data. This is probably due to the unfavourable ratio of the number of training sets versus number of parameters to be fitted. The same argument explains the relatively low recognition rate.

The above ratio is larger for larger time segments and smaller numbers of HMM states. In figure 2 we depict a 3-state PD-HMM for the same stimulus as in figure 1 but trained on 100 ms segments. This model has the best recognition rate for responses of length 500 ms (67.5%) . Thus with such a model we have a quite compact representation of the ensemble of possible responses on a given visual stimulus. From a biological point of view it is interesting to note that the relevant time scale for the information processing appears to lie in the regime around 50-100 ms.

Finally we note that we always obtained better results with left-right models than with ergodic models. This indicates that responses are well described by one prototype pattern with relatively strong noise superimposed. The latter accounts for the large variability in the responses. This picture is also confirmed by the observation that the training of ergodic models leads often to models that are essentially linear and therefore close to left-right models.

4 Summary and Discussion

We have shown that Hidden Markov Models are well suited for the classification, the coding and the analysis of neuronal responses. At the moment the small data set is the main limitation for obtaining more accurate models. Experiments with permanently implanted electrodes are currently prepared which will overcome this problem. Based on the results reported briefly in this work and in more detail in [13], we expect that HMMs are a extremely useful tool in evaluating such experimental data. It can be expected that this is even more the case if the responses in one class are more structured and differentiated due, for instance, to the presence of several features represented in the responses. In such cases it may be also worth using more sophisticated methods for quantizing the data for possible applications of VQ-HMMs.

Acknowledgements: We thank B. Dülfer and J.Krüger for their enthusiastic cooperation. This work was partially supported by the Deutsche Forschungsgemeinschaft.

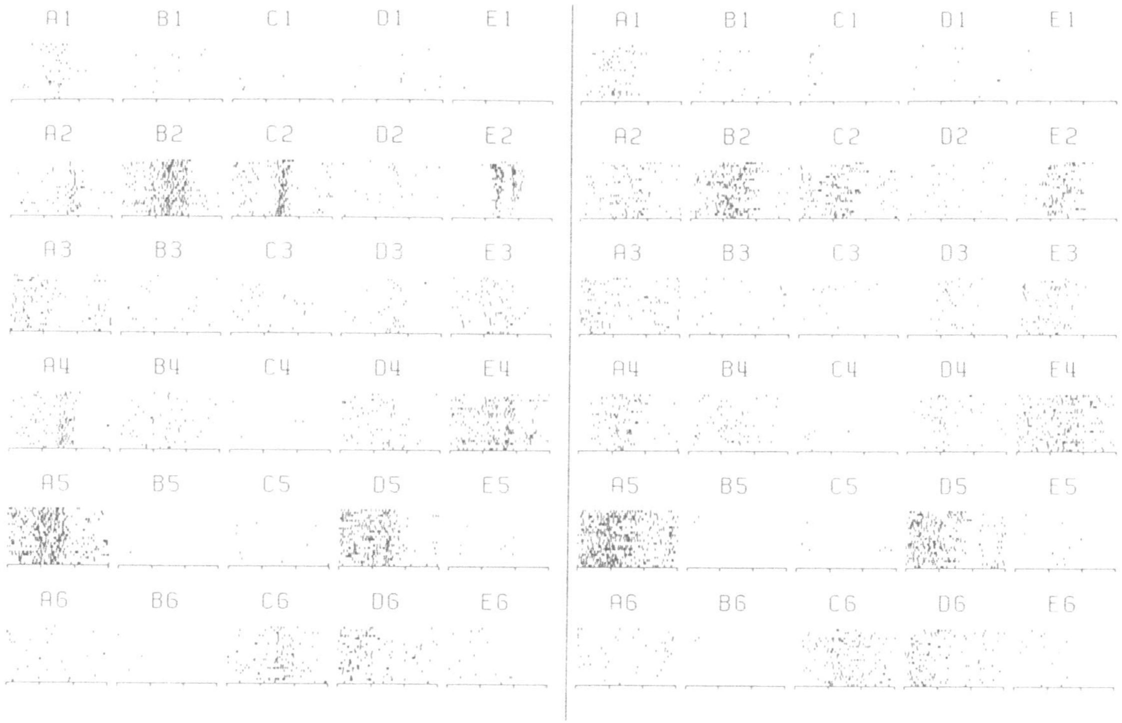

Figure 1: Left: Display of 21 repetitions of the response to the same oriented moving bar. The total length of the shown regions is 500 ms. Each vertical line represents a spike. Repetitions are shown beneath each other for every electrode. Electrodes are arranged in a 5×6 array with a spacing of 160 μm. They are labelled A–E (columns) and 1–6 (rows). The stimulus was a monocularly presented bar 7 min of arc wide, moving at 1 min of arc per 10 ms. Left: Simulation of the same response with a 12-state strict left-right PD-HMM based on 25 ms segments.

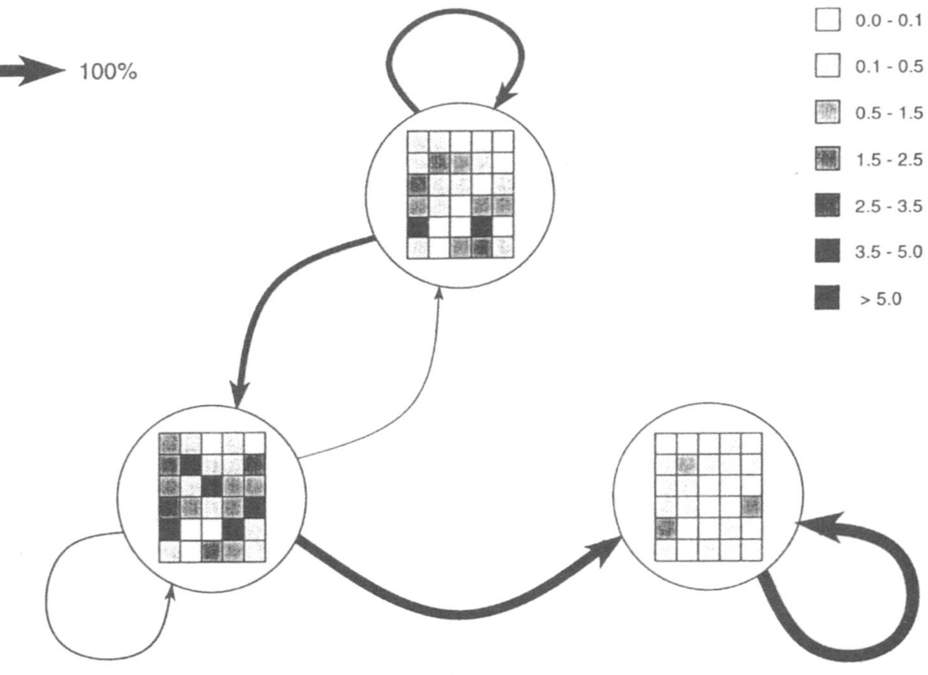

Figure 2: An example of a 3-state PD-HMM which gives almost optimal recognition rates. The transition probabilities are coded in the thickness of the arrows. The grey-tones in the squares inside each node code the mean firing rate at the corresponding electrode (arranged physically in the same order). The initial probability is concentrated in the state drawn on the top ($\pi_1 = 91\%$).

References

[1] L. E. Baum, T. Petrie, G. Soules, and N. Weiss. A maximization technique occuring in the statistical analysis of probabilistic functions of Markov chains. *Ann. Math. Stat.*, 41:164–171, 1970.

[2] J. D. Becker. Hidden Markov models for analyzing ion channels – draft of a theory. Preprint, Universität Freiburg THEP 04, 1993.

[3] J. D. Becker and B. Dülfer. A hidden Markov model for point processes. *IEEE Trans. Signal Processing*, 00:submitted, 1993.

[4] S. H. Chung, V. Krishnamurthy, and J. B. Moore. Adaptive processing techniques based on hidden Markov models for characterizing very small channel currents buried in noise and deterministic interferences. *Phil. Trans. R. Soc. Lond. B*, 334:357–384, 1991.

[5] J. Deppisch, K. Pawelzik, and T. Geisel. Uncovering the synchronization dynamics from correlated neuronal activity quantifies assembly formation. Preprint Universität Frankfurt, 1992.

[6] R. O. Duda and P. E. Hart. *Pattern Classification and Scene Analysis*. John Wiley & Sons, 1973.

[7] D. R. Fredkin and J. A. Rice. Maximum likelihood estimation and identification directly from single-channel recordings. *Proc. R. Soc. Lond. B.*, 249:125–132, 1992.

[8] R. M. Gray. Vector quantization. *IEEE ASSP Mag.*, 1:4–29, 1984.

[9] X. D. Huang, Y. Ariki, and M. A. Jack. *Hidden Markov Models for Speech Recognition*. Edinburgh University Press, 1990.

[10] J. Krüger and F. Aiple. Multimicroelectrode investigation of monkey striate cortex: Spike train correlations in the infragranular layers. *J. Neurophysiology*, 60:789–828, 1988.

[11] J. Krüger and J. D. Becker. Recognizing the visual stimulus from neuronal discharges. *TINS*, 14:282–286, 1991.

[12] L. R. Rabiner. A tutorial on hidden Markov models and selected application in speech recognition. *Proc. IEEE*, 77:257–285, 1989.

[13] G. Radons, J. D. Becker, B. Dülfer, and J. Krüger. Analysis, classification, and coding of multielectrode spike trains with hidden Markov models. *IEEE Biological Cybernetics*, 00:submitted, 1993.

[14] B. J. Richmond, L. M. Optican, M. Podell, and H. Spitzer. Temporal encoding of two-dimensional patterns by single units in primate inferior temporal cortex. I. response characteristics. *Journal of Neurophysology*, 57:132–146, 1987.

Untersuchung zur Bestimmung der Blutgeschwindigkeit und des Blutvolumens in der Aorta durch Bolus-Tracking in NMR-Schichtbildern mit Methoden der digitalen Bildverarbeitung

D. Richter[1], U. Jendrysiak[2], M. Just[2], A. Sahil[1,3], D. Sudheimer[1,4]

[1]Fachhochschule Wiesbaden, FB Informatik,
[2]Klinikum der Johannes Gutenberg-Universität Mainz,
[3]Abbot GmbH, seit 1992, [4]MCS AG, seit 1993

Die vorliegende Arbeit beschreibt die Bestimmung des während einer Herzaktion geförderten Blutvolumens durch die Aorta descendens. Das Blutvolumen wird dabei mittels Bolus-Tracking eines durch Kernspin-Resonanz vorgesättigten Blutvolumens in einer Sequenz von ca. 12 bis 15 Einzelaufnahmen ermittelt. Die Einzelaufnahmen werden in einem zeitlichen Abstand von ca. 20 - 35 ms aufgenommen und decken dabei den Zeitraum ab, in dem während einer Herzaktion Blut gefördert wird. Aus den aus den Einzelaufnahmen ermittelten Momentangeschwindigkeiten des Blutes wird das gesamte Blutvolumen berechnet. Das verwendete Aufnahmeverfahren geht auf einen Vorschlag von Edelman, Mattle, Kleefield und Silver [1] zurück.

1. Einleitung

Bei der Diagnose von Herzkammerperforationen (Septumdefekte) werden üblicherweise mit Hilfe von Sonden in unmittelbarer Nähe des Herzens die Sauerstoffkonzentrationen im arteriellen und im venösen Blut gemessen und daraus auf das Vorhandensein und auf die Größe des Defektes zwischen den Herzkammern geschlossen. Bei einer anderen gebräuchlichen Technik wird dem Patienten ein Kontrastmittel injiziert und dieses auf seinem Weg durch die Gefäße mit Röntgenstrahlen verfolgt. Diese Untersuchungsmethoden setzen den Patienten Belastungen und Risiken aus. Ein neuer Ansatz der Untersuchung ist die Bestimmung des Shuntvolumens auf nicht-invasive Art mit Hilfe der Kernspintomographie (NMR) [2]. Dabei soll das Blutvolumen im großen und kleinen Kreislauf während einer Herzaktion bestimmt werden. Aus der Differenz der Volumina kann man unter Berücksichtigung weiterer Effekte das Shuntvolumen berechnen. Die durchgeführte Arbeit dient dazu, die grundsätzliche Eignung von Methoden der digitalen Bildverarbeitung zur Messung dynamischer Prozesse in großen Gefäßen zu untersuchen. Dazu soll in einem Bereich der Aorta, in der weder Lageänderungen noch große Durchmesserschwankungen durch Druckänderungen während der Dauer einer Herzaktion zu erwarten sind, eine Blutvolumenmessung durchgeführt werden. Es bleibt weiteren Arbeiten vorbehalten, die hier gewonnenen Erkenntnisse auf den Bereich des Pulmonalarterienstammes und der Aorta ascendens anzuwenden.

2. Aufnahmeverfahren

Um dynamische Prozesse auszuwerten, müssen Ort und Zeitpunkt der Lage relevanter Objekte bestimmt werden. Hierzu wird eine Serie von 12 - 15 Einzelaufnahmen einer Herzaktion zu definierten Zeiten angefertigt. Mit dem EKG-Signal des Patienten wird die

Aufnahmeserie synchronisiert. Zusätzlich werden über einen Atemmonitor Bewegungsartefakte, die durch die Atmung des Patienten erzeugt werden, dadurch minimiert, daß bestimmte Bilddaten nur bei gleicher Lage des Brustkorbes aufgenommen werden.

Die Aufnahmeschicht der Bilder wird so gelegt, daß ein möglichst langer Teil der Achse der Aorta descendens unterhalb des Aortabogens in der Bildebene liegt. Zusätzlich wird unmittelbar vor der Aufnahme jedes einzelnen Bildes der Sequenz orthogonal zu der Aufnahmeebene und orthogonal zu dem Verlauf der Aortaachse eine Gewebeschicht vorgesättigt (Anregungsschicht, Bild 1). Die vorgesättigten Atomkerne leisten zum anschließenden Spinresonanzsignal für die Bildaufnahme keinen Beitrag; sie werden im Bild schwarz dargestellt (Bild 2).

3. Auswertung

Zur Bestimmung des während einer Herzaktion geförderten Blutvolumens ist es notwendig, den zeitlichen Verlauf der Blutgeschwindigkeit über eine Herzaktion zu ermitteln. Hierzu werden aus einer Aufnahmesequenz aus bekannten zeitlichen Parametern die Momentangeschwindigkeiten des vorgesättigten Blutvolumens durch die Messung der zurückgelegten Wegstrecke von der Vorsättigungsschicht aus gemessen. Mit dem im Messbereich bestimmten und als kreisförmig angenommenen Aortadurchmesser erhält man das geförderte Blutvolumen [3].

Hierfür sind die Anregungsschicht, die Aorta und der Bolus innerhalb der Bilddaten zu identifizieren und ihre Lage zueinander zu bestimmen. Mit dem a-priori-Wissen der Aufnahmetechnik und der Lage der Aufnahmeschicht können die Objekte modelliert werden.

Die Anregungsschicht

Die Anregungsschicht ist innerhalb der Bilder der Aufnahmesequenz kein reales Objekt, sondern beschreibt eine geometrische, physikalisch veränderte Gewebezone. Sie ist abhängig von eingestellten Parametern des NMR-Tomographen und verläuft als Linie ungefähr bekannter Breite nahezu horizontal durch das Bild mit scharfem Kontrast zu dem umliegenden, nicht vorgesättigten Gewebe. Die Grauwertgradienten in vertikaler Richtung liegen betragsmäßig über einer definierbaren Schwelle. Sie haben an der oberen und unteren Kante jeweils gleiche bzw. umgekehrte Vorzeichen und einen festen, ungefähr bekannten Abstand voneinander.

Zur Auswertung wird das Bild mit einer entsprechenden vertikalen Gradientenmaske spaltenweise abgetastet und das Auftreten des entsprechenden Gradientenverlaufs in einer zweidimensionalen Punktmatrix markiert (Schichtfilterung, Bild 3a). Aus dieser Matrix werden nun alle möglichen Linien generiert, die durch Verbindung derjeniger Punkte erzeugt werden können, die einen vorgegebenen Abstand nicht überschreiten, und deren auftretender Steigungswinkel nicht größer als 45° wird (Bild 4). Die Liste der möglichen Linien wird dann Bewertungskriterien wie die Anzahl der enthaltenen Punkte, Länge, Punktdichte, Ebenheit und Korrelationskoeffizient der Regressiongeraden unterworfen und bewertet. Diejenige

Linie, die bei der Bewertung am besten abschneidet, wird als Gerade für die Anregungsschicht verwendet (Schichtextraktion, Bild 3b).

Die Aorta

Im Gegensatz zur Anregungsschicht ist die Aorta ein real vorhandenes Objekt, deren Verlauf innerhalb des Bildes sehr stark von dem jeweiligen Patienten und von den vom Arzt eingestellten Aufnahmeparametern abhängt. Erschwerend kommt eine große strukturelle und morphologische Ähnlichkeit mit weiteren Bildobjekten wie beispielsweise den Herzgefäßen oder der subkutanen Fettschicht hinzu. Trotz dieser Schwierigkeiten kann die Aorta durch folgende Eigenschaften modelliert werden: Sie weist senkrecht zu ihrer Achse einen charakteristischen Gradientenverlauf auf, der Grauwerteverlauf ist weitgehend symmetrisch zur Aortaachse und sie weist in dem zu untersuchenden Bereich höchstens eine geringe Krümmung auf. Weitere Eigenschaften sind durch die Aufnahmetechnik festgelegt : sie verläuft nahezu senkrecht durch das Bild und senkrecht zur Anregungsschicht, sie hat einen relativ konstanten Durchmesser unterhalb der Anregungsschicht, der Durchmesser ist in den anatomisch möglichen Grenzen bekannt und sie verläuft vorzugsweise durch die Bildmitte der Aufnahmen.

Störend macht sich bei dem Modell der Aorta der bis an die Aortawand reichende Bolus bemerkbar. Daher wird zunächst eine Mittelwertbildung über die komplette Aufnahmesequenz durchgeführt. Anschließend wird das Bild ähnlich wie bei der Extraktion der Anregungsschicht, jedoch zeilenweise mit der Gradientenmaske und der Grauwertsymmetriebedingung abgetastet. In einer zweidimensionalen Matrix werden bei einer Übereinstimmung der Bilddaten die Spaltenposition der rechten und linken Aortawand und die Aortamitte gespeichert (Aortafilterung, Bild 5). In der nun folgenden Auswertung der Matrix werden wieder alle möglichen Linien durch Verbindung derjenigen Punkte generiert, deren Zeilenabstand zu einem neuen Punkt zu dem letzten in die jeweilige Liste einsortierten Punkt einen vorgegebenen Abstand nicht überschreitet und der sich durch eine Linie von maximal 45° gegen die Senkrechte mit der vorhandenen Linie verbinden lässt. Diese Liste der möglichen Mittellinien der Aorta wird dann wieder den Bewertungskriterien Anzahl der enthaltenen Punkte, Länge, Punktdichte und Ebenheit unterworfen und bewertet. Weiterhin kommen noch zwei neue Kriterien hinzu : die Zentrierung zur Bildmitte und die Durchmesserhomogenität der Aortaränder über die gefundene Länge. Diejenige Linie, die bei der Bewertung am besten abschneidet, wird als Mittellinie für die Aorta mit der dazugehörigen Aortawand verwendet (Aortaextraktion).

Die gefundene Aorta wird einer weiteren Plausibilitätsprüfung unterworfen. Sie muß in dem Bildbereich liegen, in dem bei der Differenzbildung zweier zeitlich aufeinanderfolgender Bilder einer Sequenz der Absolutbetrag der Grauwertdifferenzen größer als ein vorgegebener Schwellwert ist. Dieser Bildbereich wird durch den sich bewegenden, vorgesättigten Bolus gebildet. Diese Plausibilitätsprüfung entfällt bei einem Test an einem Flußphantom mit konstanter Fließgeschwindigkeit.

Standardverfahren zur Glättung und zum Schließen von Lücken in der gefundenen Aortawand beenden die Aortaextraktion.

Der Bolus

Der Bolus ist relativ einfach zu modellieren, da er wie die Anregungsschicht kein reales Objekt repräsentiert. Er ist in Blutflußrichtung von magnetisch ungesättigtem Blut umgeben, das ein sehr gleichmäßiges Signal erzeugt. Der Bolus liegt innerhalb der Aorta, er liegt in Blutflußrichtung meist unterhalb der Anregungsschicht (gegen Ende der Herzaktion kann er auch gering darüber liegen), die Bolushinterkante bildet einen starken Kontrast zum nachfließenden Blut und sie ist die erste signifikante Struktur innerhalb der Aorta.

Um den Bolus zu identifizieren werden von derjenigen Zeile, in der das obere Ende der Aortakontur gefunden wurde, innerhalb der Aorta die Gradienten längs der Richtung der Aortamittellinie gebildet und in eine Tabelle eingetragen. Die Bolushinterkante wird im allgemeinen gut gefunden, kann aber noch Lücken oder einzelne Ausreißer haben, die durch Standardverfahren geschlossen oder eliminiert werden können (Bild 6).

Bestimmung der Momentangeschwindigkeit

Nachdem durch die oben beschriebenen Algorithmen alle für die Untersuchung relevanten Objekte und deren Lage in den Bilddaten identifiziert wurden, wird die Strecke s_i längs der Aortamittellinie vom Schnittpunkt der Oberkante der Anregungsschicht bis zur Oberkante des Bolus berechnet, wobei aus den vom NMR-Tomographen aufgezeichneten Aufnahmeparametern die Kantenlängen der Pixel in Zeilen- und Spaltenrichtung bekannt sind.

Die Propagationszeit t_p, d.h diejenige Zeit, die dem Bolus zum Zurücklegen der oben bestimmten Strecke zur Verfügung steht, setzt sich aus zwei Anteilen zusammen : der Zeit t_s zwischen der Vorsättigung der Anregungsschicht und dem HF-Impuls zur Synchronisation der Spins im nicht vorgesättigten Gewebe und der Echozeit t_e zwischen der Anregung der Spins und der Aufnahmezeit des Spinechos. Während t_e von dem NMR-Tomographen mitprotokolliert wird und somit bekannt ist, muß t_s am Gerät durch direkte Messung im Einzelfall ermittelt werden.

Aus den Strecken s_i und der Propagationszeit t_p werden die Monentangeschwindigkeiten der Boli in den einzelnen Aufnahmem während der Herzaktion bestimmt (Bild 7).

Den Aortadurchmesser kann man wie in der vorliegenden Arbeit aus den Aufnahmen der Sequenz vorzugsweise unmittelbar oberhalb der Anregungsschicht entnehmen, da dort Bildstörungen durch die Boli am unwahrscheinlichsten sind, oder man entnimmt ihn aus einer gesonderten Aufnahme, deren Ebene senkrecht zur Aortaachse liegt.

Da Fehler in der Bestimmung des Aortadurchmessers quadratisch in die Berechnung des Blutvolumens eingehen, wurde vorgesehen, daß die Aortawand nach einer visuellen Kontrolle auch interaktiv definiert werden kann.

4. Ergebnisse

Die Algorithmen wurden an 6 Aufnahmeserien zu je 15 Bildsequenzen getestet. Bei ca. 50 % der Auswertungen konnte auf die vorgesehene Möglichkeit von interaktiven Eingriffen verzichtet werden. Die notwendigen Interaktionen bezogen sich im wesentlichen auf Korrekturen einzelner Abschnitte der gefundenen Aortakonturen.

In der Tabelle 1 sind die derzeitigen Meßergebnisse zusammengestellt. Die Echozeit und die Herzperiode wurden den vom Tomographen automatisch mitprotokollierten Geräteparametern entnommen. Die Verzögerungszeit zwischen der Vorsättigung der Anregungsschicht und dem HF-Impuls wird nicht dokumentiert und wurde nach jeder Aufnahmesequenz mit einem Oszilloskop am Tomographen gemessen. Aus den ausgewerteten Teilstrecken der einzelnen Aufnahmen und den dafür benötigten bekannten Zeiten wird die Gesamtstrecke, die das Blut während einer Herzaktion zurückgelegt hat, berechnet. Der Aortadurchmesser wurde, wie oben beschrieben, aus den in den Bildern gefundenen Aortakonturen bestimmt. Aus diesen Werten wurden die Ergebnisse für die Blutgeschwindigkeit, das Blutvolumen und den Blutfluß berechnet. Die derzeit größte Meßunsicherheit liegt vor, wenn die Aortakontur an der Stelle, an der der Durchmesser bestimmt wird, nicht exakt gefunden wurde oder wenn die Aufnahmeebene nicht mit der Aortaachse zusammenfällt.

Diese Ergebnisse sind mit bekannten Werten zu vergleichen [4]. Das Herzzeitvolumen, das ist das innerhalb einer Minute vom Herzen geförderte Blutvolumen, beträgt bei einer Herzfrequenz von 70 Schlägen/min und einem typischen Schlagvolumen von 70 ml bei einem Erwachsenen ca. 5 l/min. Bis zu dem benutzten Auswertebereich der Aorta werden für die Blutversorgung des Kopfes ca. 15 %, für die Versorgung des Herzens ca.5 % und für die Versorgung der Arme ca. 3 % des Herzzeitvolumens abgezweigt. Folglich erwartet man an dieser Stelle ein gefördertes Blutvolumen von ca. 54 ml und einen Blutfluß von ca. 3.8 l/min.

Die Algorithmen wurden an einem Flußphantom überprüft, wobei allerdings dessen Strömungsdynamik nicht ohne Einschränkungen auf den Blutstrom übertragen werden kann.

5. Geplante Weiterentwicklung

Als geplante Verbesserung des Verfahrens soll die Extraktion der Aorta und deren Durchmesser aus zwei zur ursprünglichen Aufnahmeebene orthogonalen Schichten verbessert und die Lage der ursprünglichen Aufnahmeebene längs der Aortaachse verifiziert werden (Bild 8 und 9). Weiterhin ist die Strömungsdynamik des Blutes nicht berücksichtigt. Die Ergebnisse entsprechender Arbeiten [5] sollen berücksichtigt werden.

6. Apparative Ausrüstung

Für die Bildgebung wurde der Kernspintomograph Gyroscan S15/ACS der Fa. Philips verwendet. Er wird von einer DEC MicroVAX 3400 gesteuert. Die Signalaufbereitung erfolgt durch einen Signalprozessor Analogic AP 508. Die geometrische Auflösung der vorliegenden Bilder beträgt 256^2 Pixel, die Grauwertauflösung der gelieferten Bilddaten 12 Bit. Die Bilddaten wurden bei einem Magnetfeld von 1.5 Tesla mit T_2-gewichtetem Spinecho-

Verfahren mit ca. 6 ms Echozeit aufgenommen. Die reine Aufnahmezeit für eine Bildsequenz dauert je nach Atemrhythmus und Herzfrequenz ca.7 bis 10 Minuten.

Bei den verwendeten Geräteparametern beträgt die Schichtdicke, aus der das aufgefangene Spinecho stammt, 8 mm, die Kantenlänge der Pixel in Zeilen- und Spaltenrichtung 1.17 mm.

Die Auswertung der Bilder erfolgte auf einer IBM RISC 6000-530 mit 64 MByte Hauptspeicher unter AIX 3.2. Die Anwendung wurde in C programmiert unter Verwendung von X Window X11.4, X-Toolkit und OSF/Motif. Die gesamte Auswertung einer Bildsequenz erfolgt in ca. 1 - 2 Minuten, sofern keine Interaktionen notwendig sind.

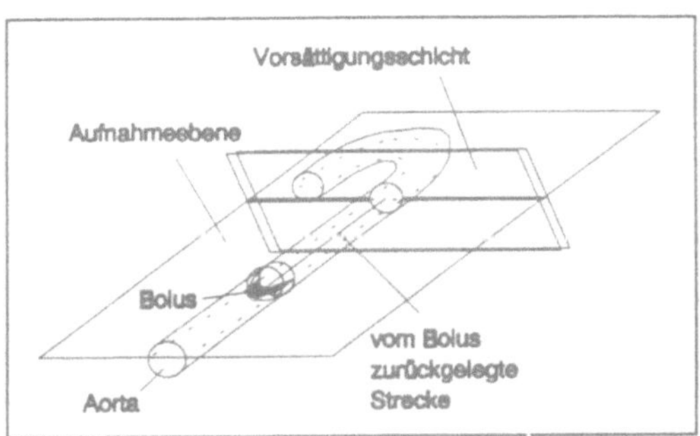

Bild 1: Schnittführung der Aufnahmeebene und der Anregungsschicht

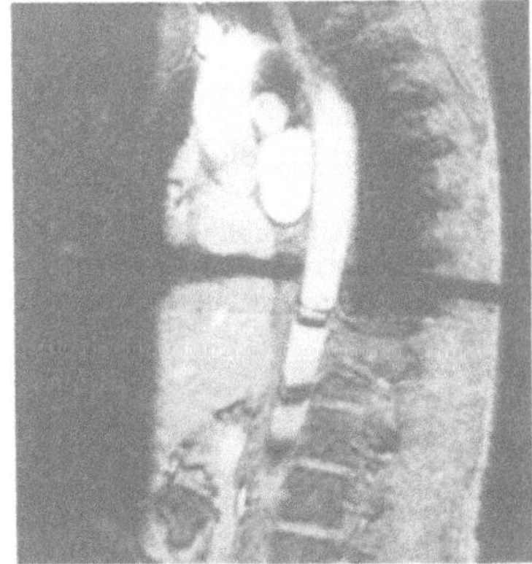

Bild 2: Orginalbild mit Herz, Aorta, Anregungsschicht, Boli der laufenden und der vorhergehenden Herzphasen, Wirbelsäule und Bandscheiben

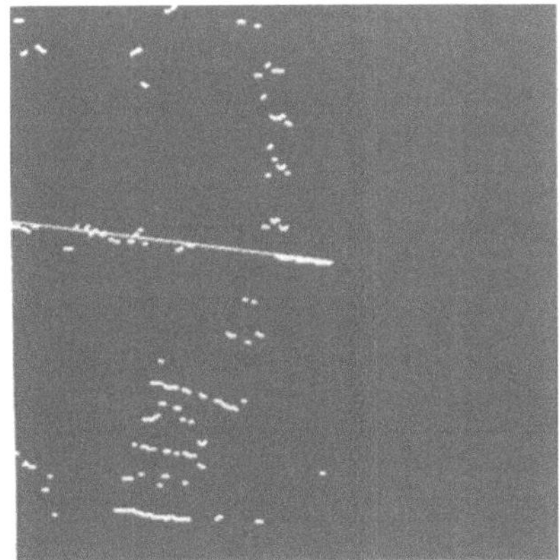

Bild 3a: Filterung der Anregungsschicht

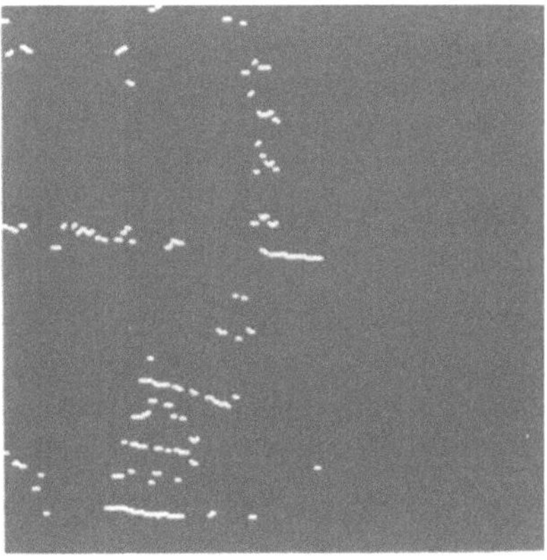

Bild 3b: Extrahierte Anregungsschicht

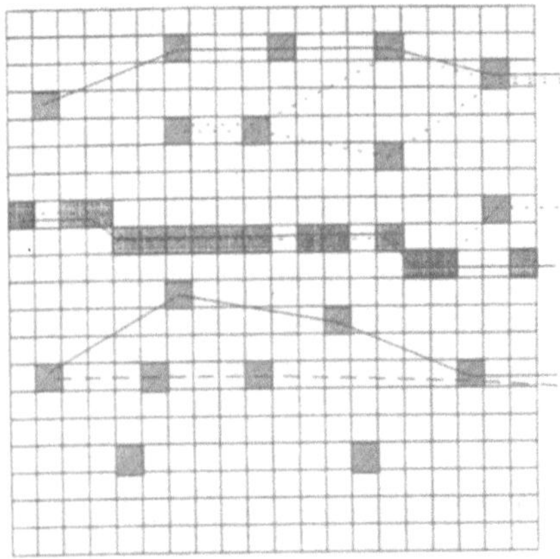

fälschlicherweise markierte Punkte

korrekt markierte Punkte

Bild 4: Schichtextraktion durch Generierung aller möglichen Linien aus der zweidimensionalen Punktematrix

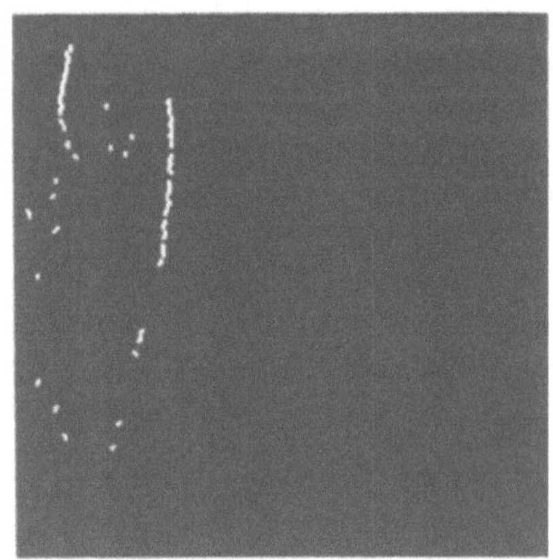

Bild 5: Aortafilterung

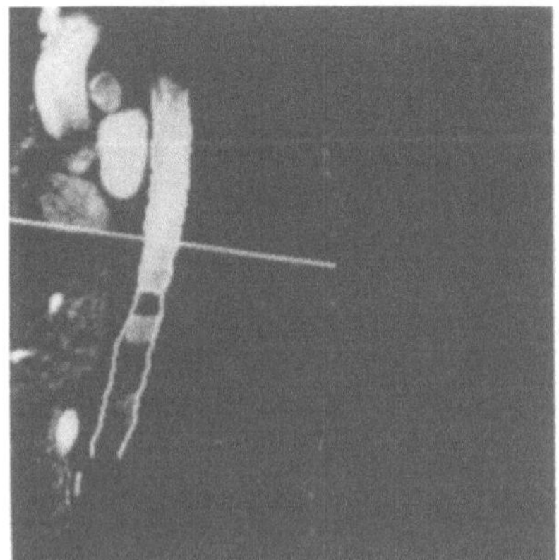

Bild 6: Markierung der Ergebnisse der Extraktion der Schicht, der Aorta und des Bolus

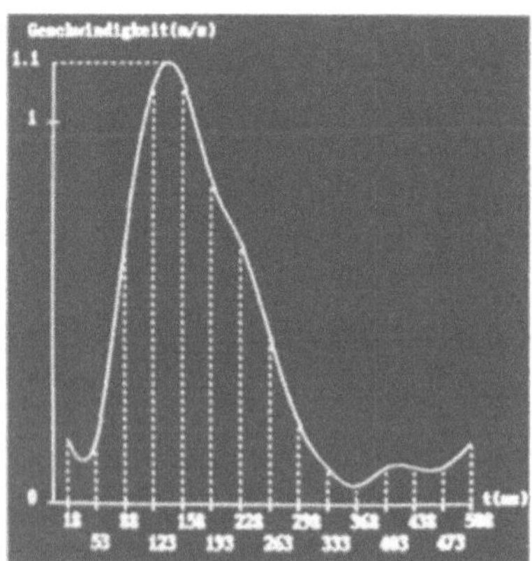

Bild 7: Das berechnete Diagramm der der Momentangeschwindigkeiten des Blutes in der Aorta

Tabelle 1: Zusammenstellung der Ergebnisse der Auswertung. Bei der mit * markierten Aufnahmeserie verlief die Aufnahmeschicht nicht durch die Aortamitte

Aufnahme / Anzahl Phasen	Echozeit (t_e) [ms]	Herzperiode (T_{rr}) [ms]	Gesamtstrecke ausgewertet [cm]	Aortadurchmesser ausgewertet [cm]	Blutgeschwindigkeit ausgewertet [cm/s]	gefördertes Volumen ausgewertet [cm³]	Fluß ausgewertet [l/min]
A1/14*	6,27	627	13,90	1,88	21,25	38,37	3,67
A2/14	6,01	691	13,83	2,23	19,53	53,84	4,68
A3/14	6,45	696	13,82	2,23	19,85	53,81	4,64
B1/15	6,43	755	20,44	1,88	26,72	56,43	4,48
B2/15	6,43	805	17,15	1,99	21,28	53,46	3,98
B3/15	6,01	814	16,11	1,88	19,77	44,48	3,28

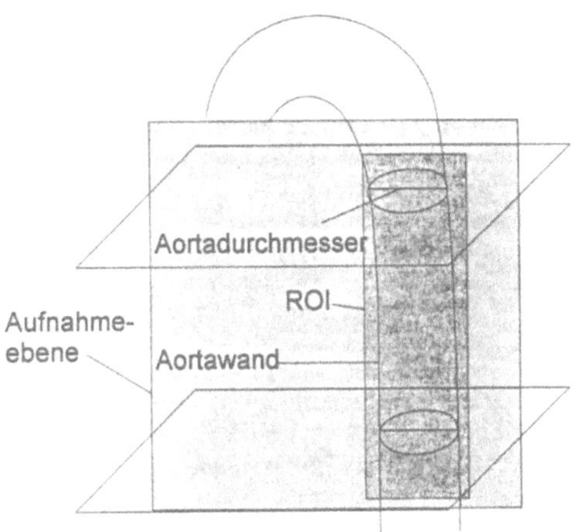

Bild 8: Schichtführung für die verbesserte Bestimmung der Aortadurchmesser

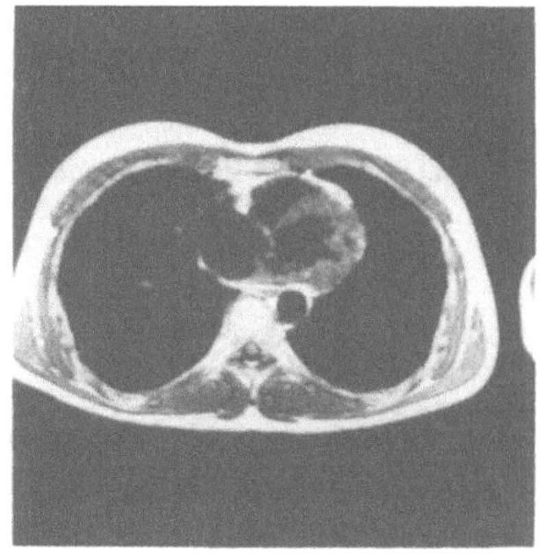

Bild 9: Darstellung der Aorta in einer zur ursprünglichen Aufnahmeebene orthogonalen Schicht

7. Literatur

[1] R.R. Edelman, H.P. Mattle, J. Kleefield, M.S. Silver : Quantification of Blood Flow with Dynamic MR Imaging and Presaturation Bolus Tracking, Radiology, Mai 1989

[2] E. Zeitler : Kernspintomographie, Deutscher Ärzteverlag, 1984

[3] A. Sahil, D. Sudheimer : Untersuchung zur Bestimmung der Blutgeschwindigkeit und des Blutvolumens in der Aorta durch Bolus-Tracking in NMR-Schichtbildern mit Methoden der digitalen Bildverarbeitung, FH Wiesbaden, 1992

[4] W.D. Keidel : Kurzgefaßtes Lehrbuch der Physiologie, 1985

[5] D. Lipsch: Pulsierende Strömung von nicht-Newtonschen Fluiden in starren und elastischen Modellen der menschlichen Arterien, Hämostaseologie 9, 1989

Klassifikation von biomagnetischen Feldmustern mit neuronalen Netzen

Martin F. Schlang[a,1], Klaus Abraham-Fuchs[b], Ralph Neuneier[a,c]
a) Siemens AG, Zentralabteilung Forschung und Entwicklung, Otto-Hahn-Ring 6, D-8000 München 83
b) Siemens AG, Geschäftsbereich Medizintechnik, D-8520 Erlangen
c) Universität Kaiserslautern, Institut für Informatik, D-6750 Kaiserslautern

1. Zusammenfassung

Bei der Auswertung von biomagnetischen Feldern spielt die dreidimensionale Quellenrekonstruktion eine bedeutende Rolle. Wir stellen ein Verfahren vor, das die Vorverarbeitung für eine solche Rekonstruktion deutlich erleichtert. Ein neuronales Netz klassifiziert gemessene Magnetoenzephalogramme in lokalisierbare bzw. nicht lokalisierbare Feldmuster. Neben den vorauszusetzenden physikalischen Modellannahmen wird die praktische Ausführung mit verschiedenen Netzwerktypen verglichen. Durch Extraktion von Regeln aus den trainierten Netzen wird zudem weiteres Wissen über die biomagnetische Quellenrekonstruktion gewonnen.

2. Einleitung

Innerhalb des menschlichen Körpers werden in Nerven oder Muskeln Informationen übertragen. Dies geschieht mit Hilfe von elektrischen Impulsen. Die daraus resultierenden Ströme erzeugen, den Maxwellschen Gleichungen zufolge, ein Magnetfeld, das auch außerhalb des Körpers meßbar ist. Mit KRENIKON®, einem Meßgerät für solche sehr kleine Felder, kann das vom Menschen erzeugte Magnetfeld mit Hilfe von SQUID's (superconducting quantum interference devices) gemessen werden /SHR 90/. Diese Felder sind typischerweise um 6 bis 8 Größenordnungen kleiner als das Erdmagnetfeld. Sie werden an bis zu 37 Aufpunkten gleichzeitig gemessen. Aufgrund der außerhalb des Körpers gemessenen Feldverteilung lassen sich die im menschlichen Körper ablaufenden elektrischen Prozesse rekonstruieren. Ein Hauptvorteil des Verfahrens besteht darin, daß es nicht invasiv und für den menschlichen Körper völlig unschädlich ist.

Bei der Messung treten viele verschiedene Probleme auf. Dank der großen Anzahl an Magnetometern wird eine enorme Flut an Daten gemessen. Nicht alle diese Daten sind aber von medizinischer Relevanz. Am Beispiel von sogenannten Extrasystolen (Extrasystolen sind krankhafte Schläge des menschlichen Herzens) konnte gezeigt werden /TLS 92/, /STA 92/, wie mit Hilfe einer speziellen Architektur von neuronalen Netzen diese Datenflut automatisch gefiltert, segmentiert und geclustert werden kann. In der vorliegenden Arbeit wird, wiederum mit neuronalen Netzen, eine weitere Möglichkeit zur Segmentierung von biomagnetischen Meßdaten, dem Magnetoenzephalogramm, gezeigt.

Ein Magnetoenzephalogramm ist das magnetische Analogon zum Elektroenzephalogramm. Es ist ein Abbild der elektrophysiologischen Prozesse, die sich im menschlichen Gehirn abspielen. Bei epileptischen Anfällen oder bei wiederholter künstlicher Stimulation und anschließender Mittelwertbildung kann es vorkommen, daß ein Prozeß erheblich dominiert. In diesem Fall ist es möglich, aus der gemessenen Magnetfeldverteilung eine genaue Aussage über den Ort dieser Prozesse zu machen.

Als aktive Prozesse werden hier näherungsweise Stromdipole angenommen /SSA 90/. Unter der Voraussetzung, daß einer der aktiven Stromdipole stark dominiert, kann aus dem Magne-

1. Tel.: +49/89/636-49408, FAX: +49/89/636-3320, e-mail: ms@leonce.zfe.siemens.de

toenzephalogramm mit Hilfe von sehr rechenaufwendigen Verfahren eine Rekonstruktion dieser Stromquelle durchgeführt werden. Neben der Stromstärke des Dipoles in den einzelnen räumlichen Dimensionen ergibt die Quellenrekonstruktion auch noch den Ort dieses Dipoles. Das Verfahren zur Rekonstruktion der Dipolquellen ist iterativ und erfordert somit einen sehr hohen Aufwand an Rechenkapazität. Daher ist es in praktischen Anwendungen nicht möglich, alle gemessenen Daten einer Quellenrekonstruktion zu unterziehen.
Deshalb ist es die Aufgabe eines Operators oder Arztes, Zeitbereiche, in denen eine Quellenrekonstruktion möglich ist, mit Hilfe seines persönlichen Expertenwissens, mittels Verfahren der digitalen Signalverarbeitung und durch seine Intuition aufzuspüren. Dies ist eine sehr zeitaufwendige und auch sehr ermüdende Angelegenheit. Mit Hilfe von neuronalen Netzen ist es uns gelungen, einen schnellen und zuverlässigen Klassifikator für die Lokalisierbarkeit aktiver Stromdipole zu realisieren. Der Klassifikator kann die Zeit, die zur Auswertung der gemessenen Daten benötigt wird, drastisch verkürzen.
Der physiologische und klinische Hintergrund zu dieser Klassifikationsaufgabe ist in /ASU 92/ genauer beschrieben.

3. Aufgabenstellung

Ein gemessenes MEG (Magnetoenzephalogramm) oder MKG (Magnetokardiogramm) soll mit Hilfe eines Algorithmus darauf untersucht werden, ob eine Quellenrekonstruktion möglich und sinnvoll ist. Diese Klassifikation muß für jeden Zeitpunkt getrennt erfolgen. Menschliche Experten lösen diese Aufgabe dadurch, daß sie für jeden Abtastzeitpunkt die zu dem Meßwerten zugehörige biomagnetische Karte beurteilen. Eine solche Karte kann durch Höhenlinien visualisiert werden (Abb. 1a). Zum einen ist es aber selbst für Experten nicht einfach, jede magnetische Karte genau zu klassifizieren. Zum anderen muß berücksichtigt werden, daß bei nur fünf Minuten Meßdauer und einer typischen Abtastrate von 1000 Hz 300000 solcher Karten zu beurteilen sind. Deshalb verwenden wir ein neuronales Netz als Klassifikator, das eine solche Aufgabe schnell und zuverlässig lösen kann. Dieses Netzwerk sagt auf Grund einer gegebenen magnetischen Karte ein Maß für die Wahrscheinlichkeit P voraus, daß diese Karte durch einen einzelnen elektrischen Stromdipol ESD erzeugt wurde (Abb. 1b).

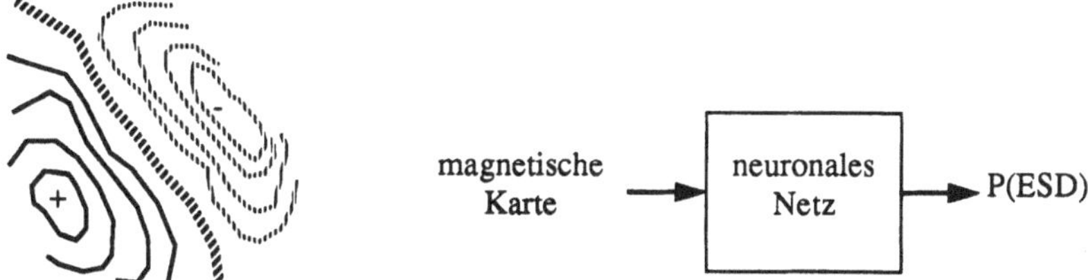

Abb. 1: Aufgabenstellung: a) Beispiel einer magnetischen Karte b) neuronaler Ansatz: das Netzwerk sagt ein Maß für die Wahrscheinlichkeit voraus, daß die Karte durch einen elektrischen Stromdipol erzeugt wurde.

4. Modellannahmen

Zur Berechnung des Magnetfeldes aus einzelnen Quellen wie auch zur Quellenrekonstruktion müssen Modellannahmen gemacht werden. Als Quellen werden hier Stromdipole angenommen, die sich innerhalb einer vorgegebenen Geometrie (Kopf, Torso) befinden. Für diese Geometrien muß ein entsprechendes elektromagnetisches Modell gebildet werden /SSA 90/. Beim Biomagnetismus werden überwiegend sehr einfache Modelle eingesetzt: bei Messungen am menschlichen Kopf wird dieser als eine homogen leitfähige Kugel angenommen, bei Messungen am menschlichen Herzen wird üblicherweise der Torso durch einen unendlichen Halbraum beschrieben.

Unter diesen Voraussetzungen ist es möglich, aus einer vorgegebenen Stromquelle (Ort, Orientierung und Stärke eines Dipoles im dreidimensionalen Raum) das zugehörige magnetische Feld am Ort des Sensorarrays zu berechnen. Die Inverse dieses Problems, also die Berechnung der Eigenschaften der Dipolquelle in Abhängigkeit eines vorgegebenen magnetischen Feldes, ist aus physikalischen Gründen nur dann eindeutig möglich, wenn vorausgesetzt wird, daß es sich bei dem Erzeuger dieses Feldes beispielsweise um nur eine einzige, punktförmige elektrische Dipolquelle handelt.

5. Erzeugung von Trainingsdaten

Ein neuronales Netz, das die Wahrscheinlichkeit von lokalisierbaren Stromdipolen voraussagen kann, muß für diese Aufgabe zunächst trainiert werden. Da dies hier durch überwachtes Training geschieht, müssen entsprechende Eingangs-/Ausgangsmuster bereitgestellt werden. Wir verwenden Simulationen zur Generierung von solchen Trainingsmustern, da somit nahezu beliebig viele Musterpaare erzeugt werden können.

Zunächst werden Dipole mit zufälligen Eigenschaften (Stärke, Orientierung, Ort) simuliert und mittels einer Vorwärtsberechnung unter den gegebenen Modellannahmen das zugehörige magnetische Feld am Ort des Sensorarrays berechnet (Abb. 2). Drei verschiedene Klassen von

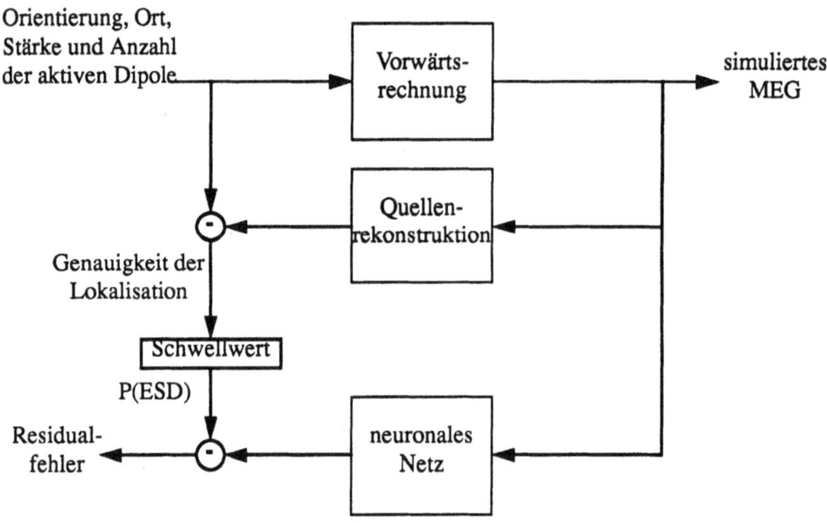

Abb. 2: Fehlerfunktion, die in der Trainingspfase minimiert wird.

Feldmustern werden für das Training verwendet: eine fokale Gehirnaktivität wird durch einen einzigen ESD, multifokale Prozesse werden durch die gleichzeitige Aktivierung von drei ESD mit zufälligen Dipolparametern und die Hintergrundaktivität des Gehirns wird durch weißes Rauschen simuliert. Den Feldmustern wird anschließend gaußförmiges Rauschen mit unterschiedlichen Pegeln überlagert. Dieses simulierte magnetische Feld wiederum verwenden wir, um eine Quellenrekonstruktion durchzuführen. Stimmen die Ergebnisse der Quellenrekonstruktion mit den Parametern der zugrundeliegenden Dipole in einem vorgegebenen Toleranzbereich überein, so wird das zugehörige magnetische Feldmuster als lokalisierbar markiert (Dipolwahrscheinlichkeit $P = 1$).

In dieser Arbeit verwenden wir zum Training der Netzwerke 3000 Beispiele, zur Überprüfung der Generalisierungsfähigkeit 2000 Muster. Es handelt sich dabei um monopolare Karten, mit einem Signal/Störverhältnis von drei.

6. Vorverarbeitung

Zur Erhöhung der Erkennungsrate werden die magnetischen Karten, bevor sie dem Netzwerk gezeigt werden, einer Vorverarbeitung unterzogen. Zunächst erfolgt eine Normierung auf konstante Energie der Karte. Dies ist erlaubt, da, ein linearer physikalischer Zusammenhang zwischen der Dipolstärke und dem magnetischen Feld besteht.
Des weiteren wurde untersucht, ob durch geeignete Transformationen die Erkennungsrate gesteigert werden kann. Es kann nämlich gezeigt werden, daß durch die praktische Anordnung des Sensorarrays und dank der zugrundeliegenden physikalischen Phänomene die magnetischen Karten Invarianzen aufweisen. Eine solche ist die Rotationsinvarianz, da das Sensorarray in einem beliebigen Winkel gegenüber dem Meßobjekt positioniert werden kann.
Erste Untersuchungen mit einer Fourier-Mellin-Transformation in Polarkoordinaten (eine solche Transformation ist rotationsinvariant) zeigten aber eine leichte Verschlechterung der Erkennungsraten. Der Grund dafür mag in der zur Fouriertransformation notwendigen Interpolation der Sensorwerte liegen. Weil beim KRENIKON® die Sensoren auf einem hexagonalen Gitter angeordnet sind, müssen für die verwendete Fouriertransformation die magnetischen Feldstärken auf ein polares Gitter umgerechnet werden. Besonders bei der Berechnung der Werte am Rand des Sensorarrays sind aber die Interpolationsfehler nicht mehr vernachlässigbar.
Eine noch stärkere Verschlechterung der Erkennungsergebnisse wurde bei der Verwendung der Momente als Merkmalsvektoren beobachtet: das Netzwerk konnte selbst auf der Trainingsstichprobe nur deutlich schlechtere Erkennungsraten erreichen.
Als die beste Vorverarbeitung erwies sich eine Rotation ohne Interpolation der Meßdaten: zunächst wird der Schwerpunkt der magnetischen Karte, also das Moment erster Ordnung, ermittelt. Die Karte wird dann, durch Symmetrien des Sensorarrays vorgegeben, modulo 60 Grad in eine Standardlage gedreht. Die Standardlage ist diejenige, in der sich der Schwerpunkt der Magnetkarte in einer definierten Richtung zum Mittelpunkt befindet.

7. Wahrscheinlichkeitstheoretischer Hintergrund

Die Aufgabe besteht darin, die Klassen ESD und $\overline{ESD}$ (nicht ESD) in Abhängigkeit eines gegebenen Magnetoencephalograms (MEG) richtig zu klassifizieren. In der Trainingsphase werden simulierte Beispiele von MEG für beide Klassen, ESD und $\overline{ESD}$ verwendet. Die Likelihood des MEG, gegeben die Klasse ESD, beträgt $P(MEG|ESD)$. Dieses Vorwissen wird in der Trainingsphase dem Netzwerk eingeprägt. In der Generalisierungsphase hingegen ist die a posteriori Klassifikationswahrscheinlichkeit $P(ESD|MEG)$, d.h. die Wahrscheinlichkeit, ob eine vorgegebene magnetische Karte (MEG) von einem elektrischen Stromdipol erzeugt wurde. Die Likelihood und die a posteriori Klassifikationswahrscheinlichkeit sind durch die Bayes-Formel verknüpft

(EQ 1)

$$P(ESD|MEG)P(MEG) = P(MEG|ESD)P(ESD)$$

Es kann gezeigt werden, daß die Aktivität am Ausgang des neuronalen Netzes gleich $P(ESD|MEG)$ ist, wenn die Trainingsdaten entsprechend normalisiert werden.
Gleichung (1) zeigt, daß die Häufigkeit des Auftretens der beiden verschiedenen Klassen für optimale Klassifikationsergebnisse berücksichtigt werden muß. In dieser Arbeit verwenden wir gleich viele Trainingsbeispiele der Klassen ESD und $\overline{ESD}$.

Zur Berechnung der Erkennungsraten in Prozent benutzen wir hier die maximale a posteriori Klassifikationswahrscheinlichkeit: jede der beiden Wahrscheinlichkeiten P(ESD) und P($\overline{\text{ESD}}$) wird durch einen eigenen Ausgangsknoten des neuronalen Netzes repräsentiert. Der Ausgangsknoten, der die stärkste Aktivität aufweist, wird als Gewinner angesehen.

8. Topologie der neuronalen Netze

In dieser Arbeit werden drei verschiedene Klassifikationsverfahren verglichen: neuronales Netz mit sigmoidalen Nichtlinearitäten /ASU 92/, neuronales Netz mit gaußförmigen radialen Basisfunktionen und ein konventioneller k-nächste-Nachbarn-Klassifikator.
Wir beschreiben hier genauer eine Architektur mit sogenannten Gaussian Mixtures, wie sie von /THA 93/ vorgeschlagen wurde. Das Netzwerk reagiert auf einen Eingangsvektor mit einer Klassifikation einer einzelnen von k Klassen (in der vorgestellten Anwendung ist k=2). Jeder Klasse ist ein Ausgangsknoten zugeordnet, die Stärke seiner Aktivität ist ein Maß für die Sicherheit, mit der sich das Netzwerk für diese Klasse entschieden hat. Am k-ten Ausgangsknoten wird die Summe der normierten Aktivität der Basisfunktionen b_i^k berechnet, wobei der Index k die Zugehörigkeit zur Klasse k angibt.

(EQ 2)

$$y_k(x) = NN_k(x) = \frac{\sum_i w_i^k b_i^k(x)}{\sum_l b_l(x)}$$

Die Ausgangsgewichte w_i^k werden zunächst als konstant eins gehalten. Die Summe im Nenner von (2) wird über alle l Knoten unabhängig von ihrer Klassenzugehörigkeit gebildet, die Summe im Zähler erstreckt sich nur über die i Basisfunktionen, die der Klasse k zugeordnet sind. Dies hat zur Folge, daß die Ausgangsaktivitäten des Netzwerkes immer positiv sind und die Summe über alle Aktivitäten eins beträgt. Wenn wir Basisfunktionen von der Form

(EQ 3)

$$b_i(x) = \kappa_i exp\{-\frac{1}{2}(x-\mu_i)^T \Sigma_i^{-1}(x-\mu_i)\}$$

verwenden, wobei die Varianzen σ_{ij} außerhalb der Diagonalen der Korrelationsmatrix Σ_i Null sind, so erhalten wir einen Gaußschen Klassifikator (Abb. 3). Dieser Typ von Klassifikator hat

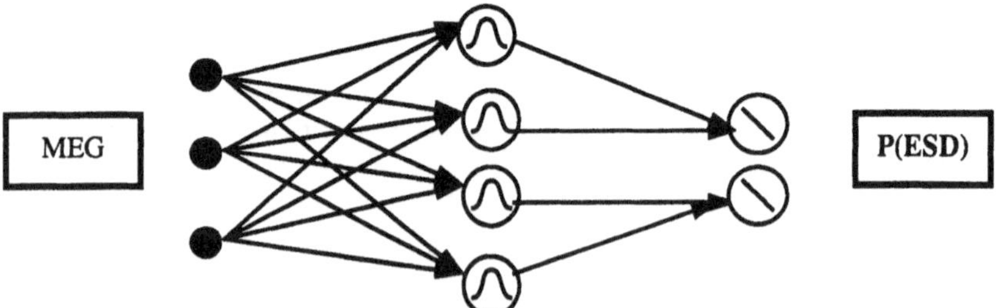

Abb. 3: Klassifikator mit gaußförmigen Basisfunktionen für zwei Klassen

drei wesentliche Vorteile: **Erstens** kann das Netzwerk mit Hilfe der Wahrscheinlichkeitstheorie interpretiert werden. Man betrachte die Summe $\Sigma_i\, b_i^k(x)$ aus den Basisfunktionen, die mit der Wahrscheinlichkeit $P(x|\,class_k)P(class_k)$ zur Klasse k gehören. Wir erhalten dann die Wahrscheinlichkeit

(EQ 4)

$$P(class_k|x) = \frac{P(x|class_k)\,P(class_k)}{\sum_l P(x|class_l)\,P(class_l)}$$

daß der Eingangsvektor x zur Klasse k gehört, was auch den Erkenntnissen aus (2) entspricht.

Zweitens gibt es viele Möglichkeiten, solche Netzwerke vorzustrukturieren. Eine sehr verbreitete Technik ist die von /DH 73/, /MD 90/, welche die Verteilung der Zentren μ der Basisfunktionen mit Hilfe von einer k-means Clusterung festlegt. Dadurch wird die Summe des euklidischen Abstandes zwischen den einzelnen Datenpunkten und den Zentren minimiert. Dieser Algorithmus kann noch erweitert werden, indem zusätzliches Wissen über den Ausgangsraum berücksichtigt wird: die Zentren μ_{ij}^k der Basisfunktionen b_i^k werden nur für den Teil der Trainingsdaten adaptiert, der zur Klasse k gehört. Dadurch wird eine bessere Verteilung der Zentren bezüglich der klassenspezifischen Datenpunkte erreicht.
Nach der Clusterung der Zentren können die Varianzen σ_{ij} mit Hilfe der k-nächste-Nachbarn Methode /MD90/ initialisiert werden. Die einzelnen σ_{ij} werden dabei aus dem Mittelwert der Abstände vom Zentrum μ_{ij} zu den nächsten k Zentren bestimmt. Dabei werden die einzelnen σ_{ij} auf Werte von größer 0.1 begrenzt. So kann eine allzu lokale Modellierung der Trainingsdaten vermieden werden. Wie bei den Zentren verwenden wir auch hier die klassenspezifische Information: es werden nur diejenigen Zentren μ_{ij}^k berücksichtigt, deren Basisfunktionen b_i^k zur gleichen Klasse gehören.
Tabelle 1 zeigt die Wirksamkeit verschiedener Verfahren zur Vorstrukturierung. Angegeben ist der Prozentsatz von richtig klassifizierten Mustern für die Trainings- und Generalisierungsstichprobe. Die Methoden, die klassenspezifisches Wissen mit berücksichtigen, sind offensichtlich den anderen bezüglich der Erkennungsgenauigkeit überlegen.
Der **dritte** Vorteil der verwendeten Architektur besteht darin, daß die Komplexität des Netzwerkes mit einer iterativen Strategie /THA 93/ reduziert werden kann. Diese Methode erlaubt auch das Vermeiden von Overfitting beim Training des Netzwerkes. Die relative Bedeutung

Table 1: Erkennungsraten nach der Initialisierung T: Trainingsdaten; G: Generalisierungsdaten

	fixierte σ_{ij}	k-nächster	Klasse-k-nächster
k-means	56% T / 53% G	53% T / 52% G	51% T / 48% G
Klasse-k-means	75% T / 72% G	75% T / 70% G	75% T / 72% G

jeder Basisfunktion $\phi_i = b_i(\mu_i) / \Sigma_i\, b_i(\mu_i)$ am Ort ihres Zentrums gibt Aufschluß darüber, welche Basisfunktionen überflüssig sind: die Basisfunktionen mit den kleinsten ϕ_i werden sukzessive entfernt und anschließend das Netzwerk entsprechend nachtrainiert. Damit ergeben sich die in Tabelle 2 gezeigten Erkennungsraten.

Table 2: Erkennungsraten nach dem Training (beste Generalisierung)

	Training	Generalisierung	Knoten	Epochen
Fixierte Ausgangsgewichte	78	74	40	400
Gelernte Ausgangsgewichte	80	75	40	550

Table 2: Erkennungsraten nach dem Training (beste Generalisierung)

	Training	Generalisierung	Knoten	Epochen
Sigmoidale Nichtlinearitäten	81	76	50	1000
Konventioneller k-nächster Nachbar Klassifikator[a]	68%	51%	3000	-

a. Auf Grund einer größeren Datenbasis sind die Erkennungsraten nicht direkt mit den anderen in dieser Arbeit vergleichbar.

Wenn die Gewichte w_i^k auch von eins verschiedene Werte annehmen dürfen, so kann die Erkennungsrate noch weiter gesteigert werden. Die geht aber auf Kosten der theoretischen Interpretation des Netzwerkes mit Hilfe der Wahrscheinlichkeitstheorie.
Der in Tabelle 2 verwendete konventionelle k-nächster-Nachbar-Klassifikator berücksichtigte k = 11 Nachbarn. Bei diesem Algorithmus müssen aber bei den vorliegenden Datensätzen sehr viele Abstände berechnet werden. Deshalb kommt er für die praktische Anwendung nicht in Frage.

9. Extraktion von Wissen aus der Gewichtsmatrix

Bei der Verwendung von Netzwerken mit radialen Basisfunktionen können aus den Gewichtsmatrizen Regeln über das modellierte System extrahiert werden /HT 92/. In Abb. 4 links ist die

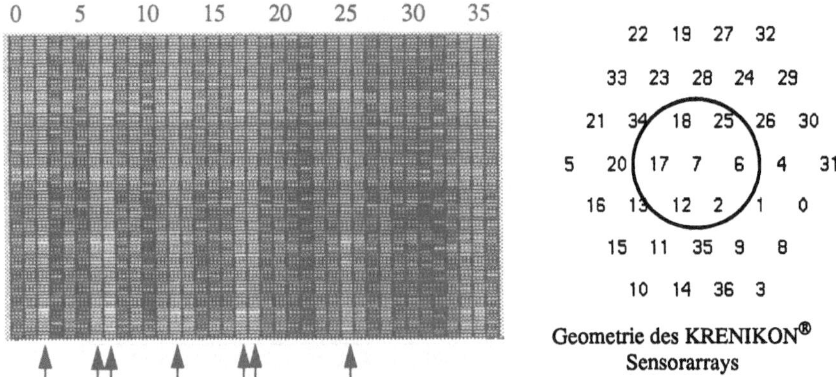

Geometrie des KRENIKON® Sensorarrays

Abb. 4: Gewichtsmatrix eines trainierten neuronalen Netzes mit gaußförmigen Basisfunktionen: Eingangsknoten, die weniger zur Klassifikation beitragen, sind hell codiert (links). Sie korrespondieren mit den magnetischen Sensoren innerhalb des rechts markierten Kreises (rechts).

Verteilung der σ_{ij} aufgetragen. Jede Zeile enthält die Varianzen einer einzelnen Basisfunktion, wobei die zu einem Eingangsknoten zugehörigen Werte jeweils spaltenweise angegeben sind. Große σ_{ij}, also weniger dominante Verbindungen, sind in der Gewichtsmatrix hell codiert. Kleine σ_{ij} deuten auf eine sehr lokale Repräsentation hin. Sie sind in der Gewichtsmatrix dunkel dargestellt. Auffällig sind die vertikalen hellen Streifen, die durch Pfeile gekennzeichnet sind. Die zu diesen Eingangsknoten zugehörigen Magnetsensoren des KRENIKON® befinden sich alle in der Mitte des Sensorarrays, innerhalb des Kreises in Abb. 4 rechts. Dies hat die praktische Bedeutung, daß bei der hier verwendeten Zusammensetzung des Trainingsmaterials diese Sensoren für die Klassifikation der Magnetkarten weniger wichtig sind als die Sensoren, die sich weiter vom Mittelpunkt des Sensorarrays entfernt befinden.

10. Schlußfolgerungen

Es kann gezeigt werden, daß neuronale Netze sehr wohl auch nicht triviale Klassifikationsaufgaben lösen können. Ihre Erkennungssicherheit ist größer als bei einem konventionellen k-nächster-Nachbar-Klassifikator. Die Erkennungsraten können durch eine der Aufgabenstellung angepaßte Vorverarbeitung deutlich erhöht werden.

In der vorliegenden Aufgabe schneidet das Netzwerk mit radialen Basisfunktionen bezüglich der Erkennungsgenauigkeit in etwa gleich gut wie ein perzeptronartiges Feedforward-Netz mit sigmoidalen Nichtlinearitäten ab. Ersteres bietet aber noch den Vorteil, daß aus den Gewichtsmatrizen durch Extraktion von Regeln noch weiteres Hintergrundwissen über den physikalisch zugrundeliegenden Prozeß gesammelt werden kann. Dies kann durch Analyse der Kovarianzmatrizen der Basisfunktionen des trainierten Netzwerkes ereicht werden. Ein weiterer Vorteil dieser Struktur besteht in der Möglichkeit der Vorstrukturierung solche Netze. Damit kann die für das Training des Netzwerkes benötigte Zeit drastisch verringert werden.

Das hier beschriebene Verfahren zur Klassifikation von magnetischen Karten wird bald auch in der Praxis eingesetzt. Zur Zeit laufen Untersuchungen, in wieweit diese Netzwerke auch zur Klassifikation von elektrischen Potentialen und zur Quellenrekonstruktion selbst verwendet werden können.

Literatur

/ASU 92/ Abraham-Fuchs, K.; Schlang, M.; Übler, J.; Hellstrand, E.; Stefan H.: "Map Classification in the Magnetoencephalogram by Neural Networks", Proc. 14. Ann. Int. Conf. IEEE Eng. in Med. & Biol. Soc, Lyon, 1992.

/DH 73/ Duda, R. O.;Hart, P. E.: "Pattern Classification and Scene Analysis", Wiley-Interscience, 1973.

/HT 92/ Hollatz, J.; Tresp, V.: "A Rule-based Network Architecture", Artifical Neural Network II, I. Aleksander, J. Taylor,eds., Elsevier, Amsterdam, 1992.

/MD 90/ Moody, J.; Darken, C.:"Fast adaptive k-means clustering: Some emperical Results", Proc. Int. Joint Conf. on Neural Networks, San Diego, 1990.

/TLS 92/ Tresp, V.; Leuthäusser, I.; Schlang, M.; Neuneier, R.; Abraham-Fuchs, K.; Härer, W.: "The Neural Impulse Response Filter", Artifical Neural Networks II, I. Aleksander, J. Taylor,eds., Elsevier, Amsterdam, 1992.

/SHR 90/ Schneider, S.; Hoenig, E.; Reichenberger, H.; Abraham-Fuchs, K.; Daalmans, G.; Moshage, W.; Oppelt, A.; Röhrlein, G. Stefan, H.; Vieth, J.; Weikl, A.; Wirth, A.: "A Multichannel Biomagnetic System for High Resolution Functional Studies of Brain and Heart", Radiology, Vol. 176, pp 825 - 830, 1990.

/SSA 90/ Stefan, H.; Schneider, S.; Abraham-Fuchs, K.; Bauer, J.; Feistel, H.; Pawlik, G.; Neubauer, U.; Röhrlein, G.; Huk, H. J.: "Magnetic source localisation in focal epilepsy: Multichannel magnetoencephalography correlated with magnetic resonance brain imaging", Brain, 113, 1990.

/STA 92/ Schlang, M.F.; Tresp, V.; Abraham-Fuchs, K.; Härer, W.; Weismüller, P.: "Neural Networks for Segmentation and Clustering of Biomagnetic Signals", Proc. IEEE Workshop on Neural Networks for Signal Processing, Copenhagen, 1992.

/THA 93/ Tresp, V.; Hollatz, J.; Ahmad, S.: "Network Structuring and Training Using Rule-based Knowledge", to appear in Advances in Neural Information Processing System 5, C. L. Giles, S. J. Hanson, J. D. Cowan, eds., Morgan Kaufmann, San Mateo, CA, 1993.

Kompensation von Rippenschatten in digitalen Thorax-Röntgenbildern

M. Schreckenberg und M. Joswig

Lehrstuhl für Meßtechnik, RWTH Aachen, D-52056 Aachen

Einleitung

Die Einführung der Speicherphosphor- oder Luminezenzradiografie brachte gegenüber der konventionellen Film-Folientechnik eine um Größenordnungen höhere Signaldynamik; ihre Kennlinie ermöglicht die lineare Abbildung des höheren Informationsgehalts im Aufnahmemedium (Tateno et al., 1987). Bei der Darstellung dieser Information für den menschlichen Betrachter muß jedoch auf die begrenzte Anzahl differenzierbarer Graustufen Rücksicht genommen werden. Deshalb kann lediglich ein kleiner Teil der vorhandenen Information gleichzeitig dargestellt werden. Praktisch bewährt hat sich die Simulation einer konventionellen Filmkennlinie, womit sich die Kontrastauflösung kleiner Intensitätsunterschiede auf Kosten der Breite des darstellbaren Intensitätsintervalls erhöhen läßt. Letztlich erhält man - mit hohem technischen Aufwand - optimal ausgeleuchtete konventionelle Röntgenaufnahmen. Die Anwendungsvorteile liegen in der in weiten Grenzen freien Belichtung z.B. für bettlägrige Patienten; prinzipiell sind jedoch keine im Vergleich zu guten konventionellen Röntgenbildern informativeren Ergebnisse zu erwarten (Blume und Jost, 1992).

Kompensation von Störstrukturen

Um den Anteil der für eine Befundung relevanten Information auf einem Bild zu erhöhen, muß zwangsläufig der Anteil der redundanten / irrelevanten Daten verringert werden. Diese Bildverarbeitungsoperation ist natürlich abhängig vom jeweiligen Untersuchungsziel. Bei der Thoraxbefundung z.B. überlagern die Schatten der Costae (Rippen) vor allem lateral wesentliche Teile der pulmonalen Strukturen. Eine Unterdrückung dieser Störsignale läßt eine verbesserte Befundbarkeit z.B. auf Konturen eines Pleuraergusses erwarten und stellt zudem die Basis für weitergehende diagnoseunterstützende Mustererkennungsverfahren dar (Lo et al., 1993; MacMahon et al., 1989; Yoshimura et al., 1992).

Zur Reduktion störender Bildstrukturen werden in der Literatur verschiedene Ansätze diskutiert. Beim Zweispektren- bzw. "Dual Energy" Verfahren werden für jede Untersuchung zwei Bilder mit unterschiedlicher physikalischer Information erzeugt. Dies geschieht entweder durch zwei aufeinanderfolgende Aufnahmen mit unterschiedlicher Beschleunigungsspannung oder bei einmaliger Belichtung durch die Verwendung einer zweiten Detektorfolie, die durch einen Metallfilter abgeschirmt wird (Shaw und Gur, 1992). Im höherenergetischen Bild ist das Durchdringungsvermögen für Weichteile sehr viel besser ist als für Knochen; durch eine geeignet gewichtete Subtraktion vom zugehörigen Niederenergiebild ist es also möglich, Knochenstrukturen im Ergebnisbild fast vollständig zu entfernen. Durch den Subtraktionsprozeß wird jedoch das Bildrauschen signifikant erhöht. Auch enthält das Hochenergiebild Signalanteile von dichteren Weichteilbereichen (vor allem im Mediastinum), woraus weitere Artefakte resultieren. Das Ziel der Entfernung der störenden Rippenstrukturen muß also mit einem verschlechterten S/N Verhältnis, welches sich nur durch eine Verringerung der Ortsauflösung wieder verbessern läßt, und erhöhtem praktischen Aufwand in der Routine erkauft werden (Ito et al., 1993).

Eine ganze Klasse von weiteren Methoden basiert auf der Information, die in lediglich einem Röntgenbild enthalten ist. Beim "Unsharp-Masking" wird durch gewichtete Subtraktion der niederfrequenten Bildanteile eine Dynamikkompression erzielt, wodurch die für den Detailreichtum verantwortlichen höherfrequenten Anteile kontrastreicher dargestellt werden können (Abe et al., 1992; Souto et al., 1992; Tahoces et al., 1991). An Grauwertsprüngen, also Knochen- oder Gefäßgrenzen und Überlagerungen treten jedoch Artefakte auf, die die diagnostische Aussagekraft stark beeinträchtigen. Das Verfahren kann daher gerade im Bereich der Pulmones nur sehr eingeschränkt Verwendung finden. Auch eine wirkungsvolle Unterdrückung der überlagernden Costaestrukturen ist nicht möglich.

Will man das Störsignal der Rippen kompensieren, ohne dabei die übrige Bildinformation zu verändern und zusätzliche physikalische Information zu verwenden, bieten sich Verfahren der Mustererkennung an. Zur Erkennung von Rippen in konventionellen Thoraxaufnahmen sind in der Literatur verschiedene Verfahren referiert worden (Persoon, 1976; Kulick et al., 1978; de Souza, 1983; Toriwaki et al., 1973, 1980; Wechsler und Sklansky, 1977; Wechsler und Fu, 1978). Trotz vielfältiger Unterschiede im Detail lassen sich alle Verfahren in das grundsätzliche Schema von Tabelle 1 einordnen. Zunächst findet eine Vorverarbeitung statt, bei der das Bild gefiltert und die örtliche Auflösung durch Reduktion der Bildmatrix verkleinert wird. Im nächsten Schritt werden mit geeigneten Extraktoren Merkmalsbilder erzeugt, die die Eigenschaft der Rippenzugehörigkeit besser vom Restbild diskriminieren sollen. Dann wird anhand eines Modells für den Konturverlauf der Rippen eine Mustererkennung durchgeführt; sie liefert eine robuste Schätzung aufgrund der globalen Anpassung. Mit diesem Ergebnis kann ein Suchraum für die nachfolgende Verfeinerung definiert werden, die die tatsächliche Lage der Rippenkanten durch lokale Anpassung bestimmt. Alle Verfahren basieren auf impliziten Modellannahmen, wie sie in der letzten Spalte von Tabelle 1 aufgelistet sind.

Die bisher erzielten Resultate sind jedoch für eine Kompensation der Rippenschatten im Thoraxbild nicht geeignet: (I) Die Rippen werden nicht als komplette Struktur modelliert, sondern lediglich als Teilbögen erkannt, was zu Übergangsfehlern zwischen dorsalen und ventralen Rippenanteilen und einer generell schlechten Erkennung ventraler Ausläufer führt. (II) Durch die signifikante Reduktion der Bildmatrix ist die Konturierung der Rippenkanten nicht hinreichend genau. (III) Ein Modell zur Kompensation der durch Rippenstruktur und deren Überlagerungen verursachten Grauwerte fehlt vollständig.

Erzeugung des Kompensationsbildes

Das hier vorgestellte Verfahren arbeitet auf den unreduzierten Daten der Original-Bildmatrix von 2560*1870 Pixel (Bild 1). Nur so kann die Erkennung und Kompensation der Rippenschatten mit der in der Diagnostik erforderlichen Genauigkeit durchgeführt werden. Ansonsten entsprechen die ersten drei Verarbeitungsschritte den in Tabelle1 skizzierten Grundfunktionen: Zuerst werden mögliche Kantensegmente mit Gradientenfiltern in horizontaler und vertikaler Richtung extrahiert (Bild 2). Dann werden im lateralen Bereich die Rippenbögen mit liegenden Parabelhälften per Hough-Transformation modelliert. Es werden nur Kantenpunkte innerhalb vorgegebener Teilbereiche des Bildes und mit Kantenrichtung innerhalb vorgegebener Winkelintervalle berücksichtigt. Die letztendliche Auswahl der Häufungspunkte im Parameterraum basiert auf Zusatzinformationen über Rippenabstand und Krümmungsradius, um Fehlzuordnungen zu vermeiden. Für die dorsalen Anteile findet ein ähnliches Verfahren Anwendung, wobei der Konturverlauf nun als stehende Parabelhälfte modelliert wird. Beide Teilbögen überlappen sich hinreichend lange, um eine eindeutige Zuordung zu vollständigen Rippenbögen vorzunehmen. Im nächsten Schritt wird die so grob bestimmte Rippenkontur verfeinert. Dazu wird eine Faltung mit einem linienförmigen Faltungskern senkrecht zum bisherigen Konturverlauf durchgeführt. Der Faltungskern realisiert ein "matched filter" für die 1. Ableitung des Grauwertprofils einer Rippenkante. Das Ergebnis einer kompletten Rippenerkennung zeigt Bild 2.

Die Kompensation der Helligkeitsunterschiede durch die komplette Rippenstruktur erfordert ein zusätzliches Modul, welches den zu subtrahierenden Anteil festlegt. Dazu werden in Rippensegmenten ohne Mehrfachüberlagerung Schwächungsprofile senkrecht zur Knochenrichtung bestimmt. Dies geschieht mit Medianstatistiken in Knochenrichtung. Basierend auf den aktuellen Meßwerten wird ein komplettes Modell mit berechneten Intensitäten für Knochenüberlagerung und den zur Seite zunehmenden Längsdurchstrahlungen der Rippenbögen erstellt und von den Originaldaten abgezogen (Bild 3). Das Ergebnis ist ein "knochenkorrigiertes" Thoraxbild mit homogener Intensitätsverteilung ohne Störschatten, welches leichtere Befundung der Lunge erlaubt. Da jedoch ein Teil der Diagnose auf anatomischer Orientierung basiert, bleibt die Rippenstruktur als rotes Strichbild einblendbar - zusätzlich zur immer geforderten Umschaltung auf die Referenz des unbearbeiteten Bildmaterials.

Ausblick

Die Kompensation von Rippenschatten ist eine Möglichkeit, den in digitalen Röntgenbildern vorhandenen "Mehrwert" an Information in neuartiger Darstellung sichtbar zu machen. Erst die erweiterte Dynamik erlaubt, nach Abziehen der Strahlschwächung durch die Knochenstruktur die tatsächlich gemessene Intensitätsunterschiede des Lungengewebes aufzulösen und sie in das Fenster visuell differenzierbarer Grauwerte abzubilden. Eine solchermaßen homogen gemachte Lungenstruktur erleichtert die ärztliche Befundung konturschwacher Symptome, allerdings um den Preis der Aufgabe des "einzig" richtigen Röntgenbildes.

Literatur

K. Abe, S. Katsuragawa, Y. Sasaki and T. Yanagisawa, "A fully automated adaptive unsharp masking technique in digital chest radiograph", *Invest. Radiol.*, **27**, 64-70, 1992.

H. Blume and R. G. Jost, "Chest imaging within the radiology department by means of photostimulable phosphor computed radiography: a review", *J. Digit. Imag.*, **5**, 67-78, 1992.

E. Persoon, "A new edge detection algorithm and its application in picture processing", *Comp. graph. and image proces.*, **5**, 425-446, 1976.

J. H. Kulick, C. M. Brace and T. W. Challis, "Automatic rib detection in chest radiographs", *Proc. 1977 int. joint conf. artific. intellig.*, 697-698, 1978.

W. Ito, K. Shimura, N. Nkajima, M. Ishida and H. Kato, "Improvement of detection in computed radiography by new single-exposure dual-energy subtraction", *J. Digit. Imag.*, **6**, 42-47, 1993.

S. B. Lo, M. T. Freedman, J. Lin and S. K. Mun, "Automatic lung nodule detection using profile matching and back-propagation neural network techniques", *J. Digit. Imag.*, **6**, 48-54, 1993.

H. MacMahon, K. Doi, M. L. Giger, S. Katsuragawa, N. Nakamori and Y. Sasaki, "Artificial intelligence in chest radiology: computer-aided detection and quantitation of disease", *Proc. CAR'89 Computer Assisted Radiol.*, H. U. Lemke, M. L. Rhodes, C. C. Jaffe and R. Felix eds., 441-451,Springer, Berlin, 1989.

C. C. Shaw and D. Gur, "Comparison of three different techniques for dual-energy subtraction imaging in digital radiography: a signal-to-noise analysis", *J. Digit. Imag.*, **5**, 262-270, 1992.

M. Souto, J. Correa, P. G. Tahoces, D. Tucker, K. S. Malagari, J. J. Vidal and R. G. Fraser, "Enhancement of chest images by autmatic adaptive spatial filtering", *J. Digit. Imag.*, **5**, 223-229, 1992.

P. de Souza, "Automatic rib detection in chest radiographs", *Comp. vision, graph. and image proces.*, **23**, 129-161, 1983.

P. G. Tahoces, J. Correa, M. Souto, C. Gonzalez, L. Gomez and J. J. Vidal, "Enhancement of chest and breast radiographs by automatic spatial filtering", *IEEE trans. Med. Imag.*, **10**, 330-335, 1991.

Y. Tateno, T. Iinuma and M. Takano eds., *Computed radiography*, Springer, Berlin, 1987.

J. Toriwaki, Y. Suenaga, T. Negoro and T. Fukumura, "Pattern recognition of chest x-ray images", *Comp. graph. and image proces.*, **2**, 252-271, 1973.

J. Toriwaki, J. Hasegawa, T. Fukumura and Y. Tagaki, "Computer analysis of chest photofluorograms and its application to automated screening", *Automedica*, **3**, 63-81, 1980.

H. Wechsler and J. Sklansky, "Finding the rib cage in chest radiographs", *Pattern recogn.*, **9**, 21-30, 1977.

H. Wechsler and K. S. Fu, "Image processing algorithms applied to boundary detection in chest radiographs", *Comp. graph. and image proces.*, **7**, 375-390, 1978.

H. Yoshimura, M. L. Giger, K. Doi, H. MacMahon and S. M. Montner, "Computerized scheme for the detection of pulmonary nodules - a nonlinear filtering technique", *Invest. Radiol.*, **27**, 124-129, 1992.

	Abtastung	Vorfilterung	Extraktion eines Merkmalsbildes	Modellierung des Konturverlaufes	Verfeinerung: Suchraum	Verfeinerung	Modellannahmen
Toriwaki	256x 256		Ableitung 2. Ordnung richtungssensitive Kantentemplates	Polynom 2. Grades	vertikales Intervall	einmalige Optimierung der Polynome	- lokale Maxima der dorsalen Rippenkanten liegen auf einer Geraden - Rippen konvex - Thorax symmetrisch - regionale Kanten-Vorzugsrichtung
Persoon	256x 256		richtungssensitives Kantenfilter	Konturverfolgung			- Rippenabstände stehen in Beziehung zueinander - Rippendicke annähernd konstant - Rippen konvex - Thorax symmetrisch
Wechsler	1024x 896	Smoothing Matrixreduktion Hochpass	Ableitung 1. Ordnung Ableitung 2. Ordnung	Polynome 2. Grades	Baum	lokale Konturanpassung	- dorsale Rippenkanten bilden lokale Maxima - Thorax symmetrisch
Kulick	256x 256	nichtlineares Smoothing	Smoothing, Ableitung 1. Ordnung	Konturverfolgung			- Rippenabstände stehen in Beziehung zueinander - Rippendicke annähernd konstant - Rippen konvex - Thorax symmetrisch
Souza	1200x 1200	Smoothing Matrixreduktion	Scharmittelung, Smoothing, Ableitung 1. Ordnung	Polynome 2. Grades	vertikales Intervall	iterative Optimierung der Polynome	- Rippenabstände stehen in Beziehung zueinander - Rippendicke konstant - Rippen konvex - Thorax symmetrisch - regionale Kanten-Vorzugsrichtungen - Rippen innerhalb der Lungenränder

Tabelle 1

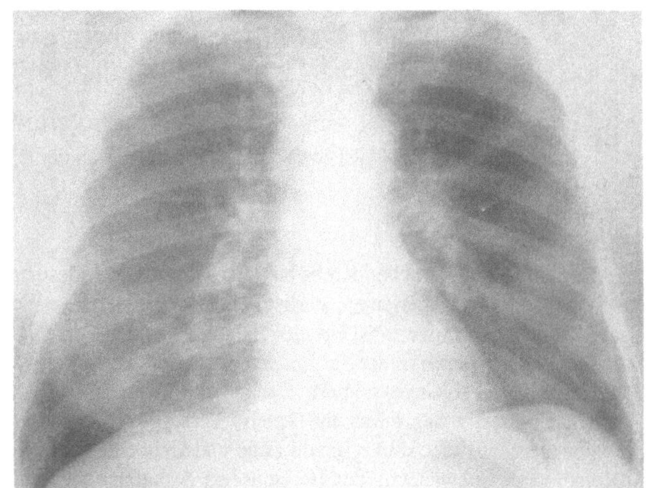

Bild 1

Originales Thorax-Röntgenbild

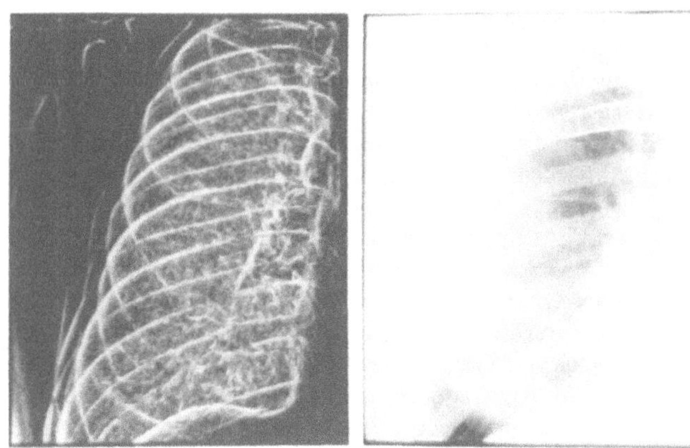

Bild 2

Kantenbild des Thorax und segmentierte Rippe

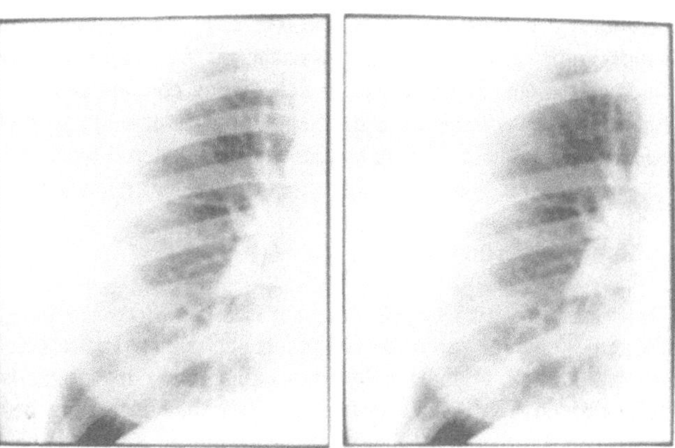

Bild 3

Kompensation eines Rippenbogens

Neue Ansätze zur HTEKG-Signalanalyse bei akutem Myocardinfarkt mittels selbstorganisierender Karten

Barbi Schulz[1], Erich Pelikan[1], Jiri Silny[2], Günther Rau[2]

1: Institut für Medizinische Informatik und Biometrie, RWTH Aachen (R. Repges)
2: Helmholtz-Institut für Biomedizinische Technik an der RWTH Aachen (G. Rau)

1. Einführung:

Der Herzinfarkt bildet neben Krebs eine der häufigsten Todesursachen. Rund 25 % aller Patienten, die einen Infarkt erleiden, sterben binnen der ersten vier Wochen an den Folgen. Die Ursache für diese Erkrankung ist ein plötzlich auftretender Verschluß (Thrombus) oder eine starke Verengung (Stenose) einer der Herzkranzarterien. Die damit verbundene Unterbrechung der Blutzufuhr führt, abhängig von der Dauer der Unterbrechung, zu einer Ischämie mit nachfolgendem Untergang des durch die Arterie versorgten Muskelgewebes und damit zu dessen funktionellem Ausfall. Abhängig von Lage und Größe des betroffenen Muskelareals erfolgt eine funktionelle Schwächung des Herzmuskels, was u.a. an einem Absinken der Auswurffraktion zu beobachten ist. Im Extremfall verstirbt der Patient im kardiogenen Schock.
Die Notfalltherapie zielt deshalb vor allem auf eine Begrenzung der Infarktgröße, wobei insbesondere binnen der ersten 2 Stunden Aussicht auf Erfolg besteht [Mat85]. Die lokale Thrombolyse mittels Herzkatheter bildet das zur Zeit wirksamste Mittel zur Wiedereröffnung verschlossener Herzkranzgefäße, ist aber als invasives Verfahren nicht ohne Risiko. Eine Therapie-Entscheidung bzw. eine Abschätzung der Erfolgsaussichten kann an folgenden Parametern orientiert werden:

- Alter des Patienten
- Infarktausmaß: die Größe des betroffenen Mycards, abhängig von der Lage des Thrombus (proximal oder distal) und der Kolateralenversorgung [Fei87],
- Infarktstatus: der bereits vorliegende Schädigungsgrad, abhängig sowohl von der zeitlichen Differenz zwischen dem Auftreten der Symptome und dem Therapiebeginn, als auch vom Grad des Verschlusses und der Kolateralenversorgung.

Wichtigste Informationsquelle in der klinischen Routine bildet bislang das 12-Elekroden-Elektrokardiogramm. Eine tendenzielle Abschätzung der Infarktgröße (IG) aus dem 12-Elekroden-Elektrokardiogramm ist 72 Stunden nach akutem Myocardinfarkt (AMI) mittels des Selvester-Scoring-Systems möglich [Ide82, Roa83, Sel85, Wag82, War84]. Zu diesem Zeitpunkt stehen aber auch die Enzymwerte CK und α-HBDH zur Verfügung, die eine gute Aussage über die Infarktgröße erlauben. Unmittelbar nach Auftreten der Symptome, wenn durch eine Intervention eine Begrenzung der Schädigung noch möglich ist, liefert das Selvester-Scoring-System keine verwertbaren Aussagen. Grund dafür sind die erheblichen Veränderungen, denen das Signal innerhalb der ersten Stunden nach akutem Myocardinfarkt unterliegt.

2. Mapping-Verfahren

Das dieser Arbeit zu Grunde liegende HTEKG-Meßverfahren (Hochauflösende Thorax Elektrokardiographie) wurde am Helmholtz Institut für Biomedizinische Technik in Aachen entwickelt [Hin85]. Grundlage der Entwicklung war der Ansatz, daß durch eine feinere, über das 6-Elektroden EKG nach Wilson herausgehende Erfassung des elektrischen Feldes auf dem Brustkorb ein deutlicher Informationsgewinn zu erzielen sein müßte. Das als *Mapping* bezeichnete Verfahren arbeitet mit 63 Brustwand- und 3 Extremitätenableitungen und ist damit ein System mittlerer Größe. Das Signal wird dabei parallel in allen Kanälen bei einer Meßdauer von 2.56 Sekunden mit einer Abtastrate von 400/s bei 12 Bit Auflösung digitalisiert.

Die in dieser Arbeit verwendeten Daten stammen aus einer Mapping-Studie, die primär die EKG-Veränderungen bei Wiedereröffung eines verschlossenen Gefäßes mittels lokaler und systemischer Thrombolyse untersuchte. Im Rahmen dieser Studie wurden die EKGs von 39 Patienten mit Vorderwandinfarkt (VWI) und 52 Patienten mit Hinterwandinfarkt (HWI) in der akuten Phase, d.h. unmittelbar nach Einlieferung, unter Thrombolyse aufgezeichnet (Lysekollekiv). Weiterhin wurden 78 gesunde Probanden erfaßt (Normalkollektiv). Zu 16 der VWI-Patienten und 28 der HWI-Patienten standen ergänzend die Verläufe der Serumwerte von CK und α-HBDH zur Infarktgrößenbestimmung zur Verfügung. Alle Patienten wurden im Abstand von 3 Tagen und drei Wochen einer Mapping-Kontrolluntersuchung unterzogen, die ebenfalls zur Auswertung herangezogen werden konnten.
Es wurden bereits zahlreiche Ansätze zur Infarktgrößenabschätzung mittels Mapping vorgestellt. Als Parameter wurden u.a. verwendet:

- Q/ST- bzw. QRS- bzw. QRST-isointegral [Hul92, Kub85, Mon84, Sta86, Iga87]*
- R-isochrone [Ike87]
- QRS-isopotential [Kor87]
- QRS/ST-T isoarea [Hor85]
- Q-Zacke - Anzahl pathologischer Q-Zacken [Awa80]*
- ΣST, ΣR, ΣQ, ΣR/(Q-S), ΣR/Q+S) [Yus79,Mar80]*
- Scoring-Systeme [Mar80, Bab84, Cow87]*

Die vorgestellten Bewertungsverfahren sind an strenge Randbedingungen hinsichtlich des Meßzeitpunktes gebunden (i.a. 72h nach AMI). Sie verwenden einzelne Zeitpunkte oder Abschnitte zur Generierung der Merkmale, aus denen die Infarktgröße ermittelt wird. Die mit * gekennzeichneten Verfahren wurden für das HTEKG-System implementiert und auf das Lysekollektiv angewendet. Für keines der Verfahren konnte ein signifikanter Zusammenhang mit der enzymatisch durch den Anstieg der Creatin-Kinase dokumentierten Infarktgröße festgestellt werden.
Qualitative Untersuchungen mit EKG's aus der akuten Phase des Infarkts wurden nur von [Kor87] vorgestellt. Eine Bewertung der Signalform im Sinne einer Ähnlichkeit zu einer von 5 EKG-Musterformen wird nur von [Mar80] vorgenommen.

3. Physiologische Signalinterpretation

Das Versagen der statistischen Ansätze bei Daten aus der akuten Phase ist auf die rapiden Veränderungen des Signals in dieser Phase zurückzuführen. Die für diese Situation unzureichende Modellbildung muß entsprechend erweitert werden.
Entsprechend den unterschiedlichen pathologischen Veränderungen (Ischämie, Läsion, Nekrose) erfährt das EKG und insbesondere der QRS-Komplex, der die Erregungsausbreitung in den Ventrikeln repräsentiert, charakteristische Veränderungen [Mul78]:

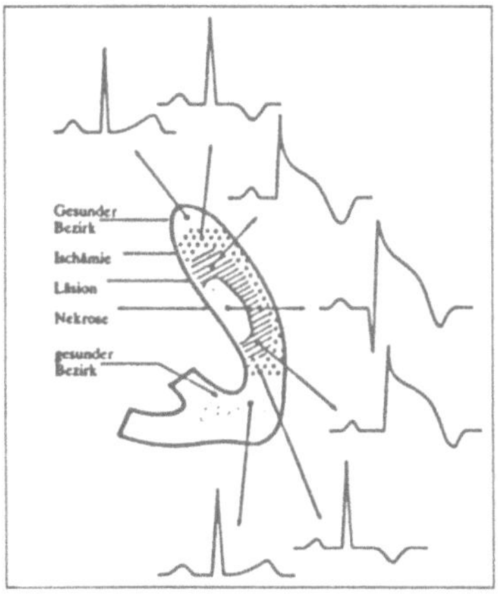

Ischämie	→	Veränderungen der T-Welle
Läsion	→	Veränderungen der ST-Strecke (Hebung/Senkung)
Nekrose	→	Einbruch von R, Ausbildung pathologischer Q-Zacken.

Abhängig von der Lage des betroffenen Areals ergeben sich Mischsignale. Bild 1 verdeutlicht diesen Effekt. Im folgenden sollen deshalb Ansätze zur ganzheitlichen Signalinterpretation vorgestellt werden.

Bild 1: Kombination der Grundaspekte von Ischämie/Läsion/Nekrose

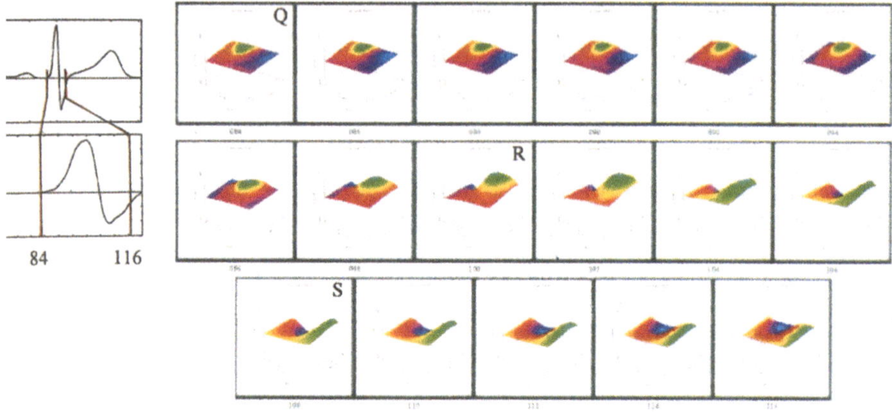

Bild 2: Potentialverlauf eines gesunden Probanden. Bild 1: Q, Bild 9: R, Bild 13: S.

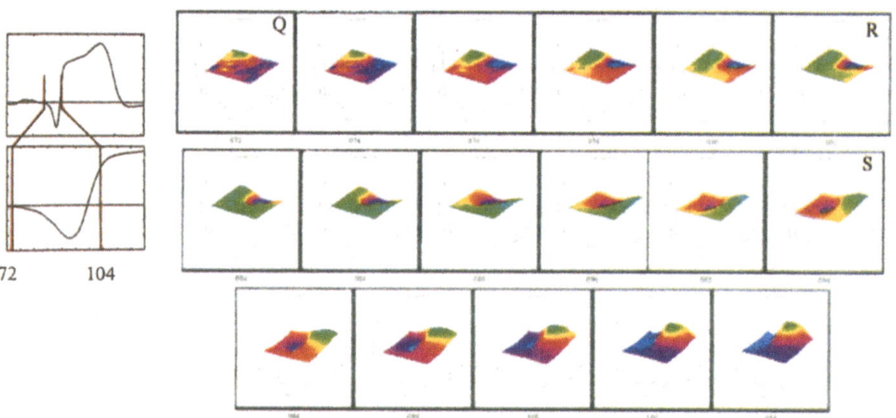

Bild 3: Potentialverlauf eines Patienten mit Vorderwandinfarkt. Bild 1: Q, Bild 6: R, Bild 12: S.

4. Untersuchung der Erregungsausbreitung

Betrachtet man nur die Phase der Erregungsausbreitung in den Ventrikeln, also den QRS-Komplex, so können qualitative Unterschiede zwischen gesunden Probanden und Infarkpatienten sichtbar gemacht werden. Die Bilder 2 und 3 verdeutlichen dies. Dargestellt ist die Potentialverteilung auf dem Thorax vom Beginn der Q-Zacke bis zum Zeitpunkt ST60 (60 ms nach der S-Zacke). Deutliche Unterschiede in der Erregungsausbreitung sind insbesondere im zentralen Bereich sichtbar. Gegen Ende des QRS-Komplexes bleibt beim Infarktpatienten eine charakteristische Potentialverteilung zurück, während beim gesunden Probanden das Potential wieder auf die Isopotentialfläche abfällt. Grund für diese Veränderungen gegenüber dem regelrechten Erscheinungsbild ist die Tatsache, daß ein Teil des Myocards durch den Infarkt "elektrisch stumm" geworden ist bzw. nur noch einen sehr kleinen Beitrag zur elektrischen Gesamtaktivität leistet. Die sich ausbreitende Erregung muß dieses Gebiet umgehen - es kommt zu Verzögerungen in der Erregungsfortleitung und Einbrüchen in den abgreifbaren Oberflächenpotentialen. So ist beispielsweise aus der Beurteilung der Potentialverteilung zum Zeitpunkt ST60 eine eindeutige Identifizierung des Infarktyps (VWI/HWI) und eine relativ genaue Bestimmung des verschlossenen Gefäßes möglich [Vog93]. Aus den Bildern 2 und 3 wird aber auch deutlich, daß die Bewertung nur eines Zeitpunktes (ST60) des EKG's nicht die gesamte zur Verfügung stehende Information nutzt.

5. Formelementanalyse

Ausgehend von den skizzierten Signalveränderungen wurde ein mehrstufiges Verfahren entwickelt, das die Form des Signals analysiert und einer von drei Klassen (geringe/mittlere/starke Schädigung) zuordnet. Zunächst werden die Grenzen des QRS-Komplexes ermittelt. Sodann wird in Anlehnung an das Verfahren von Tonooka [Ton83] für jede Elektrode das Integral, einmal unter Berücksichtigung des Vorzeichens (R-Map) und einmal unter Betragsbildung (A-

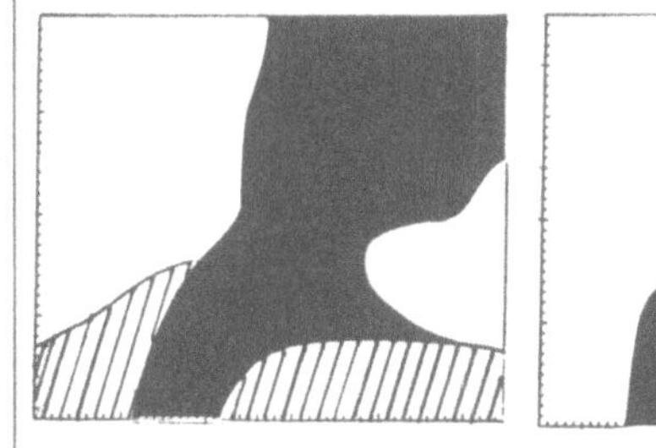
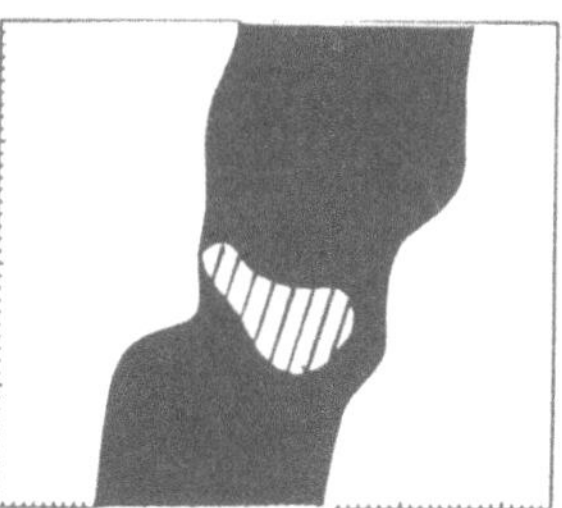
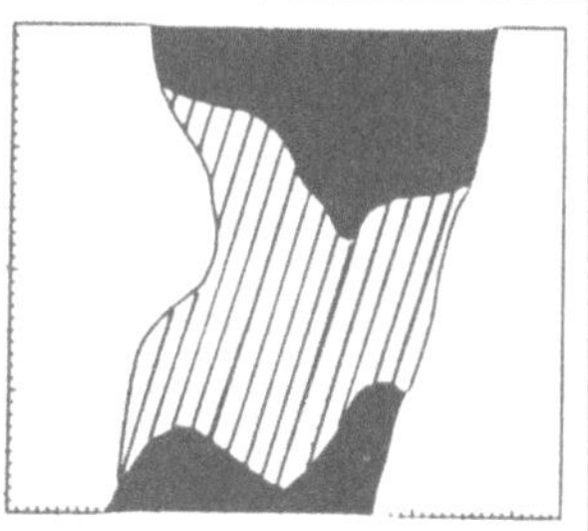

Bild 4: Veränderungen des HTEKG's unter Lyse. Dargestellt sind die Bewertungen durch die Formelementanalyse bei Therapiebeginn (links), nach 1.5 h (mitte) und nach 3 h (rechts). (weiß: geringe Schädigung, schwarz: mittlere Schädigung, schraffiert: starke Schädigung)

Map) berechnet. Mit dem A-Map erfolgt eine Bewertung hinsichtlich Potentialsenken, während das R-Map die Symmetrie des Signals bewertet. Die Werte der Maps werden bezüglich eines Normalkollektivs in Vielfachen der Standardabweichung dargestellt, wobei Werte innerhalb der einfachen Standardabweichung als nicht-pathologisch eingestuft werden. Aus dieser Vorstufe ergibt sich eine Entscheidungsmatrix mit neun Feldern, die bereits eine grobe Zuordnung der Elektroden zu den Klassen zuläßt [Pel90]. Danach erfolgt eine Klassifizierung der QRS-Komplexe aller Elektroden hinsichtlich der vorhandenen Zacken. Die Lage und Amplitude der Zacken wird dann zusammen mit den Werten aus dem R- und A-Map einem Regelwerk zugeführt, das eine abschließende Klassifizierung durchführt. Die Anwendung dieses

Verfahrens auf die HTEKG's eines Infarktpatienten bei Einlieferung sowie 1.5 Stunden und 3 Stunden danach zeigt eine deutliche Progredienz (Bild 4). Nachteil dieses Verfahrens, das dem von Selvester in [Sel85] vorgestellten Scoring-System prinzipiell ähnlich ist, ist das notwendige komplexe Regelwerk, das einem Expertensystem vergleichbar ist. Es zeigt sich aber, das die Bewertung der Ähnlichkeit von Signalen eher zum Erfolg führt als die Untersuchung einzelner Werte des EKG's mit statistischen Methoden.

6. Signalklassifikation mittels selbstorganisierender Karten

Eine mögliche, wenn auch stark reduzierte Form der Darstellung ergibt sich durch die Repräsentation des QRS-Komplexes mittels der Amplituden zu den Zeitpunkten Q, R und S, die je

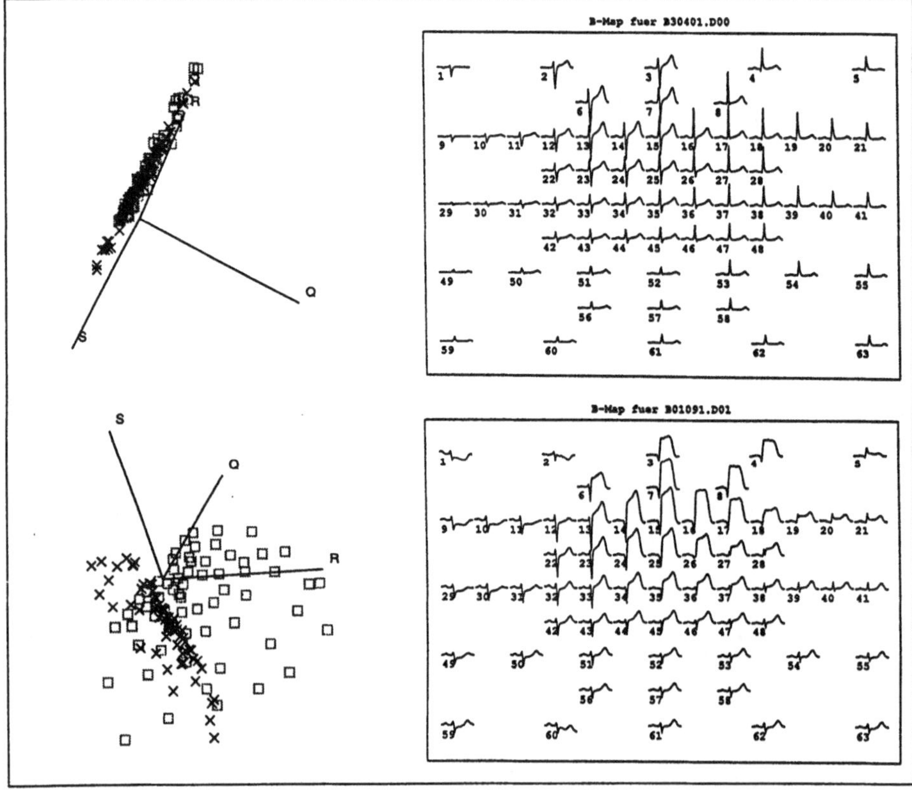

Bild 5: Darstellung der aus dem HTEKG abgeleiteten Merkmalsvektoren eines gesunden Probanden (□) und eines Vorderwandinfarkt-Patienten (x) sowie der zugehörigen HTEKG's (oben: gesunder Proband, unten: VWI)

Elektrode zu einem Vektor zusammengefaßt werden. Die skalierten Werte im dreidimensionalen Merkmalsraum zeigt Bild 5. Dargestellt sind die Elektroden eines gesunden Probanden (□) gegenüber einem Patienten mit Vorderwandinfarkt (x) sowie die zugehörigen HTEKG's. Für alle Probanden des Normalkollektives spannen die Merkmalsvektoren eine Ebene unterhalb des QRS-Koordinatensystems auf (Bild 5 oben links). Die Merkmalsvektoren der Infarktpatienten fallen ebenfalls weitgehend in diese Ebene, zeigen aber eine deutliche Clusterung bzgl. der gesunden Probanden (vergl. Bild 5 unten links). Ausgehend von diesen Ergebnissen und unter

Berücksichtigung der in 3 bis 5 geschilderten Aspekte wurde eine selbstorganisierende Karte nach Kohonen zur weiteren Signalanalyse eingesetzt [Koh84, Rit91].
Dabei wird insbesondere die Eigenschaft der topologischen Merkmalskarte, in einem hochdimensionalen Problemraum selbstständig nichtlineare Zusammenhänge aufzufinden und unter weitgehendem Erhalt der Nachbarschaftsbeziehungen eine Dimensionsreduktion durch Abbildung auf einen niederdimensionalen Raum vorzunehmen, ausgenutzt, um neue Aspekte für die Interpretation der EKG-Daten aus der Akutphase des Infarkts offenzulegen. Da die Beschreibung des QRS-Komplexes mit 3 Merkmalen nur sehr grob ist, wurde die Anzahl der Merkmale respektive Meßpunkte unter Einbeziehung des ST60-Wertes und eines Teiles der T-

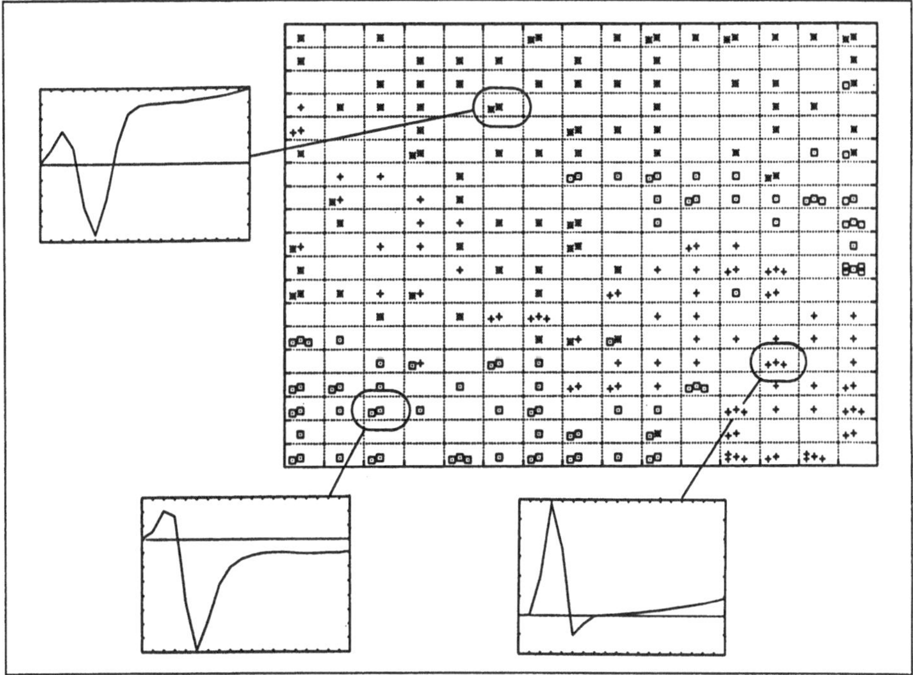

Bild 6: 15x19 Vektoren große Kohonen-Karte trainiert mit Daten von 9 Patienten/Probanden sowie exemplarische Darstellung der Kartenvektoren (□: Hinterwandinfarkte, +: gesunde Probanden, *: Vorderwandinfarkte).

Welle auf 20 erhöht. Die sich daraus je Elektrode ergebenden 20 dimensionalen Merkmalsvektoren dienen als Eingabekomponenten für die selbstorganisierende Karte. Bild 6 zeigt eine 15x19 Vektoren große Karte mit Visualisierung der 252 für das Training herangezogenen Vektoren. Das Trainingskollektiv bestand in diesem Fall aus 3 gesunden Probanden (+), 3 Patienten mit Vorderwandinfarkt (*) und 3 Patienten mit Hinterwandinfarkt (□). In das Training gingen nur Vektoren, die aus den Ableitungen der Elektroden des Zentrums gewonnen wurden, ein, da sich diese Reduktion der Eingabedaten in Vorstudien als ausreichend erwiesen hatte.
Es zeigt sich eine deutliche Clusterung unter pathologischen Aspekten, was die Annahmen aus Abschnitt 5 bestätigt. Dies wird durch die Visualisierung der Kartenvektoren, wie in Bild 6 exemplarisch gezeigt, deutlich. Die Kartenvektoren repräsentieren Prototypen unterschiedlicher QRS-Komplexe bzw. EKG-Abschnitte. Die Karte formt damit eine kontinuierliche Repräsentation unterschiedlicher pathologischer Erscheinungsbilder des EKG´s .

Zur Verifikation wurden der Karte Merkmalsvektoren aus einem Klassifikationskollektiv präsentiert und ihre Anordung auf der Karte visualisiert. Das Klassifikationskollektiv bestand aus Merkmalsvektoren von gesunden Probanden und von Infarktpatienten, die nicht im Trainingskollektiv enthalten sind. Bild 7 zeigt exemplarisch die Klassifikation eines Vorderwandinfarktpatienten. Die Vektoren wurden in die Zone abgebildet, in die die Karte beim Training die Vektoren der Vorderwandinfarkte eingeordnet hatte. Hierbei ist aber zu beachten, daß innerhalb eines als "Infarkt" eingestuften Datensatzes sowohl Komplexe mit starker als auch mit schwacher pathologischer Veränderung vorliegen. Im Sinne der Repräsentation von Signal-Prototypen sind die in der Karte scheinbar auftretenden Verwischungen zwischen den Klassen Ausdruck kontinuierlicher Veränderungen der abgeleiteten Potentiale in Abhängigkeit von der Lage der Läsion und der Lage der Ableitung. Somit ist bei einer Interpretation der Prototypen respektive Kartenvektoren unter medizinischen Gesichtspunkten, beispielsweise nach Schwere der Schädigung, die Verwendung des Verfahrens zu einer Abschätzung des Infarkttyps und des Umfangs der Schädigung denkbar. Von entscheidender Bedeutung ist dabei eine geeignete Parametrierung hinsichtlich der jeweiligen Fragestellung. Bild 8 verdeutlicht diesen Sachverhalt. Für das Training dieser Karte wurden die gleichen Datensätze wie in Bild 6 verwendet. Es erfolgte jedoch eine Beschränkung auf den QRS-Komplex bei gleicher Dimension der Merkmalsvektoren. Die damit verbundene wesentlich genauere Abtastung der Eingabesignale führt beispielsweise zu einer verstärkten Bewertung intraindividueller Unterschiede bei den gesunden Probanden. Die Karte kann dadurch die gewünschte Generalisierungsleistung nicht erbringen.

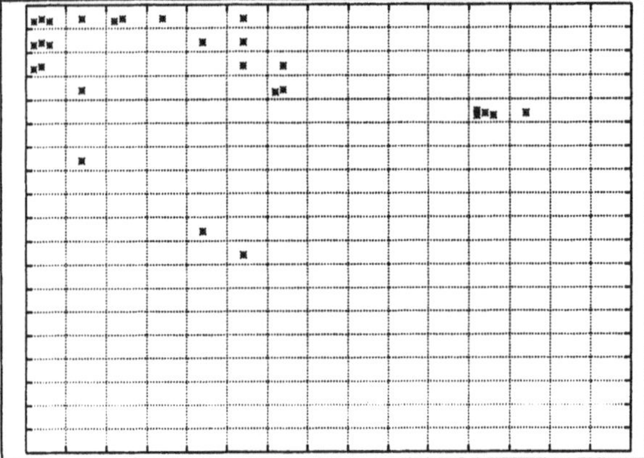

Bild 7: Klassifikation der Daten eines Patienten mit Vorderwandinfarkt auf Basis der topologischen Karte aus Bild 6

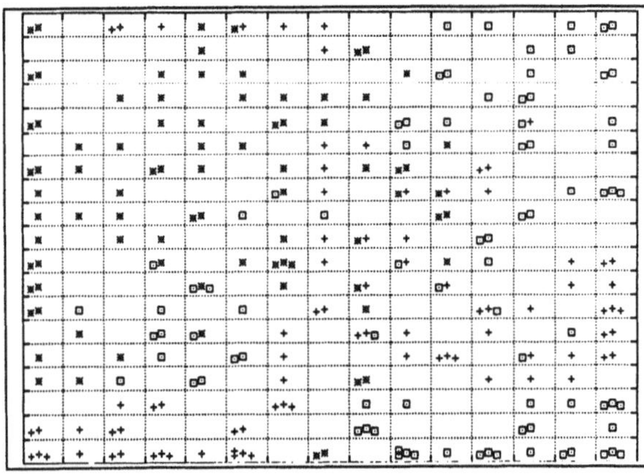

Bild 8: 15x19 Vektoren große Kohonen-Karte trainiert mit Daten von 9 Patienten/Probanden. (genauere Abtastung des QRS-Komplexes) □: Hinterwandinfarkte, +: gesunde Probanden, *: Vorderwandinfarkte

Die Anwendung des Verfahrens auf die Erkennung von Abstoßungsreaktionen zeigen die Bilder 9 und 10. Die zur Klassifikation genutzte Karte wurde mit histologisch gesicherten Referenzdaten (Abstoßung/keine Abstoßung) eines Patienten trainiert. Eine charakteristische Anordnung der Vektoren ist auch in diesem Fall zu beobachten. Die Projektion der Daten einer mit dem Referenzverfahren nicht erkannten Abstoßung zeigt Bild 10. Sie werden von der Karte in Bereiche eingeordnet, die während der Trainingsphase von Vektoren aus Abstoßungsreaktionen belegt wurden.

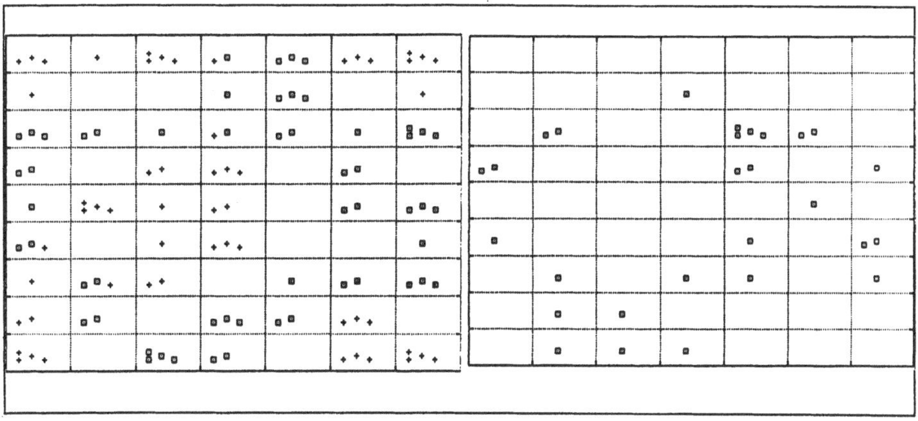

Bild 9 und 10: Links: 7x9 Vektoren große Kohonenkarte, trainiert mit Daten eines Transplantationspatienten. +: keine Abstoßung, □: Abstoßungsreaktion. Rechts: Klassifikation der Daten einer histologisch gesicherten Abstoßungsreaktion (gleicher Patient).

7. Zusammenfassung und Ausblick

Das vorgestellte Verfahren verwendet abweichend von den klassischen in der Literatur dokumentierten Ansätzen eine ganzheitliche Betrachtung des Signals als Grundlage einer Klassifikation. Diese basiert auf einem euklidischen Ähnlichkeitsmaß und führt im Zuge des Trainings der topologischen Merkmalskarte zur Ausbildung einer quasikontinuierlichen Menge von Signalprototypen, die unter medizinischen Gesichtpunkten interpretierbar bleiben. Die dem Verfahren immanente Eigenschaft der Dimensionsreduktion bei weitgehender Erhaltung von Nachbarschaftsbeziehungen führt auch in diesem Anwendungsgebiet zu einer Anordnung, die sowohl den topologischen Eigenschaften der Eingabedaten als auch den unterliegenden pathologischen Gegebenheiten Rechnung trägt.

Die bislang erzielten Ergebnisse deuten auf Vorteile des Verfahrens gegenüber einer singulären Bewertung einzelner Signalabschnitte hin, bedingt durch die Ausnutzung von *mehr* Signalinformation. Insbesondere dem Aspekt einer unscharfen Klassenzugehörigkeit der Mustersignale trägt der Ansatz Rechnung. Die sich daraus ergebenden Möglichkeiten der Datenanalyse und der Interpretation von Fremddaten mittels der topologischen Merkmalskarte sind Gegenstand weiterer Untersuchungen.

Literatur

[Awa80] Awan, N.A., Vera, Z., Mason, D.T.: Determination of MI Size and LV Performance by Precordial Q-Wave MappingQuantification of Myocardial Ischemia. In: Advances in Clinical Cardiology Vol.1 , 32, (H.W.Heiss eds.), 1980

[Bab84] Babbitt, D. Marcus, F., Wagner, G.S., Serokman, R.: Comparison of a QRS-scoring system for estimating acute infarct size with radionuclide left ventriculography. Am Heart J 108, 1426, 1984

[Cow87] Cowan, M. et al.: Estimation of Myocardial Infarct Size by Electrocardiographic and Radionuclide Techniques. J Electrocardiol - Supp. Issue, 78, 1987

[Fei87] Feiring, A.J., Johnson, M.R., Kioschos, J.M., Kirchner, P.T., Marcus, M.L., White, C.W.: The Importance of the determination of the myocardial area at risk in the evaluation of the outcome of acute MI in patients. Circulation 75, 980, 1987

[Hin85] Hinsen, R.: Ein Beitrag zur Methodik und Anwendung der Hochauflösenden Thorax-Elektrokardiographie. Dissertation, Fakultät für Elektrotechnik, RWTH Aachen, 1985

[Hor85] Horan, L.G., Sridharan, M.R., Hand, R.C., Flowers, N.C.: Variation of the Precordial QRS Transition Zone in Normal Subjects. J Electrocardiol 21, 25, 1985

[Hul92] Hulin, I., Slavkovsky, P.: Special computer graphical method for detection of myocardial infarction. Comp Methods and Programs in Biomedicine 39, 85, 1992

[Ide82] Ideker, E.R., et al.: Evaluation of a QRS Scoring System for Estimating Myocardial Infarct Size II. Am J Cardiol 49, 1604, 1982

[Iga87] Igarashi, A., Kubota, I., Ikeda, K., Tsuiki, K., Yasui, S.: Determination of the Site of Myocardial Infarction by QRST Isointegral Mapping in Patients with Abnormal Ventricular Activation Sequence. Jap Heart J 28, 165, 1987

[Ike87] Ikeda et al.: Temporal Changes in Body Surface Peak R Isochrone Maps and Left Ventricular Function in Patients with Myocardial Infarction. J Electrocardiol 20,121, 1987

[Koh84] Kohonen, T.: Self-Organisation and Associative Memory. Springer Series in Information Science 8, Heidelber, 1984

[Kor87] Kornreich, F. et al.: Qualitative and Quantitative Analysis of Characteristic BSPM Features in Anterior and Inferior MI. Am J Cardiol 60, 1230, 1987

[Kub85] Kubota, I., et al.: Noninvasive assessment of left ventricular wall motion abnormalities by QRS isointegral maps in previous anterior infarction. Am Heart J 109, 464, 1985

[Mar80] Maraoko, P.R., DeBoer, L.W.V.: Usefulness of Precordial Electrocardiographic Mapping in Assessing the Effect of Interventions in Myocardial Ischemic Damage. In: Advances in Clinical Cardiology 1, 39, (H.W.Heiss eds.), 1980

[Mat85] Mathey, D.G., Schofer, J., Bleifeld, W.: Zeitabhängigkeit der Myokarderhaltung nach Thrombolyse, Dtsch. med. Wschr. 110, 1681, 1985

[Mon84] Montague, T.J., et al.: Temporal Evolution of Body Surface Map Patterns Following Acute Inferior MI. J Electroardiol 17, 319, 1984

[Mul78] Muller,J.E., Maroko, P.R., Braunwald, E.: Precordial Electrocardiographic Mapping: A Technique to Asses the Efficacy of Interventions Designed to Limit Infarct Size. Circulation 57, 1, 1978

[Pel90] Pelikan, E.: Lineare und nicht-lineare Merkmalsextraktionsverfahren zur Infarktgrößenbestimmung aus HT-EKG-Mustern, Diplomarbeit, Fak. f. Elektrotechnik, RWTH Aachen, 1990

[Rit91] Ritter, H., Martinetz, T., Schulten, K. : Neuronale Netze. Addison Wesley, Bonn, 1991

[Roa83] Roark, S., et al.: Evaluation of a QRS Scoring System for Estimating Myocardial Infarct Size III. Am J Cardiology 51, 382, 1983

[Sel85] Selvester, R.H., Wagner, G.S., Hindman, N.C.: The Selvester QRS Scoring System for Estimating the Myocardial Infarct Size - Development and Application of the System. Arch Intern Med 145, 1877, 1985

[Sta86] Startt, R.H., Selvester, R.H., Solomon, J.C., Pearson, R.B.: Body Surface Map Criteria for Quantitating Infarct as Direved from Computer Simulations. In: Electrocardiographic Body Surface Mapping, 49, (van Dam, van Oosterom eds.), Martinius Nijhoff Publishers, 1986

[Ton83] Tonooka, I., et al: Isointegral Analysis of Body Surface Maps for the Assessment of Location and Size of Myocardial Infarction. Am J Cardiol 52, 1174, 1983

[Vog93] Vogt, L.: Ein integriertes Mustererkennungssystem zur Interpretation Hochauflösender Thorax-Elektrokardiogramme. Dissertation, Fak. f. Elektrotechnik, RWTH Aachen, 1993

[Wag82] Wagner, G.S., et al: Evaluation of a QRS Scoring System for Estimating Myocardial Infarct Size I. Circulation 65, 342, 1982

[War84] Ward, R.M., et al.: Evaluation of a QRS Scoring System for Estimating Myocardial Infarct Size IV.Am J Cardiology 53, 706, 1984

[Yus79] Yusuf, S., Lopez, R., Maddison, A., Maw, P., Ray, N., McMillan, S., Sleight P.: Value of electrocardiogramm in predicting and estimating infarctsize in man. British Heart J 42, 286, 1979

Neuronales Netzwerk und Korrelationsanalyse als Methoden der automatischen Wehenerkennung in der Geburtshilfe im Vergleich

Johannes Steffens, Falk Fallenstein und Ludwig Spätling
Universitäts-Frauenklinik Bochum - Marienhospital Herne

1 Einleitung

Die Behandlung akuter Frühgeburtsbestrebungen durch Wehenhemmung mit betamimetischen Substanzen (Tokolyse) gehört seit vielen Jahren zum geburtshilflichen Standard. Obwohl diese Therapie mit erheblichen, unerwünschten Nebenwirkungen verbunden ist, gibt es bislang keine annähernd gleichwirksame Alternativmethode. Die der hier vorgestellten Arbeit übergeordnete wissenschaftliche Zielsetzung besteht daher darin, ein klinisches Procedere zu finden, das bei drastisch verminderter Medikamentenzufuhr die Nebenwirkungsrate zu senken vermag und trotzdem eine ausreichende Wehenhemmung sicherstellt.

Wegweisend hierfür war die Entwicklung der "Bolustokolyse", bei der kleine, hochkonzentrierte Wirkstoffmengen in bestimmten Zeitabständen intravenös appliziert werden. In einer klinischen Untersuchung wurde nachgewiesen, daß ein mit der üblichen Dauerinfusion vergleichbarer therapeutischer Effekt erreicht wird bei gleichzeitiger Senkung der im Mittel pro Tokolysebehandlung benötigten Betamimetikamenge auf weniger als ein Fünftel (1).

In einem weitergehenden Ansatz wird versucht, die Dosierung der Bolustokolyse enger an die aktuelle diagnostische Situation zu koppeln, als dies in der normalen klinischen Routineüberwachung möglich ist. Hierfür wird die Wehenaktivität mit geeigneten Sensoren kontinuierlich erfaßt und die daraus ermittelte Wehenhäufigkeit als Kriterium zur Einstellung der Tokolysedosierung herangezogen (*).

Pilotmessungen haben gezeigt, daß eine solche, weitgehend automatisch von der Wehentätigkeit geregelte Bolustokolyse prinzipiell klinisch realisierbar ist (2). Ziel ist jedoch ein für den Routineeinsatz brauchbares Verfahren. Hierfür sind neben der Verbesserung der Wehensensortechnik vor allem zuverlässige, d.h. klinisch plausibel arbeitende Computeralgorithmen zur Erkennung von Wehenmustern aus den aufgenommenen Signalen zu entwickeln.

2 Material und Methoden

2.1 Zwei Lösungsansätze

Zur Lösung dieser Aufgabe bieten sich zwei Verfahren an:

a) Die Korrelationsanalyse:

Ein idealisiertes Wehensignal wird im Arbeitsspeicher bereit gestellt. Das Programm vergleicht es laufend mit Meßsignalausschnitten, indem es den Korrelationskoeffizienten als Maß für die Ähnlichkeit beider Signale berechnet.

b) Analyse durch ein neuronales Netzwerk:

Das Netzwerk ist ein Unterprogramm, das in der Lage ist, die Signalformen von Wehen zu lernen und nach Abschluß des Lernprozesses auf einen beliebigen Signalausschnitt mit der Reaktion "Wehe" bzw. "nicht Wehe" zu antworten.

(*) Mit Unterstützung der Deutschen Forschungsgemeinschaft

2.2 Meßdaten für die Programmtestung

Zur Testung der Verfahren liefen die Programme mit off-line gespeicherten Meßdaten. Diese Daten stammen aus sechsstündigen Aufzeichnungen von 50 Patientinnen, die wegen vorzeitiger Wehen stationär behandelt wurden. Die Meßdaten wurden zuvor von geburtshilflich geschultem Personal durchsucht und die darin enthaltenen Wehenkomplexe markiert.

2.3 Beurteilung der Leistungsfähigkeit

Die Evaluierung der Verfahren erfolgte durch Vergleich mit der visuellen Auswertung. Zur Vereinfachung der nachfolgenden Beschreibung soll ein maschinell (automatisch) als Wehe erkannter Signalkomplex "A-Wehe" und ein vom Menschen (manuell) als Wehe erkannter Signalkomplex "M-Wehe" genannt werden. Jedem Komplex ist zur Lokalisation ein zeitlicher Abstand zum Meßbeginn zugeordnet. A-Wehe und M-Wehe werden als gleich gewertet, wenn die Differenz korrespondierender Zeitpunkte kleiner als 30 Sekunden ist.

Aus der Gesamtheit der M-Wehen und A-Wehen eines Meßdatensatzes lassen sich zwei Kenngrößen ableiten, die für die Beurteilung eine Rolle spielen:

a) Anteil der M-Wehen, die auch A-Wehen sind. Diese Größe nennen wir Vorwärtsindex.
b) Anteil der A-Wehen, die auch M-Wehen sind. Diese Größe nennen wir Rückwärtsindex.

Der Vorwärtsindex gibt die Empfindlichkeit und der Rückwärtsindex die Zuverlässigkeit des Analyseverfahrens - gemessen an der visuellen Beurteilung - wieder.

2.4 Vektorielle Darstellung von Signalabschnitten

Beide Verfahren arbeiten grundsätzlich mit der Analyse von zeitlichen Signalausschnitten, die etwa 3 min lang sind. Ein Signalausschnitt liegt im Computer in einem Speicherfenster vor. Durch dieses Fenster wird die betrachtete Meßreihe Wert für Wert hindurchgeschoben. Zwischen zwei Schiebeoperationen arbeitet ein Unterprogramm (Analysefunktion) mit dem Speicherfenster, indem es die Inhalte bestimmter, symmetrisch zur Fenstermitte angeordneter Speicherstellen (Abtastpunkte) des Fensters übernimmt und einen Zahlenwert (Analyseantwort) zurückliefert. Das Programm nimmt durch Vergleich der Analyseantwort mit einer fest definierten Klassenschwelle eine Klassifikation des augenblicklich im Fenster "betrachteten" Signalausschnittes in die Musterklassen "A-Wehe" bzw. "nicht-A-Wehe" vor. Durchläuft die Analyseantwort ein lokales Maximum, das oberhalb der Klassenschwelle liegt, so markiert dieses Maximum eine A-Wehe (Abb. 1).

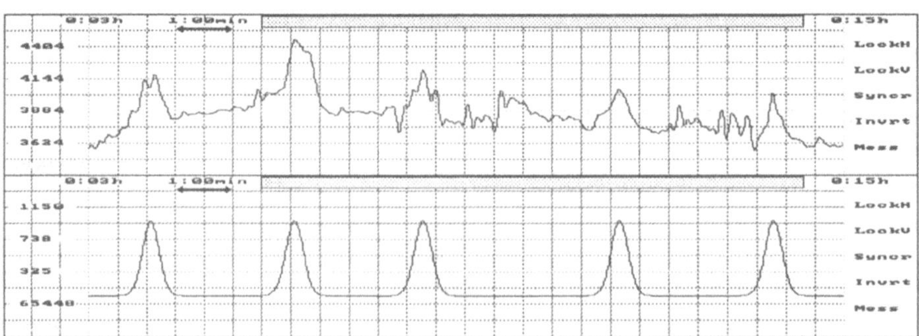

Abb. 1: Ausschnitt aus einer Wehenmessung (oben) zusammenen mit der zugehörigen idealen Analyseantwort (unten).

Im folgenden beschreiben wir die Signalausschnitte als Vektoren, deren Komponenten jeweils aus den Abtastwerten innerhalb des Speicherfensters gebildet werden. Die Vektoren seien Elemente eines euklidischen Vektorraumes E, den wir Eingaberaum nennen. n sei die Zahl der Dimensionen von E (=Zahl der Abtastpunkte). Die Elemente aus E nennen wir Eingabevektoren. Die Analysefunktion ist eine Abbildung f von E in den Raum der reellen Zahlen R. Sei T die Klassenschwelle, so entstammen alle Vektoren $\mathbf{a} \epsilon E$ mit $f(\mathbf{a}) > T$ nach Definition von f einem Signalausschnitt, das eine A-Wehe enthält. Die Gesamtheit aller Signalausschnitte, die eine A-Wehe enthalten ist demnach durch ein A-Wehen-Gebiet in E definiert. Die aus allen anderen Signalausschnitten abgeleiteten Eingabevektoren bilden das Komplement des A-Wehen-Gebietes. Prinzipiell läßt sich ein M-Wehen-Gebiet in E finden, das der Gesamtheit aller Signalausschnitte, die eine M-Wehe enthalten, entspricht. Die Leistungsfähigkeit der Analysefunktion erhöht sich durch Verminderung der symmetrischen Differenz des M-Wehen-Gebietes und des A-Wehen-Gebietes. Die Wehen-Gebiete lassen sich im Fall n=3 recht anschaulich darstellen (wenn auch für die richtige Anwendung ein wesentlich höheres n zum Einsatz kommt).

2.5 Invarianz der Analyseantwort gegenüber Verschiebung der Basislinie

Der Verschiebung der Basislinie eines Signalausschnittes entspricht die Verschiebung des Mittelwertes des abgeleiteten Eingabevektors. Da die Zugehörigkeit des Ausschnittes zu einer Musterklasse nicht vom Mittelwert des korrespondierenden Eingabevektors abhängen soll, transformieren wir die Eingabevektoren auf den Mittelwert Null. Die Transformation ist eine Projektion aller Eingabevektoren auf eine Hyperebene in E (Abb. 2). Wir nennen diese Ebene Eingabeebene. Im folgenden betrachten wir nur noch Vektoren aus der Eingabeebene.

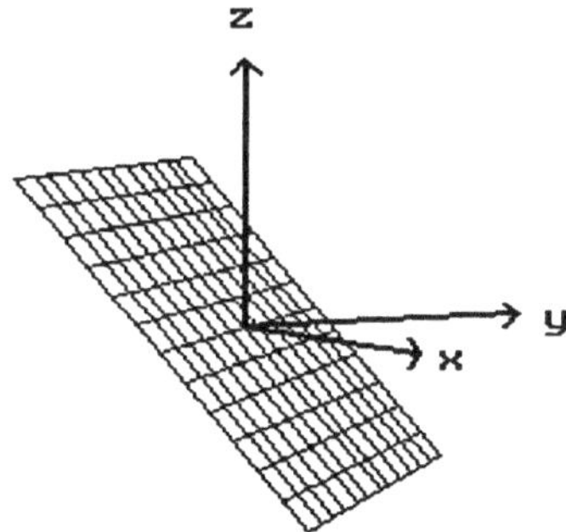

Abb. 2: Lage der Eingabeebene im Eingaberaum für n=3 (3 Abtastpunkte).

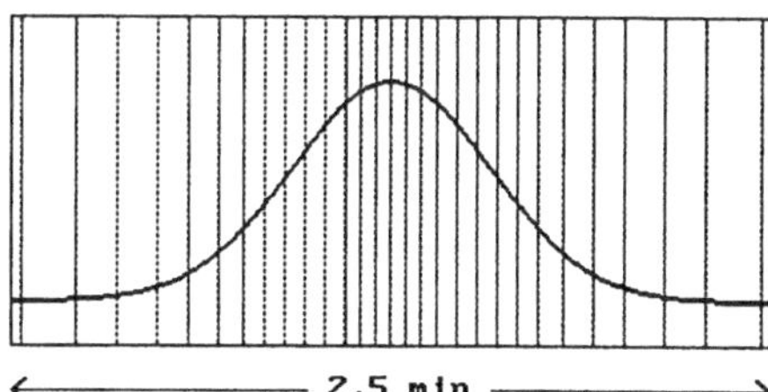

Abb. 3: Idealisiertes Wehensignal im Speicherfenster. Die senkrechten Linien geben die Lage der Abtastpunkte wieder, die wir in den getesteten Analysefunktonen verwendeten.

2.6 Geometrie der Korrelationsanalyse

Die Korrelationsanalyse beruht auf der Berechnung des Produkt-Moment-Korrelations-koeffizienten, der die Ähnlichkeit zweier Signale wiedergibt. In unserem Fall wird der Meßsignalausschnitt mit einem idealisierten Signalausschnitt, der eine Wehe darstellt (Abb. 3), verligchen. Dem idealisierten Signal läßt sich in gleicher Weise ein Eingabevektor $\mathbf{b}=(b_1,b_2,\ldots,b_n)$ entnehmen wie dem Meßsignal mit dem Vektor $\mathbf{a}=(a_1,a_2,\ldots,a_n)$. Die Analysefunktion $f:=f_{\mathbf{b}}(\mathbf{a})$ soll den Korrelationskoeffizienten zurückliefern. Sie lautet demnach:

$$f_{\mathbf{b}}(\mathbf{a}) = \frac{\Sigma a_i b_i - (\Sigma a_i \Sigma b_i)/n}{\sqrt{((\Sigma a_i^2 - (\Sigma a_i)^2/n)(\Sigma b_i^2 - (\Sigma b_i)^2/n))}}$$

ufgrund der Transformation gilt: $\Sigma a_i = \Sigma b_i = 0$. Daraus folgt:

$$f_{\mathbf{b}}(\mathbf{a}) = \frac{\Sigma a_i b_i}{\sqrt{(\Sigma a_i^2)}\sqrt{(\Sigma b_i^2)}} = \frac{\mathbf{a}\cdot\mathbf{b}}{|\mathbf{a}|\cdot|\mathbf{b}|} = \cos(\text{Winkel}(\mathbf{a},\mathbf{b}))$$

Der Korrelationskoeffizient ist also der Cosinus des Winkels, den die Vektoren **a** und **b** einschließen.

Offenbar ist das A-Wehen-Gebiet in E durch Vorgabe von **b** und der Klassenschwelle T eindeutig definiert (Abb. 4). Das A-Wehen-Gebiet ist hier ein verallgemeinerter Kegel mit dem Öffnungswinkel 2·arccos(T).

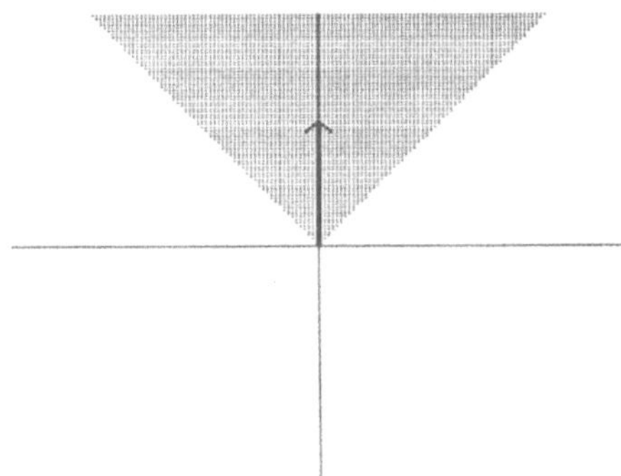

Abb. 4: Eingabeebene für den Fall n=3 (3 Abtastpunkte). Der Pfeil repräsentiert den, aus dem idealisierten Wehensignal abgeleiteten, transformierten Vektor **b**. Die hervorgehobene Fläche stellt das A-Wehen-Gebiet für die Korrelationsanalyse mit der Klassenschwelle T=0,8 dar.

Daß das Gebiet der M-Wehen anders geformt ist, ist sicherlich eine zulässige Annahme. Die Beschränkung der Freiheitsgrade des A-Wehen-Gebietes (Ausrichtung und Öffnungswinkel) verhindert demnach eine Übereinstimmung des A-Wehen-Gebietes mit dem M-Wehen-Gebiet. Uns stellte sich die Frage, ob es eine Analysefunktion gibt, deren Wehen-Gebiet besser angepaßt werden kann.

2.7 Struktur des Perzeptrons

Aus der Vielzahl der Netzwerkmodelle, die allesamt lernfähig sind, haben wir die Anwendbarkeit modifizierter Versionen des von F. Rosenblatt 1958 vorgeschlagenen Perzeptrons auf unser Problem untersucht (3).

Wie auch bei anderen neuronalen Netzwerkmodellen war hier die Nervenstruktur des Gehirns eine Grundlage, was sich in der verwendeten Terminologie widerspiegelt: Das Perzeptron setzt sich aus einer festen Anzahl Neuronen zusammen, die als mathematische Funktionen aufzufassen sind. Ein Neuron bildet Vektoren des Eingaberaumes E jeweils auf reelle Zahlen ab. Es hat demnach n Eingangsleitungen, die Synapsen heißen, und eine Ausgangsleitung. Das Neuron antwortet auf einen Vektor aus E mit einer reellen Zahl, die im allgemeinen als Aktivität des Neurons bezeichnet wird. Die Funktion $f_{s,t}$ eines Neurons ist folgendermaßen definiert: $f_{s,t}(a) = \Theta(s \cdot a - t)$; $a,s \epsilon E$; $t \epsilon R$. $\Theta(x)$ ist eine sigmoide Funktion und heißt Antwortfunktion des Neurons. Die exakte Definition der Antwortfunktion ist weniger wichtig als der qualitative Verlauf. Häufig wird die in der statistischen Physik gebräuchliche Fermifunktion $\Theta(x)=1/(1+\exp(-x))$ eingesetzt (Abb.5). Diese haben auch wir verwendet.

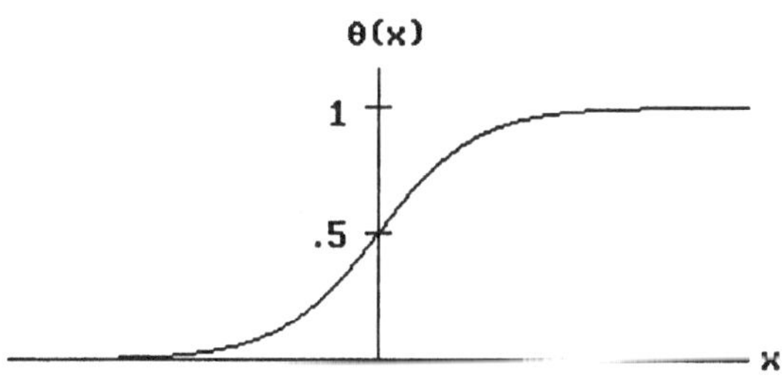

Abb. 5: Funktionsgraph der Fermifunktion $\Theta(x)=1/(1+\exp(-x))$.

Den Vektor s nennen wir Synapsenvektor, seine Komponenten heißen Synapsenstärken. s entstammt nicht einem Meßsignal (und ist deshalb auch nicht transformiert), hat aber die gleiche Struktur wie der transformierte Vektor **a**, der zu einem Signalausschnitt gehört. Die Größe t heißt Erregungsschwelle des Neurons. Überschreitet das Skalarprodukt $s \cdot \mathbf{a}$ die Erregungsschwelle, so bedeutet dies eine starke Erregung des Neurons: $f_{s,t}(a) > 0.5$. Man findet für jedes Neuron zwei Gebiete G_1, G_2 in der Eingabeebene, auf die es jeweils stark ($f_{s,t}(a) > 0,5$ für $\mathbf{a} \epsilon G_1$) oder schwach ($f_{s,t}(a) < 0,5$ für $\mathbf{a} \epsilon G_2$) aktiv ist. Die Grenze dieser beiden Gebiete ist durch $f_{s,t}(a)=0.5 \iff s \cdot \mathbf{a}=t$ definiert. Sie ist demnach eine n-2 dimensionale affine Hyperebene (Trennebene) in der Eingabeebene, deren Lage durch **s** (Normale) und $t/|s|$ (Abstand vom Ursprung) eindeutig festgelegt ist. Im Perzeptron sind die Neuronen derart zusammengeschaltet, daß alle mit demselben Eingabevektor **a** arbeiten (Abb. 6a).

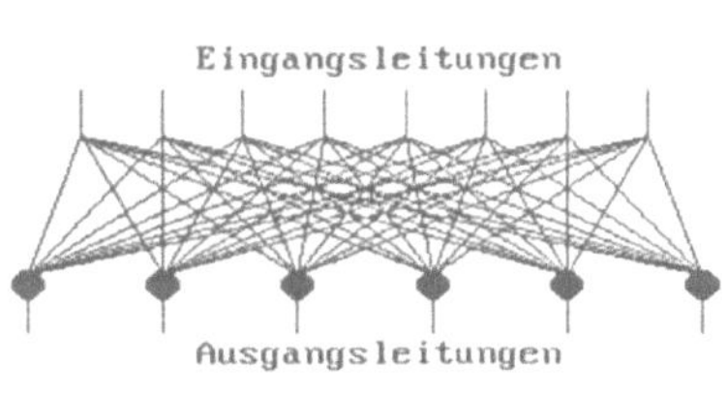

Abb. 6a: Perzeptron mit 8 Eingangsleitungen und 6 Neuronen.

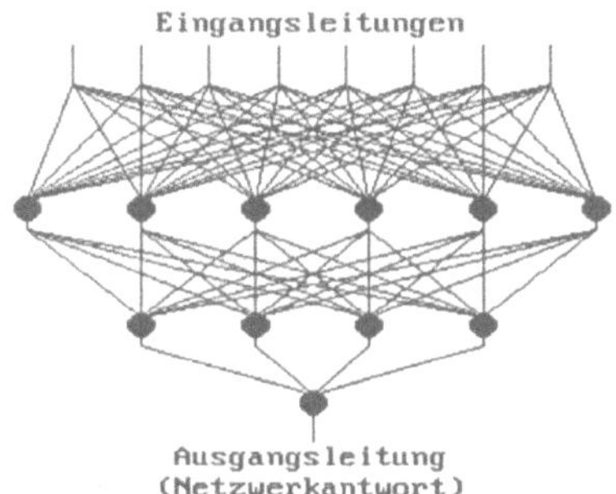

Abb. 6b: Dreilagiges Perzeptron.

2.8 Struktur des mehrlagigen Perzeptrons

Vernetzt man mehrere Neuronenreihen derart, daß die Eingangsleitungen eines Neurons in einer Reihe durch die Ausgänge der Neuronen aus der darüberliegenden Reihe gebildet werden (Abb. 6b), so erhält man ein mehrlagiges Perzeptron.

In unserem Fall besteht die letzte Neuronenreihe aus einem einzigen Neuron, dessen Ausgangsleitung die Antwort des Netzwerkes auf einen Eingabevektor darstellt. Ein solches Netzwerk mit festgelegten Synapsenstärken und Erregungsschwellen repräsentiert eine Analysefunktion.

Die Form des A-Wehen-Gebietes eines mehrlagigen Perzeptrons hängt vom Synapsenvektor und der Erregungsschwelle eines jeden Neurons ab. Sie hat gewöhnlich eine derart komplexe Struktur, daß eine elementare mathematische Beschreibung nicht möglich ist. Jedoch existiert eine Konstellation der Synapsenstärken und Erregungsschwellen, bei der das zugehörige A-Wehen-Gebiet näherungsweise durch die Trennebenen der Neuronen in der ersten Reihe eingerahmt wird (Abb. 7).

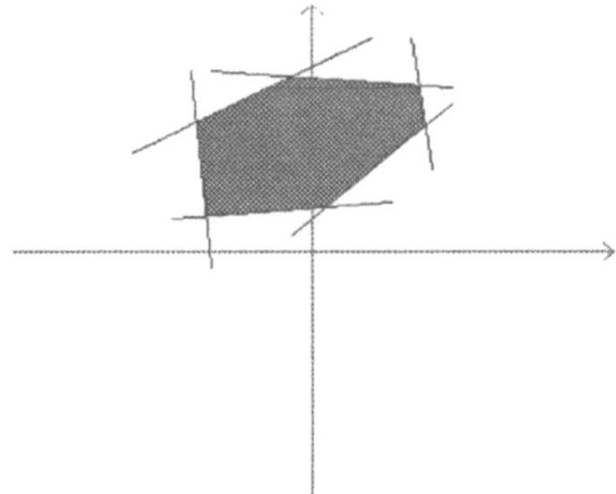

Abb. 7: Eingabeebene für den Fall n=3 (3 Abtastpunkte). Die durch angedeutete Geraden eingerahmte Fläche stellt näherungsweise das A-Wehen-Gebiet eines mehrlagigen Perzeptrons mit spezieller Verteilung der Synapsenstärken und Erregungsschwellen dar. Die Geraden repräsentieren hier die Trennebenen der Neuronen in der ersten Reihe.

2.9 Training des mehrlagigen Perzeptrons.

Der Lernvorgang des Netzwerkes besteht darin, die Synapsenstärken und Erregungsschwellen schrittweise der Konstellation zu nähern, bei der es durch seine Antwort die Eingabemuster ihren zuvor festgelegten Musterklassen mit einer minimierten Fehlerrate zuordnet.

Der Lernalgorithmus sucht nun durch Gradientenabstieg ein Minimum in dem Fehlerpotentialfeld des Raumes, der durch alle Synapsenstärken und Erregungschwellen aufgespannt wird. Er ist als Backpropagation-Algorithmus bekannt und wurde 1974 von Paul J. Werbos entwickelt (4)(5).

Die Vorbereitung der Trainingsdaten bestand darin, eine Sollantwort für visuell markierte Meßdaten zu generieren. Die Sollantwort gibt die gewünschte Netzwerkreaktion auf M-Wehen vor. Hierzu verwendeten wir Wehensignal-Aufzeichnungen, bei denen besonders viele M-Wehen (>7 Wehen/h) erkannt wurden. Das Netzwerk ist durch das Training bestrebt, die Sollantwort nachzuahmen, indem es sich an der Meßreihe orientiert (Abb. 8).

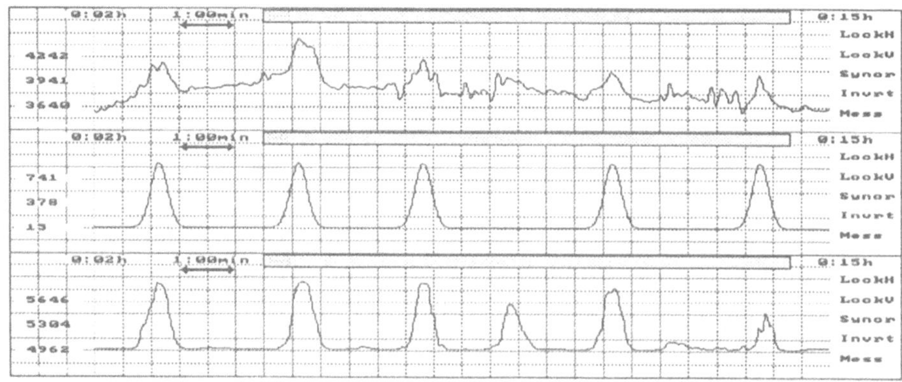

Abb. 8: Ausschnitt aus einer Wehenmessung (oben) zusammen mit vorgegebener Sollantwort (Mitte) und der Nachahmung der Sollantwort durch ein neuronales Netzwerk nach mehreren Stunden Training (unten).

3 Ergebnisse

Bei allen getesteten Analysefunktionen verwendeten wir 29 Abtastpunkte, die einen Bereich von 2,5 min überdeckten und deren Abstände in der Mitte des Speicherfensters kleiner waren als am Rand. Vor- und Rückwärtsindex wurden für verschiedene Klassenschwellen ermittelt. Besondere Beachtung gilt hier einer Einstellung, bei der Vor- und Rückwärtsindex gleich sind: Bei dieser Einstellung stimmt die Zahl der M-Wehen mit der Zahl der A-Wehen überein. Sie kommt der Forderung nach, daß die automatische Auswertung zumindest quantitativ mit der visuellen Auswertung übereinstimmen muß. Der Wert der Indizes bei dieser Einstellung ist ein Maß für die Qualität der verwendeten Funktion. Wir nennen ihn Q-Index. Die Meßdaten wurden in 3 Gruppen unterteilt:

Gruppe 1: hohe M-Wehenhäufigkeit (>7 Wehen/h) (Trainingsdaten)
Gruppe 2: mittlere M-Wehenhäufigkeit (2-7 Wehen/h)
Gruppe 3: niedrige M-Wehenhäufigkeit (<2 Wehen/h)

3.1 Korrelationsanalyse

Als Vergleichssignal verwendeten wir die Gaußfunktion mit einer Halbwertsbreite von 45 sek, wie sie im Mittel bei Wehen anzutreffen ist. Wir fanden folgende Q-Indizes:

Tab. 1: Verteilung von Q-Index und Klassenschwelle bei der Korrelationsanalyse.

	Gruppe 1	Gruppe 2	Gruppe 3	Gesamt
Q-Index	76%	57%	51%	68%
Kl.Schwelle	0,80	0,81	0,84	0,82

3.2 Netzwerkanalyse

Wir haben sechs Netzwerktypen getestet. Dabei hat sich eine Konstellation mit insgesamt neun Neuronen (acht in der 1. Reihe und eins in der letzten Reihe) als vielversprechend erwiesen. Wir fanden folgende Q-Indizes:

Tab. 2: Verteilung von Q-Index und Klassenschwelle bei der Netzwerkanalyse.

	Gruppe 1	Gruppe 2	Gruppe 3	Gesamt
Q-Index	79%	60%	30%	69%
Kl.Schwelle	0,64	0,64	0,76	0,66

4 Diskussion der Ergebnisse

Die Ergebnisse zeigen, daß grundsätzlich eine gute Selektivität sowohl durch das Netzwerk als auch durch die Korrelationsanalyse erreicht wird: etwa 70% der M-Wehen werden automatisch erkannt und umgekehrt. Da im klinischem Einsatz mit einer festen Klassenschwelle gearbeitet werden soll, muß die Varianz des Optimalwertes klein sein. Dies wird bei beiden Analyseverfaren erfüllt. Das schlechte Abschneiden bei der Datengruppe 3 (wenig Wehen) liegt z.T. daran, daß der Grundstörpegel der in jeder Messung vorhanden ist sich statistisch stärker auswirkt, wenn die Zahl der Wehen klein ist.

Folgende Argumente versprechen eine Verbesserung der Selektivität der Netzwerkanalyse:

a) Die verwendeten Trainingsdaten hatten z.T. widersprüchliche M-Wehen-Markierungen, die durch die unabhängige visuelle Auswertung von mehreren Personen korrigiert werden können. Hierdurch kann der Trainingseffekt verbessert werden.

b) Die verwendete Zahl der Neuronen, die mit der "Netzwerkintelligenz" korreliert, wurde bisher durch den erforderlichen Rechenaufwand für das Netzwerktraining begrenzt. Hier können die Beschleunigung des Lernalgorithmus und der Einsatz leistungsfähigerer Hardware in das Netzwerktraining Abhilfe schaffen.

Literatur:

(1) Spätling L, Fallenstein F, Schneider H, Dancis J (1989). Bolus Tocolysis: Treatment of preterm labor with pulsatile administration of a ß-adrenergic agonist. Am J Obstet Gynecol 160: 713-717.

(2) Fallenstein F, Spätling L, Behrens C, Abdallah A (1993). Wehengesteuerte Bolustokolyse. In: Bolustokolyse in Theorie und Praxis. Spätling L, Fallenstein F (Hrsg.), Steinkopff Verlag, Darmstadt, Seite 135-139.

(3) Ritter H, Martinez T, Schulten K, (1991). Neuronale Netze. Eine Einführung in die Neuroinformatik selbstorganisierender Netzwerke. Addison-Wesley, Bonn, Seite 27-32.

(4) Ritter H, Martinez T, Schulten K, (1991). Neuronale Netze. Eine Einführung in die Neuroinformatik selbstorganisierender Netzwerke. Addison-Wesley, Bonn, Seite 53-60.

(5) Hinton G (1992). Wie Neuronale Netze aus Erfahrung lernen. Spektrum der Wissenschaft, Heft 11; 134-139.

Anschrift des Erstautoren:

Johannes Steffens
Universitäts-Frauenklinik Bochum
Marihenhospital Herne
Hölkeskampring 40
44625 Herne 1

Ein System zur TV-Echtzeit-Detektion und -Kompensation translatorischer Objektbewegungen in Anwendung auf kapillarmikroskopische Videoaufnahmen des Nagelfalzes

G. Sträßle, U. Dreher

Institut für Physikalische Elektronik der Universität Stuttgart
in Zusammenarbeit mit dem
Mikrozirkulationslabor der Hautklinik der Universität Tübingen

1 Zusammenfassung:

Die Messung der Blutstromgeschwindigkeit in Kapillargefäßen des Nagelfalzes ist ein wichtiges Maß zur Diagnostik und Behandlungskontrolle peripherer Durchblutungsstörungen. Eine mögliche nicht-invasive Meßmethode basiert auf der bildanalytischen Verarbeitung kapillarmikroskopischer Auflicht-TV-Videoaufnahmen. Die Geschwindigkeit wird relativ zur Kapillare gemessen, damit muß die Kapillarenbewegung selbst kompensiert werdeen [CapiFlow]. Um nun ein vorhandenes von A.R. Pries vorgestelltes kapillarmikroskopisches Bildanalysesystem [Pries88] für die Meßaufgabe dennoch einsetzen zu können, wurde von uns ein Bewegungsfilter realisiert, das in TV-Echtzeit die Lageveränderung der Kapillargefäße erfaßt und kompensiert.

Das hier vorgestellte, dem Bewegungsfilter zugrunde liegende Verfahren dient zur "online"-Detektion sowie -Kompensation translatorischer Objektbewegungen in kapillarmikroskopischen TV-Videoaufnahmen des Nagelfalzes. Basierend auf einem modifizierten, geschwindigkeitsoptimierten Verfahren zur Targetkorrelation orthogonaler Projektionsschnitte werden in aufeinander-folgenden TV-Halbbildern translatorische Objektbewegungen mit bis zu ± 64 Bildelementen innerhalb des gesamten dargestellten Bildausschnittes in TV-Echtzeit erfaßt und optional kompensiert.

Eingeschleift in den TV-Signalweg zwischen Videoquelle und Analysesystem Pries wurde das Bewegungsfilter in Software auf einer "low cost" Hardware-plattform bestehend aus einem PC 486DX/50 mit Framegrabber Matrox IP-8 realisiert. Als Resultat einer klinischen Testphase hat sich bereits gezeigt, daß die Bewegungskorrektur mit Projektionsschnitten für die von der Anwendung vorgegebenen Szenen gute Korrekturergebnisse liefert.

2 Einleitung:

Zur Diagnostik und Behandlungskontrolle peripherer Durchblutungsstörungen und weiterer Erkrankungen, die zu derartigen Störungen führen (z.B. Morbus Renot, Sklerodermie, Diabetes mellitus), werden in neuerer Zeit u.a. kapillarmikroskopische Messungen durchgeführt.
In der speziellen Anwendung werden kapillarmikroskopische Videoaufzeichnungen des Nagelfalzes angefertigt, die nach der eigentlichen Messung "off-line" ausgewertet werden (u.a. Messung der Flußgeschwindigkeit der Erythrozyten, Gefäßdurchmesser). Die Kapillaren sind in diesen Sequenzen trotz Fixierung des Patienten nicht

ortskonstant, sondern zeigen zum Teil erhebliche Lageveränderungen, die u.a. durch Einflüsse des Pulses auftreten.

Für die eigentliche Analyse soll auf ein bereits existierendes PC-basiertes System [Pries88] zurückgegriffen werden, das manuelle und automatische Messungen erlaubt. Die automatischen Meßverfahren zur Ermittlung der Flußgeschwindigkeit liefern jedoch nur bei ortskonstanten Kapillaren verwertbare Meßergebnisse, während die manuelle Messung ("Flying spot"-Verfahren) geschultes Personal erfordert und systembedingt nur zur Ermittlung des Mittelwertes der Flußgeschwindigkeit geeignet ist.

Aus diesem Grund wurde in einer Machbarkeitsstudie [Burger91] untersucht, ob es möglich ist, eine 2-dimensionale Bewegungskorrektur in Form eines Bildvorverarbeitungsprozesses durchzuführen, die zum einen hinreichend genaue Ergebnisse des Bildversatzes liefert, zum anderen die Ausführung der Bewegungskorrektur übernimmt und eine nachgeschalteten Messung der Blutstromgeschwindigkeit in TV-Echtzeit (Halbbild-synchron mit 50 Hz) ermöglicht. Basierend auf den Resultaten dieser Arbeit konnte jetzt ein Bewegungsfilter realisiert werden, das den oben genannten Systemanforderungen entspricht.

Das dem Bewegungsfilter zugrunde liegende Verfahren wurde zur "online"-Detektion sowie -Kompensation translatorischer Objektbewegungen in kapillarmikroskopischen TV-Videoaufnahmen des Nagelfalzes ausgelegt. Basierend auf einem modifizierten, geschwindigkeitsoptimierten Verfahren zur Targetkorrelation orthogonaler Projektionsschnitte werden in aufeinanderfolgenden TV-Halbbildern translatorische Objektbewegungen mit bis zu ± 64 Bildelementen innerhalb des gesamten erfaßten Bildausschnittes in TV-Echt-zeit erfaßt und optional kompensiert.

3 Systemkonfiguration:

Als Hardwarebasis wird ein PC 486 DX/50 mit ISA-Bussystem eingesetzt. Die Bildverarbeitung erfolgt als Grauwertverarbeitung mit einer IP-8-Karte von Matrox. Diese Karte ist weitgehend frei konfigurierbar und erlaubt Datenerfassung und Bildwiedergabe mit der hier erforderlichen Halbbildfrequenz von 50 Hz. Wichtig ist, daß für die synchrone Bewegungsdetektion und -Filterung mindestens mit zwei getrennten Vollbildspeichern gearbeitet werden muß:

- Speicher 1: Bildaufnahme,
- Speicher 2: Bildwiedergabe, bewegungskorrigiert.

Da Datenerfassung und Bildausgabe über Genlock synchronisiert sind, erfolgt die Bewegungsdetektion mit einem Halbbild Verzögerung. Damit stehen jeweils am Ende eines Halbbilds die Korrekturdaten für das nächste Halbbild zur Verfügung. Durch dieses Verfahren der Synchronisation erfolgt die Bildausgabe der bearbeiteten Sequenz gegenüber dem Eingangssignal um ein Vollbild verzögert.

Diese Lösung zeichnet sich durch eine ausreichend hohe Rechenkapazität bei relativ niedrigen Hardwarekosten aus und ist flexibel, da der nachfolgend beschriebene Algorithmus zur Bewegungsdetektion in Software realisiert werden kann. Abbildung 1 zeigt ein Blockschaubild des Bewegungsfiltersystems mit dem nachgeschalteten Auswertesystem Pries.

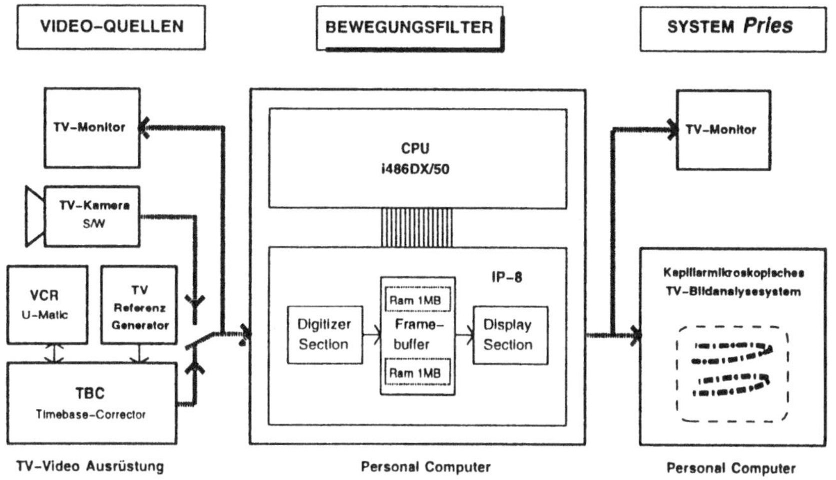

Abbildung 1: Blockschaubild des Bewegungsfilters mit nachgeschaltetem Auswertungssystem (System Pries).

4 Bewegungsdetektion

4.1 Beschreibung des Verfahrens

Die Bewegungsdetektion wurde als Targetkorrelationsprozeß innerhalb eines zeitinvarianten Suchbereiches realisiert. Das Target entspricht einer "area of interest" innerhalb eines Videohalbbildes und wird im Standbild zu Beginn einer Meßreihe interaktiv definiert (siehe Abb. 2). Die Translationsparameter werden durch die Lage des globalen Maximums der Kreuzkorrelationsfunktion bestimmt.

Die Forderung nach Detektion und Kompensation translatorischer Bewegungen der im Videosignal abgebildeten Kapillargefäße mit TV-Halbbild-frequenz (20msec) kann mit der beschriebenen Systemkonfiguration nur durch eine drastische Datenreduktion erreicht werden:

a) Repräsentation des zweidimensionalen Target- und Suchbereiches durch zwei eindimensionale orthogonale Projektionschnitte $P_y(x)$ und $P_x(y)$.

b) Unterabtastung des Targetbereichs bei der Berechnung der Projektionsschnitte mit anschliessender Rekonstruktion durch lineare Interpolation.

c) Die Bestimmung der Translationsparameter wird in einem 2-stufigen Prozeß, einer Grobdetektion mit nachgeschalteter Feindetektion, durchgeführt. Danach wird innerhalb des Suchbereichs, getrennt für $P_y(x)$ und $P_x(y)$, die Korrelationsfunktion zunächst nur an jeder n-ten Stützstelle berechnet und das Maximum detektiert. Aufgrund der Eigenschaften der Autokorrelationsfunktion der Projektionsschnitte bezüglich der Kapillargefäßszenen - sie ist stetig und lokal monoton - liegt das Maximum näherungsweise in einer Nachbarschaft von ±(n+2)/2 Bildelementen vom globalen Maximum entfernt. Die Feindetektion mit Berechnung der Kreuzkorrelationsfunktion bei voller Auflösung erfolgt somit nur noch in einem Suchbereich von ±(n+2)/2.

4.2 Berechnung der Projektionsschnitte

Die Länge der Suchbereiche innerhalb eines Videohalbbildes mit 512*256 Bildelementen wurde mit 256 Pixel in x-Richtung und 128 Pixel in y-Richtung (Halbbildverarbeitung!), die maximale Größe des Targets mit 128*64 Pixel festgelegt (siehe Abb.2). Damit ergibt sich, daß zur Ermittlung der Projektionsschnitte des Targets je Halbbild 8 kPixel ausgelesen werden müssen. Da dies die Möglichkeiten des Systems überfordern würde (Datentransferrate aus dem Bildspeicher <1,2 MByte/Sek.), wird das Target in x- und y-Richtung 2:1 unterabgetastet. Anschließend werden die fehlenden Zwischenwerte im Projektionsschnitt als arithmetisches Mittel der Nachbarn interpoliert. Somit reduziert sich der Berechnungsaufwand für die Projektionsschnitte mit 2,7 ms auf ¼ des ursprünglichen Wertes.

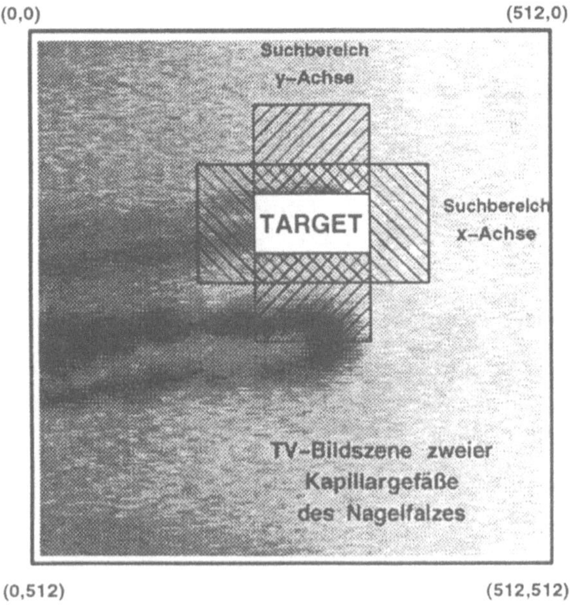

Abbildung 2: Definition und Lage des Targets und der daraus abgeleiteten Suchbereiche für x und y.

4.3 Korrespondenzanalyse

4.3.1 Korrelationsverfahren

Die Korrespondenzanalyse zur paßlagengenauen Kompensation der Objektbewegung wird mit Hilfe der normierten Kreuzkorrelationsfunktion [Lenz86] gelöst.

Zur weiteren Verringerung des numerischen Rechenaufwandes wird die Korrelation des Targets innerhalb des Suchbereiches 2-stufig vorgenommen:

a) In einem ersten Schritt wird eine Korrelation mit 2:1 unterabgetasteten Projektionsschnitten des Targets und des Suchbereichs durchgeführt. Diese Korrelation wird ebenfalls nur für jede zweite Koordinate ermittelt, so daß - bezogen auf die Originaldaten - nur jeder vierte Korrelationsschritt durchgeführt wird. In einem zweiten Schritt wird dann zur genauen Findung der Paßlage mit den Originaldaten in einer Umgebung von ±3 Pixeln um das gefundene Maximum erneut die Korrelation durchgeführt. Als Kriterium für eine erfolgreiche Bewegungsdetektion dient der normierte Korrelationskoeffizient (Wertebereich -1 .. 1).

Diese Maßnahmen reduzieren insgesamt die effektive Verarbeitungszeit um ca. eine Größenordnung. Erkauft wird dies jedoch durch eine Zunahme der Detektionsunsicherheit von ±1 Pixel. Bezogen auf unseren Anwendungsfall liegt dieser Wert noch innerhalb des Toleranzbereiches.

4.3.2 Targetverfolgung

Es ist einleuchtend, daß mit zunehmender Verschiebung des Eingangsbildes das Target nachgeführt werden muß, um die Struktur, mittels derer die Bewegungsdetektion erfolgt, nicht aus dem Targetbereich zu verlieren.
Hierbei wurde keine Bewegungsprädiktion implementiert, vielmehr wird das Target einfach nachgeführt. Dieses Verfahren hat sich bewährt, da große Sprünge in aufeinanderfolgenden TV-Halbbildern ggf. durch eine Vergrößerung des Targets kompensiert werden können. Die maximal zulässige Targetgröße beträgt immerhin 1/16 der gesamten Bildfläche.

4.4 Anforderungen an die Videoszenen

Das vorgestellte Verfahren zur Bewegungsdetektion mittels Projektionsschnitten setzt voraus, daß folgende Vorbedingungen an die Bildszenen erfüllt werden:
a) Die Struktur, mittels derer die Bewegungsdetektion durchgeführt wird, muß sich auf den Projektionsschnitt abbilden lassen. Dies bedeutet insbesondere, daß in horizontaler und vertikaler Richtung im Targetbereich ausgeprägte Hell-Dunkel-Übergänge vorhanden sein müssen.
b) Der Kontrast zwischen hellen und dunklen Stellen im Targetbereich muß ausreichend groß sein, um das Signal/Rausch-Verhältnis durch Rundungsrauschen nicht zu verschlechtern.
c) Innerhalb einer Meßreihe sollte die Objektinformation bezogen auf das Target näherungsweise stationär sein.

5 Bewegungsfilter

Das Bewegungsfilter wird, als autonome Einheit realisiert und dem kapillarmikroskopischen Bildanalysesystem Pries vorgeschaltet (siehe Abb. 1). Es hat die Aufgabe translatorische Lageveränderungen der abgebildeten Kapillargefäße in TV-Echtzeit im Videosignal zu kompensieren, da System Pries für die Geschwindigkeitsanalyse des Blutstromes bzgl. der Kapillargefäße deckungsgleiche TV-Bildsequenzen an seinem Videoeingang erwartet. Die Verarbeitung der Videobilder erfolgt in einer Pipeline mit einer zeitlichen Verzögerung zwischen Akquisition und Display um 1 TV-Vollbild.

Das Bewegungsfilter besteht aus zwei Modulen:
a) Bewegungsdetektion und
b) Bewegungskompensation.
Dabei ist das in Abschnitt 4 beschriebene Verfahren zur Bewegungsdetektion elementarer Bestandteil des Filters.
Im Gegensatz zur Detektion ist die Kompensation einer Objektbewegung die bei weitem einfachere Aufgabe. Sie wird mit Hilfe des Videocontrollers der TV-Framegrabber-Karte durchgeführt, indem der Offset des Displayfensters die Objektbewegung kompensiert.

Aufgrund von Einschränkungen im Videocontrollerteil der Framegrabber-Karte erreicht die Bewegungskompensation jedoch nicht ganz den Bereich der Bewegungsdetektion.

6 Experimentelle Resultate

Versuche an vielen verschiedenen Sequenzen haben ergeben, daß das Verfahren tadellos arbeitet, solange die vorliegenden Bilddaten den zuvor beschriebenen Anforderungen hinreichend genügen. Die visuelle Kontrolle erfolgte durch Dunkeltasten derjenigen TV-Halbbilder, die eine untere Schranke des Korrelationskoeffizienten nicht erreichten. Der Grenzwert für die Zuverlässigkeit der Detektion wurde mit 0.9 sehr hoch angesetzt. Auch größere Objektsprünge wurden sicher detektiert. Zurückweisungen hatten ihre Ursache häufig in zeitlichen Veränderungen der Objektinformation, so z.B. dem Auftreten neuer Objekte und stark schwankenden Fokuslagen.

7 Diskussion und Ausblick

Das von uns realisierte Bewegungsfilter ist in der Lage translatorische Bewegungen von Kapillargefäßen des Nagelfalzes in TV-Bildsignalen in TV-Echtzeit über den nahezu vollständigen Bildbereich zu erfassen und als bewegungskompensiertes TV-Signal mit einem TV-Bild Verzögerung live wiederzugeben. Zuvor muß jedoch interaktiv der Target- und Suchbereich definiert werden. Eine Qualitätskontrolle steht mit dem berechneten normierten Korrelationskoeffizienten für jedes TV-Halbbild zur Verfügung. Bedingt durch die Repräsentation der Bildinformation durch orthogonale Projektionsschnitte verhält sich das Bewegungsfilter robust gegenüber Aufnahmen mit schlechtem Signal/Rausch-Verhältnis.

Nicht unerwähnt bleiben soll, daß das Bewegungsfiltersystem, das in der konkreten Anwendung mit einem Videorekorder arbeitet, nur mit einem Timebase-Korrektor arbeitsfähig ist, da der durch den Rekorder bedingte Phasenjitter die Bildausgabe stark stört. Bei TV-Bildquellen mit konstantem Zeitverhalten (z.B. Videokamera) treten derartige Effekte nicht auf.

Das Filter wurde als autonomes System realisiert und besitzt damit den Vorteil, daß nachgeschaltete bestehende Analysesysteme dadurch mit der Qualität einer lageinvarianten Verarbeitung (nur Translation!) ausgestattet werden können, ohne

Eingriffe in diese Systeme vornehmen zu müssen, die auch häufig überhaupt nicht möglich sind. Das Filtersystem basiert auf einer Standard-Hardwareplattform mit einer Softwareimplementierung der Algorithmen, was die Adaption des Filters an andere Fragestellungen sehr erleichtert. Getestet wurde in diesem Zusammenhang das Filter auch in Anwendung auf künstliche Laborszenen (schwingendes Pendel).

8 Literatur

[Burger91] R. Burger: "Schnelle Bewegungskompensation für Kapillarbildsequenzen auf einem PC". Diplomarbeit am Institut für Physikalische Elektronik, Universität Stuttgart 1991.

[CapiFlow] Capiflow: "A multi purpose computerized image analysis system for experimental and clinical scientists". Informationsmaterial der Fa. IM CapiFlow, Kista, Schweden.

[Lenz86] R.Lenz: "Ein Verfahren zur Schätzung der Parameter geometrischer Bildtransformationen". Dissertation, Technische Universität München, 1986.

[Nagel85] H-H.Nagel: "Analyse und Interpretation von Bildfolgen". Informatik Spektrum 8, 1985, p. 178-200. Springer-Verlag 1985.

[Pries88] A. R. Pries: "A versatile video image analysis system for microcirculatory research". Int. Journal Microcirc: Clin. Exp. 7: p. 327-345. Kluwer Academic Publishers, 1988.

Ein Verfahren zur effizienten Merkmalsauswahl für die Texturfehleranalyse

Jörg Amelung
Technische Hochschule Darmstadt
Fachgebiet Regelsystemtheorie und Robotik
Landgraf-Georg-Str.4, 64283 Darmstadt

Zusammenfassung

Die maschinelle Erkennung von Fehlern in texturierten Oberflächen durch Bildverarbeitungssysteme gewinnt für die Qualitätssicherung in der modernen Fertigung zunehmend an Bedeutung. Gerade problematische Aufgabenstellungen, wie die Analyse statistischer Texturen, kann man in einen Aufgabenteil mit hoher Priorität, nämlich die vollständige Erkennung der Fehler, und einen mit niedrigerer Priorität, nämlich die Unterscheidung der einzelnen Fehlerklassen, unterteilen. Es wird ein Ansatz zur Merkmalsauswahl vorgestellt, der dieser Auffassung entsprechend besonders die Signifikanz der Merkmale für die Unterscheidung zwischen fehlerfreier und fehlerhafter Textur bewertet. Als Hilfsmittel für die Bewertung werden sogenannte Flexibilitätsmatrizen berechnet. Die erreichte Entscheidungssicherheit wird anhand experimenteller Ergebnisse gezeigt.

1 Einleitung

Zur Einführung einer computerunterstützten Qualitätssicherung (CAQ) werden Bildverarbeitungssysteme benötigt, die die Sichtinspektion gemusterter (texturierter) Oberflächen (z.B. Abb.1(a) Stoff, Abb.1(b) Teppich) ermöglichen.

Bei der Vielzahl der zur Auswahl stehenden Texturanalyseverfahren ([TATA87]) ist dabei von entscheidender Bedeutung, daß für die vorliegende Aufgabe eine effiziente Algorithmenauswahl erfolgt.

Ein periodisches Muster wie in Abbildung 1(a) kann man z.B. sehr effizient mit Eigenfilterverfahren ([DEWA88]) bearbeiten, für ein relativ unregelmäßiges Muster wie den in Abbildung 1(b) dargestellten Teppich kann man mit statistischen Methoden Erfolge erzielen. Dabei ist jedoch die Eigenfilterung mit sehr viel weniger Aufwand verbunden als die Berechnung von Texturmerkmalen auf der Basis von Bildpunktstatistiken.

Das Fehlererkennungssystem AST (Analyse Statistischer Texturen) stellt eine Verfahrensgruppe bildpunktbezogener Texturanalyseverfahren zur Verfügung. Mit diesen werden die relativen Häufigkeiten der Eigenschaften lokaler Bildpunktgruppen eines Texturausschnittes (Bildfenster) berechnet und anschließend klassifiziert. Um gute Ergebnisse zu erzielen, müssen zum einen die Parameter der Texturanalyseverfahren eingestellt werden ([AMEL92]), zum anderen aber auch aus der Vielzahl der vorhandenen Merkmale (AST berechnet bis zu 78 Merkmale) die für das vorliegende Problem signifikanten ausgewählt werden.

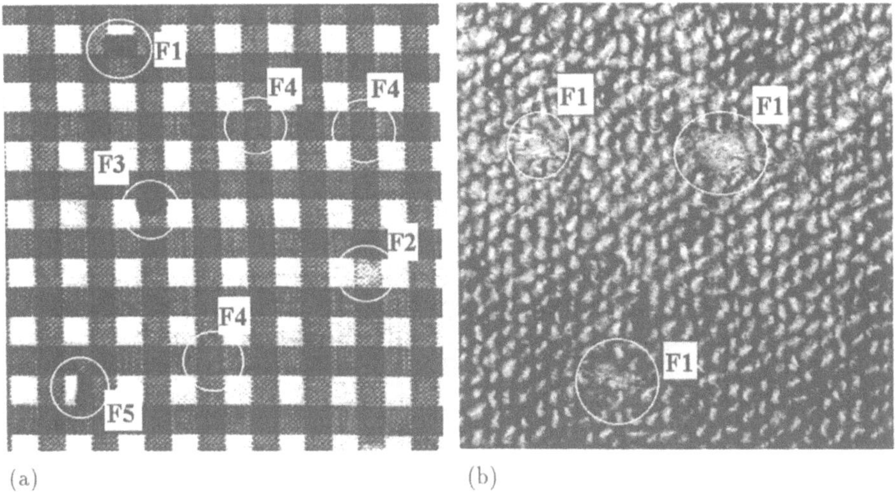

(a) (b)

Abbildung 1: Beispiele für Texturmuster. In (**a**) ist ein Stoff und in (**b**) ein Teppich abgebildet. Die jeweiligen Fehler sind markiert.

Im folgenden stellen wird eine Vorgehensweise zur Merkmalsauswahl vorgestellt, die besonders auf die Problemstellung der Fehlererkennung für die Qualitätssicherung abzielt. Es wird davon ausgegangen, daß eine Erkennung der vorhandenen Fehler absolute Priorität hat, eine Unterscheidung der einzelnen Fehler zwar sehr wünschenswert, aber von zweitrangiger Bedeutung ist. Bei der Merkmalsauswahl wird deshalb besonderer Wert auf eine sichere Unterscheidung zwischen fehlerfreier und fehlerhafter Textur gelegt und dabei eine mögliche Verschlechterung der Unterscheidungsfähigkeit für die verschiedenen Fehlerklassen in Kauf genommen.

2 Merkmalsauswahl für die Texturfehlererkennung

Das Fehlererkennungssystem AST benutzt zur Berechnung von Texturmerkmalen die folgenden Verfahren :

- Histogrammanalyse,
- Gradientenhistogrammanalyse,
- Co-occurrence-Matrix (SGTDM) [HARA73],
- Grauwertlauflängenmatrix (GLRLM) [GALL75],
- Nachbarschafts-Grauwertabhängigkeitsmatrix (NGLDM) [SUN82]),
- Grauwertdifferenz-Abhängigkeitsmatrix (GLDDM) [DAPE86],
- Nachbarschafts-Grauwertdifferenzmatrix (NGTDM) [AMAD89].

Diese Verfahren liefern 78 Texturmerkmale, die für verschiedene Aufgabenstellungen unterschiedlich signifikant sind. Die Signifikanz der Merkmale hängt dabei sowohl von der vorliegenden Textur als auch von der Art der Fehler ab. Merkmale können also nur für einzelne Fehlerklassen signifikant oder aber für die vorliegende Textur insgesamt nicht signifikant sein.

Um den Rechenaufwand sowohl zur Berechnung der Merkmale als auch der Klassifikation so gering wie nötig zu halten und Klassifikationsfehler bedingt durch nicht signifikante Merkmale zu vermeiden, ist es notwendig, eine geeignete Merkmalsauswahl zu treffen.

Die ausgewählten und zum Merkmalsvektor zusammengefaßten Merkmale werden jeweils für ein über das aquirierte Bild wanderndes Bildfenster berechnet und für jedes Fenster einzeln klassifiziert (siehe Abb.2).

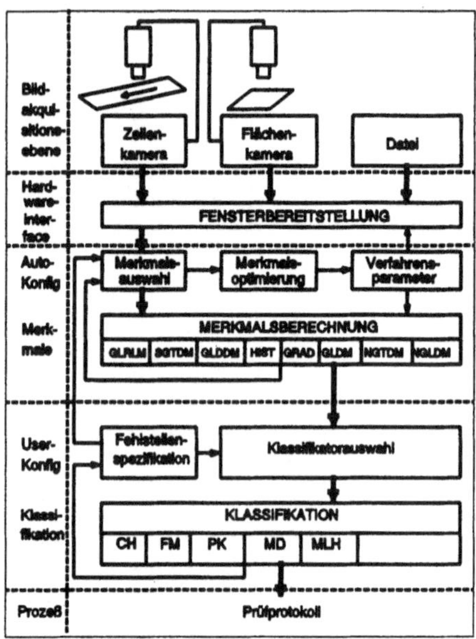

Abbildung 2: Die Struktur des Fehlererkennungssystems AST. Man erkennt die klassischen Arbeitsschritte mit Datenaquisition, Merkmalsberechnung und Klassifikation. Eingearbeitet ist eine automatische Parameterkonfigurierung, mit der eine hohe Signifikanz der Merkmale erzielt wird. Für die Durchführung dieser Konfigurierung muß der Benutzer lediglich die Fehler in einer Texturvorlage, anhand derer das System seine Parameter adaptiert, markieren.

3 Flexibilitäts-Matrizen zur Merkmalsauswahl

3.1 Verfahren zur Merkmalsauswahl

Die Auswahl geeigneter Merkmale zählt zu den Grundproblemen der Mustererkennung. Ziel der Merkmalsauswahl ist es, eine Untermenge M' der Merkmalsmenge M zu finden, die die Fehlerwahrscheinlichkeit bei der Klassifikation minimiert.

Da eine Erprobung aller möglichen Untermengen der Merkmalsmenge M in der Regel aus Aufwandsgründen nicht möglich ist, findet man immer nur eine suboptimale Lösung des Problems.
In der Regel werden Gütemaße berechnet, um den Beitrag der Merkmale zum Klassifikationsergebnis zu beurteilen. Diese Gütemaße werden verwendet, um die Merkmale einer Rangbildung zu unterziehen, wobei die m' ersten Merkmale der Rangliste für die Klassifikation benutzt werden. In der Regel werden für die Rangbildung heuristischen Vorgehensweisen zur Auswertung der Gütemaße verwendet.

Zu diesen heuristischen Vorgehensweisen gehören z.B. die direkte Rangbildung aus der jeweiligen Güte der Merkmale oder die Bewertung des Beitrages eines Merkmales zur Minimierung des Klassifikationsfehlers. Bei den Gütemaßen muß man unterscheiden zwischen Maßen, die die statistischen Abhängigkeiten der Merkmale berücksichtigen (vektorbasiert) und Maßen, die jeweils nur ein Merkmal berücksichtigen ([NIEM83]). Die einfache Berechnung der Merkmalsgüten und die direkte Rangbildung führen durchaus zu zufriedenstellenden Resultaten ([NICO90]), jedoch in den seltensten Fällen tatsächlich zu einer optimalen Merkmalsauswahl ([NIEM83]).

Das im folgenden beschriebene Auswahlverfahren basiert ebenso auf einer einfachen Bewertung der Merkmalsgüten, die jedoch nach einer auf die Anforderungen der Qualitätssicherung angepaßten Auswahlheuristik ausgewertet werden. Dabei kann man davon ausgehen, daß eine Erkennung der vorhandenen Fehler absolute Priorität hat und eine Unterscheidung der einzelnen Fehler zwar von zweitrangiger Bedeutung, trotzdem jedoch sehr wünschenswert ist.

3.2 Beurteilung der Merkmalssignifikanz

Die Beurteilung der Signifikanz der Merkmale durch die Berechnung eines Gütemaßes wird anhand einer Texturstichprobe durchgeführt, die Texturausschnitte (Fenster) von Fehlern aller Fehlerklassen und fehlerfreier Textur enthält. Berechnet man die Werte eines Merkmals für die Fenster der Texturstichprobe, so erhält man die Verteilungsdichten des Merkmals für die verschiedenen, in der Texturstichprobe enthaltenen Fehlerklassen [1]. Um die Signifikanz des Merkmals zu beurteilen, werden die Verteilungsdichten des Merkmals für die einzelnen Klassen durch Gaußdichten approximiert und die resultierenden Parameter, Mittelwert m und Standardabweichung σ, jeweils für zwei Klassen i, j in Beziehung gesetzt. Die Güte q des Merkmals für die Klassenkombination i, j ergibt sich zu (siehe auch Abb.3(a),(b))

$$q_{ij} = \frac{|m_i - m_j|}{\sigma_i + \sigma_j} \tag{1}$$

Tabelle 1 zeigt, wie sich die Güte q_{ij} auf das Klassifikationsverhalten auswirkt. Man kann beobachten, daß sich die Ergebnisse mit steigender Güte q_{ij} verbesseren, also weniger fehlerfreie Fenster der entsprechenden Fehlerklasse zugeordnet werden. Ab einer Güte von $q_{ij} \geq 3$ können sehr gute Ergebnisse erzielt werden.

Um eine Texturstichprobe mit allen vorhandenen Fehlerklassen zu erstellen, wird eine vom Benutzer mit einem Fehler-Editor anzugebende Liste der Lage der vorhandenen Texturfehler mit einem speziellen Match-Verfahren ausgewertet, so daß eine automatische Zuordnung eines Bearbeitungsfensters zu einer der Fehlerklassen vorgenommen werden kann. Dieses Match-Verfahren stellt fest, wieviel Prozent der Fläche eines Bearbeitungsfensters von einem Fehler überdeckt werden, bzw. wieviel Prozent eines Fehlers im Bearbeitungsfenster liegen. So kann man über Schwellen eine „Autoklassifikation" vornehmen, um die Klassenzuweisung für die einzelnen Fenster der Texturstichprobe zu automatisieren. Dies läßt sich beispielsweise folgendermaßen in Worten ausdrücken: „95% des Bearbeitungsfensters sind vom Fehler überdeckt, also wird das Fenster als Fehler klassifiziert" bzw. „100% des Fehlers liegen im Bearbeitungsfenster, also muß das Fenster als Fehler klassifiziert werden" (s.Abb.3(c)).

[1] Wir betrachten die fehlerfreie Textur als Fehlerklasse F0 und die Fehler als Fehlerklassen F1, F2 usw., um die Behandlung zu vereinheitlichen.

NGTDM_COA	
Klassen-kombination	q_{ij}
$0 \Longleftrightarrow 1$	0.59
$0 \Longleftrightarrow 2$	2.12
$0 \Longleftrightarrow 3$	3.22
$0 \Longleftrightarrow 4$	3.84
$0 \Longleftrightarrow 5$	0.66

(a)

HIST_GLA	
Klassen-kombination	q_{ij}
$0 \Longleftrightarrow 1$	3.70
$0 \Longleftrightarrow 2$	5.08
$0 \Longleftrightarrow 3$	1.70
$0 \Longleftrightarrow 4$	2.96
$0 \Longleftrightarrow 5$	7.34

(b)

Klasse:	0	1	2	3	4	5
	79	76	16	3	0	58

(c)

Klasse:	0	1	2	3	4	5
	159	0	0	67	6	0

(d)

Tabelle 1: Die Signifikanz q_{ij} der Merkmale **(a)** NGTDM_COARSENESS und **(b)** HISTOGRAMM_GRAUWERTMITTELWERT für die Klassenkombinationen $i \Longleftrightarrow j$ der Textur Stoff (Abb.1(a)). In **(c)** und **(d)** wird dargestellt, wie 232 fehlerfreie Fenster (Klasse 0) mit den beiden Merkmalen aus (a) und (b) klassifiziert wurden. Die erste Spalte gibt an, wieviele Fenster richtig klassifiziert wurden. Man erkennt, daß sich niedrige q_{ij} durch entsprechend viele Fehlklassifikationen (Spalten 1 bis 5) niederschlagen.

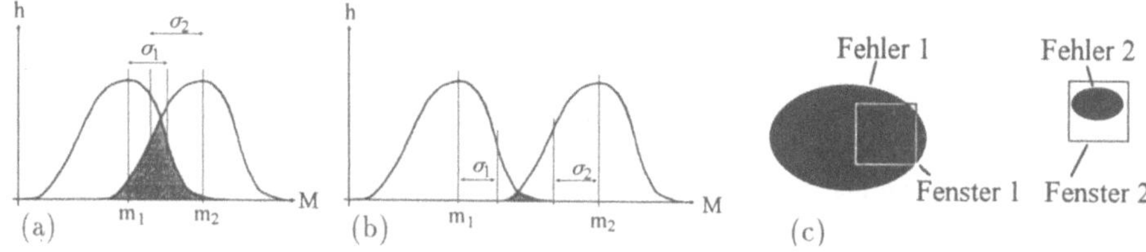

Abbildung 3: **(a)** Das Merkmal M mit der dargestellten Verteilung für die Klassen 1 und 2 hat für die Unterscheidung der Klassen 1 und 2 eine Güte von $q_{12} = 0,8$, in **(b)** ist die Güte $q_{12} = 1,8$. **(c)** Darstellung von zwei Texturfehlern und zwei Bearbeitungsfenstern. Fenster 1 ist zu 95% mit dem Fehler 1 bedeckt, aber nur 30% von Fehler 1 liegen im Fenster 1. Entsprechend sind nur 50% von Fenster 2 mit Fehler 2 bedeckt, obwohl dieser ganz in Fenster 2 liegt.

3.3 Berechnung der Flexibilitätsmatrizen

Strebt man eine Merkmalsauswahl auf der Basis der Signifikanzbeurteilung nach Gl.(1) an, dann werden die Merkmalsgüten q_{ij} zur Bildung einer Rangliste herangezogen. Betrachtet man die Ranglistenbildung nach dem Verfahren der Summation der Güten q_{ij} ([NIEM83], [NICO90]), so tritt das Problem auf, daß Merkmale, die für einzelne Klassenkombinationen eine hohe Signifikanz aufweisen, ausgewählt werden, obwohl viele nicht signifikante Klassenkombinationen vorhanden sind.

Deshalb werden durch eine Schwellwertbildung der Güten sogenannte Flexibilitätsmatrizen FM gebildet, so daß nur noch zwischen *signifikant* und *nicht signifikant* unterschieden wird :

$$FM_m(i,j) = \begin{cases} 1 & \text{wenn } q_{ij_m} > q_s \\ 0 & \text{sonst} \end{cases} \qquad (2)$$

Für den Parameter q_s gilt $q_s \geq 3$. Die Flexibilitätsmatrix FM wird für jedes Merkmal m gebildet und erlaubt eine Aussage darüber, wie flexibel das jeweilige Merkmal ist, d.h. für wieviele und welche Klassenkombinationen das Merkmal signifikant ist. In Tabelle 2 wird eine Flexibilitätsmatrix dargestellt.

$FM_m(i,j)$:

$i \downarrow j \rightarrow$	1	2	3	4	5
0	0	0	1	1	0
1		1	1	1	0
2			1	1	1
3				1	1
4					1

Tabelle 2: Die Flexibilitätsmatrix des Merkmales NGTDM_COA für die Textur „Stoff" ($q_s = 3.0$).

3.4 Auswertung der Flexibilitätsmatrizen

Um aus den Flexibilitätsmatrizen eine Rangliste für die Auswahl der Merkmale zu erhalten, kann man für die jeweilige Aufgabe gezielte Strategien einsetzen. Für den Anwendungsfall in der Qualitätssicherung, bei dem man davon ausgehen kann, daß eine Erkennung der vorhandenen Fehler absolute Priorität hat, eine Unterscheidung der einzelnen Fehler zwar sehr wünschenswert, aber von zweitrangiger Bedeutung ist, wird vor allem die Flexibilität der Merkmale für die Klassenkombinationen $0 \Longleftrightarrow j$, mit $j = 1 \ldots K$, gewichtet.

Um die Rangliste der Merkmale zu bilden, wird zunächst für jedes Merkmal m als Kennwert die Anzahl $q^{\Sigma}_{0j_m}$ der signifikanten Klassenkombinationen $0 \Longleftrightarrow j$ berechnet:

$$q^{\Sigma}_{0j_m} = \sum_{j=1}^{K} FM_m(0,j) \tag{3}$$

Da dieser Kennwert zur Bildung einer Rangliste in der Regel nicht eindeutig ist, wird ein zweiter Kennwert, die Anzahl $q^{\Sigma}_{ij_m}$ der signifikanten Klassenkombinationen $i \Longleftrightarrow j$, mit $i,j \geq 1$, gebildet:

$$q^{\Sigma}_{ij_m} = \sum_{j=1}^{K} \sum_{j=i+1}^{K} FM_m(i,j) \tag{4}$$

Die Ranglistenbildung erfolgt nun durch Sortieren der Merkmale nach ihrem Kennwert $q^{\Sigma}_{0j_m}$, bei Gleichheit entscheidet der zweite Kennwert $q^{\Sigma}_{ij_m}$. Eine weitere Unterscheidung ist möglich, aber nicht vorgesehen.

Durch diese Auswahlstrategie wird die Priorität auf die Signifikanz der Merkmale für die Kombinationen $0 \Longleftrightarrow j$ gesetzt, die allgemeine Flexibiliät der Merkmale aber durch den Kennwert $q^{\Sigma}_{ij_m}$ berücksichtigt.

Um zu gewährleisten, daß für alle Klassenkombinationen $0 \Longleftrightarrow j$ signifikante Merkmale ausgewählt werden, ermittelt man für jede Kombination $0 \Longleftrightarrow k$, mit $k = 1 \ldots K$, die t ersten Merkmale aus der Rangliste, die für die jeweilige Klassenkombination signifikant sind, d.h. deren $FM_m(0,k) = 1$ ist (Tabelle 3). Da in der Regel Überschneidungen vorhanden sind, ist die zu erwartende Gesamtanzahl der ausgewählten Merkmale kleiner als $t \cdot (K-1)$.

Rang-liste	Merkmal	$q^{\Sigma}_{0j_m}$	$q^{\Sigma}_{ij_m}$	$FM_m(0,j)$ $j=1$	2	3	4	5
1.	SGTDM_ENT_045	5	6	1	1	1	1	1
2.	SGTDM_CON_090	4	8	0	1	1	1	1
3.	SGTDM_IDM_090	4	8	0	1	1	1	1
4.	SGTDM_CON_045	4	7	0	1	1	1	1
5.	SGTDM_IDM_090	4	7	0	1	1	1	1
6.	NGLDM_LNE	4	7	1	0	1	1	1
7.	GLDDM_GLA	4	6	1	1	0	1	1
8.	SGTDM_ASM_000	3	7	1	1	0	0	1
9.	NGLDM_COA	3	7	0	1	1	1	0

(a)

Klassenkombination $0 \Longleftrightarrow 1$:
SGTDM_ENT_045
GLDDM_GLA
NGLDM_LNE
Klassenkombinationen $0 \Longleftrightarrow 2$, $0 \Longleftrightarrow 3$, $0 \Longleftrightarrow 4$, $0 \Longleftrightarrow 5$:
SGTDM_ENT_045
SGTDM_CON_090
SGTDM_IDM_090

(b)

Tabelle 3: Die Auswahl der Merkmale anhand der Kennwerte $q^{\Sigma}_{0j_m}$ und $q^{\Sigma}_{ij_m}$. In **(a)** ist die Rangliste dargestellt und in **(b)** die für die Klassenkombinationen $0 \Longleftrightarrow j$ jeweils ausgewählten Merkmale für $t = 3$.

4 Ergebnisse

Bei der Erkennung von Fehlern in texturierten Oberflächen handelt es sich um die Erkennung von Abweichungen in der Textur eines Bearbeitungsfensters. Die Aufgabenstellung hat jedoch ein spezielles Ziel. Zunächst sollen alle fehlerfreien Bereiche als solche erkannt werden, es sollen aber auch alle Fehler erkannt werden, wobei die Unterscheidung zwischen verschiedenen Fehlerklassen zunächst zweitrangig erscheint, insgesamt aber doch wünschenswert ist. Damit hat in der Tabelle 4 jeweils die erste Zeile höchste Bedeutung, da sie das Ergebnis der Unterscheidung fehlerfrei–fehlerhaft dokumentiert.

Die Untersuchungen wurden anhand der beiden in Abb.1 dargestellten Texturen „Stoff“ und „Teppich“ durchgeführt. Die Textur Teppich enthält drei Fehler, die von gleicher Art sind und deshalb nicht unterschieden werden, d.h. es gibt nur die Klassen 0 (fehlerfrei) und 1 (Fehler). Die Textur Stoff enthält 5 Fehler, die unterschiedliche Gestalt haben und unterschieden werden.

Klasse:	0	1	2	3	4	5	richtig
0	3434	26	0	0	0	30	98,4%
1	0	53	0	0	0	0	100,0%
2	0	0	53	5	0	0	91,4%
3	2	0	0	58	0	0	96,7%
4	31	0	0	0	126	0	80,2%
5	0	1	0	0	0	53	98,1%

(a)

Klasse:	0	1	richtig
0	3470	12	99,6%
1	3	118	97,5%

(b)

Tabelle 4: Klassifikationsergebnis :**(a)** Stoff. Es wurden die in Tabelle 3(b) dargestellten Merkmale ausgewählt. **(b)** Teppich. Es wurden die Merkmale SGTDM_ASM_000, SGTDM_ASM_090, SGTDM_ASM_135, SGTDM_ENT_000 und SGTDM_ENT_135 ausgewählt.

Für die jeweils untersuchte Textur wurde von AST eine automatische Parameterkonfiguration durchgeführt und anschließend aus der Texturvorlage eine Lernstichprobe berechnet. Diese Lernstichprobe wurde herangezogen, um einen Minimum-Distanz-Klassifikator zu initialisieren, wobei die Auswahl der Merkmale mit der beschriebenen Flexibilitätsauswahl vorgenommen wurde. Die so ermittelten Ergebnisse werden in Tabelle 4 dargestellt.

5 Zusammenfassung

Im Bildverarbeitungssystem AST zur Erkennung von Fehlern in unregelmäßig gemusterten Oberflächen wurde ein Verfahren zur Merkmalsauswahl realisiert, das durch besondere Beachtung der Merkmalssignifikanzen für die Unterscheidung fehlerfrei—fehlerhaft eine hohe Erkennungssicherheit für Aufgaben in der Qualitätsicherung erreicht. Für die Merkmalsauswahl wurden Flexibilitätsmatrizen eingeführt und eine Strategie zur Auswertung hergeleitet. Entsprechende Ergebnisse wurden dargestellt.

Die vorgestellte Strategie zur Auswertung der Flexibilitätsmatrizen ist offen für Erweiterungen. Insbesondere kann durch Hinzunahme weiterer Ranglisten eine Bewertung anderer Auswahlkriterien erreicht werden. Hier ist z.B. eine Bewertung der Rechenzeit (schnelle ungenaue oder langsame genaue Bildauswertung) oder ein Einbeziehen der statistischen Abhängigkeit der Merkmale denkbar.
Die Untersuchungen wurden mit einem MC68020-VME-Bus-System durchgeführt. Es ergaben sich hierbei Rechenzeiten von durchschnittlich 0.15 s pro Merkmal bei einer Fenstergröße von 40×40 Bildpunkten. Dies entspricht einer Zeit von ca. 1 Minute für $10 \times 10cm^2$ Oberfläche bei den untersuchten Texturen. Durch den Einsatz modernerer Prozessoren (MC68040) oder Transputerhardware läßt sich die Bearbeitungszeit entsprechend verkürzen.

Literatur

[AMAD89] Moses Amadasun and Robert King: Textural Features Corresponding to Textural Properties; IEEE Transactions on Systems, Man and Cybernetics, Vol.SMC-19 (1989), No.5, pp.1264-1274.

[AMEL92] Jörg Amelung: Automatische Konfiguration für bildpunktbezogene Texturanalyseverfahren;Proc.37.Int.wiss.Kolloquium,Ilmenau,1992, pp.533-538.

[DAPE86] Zhang Dapeng and Li Zhongrong: Digital image texture analysis using gray level and energy cooccurrence; Proc. SPIE Int. Soc. Opt. Eng. Conference on Applications of Artificial Intelligence IV, Innsbruck, Austria, 15-16 April 1986, Vol.657 (1986), pp.152-156.

[DEWA88] P.Dewaele, L.Van Gool, P.Wambacq and A.Oosterlinck: Texture Inspection with Self-adaptive Convolution Filters; Proc.IEEE 9th Intern.Conf.on Pattern Recognition, Washington DC, 1988, pp.56-60.

[GALL75] Mary M. Galloway: Texture Analysis Using Gray Level Run Lengths; Computer Graphics and Image Processing, Vol.CGIP-4 (1975), pp.172-179.

[HARA73] Robert M.Haralick, K.Shanmugam and I.Dinstein: Textural Features for Image Classification; IEEE Transactions on Systems, Man and Cybernetics, Vol.SMC-3 (1973), No.6, pp.610-621.

[HE88] Dong-Chen He, Li Wang und Jean Guibert: Textural Discrimination Based On An Optimal Utilization Of Texture Features; Pattern Recognition, Vol.21 (1988), No.2, pp.141-146.

[NICO90] Bertram Nicolay: Überwacht lernendes Bildauswertungssystem zur Erkennung von Oberflächenfehlern; Hanser-Verlag, München, Wien, 1990.

[NIEM83] Heinrich Niemann: Klassifikation von Mustern; Springer Verlag, Berlin, Heidelberg, New York, Tokio, 1983.

[SUN82] Sun,C.and Wee,W.G.: Neighboring Gray Level Dependence Matrix for Texture Classification; Computer Vison,Graphics and Image Processing, Vol.23 (1982), pp.341-352.

[TATA87] Tatari, S.: Auswahl von Texturanalyseverfahren zur Automatisierung industrieller Sichtprüfungen, Fortschr.Ber.VDI Reihe 10 Nr.67, VDI-Verlag, Düsseldorf, 1987.

Depth-from-Focus zur Bestimmung der Konzentration und Größe von Gasblasen

Peter Geißler[1] und Bernd Jähne[2]

[1] Institut für Umweltphysik der Universität Heidelberg
Im Neuenheimer Feld 366, 69120 Heidelberg
[2] Scripps Institution of Oceanography, PORD
La Jolla, CA 92093-0230 USA

Zusammenfassung

Es wird ein Depth-from-Focus Verfahren vorgestellt, mit dem die Tiefeninformation einer 3D-Szene durch die Auswertung eines einzigen 2D-Bildes gewonnen werden kann. Dies wird durch die konsequente Ausnutzung bekannter Objekteigenschaften ermöglicht. Das Verfahren wird zur Untersuchung von Gasblasen im Wasser angewendet. Mit Hilfe der Tiefeninformation wird die genaue Objektgröße ermittelt. Darüber hinaus begrenzt ein Unschärfekriterium das Meßvolumen in Abhängigkeit von der Teilchengröße und ermöglicht so die Bestimmung der Konzentration.

1 Einleitung

Ein grundlegendes Problem bei der Auswertung zweidimensionaler Bilddaten ist der Verlust der 3D-Ortsinformation durch die Projektion der dreidimensionalen Szene auf eine zweidimensionale Sensorfläche. Zur Rekonstruktion der Tiefeninformation kann die Unschärfe der Abbildung herangezogen werden, die mit der Entfernung der Objekte von der Fokusebene wächst. Die meisten bisher entwickelten Depth-from-Focus Verfahren bestimmen die Entfernung der Objekte senkrecht zur Bildebene anhand mehrerer, mit unterschiedlichen Kameraparametern aufgenommener Bilder derselben Szene. Beispielsweise bestimmt [Darell, 1990] die Position der Objekte durch Auffinden des Maximums der Schärfe in einer Fokusserie. [Ens, 1993] bestimmt die Defokussierung aus der Aufnahme zweier Bilder mit unterschiedlicher Kamerablende. Für die Analyse bewegter Szenen ist es aber nicht möglich, die benötigten Bilder nacheinander mit nur einer Kamera aufzunehmen. Durch Verwendung eines geteilten Strahlengangs ist die gleichzeitige Aufnahme zweier Bilder prinzipiell möglich, jedoch werden hierzu zwei Kameras benötigt [Pentland, 1987]. [Bove, 1989] verwendet ein ähnliches Verfahren mit einer Lochkamera und einer Kamera endlicher Blende.

In vielen Anwendungsfällen treten jedoch nur bestimmte Objekte auf. Damit steht eine weitere Informationsquelle zur Verfügung, die zur Bestimmung der Defokusierung herangezogen werden kann: die a priori bekannten Objekteigenschaften.

Im vorliegenden Beitrag wird ein neues Verfahren zur Bestimmung der Entfernung aus der Unschärfe vorgestellt. Es basiert auf der konsequenten Ausnutzung der Eigenschaften der zu untersuchenden Objekte und erfordert deshalb nur eine Aufnahme jeder Szene zur Auswertung. Das Verfahren wurde zur Untersuchung von Luftblasen im Wasser entwickelt. Die Verwendung einer Depth-from-Focus Methode bringt dabei einen entscheidenden Vorteil: das Meßvolumen wird durch ein Unschärfekriterium begrenzt. Damit wächst es mit der Blasengröße linear an und es wird eine verbesserte Statistik bei großen Blasen erzielt, deren Anzahl erheblich geringer als die der kleinen Blasen ist. Darüber hinaus vermeidet diese Art der Volumenbegrenzung Störungen der zu untersuchenden Prozesse, wie sie bei mechanischer Begrenzung auftreten.

Luftblasen werden durch große, brechende Wellen aus der Atmosphäre ins Wasser geschlagen. Sie besitzen eine große Bedeutung für die klimarelevanten Austauschprozesse zwischen Ozean und Atmosphäre [Merlivat, 1983], [Farmer, 1993] sowie für die Dynamik des Wellenbrechens [Lamarre, 1991].

Depth-from-Focus Verfahren können auch in anderen Bereichen zur Konzentrationsbestimmung eingesetzt werden. In Zusammenarbeit mit dem Konzernforschungszentrum der Asea Brown Boveri AG und der Universität Hannover arbeitet das Institut für Umweltphysik an einem solchen Verfahren zur on-line-Messung von Zellkonzentrationen [Suhr et al., 1990].

Im folgenden wird zunächst die verwendete Visualisierungstechnik kurz beschrieben, dann die Depth-from-Focus Methode und das Bildanalyseverfahren erläutert.

2 Bildaufnahme

Zur berührungslosen Messung von Blasenverteilungen werden sowohl optische [Johnson, 1979], [Jähne, 1984] als auch akustische Verfahren [Melville, 1993] eingesetzt. Die hier verwendete optische Meßmethode beruht auf der Visualisierung

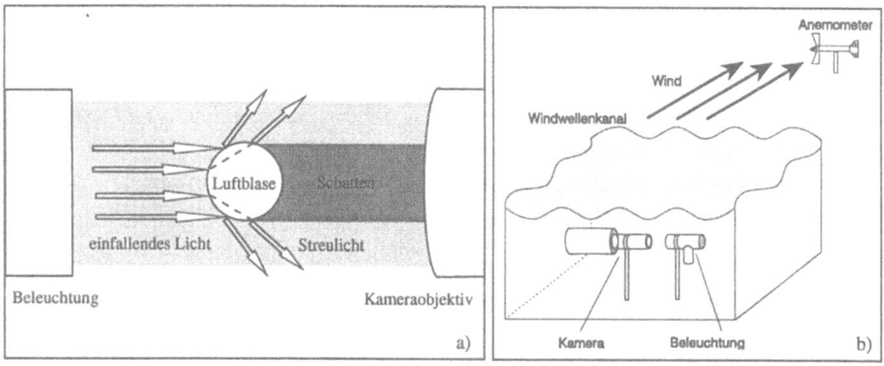

Abb. 1. *a) Prinzip der Durchlichtbeleuchtung: Die Visualisierung erfolgt durch Lichtstreuung an den Blasen (Die Darstellung ist nicht maßstabgerecht). b) Versuchsanordnung: Die der Beleuchtung gegenüberliegende Kamera beobachtet die durch Wellen eingetragenen Gasblasen. Die aufgenommenen Bilder werden analog auf einem Videorekorder gespeichtert. Gleichzeitig mißt ein Anemometer die Windgeschwindigkeit.*

der Luftblasen durch Lichtstreuung an der Blasenoberfläche sowie der Begrenzung des Meßvolumens durch die Unschärfe der Abbildung bei Verwendung einer Optik geringer Tiefenschärfe. Beleuchtung und Kamera liegen sich dabei gegenüber (Abb. 1b). Dazwischen befindliche Blasen streuen das Licht aus seiner ursprünglichen Richtung heraus, so daß es nicht mehr ins Objektiv gelangen kann. Luftblasen erscheinen daher im Bild dunkel vor dem hellen Hintergrund der unbeeinflußten Beleuchtung (Abb. 1a). Durch die geringe Tiefenschärfe der Optik werden Blasen nur in der Nähe der Schärfenebene, die sich genau in der Mitte zwischen Beleuchtung und Kamera befindet, erkennbar abgebildet. Das Meßvolumen befindet sich dann zwischen den Geräten und reicht nicht bis an diese heran, ist also von deren störenden Einflüssen entfernt. Gleichzeitig führt die Tatsache, daß große Blasen bei gleicher Unschärfe, d.h. gleicher Entfernung von der scharfen Ebene, wesentlich besser erkennbar sind als kleinere Blasen und erst in größerer Entfernung nicht mehr detektiert werden, zu einem mit der Blasengröße anwachsenden Meßvolumen. Die kurze Belichtungszeit von 63 μs dient zur Vermeidung zusätzlicher Bewegungsunschärfe. Abb. 2 zeigt zwei mit der beschriebenen Anordung aufgenommene Bilder.

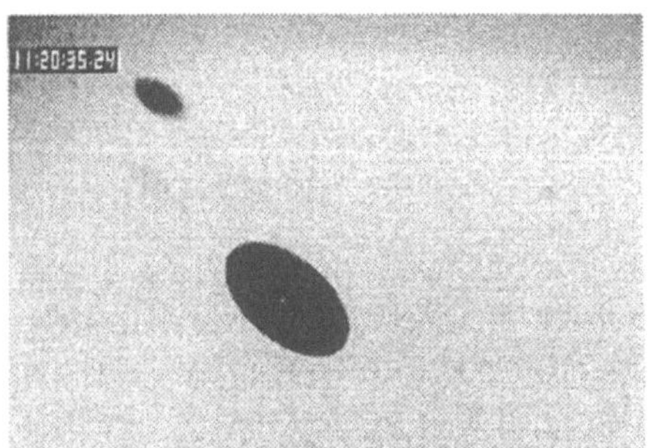

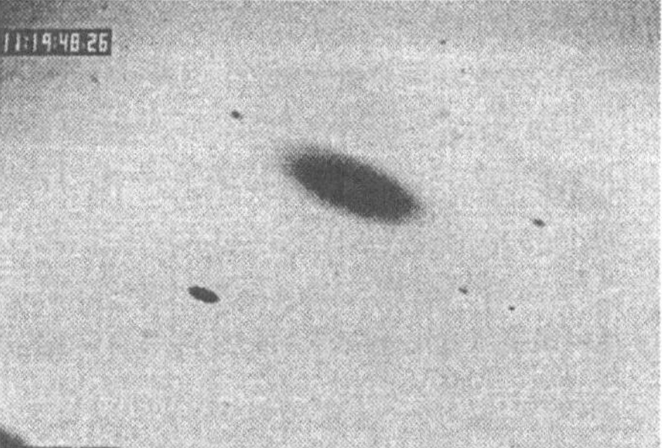

Abb. 2. *Zwei typische, mit dem beschriebenen Aufnahmeverfahren erhaltene digitale Bilder mit 240 x 512 Bildpunkten (Halbbild). Die Ortsauflösung beträgt etwa 15 μm in horizontaler und 27 μm in vertikaler Richtung.*

Durch den Vorgang des Digitalisierens geht die Information über die absolute Helligkeit verloren. Die Grauwerte $g(\mathbf{x})$ des digitalisierten Bildes hängen linear von der Lichtintensität $I(\mathbf{x})$ auf dem CCD ab:

$$g(\mathbf{x}) = aI(\mathbf{x}) + b \tag{1}$$

Dabei sind die Parameter a und b von den Eigenschaften des verwendeten Aufnahmesystems abhängig. Deshalb wird eine von diesen Parametern sowie der absoluten Lampenhelligkeit unabhängige Information benötigt. Man erhält sie durch die Aufnahme zweier zusätzlicher Eichbilder. Bei ausgeschalteter Beleuchtung ($I(\mathbf{x}) = 0$) wird ein Schwarzbild aufgenommen, dessen Grauwert durch $g_s(\mathbf{x}) = b$ gegeben ist. Bei eingeschalteter Beleuchtung, aber ohne daß Objekte im Meßvolumen vorhanden sind ($I(\mathbf{x}) = I_0(\mathbf{x})$), wird ein Nullbild $g_n(\mathbf{x}) = aI_0(\mathbf{x}) + b$ aufgenommen. Durch die Normierung

$$g'(\mathbf{x}) = 1 - \frac{g(\mathbf{x}) - g_s(\mathbf{x})}{g_n(\mathbf{x}) - g_s(\mathbf{x})} = 1 - \frac{I(\mathbf{x})}{I_0(\mathbf{x})} \tag{2}$$

erhält man ein von den erwähnten Parametern unabhängiges Signal.

3 Punktantwort

Durch die verwendete Aufnahmetechnik erscheinen die einzelnen Objekte lichtundurchlässig. Sie können daher durch lichtabsorbierende Scheibchen beschrieben werden. Allgemeiner wird zur Darstellung der Objekte der Verlauf des Lichtabsorptionskoeffizienten $\tau(\mathbf{x})$ in der Objektebene verwendet. Zwei Faktoren, die beide vom Abstand z zwischen Objekt- und Fokusebene abhängen, bestimmen das Bild des Objekts. Die *geometrische Vergrößerung* $V_g(z)$ bedingt eine Veränderung der Bildgröße. Bei ideal scharfer Abbildung wäre das Bild[3] des Objekts durch $1 - \tau(\frac{1}{V_g}\mathbf{x})$ gegeben. Die Unschärfe wird durch die Faltung des scharfen Bildes mit der *point spread function* $PSF_z(\mathbf{x})$ des optischen Systems beschrieben. Dabei wird die Form der PSF durch die wirksame Blende des optischen Systems bestimmt und ist daher von der Lage der Objektebene bezüglich der Fokusebene unabhängig. Diese bestimmt nur die Ausdehnung der PSF. Sie ist somit durch eine Blendenfunktion $B(\mathbf{x})$ und einen entfernungsabhängigen Vergrößerungsfaktor $V_p(z)$ gegeben:

$$PSF_z(\mathbf{x}) = kB\left(\frac{1}{V_p(z)}\mathbf{x}\right) \tag{3}$$

wobei der Normierungsfaktor k durch $k^{-1} = \int B\left(\frac{\mathbf{x}}{V_p(z)}\right) d\mathbf{x}$ gegeben ist. Oftmals

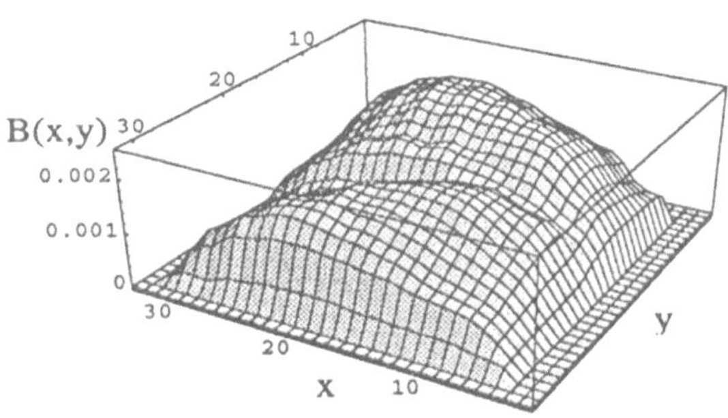

Abb. 3. *3D Darstellung der Blendenfunktion des optischen Aufbaus*

[3] Unter dem Bild wird hier die Intensitätsverteilung auf der Bildebene verstanden

wird für die Blendenfunktion eine einfache Kastenfunktion oder ein gaußförmiger Verlauf angenommen. Für das hier eingesetzte System wurde sie exakt bestimmt. Bei dieser Optik dient das Bild der Lampenleuchtfläche als effektive Blende. Die Blendenfunktion wird daher von der Helligkeitsverteilung der Leuchtfläche bestimmt (Abb. 3).

Insgesamt wird das Bild eines Objekts in der Entfernung z von der Fokusebene durch

$$I(\mathbf{x}) = n(\mathbf{x}) \left[1 - \tau(\frac{1}{V_g(z)}\mathbf{x}) * PSF_z(\mathbf{x}) \right] \tag{4}$$

beschrieben, wobei die Vignettierungsfunktion $n(\mathbf{x})$ insbesondere auch die absolute Lampenhelligkeit enthält. An dieser Stelle wird deutlich, daß die oben erwähnte Normierung gerade die relevante Information $\tau(\frac{1}{V_g(z)}\mathbf{x}) * PSF_z(\mathbf{x})$ aus dem Bild extrahiert.

4 Depth from Focus

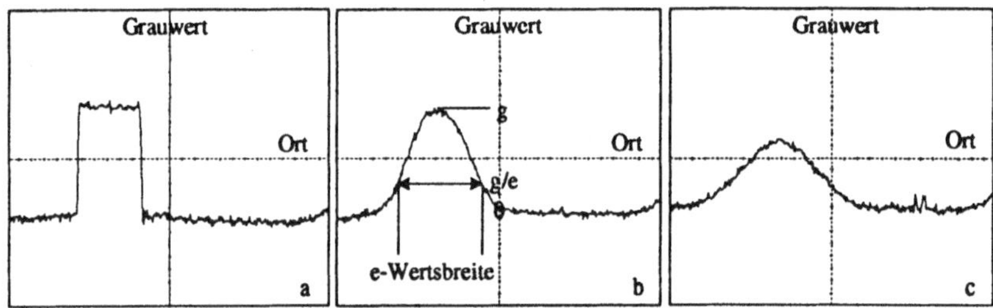

Abb. 4. *Grauwertverläufe von Blasen in zunehmender Entfernung zur Fokusebene. Die Entfernung nimmt von a) nach c) zu.*

Abb. 4 zeigt den Grauwertverlauf für Blasen, die sich in unterschiedlicher Entfernung von der Fokusebene befinden. Die zunehmende Unschärfe bei immer größer werdendem Abstand verursacht neben flacher werdenden Kanten auch abnehmende Helligkeit. Da keine scharfe Objektbegrenzung vorhanden ist, wurde die e-Wertsfläche (Abb. 4b) als Maß für die Größe definiert.

Zur Bestimmung der Größe und Entfernung ist zusätzlich noch ein Maß für die Unschärfe des Objekts notwendig. Aufgrund der starken Abhängigkeit von der Entfernung ist der mittlere Grauwert auf der e-Wertsfläche als Unschärfemaß geeignet. Abb. 5 zeigt die Abhängigkeit der beiden Kenngrößen von der Entfernung für ein kreisförmiges Objekt von 0.5 mm Durchmesser.

Das Depth-from-Focus Verfahren beruht auf folgendem Prinzip:
Aus der e-Wertsfläche und dem mittleren Grauwert kann die Entfernung von der Fokusebene sowie die tatsächliche Größe eines Objekts bestimmt werden. Anhand von Gleichung (4) kann das Bild einer Blase berechnet werden, die sich in beliebiger Entfernung zur Fokusebene befindet. Die Blendenfunktion des optischen Aufbaus wird dabei gemessen, die Vergrößerungsfaktoren $V_g(z)$

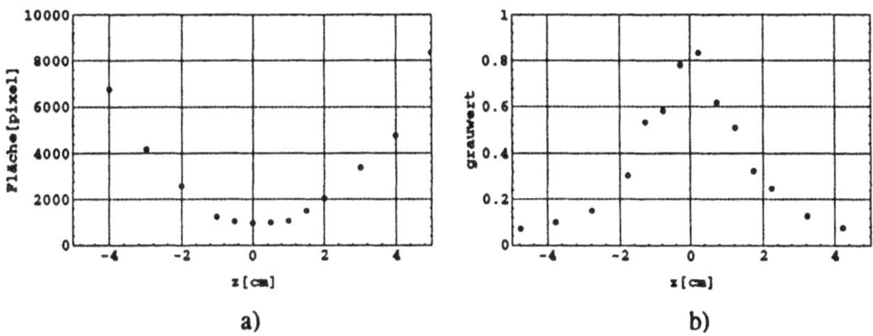

Abb. 5. *Verlauf der e-Wertsfläche a) und des mittleren Grauwerts b) mit der Entfernung z von der Fokusebene*

und $V_p(z)$ berechnet. Damit wird der Verlauf der e-Wertsfläche und des Grauwerts mit der Entfernung für eine Reihe von Blasen verschiedener Radien bestimmt. Diese Daten werden in *Look-Up-Tabellen* zusammengefaßt, deren Eingangsdaten die beiden Kenngrößen und deren Ausgangsdaten die Entfernung bzw. die wahre Größe sind. Die LUTs für positive Werte von z sind in Abb. 6 gezeigt. Für negative z-Werte ergeben sich ähnliche, aber nicht identische Daten. Mit zunehmendem Abstand führt nämlich der Einfluß der PSF zu wachsender e-Wertsfläche, während die Vergrößerung nur bei abnehmendem Abstand zum Objektiv ($z < 0$) zum Anstieg der Fläche führt, in der umgekehrten Richtung ($z > 0$) dagegen die Fläche verkleinert. Zur Bestimmung der Konzentration muß neben der Größe und Anzahl der Blasen auch das Meßvolumen bekannt sein. Aufgrund der Abnahme des mittleren Grauwerts mit der Entfernung ist eine Volumenbegrenzung durch eine untere Grauwertgrenze möglich. Die Grauwertabnahme mit der Entfernung ist bei großen Blasen geringer als bei kleinen Blasen. Diese Tatsache bedingt ein mit der Blasengröße anwachsendes Meßvolumen. Das

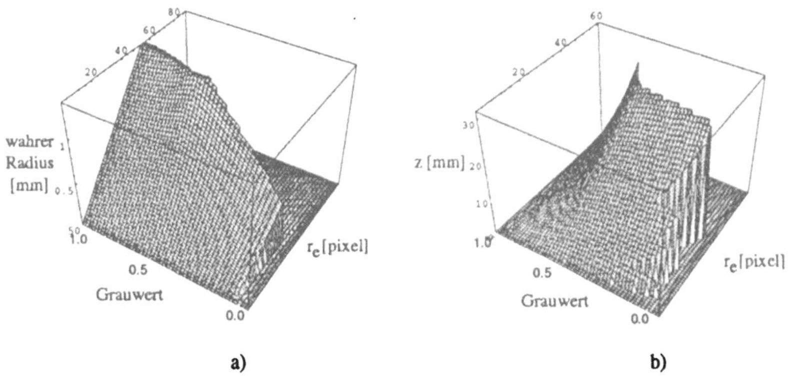

Abb. 6. *Ausschnitte aus der Radius- und Entfernungs-LUT für $z > 0$*

ist insofern wünschenswert, als die Anzahl der großen Blasen erheblich kleiner als die der kleinen Blasen ist. Durch das größenabhängige Meßvolumen erzielt man eine verbesserte Statistik bei großen Blasen. Die Bilder zweier Blasen besitzen denselben mittleren Grauwert g_e, wenn in beiden Fällen das Verhältnis der Ausdehnung von PSF zu geometrisch vergrößertem Objekt dasselbe ist:

$$\frac{rV_g(z)}{V_p(z)} = const \Leftrightarrow g_e = const \tag{5}$$

Bezeichnet man die Konstante mit $\gamma(g_e)$, so werden die Volumengrenzen für den minimalen Grauwert $g_e^\star$ durch

$$z_{gr} = \gamma(g_e^\star)\,\nu^{-1}(r) \quad mit \quad \nu(r) = \frac{V_p(z)}{V_g(z)} \tag{6}$$

beschrieben. Diese Gleichung besitzt im Allgemeinen verschiedene Lösungen für positive und negative z_{gr}. Bei der verwendeten Optik ist $\nu(z) \sim |z|$ und damit $|z_{gr}| = \gamma(g_e^\star)r$, so daß die Volumengrenzen symmetrisch sind. Das Volumen ist dann proportional zum Blasenradius r.

5 Ergebnisse

Mit dem beschriebenen Verfahren kann die Größenverteilung von Gasblasen im Wasser gemessen werden. Ein Beispiel ist in Abb. 7 dargestellt. Sie zeigt die Abhängigkeit der Anzahl der pro Radiusintervall und Volumeneinheit vorhandenen Luftblasen vom Radius bei einer hoher Windgeschwindigkeit. Deutlich ist die starke Abnahme der Blasenhäufigkeit mit wachsenden Radius zu erkennen.

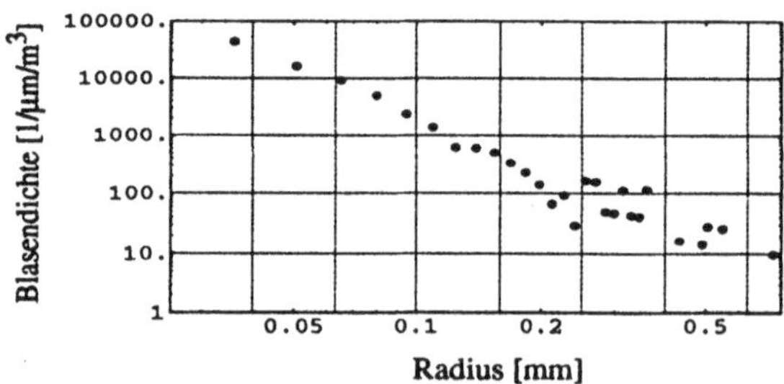

Abb. 7. *Größenverteilung von Luftblasen im Wasser. Die Daten wurden bei einer Windgeschwindkeit von 14 m/s aufgenommen.*

Die Werte ergaben sich aus der Analyse von 600 Bildern. Da die LUTs nur einmal berechnet werden müssen, reduziert sich die Depth-from-Focus Analyse während der Auswertung auf das Nachschauen der Werte in den Tabellen und nimmt daher kaum Zeit in Anspruch. Insgesamt ist die Segmentierung der Blasen der zeitaufwendigste Bearbeitungsschritt. Dazu wurde ein kombiniertes Schwellwert/Regionenwachstumsverfahren eingesetzt, das mit nur wenigen Keimpunkten pro Objekt auskommt und daher eine hohe Geschwindigkeit erreicht. So konnte eine Rechenzeit von nur etwa 1 bis 2 Sekunden pro Bild erreicht werden[4].

Mit dem beschriebenen Verfahren konnte Depth-from-Focus erfolgreich zur Messung physikalischer Größen eingesetzt werden. Die Volumenbegrenzung durch gezielte Unschärfe weist gegenüber anderen Verfahren deutliche Vorteile auf. Da das Verfahren mit nur einem Bild jeder Szene auskommt, konnte ein kompakter Meßaufbau mit einer Kamera realisiert werden. Eine Weiterentwicklung der Blasensonde soll auch zu Feldmessungen auf einer auf dem Ozean treibenden Boje eingesetzt werden. Angestrebt wird auch eine on-line-Auswertung ohne Zwischenspeicherung der Bilder auf Videoband.

Danksagung: Wir bedanken uns bei der National Science Foundation, dem Office of Naval Research und der Europäischen Gemeinschaft (Large Installation Program) für die finanzielle Unterstützung des Projekts. Weiterhin danken wir dem Forschungszentrum der Asea Brown Boveri AG für die im Rahmen des Projekts 'In Situ Mikroskopie' erfolgte Unterstützung.

Literaturverzeichnis

Bove,V.M, Jr.: *Discrete Fourier transform based depth-from-focus.* Image Understanding and Machine Vision 1989. Tech. Dig. Series, Opt. Soc. America, Vol. 14, 1989

Darell,T., Wohn, K.: *Depth from Focus using a pyramid architecture.* Pattern Recognition Letters 11 pp.787-796, 1990

Ens, J., Lawrence, P.: *An Investigation of Methods for Determining Depth from Focus.* PAMI Vol. 15, No. 2 , 1993

Farmer, D. M, McNeil, C. L., Johnson, B. D.: *Evidence for the importance of bubbles in increasing air-sea gas flux.* Nature Vol. 361, pp.620 - 623, Feb.1993

Jähne, B., Wais, T., Barabas, M.: *A New Optical Bubble Measuring Device; A Simple Model for Bubble Contribution to Gas Exchange.* in: Gas Transfer at Water Surfaces 1984

Johnson, D., Cooke, R. C.: *Bubble Populations and Spectra in coastal Waters: A photographic Approach.* Jour. of Geophysical Res. Vol.84 , pp.3761 - 3766, 1979

Lamarre, E., Melville,W. K. : *Air Entrainment and Dissipation in breaking waves.* Nature Vol.351, pp.469 - 742, 1991

Melville, W. K. , M.R. Loewen, Eric Lamare, E.: *Bubbles, Noise and breaking waves; a review of laboratory experiments.* Natural Physical Scources of Underwater Sound, pp.483-501, Kluwer Academic Publisher, 1993

Merlivat, L., Memery, L.: *Gas exchange across an air-water interface: Experimental results and modelling of bubble contribution to transfer.* Jour. of Geophysical Res. Vol.88, pp.707 - 724, 1983

Pentland, A. P.: *A new Sense for Depth of Field.* PAMI Vol. 9, 1987

Suhr, H., Speil, P., Wehnert, G., Sorhas, H.: *In Situ Mikroskopsonde und Meßverfahren zur Analyse bewegter und ruhender Partikel in Bioreaktoren.* Deutsche Offenlegschrift 4032002, 1990

---

[4] Das Verfahren wurde auf einem i860 Risc Prozessorboard mit integriertem Framegrabber implementiert

THE MHD MEMORY

B. Klauer
J.W.Goethe - Universität Frankfurt
Technische Informatik, Prof. Dr. K. Waldschmidt
Robert Mayer Straße 11 - 15
D - 60054 Frankfurt am Main

1. Abstract

This paper introduces a full-parallel associative pattern recognition engine. It uses the Masked Hamming-Distance as classifier to identify an unknown pattern by locating the most similar pattern stored in it's pattern database. The Masked Hamming-Distance function (MHD) has been introduced in [KL1]. It has then been implemented in C++ and prooven to be suitable for handprinted character recognition [KL2]. A special version has been prepared for the requirements of pen computers [KL3]. It is distributed as recognition engine (PenStar HWR®) for the INFOS 386-SX notepad computer®. The MHD-based character recognition method is expected to be suitable for a varity of other pattern classification problems. A speech recognition method using the MHD function is currently under investigation. The full-parallel MHD memory architecture is the final result of a migration process of the pure software implementation into hardware. The hardware architecture will be shown in this paper preceded by a brief introduction into the MHD pattern recognition method. The MHD memory is a storage device for boolean patterns combined with a special retrieval mechanism to identify unknown patterns. It will be shown how to perform the classification as well as the computation of the best matching database item in parallel. An emulation of the MHD memory has been implemented with TMS320C31 RISC processors. A heterogeneous analog/digital approach for a full parallel implementation of the memory will be proposed in this paper.

2. The MHD pattern recognition method

The MHD pattern recognition method has been discribed in [KL1] in detail. This chapter summarizes the method proposed in that paper to give a background for the hardware architecture in the next chapter.

2.1 The Hamming Distance

The Hamming Distance of two patterns has originally been proposed by Hamming [HAM] as error-correction mechanism for binary codes. Let A and B be boolean vectors. HD(A,B) is the total of all bits a_i,b_i with $a_i \neq b_i$. Within a discussion of similarity measures Kohonen discribes it to be "perhaps the best known measure between digital representations" [KOH].

$$HD(A,B) = \Sigma\, (a_i, \oplus b_i)$$

$\oplus$: boolean EXOR

2.2 The Masked Hamming Distance

The Masked Hamming Distance MHD of two patterns A,B and a mask M is defined as follows:

$MHD(A,B,M) = \Sigma (a_i \oplus b_i \bullet \neg m_i)$
$\bullet$: boolean AND
$\neg$: boolean NOT

MHD behaves as HD if $M = \underline{0}$. If $M \neq \underline{0}$ MHD computes the total of all different elements a_i,b_i with $a_i \neq b_i$ which are not hidden by m_i. HD as well as MHD can be used as similarity indicators in recognition systems. For the recognition of handprinted characters we have shown that MHD is much more suitable than HD [KL1].

2.3 The MHD Method

An image I = (BP,s) be a tupel consisting of a boolean pattern BP and the semantics s of the pattern. All well known symbols in a database can be represented as shown in the example. ϕ is used to indicate that the semantics of a symbol is unknown. The following method can be used to find the semantics of an unknown image $U=(X,\phi)$:

Compute the mask argument M
Compute $MHD(X,P_i,M)$ for all i
Replace ϕ with s_i if $MHD(X,P_i,M)$ is minimal for all i

2.4 The masking argument for the MHD method

The Masked Hamming-Distance distinguishes from the pure Hamming-Distance in the simple fact that specific areas of the patterns to be compared can be hidden by the masking argument M. The MHD classifier computes the Hamming-Distance of all bits which are not masked. This special feature can be used for fault-tolerant classification. An efficient method to compute the mask argument for handprinted characters has been shown in [KL1]. It has been prooven suitable for character-recognition. Methods for speech recognition are currently under investigation.
Example: Figure 1 shows two "+" symbols which are similar but not equal. A recognition system holding one of both symbols in it's database should classify them to be similar. The Hamming-Distance of pattern A and B is 28, indicating that both patterns are different - even if they are not from a human point of view. The masked Hamming-Distance of A and B and a mask argument as shown in figure 1 is 13, indicating that both patterns are similar (not equal). This result is more adequate to the human impression. The next example (figure 2) shows that the MHD classifier works also in case that both symbols are different. In this case MHD(A,C,Mask) = 24, indicating that both symbols are different.

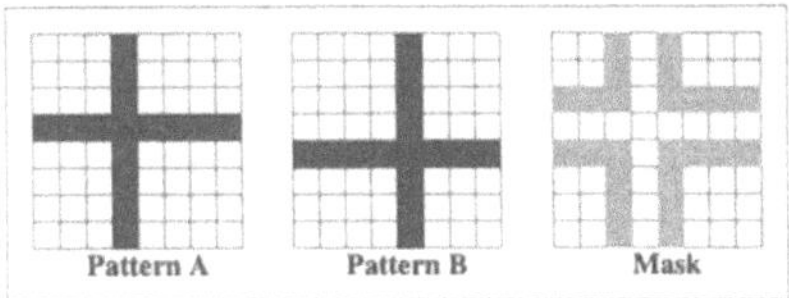

figure 1

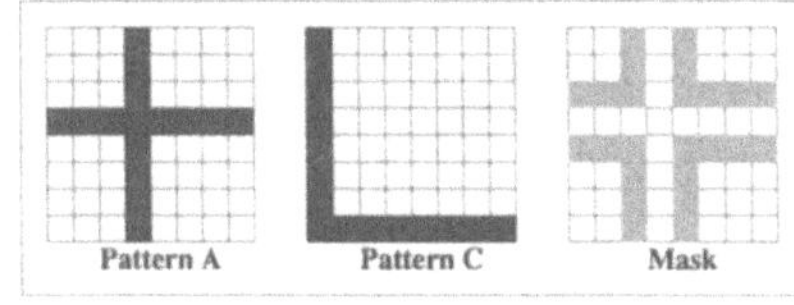

figure 2

3. The MHD memory

3.1 Migration into hardware

The first step in the migration process was the evaluation of the MHD-method on a 33MHz 486 computer. The next step was the developement of a special coprocessor-board for PCs based on the TMS 32C031 RISC processor running at 40 MHz [HEN]. A pattern recognition engine supported by the TMS processor performs approximately 10 times faster than a 33MHz 486 machine. It has been used as fast sequential pattern-matcher to proove the hardware interface and to decide:

3.1.1 Is full-parallel hardware really neccessary ?

The recognition time within the sequential implementations grows linear with the size of the pattern database. The size of the database depends on the size of the patterns and on the count of patterns. The database size is tightly related with the recognition accuracy. High recognition quality reqires in most cases large databases. High speed requires small databases - or high performance. Performance can be increased by physical acceleration (clock frequency) or architectural speed-up (parallelism). The MHD memory has been developed to combine a short physical memory access time with high grade parallelism. The performance of the sequential implementation was adequate for the recognition of hand or machine printed characters with high accuracy (>90% with the pure pattern matcher without ambiguity resolver, >97% together with the ambiguity-resolver) in reasonable time. Due to the high sample rate required for speech recognition [RAS,RUS] and large database sizes the performance of the sequential implementations is insufficient to gain a reasonable response time.

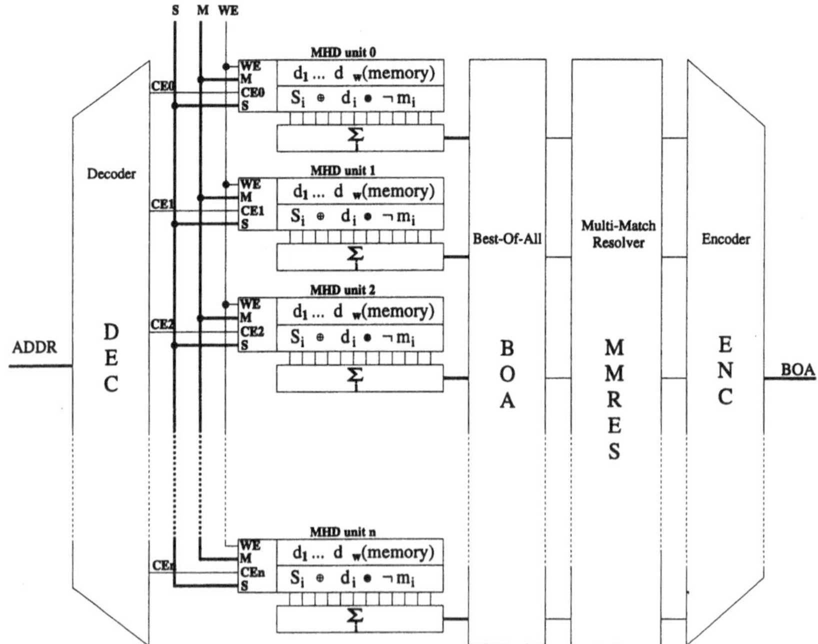

figure 3: Architecture of the MHD memory

3.2 Global architecture of the MHD memory

The architecture of the MHD memory is shown in figure 3. It is based on a standard static RAM. Each RAM cell is embedded into a so-called MHD unit. An MHD unit consists of a RAM memory cell and a logical unit to compute ($d_i \oplus s_i \bullet \neg m_i$) for all i. d_i is one bit within the RAM-cell D. s_i and m_i are bits from the search and mask arguments S and M. The Σ unit computes the Hamming-weight of the result computed by the MHD unit. One MHD unit together with one Σ unit is provided for each RAM cell instead of a sequential computation on a single arithmetical logical unit. The BOA (**B**est **O**f **A**ll) unit evaluates the totals computed by the Σ units. It retrieves the highest of all totals. A multi match resolver is provided to separate matches in case that two patterns in the memory have the same Masked Hamming Distance to a search pattern. An encoder is then used to compute the binary address of the best matching memory cell. This address corresponds with the semantics of a pattern.

3.2.1 The MHD unit

The MHD unit consists of a standard static RAM cell and the MHD evaluation logic. The MHD function $MHD(D,S,M) = \Sigma((d_i \oplus s_i) \bullet \neg m_i)$ is disassembled into a logical part $((a_i \oplus s_i) \bullet \neg m_i)$ and an arithmetical part Σ computing the total for all i. The logical part of the MHD function between a_i coming from the storage cell, s_i provided by the S BUS (Search argument) and m_i provided by the M BUS (Mask argument) is then computed using a XOR gate and an AND gate. An OR gate is used to evaluate the OE signal. The output of the logical part is propagated only if OE = 0. The total for all i is then computed in an analog circuit. A MHD unit is shown in figure 4.

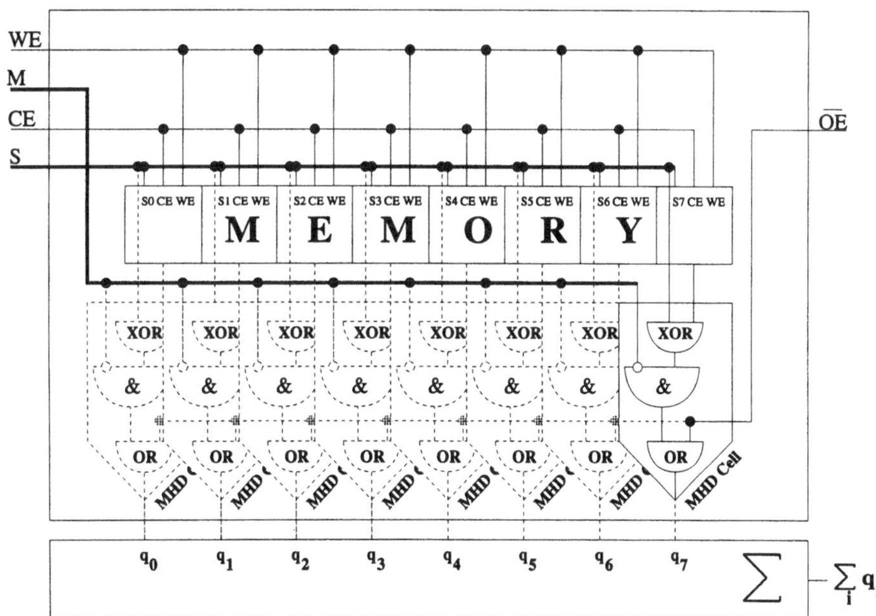

figure 4: The MHD unit

3.2.2 The BOA-unit

The BOA unit finds the lowest of all totals from the MHD units. It has one input for each MHD unit and one corresponding output. The output indicates whether the corresponding input from the MHD unit is the lowest of all inputs. The BOA circuit could be implemented as network of comparators. A proposal for an analog implementation of the BOA circuit with a significantly lower complexity will be given in the next chapter.

3.2.3 The Multi-Match-Resolver

The task of the Multi-Match resolver provides a mechanism to separate multiple matches for sequential retrieval and to support the computation of the best n matches. Multiple matches can occur if the MHD function returns the same value for two different patterns or in case of resolution errors in the Σ-BOA unit. The architecture of a suitable multi-match resolver is given in [TAV]. Tavagarian uses the multi-match resolver in the ARAM Architecture for separating multiple matches computed by the masked identity or relational search functions. A match signal is propagated to an output only if there is no match with higher priority. The match with the highest priority is computed using AND gates as shown in figure 5.

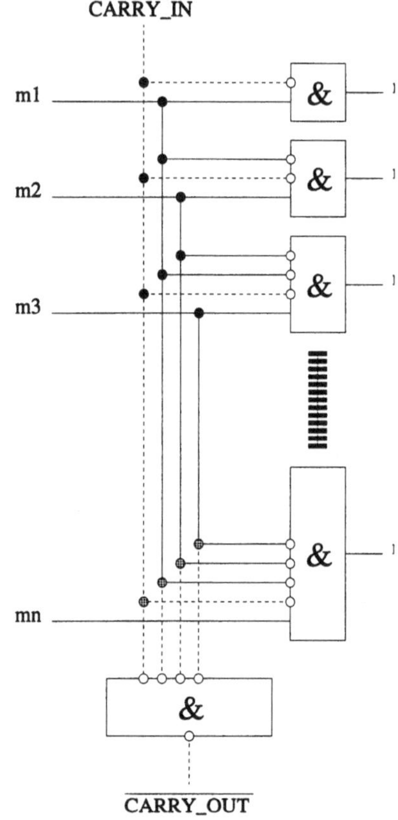

figure 5: The Multi-Match-Resolver

4. Implementations of the MHD memory

After a performance evaluation on a behavioural simulation the MHD-memory was implemented as sequential emulation based on the TMS 320C31 signal processor. A concept for a full-parallel VLSI implementation of the MHD memory has been developed.

4.1 The TMS 320C31 implementation

The TMS 320C31 implementation is a subsystem consisting of a processor with RAM and ROM. It is mapped into the I/O space of the PC. The bus systems of the coprocessor and the PC are connected via fast FIFOs. Up to four coprocessors with a storage capacity of 256KB each can be used in a PC. This mixed Sequential/parallel approach has the following advantages:

- Up to four coprocessors can work concurrently.
- Up to 1MB storage space can be provided for a pattern database without ocupying internal RAM of the PC.
- Of-the-shelf components can be used.

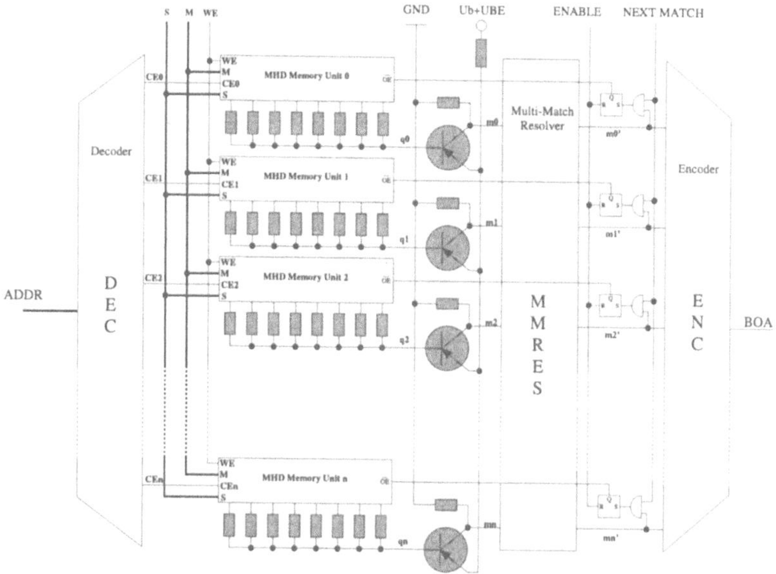

figure 6: The full parallel MHD memory
The components of the Σ-BOA-unit are shaded grey

4.2 The concept for a full-parallel MHD memory implementation

The concept for a full-parallel implementation of the MHD memory is based on the following idea: The Σ unit and the BOA unit are replaced by an analog circuit performing both functions. This so-called Σ-BOA unit is an analog network computing the total of matches within the MHD units and making the decision which of the cells contains the best matching pattern.

4.2.1 The analog Σ-BOA-unit (Sum and Best OF ALL)

Figure 6 shows the full parallel MHD memory. The analog Σ-BOA unit (shaded grey) replaces the Σ units and the BOA circuit. It totals all matching bits for a data word in the memory and selects the best of all database items as in the digital approach. The resistor network is used to compute the total of matching bits.The ECL circuit is used to determine the memory cell with the best matching contents concurrently with the computation of the total of all matching bits. The voltage at the nodes q_0 ... q_n depends on the ratio of the high and low outputs of the MHD units. MHD unit l be the unit containing the best matching pattern due to a search/mask argument. q_l is then the node with the lowest voltage U_l. In this case U_l causes a base-emitter current I_l in T_l . The current causes the transistor to be open and a voltage U_E at the coupled emitters. Since U_l was assumed to be the lowest of all base-voltages of the transistors all other base-voltages remain higher than: U_E - U_{BE}. This means that all transistors - but T_l remain in an high impedance state. The index l indicated by a high level at ouput m_l is then filtered by a muti-match resolver and then directly encoded into it's binary equivalent, corresponding to the address of the MHD unit holding the best matching data.

4.2.2 A write cycle

A write cycle to store data in the MHD memory is similar to a write cycle for a standard static RAM. The data to be written into the memory is applied to the S-BUS (DATA-BUS in standard static RAMs). The address is applied to the ADDR-BUS. The MHD-unit with the corresponding address is then activated by the decoder. The data is transferred into the storage cells with a WE (write enable) signal. Due to the static behaviour of the storage cells their contents is permanently applied to the MHD cells within the MHD unit.

4.2.3 A best-match cycle

The match cycle of the MHD memory corresponds with the read cycle in a standard RAM with the difference that it selects the data in the storage-cells by evaluating their contents instead of address related information. The data of all storage cells is evaluated in the MHD units and the Σ-BOA unit in parallel. To match an unknown pattern with the data stored in the memory all the RS latches controlled by the MMRES unit must be reset by triggering the ENABLE ALL signal to enable all MHD units. The unknown pattern must then be applied to the S-BUS and the corresponding mask to the M-BUS. The contents of the static memory cells, the unknown pattern and the mask are now matched concurrently in all MHD units. The output of all MHD units is evaluated in the S-BOA circuit concurrently for all units. The result of the Σ-BOA unit is filtered by the multi-match-resolver MMRES to be shure that only one match is generated as input for the encoder. The output of the encoder is the binary encoded signal BOA (Best Of All) representing the address of the MHD unit holding the best matching data due to the Masked Hamming Distance criterion for similarity. The address contained in the BOA signal can then be used as a pointer into a lookup table to find the semantics of the stored data.

4.2.4 A next-match cycle

The MHD memory is not restricted to retrieving the best matching pattern only. It can also support the retrieval of the next-best pattern. After retrieving the best matching pattern the latches at the outputs of the MMRES unit can be triggered by the NEXT MATCH signal to disable the best matching MHD unit. The S and D input must be left unchanged. The Σ-BOA unit repeats the best-match cycle as mentioned above, ignoring the previously identified best-matching unit, which has been disabled. This procedure can be repeated to compute a sorted list of the n best-matching patterns in descending order.

5. The architecture of a general pattern recognizer using the MHD method

The following architecture has been prooven suitable and very efficient in speed and quality for the recognition of handprinted characters. It is currently under investigation for cursive character recognition and speech recognition. It demonstrates how the pattern-matcher can be embedded into pattern recognizer and into applications. The proposed architecture consists of three modules a preprocessor, a pattern-matcher and an ambiguity resolver as shown in figure 7. The preprocessor module contains an extractor to provide the pattern matcher and the ambiguity resolver with significant information on the symbol to be recognized. It works "online"

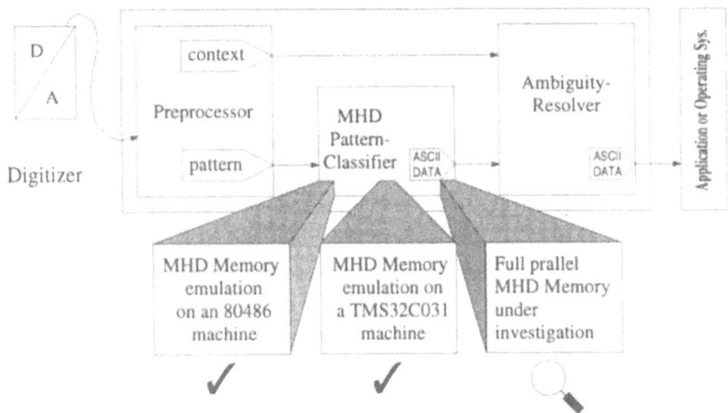

figure 7: A pattern matcher using the MHD method

during the aquisition phase. It passes a boolean pattern to the pattern matcher and context-information to the ambiguity resolver. The pattern matcher is a very fast pre-classifier but in case of similar patterns e.g. the phonetic equivalent of the words "try" and "dry" it might be insufficient in recognition accuracy. It passes it's result to the ambiguity resolver module to detect - and to resolve ambiguities as mentioned above. How ambiguities are resolved depends on the specific recognition problem. An ambiguity resolver for speech recognition could for example use the phonetic context. In the sentences: "The desert is _ry" and "_ry to write legible" the substitution of the _ can be decided by the context. Specific pattern recognition modules using this model distinguish in their preprocessors and ambiguity resolvers.

6. A view to the future

The full-parallel MHD memory as proposed in this paper is currently under simulation to proove the functions of the MHD memory and to find optimal parameters for the transistors and resistors. The VLSI realisation of the memory is planned after the evaluation of the simulation results.The sequential pattern matcher currently used for character- and speech recognition can then be replaced by the full-parallel hardware component.

References:

[HEN] Entwicklung eines Assoziativspeichers zur Mustererkennung, Diplomarbeit, J.W. Goethe-Universität Frankfurt, Technische Informatik, 1993

[KL1] Pen Based Recognizing of Handprinted Characters, Proc. of the EUROMICRO '93, Elsevier, 1993

[KL2] An Object-Oriented Pen-Based Recognizer for Handprinted Characters, Proc. of the CAIP '93, Springer 1993

[KL3] An Object-Oriented Character Recognition Engine for Pen-Based computers, Proc. of the EuroARCH '93, Springer, 1993

[KOH] T. Kohonen, Content-Addressable Memories,Springer, 1987

[TAV] D. Tavangarian, Flagorientierte Assoziativspeicher und Prozessoren, Springer, 1990

[RAS] L.R. Rabiner W. Schafer, Digital processing of speech signals, Prentice-Hall, 1978

[RUS] G. Ruske, Automatische Spracherkennung, Oldenbourg, 1988

Texturklassifikation mit Polynomialverteilungen

Gabriele Lohmann

Technische Universität München
Institut für Informatik – Lehrstuhl Prof. Radig
e-mail: lohmanng@informatik.tu-muenchen.de

Abstract. In diesem Beitrag wird ein neues Verfahren zur überwachten Klassifikation von Texturen vorgestellt. Bei diesen Verfahren werden Cooccurrence Matrizen als Merkmalsvektoren verwendet. Es wird gezeigt, daß diese Merkmalsvektoren als polynomialverteilte Zufallsvariablen aufgefasst werden können. Die polynomialen Dichtefunktionen werden eingesetzt, um Gewichtsfunktionen für die Klassifizierung zu erhalten.

1 Einleitung

In diesem Beitrag wird ein neues Verfahren zur überwachten Klassifikation von Texturen vorgestellt. Die Idee, die diesem Verfahren zugrundliegt, beruht darauf, Cooccurrence Matrizen als polynomialverteilte Zufallsvariablen aufzufassen. Wir verwenden diese Verteilungsfunktionen, um die Stärke der Evidenz zu gewichten, die durch die Cooccurrence Matrizen (im folgenden "CM" genannt) gegeben ist.

Bisherige CM-basierte Verfahren führen üblicherweise eine erhebliche Informationsreduktion auf den CMs durch, so daß nicht die CM selber, sondern sekundäre, aus der CM abgeleitete Merkmale die Klassifizierung bestimmen. Im Gegensatz dazu wird hier die CM selber als Merkmalsvektor verwendet. Der Vorteil dabei ist, daß erheblich mehr Information in die Klassifizierungsentscheidung eingeht. Trotzdem ist aber der Rechenaufwand insgesamt sogar geringer als bei der sonst üblichen Verwendung der CM. Ein weiterer Vorteil dieses neuen Verfahrens ist darüberhinaus die erheblich einfachere Parametrisierung.

2 Cooccurrence Matrizen

Cooccurrence Matrizen werden häufig zur Beschreibung und Klassifikation von Texturen verwendet. In diesem Abschnitt werden die wichtigsten Begriffe, die im folgenden verwendet werden, zusammengestellt. Ausführlichere Darstellungen über CMs finden sich z.B. in [2], [3].

Eine CM ist eine Art Histogramm, in dem nicht die Häufigkeiten von einzelnen Grauwerten dargestellt sind, sondern die Häufigkeiten von Paaren von Pixeln, die in einer bestimmten räumlichen Beziehung zueinander stehen. Man betrachte Paare von Pixeln, die im Abstand d und Winkel α voneinander getrennt sind. Dabei sind die $(d = 1, \alpha = 0)$-Paare horizontal benachbart, die $(d = 1, \alpha = 90)$-Paare sind vertikal benachbart, die $(d = 1, \alpha = 45)$-Paare sind links-diagonal benachbart und die $(d = 1, \alpha = 0)$-Paare sind rechts-diagonal benachbart.

Ferner sei m die Anzahl der verschiedenen Graustufen des Bildes. Eine (d, α)-CM C ist dann eine $m \times m$-Matrix, deren Einträge c_{ij} die Anzahl derjenigen Pixelpaare enthalten, deren Pixel mit Abstand d und Winkel α voneinander entfernt sind und die die Grauwerte i und j haben.

Von Interesse ist in unserem Kontext auch die normalisierte Version der Cooccurrence Matrix. Da die CMs symmetrisch sind, ist der Normalisierungsfaktor hier die Summe über die untere Dreiecksmatrix, so daß

$$\sum_{i=0}^{m-1} \sum_{j=0}^{i} c_{ij} = 1.$$

Folgendes Beispiel einer $(d = 1, \alpha = 0^0)$-CM illustriert die obigen Definitionen:

Bild		CM				normalisierte CM		
			0	1			0	1
○ ● ○								
● ● ●		0	2	3		0	1/3	1/2
○ ○ ●		1	3	1		1	1/2	1/6

3 Die Polynomialverteilung

Im folgenden wird dargestellt, daß Texturen mit Hilfe von Polynomialverteilungen und Cooccurrence Matrizen modelliert werden können.

Es sei ein Bild beliebiger Größe gegeben, das mit m Graustufen digitalisiert ist. Wir betrachten alle Paare von Pixeln in diesem Bild, die in einer (d, α)-Nachbarschaft stehen, also z.B. horizontal benachbarte Paare. Ferner sei q_{ij} der Anteil dieser Pixelpaare, die Grauwerte i und j haben.

Wir wählen nun n solcher Pixelpaare zufällig aus diesem Bild aus. Wir interessieren uns dabei für die Wahrscheinlichkeit, daß unter diesen n benachbarten Paaren genau u_{ij} Paare sind, die die Grauwerte i und j haben. Situationen dieser Art lassen sich durch Polynomialverteilungen modellieren. In diesem Kontext erhalten wir damit folgende Wahrscheinlichkeitsdichte:

$$p(u) = n! \prod_{i=0}^{m-1} \prod_{j=0}^{i} \frac{q_{ij}^{u_{ij}}}{u_{ij}!}$$

wobei

$$\sum_{i=0}^{m-1} \sum_{j=0}^{i} q_{ij} = 1,$$

und

$$\sum_{i=0}^{m-1} \sum_{j=0}^{i} u_{ij} = n.$$

Beispiel:
Es sei ein Bild mit den 2 Graustufen "weiß" und "schwarz" gegeben, in dem 70 Prozent aller horizontal benachbarten Pixel unterschiedliche Grauwerte haben, und 30 Prozent aller horizontal benachbarten Pixel schwarz sind. Die Wahrscheinlichkeit, bei 10 zufällig ausgewählten Paaren 7 Paare mit unterschiedlichen Grauwerten anzutreffen und 3 Paare, bei denen beide Pixel schwarz sind, ist:

$$p(u_{00} = 3, u_{10} = 7, u_{11} = 0) = \frac{10!}{3!\,7!\,0!}\, 0.3^3\, 0.7^7\, 0.0^0 \approx 0.2668.$$

Im folgenden werden wir diese Grundidee verwenden, um einen neuen Texturklassifikator zu konstruieren.

4 Gewichtsfunktionen

In diesem Abschnitt werden Gewichtsfunktionen eingeführt, die in vielen Klassifizierungsproblemen anwendbar sind. Es stellt sich heraus, daß diese Gewichtsfunktionen für die Texturklassifikation besonders gut geeignet sind. Diese Funktionen sind bereits in [4],[5] dargestellt worden, so daß wir uns hier auf eine kurze Zusammenfassung beschränken.

Sei x ein Merkmalsvektor und ω eine zu identifizierende Objektklasse. Der Vektor x kann eine Klassifizierungshypothese stützen oder sie schwächen. Wenn die Hypothese durch x unterstützt wird, so erwarten wir, daß $p(\omega|x) > p(\omega)$.

Wenn umgekehrt x die Hypothese weniger wahrscheinlich erscheinen läßt, so ist $p(\omega|x) < p(\omega)$. Der Logarithmus der Ratio von $p(\omega|x)$ und $p(\omega)$

$$I(x, \omega) = \log \frac{p(\omega|x)}{p(\omega)}$$

ist also ein Maß für die Stärke der Evidenz, die der Merkmalsverktor x in Bezug auf die Hypothese ω darstellt. Dieser Ausdruck ist in der Informationstheorie unter der Bezeichnung "gegenseitige Information" (engl. "mutual information") bekannt (cf.[1]).

Wenn $I(x, \omega) > 0$, dann wird die Hypothese ω durch x gestützt. Wenn $I(x, \omega) < 0$, dann wird die Hypothese ω durch x geschwächt. Wenn $I(x, \omega) = 0$, so liefert x keine Information bzgl. ω. Man beachte, daß

$$\log \frac{p(\omega|x)}{p(\omega)} = \log \frac{p(\omega, x)}{p(\omega)p(x)} = \log \frac{p(x|\omega)}{p(x)}.$$

Die Berechnung von $I(x, \omega)$ erfordert also nicht die Kenntnis von $p(\omega|x)$ oder $p(\omega)$. Statt dessen muß aber eine Approximation an $p(x)$ und an $p(x|\omega)$ bekannt sein. Im folgenden werden wir zeigen, daß $I(x, \omega)$ sehr einfach berechnet werden kann, wenn x ein Merkmalsvektor ist, der auf einer Cooccurrence Matrix basiert.

5 CM-basierte Gewichtsfunktionen

In diesem Abschnitt setzen wir die oben eingeführten Gewichtsfunktionen zur Texturklassifikation ein. Die Merkmalsvektoren x sind hierbei Cooccurrence Matrizen, deren Wahrscheinlichkeitsdichten durch Polynomialverteilungen modelliert werden können. Wir gelangen damit zu einem neuen Verfahren zur Texturklassifikation.

Sei ω eine zu identifizierende Objektklasse, die Teile eines größeren Bildes bedeckt. Ferner sei p_{ij} die normalisierte Cooccurrence Matrix von ω, und sei q_{ij} die normalisierte Cooccurrence Matrix des gesamten Bildes. Wir betrachten nun kleine Bildausschnitte der Größe $s \times s$ Pixel, wobei im allgemeinen $5 \leq s \leq 15$.

In dieser Umgebung betrachten wir nun n Pixelpaare, die (d, α)-benachbart sind, und die sich nicht überlappen. Zum Beispiel ergeben sich in einer 5×5-Umgebung $n = 10$ horizontal benachbarte, nicht-überlappende Paare (siehe Fig. 1). Man kann zwei verschiedene Gruppen C_1 und C_2 von Pixelpaaren, in denen die Paare jeweils unabhängig, d.h. nicht überlappend sind, identifizieren (siehe Fig. 1).

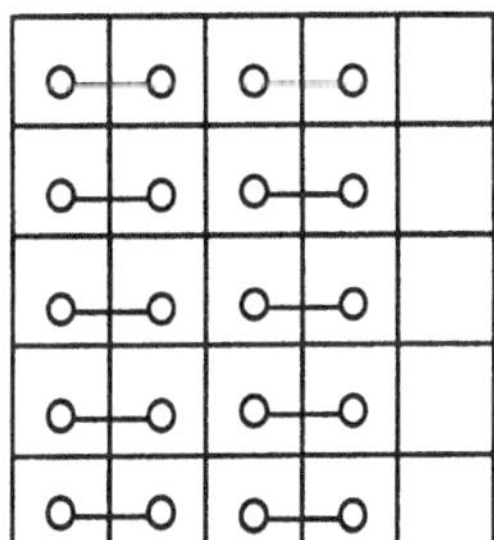
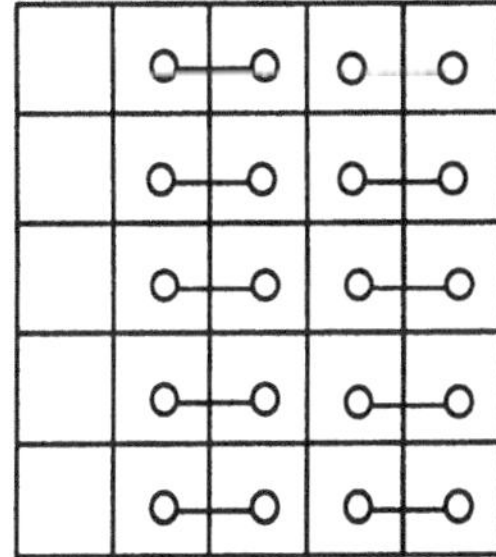

Fig. 1: Die Gruppen C_1 und C_2 in einer 5×5-Umgebung

In jeder dieser beiden Gruppen C_1 und C_2 berechnen wir nun die nicht-normalisierten CMs u_1 und u_2. Da die Paare innnerhalb jeder Gruppe unabhängig sind, können wir nach unseren Überlegungen aus Abschnitt 3 damit sowohl $p(u_k)$ als auch $p(u_k|\omega), k = 1, 2$ wie folgt berechnen:

$$p(u_k|\omega) = n! \prod_{i=0}^{m-1} \prod_{j=0}^{i} \frac{p_{ij}^{u_{k,ij}}}{u_{k,ij}!}$$

$$p(u_k) = n! \prod_{i=0}^{m-1} \prod_{j=0}^{i} \frac{q_{ij}^{u_{k,ij}}}{u_{k,ij}!}$$

Und somit:

$$\log \frac{p(u_k|\omega)}{p(u_k)} = \log \frac{\prod_{i=0}^{m-1} \prod_{j=0}^{i} p_{ij}^{u_{k,ij}}}{\prod_{i=0}^{m-1} \prod_{j=0}^{i} q_{ij}^{u_{k,ij}}}$$

$$= \sum_{i=0}^{m-1} \sum_{j=0}^{i} u_{k,ij} \left(\log p_{ij} - \log q_{ij}\right).$$

Wir fassen nun die Evidenz, die aus den beiden Gruppen stammt, durch Mittelwertbildung wie folgt zusammen:

$$0.5\,(\log\frac{p(u_1|\omega)}{p(u_1)}+\log\frac{p(u_2|\omega)}{p(u_2)}) \;=\; 0.5\,\log\frac{p(u_1,u_2|\omega)}{p(u_1,u_2)} \;=\; 0.5\,\log\frac{p(u|\omega)}{p(u)}$$

Die Berechnung dieses Mittelwertes ist hier etwas verkürzt dargestellt. Die Details lassen sich aber sehr leicht nachrechnen. Man beachte, daß u die nicht-normalisierte CM der gesamten $s \times s$-Umgebung ist, und $u_{ij} = u_{1,ij} + u_{2,ij}$. Insgesamt erhalten wir also den Term

$$0.5\,\log\frac{p(u|\omega)}{p(u)} \;=\; 0.5\sum_{i=0}^{m-1}\sum_{j=0}^{i} u_{ij}\,(\log p_{ij} - \log q_{ij})$$

als ein Maß für die Stärke der Evidenz, die die nicht-normalisierte CM u_{ij} bezüglich der Hypothese ω liefert.

Dieser Term kann sehr einfach berechnet werden. Die normalisierten CMs p_{ij} und q_{ij} werden in einer Trainingsphase zu Beginn bestimmt. Die nicht-normalisierte CM u_{ij} wird jeweils in einem $s \times s$ Fenster berechnet, das im Laufe der Verarbeitung über das Bild geschoben wird. Je nach Körnigkeit der Textur sind dabei Fenstergrößen von 5×5 bis 15×15 sinnvoll.

Man beachte, daß $\log p_{ij}$ für $p_{ij} = 0$ nicht definiert ist. Um dieses Problem zu vermeiden, werden die CMs p_{ij} und q_{ij} mit einem Gaussfilter geglättet, so daß $p_{ij}, q_{ij} > 0$ für alle i, j.

6 Addition von Gewichten

Im allgemeinen wollen wir für die Klassifizierung nicht nur eine, sondern mehrere Arten von Nachbarschaftsrelationen verwenden, z.B. horizontale und vertikale Nachbarschaften. In diesem Fall muß mehr als eine Cooccurrence Matrix in die Berechnung der Gewichtsfunktion eingehen. In diesem Abschnitt werden wir zeigen, daß es ausreicht, die Gewichte der einzelnen CMs zu addieren.

Seien also x und y zwei verschiedene CMs. Dann können wir x und y als Realisierungen zweier Zufallsvariablen X und Y auffassen. Falls X und Y voneinander unabhängig sind, und auch konditionional unabhängig bzgl. ω sind, so gilt:

$$\begin{aligned} I(x,y,\omega) &= \log\frac{p(x,y|\omega)}{p(x,y)} \\ &= \log\frac{p(x|\omega)p(y|\omega)}{p(x)p(y)} \\ &= \log\frac{p(x|\omega)}{p(x)} + \log\frac{p(y|\omega)}{p(y)} \\ &= I(x,\omega) + I(y,\omega) \end{aligned}$$

Die Annahme der Unabhängigkeit ist zwar i.a. unrealistisch, die Praxis zeigt jedoch, daß der dadurch verursachte Fehler tolerabel ist..Insgesamt wird hierbei der Absolutbetrag der Gewichte zu groß angenommen. Jedoch betrifft dieser Effekt alle Klassen in etwa gleichermaßen, so daß das Klassifizierungsergebnis nur geringfügig verfälscht wird.

7 Der Algorithmus

Aus den Überlegungen der vorangegangenen Abschnitte ergibt sich folgender Algorithmus:

Trainingsphase:

1. für alle Nachbarschaftsrelationen $r = 1, ..., l$:
   berechne normalisierte CM q_r des gesamten Bildes;
2. für alle Klassen ω und Relationen $r = 1, ..., l$:
   berechne normalisierte CM p_r^ω;

Klassifizierungsphase:

für alle Pixel :
1. für alle Nachbarschaftsrelationen $r = 1, ..., l$:
   berechne die nicht-normalisierten CMs in einer $s \times s$-Umgebung des Pixels;
   bezeichne die CM der Relation r als u_r;
2. für alle Klassen ω :
   für alle CMs u_r:
      berechne Gewichtsfunktion $I(u_r, \omega) = 0.5 \sum_{i,j} u_{r,ij}(\log p_{r,ij}^\omega - \log q_{r,ij})$;
   addiere Gewichte $\sum_r I(u_r, \omega)$;
3. bestimme Klasse ω_{max} mit dem größten Gesamtgewicht;
4. falls $\sum_r I(u_r, \omega_{max}) > treshhold$, so klassifiziere das Pixel als ω_{max},
   andernfalls als *NIL*;

8 Tests

Der hier eingeführte Algorithmus wurde implementiert und an einigen Beispielen getestet. Die Tests verliefen durchweg vielversprechend.

Die ersten Tests wurden mit Brodatz-Testmustern durchgeführt. Die Ergebnisse sind in Fig. 2a,2b dargestellt.

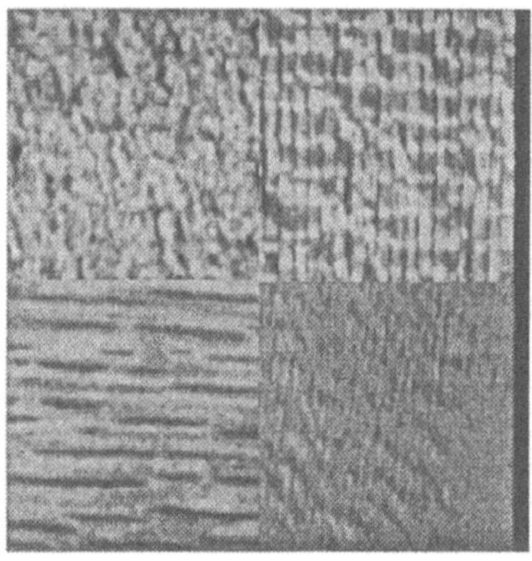

Fig. 2a: Brodatz-Bild

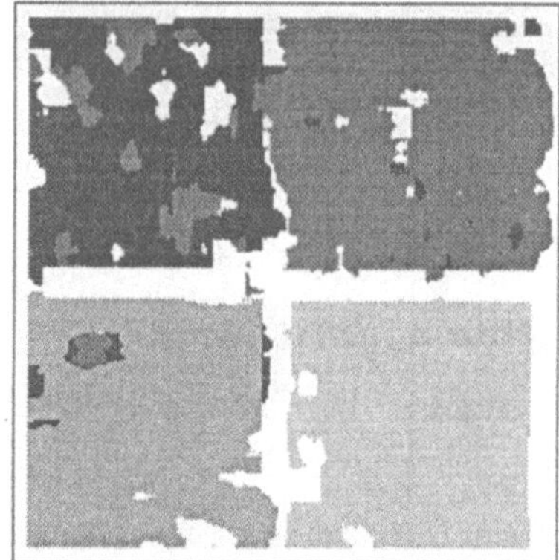

Fig. 2b: Klassifikation

Darüberhinaus wurden Tests mit Tomographie-Bildern des menschlichen Gehirns durchgeführt. In Fig. 3a,3b sind Tests zur Identifizierung eines Gehirntumors dargestellt.

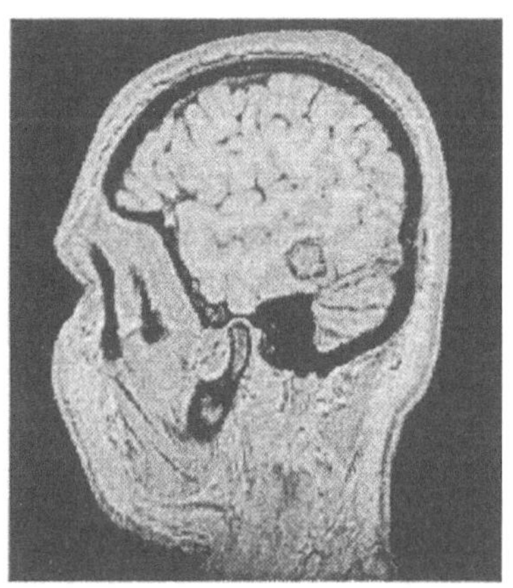

Fig. 3a: CT-Bild mit Tumor

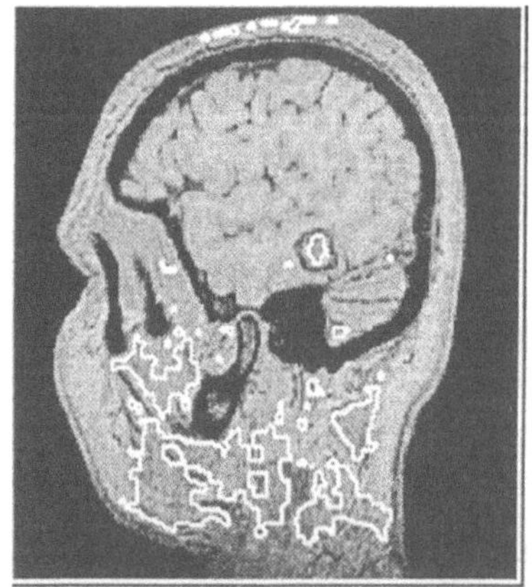

Fig. 3b: Identifizierung des Tumors

Der Tumor wurde innerhalb des Gehirns sicher erkannt. Fehlklassifizierungen gibt es allerdings in einigen Bereichen der Schädeldecke und dem unteren Teil des Kopfes. Die hier verwendete Fenstergröße betrug 5×5 Pixel, die Anzahl der Graustufen war $m = 8$. Es wurde sowohl die horizontale als auch die vertikale Nachbarschaftsrelation verwendet.

9 Zusammenfassung und Diskussion

In diesem Beitrag wurde gezeigt, daß Cooccurrence Matrizen als Merkmalsvektoren aufgefaßt werden können, die polynomialverteilt sind. Mit Hilfe von Gewichtsfunktionen, die auf dem informationstheoretischen Bergriff der "gegenseitigen Information" beruhen, wurde damit ein neuer Texturklassifikator definiert.

Ein großer Vorteil des hier dargestellten Klassifikators ist die Tatsache, daß er sehr einfach parametrisiert ist. Die einzigen Parameter sind die Fenstergröße, die Anzahl der Graustufen und der Typ der Cooccurrence Matrix. Für alle drei Parameter lassen sich Wertebereiche angeben, die in fast allen Fällen anwendbar sind, nämlich $5 \leq s \leq 15$, $8 \leq m \leq 16$ und $(d = 1, \alpha = 0^0)$ und $(d = 1, \alpha = 90^0)$.

Üblicherweise wurden CMs bislang nicht unmittelbar eingesetzt, sondern es wurden sekundäre Merkmale wie z.B. Energie oder Kontrast aus den CMs abgeleitet [2],[3],[6]. Nur diese abgeleiteten Merkmale wurden dann für die Klassifikation verwendet. Diese abgeleiteten Merkmale enthalten wesentlich weniger Information als die ursprüngliche CM. Insbesondere geht die Grauwert-Information im wesentlichen verloren. Zum Beispiel ist der Kontrast von kleinen dunklen Flecken auf hellen Hintergrund derselbe wie der Kontrast von kleinen hellen Flecken auf dunklem Hintergund. Bei traditioneller Verwendung der CM sind diese beiden Texturen nicht unterscheidbar.

Bei der hier eingeführten Verwendungsweise der Cooccurrence Matrizen dagegen bleibt die in den CMs enthaltene Information vollständig erhalten, so daß sowohl Textur-Information als auch Grauwert-Information zum Tragen kommt. Der Rechenaufwand ist hier sogar geringer als bei üblicher Verwendung vom CM-Matrizen, da die zusätzliche Berechnung der sekundären Merkmale entfällt.

Der hier vorgestellte Klassifikator ist also besonders dann gut einsetzbar, wenn sowohl Textur- als auch Grauwert-Information relevant sind. Dies ist bei vielen Klassifikationsproblemen der Fall. Zukünftige Einsatzmöglichkeiten bieten sich vermutlich u.a. bei der Verarbeitung von Satellitenbilddaten an, insbesondere bei stark texturierten Bildern wie etwa des Satelliten ERS-1.

References

1. Robert G. Gallager, *Information Theory and Reliable Communication*, John Wiley, 1968.
2. R.M. Haralick, "Statistical and Structural Approaches to Texture", *Proceedings IEEE-67*, 1979, pp.786-804.
3. R.M. Haralick, L. Shapiro, *"Computer and Robot Vision"*, Vol. 1, Addison-Wesley, 1992.
4. G. Lohmann, "An Evidential Reasoning Approach to the Classification of Satellite Images", *DLR-Forschungsbericht 91-29*, Deutsche Forschungsanstalt für Luft- und Raumfahrt, Köln, 1991.
5. G. Lohmann, "Evidenzbasierte Interpretation von Satellitenbilddaten", in: *Mustererkennung 1991*, 13. DAGM-Symposium, München 1991, Springer Verlag, Informatik-Fachberichte 290.
6. Philippe P. Ohanian, Richard C. Dubes, "Performance Evaluation for Four Classes of Textural Features", *Pattern Recognition*, Vol. 25, No. 8, pp. 819-833, 1992.

Shape from Shading (SFS) und Integration von SFS und Stereo unter perspektivischer Projektion

An Luo und Hans Burkhardt
Technische Informatik I, TU Hamburg-Harburg
Postfach 90 10 52, 21071 Hamburg, Germany

Zusammenfassung

In diesem Beitrag werden neue vereinheitlichte globale Methoden von monokularem bzw. binokularem SFS unter perspektivischer Projektion entwickelt, mit denen das Tiefenfeld und die Oberflächenneigung (wofür angenommen wird, daß die Integrabilitätsbedingung erfüllt ist) zugleich aus einem Graubild oder aus Stereobildern rekonstruiert werden können. Die Annahme perspektivischer Projektion ist bei relativ starker Tiefenänderung notwendig und führt auch zur leichten Integration mit der Stereoauswertung. Es ergibt sich dadurch eine Erhöhung der Zuverlässigkeit der Schätzung.

1 Einführung

Aus der Schattierung ergeben sich wichtige Informationen im menschlichen und künstlichen Sehsystem. In den letzten Jahren wurden verschiedene Ansätze zur Ermittlung der *Form aus der Schattierung* (SFS) vorgestellt [HB89], und die meisten setzen orthogonale Projektion, Lambert'sche Oberflächen und unendlich entfernte Beleuchtung voraus [Hor86]. In [IH81] wurden die Algorithmen zur globalen Ermittlung der Oberflächenneigung aus einem einzigen Graubild unter der Annahme einer kontinuierlich verlaufenden Neigung, jedoch ohne Integrabilitätsbedingung vorgestellt. Die hier vorgestellte Lösung erweitert die in [HB86] [FC88] vorgestellten Ergebnisse von der orthogonalen auf die perspektivische Projektion und behandelt gleichzeitig die Stereobildauswertung.

Weil im allgemeinen die Repräsentation einer 3-D Szene aus einem einzigen 2-D Grauwertbild nicht eindeutig festgelegt wird, werden zur Verbesserung der Schätzung die Methode des SFS und das Stereoverfahren integriert, wobei die 3-D Form der Objekte aus zwei oder mehreren Schattierungsbildern berechnet wird [BM88] [SCS91] [Hei92]. Mit Stereoverfahren läßt sich das absolute Tiefenfeld aus zwei Grauwertverläufen bestimmen. Dies setzt jedoch die Kenntnis korrespondierender Merkmale oder Texturen voraus. Beim SFS wird andererseits eine Oberfläche gleichartiger Reflektion zur Extraktion der Neigung verwendet, wodurch nur eine relative Tiefeninformation gewonnen werden kann. Durch die Kombination beider Verfahren (binokulares SFS) kann ein eindeutiges Tiefenbild ohne die Notwendigkeit der Auswertung korrespondierender Merkmale gewonnen werden.

In dem vorliegenden Beitrag wird eine neue, vereinheitlichte globale Methode des monokularen und des binokularen SFS entwickelt, mit der das Tiefenfeld und die Oberflächenneigung gleichzeitig aus einem Graubild oder Graubildpaar unter Beachtung der Geometrie der perspektivischen Projektion rekonstruiert werden. Die perspektivische Projektion des Bildaufnahmeprozesses wird bei den meisten existierenden Methoden aus dem Bereich SFS unzulässigerweise vernachlässigt. Bei dem hier vorliegenden Ansatz werden keine Lambert'schen Oberflächen vorausgesetzt, aber die Beleuchtung und das reflektorische Modell der Objekte sollen bekannt sein. Zur Minimierung der a-posteriori Energiefunktion aus der Strahlungsgleichung, dem a-priori Modell und evtl. der Stereo-Geometrie ist das iterative Gradientenverfahren zu verwenden, so daß das Tiefenfeld und die Oberflächenneigung (unter Beachtung der Integrabilitätsbedingung) global optimal geschätzt werden können. Die Vorteile dieser Methode sind, daß das Tiefen- und Neigungsfeld in einer einheitlichen Stufe ermittelt werden können und die Methode des SFS wegen der perspektivischer Projektion leicht mit dem Stereoverfahren kombiniert werden kann. Während beim monokularen SFS immer die Skalierungsunsicherheit besteht, wird beim binokularen SFS ein absolutes Tiefenfeld genauer und zuverlässiger geschätzt.

Nach der Modellierung der Bildintensität und der Geometrie der Oberfläche werden zugleich diese beiden neuen Ansätze zur Ermittlung sowohl des Tiefenfeldes als auch der Oberflächenneigungen vorgestellt. Dann folgt die Schlußfolgerung mit einigen Beispielen.

2 Grundlagen für SFS unter perspektivischer Projektion

Ein Objekt wird von einer künstlichen oder natürlichen Lichtquelle beleuchtet, und die reflektierte Intensität wird in die Bildebene projiziert. Eine unendlich entfernte Lichtquelle mit konstanter Strahlungsstärke E_s aus Richtung $\vec{s}$ erzeugt durch Reflektion auf der Oberfläche eines Objektes mit den Raumkoordinaten (X, Y, Z) in der Bildebene $g(x, y)$ eine Intensität gemäß:

$$g(x,y) = \rho\lambda R(p,q) \tag{1}$$

mit $\lambda = E_s\tau(\frac{d}{f})^2\frac{cos^4\alpha}{4}$, wobei d, f und τ Konstanten der Kamera sind, der Winkel α zwischen der optischer Achse und Betrachtungsrichtung $\vec{v}$ nur von der Bildkoordinate (x, y) abhängt. Ohne Einschränkung der Allgemeingültigkeit wird der Faktor $\rho\lambda$ nachher immer auf 1 normiert.

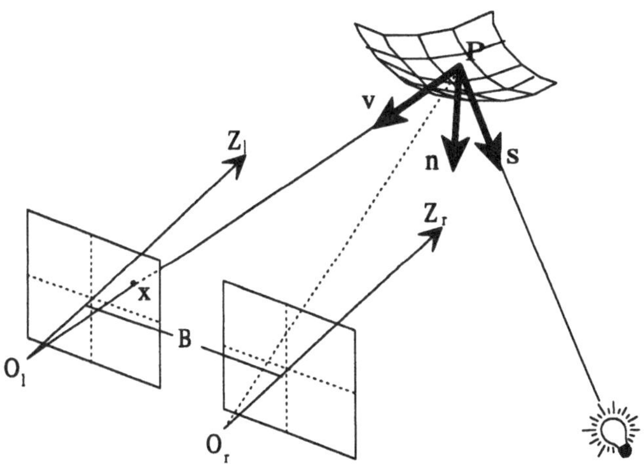

Abbildung 1: Das Modell für SFS

Die Oberflächengeometrie kann mit den Oberfächenneigungen $p = Z_X(X, Y)$ und $q = Z_Y(X, Y)$, die sich direkt auf Tiefenwerte beziehen, beschrieben werden:

$$\vec{n} = (\frac{-p}{\sqrt{p^2+q^2+1}}, \frac{-q}{\sqrt{p^2+q^2+1}}, \frac{1}{\sqrt{p^2+q^2+1}})^T.$$

Beispielsweise besitzt eine ideale matte Oberfläche das Lambert'sche Reflexionsverhalten $R(p, q)$, das nur vom Winkel zwischen der Lichteinfallsrichtung $\vec{s}$ und Oberflächennormale $\vec{n}$ abhängt:

$$R(p,q) = \frac{p_s p + q_s q + 1}{\sqrt{p_s^2 + q_s^2 + 1}\sqrt{p^2 + q^2 + 1}}. \tag{2}$$

Für das Tiefenfeld im Bildkoordinatensystem $Z(x,y)$ gelten $x = f\frac{X}{Z}$ und $y = f\frac{Y}{Z}$ nach perspektivischer Projektion. Aus diesem Tiefenfeld kann man die Oberflächenneigung p und q berechnen:

$$\begin{aligned} p(x,y) = Z_X &= \frac{fZ_x(x,y)}{Z(x,y) + xZ_x(x,y) + yZ_y(x,y)} \\ q(x,y) = Z_Y &= \frac{fZ_y(x,y)}{Z(x,y) + xZ_x(x,y) + yZ_y(x,y)}. \end{aligned} \tag{3}$$

Offensichtlich erfüllen die Neigungsfelder $\mathbf{p} = \{p(x,y)\}$ und $\mathbf{q} = \{q(x,y)\}$ in der obigen Gleichung die Integrabilitätsbedingung $p_Y(x,y) = q_X(x,y)$. Zur Ermittlung von $\mathbf{p}$ und $\mathbf{q}$ muß diese Bedingung erfüllt sein, da sonst aus einem Schattierungsbild Oberflächenneigungen rekonstruiert werden könnten, die keiner tatsächlichen Oberfläche entsprechen. Bei vielen Ansätzen des SFS wird diese Einschränkung vernachlässigt und die Oberflächenneigungen werden allein ohne Integrabilitätsbedingung geschätzt. Bei anderen Methoden wird zwar diese Nebenbedingung verwendet, aber es wird eine orthogonale Projektion des Bildaufnahmeprozesses vorausgesetzt, so daß dabei eigentlich $Z_x = p$ und $Z_y = q$ angenommen wird.

3 Ein neuer vereinheitlichter Ansatz von monokularem oder binokularem SFS

Anders als bei Stereoverfahren oder "Shape from Motion" (SFM), die mehrere Bilder zur Tiefenschätzung der Szene ausnutzen, wird beim SFS nur die Grauwertveränderung eines einzigen Bildes zur Modellierung der Szene-Geometrie verwendet, wobei nur im begrenzten Maße die Oberflächenneigung und das relative Tiefenfeld zu rekonstruieren sind. Aus zwei Schattierungsbildern kann man das echte Tiefenfeld und die Oberflächenneigung besser bestimmen, und zwar durch Integration von SFS und Stereoverfahren, wobei die Nachteile der einzelnen visuellen Lösungsansätze gegenseitig zu überwinden sind.

Wird das Oberflächenneigungsfeld mit einem MRF modelliert, so kann man die Oberflächenneigungen $p(x,y)$ und $q(x,y)$ einschließlich des Tiefenfeldes $Z(x,y)$ aus einem Schattierungsbild $g(x,y)$, oder Stereobildpaar, und zwar linkem $g(x,y)$ und rechtem $g_r(x,y)$, gemäß MAP-Kriterium optimal dadurch schätzen, daß eine a-posterior Energiefunktion bezüglich der Oberflächenneigungen und des Tiefenfeldes minimiert wird. Aus dem a-priori Modell für die Oberflächenneigungen, dem Strahlungsmodell für ein monokulares Bild oder binokulare Bilder, der Integrabilitätsbedingung und evtl. Stereo-Geometrie kann man die a-posteriori Energiefunktion formulieren ($S = 0$ für monokulares SFS bzw. $S = 1$ für binokulares SFS):

$$\begin{aligned} U_p(\mathbf{p},\mathbf{q},\mathbf{Z}) \;=\; \int\int_{\mathbf{X}} & [(g(x,y) - R(p,q))^2 + S(g_r(x - \frac{Bf}{Z(x,y)}, y) - R(p,q))^2 + \\ & \mu(fZ_x(x,y) - p(x,y)Z_s(x,y))^2 + \\ & \mu(fZ_y(x,y) - q(x,y)Z_s(x,y))^2 + \\ & \lambda(V_{\mathbf{X}}(\mathbf{p}) + V_{\mathbf{X}}(\mathbf{q}))]dxdy \end{aligned} \tag{4}$$

mit Abkürzung

$$Z_s(x,y) = Z(x,y) + xZ_x(x,y) + yZ_y(x,y), \tag{5}$$

wobei B und f die Basisbreite bzw. Brennweite der Stereokameras, $V_{\mathbf{x}}(\mathbf{p})$ und $V_{\mathbf{x}}(\mathbf{q})$ die lokalen Potentialfunktionen der Gibbs'schen Verteilung sind. Hier sind λ und μ Kopplungskonstanten für das a-priori Modell und die Integrabilitätsbedingung. Während für die global glatte Oberfläche $V_{\mathbf{x}}(\mathbf{p}) = p_x^2(x,y) + p_y^2(x,y)$ gilt, entspricht eine neue lokale Potentialfunktion $V_{\mathbf{x}}(\mathbf{p}) = (p(x,y) - p_{(med)}(x,y))^2$ der Oberfläche, deren Neigung schwach kontinuierlich ist. In diesem Beitrag wird jedoch nur die global glatte Oberfläche begrenzt. Um eine konvergierende Lösung zur Mimimierung der Energiefunktion zu erzielen, wird auf der strengen Integrabilitätsbedingung der Oberflächenneigungen, die zu einer sehr komplizierten nichtlinearen Energiefunktion bezüglich des Tiefenfeldes führt, verzichtet,

Mit Hilfe der Variationsrechnung sind die folgenden Euler-Lagrange'schen Gleichungen für diese a-posteriori Energiefunktion bezüglich des Tiefenfeldes und der global glatten Neigungen zu ermitteln:

$$
\begin{aligned}
\lambda\Delta p(x,y) + \mu(fZ_x(x,y) - p(x,y)Z_s(x,y))Z_s(x,y) + (g(x,y) - R(p,q))R_p(p,q) &\quad + \\
S(g_r(x - \frac{Bf}{Z(x,y)}, y) - R(p,q))R_p(p,q) &\quad = \quad 0 \\
\lambda\Delta q(x,y) + \mu(fZ_y(x,y) - q(x,y)Z_s(x,y))Z_s(x,y) + (g(x,y) - R(p,q))R_q(p,q) &\quad + \\
S(g_r(x - \frac{Bf}{Z(x,y)}, y) - R(p,q))R_q(p,q) &\quad = \quad 0 \\
a_{20}Z_{xx}(x,y) + 2a_{11}Z_{xy}(x,y) + a_{02}Z_{yy}(x,y) + a_{10}Z_x(x,y) + a_{01}Z_y(x,y) + a_{00}Z(x,y) &\quad - \\
\frac{SBf}{\mu Z^2(x,y)}(g_r(x - \frac{Bf}{Z(x,y)}, y) - R(p,q)(g_{rx}(x - \frac{Bf}{Z(x,y)}, y) &\quad = \quad 0
\end{aligned}
$$

mit Abkürzungen $Z_s(x,y)$ von Gleichung (5) und:

$$
\begin{aligned}
a_{20}(p,q) &= (f - xp)^2 + (xq)^2 \\
a_{11}(p,q) &= xy(p^2 + q^2) - f(yp + xq) \\
a_{02}(p,q) &= (yp)^2 + (f - yq)^2 \\
a_{10}(p,q) &= 3x(p^2 + q^2) - f(3p + 2xp_x + yp_y + xq_y) + 2x^2(pp_x + qq_x) + 2xy(pp_y + qq_y) \\
a_{01}(p,q) &= 3y(p^2 + q^2) - f(3q + yp_x + xq_x + 2yq_y) + 2xy(pp_x + qq_x) + 2y^2(pp_y + qq_y) \\
a_{00}(p,q) &= (p^2 + q^2) - f(p_x + q_y) + 2x(pp_x + qq_x) + 2xy(pp_y + qq_y). \qquad (6)
\end{aligned}
$$

Wegen der nichtlinearen Eigenschaft kann man diese Gleichungen nur mit iterativen Methoden, z.B. Gradientenverfahren, lösen, wobei die Randbedingungen erforderlich sind. Die Randbedingung für die Oberflächenneigungen kann man durch die Kontur der Objekte bestimmen [IH81]. Weiterhin kann man mit dem Stereoverfahren die Randbedingung für das Tiefenfeld einschließlich der Neigungen entlang der Kontur bestimmen. Das konkrete Rechenschema der Algorithmen für monokulares (**S=0**) bzw. binokulares (**S=1**) SFS wird wie folgt dargestellt:

- *Initialisierung für $Z(\mathbf{x})$, $p(\mathbf{x})$ und $q(\mathbf{x})$*
- *Wiederhole die folgende Iteration bis zur Konvergenz:*

Iteration für $Z(\mathbf{x})$:

$$\begin{aligned} Z(x,y) \quad \leftarrow \quad & \frac{1}{\beta}[a_{20}(Z(x+1,y)+Z(x-1,y))+a_{02}(Z(x,y+1)+ \\ & Z(x,y-1))+a_{11}Z_{xy}(x,y)+a_{10}Z_x(x,y)+a_{01}Z_y(x,y)- \\ & \frac{SBf}{\mu Z^2(x,y)}(g_r(x-\frac{Bf}{Z(x,y)},y)-R(p,q))g_{rx}(x-\frac{Bf}{Z(x,y)},y)] \\ & \textit{wobei } \beta = 2a_{20}+2a_{02}-a_{00} \\ & \textit{mit } a_{20}, a_{02}, a_{11}, a_{10}, a_{01} \textit{ und } a_{00} \textit{ der Gleichungen } (6) \end{aligned}$$

Normalisierung für $Z(\mathbf{x})$ (entfällt bei binokularem SFS): *Bei monokularem SFS allein sollte das Tiefenfeld mit einem Skalierungsfaktor aus Vorwissen normalisiert werden.*

$$Z(x,y) \leftarrow kZ(x,y)$$

Iteration für $p(\mathbf{x})$ und $q(\mathbf{x})$:

$$\begin{aligned} p(x,y) \quad \leftarrow \quad & p_{(ave)}(x,y)+\frac{1}{\lambda}[\mu(fZ_x(x,y)-p(x,y)Z_s)Z_s+(g(x,y)- \\ & R(p,q))R_p(p,q)+S(g_r(x-\frac{Bf}{Z(x,y)},y)-R(p,q))R_p(p,q)] \\ q(x,y) \quad \leftarrow \quad & q_{(ave)}(x,y)+\frac{1}{\lambda}[\mu(fZ_y(x,y)-q(x,y)Z_s)Z_s+(g(x,y)- \\ & R(p,q))R_q(p,q)+S(g_r(x-\frac{Bf}{Z(x,y)},y)-R(p,q))R_q(p,q)] \\ & \textit{mit } Z_s = Z(x,y)+xZ_x(x,y)+yZ_y(x,y) \end{aligned}$$

Zur iterativen Lösung wird bei obigem Rechenschema normalerweise der Gauß-Seidel-Algorithmus verwendet, so daß eine nicht nur schnellere sondern auch stabilere Konvergenz gewährleistet wird. Mit dem Jacobi-Algorithmus sollte dagegen zur Stabilisierung der Konvergenz ein Glättungsoperator zwischen Iterationen verwendet werden.

Um schnelle und stabile Konvergenz zur global-optimalen Schätzung zu garantieren, muß man die Oberflächenneigungen und das Tiefenfeld gut initialisieren. Eine geeignete Methode dafür ist das Mehrgitterverfahren, welches auch moderaten Rechenaufwand besitzt.

Weiterhin kann eine klassische Methode von SFS in [IH81] eine Schätzung der Oberflächenneigungen ohne Integrabilitätsbedingung und somit ein approximiertes Tiefenfeld als Anfangswerte liefern.

Wenn zwei stereoskopische Schattierungsbilder verfügbar sind, kann man allein mit dem Stereoverfahren das Tiefenfeld annähernd gut schätzen, aus dem sich die Anfangsschätzung der Oberflächenneigungen direkt ergibt.

An Gleichung (3) sieht man offensichtlich, daß allein aus einem Schattierungsbild kein absolutes Tiefenfeld rekonstruiert werden kann, denn alle Tiefenfelder $kZ(x,y)$ bei beliebigem Faktor k entsprechen identischen Oberflächenneigungsfeldern $p(x,y)$ und $q(x,y)$ und daher auch einem identischen Schattierungsbild. Um eine eindeutige Lösung für das Tiefenfeld zu garantieren, muß man den Tiefenwert $Z^*(x_0,y_0)$ in irgendeinem Bildpunkt

(x_0, y_0) festlegen, und es ist erforderlich in jeder Iteration einen Normalisierungsprozeß mit:

$$k = Z(x_0, y_0)/Z^*(x_0, y_0)$$

für monokulares SFS vorzusehen.

Bei binokularem SFS wird zur Erzielung einer eindeutigen Lösung kein Normalisierungsvorgang benötigt, da aus beiden Grauwertstereobildern ein absolutes Tiefenfeld rekonstruiert werden kann. Aus der Schattierungsinformation sind die Oberflächenneigungen zu berechnen, welche die nur aus Stereo ermittelte Tiefenschätzung erheblich verbessern können. Weiterhin kann das Tiefenfeld wieder ein Oberflächenneigungsfeld erzwingen, welches die Integrabilitätsbedingung erfüllt.

Zusammenfassend kann mit diesen neuen Ansätzen des monokularen oder binokularen SFS nach dem MAP-Kriterium das relative bzw. echte Tiefenfeld und die Oberflächenneigungen unter der Integrabilitätsbedingung global optimal geschätzt werden, so daß die daraus erzeugten Schattierungsbilder mit den tatsächlich aufgenommenen Graubildern in jedem Bildpunkt gute Übereinstimmung zeigen. Bei diesem Ansatz wird die perspektivische Projektion vorausgesetzt, so daß das Aufnahmemodell in jedem Fall der Wirklichkeit besser entspricht und der daraus gewonnene Lösungsansatz einfacher mit dem Stereoverfahren kombiniert werden kann. Ein weiterer Vorteil dieser Methode ist, daß das Tiefen- und Neigungsfeld in einer einheitlichen Stufe ermittelt werden.

4 Experimentelle Ergebnisse und Schlußfolgerung

Wir haben zwei neue vereinheitlichte globale iterative Ansätze von monokularem und binokularem SFS entwickelt, wobei der Bildaufnahmeprozeß mit der perspektivischen Projektion modelliert wird. Unter der Annahme der orthogonalen Projektion muß allerdings das Objekt weit genug entfernt von der Kamera bleiben. Zur Demonstration dieser beiden Ansätze wurden die Schattierungsbilder künstlich erzeugt, wobei die Stereokameras die Brennweite $f = 225$ Pixel, das Format 128×128 Pixel und die Basisbreite $B = 4$ cm haben. Als Objekt wird eine matte Kugel mit einem Durchmesser von 80 cm angenommen, deren Zentrum 160 cm weit von der Basislinie entfernt liegt. Zusätzlich wird eine parallel zur Z-Achse ausgerichtete, unendlich entfernte Punktlichtquelle mit $\lambda\rho = 1$ angenommen. Diese Annahmen wurden in den meisten Aufsätzen über SFS ohne Einschränkung der Allgemeingültigkeit verwendet.

Die Profile beider gestörten Schattierungsbilder entlang der x-Achse, und zwar linkes $g(x,0)$ und rechtes $g_r(x,0)$, werden in den Abbildungen 2 (a) und (b) dargestellt. Die Abbildungen 2 (c) und (d) zeigen die Profile des idealen Tiefenfeldes $Z(x,0)$ und einer idealen Neigungskomponente $p(x,0)$ entlang der x-Achse im linken Kamerakoordinatensystem. Allein mit einem einfachen Stereoverfahren kann man eine Schätzung des Tiefenfeldes ermitteln, dessen Profil entlang der x-Achse in Abbildung 2 (e) dargestellt wird. Weiterhin kann man allein die Methode von SFS in [IH81] zur Schätzung der Oberflächenneigungen ohne Integrabilitätsbedingung aus dem linken Schattierungsbild verwenden, dessen Ergebnisse von $p(x,0)$ in Abbildung 2 (f) zu sehen sind. Unter Verwendung unserer neuen Ansätze wird die Schätzung des Tiefenfeldes und der Oberflächenneigungen erheblich verbessert:

Wird der Tiefenwert $Z(0,0) = 120$ cm in einem Bildpunkt zur Vermeidung der Skalierungsmehrdeutigkeit festgesetzt, werden das Tiefenfeld mit einem durchschnittlichen Fehler von 0.244 cm und die Oberflächenneigungen mit einem Fehler von 0.046 nur aus dem linken Bild geschätzt. Die entsprechenden Profile von $Z(x,0)$ und $p(x,0)$ werden in den Abbildungen 2 (g) und (h) dargestellt.

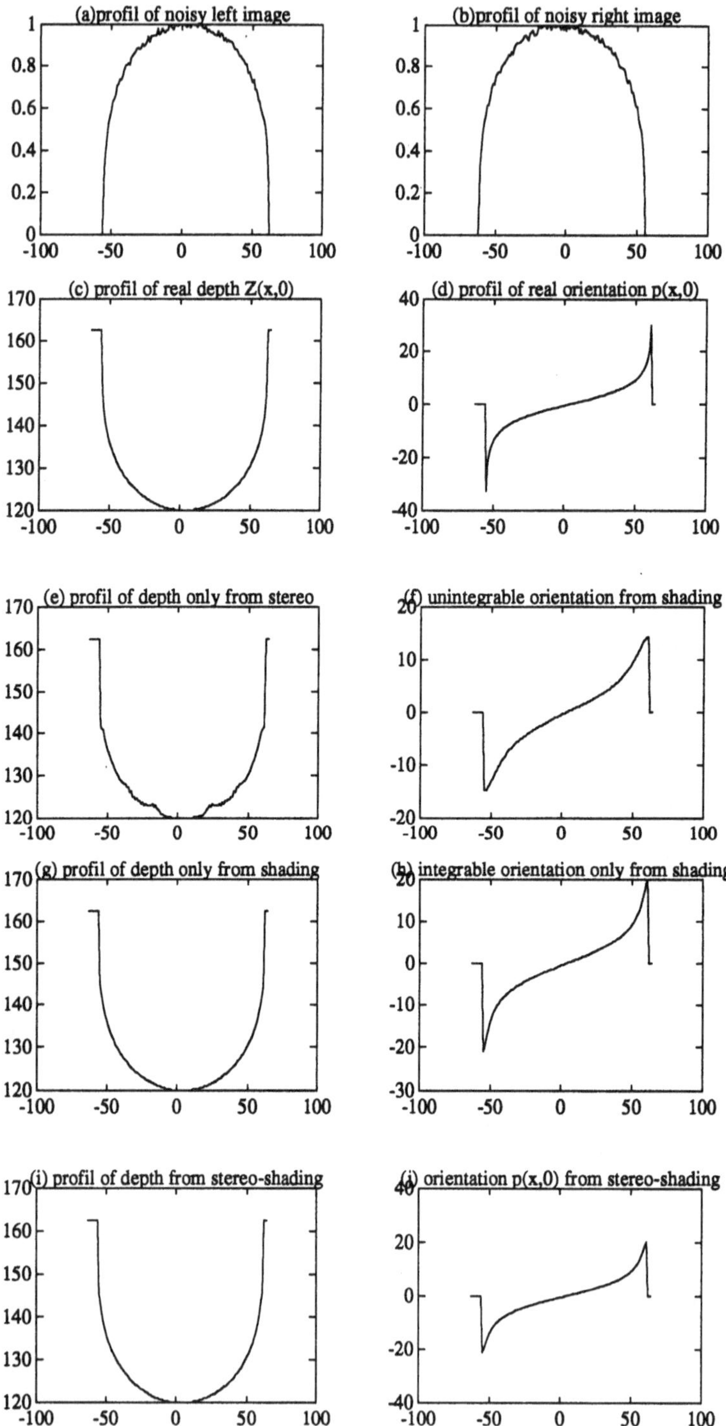

Abbildung 2: Schattierungsbilder und Ergebnisse der Experimente

Aus zwei binokularen Schattierungsbildern $g(x, y)$ und $g_r(x, y)$ werden das absolute Tiefenfeld mit einem Fehler von 0.205 cm und die Oberflächenneigung mit einem Fehler von 0.047 optimal geschätzt. Die Abbildungen 2 (i) und (j) zeigen die zugehörigen Profile $Z(x, 0)$ und $p(x, 0)$ im linken Kamerakoordinatensystem. Im Vergleich zu dem Tiefenfeld und der Neigung in den Abbildungen 2 (g) und (h) wird hierbei nur eine kleine Verbesserung erreicht, denn die Schätzung mit monokularem SFS liefert bereits gute Werte. Entscheidend ist hierbei, daß das echte Tiefenbild hier eindeutig geschätzt wird.

Die obigen Experimente zeigen, daß die neuen Ansätze des monokularen oder binokularen SFS für die Schätzung des Tiefenfeldes und der Oberflächenneigungen aus Schattierungsbildern gut geeignet sind. Besonders wenn bei starker Tiefenänderung der Oberfläche relativ zu den absoluten Tiefen die orthogonale Projektion zur Modellierung des Bildaufnahmeprozesses nicht geeignet ist, ist der perspektivische Einfluß entscheidend und hier vorgestellten Lösungen liefern gute Ergebnisse. Weitere Vorteile dieser Ansätze sind, die Schätzung der Oberflächenneigungen unter der Integrabilitätsbedingung zu garantieren, und das Tiefenfeld damit zugleich in einer einheitlichen Stufe nach dem MAP-Kriterium optimal zu schätzen. In diesem Rechenschema sind die beiden visuellen Informationen, und zwar Schattierung und Stereoparallaxe, mit einem Bayes'schen Modell leicht zu koppeln, so daß der Ansatz des monokularen SFS unmittelbar auf die Integration von SFS und Stereo erweitert werden kann. Durch Ausnutzung der Vorteile beider visuellen Lösungansätze, wird eine Lösung erreicht, welche die Nachteile der ursprünglichen Verfahren überwindet, und damit in der Lage ist, eine eindeutige optimale Schätzung des Tiefenfeldes und der Oberflächenneigung mit guter Zuverlässigkeit zu erzielen.

Literatur

[BM88] H. H. Bülthoff and H .A. Mallot, Integration of depth modules: stereo and shading, Journal of Optical Society of America A, Vol. 5, 1988, pp. 1749-1758

[FC88] R. T. Frankot and R. Challapa, A method for enforcing integrability in shape from shading algorithms, IEEE Trans. Pattern Anal. Mach. Intelligence, Vol. 10, 1988, pp. 439-451

[Hei92] C. Heipke, Integration of digital image matching and multi image shape from shading, in S. Fuchs and R. Hoffmann (eds.), Mustererkennung 1992, 14. DAGM-Symposium, Springer Verlag 1992, pp. 186-198

[Hor86] B. K. P. Horn, Robot Vision, MIT Press, Cambridge, MA, 1986

[HB86] B. K. P. Horn and M. J. Brooks, The variational approach to shape from shading, Computer Vision, Graphics and Image Processing, Vol. 33, 1983, pp. 174-208

[HB89] B. K. P. Horn and M. J. Brooks (eds.), Shape from Shading, MIT Press, 1989

[IH81] K. Ikeuchi and B. K. P. Horn, Numerical Shape from Shading and Occluding Boundaries, Artif. Intell., Vol. 17, pp141–184, 1981

[SCS91] M. Shao, R. Chellappa and T. Simchony, Reconstructing a 3-D depth map from one or more images, CVGIP: Image Understanding, Vol. 53, 1991, pp. 219-226

Gewinnung von Tiefeninformation mit Moiré-Techniken

W. Ortmann und T. Leditzky
Friedrich-Schiller-Universität Jena,
Fakultät für Mathematik und Informatik
07740 Jena

1. Einleitung

Moiré-Effekte entstehen überall dort, wo sich Gitterstrukturen überlagern. In einer geeigneten Anordnung können Moiré-Streifen Tiefeninformation liefern, die bei der Projektion in die zweidimensionale Ebene verlorengehen. In der vorliegenden Arbeit wird die Entstehung der Moiré-Streifen mit Hilfe der Autokorrelation erklärt. Es wird ein Verfahren angegeben, bei dem unter Ausnutzung der Struktur der CCD-Sensoren moderner Video-Kameras Moiré-Effekte mit einfachen Mitteln erzeugt werden können. Die Eignung des Verfahrens zur Gewinnung von Tiefeninformation wird nachgewiesen.

2. Probleme bei der Gewinnung von Tiefeninformation

Zahlreiche Verfahren zur Gewinnung von Tiefeninformation beruhen auf der Triangulation (Bild 1). Dabei werden von zwei Referenzpunkten P1 und P2 aus Strahlen in Richtung zum Objektpunkt vermessen. Die Tiefeninformation ergibt sich aus den Strahlrichtungen und dem Abstand der beiden Referenzpunkte.

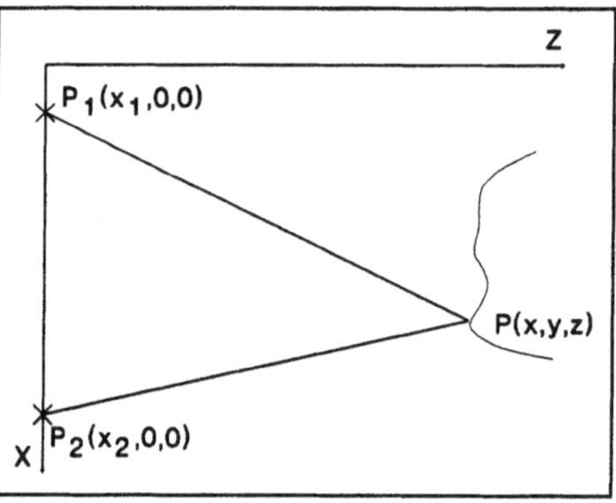

Bild 1: Prinzipdarstellung der Triangulation

Bei den "passiven" Verfahren befinden sich in den beiden Punkten Kameras. Die Strahlrichtungen zu einem Punkt ergeben sich aus den Kamerabildern und den bekannten Kameraparametern. Ein schwieriges Problem besteht in der Erkennung von korrespondierenden Punkten in den Bildern beider Kameras. Zur eindeutigen Zuordnung werden oft die Bilder weiterer Kameras herangezogen. Die Lösung dieses Problems ist unsicher und zudem zeitaufwendig. Auf glatten Oberflächen sind korrespondierende Punkte praktisch nicht zu finden.

"Aktive" Verfahren beruhen darauf, daß sich in einem Referenzpunkt eine Lichtquelle, im anderen eine Kamera befindet. Durch eine Strukturierung des Lichtes der Lichtquelle lassen sich im Bild der Kamera die Punkte den jeweiligen Strahlrichtungen der Lichtquelle zuordnen. Bei bekannten angewendeten Verfahren wird das Licht als Streifen (Lichtebene) über die Szene bewegt, entsprechend des Raumwinkels in zeitlich aufeinanderfolgenden Phasen hell oder dunkel geschaltet (codiert) oder auch in verschiedenen spektralen Farben über den Raum verteilt [Ta90,St92]. Lichtquellen dieser Art sind sehr aufwendig. Längere Aufnahmezeiten ergeben sich entweder durch die aufeinanderfolgende Abtastung in mehreren Schritten bzw. durch die geringe Lichtstärke der Lichtquelle.

3. Moiré-Methodik

Ein aktives Verfahren, welches bisher vorrangig mit fotografischen Methoden verwirklicht wurde, ist die Moiré-Technik [Xe79]. Sie beruht auf der Überlagerung von Gitterstrukturen. In einem Referenzpunkt der Anordnung befindet sich die Kamera oder auch ein Beobachter, im zweiten Referenzpunkt eine Lichtquelle, die unstrukturiertes Licht aussendet (Bild 2). Zwischen den Referenzpunkten und dem Objekt befindet sich ein senkrechtes Gitter mit einer von x abhängigen optischen Durchlässigkeit $f(x)$. Die Gitterfunktion ist periodisch mit einer Periode Δg:

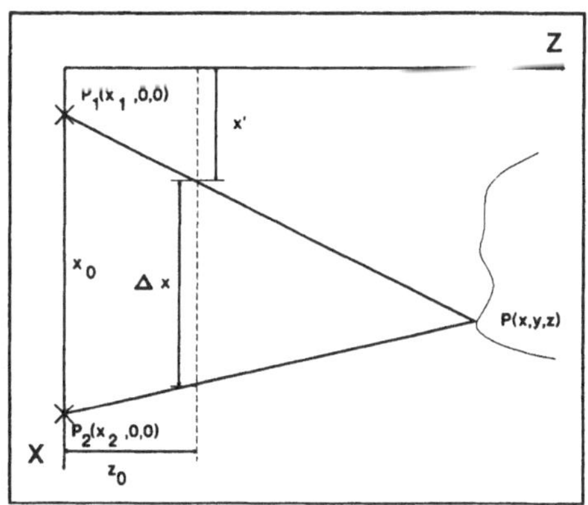

Bild 2: Prinzip der Erzeugung von Moiré-Bildern

$$f(x) = f(x+\Delta g) \quad .$$

Gegenüber dem Kamerabild ohne Gitter modifiziert sich die Helligkeit, mit der ein Objekt abgebildet wird. Das Licht durchläuft das Gitter zweimal: einmal auf dem Weg von der Lichtquelle zum Objekt und zum zweiten Mal auf dem Weg vom Objekt zur Kamera. Diese Überlagerung des Gitters "mit sich selbst" garantiert identische Gitterkonstanten und gleiche Ausrichtung der überlagerten Gitter.

Berechnet man die relative Helligkeit $i(x,y,z)$, mit der der Punkt $P(x,y,z)$ beobachtet wird, als Verhältnis der Lichtintensitäten für den Fall der Anordnung mit Gitter $I(x,y,z)$ im Vergleich zum Fall ohne Gitter $I_0(x,y,z)$, so ergibt sich (Bild 2):

$$i(x,y,z)=I(x,y,z)/I_0(x,y,z)$$
$$=f(x')\cdot f(x'+\Delta x)$$

mit

$$x'(x,z)=\frac{(x-x_1)\cdot z_0}{z}+x_1$$

und

$$\Delta x(z)=\frac{(z-z_0)}{z}\cdot x_0 \quad .$$

4. Tiefeninformation

Die Helligkeit eines Punktes hängt wie gewünscht von der Tiefe z ab. Leider ist aber für eine konstante Tiefe z auch noch eine periodische Abhängigkeit von x vorhanden, die durch das Gitter gegeben ist. Um die Abhängigkeit von x zu eliminieren, gibt es zwei Möglichkeiten:

- Bewegung des Gitters in x-Richtung und Integration über eine Gitterperiode,
- Integration über einen Objektbereich, der einer Gitterperiode entspricht.

Ist die Gitterkonstante klein gegenüber den anderen geometrischen Abmessungen, so sind die beiden Fälle durch das gleiche Modell zu beschreiben. Es werde der Einfachheit halber die Integration über die Verschiebung des Gitters betrachtet:

$$i_i(x,y,z)=\int_0^{\Delta g} f(x'+s)\cdot f(x'+\Delta x+s)\,ds$$

$$=\int_0^{\Delta g} f(s)\cdot f(s+\Delta x)\,ds \quad .$$

Da diese Integration über eine Periode erfolgt, ist das Ergebnis vom Startpunkt unabhängig und deshalb ist auch die integrale Intensität von x' unabhängig. Dieses Integral über eine Periode des Gitters stellt die periodische Autokorrelationsfunktion $F(\Delta x)$ der Gitterfunktion $f(x)$ dar. Damit erhält man in vereinfachter Schreibweise:

$$i_i(x,y,z)=F(\Delta x)=F\left(\frac{z-z_0}{z}\cdot x_0\right)$$

Die resultierende relative Helligkeit ist nur noch von der Tiefe z abhängig. Für ein einfaches Rechteckgitter mit gleich breiten hellen und dunklen Streifen der Breite $\Delta g/2$ ergibt sich z.B.

$$F_r(\Delta x)=\begin{cases}\Delta x, & \Delta x\leq\Delta g/2\\ \Delta g-\Delta x, & \Delta x>\Delta g/2\end{cases}$$

Im Bild auftretende helle und dunkle Streifen verbinden Punkte konstanter Tiefe. Die Maxima der relativen Helligkeit liegen bei den Werten von $\Delta x = n*\Delta g$. Der n-te helle Streifen entspricht damit einer Tiefe von

$$z_n=\frac{z_0\cdot n\cdot\Delta g}{x_0-n\cdot\Delta g}$$

Gelingt eine Zuordnung der Streifennummern zu den Streifen, ist eine Tiefenbestimmung möglich. Die hier vorgestellte Herleitung dieser Formeln über die Autokorrelationsfunktion ist wesentlich einfacher als die üblicherweise durchgeführte explizite Integration [Xe79].

5. Praktische Realisierung

Die praktische Realisierung dieses Prinzips stößt auf eine Reihe von Schwierigkeiten:

- Die Verschiebung des Gitters und Integration über verschiedene Stellungen bedingt relativ lange Belichtungszeiten.

- Da das Gitter nicht unmittelbar in der Objektebene positioniert werden kann, gibt es ein Schärfeproblem. Zur Erzielung der notwendigen Tiefenschärfe sind relativ kleine Blendenwerte notwendig.

Selbst bei Verwendung von hellen Farben für die Objekte zur Erhöhung der Lichtausbeute werden Belichtungszeiten von 1/8 Sekunde und mehr angewendet [Xe79,Ta73].

Eine Lösung des Schärfeproblems wäre möglich, wenn das Gitter in einer zweiten Schärfeebene des optischen Systems der Kamera und der Beleuchtungseinrichtung angeordnet würde. Dann kann allerdings nicht mehr dasselbe Gitter für beide Teilstrahlengänge Anwendung finden. Dies ist aber auch gar nicht notwendig, wenn nur die effektiven Gitterkonstanten und die Orientierung der beiden Gitter die gleichen sind. Geeignete Lichtquellen sind Dia-Projektoren bzw. Overhead-Projektoren für Folien. Will man hingegen ein Gitter in den Strahlengang einer Kamera einbringen, benötigt man ein speziell konstruiertes optisches System. Bei genauerer Betrachtung erweist sich auch das jedoch als nicht notwendig. Moderne Videokameras mit einer CCD-Sensormatrix verfügen bereits über ein Gitter. Die Sensoren der CCD-Matrix sind zeilenweise (oder spaltenweise) angeordnet, zwischen den Zeilen liegen die lichtunempfindlichen Bereiche des Ausleseschieberegisters. Bei der Nutzung dieses "natürlichen" Gitters einer Kamera kommt es zu einem zweiten positiven Effekt: Da sich die einzelnen Sensoren und der zugehörige Zwischenraum genau über eine Streifenbreite erstrecken, erfolgt damit sofort eine Integration über eine Periode dieses Gitters.

Mit diesen Voraussetzungen wurde versucht, zunächst eine einfache experimentelle Anordnung aufzubauen und den grundlegenden Effekt nachzuweisen. Dazu wurden vorhandene Kameras und Projektoren auf ihre Eignung untersucht. Da eine nachträgliche Justierung der Gitterkonstanten unvermeidbar ist, mußte für eines der Objektive ein Objektiv variabler Brennweite angewendet werden. Damit ist es möglich, die wirksame Gitterkonstante zu verändern, indem der Abbildungsmaßstab geändert wird. Deshalb wurde als Kamera ein Camcorder S-VS 180 (Grundig) mit einem Zoom-Objektiv eingesetzt. Als Lichtquelle stand ein Streifenlichtprojektor LCD-320 (ABW) zur Verfügung. Dieser ist in der Lage, Streifenmuster für die Tiefenbestimmung mittels codierten Lichtansatz zu projizieren [St92]. Dies wäre für einen normalen Einsatz natürlich ein unnötig hoher Aufwand. Aus praktischen Gründen wurde ein horizontales Gitter verwendet und Kamera und Projektor übereinander angeordnet.

Der Streifenlichtprojektor kann maximal 320 Streifen erzeugen, das heißt mit 160 hellen und 160 dunklen Streifen genau 160 Perioden des Gitters. Eine Anpassung der effektiven Gitterkonstanten der Kamera mit Hilfe des Zoom-Objektives war nicht möglich: im besten

Fall entsprach eine Periode des Gitters des Streifenlichtprojektors vier Zeilen im Fernsehbild. Die 160 Gitterperioden des Streifenlichtprojektors füllen damit das Fernsehbild recht gut aus. Eine Überlagerung der Gitter mit der Folge der Bildung von Moiré-Streifen findet damit nicht statt, die Gitterstruktur des projizierten Gitters ist im aufgenommenen Bild direkt sichtbar. Es wurde deshalb der Versuch unternommen, das abgetastete Bild rechnerisch mit einem geeigneten Gitter zu überlagern und die Helligkeitswerte über eine Periode zu integrieren. Dazu wurden jeweils 4 Zeilen gemeinsam ausgewertet, indem die Helligkeitswerte der ersten zwei Zeilen addiert (Multiplikation mit 1) und die Helligkeitswerte der nächsten zwei Zeilen subtrahiert wurden (Multiplikation mit -1). Da sich hiermit die Auflösung in y-Richtung um den Faktor 4 verringert, wurden auch in x-Richtung nachfolgend jeweils 4 Bildpunkte gemittelt. Gegenüber dem optischen Gitter besitzt diese Lösung den Vorteil, daß hier auch "negative Durchlässigkeiten" des Gitters möglich sind: durch die Subtraktion benachbarter Punkte ergibt sich eine relative Unabhängigkeit von der Umgebungshelligkeit und dem Absorptionsvermögen des Objektes. Nachteilig ist jedoch, daß die Moiré-Streifen erst nach der Bearbeitung der Bilder im Computer (d.h. nach wenigen Sekunden) sichtbar werden, was die Justierung der Anordnung deutlich erschwert.

6. Ergebnisse

Der Aufbau mit den genannten verfügbaren Komponenten gestaltete sich relativ einfach. Die Justierung war etwas zeitaufwendig, da nach jedem Abgleich zunächst die Abtastung und Berechnung des Moiré-Bildes erfolgen mußte. Zur Ermittlung der Eigenschaften der Anordnung wurde eine Serie von Moiré-Bildern aufgenommen, bei denen eine zur x-y-Ebene parallele Platte in Richtung der z-Achse verschoben wurde. Im Idealfall hätten sich hier Bilder einheitlicher Helligkeit zeigen müssen, da ja alle Punkte die gleiche Tiefe aufweisen. Diese Helligkeit hätte sich mit der Verschiebung in z-Richtung periodisch ändern müssen. Bei der praktischen Realisierung zeigten sich Abweichungen von diesem Idealbild, die durch ungenaue Justierung erklärt werden können und somit Anhaltspunkte für die weitere Justierung liefern.

- Streifen in Richtung des Gitters (waagerecht):

  Der Abgleich auf gleiche Gitterkonstanten ist unexakt.
- Streifen in Richtung senkrecht zum Gitter (senkrecht):

  Die beiden Gitter sind zueinander verdreht.

- Bei einem Schnitt in Richtung der y-z-Ebene sind die Streifen gerade, aber nicht parallel:
  Die Hauptpunkte der beiden Objektive liegen nicht exakt in einer x-y-Ebene.
- Bei einem Schnitt in Richtung der y-z-Ebene sind die Moiré-Streifen gewölbt:
  Die Ebenen der Gitter in Kamera und Projektor sind gegeneinander verkippt.

Nach einer Justierung verbleiben Abgleichfehler, die in der Größenordnung eines Streifens liegen. Das läßt sich auf verbleibende Ungenauigkeiten in der Justierung sowie weitere Fehlerquellen wie z.B. Verzeichnungen der Objektive zurückführen. Werden diese Fehler jedoch einmal ermittelt, sind sie auch numerisch korrigierbar. Der Streifenabstand entspricht in der verwendeten Anordnung einer Tiefendifferenz von 12 mm. Ein Tiefenbild läßt sich dementsprechend mit einer Auflösung vom mindestens 6 mm aufbauen (Abstand heller zu dunkler Streifen) wenn A-priori-Wissen genutzt wird, um die Streifen zuzuordnen.

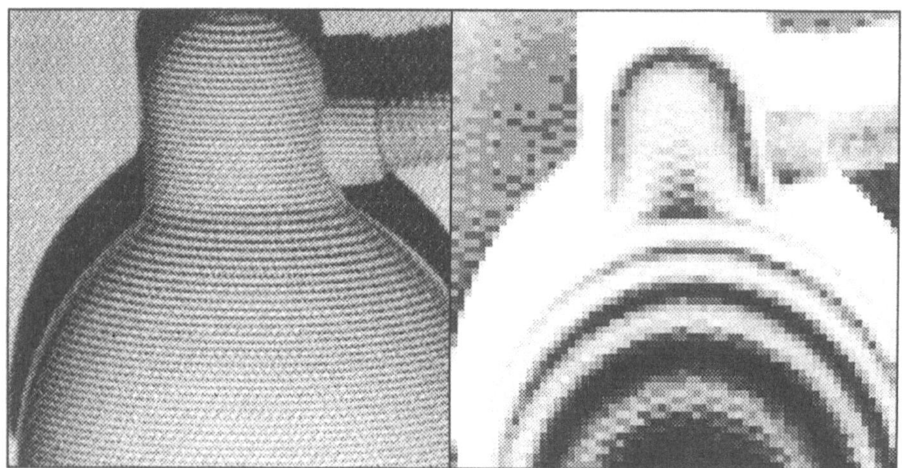

Bild 3: Lampe mit Moiré-Streifen
links - Kamerabild, rechts - mit Referenzgitter überlagertes und integriertes Bild

Ein einfaches Moiré-Bild zeigt Bild 3. Die Streifenstruktur ist auf der glatten (im Original weißen) Oberfläche des Lampenschirmes gut zu erkennen. Der Hintergrund, eine glatte weiße Holzplatte, die ca. 20 cm hinter der Lampe stand, weist wegen der ungenügenden Tiefenschärfe jedoch keine Moiré-Streifen mehr auf.

7. Zusammenfassung und Ausblick

Weitere Experimente sollen dazu dienen, die Grenzen der meßtechnischen Möglichkeiten auszuloten. Dazu soll die Lichtquelle mit einer feineren Gitterstruktur versehen werden, um an die Gitterstruktur der Kamera angepaßt zu sein. Da dann der Rechenaufwand zur Bildung der Moiré-Streifen entfällt, liegt eine Messung in Echtzeit im Rahmen der Möglichkeiten. Im Rahmen der Software-Lösung muß geprüft werden, wie durch Überlagerung mit einem verschobenen Gitter die Auflösung im Bereich einer Streifenperiode verbessert werden kann (Phasenshift-Methode, siehe [Di88,Ja92]).

Literatur

[Di88] J.J.J.Dirckx, W.F.Decraemer, G.Dielis: Phase shift method based on object translation for full field automatic 3-D surface reconstruction from moiré topograms. Applied Optics 27 (1988) 1164-1169

[Ja92] G.Jansen: Topometrische Verfahren in der dreidimensionalen Oberflächen-Meßtechnik. Bild & Ton 45 (1992) 166-171

[St92] T.Stahs, F.Wahl: Object recognition and pose estimation with a fast and versatile 3D robot sensor. Proc. 11th IAPR Int.Conf.Pattern Recognition, Den Haag 1992, Vol. I, pp. 684-687

[Ta73] H.Takasaki: Moiré topography. Applied Optics 12 (1973) 845-850

[Ta90] J.Tajima, M.Iwakawa: 3-D data acquisition by rainbow range finder. Proc. IEEE 1990, pp. 309-313

[Xe79] S.S.Xenofos, C.H.Jones: Theoretical aspects and practical applications of moiré topography. Phys. Med. Biol. 24 (1979) 250-261

S.S.Xenofos, C.H.Jones, D.R.Dance: Evaluation of a moiré imaging system. Phys. Med. Biol. 24 (1979) 262-270

Adaptive Ausgleichsrechnung und Ausreißerproblematik für die digitale Bildverarbeitung

H. Suesse und K. Voss
Friedrich-Schiller-Universität Jena,
Fakultät für Mathematik und Informatik
07743 Jena

1 Einleitung

Verfahren der robusten Schätzung und der Behandlung von Ausreißern sind in der Literatur ausführlich beschrieben [Hu81, Ha86, Ro87, Fö86, Fö91, Fö92]. Die Ausgleichsrechnung findet in der Bildverarbeitung eine breite Anwendung (Kurven- und Flächenfitting, Filterentwurf [Vo90], photogrammetrische Messungen usw.). Ein wesentliches Problem besteht dabei im Einfluß von Ausreißern auf die Ergebnisse, die in kritischen Fällen völlig unbrauchbar werden können. In der vorliegenden Arbeit wird eine adaptive Anpassung der Parameter eines Ausgleichsmodells vorgestellt, wobei der Einfluß von Ausreißern systematisch verringert wird. Dabei ist keine Schwellwertentscheidung zur Detektion von Ausreißern nötig (parameterfreie Eliminierung der Ausreißer).

Das allgemeine Modell der Ausgleichsrechnung geht davon aus, daß eine Funktion $F(\mathbf{a},\mathbf{x})$ der im Tupel $\mathbf{x}=(x_1,x_2,\ldots,x_n)$ zusammengefaßten Meßgrößen x_i die Gleichung $F(\mathbf{a},\mathbf{x})=0$ erfüllen soll. Die im Tupel $\mathbf{a}=(a_1,a_2,\ldots,a_m)$ zusammengefaßten Parameter sollen anhand gemessener Werte $\mathbf{x}^{(k)}$ nun so bestimmt werden, daß ein Zielfunktional $Z_F(\mathbf{a})$ zu einem Minimum wird. Die am häufigsten angewendete Gaußsche Ausgleichung geht von einer linearen Funktion

$$F(\mathbf{a},\mathbf{x}) = \sum_{i=1}^{n} a_i x_i$$

und einem quadratischen Zielfunktional

$$Z_F(\mathbf{a}) = \sum_{k=1}^{K} (F(\mathbf{a},\mathbf{x}^{(k)}))^2 \rightarrow \mathit{Minimum!}$$

aus [Ro87]. Es ist bekannt, daß die für dieses Zielfunktional verwendete L_2-Norm sehr empfindlich auf Ausreißer reagiert [Fö92, Ro87]. In der Bildverarbeitung, wo die Anpassung von Geraden eine große Rolle spielt, macht sich das unangenehm bemerkbar: Geraden, die fast Parallelen zur y-Achse sind und deshalb nur bedingt als "vernünftige" lineare Funktion $y=a_1+a_2x$ darstellbar sind, reagieren sehr empfindlich auf x-Ausreißer (Abbildung 1).

Das Problem kann auch nicht durch eine Regression von x bezüglich y gelöst werden, da sich dann y-Ausreißer störend bemerkbar machen. Eine Lösung für die Bildverarbeitung ist die "orthogonale Regression", bei der die geometrischen Abstände der Punkte zur Geraden minimiert werden [Vo92]. Aber nicht nur die konkrete Funktion $F(\mathbf{a},\mathbf{x})$, sondern auch die Art des Zielfunktionals (z.B. L_1-Norm, Median-Kriterien usw.) bestimmen die Robustheit des jeweiligen Verfahrens gegenüber Ausreißern.

Zur Analyse und zur Detektion von Ausreißern werden oft die *Defekte* δ_k (*Residuen*) herangezogen:

$$\delta_k = F(\boldsymbol{a}, \boldsymbol{x}^{(k)}) \quad ,$$

wobei δ_k für das k-te Meßwerttupel $\boldsymbol{x}^{(k)}$ die Abweichung der Funktion $F(\boldsymbol{a},\boldsymbol{x}^{(k)})$ vom "eigentlich" erwarteten Wert Null ist. Letztendlich ist die Verteilung der Residuen über eine Meßserie der Maßstab für die Entscheidung, ob ein einzelnes Tupel $\boldsymbol{x}^{(k)}$ als Ausreißer angesehen werden muß oder nicht [Fö86, Ro87].

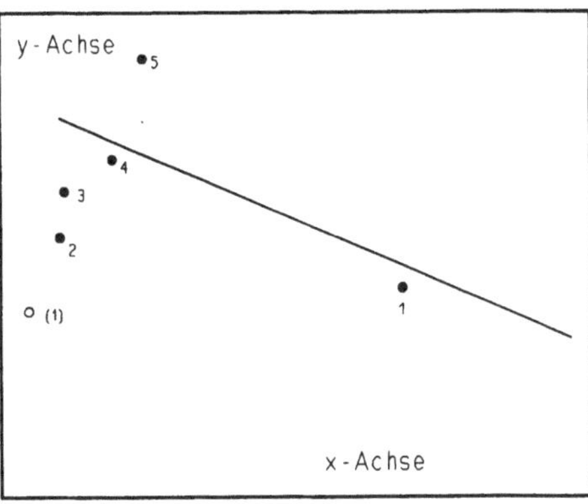

Abb. 1: Wirkung eines x-Ausreißers (1) → 1 auf die Lage der Ausgleichsgeraden

Die Robustheit eines Ausgleichsverfahrens wird oft durch den Anteil p der sogenannten "break-down-points" gemessen: Man untersucht, wieviel Meßwerttupel beliebig stark in ihren Daten x_i verändert werden können, ohne daß die zu ermittelnden Parameter a_j völlig verfälscht werden [Hu81, Ro87]. Ist diese Zahl z, so wird bei K Meßwerttupeln der "Break-down-points"-Index p durch das Verhältnis $p=(z+1)/K$ bestimmt.

Für die quadratische Ausgleichsrechnung ist $p=1/K$, weil kein einziger Meßwert bei beliebiger Veränderung die zu ermittelnden Parameter unbeeinflußt läßt. Andererseits hat z.B. der "Repeat-median"-Schätzer [Si82] einen "Break-down-points"-Index von 50%.

Häufig wird zur Herleitung von robusten Ausgleichsverfahren die sogenannte "Einflußfunktion" herangezogen [Hu81, Ha86]: Mit $Z=Z[F(\boldsymbol{a},\boldsymbol{x})]$ als Zielfunktional bewertet die Ableitung $\partial Z/\partial F$ die Wirkung von Veränderungen des Meßwerttupels $\boldsymbol{x}$ auf die Defekte und damit auch die Wirkung von Ausreißern auf das Zielfunktional. Deshalb kann man bezüglich der Robustheit Forderungen an $\partial Z/\partial F$ stellen und damit das Zielfunktional selbst charakterisieren.

Für die Methode der kleinsten Quadrate spielt das Skalarprodukt $<\boldsymbol{a},\boldsymbol{x}>$ zwischen den Parametern und den Meßwertkomponenten die entscheidende Rolle, so daß mit $Z= Z[F(\boldsymbol{a},\boldsymbol{x})]=(<\boldsymbol{a},\boldsymbol{x}>)^2$ und $\partial Z/\partial F=2<\boldsymbol{a},\boldsymbol{x}>$ die Ausreißer einen linearen Einfluß auf die Änderung des Zielfunktionals ausüben. Dadurch wird die große Sensitivität dieser Ausgleichsmethode unterstrichen.

In [Hu81] werden ausführlich Klassen von Funktionen und deren Eigenschaften in Hinblick auf Robustheit untersucht. Dabei wird oft die Idee verfolgt, eine "gewichtete Ausgleichung" durchzuführen, indem man die einzelnen Meßwerttupel unterschiedlich stark in das Zielfunktional eingehen läßt. Als Gewichtsfunktion wird beispielsweise die Funktion $w(\mathbf{x})=(\partial Z/\partial F)/|F|$ benutzt. Eine weitere Methode zur Erhöhung der Robustheit von Schätz- und Ausgleichsverfahren besteht in der "iterativen Elimination" von Ausreißern. Diese bekannte Methode funktioniert im wesentlichen wie folgt:

1. Man bestimme eine globale Lösung mit allen Beobachtungen.
2. Man ermittle die Verteilung der Residuen.
3. Man extrahiere die "glaubwürdigsten" Punkte.
   Keine weitere Änderung → Stop.
4. Man bestimme erneut die globale Lösung mit den "glaubwürdigsten" Punkten. Gehe zu 2.

Im folgenden soll ein robustes Ausgleichsverfahren derart entwickelt werden, daß die Methode der iterativen Elimination mit einer adaptiven Gewichtsanpassung gekoppelt wird. Das hat den entscheidenden Vorteil, daß man nicht entscheiden muß, welche Punkte "glaubwürdig" sind oder nicht - sie werden stets alle mit entsprechenden Gewichten in die Iteration einbezogen. Die Gewichte werden aus der Residuenverteilung bestimmt und sind demnach in jedem Iterationsschritt neu zu bestimmen.

2 Adaptive Gewichtsberechnung

Es sei folgendes allgemeines Modell gegeben:

$$F(\mathbf{a},\mathbf{x}) = 0 \ .$$

Bei Vorgabe von K Meßwerttupeln $\mathbf{x}^{(k)}$ ist das Parametertupel $\mathbf{a}$ so zu bestimmen, daß

$$\sum_{k=1}^{K} Z\big(F(\mathbf{a},\mathbf{x}^{(k)})\big)\cdot w_k \rightarrow \mathit{Minimum}$$

erfüllt ist. Die w_k stellen zunächst beliebige Gewichte dar. Weiterhin bezeichne

$$F\big(\mathbf{a},\mathbf{x}^{(k)}\big) = \delta_k$$

das Residuum oder den Defekt der k-ten Messung.

Algorithmus:

1. Man setze den Iterationsindex $s=1$ und die Gewichte $w_k^{[s]}=1$ für $k=1,\ldots,K$.

2. Man ermittle eine Lösung $\mathbf{a}^{[s]}$ der Ausgleichsaufgabe $Z[F(\mathbf{a},\mathbf{x})] \rightarrow$ Minimum, d.h. im speziellen Fall eine Lösung des linearen Ausgleichsproblems

$$\sum_{k=1}^{K} w_k^{[s]} \cdot Z(F(\mathbf{a},\mathbf{x}^{(k)})) \rightarrow \mathit{Minimum}$$

   durch Lösung der Gaußschen Normalengleichungen.

3. Wenn keine Änderung der Lösung $\rightarrow$ Stop.

4. Man berechne die Residuen

$$\delta_k^{[s]} = Z\big(F(\mathbf{a}^{[s]},\mathbf{x}^{(k)})\big)$$

   im s-ten Schritt für alle $k=1,\ldots,K$.

5. Nach dem Gesetz der Großen Zahlen nehmen wir an, daß die Defekte normalverteilt sind und schätzen daraus den Erwartungswert und die Streuung:

$$\mu^{[s]} = \frac{1}{K} \sum_{k=1}^{K} \delta_k^{[s]} ,$$

$$\sigma^{[s]} = \sqrt{\frac{1}{K-1} \sum_{k=1}^{K} \left(\delta_k^{[s]} - \mu^{[s]}\right)^2} .$$

6. Aus der Dichtefunktion

$$\varphi(\delta ; \mu^{[s]}, \sigma^{[s]}) = \frac{1}{\sqrt{2\pi}} e^{-\frac{(\delta-\mu^{[s]})^2}{2(\sigma^{[s]})^2}}$$

bestimmen wir die Gewichte der einzelnen Meßwerttupel $\mathbf{x}^{(k)}$ zu

$$w_k^{[s]} = \frac{\varphi\left(\delta_k^{[s]} ; \mu^{[s]}, \sigma^{[s]}\right)}{\sum_{l=1}^{K} \varphi\left(\delta_l^{[s]} ; \mu^{[s]}, \sigma^{[s]}\right)} .$$

7. Setze $s+1 \rightarrow s$ und gehe zu 2.

Die angegebene Methode läßt sich in den verschiedendsten Gebieten der Bildverarbeitung und Mustererekennung anwenden. Einige interessante Beispiele werden im folgenden angegeben.

3 Beispiel Geradenanpassung

Zur Geradenanpassung wählen wir die Gleichung $F(\boldsymbol{a},\boldsymbol{x}) = a_1 + a_2x - y = 0$, um den in Abschnitt 2 beschriebenen Algorithmus durchzuführen. In Abbildung 2 ist zu sehen, wie nach wenigen Iterationen ($s_{max}=6$) eine Ausgleichsgerade so angepaßt wird, wie es etwa auch durch nutzergesteuerte Interaktion geschehen würde.

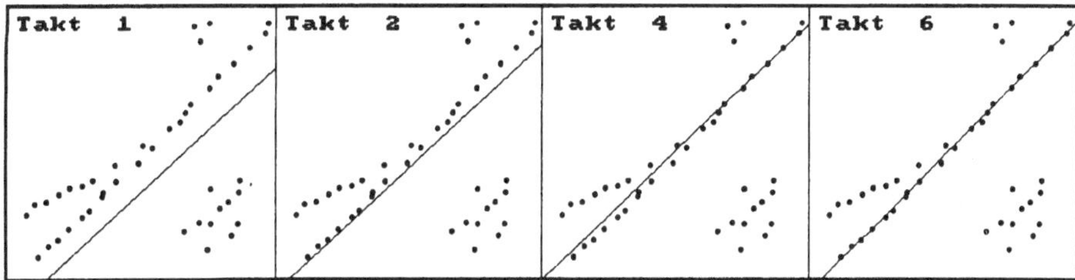

Abb. 2: Adaptive Geradenanpassung

Natürlich kann der Algorithmus bezüglich des Break-down-points-Index nicht besser sein als die normale Regressionsmethode. Liegt aber die Anfangslösung ($s=1$) im "Einzugsbereich" der zu erwartenden Lösung, dann wird auf verblüffend rasche und zuverlässige Weise die "richtige" Lösung ermittelt.

In der Bildverarbeitung sind in der Regel Geraden als zweidimensionale Kurven $f(x,y)=0$ zu interpretieren und nicht als Funktionen $y=f(x)$. Die Ausgleichsproblematik sollte dann mit Hilfe der "orthogonalen" Regression gelöst werden, d.h. mit Hilfe des Ansatzes

$$F(\mathbf{a},\mathbf{x}) = a_1 + x \cdot \cos(a_2) + y \cdot \sin(a_2) = 0 \ .$$

Die Lösung für $s=1$ ist bei [Vo92] nachzulesen. Die entsprechende Modifikation für die gewichtete Ausgleichung wird durch die folgenden Formeln geliefert:

$$G(f(x,y)) =_{def} \sum_{k=1}^{K} w_k \cdot f(x_k, y_k) \ ,$$

$$\tan(2a_2) = \frac{2\left(G(xy) - G(x) \cdot G(y)\right)}{[G(x^2) - G^2(x)] - [G(y^2) - G^2(y)]} \ ,$$

$$a_1 = G(x) \cdot \cos(a_2) + G(y) \cdot \sin(a_2) \ .$$

Wie in Abbildung 3 zu sehen ist, wird bei diesem Vorgehen nicht nur ein stabiles Resultat erhalten, sondern auch das Ausreißerproblem aus Abbildung 1 ist nach wenigen Iterationen gelöst.

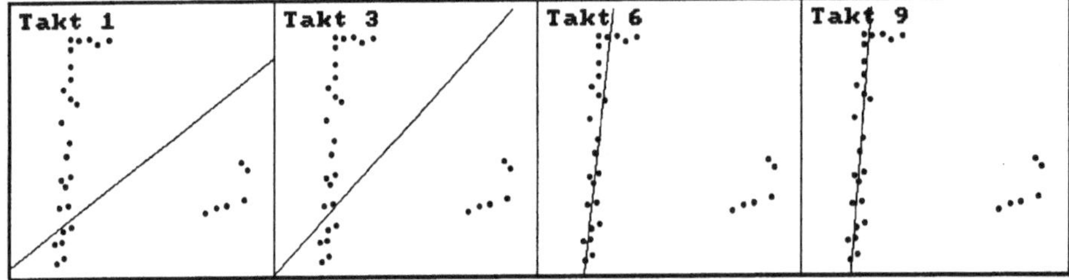

Abb. 3: Geradenanpassung mit adaptiver orthogonaler Regression

Man kann die Ermittlung von Ausgleichsgeraden auch auf völlig andere Art und Weise angehen: Wir betrachten die gegebene Punktmenge als Realisierungen einer zweidimensionalen Normalverteilung. Demzufolge haben wir die Erwartungswerte für x und y, die Varianzen und die Kovarianz zu schätzen. Diese Parameter werden im s-ten Iterationsschritt gewichtet geschätzt, wobei die Gewichte die auf 1 normierten Funktionswerte der im (s-1)-ten Iterationsschritt geschätzten Dichtefunktion der zweidimensionalen Normalverteilung darstellen. In Abbildung 4 ist zu sehen, wie sich nach wenigen Iterationsschritten das "Grauwertgebirge" der Normalverteilung auf die zur Geradenanpassung wesentlichen Punkte "zusammenzieht". Durch den in die Normalverteilung eingehenden Mahalanobis-Abstand paßt sich die Dichtefunktion der Topologie der "ungefähr eine Gerade bildenden" Punkte an.

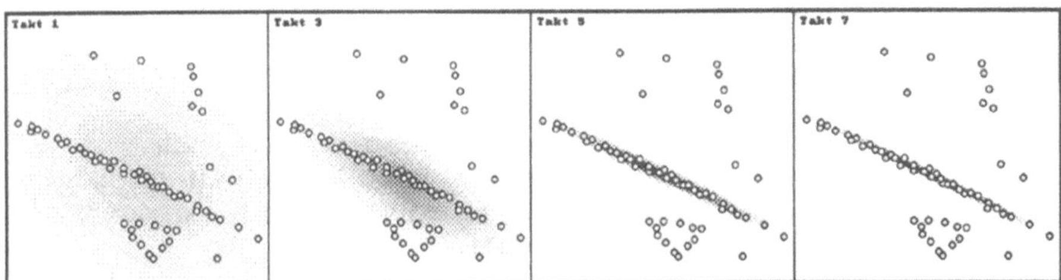

Abb. 4: Korrelationsellipsen zur Geradenanpassung

4 Beispiel Kreis- und Ellipsenanpassung

Zur Kreis- und Ellipsenanpassung benutzen wir

$$F(\mathbf{a},\mathbf{x}) = a_1 + a_2x + a_3y + a_4x^2 + a_5xy + a_6y^2 = 0,$$

wobei ein Koeffizient auf 1 zu normieren ist. Diese Vorgehensweise hat den entscheidenden Vorteil, daß die Normalengleichungen linear sind. Aber durch die Normierung eines Koeffizienten gibt es immer Fälle, die nicht lösbar sind (beispielsweise kann mit $a_1 = 1$ keine durch den Nullpunkt gehende Ellipse beschrieben werden). Man kann dieses Problem behandeln, wenn für $F(\boldsymbol{a},\boldsymbol{x})$ die geometrischen Abstände der Punkte von der zu bestimmenden Kurve benutzt werden. Dann sind aber die Normalengleichungen nichtlinear und lassen sich nur aufwendig lösen.

In Abbildung 5 ist einer Punktmenge ein Kreis anzupassen. Die Anfangslösung ($s = 1$) ist unbefriedigend. Aber schon nach wenigen Iterationen wird der gesuchte Kreis gefunden. Die Grauwertverteilung beschreibt hier die Gewichtung der Residuen im jeweiligen Iterationsschritt. Punkte, die im hellen Bereich liegen, werden bei der entsprechenden gewichteten Ausgleichung kaum noch berücksichtigt (beispielsweise Gewichte kleiner als e^{-20}).

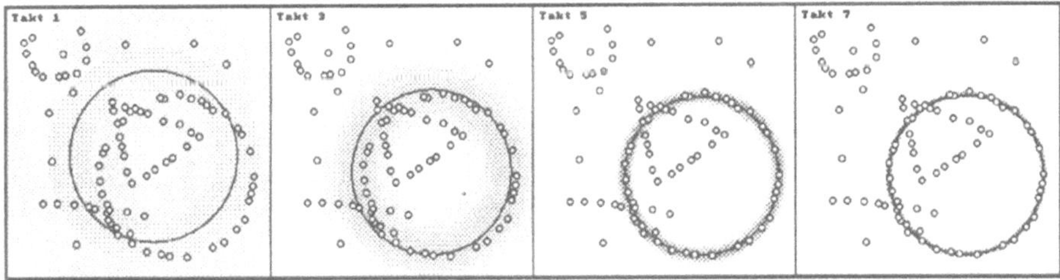

Abb. 5: Adaptive Kreisanpassung

Auch bestimmte Ausreißersituationen werden mit Hilfe des hier vorgestellten Algorithmus robust beherrscht. In Abbildung 6 ist eine typische Ausreißer-Situation gegeben. Es zeigt sich, daß Ausreißer in einer Vielzahl von Fällen eliminiert werden können, so daß nunmehr eine robuste Schätz- und Ausgleichsmethode zur Verfügung steht.

5 Beispiel lokal adaptive Filter

Wenn man die Grauwerte eines Bildes lokal durch analytisch vorgegebene Funktionen approximiert ([Vo90]), so kann man beliebige Filter entwerfen, indem man die Filteroperation aus dieser Funktion ableitet. So lassen sich Mittelwertfilter, Ableitungsfilter, Laplacefilter, Krümmungsfilter usw. entwerfen. Diese Filter reagieren oft sehr sensibel auf Rauschen. Betrachtet man Rauschen als Ausreißerproblematik, so kann man die Koeffizienten der gesuchten Funktion mit der angegebenen adaptiven Ausgleichsmethode bestimmen, wobei sich der Rechenaufwand natürlich erhöht. Abbildung 7 zeigt die Wirkung durch Anpassung einer Ebene in einer 3 x 3 Umgebung eines Gradientenfilters (3 Iterationen). Speziell die empfindlichen Ableitungsfilter zeigen somit eine wesentlich robustere Wirkung.

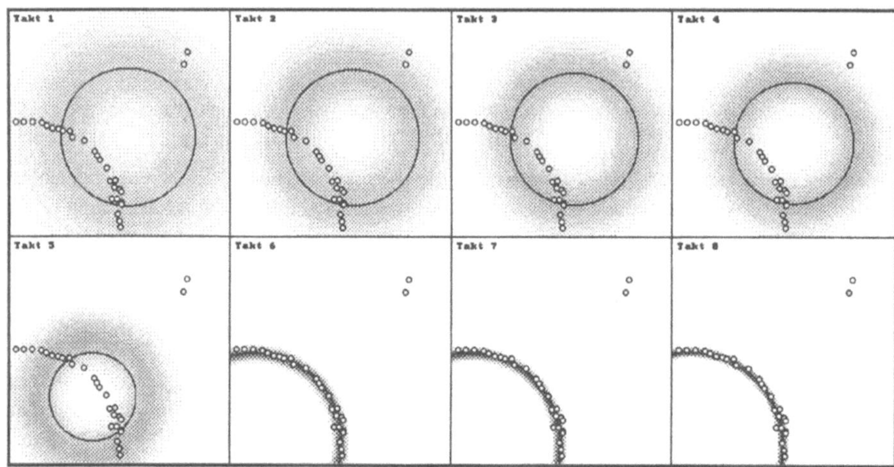

Abb. 6: Kreisanpassung mit Ausreißer

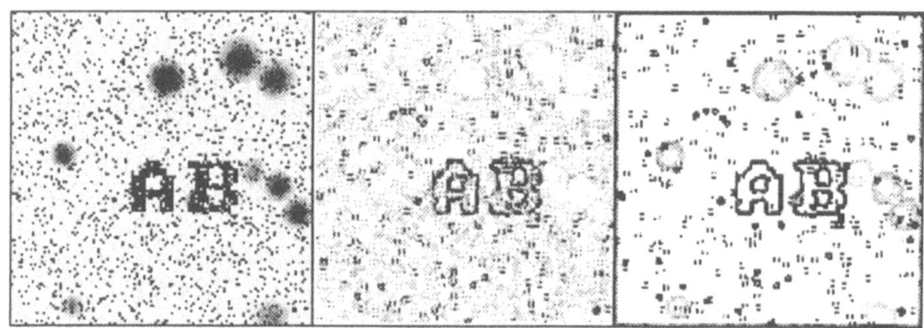

Abb. 7: Originalbild - Gradientenfilter - adaptiver Gradientenfilter

6 Vektorielle Ausgleichsrechnung

Liegen mehrere zu erfüllende Gleichungen $F_1(\boldsymbol{a},\boldsymbol{x})=0,\dots,F_R(\boldsymbol{a},\boldsymbol{x})=0$ vor, dann suchen wir eine Lösung $\boldsymbol{a}^{[s]}$ derart, daß alle Gleichungen F_i für alle Meßwerttupel näherungsweise erfüllt sind. Zu diesem Zweck muß ein gemeinsames Zielkriterium formuliert werden.

Betrachten wir den Vektor

$$\boldsymbol{v}^{\mathrm{T}} = \left(F_1(\boldsymbol{a},\boldsymbol{x}^{(1)},\dots,F_1(\boldsymbol{a},\boldsymbol{x}^{(K)}),\dots,F_R(\boldsymbol{a},\boldsymbol{x}^{(1)}),\dots,F_R(\boldsymbol{a},\boldsymbol{x}^{(K)})\right) ,$$

so ist für eine beliebige Norm das Kriterium

$$\| \boldsymbol{v} \|^2 \rightarrow Minimum$$

evident. Für die Euklidische Norm führt dies dann zur üblichen Bedingung

$$\sum_{r=1}^{R} \sum_{k=1}^{K} \left(F_r(\boldsymbol{a},\boldsymbol{x}^{(k)})\right)^2 \rightarrow Minimum .$$

Nun kann man aber nicht mehr so einfach wie im eindimensionalen Fall die Defekte bestimmen, da die Defekte als Defektvektoren zu behandeln sind:

$$\begin{pmatrix} F_1(\mathbf{a}^{[s]}, \mathbf{x}^{(k)}) \\ F_2(\mathbf{a}^{[s]}, \mathbf{x}^{(k)}) \\ \cdot \\ \cdot \\ \cdot \\ F_R(\mathbf{a}^{[s]}, \mathbf{x}^{(k)}) \end{pmatrix} = \begin{pmatrix} \delta_k^{(1)} \\ \delta_k^{(2)} \\ \cdot \\ \cdot \\ \cdot \\ \delta_k^{(R)} \end{pmatrix} .$$

Wir nehmen wieder an, daß diese Defektvektoren einer R-dimensionalen Normalverteilung unterliegen. Daraus schätzen wir den Erwartungswertvektor und die Kovarianzmatrix und bilden normierte Gewichte w_k durch Einsetzen von $\mathbf{x}_{(k)}$ in die R-dimensionale Dichtefunktion. Damit haben wir das Problem

$$\sum_{k=1}^{K} w_k \sum_{r=1}^{R} \left(F_r(\mathbf{a}, \mathbf{x}^{(k)}\right)^2 \rightarrow Minimum$$

zu lösen. Nun können wir wie im eindimensionalen Fall durch iterative Gewichtsbestimmung das Ausgleichsproblem lösen. Sind die Defekte dekorreliert, dann kann für jeden Defekt eine eindimensionale Normalverteilung benutzt werden.

Ein typisches Beispiel zur vektoriellen Ausgleichsrechnung ist die Bestimmung der perspektivischen Kameraparameter zur Kamerakalibrierung.

7 Literaturverzeichnis

Fö86 Reliability Analysis of Parameter Estimation in Linear Models with Applications to Mensuration Problems in Computer Vision, Computer Vision, Graphics, and Image Processing 40, 273-310, 1987.

Fö91 W.Förstner: Statistische Verfahren für die automatische Bildanalyse und ihre Bewertung bei der Objekterkennung und -vermessung. Habilitationsschrift, Reihe C, Heft Nr. 370, Verlag der Bayerischen Akademie der Wissenschaften 1991

Fö92 W.Förstner: Robust Methods for Computer Vision. Tutorial Notes, 2nd Intern. Workshop on Robust Computer Vision, Bonn 1992

Ha86 F.R.Hampel, E.M.Ronchetty, P.J.Rousseeuw, W.A.Stahel: Robust Statistics - The Approach Based on Influence Functions. John Wiley & Sons, New York 1986

Hu81 P.J.Huber: Robust Statistics. John Wiley & Sons, New York 1981

Ma92 A.D.Marshall, R.R.Martin: Computer Vision, Models and Inspection. World Scientific, Singapore 1992, pp. 395/396

Ro87 P.J.Rousseeuw, A.M.Leroy: Robust Regression and Outlier Detection. John Wiley & Sons, New York 1987

Si82 A.F.Siegel: Robust regression using repeated medians. Biometrika 69 (1982) 242-244

Vo90 K.Voss: Differentialgeometrie und digitale Bildverarbeitung. Bild und Ton 43 (1990) 165-170

Vo91 K.Voss, H.Süße: Praktische Bildverarbeitung. Carl Hanser Verlag, München 1991

Vo92 K.Voss: Kontursegmentierung durch Anpassung grafischer Elemente. Proc. 14. DAGM-Symposium, Dresden 1992, Springer-Verlag Berlin/Heidelberg 1992, S. 274-281

Ein Verfahren zu monokularen Tiefenbestimmung in Grauwertbildern

Thomas Wieland
Fraunhofer Institut für
Produktionsanlagen und Konstruktionstechnik (IPK)
Abteilung für Mustererkennung
Pascalstraße 8-9
D-10587 Berlin

Zusammenfassung

Die Berechnung des Abstandes zwischen Kamerakoordinatensystem und dem mittels Bildverarbeitung zu analysierenden Objekt ist eine häufig auftretende Aufgabenstellung innerhalb der 3-D Bildverarbeitung. Ist zusätzlich der Raumwinkel zwischen der optischen Achse der Kameralinse und dem Normalenvektor der zu analysierenden Oberfläche zu bestimmen, so wird dieses Problem bisher unter Einsatz zusätzlicher Hardware (Abstandssensoren) oder mit der Technik des Stereosehens (2. Kamera) gelöst, d.h. es entstehen Mehrkosten und eine erhöhte Rechenzeit.

Im folgenden Beitrag wird ein Verfahren vorgestellt, das die Abstandsinformation aus einem monokular aufgenommenen Videobild berechnet, in dessen Objektkoordinatensystem ein Laserspot eingeblendet wird. Über eine zuvor einmalig durchzuführende Kalibrationsprozedur sind die 12 Parameter eines linearen Kameramodells sowie die Geradengleichung des Lasermodells bekannt. Der Sehstrahl des im Rechnerkoordinatensystem abgebildeten Laserspots und die Gerade des Laserstrahls werden windschief geschnitten und somit die z Koordinate des Bildpunktes bestimmt. Durch Einblenden dreier Laserspots kann eine Objektebene definiert werden und der Raumwinkel zwischen dem Normalenvektor der Ebene und der optischen Achse der Kameralinse berechnet werden.

Die Einsatzgebiete dieses Verfahrens sind u.a. die Roboterkalibration, sowie die in diesem Beitrag beschriebene Entzerrung einer Texturfläche zu deren Anlalyse.

Motivation und Problemstellung

Die Abstandsbestimmung zwischen Kamera- und Objektkoordinatensystem ist eine häufig zu lösende Aufgabenstellung innerhalb der 3-D Bildverarbeitung. Anwendungsgebiete sind die Steuerung und Kalibration von Handhabungsautomaten (Robotern) mit Standard Bildverarbeitungshardware und die Analyse von geneigten bzw. gekrümmeten Oberflächen (3-D Texturanalyse) [1]. Die dazu nötige Information soll aus einem Videobild berechnet werden, um der Echtzeitbedingung möglichst nahe zu kommen.

Die dazu nötigen Modellparameter werden mit einer zuvor einmalig durchgeführten Kalibrationsprozedur berechnet. Diese Kalibrationsprozedur wird in zwei Teilbereiche unterteilt; in den der Kamera- und der Laserkalibration.

Sind sämtliche Modellparameter bestimmt, so können der Sehstrahl des im Rechnerkoordinatensystem abgebildeten Laserspots und die Gerade des Lasersstrahls windschief geschnitten werden.

Kamerakalibration

Das verwendete Kameramodell von Lenz & Tsai [2] besitzt 12 Parameter; 6 externe, sie beschreiben die Abbildung vom Weltkoordinatensystem (WKS) in das Kamerakoordinatensystem (KKS) und 6 interne sie, beschreiben die Abbildung vom KKS in das Rechnerkoordinatensystem (RKS) (Bild 1).

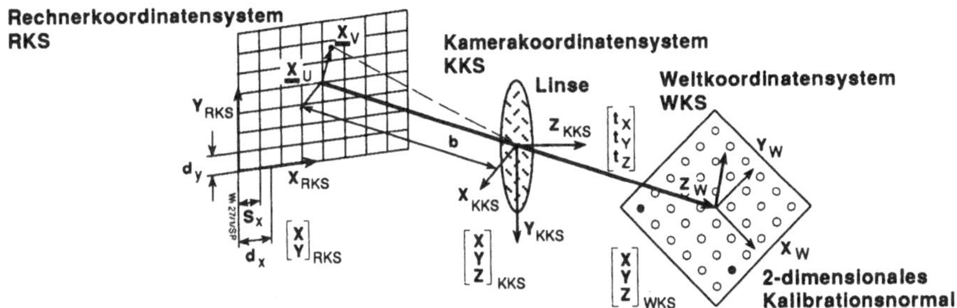

Bild 1: Das geometrische Kameramodell [2]

Die Bedeutung der Parameter ist wie folgt, ihre Berechnung erfolgt nach dem von Lenz & Tsai vorgeschlagenen Verfahren [2].

Die Abbildung vom Weltkoordinatensystem (WKS) in das KKS beschreiben die 6 externen Parameter

$$\underline{x}_{KKS} = \underline{R} \cdot \underline{x}_{WKS} + \underline{t} \tag{1}$$

Die Abbildung vom KKS in das unverzeichnete Sensorkoordinatensystem wird durch eine idealisierte Lochkamera beschrieben

$$\underline{x}_u = \frac{b}{z_{KKS}} \underline{D} \cdot \underline{x}_{KKS} \tag{2}$$

Um der Linsenverzeichnung 3. Ordnung Rechnung zu tragen, wird das unverzeichnete Koordinatensystem in ein radial verschobenes transformiert

$$\underline{x}_v = \frac{2}{1 + \sqrt{1 - 4k_3(x_u^2 + y_u^2)}} \underline{x}_u \tag{3}$$

Die örtliche Abtastung eines Punktes im verzeichneten Sensorkoordinatensystem durch einen CCD-Flächensensor legt die Skalierung im RKS fest.

$$\underline{x}_{RKS} = \begin{pmatrix} \frac{f_{ADU}}{f_p d_x} & 0 \\ 0 & \frac{1}{d_y} \end{pmatrix} \cdot \underline{x}_v + \begin{pmatrix} c_x \\ c_y \end{pmatrix} \tag{4}$$

Die Gesamttransformation ist in Bild 2 dargestellt.

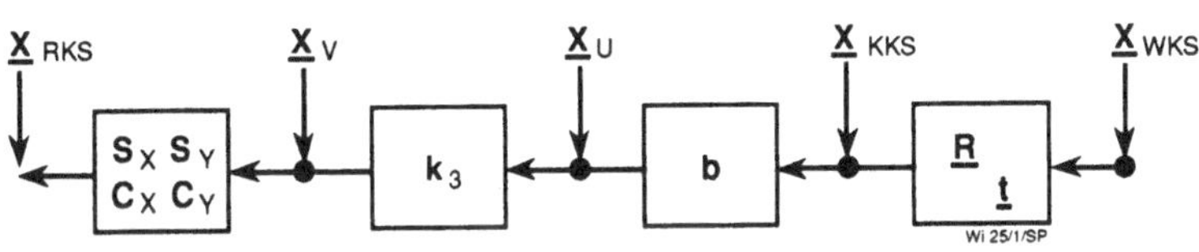

Bild 2: Die Gesamttransformation [3]

Dieses Kameramodell hat 3 Vorteile; erstens entsprechen sämtliche Parameter physikalischen Größen; zweitens werden die Parameter direkt d.h. nicht iterativ berechnet; drittens wird zur Berechnung der Parameter nur ein zweidimensionales Kalibrationsnormal benötigt.

Laserkalibration

Bei dem hier vorgestellten Lasermodell wird davon ausgegangen, daß der Laserstrahl durch eine Geradengleichung beschrieben werden kann

$$\underline{x}_L = \underline{p}_1 + \alpha(\underline{p}_2 - \underline{p}_1) \qquad (5)$$

$$= \underline{p}_1 + \alpha \underline{l}_1$$

Zur Bestimmung des Richtungsvektors $\underline{l}_l$, sowie des Startpunktes $\underline{p}_l$ wird wie folgt vorgegangen.

Der Laserspot wird auf eine Mattscheibe projeziert. Diese Projektion wird von zwei Kameras aufgenommen. Diese Kameras sind auf dasselbe WKS kalibriert. Danach werden die Schwerpunkte der abgebildeten Laserspots mit einer Grauwertinterpolation bestimmt. Diese Schwerpunkte sind Ausgangspunkte zweier Sehstrahlen, die windschief geschnitten werden. Somit erhält man die Koordinate von $\underline{p}_l$ und durch Verschieben der Mattscheibe um einen beliebigen Vektor $\underline{\Delta v}$ die Koordinate von $\underline{p}_2$ (Bild 3).

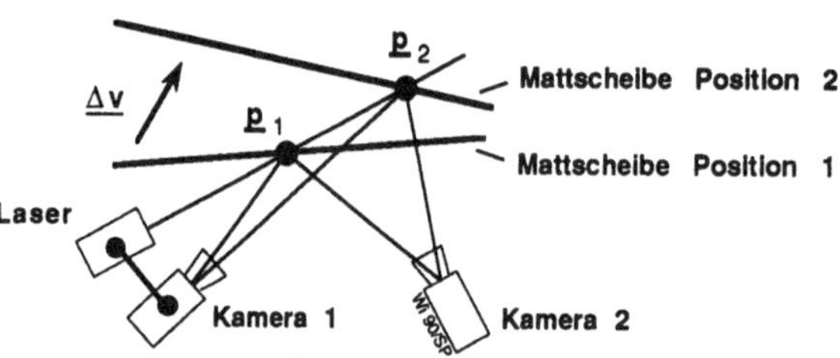

Bild 3: Die Laserkalibration

Sämtliche Modellparameter zur monokularen Abstandsbestimmung sind nach Durchführung dieser Kalibrationsprozedur bekannt.

Abstandsmessung

Zur Abstandsmessung wird der Schwerpunkt des abgebildeten Laserspots mit einer Grauwertinterpolation bestimmt. Der Sehstrahl ergibt sich mit den Gleichungen (1) und (2) zu

$$\underline{x}_{WKS} = \underline{R}^{-1} \cdot \underline{x}_{KKS} - \underline{R}^{-1} \cdot \underline{t} \qquad (6)$$

$$= \underline{R}^{-1} \cdot \begin{pmatrix} \frac{x_u}{b} \\ \frac{y_u}{b} \\ 1 \end{pmatrix} z_{KKS} - \underline{R}^{-1} \cdot \underline{t}$$

$$= \underline{\Gamma}_{KKS} z_{KKS} - \underline{R}^{-1} \cdot \underline{t}$$

Dieser wird windschief mit der Geraden des Laserstrahls geschnitten (Bild 4).

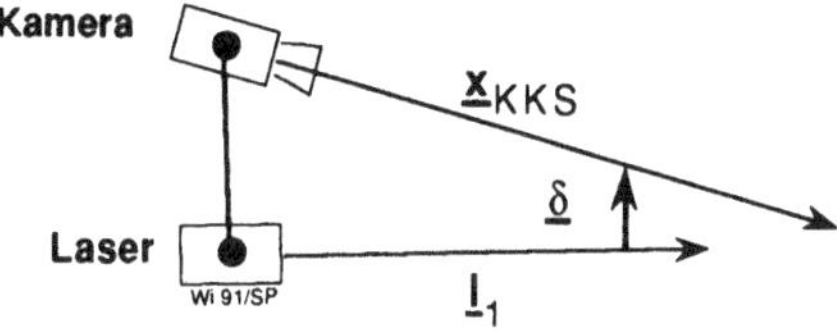

Bild 4: Die Abstandsmessung

Der Abstand der windschiefen Geraden ergibt sich zu

$$d = \frac{\left\| (\underline{p}_1 + \underline{R}^{-1} \cdot \underline{t}) \cdot (\underline{l}_1 \times \underline{\Gamma}_{KKS}) \right\|}{\left\| \underline{l}_1 \times \underline{\Gamma}_{KKS} \right\|} \tag{7}$$

Der Schnittpunkt wird wie folgt bestimmt

$$\underline{l}_1 \alpha + \underline{p}_1 + \frac{(\underline{l}_1 \times \underline{\Gamma}_{KKS})}{\left\| \underline{l}_1 \times \underline{\Gamma}_{KKS} \right\|} d = \underline{\Gamma}_{KKS} z_{KKS} - \underline{R}^{-1} \cdot \underline{t} \tag{8}$$

$$\underline{l}_1 \alpha + \underline{p}_1 + \underline{\delta} = \underline{\Gamma}_{KKS} z_{KKS} - \underline{R}^{-1} \cdot \underline{t}$$

$$\begin{pmatrix} \frac{x_u}{b} \\ \frac{y_u}{b} \\ 1 \end{pmatrix} z_{KKS} - (\underline{R} \cdot \underline{l}_1)\alpha = \underline{t} + \underline{R} \cdot (\underline{p}_1 + \underline{\delta})$$

Für die unbekannten Parameter z_{KKS} und α ergibt sich das lineare überbestimmte Gleichungssystem

$$\begin{pmatrix} \frac{x_u}{b} & -(r_{11} l_{1x} + r_{12} l_{1y} + r_{13} l_{1z}) \\ \frac{y_u}{b} & -(r_{21} l_{1x} + r_{22} l_{1y} + r_{23} l_{1z}) \\ 1 & -(r_{31} l_{1x} + r_{32} l_{1y} + r_{33} l_{1z}) \end{pmatrix} \cdot \begin{pmatrix} z_{KKS} \\ \alpha \end{pmatrix} = \begin{pmatrix} \epsilon_1 \\ \epsilon_2 \\ \epsilon_3 \end{pmatrix} \tag{9}$$

mit

$$\begin{pmatrix} \gamma_{11} & \gamma_{12} \\ \gamma_{21} & \gamma_{22} \\ \gamma_{31} & \gamma_{32} \end{pmatrix} \cdot \begin{pmatrix} z_{KKS} \\ \alpha \end{pmatrix} = \begin{pmatrix} \epsilon_1 \\ \epsilon_2 \\ \epsilon_3 \end{pmatrix} \tag{10}$$

Zur Lösung dieses überbestimmten linearen Gleichungssystems wird die Pseudoinverse verwendet [4]. Die Anwendung dieser minimiert den Fehler im Sinne einer linearen Regression.

$$\begin{pmatrix} (\gamma_{11}\gamma_{12} + \gamma_{21}\gamma_{22} + \gamma_{31}\gamma_{32}) & (\gamma_{12}^2 + \gamma_{22}^2 + \gamma_{32}^2) \\ (\gamma_{11}^2 + \gamma_{21}^2 + \gamma_{31}^2) & (\gamma_{11}\gamma_{12} + \gamma_{21}\gamma_{22} + \gamma_{31}\gamma_{32}) \end{pmatrix} \cdot \begin{pmatrix} z_{KKS} \\ \alpha \end{pmatrix} = \begin{pmatrix} \epsilon_1\gamma_{12} + \epsilon_2\gamma_{22} + \epsilon_3\gamma_{32} \\ \epsilon_1\gamma_{11} + \epsilon_2\gamma_{21} + \epsilon_3\gamma_{31} \end{pmatrix} \tag{11}$$

Mit der Substitution

$$\begin{pmatrix} \beta_{11} & \beta_{12} \\ \beta_{21} & \beta_{11} \end{pmatrix} \cdot \begin{pmatrix} z_{KKS} \\ \alpha \end{pmatrix} = \begin{pmatrix} \tau_1 \\ \tau_2 \end{pmatrix} \tag{12}$$

und der Inversion

$$\begin{pmatrix} z_{KKS} \\ \alpha \end{pmatrix} = \frac{1}{\beta_{11}^2 - \beta_{12}\beta_{21}} \begin{pmatrix} \beta_{11} & -\beta_{12} \\ -\beta_{21} & \beta_{11} \end{pmatrix} \cdot \begin{pmatrix} \tau_1 \\ \tau_2 \end{pmatrix} \tag{13}$$

ergibt sich

$$z_{KKS} = \frac{\tau_1\beta_{11} - \tau_2\beta_{12}}{\beta_{11}^2 - \beta_{12}\beta_{21}} \tag{14}$$

Der zu bestimmende Abstand im WKS wird definiert durch

$$\underline{x}_{WKSL} = \underline{R}^{-1} \cdot \left(\begin{pmatrix} \frac{x_u}{b} \\ \frac{y_u}{b} \\ 1 \end{pmatrix} z_{KKS} - \underline{t} \right) - \frac{(\underline{l}_{1x} \times \underline{r}_{KKS})}{\| \underline{l}_{1x} \times \underline{r}_{KKS} \|} \frac{d}{2} \tag{15}$$

Raumwinkelmessung

Zur Bestimmung des Raumwinkels zwischen der optischen Achse der Kameralinse und dem Normalenvektor der zu analysierenden Oberfläche werden drei Laserspots projeziert. Die einzelnen Laser werden wie beschrieben kalibriert. Die drei Weltkoordinatenpunkte werden zu einem Schwerpunkt zusammengefasst

$$\underline{x}_{SL} = \frac{1}{3} \sum_{i=1}^{3} \underline{x}_{WKSL(i)} \tag{16}$$

Der Raumwinkel berechnet sich dann zu

$$\phi = \arccos\left(\frac{\underline{x}_{SL} \cdot \underline{o}}{\| \underline{x}_{SL} \| \, | \underline{o} |} \right) \tag{17}$$

wobei $\underline{o}$ der Sehstrahl des Sensortargetmittelpunktes ist.

Experimentelle Ergebnisse

Die geometrische Entzerrung einer texturierten Oberfläche zu deren Analyse wurde im Folgenden untersucht. Ziel dieser Untersuchung war es, die Abhängigkeit statistischer Merkmale zur Texturanalyse in Bezug auf die Beobachtungsrichtung quantitativ zu messen. Die eingesetzte CCD Kamera war eine Panasonic WV-CD50 mit einem Standard TV 35mm Objektiv mit 3mm Zwischenring. Die Laser waren Standard Laser 2mW / 670nm.

Die Parameter des Kameramodells waren

Rotationsvektor in Rodriguez Notation [dimensionslos]
wx = -0,235951 wy = -0,136936 wz = 0,000766
Translationsvektor in [mm]
tx = 3,876775 ty = 3,259869 tz = 537,152489
Kamerahauptpunkt in [Pixel]
cx = 254,653102 cy = 241,482699
Skalierungsfaktoren in [μm]
sx = 10,469 sy = 11,0
Kammerkonstante in [mm]
b = 39,330465
Linsenverzeichnungskoeffizient 3. Ordnung in $1/mm^2$
k_3 = 0,000009

die Entfernung des Schwerpunktes der Laserspots in [mm] ergab sich zu

$x_{SL} = -13{,}999879$ $y_{SL} = 11{,}438947$ $z_{SL} = -57{,}697900$

der Raumwinkel war $\phi = 22{,}3°$

In Bild 5 ist die unter dem Winkel $\phi = 22{,}3°$ aufgenommene Texturfläche zu sehen.

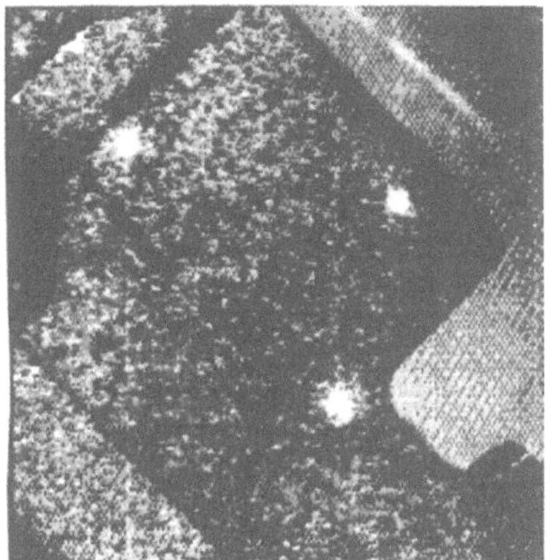

Bild 5: Texturfläche unter Raumwinkel $\phi = 22{,}3°$

In Bild 6a ist die rechnerisch zurückgedrehte Texturfläche und in Bild 6b die unter einem Raumwinkel $\phi \approx 0°$ aufgenommene Texturfläche dargestellt.

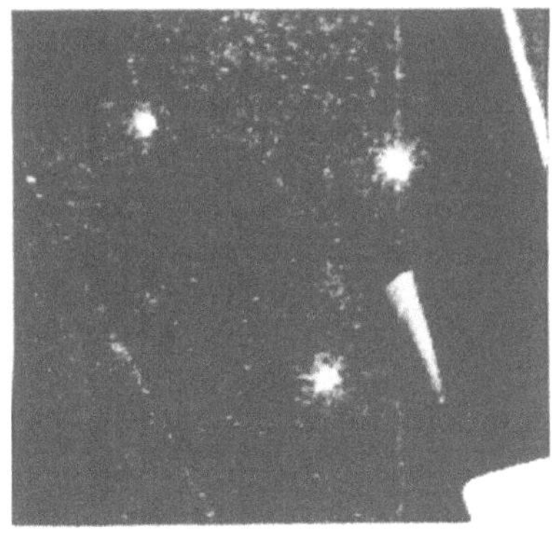

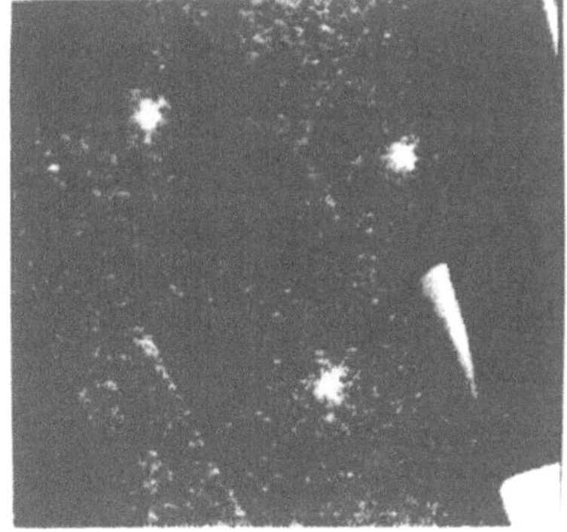

Bild 6a: Texturfläche unter Raumwinkel $\phi \approx 0°$

Bild 6b: Rechnerisch gedrehte Texturfläche

Die untersuchten statistischen Merkmale (Co-Occurrencematrix) [5] sind zwischen Bild 6a und 6b equivalent, während Bild 5 erhebliche Abweichungen liefert. Eine Klassifikation der Merkmale lieferte das Ergebnis einer "fehlerhaften" Textur, obwohl die Texturverzerrungen nur auf Abbildungsfehlern beruhen, nicht auf Fertigungsfehlern.

Ausblicke

Diese Methode kann nicht nur zur Entzerrung von Oberflächen eingesetzt werden, sondern in jeder Applikation, bei der die Abstandsbestimmung zwischen Kamera- und Objektkoordinatensystem notwendig ist.

Weitere Untersuchungen zur Einsetzbarkeit bei der Kalibration (online) von Handhabungsautomaten werden derzeit durchgeführt.

Dank

An dieser Stelle möchte ich dem Direktor des Bereiches Prozeßtechnik am IPK-Berlin, Dr.-Ing. W.Adam, sowie dem Leiter der Abteilung für Mustererkennung, Dr.-Ing. B.Nickolay, meinen Dank für die wohlwollende Unterstützung dieser Arbeit ausprechen.

Verwendete Symbole

Symbol	Bedeutung
$\underline{t} \in R^3$	Translationsvektor zwischen WKS und KKS im [mm]
$\underline{R} \in R^{3x3}$	Rotationsmatrix zwischen WKS und KKS
$\underline{D} = \begin{pmatrix} 1 & 0 & 0 \\ 0 & 1 & 0 \end{pmatrix}$	Matrix zur Dimensionsreduktion
$b \in R$	Kamerkonstante der idealisierten Lochkamera in [mm]
$\underline{x}_{WKS} \in R^3$	Vektor im WKS in [mm]
$\underline{x}_{KKS} \in R^3$	Vektor im KKS in [mm]
$\underline{x}_u \in R^2$	Vektor im unverzeichneten Sensorkoordinatensystem in [mm]
$\underline{x}_v \in R^2$	Vektor im verzeichneten Sensorkoordinatensystem in [mm]
$k_3 \in R$	Linsenverzeichnungskoeffizient 3. Ordnung in $[1/mm^2]$
$f_{ADU} \in R$	Abtastfrequenz des Analog/Digitalumsetzers in [MHz]
$f_p \in R$	Pixelfrequenz des Kamerataktes in [MHz]
$\{d_x, d_y\} \in R$	Abstand der Sensorelemente in [μm]
$c_x, c_y \in R$	Sensormittelpunkt in [Pixel]
$\underline{x}_{RKS} \in R^2$	Vektor im RKS in [Pixel]
$\{\underline{x}_L, \underline{p}_1, \underline{p}_2\} \in R^3$	Laserspotkoordinaten im WKS in [mm]
$\alpha \in R$	Skalar des Richtungsvektors
$\underline{l}_l \in R^3$	Richtungsvektor des Laserstrahls im WKS in [mm]
$\underline{\Delta v} \in R^3$	Verschiebungsvektor der Mattscheibe in [mm]
$\{\underline{\Gamma}_{KKS}, \underline{\delta}\} \in R^3$	Hilfsgrößen zur Berechnung
$\{\epsilon_i, \gamma_{ii}, \tau_i\} \in R$	
$d \in R$	Abstand zwischen Seh- und Laserstrahl in [mm]
$\underline{x}_{SL} \in R^3$	Schwerpunkt dreier Laserspots in [mm]
$\phi \in \{x \mid 0{,}0 \leq x \leq 180{,}0\}$	Raumwinkel zwischen Oberflächennormalen und optischer Achse der Kameralinse in [DEG]

Literaturhinweise

[1] Nickolay, B.; Wieland, T.J. 1992: Texturanalyse an leicht gekrümmten Oberflächen. Zeitschrift für wirtschaftliche Fertigung und Automatisierung, Vol.87, No.10. München : Carl Hanser Verlag, S.600-602.

[2] Lenz, R.K.; Tsai, R.Y. 1988: Techniques for Calibration of Scale Factor and Image Center for High Accuracy 3-D Machine Vision Metrology. IEEE Transaction on Pattern Analysis and Machine Intelligence, Vol.10, No.5. Washington : IEEE Press, S.713-720.

[3] Wieland, T.J. 1992: Untersuchung eines linearen direkten Verfahrens zur Kamerakalibration. Vision & Voice Magazine, Vol.6 , No.1. Coburg : Sprechsaal Verlag, S.31-36.

[4] Maess, G. 1985: Vorlesungen über numerische Mathematik I Lineare Algebra. Basel und Stuttgart : Birkhäuser Verlag.

[5] Nickolay, B.; Vollmerhaus, D. 1988: Texturanalyse mittels Co-Occurrence Matrizen zur automatischen Oberflächenprüfung. Zeitschrift für wirtschaftliche Fertigung und Automatisierung, Vol.83, No.10. München : Carl Hanser Verlag, S.90-94.

Ein Verfahren zur Erkennung linienhafter Objekte in digitalen Bildern *

Andreas Busch

Institut für Angewandte Geodäsie
Abt. Photogrammetrische Forschung
Richard-Strauß-Allee 11
60598 Frankfurt am Main
e-mail: busch@p9.ifag.de

Zusammenfassung: Es wird ein Verfahren zur Erkennung von Linien vorgestellt, das auf der lokalen Anpassung eines Polynoms 2. Grades an die Bildfunktion und der Anwendung von Markoff-Zufallsfeldern beruht. Markoff-Zufallsfelder ermöglichen die Beschreibung von Beziehungen zwischen benachbarten Linienpixeln, um z.B. Lücken in Linien zu überbrücken und breite Linien zu skelettieren. Die Anwendung des Verfahrens wird an einem Beispiel aus der Fernerkundung demonstriert.

1 Einführung

Die automatische Erkennung von Linien ist in vielen Anwendungsgebieten der digitalen Bildverarbeitung von Bedeutung, z.B. in der Fernerkundung und Kartographie (Erkennung von Straßen, Eisenbahnstrecken, Wasserstraßen) und in der Medizin (Erkennung von Blutgefäßen, Teilen des Skelettes und bei Aufnahmeverfahren, die Schnittbilder von Objekten liefern). Vor dem Hintergrund der erwähnten Objekte kann eine Linie als ein gekrümmtes oder gerades Objekt definiert werden, das bei näherungsweise konstanter Breite eine im Vergleich zur Breite große Länge besitzt. Linien sind demnach von zwei Kanten begrenzt. Deshalb sind prinzipiell zwei Vorgehensweisen zur Erkennung von Linien denkbar. Eine Möglichkeit liegt darin, erst eine Kantenextraktion durchzuführen und Linien dann durch die Zuordnung benachbarter paralleler Kanten zu detektieren. Bei der anderen Vorgehensweise wird ein Operator auf das digitale Bild angewandt, der als Ergebnis direkt Linienpixel oder zumindest Linienpixelkanditaten liefert. Im folgenden wird dem letztgenannten Ansatz nachgegangen. Dies ist sinnvoller, da er ohne nacheinander ablaufende Bearbeitungsschritte auskommt.

Bei bestehenden Verfahren zur Linienextraktion, z.B. Wang et al. (1992) und die dort aufgeführten Literaturstellen, werden im allgemeinen zunächst mit Hilfe eines geeigneten Operators Linienpixel klassifiziert. Das Ergebnis dieses Prozesses ist ein Binärbild, das dann eine Reihe weiterer Bearbeitungsschritte durchläuft. So werden einzelne Linienpixel und sehr kurze Linien eliminiert, breite Linien mit Hilfe von Skelettierungsalgorithmen (Rosenfeld und Kak 1982, Bd.2,

* Mitteilung aus dem Institut für Angewandte Geodäsie

S.232) verdünnt und Lücken durch Linienverfolgung (Haberäcker 1989, S.349) geschlossen. Dagegen wird in diesem Beitrag der Weg beschritten, die Bearbeitung der genannten Einzelaufgaben in einem geschlossenen Modell zu integrieren. Dazu wird zunächst ein Verfahren vorgestellt, das Pixel eines Grauwertbildes als Linienpixel klassifiziert. Es wird aus einem allgemeineren Operator (Haralick et al. 1983) entwickelt, der vor allem zur Kantendetektion (Haralick 1984) verwendet wird. Beziehungen zwischen benachbarten Linienpixeln, die bei diesem Operator nicht berücksichtigt werden, werden dann als Markoff-Zufallsfelder formuliert. Beide Ansätze werden zum vollständigen Modell zur Linienerkennung zusammengefaßt. Schließlich folgt ein Anwendungsbeispiel aus dem Bereich der Fernerkundung.

2 Linienextraktion durch lokale Polynomanpassung

Die Berechnung eines lokal an die Grauwerte angepaßten Polynoms zur Beschreibung eines digitalen Bildes geht auf Haralick et al. (1983) zurück. Sie benutzen ein Polynom 3. Grades, um anhand der Ableitungen und der Ableitungen in bestimmten Richtungen, z.B. in Richtung des größten Gradientenbetrages, Pixel mit besonderen Eigenschaften, z.B. Spitzen, Einkerbungen und Grate, zu klassifizieren. Der Grundgedanke dieses Abschnittes liegt darin, für die Extraktion von Linienpixeln ein Polynom 2. Grades zu verwenden, da sich eine Parabel sehr gut an den Querschnitt einer Linie anpaßt. Die Entscheidung über den Zustand Linien-Pixel oder Nicht-Linien-Pixel wird dann mit Hilfe der zweiten Ableitung, also der Krümmung, in Hauptkrümmungsrichtung (= Richtung stärkster Krümmung) getroffen.

Ein Polynom 2. Grades in Abhängigkeit der Zeilenkoordinate x und der Spaltenkoordinate y hat die Form

$$g(x,y) = k_0 + k_1 x + k_2 y + k_3 x^2 + k_4 xy + k_5 y^2 \quad . \tag{1}$$

Die sechs Koeffizienten $k_0, \ldots, k_5$ des Polynoms werden für jedes Pixel anhand eines Fensters der Bildfunktion nach der Methode der kleinsten Quadrate geschätzt. Zur Bestimmung der Koeffizienten reicht ein 3×3 Fenster aus. Sinnvoller ist aber die Verwendung größerer Fenster, also 5×5, 7×7, usw. Die Koeffizienten ergeben sich als Matrizenprodukt einer $6 \times n$ Matrix $\mathbf{A}$ und des $n \times 1$ Vektors der Grauwerte im Fenster. Die Matrix $\mathbf{A}$ ist für alle Bildpunkte identisch, wenn die Koordinaten x und y auf das Fenster bezogen werden.

Für die ersten und zweiten Ableitungen des Polynoms (1) gilt

$$\begin{aligned}
\frac{\partial g(x,y)}{\partial x} &= k_1 + 2k_3 x + k_4 y & \frac{\partial g(x,y)}{\partial y} &= k_2 + k_4 x + 2k_5 y \\
\frac{\partial^2 g(x,y)}{\partial x^2} &= 2k_3 & \frac{\partial^2 g(x,y)}{\partial y^2} &= 2k_5 \\
\frac{\partial^2 g(x,y)}{\partial x \partial y} &= k_4 & \frac{\partial^2 g(x,y)}{\partial y \partial x} &= \frac{\partial^2 g(x,y)}{\partial x \partial y} \quad .
\end{aligned} \tag{2}$$

Die zweiten Ableitungen werden zur symmetrischen Hesse-Matrix zusammengefaßt.

$$\mathbf{H} = \begin{pmatrix} \frac{\partial^2 g(x,y)}{\partial x^2} & \frac{\partial^2 g(x,y)}{\partial x \partial y} \\ \frac{\partial^2 g(x,y)}{\partial y \partial x} & \frac{\partial^2 g(x,y)}{\partial y^2} \end{pmatrix} = \begin{pmatrix} 2k_3 & k_4 \\ k_4 & 2k_5 \end{pmatrix} \tag{3}$$

Aus der Hesse-Matrix ergibt sich die Richtung α stärkster Krümmung als Richtung des zum betragsgrößten Eigenwert gehörenden Eigenvektors. Die Lösung des Eigenwertproblems ist bei einer 2×2 Matrix unproblematisch (Faddejew und Faddejewa 1976, S.277). Zur Bestimmung der Parabel, die in der Schnittebene in Richtung dieses Eigenvektors liegt, wird die Richtungsableitung benötigt. Sie ist definiert als (Haralick et al. 1983)

$$g'_\alpha(x,y) = \lim_{h\to\infty} \frac{g(x+h\sin\alpha, y+h\cos\alpha) - g(x,y)}{h} \quad . \tag{4}$$

Hieraus folgt

$$\begin{aligned} g'_\alpha(x,y) &= \frac{\partial g(x,y)}{\partial x}\sin\alpha + \frac{\partial g(x,y)}{\partial y}\cos\alpha \\ g''_\alpha(x,y) &= \frac{\partial^2 g(x,y)}{\partial x^2}\sin^2\alpha + 2\frac{\partial^2 g(x,y)}{\partial x \partial y}\sin\alpha\cos\alpha + \frac{\partial^2 g(x,y)}{\partial y^2}\cos^2\alpha \quad . \end{aligned} \tag{5}$$

Die Parameter der Schnittparabel

$$f(r) = c_0 + c_1 r + c_2 r^2 \tag{6}$$

werden durch Koeffizientenvergleich der Ableitungen (5) mit den Ableitungen

$$\frac{\partial f(r)}{\partial r} = c_1 + 2c_2 r \ , \qquad \frac{\partial^2 f(r)}{\partial r^2} = 2c_2 \tag{7}$$

an der Stelle $x = 0, y = 0$ bzw. $r = 0$ erhalten, so daß

$$c_1 = k_1 \sin\alpha + k_2 \cos\alpha \tag{8}$$

und

$$c_2 = k_3 \sin^2\alpha + k_4 \sin\alpha\cos\alpha + k_5\cos^2\alpha \tag{9}$$

gilt. Der Parameter c_0 wird nicht benötigt.

Ein Pixel gilt als Linienpixel, wenn der Scheitelpunkt der Parabel (7) innerhalb des Pixels liegt und der Betrag der Krümmung der Parabel einen vorzugebenden Schwellwert T überschreitet. Der Scheitelpunkt der Parabel ergibt sich durch gleich Null setzen von $\partial f(r)/\partial r$ zu

$$r_0 = -\frac{c_1}{2c_2} \quad . \tag{10}$$

Ein ähnlicher Ansatz ist bei Wang et al. (1992) zu finden. Dort werden jedoch anstelle der flächenhaften Anpassung eines Polynoms zweier Variablen vier Parabeln einer Variablen in Richtung der beiden Koordinatenachsen und der Diagonalen berechnet. Dies führt insbesondere bei großen Nachbarschaften zu schlechteren Isotropieeigenschaften.

3 Anwendung von Markoff-Zufallsfeldern

Das bisher vorgestellte Verfahren ist noch nicht in der Lage, Beziehungen zwischen benachbarten Linienpixeln herzustellen, was notwendig ist, um Lücken zu schließen, breite Linien zu skelettieren und kurze Linien zu eliminieren. Geeignet für diese Aufgaben ist die Anwendung von Markoff-Zufallsfeldern, die seit der Arbeit von Geman und Geman (1984) große Beachtung in der digitalen Bildverarbeitung gefunden haben. Sie sind dadurch charakterisiert, daß die bedingte Wahrscheinlichkeit einer einzelnen Variablen des Zufallsfeldes nur von den Variablen der Nachbarschaft abhängt. Es gilt also

$$p(\beta_i \mid \beta_1, \ldots, \beta_{i-1}, \beta_{i+1}, \ldots, \beta_n) = p(\beta_i \mid \beta_j \in \partial\beta_i) \ , \tag{11}$$

wobei $\boldsymbol{\beta} = (\beta_1, \ldots, \beta_n)^T$ ein $n \times 1$ Zufallsvektor ist und $\partial\beta_i$ die Nachbarschaft der Variablen β_i bezeichnet, z.B. eine 4er-, 8er- oder 12er-Nachbarschaft. Äquivalent zu Markoff-Zufallsfeldern sind Nachbarschafts-Gibbsfelder, deren Dichtefunktion durch

$$p(\boldsymbol{\beta}) \propto \exp\Big(-\sum_C U_C(\boldsymbol{\beta})\Big) \tag{12}$$

definiert ist. Die Summation innerhalb der Exponentialfunktion wird über sogenannte Cliquen durchgeführt. Dies sind Teilmengen des Zufallsfeldes, bei denen alle Elemente gegenseitig benachbart sind. Das sogenannte Cliquenpotential $U_C(\boldsymbol{\beta})$ ist eine Funktion, die wahrscheinlichen Zuständen kleine Funktionswerte und unwahrscheinlichen Zuständen große Funktionswerte zuweist. Wegen (12) sind kleine Potentiale mit großen Wahrscheinlichkeiten verbunden. Eine wichtige Beziehung, die aus der Äquivalenz von Markoff-Zufallsfeldern und Nachbarschafts-Gibbsfeldern erhalten wird (Besag 1974, Busch 1992, S.13), ist

$$p(\beta_i \mid \beta_j \in \partial\beta_i) \propto \exp\Big(-\sum_{C \mid \beta_i \in C} U_C(\boldsymbol{\beta})\Big) \ . \tag{13}$$

Sie besagt, daß die bedingte Dichte von β_i anhand aller Cliquen, die β_i enthalten, berechnet werden kann.

Nachfolgend wird ein Cliquenpotential vorgestellt, das auf der Grundlage eines Binärbildes Konstellationen von Linienpixeln bewertet. Es ist für Cliquen der Größe 4×4 und größer geeignet. Die Zufallsmatrix des Linienattributes wird mit $\boldsymbol{\Lambda} = (\lambda_{ij})$ bezeichnet. Die λ_{ij} nehmen die Werte 0, d.h. das Pixel ist kein Linienpixel und 1, d.h. das Pixel ist ein Linienpixel, an. Das Potential beruht auf einfachem, elementaren Wissen über Linien. Ihm liegt die Überlegung zugrunde, daß ein Linienpixel in seiner 8er-Nachbarschaft in der Regel zwei Nachbarlinienpixel besitzt. Hat es weniger Nachbarn, ist es entweder ein Einzelpixel (kein Nachbarlinienpixel) oder ein Linienende (ein Nachbarlinienpixel). Pixel mit mehr als zwei Nachbarlinienpixeln sind in einer Clique immer dann zu erwarten, wenn mehr als zwei Linien in die Clique hineinlaufen. Dies wird anhand des Cliquenrandes überprüft. So sind keine Verzweigungen zu erwarten, wenn an zwei Stellen des Randes Linienpixel vorhanden sind. Eine bzw. zwei Verzweigungen treten auf, wenn drei bzw. vier Linien am Rand der Clique gezählt werden. Ein sinnvolles Cliquenpotential zur Bewertung der Linienkonstellationen ist deshalb

$$U_C(\boldsymbol{\Lambda}) = n_{<2} + n_{>2} - n_{\mathrm{V}} \ . \tag{14}$$

Darin bezeichnet $n_{<2}$ die Anzahl der Linienpixel der Clique, die weniger als zwei Liniennachbarn in ihrer 8er-Nachbarschaft besitzen. $n_{>2}$ ist die Anzahl der Linienpixel mit mehr als zwei Liniennachbarn und n_{V} ist die Anzahl der in der Clique zu erwartenden Verzweigungen.

4 Vollständiges Modell zur Linienerkennung

Die in den Abschnitten 2 und 3 vorgestellten Ansätze werden nun in einem vollständigen Modell integriert, das die Eigenschaften beider Ansätze vereinigt. Anhand dieses Modells wird dann die Matrix $\boldsymbol{\Lambda}$ geschätzt. Dazu wird als Verteilung

$$p(\boldsymbol{\Lambda}\,|\,\mathbf{Y}) \propto \exp\Big(-c_{\mathrm{P}}\sum_{i,j}\lambda_{ij} - c_{\mathrm{G}}\sum_{C}(n_{<2}+n_{>2}-n_{\mathrm{V}}) - \sum_{i,j}\begin{cases} -\lambda_{ij}c_2^2 & \text{falls} \quad -0.5 \le x_0, y_0 \le 0.5 \ \wedge \ |c_2| \ge T \\ 0 & \text{sonst} \end{cases}\Big) \tag{15}$$

eingeführt, wobei $x_0 = r_0 \sin\alpha, y_0 = r_0 \cos\alpha$ mit r_0 entsprechend (10). Der Parameter c_{G} dient zur Gewichtung des Anteils der Linienkonstellationen gegenüber dem Anteil aus der Polynomanpassung. Der Term $c_{\mathrm{P}}\lambda_{ij}$ bewirkt eine „Bestrafung" des Auftretens von Linienelementen. Dadurch wird vermieden, daß sowohl Cliquen mit bestimmten Linienkonstellationen als auch Cliquen, in denen keine Linien auftreten, das Potential 0 zugewiesen wird. $\mathbf{Y}$ bezeichnet die Bildmatrix des digitalen Grauwertbildes, die über die lokale Anpassung der Polynome nach der Methode der kleinsten Quadrate funktional in $p(\boldsymbol{\Lambda}\,|\,\mathbf{Y})$ enthalten ist. Gl. (15) ist als Posteriori-Dichte des Bayes-Theorems (s. z.B. Koch 1990), also als bedingte Verteilung der unbekannten Parameter $\boldsymbol{\Lambda}$ bei gegebenen Beobachtungen $\mathbf{Y}$, interpretierbar. $p(\boldsymbol{\Lambda}\,|\,\mathbf{Y})$ wird hier jedoch per Definition eingeführt. Ansätze zur Herleitung mit Hilfe des Bayes-Theorems werden von Busch (1992) im Zusammenhang mit der Bildrestaurierung beschrieben.

Zur Bestimmung des MAP-Schätzers $\hat{\boldsymbol{\Lambda}}$ muß $\boldsymbol{\Lambda}$ so gewählt werden, daß $p(\boldsymbol{\Lambda}\,|\,\mathbf{Y})$ maximal wird. Da dieses Problem analytisch nicht lösbar ist, werden numerische Methoden angewandt, die aufgrund des großen Suchraumes i.a. nur Näherungslösungen finden. Die Berechnungsverfahren können unterteilt werden in stochastische Algorithmen, z.B. Simulated Annealing (Geman und Geman 1984) und deterministische Algorithmen, z.B. Iterated Conditional Modes (Besag 1986). Eine Übersicht und ein Vergleich der Verfahren ist bei Busch (1992, S.43), Busch und Koch (1990) zu finden. Den Verfahren ist gemeinsam, daß sie die Beziehung (13) nutzen, um lokal mit Hilfe der Nachbarschaft verbesserte Schätzwerte zu erhalten. Zur Berechnung der nachfolgend vorgestellten Ergebnisse werden in zufälliger Reihenfolge alle Pixel des Bildes aufgesucht. In einem 3×3 Fenster um das jeweilige Pixel wird durch Bestimmung der Potentiale aller $2^9 = 512$ Zustände der wahrscheinlichste Zustand ausgewählt.

5 Anwendungsbeispiel

Das beschriebene Verfahren wird auf einen Ausschnitt einer Szene des Fernerkundungssatelliten SPOT vom 25.9.1986 angewandt. Der Ausschnitt zeigt ein Autobahndreieck im Raum Frankfurt (s. Abb. 1). Es werden zwei Varianten des Verfahrens durchgeführt. Bei der ersten Variante wird für die lokale Anpassung des Polynoms 2. Grades ein 5×5 Fenster verwendet, bei der zweiten ein 7×7 Fenster. Die Cliquengröße bei der Berücksichtigung der Linienkonstellationen beträgt in beiden Fällen 5×5 Pixel. Die Steuerparameter des Verfahrens sind bei der ersten Variante $T = 1.0, c_G = 0.2, c_P = 0.5$ und bei der zweiten Variante $T = 0.4, c_G = 0.075, c_P = 0.1$. Die Parameter wurden in mehreren Durchläufen interaktiv festgelegt. Wegen der Anwendung von $c_2 \leq -T$ anstelle von $|c_2| \geq T$ in (15) berücksichtigt das Verfahren nur die Linien, die heller als ihre Umgebung sind. Die Ergebnisse sind in Abb. 2 zu sehen.

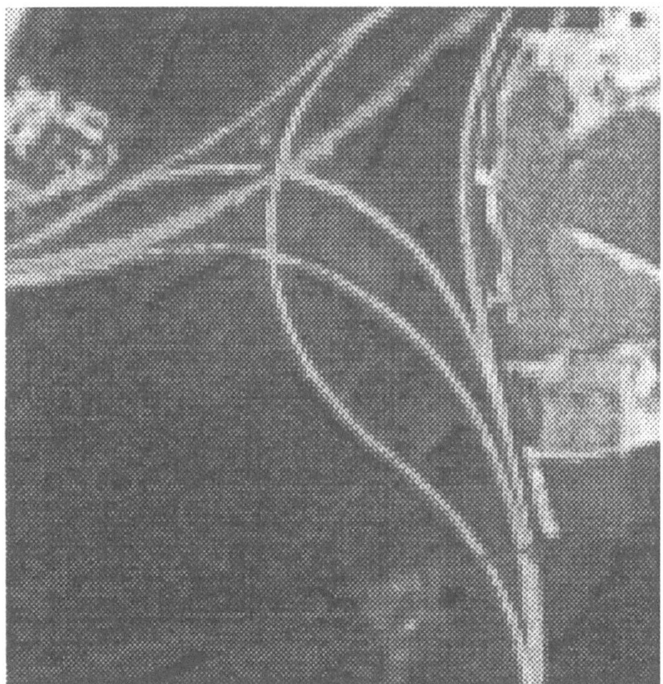

Abb. 1: Ausschnitt einer SPOT-Szene, 128×128 Pixel

Die Ergebnisse zeigen, daß es gelungen ist, Lücken zu schließen, Linien auf die Breite eines Pixels zu reduzieren und kurze Linien zu eliminieren. Es werden

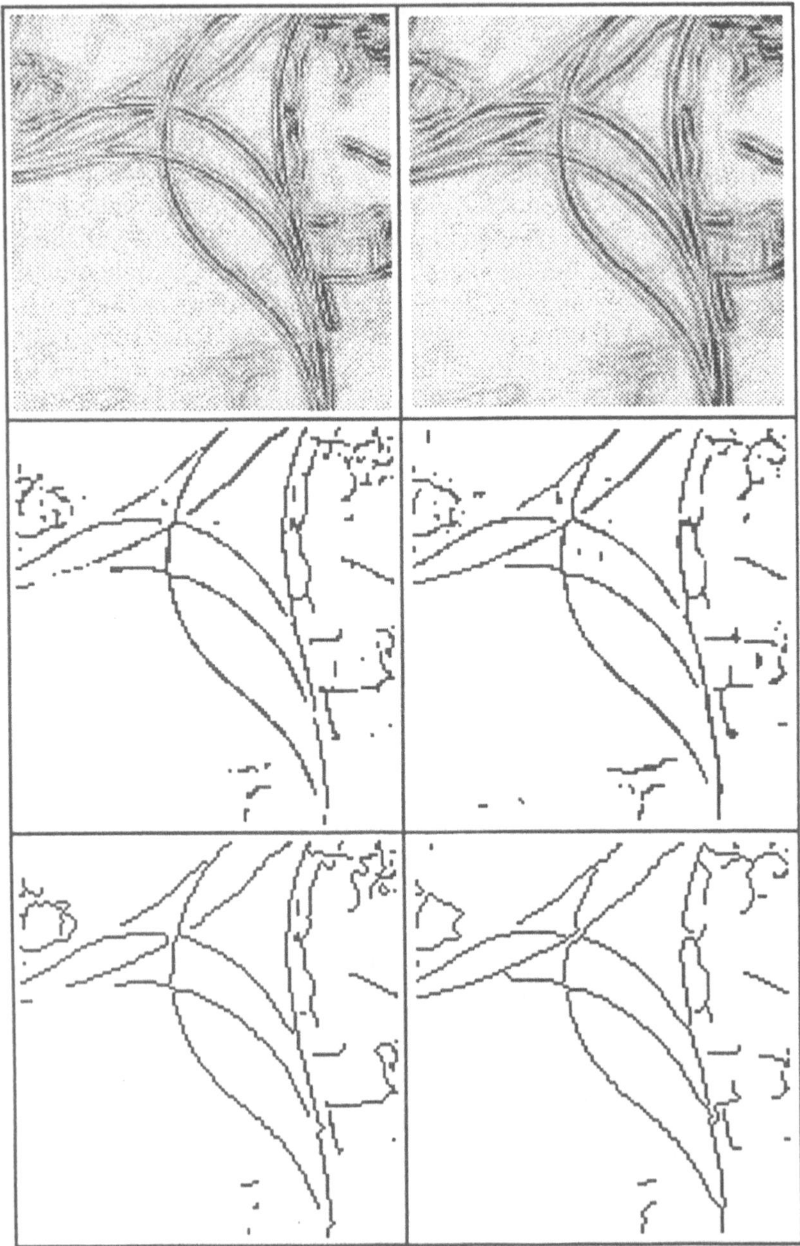

Abb. 2: Ergebnisse von Variante 1 (links) und Variante 2 (rechts), Betrag des Parameters c_2 aus (6) (oben), Ergebnisse mit Verfahren aus Abschnitt 2 (Mitte), Ergebnisse mit Verfahren aus Abschnitt 4 (unten)

aber auch Schwächen deutlich, die in stark strukturierten Gebieten, z.B. bei Kreuzungen, auftreten. Der Vergleich beider Varianten verdeutlicht, daß bei der Wahl der Fenstergröße für die Polynomanpassung die Breite und der Abstand der Linien im Bild berücksichtigt werden müssen.

6 Schlußfolgerungen und Ausblick

Das vorgestellte Verfahren ist in der Lage, in nicht zu stark strukturierten Gebieten gute Ergebnisse bei der Linienerkennung zu liefern. Verbesserungen sind zu erwarten, wenn z.B. die Richtung einer Linie, die orthogonal zu der in Abschnitt 2 verwendeten Richtung ist, für die Linienverknüpfung genutzt wird. Auch ein Modell, das in der Lage ist, Knotenpunkte bei Linienkreuzungen und Verzweigungen zu erkennen, würde eine bessere Verknüpfung der Linien ermöglichen. Unbefriedigend ist, daß dem Verfahren drei Steuerparameter vorgegeben werden müssen. Sinnvoll ist deshalb die Reduzierung der Anzahl der Steuerparameter oder die Entwicklung von Modellen zur Bestimmung der Parameter aus dem Datenmaterial.

Literatur

Besag, J.: Spatial interaction and the statistical analysis of lattice systems. J. Roy. Stat. Soc. Series B **36** (1974) 192–236

Besag, J.: On the statistical analysis of dirty pictures. J. Roy. Stat. Soc. Series B **48** (1986) 259-302

Busch, A.: Bayes-Statistik und Markoff-Felder für die Restaurierung digitaler Bilder. Deutsche Geodätische Kommission, Reihe C: Dissertationen, Nr.396, Verl. des Inst. f. Angew. Geodäsie, Frankfurt/Main 1992

Busch, A., Koch, K.R.: Reconstruction of digital images using Bayesian estimates. Z. Photogrammetrie Fernerkundung **58** (1990) 173–181

Faddejew, D.K., Faddejewa, W.N.: Numerische Methoden der linearen Algebra. 4.Aufl., R. Oldenbourg Verlag, München 1976

Geman, S., Geman, D.: Stochastic relaxation, Gibbs distributions, and the Bayesian restoration of images. IEEE T-PAMI **6** (1984) 721–741

Haberäcker, P.: Digitale Bildverarbeitung. 3. Aufl., Carl Hanser Verlag, München 1989

Haralick, R.M., Watson, L.T., Laffey, T.J.: The topographic primal sketch. Int. J. Robotics Research **2** (1983) 50–72. Reprint in: Kasturi, R., Jain, R.C. (Eds.): Computer Vision: Principles. IEEE Computer Society Press, Los Alamitos 1991

Haralick, R.M.: Digital step edges from zero crossing of second directional derivatives. IEEE T-PAMI **6** (1984) 58-68

Koch, K.R.: Bayesian Inference with Geodetic Applications. Springer-Verlag, Berlin 1990

Rosenfeld, A., Kak, A.C.: Digital Picture Processing. 2.Aufl., Academic Press, San Diego 1982

Wang, J., Treitz, P.M., Howarth, P.J.: Road network detection from SPOT imagery for updating geographical information systems in rural-urban fringe. Int. J. Geographical Information Systems **6** (1992) 141–157

Neuronales System zur Objektsegmentierung in der Mikrostrukturvermessung*

B. Flach *E. Kask* *R. Osterland*

Projektgruppe "Technische Kognition" (WIP-KAI e.V.)
Technische Universität Dresden
Institut für Künstliche Intelligenz
01069 Dresden, Mommsenstr. 13

1 Einführung

In dem vorliegenden Beitrag wird ein neuronales System zur Segmentierung von Objekten für die automatisierte Vermessung von Mikrostrukturen[1] aus Resistwerkstoffen vorgestellt.

Für die Gütekontrolle solcher Strukturen werden mittels eines Auflichtmikroskops und einer CCD-Kamera Grauwertbilder aufgenommen und digitalisiert. Durch einen automatisch positionierbaren Objekttisch werden die Strukturen, die sich auf einem Substrat befinden, in mehreren Teilbildern aufgenommen. Entsprechend des verwendeten Materials und Herstellungsverfahrens weisen die Bilder große Unterschiede in der Stärke der Strukturierung und der Ausprägung der Kanten auf. So sind die Objektoberflächen von Strukturen aus Resistwerkstoffen stark strukturiert.

Bei diesen Bildern ist es schwierig zwischen realen Objekten und der Oberflächenstrukturierung zu unterscheiden (siehe Abb. 2, 10, 12). Andererseits erleichtert das Hinzuziehen der CAD-Daten für die Sollstruktur das Auffinden des Randes der realen Struktur. Dabei ist allerdings die Position der Objekte in den Bildern wegen systematischer Fehler der Objekttischpositionierung nicht genau bekannt.

Zur Lösung des Problems wurde ein modular aufgebautes neuronales Netz zur Objektsegmentierung entworfen. Dieses Netz liefert als Resultat ein Binärbild in dem die Randlinien markiert sind.

Das System besteht zum einen aus einer Vorverarbeitungskomponente, die ähnlich wie die Region V1 des visuellen Kortex aus identischen Subnetzen mit topologisch angeordneten rezeptiven Feldern besteht. Die Subnetze detektieren Kanten (Stärke und Richtung) in den jeweiligen rezeptiven Feldern. Durch einen Vergleich der CAD-Daten für die Sollkanten mit den Resultatbildern der Vorverarbeitung wird die exakte Position der Objekte im Bild bestimmt. Die weitere Verarbeitung erfolgt dann in einem schlauchförmig um die Sollkante liegenden Aufmerksamkeitsgebiet.

Die zweite Komponente besteht wiederum aus identischen Subsystemen mit rezeptiven Feldern, die entlang des Schlauches angeordnet sind. Jedes dieser Subsysteme bestimmt ein Pixel als Stützstelle für den Rand.

*Partiell finanziert durch das Kernforschungszentrum Karlsruhe. Wir danken H. Guth für die Aufgabenstellung und die anregenden Diskussionen.

[1] Dreidimensionale Strukturen beliebiger lateraler Ausbildung auf planaren Substraten (Auflösung $< 1\mu m$).

Die Liste dieser Pixel bildet schließlich den Input für ein elastisches Netz (ringförmiges Feature Map), welches als Output die Randkontur liefert.

2 Vorverarbeitung

Detektion primärer Merkmale

Die Vorverarbeitung erfolgt in zwei Ebenen. In der ersten werden einfache lokale Merkmalsdetektoren in Form linearer Filter auf das Inputbild (256 Grauwerte) angewendet. Diese Merkmalsdetektoren entsprechen den "einfachen Zellen" der Region V1 des visuellen Kortex. Ihre Herausbildung kann in einem einfachen zweischichtigem Netz mit lateraler Wechselwirkung simuliert werden, das sich mittels Hebb'schen (bzw. anti-Hebb'schen) Lernen selbst organisiert ([4], [5], [2]). Die dabei entstehenden Merkmalsdetektoren entsprechen den Eigenvektoren der Kovarianzmatrix der für das Lernen benutzten Bilder.

Da in der folgenden Verarbeitungsebene die Ausprägung und Orientierung der in dem Bild vorkommenden Kanten detektiert werden soll, wurden nur die beiden Merkmalsdetektoren verwendet, die auf Kanten reagieren. Die erste Verarbeitungsebene besteht aus einem Satz dieser Detektoren, die topologisch geordnete überlappende rezeptive Felder der Größe 3×3 haben (siehe Abb. 1). Die Resultate dieser Vorverarbeitung für das Inputbild in Abb. 2 sind in den Abb. 3, 4 dargestellt.

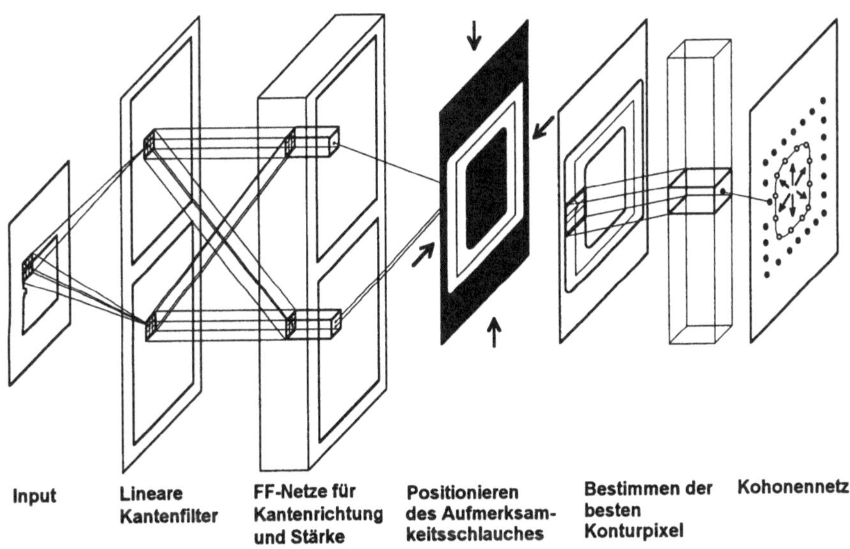

Abb. 1: Aufbau des Systems

Detektion der Ausprägung und Orientierung der Kanten

Die zweite Ebene der Vorverarbeitung besteht aus einem Satz identischer Paare von Feed-Forward-Netzen deren rezeptive Felder wiederum topologisch geordnet sind (siehe Abb. 1). Während der eine Netztyp (3×3 Inputs und 4, 2, 1 Neuronen in den entsprechenden

Schichten) auf die Kantenausprägung reagiert, detektiert der andere (3×3; 4/2/2) die Kantenrichtung. Beide Netze wurden mittels Error-Back-Propagation mit folgendem Mustersatz angelernt: Im Template der Inputebene wurden Grauwertmuster mit

$$g(x,y) = \Delta \tanh(\beta\,(x\ \cos\alpha + y\ \sin\alpha)) + \Delta_0 \tag{1}$$

angelegt. Hierbei sind x, y die auf das Zentrum des Templates bezogenen Koordinaten der Pixel. Diese Muster werden durch die Primärmerkmalsfilter verarbeitet, deren Outputs dann als Input für die Feed-Forward-Netze dienen. Die gewünschten Outputs der Netze sind dann entsprechend

$$s = \beta\,\Delta \tag{2}$$

und

$$s_1 = \cos(\alpha + \frac{\pi}{2}) \tag{3}$$

$$s_2 = \sin(\alpha + \frac{\pi}{2})\,. \tag{4}$$

Die Parameter werden dabei zufällig in den Bereichen $-1 < \Delta_0 < 1$, $0 < \Delta < 1 - \Delta_0$, $0 \leq \alpha \leq 2\pi$ und $0 < \beta < 1$ ausgewählt. Während des Lernens bleiben die vorher eingestellten Primärmerkmalsfilter unverändert. Das Resultat der Anwendung der gesamten Vorverarbeitungsstufe auf das Bild in Abb. 2 ist in den Abb. 5, 6, 7 dargestellt.

3 Extraktion des Randes

Bestimmung der Objektposition

Die Extraktion des Randes erfolgt in einem um die zu erwartende Kontur schlauchförmig gelegten Aufmerksamkeitsgebiet (siehe Abb. 1). Dazu muß zunächst die genaue Position des Objektes im Bild gefunden werden. Dafür wird die aus den CAD-Daten gewonnene Sollkontur in allen Positionen innerhalb des durch die Positionstoleranz des Objekttisches vorgegebenen Bereichs mit dem Resultat der Vorverarbeitung verglichen. Für jede Position wird ein Übereinstimmungsmaß gebildet und anschließend die Position mit dem besten Maß gewählt. Das Gesamtmaß ist dabei die Summe lokaler Maße für die Umgebung jedes Pixels p_i der Sollkontur K:

$$M = \sum_{p_i \in K} \min_{p \in U(p_i)} \left[\frac{(s_p - s_{\max})^2}{s_{\max}^2} + \frac{(\alpha_p - \alpha_{p_i}^k)^2}{(2\pi)^2} \right] \tag{5}$$

Hierbei sind s_p, α_p die in p detektierten Kantenausprägung bzw. Winkel und $\alpha_{p_i}^k$ der Winkel der Sollkontur in p_i.

Lokale Bestimmung von Konturpixeln

An die Vorverarbeitung schließt sich ein Modul an, welcher Stützstellen für den Rand liefern soll. Er besteht aus einem linear angeordneten System von unabhängig arbeitenden Prozeßeinheiten. Diese „Prozessoren“ sind funktional identisch und haben sich überlappende rezeptive Felder, die entlang des Aufmerksamkeitsschlauches angeordnet werden. Die parallel arbeitenden Prozeßeinheiten können durch einen nichtadaptiven Algorithmus oder neuronal realisiert werden. Wir haben zunächst die erste Methode gewählt. Im weiteren bezeichnen wir das Bild mit den Kantenausprägungen als A-Bild (s. Abb.5) und das Bildpaar mit den Kanterichtungen als W-Bild (s. Abb.6 und 5).

In jedem rezeptiven Feld wird zunächst die Linie markiert, die durch dessen Zentrum verläuft und die Sollkontur im rechten Winkel schneidet. Jede Prozeßeinheit wählt nun anhand der Resultate der Vorverarbeitung ein Pixel dieser Linie als Stützstelle für den Rand.

Zunächst werden die Pixel der Linie bestimmt, in denen das A-Bild Maxima entlang der Linie aufweist. In diesen Pixeln werden auf das A-Bild Balkendetektoren angewendet deren Richtung sich aus dem W-Bild ergibt. Bewegt man sich auf der Linie von außen nach innen, so wird das erste dieser Pixel, in dem die Balkenintensität eine Schwelle überschreitet, als Stützstelle für den Rand gewählt. Wir erhalten somit eine Menge $P = \{p_i\}_{i=1,\dots,N}$ von Pixeln, die als Randpixel in Frage kommen (Abb. 8).

Elastisches Netz zur Erzeugung der Randkontur

Um nun eine geschlossene Kontur zu erhalten, die sich den Pixeln aus P "anpaßt", haben wir ein Feature Map verwendet, dessen Neuronen topologisch als Ring angeordnet sind. Jedes Neuron des Netzes hat zwei Inputplätze. Die Koordinaten der Pixel aus P bilden die Mustermenge mit der sich das Netz organisiert.

Die Vorgehensweise ist ähnlich der zur Nutzung von Feature Maps [3] bzw. des in [1], [6] vorgeschlagenen elastischen Netzes zur Lösung des Traveling Salesman Problems. Der Unterschied besteht nur darin, daß in unserem Fall die Randkontur als Abweichung der Sollkontur aufgefaßt werden kann. Somit können die Gewichte des Netzes gleichmäßig verteilt auf der Sollkontur initialisiert werden. Das beschleunigt das Lernen, da die Phase der Groborganisation übersprungen wird.

Allerdings muß auch hier am Anfang die Lernumgebung genügend groß gewählt werden, um eine gute Anpassung insbesondere in den Objektecken zu erreichen. Sind die Abweichungen der Randkontur von der Sollkontur nicht zu groß, so genügen i.A. ca 50 Lernzyklen zur Organisation des Netzes.

4 Resultate und Ausblick

In den Abb. 9, 11 und 13 sind erzeugte Randkonturen dargestellt. Es zeigt sich, daß ein Vorteil der fast durchgängig neuronalen Realisierung darin liegt, daß erst am Ende der Verarbeitung scharfe Entscheidungen getroffen werden. Ein weiterer Vorteil ist natürlich die inherente Parallelität des Systems.

Ein Nachteil besteht darin, daß in der Vorverarbeitung nur Merkmalsfilter zur Kantendetektion verwendet werden. Es zeigt sich, daß nahe am Objektrand verlaufende Oberflächenstrukturen dazu führen, daß der Rand nur noch als 1–2 Pixel breite Balken ausgeprägt ist. Dieser Nachteil kann durch das Hinzuziehen von Balkendetektoren und entsprechendes Neuanlernen der Feed-Forward-Netze behoben werden.

Ein weiterer Nachteil ergibt sich daraus, daß die Randkontextauswertung nicht neuronal realisiert wurde. Ein entsprechendes neuronales Netz müßte Kurvenstücke detektieren können und besäße somit erste Ansätze von kognitiven Fähigkeiten.

Literatur

[1] R. Durbin and D. Willshaw. An Analogue Approach to the Travelling Salesman Problem. *Nature*, 326:689–691, 1987.

[2] H. Kühnel and P. Tavan. The Anti-Hebb Rule Derived from Information Theory. In R. Eckmiller, G. Hartmann, and G. Hauske, editors, *Parallel Processing Systems in Neural Systems and Computers.* North-Holland, 1990.

[3] H. Ritter, T. Martinetz, and K. Schulten. *Neuronale Netze: Eine Einführung in die Neuroinformatik selbstorganisierender Netzwerke*, pages 95–98. Addison-Wesley, 1990.

[4] J. Rubner and K. Schulten. Development of Feature Detectors by Self-Organization. *Biological Cybernetics*, 62:193–199, 1990.

[5] J. Rubner and P. Tavan. A Self-Organizing Network for Principal-Component Analysis. *Europhys. Lett.*, pages 694–698, 1989.

[6] J. V. Stone. The Optimal Elastic Net: Finding Solutions to the Travelling Salesman Problem. In I. Aleksander and J. Taylor, editors, *Artifical Neural Networks, 2.* North Holland, 1992.

Anhang

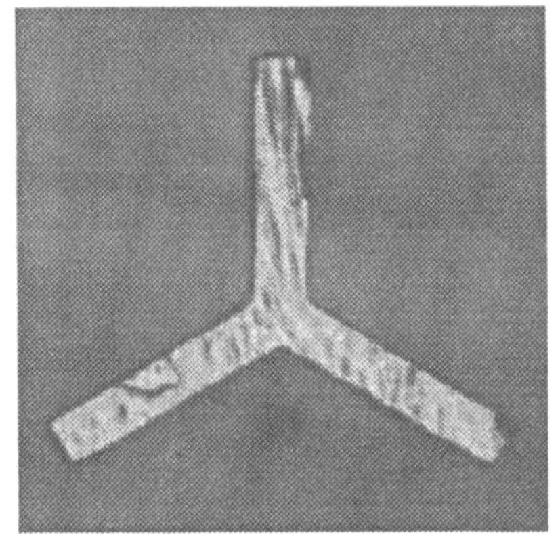

Abb. 2 Inputbild

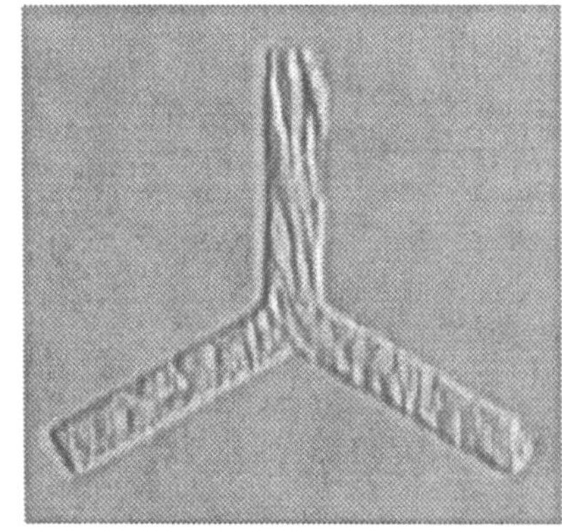

Abb. 3

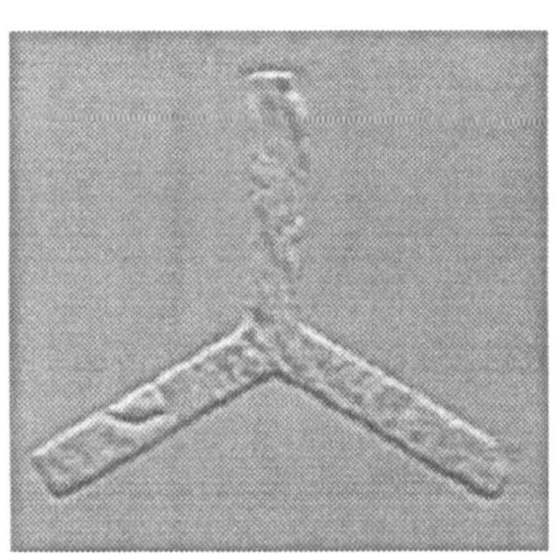

Abb. 4

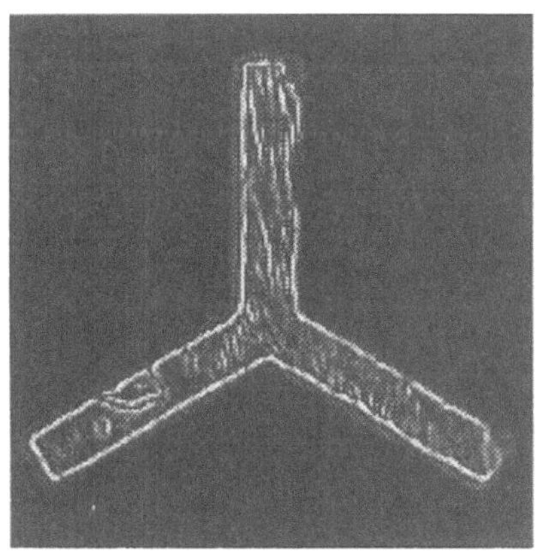

Abb. 5

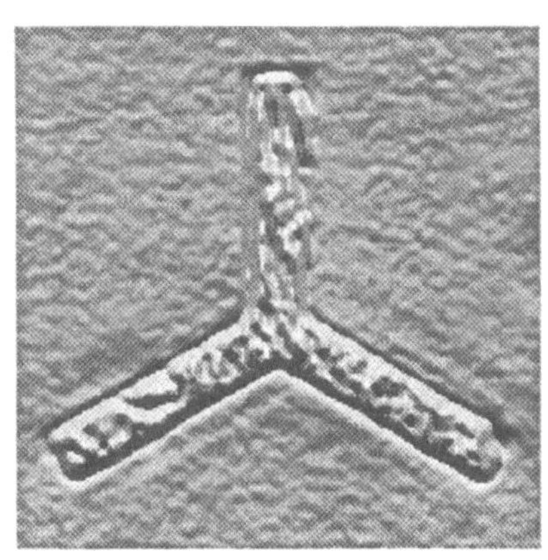

Abb. 6

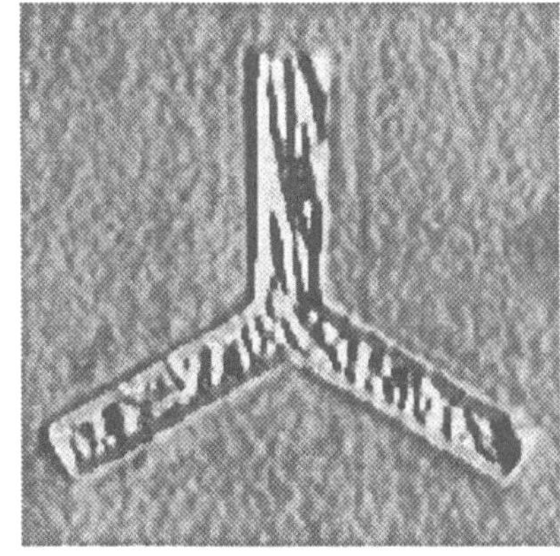

Abb. 7

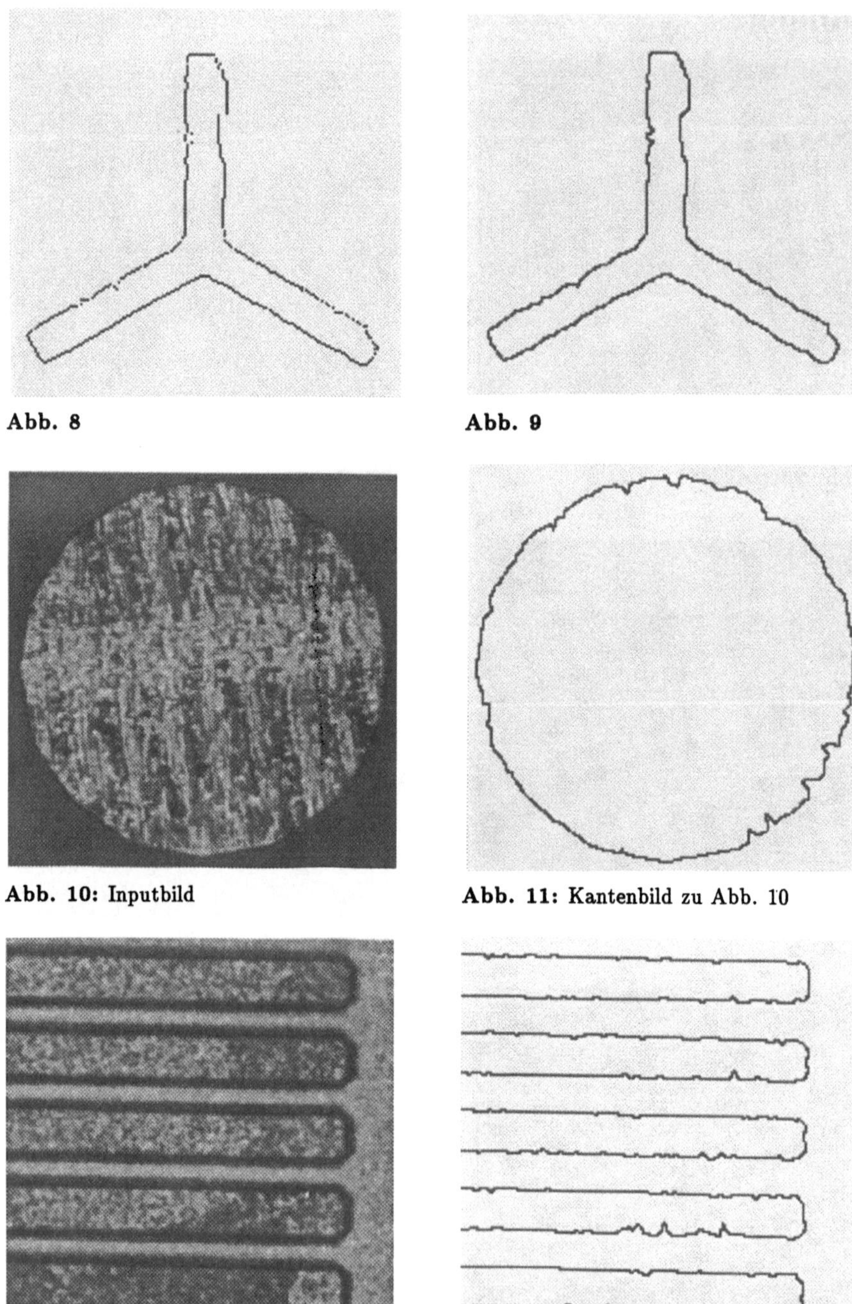

Abb. 8

Abb. 9

Abb. 10: Inputbild

Abb. 11: Kantenbild zu Abb. 10

Abb. 12: Inputbild

Abb. 13: Kantenbild zu Abb. 13

Ein Kalman-Filter zur Konturverfolgung und Parameterschätzung

R. Gerdes, R. Kammüller

Fachbereich Elektrotechnik und Informatik, Universität-GH-Siegen

Hölderlin-Str. 3, 57068 Siegen

Zusammenfassung

Dieser Beitrag beschreibt ein Verfahren zur Konturverfolgung und suboptimalen Parameterbestimmung implizit vorgegebener Konturen. Über die Formulierung eines statischen Zustandsraummodells mit nichtlinearen Meßfunktionen wird die Anwendung eines erweiterten Kalman-Filters möglich. Hiermit ist die modellgesteuerte Kantenextraktion sowie eine gleichzeitige Parameterschätzung möglich. Die Genauigkeit der bestimmten Parameter ist bereits nach einem Durchlauf der Meßdaten hoch im Vergleich zur Methode der gewichteten, kleinsten Fehlerquadrate.

1 Einleitung

Die exakte Positionsbestimmung von Objekten, ausgehend von einer Lagehypothese, ist ein wesentliches Problem der Kamerameßtechnik. Korrelationsmethoden liefern unter der Annahme weißen Rauschens zwar optimale Ergebnisse [1], ihre Implementierung scheitert aber oft an der benötigten Rechenzeit. Zudem sind vielfach neben den Lageparametern auch die Objektparameter selbst (Radius, Halbachsen, ...) mit Unsicherheiten behaftet. Solche Aufgabenstellungen werden durch die Theorie der Parameterschätzung beschrieben [2]. Üblich sind Lösungsansätze nach der Methode der kleinsten Fehlerquadrate [3], welche, abhängig von der Güte der Hypothese, mehrere Iterationen bis zum Erreichen einer ausreichend genauen Lösung benötigen. Die Parameterschätzung verläuft nicht rekursiv und ist daher für eine Schätzung der Konturparameter während der Extraktion der Kantenpunkte zu aufwendig.

In [4] wurde ein Konturverfolger auf Basis eines linearen Kalman-Filters vorgestellt, welcher Konturen konstanter Krümmung, also Kreise und Geraden, auf der Basis einer natürlichen Konturbeschreibung extrahiert. Bedingt durch die Verwendung der Bogenlänge zur Konturbeschreibung ist eine exakte Parameterschätzung jedoch nicht möglich.

Ausgehend von einem Zustandsraummodell mit konstanten Systemvariablen wird ein alternativer Ansatz unter Verwendung eines erweiterten Kalman-Filters vorgestellt. Das Filter arbeitet modellgestützt und erzeugt eine exakte Parameterschätzung. Aufgrund der sequentiellen Abarbeitung der Meßpunkte durch das Kalman-Filter können die fortlaufend berechneten Schätzwerte zur Einschränkung des Suchbereichs und damit zur Konturverfolgung herangezogen werden.

Als Eingangsdaten des Kalman-Filters werden aus Gradientenbildern gewonnene Koordinaten von Konturpunkten verwendet sowie die lokale Orientierung des Gradienten in den Kantenpunkten.

2 Zustandsraummodell

Ausgangspunkt ist die Darstellung einer Kontur in impliziter Form:

$$F(x,y,\underline{\beta})=0$$

mit $\underline{\beta}$, dem zu bestimmenden Parametervektor. Diese Parameter können als zufällige Systemvariablen $\underline{x}$ direkt geschätzt werden. Startwerte für Erwartungswert $E\{\underline{x}\}$ und Varianz werden aufgrund von Hypothesen bzgl. der Lage der Kontur angegeben. Die Systemgleichung nimmt wegen der Konstanz der Parameter die einfache, lineare Form an:

$$\dot{\underline{x}}=\underline{0}$$

Wegen der Art des vorliegenden Problems wird nicht nach der Zeit, sondern nach dem Winkel α zum Koordinatensystem-Nullpunkt abgeleitet. Im allgemeinen ist der Zusammenhang zwischen Meßwerten und Systemvariablen nichtlinear, weshalb zur Nachbildung des Systems ein erweitertes Kalman-Filter notwendig ist.

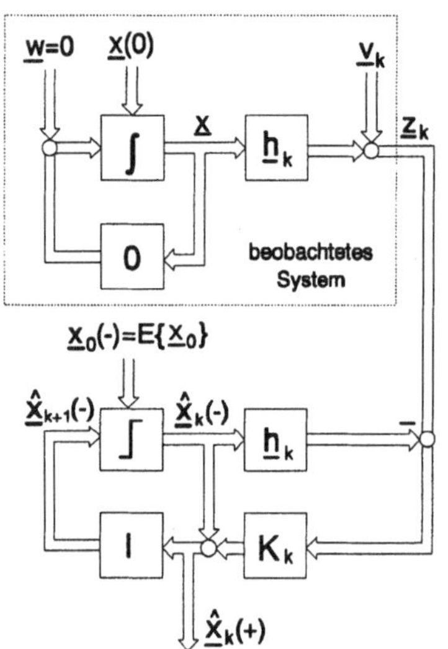

mit:

$\hat{\underline{x}}_k(+)$: durch die Messung $\underline{z}_k$ aktualisierter Schätzwert der Systemvariablen an der Stelle k.

$\hat{\underline{x}}_k(-)$: Prädiktion der Systemvariablen an der Stelle k aufgrund der Messungen $\underline{z}_0$, $\underline{z}_1$, .. $\underline{z}_{k-1}$.

$\underline{h}_k$: Vektor, welcher die nichtlinearen Meßfunktionen umfaßt.

$\mathbf{K}_k$: Kalman-Gain-Matrix

I: Einheitsmatrix

$\mathbf{P}_k$: Kovarianz-Matrix der Systemvariablen ($\mathbf{P}_k(+)$, $\mathbf{P}_k(-)$ analog zu $\hat{\underline{x}}_k(+)$, $\hat{\underline{x}}_k(-)$).

$\mathbf{R}_k$: Kovarianz-Matrix des Meßrauschens $\underline{v}_k$.

$\mathbf{H}_k$: Jakobi-Matrix der Meßfunktionen an der Stelle $\hat{\underline{x}}_k(-)$

Bild 1: Zustandsmodell des erweiterteten Kalman-Filters

Die rekursive Vorschrift zur Bestimmung der Systemvariablen lautet allgemein [5]:

$$\hat{\underline{x}}_k(+)=\hat{\underline{x}}_k(-)+\mathbf{K}_k\cdot(\underline{z}_k-\underline{h}_k\cdot\hat{\underline{x}}_k(-))$$

$$\mathbf{P}_k(+)=\left[\mathbf{I}-\mathbf{K}_k\cdot\mathbf{H}_k(\hat{\underline{x}}_k(-))\right]\cdot\mathbf{P}_k(-)$$

$$\mathbf{K}_k=\mathbf{P}_k(-)\cdot\mathbf{H}_k{}^T(\hat{\underline{x}}_k(-))\cdot\left[\mathbf{H}_k(\hat{\underline{x}}_k(-))\cdot\mathbf{P}_k(-)\cdot\mathbf{H}_k{}^T(\hat{\underline{x}}_k(-))+\mathbf{R}_k\right]^{-1}$$

$$\mathbf{H}_k(\hat{\underline{x}}(-))=\frac{\partial\underline{h}_k(\underline{x}(\alpha_k))}{\partial\underline{x}(\alpha_k)}\bigg/_{\underline{x}(\alpha_k)=\hat{\underline{x}}(-)}$$

Wegen der Parameterkonstanz ergibt sich die Einheitsmatrix **I** zur Prädiktion des nächsten Systemzustands $\underline{x}_{k+1}(-)$:

$$\underline{x}_{k+1}(-) = \underline{x}_k(+) \qquad \text{und} \qquad \mathbf{P}_{k+1}(-) = \mathbf{P}_k(+)$$

Das Systemrauschen $\underline{w}$ wird zu $\underline{0}$ gesetzt, da die Kontur ideal durch $F(x, y, \underline{\beta}) = 0$ beschrieben wird. Im Gegensatz zu einer Konturbeschreibung in Abhängigkeit der Bogenlänge entstehen keine Fehler in der Bestimmung der freien Variablen α. Die Messungen sind hingegen durch das überlagerte weiße, gaußverteilte Meßrauschen $\underline{v}$ gestört.

3 Algorithmus zur Parameterschätzung

Bedingt durch die Auswertung von Gradientenbildern ergeben sich Meßwertpaare bestehend aus

$$\underline{z}_k = \begin{pmatrix} x_k \\ y_k \\ \varphi_k \end{pmatrix}$$

mit φ_k der Gradientenrichtung in dem Kantenpunkt (x_k, y_k). Es wird eine Referenzlage der Kontur definiert, welche eine minimale Anzahl von Parametern zur Beschreibung der Konturform benötigt (z. B. die Normalform einer Kurve 2. Ordnung). Damit kann der Parametervektor $\underline{\beta}$ unterteilt werden in Parameter der Konturform $\underline{p}$ sowie einen Transformationsvektor $\underline{t}$, bestehend aus Rotationswinkel γ und den Verschiebungen t_x und t_y.

$$F(x, y, \underline{p}, \underline{t}) = 0 \quad \text{mit} \quad \underline{t} = \begin{pmatrix} t_x \\ t_y \\ \gamma \end{pmatrix}$$

Vor Eintritt in eine für jeden Meßwert zu durchlaufende Schleife wird eine Transformation der Kontur in ihre Referenzlage durchgeführt. Der Vorteil der wiederholten Transformation liegt darin, daß lediglich eine Routine zur Bestimmung eines Punktes auf der Kontur in Referenzlage bei gegebenem Winkel α zur Verfügung gestellt werden muß. Zudem können durch geschickte Wahl der Referenzlage Mehrdeutigkeiten vermieden werden.

Der Algorithmus ist zusammenhängend in Bild 2 wiedergegeben. Es werden die folgenden Bezeichnungen verwendet: $\mathbf{T}_k$ bezeichnet die dem Transformationsvektor $\underline{t}_k$ zugehörige homogene 2D-Transformationsmatrix:

$$\mathbf{T}_k = \begin{pmatrix} \cos\gamma_k & -\sin\gamma_k & t_{xk} \\ \sin\gamma_k & \cos\gamma_k & t_{yk} \\ 0 & 0 & 1 \end{pmatrix} = \mathbf{T}\{\underline{t}_k\}$$

$K\{\underline{x}\}$ steht für die Kontur, welche durch den Parametervektor $\underline{x}$ beschrieben wird. $\underline{e}$ bezeichnet den Einheitsvektor in Richtung der Achse i.

Die Meßpunktkoordinaten P_k werden mit $\mathbf{T}_k^{-1}$ in das aktuelle Koordinatensystem transformiert, während man die zugehörige Gradientenrichtung φ_k durch Subtraktion von γ_k erhält. Um die nächsten Schätzwerte für den Systemvektor $\hat{\underline{x}}_k(+)$ und die zugehörige Kovarianz $\mathbf{P}_k(+)$ berechnen zu können, muß

Hauptprogramm:

1. Variableninitialisierung: $\hat{\underline{x}}_0 = E\{\underline{x}\}$, $\mathbf{P}_0(-) = E\{\underline{x} \cdot \underline{x}^T\}$, $\mathbf{R}_k = E\{\underline{v} \cdot \underline{v}^T\}$

   Berechnung der Maschinengenauigkeit ε

   k=0

2. Transformation der Kontur in Referenzlage $F(x, y, \underline{p}, \underline{0}) = 0$, $\mathbf{T}_0 = \mathbf{T}\{\underline{t}_0\}$

3. Meßpunkt $\underline{P}_k$ in aktuelles Koordinatensystem transformieren:

   Meßpunktkoordinate: $\underline{P}_k = \begin{pmatrix} x_k \\ y_k \end{pmatrix} = \mathbf{T}_k^{-1} \cdot \underline{P}_k$

   Gradientenrichtung: $\varphi_k = \varphi_k - \gamma\{\mathbf{T}_k\}$

   } Meßwert $\underline{z}_k$

4. Winkel zu Meßpunkt bestimmen: $\alpha_k = atan(y_k/x_k)$

5. Ermittlung der Jakobi-Matrix $\mathbf{H}_k$:

   für alle Variablen $x_i \in \underline{x}$

   bestimme Punkt P_i und Normalenwinkel auf Kontur $K\{\hat{\underline{x}}_k(-) + \varepsilon \cdot \underline{e}_i\}$ bei gegebenem Winkel α_k, zusammengenommen $\underline{z}_i$

   $$\text{Spalte i von } \mathbf{H}_k = \frac{residuum(\underline{z}_i, \hat{\underline{x}}_k(-))}{\varepsilon} \approx \left.\frac{\partial \underline{h}}{\partial x_i}\right/_{\hat{\underline{x}}(\alpha_k) = \underline{x}(-)}$$

6. Berechne Kalman-Gain und Kovarianz:

   $$\mathbf{K}_k = \mathbf{P}_k(-) \cdot \mathbf{H}_k^T \cdot \left[\mathbf{H}_k \cdot \mathbf{P}_k(-) \cdot \mathbf{H}_k^T + \mathbf{R}_k\right]^{-1}$$

   $$\mathbf{P}_k(+) = \left[\mathbf{I} - \mathbf{K}_k \cdot \mathbf{H}_k\right] \cdot \mathbf{P}_k(-)$$

7. Bestimme Residuum $\underline{z}_k - h_k(\hat{\underline{x}}_k(-)) = residuum(\underline{z}_k, \hat{\underline{x}}_k(-))$

8. $$\hat{\underline{x}}_k(+) = \hat{\underline{x}}_k(-) + \mathbf{K}_k \cdot \left[\underline{z}_k - \underline{h}_k(\hat{\underline{x}}_k(-))\right]$$

   $$\hat{\underline{x}}_{k+1}(-) = \hat{\underline{x}}_k(+)$$

9. Transformationsmatrix aufstellen: $\mathbf{T}_{k+1} = \mathbf{T}\{\underline{t}_k\}$

   Transformationen zusammenfassen: $\mathbf{T}_{k+1} = \mathbf{T}_k \cdot \mathbf{T}_{k+1}$ k=k+1

10. alle Meßpunkte abgearbeitet? wenn nein, dann weiter mit 3.

    sonst, Parameter $\underline{p} = \hat{\underline{x}}_k$, $\underline{t}_k = \underline{t}\{\mathbf{T}_k\}$

Routine *residuum*($\underline{z}$, $\underline{x}$):

1. Bestimme Punkt auf Kontur $K\{\underline{x}\}$ mit minimalem Abstand zu Meßwert $\underline{z}$
2. Berechne Differenzen zu $\underline{z}$, also die Residuen
3. Rückgabe der Residuen

Bild 2: Algorithmus

1. der neue Meßvektor $\underline{z}_k$,

2. eine Schätzung für den neuen Meßvektor $\underline{h}_k(\hat{\underline{x}}_k(-))$ und

3. die zu den Meßfunktionen gehörige Jakobi-Matrix $\mathbf{H}_k$

bestimmt werden.

Der Winkel α_k gibt den Punkt auf der Kontur in Referenzlage

$$F(x,y,\underline{\beta} = \hat{\underline{x}}_k(-)) = \underline{0} \qquad \text{mit} \qquad \underline{t} = \underline{0}$$

an, in dem die Jakobi-Matrix $\mathbf{H}_k$ zu bestimmen ist. Im allgemeinen kann $\mathbf{H}_k$ nicht geschlossen bestimmt werden und muß daher durch Differenzenquotienten, entsprechend dem in [6] angegebenem Verfahren, angenähert werden. Als Ergebnis der Kalman-Filterung für einen Meßwert liegt ein modifizierter Parametervektor $\underline{p}$ sowie ein Transformationsvektor $\underline{t}_{k+1} \neq \underline{0}$ vor. Zu dieser Transformation muß die entsprechende Transformationsmatrix $\mathbf{T}_{k+1}$ bestimmt werden. Nach einer Multiplikation mit $\mathbf{T}_k$ ergibt sich die Gesamttransformation von der ersten Lagehypothese der Kontur bis zur aktuellen Referenzlage.

Der Parametervektor $\underline{p}$ enthält nach Durchlaufen aller Meßwerte die Schätzwerte, welche die Konturform beschreiben. Die resultierende Rotation und Translation der Kontur wird aus der Matrix $\mathbf{T}_k$ berechnet. Die Bestimmung des Konturpunktes mit minimalem Abstand zum Meßpunkt erfordert im allgemeinen ein iteratives Lösungsverfahren. Für die Spezialfälle Kreis und Gerade können die Meßfunktionen direkt angegeben werden. Damit ist zum einen die direkte Berechnung der Jakobi-Matrix möglich, zum anderen vereinfacht sich die Routine zur Bestimmung des Residuums wegen des einfachen geometrischen Zusammenhangs.

4 Modellgestützte Konturverfolgung

Die von dem Kalman-Filter gelieferten Schätzwerte für Position und Orientierung der Objektkontur werden verwendet, um den Vorgang der Kantenpunktbestimmung zu steuern. In einem der Varianz der Zustandsgrößen angepaßten Bereich um den erwarteten Kantenpunkt wird der Gradient entlang einer Gerade in Richtung des erwarteten Normalenwinkels φ gebildet. Zur Eliminierung von Störungen wird lediglich die Komponente des Gradienten in Richtung der Normalen an die Kontur ausgewertet:

$$g_d = \frac{\partial g(x,y)}{\partial x} \cdot \cos\varphi + \frac{\partial g(x,y)}{\partial y} \cdot \sin\varphi$$

Die exakte Lokalisierung der Kante und die Ermittlung der zugehörigen Varianz können nach dem in [7] angegebenem Verfahren mittels einer Korrelation mit einem Modellkantenverlauf durchgeführt werden. Über die Vorgabe der Kovarianzmatrix der Meßstörungen $\mathbf{R}_k$ wird die Zuverlässigkeit der Kantenpunktbestimmung berücksichtigt. Abhängig von der Anwendung kann zusätzliches Modellwissen in gleicher Weise einbezogen werden.

5 Ergebnisse

Vergleiche zur Methode der kleinsten Fehlerquadrate wurden anhand von simulierten Konturdaten durchgeführt. Die Möglichkeit der Konturverfolgung wurde in diesem Fall nicht genutzt.

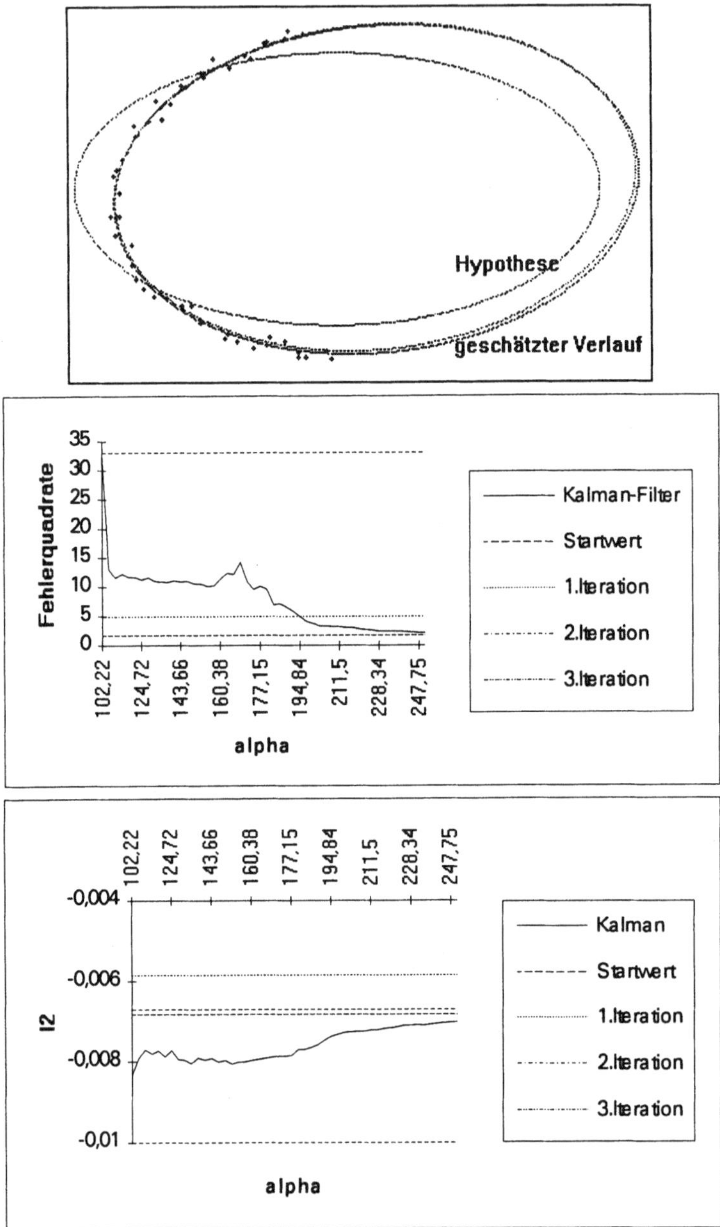

Bild 3: Vergleich des Kalman-Filters und der Methode der kleinsten Quadrate anhand simulierter Konturdaten für einen Ellipsenteilbereich

Die Genauigkeit der Parameterbestimmung kann am Beispiel der in Bild 3 dargestellten Ellipse demonstriert werden. Ausgangspunkt ist die Normalform einer Kurve 2.Ordnung:

$$l_1 \cdot x^2 + l_2 \cdot y^2 + 1 = 0$$

Im oberen Bild ist die Kontur (s. auch Tabelle 1) entsprechend ihrer Hypothese, ihre tatsächliche Lage und die mit dem Kalman-Filter bestimmte Kontur dargestellt. Das erste

Diagramm zeigt die Summe der Fehlerquadrate zwischen Konturschätzung und den 50 extrahierten Konturpunkten. Im zweiten Diagramm ist der Konturparameter l_2 während des Verlaufs der Parameterschätzung aufgetragen.

Das Kalman-Filter liefert trotz gleicher Startwerte nach der ersten Iteration einen besseren Schätzwert als die Methode der kleinsten Fehlerquadrate. Der Grund hierfür liegt in der fortlaufend verbesserten Schätzung. Damit werden die Fehler durch falsche Vorgabe des Punktes, um den die Meßfunktionen linearisiert werden, schon während des ersten Durchgangs verringert. In den meisten Fällen reicht die Genauigkeit des ersten Schätzwerts bereits aus. Analog zur Methode der kleinsten Quadrate liegt der wesentliche Rechenaufwand in der Bestimmung der linearisierten Meßfunktionen um den aktuellen Schätzwert sowie in der Berechnung des Abstands eines Punkts zu der Kontur. Die Rechenzeiten unterscheiden sich daher nur unwesentlich.

	vorgegebene Kontur (Startwert)	tatsächliche Kontur	Kalman-Filter nach 1. Durchlauf	1 Iteration mit Meth. der Fehlerquadr.	2 Iterationen mit Meth. der Fehlerquadr.
l_1	-0,002500	-0,002500	-0,002543	-0,002394	-0,002447
l_2	-0,010000	-0,006944	-0,007027	-0,005854	-0,006724
t_x	20	23	22,9041	24,0930	23,2150
t_y	0	0	0,1751	-0,2143	0,0554
γ [rad]	0	0,0872 (5°)	0,0927	0,0412	0,0873
ΣFehlerquadr	33,059	1,709	2,224	4,592	1,713
Rechenzeit [sek] (´486)			3,8	3,5	7,0

Tabelle 1: Vergleichswerte zu Kalman-Filter und Methode der kleinsten Fehlerquadrate

Bild 4: Bildausschnitt (64 x 64 Pixel) mit Rohrstutzen

Weiterhin wird das Kalman-Filter in einer Schweißtechnik-Applikation zur Bestimmung der Position und des Radius von Rohrstutzen (s. Bild 4) eingesetzt. Da die aufgenommenen Bilder aufgrund des Materials sehr kontrastarm sind, ist durch die modellgestützte Konturverfolgung eine erhebliche Genauigkeitssteigerung möglich. Die Vermessung kann auf 0,1 Pixel genau ($< 0{,}1$mm) durchgeführt werden.

6 Diskussion und Ausblick

Das vorgestellte Kalman-Filter führt parallel zur Extraktion der Kantenpunkte eine Parameterbestimmung der Konturparameter durch. Hiermit ist die modellgestützte Steuerung der Bildauswertung durch die Schätzwerte des Kalman-Filters möglich. Die Parameterschätzung liefert im Vergleich zu der Methode der kleinsten Quadrate bereits nach dem ersten Durchlauf gute Ergebnisse, wodurch zeitaufwendige weitere Iterationen überflüssig werden. Der Algorithmus setzt lediglich eine implizite Konturbeschreibung voraus.

Eine Erweiterung zur Parameterbestimmung dreidimensionaler Konturen oder Flächen ist möglich. Zu untersuchen ist, inwieweit Hypothesentests verwendet werden können, um Segmentationspunkte exakt zu lokalisieren.

7 Literatur

[1] Pratt, W.: Digital Image Processing. John Wiley & Sons, New York 1978

[2] Koch, K.-R.: Parameterschätzung und Hypothesentests in linearen Modellen. Dümmler-Verlag, Bonn 1987

[3] Shirai, Y.: Three-Dimensional Computer Vision. Springer Verlag 1978

[4] Raubenheimer, H. R.: Konturverfolgung und -segmentation mit Kalman-Filtern. Dissertation, Clausthal 1984

[5] Gelb, Arthur: Applied Optimal Estimation. The MIT-Press, Cambridge 1974

[6] Engeln-Müllges, G.; Reutter, F.: Formelsammlung zur numerischen Mathematik mit C-Programmen. BI Wissenschaftsverlag, Mannheim 1987

[7] Förstner, W.: Modelle intelligenter Bildsensoren und ihre Qualität. DAGM 90, Informatik-Fachberichte 254, Springer-Verlag 1990

Kantenorientierte Farbsegmentation im CIE-Lab Raum

Stefan Lanser
Technische Universität München
Institut für Informatik - Lehrstuhl Prof. Radig*)
Orleansstraße 34, 81667 München
e-mail: lanser@informatik.tu-muenchen.de

Die Segmentation eines ein- oder mehrkanaligen Eingabebildes stellt die wohl wichtigste Phase in der Low- bzw. Medium-Level Bildverarbeitung dar. Eine "gute" Segmentation kann die anschließende Bildinterpretation erheblich vereinfachen, ja in vielen Fällen sogar überhaupt erst ermöglichen. Die Ausnutzung von Farbinformation führt dabei häufig zu "besseren" Ergebnissen als sie unter Verwendung von reiner Luminanz- oder auch Texturinformation erzielbar wären. In diesem Beitrag wird zunächst ein robuster, parameterfreier Ansatz zur Segmentation eines einkanaligen Grauwertbildes vorgestellt, der dann zu einem kantenorientierten Farbsegmentations-Algorithmus erweitert wird. Besonders bewährt hat sich dabei der CIE-Lab Farbraum.

1. Einleitung

Segmentationsverfahren lassen sich grob in zwei Kategorien unterteilen: Regionen- und kantenorientierte Ansätze. Beide haben die Partitionierung eines Eingabebildes in disjunkte, bezüglich eines bestimmten Kriteriums homogene Bildbereiche (Segmente) zum Ziel. Die eine Gruppe von Verfahren versucht dabei, die Segmente direkt zu bestimmen (regionenorientiert), die andere sucht hingegen nach den Diskontinuitäten (Kanten), die die Segmente voneinander trennen.
Regionenorientierte Segmentationsverfahren gliedern sich weiter in globale und lokale Verfahren. Globale Verfahren, wie etwa Schwellwert-Techniken ([4]) oder Cluster-Ansätze ([5]) zeigen dabei eine klare Tendenz zur Übersegmentation in Bildbereichen mit allmählichen Übergängen in der Bildinformation (sei dies nun Luminanz, Chrominanz, Textur o. ä.). Demgegenüber neigen lokale Ansätze, wie bekannte Regionenwachstumsverfahren, dazu, Segmente schon wegen einzelner z.B. rauschbedingter Störungen entlang der gemeinsamen Konturlinie zu verschmelzen. Es ist zumeist nicht trivial, zu einem späteren Zeitpunkt im Bildanalyseprozeß eine solche Segmentverschmelzung zu erkennen und zu korrigieren. Dieses Problem läßt sich durch Verwendung eines segmentbezogenen Merging-Kriteriums teilweise lösen. Beispielsweise werden Punkte nur zu einem Segment hinzugenommen, deren Abweichung vom aktuellen Mittelwert (bzgl. des Bildsignals) des Segments einen bestimmten Toleranzwert nicht übersteigt. Hiermit wird das Verfahren allerdings stark abhängig von der Wahl der Anfangspunkte für das Regionwachstum.
Die angesprochenen Probleme treten bei einem kantenorientierten Ansatz nicht auf. Kanten markieren Diskontinuitäten, also mehr oder weniger sprunghafte Änderungen, im Bildsignal und trennen somit homogene Bildbereiche voneinander. Bereiche mit allmählicher Ände-

*) Die vorliegende Arbeit wurde mit Unterstützung der DFG im Rahmen des Sonderforschungsbereiches 331 "Informationsverarbeitung in autonomen, mobilen Handhabungssystemen", Teilprojekt L5 angefertigt.

rung der Bildinformation werden somit nicht unerwünscht in Segmente zerlegt. Sofern es gelingt, weitgehend *geschlossene* Kantenzüge zu gewinnen, werden benachbarte Segmente allenfalls durch schmale Brücken miteinander verbunden, die sich durch eine geeignete Nachverarbeitung wieder auftrennen lassen.
Drei Gründe sprechen allerdings dafür, kantenorientierte Ansätze um regionenorientierte Techniken zu erweitern: Erstens lassen sich so isolierte, z.B. rauschbedingte, Kanten(stücke), die zu keiner Objektkontur gehören, einfacher eliminieren. Zweitens ist das Verschmelzen zu kleiner Segmente mit geeigneten benachbarten Segmenten regionenorientiert wesentlich einfacher. Schließlich ergeben sich bei kantenorientierten Ansätzen häufig Probleme in Bildbereichen, wo mehrere Kanten aufeinandertreffen oder Kanten sprunghaft ihre Richtung ändern ("abgerundete" Ecken etc.). Ein Regionenwachstumsverfahren liefert hier meist etwas "detailgetreuere" Objektkonturen.
Ein möglicher Algorithmus für eine einkanalige, kantenorientierte Segmentation mit regionenorientierten Erweiterungen ist, grob skizziert, folgender:

A. Kantendetektion (Eingabebild, Kanten),
   Konturfindung (Kanten, Konturen),
B. Regionenextraktion (Kanten, Segmente),
   Segmentverbesserung (Segmente).

Hier, wie im folgenden auch wird dabei eine PROLOG-ähnliche Notation verwendet. In Abschnitt 2 wird obiger Algorithmus verfeinert und in Abschnitt 3 auf mehrere Kanäle erweitert. Außerdem wird die Wahl des CIE-Lab Raumes zur Farbsegmentation motiviert. Abschnitt 4 präsentiert die Ergebnisse der Anwendung des resultierenden Farbsegmentations-Algorithmuses auf zwei Bilder.

2. Einkanalige Segmentation

2.1 Kantendetektion

Die im vorigen Abschnitt präsentierte Grobskizze eines kantenorientierten, einkanaligen Segmentationsverfahrens läßt bereits erahnen, daß hier recht hohe Anforderungen an die Kantendetektion zu stellen sind: Natürlich sollte der verwendete Kantenoperator eine hohe Detektions- und Lokalisierungsgüte aufweisen, d.h. er sollte alle relevanten Kanten und nur diese möglichst genau im Zentrum der Diskontinuitäten im Eingangsbild markieren. Deshalb ist der modifizierte Deriche-Operator, der auf eindimensionalen rekursiven Filtern beruht und darum auch effizient implementierbar ist, eine sehr gute Wahl ([1]). Im folgenden werden zwei Ansätze zur Kantendetektion präsentiert, die beide diesen Operator verwenden.

2.1.1 Kantendetektion mittels erster und zweiter Ableitung

Es ist von elementarer Bedeutung für den vorgestellten Segmentations-Algorithmus, daß der Kantendetektor (zumindest weitgehend) *geschlossene* Konturen liefert. Sonst würden benachbarte Segmente miteinander verschmelzen. Dies legt die Verwendung eines Operators nahe, der auf der Bestimmung von Nulldurchgängen der zweiten Ableitung beruht. Tatsächlich liefern solche Kantendetektoren (naturgemäß) geschlossene Kantenzüge. Leider haben die so gefundenen Konturen nicht immer etwas mit den gesuchten Objektkonturen im Bild zu tun. Teilweise ergeben sich deutliche Verästelungen der Konturen, die eine Weiterverarbeitung der gefundenen Segmente sehr erschweren (siehe Abbildung 1). Zudem sind Opera-

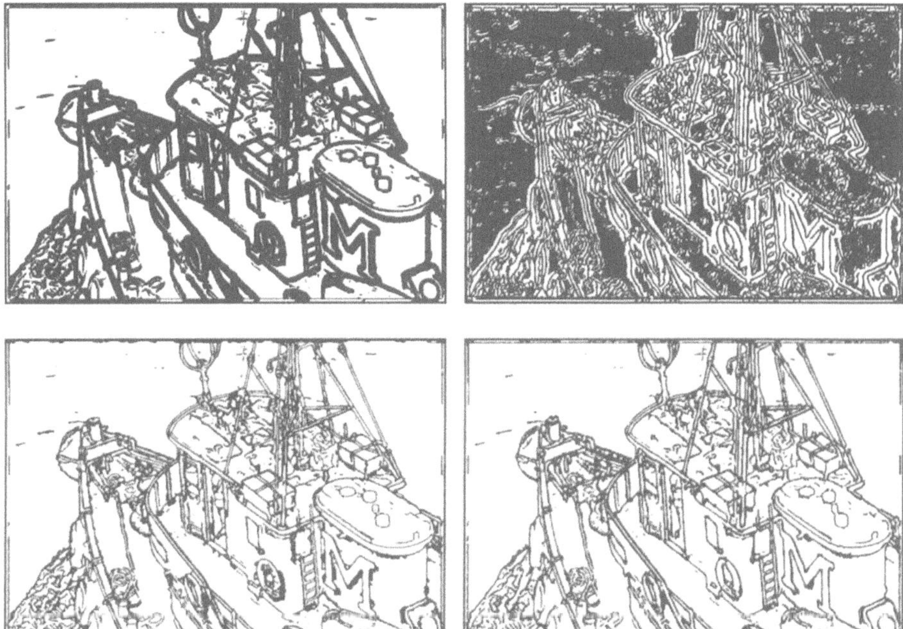

Abbildung 1. Kantendetektion in den L- und b-Komponenten eines in den CIE-Lab Farbraum transformierten Bildes aus Abbildung 2 (nach 3.1).
links oben: 1. Ableitung (Anwendung des mod. Deriche-Operators) + Schwellwertoperation.
rechts oben: Nulldurchgänge der 2. Ableitung (neuerliche Anwendung des Operators).
links unten: Maxima der 1. Ableitung + Verdünnung (Variante A').
rechts unten: Kombination der 1. und 2. Ableitung (Variante A).

toren, die die zweite Ableitung bestimmen, sehr rauschempfindlich, was zu einer großen Zahl von Scheinkanten im Bild führt.
Beide Probleme lassen sich durch Hinzunahme der ersten Ableitung beseitigen. Kanten zeigen sich hier als Maxima. In der Praxis verzichtet man aber häufig auf eine exakte Maximumssuche und führt statt dessen eine einfache Schwellwert-Operation durch. Schneidet man die Mengen der Kantenpunkte resultierend aus der ersten und der zweiten Ableitung miteinander, erhält man ein qualitativ hochwertiges Kantenbild (Abbildung 1):

Variante A:
A.1 edges (EingabeBild, ErsteAbleitung),
A.2 edges (ErsteAbleitung, ZweiteAbleitung),
A.3 threshold (ErsteAbleitung, Kanten1),
A.3 zero_crossings (ZweiteAbleitung, Kanten2),
A.4 intersection (Kanten1, Kanten2, Kanten).

Insbesondere sind die ermittelten Kanten ausreichend dünn. Dies ist von Bedeutung für den angestrebten Segmentationsalgorithmus, um die Lücken zwischen den Segmenten, die durch die Komplementbildung in Schritt B entstehen, möglichst klein zu halten. Allerdings sind die Konturen nicht mehr geschlossen, da allzu weite Verästelungen der Nulldurchgänge der zweiten Ableitung durch die Schnittbildung gekappt werden.

2.1.2 Kantendetektion mittels erster Ableitung und gerichteter Maximasuche

Da also der große Vorteil der Verwendung der zweiten Ableitung - geschlossene Konturen - doch nicht ohne zusätzlichen Aufwand zu halten ist, bietet sich alternativ dazu auch folgendes effizienteres Verfahren an:

Variante A':

A'.1 edges (Eingabebild, ErsteAbleitung),
A'.2 nonmax_suppression (ErsteAbleitung, Kanten1),
A'.3 threshold (Kanten1, Kanten2),
A'.4 skeleton (Kanten2, Kanten).

Hier werden Kanten als gerichtete Maxima (normal zur Kantenrichtung) in der ersten Ableitung bestimmt ([2]). Etwaige Plateaus eliminiert ein - heuristisch gewählter - nachfolgender Skelettierungsschritt. Die vom modifizierten Deriche-Operator gelieferten Amplitudenbilder wiesen allerdings in den durchgeführten Experimenten nur selten solche Plateaus auf. Dieser Ansatz ist um beinahe Faktor zwei schneller, da nur eine Ableitung zu bestimmen ist, erweist sich allerdings bei sehr schlechtem Bildmaterial als qualitativ unterlegen.
Die in A.3 bzw. A'.3 benötigte Schwelle läßt sie sich gemäß einer von Canny vorgeschlagenen Heuristik einfach aus dem Histogramm der Kantenamplituden abschätzen ([2]).

2.1.3 Schließen der Konturen

Beide vorgestellten Verfahren zur Kantendetektion müssen durch ein Verfahren zur Schließung der Konturen ergänzt werden, um hier verwendet werden zu können:

A.5 close_edges (Kanten, Konturen).

Ausgehend von den gefundenen Kantenendpunkten werden jeweils die beiden amplitudenstärksten Nachbarpunkte iterativ zur Kante hinzugenommen, falls es sich nicht ohnehin schon um Kantenpunkte handelt oder ihre (Kanten-)Amplitude unter einer gewissen Schwelle liegt. Der Konturverlauf wird dabei nicht weiter berücksichtigt - in Gebieten mit Amplitudenplateaus beginnt die Kontur daher i.a. zu schwingen. Plateaus waren aufgrund der hohen Qualität des vom modifizierten Deriche-Operators gelieferten Amplitudenbildes in den durchgeführten Experimenten selten bzw. nicht sehr breit, so daß sich das skizzierte Verfahren zur Konturschließung als ausreichend erweist.

2.2 Der Übergang zu Regionen

Der Übergang von den gefundenen Konturen zu Regionen erfolgt durch Komplementbildung. Geschlossene Konturen vorausgesetzt, erhält man so die angestrebten Segmente - getrennt durch feine Linien, die von den Kanten belegten Bildpunkte. Isolierte, z.B rauschbedingte, Kanten(stücke) werden hingegen eliminiert. Es bleiben an ihrer Stelle Löcher in den Segmenten, die durch die weiter unten skizzierte Nachverarbeitung geschlossen werden. In der Praxis kann es jedoch vorkommen, daß benachbarte Segmente durch schmale Brücken miteinander verbunden sind. Lücken entlang der gemeinsamen Konturlinie führen allerdings nicht wie beim Regionenwachstum zu einer nicht mehr korrigierbaren Verschmelzung von Segmenten. Vielmehr können besagte Brücken leicht durch Anwendung einer morphologischen Operation - eines Openings mit einer sehr kleinen Kreismaske (ein oder zwei Pixel Radius) - wieder aufgetrennt werden. Als letzter Schritt der Regionenextraktion werden die ermittelten Regionen in Zusammenhangskomponenten zerlegt und zu kleine Segmente unterdrückt.

B.1 complement (Konturen, RegionenKanditaten),
B.2 opening (RegionenKanditaten, Regionen),
B.3 connection (Regionen, Segmente),

Am Ende dieses Schrittes gibt es natürlich noch Bildpunkte, die keinem Segment zugeordnet sind (Lücken zwischen den Segmenten, Löcher in den Segmenten und Bildbereiche mit unterdrückten Segmenten). Daher werden die Segmente noch durch Regionenwachstum kontrolliert expandiert und damit insbesondere auch die Segmentkonturen verfeinert. Noch nicht zugeordnete Bildpunkte werden dabei in das benachbarte Segment aufgenommen, zu dem sie die kleinste Differenz bezüglich des Bildsignals aufweisen. Die Entscheidung, zu welchem Segment die Zuordnung erfolgen soll, ist dabei unkritisch, da es sich zumeist um Kantenbereiche handelt, also um Bildbereiche mit deutlichen Diskontinuitäten. Somit ergibt sich als letzter Teilschritt:

B.4 expand_segments (Segmente).

3. Mehrkanalige Segmentation

3.1 Der Übergang zu mehreren Kanälen

Der im vorigen Abschnitt vorgestellte einkanalige Segmentationsalgorithmus läßt sich im wesentlichen auf zwei Arten auf mehrere Kanäle (Farbinformation oder zusätzliche Texturinformation, etc.) verallgemeinern. Der naheliegendste Ansatz ist, die Kanäle sequentiell abzuarbeiten: Es erfolgt erst eine (einkanalige) Segmentation im ersten Kanal. Die Ergebnissegmente bilden dann die Eingabe für die Segmentation bezüglich des zweiten Kanals und so weiter. Hier soll jedoch ein anderer Weg vorgestellt werden - die simultane Kantensuche in allen Kanälen. Dazu wird Schritt A für jeden einzelnen Kanal durchgeführt und die gefundenen Kanten vereinigt:

```
multi_channel_edges ([ErsterKanal | RestlicheKanäle], Kanten) :-
    A (ErsterKanal, Kanten1),
    multi_channel_edges (RestlicheKanäle, Kanten 2),
    union (Kanten1, Kanten2, Kanten).
```

Das so ermittelte Gesamtkantenbild wird dann wie im vorigen Abschnitt diskutiert weiterverarbeitet. Alternativ dazu wäre natürlich auch eine mehrdimensionale Kantensuche vorstellbar.

3.2 Der CIE-Lab Farbraum

Im vorigen Abschnitt wurde ein Segmentationsverfahren präsentiert, das prinzipiell für beliebige mehrkanalige Eingabeobjekte geeignet ist. Bei der Farbsegmentation als speziellem dreikanaligen Segmentationsproblem ist dabei insbesondere noch die Wahl des Farbraumes zu klären (eine vergleichende Übersicht findet sich z.B in [6]). Der technische Standard-Farbraum, der RGB-Raum, ist weder für die Segmentation noch für die anschließende Bildinterpretation besonders geeignet. Auch der Mensch sieht und denkt nicht in RGB-Tripeln (sondern in Farbton, Sättigung und Helligkeit). Doch auch wahrnehmungsorientierte Farbräume, wie der HSI- oder der HSV-Raum, sind nicht vollständig an das menschliche Farb-Wahrnehmungsvermögen angepaßt: Die geometrischen Abstände der Farben im Farbraum

entsprechen nicht immer den empfindungsgemäßen - manche Farbbereiche sind stärker untergliedert als andere.
Beim CIE-Lab Raum ([3]) handelt es sich hingegen um einen bezogen auf das menschliche Wahrnehmungsvermögen *äquidistanten* Farbraum. Farben, die für den normalsichtigen Betrachter ähnlich unterschiedlich wirken, haben denselben euklidischen Abstand im CIE-Lab Raum.
Hinzu kommt noch ein pragmatischer Vorteil: Die Komponenten des CIE-Lab Raumes sind nicht zyklisch (im Gegensatz beispielsweise zur H-Komponente des HSI- oder des HSV-Raumes). Damit sind Standardoperationen zur Glättung, Kantendetektion etc. ohne Änderungen auf die einzelnen Komponenten anwendbar.
Experimente haben gezeigt, daß zur vollständigen Farbsegmentation eines Bildes bereits die L-Komponente (Helligkeit) und die b-Komponente (Gelb-Blau-Komponente) ausreichen. Die a-Komponente (Rot-Grün-Komponente) ist hingegen in vielen Fällen nur schwach strukturiert und liefert einen eher geringen Beitrag zur Segmentation.
Der Vollständigkeit halber sei hier noch die Berechnung der drei CIE-Lab Kanäle (auch als L, a^* und b^* bezeichnet) aus den CIE XYZ-Grundvalenzen angegeben:

$$L = 25 \sqrt[3]{\frac{100Y}{Y_w}} - 16,$$

$$a = 500 \left(\sqrt[3]{\frac{X}{X_w}} - \sqrt[3]{\frac{Y}{Y_w}} \right),$$

$$b = 200 \left(\sqrt[3]{\frac{Y}{Y_w}} - \sqrt[3]{\frac{Z}{Z_w}} \right)$$

$[X_w, Y_w, Z_w]$ ist dabei reines Weiß.

3.3 Effizienzsteigerung durch Auflösungsreduktion

Das in diesem Beitrag vorgestellte Segmentationsverfahren läßt sich durch Verwendung der Gauß'schen Auflösungspyramide deutlich beschleunigen, indem die Kantendetektion, die rund die Hälfte der Rechenzeit in Anspruch nimmt, in die nächst höhere Ebene der Auflösungspyramide verlegt wird, d.h. die Kanten werden im horizontal und vertikal um den Faktor zwei unterabgetasteten Bild gesucht. Die gefundenen Kanten werden dann in die Originalauflösung zurück transformiert und wie oben beschrieben weiterverarbeitet. Insbesondere wird die Segmentexpansion (Schritt B.4) im Originalbild ausgeführt. Dadurch bleiben die Originalkonturen im Bild weitgehend erhalten (bei einer Laufzeitreduktion um etwa Faktor zwei). Allerdings gehen naturgemäß feine Details verloren.

4. Experimente

Das vorgestellte Farbsegmentationsverfahren wurde auf eine Reihe von Testbildern angewandt, von denen hier zwei vorgestellt werden. Bei der Aufnahme eines Hafenschleppers handelt es sich um ein 24-Bit Farbbild von einer Photo-CD (Auflösung 735 x 485 Bildpunkte) mit sehr guter Bildqualität. Das Bild mit dem Bajazzo wurde hingegen mit einem S-VHS Camcorder aus einem Buch abgefilmt und mit einem S-Bus VideoPix-Framegrabber digitalisiert (Auflösung 768 x 575 Bildpunkte, 24-Bit Farbtiefe). Dieses Bild ist stark verrauscht. Insbesondere sind die Farbkanten schlecht ausgeprägt, sie "flattern".
Das beschriebene Verfahren wurde in C implementiert und in das Bildverarbeitungssystem HORUS integriert, das von Dr. Eckstein an der TU München am Lehrstuhl von Prof. Radig

Bild	Originalauflösung	Reduzierte Auflösung bei der Kantendetektion
Boot	A: 31.1 Sekunden A´: 21.2 Sekunden	A: 16.2 Sekunden A´: 11.9 Sekunden
Bajazzo	A: 40.5 Sekunden A´: 26.9 Sekunden	A: 19.9 Sekunden A´: 15.3 Sekunden

Tabelle 1. Laufzeiten des vorgestellten Farbsegmentations-Verfahrens auf einer HP 9000/720 Workstation (inkl. der Farbtransformation vom RGB- in den CIE-Lab Raum).

entwickelt wurde ([7]). Tabelle 1 gibt eine Übersicht über die Laufzeiten auf einer HP 9000/720 Workstation. Zur Segmentation wurden jeweils nur die L- und b-Komponenten der Bilder verwendet.

Abbildung 2 zeigt die Originalbilder und die Segmentationsergebnisse. Segmente mit weniger als 60 Pixeln wurden unterdrückt und die Schwellwert-Parameter für die Kantendetektion zur Laufzeit geschätzt. Somit erfolgte die Segmentation völlig parameterfrei! Die gefundenen Segmente wurden für die Darstellung mit ihrem mittleren Farbwert eingefärbt und zusätzlich der Deutlichkeit halber die Segmentränder schwarz eingezeichnet.

5. Zusammenfassung

In diesem Beitrag wurde ein robustes, parameterfreies Farbsegmentationsverfahren vorgestellt, das auf einer qualitativ hochwertigen Kantendetektion in den einzelnen Farbkanälen und der anschließenden Kombination der gefundenen Kanten zu einem Gesamtkantenbild beruht. Als besonders geeignet hat sich dabei der bezogen auf das menschliche Farb-Wahrnehmungsvermögen äquidistante CIE-Lab Farbraum erwiesen. Das Verfahren wird im SFB 331 "Informationsverarbeitung in autonomen, mobilen Handhabungssystemen", Teilprojekt L5 zur Szenenanalyse eingesetzt.

Bedanken möchte ich mich bei Herrn Dr. W. Eckstein und Herrn O. Munkelt für einige hilfreiche Anmerkungen und Diskussionen.

Literatur:

[1] S. Lanser, W. Eckstein: "Eine Modifikation des Deriche-Verfahrens zur Kantendetektion"; *13. DAGM-Symposium*, München 1991; Informatik Fachberichte, Bd. 290; S. 151 - 158; 1991.

[2] J. Canny: "Finding Edges and Lines in Images"; Report, AI-TR-720, M.I.T. *Artificial Intelligence Lab.*, Cambridge; 1983.

[3] A.K. Jain: "Fundamentals of Digital Image Processing"; S. 66 - 73; *Prentice-Hall International, Inc.*; 1989.

[4] S. U. Lee, S. Y. Chung, R. H. Park: "A comparative performance study of several global thresholding techniques for segmentation"; *Comp. Vision, Graphics and Image Processing*; Vol. 52; S. 171 - 190; 1990.

[5] H. Celenk: "A color clustering technique for image segmentation"; *Comp. Vision, Graphics and Image Processing*; Vol. 52; S. 145 - 170; 1990.

[6] Y. Ohto, T. Kanade, T. Sakai: "Color information for region segmentation"; *Computer Graphics and Image Processing*; Vol. 13; S. 222 - 241; 1980.

[7] W. Eckstein et al: "Benutzerfreundliche Bildanalyse mit HORUS: gegenwärtiger Stand und Weiterentwicklungen"; *15 DAGM-Symposium*, Lübeck, 1993.

Abbildung 2. Testergebnisse.

links oben:	"Schlepper": 24-Bit Farbbild, 735 x 485 Bildpunkte, gute Qualität.
rechts oben:	Segmentationsergebnis von "Schlepper": Variante A, Originalauflösung.
links, Mitte:	"Bajazzo": 24-Bit Farbbild, 768 x 575 Bildpunkte, geringe Qualität.
rechts, Mitte:	Segmentationsergebnis von "Bajazzo": Variante A, Originalauflösung.
links unten:	Segmentationsergebnis von "Bajazzo": Variante A', Originalauflösung.
rechts unten:	Segmentationsergebnis von "Bajazzo": Variante A, reduzierte Auflösung.

Lokalisierungseigenschaften direkter Ansätze zur Ermittlung von Grauwertecken

Karl Rohr

Arbeitsbereich Kognitive Systeme, FB Informatik, Universität Hamburg
Bodenstedtstr. 16, 22765 Hamburg, FRG

Zusammenfassung

Die Leistungsfähigkeit bestehender Ansätze zur direkten Bestimmung von Grauwertecken wurde bisher weitgehend durch experimentellen Vergleich demonstriert. In diesem Beitrag werden Lokalisierungseigenschaften verschiedener Ansätze anhand eines analytischen Modells von Grauwertecken untersucht. Die jeweils lokalisierten Eckenpositionen werden in Abhängigkeit der Modellparameter berechnet und miteinander verglichen. Für praktische Anwendungen, bei denen Grauwertecken als Bildmerkmale verwendet werden sollen, können die Ergebnisse dieser Arbeit als Entscheidungshilfe bei der Auswahl eines bestimmten Verfahrens nützlich sein.

1 Einleitung

Direkte Ansätze zur Ermittlung von Grauwertecken sind dadurch gekennzeichnet, daß Ergebnisse von direkt auf das Grauwertbild angewandten (lokalen) Operatoren unmittelbar zur Eckenbestimmung verwendet werden und keine vorherige Segmentierung, z.B. durch Verkettung von Kantenpunkten, durchgeführt wird. Entscheidend für den Einsatz bestimmter Verfahren ist der benötigte Aufwand, die Robustheit und die Lokalisierungsgenauigkeit. Bei der Ermittlung von Positionen dreidimensionaler Objekte können schon kleine Meßfehler im Bildbereich zu großen Ungenauigkeiten im Szenenbereich führen. Daher ist eine genaue Lokalisierung von Grauwertecken wichtig. Auch bei der Ermittlung einer qualitativen Bildbeschreibung verringert eine genaue Lokalisierung darüber hinaus i.a. die Anfälligkeit gegenüber Fehlinterpretationen (beispielweise dann, wenn Grauwertecken nahe beieinanderliegen).

Unter Zugrundelegung einer bestimmten Klasse von Grauwertverläufen wird in der vorliegenden Arbeit die Lokalisierung von L-Ecken durch verschiedene in der Literatur vorgestellte differentialgeometrische Verfahren analysiert. Durch Verwendung des quantitativen Modells in Rohr *[16]*, *[17]* sind die hier erhaltenen Ergebnisse für Öffnungswinkel der L-Ecke im Bereich $0^o < \beta < 180^o$ gültig. Die von den jeweiligen Verfahren gefundenen Positionen werden für diesen Bereich numerisch berechnet und mit den eigentlich gesuchten Eckpunkten verglichen. Im Unterschied hierzu bestimmen Deriche&Giraudon *[6]* die Eckpunkte nur für bestimmte Werte von β für einen Teil der hier untersuchten Eckendetektoren.

Es wird sich zeigen, daß die untersuchten Verfahren für die hier zugrundegelegten Grauwertflächen zu unterschiedlichen Ergebnissen führen und ohne zusätzliche Maßnahmen nicht die gewünschten Eckpunkte liefern. Schon für allgemeine (generische) Flächen wurde in Rieger *[14]* gezeigt, daß beispielsweise der von Kitchen&Rosenfeld *[10]* vorgestellte Operator nicht die gewünschten Eckpunkte detektiert und daß die von Nagel *[11]* geforderten Eckenbedingungen an keinem Punkt erfüllt sind.

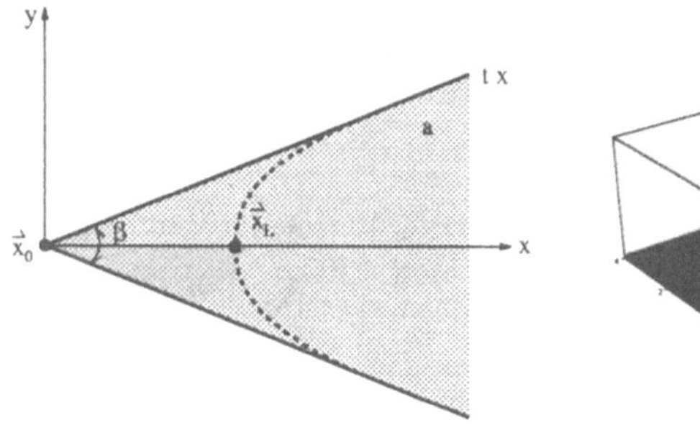

Abbildung 1: L-Ecke

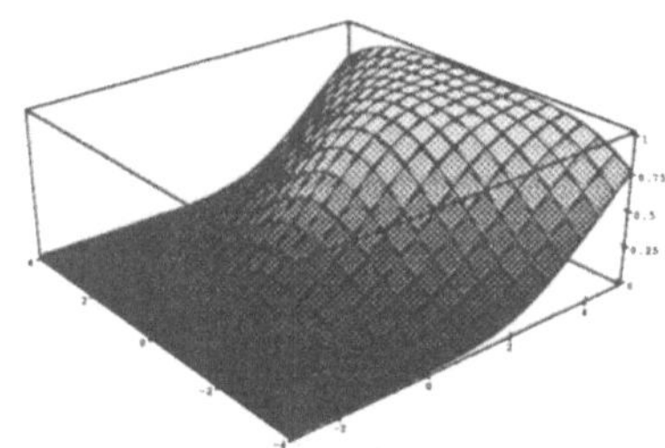

Abbildung 2: 3D-Darstellung einer L-Ecke

Im folgenden wird die übliche Definition von Eckpunkten diskutiert. Ausgehend von einer analytischen Beschreibung der Grauwertstruktur werden dann die Ansätze von Beaudet *[1]*, Dreschler&Nagel *[8]*, Kitchen&Rosenfeld *[10]*, Zuniga&Haralick *[20]*, Nagel *[11]*, Förstner *[9]*, Harris *[13]*, Rohr *[15]*, Blom et al. *[3]* und Brunnström et al. *[4]* zur direkten Bestimmung von Eckpunkten untersucht und miteinander verglichen.

2 Definition von Eckpunkten

Unter der Annahme, daß die 3D-Kanten von polyedrischen Szenenobjekten ideal scharf sind und die aufnehmende Kamera beobachtete Intensitäten nicht beeinflußt, bildet sich beispielsweise eine 3D-Ecke, von der zwei 3D-Kanten zu sehen sind, auf die durchgezogenen Linien der L-Ecke in Abb. 1 ab. Als Eckpunkt ist dann der Punkt $\vec{x}_0$ ausgezeichnet. Berücksichtigt man zusätzlich eine Verschmierung von Grauwertverläufen durch eine (reale) bandbegrenzte Kamera und bestimmt jetzt diejenigen Punkte, für die der Betrag des Grauwertgradienten in dessen Richtung maximal wird (z.B. mit dem Kantendetektionsverfahren in Canny *[5]*), so ergibt sich qualitativ die gestrichelte Linie. In diesem Fall bietet es sich an, den Punkt maximaler Krümmung entlang dieser Linie ($\vec{x}_L$) als Eckpunkt zu definieren. Sowohl indirekte Verfahren, die entlang von verketteten Kantenpunkten die Stelle maximaler Krümmung bestimmen, als auch direkte Verfahren versuchen, diesen Punkt zu ermitteln.

Diese Gegenüberstellung macht deutlich, daß bei ideal scharfen Kanten im Szenenbereich der Punkt $\vec{x}_0$ der gesuchte Eckpunkt ist und daher dessen Position für eine 3D-Interpretation verwendet werden sollte. Ansätze zur Bestimmung dieses Punktes stellen beispielsweise Rohr *[16]*, *[17]* und Deriche&Giraudon *[6]* vor. Bei den üblichen direkten Eckendetektoren wird stattdessen i.a. der Punkt $\vec{x}_L$ gesucht, wodurch sich offensichlich ein systematischer Fehler ergibt, der bei einer Weiterverarbeitung berücksichtigt werden sollte. Da direkte Eckendektoren i.a. versuchen, den Punkt $\vec{x}_L$ zu lokalisieren, werden die im weiteren erhaltenen Ergebnisse mit dieser Position verglichen.

3 Analytische Beschreibung einer L-Ecke

Die Grauwertfunktion einer L-Ecke in Rohr *[16]* , *[17]* gilt für beliebige Öffnungswinkel $0^o < \beta < 180^o$ und stellt eine Überlagerung von einer Modellfunktion in Berzins *[2]* und der durch Spiegelung dieser Modellfunktion an der x-Achse entstandenen Funktion dar (Abb. 1). Dieses quantitative Modell entspricht dem qualitativen Modell in Dreschler *[7]* und Nagel *[11]* . Mit den Ortskoordinaten $\vec{x} = (x, y)$, der Höhe des Grauwertkeils a und der Stärke der Verschmierung $\sigma = 1$ ergibt sich

$$g_{ML}(\vec{x}, \beta, a) = a\left(\phi(x) - m(\vec{x}, \beta) - m(\vec{x}^{\#}, \beta)\right), \quad 0^o < \beta < 180^o, \tag{1}$$

wobei $m(\vec{x},\beta) = \int_{-\infty}^{x} D(\xi, t\xi - \zeta_2)d\xi$, $D(\vec{x}) = G(x)\phi(y)$, $G(x)$ die Gaußfunktion, $\phi(x)$ die Gaußsche Fehlerfunktion, $t = \tan(\beta/2)$, $\zeta_2 = tx - y$ und $\vec{x}^{\#} = (x, -y)$ bezeichnen. Eine 3D-Darstellung dieser L-Ecke mit $\beta = 90^o$, $a = 1$ und $\sigma = 1$ ist in Abb. 2 wiedergegeben. Aufgrund der Symmetrie des Eckenmodells zur x-Achse kann davon ausgegangen werden, daß die zu ermittelnden Eckenpositionen auf der Geraden $y = 0$ liegen. Daher wird im weiteren nur die x-Koordinate der Eckenpositionen angegeben.

Die Linie, für die der Betrag des Grauwertgradienten in dessen Richtung maximal wird, ist mit der Hessematrix $\underline{H}$ und dem Grauwertgradienten ∇g durch $(\nabla g)^T \underline{H}\, \nabla g = 0$ festgelegt (z.B. Canny *[5]*). Schneidet man zur Bestimmung der Position $\vec{x}_L$ (Abb. 1) diese Linie mit der Geraden $y = 0$, so ergibt sich mit $q = \sqrt{1+t^2}$ und $x' = x/q$ folgende implizite Gleichung.

$$G(x') - t^2 x' \phi(x') = 0 \tag{2}$$

Die Eckenposition ist daher unabhängig von der Höhe des Grauwertkeils a. Aufgrund unseres Modells ist obiger Zusammenhang für Öffnungswinkel im Bereich $0^o < \beta < 180^o$ gültig. Die aus (2) numerisch bestimmten Positionen x_L in Abhängigkeit von β sind in Abb. 5 aufgetragen (fettgedruckte Kurve). Für $\beta = 90^o$ erhält man beispielsweise den Wert $x_L = 0.71567$. Positionen von x_L für unterschiedliche Verschmierungsstärken lassen sich durch Skalierung mit x/σ gewinnen.

4 Direkte Ansätze zur Ermittlung von Grauwertecken

4.1 Beaudet 78

Zur direkten Bestimmung von Grauwertecken wird in Beaudet *[1]* die Determinante der Hesse Matrix $\underline{H}$ ausgewertet, die große Werte in der Nähe von Ecken liefert. Zu suchen sind daher lokale Extrema

$$DET(\vec{x}) = det\underline{H} = g_{xx}g_{yy} - g_{xy}^2 \quad \rightarrow \quad Extremum, \tag{3}$$

die den notwendigen Bedingungen $DET_x = DET_y = 0$ genügen müssen. Angewandt auf unser L-Eckenmodell führen diese Bedingungen für $y = 0$ und beliebige Öffnungswinkel β auf die Gleichung $g_{xx}g_{xyy} + g_{yy}g_{xxx} = 0$, die zwei Lösungen besitzt, ein negatives und ein positives lokales Extremum (x_{Bn} und x_{Bp}). Für unterschiedliche Öffnungswinkel und Verschmierungsstärken verschieben sich diese beiden Extrema (siehe die durchgezogenen Kurven in Abb. 5). Für $\beta = 90^o$ ergibt sich $x_{Bn} = 0$ und $x_{Bp} = 1.65653$.

4.2 Dreschler&Nagel 81

Dreschler&Nagel *[8]* beschreiben einen Ansatz, der die lokalen Extrema der Determinante der Hesse Matrix DET nach Beaudet *[1]* heranzieht (siehe auch Nagel *[11]* und Shah&Jain *[19]*). An der Stelle x_L ist $DET = 0$. In *[8]* werden lokale Extrema von DET interpoliert, um die Position zu finden, für die $DET = 0$ ist (siehe Dreschler *[7]*). Bei exakter Interpolation würde sich an L-Ecken tatsächlich die gesuchte Stelle $x_{DN} = x_L$ ergeben. Bei linearer Interpolation liefert diese Vorgehensweise für $\beta = 90^o$ die Position $x_{DN} = (x_{Bn} + x_{Bp})/2 \approx 0.82826$. Die Positionen in Abhängigkeit von β ebenfalls bei linearer Interpolation zeigt Abb. 5.

4.3 Kitchen&Rosenfeld 82

Der von Kitchen&Rosenfeld *[10]* vorgestellte Differentialoperator besteht mit $\nabla g^{\perp} = (-g_y, g_x)^T$ und dem Betrag des Grauwertgradienten $|\nabla g|$ aus

$$KR(\vec{x}) = \frac{(\nabla g^{\perp})^T \underline{H}\, \nabla g^{\perp}}{|\nabla g|^2} \quad \rightarrow \quad Extremum \tag{4}$$

und ergibt sich durch Multiplikation der Krümmung einer ebenen Kurve (Isointensitätslinie) mit $|\nabla g|$. Für $y = 0$ führen die Bedingungen $KR_x = KR_y = 0$ auf die Bestimmungsgleichung $g_{xxy} = 0$, womit sich die Eckenpositionen wie in Abb. 5 ergeben (fettgedruckte gestrichelte Linie). Für $\beta = 90^o$ ist $x_{KR} = 1.18783$. Zur besseren Lokalisierung wird in *[10]* vorgeschlagen als Eckpunkte nur Kantenpunkte zu berücksichtigen (maximaler Grauwertgradient in dessen Richtung). Durch diese zusätzliche Maßnahme erhält man bei exakter Lage der Kantenpunkte tatsächlich den gesuchten Eckpunkt $x_{KR} = x_L$.

4.4 Zuniga&Haralick 83

Zuniga&Haralick *[20]* verwenden im Unterschied zu Kitchen&Rosenfeld *[10]* direkt die Krümmung einer ebenen Kurve

$$\kappa(\vec{x}) = KR(\vec{x})/|\nabla g| \quad \rightarrow \quad Extremum \tag{5}$$

unter der Voraussetzung, daß nur Kantenpunkte als Eckpunkte zugelassen werden. Erstaunlicherweise liefert $\kappa(\vec{x})$ angewandt auf das Eckenmodell für $y = 0$ in x-Richtung kein lokales Extremum (siehe Abb. 3 für $\beta = 90^o$), d.h. die aus (5) folgende Bedingung $g_x g_{xyy} - g_{xx} g_{yy} = 0$ ist an keiner Stelle erfüllbar. Zur Bestimmung von Eckpunkten wird demzufolge die Extremaleigenschaft von $\kappa(\vec{x})$ nur in y-Richtung ausgenutzt. Die Positionierung in x-Richtung geschieht ausschließlich durch Bestimmung von Kantenpositionen.

4.5 Nagel 83

In Nagel *[11]* wird vorgeschlagen, Grauwertecken aufgrund der Bedingungen

$$\begin{aligned} g_x &\rightarrow Extremum & g_y &= 0 \\ g_{xx} &= 0 & g_{yy} &\rightarrow Extremum \end{aligned} \tag{6}$$

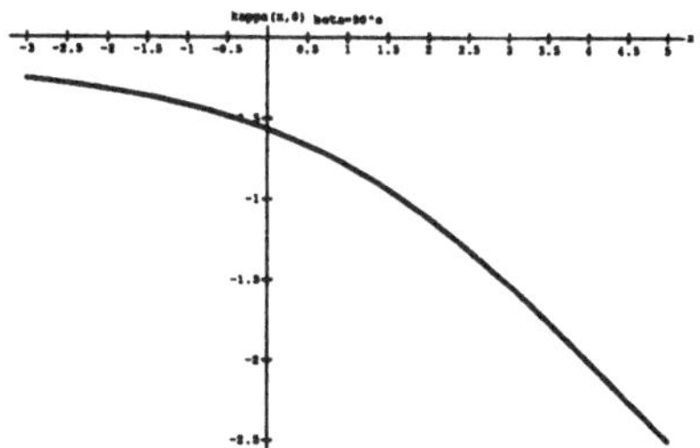

Abbildung 3: $\kappa(x,0)$ für $\beta = 90^o$

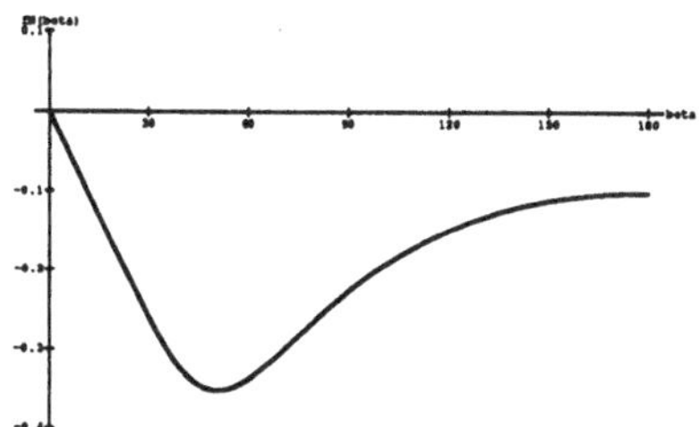

Abbildung 4: $f_N(x,0)$ für $\beta = 90^o$

zu ermitteln, wobei das lokale Koordinatensystem in der Weise gedreht sein sollte, daß $g_{xy} = 0$ ist. Aus diesen Bedingungen lassen sich die folgenden Gleichungen ableiten.

$$1.)\ g_y = 0 \qquad 2.)\ g_{xx} = 0 \qquad 3.)\ g_{xyy} = 0 \qquad 4.)\ g_{yyy} = 0 \tag{7}$$

Nachdem das Koordinatensystem festliegt (hier $g_{xy} = 0$) müßten an Eckpunkten diese 4 Bedingungen erfüllt sein (und darüber hinaus alle anderen Punkte aufgrund dieser Bedingungen ausgeschlossen werden). Allerdings legen schon 2 Bedingungen einen Punkt auf einer hinreichend allgemeinen zweidimensionalen Fläche eindeutig fest. Für allgemeine (generische) Flächen werden daher in einem Punkt nie alle 4 Bedingungen erfüllt sein. Darüber hinaus sind i.a. partielle Ableitungen bis zur 5. Ordnung zu berücksichtigen, um den gesuchten Eckpunkt x_L zu finden (siehe Rieger *[14]*).

Für die in der vorliegenden Arbeit zugrundegelegten Modelle gilt für die Symmetrielinie der L-Ecke, daß $g_{xy} = 0$, also das lokale Koordinatensystem in der von Nagel *[11]* geforderten Weise gedreht ist. Ebenso gilt für alle Punkte auf $y = 0$, daß $g_y = 0$ und $g_{yyy} = 0$ ist. Die Bedingungen 1.) und 4.) in (7) treffen daher zu. Sollen zusätzlich die Bedingungen 2.) und 3.) erfüllt sein, müßte

$$f_N(\beta) = G(\frac{1}{tq}) - \frac{t}{q}\phi(\frac{1}{tq}) = 0 \tag{8}$$

gelten. Trägt man $f_N(\beta)$ auf (Abb. 4), so wird deutlich, daß (8) nur für $\beta = 0^o$ erfüllt sein kann. Es gibt daher unter der hier betrachteten Klasse von Grauwertverläufen für L-Ecken keinen Punkt auf $y = 0$, der den in *[11]* geforderten Bedingungen genügt.

4.6 Förstner 86, Harris 87, Rohr 87

Die Eckendetektoren von Förstner *[9]* und Harris (Plessey Eckendetektor; siehe Noble *[13]*) nutzen die über eine lokale Umgebung gemittelten Komponenten des Grauwertgradienten aus. Die dabei verwendete Matrix besteht aus

$$\underline{C} = \begin{pmatrix} \overline{g_x^2} & \overline{g_x g_y} \\ \overline{g_x g_y} & \overline{g_y^2} \end{pmatrix} \tag{9}$$

Entwickelt man die in $\underline{C}$ auftretenden Gradientenkomponenten bis zur 1. Ordnung, so folgt daraus die Matrix

$$\underline{C} = \nabla g\,(\nabla g)^T + c\underline{H}^2 \tag{10}$$

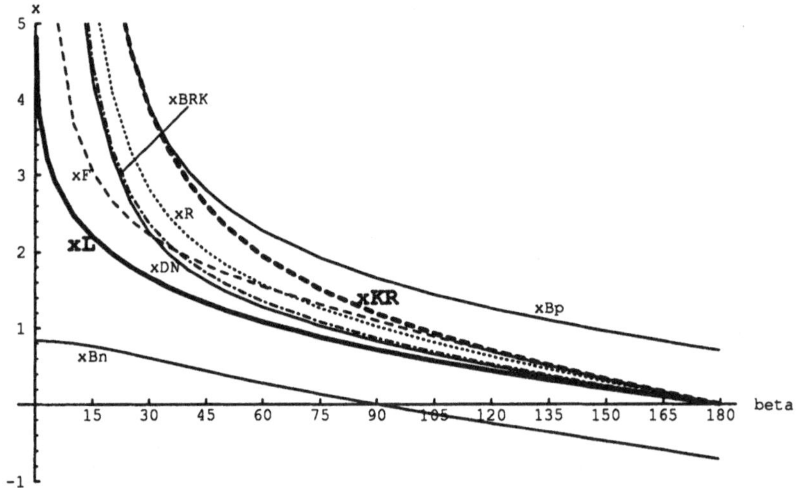

Abbildung 5: Lokalisierung verschiedener Verfahren

in Nagel *[12]*, die in Rohr *[15]* zur Eckenbestimmung verwendet wurde. Die Konstante c stellt dabei ein Maß für die Bereichsgröße dar, über der die Mittellung stattfindet.

Zur Bestimmung von Eckpunkten maximieren Förstner *[9]* und Harris *[13]* den Quotienten (bzw. minimieren dessen Kehrwert) aus der Determinante und der Spur von $\underline{C}$

$$\frac{det\underline{C}}{trace\underline{C}}(\vec{x}) \quad \rightarrow \quad Extremum \tag{11}$$

Für $y = 0$ lassen sich hieraus die Eckenpositionen x_F in Abb. 5 gewinnen (gestrichelte Linie). Angenommen wurden dabei 3×3-Operatoren, für die sich $c = 2/3$ ergibt. Für $\beta = 90^o$ ist $x_F = 1.09673$.

Der Eckendetektor in Rohr *[15]* wertet

$$det\underline{C}(\vec{x}) \quad \rightarrow \quad Extremum \tag{12}$$

aus, wobei sich in diesem Fall die Eckpunkte x_R in Abb. 5 ergeben (punktierte Linie). Die Position für $\beta = 90^o$ ist $x_R = 1.01947$.

4.7 Blom et al. 92, Brunnström et al. 92

Blom et al. *[3]* und Brunnström et al. *[4]* verwenden zur Eckendetektion eine modifizierte Version des Ansatzes von Kitchen&Rosenfeld *[10]* :

$$BRK(\vec{x}) = -(\nabla g^{\perp})^T \underline{H}\, \nabla g^{\perp} \quad \rightarrow \quad Extremum \tag{13}$$

Damit erhält man die in Abb. 5 durch die gestrichelt-punktierte Linie wiedergegebenen Positionen x_{BRK}.

5 Vergleich und Diskussion

In Abb. 5 sind die lokalisierten Positionen der verschiedenen Ansätze in Abhängigkeit des Öffnungswinkels β der L-Ecke (bei einer Gaußverschmierung von $\sigma = 1$) dargestellt. Die Positionen bei anderen Verschmierungsstärken lassen sich hieraus durch Skalierung mit x/σ gewinnen. Die Höhe a der L-Ecke hat keinen Einfluß auf die Positionen.

Mit Ausnahme des negativen Extremums x_{Bn} des Ansatzes nach Beaudet *[1]* liegen alle 'Eckpunkte' oberhalb der Positionen x_L (Krümmungsextrema enlang von Kanten). Die durch lineare Interpolation der beiden Extrema nach Beaudet *[1]* bestimmten Positionen x_{DN} (Dreschler&Nagel *[8]*) liegen für $\beta >\approx 30^o$ am nächsten bei x_L. Bei exakter Interpolation oder Einbeziehung von Kantenpunkten ergibt sich $x_{DN} = x_L$. Auch für alle anderen Ansätze (bis auf Nagel *[11]*) führt die Einbeziehung von Kantenpunkten zu den gewünschten Positionen x_L.

Sieht man von solchen zusätzlichen Maßnahmen (Interpolation; Berücksichtigung von Kanten) ab und betrachtet ausschließlich Maße, die zu maximieren bzw. zu minimieren sind, so läßt sich folgendes sagen. Die besten Ergebnisse, betrachtet über den gesamten Bereich $0^o < \beta < 180^o$ im Vergleich zu x_L, liefern die Ansätze von Förstner *[9]* (x_F) und Blom et al. *[3]* (x_{BRK}). For $\beta <\approx 35^o$ sind die Abstände von x_F zu x_L am geringsten, während dies für $\beta >\approx 35^o$ für x_{BRK} der Fall ist. Für $\beta > 20^o\,(10^o)$ sind die Abweichungen von x_F zu x_L kleiner als 0,7 (1,3) Bildpunkte. Für $\beta >\approx 65^o$ ist dic Lokalisierung x_R nach Rohr *[15]* etwas besser als x_F. Die von Kitchen&Rosenfeld *[10]* gewonnenen Positionen x_{KR} sind nie besser als x_F, x_{BRK} und x_R. Das positive Extremum x_{Bp} nach Beaudet *[1]* ist am weitesten von x_L entfernt und verläuft für abnehmende Werte $\beta <\approx 35^o$ asymptotisch gegen die durch x_{KR} festgelegte Kurve. Für nicht allzu große Winkel β bietet x_{Bn} eine Alternative zur Lokalisierung. Für $\beta <\approx 65^o$ (bzw. 35°, 20°) ist x_{Bn} jeweils näher bei x_L als x_{KR} (bzw. x_{BRK}, x_R). Die Positionierung für $\beta <\approx 7^o$ ist besser als bei allen anderen Ansätzen. Mit dem Maß in Zuniga&Haralick *[20]* lassen sich ohne Kantenpunkte keine Eckpunkte lokalisieren. Es wurde außerdem gezeigt, daß die von Nagel *[11]* geforderten Bedingungen für die hier zugrundegelegten Modelle nicht zur Eckenbestimmung verwendet werden können.

Weiterhin ist erwähnenswert, daß für kleine Öffnungswinkel enorme Abweichungen auftreten können. Für $\beta = 1^o$ ist beispielsweise der Abstand von x_{Bp} und x_{KR} zu x_L größer als 100. Für eher realistische Winkelwerte, z.B. $\beta = 10^o$, ist $x_{Bp} \approx x_{KR} \approx 11.5$, d.h. bei einer angenommenen Verschmierung des Originalbildes von $\sigma_0 = 1$ führt eine Gaußfilterung mit $\sigma_F = 1\,(2)$ zu einer Verschiebung der Position um $\Delta x = (\sqrt{\sigma_0^2 + \sigma_F^2} - \sigma_0)\, x_{Bp} \approx 4.8\,(14.2)$ Bildpunkte. In diesem Fall könnten solche Verfahren Schwierigkeiten haben, die die mit diesen Ansätzen gewonnenen Ecken bei unterschiedlich starker Filterung einander zuordnen (z.B. Deriche&Giraudon *[6]*).

Da reale Bilder diskretisiert, quantisiert und durch Rauschen gestört sind, sollte in weiteren Untersuchungen der Einfluß dieser Effekte auf die Leistungsfähigkeit der verschiedenen Eckendetektoren analysiert werden.

Dank

Für Diskussionen und hilfreiche Kommentare danke ich J. Rieger, C. Schnörr, B. Neumann, G. Winkler, D. Bister, K. Daniilidis, D. Koller und H. Neumann. Gefördert wurde diese Arbeit von der EG, Esprit-Projekt VIVA.

Literatur

[1] P.R. Beaudet, "Rotationally invariant image operators", *Proc. Intern. Joint Conf. on Pattern Recognition,* Kyoto/Japan, Nov. 7-10, 1978, 579-583

[2] V. Berzins, "Accuracy of Laplacian Edge Detectors", *Computer Vision, Graphics, and Image Processing* 27 (1984) 195-210

[3] J. Blom, B.M. ter Haar Romeny, and J.J. Koenderink, "Affine invariant corner detection", submitted for publication, see "Topological and Geometrical Aspects of Image Structure", Blom, J., doctoral dissertation, University of Utrecht, May 1992

[4] K. Brunnström, T. Lindeberg, and J.-O. Eklundh, "Active Detection and Classification of Junctions by Foveation with a Head-Eye System Guided by the Scale-Space Primal Sketch", *Proc. 2nd ECCV,* S. Margherita/Italy, May 18-23, 1992, Lecture Notes in Computer Science 588, G. Sandini (Ed.), Springer-Verlag Berlin Heidelberg, 701-709

[5] F. Canny, "A computational approach to edge detection", *IEEE Trans. on Pattern Anal. and Machine Intell.* 8 (1986), 679-698

[6] R. Deriche, G. Giraudon, "Accurate Corner Detection : An Analytical Study", *Proc. Third Intern. Conf. on Computer Vision,* Dec. 4-7, 1990, Osaka/Japan, 66-70

[7] L. Dreschler, "Zur Reproduzierbarkeit von markanten Bildpunkten bei der Auswertung von Realwelt-Bildfolgen", *Modelle und Strukturen,* DAGM - Symposium, Hamburg, Oktober 1981, Informatik-Fachberichte 49, B. Radig (Hrsg.), Springer-Verlag Berlin Heidelberg 1981, 76-82

[8] L. Dreschler, H.-H. Nagel, "Volumetric Model and 3D-Trajectory of a Moving Car Derived from Monocular TV-Frame Sequences of a Street Scene", *Proc. IJCAI,* Vancouver, BC (1981) 692-697, see also: *Computer Graphics and Image Processing* 20 (1982) 199 - 228

[9] W. Förstner, "A Feature Based Correspondence Algorithm for Image Matching", *Intern. Arch. of Photogrammetry and Remote Sensing* 26-3/3, 1986, 150-166

[10] L. Kitchen and A. Rosenfeld, "Gray-level corner detection", *Pattern Recogn. Letters* 1 (1982) 95-102

[11] H.-H. Nagel, "Displacement Vectors Derived from Second-Order Intensity Variations in Image Sequences", *Computer Vision, Graphics, and Image Processing* 21 (1983) 85-117

[12] H.-H. Nagel, "Constraints for the Estimation of Displacement Vector Fields from Image Sequences", *Proc. IJCAI,* Karlsruhe, 8.-12. August 1983, 945-951

[13] J.A. Noble, "Finding Corners", *Proc. Third Alvey Vision Conf.* , University of Cambridge, Cambridge/UK, 15.-17. Sept. 1987, 267-274

[14] J.H. Rieger, "Generic properties of edges and "corners" on smooth greyvalue surfaces", *Biol. Cybern.* 66 (1992) 497-502

[15] K. Rohr, "Untersuchung von grauwertabhängigen Transformationen zur Ermittlung des optischen Flusses in Bildfolgen", Diplomarbeit, Institut für Nachrichtensysteme, Universität Karlsruhe, 1987

[16] K. Rohr, "Über die Modellierung und Identifikation charakteristischer Grauwertverläufe in Realweltbildern", *12. DAGM - Symposium Mustererkennung,* 24.-26. Sept. 1990, Oberkochen-Aalen, Informatik-Fachberichte 254, R.E. Großkopf (Hrsg.), Springer-Verlag Berlin Heidelberg, 217-224

[17] K. Rohr, "Modelling and Identification of Characteristic Intensity Variations", *Image and Vision Computing* 10 (1992) 2, 66-76

[18] K. Rohr, "Lokalisierungseigenschaften direkter Ansätze zur Ermittlung von Grauwertecken", Techn. Bericht FBI-HH-M-217/92, FB Informatik, Universität Hamburg, 1992

[19] M.A. Shah, R. Jain, "Detecting Time-Varying Corners", *Proc. 7th Intern. Conf. on Pattern Recognition,* Montreal/Canada, July 30 - August 2, 1984, 2-5

[20] O.A. Zuniga and R.M. Haralick, "Corner detection using the facet model", *Proc. IEEE Conf. on Computer Vision and Pattern Recognition,* Washington/D.C., June 19-23, 1983, 30-37

Preserving Topology in the Irregular Curve Pyramid*

Dieter Willersinn and Walter G. Kropatsch

Technical University of Vienna, Institute for Automation 183/2, Department for Pattern Recognition and Image Processing, Treitlstr. 3, A-1040 WIEN
Email: wil@prip.tuwien.ac.at

Abstract. The irregular curve pyramid processes symbolic representations of the content of an image, binary curve relations, in logarithmic time. It is flexible enough to adapt to the structure of the data. The process to construct the pyramid may corrupt the topology of levels, introducing holes in the cells that hold image data. We solve the problem by using distance information to control the construction process. The (dual) concept of the irregular curve pyramid can be generally helpful in applications, where the rigidity of classical pyramidal concepts causes instabilities due to shift variance and logarithmic complexity needs to be preserved.

1 Introduction

Pyramids are efficient, parallel processing schemes that permit to cope with the large amount of data in image processing. In a large class of images, e.g. aerial photographs, important structures are elongated and thin, e.g. highways, rivers. To extract these objects from an image, we represented them by symbolic binary relations between sides of observation windows and built a pyramid on top of this representation, the regular, $2 \times 2/2$ curve pyramid [4].

Shift variances of this pyramid scheme motivated our research in view of a more flexible structure [7], which gave rise to the irregular curve pyramid [6]. It is based on the principle of the stochastic pyramid [8], which permits to select image cells that will be represented at the next level. This selection has been firstly related to the image content in [3], and to symbolic curve relations in [7]. The flexibility of the structure may also represent a drawback, when it leads to a modification in the topology of the pyramidal cells, as we will show in Section 3.

The structure of this paper is as follows. In Section 2 we summarize basic notions of the irregular curve pyramid and outline the process of constructing it. Section 3 contains a precise statement of the problem addressed in this paper, and Section 4 introduces the measure and the solution which is based on this measure. Section 5 contains concluding remarks.

* This work was supported by the Austrian Science Foundation under grant P 8785

2 The irregular curve pyramid

Definition 1. A level of the *irregular curve pyramid* consists of the following elements:

1. A *neighborhood graph* $G(V, E)$ represents *vertices* $v \in V$ that are connected by *edges* $e \in E$.
2. *Cycles* in $G(V, E)$ are ordered sequences of vertices, $(v_0,, v_n), v_i \in V; i = 0, \ldots, n, v_0 = v_n, (v_i, v_{i+1}) \in E$. We call a cycle a *primitive cycle* if $(v_i, v_j) \notin E; i = 0, \ldots, n, j = 0, \ldots, n, j \notin \{i-1, i, i+1\}$. We refer to primitive cycles of $G(V, E)$ as *faces* and represent them in the dual of $G(V, E)$, $\overline{G}(F,\overline{E})$, with faces $f \in F$ and sides $\overline{e} \in \overline{E}$ that connect faces $f \in F_n$.
3. Curves are represented with respect to faces, as binary relations between sides.

We formally distinguish between the edges $e \in E$ and sides $\overline{e} \in \overline{E}$. However, there is a one-to-one match between sides and edges, and their graphical representation is the same line segment. The *levels of the pyramid* are counted bottom up, with the base being level 0 of the pyramid. For graphs representing a level n of the pyramid we use a notation with an index n: $G_n(V_n, E_n)$, $\overline{G_n}(F_n,\overline{E_n})$, or, for convenience: G_n, $\overline{G_n}$.

2.1 Selecting cells of a new pyramid level

We start building a new pyramid level $n+1$ by selecting vertices $v \in V_n$ and faces $f \in F_n$ that will be represented in V_{n+1} and F_{n+1} respectively. Selection of vertices divides the vertices in two classes (non-survivors and survivors) who satisfy the following rules:

1. No two vertices must survive.
2. Every non-survivor must have at least one survivor in its neighborhood.

The process to construct G_{n+1} from G_n is described in Definition 2.

Definition 2. *Stochastic decimation* of a graph is a process that is executed in the following steps:

1. Random numbers are assigned to the vertices.
2. Vertices with a local maximum of this variable (*surviving vertices*) are selected.
3. All *non-surviving vertices* are assigned to one survivor out of their neighborhood.
4. Repeat steps 1 ...3 for all non-survivors who do not have a survivor in their neighborhood.
5. The *receptive field* $RF(v_s)$ of a survivor v_s is formed by all non-survivors (*sons*) that are assigned to v_s (*father*). It also includes v_s itself.
6. Vertices of the reduced graph are connected if vertices of their receptive fields are connected in the original graph.

At a level $n; n \geq 0$, of the pyramid, we decimate the vertices of $G_n(V_n, E_n)$ to obtain the vertices of level $n+1$.

After the decimation of the neighborhood graph, we can now identify surviving faces. We therefore define two two notions corresponding to primitive cycles in the neighborhood graph.

Definition 3. The *receptive field set* $RS(f)$ of face f is the set of fathers of all different receptive fields that are represented by the vertices of cycle $C(f)$ corresponding to f. The *valence* $val(f)$ of a face f is defined as the number of different receptive fields represented in $C(f)$,

$$val(f) = card(RS(f)).$$

In [5] we proved the following helpful property for the definition of F_{n+1}.

Theorem 4. *Faces f with $val(f) \geq 3$ at a level n of an irregular curve pyramid survive to level $n+1$. At level $n+1$, the cycle corresponding to f is made of the father vertices in $RS(f)$. Faces f with $val(f) \leq 2$ do not survive.*

2.2 Connecting new pyramid cells

Figure 1 illustrates the construction of a new level $n+1$ of a irregular curve pyramid. Survivors are marked with black spots (•), non-survivors with circles (o),

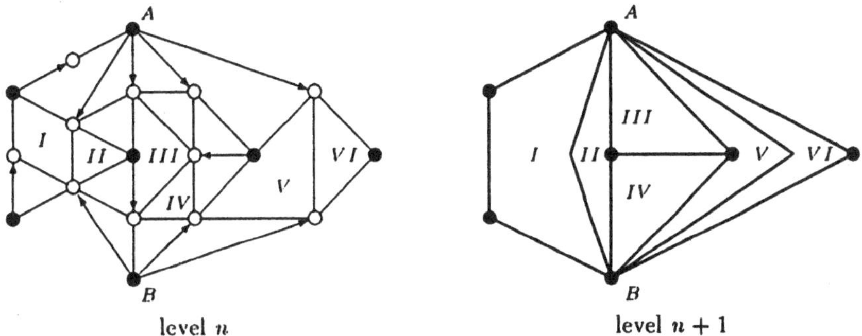

Fig. 1. A level of an irregular curve pyramid may contain multiple connections between vertices.

receptive field assignments with arcs ($\longrightarrow$), and surviving faces with $(I \ldots VI)$. At level $n+1$, faces I and II, as well as faces V and VI are connected via a side that is spanned by vertices A and B.

With respect to G_{n+1}, this means that two edges $(A, B)_1, (A, B)_2 \in E_{n+1}$ have to be created to satisfy polyhedron formula (cf. Equation 1), which we cite from [2] without counting the exterior face.

$$card(V) - card(E) + card(F) = 1. \tag{1}$$

However, $(A,B)_1$ and $(A,B)_2$ are not identical since the corresponding sides connect different faces of the dual graph. To create all connections between faces and vertices of a new pyramid level, it is therefore not sufficient to establish all connections between survivors, the receptive fields of which are connected in G_n. For level $n+1$ of the example in Figure 1, polyhedron formula is satisfied with

$$card(V) = 7, card(F) = 6, card(E) = 12.$$

In [5] we solved the problem of finding all connections between faces by searching paths in F_n between surviving faces, which is equivalent to growing disjoint face regions starting with the surviving faces.

Definition 5. A face $f \in F_n$ belongs to the *face region* $FR(g) \subset F_n$ of a surviving face $g, g \in F_n \cap F_{n+1}$ if

1. $f \notin FR(h); h \neq g; g, h \in F_n \cap F_{n+1}$,
2. $\exists (f,k) \in \overline{E_n}, k \in FR(g)$,
3. $RS(f) \subset RS(g)$.

Every face in FR(g) carries as a label of $g \in F_{n+1}$. The *boundary* of a face region $FR(g)$ is formed by sides $(k,l); k,l \in F_n$ separating faces $k \in FR(g)$ from faces $l \notin FR(g)$. Surviving face g is called the *core face* with respect to $FR(g), g \in FR(g)$.

We obtain a complete partitioning of F_n into connected regions.

Definition 6. We assume a partitioning of a level n of an irregular curve pyramid into face regions. Assuming one processing unit representing every edge/side, we obtain *primary links* (sides, edges) between faces and vertices at level $n+1$ in two processing steps for the whole graph:

1. sides $(k,l) \in \overline{E_n}$ of the boundary between face regions, i.e. $k \in FR(f), l \in FR(g); f, g \in F_n, F_{n+1}$, build the intersection between the regions' receptive field sets, $RS(f) \cap RS(g)$.
2. If this "RS intersection" contains more than one vertex, then theoretically $\binom{n}{2}; n = card(RS(f) \cap RS(g))$ sides may be shared by the cycles corresponding to f and g, $\binom{n}{2}$ being the number of different edges to connect the n vertices of the RS intersection. All $\binom{n}{2}$ new sides, as well as the edges corresponding to them are created, provided that they do not exist already.

3 The problem of holes in cells

Problems may occur if two surviving faces have identical receptive field sets. Figure 2 a) shows such a situation, $RS(II) = RS(III) = \{A,B,C\}$: the face region of one of the two survivors can surround completely the other's, as illustrated in Figure 2 a). Face III is surrounded by faces assigned exclusively to $FR(II)$, and links to III can only be established from II. On the other hand, II will be connected to all other faces of the configuration, cf. Figure 2 b). We

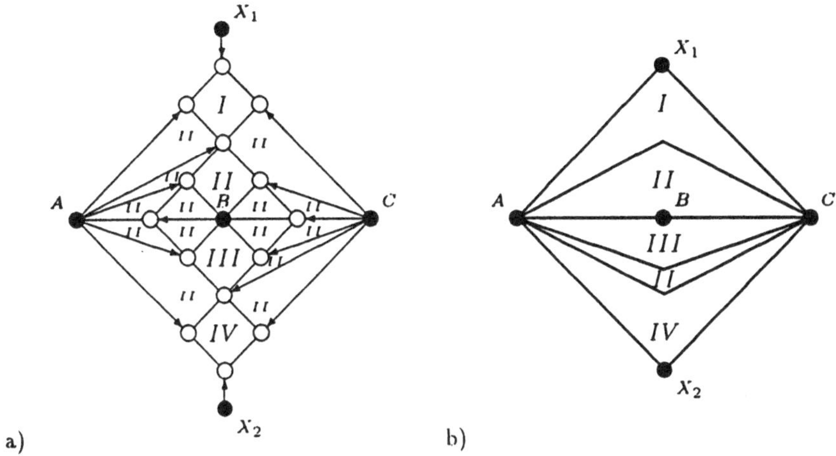

Fig. 2. A face incorporating a face with an identical receptive field set.

obtain a configuration with two faces nested into one another. Such a topology has been excluded in the definition of the irregular curve pyramid because it contains a primitive cycle with two vertices (cf. [5]).

4 Controlling the growth of face regions

The unstable topology of pyramid levels (see Figure 2) is due to a lack of control in the region growing process. Definition 5 does not provide a means to modify a face region. To remedy the situation, we introduce the notion of distance between faces.

Definition 7. The distance $d(f, g)$ between two faces f and g; $f, g \in F_n$, is the minimum number of sides that separates f from g.

We now have a measure that can be used to control the growth of face regions. During the formation of a face region $FR(g)$, every face associates with its region label the distance to the core face g. We start with initializing this distance value to 0 for all surviving faces. Every face f that overtakes a region label from one of its neighbors, e.g. $k \in FR(g)$, assigns its own distance value the distance value of k, incremented by one:

$$d(f, g) = d(k, g) + 1.$$

For the example in Figure 2, Figures 3 a) and b) show distances from faces II and III, respectively. In both subfigures it is assumed that the face region has grown without concurrency to its maximum extent, the respective core face

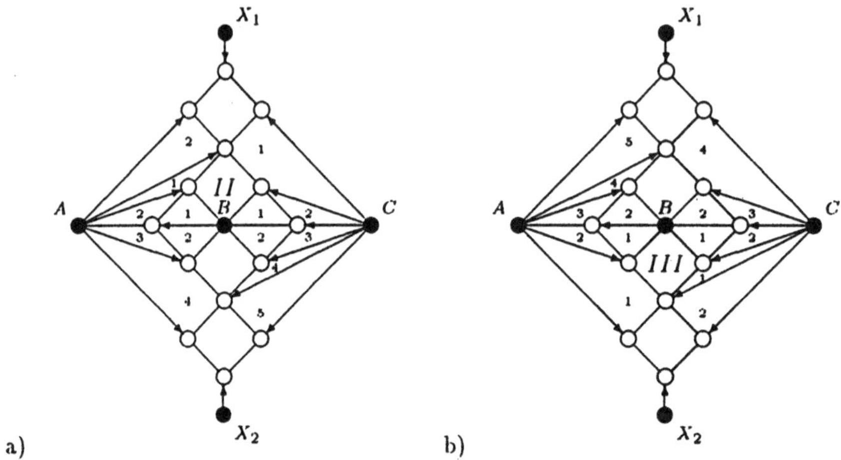

Fig. 3. Distances to faces *II* and *III*

is marked with its number. For all faces of these face regions the distance to the core face is given.

With the distance value on every face $f \in F_n$ we can extend Definition 5 to allow to modify the FR-assignment of a face.

Definition 8. The *reassignment* of a face $f \in FR(h); h \in F_n \cap F_{n+1}$ to a face region is done in favour of a face region $FR(g)$ if

1. $\exists (f,k) \in \overline{E_n}, k \in FR(g)$.
2. $RS(f) \subset RS(g)$,
3. $d(f,h) > d(f,g); h \neq g; g,h \in F_n \cap F_{n+1}$,

The face region growing, weighted with the distance information, can solve the problem of one surviving face being "submerged" into the region of another face with an identical receptive field set. However, there are still ambiguities, where faces have the same minimum distance to two or more surviving faces. In Figure 4, two possible configurations of face regions are shown. Table 1 reflects this situation. It contains all primary links between faces (and vertices) that can possibly be established by sides on face region boundaries for the example of Figure 4. The remaining ambiguities consist in multiple connections from one surviving face to other survivors. The following local correction process removes inconsistent connections:

1. Consider a surviving face having more than one primary link to another surviving face g, $L_g = \{\overline{e_1}, \dots, \overline{e_k}\} = \{(\overline{v_0, v_1}), \dots, (\overline{v_{k-1}, v_k})\}$, but only one connection to another face h, $L_h = \{\overline{e_i}\} = \{(\overline{v_{i-1}, v_i})\}$, with $\overline{e_i} \in L_g$. Remove $\overline{e_i}$ from L_g.

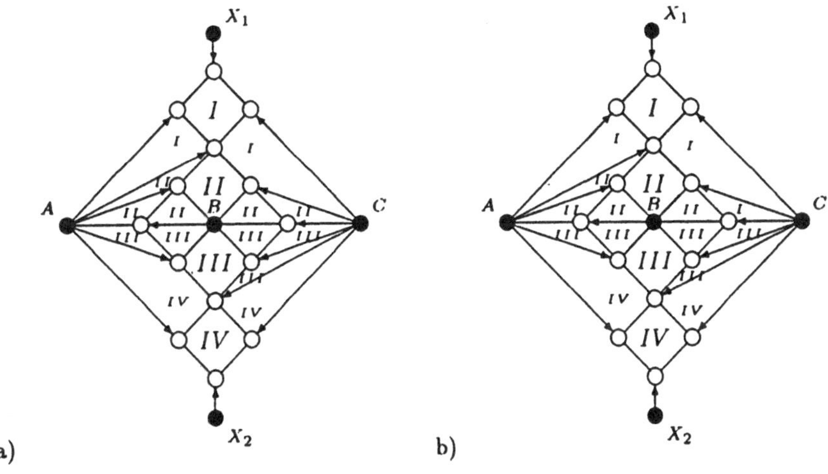

Fig. 4. Two examples for region formations using distances

face	neighbor	RS intersection	possible connections	distance
I	II	$\{A,C\}$	$\overline{(A,C)}$	2
	III	$\{A,C\}$	$\overline{(A,C)}$	6
II	I	$\{A,C\}$	$\overline{(A,C)}$	2
	III	$\{A,B,C\}$	$\overline{(A,B)},\overline{(A,C)},\overline{(B,C)},$	3
III	I	$\{A,C\}$	$\overline{(A,C)}$	6
	II	$\{A,B,C\}$	$\overline{(A,B)},\overline{(A,C)},\overline{(B,C)},$	3
	IV	$\{A,C\}$	$\overline{(A,C)}$	2
IV	III	$\{A,C\}$	$\overline{(A,C)}$	2

Table 1. Primary links between faces at level $n+1$

2. Consider a face f having primary links to potential neighbors $g_1,\ldots,g_k$ with $\overline{(f,g_1)} = \ldots = \overline{(f,g_k)} = (v_1,v_2); v_1,v_2 \in V_n$.
Remove all primary links except

$$(f,g_j)|d(f,g_j) = min\{d(f,g_i), g_i \in \{g_1,\ldots,g_k\}\};$$

Table 2 shows the result of the correction process for the example in Figure 4.

5 Conclusion

We have shown that, by using local distance measurements, topology is preserved in the irregular curve pyramid. The construction of the pyramid is done by a cooperation between the neighborhood graph and its dual.

This principle avoids a drawback of stochastic decimation: the vertex degree in the neighborhood graph does not remain bounded [5], and hence the com-

face	neighbor	RS intersection	connections
I	II	$\{A,C\}$	$\overline{(A,C)}$
II	I	$\{A,C\}$	(A,C)
	III	$\{A,B,C\}$	$\overline{(A,B)},\overline{(B,C)}$
III	II	$\{A,B,C\}$	$\overline{(A,B)},\overline{(B,C)}$
	IV	$\{A,C\}$	$\overline{(A,C)}$
IV	III	$\{A,C\}$	(A,C)

Table 2. New sides after elimination of inconsistent edges

putational complexity of local processes. However, also in [5], we proved that the degree of faces does not increase. The dual concept of the irregular curve pyramid combines therefore the flexibility of the structure with neighborhood operations that can be executed in constant time ($O(1)$). Logarithmic computational complexity of processes is therefore preserved.

The concept of irregular pyramids, extended by the concept of dual graphs can generally be helpful in applications where rigidity of classical pyramid structures causes instabilities due to shift variances [1], and where logarithmic complexity needs to be preserved.

References

1. M. Bister, J. Cornelis, and A. Rosenfeld. A critical view of pyramid segmentation algorithms. *Pattern Recognition Letters*, Vol. 11(No. 9):pp. 605–617, September 1990.
2. F. Harary. *Graph Theory*. Addison-Wesley, Reading, Mass., 1972.
3. J. M. Jolion and A. Montanvert. The adaptive pyramid, a framework for 2D image analysis. *Computer Vision, Graphics, and Image Processing: Image Understanding*, 55(3):339–348, May 1992.
4. W. G. Kropatsch. Kurvenrepräsentation in Pyramiden. In W. G. Kropatsch and P. Mandl, editors, *Mustererkennung'86*, OCG-Schriftenreihe, Österr. Arbeitsgruppe für Mustererkennung, pages 16–51. Oldenbourg, 1986. Band 36.
5. W. G. Kropatsch and A. Montanvert. Irregular versus regular pyramid structures. In U. Eckhardt, A. Hübler, W. Nagel, and G. Werner, editors, *Geometrical Problems of Image Processing*, pages 11–22, Georgenthal, Germany, March 1991. Akademie Verlag, Berlin.
6. W. G. Kropatsch and D. Willersinn. Irregular curve pyramids. In *Proceedings of the NATO Workshop Shape in Picture (in press)*, Driebergen, (The Netherlands), September 1992.
7. W. G. Kropatsch and D. Willersinn. Representing curves in irregular pyramids. In W. G. Kropatsch and H. Bischof, editors, *Pattern Recognition 1992*, pages 333–348, Vienna, Austria, May 1992. Oldenbourg.
8. P. Meer. Stochastic image pyramids. *Computer Vision, Graphics, and Image Processing*, Vol. 45(No. 3):pp.269–294, March 1989.

Sprachverarbeitung in neuronaler Architektur

A.Grauel[1], H.-G.Grundmann[2], M.Stork[1]
Universität Paderborn, Abt.Soest, FB 16
Elektrische Energietechnik / Automatisierungstechnik
[1]Fachgebiet Mathematische Methoden und Systemtheorie
[2]Rechenzentrum
Steingraben 21, 4770 Soest

Abstract: We propose in this investigation the speech recognition of long words in neural architecture. To that we consider a Multilayer-Perceptron where the input vectors are normalized with respect to time. In this consideration we have used different distance measures to study their influence. We obtain a highly accuracy about few learning steps with respect to the memorized states. Moreover the speech recognition can be obtained with a satisfactory accuracy.

1 Einleitung

Es wird eine Methode zur Spracherkennung langer Worte in neuronaler Architektur vorgeschlagen. Mit Hilfe der Zeitnormierung (Ortskurvensegmentierung) wurden die Sprachvektoren als Inputvektoren für ein Multilayer-Perceptron zunächst auf einheitliche Länge gebracht. Dabei wurden unterschiedliche Abstandsmaße betrachtet. Es ergab sich, daß bereits mit wenigen Lernschritten eine hohe Genauigkeit bezüglich der Memorisierung erreicht werden konnte. Für die Spracherkennung ergab sich eine zufriedenstellende Erkennungsrate.

2 Spracherkennung

In den letzten Jahren wurden große Anstrengungen im Bereich der Mensch-Maschine-Kommunikation, z.B. der automatischen Spracherkennung, der Robotertechnik etc. unternommen. Trotz der großen Fortschritte, die auf dem Gebiet der Spracherkennung und Sprachverarbeitung erzielt wurden, sind noch erhebliche Anstrengungen nötig. Durch die neuronalen Netze erfuhr die Sprachverarbeitung neben den Hidden-Markov-Modellen[1-3] zusätzlich weiteres Interesse. Mindestens zwei Netzwerkarchitekturen im Bereich der neuronalen Netze dürften dafür von besonderem Interesse sein: Rückgekoppelte Multilayer-Perceptrons und Time-delayed Networks[4,5]. Time-delayed Netzwerke besitzen asymmetrisches Verhalten, simuliert durch eine asymmetrische Kopplung zwischen den Neuronen. Untersuchungen zeigen, daß mit asymmetrischen Netzwerken Zyklen von Mustern und Mustersequenzen erzeugt werden können[6-11]. In einer vorangegangenen Arbeit[12] haben wir musterinduzierte Übergänge in einer seriellen Assoziation untersucht, nämlich die Generierung einfacher geometrischer Muster und Sprachmuster von Lauten und Umlauten. Solche Mustersequenzen sind für die Sprachverarbeitung wichtig, um z.B. Laute zu Worten und Worte zu Sätzen zu verketten. Es konnte gezeigt werden[13], daß die Multilayer-Perceptron (MLP) Architektur für die Sprachdatenverarbeitung bzw. Spracherkennung geeignet ist. Der Vorteil der Multilayer-Architektur liegt in der Anwendung eines einfachen Lernalgorithmus. Als Sprachdaten standen Merkmalsvektoren von Lauten /Umlauten zur Verfügung, die aus einer Ceptral-Analyse gewonnen wurden. Sprecherunabhängigkeit wurde durch vier Versionen der Sprachdaten simuliert. Für die Erkennung wurde ein Multilayer-Perceptron optimiert, wobei als Lernalgorithmus zwei unterschiedliche Gradientenverfahren benutzt wurden. Die Resultate zeigten, daß sich bessere Resultate mit einem konjugierten Gradientenabstiegsverfahren gegenüber dem üblichen Gradientenverfahren erreichen lassen[12].

In diesem Beitrag wird die Einzelworterkennung von langen Wörtern in neuronaler Architektur diskutiert. Das hier verwendete Multilayer-Perceptron besteht aus drei Neuronenschichten, der Input- und Output-Schicht sowie einer Hidden-Layer-Schicht (siehe Fig.1) mit variabler Neuronenzahl in Form eines feedforward Netzes. Die Zahl der Input- bzw. Output-Neuronen richtet sich nach der Größe des zu betrachtenden Zeitfensters.

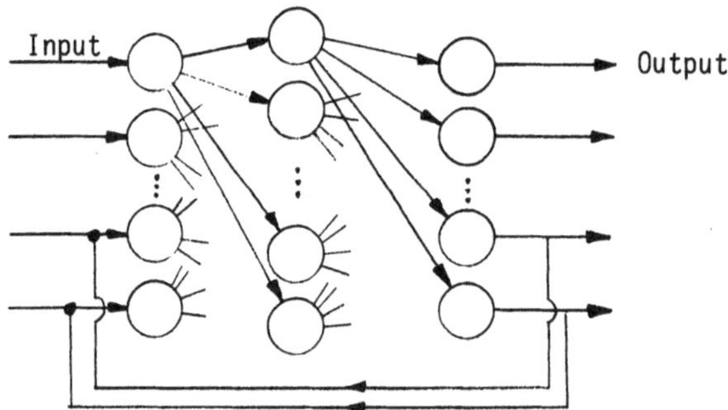

Fig.1 Architektur eines rückgekoppelten Multilayer-Perceptrons. Verbindungstopologie: Feedforward-Verknüpfung und keine laterale Verkopplung.

Bei unseren Untersuchungen wurde die Zahl der Inputneuronen auf 240 begrenzt. Als Aktivierungsfunktion für die formaliserten Neuronen haben wir eine sigmoide Funktion benutzt. Als Lernregel wurde die Methode Backpropagation benutzt, wobei die Abänderung der Gewichte wie folgt vorgenommen wurde

$$\Delta_p w_{ij} = \eta \cdot \delta_{pi} \cdot a_{pj}$$

Für eine sigmoide Aktivierungsfunktion gilt:

$$\delta_{pi} = \begin{cases} a_{pi} \cdot (1 - a_{pi}) \cdot (t_{pi} - a_{pi}) & i \in O \\ & \text{für} \\ a_{pi} \cdot (1 - a_{pi}) \cdot \sum_k w_{ki} \cdot \delta_{pk} & i \in H \end{cases}$$

O kennzeichnet den Outputlayer und H die Hiddenlayer-Schicht. Die numerischen Ergebnisse mit dem Backpropagation-Algorithmus werden mit einem modifizierten Gradientenabstiegsverfahren, dem sogenannten konjugierten Gradientenverfahren, verglichen. Die Gewichtsänderung zum Zeitpunkt t berechnet sich aus der Summe der partiellen Ableitungen (Zeitpunkt t) und der im vorhergehenden Zeitschritt t-1 berechneten Gewichtsänderung:

$$\Delta_p w_{ij}(t) = -\eta \cdot \frac{\partial E_p}{\partial w_{ij}(t-1)} + \alpha \cdot \Delta_p w_{ij}(t-1)$$

Der Trägheitssummand α verhindert ein quasi-steckenbleiben bei der Methode des steilsten Abstieges bei konstanter Schrittweite. Er bewirkt, daß die "Bewegung" in weiten Plateaus der Fehlerfläche E beschleunigt und bei starken Zerklüftungen gehemmt wird[13].

Da die Sprachmerkmalsvektoren stets unterschiedliche Länge besitzen, die neuronalen Netze jedoch stets gleich lange Eingangsvektoren benötigen, sind die Daten zunächst auf einheitliches Datenformat zu bringen. Eine lineare Anpassung der Sprachvektoren ist im allgemeinen nicht möglich, dieses hängt damit zusammen, daß die verschiedenen Lautgruppen bei schnellen und langsamen Sprechen unterschiedlich gedehnt bzw. gestaucht werden. Deshalb wird für dic Merkmalsextraktion die Fourier-Cepstralanalyse benutzt. Die mit 8 Bit quantisierten und mit 10 kHz abgetasteten Signale werden segmentweise mit einem Hamming-Fenster gewichtet und fouriertransformiert. Es werden dabei Segmente bestehend aus 256 Werten (Länge 32 ms) betrachtet und das logarithmierte Leistungsdichtespektrum gebildet, aus dem durch inverse Fourier-Transformation letztlich das Cepstrum gebildet wird. Zur parametrischen Charakterisierung eines Segmentes werden zwölf Cepstralkoeffizienten niedrigster Ordnung benutzt. Mit Hilfe einer Ortskurvensegmentierung (Zeitnormierung) werden die Merkmalsvektoren auf eine konstante Länge gebracht. Wir erhalten für eine Ortskurvensegmentierung mit 20 charakterisierenden Segmenten und 12 Cepstralwerten insgesamt einen Merkmalsvektor mit 240 Komponenten.

Zur Veranschaulichung der Dynamik der Sprachprobe wurde der Abstand zwischen aufeinanderfolgenden Merkmalsvektoren benutzt. Dabei wurden als Abstandsmaße untersucht

Euklidischer Abstand:

$$d_{1,2} = \left|\underline{X}_1 - \underline{X}_2\right| = \sqrt{\sum_{i=1}^{12}\left|X_{1,i} - X_{2,i}\right|^2}$$

LPC-Klassifikator:

$$d_{x_1,x_2} = \left|\frac{\underline{x}_j \underline{\underline{R}}_i(|1-k|)\underline{x}_j^T}{\underline{x}_{j+1} \underline{\underline{R}}_i(|1-k|)\underline{x}_{j+1}^T}\right|$$

Mahalanobis-Klassifikator:

$$d_i(\underline{x}) = \left((\underline{x} - \underline{m}_i)^T \underline{\underline{C}}_i^{-1} (\underline{x} - \underline{m}_i)\right)^{\frac{1}{2}}$$

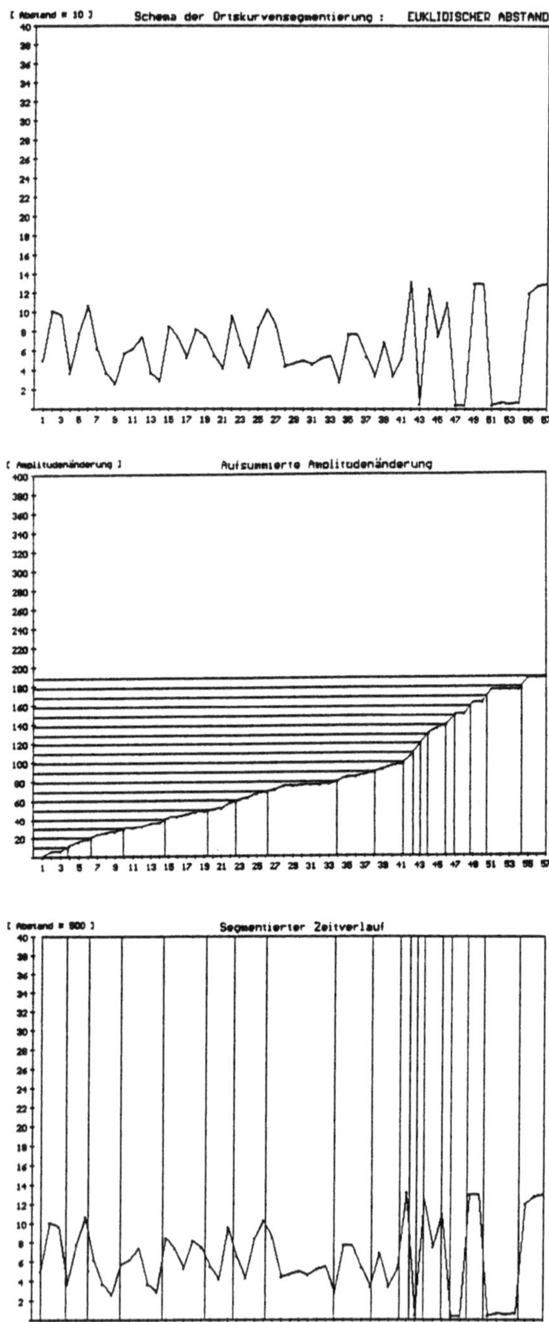

Fig.2 Zeigt das Schema der Ortskurvensegmentierung (Wahl: Euklidischer Abstand), dabei werden die Merkmalsvektoren auf einheitliche Länge gebracht. Abbildungen: a) aufgenommener Zeitverlauf für das Wort "Programmiersprache", b) zeigt die aufsummierte Amplitudenänderung und c) der segmentierte Zeitverlauf.

Betrachtet man die zeitliche Änderung der Sprachparameter zwischen zwei aufeinanderfolgenden Merkmalsvektoren, so sind gewisse Unterschiede im Dynamikverlauf erkennbar. Es ist bekannt, daß sich für Vokale relativ breite nahezu stationäre Bereiche ergeben, während Lautübergänge und Konsonanten schnell veränderlich sind und demzufolge in den Dynamikbereichen zu finden sind, in den größere Amplitudenänderungen vorkommen.
In den Figuren 2, 3 und 4 ist zu erkennen, daß der segmentierte Zeitverlauf für die gleiche Sprachprobe durchaus unterschiedlich ausfallen kann. Als Beispiel wurde aus einem Codebook das Wort Programmiersprache ausgewählt.

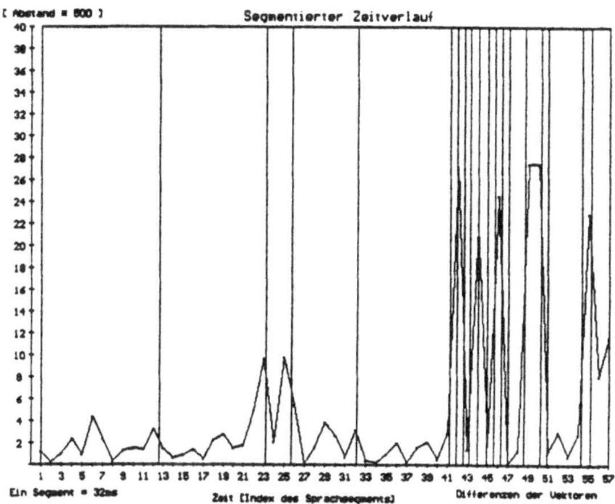

Fig.3 Aufnahme des segmentierten Zeitverlaufs mit dem LPC-Klassifikator als Abstands maß für das Wort "Programmiersprache".

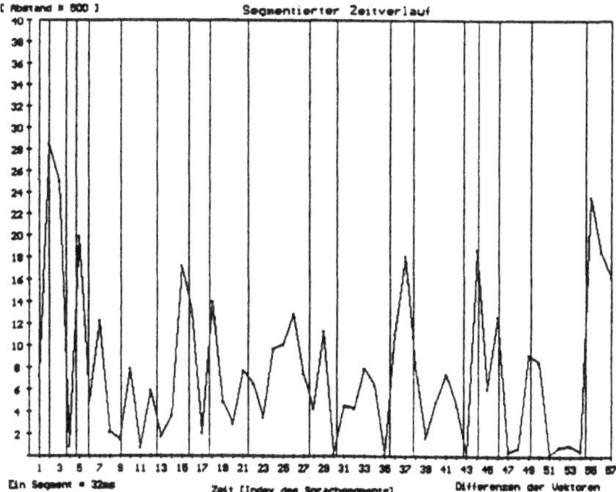

Fig.4 Zeigt den segmentierten Zeitverlauf für das Wort "Programmiersprache", dabei wurde der Mahalanobis-Klassifikator als Abstandsmaß benutzt.

Bei der Optimierung des Netzwerkes (siehe Fig.1) stellte sich heraus, daß sowohl beim Backpropagation-Verfahren als auch beim konjugierten Gradientenabstiegsverfahren die beste Genauigkeit mit einem Lernparameter $\eta = 0{,}5$ zu erreichen war. Parameter α hat üblicherweise Werte zwischen 0,2 und 0,99. Der optimale Wert lag ebenfalls bei $\alpha = 0{,}5$. Numerische Resultate für die Anzahl der Zwischenschichtneuronen zeigen für eine optimale Parametereinstellung, daß bereits bei 150 Lernschritten mit 260 Neuronen in der Zwischenschicht eine Genauigkeit von 97,7 % mit dem konjugierten Gradientenabstiegsverfahren erreichbar sind (Fig.5). Lernkurven für das Backpropagation-Verfahren zeigen ein analoges Verhalten, allerdings mit geringerer Genauigkeit.

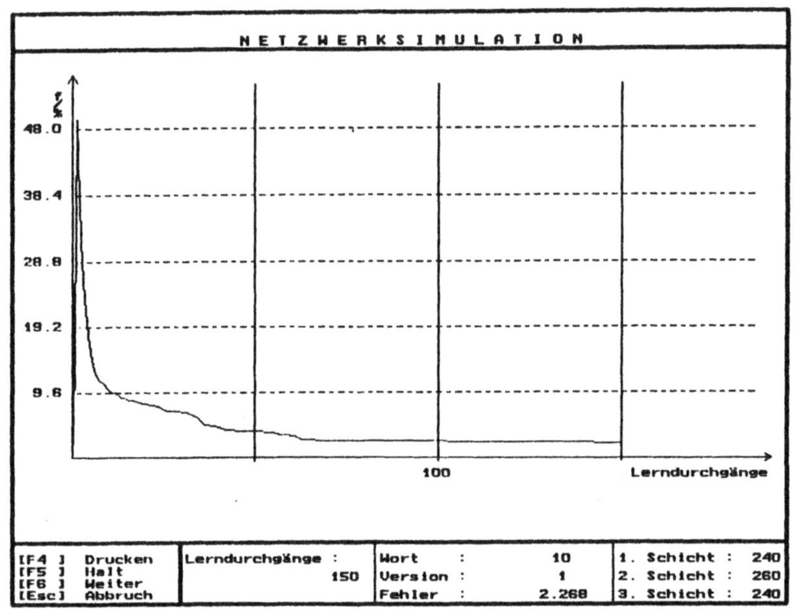

Fig.5 Ergebnis einer Netzwerksimulation mit dem oben beschriebenen optimierten MLP. Es ist zu erkennen, daß bereits bei einer geringen Zahl von 150 Lerndurchgängen der Fehler nur noch 2,3% beträgt. Die Parameter wurden wie folgt gewählt: $\eta = 0{,}5$, $\alpha = 0{,}5$. Bei der Sprachprobe handelt es sich um das Wort Programmiersprache.

Die Untersuchungen mit dem optimierten MLP zeigen (Fig.5), daß die Einspeicherung der zeitnormierten Vektoren (Merkmalsvektoren) bereits nach relativ wenigen Lernschritten mit brauchbaren Genauigkeiten erreicht werden kann (Rechenzeit für 150 Lernschritte, ca. 30 min auf 386'er PC, Programmiersprache Turbo Pascal). Insgesamt wurden aus einem Codebook zehn lange Worte im Recall getestet. Bei diesen Voruntersuchungen zur Spracherkennung ergab sich, daß die Erkennungsrate beim Selbstaufruf bei ca. 97,5% lag. Die Erkennungsrate allgemein bei Verwendung unterschiedlicher Abstandsmaße lag in der gleichen Größenordnung. Insgesamt können die Recall-Eigenschaften und somit auch die

Spracherkennung durchaus als zufriedenstelled beurteilt werden. Für höhere Genauigkeitsansprüche muß die Anzahl der angegebenen Lernschritte erhöht werden. Eine Performance-Verbesserung ist sicherlich möglich, dabei dürfte die Trainingszeit ein unkritischer Parameter sein.

Die Rückkopplungen in der Multilayer-Perceptron-Architektur erscheinen biologisch gerechtfertigt. Unsere Untersuchungen jedoch zeigten, daß bei Benutzung des Feedback-Mechanismus keine signifikaten Verbesserungen z.B. beim Lernen erreicht werden konnten. Weitere Analysen und Untersuchungen erscheinen im Hinblick auf die Biologie sinnvoll und notwendig, um die Performance der neuronalen Netze zu verbessern.

3 Literaturverzeichnis

1. L.R.Rabiner, S.E.Levinson: A Speaker-Independent, Syntax-Directed, Connected Word Recognition System Based on Hidden Markov Models and Level Buildung, IEEE Trans. on Acoustics, Speech, and Signal Processing 33,561(1985).
2. L.R.Rabiner: A Tutorial on Hidden Markov Models and Selected Applications in Speech Recognition, Proceedings of the IEEE, 77,257(1989).
3. H.A.Bourlard: How Connectionist Models Could Improve Markov Models for Speech Recognition, in Advanced Neural Computers (R.Eckmiller, Editor), Elsevier Science Publ.B.V., North-Holland, Amsterdam (1990).
4. R.Pels: Diplomarbeit, Universität Paderborn, Abt. Soest (1992).
5. S.M.Peeling, R.K.Moore: Isolated Digit Recognition Experiments Using the Multilayer Perceptron, Speech Communication, 7,403(1988).
6. D.Kleinfeld: Sequential State Generation by Model Neural Networks, Proc. Natl. Acad. Sci. USA 83,9469(1986).
7. H.Sompolinsky, I.Kanter: Temporal Association in Asymmetric Neural Networks, Phys. Rev. Lett. 57,2861(1987).
8. S.Dehaene, J.P.Changeux, J.V.Nadal: Neural Networks That Learn Temporal Sequen ces by Selection, Proc. Natl.Acad.Sci. USA 84,2767(1987).
9. A.Herz, B.Sulzer, R.Kühn, J.L.van Hemmen: The Hebb Rule: Storing Static and Dynamic Objects in an Associative Neural Network, Europhys. Lett 7,663(1988).
10. A.Herz, B.Sulzer, R.Kühn, J.L.van Hemmen: Hebbian Learning Reconsidered, Biol. Cybern. 60, 457(1989).
11. R.Kühn, J.L.van Hemmen: Temporal Association,in Models of Neural Networks (E. Domany, J.L.van Hemmen, K.Schulten, Eds.), Springer-Verlag, Heidelberg (1991).
12. A. Grauel, H.-G.Grundmann, R.Pels: Muster-Assoziation mit Time-delayed Networks, Informatik aktuell (S.Fuchs, R.Hoffmann eds.), Springer-Verlag, Heidelberg (1992).
13. A.Grauel: Neuronale Netze, Grundlagen und mathematische Modellierung, B. I. Wis senschaftsverlag, Mannheim (1992)

Modulares Neuronales Phonemerkennungskonzept mit Radialbasisfunktionen

Anke Meyer-Bäse[1], W. Endres[2], W. Hilberg[3], H. Scheich[4]

[1]Institut für Flugmechanik und Regelungstechnik, Technische Hochschule Darmstadt
[2]Institut für Übertragungstechnik und Elektroakustik, Technische Hochschule Darmstadt
[3]Institut für Datentechnik, Technische Hochschule Darmstadt
[4]Institut für Neurobiologie, Magdeburg

Dieser Beitrag beschäftigt sich mit der Erkennung von Phonemen durch spezialisierte modulare neuronale Netze. Die Lautmerkmale liegen in Form von Cepstralkoeffizienten vor und durchlaufen mehrere neuronale Netze, die als Radialbasisklassifikatoren implementiert sind und je nach Schwierigkeitsgrad der Entscheidungsstufe die Klassenzugehörigkeit mehr oder weniger berücksichtigen. Der hier gewählte Ansatz ergab gute Erkennungsergebnisse.

1 Einleitung

Phonemerkennung erfolgt auf dem Gebiet der neuronalen Netze fast ausschließlich durch das Multilayer-Perzeptron [End92] und dessen erweiterten Ableitung, dem Time-Delay Neural Network. Beim implementierten Lernverfahren handelt es sich um den Backpropagation-Algorithmus — eine überwachte Lernmethode —, bei dem die Verbindungsgewichte zwischen den Neuronen in Richtung des negativen Gradienten einer Fehlerfunktion verändert werden.

Radialbasisklassifikatoren wurden bisher relativ wenig in der Phonemerkennung [RR88] getestet, obwohl sie eigentlich sowohl die Vorzüge des Perzeptrons als auch die der Kohonen-Karten verknüpfen. Demnach ist der Lernalgorithmus hybrid, denn er implementiert einerseits ein Gradientenverfahren zur Minimierung des Lernfehlers, andererseits wird die Verteilung der Merkmale im Raum, die Clusteranalyse, vorgenommen. Ein dreischichtiger Aufbau des neuronalen Netzes ist vollkommen ausreichend, die Anzahl der Neurone stellt sich von allein ein und ist datenabhängig. Bei der Entstehung neuer Neuronen in der mittleren Schicht wird sowohl nach klassenspezifischen als auch nach fehlerminimierenden Kriterien vorgegangen.

Radialbasisklassifikatoren lassen die Implementierung verschiedener Arten neuronaler Netze zu, wobei die Rechengeschwindigkeit und die genaue Klassifizierung selbst bei sich stark überlappenden Klassen mit einer im Vergleich zum Perzeptron minimalen Anzahl an Neuronen möglich ist.

2 Neuronale Netze mit Radialbasisfunktionen

Neuronale Netze mit Radialbasisklassifikatoren stellen einen universellen Approximator bei einem dreischichtigen Aufbau [HKK90] dar. Die Input- und Outputschicht bestehen aus linearen Einheiten, während die Hiddenschicht meistens Kernelfunktionen enthält. Die Hiddenschichtneuronen bestimmen den euklidischen Abstand zwischen Input- und Referenzvektor und bewerten diesen schließlich über eine Kernelfunktion, die meistens eine Radialbasisfunktion ist. Die Outputschicht realisiert die Überlagerung der gewichteten Radialbasisneurone der mittleren Schicht und damit die Approximation der geforderten Funktion.

Die allgemeine mathematische Beschreibung des Radialbasisansatzes [PG90] lautet:

$$F(\mathbf{x}) = \sum_{i=1}^{N} c_i * h(||\mathbf{x} - \mathbf{x_i}||) + \sum_{i=1}^{m} d_i * p_i(\mathbf{x}) \qquad (1)$$

mit $m \leq N$, wobei N die Anzahl der Hiddenschichtneurone, m die Basis des Linearraumes, $\mathbf{x}$ den Eingangsvektor, $\mathbf{x_i}$ die Referenzvektoren, $h(||\mathbf{x} - \mathbf{x_i}||)$ die Radialbasisfunktion und c_i die Gewichte, mit denen diese zum Ausgang beitragen, darstellen. Der Radialbasisansatz (1) enthält zusätzlich auch eine Überlagerung von gewichteten Polynomen $p_i(\mathbf{x})$ als sogenannte Defaultterme und erfüllt in dieser Form nun vollständig das Approximationsproblem aus mathematischer Sicht. Der Defaultterm kann gleich Null gesetzt werden, wenn die Radialbasisfunktion h stetig auf $[0, \infty)$ und positiv auf $(0, \infty)$ ist und die erste Ableitung dieser monoton und nicht konstant auf $(0, \infty)$ ist.

Vom allgemeinen Ansatz (1) werden nun zwei verschiedene Formen von neuronalen Netzen abgeleitet, die mit einer minimalen Anzahl an Neuronen in der Hiddenschicht auskommen und diese im Laufe des Lernverfahrens an den Merkmalsraum angepaßt werden.

Neuronale Netze mit Gaußschen Potentialfunktionen

Die Radialbasisfunktionen werden als Gaußsche Potentialfunktionen [LK91] angesetzt, die einen Mittelpunkt m^i und eine Konturmatrix K^i — eine inverse Kovarianzmatrix — besitzen.

Der spezielle Ansatz lautet nun folgendermaßen:

$$F(\mathbf{x}) = \sum_{i=1}^{N} c_i * e^{-d(\mathbf{x},\mathbf{m_i})/2} \tag{2}$$

mit

$$d(\mathbf{x}, \mathbf{m_i}) = (\mathbf{x} - \mathbf{m^i})^t * \mathbf{K^i} * (\mathbf{x} - \mathbf{m^i}). \tag{3}$$

Die Konturmatrix $\mathbf{K^i}$ ist positiv semidefinit und ihre Elemente k^i_{jk}

$$k^i_{jk} = \frac{h^i_{jk}}{\sigma^i_j * \sigma^i_k} \tag{4}$$

bestehen aus den Korrelationskoeffizienten h^i_{jk} und den Standardabweichungen σ^i_j und σ^i_k. Dabei ist $h^i_{jk} = 1$ für $j = k$ und $|h^i_{jk}| \leq 1$ sonst.

Wählt man einen einfacheren Ansatz für die Konturmatrix [LK91], so wird man die einzelnen Elemente dieser Matrix nach

$$k^i_{jk} = \begin{cases} \frac{1}{\sigma_i^2} & j = k \\ 0 & \text{sonst} \end{cases} \tag{5}$$

beschreiben. Das so entstandene neuronale Netz mit nichtorientierten Gaußschen Potentialfunktionen wird im Laufe dieses Beitrags als GPF1-Netz bezeichnet.

Eine genauere Klassifizierung erhält man durch die Wahl der Konturmatrix in ihrer allgemeinen Form als nicht dünnbesetzte Matrix. Als Gebiete konstanter Wahrscheinlichkeitsdichte werden orientierte und um ihren Mittelpunkt drehbare Ellipsen erzeugt. Dieses Netz mit orientierten Gaußschen Potentialfunktionen wird als GPF2 bezeichnet und ermöglicht eine genaue Klassifizierung bei einer geringeren Anzahl an Neuronen in der mittleren Schicht.

Sowohl beim GPF1 als auch beim GPF2 werden zusätzliche Neuronen nach einem hierarchisch selbstorganisierten Lernalgorithmus [LK91] generiert, der deren Anzahl langsam anwachsen läßt und sowohl eine Fehlerminimierung am Ausgang, als auch eine Zuordnung der Neuronen zu den vorgegebenen Klassen vornimmt.

Resource Allocating Neural Network (RAN)

Anders als bei den GPF-Netzen steht bei diesem Netz nicht die extrem genaue Klassifizierung sich stark überlappender Klassengebiete durch besonders spezialisierte Neuronen im Vordergrund, sondern die Einsparung von Rechenzeit und von Speicherplatz [Pla91].

Die rechenzeitaufwendige Exponentialfunktion mit zahlreichen Matrizenoperationen im Exponenten wird durch eine gestürzte Parabel z_j approximiert, die eine kontinuierliche C^1 Polynomapproximation darstellt. Der allgemeine Ansatz lautet:

$$F(\mathbf{x}) = \sum_j \mathbf{c_j} * z_j + \gamma \tag{6}$$

mit

$$z_j = \begin{cases} [1 - (d/q * w_j^2)]^2 & z_j < q * w_j^2 \\ 0 & \text{sonst} \end{cases} \tag{7}$$

und mit

$$d = ||\mathbf{x} - \mathbf{m_i}||^2. \tag{8}$$

γ ist der Defaultterm am Ausgang des Netzes, wenn kein Eingangsneuron aktiv ist. q wird gleich 2,67 gesetzt und erzielt die beste Nachbildung der Exponentialfunktion. w_j beschreibt den Abstand zwischen dem nächsten Nachbarn eines Inputvektors und diesem Inputvektor selbst und wird entsprechend gewichtet, um eine mehr oder weniger starke Überlappung der Klassengebiete zu erzielen [Pla91].

Die Neurone der Hiddenschicht bilden sich im Laufe des Lernvorganges aus und sind auf eine rasche Approximation der zu erlernenden Funktion spezialisiert, ohne mit der Eigenschaft der Klassenzugehörigkeit versehen zu sein. Auch bei diesem Netz wird eine Minimierung der Fehlerfunktion am Ausgang angestrebt.

3 Sprachdatenmaterial und Vorverarbeitung

Das Sprachdatenmaterial umfaßt Zahlwörter und einzelne Wörter von je insgesamt 25 männlichen und weiblichen Sprechern, gesprochen. Es wurde telefonbandbegrenzt, mit 8kHz abgetastet und mit 12 Bit quantisiert. Die Vorverarbeitung anhand einer Prädiktionsanalyse liefert 8-dimensionale Cepstralkoeffizienten zu den einzelnen vorhandenen Phonemen. Es wurden insgesamt ca. 100 Lern- und Testmustervektoren pro Lautklasse ausgewählt. Die Untersuchungen in diesem Beitrag zur Phonemerkennung umfassen Vokale, Nasale, Frikative und den Lateral l. Ziel der hiesigen Spracherkennung ist es, ein modulares System unterschiedlicher neuronaler Architekturen mit Radialbasisfunktionen zur Unterscheidung von ca. 22 verschiedenen Phonemen zu implementieren.

4 Modulares Spracherkennungskonzept

Die gleichzeitige Erkennung aller 22 Phoneme durch ein einziges neuronales Netz erweist sich hinsichtlich der damit erzielten Erkennungsrate als nicht zufriedenstellend [Vas92]. Die erheblichen artikulatorischen Unterschiede zwischen den einzelnen Lautgruppen legen einen modularen Aufbau und eine Unterscheidung nach Lautgruppen durch spezialisierte neuronale Netze nahe.

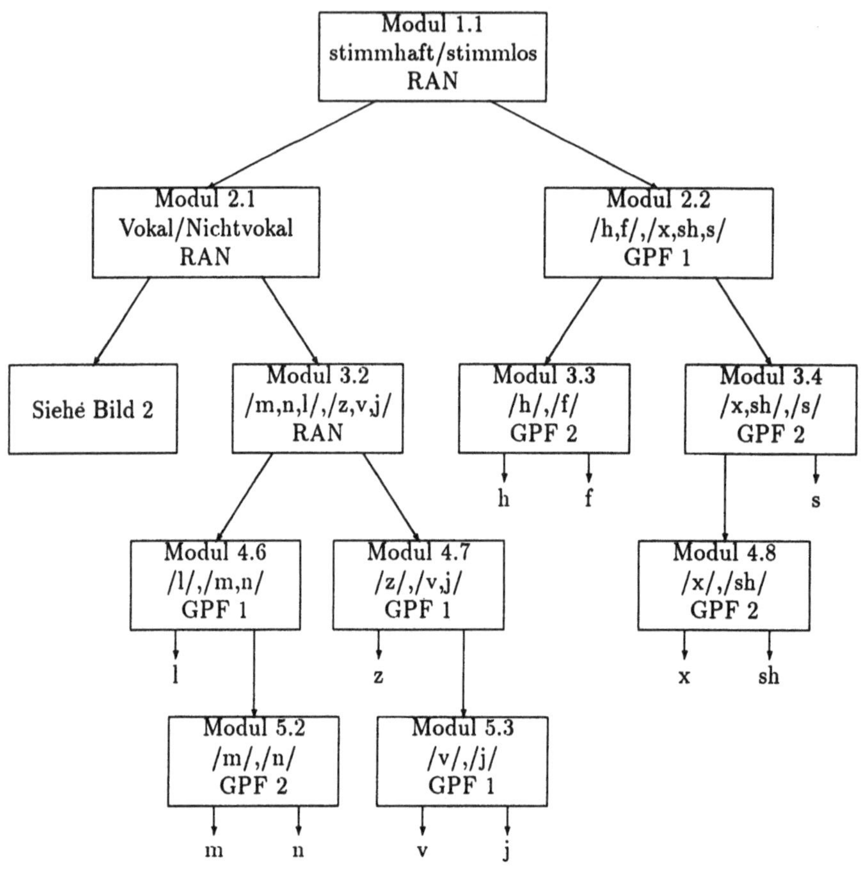

Bild 1 Modularer neuronaler Aufbau mit Radialbasisfunktionen zur Phonemerkennung.

Das hiesige Phonemerkennungskonzept hat eine Baumstruktur und verwendet je nach Schwierigkeitsgrad der Phonemerkennung die vorhin beschriebenen Netze mit Radialbasisfunktionen GPF1, GPF2 und RAN [MB93]. In einer ersten Ebene werden stimmhafte und stimmlose Laute erkannt, anschließend erfolgt bei den stimmhaften Lauten eine Unterscheidung zwischen Vokalen und Nasalen, dem Lateral l und stimmhaften Frikativen.

Bei den stimmlosen Lauten erfolgt eine Erkennung ausschließlich innerhalb der Lautgruppe der Frikative.

Den dabei entstehenden modularen Aufbau kann man Bild 1 entnehmen. Die eingeführten Module enthalten die Netzbezeichnung und die Phonemgruppen, die erkannt werden sollen. Das Modul zur Vokalerkennung mußte aus Platzgründen getrennt aufgeführt werden und ist in Bild 2 aufgeführt.

Den untenstehenden Tabellen kann man die Netzkonfiguration und die erzielten Erkennungsraten durch die einzelnen Module entnehmen. Die Bezeichnung "8/11/2" bedeutet,

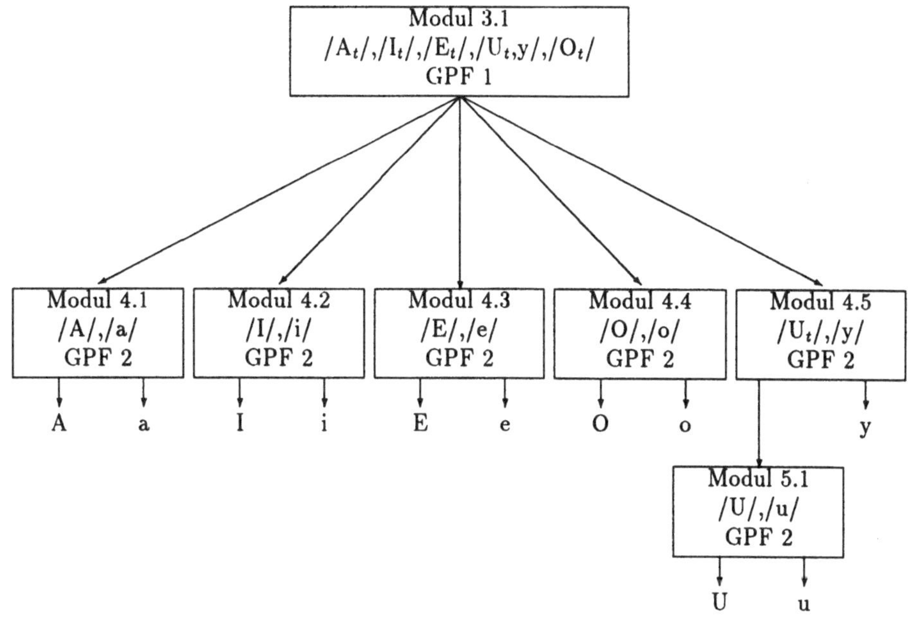

Bild 2 Submodul Vokalerkennung.

daß man einen 8-dimensionalen Eingangsvektor, ein Netz mit 11 Neuronen in der Hiddenschicht und 2 Ausgangsneurone zur Erkennung zweier verschiedener Lautgruppen hat.

Die Erläuterungen zum Gesamtmodul entnimmt man der Tabelle 1.

	Netzkonfiguration	Erkennungsrate
Modul 1.1	8/ 6/2	92,8%
Modul 2.1	8/ 8/2	92,2%
Modul 2.2	8/12/2	86,6%
Modul 3.2	8/ 3/2	91,5%
Modul 3.3	8/ 5/2	85,0%
Modul 3.4	8/ 8/2	77,5%
Modul 4.6	8/ 3/2	92,0%
Modul 4.7	8/ 4/2	91,5%
Modul 4.8	8/16/2	70,0%
Modul 5.2	8/ 6/2	91,0%
Modul 5.3	8/ 3/2	92,0%

Tabelle 1 Erläuterungen zum Gesamtmodul

Die Erläuterungen zum Modul Vokalerkennung entnimmt man der Tabelle 2.

	Netzkonfiguration	Erkennungsrate
Modul 3.1	8/12/5	90,6%
Modul 4.1	8/18/2	68,0%
Modul 4.2	8/12/2	86,0%
Modul 4.3	8/ 7/2	89,0%
Modul 4.4	8/11/2	90,0%
Modul 4.5	8/ 2/2	89,7%
Modul 5.1	8/ 3/2	86,0%

Tabelle 2 Erläuterungen zum Submodul der Vokalerkennung

Allen Erkennungsstufen ist die ökonomische Arbeitsweise der neuronalen Netze mit Radialbasisfunktionen gemeinsam. Die damit erzielten hohen Erkennungsraten gehen nicht zu Lasten der Komplexität des neuronalen Netzes. Dieses stellt seine Neurone adaptiv ein und hält deren Anzahl so gering wie möglich.

In den oberen Entscheidungsstufen reichen die schnellen RAN-Netze vollkommen zur Klassifizierung aus, während in den unteren Entscheidungsstufen komplexere GPF-Netze erforderlich sind. Die Erfassung der Intra- und Interklassen-Korrelationen innerhalb des Lernvorganges bewirkt eine minimale Anzahl an Neuronen, die zugleich aber problembezogen und vollkommen ausreichend ist.

Deutlich werden auch die Grenzen der erreichten Vokalerkennung bei Anwendung dieser neuronalen Netze vor allem bei der Unterscheidung von langem a wie „lahm“ und kurzem A wie „hatte“ sichtbar. Die beiden Laute haben den gleichen Artikulationsort und einen ähnlichen Frequenzverlauf. Zeitvariante Verfahren werden zur Zeit zum Lösen dieses Problems erarbeitet.

5 Zusammenfassung und Ausblick

Bei der Phonemerkennung mit neuronalen Netzen unter Zuhilfenahme von Radialbasisfunktionen lassen sich für bestimmte Lautgruppen in den einzelnen Erkennungsstufen hohe Erkennungsraten erzielen. Die Struktur dieser Netze ist problembezogen und so ausgelegt, daß sie die Merkmalsverteilung im Laufe des Lernvorganges erfaßt. Die Komplexität des Netzes wird durch eine minimale und gut ausgewählte Neuronenanzahl gering gehalten. Die Neurone der Hiddenschicht sind orientierte und formveränderliche Cluster, die mit dem zusätzlichen Merkmal der Klassenzugehörigkeit versehen sind. Die Hardwareunterstützung ist bei diesen neuronalen Netzen besonders effektiv und ermöglicht den Aufbau eines Spracherkennungssystems auch in Echtzeit [Sch92] und [SK92].

Im weiteren soll die modulare Erkennung auch auf andere Lautgruppen ausgeweitet werden, wobei zeitvariante Verfahren einzusetzen sind.

Die vorliegende Arbeit wurde im Rahmen des DFG-Sonderschwerpunktprogramms „Theorie und Physiologie von neuronalen Netzen" erstellt.

Literaturverzeichnis

[End92] W. Endres. *Nachrichtenübertragung durch Sprache.* Vorlesungsskript der TH Darmstadt, 1992.

[HKK90] Eric J. Hartman, James D. Keeler und Jacek M. Kowalski. *Layered Neural Networks with Gaussian Hidden Units as Universal Approximations. Neural Computation*, 2:210–215, 4 1990.

[LK91] Sukhan Lee und Rhee M. Kil. *A Gaussian Potential Function Network with Hierarchically Self- Organizing Learning. Neural Networks*, 4:207–224, 9 1991.

[MB93] A. Meyer-Bäse. *Sprachverarbeitung mit Radialbasisklassifikatoren.* Institutsbericht 1, Institut für Datentechnik, 1993.

[PG90] Tomaso Poggio und Federico Girosi. *Networks for Approximation and Learning. Proceedings of the IEEE*, 78:1481–1497, 9 1990.

[Pla91] John Platt. *A Ressource-Allocating Network for Function Interpolation. Neural Computation*, 3:213–225, 6 1991.

[RR88] S. Renals und R. Rohwer. *Phoneme Classification Experiments Using Radial Basis Functions. Proceedings of the International Joint Conference on Neural Networks*, 1:461–467, 6 1988.

[Sch92] R. Schimpf. *Implementierung eines neuronalen Netzes zur Sprachvorverarbeitung auf einem Fließkommasignalprozessorsystem.* Diplomarbeit DT 553, Institut für Datentechnik, TH Darmstadt, 1992.

[SK92] Thomas Häuser Stefan Kenne. *Entwicklung einer vollparallelen neuronalen CORDIC-Architektur Fließkommasignalprozessor und programmierbaren Gate-Arrays.* Studienarbeit ST 570, Institut für Datentechnik, TH Darmstadt, 1992.

[Vas92] J. Vassiliadis. *Sprach- und Schrifterkennung mit Hilfe von selbstorganisierten neuronalen Netzen.* Diplomarbeit DT 548, Institut für Datentechnik, TH Darmstadt, 1992.

Automatische Erkennung und Ausführung von Korrekturanweisungen in Textdokumenten

D. Möri, H. Bunke

Institut für Informatik und angewandte Mathematik
Universität Bern
Länggassstr. 51, CH-3012 Bern
moeri@iam.unibe.ch, bunke@iam.unibe.ch

1 Einführung und Übersicht

Die Dokumenterkennung ist ein Teilgebiet der Mustererkennung, welches in den letzten Jahren rapide Fortschritte zu verzeichnen hat [1–6]. So werden heute z.B. für die automatische Konversion von gedruckten Textdokumenten in ASCII-Dateien zahlreiche kostengünstige Systeme kommerziell angeboten. Auf der anderen Seite beinhaltet die Dokumenterkennung aber auch noch eine Reihe ungelöster Aufgaben, sowohl aus theoretischer als auch aus praktischer Sicht. Im vorliegenden Beitrag beschäftigen wir uns mit der automatischen Erkennung und Ausführung von handschriftlichen Korrekturanweisungen in Textdokumenten. Die Erzeugung von Textdokumenten geschieht heute typischerweise computerbasiert mithilfe von Textverarbeitungssoftware. Nach Erstellung einer Textdatei auf dem Rechner erfolgt das Korrekturlesen jedoch häufig anhand eines Ausdrucks des Textes auf Papier. Hierbei markiert der Autor handschriftlich verschiedene Korrekturen auf der Papiervorlage. Anschliessend wird die Textdatei entsprechend verändert. Die Erfahrung zeigt, dass gerade bei Verwendung von sogenannten Non-WYSIWYG-Textverarbeitungssytemen[1] — ein Beispiel ist das gerade bei Informatikern sehr beliebte LaTeX-System [7] — häufig mehrere Iterationen nötig sind, bis das Textdokument in der gewünschten Form erstellt ist.

Im vorliegenden Beitrag beschreiben wir ein in Form eines Prototypen realisiertes System, welches in der Lage ist, Korrekturanweisungen auf einer Textvorlage zu erkennen, die Korrekturanweisungen in Editierbefehle zur Änderung der zugrunde liegende Textdatei umwandelt und diese schliesslich ausführt. Das System erhält als Eingabe einerseits ein mithilfe eines Dokumentscanners digitalisiertes Dokument mit Korrekturanweisungen und andererseits die zum Dokument gehörige Textdatei. Als Ausgabe wird die korrigierte Version der Textdatei erzeugt. Ein Ausschnitt eines Beipieldokuments mit manuell eingetragenen Korrekturen ist in Fig. 1 dargestellt. Das Ergebnis der Verarbeitung — genauer, ein Ausdruck der modifizierten Textdatei — zeigt Fig. 2. Bis auf die Erkennung isoliert geschriebener handschriftlicher alphanumerischer Zeichen am Rand des Dokuments laufen alle Schritte im gegenwärtigen Protoypen vollautomatisch ab.

[1] WYSIWYG = What you see is what you get

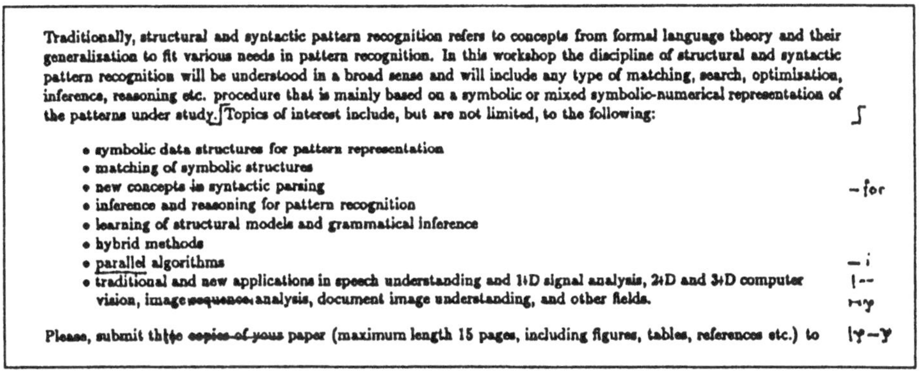

Traditionally, structural and syntactic pattern recognition refers to concepts from formal language theory and their generalization to fit various needs in pattern recognition. In this workshop the discipline of structural and syntactic pattern recognition will be understood in a broad sense and will include any type of matching, search, optimisation, inference, reasoning etc. procedure that is mainly based on a symbolic or mixed symbolic-numerical representation of the patterns under study. Topics of interest include, but are not limited, to the following:

- symbolic data structures for pattern representation
- matching of symbolic structures
- new concepts in syntactic parsing
- inference and reasoning for pattern recognition
- learning of structural models and grammatical inference
- hybrid methods
- parallel algorithms
- traditional and new applications in speech understanding and 1+D signal analysis, 2+D and 3+D computer vision, image sequence analysis, document image understanding, and other fields.

Please, submit three copies of your paper (maximum length 15 pages, including figures, tables, references etc.) to

Fig. 1. Beispiel eines Ausschnittes aus einem mit Korrekturanweisungen versehenen Textdokumentes; die Auflösung beträgt 350 dpi

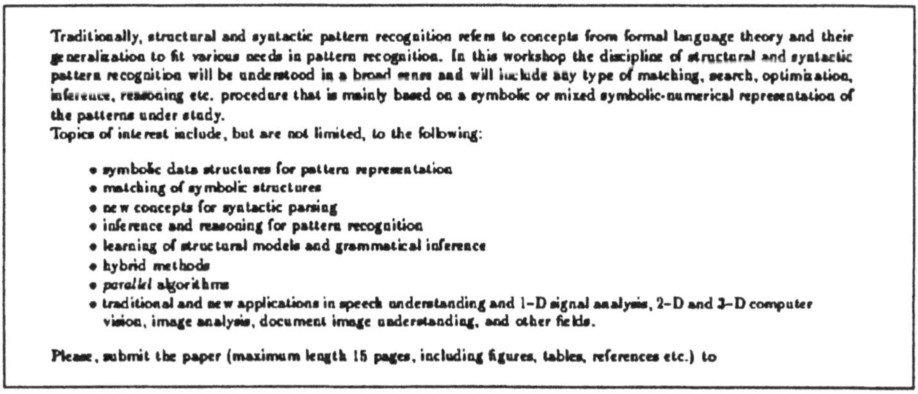

Traditionally, structural and syntactic pattern recognition refers to concepts from formal language theory and their generalization to fit various needs in pattern recognition. In this workshop the discipline of structural and syntactic pattern recognition will be understood in a broad sense and will include any type of matching, search, optimisation, inference, reasoning etc. procedure that is mainly based on a symbolic or mixed symbolic-numerical representation of the patterns under study.
Topics of interest include, but are not limited, to the following:

- symbolic data structures for pattern representation
- matching of symbolic structures
- new concepts for syntactic parsing
- inference and reasoning for pattern recognition
- learning of structural models and grammatical inference
- hybrid methods
- *parallel* algorithms
- traditional and new applications in speech understanding and 1-D signal analysis, 2-D and 3-D computer vision, image analysis, document image understanding, and other fields.

Please, submit the paper (maximum length 15 pages, including figures, tables, references etc.) to

Fig. 2. Das Textdokument aus Fig. 1 nach Erkennung und Ausführung der Änderungen

Die Klassifikation der Zeichen erfolgt in der momentanen Version des Systems noch interaktiv.

Unser Prototyp unterstützt bis jetzt nur die automatische Bearbeitung von reinen Textdokumenten in LaTeX-Form. Das auf einem Laserdrucker ausgedruckte Dokument mit den angebrachten Korrekturen wird von einem Farb-Scanner mit einer Auflösung von mindestens 300 dpi (dots per inch) eingelesen und in eine RGB-Datei mit 3 bit pro Pixel und Farbkanal konvertiert. Wir verlangen, dass der Korrektor die Korrekturen mit einem roten Farbstift anbringt. Dies ist aber keine echte Einschränkung, da mit Änderung von Parameterwerten in der Vorverarbeitung auch eine andere Farbe verwendet werden kann. Die Korrekturen müssen den Vorschriften entsprechen, welche von der Dudenredaktion und dem Deutschen Normenausschuss festgehalten wurden [8]. Die Hauptregel dabei lautet, dass jedes Korrekturzeichen auf dem Rand zu wiederholen ist.

Der gegenwärtig implementierte Umfang an möglichen Korrekturen umfasst die Löschung, Einfügung und Ersetzung einzelner Zeichen oder Zeichenfolgen, den Beginn einer neuen Zeile sowie die Änderung des Typenfonts. Die in Fig. 1 auftretenden Korrekturen pro Zeile von oben nach unten sind: Neuer Absatz vor "Topics"; Ersetzung der Zeichenfolge "in" durch die Zeichenfolge "for"; Änderung des Typenfonts von "normal" nach "italic"; Ersetzung des Zeichens "kurzer Bindestrich" durch "mittellanger Bindestrich" an drei Stellen in der gleichen Textzeile; Streichen der Zeichenfolge "sequence"; Streichen der Zeichen "r" und "e" sowie der Zeichenfolge "copies of your".

2 Beschreibung des Prototypen

Der Protoyp lässt sich in drei Module unterteilen:

1. **Vorverarbeitung**: Ausgehend von dem RGB–Bild werden ein Korrekturbild und ein Textbild als binäre Bilder mit den Korrekturen respektive dem reinen Text erzeugt.
2. **Erkennung**: Es erfolgt die Erkennung eines jeden Korrektursymbols im Korrekturbild und die Generierung einer Liste mit dem erkannten Korrekturtyp und den zugehörigen Koordinaten.
3. **Änderung**: Für jede vorzunehmende Korrektur wird die zu editierende Textstelle bestimmt und die originale LaTeX–Datei entsprechend verändert.

Im folgenden werden die einzelnen Module genauer beschrieben.

2.1 Vorverarbeitung

Beim Laden des eingelesenen Bildes (Fig. 1) werden sämtliche auftretenden RGB–Werte in einer Farbtabelle festgehalten. Für jeden Eintrag werden dann mit einer Farbraumtransformation die Werte für die Intensität (I), den Buntton (Hue) und die Sättigung (Saturation) berechnet [9, 10]. Mit Hilfe vordefinierter Bereiche für H, S und I erfolgt die Zuordnung eines jeden Pixels zu einer der Klassen `Korrektur`, `Hintergrund`, `Text` und `ev_Text` (eventueller Text). Bezeichnen I_P, H_P und S_P Intensität, Buntton und Sättigung des Pixels P, dann lautet die detaillierte Klassifikationsvorschrift

$$
\begin{aligned}
&P \in \texttt{Korrektur} \Leftrightarrow H_P \in [H_{min}, H_{max}] \ \wedge S_P \in [S_{min}, S_{max}] \ \wedge I_P > I_{min} \\
&P \in \texttt{Text} \Leftrightarrow I_P \leq I_{max}/2 \ \wedge P \notin \texttt{Korrektur} \\
&P \in \texttt{ev\_Text} \Leftrightarrow I_{max}/2 \ < I_P \leq I_{max} \ \wedge P \notin \texttt{Korrektur} \\
&P \in \texttt{Hintergrund} \Leftrightarrow I_P > I_{max} \ \wedge P \notin \texttt{Korrektur}
\end{aligned}
$$

Die Schwellwerte $H_{min}, H_{max}, S_{min}, S_{max}, I_{min}, I_{max}$ sind dabei in erster Linie von der gewählten Farbe des Korrekturstiftes — in der gegenwärtigen Version unseres Systems rot — abhängig und wurden anhand einer Stichprobe bestimmt. Pixel, die nach dieser Vorschrift zur Klasse `ev_Text` gehören, werden in einem

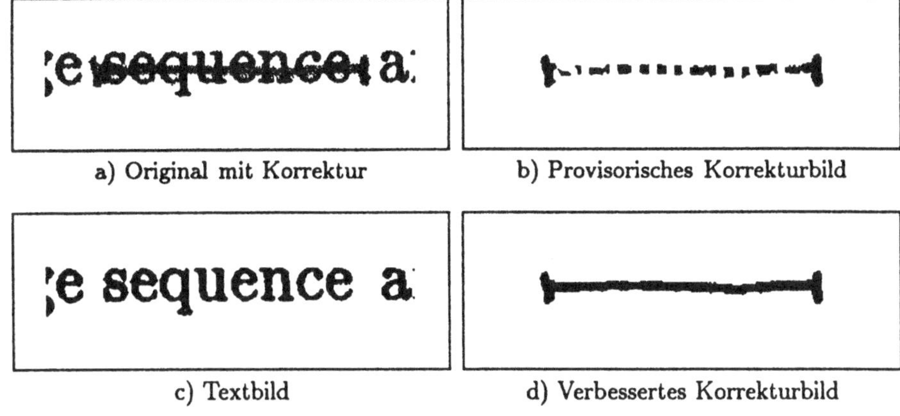

a) Original mit Korrektur

b) Provisorisches Korrekturbild

c) Textbild

d) Verbessertes Korrekturbild

Fig. 3. Ausschnitt aus Fig. 1

zweiten Schritt aufgrund eines Nachbarschaftstests entweder dem Typ `Text` oder `Hintergrund` zugeordnet. Hat ein Pixel $P \in$ `ev_Text` mindestens einen Nachbarn $Q \in$ `Text`, so wird P aus `ev_Text` entfernt und nach `Text`, andernfalls nach `Hintergrund` eingeordnet.

Mithilfe des geschilderten Klassifikationsverfahrens werden aus dem ursprünglichen RGB-Bild zwei Binärbilder, das sogenannte provisorische Korrekturbild und das Textbild erzeugt. Ein Beispiel ist in Fig. 3 zu sehen. Es handelt sich hier um einen Ausschnitt aus Fig. 1.

Wie in Fig. 3.b) zu sehen ist, sind viele Punkte der Korrektur im provisorischen Korrekturbild nicht enthalten, da sie als Text detektiert wurden. Mit einer Nachbearbeitung werden diese Lücken im provisorischen Korrekturbild nun gefüllt. Dabei werden alle Punkte nachträglich der Klasse `Korrektur` zugeordnet, falls in horizontaler, d.h. in linker und in rechter, Richtung je ein Nachbar aus der Klasse `Korrektur` vorhanden ist. Die Suche nach einem Nachbar wird abgebrochen, sobald ein Punkt aus der Klasse `Hintergrund` auftaucht. Analog erfolgt die Suche in vertikaler Richtung. Das resultierende verbesserte Korrekturbild ist in Fig. 3.d) zu sehen.

Zur Weiterverarbeitung werden das verbesserte Korrekturbild sowie das Textbild verwendet. Zusätzlich halten wir die Anzahl der Textpunkte pro Bildzeile und die Position des rechten Randes des Textes fest.

2.2 Erkennung der Korrektursymbole

Auf das verbesserte Korrekturbild wird der Algorithmus von Hilditch [11] zur Verdünnung angewendet. Die dadurch entstehenden pixel-breiten Korrektursymbole sind 8-Nachbarschaft-verbunden. Anschliessend werden die Symbole durch gerade Liniensegmente in horizontaler, vertikaler oder diagonaler Richtung approximiert. Gemäss den Korrekturvorschriften muss jedes Symbol auf dem rechten Textrand wiederholt werden. Deshalb wird für jedes Korrektursymbol innerhalb des Textbereichs ein entsprechendes auf dem Textrand gesucht. Die Zuordnung sich entsprechender Korrektursymbole erfolgt durch einen Strukturvergleich, bei dem zwei Korrektursymbole - eines im Textbereich und eines im Randbereich - hinsichtlich der Anzahl, des Typs (horizontal, vertikal oder diagonal) und der gegenseitigen Lage ihrer Liniensegmente überprüft werden. Dieses Erkennungsverfahren ist sehr einfach, da Korrekturzeichen aus wenigen Liniensegmenten bestehen.

Die auf dem rechten Rand direkt nach einem Korrektursymbol auftretenden alphanumerischen Zeichen oder Sonderzeichen dienen der genaueren Spezifikation der durchzuführenden Korrekturen. So gibt z.B. bei einer Änderung des Typenfonts der auf das Korrekturzeichen folgende Buchstabe den neuen Font an (vgl. die vierte Korrekturzeile von unten in Fig. 1). Löschungen einzelner Textzeichen oder Zeichenfolgen werden durch ein Spezialzeichen (Deleatur) bezeichnet (vgl. die letzte und vorletzte Korrekturzeile in Fig. 1). Bei Ersetzungen oder Einfügungen sind es die neu einzusetzenden Zeichen, die auf das Korrektursymbol folgen. Alle alphanumerischen Zeichen und das Sonderzeichen Deleatur werden im Moment interaktiv klassifiziert, da in der gegenwärtigen Version des Systems noch keine Handschriftzeichenerkennung implementiert ist.

Wurde die Bedeutung eines jeden Korrektursymbols im Textbereich unter Zuhilfenahme des korrespondierenden Korrektursymbols und eventuell zugehöriger alphanumerischer Zeichen oder Sonderzeichen im Randbereich erkannt, wird die durchzuführende Änderung am Textfile in einer Liste gespeichert. In diese Liste werden auch die Bildkoordinaten des Korrektursymbols im Textbereich aufgenommen.

2.3 Änderung der Textdatei

Anhand der von der Erkennung erzeugten Liste von Korrekturen und des Textbildes wird pro Korrektursymbol das umschreibende Rechteck (Bounding Box) bestimmt. Die Bounding Box wird mit Hilfe einer Konturverfolgung des durch die Korrektur berührten Textes ermittelt. Das Rechteck umfasst dabei immer mindestens ein ganzes Wort. Falls z.B. nur ein Buchstabe gelöscht oder ersetzt werden soll, so wird nach dem nächsten links stehende Buchstaben gesucht, ein Zähler um eins erhöht und die Bounding Box um dessen umschreibendes Rechteck erweitert. Die Erweiterung stoppt, falls der Abstand zum nächsten Buchstaben grösser ist als ein Schwellwert. Dieser Schwellwert wird abhängig von der Höhe der Textzeile berechnet. Die Höhe kann dabei aufgrund der Information über die Anzahl Textpixel pro Bildzeile ermittelt werden und ist somit unabhängig von der verwendeten Bildauflösung und der Textgrösse. Eine analoge Erweiterung der Bounding Box nach rechts liefert das ganze Wort. Der Zähler weist auf den Buchstaben innerhalb des Wortes hin.

Nachdem für jedes Korrektursymbol im Textbereich das umschreibende Rechteck ermittelt wurde, ist das Ziel des nächsten Verarbeitungsschrittes, sämtliche Schriftzeichen des Dokumentes innerhalb des Rechtecks zu klassifizieren. Wir übergeben hierfür den Inhalt jeder Bounding Box als Teilbild dem kommerziellen OCR-Softwarepaket ScanWorX von Xerox. Dieses Softwarepaket liefert uns den in der Bounding Box enthaltenen Text als Kette von ASCII-Zeichen. Diese Zeichenkette wird dann in der LaTeX-Datei gesucht und ihre Startposition in einer Liste eingetragen. Tritt der in der Bounding Box erkannte Text mehrfach in der LaTeX-Datei auf, so wird der Kontext vergrössert. D.h. es wird die Bounding Box ausgedehnt und der Prozess mit der Texterkennung und Textsuche solange wiederholt bis der Inhalt der Bounding Box eindeutig in der LaTeX-Datei lokalisiert werden kann. Tritt ein Fehler bei der Texterkennung auf, so führt das u.U. dazu, dass die zu suchende Textkette nicht in der LaTeX-Datei gefunden wird. In diesem Fall kommt ein fehlertoleranter Suchalgorithmus zur Anwendung [12]. Sind schliesslich alle Korrekturen behandelt, so wird die Liste nach den Startpositionen sortiert und eine neue LaTeX-Datei mit den angewandten Korrekturen erzeugt.

2.4 Weitere Bemerkungen

Jedes der vorgestellten Module ist ein selbständiges Programm mit einer graphischen Benutzeroberfläche, wobei die einzelnen Vorgänge anschaulich angezeigt werden. Als Schnittstelle zwischen den Modulen dienen Dateien in Pixelformat (Korrektur- und Textbilder) sowie ASCII-Dateien (Liste der Korrekturen). Die Verwendung von ASCII-Dateien für die Korrekturliste hat den Vorteil, dass das dritte Modul (Änderung der Textdatei) getestet werden kann, ohne dass die Handschrifterkennung bereits implementiert ist.

3 Resultate und Ausblick

In diesem Abschnitt möchten wir einige Zwischenschritte bei der Verarbeitung von Fig. 1 anschaulich präsentieren. Die Fig. 4–7 entsprechen aufgenommenen Bildschirmausschnitten, wobei die Abbildungen unter a) das Original mit den angebrachten Korrekturen und die Abbildungen unter b) das Textbild mit den darauf eingezeichneten stilisierten Korrekturzeichen, den umschreibenden Rechtecken, den von der Texterkennung zurückgegebenen Text sowie die in der Textdatei vorzunehmenden Änderungen zeigen.

Der gegenwärtige Prototyp ist in der Sprache C auf einer Sun Workstation implementiert. Der Speicherbedarf des Laufzeitsystems entspricht der dreifachen Anzahl Bytes der im digitalisierten Bild vorhandenen Pixel. Die komplette Bearbeitung des Beipiels in Fig. 1 beträgt etwa 40s (ohne Bildaufnahme, aber mit Unterhalt der Graphik).

Der Prototyp wurde an einer Reihe von Beispielen mit einem ähnlichen Komplexitätsgrad wie in Fig. 1 getestet und hat korrekte Resultate geliefert. Um die gegenwärtige Version des Systems zu einem universell anwendbaren und robusten Werkzeug weiterzuentwickeln, bedarf es allerdings noch zahlreicher Anstrengungen. Ein momentan ungelöstes Problem, welches zwar bei den bisherigen Tests nie auftrat, theoretisch aber durchaus vorkommen kann, sind OCR Fehler, die ein Wort in ein anderes im Dokument vorhandenes Wort überführen. Derartige OCR–Fehler führen i.a. zu falschen Korrekturen. Weitere ungelöste Probleme haben ihre Ursache im LaTeX–System. So haben z.B. spezielle Umgebungen (tabular, itemize, u.a.) Auswirkungen auf das Layout einer Seite. Die Editierbefehle sind demzufolge abhängig von der aktuellen Umgebung der Startposition. Ganz allgemein verlangen alle Formatanweisungen fixe Angaben, welche vom Korrektor angegeben werden müssen. Da aber LaTeX das Seitenlayout festlegt, können gewisse Formatanweisungen eventuell ignoriert oder verändert interpretiert werden. Eine genauere Untersuchung dieser Probleme ist Gegenstand unserer momentanen Arbeiten.

4 Literatur

References

1. T. Pavlidis, S. Mori: *Optical Character Recognition*, Special Issue of Proceedings of the IEEE, Vol. 80, No. 7, July 1992.
2. L. O'Gorman, R. Kasturi: *Document Image Analysis Systems*, Special Issue of IEEE Computer, Vol. 25, No. 7, July 1992.
3. R. Kasturi, L. O'Gorman: *Document Image Analysis Systems: A Bibliography*, Special Issue of Machine Vision and Application, Vol. 5, No. 3, Summer 1992.
4. H. Baird, H. Bunke, K. Yamamoto (eds.): *Structured Document Image Analysis*, Springer Verlag, 1992.
5. R. Plamondon (ed.): *Handwriting Processing and Recognition*, Pattern Recognition, Vol. 26., No. 3, March 1993.
6. H. Tominaga (ed.): *Special Issue on Postal Processing and Character Recognition*, Pattern Recognition Letters. Vol. 14, No. 4, April 1993.
7. L. Lamport: LaTeX, *A Document Preparation System*, Addison–Wesley Publishing Company.
8. Duden, Band 1: *Die deutsche Rechtschreibung*, Meyers Lexikonverlag, Mannheim, 1991, pp 77–80.
9. A.K. Jain: *Fundamentals of Digital Image Processing*, Prentice Hall, 1989.
10. W.K. Pratt: *Digital Image Processing (second edition)*, J. Wiley & Sons, Inc., 1991.
11. C.J. Hilditch: *Linear Skeleton from Square Cupboards*, in Machine Intelligence 6, Edinburgh Univ. Press, 1969, pp 403–420.
12. Z. Galil, K. Park: *An Improved Algorithm for Approximate String Matching*, SIAM J. COMPUT. 19, December 1990, pp 989–999.

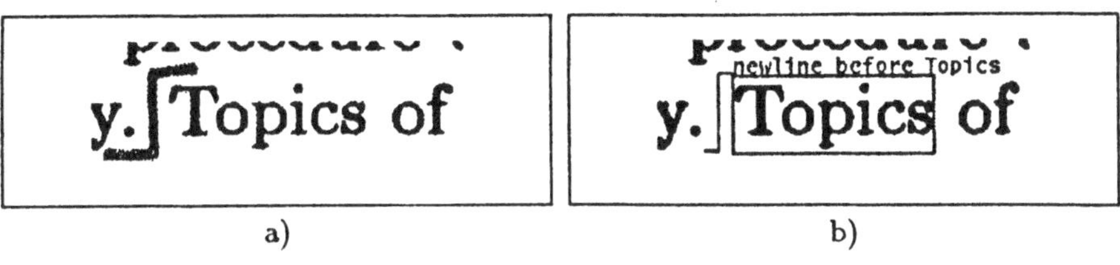

Fig. 4. Neuer Absatz

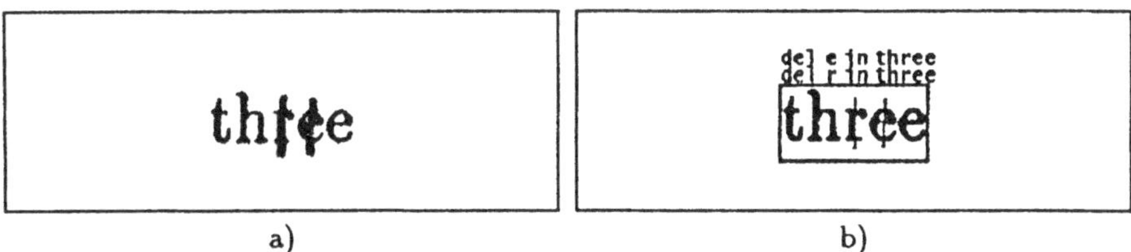

Fig. 5. Löschen mehrerer Zeichen innerhalb eines Wortes

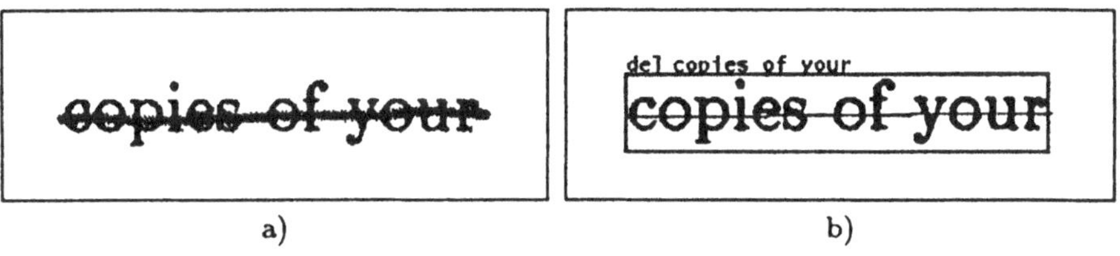

Fig. 6. Löschen mehrerer Wörter

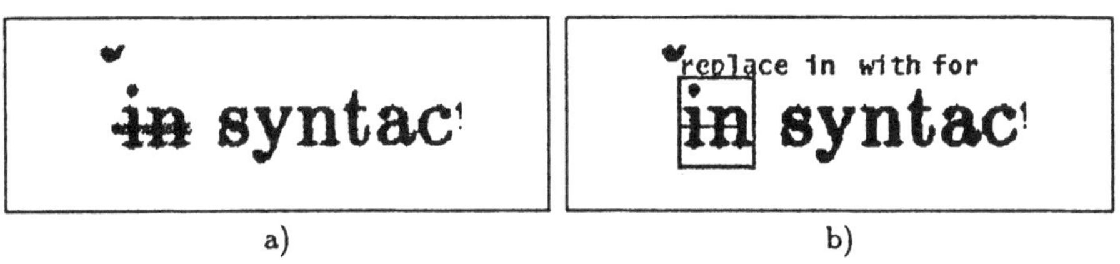

Fig. 7. Löschen eines Wortes mit Unterbrechung in horizontaler Projektion

A Font-Classifier for Printed Chinese Characters Based on Possibility Theory

V. Müller
MAZ Mikroelektronik Anwendungszentrum Hamburg GmbH
Karnapp 20, 21079 Hamburg[1]

Abstract

Optical Character recognition of single fonts yields much higher recognition accuracy than recognition of mixed fonts. In this paper a font-classifier for Chinese characters is proposed, that is applied prior to single-font OCR. The font-classifier is designed as a self-learning system based on possibility theory. Several features especially developed for font-recognition are introduced. Font-classification of Chinese characters is seen as an example, that tendencies in complex scenes, so-called secondary properties, can be detected without classifying single objects.

1. Introduction

Font-Classification of Chinese Characters is a challenging task both from the theoretical and practical point of view. Research on Chinese Character Recognition started 1966 in New York [1]. At that time efforts were not aimed at developing a commercial usable OCR-device, but Chinese Characters were taken as an example, that it is possible to solve a recognition task involving several thousand classes. In this paper font-classification is taken as an example, that certain tendencies - we call them secondary properties - can be detected in a very complex scene without classifying single objects. The clear, obvious property of a character is its meaning. On the other hand, the font creates a more general impression of the whole text. There are ample potential applications demanding to analyse and classify secondary properties, for example in surface inspection, when the surface has such a complex pattern, that standard texture analysis does not provide reliable results.
Since all common classification methods have been applied to Latin Character OCR, the focus of research efforts has shifted to Chinese character recognition (surveys of the stand-of-the-art are given in [2] - [5]). But up to now, recognition rates are far away from being satisfactory, the minimum error rates reported are still around 1% [6], that means about one recognition errors every two lines.

---

[1] The research resulting in this paper was carried out at Chongqing University, Dept. of Optoelectronics, 630044 Chongqing, P.R. China

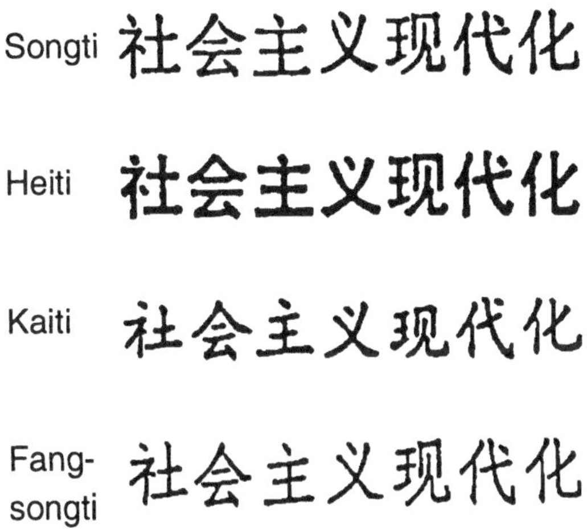

Songti 社会主义现代化

Heiti 社会主义现代化

Kaiti 社会主义现代化

Fang-songti 社会主义现代化

Fig1: Standard fonts of Chinese Characters

Latin character OCR gradually developed from single-font to multi-font recognition. This approach seems to be appropriate to Chinese character recognition, too. But most authors want to skip single-font recognition and develop multi-font Chinese OCR - up to now with limited success. However the precondition to use single-font classification is the automatic classification of the font, a gap, that may be closed with the method introduced in this paper.

2. Feature Extraction for Font-Classification

The huge number of classes in Chinese creates enormous difficulties for character recognition, but there is definitely one property, that is beneficial for character recognition: in standard printed text only four distinct fonts[2] are used (Fig. 1), compared with hundreds of fonts available for latin characters. However there may be considerable modifications of the character style within one font [7].
To determine and extract adequate features is essential for successful font-recognition. Some of the implemented features are described below:

Average Stroke Width

Width of the strokes and especially the relation between the width of horizontal strokes and the width of vertical strokes depends on the font. The width of a stroke is measured from one white-black transition to the next black-white transition (fig. 2a). To measure the width of vertical strokes, the character is scanned line by line. In order to avoid interference of horizontal strokes, only strokes, which width is smaller than a threshold $\varepsilon * h$ (h = hight of the character) are taken into account. The average width of vertical strokes w_V is calculated according to

[2] This statement is only true for the simplified characters used in the P.R.China. Different fonts of traditional characters are used in Hongkong and Taiwan

$$w_v = \frac{\sum_{i=0}^{s} w_i}{s*h} \qquad \text{for all } w < \varepsilon*h \qquad (1)$$

s - number of measure points, that satisfy $w < \varepsilon*h$

ε - threshold avoiding that other than vertical strokes are measured

The hight h is used in (1) to make feature extraction independent of the size of the character. The same method is used to measure the width of horizontal strokes w_v. Relation w_v/w_h is used as a third feature.

Distance between boarder of character and first stroke

Songti (see figure 1) has distinct serifs, the other fonts have not. The distance d between the upper boarder of the character and the first horizontal stroke is measured in order to detect the existence of serifs (fig 2b):

$$d = \begin{cases} x_{\min}/h & \text{if } x_{\min} < \varphi*h \\ \text{invalid} & \text{else} \end{cases} \qquad (2)$$

φ - threshold necessary to handle characters without horizontal stroke at the top

A similar feature is extracted measuring the distance between the right boarder of the character and the rightmost vertical stroke.

Slant of horizontal strokes

Kaiti and Fangsongti have slanted vertical strokes, Songti and Heiti have not. The slant α_s can reliably be detected by measuring the length l of nearly-horizontal strokes in dependence on the angle of spectation α (fig. 2c). 0° and ϕ are thresholds defining 'nearly-horizontal strokes':

$$\alpha_s = \alpha \quad (l = maximum) \qquad 0^\circ < \alpha < \phi \qquad (3)$$

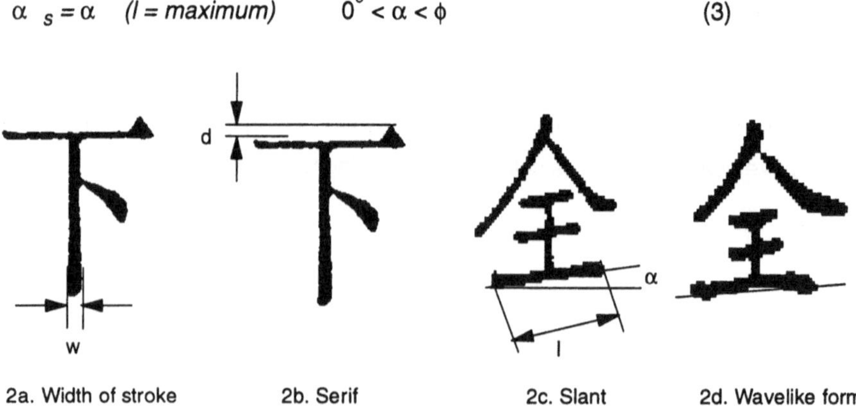

2a. Width of stroke 2b. Serif 2c. Slant 2d. Wavelike form

Wavelike shape of horizontal strokes

The shape of horizontal strokes can help to distinguish Kaiti and Fangsongti. Strokes of Kaiti have wavelike form, Fangsongti has straight strokes.

The lower black-white transition is taken to represent the shape of a vertical stroke (fig. 2d). This boarder is approximated by a straight line using standard regression analysis:

$$y = a + b * x \tag{4}$$

$$a = \overline{y}_n - b * \overline{x}$$

$$b = \frac{\sum_{i=1}^{n} (x_i - \overline{x})(y_i - \overline{y})}{\sum_{i=1}^{n} (x_i - x)^2}$$

The variance of wavelike strokes $\overline{d}_w^2$ is much larger than that of straight strokes $\overline{d}_s^2$:

$$\overline{d}_w^2 >> \overline{d}_s^2$$

$$\overline{d}^2 = \frac{1}{n} * \sum_{i=1}^{n} \left(y_i - a + b * x_i \right)^2 \tag{5}$$

The quadratic distance emphasizes boarder points in greater distance from the straight line, avoiding strong influence from rugged boarders.

Feature extraction provides a feature set **s** for each character. Because of the large variety of different characters, it is in general not possible to determine the font analysing only single characters. Therefore the mean-value of the features is calculated:

$$\overline{s}_m = \frac{1}{n'_m} * \sum_{i=0}^{n'_m} s_i \qquad n'_m \leq n \tag{6}$$

n is the number of characters forming a portion of text, which font is classified. n'_m is the number of characters, which provide a valid feature s_m. n'_m may be less then n, since some features can only be extracted, if certain conditions are fulfilled.

3. Classification by Methods of Possibility Theory

The classification method is choosen according to the following criteria

- because of the large number of different characters, it is nearly impossible to create statistical representative samples and to use statistical methods
- neural networks require too much memory
- since each single feature is not very reliable, decisions must been made combining several features. Therefore decision trees are not appropriate
- because of the huge variaty of characters, a self-learning method is required

The solution is a self-learning system, that automatically creates possibility distributions (which also may be interpreted as fuzzy sets [9], [10]). One possibility function $\pi_{m,f}$ is calculated for each feature (index m) and each font (index f), resulting in an matrix π, with F*M elements (M - number of features, F - number of fonts):

$$\pi = \begin{matrix} \pi_{1,1} & \dots & \pi_{1,f} & \dots & \pi_{1,F} \\ \dots & \dots & \dots & \dots & \dots \\ \pi_{m,1} & \dots & \pi_{m,f} & \dots & \pi_{m,F} \\ \dots & \dots & \dots & \dots & \dots \\ \pi_{M,1} & \dots & \pi_{M,f} & \dots & \pi_{M,F} \end{matrix} \tag{7}$$

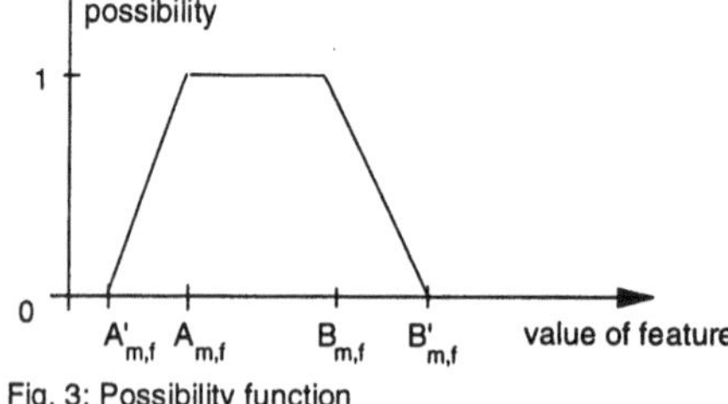

Fig. 3: Possibility function

Each possibility function $\pi_{m,f}$ takes values between 0 (impossible) and 1 (definitely possible). The possibility functions according to fig. 3 are represented by four points each: A_{mf}, B_{mf}, A'_{mf}, B'_{mf}. A_{mf} and B_{mf} are directly gained from the training set: A_{mf} is the smallest occuring value of the training set, B_{mf} the largest. Values outside $\{A_{mf}, B_{mf}\}$ may exist, but with a smaller possibility. A'_{mf} and B'_{mf} are determined empirically, there is no way to derive them.

$$\begin{aligned} A'_{mf} &= A_{mf} - 0.8 * (B_{mf} - A_{mf}) \\ B'_{mf} &= B_{mf} + 0.8 * (B_{mf} - A_{mf}) \end{aligned} \tag{8}$$

yields the best results. Each possibility function can be stored requiring just 4 bytes.

During recognition, feature extraction provides a feature vector $\vec{x}$ with M elements (fig 4.). Calculation of possibilities results in a matrix **p** with F columns and M lines, equivalent to π in (7).

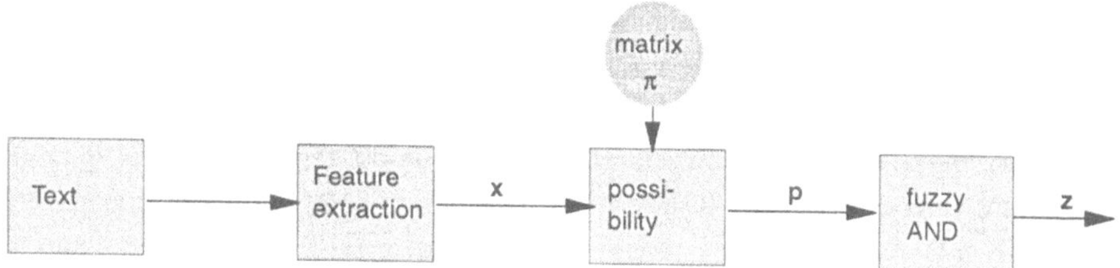

Fig. 4: classification

The next step is to combine the possibilities belonging to one font, leading to one single recognition result for each font. This combination of possibilities is nothing else but a 'fuzzy AND'. Many different implementations for 'fuzzy AND' have been proposed [8]. The best classification results were gained by applying the arithmetic mean value:

$$z_f = p_{f,1} \oplus p_{f,2} \oplus \ldots \oplus p_{f,M} = \frac{1}{m}\sum_{i=1}^{m} p_{f,i} \tag{9}$$

The font f, that has the highest possibility z_f is the 'winner' of the classification.

$$z = \max\{z_1,..,z_f,...,z_F\} \quad \Rightarrow \; f \tag{10}$$

There are two thresholds to check the reliability of the recognition result: the maximum possibility z_f, e.g. the possibility of the 'winner', must be greater than a threshold γ:

$$z_f > \gamma \tag{11}$$

and the difference of the possibility of the choosen font and the second largest possibility
z_g must be greater than threshold η:

$$z_f - z_g > \eta \tag{12}$$

If one of the conditions (11) or (12) is violated, the recogniton result is 'reject'.

4. System Design

Fig. 5 shows the design of the complete Chinese character recognition system. During preprocessing headlines, body text, footnotes and so on are detected. According to this results the areas are defined, in which the font is classified separately.

The background knowledge consists of some rules about the possibility of fonts at certain positions of a text. For example long portions of body text will in general not be written in Heiti, Fangsongti rarely appears in headlines. These rules are used to confirm or refute the recognition results of the font classifier. One single-font classifier is choosen to classify the Chinese characters. If the recognition result is unreliable (including context analysis), e.g. the reject rate is high, the decision of the font classifier may be checked again and a second single-font classifier may be applied.

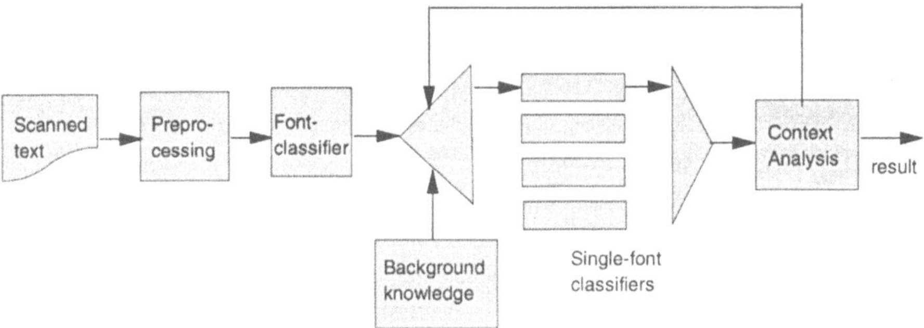

Fig. 5: recogniton system

5. Recognition Results

The recognition result mainly depends on the number of characters in the portion of text, which font has to be classified (fig. 6). The font classification is absolutly reliable for longer portions of text (more than 30 characters) and nearly without problems for standard headlines (10 to 20 characters). However it is not reliable for short marked parts of a text with less than 10 characters. In this case, the decision must interactively be made by the user.

The character error-rate of the system could be reduced by the factor 10, compared with a multifont-classification system without font-classifier [7].

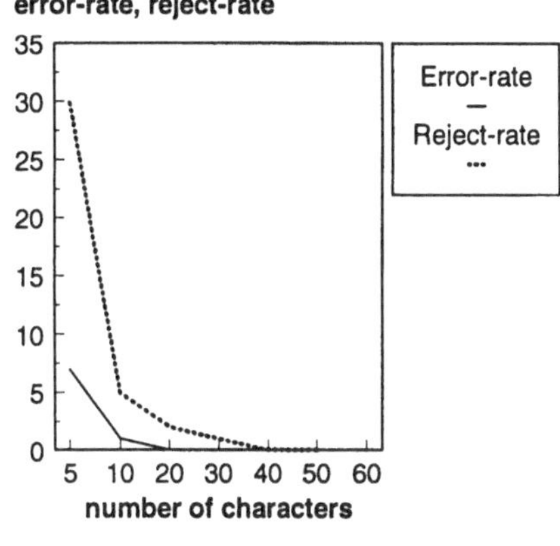

Fig. 6: Error- and reject-rate depending on the number of characters

Conclusions

Font-classification allows the use of single-font classification and therefore improves the recognition performance of the whole system.

It has been shown, that in an extremly complex scene of more than thousand classes secondary properties can reliably be classified. This may be the base for further research on a more general subject: detecting tendencies in natural scenes without classifying single objects.

References

[1] R.G. Casey and G. Nagy: automatic Recognition of Machine Printed Chinese Characters, IEEE-TEC, 1966

[2] R. Nagy: Chinese Character Recognition: A Twenty-Five-Year Retrospective, Proceedings of ICPR 1988, pp. 163-167

[3] Tai J.-W.: Some Research Achievements on Chinese Character Recognition in China, International Journal of Pattern Recognition and Artificial Intelligence, 1&2:199-206, 1991

[4] Cheng F.-H. and Xu W.-H.: Research on Chinese OCR in Taiwan, International Journal of Pattern Recognition and Artificial Intelligence, 1&2:139-164,1991

[5] Ye N.-F.: Chinese Characters on Microcomputers and Chinese Character Recognition (in Chinese), Beijing 1989

[6] Zhang X.-Z.: Chinese Character Recognition Develops towards Practical Systems (in Chinese) Computerworld 2:17, 1992

[7] V.Müller: A self-learning Recognition System for Chinese Characters (in Chinese), Pattern Recognition and Artificial Intelligence (to be published)

[8] R. Suchenwirth: Optical Recognition of Chinese Characters, Braunschweig 1989

[9] H.-J. Zimmermann: Fuzzy Set Theory and its Applications, Boston 1991

[10] D. Dubois and H. Prade: Possibility Theory, New York 1988

Verkehrsszenenanalyse in hierarchisch codierten Bildern

Ulrich Büker, Hubert Austermeier, Georg Hartmann, Bärbel Mertsching
Universität-Gesamthochschule-Paderborn
Fachbereich Elektrotechnik
Pohlweg 47-49
33098 Paderborn
e-mail: getbuek@get.uni-paderborn.de

Zusammenfassung

Eine schnelle Objekterkennung in natürlichen Szenen ist eines der herausragenden Ziele der digitalen Bildverarbeitung. Im Rahmen des PROMETHEUS-Projektes[1] wurden Untersuchungen zur Verkehrszeichenerkennung in hierarchisch codierten Bildern vorgenommen. Verfahren zur effizienten und robusten Erkennung von Verkehrszeichen wurden hierzu entwikkelt. Die Implementation des Systems erfolgte auf einem Transputer-System und einer SUN-Workstation für die grafische Ausgabe der Ergebnisse. Das System konnte bereits an einem Satz von über 300 Verkehrszenen erfolgreich getestet werden. Die angewandten Strategien sowie quantitative Untersuchungsergebnisse werden im weiteren vorgestellt.

1 Einleitung

Eine Fahrzeugführung basierend auf Bildverarbeitungskomponenten kann auf verschiedenen Komplexitätsstufen durchgeführt werden. Erstens sei hier die Straßenranderkennung aufgeführt, die in Zusammenhang mit der Position des Fahrzeuges sowie mit geeigneten Orts-Zeit-Modellen eine Spurhaltung des Fahrzeugs erlaubt [Graefe 91], [Franke 91], [Sowers 89]. Die Bildverarbeitungskomponente steuert in diesem Prozeß Informationen über die Orientierung des Straßenrandes bei. Übergeordnete Informationen wie Form und Symmetrie werden bei der Erkennung vorrausfahrender Fahrzeuge benötigt [Kühnle 91], während die Erkennung von Kreuzungen oder Verkehrszeichen Bildverarbeitungsverfahren auf einer symbolischen Ebene benötigt [Seitz 91], [Austermeier 92].

Unser Ansatz zur Verkehrsszenenauswertung besteht nun aus einem zweistufigen System. In der ersten Stufe werden die Bilder dazu in eine hierarchische Datenstruktur, den Hierarchischen Strukturcode, transformiert. Diese Stufe ist applikationsunabhängig und erste Hardware-Implementationen [Bilau 93] stehen hierzu zur Verfügung. Der Erkennungsvorgang findet in der zweiten Stufe statt und beinhaltet sowohl die Fahrbahnranderkennung als auch eine Verkehrszeichenanalyse.

---

[1]In Zusammenarbeit mit der Universität Koblenz, Koblenz und der Daimler-Benz AG (Forschungszentrum, Ulm und Mercedes-Benz Forschungszentrum, Esslingen)

2 Hierarchische Repräsentation von Verkehrsszenen

Die rechtzeitige Erkennung von Verkehrszeichen in einer natürlichen Straßenszene erfordert eine ausreichend gute Auflösung des Kamerabildes. Dies hat aber gleichzeitig zur Folge, daß ebenso Strukturinformationen durch Bäume und andere Details der Umgebung sowie durch Rauschen im Bild entstehen. Der Mensch ist nun trotz dieser Vielfalt von Informationen in der Lage, sich sehr schnell einen Überblick in derart komplexen Situationen zu verschaffen, da er rasch die bildbestimmenden kontinuierlichen Strukturen wahrnehmen kann [Hartmann 87].

Abb 1-4.: Grauwertbild (1) einer Verkehrsszene in hoher Auflösung. Während in (2) alle Kantenelemente dargestellt sind, werden in (3) nur Kantenelemente von Codebäumen $n \geq 4$ gezeigt. Bildrauschen und kleine Detailstrukturen sind bereits verschwunden. Nur die bilddominierenden Strukturen sind in der Lage, Codebäume einer Größe $n \geq 7$ aufzubauen (4).

Im algorithmischen Sinne ist es also wünschenswert, gezielt auf Bildstrukturen einer vorgegebenen Größe zugreifen zu können. Der Hierarchische Strukturcode (HSC) bietet eine solche Möglichkeit. Kontinuierliche Strukturen wie Kanten, Linien oder Flecken werden auf sogenannte Codebäume abgebildet. Diese Codebäume sind umso höher, je ausgedehnter die Strukturen im Bild sind. Die Datenstruktur erlaubt weiterhin einen gezielten Zugriff auf die Wurzelknoten der Bäume einer gewünschten Höhe. Eine detaillierte Beschreibung des Hierarchischen Strukturcodes und seiner Erzeugung findet sich in [Mertsching 91], [Bilau 92]. In Abbildung 1-4 wird beispielhaft an einer Autobahnszene die Leistung des HSC zur Trennung von bildbestimmenden und unwichtigen Strukturen gezeigt.

3 Die Auswertung von Verkehrsszenen

Nachdem im ersten Schritt der HSC des aufgenommenen Bildes erzeugt wurde, erfolgt in der zweiten Systemstufe die applikationsspezifische Auswertung der Szene. Hierzu wird zunächst die Fahrbahn lokalisiert. Mit Hilfe der Fahrbahnränder wird der Fluchtpunkt bestimmt, der zur Festlegung einer RoA (Region-of-Attention), in der die Verkehrszeichen erwartet werden, herangezogen wird.

Zur Extraktion der relevanten Informationen aus dem HSC wurde ein Satz von applikationsunabhängigen Operationen entwickelt, die ausführlich in [Mertsching 91] beschrieben sind. Hiervon werden für die Verkehrsszenen-Analyse die Operationen ALLROOT zur Wurzelknotensuche, SEQU zum Aufspannen der zu einem Wurzelknoten gehörenden Kantensequenz, einer geordneten Liste der Blätter eines Codebaumes, sowie SHAPE zur Formanalyse von Sequenzen genutzt. Dieser Satz wurde speziell für die Bestimmung der Fahrbahnränder um die Operation CUSTL (Cut Straight Lines) erweitert.

3.1 Fahrbahnrandbestimmung und Region-of-Attention

Zur Fahrbahnbestimmung werden in einem ersten Schritt alle Kanten-Wurzelknoten der Verknüpfungsebene $n \geq 5$ eingesammelt. Aus den mit Hilfe von SEQU aufgespannten Sequenzen werden mit Hilfe von CUSTL alle Geradenstücke einer bestimmten Mindestlänge herausgeschnitten. Eine detaillierte Beschreibung des Algorithmus befindet sich in [Austermeier 92].

In einem nächsten Schritt werden die Geraden, die die Fahrbahnbegrenzung codieren, unter Verwendung gewisser Annahmen herausgefiltert (das Auto fährt z.Z. geradeaus auf der rechten Spur, die Kamera blickt in Fahrtrichtung auf die Fahrbahn, etc.). Die Geraden müssen demnach bestimmte Anforderungen erfüllen, was ihren Neigungswinkel und ihre absolute Position im zweidimensionalen Bildarray anbelangt (Abb. 5 und 6).

Die verbliebenen Geraden werden in Gruppen zusammengefaßt, von denen jede eine Ausgleichsgerade liefert. Es gehören solche Geraden in eine Gruppe, die kollinear zueinander sind (innerhalb gewisser Toleranzbereiche). Unter den Ausgleichsgeraden werden nun wiederum unter Annahme der oben genannten Voraussetzungen die zwei Geraden herausgesucht, die mit der größten Wahrscheinlichkeit die Fahrbahnbegrenzungen markieren. Den Schnittpunkt dieser beiden Ausgleichslinien definieren wir nun als den Fluchtpunkt.

Ausgehend vom Fluchtpunkt ziehen wir nun eine obere Begrenzungslinie zum rechten Rand des Bildes bzw. eine untere Begrenzungslinie zum rechten Rand. Das entstehende Dreieck (die beiden Begrenzungslinien und der rechte Rand) stellt den eingeschränkten Suchbereich RoA dar, der häufig nur 15% der Bildfläche ausmacht.

Eine weitere RoA wird analog auf der linken Straßenseite ermittelt. Desweiteren können Verkehrszeichen auch an Schilderbrücken befestigt sein, sodaß eine dritte RoA in einem Bildstreifen oberhalb des Fluchtpunktes festgelegt wird.

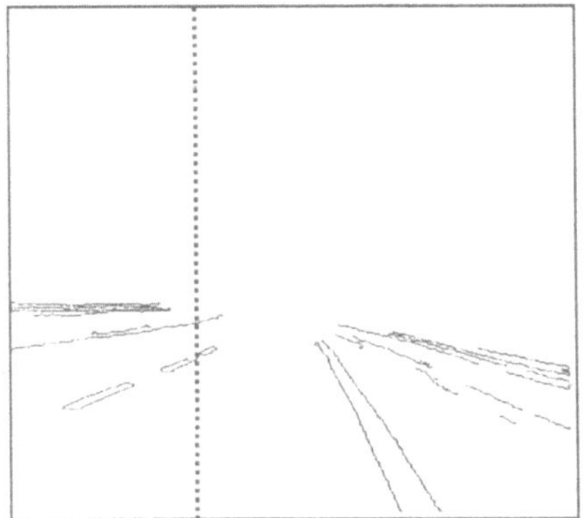

Abb. 5: Die interessanten Kandidaten aus dem Geradenpool zusammen mit einer imaginären Teilungslinie.

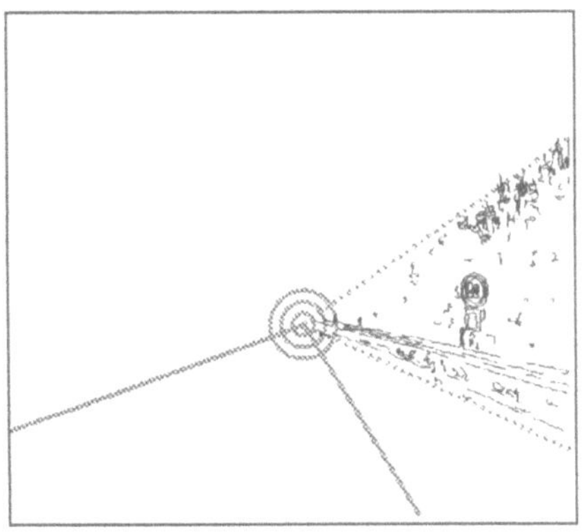

Abb. 6: Die resultierende Fahrbahnbegrenzung, der Fluchtpunkt und die RoA mit dem darin befindlichen Code.

3.2 Verkehrszeichenerkennung

Die eigentliche Suche nach Verkehrszeichen startet nun mit dem Aufsammeln aller Wurzelknoten von Codebäumen, deren Zeile/Spalte-Koordinaten innerhalb des RoA liegen und beschränkt sich auf Verknüpfungsebenen $n \geq 4$ (Operation ALLROOT, [Mertsching 1991]). Die zugehörige Sequenz von Kantenelementen wird dann mit der formbestimmenden Operation SHAPE behandelt, die eine Untersuchung auf die geometrischen Grundformen Kreis, Dreieck und Rechteck durchführt, da der Großteil aller VZ eine dieser Formen haben. Als Ergebnis dieser Operation erhalten wir die gefundene Form (eventuell auch mehrere) zusammen mit einer Maßzahl für die Güte der Übereinstimmung.

Um die Robustheit des zuvor beschriebenen Verfahrens zu untersuchen, wurden ca. 300 Bilder in Form von Einzelbildern und Bildserien, die vorwiegend auf der Autobahn aus einem mit konstanter Geschwindigkeit fahrenden Fahrzeug aufgenommen wurden, ausgewertet. Mit Hilfe geeigneter Werkzeuge wurden die Bilder nun einzeln untersucht und alle zu Verkehrszeichen gehörenden, relevanten Informationen in Tabellen systematisch eingetragen. Interessant waren hier vor allen Dingen die Konnektivität von am VZ beteiligten Kantensequenzen (Wie häufig reißen Kanten auf?), ihre Längen und Ausdehnungen (zur Ermittelung der Schildgröße im Bild), geometrische und statistische Werte über die Elemente einer Sequenz (Lage des Schwerpunktes, mittlerer Abstand der Elemente zum Schwerpunkt, Varianz), sowie Aussagen über die Codebaumhöhe (Verknüpfungsebene des Wurzelknotens) und Anzahl der Sequenzelemente, um nur einige zu nennen.

Die Auswertung der Tabellen führte schließlich zu einer Bereichseinteilung nach Entfernungen zwischen Kamera und VZ. Im Bereich 1 (bis zu 30 m) gab es kaum Probleme, und die wichtigen Kanten waren ungestört vorhanden. Mit zunehmender Entfernung läßt die Qualität der Sequenzen dann nach (Bereich 2 bis 40 m, Bereich 3 bis 70 m), und es kommt dort häufiger zu Verschmelzungen. Im Bereich über 70 m trat häufig der Fall auf, daß vom VZ genau eine dominate Kante vorhanden war und kleinere Details, die vorher hauptsächlich zu den Verschmelzungen führten, wegen der großen Entfernung nicht mehr im Bild vorhanden waren. Die Bereichsangaben sind als relativ zu betrachten, sie hängen direkt von der verwendeten Kamera und dem zugehörigen Objektiv ab.

	Ber. 1 <30m	Ber. 2 30-40m	Ber. 3 40-70m	Ber. 4 70-90m	Ber. 5 >90m	total
untersucht	29	12	33	8	7	89
erkannt	29	10	12	6	4	61

Tabelle 1: Auszug einer Untersuchungsreihe aufgeteilt in die unterschiedlichen Entfernungsbereiche. Es ist die Anzahl der untersuchten, sowie der mit HSC-Methoden erkannten Verkehrszeichen eingetragen. In dieser Tabelle sind nur Vorschriftzeichen und Gefahrzeichen mit ähnlicher Schildausdehnung berücksichtigt.

Die folgenden Abbildungen zeigen den Kantencode von einem exemplarischen Verkehrszeichen in vier verschiedenen Entfernungsbereichen. Es handelt sich dabei um vergrößerte Bildausschnitte mit jeweils 50x125 Bildpunkten.

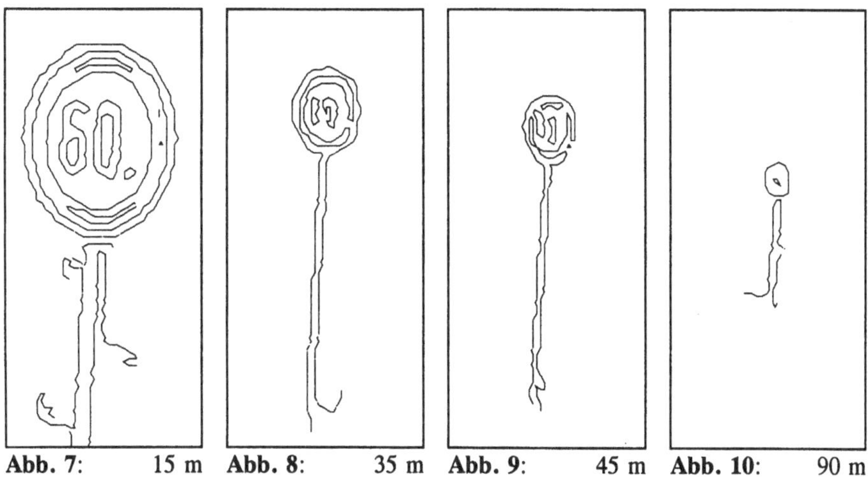

Abb. 7: 15 m **Abb. 8**: 35 m **Abb. 9**: 45 m **Abb. 10**: 90 m

4 Projektkooperationen

In einer Zusammenarbeit mit der Daimler-Benz AG und der Universität Koblenz wurde ein Gesamtsystem aufgebaut, das in der Lage ist, in Verkehrsszenen sowohl eine Fahrbahnbestimmung als auch eine genaue Analyse der vorhandenen Verkehrszeichen durchzuführen. Hierbei werden die im HSC erkannten Verkehrszeichen vorklassifiziert (Gefahrzeichen, Vorschriftszeichen, etc.) und die Koordinaten des umschließenden Rechtecks an einen Pixelklassifikator übertragen. Dieser übernimmt dann eine detaillierte Auswertung des Piktogrammes innerhalb des Verkehrszeichens [Ritter 93]. Alternativ zum HSC kann für die Verkehrszeichensuche auch der Color Structure Code (CSC) eingesetzt werden. Der CSC wurde von unserem Projekt-Partner an der Universität Koblenz entwickelt. Farbinformationen werden hierbei auf ähnliche Weise wie Strukturinformationen im HSC hierarchisch codiert [Priese 93]. Die durchgeführten Tests haben gezeigt, daß sich HSC und CSC in Situationen unter schwierigen Lichtverhältnissen gegenseitig sehr gut unterstützen. Abbildung 11 zeigt die Benutzerschnittstelle zum Gesamtsystem. Auf der linken Seite ist das auszuwertende Farbbild dargestellt, rechts die erkannten Verkehrszeichen.

5 Zusammenfassung und Ausblick

Ein einheitlicher Ansatz sowohl für die Fahrbahnrandbestimmung als auch für eine Vorklassifizierung von Verkehrszeichen wurde vorgestellt. Die Funktionsfähigkeit und Robustheit des Systems wurde an einem Satz von über 300 Testbildern nachgewiesen. Eine weitere Steigerung der Robustheit kann durch eine engere Kopplung des HSC und des CSC erreicht werden. Die Laufzeit des momentan sequentiell implementierten Systems beträgt zur Zeit einige Sekunden für die Erzeugung und Auswertung des HSC. Zukünftige Arbeiten werden

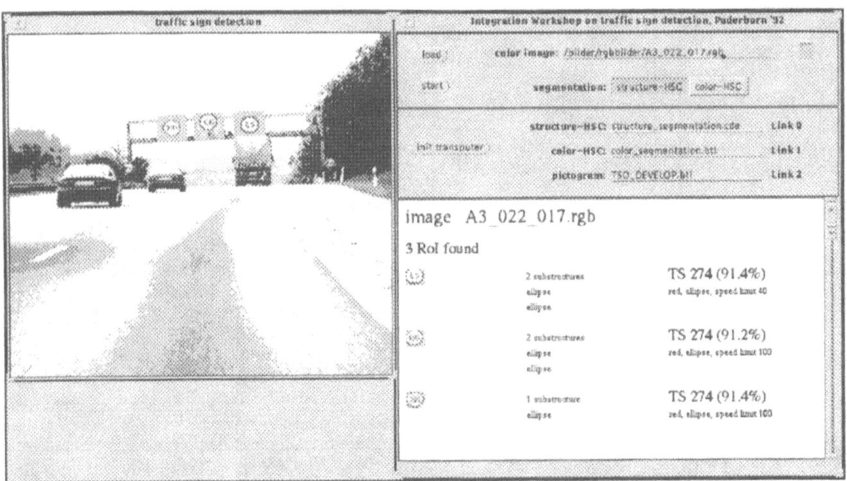

Abb. 11: Grafische Benutzerschnittstelle des Gesamtsystems

sich daher mit der Integration eines ebenfalls im Rahmen eines PROMETHEUS-Projektes entstandenen VLSI-Chip zur HSC-Generierung in unsere Transputer-Umgebung sowie mit einer Parallelisierung der Auswertung beschäftigen.

Literatur

[Austermeier 92]
Austermeier, H.; Büker, U.; Mertsching, B.; Zimmermann, S.: Analysis of Traffic Scenes by Using the Hierarchical Structure Code. In: Bunke, H. (Hg.): Advances in Structural and Syntactic Pattern Recognition. Series on Machine Perception and Artificial Intelligence. Bd. 5. Singapore (World Scientific Publishing Co.) 1993

[Bilau 92]
Bilau, N.; Schnusenberg, J: Ein schneller Codierungsprozessor für ein System zur echtzeitnahen Generierung des Hierarchischen Strukturcodes (HSC) mit Schnittstelle zum Erkennungssystem PANTER. In: Fuchs, S.; Hoffmann, R. (Hg.): Mustererkennung 1992. Informatik aktuell. Berlin u. a. (Springer-Verlag) 1992, S. 310-315

[Bilau 93]
Bilau, N.; Schnusenberg, J.; Hartmann, G.; Siggelkow, A.; Schwederski, T.: A VLSI-Processor for the Generation of the Hierarchical Structure Code in Real-Time. Eingereicht zur CAMP 1993

[Franke 91]
Franke, U.; Fritz, H.; Mehring, S.: Long Distance Driving with the Daimler-Benz Autonomous Vehicle VITA. PROMETHEUS Workshop, Grenoble, 1991

[Graefe 91]
Graefe, V.; Blöchl, B.: Visual Recognition of Traffic Situations for an Intelligent Automatic Copilot, PROMETHEUS, Proceedings of the 5th workshop, München, 1991, S.98-108

[Hartmann 87]
G. Hartmann: Recognition of Hierarchically Encoded Images by Technical and Biological Systems. In: Biological Cybernetics 57, 1987, 73-84

[Kühnle 91]
Kühnle, A.: Symmetry-based recognition of vihicle rears. In: Pattern Recognition Letters 12, 1991, S.249-258

[Mertsching 91]
Mertsching, B.; Zimmermann, S.; Büker, U.: Ein Satz von merkmalsbestimmenden Basisoperationen zur Auswertung von Bildpyramiden. In: Radig, B. (Hg.): Mustererkennung 1991. Informatik-Fachberichte 290. Berlin u. a. (Springer-Verlag) 1991, S. 180-186

[Priese 93]
Priese, L.; Rehrmann, V.; Schian, R.; Lakmann, R.: Traffic Sign Recognition Based on Color Image Evaluation. In: Proc. of the Intelligent Vehicles Symposium 1993

[Ritter 93]
Ritter, W.; Janssen, R.; Ott, S.: Hybrid Approach for Traffic Sign Recognition. In: Proc. of the Intelligent Vehicles Symposium 1993

[Sowers 89]
Sowers, J.P.; Mehrotra, R.: Road boundary detection for autonomous navigation. In: Proc. of 3rd Int. Conf. on Image Processing and its Applications, 1989, S.68-71

Kombination seismischer Verfahren mit Methoden aus der Bildverarbeitung zur Auswertung von Radargrammen

MARTIN FRITZSCHE
Daimler-Benz AG, Forschung und Technik, Wilhelm-Runge-Str.11, D-7900 Ulm

Zusammenfassung

Vorgestellt wird ein Verfahren mit dem der für die Verarbeitung von Bodenradargrammen erforderliche Zeitaufwand drastisch gesenkt werden kann. Unter Verwendung von konturverfolgenden Algorithmen, angewendet auf ein binarisiertes Kantenbild, gelingt es die wichtigsten Bildausschnitte, solche in denen sich vergrabene Gegenstände abzeichen, zu identifizieren. Mit Hilfe der Hough-Transformation können diese weiterverarbeitet werden. Konturverfolgende Verfahren können auch zur weiteren Untersuchung von Radargrammen eingesetzt werden, welche bereits mit seismischen Verfahren prozessiert wurden.

Einleitung

Bei der Untersuchung oberflächennaher Bodenschichten gelangen in den letzten Jahren vermehrt Georadare zum Einsatz. Sie stellen ein kostengünstiges Verfahren dar mit dem z.B. Inspektionen von Straßen, Deponien oder vergrabenen Kanälen bzw. Leitungen durchgeführt werden können, um anschließend eventuell erforderliche bauliche Maßnahmen zu ergreifen. Das Meßprinzip wird anhand von Abb. 1 erläutert. Zur Gewinnung der Meßdaten wird eine Sende-Empfangseinheit längs einer sogenannten Profillinie entlang der Oberfläche verschoben. Fortlaufend ausgestrahlte Radarpulse breiten sich im Untergrund aus und werden an geologischen Schichtgrenzen oder vergrabenen Gegenständen reflektiert. Die Echolaufzeiten der reflektierten Signale sind proportional zum zurückgelegten Signallaufweg. Aufgrund der Antennenbewegung entlang des Meßprofils zeichnet sich die Streuantwort eines Diffraktors im Boden als Hyperbel ab.

Zur Gewinnung von Versuchsdaten wurden Objekte in einem Sandkasten mit drei auf sechs Meter Grundfläche und etwa eineinhalb Meter Tiefe vergraben, Abbildung 2 zeigt die im folgenden verwendeten Rohdaten. Sie stellen einen Tiefenschnitt von etwa 3 m Breite und 20 ns Reflexionslaufzeit dar, d.h. die vertikale Achse ist eine Zeitachse, sie entspricht einer maximalen Tiefe von etwa 1.5 m. Die linke Hyperbel beruht auf der Diffraktion an einem Rohr (25 cmDurchmesser), die rechte wird verursacht durch Streuung an einer Metallkugel (7 cm Durchmesser).

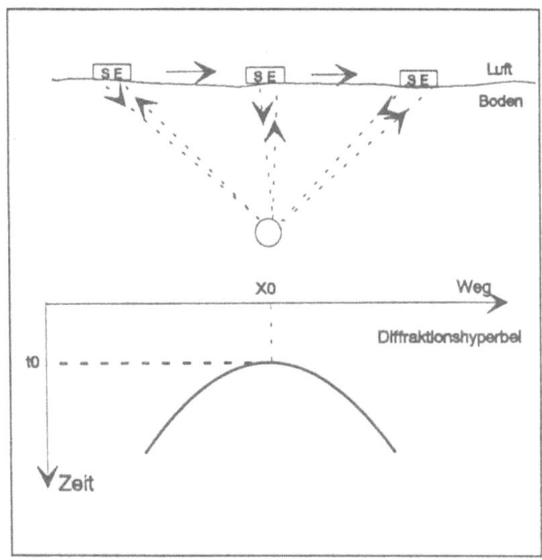

Abbildung 1 : Meßanordung

Der verwendete Frequenzbereich liegt je nach Sendeantenne zwischen 50 MHz und 900 MHz. Die Eindringtiefe der Signale beträgt je nach Bodenfeuchte zwischen 2 m in feuchten Böden, bis zu 20 m in trockenem Sandboden. Die Wellenlängen liegen im Bereich von Dezimetern, das zu erzielende Auflösungsvermögen liegt zwischen $\frac{\lambda}{4}$ und $\frac{\lambda}{2}$. Der Abstand zwischen zwei Messungen ist in der Regel < 1 cm bis hin zu mehreren Dezimetern (z.B. bei Straßenmessungen).

Konventionelle seismische Verarbeitung

Die Datenverarbeitung hat zum Ziel das vertikale Schnittbild durch den Untergrund, welches im Orts-Zeitbereich gegeben ist, in eine Orts-Orts-Abbildung überzuführen, welche der tatsächlichen Struktur des Untergrundes entspricht. In der Seismik werden hierzu sogenannte Migrationsverfahren [YIL87] eingesetzt, welche die gestreute Signalenergie entlang der Diffraktionshyperbel in deren Scheitelpunkt aufakkumulieren. Hierzu ist die ungefähre Kenntnis der Bodenausbreitungsgeschwindigkeit elektromagnetischer Wellen erforderlich. Abb. 3 zeigt das Resultat der Migration, nachdem einige Filteroperationen

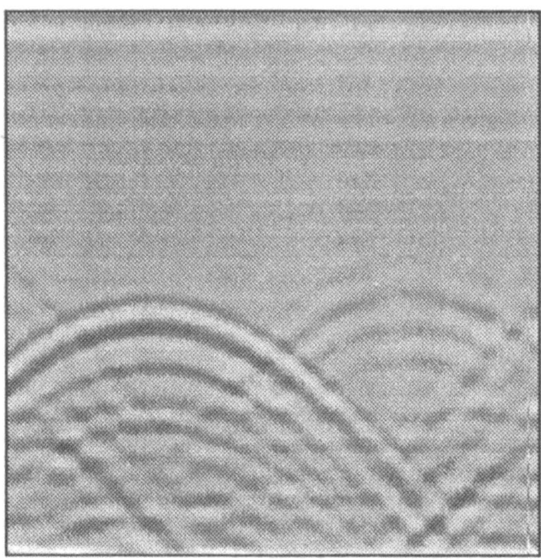

Abbildung 2 : Bei Meßung im Sandkasten aufgezeichnete Rohdaten. Die horizontale Achse gibt die Position entlang des Meßprofils wieder, die vertikale Achse die Reflexionslaufzeit.

durchgeführt wurden. Das resultierende Bild zeigt eine deutliche Fokussierung der Hyperbeln, die vertikale Achse ist nun in eine Längenachse überführt worden.

Alternativer Verarbeitungsweg

Während in der oben beschrieben Vorgehensweise jeweils das ganze Radargramm verarbeitet werden muß, soll nun gezeigt werden wie man eine Eingrenzung auf wesentliche Bildbereiche erreichen kann.

Mit der Hough-Transformation [BAL82] können hyperbelförmige Objekte auf Punkte in einem dreidimensionalen *Houghraum* abgebildet werden. Die drei Parameter Horizontal- und Tiefenposition des Streuers, sowie die Bodenausbreitungsgeschwindigkeit spannen den Raum auf. Die Durchführung der Hough-Transformation ist von wesentlich geringerem Aufwand, wenn der Wert des Grauwertgradienten in jedem Bildpunkt bekannt ist. Mit dem weiter unten erläuterten *BCC-Verfahren* gelingt es aber, die Anzahl der zu berücksichtigenden Bildpunkte so zu vermindern, daß eine Bestimmung der Grauwertgradienten mithilfe von zweidimensionalen Filteroperatoren entfallen kann.

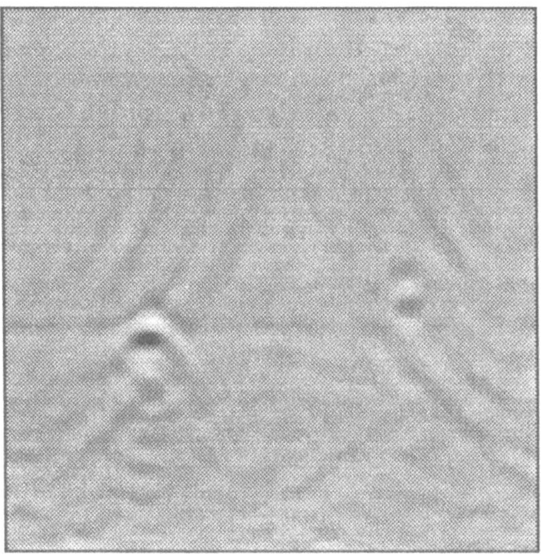

Abbildung 3 : Migration der gefilterten Daten. Die Hyperbeläste *wandern* in den Scheitelpunkt zurück, die vertikale Achse ist nun eine Tiefenachse.

Zuerst wird aus dem Ausgangsbild ein Kantenbild abgeleitet, wobei mit Kante im folgenden lediglich eine steil ansteigende oder abfallende Signalamplitude gemeint ist. Hierfür werden die *Hilbert-Transformation* und die aus ihr berechenbaren *komplexen Attribute* benutzt. Ein analytisches Signal $u(t)$ ist definiert durch:

$$u(t) = x(t) + \jmath y(t) = R(t) \exp[\jmath \Phi(t)]$$

mit $x(t)$ als dem (reellen) Signal, welches jeweils einer Bildspalte entspricht und $y(t)$ der Hilberttransformierten von $x(t)$. $R(t)$ wird als Magnitude bezeichnet, $\Phi(t)$ als momentane Phase, welche zwischen $-\pi$ und $+\pi$ oszilliert. Der Phasensprung erfolgt genau an den Stellen, an denen das Signal extremal wird, d.h. an vorhandenen Peaks. Durch Differentiation in Zeitrichtung erhält man große Werte vor allem an den Sprungstellen der Phase. Nach einer Schwellwertbildung ergibt sich das binäre Kantenbild, welches in Abb. 4 in der Mitte dargestellt ist. In der Folge können nun mit einem konturverfolgenden Algorithmus, dem BCC-Verfahren (*binary connected components*) Objekte gebildet werden [MAN90]. Mit einer einmaligen Dilatationsoperation [BAL82] können getrennte Objekte mit nur einem Pixel Abstand verbunden werden. Zusammenhängende Gebiete mit gleicher Farbe (schwarz oder weiß, da ein Binärbild vorausgesetzt wird) werden als *ein* Objekt erkannt und mit dem sog. Randliniencode beschrieben. Zusätzlich werden geometrische Attribute wie Lage des Schwerpunktes, Umfang, Fläche, umschreibendes Rechteck usw. zugeordnet.

Erweisen sich die Kanten als nicht zusammenhängend, kann versucht werden sie mit den aus der Bildverarbeitung bekannten Dilatationsalgorithmen zu verbinden, um die Anzahl der entstehenden Objekte zu vermindern und eine kritische Mindestgröße derselben zu erreichen. Anhand des Umfangs und der Fläche eines Objektes kann eine Auswahl getroffen werden, d.h. sehr kleine und formuntypische Objekte, wie z.B. Linien, können erkannt und gelöscht werden. Damit ist, ausgehend vom Originalbild, bereits eine große Reduktion zu verarbeitender Pixel erreicht worden. Abb. 4 zeigt im linken Bildteil die Rohdaten, in der Mitte die detektierten Kanten und rechts die via Houghtransformation gefundenen Häufungspunkte. Der horizontale Maßstab entspricht dem des Ausgangsbildes, die vertikale Achse ist eine Längenachse. Man erkennt deutlich die den Hyperbeln zugeordneten Häufungspunkte. Da sich bei der Kantenbildung mehrere Kanten pro Hyperbel ergeben hatten, kommen die zugehörigen Häufungspunkte senkrecht untereinander zu liegen.

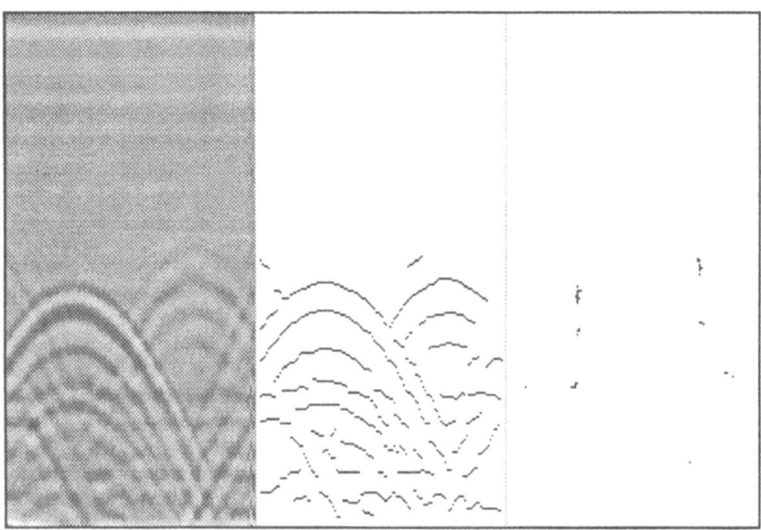

Abbildung 4 : Hough-Transformation, angewendet auf Sandkastendaten, links: Rohdaten, Mitte: detektierte Kanten, rechts: zugehörige Houghtransformierte (vertikale Achse ist die Objekttiefe).

Ein Problem bei der Bestimmung der Hyperbelparameter tritt vor allem dann auf, wenn sich zwei Hyperbeln berühren oder schneiden, d.h. zusammen ein Objekt bilden. In diesem Fall würde der Algorithmus versuchen eine einzige Hyperbel durch das Objekt zu legen. Am Schnittpunkt zweier Hyperbeln weist die Orientierung der Randkurve eines

Objekts im allgemeinen einen Vorzeichenwechsel auf. Durch Verfolgung der Kontourlinie und Untersuchung des Anstieges innerhalb eines Bereichs konstanter Größe können Stellen, an denen sich die Orientierung der Randkurve sprunghaft ändert, gefunden werden. Der implementierte Algorithmus lieferte, angewendet auf die Meßdaten, korrekte Ergebnisse.

Anhand der Lage der Cluster im Houghraum kann auf die Hyperbeln im Ausgangsbild zurückgeschlossen werden. Die Größe der so erhaltenen Datenfenster beträgt nur noch einen Bruchteil der Größe des Ausgangsbildes. Da die umschreibenden Rechtecke jedes Objekts bekannt sind, kann nun eine Migration des so erhaltenen Bildausschnitts (s. Abb. 5 durchgeführt werden. Die hierfür erforderliche Geschwindigkeit wurde über die Hough-Transformation als diejenige ermittelt, bei der die größte Häufung im Houghraum auftrat.

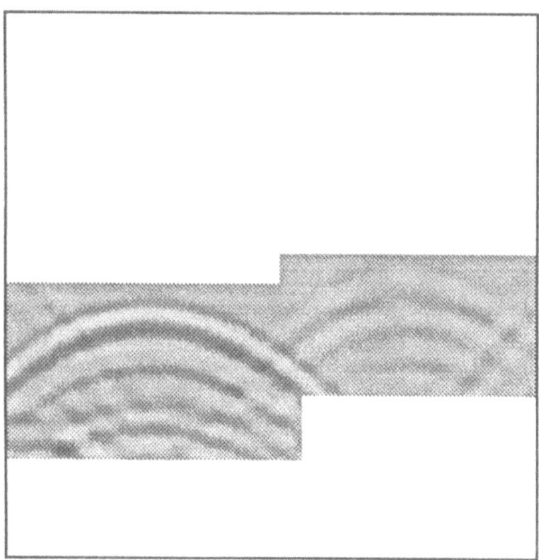

Abbildung 5 : Mit Hilfe der Hough-Transformation ermittelte Datenfenster mit den wichtigsten Bildausschnitten.

Mit einer Modifikation des üblichen Migrationsalgorithmus ist es möglich, durch wiederholte Migration desselben Datenfensters, den Radius eines Rohres abzuschätzen. Der erforderliche Zeitaufwand ist gering, da nur die in Abb. 5 gezeigten Fenster verarbeitet werden.

Wendet man das BCC-Verfahren auf bereits im konventionellen Sinne verarbeitete Bilder an, können vorhandene Objekte (vergrabene Gegenstände) wesentlicher besser herausgestellt werden. Abb. 6 zeigt in der oberen Hälfte eine Straßenmeßung, unten das verarbeitete Bild. Das Radargramm entspricht einem vertikalen Schnitt von 6 m Breite und 4 m Tiefe. Die gesuchten Objekte, in diesem Fall handelte es sich um drei Leitungen, sind deutlich zu erkennen. Der Clutter im linken oberen Bildteil ist auf Stördiffraktionen zurückzuführen.

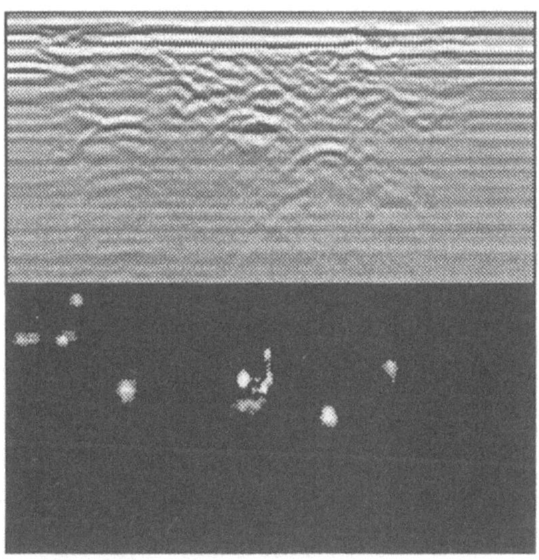

Abbildung 6 : Verarbeitung von Straßendaten. Der obere Bildteil zeigt die Rohdaten, der untere das prozessierte Objektbild mit drei deutlich zu erkennenden Objekten.

Literatur

[BAL82] **Ballard, D. H. , Brown, Ch. M.** 1982, Computer Vision, Prentice Hall, New Jersey

[MAN90] **Mandler, E. , Oberländer, M. F.** 1990, Proc. 10th Intern. Conf. on Pattern Recognition, Atlantic City, New Jersey, June 16-21, 1990

[YIL87] **Yilmaz, Ö.** 1987, Seismic Data Processing, Society of Exploration Geophysics, Series: Investigations in Geophysics, Vol. 2

Layout Retrieval
Suche von Dokumentenbildern mittels Layoutwissen

Per Herrmann[1], Gunther Schlageter[2]
[1] BASF AG, Abt. ZX/ZT, D-67056 Ludwigshafen
[2] FernUniversität, FB Praktische Informatik I, D-58084 Hagen

Zusammenfassung

Systeme zur Erfassung und Speicherung von gescannten Dokumenten ersetzen zunehmend Papier- und Mikrofilmablagen. Üblicherweise kombiniert man solche Dokumenten-Speicher-Systeme (DSS) mit einer Stichwortdatenbank oder mit Volltextsuche, um die Dokumentenbilder recherchierbar zu machen. Dieses Papier stellt ein zusätzliches Suchverfahren für bereits bekannte Dokumente in DSS vor. Dazu wird davon ausgegangen, daß der Mensch während der Beschäftigung mit den Inhalt eines Dokuments auch Layoutinformationen aufnimmt. Man erinnert sich beispielsweise an die Position von Stichworten, an Layoutobjekte und deren Anordnung, Größe und Farbe. Dieses bildliche Wissen läßt sich zur Unterstützung der textuellen Suche verwenden. Ein entsprechendes System zur Layoutsuche besteht aus Komponenten zur Layouterkennung, zur Eingabe des gesuchten Layouts und aus der eigentlichen Layoutsuche. Während für die Layouterkennung genügend bekannte Verfahren existieren, besteht bei der Eingabe und Suche einer Layoutbeschreibung noch Forschungsbedarf. Es wird ein Layouteditor vorgeschlagen, mit dem sich das gesuchte Layout zusammen mit Textinhalten interaktiv aus Layoutobjekten bildlich erzeugen läßt. Die Layoutsuche ist mit alphanumerischen Datenbanken eingeschränkt möglich, Verbesserungen lassen sich mit speziellen Suchstrukturen für bestimmte Layout- und Textinformationen erzielen. Verwendet werden ein Quad-Tree für die absolute räumliche Suche von Layoutobjekten, Nachbarschaftstabellen für relative Objektanordnungen und Volltextsuche an Seiten- oder Objektpositionen.

1 Einführung

Im Laufe der letzten 5 Jahre haben sich Systeme zur Verwaltung von Dokumentenbildern von teueren Spezialsystemen zu erschwinglichen Standardlösungen entwickelt. Sie bestehen aus zwei Hauptkomponenten: dem Bildspeicher und einer Anwendung (siehe Abbildung 1). Der Bildspeicher erlaubt die digitale Speicherung und Reproduktion von schwarzweißen Dokumenten mit hoher Qualität. Dazu dienen Scanner mit hohem Durchsatz, schnelle Kompression und Dekompression nach CCITT Rec. T.4 oder T.6, Massenspeicher mit hoher Kapazität (meist optische Medien) und hochauflösende Bildschirme und Drucker. Heute arbeiten alle diese Komponenten zuverlässig, wodurch Dokumentenbilder als neuer Multimedia-Datentyp neben Farbbildern oder Audio zur Verfügung stehen. Typische Anwendungen für Dokumentenbilder sind Retrievalsysteme, elektronische Post (Fax) und Vorgangsbearbeitung [Avedon92].

Retrievalsysteme zum Ersatz von Papier- und Mikrofilmarchiven bilden die häufigste Anwendung. Man erhofft sich verringerten Raumbedarf, lange Datenhaltbarkeit (mit optischen Medien), schnelle und treffsichere Suche sowie raschen Informationstransport. Existierende

System zeigen aber, daß die Suche problematisch ist. Für strukturierte Dokumente genügt es in den meisten Fällen, einige beschreibende Stichworte in einem Datenbank-Managementsystem abzulegen. Für unstrukturierte Dokumente (z. B. Bücher, Zeitschriften, Briefe) sind dagegen anspruchsvollere Suchverfahren notwendig. Hierzu gehören die automatische Indexierung oder die Volltextsuche in Kombination mit Textleseverfahren, um die textuelle Information automatisch aus den Dokumentenbildern zu gewinnen.

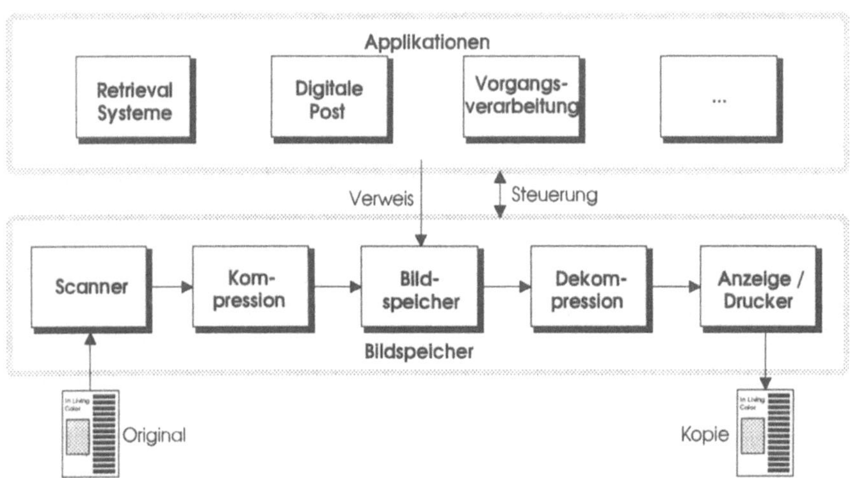

Abbildung 1 : Komponenten eines Dokumenten-Speicher-Systems

Da Dokumente in DSS als Bilder gespeichert werden, bietet sich die Ergänzung der textuellen Suche durch zusätzliche Berücksichtigung von bildhafter Erinnerung an. Die Erfahrung zeigt, daß solches Wissen für bekannte Dokumente, wie man sie überwiegend in privaten Archiven findet, oftmals vorhanden ist. Allerdings ist die Suche nach bildlichem Inhalt (d.h. nach beliebigen Objekten in Photos und Darstellungen) mit den heute verfügbaren Verfahren zur Bilderkennung nicht durchführbar. Statt dessen soll nach einer anderen Form bildlichen Wissens gesucht werden: dem Dokumentenlayout. Dazu gehören die Position und Größe von Textblöcken, Bildern und anderen Layoutobjekten, deren Farbe und weitere Eigenschaften. In der vorliegenden Arbeit wurde untersucht, ob solches Layoutwissen überhaupt vorliegt und wie es zur Verbesserung der textuellen Suche eingesetzt werden kann.

2 Layoutwissen

Grundlegende Untersuchungen zur Erinnerung von bildlicher Information finden sich in der kognitiven Psychologie. Es gilt als gesichert, daß Bilder besser erinnert werden als Textinformationen ("Bildüberlegenheitseffekt") [Ande88]. Weiterhin wurde gezeigt, daß die Erinnerungsleistung beim Wiedererkennen besser ist als beim freien Reproduzieren und daß die Abfrage von Informationen in ihrer Originalkodierung erfolgen sollte (d. h. Bilder in Bildform anstatt textuell) [Enge91]. Spezielle Befunde zur unterbewußten Speicherung von Layoutinformationen während der Beschäftigung mit dem textuellen Inhalt eines Dokuments sind dem Autor allerdings nicht bekannt.

Um das Vorhandensein von Layoutwissen nachzuweisen und festzustellen, welche Layoutelemente und -attribute erinnert werden, wurden einige Experimente durchgeführt. 24 Personen erhielten 4 verschiedene Dokumente zum Lesen mit dem einzigen Hinweis, daß sie an einem Erinnerungsexperiment teilnehmen. Nach unterschiedlichen Zeitabständen sollten sie das Dokumentenlayout beschreiben, entweder verbal mit einem rechnerkontrollierten Fragebogen oder bildlich mit einem einfachen Zeichenprogramm (siehe Abschnit 3.1). Die resultierenden Layoutbeschreibungen wurden mit denen einiger Referenzkandidaten verglichen, welche die Dokumente während der Eingabe vorliegen hatten. Die Auswertung ergab folgendes:

- Es existiert genügend Layoutwissen, um die Idee der Layoutsuche zu rechtfertigen.
- Die Reproduktionen mit dem Zeichenprogramm erfolgten besser und schneller als mit dem Fragebogen. Dies stimmt mit den o. g. Untersuchungen überein, evtl. wird die bildliche Reproduktion durch einen Wiedererkennenseffekt unterstützt.
- Korrelationen der Ergebnisse mit dem Dokumenteninhalt oder mit den Erinnerungszeiträumen ließen sich nicht nachweisen. Für eine Auswertung nach einzelnen Layoutobjekten oder -attributen ist die Anzahl der Abfragen ebenfalls zu gering.
- Auffällig war die bessere Erinnerung von räumlichen Anordnungen der Objekte untereinander im Vergleich zu absoluten Positionen und Abmessungen.

Zusätzliche und geeignetere Untersuchungen sind notwendig, um stichhaltige Ergebnisse zu erhalten (z. B. mit synthetischen Dokumenten zur gezielten Überprüfung einzelner Fragestellungen).

3 Layoutsuche

Abbildung 2 faßt alle Komponenten eines DSS zusammen, welches die Suche nach Dokumentenbildern aufgrund ihres Layouts als auch ihres textuellen Inhalts ermöglicht:

- Die Komponenten des Bildspeichers wurden in Abschnitt 1 beschrieben.
- Die Layoutanalyse separiert und klassifiziert die Layoutobjekte in einer Dokumentenseite. Dabei wird zwischen vier Objekttypen unterschieden: Textblock, Textzeile (Überschrift), Grafik und Photo. Objektattribute sind Position, Abmessungen und die vorherrschende Farbe.
- Die Texterkennung wandelt bildliche Textinhalte in Textobjekten in kodierte Informationen um. Der erkannte Text wird als zusätzliches Attribut zum Layoutobjekt hinzugefügt, ergänzt um Angaben zur Textsprache und zum verwendeten Zeichensatz.
- Die Speicherung der erkannten Layout- und Textinformationen erfolgt für jede Seite in einer Layoutdatei. Die Datei enthält auch Angaben zur Seite selbst, wie Größe, Lage, Farbe, u. a. . Das Datenformat basiert auf der SGML-Notation (ISO 8879).
- Die Eingabe einer Suchanfrage wird durch Skizzieren des gesucheten Layouts (inklusive Text) mit einem "Layouteditor" vorgenommen.

- Für die Suche stehen unterschiedliche Suchmodule für die verschiedenen Informationstypen (z. B. Layout und Text) zur Verfügung. Die gefundenen Dokumente werden in einer Trefferliste präsentiert, sortiert nach der Ähnlichkeit der Seiten bezüglich der Suchanfrage.

Verfahren zur Layoutanalyse und Texterkennung werden in der Literatur ausführlich besprochen, oft im Zusammenhang mit OCR und Dokumentanalyse [Nagy89]. Deshalb wird auf diese Komponenten hier nicht weiter eingangen. Eine Besonderheit stellt allerdings die Berücksichtigung von Farben im Layout als wichtiger Teil menschlicher Wahrnehmung dar. Dies impliziert die Verwendung eines Farbscanners zur Dokumentabtastung und ermöglicht die Auswertung der Farbinformationen während der Erkennungs- und Analyseschritte. Die folgenden Abschnitte beschäftigen sich genauer mit der Layouteingabe- und suche, da diese den innovativen Anteil der Arbeit repräsentieren.

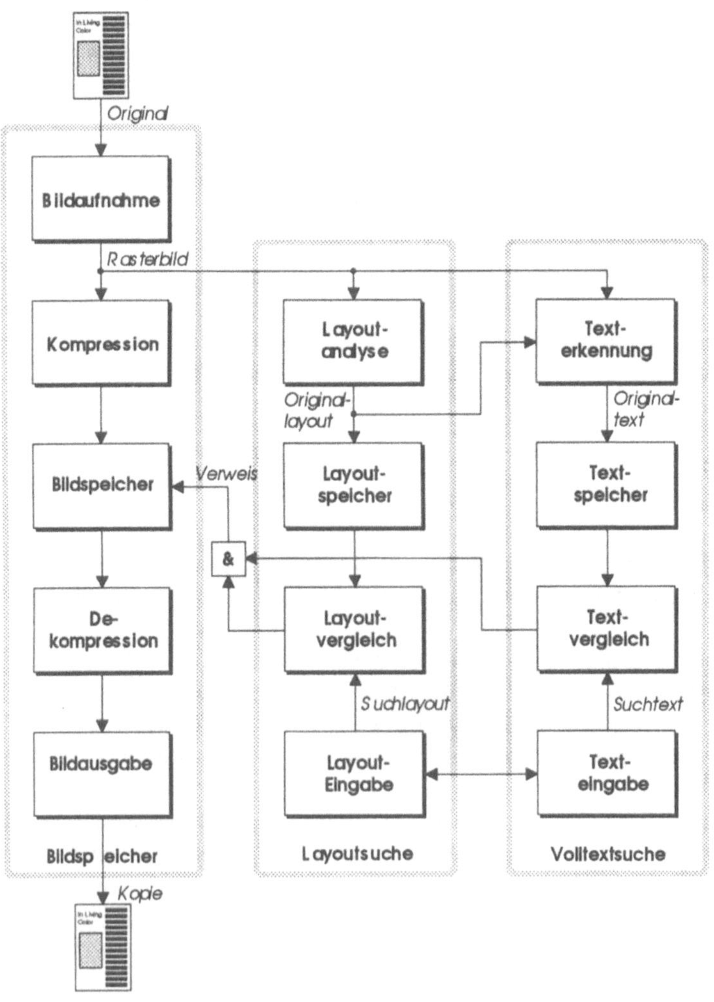

Abbildung 2 : Komponenten der Layoutsuche

3.1 Layouteditor

Die Eingabe von Suchanfragen erfolgt unter Berücksichtigung der unter 2 beschriebenen Beobachtungen mit einem "Query by Example"-Verfahren. Dazu wurde ein spezielles Zeichenprogramm namens Layouteditor entwickelt. Unter Ausnutzung der Konzepte grafischer Benutzerschnittstellen erlaubt es die schrittweise Erstellung des gesuchten Layouts:

- Als erstes werden die Seitenattribute in einem Dialogfenster ausgewählt. Das Programm stellt die Seite entsprechend dieser Angaben als Rechteck auf dem Bildschirm dar.
- Im nächsten Schritt lassen sich Layoutobjekte mit einem grafischen Zeigegerät auf der Seitenfläche positionieren und skalieren. Die dazugehörigen Objektattribute werden in einem Dialogfenster selektiert und anschließend möglichst realistisch dargestellt. Alle Attribute sind optional (auch die Objektposition und -abmessungen); ein Sicherheitsparameter dient zur Angabe, wie zuverlässig ein Objekt erinnert wurde.
- Schließlich kann man Suchtext als spezielles Layoutobjekt eingeben. Die Textsuche erfolgt innerhalb des Objektbereichs, wobei dieser Bereich nicht mit dem eines bereits gezeichneten Textobjekts übereinstimmen muß.

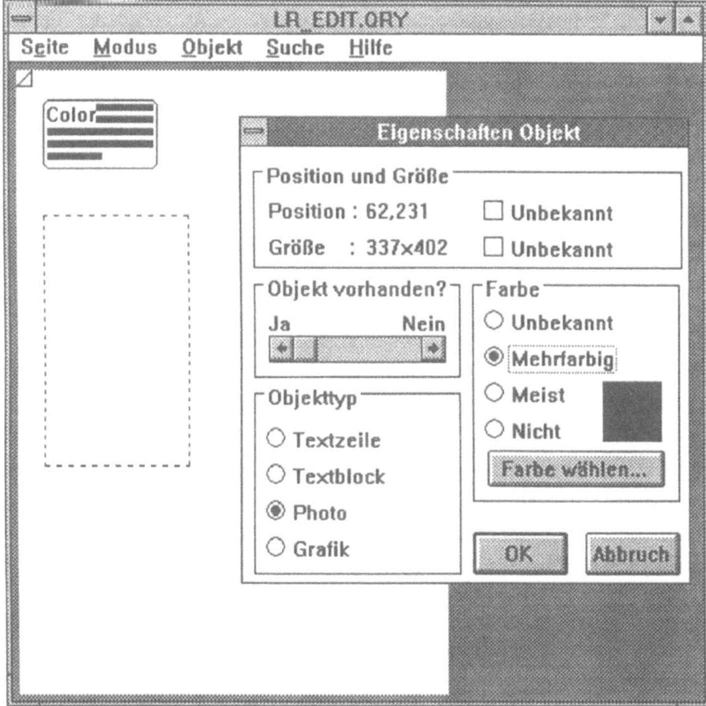

Abbildung 3 : Definition eines Layoutobjekts während der Eingabe eines Suchlayouts

Der gesamte Eingabevorgang sollte so einfach und intuitiv wie möglich ablaufen, um den Erinnerungsprozeß nicht unnötig zu stören. Dazu gehört auch eine einfache Form der Farbauswahl durch Verwendung eines 2-dimensionalen HSL-Diagramms, in dem der Sättigungsparameter nur zur Anzeige von Grauwerten dient. Das Suchlayout wird im gleichen Format wie die Dokumentenlayouts als SGML-Datei gespeichert. Abbildung 3 zeigt ein Bildschirmphoto des Layouteditors während der Eingabe einer Suchanfrage.

3.2 Suchverfahren

Ein Suchlayout enthält eine Vielzahl von Einzelinformationen für die Layoutsuche, die den Einsatz unterschiedlicher Suchverfahren erfordern. Allgemein betrachtet handelt es sich um Verfahren zur Suche von attributierten Bildobjekten, die sich auch für andere Anwendungen von objektbasierten Bilddatenbanken ("Iconic Image Database") eignen:

- Für den boole'schen Vergleich der numerisch kodierten Seiten- und Objekteigenschaften bieten sich die üblichen tabellen- oder zeigerorientierten Datenbank-Managementsysteme (DBMS) an.
- Der Vergleich von Farbattributen erfordert eine Sonderbehandlung, wenn man auch Farbähnlichkeiten berücksichtigen möchte. Bei Verwendung geeigneter Farbräume wie CIELAB, CIELUV oder HSL lassen sich die geometrischen Abstände der zu vergleichenden Farben im Farbraum als Ähnlichkeitsmaß verwenden. Aufgrund der Subjektivität von Farbwahrnehmungen sind solche Berechnungen allerdings nur für geringe Farbabstände zulässig [Richt76].
- Die Suche nach Layoutobjekten aufgrund ihrer Position bzw. Abmessungen ist ebenfalls nur unter Zulassung bestimmter Toleranzen sinnvoll. Dazu kann die Suche mit DBMS innerhalb festlegbarer Ober- und Untergrenzen stattfinden. Eine Alternative ist die Quantisierung der räumlichen Attribute durch Zuordnung zu einem Positionsraster aus n x m Feldern bzw. zu Größen- oder Flächenintervallen. Dann genügt ein Vergleich auf Übereinstimmung, allerdings erfordern Grenzfälle (beim Quantisieren) eine besondere Behandlung. Verschiedene Autoren schlagen für die räumliche Suche in Bilddatenbanken spezielle Indexstrukturen vor, u. a. Quad-Trees und R-Trees [RouLei85]. Für die Layoutsuche wurde eine modifizierte Form des Quad-Trees erprobt.
- Ein anderes Suchverfahren berücksichtigt die Tendenz, Layoutobjekte in der Suchanfrage gruppenweise falsch zu positionieren oder zu skalieren. In diesem Fall führt die absolute räumliche Suche zu falschen Ergebnissen, was sich durch einen Vergleich relativer Objektanordnungen vermeiden läßt. Auch hierfür findet man in der Literatur verschiedene Suchverfahren, z. B. 2D-Strings [CSYa87] und Nachbarschaftstabellen [ChaLee91]. Letztere wurden für die Layoutsuche adaptiert: Die Tabelle enthält die Fundstellen aller Objektpaare je nach (quantisierter) Richtung und Länge der dazugehörigen Verbindungsvektoren.
- Zur Realisierung der Volltextsuche hat man die Wahl zwischen sequentieller Suche, invertierten Indextabellen und Signaturmethoden [Salt89]. Die Layoutsuche arbeitet mit einer Indextabelle, die für jedes Wort auf alle Layoutobjekte verweist, in denen das Wort enthalten ist. Im Gegensatz zur üblichen Textsuche kann im Suchlayout der Suchbereich innerhalb einer Seite eingeschränkt werden (siehe Abschnitt 3.1). Dabei sind zwei Interpretationen möglich: Entweder man sucht den Text direkt im vorgegebenen Seitenbereich oder

aber man sucht den Text in solchen Layoutobjekten, die im Suchbereich des Suchlayouts liegen, wobei die tatsächliche Position der Layoutobjekte von der vermuteten abweichen kann. Diese Unterscheidung entspricht der o. g. Trennung von absoluter und relativer räumlicher Suche.

Das Ergebnis der einzelnen Teilsuchen ist jeweils eine Menge von Seiten, welche die gesuchten Eigenschaften, Objekte oder Wörter enthalten. Durch Abzählen läßt sich die Häufigkeit s(p) ermitteln, mit der eine Seite p in allen Mengen L enthalten ist. Die Relevanz einer Teilsuche i kann man durch einen Gewichtswert g_i angeben (mit $0 \leq g_i \leq 1$), analog werden die Sicherheitsfaktoren c_j für die Layoutobjekte O_j berücksichtigt (wobei $0 \leq c_j \leq 1$). Das Abzählgewicht einer Menge L_{ij} ergibt sich dann in Anlehnung an die Fuzzy Set Theorie durch Multiplikation von g_i mit c_j:

$$s(p) = \sum_i \sum_j \begin{cases} g_i \cdot c_j, & \text{falls } p \in L_{ij} \\ 0, & \text{sonst} \end{cases}$$

Als Gesamtergebnis einer Suche entsteht auf diese Weise letztlich eine Liste von Seitenzahlen, die mit Häufigkeitswerten versehen sind. Je höher dieser Wert ist, umso genauer stimmen das Suchlayout und das Layout der gefundenen Seite überein.

Die bisher beschriebenen Vergleichsfunktionen sind boole'scher Art, d. h. eine Seite wird entweder in die Ergebnismenge aufgenommen oder nicht. Für Attribute mit kontinuierlichen Wertebereichen besteht auch die Möglichkeit, die Übereinstimmung zwischen gesuchter und der in einer Seite p enthaltenen Information mit einem Ähnlichkeitswert a im Intervall [0..1] zu bewerten. Dies bietet sich besonders an für Farben und räumliche Angaben (wobei die Ähnlichkeit mit zunehmenden Abweichungen abnimmt) sowie für die Textsuche (hier werden Worthäufigkeiten und Wortähnlichkeiten ausgewertet). Suchergebnisse, deren Ähnlichkeitswert a einen festgelegten Grenzwert übersteigt, werden mit der dazugehörigen Seite p als Paar (p,a) in die dafür vorgesehene Treffermenge L_{ij} aufgenommen. Die Zusammenfassung der Teilsuchen zu einem Gesamtergebnis erfolgt analog zur Auswertung der boole'schen Vergleiche:

$$s(p) = \sum_i \sum_j \begin{cases} a \cdot g_i \cdot c_j, & \text{falls } (p,a) \in L_{ij} \\ 0, & \text{sonst} \end{cases}$$

3.3 Auswertung der Suchergebnisse

Anstatt einer alphanumerischen Anzeige der Trefferliste nach der Layoutsuche erfolgt die Präsentation aller gefundenen Seiten in Form von verkleinerten Abbildungen (sog. "Thumbnails"). Auf diese Weise besteht die Möglichkeit, die tatsächlich gesuchte Seite direkt wiederzuerkennen. Um den hohen Zeitaufwand für die Datenübertragung, Dekompression, Verkleinerung und Anzeige aller Dokumentenbilder zu verringen, kann man die Thumbnails in der benötigten geringen Auflösung zusätzlich speichern. Eine interessante Alternative ist die

Synthese eines Thumbnails aus der dazugehörigen Layoutbeschreibung, was allerdings eine möglichst umfassende und korrekte Layoutanalyse voraussetzt.

Zur Verfeinerung einer Suchanfrage bieten sich Feedback-Verfahren an, die das Suchergebnis als Ausgangspunkt einer weiteren Suche benutzen. Beispielsweise kann man in Anlehnung an das aus der Textsuche bekannte "Relevanz-Feedback" die Relevanz gefundener Seiten bewerten [Salt89]. Die Eigenschaften und Objekte der demnach relevantesten Seiten lassen sich dann automatisch zu einem neuen Suchlayout kombinieren und im Layouteditor als Vorschlag für die nächste Suchanfrage vorgeben. Eine einfachere Lösung ist die manuelle Selektion und Übernahme von relevanten Objekten und Eigenschaften gefundener Seiten in den Layouteditor.

4 Ergebnisse und Ausblick

Unter Zuhilfenahme von käuflichen Hard- und Softwarekomponenten (für Bildspeicher, Layoutanalyse, Zeichenerkennung und DBMS) wurde ein Prototyp realisiert. Sein Ziel ist die praktische Erprobung der beschriebenen Suchverfahren, um deren Leistungsfähigkeit einschätzen und verbessern zu können. Statistiken über alle Suchanfragen dienen der Feststellung, welche Teilsuchen wie oft das gesuchte Dokument ermittelten. Niedrige Trefferraten deuten entweder auf prinzipiell schlecht erinnerte Layouteigenschaften oder auf unbrauchbare Suchverfahren hin.

Erste Versuche bestätigen, daß die Leistungsfähigkeit herkömmlicher textueller Suche in DSS für bekannte Dokumente durch die Layoutsuche verbessert werden kann. Die weiteren Arbeiten gelten der Optimierung der Eingabeschnittstelle und insbesondere der Suchverfahren.

Literatur

[Ande88] John R. Anderson: Kognitive Psychologie. Spektrum der Wissenschaft, 1988

[Avedon92] Don M. Avedon: Introduction to Electronic Imaging. Association for Information and Image Management (AIIM), 1992

[ChaLee91] Chin-Chen Chang, Suh-Yin Lee: Retrieval of similar pictures on pictorial databases. Pattern Recognition, Vol. 24, No. 7, 1991, S. 675-680

[CSYa87] Shi-Kuo Chang, Qing-Yun Shi, Cheng-Wen Yan: Iconic indexing by 2D strings. IEEE Transactions on Pattern Recognition and Machine Intelligence Vol. PAMI-9, No. 3, 1987, S. 413-428

[Enge91] Johannes Engelkamp: Das menschliche Gedächtnis. Verlag für Psychologie, 1990 und 1991

[Nagy89] George Nagy: Document Analysis and Optical Character Recognition. Proceedings of the 5th International Conference on Image Analysis and Processing 1989, S.511-529

[Richt76] Manfred Richter: Einführung in die Farbmetrik. De Gruyter, 1976

[RouLei85] Nick Roussopoulos, Daniel Leifker: Direct spatial search on pictorial databases using packed R-Trees. Proc. ACM-SIGMOD 1985 Int. Conf. on Managment of Data, May 1985, S. 17-31

[Salt89] Gerard Salton: Automatic Text Processing. Addison-Wesley, 1989

Von der 3D-Digitalisierung über die Mustererkennung zur Fertigung

Marinos Ioannides
Institut für Steuerungstechnik der Werkzeugmaschinen und Fertigungseinrichtungen der Universität Stuttgart (ISW)
Seidenstr. 36; D-70174 Stuttgart
Aloysius Wehr
Institut für Navigation der Universität Stuttgart
Keplerstr. 11
D-70174 Stuttgart

1 Einleitung

Die dreidimensionale Digitalisierung in Design, Modellbau, Architektur und Maschinenbau gewinnt zunehmend an Bedeutung. Insbesondere für den Designbereich und für den Modellbau werden bereits eine Reihe von 3D-Digitalisiergeräten angeboten [1]. Diese arbeiten entweder optisch berührungslos nach dem Triangulations- oder nach dem Phasendifferenzmeßprinzip [2, 3] oder mechanisch berührend als sogenannte tastende Koordinatenmeßmaschinen [2]. Die optischen Meßverfahren zeichnen sich durch eine hohe Aufzeichnungsrate aus und gestatten auch komplexe Körper bzw. Objektoberflächen in einer akzeptablen Zeit zu digitalisieren, Bild 1 und Bild 2. Diese optischen Meßgeräte gestatten, bedingt durch die hohe Tast- und Meßrate, das Meßobjekt bildhaft aufzunehmen. Hierdurch können auch kleine Details erfaßt und ausgewertet werden. Diesem großen Vorteil steht jedoch als Nachteil die große Datenmenge der digitalen Bilder entgegen. Diese Daten beinhalten sehr viel redundante Objektinformationen und erforden viel Speicherplatz sowie große Rechner mit hoher Rechnerleistung. Im Rahmen dieser Präsentation soll gezeigt werden, wie durch den gezielten Einsatz der Mustererkennung die Datenflut von optischen 3D-Digitalisiersystem drastisch reduziert werden kann und eine effektive Bearbeitung der Daten in CAD/CAM-Systemen ermöglicht wird. Mit Hilfe dieser Daten können die digitalisierten Objekte mit CNC-Maschinen gefertigt werden.

Der Arbeitsablauf von der 3D-Digitalisierung bis hin zur Fertigung wird an ausgewählten Beispielen erläutert und die einsetzbaren Mustererkennungsmethoden, wie z.B. Konturverfolgung, Flächengenerierung, Höhenlinien, Triangulierung und Volumenbildung werden präsentiert. Die in den Beispielen dargestellten Objekte wurden mit einem 4D-Laser-Scanner digitalisiert [3].

2 Datenaufbereitung

Nach der Digitalisierung des Objektes mit Hilfe des 4D-Laser Scanners liegen die Meßdaten in Objektkoordinaten vor. Diese werden durch den Einsatz verschiedener Algorithmen, die interaktiv vom Benutzer ausgewählt werden können, reduziert, geglättet und segmentiert [4]. Diese Verfahren arbeiten auf der Basis von 3D-Punkten und stellen als Ergebnis eine im wesentlichen gerasterte Punktemenge zur Verfügung. Diese beschreibt unter Einhaltung einer vorgegebenen Genauigkeit die Geometrie des Objektes. Im folgenden Kapitel werden die auf unseren Rechnern implementierten Verfahren zur Datenaufbereitung vorgestellt.

2.1 Datenreduktion

Folgende Verfahren sind zur Datenreduktion implementiert:

1. Das Tangentiale Distanzschrittverfahren.
   Hierbei wird überprüft, ob Digitalisierdaten innerhalb eines um die Punktekette gelegten tangentialen Toleranzzylinders liegen. Der Zylinder hat eine beliebige Länge und einen Durchmesser, der der Toleranz entspricht. Die Mittellinie des Zylinders ist eine Gerade.

2. Datenreduktion mit dem Kurven-Näherungsverfahren.
   Diese Methode stellt eine Erweiterung des vorherigen Algorithmus dar. Hierbei wird der Toleranzzylinder als Toleranzschlauch gebildet. Die Mittellinie des Zylinders ist im Gegensatz zum vorherigen Verfahren ein Spline.

Nach erfolgreicher Datenreduktion und Glättung der 3D-Meßdaten liegt eine geordnete Punktmenge vor. Mit ihr können Splines approximiert oder interpoliert werden. Diese Splines, die als Eingabe für die Segmentierung dienen, ermöglichen sowohl eine zusätzliche Datenreduktion als auch eine Krümmungsabhängige Segmentierung.

3 3D-Mustererkennung

Der 4D-Laser-Mapper tastet die Objektoberfläche mit einem scharf gebündelten Laserstrahl ab, der mit Hilfe eines Spiegelablenkmechanismus in vertikaler und horizontaler Richtung abgelenkt wird. Zu jedem Bildpunkt werden gleichzeitig die beiden Ablenkwinkel α und β, die Schrägentfernung R zwischen der Apertur des 4D-Laser-Mappers und dem Beleuchtungspunkt auf der Objektoberfläche und die Intensität I des von der Objektoberfläche rückgestreuten Laserlichtes gemessen. Der

Meßvektor für jeden Bildpunkt beinhaltet also vier Komponenten, d.h. er hat die Dimension 4 (4D), Bild 1 und Bild 2:

$$\vec{m} = [\alpha, \beta, R, I]^T$$

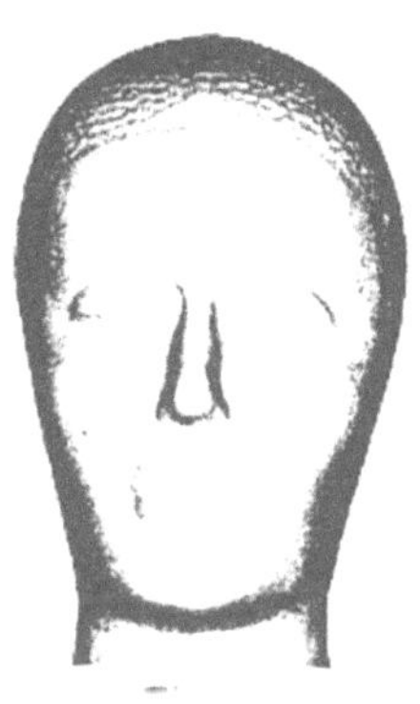

Bild 1: Laserintensitätsbild

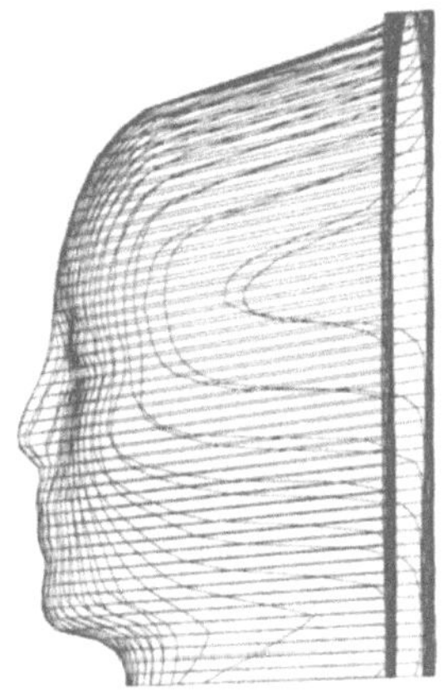

Bild 2: Entfernungsdaten

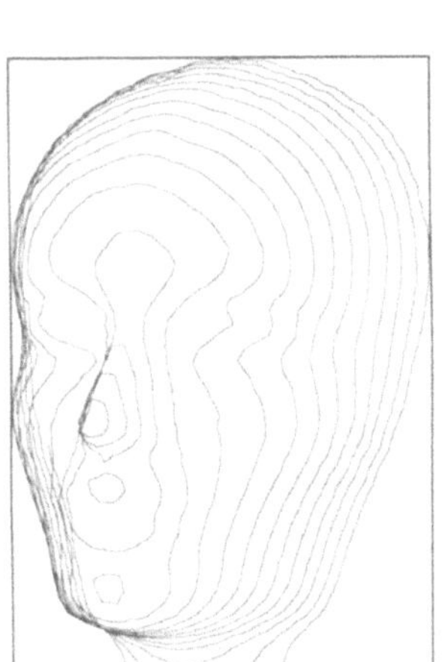

Bild 3: Höhenlinien

Normalerweise wird der Vektor durch Anwendung eines Raytracing-Algorithmus in kartesische Koordinaten umgerechnet. Ordnet man aber der Einfachheit halber die Scanwinkel α und β einem zweidimensionalen Raster zu, ergeben sich zwei Rasterbilder, die zum einen in jedem Bildpunkt die Schrägentfernung und zum anderen die Intensität beinhalten. Auf diese Bilder können direkt die bekannten Algorithmen aus der Bilddatenverarbeitung angewendet werden. Sie erlauben eine effektive Vorverarbeitung der Meßdaten und eine schnelle Segmentierung. Die zusätzliche Auswertung des Intensitätsbildes gestattet, zusammenhängende Bereiche des Objektes (wie z.B. Linien, Flächen usw.) leichter und eindeutiger zu erkennen und somit die Zuverlässigkeit der Segmentierung zu verbessern. Ferner ist mit Hilfe der Bilddatenverarbeitungsalgorithmen eine Konturverfolgung möglich. Im folgenden wird nun ein bereits an unserem Institut auf Rechnern implementierter Programmablauf beschrieben, wo mit Hilfe von Bildverarbeitungsmethoden die 4D-Laser-Mapper-Bilder segmentiert werden.

Die gerasterten Intensitäts-, Bild 1 bzw. Entfernungsdaten, Bild 2 können Grau- oder Farbwerten zugeordnet und auf entsprechende Farbpaletten abgebildet werden. Nach Auswahl einer Farbpalette wird ein Histogramm berechnet, das über den gesamten Paletten-Bereich die Häufigkeiten der in den Bildern auftretenden einzelnen Farben enthält. Anschließend kann die Verteilungsfunktion mit Hilfe eines speziell entwickelten Softwaremoduls als Spline berechnet werden. Diese enthält die Häufigkeit der Farbwerte von Intensität oder Entfernung im kompletten Bereich der digitalisierten Aufnahme. Die berechnete Funktion (Spline) kann entweder mit Hilfe von Approximations- oder Interpolationsverfahren berechnet werden [4]. Durch Kurvendiskussion dieser Splines können folgende Aussagen über das digitalisierte Bild getroffen werden:

1) Nullstellen der Verteilungsfunktion
An den Nullstellen sind die aufgenommenen Oberflächen unstetig, d.h. dieser Farb- oder Helligkeitswert eignet sich optimal als Stelle für eine Konturverfolgung (d.h. Trennung zwischen Vordergrund und Hintergrund; Vordergrund ==> Objekt; Hintergrund ==> unnötige Information)

2) Wendepunkte der Verteilungsfunktion
An Wendepunkten, die keine Nullstelle bilden, lassen sich Unstetigkeitsstellen im Objekt sehr leicht herauskristallisieren. Diese Information dient zur Segmentierung der Aufnahme für die nachfolgende Weiterverarbeitung in der Freiformflächengenerierung.

3) Beliebige Punkte der Verteilungsfunktion ohne besonderes Merkmal
An jeder beliebigen anderen Stelle der Funktion lassen sich Höhenlinien (Schnitte zur Aufnahmeebene) bilden. Diese Höhenlinien beschreiben direkt die Fräsbahnen, die ein Fräser bei der spanenden Fertigung fahren muß, um das Objekt zu fräsen. Dieser Vorgang wird auch als Kopierfräsen bezeichnet.

4 Höhenlinien und Triangulierung

Aus den in Objektkoordinaten vorliegenden Daten lassen sich ohne die bei einem CAD/CAM- oder CAP-System notwendige komplexe Programmierung direkt NC-Sätze gewinnen.
Zunächst werden aus den reduzierten und geglätteten Daten Höhenlinien berechnet, Bild 3. Zur Erzeugung der Höhenlinien muß der Benutzer erstens den Abstand zwischen den Höhenebenen, in die das Objekt diskretisiert wird und zweitens den Begrenzungsbereich, in dem ein Teil oder das gesamte Objekt liegen kann, wählen, Bild 4 und Bild 5. Diese Höhenlinien sind besonders für eine zweidimensionale

Berechnung eines äquidistanten Aufmaßes geeignet. Ihr Datenformat ist mit dem Schichtenmodell einer medizinischen Röntgen-Aufnahme identisch.

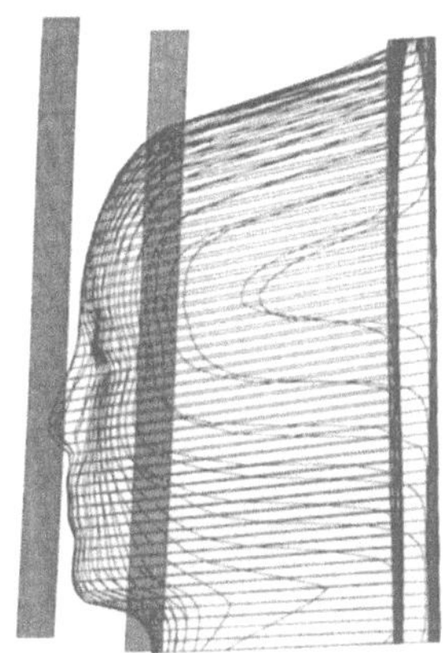

Bild 4: Begrenzungsbereiche

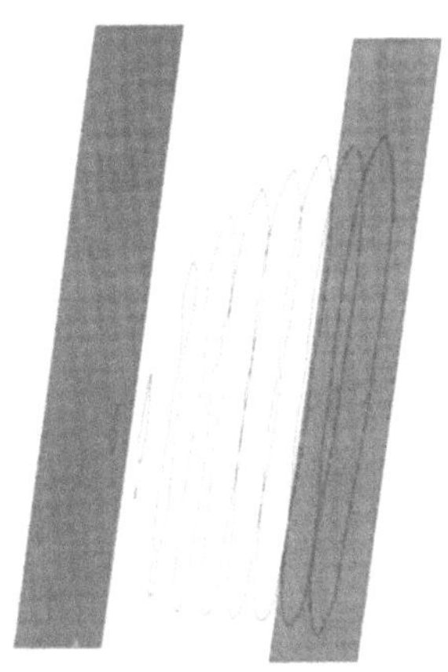

Bild 5: Begrenzungsbereiche mit Höhenlinien

Ferner können aus den Höhenlinien durch eine spezielle Triangulierung Dreiecksflächen erzeugt werden. Hierdurch wird das im Begrenzungsbereich befindliche Objekt in eine stereolithographische Struktur umgewandelt. Diese ist konsistent und eindeutig. Konsistent bedeutet hier, daß keine Löcher oder Überschneidungen existieren. Das reale Objekt oder der zu bearbeitende Teil des Objektes wird vollständig abgebildet. Eindeutigkeit bedeutet in diesem Zusammenhang, daß das Objekt mit allen seinen Elementen zweifelsfrei identifiziert werden kann. Darüberhinaus kann anschließend das Objekt in die Tetraeder-Struktur umgewandelt werden, die für die Finite-Elemente Simulation als Schnittstellenformat gilt. Ein weiterer Vorteil dieser rechnerinternen Darstellung ist, daß sie direkt Datensätze für die dreidimensionale graphische Simulation von Bearbeitungsvorgängen zur Verfügung stellt. Das eigenentwickelte Simulationssystem, das zur Grundlage ein Facettenmodell hat, erlaubt es, ein NC-Programm vor dem tatsächlichen Bearbeitungsvorgang an der CNC-Maschine zu überprüfen.

Darüberhinaus bildet dieses Datenformat die Basis für eine Echtzeitanimation "Virtual reality" (Die Reise durch das 3D-rechnerunterstützte Gebilde), die sehr häufig in der Vermessungstechnik, Bauwesen, Architektur, Medizin usw. stattfindet. Durch dieses einheitliche Format ist eine Schnittstelle geschaffen worden, die sehr effizient und ohne Übertragungs- oder Transformationsfehler Daten an verschiedene Bearbeitungssysteme übergeben kann.

5 Flächengenerierung und Anbindung an CAD/CAM-Systemen

Sollen die hier zu bearbeitenden Daten in den Fertigungsprozeß des allgemeinen Werkzeug- und Formenbaus eingespeist werden, müssen sie für die typischen Datenformate von CAD/CAM-Systemen aufbereitet werden. Im Werkzeug- und Formenbau müssen in den gängigen CAD/CAM-Systemen die Objekte mathematisch über Freiformflächen modelliert werden.

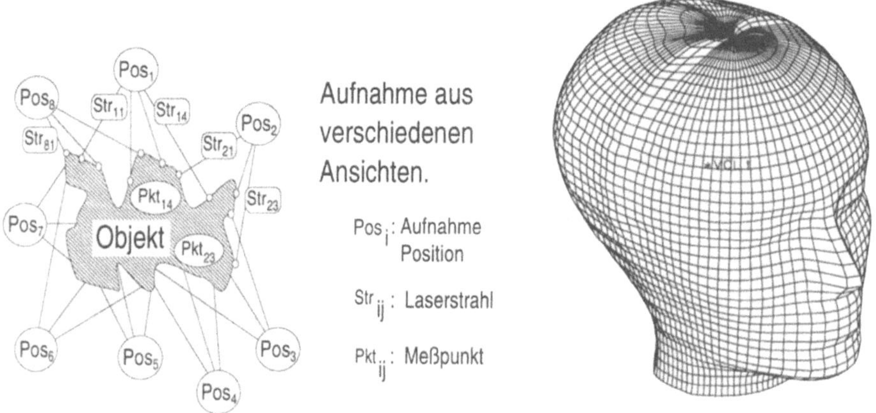

Bild 6: Rekonstruktionsverfahren

Bild 7: Volumenmodell

Diese rechnerinterne Darstellung solcher freigeformter Körper ist analytisch nicht beschreibar [4]. Um solche Flächen bearbeiten zu können, wurden folgende mathematische Verfahren implementiert:

- bikubische Bézierflächen,
- polynomielle Flächen,
- B-Spline-Flächen und
- Flächen in Non Uniform Rational B-Splines Darstellung (NURBS).

Die mit Hilfe dieser Verfahren erzeugten Kurven- und Flächendaten können dann von allen 3D-Freiformflächenorientierten CAD/CAM-Systemen eingelesen und verarbeitet werden. Für die Datenübertragung zu CAD-Systemen stehen die Schnittstellenformate

VDAFS, IGES und STEP zur Verfügung.

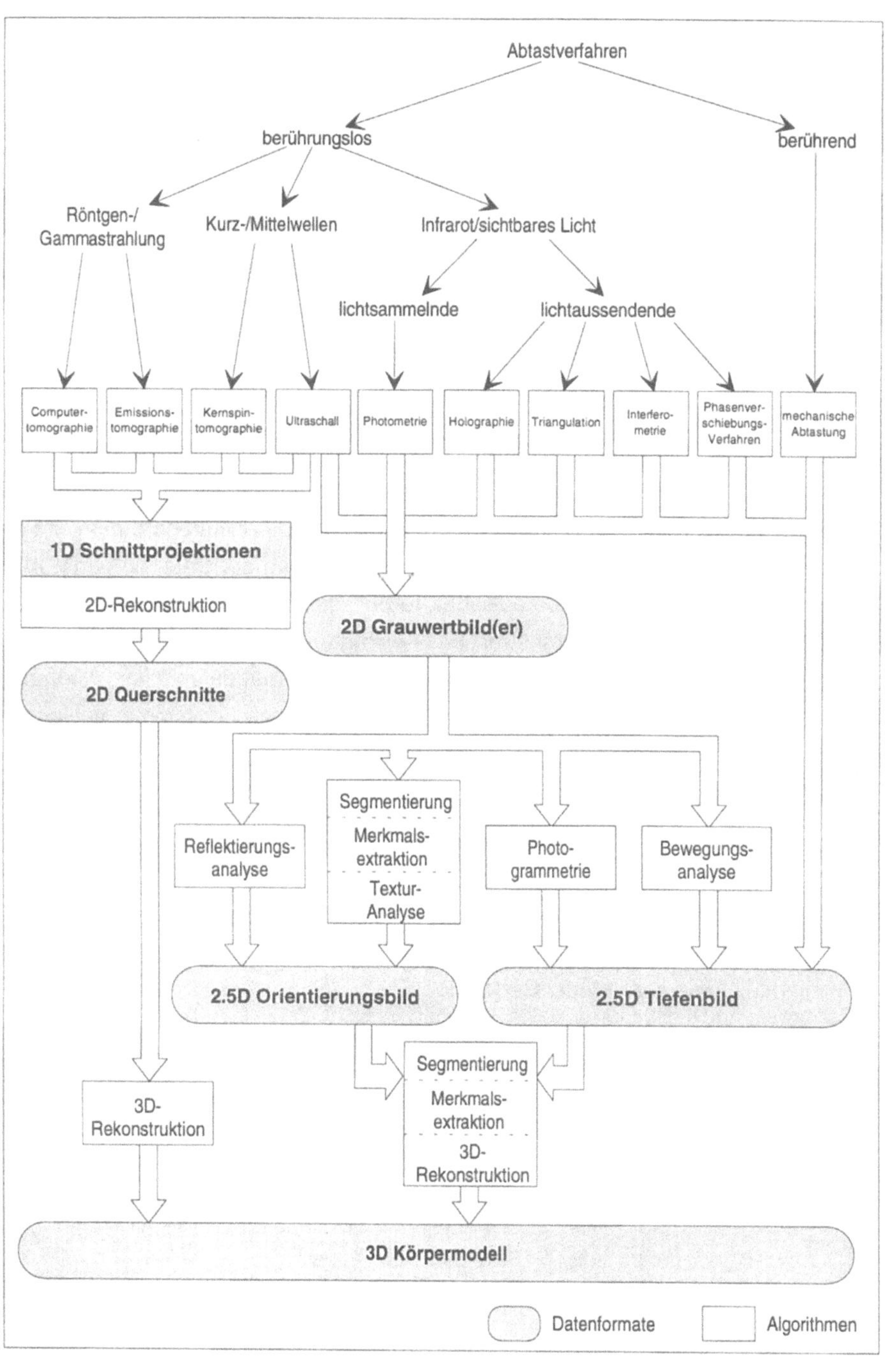

Bild 8: Klassifizierung der Abtastverfahren

Wird das digitalisierte Objekt nicht nur aus einer Ansicht, sondern aus verschiedenen überlappenden Seiten abgetastet, dann ist es möglich durch speziell entwickelten Algorithmen das gesamte Objekt (Volumen) zu rekonstruieren, Bild 6 illustriert wie eine Kontur aus verschiedenen Laserpositionen (Pos) abgetastet wird. Durch ein Strahlungszuordnungsproblem wird aus den Laserstrahlen (Str) und den zugeordneten digitalisierten Punkten (Pkt), die gesuchte eindeutige Kontur berechnet. Die Abbildung 7 zeigt beispielhaft das rekonstruierte Volumenmodell eines Glaskopfes, der aus mehreren Ansichten digitalisiert wurde.

6 Zusammenfassung

Die vorgestellten Forschungsergebnisse zeigen, daß der 4D-Laser Scanner direkt Daten liefert, die für den Einsatz von Mustererkennungsmethoden geeignet sind. Erst die Mustererkennung erlaubt eine Reduzierung der Datenflut und ermöglicht die in der Praxis geforderte schnelle und effektive Bearbeitung der Daten in CAD/CAM-Systemen. Die Ausführungen verdeutlichen ferner, daß durch die Mustererkennung auch die eindeutige Volumenrekonstruktion erleichtert und beschleunigt wird.
Obwohl die vorgestellten Algorithmen mit Hilfe der Meßdaten des 4D-Laser-Scanners erprobt wurden, lassen sie sich auf alle 3D-Digitalisiersysteme, anwenden Bild 8, die Entfernungsinformationen bereitstellen.

7 Literatur:

[1] C. M. Schuster: Designing with Laser Technology, Photonics Spectra, Jan. 1991

[2] H. J. Tiziani: Optische Verfahren: Abtastmessung berührungslos. Industrie-elektrik+Elektronik, 34. Jahrgang Nr. 4, 1989.

[3] Pritschow, G.; Hartl, Ph.; Ioannides, M. Wehr, Al.: 4D-Laser Scanning Followed by internal Computer Model Generation. Proceedings of the XVIIth ISPRS Congress, XXIX, B5 Commission V, 1992, S. 522-527.

[4] Hoschek und Lasser: Grundlagen der geometrischen Datenverarbeitung. B. G.Teubner Verlag, Stuttgart, 2. Auflage. 1992

Interpretationsgetriebene Parameteradaption und Verfahrenswahl in der Szenenanalyse

M. Kilger
Siemens AG, ZFE ST SN 3
Otto-Hahn-Ring 6
81739 München

Eine hierarchisch gegliederte Architektur zur Analyse von sehr langen Bildfolgen aus natürlichen Szenen wird vorgestellt. Parallel zur Auswertung findet eine ständige Bewertung der Güte der Ergebnisse statt. Das Resultat wird zu den niederen Verarbeitungsebenen zurückgekoppelt und zur Anpassung der Parameter bzw. Algorithmen an die gegebenen Rahmenbedingungen genützt. Durch diese Maßnahme können sehr lange Bildszenen, die auch sehr starken Änderungen unterworfen sein können (z.B. rapiden Beleuchtungsänderungen), bearbeitet werden. Dieses Prinzip wird anhand der videobasierten Verkehrserfassung mit einem in Echtzeit laufenden low-cost System gezeigt.

1. Einleitung

Zur Auswertung von langen Szenen ist eine Anpassung der Auswertealgorithmen und deren Parameter an die entsprechenden Rahmenbedingungen nötig. Dazu gibt es grundsätzlich zwei verschiedene Ansätze: Auswertung des Eingangsbildes hinsichtlich dessen Rauscheigenschaften und Benutzung der Ergebnisse aus der Szeneninterpretation.

[Vögtle & Jäger 90] nehmen für den ersten Fall an, daß es sich um ein normal verteiltes Rauschen handelt. Zur Schätzung des Binarisierungsschwellwertes verwenden sie dann ein Gauß-Markov Modell. Dieses Verfahren zur Parameterschätzung basiert auf der low-level Signalverarbeitung. Es ist rechentechnisch teuer und hängt stark von der Güte der Schätzung ab.

Die zweite Methode zur Adaption an die betrachtete Szene wenden z.B. [Kohl et al 87] an. Sie entwickelten dazu ein System GOLDIE (Goal Directed Intermediate-Level Executive), in dem ein "low-level processing controller" definiert wird. Nach Segmentierung eines Einzelbildes wird eine Hypothese aufgestellt. Mit einer Resegmentierung des Bildes wird dann versucht, diese Hypothese zu bestätigen (z.B. einen Strauch von einem Baum zu trennen).

In dieser Arbeit wird in gewisser Weise dieser Ansatz auf lange Sequenzen erweitert. Zur Verdeutlichung wird die Analyse von Verkehrsszenen verwendet.

In diesem Bereich ist eine Analyse der Szenen über sehr lange Zeiträume nötig, in denen sich durch äußere Einflüsse (z.B. ändernde Beleuchtungs- oder Verkehrsverhältnisse) die Rahmenbedingungen für die Auswertung der Bilder stark verändern können. Um zu guten Ergebnissen zu gelangen muß ein entsprechendes Anpassen an diese Bedingungen stattfinden. Es wird das Prinzip der interpretationsbezogenen Anpassung gewählt. Mit diesem Prinzip können nicht nur Parameter angepaßt werden, sondern szenenabhängig die bestpassenden Algorithmen gewählt werden. Ferner werden nicht nur Einzelbilder ausgewertet, sondern ganze Bildsequenzen müssen bearbeitet werden. Die Interpretation der bisherigen Sequenz wird dazu benutzt, künftig effizienter zu analysieren.

Videobasierte Verkehrserfassung wurde bereits von mehreren Autoren untersucht. Eine Gruppe von Autoren [Baker & Sullivan 92], [Bielik & Abramczuk 89], [Blosseville et al 89], [Dickinson & Wan 89], [Hoose & Willumsen 87], [Hoose 89], [Houghton et al 89], [Koller 92], [Leutzbach et al 87] wertete (wohl auf Grund des hohen Rechenaufwands) nur sehr kurze Videosequenzen aus. Die Analyse dieser Sequenzen erfordert keine Parameteranpassung oder

Verfahrenswechsel, da für diese kurzen Sequenzen die Parameter so eingestellt werden können, daß sie über den betrachteten Zeitraum gute Ergebnisse liefern. Nur in [Michalopoulos 91] und [Kelly 92] werden lange Szenen an einem Stück ausgewertet. Diese Autoren erwähnen auch die Wichtigkeit der Parameteranpassung. Es werden aber keinerlei Algorithmen für deren Anpassung vorgestellt.

In der hier vorgestellten Arbeit wurde das bestehende Bildverarbeitungssystem [Feiten et al 91], [Kilger 92a] zu einem adaptiven System erweitert, das on-line zwischen den bestpassenden Algorithmen wechseln und low-level Parameter an die Szene anpassen kann. Es wird high-level Szenenwissen benutzt, um diese Algorithmen bzw. Parameter effizient steuern zu können.

In Kapitel 2 wird die Systemarchitektur mit den Rückkopplungen erklärt. Anhand der Verfahren (Kapitel 3) werden in Kapitel 4 die Adaptionsmechanismen an Beispielen gezeigt. Die verwendete Implementierung wird in Kapitel 5 vorgestellt, und in Kapitel 6 werden die Ergebnisse dieser Verfahren gezeigt.

2. Systemarchitektur

Die Systemarchitektur (Fig. 2.1) ist hierarchisch gegliedert und beinhaltet sowohl Prozeßmodule als auch Parametermodule. Sie ist in fünf Hierarchiestufen eingeteilt.

Die niedrigste Stufe H1 beinhaltet alle low-level Bildverarbeitungsroutinen.

In der Stufe H2 sind links die drei Ergebnismodule (*Spuren*, *Evaluierung* und *Hintergrund*) der low-level Verarbeitung angeordnet. Diese Module stellen die Ergebnisse des augenblicklich ausgewerteten Bildes dar, d.h. die Ergebnisse einer Einzelbildauswertung.

- Das Modul *Spuren* beinhaltet alle z.Zt. im Bild befindlichen Fahrzeuge, deren Position und Geschwindigkeit.
- Das Modul *Evaluierung* beinhaltet die Plausibilität der detektierten bewegten Objekte. Diese Bewertung ist die Grundlage für eine flexible und intelligente Steuerung der Parameter und der Algorithmen aus Stufe H1.
- Als weiteres Ergebnis liegt das aktuelle *Hintergrundbild* vor. Dieses wird durch den Bewegungsdetektionalgorithmus berechnet und für den nächsten Bearbeitungszyklus verwendet. Ferner kann dieses Hintergrundbild für weitere Algorithmen zur Szeneninterpretation verwendet werden (siehe z.B. Schattenanalyse).

Auf der rechten Seite sind die Steuerungsmodule *Arbeitspunkt* und *Verfahrenswahl* angezeigt.

- Das Modul *Arbeitspunkt* beinhaltet die Werte aller zu diesem Zeitpunkt gültigen Parameter der low-level Algorithmen (z.B. Schwellwert der Bewegungsdetektion).
- Das Modul *Verfahrenswahl* gibt die zu verwendenden Algorithmen an.

Mit diesen beiden Modulen können zum aktuellen Zeitpunkt die bestpassenden Parameter der einzelnen Routinen eingestellt werden, aber auch aufgrund von wechselnden Rahmenbedin-

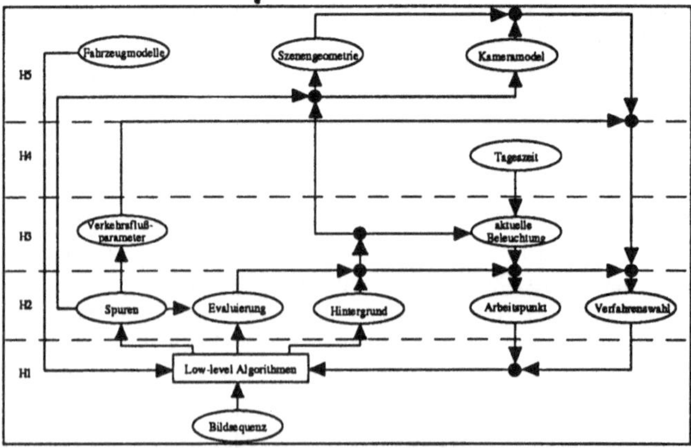

Fig. 2.1: Systemarchitektur

gungen ganze Routinen ausgetauscht werden. Somit wird die Systemperformance hinsichtlich äußerer Einflüsse wie z.B. wechselnden Beleuchtungsverhältnissen, aber auch wechselnden Verkehrsbedingungen optimiert. Allgemein befinden sich auf dieser Ebene sehr schnell zu verändernde Parameter. Sie können ihren Wert von einem Bild zum nächsten Bild ändern.

Module mit Parametern, die sich langsamer ändern, befinden sich auf Hierarchieebene H3. Links ist das Ergebnis der Spuren über einen längeren Zeitraum in die *Verkehrsflußparameter* eingetragen. Zu den Verkehrsflußparametern zählen z.B. mittlere Fahrzeugdichte, mittlere Geschwindigkeit, etc. Die Verkehrsflußparameter können aber auch die aktuellen Parameter oder Verfahren beeinflussen. Bei sehr dichtem Verkehr oder sogar Stauungen ist ein wesentlich verschiedener Parametersatz, aber auch ein anderer Detektionsalgorithmus nötig.

Das Modul *aktuelle Beleuchtung* bestimmt ebenfalls signifikant die Wahl der Algorithmen und deren Parameter. Bei Sonnenschein sind andere Algorithmen nötig, um zufriedenstellende Ergebnisse zu erzielen.

In der nächsten Hierarchieebene H4 ist die *Tageszeit* eingetragen. Sie wird verwendet, um die *aktuelle Beleuchtung* zu steuern. Man denke hier nur an die Tages-Nacht-Übergänge.

In der höchsten Hierarchiestufe H5 sind die statischen oder quasi-statischen Parameter angesiedelt. Die *Fahrzeugmodelle* sowie die *Szenengeometrie* und das *Kameramodell* werden sich relativ wenig oder fast nie ändern.

Die Pfeile in Fig. 2.1 stehen für verschiedene Abhängigkeiten zwischen den einzelnen Modulen. Sie werden für Adaptionsmechanismen verwendet. In Abhängigkeit der verschiedenen Situationen werden die bestpassenden Parameter bzw. Verfahren ausgewählt.

3. Verfahren

3.1. Bewegungsdetektion

Objektbildgenerierung

Zur Bewegungsdetektion sind zwei Algorithmen implementiert. Sie sind in ihrer Struktur ähnlich; beide basieren auf einem Hintergrundbild, das laufend an den wechselnden Lichtverhältnissen angepaßt wird.

Der erste Algorithmus (Objektbildgenerierung 1) [Karmann & Brandt 90] liefert gute Ergebnisse für leichten bis mittleren Verkehr, während der zweite (Objektbildgenerierung 2) für dichten Verkehr gute Ergebnisse liefert. Im Laufe der Arbeit wurde festgestellt, daß ein Schalten zwischen den beiden Algorithmen zu den besten Ergebnissen führt.

Objektbildgenerierung 1:

Zur Bewegungsdetektion wird ein binäres Objektbild erzeugt. Dies geschieht durch die Subtraktion des Hintergrundbildes H (Fig. 3.2) vom aktuellen Eingangsbild E (Fig. 3.1). Der Absolutwert dieser Differenz wird einer Schwellwertentscheidung mit der Schwelle S unterworfen, daraus wird das binäre Objektbild O (Fig. 3.3) erzeugt:

Fig. 3.1: Eingangsbild

Fig. 3.2: Hintergrund

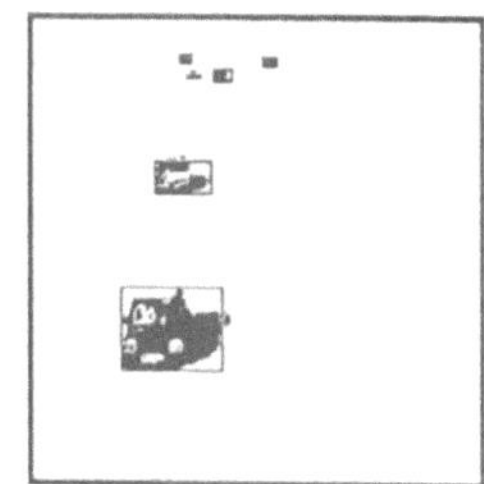

Fig. 3.3: Objektbild

$$\begin{aligned} |E_k(p) - H_k(p)| &> S_k(p) \quad \Rightarrow \quad O_k(p) = 1 \\ |E_k(p) - H_k(p)| &\leq S_k(p) \quad \Rightarrow \quad O_k(p) = 0 \end{aligned} \tag{3.1}$$

Dabei bedeutet k den Zeitpunkt und p die Position.

Objektbildgenerierung 2:
Der zweite Detektionsalgorithmus [Kilger 93] funktioniert ähnlich. Es wird wieder das Hintergrundbild vom aktuellen Eingangsbild subtrahiert. Allerdings wird nicht der Absolutwert der Differenz der Schwellwertentscheidung unterworfen, sondern nur der dunkle Anteil der Differenz als signifikant betrachtet.

$$\begin{aligned} H_k(p) - E_k(p) &> S_k(p) \quad \Rightarrow \quad O_k(p) = 1 \\ H_k(p) - E_k(p) &\leq S_k(p) \quad \Rightarrow \quad O_k(p) = 0 \end{aligned} \tag{3.2}$$

Dieser Ansatz beruht auf der Tatsache, daß die Fahrzeuge in der Regel einen dunklen Schatten werfen, selbst bei diffusen Lichtverhältnissen. Dieser Schatten befindet sich auf der Straße zwischen den vorderen Reifen. Dieser und die Reifen selbst, bilden ein dunkles Gebiet, das sich signifikant vom Hintergrund unterscheidet. Ferner kann dieser Bereich genau einem Fahrzeug zugeordnet werden. Der Vorteil dieses Detektionsverfahrens liegt darin, daß selbst bei dichtem Verkehr die Fahrzeuge noch einzeln detektiert werden können (zumindest im vorderen Bildbereich), obwohl sie sich bereits gegenseitig überlappen. Somit können Fahrzeuge und Segmente eindeutig einander zugeordnet werden. Das spätere Verfolgen der Fahrzeuge wird somit erheblich erleichtert. Nachteil dieses Verfahrens ist, daß nicht das komplette Fahrzeug detektiert wird.

Zusammenfassen der Bewegtbereiche
Im nächsten Bearbeitungsschritt werden die Zusammenhangskomponenten (Segmente) des Objektbildes O berechnet und durch ihr umschreibendes Rechteck dargestellt.

Segmentfilterung
Die Segmente werden entsprechend ihre Größe und Breite gefiltert. Dabei wird die Breite in Weltkoordinaten berechnet. Dies geschieht mittels der inversen perspektivischen Transformation unter der Annahme, daß sich die Fahrzeuge auf der Höhe Z=0 befinden.

Hintergrundadaption
Das Hintergrundbild wird laufend durch rekursive Filterung adaptiert [Karmann & Brandt 90]. Die Bereiche, die als bewegt erkannt wurden (O(p)=1), und die Bereiche, die als unbewegt erkannt wurden (O(p)=0), werden verschieden behandelt. Das geschätzte Hintergrundbild für den nächsten Bearbeitungszyklus erhält man durch eine gewichtete Addition des Differenzbildes (von Eingangsbild und altem Hintergrundbild) und des alten Hintergrundbildes:

$$H_{k+1} = H_k + \alpha * (E_k - H_k) \tag{3.3}$$

wobei α der Updatekoeffizient ist.

3.2. Schattenabspaltung

Fig. 3.4: Segment mit Kantenhistogramm

Zur Klassifikation der Fahrzeuge nach Breite (siehe Kap. 3.4) ist eine Detektion der Fahrzeuge ohne seitlichen Schatten notwendig, um Fehlklassifikationen zu minimieren. Dazu wurde ein Algorithmus [Kilger 92b] entworfen, der diesen Schattenbereich vom Fahrzeug erkennt und zur weiteren Verarbeitung vom Fahrzeug trennt. Zur Detektion des prinzipiellen Auftretens von Schatten und dessen Richtung siehe Kap. 4.2.1.

Bei der Schattenabspaltung wird davon ausgegangen, daß Fahrzeuge signifikante Kanten haben, während das Schattenbild im Innern in der Regel kantenfrei ist. Je nach Schattenrichtung (lateral oder vor dem Fahrzeug) wird ein vertikales bzw. horizontales Kantenbild des Eingangsbildes und des Hintergrundbildes berechnet. Diese Kantenbilder werden zu Histogrammen aufsummiert, diese dann voneinander subtrahiert. Durch die Differenz werden die Kanten aus dem sichtbaren Bereich des Hintergrundbildes (z.B. Fahrbahnmittelstreifen) eliminiert. Im Bereich des Schattens sind die Kanten weit weniger stark ausgeprägt als die Kante zwischen Schatten und Fahrzeug (siehe Fig. 3.4). Die Position dieser Kante wird berechnet und als Grenze Fahrzeug - Schatten interpretiert.

3.3. Verfolgung

Das Ziel der Verfolgung ist, den aktuell detektierten Segmenten eine Spur zuzuordnen, d.h. eine Information über das Verhalten der einzelnen Fahrzeuge über die Zeit zu erhalten.
Der ganze Verfolgungalgorithmus basiert auf Kalman-Filterung mittels des Bewegungsmodells der unbeschleunigten Bewegung. Der Zustandsvektor beinhaltet somit X- und Y-Position des Fahrzeuges in Weltkoordinaten und deren Geschwindigkeiten.
Das Zuordnungsmerkmal für die Verfolgung ist die Mitte der vorderen Kante des umschreibenden Rechtecks. Dieses Merkmal ist einfach und robust zu detektieren und gut zu verfolgen. Nach einigen Messungen ist die Differenz zwischen der prädizierte Position und der gemessene Position sehr gering.

3.4. Klassifikation

Die Fahrzeuge werden in drei Klassen eingeteilt:
- Lastkraftwagen, Busse,
- Personenkraftwagen und
- Motorräder, Fahrräder.

Die Fahrzeuge werden anhand der detektierten Breiten klassifiziert. Diese Breite wird über mehrere Detektionen gefiltert. In den untersuchten Sequenzen reichte die Klassifikation nach der Breite aus.

4. Adaptionsmechanismen

Bei der Analyse von Verkehrszenen über längere Zeiträume müssen die Parameter der einzelnen Algorithmen den äußeren Rahmenbedingungen angepaßt werden, um gute Ergebnisse zu erzielen [Kilger & Dietl 93]. Es muß sogar die Auswahl der Verfahren bestimmt werden, um den verschiedenen Rahmenbedingungen gerecht zu werden [Kilger 92b], [Kilger 93].

4.1. Parameteradaption

Die Güte des detektierten binären Objektbildes hängt signifikant von der Binarisierungsschwelle S und dem Hintergrund H ab. Liegt z.B. die Schwelle S zu tief, werden sehr viele uninteressante Regionen mit signifikanter Bewegung erkannt (z.B. Blätterrauschen), die dann im nachfolgenden Verarbeitungsschritt wieder eliminiert werden müssen, weil sie zu klein für ein relevantes Objekt sind. Ist das Hintergrundbild verrauscht, wird auch das Objektbild verrauscht sein, d.h. sowohl die Schwelle S als auch die Güte des Hintergrundbildes H, das wiederum durch den Koeffizienten α adaptiert wird, sind ausschlaggebend für die Qualität des Objektbildes O.

Adaption der Binarisierungsschwelle S
Zur Regelung der Schwelle S wird die Szeneninterpretation herangezogen, d.h. es werden die detektierten bewegten Objekte zu den relevanten bewegten Objekten in Bezug gesetzt. Diese beiden Ergebnisse sind vorhanden: zum einen das direkte Ergebnis aus der Segmentierung und zum anderen das Ergebnis nach der Filterung der Segmente. Der Quotient Q der beiden Größen ist ein Qualitätsmaß für die Güte der Schwelle S:

$$Q = \frac{\textit{Anzahl der relevanten Segmente}}{\textit{Anzahl der ungefilterten Segmente}} \qquad (4.1)$$

Ist die Schwelle zu niedrig, wird der Quotient sehr klein, da sehr viele kleine Segmente detektiert wurden und wieder gefiltert werden müssen. Ist die Schwelle sehr hoch, werden die Fahrzeuge nur teilweise oder gar nicht mehr detektiert. Es wird deshalb in einer langsamen Kontrollschleife der Schwellwert S laufend reduziert, bis der Quotient beginnt, zu klein zu werden. Danach wird er wieder erhöht. Dies ist eine einfache, aber auch eine sehr leistungsfähige Methode, den Schwellwert S abhängig von den äußeren Einflüssen stabil zu regeln.

Adaption der Hintergrundadaptionskoeffizienten
Die Hintergrundadaption findet mit zwei verschiedenen Parametern statt. Die Regionen mit signifikanten Bewegungen werden mit einem Koeffizienten $\alpha=\alpha_1$ gewichtet, während die Regionen mit langsamer bzw. keiner Bewegung mit einem Koeffizienten $\alpha=\alpha_2$ gewichtet werden.

Es wurde festgestellt, daß bei Kopplung der Adaptionskoeffizienten α_1 und α_2 an die Binarisierungsschwelle S (also ebenfalls an die Szeneninterpretation) die besten Ergebnisse erzielt werden können. Es ausführliche Beschreibung findet man in [Kilger & Dietl 93].

4.2. Verfahrenswechsel

Aufgrund der notwendigen Analyse von Szenen sehr langer Dauer ist es sogar nötig, einzelne Verfahren zu wechseln, da sich die äußeren Umstände drastisch verändert haben. Das Problem liegt darin, diese Veränderungen zu erfassen, entsprechend zu reagieren, die richtigen Verfahren auszuwählen und dann einzusetzen. Als Beispiel werden hier zwei relevante Veränderungen diskutiert. Zum einem geht es um den Beleuchtungswechsel verursacht vom Übergang von diffusem Licht zu Sonnenschein. Zum anderen werden hier die Einflüsse des Verkehrsaufkommens auf den Einsatz verschiedener Detektionsverfahren aufgeführt.

Beleuchtungswechsel
Um die Fahrzeuge richtig klassifizieren zu können (siehe Kap. 3.4), muß das Auftreten von Schatten schnell und robust erkannt werden. Dazu wurde ein hierarchisch gegliederter Ansatz

gewählt. In der ersten Verarbeitungsebene wird das generelle Auftreten von Schatten untersucht. Ist Schatten ausgeschlossen (z.B. während der Nacht), wird auch keine Schattenanalyse vorgenommen. Setzen Bedingungen ein, die zum möglichen Auftreten von Schatten führen (z.B. Sonnenaufgang), wird ein Algorithmus eingesetzt, der laufend bestimmte Attribute überwacht. In unserer Anwendung ist das die Breite des detektierten Fahrzeuges. Wird eine Auffälligkeit festgestellt, wie z.B. eine übergroße Breite oder eine laufende Klassifikation aller bewegten Objekte als Lkws, wird der Schattenabspaltungsalgorithmus angewendet. In diesem Algorithmus wird laufend die Plausibilität der erhaltenen Resultate analysiert. Da die Richtung des Schattenwurfs zu diesem Zeitpunkt noch nicht bekannt ist, wird versucht, Schatten sowohl in den beiden lateralen Richtungen als auch nach vorne abzuspalten. Sind, nach einer definierten Observierungszeit, die Plausibilität über das Auftreten eines Schattens und dessen Richtung über einem Schwellwert, wird davon ausgegangen, daß es sich um eine Schattenszene handelt. Es wird versucht, aus der nun bekannten Schattenrichtung das Fahrzeug ohne den Schatten zu extrahieren. Es wird aber weiterhin die Plausibiltät der Resultate mitberechnet, um bei einem evtl. Verschwinden des Schattens wieder in die ursprüngliche Verarbeitungsebene zurückzuschalten. Gelten wieder die Grundvoraussetzungen von Schattenwurf nicht mehr (z.B. Sonnenuntergang), wird in die Stufe 1 zurückgeschaltet.

Verkehrsaufkommen

Wurde aus dem Modul Verkehrsflußparameter ein sehr hohes Verkehrsaufkommen registriert, wird on-line auf die Objektbildgenerierung 2 umgeschaltet, da diese wesentlich bessere Resultate für dichteren Verkehr liefert. Da diese Detektionsmethode aber Nachteile in der Hintergrundadaption hat, wird bei leichtem bis mittleren Verkehr auf die Objektbildgenerierung 1 zurückgeschaltet.

5. Implementation

Die Algorithmen wurden einem low-cost System implementiert. Als Hardware wird ein herkömmlicher PC mit i486 Prozessor mit zwei Einsteckplatinen, einem Frame-Grabber und einem Signalprozessorboard (Motorola DSP96002 Prozessor) verwendet. Die low-level Bildverarbeitungsroutinen (Detektion, Segmentierung und Hintergrundadaption) werden auf dem Signalprozessorboard berechnet, während der PC für die high-level Routinen (Verfolgung, Klassifikation und Bewertung) verantwortlich ist. Bis auf die zeitkritischen Algorithmen (Detektion, Hintergrundadaption und Teile der Segmentierung) in Assembler des DSP96002 sind alle Algorithmen in C implementiert. Die Bildauflösung beträgt 256x256 Pixel mit 256 Graustufen. Mit dieser Hardware und den oben erwähnten Algorithmen konnten Bildfolgefrequenzen von 3 bis 5 Hz erzielt werden (ohne harte Codeoptimierungen in C).

6. Ergebnisse

Die beschriebenen Algorithmen wurden ausführlich getestet. Aufnahmen von mehr als 10 verschiedenen Standorten wurden jeweils ca. 30 Minuten aufgezeichnet. Dabei wurden verschiedene Beleuchtungsverhältnisse (Schnee, Sonnenschein, Nebel und diffuses Licht) untersucht. Ferner wurden verschiedene Verkehrsverhältnisse, vom leichten bis dichtem Verkehr, betrachtet.

Die Parameter konnten on-line den verschiedenen Szenen angepaßt werden. Es konnten, abgesehen von der völlig verschiedenen Wahl der Parameterwerte bei verschiedenen Szenen, folgende Verbesserungen hinsichtlich einer festen Parameterwahl festgestellt werden:

- Die Spurlängen wurden um ca. 10% verlängert;

- Die Abweichung des Hintergrundes bei dichtem Verkehr gegenüber dem Hintergrund ohne Verkehr wurde um 60% reduziert;
- Es wurden 20% - 30% der Rechenleistung eingespart.

Durch die Wahl der verschiedenen Verfahren konnte bei Sonnenschein durch Abspalten des Schattens die Klasse des Fahrzeuges in 95% der Fälle richtig geschätzt werden. Durch Wechseln zwischen den beiden Detektionsalgorithmen wurde bei dichtem Verkehr die Detektion der Fahrzeuge sichtlich erhöht. Bei einem Verkehrsaufkommen von ca. 100 Fahrzeugen pro Minute konnten 98% der Fahrzeuge gezählt (nach Klasse geordnet) werden. Beim Einsatz des Detektionsalgorithmus 1 waren es nur 80%.

Die Effizienz unseres Ansatzes einer interpretationsgetriebenen Parameter- und Verfahrensadaption konnte also in der Verkehrsanalyse anschaulich gezeigt werden.

7. Literaturverzeichnis

[Baker & Sullivan 92] K.D. Baker, G.D. Sullivan, Performance Assessment of Model-Based Tracking, *IEEE Workshop on Applications of Computer Vision*, Palm Springs, CA, USA, Nov. 30 - Dec. 2, 1992.

[Bielik & Abramczuk 89] A. Bielik, T. Abramczuk, Real-time wide-traffic monitoring: information reduction and model-based approach, *Proc. 6th Scandinavian Conference on Image Analysis, pp. 1223-1230*, Oulu, Finland, 1989.

[Blosseville et al 89] J.M. Blosseville, C. Krafft, F. Lenoir, TITAN: a traffic measurement system using image processing techniques, *Second International Conference on Road Traffic Monitoring*, London, UK 7-9 Feb. 1989.

[Dickinson & Wan 89] K.W. Dickinson, C.L. Wan, Road traffic monitoring using the TRIP II system, *Second International Conference on Road Traffic Monitoring*, London, UK 7-9 Feb. 1989.

[Feiten et al 91] W. Feiten, A.v. Brandt, G. Lawitzky, I. Leuthäusser, Ein videobasiertes System zur Erfassung von Verkehrsdaten, *Proc. 13th DAGM 1991*, pp. 507-514, München, 1991.

[Hoose & Willumsen 87] N. Hoose, L.G. Willumsen, Automatically extracting traffic data from videotape using CLIP4 parallel image processor, *Patt. Recog. Letters, 6, pp. 199-213*, 1987.

[Hoose 89] N. Hoose, Queue detection using computer image processing, *Second International Conference on Road Traffic Monitoring*, London, UK 7-9 Feb. 1989.

[Houghton et al 89] A.D. Houghton, G.S. Hobson, N.L. Seed, R.C. Tozer, Automatic vehicle recognition, *Second International Conference on Road Traffic Monitoring*, London, UK 7-9 Feb. 1989.

[Karmann & Brandt 90] K.P. Karmann, A.v. Brandt, Moving object segmentation based on adaptive reference images, *Proc. of EUSIPCO 1990*, Barcelona, 1990.

[Kelly 92] D.M. Kelly, Results of a field trial of the IMPACTS image processing system for traffic monitoring, *Proc. of 6th International Conference on Road Traffic Monitoring and Control*, 28-30 Apr., 1992.

[Kilger 92a] M. Kilger, Video-based traffic monitoring, *Proc. of 4th International Conference on Image Processing and Its Applications*, pp 89-92, Maastricht/The Netherlands, Apr. 1992.

[Kilger 92b] M.Kilger, A Shadow Handler in a Video-based Real-time Traffic Monitoring System, *IEEE Workshop on Applications of Computer Vision*, Palm Springs / CA (USA), Nov-Dec 1992.

[Kilger & Dietl 93] M. Kilger, T. Dietl, Interpretation-Driven Low-Level Parameter Adaptation in Scene Analysis, *Computer Aided Systems Theory - EUROCAST '93. Third International Workshop. Proceedings*, Las Palmas de GC, Spain, Feb. 1993.

[Kilger 93] M. Kilger, Heavy Traffic Monitoring in Real-Time, *Proc. 8th Scandinavian Conference on Image Analysis*, Tromsoe, Norway, May 1993.

[Kohl et al 88] C.A. Kohl, A.R. Hanson, E.M. Riseman,Goal Directed Control of Low-Level Processes for Image Interpretation, *IU Proc. DARPA, Vol.2*, pp. 538-551, 1987.

[Koller 92] D. Koller, Detektion, Verfolgung und Klassifikation bewegter Objekte in monokularen Bildfolgen am Beispiel von Straßenverkehrsszenen, *Dissertation, Universität Karlsruhe*, 1992.

[Leutzbach et al 87] W. Leutzbach, H.P. Bähr, U. Becker, T. Vögtle: Entwicklung eines Systems zur Erfassung von Verkehrsdaten mittels photogrammetrischer Aufnahmeverfahren und Verfahren zur automatischen Bildauswertung, *Technischer Schlußbericht*, Universität Karlsruhe, 1987.

[Michalopoulos 91] P.G. Michalopoulos, Vehicle detection video through image processing: the Autoscope system, *IEEE Trans. Veh. Technol. (USA) vol. 40, no. 1*. Feb 1991.

[Vögtle & Jäger 90] T. Vögtle, R. Jäger, Detektion bewegter starrer Objekte in Realwelt-Bildfolgen mit Hilfe von Differenzbildern, *Proc. 12th DAGM 1990, pp.625-633*, Oberkochen-Aalen, Germany, 1990.

Segmentierung farbiger kartographischer Vorlagen in empfindungsgemäßen Farbräumen

B. Lauterbach und W. Anheier

Universität Bremen, FB1, Arbeitsgruppe Digitale Systeme
Postfach 330440, 28334 Bremen

1. Einleitung

In den letzten Jahren haben geographische Informationssysteme in Bereichen wie Raum-, Stadt- und Umweltplanung zunehmend an Bedeutung gewonnen. Ein wichtiger Aspekt des Einsatzes dieser Systeme ist die Akquisition der notwendigen Daten. Diese liegen häufig in Form gedruckter Karten vor und müssen in langwieriger Arbeit per Hand digitalisiert werden.

Um die Überführung der in der Karte enthaltenen Information in eine Datenbank effektiver zu machen, werden in letzter Zeit häufig halb- bzw. vollautomatische Methoden zur Vektorisierung und Analyse gescannter Karten eingesetzt. Für die Anwendung im Zusammenhang mit einfachen, schwarz-weißen Karten (z.B. Katasterkarten, Netzpläne) existieren bereits kommerzielle Systeme [Maye1991]. Weitere Systeme zur Verarbeitung binärer Kartentypen wurden in [Ille1990], [Suzu1990] und [Maye1992] vorgestellt.

Ein System zur automatischen Analyse farbiger Karten hoher Komplexität (topographische Karten TK25) wird in [Laut1992] und [Ebi1992] beschrieben. Ein essentieller Bestandteil eines solchen Systems ist eine leistungsfähige Methode zur Separierung der einzelnen Farbschichten der farbig abgetasteten Karte. Diese Segmentierung ist i. allg. der erste Schritt in einer Reihe anzuwendender Verfahren. Schlechte Ergebnisse in diesem Verarbeitungsschritt wirken sich drastisch auf den Erfolg der gesamten Kartenanalyse aus.

Im folgenden wird eine neues Verfahren zur Separierung der Farbschichten einer topographischen Karte vorgestellt, das auf dem empfindungsgemäßen CIE-L*u*v*-Farbraum und dessen u'v'-Farbtafel basiert [Laut1993]. Mit der vorgestellten Methode werden sehr saubere Segmentierungsergebnisse erreicht, die eine gute Ausgangsbasis für eine nachfolgende Vektorisierung und Analyse darstellen.

2. Auswahl eines geeigneten Farbsystems für die Segmentierung

Die Separierung der Farbschichten einer topographischen Karte ist ein Problem der Segmentierung von Farbbildern. Für diese Art von Aufgabenstellung ist die Auswahl eines geeigneten Farbsystems, in dem die zu segmentierenden Daten repräsentiert werden von großer Bedeutung [Bumb1987]. Ein geeignetes Farbsystem kann z.B. durch die Betrachtung des menschlichen visuellen Systems gefunden werden. Farb-Bildverarbeitungssysteme sollten ähnlich gute Eigenschaften zur Diskriminierung von Farben aufweisen, wie das menschliche Auge [Ohta1980].

Diese Forderung läßt sich nur in Ansätzen verwirklichen, da das menschliche Auge kein Farbreizempfänger mit festen Parametern ist, wie z.B. ein CCD-Sensor. Ungleichmäßigkeiten bzgl. Helligkeit und Farbtemperatur vermag das adaptive Auge problemlos auszugleichen [Bumb1987]. Für ein Farb-Bildverarbeitungssystem bedeutet der Ausgleich von sich ändernden Umfeldverhältnissen jedoch einen hohen algorithmischen Aufwand.

Ungeachtet der angesprochenen Adaptivität kann die Forderung nach einer möglichst dem menschlichen visuellen System entsprechenden Farbunterscheidungsfähigkeit aufrecht erhalten werden. Ein idealer Farbraum wäre demnach charakterisiert als ein Raum, in dem die Farbdistanz zweier Farbvalenzen proportional zu der vom Menschen empfundenen Farbdifferenz zwischen den durch die Farbvalenzen repräsentierten Farben ist [Mukh1986]. Diese wichtige

Eigenschaft wird von keinem der üblicherweise verwendeten Farbräume erfüllt (RGB, IHS, YIQ usw.)

Eine weitere wichtige Eigenschaft des Farbraums ist, daß im verwendeten Farbraum die Gesetze der additiven Farbmischung gelten müssen. Dieses kommt dann zum Tragen, wenn Berechnungen durchgeführt werden, die mehrere Pixel mit unterschiedlichen Farbvalenzen einbeziehen (z.B. Filterung). Nichtbeachtung der Additivität führt zu falschen Ergebnissen [Mahy1991].

Die in einer Reihe von Untersuchungen erzielten Ergebnisse [Ohta1980] [Gent1990] [Kass1991] [Mahy1991] zeigen, daß es kein in jeder Hinsicht optimales Farbsystem für die Segmentierung gibt. Ein besseres Verhalten in bezug auf Empfindungsgemäßheit, Luminanz-Kontamination oder weitere Eigenschaften wird mit einem erhöhten Rechenaufwand erkauft. Als günstigste Alternative erscheint die Verwendung des CIE-L*u*v*-Farbsystems zur Segmentierung, da dieses die genannten Forderungen erfüllt und mit einigermaßen tragbarem Rechenaufwand implementierbar ist. Das L*u*v*-System 1976 dient deswegen als Grundlage für die in den folgenden Abschnitten beschriebenen Algorithmen.

3. Umrechnung der vom Scanner erzeugten RGB-Daten in die Farbtafelkoordinaten u' und v'

Die vom Scanner erzeugten Daten liegen als 8-Bit-Farbvalenzen im System RGB vor. Da diese Daten keinerlei absoluten Bezug aufweisen, das L*u*v*-Farbsystem jedoch genormt ist, muß eine Umrechnung von RGB nach u'v' sowie eine Farbkorrektur durchgeführt werden:

1) Umrechnung von RGB in das Normvalenzsystem XYZ,
2) Korrektur der Normfarbwerte,
3) Umrechnung der korrigierten XYZ-Werte in das System *u'v'*.

3.1 Umrechnung in das Normvalenzsystem XYZ

Die von einem Scanner erzeugten Farbtripel besitzen keinen festen Bezug zu einem genormten Farbsystem, so daß sich keine wirklich sinnvolle Aussage über die Bedeutung eines Tripels mit bestimmten Grauwerten machen läßt. In diesem Zusammenhang ist es von großer Wichtigkeit, die Primärvalenzen des verwendeten Scanners farbmetrisch genau zu bestimmen, um damit einen Bezug zu einem genormten Farbsystem zu schaffen. Als geeignetes System bietet sich das international anerkannte XYZ-Normvalenzsystem der CIE an [CIE1986].

Für die Berechnung der Primärvalenzen ist die Bestimmung der spektralen Eigenschaften der einzelnen Systemkomponenten notwendig. Diese sind

- die Beleuchtungseinheit des Scanners mit ihrer spektralen Strahlungsverteilung,
- die drei Farbfilter des Scanners mit ihren jeweiligen Transmissionsgraden und
- der CCD-Sensor mit seinem Transmissionsgrad (spektrale Empfindlichkeit).

Aus den Eigenschaften der Einzelkomponenten können das Verhalten des Gesamtsystems und damit die benötigten Primärvalenzen X_F, Y_F und Z_F jedes Farbkanals F (R,G,B) berechnet werden [Rich1981]:

$$X_F = \frac{1}{k}\int_K^L \varphi_\lambda \cdot \bar{x}(\lambda)\,\mathrm{d}\lambda; \quad Y_F = \frac{1}{k}\int_K^L \varphi_\lambda \cdot \bar{y}(\lambda)\,\mathrm{d}\lambda; \quad Z_F = \frac{1}{k}\int_K^L \varphi_\lambda \cdot \bar{z}(\lambda)\,\mathrm{d}\lambda, \tag{1}$$

Für die Integrationsgrenzen gilt K=380 nm und L=780 nm. Dabei wird φ_λ aus der Transmissionsfunktion des Sensors $\tau_S(\lambda)$, der Transmissionsfunktion des jeweiligen Farbfilters $\tau_F(\lambda)$ der Farbe F und der spektralen Strahlungsverteilung der Lampe S_λ berechnet:

$$\varphi_\lambda = S_\lambda \cdot \tau_S(\lambda) \cdot \tau_F(\lambda). \tag{2}$$

Der Maßstabsfktor k dient zur Normierung, so daß für eine ideal weiße Fläche die Normfarbvalenz Y=100 wird. Unter dieser Voraussetzung ergibt sich

$$k = \frac{1}{100} \int_K^L S_\lambda \cdot \bar{y}(\lambda)\, \mathrm{d}\lambda. \tag{3}$$

Die Transformation der vom Scanner erzeugten Daten in das Normvalenzsystem kann mit Hilfe einer linearen Abbildung durchgeführt werden. Jeder Normfarbwert ergibt sich dabei aus den Komponenten der Scanner-Primärvalenzen der entsprechenden Normvalenz, die mittels der auf normierten Grauwerte der Bildkanäle skaliert werden:

$$\begin{pmatrix} X \\ Y \\ Z \end{pmatrix} = \begin{pmatrix} X_R & X_G & X_B \\ Y_R & Y_G & Y_B \\ Z_R & Z_G & Z_B \end{pmatrix} \cdot \begin{pmatrix} R/255 \\ G/255 \\ B/255 \end{pmatrix} \tag{4}$$

3.2 Farbkorrektur der Normfarbwerte

Nachdem die Farbvalenzen der Bildpunkte in den Koordinaten XYZ vorliegen, müssen sie einer Farbkorrektur unterzogen werden. Die Korrektur der durch den Scanner bedingten Farbfehler kann aufgrund der nicht vollständig bekannten Parameter des Systems nicht direkt berechnet werden. Vielmehr ist eine Umrechnungsvorschrift zu finden, indem die vom Scanner aufgrund der Abtastung genormter Vorlagen gelieferten Farbvalenzen mit den bekannten Farbvalenzen der Vorlage verglichen werden. Die Umrechnungsvorschrift ist dann so zu modifizieren, daß der Fehler möglichst klein wird. Als Normvorlagen kamen 22 NCS-Farbmuster zum Einsatz [SS019100].

Das verwendete Verfahren stammt aus dem Bereich *Remote Sensing* und dient dort zur zweidimensionalen geometrischen Korrektur von Satelliten-Bilddaten [Rich1986]. Zur Anwendung für die Farbkorrektur ist eine Erweiterung auf den dreidimensionalen Fall notwendig. Es werden hier Normfarbwerte im XYZ-Farbraum aufeinander abgebildet, wobei die Abbildung für die Koordinaten X, Y und Z separat zu berechnen ist. Die Berechnungsvorschriften lauten allgemein formuliert

$$\mathbf{x} = \mathbf{F} \cdot \mathbf{a}, \quad \mathbf{y} = \mathbf{F} \cdot \mathbf{b} \quad \text{und} \quad \mathbf{z} = \mathbf{F} \cdot \mathbf{c}, \tag{5}$$

wobei **x**, **y** und **z** die Vektoren mit den Normfarbwerten der Testmuster sind, **a**, **b** und **c** die Vektoren mit den Korrekturkoeffizienten und **F** eine Matrix, die sich aus den Normfarbwerten ergibt, die aus den Rasterbildern der abgetasteten Testmuster errechnet wurden. Mit der Einschränkung, daß das Gleichungssystem stets lösbar bleiben muß, kann die Anzahl der Gleichungen frei gewählt werden. Somit entsteht ein überbestimmtes Gleichungssystem, falls die Anzahl der Gleichungen größer ist als die Anzahl der Spalten der Matrix **F**. Bei der vorliegenden Anwendung können somit alle 22 Farbmuster zur Bestimmung der Koeffizienten einbezogen werden. Um aus den Werten der Testmuster die besten Koeffizienten zu ermitteln, wird das überbestimmte Gleichungssystem mit Hilfe der Methode der kleinsten Quadrate in ein m x m-Gleichungssystem überführt. Anschließendes Lösen des Systems führt zu den gesuchten Korrekturkoeffizienten.

3.3 Umrechnung in das System u'v'

Die Berechnung der Farbwerte u' und v' aus den Normvalenzen X, Y und Z erfolgt mittels der Gleichungen

$$u' = \frac{4 \cdot X}{X + 15 \cdot Y + 3 \cdot Z} \quad \text{und} \quad v' = \frac{9 \cdot Y}{X + 15 \cdot Y + 3 \cdot Z}. \tag{6}$$

Die in Abschnitt 3. beschriebenen Schritte 3) und 4) erfordern jeweils das Einsetzen der Werte in ein lineares Gleichungssystem. Durch Zusammenfassen der diesen beiden Schritten zugrundeliegenden Gleichungen sowie der Umrechnung XYZ-u'v' kann eine einfache Beziehung zwischen den vom Scanner erzeugten Grauwerten und den Farbartkoordinaten u' und v' hergestellt werden. Für die direkte Berechnung der Werte u' und v' aus den Scanner-Rohdaten R, G und B

ergibt sich unter Einbeziehung der für den verwendeten Scanner OPTOTECH Optoscan 2000 berechneten Koeffizienten

$$u' = \frac{(0{,}6964 \cdot R + 0{,}462 \cdot G + 0{,}1812 \cdot B)}{(1{,}3531 \cdot R + 4{,}2294 \cdot G + 1{,}1052 \cdot B)} \quad \text{und} \quad v' = \frac{(0{,}7308 \cdot R + 2{,}4336 \cdot G - 0{,}072 \cdot B)}{(1{,}3531 \cdot R + 4{,}2294 \cdot G + 1{,}1052 \cdot B)}. \tag{7}$$

In diesen beiden Gleichungen können jeweils der Zähler und der Nenner durch drei Zugriffe auf Look-Up-Tabellen und zwei Additionen gebildet werden, wobei der Nenner nur einmal zu bilden und dann in beiden Gleichungen zu verwenden ist. Für die hierzu erforderlichen Operationen sind sechs eindimensionale Tabellen mit jeweils 256 Einträgen erforderlich. Das gewählte Verfahren erwies sich als schnell und verbraucht wenig Speicherplatz.

4. Segmentierung - Separierung der Farbschichten der gescannten Karte

Die in der Literatur beschriebenen Verfahren zur Segmentierung von Farbbildern weisen mehr oder weniger große Nachteile auf. Dies sind z.B. die Verwendung nicht empfindungsgemäßer Farbräume [Mass1990] [Andr1990], notwendige manuelle Vorgabe der Clusteranzahl [Bala1991] [Kuma1989] oder Verwendung nicht additiver Farbräume [Cele1990] [Tomi1992].

Das hier beschriebene Verfahren zur Segmentierung der abgetasteten topographischen Karten hat keinen dieser Nachteile. Als wesentliche Vorteile sind zu nennen:

- Das Verfahren arbeitet vollautomatisch ohne Vorgabe von Clusterzentren oder -anzahl,
- Berücksichtigung von additiv entstandenen Mischfarben an Farbübergängen,
- Farben geringer Häufigkeit werden nicht unterdrückt, sondern können auch separate Cluster bilden,
- stark streuende Bereiche werden nicht in unnötig viele Cluster unterteilt.

Das vorgestellte Verfahren ist jedoch nicht universell zur Segmentierung von Farbbildern einsetzbar. Es ist speziell für die Verwendung im Zusammenhang mit Rasterbildern vollfarbig gedruckter Dokumente entwickelt worden (z.B. topographische Karten). In der hier beschriebenen Version können Helligkeitsunterschiede nur bei den unbunten Farben festgestellt werden. Prinzipiell ist jedoch eine Erweiterung des Verfahrens für eine helligkeitsmäßige Separierung bunter Farben denkbar. Die Methode läßt sich in vier Schritte unterteilen:

- Detektion von Maxima (Clusterzentren) und Graten (Farbverläufe durch additive Mischung) im zweidimensionalen Histogramm auf Basis der u'v'-Farbtafel,
- Segmentierung des Bildes nach dem Farbort unter Berücksichtigung der additiven Mischgesetze,
- Segmentierung des unbunten Clusters nach der Helligkeit der Bildpunkte,
- Korrektur evtl. vorhandener Fehler in den separierten Farbschichten.

4.1 Detektion von Maxima und Graten im u'v'-Histogramm

Ein in [Seza1990] vorgestellter Algorithmus zur Detektion von Maxima in 1D-Histogrammen wurde für zweidimensionale Histogramme erweitert. Für die zweidimensionale Summenhistogrammfunktion $s(u'_q,v'_q)$ ergibt sich

$$s(u'_q,v'_q) = \sum_{i=0}^{u'_q} \sum_{j=0}^{v'_q} h(i,j). \tag{8}$$

Das zweidimensionale Maxima-Detektions-Signal $r_N(u'_q,v'_q)$ ergibt sich als Differenz des Summenhistogrammwertes $s(u'_q,v'_q)$ und des Mittelwertes eines ihn umgebenden Fensters:

$$r_N(u'_q,v'_q) = s(u'_q,v'_q) - \frac{1}{N^2} \cdot \sum_{i=u'_q-\frac{q\cdot N}{2}}^{u'_q+\frac{q\cdot N}{2}} \; \sum_{j=v'_q-\frac{q\cdot N}{2}}^{v'_q+\frac{q\cdot N}{2}} s(i,j). \tag{9}$$

Die Auswahl der Seitenlänge N des Fensters ist in der Praxis unkritisch. Für die aufgeführten Beispiele wurde N jeweils auf 1/4 der längeren Seite des belegten Histogrammbereiches gesetzt, wobei darauf zu achten ist, daß N ungerade ist. Anhand der nun erhaltenen Funktion $r_N(u'_q,v'_q)$ lassen sich die Maxima folgendermaßen erkennen:

$$f\left(r_N\left(u'_q,v'_q\right)\right)=\begin{cases}1 & \text{wenn } \left(r_N\left(u'_q,v'_q\right)\geq 0\right)\wedge\left(r_N\left(u'_q,v'_q-q\right)<0\right)\wedge \\ & \left(r_N\left(u'_q-q,v'_q\right)<0\right)\wedge\left(r_N\left(u'_q-q,v'_q-q\right)<0\right). \\ 0 & \text{sonst}\end{cases} \tag{10}$$

Wenn gilt $f(r_N(u'_q,v'_q))=1$, entspricht der Histogrammeintrag einem Maximum.

Zur Detektion von Gratpunkten, die ein Histogrammaximum in nur einer Richtung darstellen, muß die zweidimensionale Funktion $s(u'_q,v'_q)$ durch eine Berechnungsvorschrift $s'(u'_q,v'_q)$ ersetzt werden, die sich aus dem Produkt der eindimensionalen Summenhistogrammfunktionen jeder Zeile bzw. Spalte des Histogramms ergibt. Diese werden folgendermaßen berechnet:

$$s'_u\left(u'_q,j\right)=\sum_{j=0}^{u'_q}h(i,j) \quad \text{und}\; s'_v\left(i,v'_q\right)=\sum_{i=0}^{v'_q}h(i,j). \tag{11}$$

Dabei kennzeichnen i und j jeweils die Histogrammzeile bzw. -spalte. Weiterhin ergibt sich

$$s'\left(u'_q,v'_q\right)=s'_u\left(u'_q,v'_q\right)\cdot s'_v\left(u'_q,v'_q\right). \tag{12}$$

Der wesentliche Unterschied der hier definierten Funktion, im Gegensatz zu der zuvor beschriebenen 2D-Summenhistogrammfunktion, ist der, daß $s'(u'_q,v'_q)$ vom Nullpunkt des Koordinatensystems gesehen nicht zwingend in jeder Richtung monoton steigend ist.

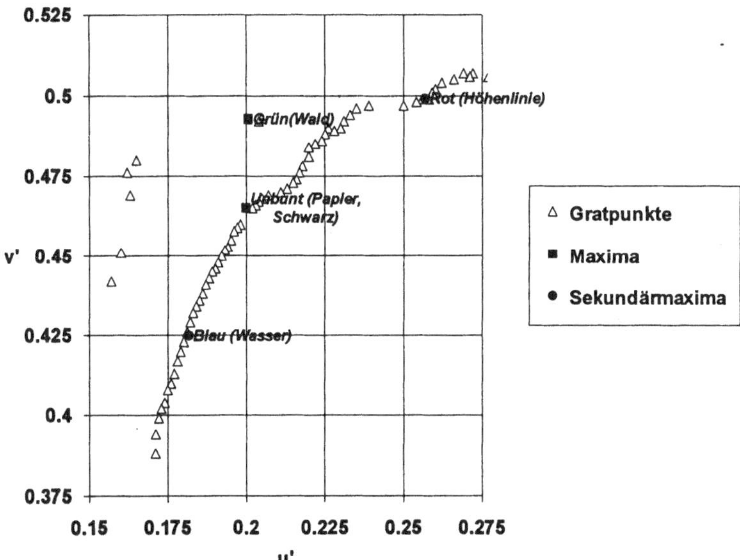

Abbildung 1: Maxima und Gratpunkte im u'v'-Histogramm des Kartenausschnitts aus Abbildung 3.

Zusätzlich zu den extrahierten Maxima können Sekundärmaxima gewonnen werden, indem durch die Punktscharen der extrahierten Gratpunkte Geraden gelegt werden, die durch die Maxima laufen. Über die Positionen der zur Gerade benachbarten Gratpunkte wird eine gewichteter Mittelwert berechnet, aus dem sich die Position des Sekundärmaximums ergibt.

Abbildung 1 zeigt die aus dem Histogramm des in Abbildung 3 dargestellten Kartenausschnitts extrahierten Maxima und Grate.

4.2 Segmentierung des Bildes auf der Basis der additiven Farbmischgesetze

Nachdem die Lage der Merkmalszentren in der u'v'-Farbtafel bestimmt wurde, kann die Segmentierung des Rasterbildes durchgeführt werden. Der entwickelten Methode ist zu eigen, daß sie sich primär an einem Abstandsmaß von den additiven Farb-Mischlinien orientiert anstatt ausschließlich die Distanz von den Merkmalszentren zur Bestimmung der Klassenzugehörigkeit zu verwenden. Dadurch können Fehlklassifikationen von Bildpunkten, die additiv gemischte Farbvalenzen repräsentieren, weitestgehend verhindert werden.

Im ersten Schritt der Segmentierung erfolgt die Trennung der einzelnen Farbvalenzen in Abhängigkeit von ihrem Farbort in der u'v'-Farbtafel. Dadurch werden helligkeitsmäßige Unterschiede der Bildpunkte jedoch nicht berücksichtigt. Abbildung 2 veranschaulicht das Prinzip der Segmentierung durch Bestimmung der nächsten additiven Farb-Mischlinie.

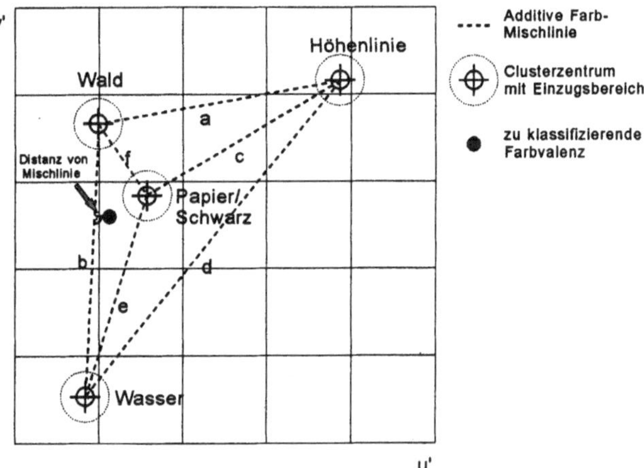

Abbildung 2: Prinzip der Segmentierung durch Bestimmung der nächsten additiven Farb-Mischlinie.

Zunächst wird um jedes Merkmalszentrum ein fester, kreisförmiger Einzugsbereich definiert. Dieser verhindert, daß Farben, die eindeutig einem Cluster zugewiesen werden können versehentlich einer dicht benachbarten Farb-Mischlinie zugeordnet werden. Die additiven Farb-Mischlinien sind Stücke von Geraden, die in der u'v'-Farbtafel durch jeweils zwei Merkmalszentren verlaufen. Für jede zu klassifizierende Farbvalenz wird der euklidische Abstand zu allen Geraden bestimmt. Die Zuordnung der Farbvalenz zu einem Cluster-Paar erfolgt dann, wenn der Abstand zu der entsprechenden Farb-Mischlinie kleiner ist als zu allen anderen Farb-Mischlinien und wenn die Distanz von der Farbvalenz zu den beiden Clustern kleiner ist als die Distanz der beiden Cluster zueinander. Das Fortlassen der zweiten Bedingung hätte zur Folge, daß auch Farben als additive Mischfarben akzeptiert würden, die nicht auf der Verbindungsstrecke zwischen den Merkmalszentren liegen.

Nachdem die an der Mischung beteiligten Cluster bestimmt sind, wird durch Anwendung eines Minimum-Distance-Kriteriums auf die Abstände zwischen Farbvalenz und den beiden Merkmalszentren, die endgültige Zuordnung des Bildpunkts zu einem Cluster durchgeführt.

Bei der Segmentierung einer topographischen Karte entstehen mehrere bunte sowie ein unbunter Cluster. In diesem sind die Bildpunkte mit den Kartenfarben *Weiß* (Kartenpapier) und *Schwarz* enthalten. Zwischen den in diesem Cluster enthaltenen Farbvalenzen muß nun eine Diskriminierung nach dem Kriterium *Helligkeit L** durchgeführt werden. Als Segmentierungsmethode für dieses eindimensionale Problem wird das bereits erwähnte Verfahren nach [Seza1990] eingesetzt.

5. Ergebnisse

Die Segmentierungsergebnisse des zuvor beschriebenen Verfahrens sollen anhand eines Kartenbeispiels veranschaulicht werden. Abbildung 3 zeigt einen Ausschnitt aus einer topographischen Karte TK25, Blatt 2717, Schwanewede, in schwarz-weißer Darstellung (Abtastauflösung 1020dpi, 24 Bit RGB). Die Original-Kartenfarben sind Rot-Braun (Höhenlinien), Blau (Wasser), Grün (Wald), Schwarz (Symbole) und Weiß (Papier). Abbildung 4 zeigt die mit dem beschriebenen Verfahren extrahierten Farbschichten *Rot-Braun* und *Blau* (links) im Vergleich mit dem Segmentierungsergebnis eines fest dimensionierten Minimum-Distance-Klassifikators (rechts).

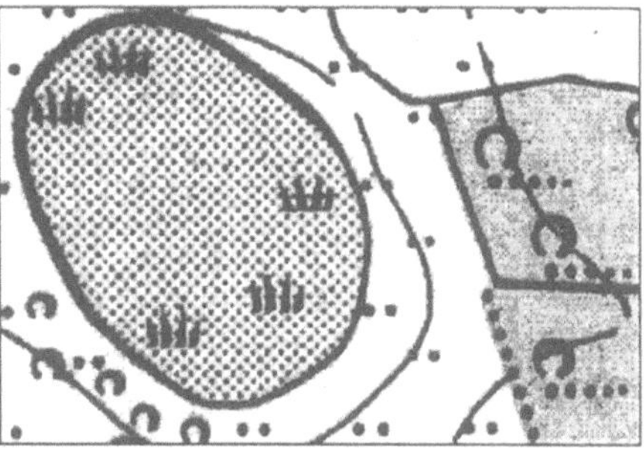

Abbildung 3: Kartenausschnitt aus der TK25, Blatt 2717, Schwanewede.

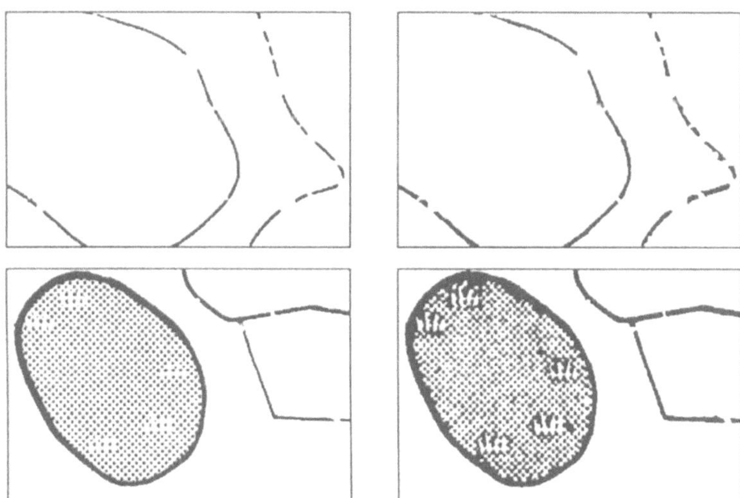

Abbildung 4: Segmentierungsergebnisse der Farbschichten *Rot-Braun* und *Blau* für das vorgestellte Verfahren (links) und einen fest dimensionierten Minimum-Distance-Klassifikator (rechts).

Literatur

[Andr1990] I. Andreadis, M. Browne und J.A. Swift, "Image pixel classification by chromaticity analysis", in *Pattern Recognition Letters*, Elsevier Science Publishers, North-Holland, Vol. 11, Nr. 1, Januar 1990, pp. 51-58.

[Bala1991] R. Balasubramanian und J. Allebach, "A New Approach to Palette Selection for Color Images", in *Journal of Imaging Technology*, The Society for Imaging Science and technology, Vol. 17, Nr. 6, Dezember 1991, pp. 284-290.

[Bumb1987] F. Bumbaca und K. Smith, "Design and Implementation of a Color Vision Model for Computer Vision Applications", in *Computer Vision, Graphics, and Image Processing*, Academic Press, Vol. 39, 1987, pp. 226-245.

[Cele1990] M. Celenk, "A Color Clustering Technique for Image Segmentation", in *Computer Vision, Graphics, and Image Processing*, Academic Press, Vol. 52, 1990, pp. 145-170.

[CIE1986] Commission Internationale de L'Éclairage - Internationale Beleuchtungskommission, "Colorimetry", *CIE Publikation Nr. 15.2*, Zentrales Büro der CIE, Wien, 1986.

[Ebi1992] N. Ebi, B. Lauterbach und P. Besslich, "Automatic Data Acquisition from Topographic Maps Using a Knowledge-Directed Image Analysis System", in *The International Archives of Photogrammetry and Remote Sensing*, International Society for Photogrammetry and Remote Sensing, Bethesda, Vol. 29, Part B4, 1992, pp. 655-663.

[Gent1990] R.S. Gentile, E. Walowit und J.P. Allebach, "Quantization and multilevel halftoning of color images for near-original image quality", in *Journal of the Optical Society of America, Part A*, Optical Society of America, Vol. 7, Nr. 6, Juni 1990, pp. 1019-1026.

[Ille1990] A. Illert, "Automatische Erfassung von Kartenschriften, Symbolen und Grundrißobjenkten aus der Deutschen Grundkarte 1:5000", Dissertation, Wissenschaftliche Arbeiten der Fachrichtung Vermessungswesen der Universität Hannover, Nr. 166, 1990.

[Kass1991] J.M. Kasson und W. Plouffe, "Subsampled device-independent interchange color spaces", in *Image Handling and reproduction Systems Integration, Proceedings of SPIE*, Vol. 1460, Hrsg. W. Bender und W. Plouffe, Society of Photo-Optical Instrumentation Engineers, 1991, pp. 11-19.

[Kuma1989] H. Kumazawa, S. Tomita, S. Nishikawa und M. Nakada, "Quantization of Color Space Using the Merging Method of Color Sub-Spaces", in *Transactions of the Institute of Electronics, Information, and Communication Engineers D-II, Japan*, Vol. J72-D-II, Nr. 5, Mai 1989, pp. 715-723, (auf Japanisch).

[Laut1992] B. Lauterbach, N. Ebi und P. Besslich, "PROMAP - A System for Analysis of Topographic Maps", in *Proceedings of IEEE Workshop on Computer Vision"*, IEEE Computer Society Press, Los Alamitos, 1992, pp. 46-55.

[Laut1993] B. Lauterbach, "Extraktion attributierter Strukturprimitiven aus multispektral abgetasteten kartographischen Vorlagen", Dissertation, Institut für Theoretische Elektrotechnik und Digitale Systeme, Universität Bremen, 1993.

[Mahy1991] M. Mahy, B. Van Mellaert, L. Van Eycken und A. Oosterlinck, "The Influence of Uniform Color Spaces on Color Image Processing: A Comparative Study of CIELAB, CIELUV, and ATD", in *Journal of Imaging Technology*, IS&T-The Society for Imaging Science and Technology, Vol. 17, Nr. 5, Nov. 1991, pp. 232-243.

[Mass1990] R. Massen, J. Gässler, P. Böttcher und W. Reichelt, "Trainable Look-Up-Tables versus Neural Networks for Real-Time Colour Classification", in *Mustererkennung 1990, Proceedings des 12. DAGM-Symposiums, Informatik Fachberichte 254*, Springer Verlag, Berlin, Heidelberg, New York, 1990, pp. 377-384.

[Maye1991] H. Mayer, "Systemtests zur Raster-/Vektor-Konvertierung", Fachzentrum Geo-Informationssysteme, Firma Siemens, München, 1991.

[Maye1992] H. Mayer, C. Heipke und G. Maderlechner, "Knowledge-Based Interpretation of Scanned Large-Scale Maps", in *The International Archives of Photogrammetry and Remote Sensing*, International Society for Photogrammetry and Remote Sensing, Bethesda, Vol. 29, Part B3, 1992, pp. 578-585.

[Mukh1986] A. Mukherjee und Y.V. Venkatesh, "Digital Color Reproduction on Color Television Monitors", in *Computer Vision, Graphics, and Image Processing*, Academic Press, Vol. 36, 1986, pp. 114-132.

[Ohta1980] Y. Ohta, T. Kanade und T. Sakai, "Color Information for Region Segmentation", in *Computer Graphics and Image Processing*, Academic Press, Vol. 13, 1980, pp. 222-241.

[Rich1981] M. Richter, "Einführung in die Farbmetrik", de Gruyter, Berlin, New York, 1981.

[Rich1986] J.A. Richards, "Remote Sensing Digital Image Processing", Springer Verlag, Berlin, Heidelberg, New York, 1986.

[Seza1990] M.I. Sezan, "A Peak Detection Algorithm and Its Application to Histogram-Based Image Data Reduction", in *Computer Vision, Graphics, and Image Processing*, Academic Press, Vol. 49, 1990, pp. 36-51.

[SS019100] SIS - Swedish Standards Institution, "Swedish Standard SS 01 91 00 - Colour notation system", 1979.

[Suzu1990] S. Suzuki und T. Yamada, "MARIS: Map Recognition Input System", in *Mapping and Spatial Modelling for Navigation*, Hrsg. L.F. Pau, Springer Verlag, Berlin, Heidelberg, 1990, pp. 95-116.

[Tomi1992] S. Tominaga, "Color Classification of Natural Color Images", in *Color Research and Application*, John Wiley & Sons, New York, Vol. 17, Nr. 4, 1992, pp. 230-239.

Farbgestützte Verfolgung von Objekten mit dem PC-basierten Multiprozessorsystem BVV 4

L. Tsinas, H. Meier, W. Efenberger
Institut für Meßtechnik
Fakultät für Luft- und Raumfahrttechnik
Universität der Bundeswehr München
85577 Neubiberg

Zusammenfassung

Im folgenden werden Algorithmen zum Entdecken und Verfolgen farbiger Objekte in Straßenszenen vorgestellt. Die Implementierung erfolgte ursprünglich mit einer *framegrabber*-Karte in einer AT-Umgebung. In einem zweiten Schritt wurden die Algorithmen auf das farbfähige Echtzeit-Bildverarbeitungssystem BVV 4 portiert. Die Struktur dieses PC-basierten Multiprozessorsystems wird ebenfalls beschrieben. Die Algorithmenlaufzeiten mit dem BVV 4 stellen überraschend geringe Ansprüche an die Rechenzeit (ca. 8 ms für die einfache Verfolgung eines Objektes). Dies erlaubt die Annahme, daß die Hinzunahme von Farbinformation (HSI-Daten) bei der Objektentdeckung und -verfolgung - zur Unterstützung anderer bewährter Verfahren - von Nutzen sein wird.

1 Einleitung

Für die Führung autonomer mobiler Roboter haben sich sichtbasierte Systeme, die evtl. durch weitere passive oder aktive Sensoren (Laser, Radar, Ultraschall o.ä.) unterstützt werden, bewährt. Auf der *hardware*-Ebene sind modular strukturierte Mehrprozessorsysteme für den Einsatz bei solchen sichtgesteuerten Robotern, die unter Echtzeitbedingungen arbeiten, besonders geeignet [Graefe 91], [Lang 92]. Eine bewährte Repräsentation dieses *hardware*-Konzepts stellt die BVV-Familie [Graefe 90] dar. Verschiedene universitäre und industrielle Bildverarbeitungsansätze zur Fahrzeugführung - vor allem aus den USA, Japan und Deutschland - werden in [Dickmanns, Graefe 88], [Mysliwetz 90], [Ichiro 91] und [Tsugawa 93] behandelt.

Zur besseren Erkennung von Objekten und zur Interpretation von Straßenszenen wurden die bisherigen auf gesteuerter Korrelation basierenden Verarbeitungsansätze [Kuhnert 88] für Grauwertbilder um Algorithmen erweitert, welche die Farbinformation in Bildfolgen (50 Halbbilder/s) auswerten. In dieser Arbeit werden zwei modulare Ansätze beschrieben, um Objekte oder Objektteile aufgrund ihrer Farbeigenschaften selektiv zu erkennen und zu verfolgen. Die in den HSI-Farbraum vortransformierten Bilddaten dienen für beide Ansätze als Eingangsgrößen. Das erste Verfahren nimmt ausgehend von (zunächst) vorgegebenen Farbbereichen eine Farbsegmentierung vor. Abhängig von der Position eines erkannten Objektes im Bild und dessen Farbeigenschaften werden für die Verfolgung (im nächsten Bild) die Suchkriterien ermittelt bzw. adaptiv angepaßt. Der zweite Ansatz zielt darauf ab, die Farbeigenschaften eines Bereichs zu überwachen, darin auftretende Objekte zu detektieren und evtl. zu verfolgen. Die Grenzen des betrachteten Bereichs werden von anderen Modulen vorgegeben (Initialisierung) bzw. möglichst für jedes Bild neu ermittelt.

2 Auswertung der Farbinformation in Bildfolgen

Durch die verbesserte Leistungsfähigkeit der Mikroelektronik (= Rechenleistung der bildverarbeitenden Prozessoren und Durchsatz der Bildspeicherstrukturen) kann die Farbinformation zur Interpretation von Bildfolgen auch unter Echtzeitbedingungen genutzt werden [Turk, Marra 86], [Thorpe et al. 88], [Münkel, Welz 91], [Priese, Rehrmann 92], [Ritter 92]. Dafür bietet sich insbesonders der HSI-Farbraum an, der dem menschlichen Farbempfinden näher liegt als der RGB-Farbraum und somit das Vorstellungsvermögen beim Algorithmenentwurf und -test positiv unterstützen kann: Die Werte für Buntton H (*hue*), Sättigung S (*saturation*) und Intensität I (*intensity*) liegen als getrennte Daten vor. Die Transformation der RGB-Daten, wie sie z.B. von einer Farbkamera geliefert werden, in HSI-Werte wird von der Bildverarbeitungskarte im Videotakt vorgenommen und verursacht keinen zusätzlichen Zeitverlust.

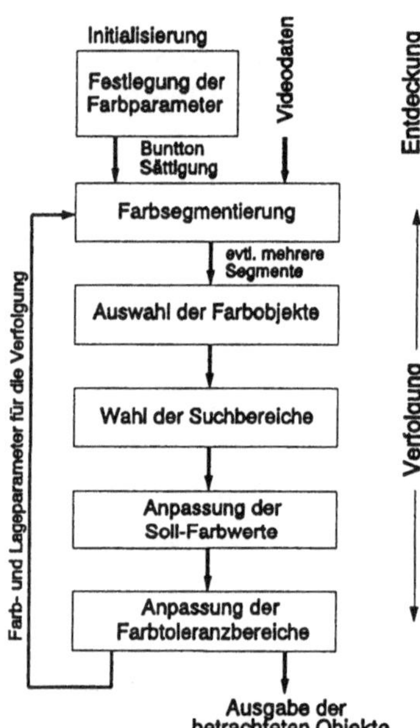

Abbildung 1: Ablauf der Entdeckung und Verfolgung farbiger Objekte in Bildfolgen

2.1 Entdeckung und Verfolgung von Objekten aufgrund ihrer Farbe

Bei der Entdeckung und Verfolgung bunter (Teil-)Objekte kann es für die Klassifikation hilfreich sein, neben typischen Gestaltmerkmalen (Kanten, Ecken, Symmetrie, Textur o.ä.) ihre jeweiligen Farbeigenschaften zu nutzen. Bei der Rückansicht eines Straßenfahrzeugs bieten sich beispielsweise die roten Rückscheinwerfer oder die gelben Blinker an. Abbildung 1 zeigt die einzelnen Module zur Verarbeitung von Farbinformation, um ein oder mehrere farbige Objekte zu erkennen und anschließend zu verfolgen.

2.1.1 Farbsegmentierung - Auswahl der Objekte

Für die Farbsegmentierung wird zuerst anhand der erwarteten Farbeigenschaften im gesamten Bild nach den gewünschten (Farb-)Merkmalen gesucht. Kleine Felder mit 2*2 bzw. 4*4 Pixeln erfüllen dann die Suchkriterien, wenn eine Mindestanzahl von Pixeln (> 50%) der Buntton- oder der Sättigungsbedingung entsprechen. Dieses Verfahren erwies sich im Vergleich zu einer zweiten Vorgehensweise, bei der eine Mindestanzahl von Pixeln gleichzeitig sowohl die Buntton- als auch die Sättigungskriterien erfüllen mußte, als robuster. Die Intensität wird nicht ausgewertet, da ausschließlich nach bunten Objekten gesucht wird.

Benachbarte Felder, welche die Suchkriterien erfüllen, werden zu einem Segment zusammengefaßt. In jedem bearbeiteten Bild können mehrere Segmente gefunden werden. Jedes wird vorerst als eigenständiges Objekt aufgefaßt und verfolgt. In den folgenden Bildern verschmelzen benachbarte Segmente zu einem Objekt. Sollten dabei zwei Objekte zusammentreffen, so werden sie, falls sie gleiche Farbeigenschaften haben, zu einem Objekt vereinigt.

2.1.2 Wahl des Suchbereichs

Zur Verfolgung eines Objekts (im Bild) wird ein rechteckiger Suchbereich um das Objekt eingestellt. Der Suchbereich ist (in horizontaler und vertikaler Richtung) um eine Feldbreite (also 2 bzw. 4 Pixel) größer als das bisher gefundene Objekt.

2.1.3 Anpassung der Soll-Farbwerte

Die Farbsensoren einer Drei-Chip-Farbkamera (z.B. Sony DXC-3000 P) weisen unterschiedliche Kennlinien für die Beleuchtungsempfindlichkeit auf, was bei Intensitätsschwankungen zu Änderungen von Buntton und Sättigung führt, obwohl diese Größen - nach den Transformationsgleichungen [Tenenbaum, et al. 74], [Frey 88]) - unabhängig von der Intensität sein sollten. Um die Auswirkungen dieses Effektes zu reduzieren schlägt [Frey 91] nach der RGB-HSI-Transformation eine sog. Linearisierung mit Unbuntabgleich vor. Bei der Verfolgung farbiger Objekte wird dieser kameraspezifische Abgleich durch eine automatische Nachführung sowohl der Soll-Farbwerte als auch ihrer Toleranzbereiche ersetzt. Die ermittelten Farbkomponenten und die Toleranzbereiche werden von Bild zu Bild adaptiv angepaßt, um beleuchtungsbedingten Variationen der Farbeigenschaften (der verfolgten Objekte) entgegenzuwirken.

Zur Anpassung der Soll-Farbwerte für das jeweils nächste Bild werden die Farbwerte der als Objektteile klassifizierten Felder des aktuellen Bildes ausgewertet. Mittels Histogramm-Bildung wird in jedem Segment das Häufigkeitsmaximum der jeweiligen Farbkomponente bestimmt, wodurch man einen neuen Soll-Farbwert gewinnt. Dieser kann maximal um die bisherige Toleranzbreite in Richtung dieses Häufigkeitsmaximums verschoben werden.

2.1.4 Anpassung der Farbtoleranzbereiche

Ein möglichst präzis eingestellter Toleranzbereich der H- bzw. S-Farbkomponente wird angestrebt, um nicht irrtümlich (andersfarbigen) Hintergrund als Teil des Objektes zu interpretieren. Wenn mehr Pixel als eine vorgegebene (empirisch ermittelte) Anzahl von Pixeln (> 50%) die Farbbedingungen erfüllen, wird der Toleranzbereich verkleinert; sind es weniger, wird er ausgedehnt. Die unabhängig voneinander behandelten Toleranzbereiche der Farbkomponenten sollten nach Möglichkeit sehr exakt nachgeführt werden, um die Qualität der Verfolgung zu erhöhen und eine Anpassung an sich ständig wechselnde Sichtverhältnisse zu ermöglichen. Zudem ist eine hinreichende Größe der Toleranzbereiche notwendig, um das verfolgte Objekt nicht zu verlieren. Es wurden Grenzwerte für die Größe des Toleranzbereichs festgelegt.
In der Verfolgungsphase werden sowohl die Farb- als auch die Lagekoordinaten der entdeckten Objekte (wie beschrieben), zyklisch überwacht.

2.2 Automatische Entdeckung und Verfolgung von Objekten auf der eigenen Fahrspur

Ein zweites Verfahren, das als Fahrzeugentdecker implementiert wurde, sollte automatisch alle auf der eigenen Fahrspur (= Suchbereich) auftretenden Objekte (z.B. Fahrzeuge) erkennen. Zuerst wird die Lage der Fahrspur im Bild ermittelt und darauf aufbauend der Suchbereich eingestellt. Als nächstes werden die Farbeigenschaften der Fahrspur bestimmt. Alle davon abweichenden Regionen innerhalb des Suchbereichs deuten auf potentielle Objekte auf der Fahrspur hin.

Für die Entdeckung der eigenen Fahrbahn kommt ein korrelationsbasiertes Verfahren zum Einsatz [Tsinas, Graefe 92], wobei aus Laufzeitgründen eine vereinfachte Version implementiert wurde. Ausgehend von den über die Straße gewonnenen Informationen kann der Suchbereich eines Objektentdekkers initialisiert werden. Anschließend erfolgt die Bestimmung der (über mehrere Bilder gemittelten) Farbwerte - diesmal aller drei Farbkomponenten - der verschiedenen Regionen der Fahrspur. Für die Initialisierung wird vorausgesetzt, daß während der ersten 10 Bilder keine Objekte auftauchen oder sich bereits auf der Fahrspur befinden.

Nun werden alle Felder (2*2 bzw. 4*4 Pixel) des Suchbereichs, die abweichende Farbmerkmale zur Fahrbahn aufweisen, ermittelt. Dabei reicht es aus, daß einer der Farbwerte Buntton, Sättigung oder Intensität außerhalb des Toleranzbereichs liegt. Im Gegensatz zum ersten Verfahren werden hier alle drei Farbkomponenten untersucht, denn neben bunten sollen auch unbunte Objekte erkannt werden. Der Buntton wird insbesondere nur dann kontrolliert, wenn die Sättigung S ausreichend groß ist ($S > 30$), so daß der Buntton signifikante Aussagekraft besitzt.
Prinzipiell könnten die gefundenen Felder, wie im ersten Verfahren, zu einem Objekt zusammengefaßt und weiter verfolgt werden. Dies wurde bislang nicht realisiert.

Bei den Versuchen hat es sich als problematisch erwiesen, daß die Straße häufig keine einheitliche Farbeigenschaften besitzt, sondern vor allem bei Schatteneinfluß die Intensität stark schwankt. Auf die Auswertung der Intensität kann hier jedoch nicht verzichtet werden, da sich unbunte Objekte (Hindernisse), die sich in der Sättigung nicht von der Fahrbahn unterscheiden, auf der Fahrbahn befinden können.

3 Beschreibung der *hardware*

Die Abbildung 2 gibt einen Überblick über das Bildverarbeitungssystem BVV 4 [Meier 93]. Dieses ist aus mehreren PC-Einsteckkarten aufgebaut: *framegrabber*-Karte, Bildverarbeitungskarte und Markiereinheit. Die analogen Videosignale werden mittels einer *framegrabber*-Karte [Leutron 92] digitalisiert -verstärkt- und allen weiteren Einsteckkarten parallel zur Verfügung gestellt. Es können entweder bis zu drei monochrome Videoquellen oder eine Farbkamera angeschlossen werden können.

Wie alle Mitglieder der BVV-Familie besitzt das BVV 4 zwei unabhängige Bussysteme. Ein Videobus mit hoher Bandbreite (45 MByte/s, d.h. drei Kanäle mit je 15 MByte/s) verteilt die digitalisierten Bilddaten an die parallelen Verarbeitungseinheiten. Der Systembus (ISA bzw. EISA) mit wesentlich kleinerer Bandbreite, erlaubt den Parallelprozessoren, Nachrichten untereinander auszutauschen. Als weitere Kommunikationskanäle besitzt jede Einheit zwei Transputerlinks TL mit wählbarer Betriebsart (INMOS- bzw. Parsytec-Standard) und einer Übertragungsrate von 10 Mbit/s oder 20 Mbit/s. Die Links ermöglichen bei Bedarf die Ankopplung an ein Transputer-*cluster* und/oder den direkten Datenaustausch zwischen Parallelprozessoren, falls z.B. die Bandbreite bzw. die Zeitbedingungen am PC-Bus nicht ausreichen. Eine parallele Schnittstelle (als EXT bezeichnet) erlaubt den Zugang zu den wichtigen Signalen, z.B. für Testzwecke, und ermöglicht zudem den Anschluß weiterer Baugruppen (z.B. direkte Ansteuerung einer *shutter*-Kamera, Speichererweiterung, spezielle Bildverarbeitungs-Coprozessoren etc.).

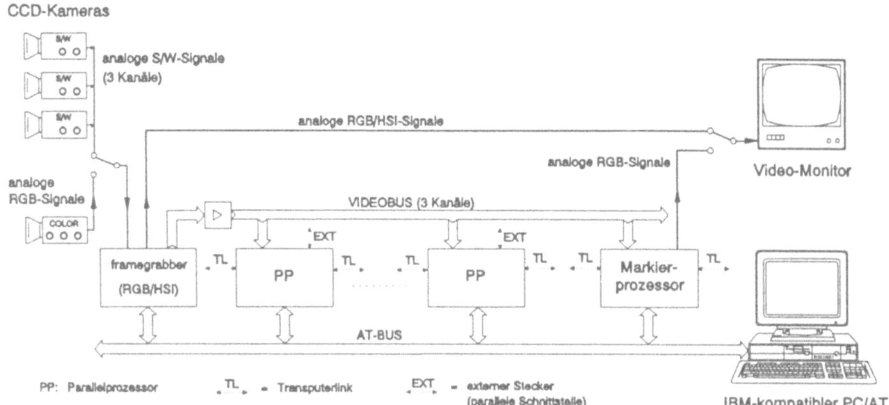

Abbildung 2: Struktur des Echtzeit-Bildverarbeitungssystems BVV 4

3.1 Struktur eines Parallelprozessors PP

Die Abbildung 3 zeigt die Struktur eines einzelnen Parallelprozessors mit lokalem Bildspeicher (= Teil des Hauptspeichers). Als Recheneinheit wurde der RISC-Prozessor Intel 80960CA mit 32 MHz Taktrate gewählt, der lt. Hersteller eine Rechenleistung von max. 66 MIPS besitzt und eine Reihe von Einheiten integriert:
Neben vier 32 Bit breiten DMA-Kanälen für verschiedene Betriebsarten (*two-cycle*-Modus mit 32 MByte/s und *fly-by*-Modus mit 59 MByte/s) verfügt er über 1 KByte *onboard*-RAM, *register*- und *instructioncache*, sowie einen komfortablen programmierbaren *buscontroller*. Dieser erlaubt es, 16 je 256 MByte große *memory regions* (= 4 GByte Adreßraum) per *software* unterschiedlich zu konfigurieren: er sieht mehrere Busbreiten (8, 16, 32 Bit) vor, paßt diese bei Datentransfers automatisch an, legt mittels einer programmierbaren *wait-state*-Logik das Bustiming für statische und/oder dynamische Speicher fest und unterstützt schnelle Speicherzugriffsarten (*burst*- und *pipelined*-Zugriff). Von den 248 möglichen externen Interrupts mit 32 programmierbaren Prioritätsebenen, werden nur wenige zum Zählen von Bild- oder Zeilenimpulsen, zur Unterscheidung zwischen Halbbildern mit geraden und ungeraden Zeilennummern (für die Vollbildverarbeitung), zur Steuerung der Anwenderprogramme etc. genutzt. Aufgrund der superskalaren Prozessorarchitektur mit mehreren chipinternen, parallelen Verarbeitungseinheiten, eignet sich der Prozessor besonders für *integer*-Operationen, wie sie häufig für Grundoperationen der Bild(vor)verarbeitung benötigt werden.

Das Einlesen digitalisierter Videodaten in den Hauptspeicher geschieht mit Hilfe des integrierten DMA-*controller*. Dank seiner freien Programmierbarkeit ist eine flexible (Bild-)Datenerfassung möglich. Bilder einer Bildfolge können abwechselnd in eine Wechselpufferstruktur aus zwei oder mehreren (Bild-)Speicherbänken abgelegt, Vollbilder aus aufeinanderfolgenden Halbbildern erfaßt und bearbeitet werden.

Zum Einlesen der Bilddaten vom Videobus in den Hauptspeicher (= lokaler Bildspeicherbereich) des Prozessors, wird ein FIFO-Zwischenspeicher (1K*32) eingesetzt. Die Bilder werden zeilenweise per DMA in den entsprechenden (Bild-)Speicherbereich übertragen. Der Nachteil dieser *software*-Lösung – gegenüber einer reinen *hardware*-Implementierung eines Bildspeichers – sind Rechenzeitverluste, die auf die Busbelegung durch die DMA-Transfers zurückzuführen sind. Aufgrund der hohen Rechenleistung des Prozessors, des prozessorinternen RAMs sowie der Übertragungsraten des DMA-*controllers* ist dies jedoch akzeptabel. Zum Einlesen eines typischen Bildformats mit 256*256 Bildpunkten wird während der Halbbilddauer von 20 ms der Prozessorbus nur zu einem Anteil von 5,5 % für den Bilddatentransfer (= maximaler Rechenzeitverlust im ungünstigsten Fall) beansprucht. Um eine hohe Rechenleistung bei Algorithmen mit häufigen Speicherzugriffen zu erhalten, ist der Hauptspeicher mit statischen RAMs mit einer Kapazität von 2 MByte (512K*32) und einem *wait state* realisiert.

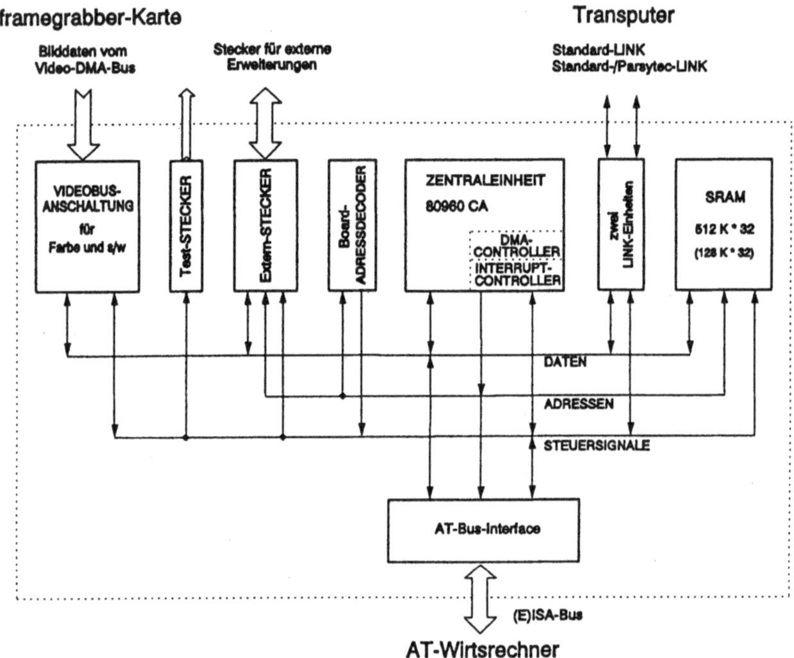

Abbildung 3: Struktur eines Parallelprozessors PP

Jede Einsteckkarte (Parallel- und Markierprozessor) besitzt mehrere serielle und parallele Kommunikationskanäle. Die Verbindung zum Systembus des Wirtsrechners erfolgt über einen *dualport*-Speicher von 4 KByte (2K*16), der zur Abkopplung und Synchronisation der Kommunikationspartner eingesetzt wird. Jeder *dualport*-Speicher wird per *software* in zwei Ringpuffer aufgeteilt, jeweils einer für Nachrichten, die von der Karte verschickt werden, und einer für Nachrichten, die von anderen Teilnehmern an die Karte übertragen werden. Die AT-Komponente des Betriebssystems hat Zugriff auf alle *dualport*-Speicher. Sie fragt zyklisch die Lese- und Schreibzeiger der Ausgabepuffer ab und transportiert nötigenfalls die Nachricht zu anderen Eingabepuffern. Die Zielkarte ist im Nachrichtenkopf angegeben.

3.2 Markierung von Ergebnissen

Einfache Markierungen zum Anzeigen der Suchbereiche und der gefunden Objekte werden bislang mit dem *overlay*-Speicher der *framegrabber*-Karte ermöglicht, der vom AT-Wirtsrechner über den Systembus bedient werden muß. Die Markiereinheit wurde inzwischen in Betrieb genommen und deren Funktionsfahigkeit demonstriert [Fischer 93]. Sie erlaubt ein erheblich schnelleres und komfortableres Markieren als die *framegrabber*-Karte.

4 Ergebnisse

Der modulare Ansatz beider Verfahren und der Einsatz des farbfähigen Bildverarbeitungssystems BVV 4 haben sich in dieser Implementierung bewährt. Dies spiegelt sich auch in den ermittelten Zeiten für die Ausführung der (nicht rechenzeitoptimierten) Algorithmen wieder. Die in Abschnitt 2 vorgestellten Verfahren wurden zunächst mit einer *framegrabber*-Karte auf einem PC (i486 mit 50 MHz) implementiert. Die Zykluszeiten beider Verfahren - ohne den Straßenfinder zu berücksichtigen - sind nahezu identisch und liegen für das erstgenannte zur Detektion von Objekten bei 1 s sowie bei 300 ms zu ihrer Verfolgung. Solche Zykluszeiten reichen bei autonomen Straßenfahrzeugen nicht aus.
Im Mittel werden bei Einsatz des BVV 4 für die Entdeckung eines Hindernisses - wenn über das ganze Bild gesucht wird - ca. 300 ms benötigt. Für die weitere Verfolgung sind jedoch nur ca. 8 ms notwendig. Dieser zeitliche Rahmen ist mehr als akzeptabel.
Im Vergleich dazu benötigen die in [Efenberger et al. 92] beschriebene Entdecker und Verfolger für monochrome Bildfolgen - implementiert auf dem leistungsfähigen Bildverarbeitungssystem BVV 3 - je 40 ms bzw. 20 ms. Das erste Verfahren wurde u.a. dazu eingesetzt, das Ansprechen der Bremslichter von Rücklichtern zu unterscheiden. Darüber hinaus konnten in Laborumgebung gleichzeitig bis zu 12 Objekte verschiedener Farbe verfolgt werden.

5 Schlußfolgerungen

Das farbfähige Multiprozessorsystem BVV 4 besitzt eine ähnliche Architektur wie die anderen Mitglieder der BVV-Familie. Zur Entdeckung und Verfolgung von Farbobjekten konnte ein Algorithmensatz auf diesem System implementiert werden. Die genannten Ergebnisse zeigen, daß die Hinzunahme der Farbinformation bei der Interpretation von Bildfolgen sehr hilfreich sein wird und ein weites Feld für künftige Forschungsaufgaben bietet. Mit dem BVV 4 steht ein leistungsfähiges Echtzeit-Bildverarbeitungssystem bereit, das gleichzeitig in die gewohnte Arbeitungsumgebung eines AT-Wirtsrechners eingebettet ist.

Anmerkungen

Teile dieser Arbeit sind im Rahmen des europäischen EUREKA-Projektes PROMETHEUS vom Bundesministerium für Forschung und Technologie, sowie der Daimler-Benz AG finanziert worden. Wir danken Hrn. Lopau für seine Unterstützung bei der Entwicklung und Implementierung zahlreicher Algorithmen [Lopau 92] sowie Hrn. Müllerschkowski für die Realisierung verschiedener *hardware*-Komponenten [Müllerschkowski 91].

Literatur

Dickmanns, E.D.; Graefe, V. (1988a,b): Dynamic Monocular Machine Vision and Application of Dynamic Monocular Machine Vision. In Machine Vision and Applications 1. Springer Verlag, November 1988, pp 223-241 and 241-261.

Efenberger, W.; Ta, Q.; Tsinas, L.; Graefe, V. (1992): Automatic Recognition of Vehicles Approaching from Behind. Proceedings, Intelligent Vehicles. Detroit, pp 57-62.

Frey, H. (1988): Digitale Bildverarbeitung in Farbräumen. Diss., Technische Universität München.

Frey, H. (1991): Über den Vorteil eines linearen Sensors bei der HSI-Farbbildverarbeitung. Proceedings, 13. DAGM Symposium. München, pp 337-342.

Fischer, H. (1993): Konzeption, Aufbau und Inbetriebnahme einer PC-Einsteckkarte mit dem RISC-Prozessor Intel 80960CA zum Einblenden graphischer Symbole in monochrome und mehrkanalige (Video-)Bildfolgen als Komponente eines Echtzeit-Bildverarbeitungssystems. Diplomarbeit am Institut für Meßtechnik, Universität der Bundeswehr München.

Graefe, V. (1990): The BVV-Family of Robot Vision Systems. In O. Kaynak (ed.): Proceedings of the IEEE Workshop on Intelligent Motion Control. Istanbul, pp IP55-IP65.

Graefe, V. (1991): Robot Vision Based on Coarsely-Grained Multiprocessor Systems. R. Vichnevetzky, J.J.H. Miller (eds.): Proceedings, IMACS World Congress. Dublin, pp 755-756.

Ichiro, M. (1991): Vision-based Vehicle Guidance. Springer-Verlag Berlin.

Kuhnert, K.-D. (1988): Zur Echtzeit-Bildfolgenanalyse mit Vorwissen. Diss., Fakultät für Luft- und Raumfahrttechnik der Universität der Bundeswehr München.

Lang, B. (1992): Digitale Bildverarbeitung mit kombinierter Pipeline- und Parallel-Architektur. Diss., Universität Hamburg-Harburg.

Leutron (1992): Echtzeit Bildverarbeitungskarte CFP-AT. Mai 1992.

Lopau, R. (1992): Entdeckung und Verfolgung von Objekten durch Farbbildverarbeitung. Diplomarbeit am Institut für Meßtechnik, Universität der Bundeswehr München.

Meier, H. (1993): Zum Entwurf eines PC-basierten Multiprozessorsystems für die Echtzeit-Verarbeitung monochromer und farbiger Videobildfolgen. Diss., Fakultät für Luft- und Raumfahrttechnik der Universität der Bundeswehr München.

Müllerschkowski, U. (1991): PC-Einsteckkarte zur monochromen Echtzeit-Bildverarbeitung mit dem RISC-Prozessor Intel 80960CA. Diplomarbeit am Institut für Meßtechnik, Universität der Bundeswehr München.

Münkel, H.; Welz, K. (1991): Verfolgung des Straßenverlaufes in einer Farbbildfolge. Proceedings, 13. DAGM Symposium. München, pp 515-520.

Mysliwetz, B. (1990): Parallelrechner-basierte Bildfolgen-Interpretation zur autonomen Fahrzeugsteuerung. Diss., Fakultät für Luft- und Raumfahrttechnik der Universität der Bundeswehr München.

Priese, L.; Rehrmann, V. (1992): A Fast Hybrid Color Segmentation Method. Fachberichte Informatik 12/92, Universität Koblenz-Landau.

Ritter, W. (1992): Traffic Sign Recognition in Color Image Sequences. Symposium on Intelligent Vehicles 1992. Detroit, pp 12-17.

Tenenbaum, J.M.; et al. (1974): An interactive facility for scene analysis research. Technical Note 87. SRI Project 1187, Artificial Intelligence Center, Stanford Research Institute, Menlo Park, California, January 1974.

Thorpe, C.; Hebert, M. H.; Kanade, T.; Shafer, S. A. (1988): Vision and Navigation for the Carnegie-Mellon Navlab. IEEE Transactions on Pattern Analysis and Machine Intelligence, Vol. 10, No. 3, pp 362-373.

Tsinas, L.; Graefe, V. (1992): Automatic Recognition of Lanes for Highway Driving. IFAC Conference on Motion Control for Intelligent Automation. Perugia, pp 295-300.

Tsugawa, S. (1993): Vision-Based Vehicles in Japan: The Machine Vision Systems and Driving Systems. IEEE International Symposium on Industrial Electronics, ISIE '93. Budapest, pp 278-285.

Turk, M. A.; Marra, M. (1986): Color Road Segmentation and Video Obstacle Detection. Proceedings of the SPIE, Vol. 727 Mobile Robots (1986). Cambridge, pp 136-142.

Ein Verfahren zur Farbraumanpassung CCD–basierter Bilderfassungssysteme

Wolfgang Wölker

Universität Hannover
Institut für Theoretische Nachrichtentechnik und Informationsverarbeitung
Appelstr. 9a, 30167 Hannover
e–mail: woelker@tnt.uni–hannover.de

Dieser Beitrag beschreibt ein Verfahren zur Abbildung gerätespezifischer Farbinformation auf einen neutralen Referenzfarbraum, wodurch die Farbbilddaten in diesem Referenzfarbraum verarbeitet und gespeichert werden können. Der mit der gerätespezifischen Korrektur verbundene Aufwand wird durch eine von der Verarbeitung entkoppelte Berechnung der Korrekturinformation wesentlich verringert. Damit ist es möglich die Farbraumanpassung, ohne nennenswerten Zeitverlust, z.B. bereits während des Bildeinzugs vorzunehmen. Das Verfahren wurde an verschiedenen CCD–Erfassungssystemen getestet.

1. Einleitung

CCD–Sensoren haben inzwischen einen nennenswerten Anteil bei der Bilderfassung erreicht. Die hohe Lebensdauer, die hohe Linearität, das gutmütige Verhalten bei Übersteuerungen und die oft sehr kostengünstigen Lösungen haben die Röhrentechnik in den Hintergrund gedrängt. Im gleichen Maße ist der Anteil an Farbkameras bei der Bilderfassung in vielen Bereichen gestiegen. Hingegen spiegelt sich die Bedeutung der Farbe für die Bildanalyse und Mustererkennung bisher nur teilweise in den bildverarbeitenden Verfahren wieder. Oft wird die Farbe aufgrund der mit der Erfassung verbundenen Probleme und des hohen Aufwandes nicht verwendet. Der in der Farbe enthaltene Informationswert gelangt damit nur selten zur Anwendung.

Ein Grund für diese Vernachlässigung könnte darin bestehen, daß eine farbgetreue Erfassung und Ausgabe der Farbinformation nur selten realisiert ist. Das Signal enthält oft gerätetypische Farbabweichungen, die eine Nutzung der Farbinformation für quantitative Messungen behindern. Trotzdem ist die Farbe in vielen Bereichen in zunehmendem Maße in den Mittelpunkt gerückt. Beispiele hierfür sind im medizinischen Bereich in der Zytologie zu finden.

Dort stellen moderne Präparationsverfahren wesentliche Farbinformation zur Verfügung. Die Immunozytologie verlangt quantitative Farbmessungen, um z.B. Hauttumorzellen klassifizieren zu können. Weitere Anwendungen im industriellen Bereich sind z.B. in der Reproindustrie zu finden, bei der die exakte Farbreproduktion eine zentrale Aufgabe darstellt. Hier müssen die typischerweise als Dia gelieferten Vorlagen derart aufgearbeitet werden, daß eine farbgetreue Reproduktion im anschließenden Druckprozeß gewährleistet ist. In allen Fällen ist eine farbgetreue Erfassung der Bildinformation unabdingbare Voraussetzung für die korrekte Weiterverarbeitung und Darstellung der Farbe.

Die derzeit verfügbaren Systeme weisen jedoch nahezu ausnahmslos gerätespezifische Farbabweichungen auf, die vor der Verarbeitung und Nutzung korrigiert werden müssen. Es ist daher zu untersuchen inwieweit es möglich ist, diese Abweichungen zu kompensieren.

2. Das Korrekturverfahren

Ausgehend von einem allgemeinen Farbbildverarbeitungssystem ist zu prüfen, an welcher Stelle im System derartige Korrekturen eingesetzt werden können und welche Darstellungsformen des Farbraums sinnvoll anwendbar sind. Das Blockdiagramm eines allgemeinen Farbbildverarbeitungssystems sieht üblicherweise wie folgt aus:

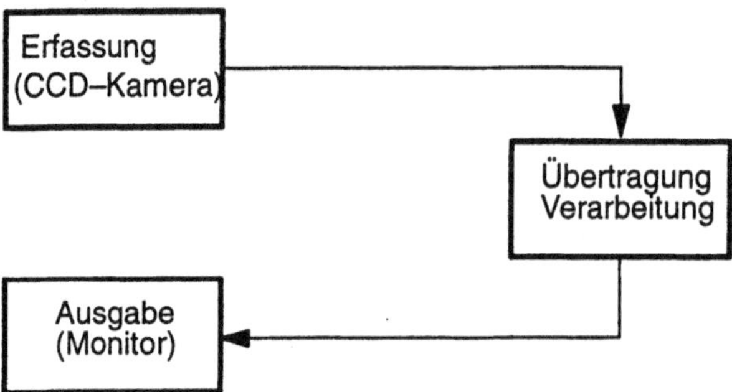

Bild 1 Allgemeiner Aufbau eines Bildverarbeitungssystems

Ausgehend von einer digitalen, verlustfreien Übertragung treten jedoch Fehler (Farbabweichungen) bei der Bilderfassung und Bildausgabe auf. Jede CCD–Kamera sowie jeder Monitor weisen gerätetypische Verfälschungen auf, die die Forderung nach einer farbgetreuen Reproduktion verletzen.

Es liegt daher nahe, einen weiteren Block zur Farbkorrektur vorzusehen, der diese Abweichungen kompensiert. Ein solcher Funktionsblock könnte beispielsweise wie folgt in den Datenfluß eingesetzt werden.

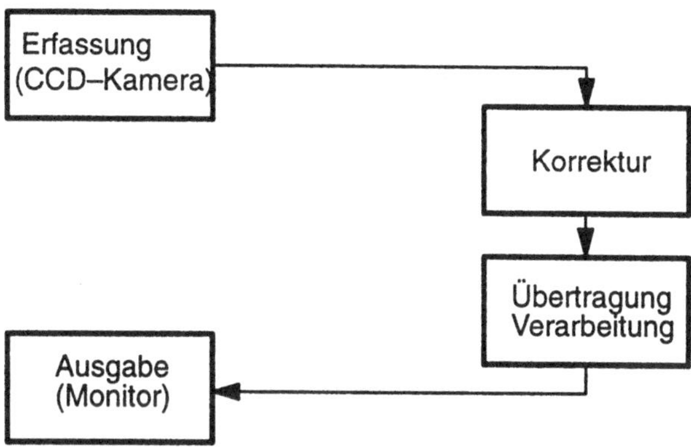

Bild 2 Aufbau eines Bildverarbeitungssystems mit globaler Korrektur

Die Korrektur kompensiert dabei sowohl die Abweichungen der CCD–Kamera, als auch die des Monitors, wodurch die Forderung der farbgetreuen Übertragung für den Betrachter im Rahmen der physikalischen Grenzen der verwendeten Geräte realisiert werden kann. Wesentlich ist hierbei bereits, daß Korrektur und Verarbeitung getrennt erfolgen. Anderntalls könnte eine falsch eingebundene Korrektur das Bildverarbeitungsergebnis zerstören. Nachteilig wirkt sich bei dieser Struktur jedoch aus, daß die Korrektur nur für eine festgelegte Gerätekombination wirksam ist. Probleme treten auf, sobald z.B. die Kamera oder der Monitor ausgetauscht werden muß, oder die Ausgabe auf mehreren parallel angeschlossenen Geräten (Bsp.: Monitor und Drucker) erfolgen soll. Diese geschlossene Lösung, die auch in vielen industriellen Produkten Anwendung findet, ist daher nicht universell verwendbar.

Die naheliegende Lösung für dieses Problem ist eine gerätespezifische Korrektur mit folgendem Blockschaltbild.

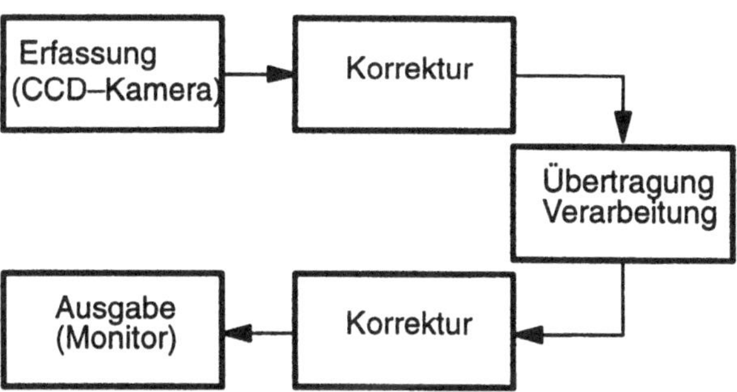

Bild 3 Aufbau eines Bildverarbeitungssystems mit gerätespezifischer Korrektur

Diese Struktur erlaubt die Transformation in und von einem definierten Referenzfarbraum. Inzwischen werden auch zumindest im Bereich der Bildverarbeitungssysteme für die Druckvorstufe (Reproindustrie) erste Lösungen dieser Art angeboten. Während einige Hersteller Monitore anbieten, die intern und somit gerätespezifisch auf einen definierten Farbraum kalibriert werden können, gehen die Möglichkeiten bei einer CCD–Kamera üblicherweise nicht über einen Weiß– und Schwarzabgleich hinaus, der jedoch in erster Linie dazu dient,

Beleuchtungsänderungen auszugleichen. Eine Transformation in einen definierten Farbraum ist bei CCD–Kameras typischerweise nicht anzutreffen. Es ist daher zu prüfen, inwieweit diese Anpassung rechnerseitig vor der Weiterverarbeitung erfolgen kann.

Mit der Darstellung der Farbinformation im Referenzfarbraum und der beschriebenen Struktur ist es möglich, z.B. unterschiedliche Ausgabegeräte mit einem jeweils gerätespezifischen Korrekturverfahren zu versorgen, um damit eine farblich einheitliche Bildausgabe mit diesen Geräten zu erreichen. Die Bildverarbeitung im geräteunabhängigen Referenzfarbraum bietet z.B. bei Segmentierungsaufgaben oder Messungen im Farbraum erhebliche Vorteile, da der Referenzfarbraum quantitative Farbmessungen erlaubt. Nachteilig hingegen ist der, mit dieser Struktur verbundene, erhöhte Korrekturaufwand. Hier sind gesonderte Betrachtungen erforderlich, um den Rechenaufwand möglichst klein zu halten.

Die möglichen Farbfehlerarten der verschiedenen Kameratypen lassen sich grob unterteilen in bildglobale und lokale Abweichungen. Die für Röhrenkameras signifikanten Konvergenz– und Farbreinheitsprobleme stellen den größten Anteil der bildlokalen Probleme dar. Bei CCD–Kameras tritt diese Fehlerart bedingt durch die fest auf dem Sensor angebrachten Filtermasken und die vorgegebene Aperturenanordnung hingegen nicht auf. Daher ist es sinnvoll die Farbkorrekturen für CCD–Sensoren auf bildglobale Korrekturen zu beschränken. Lokale Abweichungen werden bei dem hier beschriebenen Verfahren nicht berücksichtigt.

Um nun die Farbfehler erfassen zu können, sind zunächst Soll– und Istwerte zu definieren. Wird der Kamera eine bestimmte Farbvalenz angeboten, so erhält man einen aus drei Komponenten (Rot, Grün, Blau) bestehenden Farbwert, der hier als Vektor $\vec{x}$ bezeichnet ist. Den Sollwert $\vec{r}$ hingegen liefert ein Meßgerät, dem die gleiche Farbvalenz zugeführt wird. Der vom Meßgerät verwendete Farbraum muß dabei dem des Referenzfarbraums entsprechen bzw. in diesen umzurechnen sein. Wird als Referenzfarbraum beispielsweise der CIE–XYZ Farbraum gewählt, so ist eine Umrechnung in andere Farbräume leichter möglich [1, 2]. Der Fehler $\vec{e}$ für eine Farbvalenz kann anschließend aus

$$\vec{e} = \vec{x} - \vec{r} \tag{1}$$

bestimmt werden. Ermittelt man den Fehlervektor für genügend viele verschiedene Farbvalenzen, so ergibt sich kameraabhängig für jede Farbvalenz ein anderer Fehlervektor. $\vec{e}$ ist damit abhängig von $\vec{x}$. Möchte man die korrigierten Kamerawerte $\vec{y}$ über die Beziehung

$$\vec{y} = \vec{x} - \vec{e} \tag{2}$$

errechnen, so ist hierzu der Fehlervektor $\vec{e}$ für die jeweils zugeführte Eingangsvalenz zu verwenden. Da es nicht durchführbar ist, per Messung dieses für alle zulässigen Valenzen zu erfassen, sind geeignete Verfahren nötig, mit denen die nicht erfaßten Werte errechnet werden können.

Um nun das Verhalten des zu korrigierenden Systems zu erfassen, indem das Fehlervektorfeld des Farbraums bestimmt wird, sind über den gesamten Farbraum verteilte Stichproben auszuwerten. Hierzu muß zunächst der Fehlervektor für genügend viele im Farbraum verteilte Farbvalenzen bestimmt werden. Mit einer geeigneten Vorlage, entsprechend dem nachfolgenden Muster,

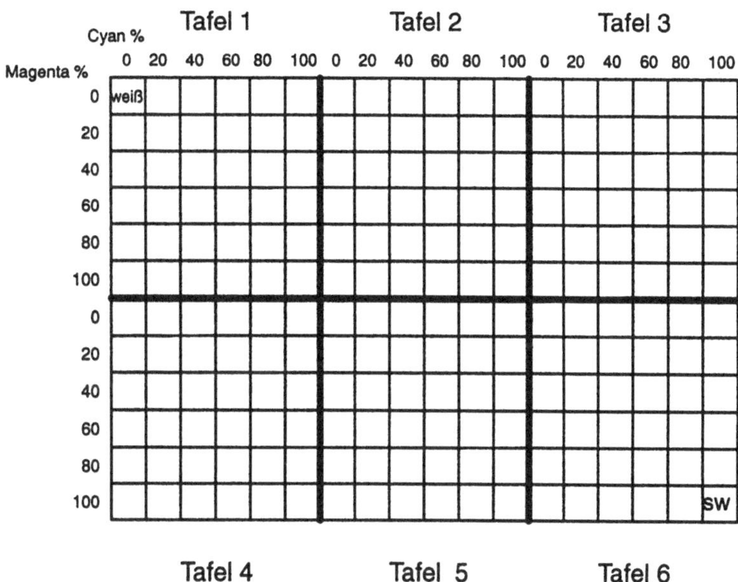

Bild 4 Testvorlage zur Erfassung des Farbraums

kann ein großer Teil des Farbraums abgedeckt werden. Die Vorlage enthält 6 Tafeln à 36 Felder, deren Intensitätsstufen der Farben Cyan, Magenta und Yellow in 20% Schritten anwachsen. Der Yellow–Anteil ist über jeweils eine Tafel konstant und steigt von Tafel zu Tafel um 20 % (Tafel 1: Yellow = 0%; Tafel 6 : Yellow = 100%). Damit sind mit den 216 Fehlervektoren, vom Weißfeld (0% Cyan, 0% Magenta, 0% Yellow) bis zum Schwarzfeld (alle Farben 100%), sämtliche Mischfarben in 20 % Stufen abgedeckt.

Die so gewonnenen 216 Fehlervektoren reichen allerdings nicht aus, um die Korrekturvektoren für alle zulässigen Eingangsfarbvalenzen abzudecken. Die Korrekturvektoren dieser Farbvalenzen lassen sich jedoch mit einer dreidimensionalen Interpolation, aus den mit der Testvorlage ermittelten benachbarten Korrekturwerten errechnen. Allerdings schränkt der damit verbundene Rechenaufwand die unmittelbaren Einsatzmöglichkeiten dieser Vorgehensweise erheblich ein. Abweichungen der Testvorlage von den Sollwerten verschärfen das Problem zusätzlich. Bedingt durch den chemischen Produktionsprozeß lassen sich keine ausreichend äquidistanten Raster herstellen, so daß dies bei der Interpolation berücksichtigt werden muß.

Zur Lösung dieses Problems sind daher Maßnahmen zu treffen, um die benötigte Rechenzeit zu reduzieren, ohne an Korrekturgenauigkeit zu verlieren. Eine Möglichkeit dazu entsteht, indem Interpolation und Korrektur entkoppelt werden. Überführt man das vorlagenbezogene, nicht äquidistante Korrekturraster durch Interpolation in ein äquidistantes Raster, so läßt sich zunächst der verbleibende Interpolationsprozeß stark vereinfachen. Ist das neue Raster zusätzlich genügend fein, so kann die Interpolation ggf. auch völlig entfallen, indem stattdessen ein Nearest–Neighbour–Zugriff auf den Korrekturvektor realisiert wird. Die Korrekturvektoren können dann in Form einer Lookup Table abgelegt werden, so daß zur Durchführung der Korrektur ein einfacher Zugriff auf diese Tabelle mit anschließender Vektorsubtraktion entsprechend (2) genügt. Der wesentliche Vorteil dieser Lösung ist darin zu sehen, daß die Erstellung der Korrekturinformation vorab geschieht. Der Berechnungsprozeß kann damit zugunsten einer optimalen Interpolation nahezu beliebig aufwendig gestaltet sein, da kaum Rücksicht auf die zur Verfügung stehende Rechenzeit genommen werden muß. Organisiert man die so berechnete Tabelle zusätzlich in Form einer 3D–LUT, so gestaltet sich auch der Zugriff während der Korrekturphase besonders effizient.

3. Ergebnisse

Das zuvor beschriebene Verfahren wurde für 3 CCD–Bilderfassungssysteme einem Test unterzogen. Zur Verfügung standen dabei:

- **Kontron ProgRes 3012**: Hochauflösende Kamera mit einem auf dem Sensor angebrachtem RGB–Mosaikfilter. Bedingt durch die zweidimensionale Mikroverschiebung des Sensors lassen sich die RGB–Aperturen derart positionieren, daß die Erfassung der drei Farbkomponenten an einem Bildpunkt und nicht wie bei Streifen– bzw. Mosaikfiltern üblich, an verschiedenen Bildpunkten erfolgt.
- **Sony DXC–325**: 3 Chip CCD–Kamera aus dem Sony Broadcast Bereich. Farbseparation über Strahlteiler und getrennte RGB–Filter.
- **Howtek Scanmaster**: DIN–A3 Flachbettscanner mit Zeilen–CCD; baugleich mit Sharp JX–450. Farbseparation durch Beleuchtung der Vorlage mit gefiltertem Blitzlicht.

Das dabei beobachtete Verhalten der Systeme war in den wesentlichen Punkten ähnlich. Stellvertretend sind hier die Daten der Kontron ProgRes 3012 Kamera gezeigt. Um die Differenzen zwischen Soll– und Istwerten zu verdeutlichen sind in Bild 5 die Farbabweichungen der Rot– und Grünkomponente der Tafel 1 dargestellt.

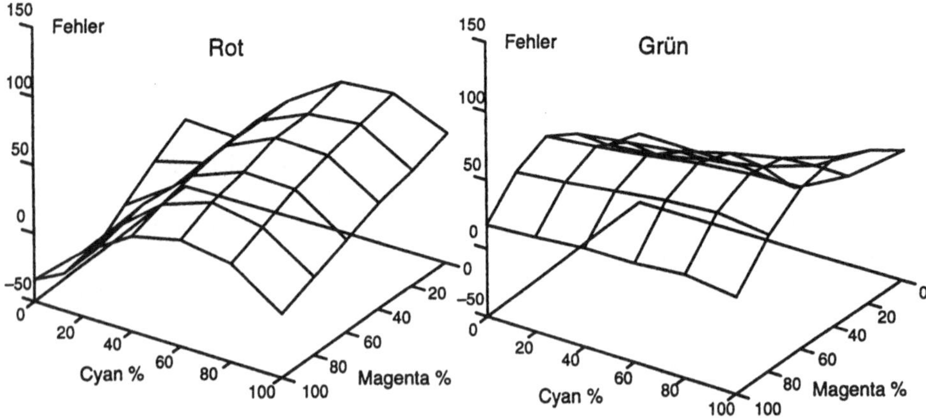

Bild 5 Farbwertabweichung (Soll – Istwert) des Rot– und Grünanteils der ProgRes3012, bezogen auf 8 Bit Daten (Wertebereich 0 bis 255) am Beispiel von Tafel 1

Die 12 Bit Ausgangsdaten der Kamera waren dabei auf 8 Bit skaliert, so daß der Wertebereich von Soll– und Istdaten auf 0 bis 255 begrenzt war. Als Sollwerte sind hier die für eine korrekte

Offset–Reproduktion erforderlichen Daten vorgegeben. Der Fehlerwert für den Blauanteil ist nicht dargestellt, da dieser in jeder Tafel konstant ist. (Die Komplementärfarbe Gelb ist für jede Tafel konstant vorgegeben.)

Diese Daten wurden der beschriebenen Korrektur zugeführt und anschließend in gleicher Weise ausgewertet. Der verbleibende Restfehler lag dabei bei allen untersuchten Systemen im Arbeitsbereich (Farbraum der Testvorlage) unter 2 % des ursprünglichen Fehlers und im Randbereich des Korrekturfarbraums unter 5 %. Auf eine grafische Darstellung des Restfehlers wurde aus diesem Grunde verzichtet.

Werden der Kamera Farbvalenzen zugeführt, die außerhalb des Farbraums der Testvorlage liegen, so steigt der Fehler an. Es liegen dann keine Korrekturinformationen vor, so daß die Daten ggf. unkorrigiert weitergereicht werden. Wenn das zur Erstellung der Korrekturvektoren verwendete Interpolationsverfahren auch eine leichte Extrapolation an den Randbereichen zuläßt, kann der Übergang vom korrigierten Farbraum zum nichtkorrigierten entsprechend sanft gestaltet werden. Hierdurch werden "harte" Umschalteffekte zwischen korrigierten und nichtkorrigierten Bereichen abgeschwächt. Sofern die Testvorlage den Farbraum vollständig abdeckt, sind diese Effekte hingegen nicht zu beobachten. Werden der Kamera Farbvalenzen zugeführt, die außerhalb des von der Kamera vorgegebenen Farbraumes liegen, so werden diese auf den Rand des Raumes abgebildet [3]. Der Kamerafarbraum wird dabei durch die RGB–Farborte der Separationsfilter vorgegebenen.

4. Diskussion

Gegenüber einer gemeinsamen, globalen Korrektur für Ein– und Ausgabegeräte bietet die Realisierung eines Referenzfarbraumes deutliche Vorteile im Hinblick auf den Austausch einzelner Geräte oder der Bildverarbeitung in einem geräteunabhängigen Farbraum. Der damit verbundene Nachteil des erhöhten Rechenaufwandes kann mit der Erstellung einer Lookup Table für die Korrekturvektoren wesentlich reduziert werden. Die untersuchten Interpolationsverfahren machten deutlich, daß auch mit relativ einfachen, linearen Verfahren die angegebenen Restfehler erreicht werden konnten [4]. Weiterhin reduzierten derartige Interpolationsverfahren die zur Erstellung einer aus $64^3 = 262144$ RGB–Fehlervektoren bestehenden Tabelle benötigte Rechenzeit auf ca. eine Stunde (SUN Sparc 10). Andere Interpolationsverfahren können jedoch problemlos eingesetzt werden. Geeignete Scattered–Data–Interpolationsverfahren [5,6] lassen eine noch höhere Genauigkeit erwarten. Ein diesbezüglicher Test, auch im Hinblick auf die damit verbundenen Anforderungen an die Rechenzeit, steht jedoch noch aus.

Die spektralen Eigenschaften des Systems werden durch die physikalischen Randbedingungen der Geräte vorgegeben. Im Falle der CCD–Kamera sind hierbei das Verhalten des Sensors sowie die zur Farbseparation eingesetzten Filter bestimmend. Die Kennlinien der RGB–Farbfilter haben dabei einen nahezu gaußförmigen Verlauf mit teilweiser Überlappung. Daraus folgt, daß unterschiedliche spektrale Erregungen die gleichen Farbwerte erzeugen können; eine Eigenschaft, die sich die CCD–Kamera mit dem menschlichen Auge teilt [2]. Da bedingt durch die Filterbandbreite keine spektral feine Auflösung erfolgt, lassen sich nur Korrekturverfahren anwenden, die auf vorhandene RGB–Information aufbauen.

Einige Sorgfalt erfordert daher die Erstellung der Testvorlage. Insbesondere ist darauf zu achten, daß der später erforderliche Farbraum möglichst umfassend dargestellt wird. Um den zulässigen Eingangsfarbraum nicht unnötig einzuschränken, kann z.B. außerhalb des von der Testvorlage vorgegebenen Farbraums auf eine Farbkorrektur verzichtet werden. Allerdings ist zu beachten, daß der Übergang vom korrigierten zum nichtkorrigierten Bereich möglichst stetig geschieht. Andernfalls können deutlich sichtbare Farbsprünge beim Übergang vom korrigierten zum nichtkorrigierten Bereich verursacht werden. Wird dies berücksichtigt, so kann das Verfahren mit dieser Einschränkung auch für Bildmaterial verwendet werden, dessen Farbraum nicht auf den Farbraum der Testvorlage beschränkt ist.

5. Literaturverzeichnis

[1] G. Wyszecki, W. S. Stiles:
Color Science: Concepts and Methods, Quantitative Data and Formulae.
2nd Edition, New York, Chichester, Brisbane, Toronto, Singapore: John Wiley & Sons 1982.

[2] H. Lang:
Farbmetrik und Farbfernsehen.
München, Wien: Oldenburg 1978.

[3] M. Richter:
Einführung in die Farbmetrik.
2. Auflage, Berlin, New York 1981.

[4] M. Schuschk:
Untersuchungen zur Farbraumanpassung an Bildein– und ausgabegeräten.
Studienarbeit. Institut für Theoretische Nachrichtentechnik und Informationsverarbeitung, Universität Hannover, Hannover 1991

[5] R. Franke:
Scattered Data Interpolation: Test of Some Methods.
Mathematics of Computation, Vol. 38, Number 157, Jan. 1982, pp: 181–200

[6] T. A. Foley, H. Hagen, G. M. Nielson:
Recent Methods for Visualizing and Modeling Unstructured Data.
Tagungsband: "Visualisierung – Rolle von Interaktivität und Echtzeit"
Workshop 2–3 Juni 1992, GMD Gesellschaft für Mathematik und Datenverarbeitung mbH, St. Augustin Schloß Birlinghoven.

Speicherung und Übertragung von stereoskopischen Bildsequenzen

Manfred Ziegler
Siemens AG, ZFE ST SN 62, 81730 München
Tel.: +49 89 636 41225, Fax: +49 89 636 2393
email: zie@bsun2.zfe.siemens.de

Zusammenfassung

Speicherung und Übertragung von Bildsequenzen sind entscheidend für Anwendungen in der Telekommunikation. Bei der Verwendung von stereoskopischen Bildsequenzen (= 3D TV) werden im Vergleich zu "normalen" 2D-Bildern die auftretenden Probleme durch die doppelte Datenrate noch verschärft. Im vorliegenden Artikel wird ein neues Bilddatenkompressionsverfahren vorgestellt, das stereoskopische Bildsequenzen kompatibel zu bestehenden "2D-Standards" übertragen kann. Damit können die für digitale TV-Signale verfügbaren Übertragungskanäle und Speichermedien benutzt werden. Ein Vorteil des beschriebenen Verfahrens besteht darin, daß ein Empfänger ohne Stereodisplay nach wie vor ein hochqualitatives 2D-Bild aus dem Gesamtdatenstrom erzeugen kann, ein Stereoempfänger aber aus dem gleichen Datenstrom alle beiden Kanäle decodieren kann, ohne die geplante Gesamtdatenrate zu erhöhen. Das beschriebene Verfahren basiert auf einer Kombination von Bewegungs- und Disparitätsschätzung.

Einleitung

Stereobilder bieten dem Menschen den dreidimensionalen Eindruck den er von seiner natürlichen Umgebung her kennt. Stereoskopische Darstellungen werden daher für viele Anwendungen - von Einzelbildern für die räumlichen Darstellung komplexer Zusammenhänge bis hin zu 3D-Spielfilmen - eingesetzt.

Bei der Wiedergabe von Stereobildern auf einem Fernsehsystem müssen im Gegensatz zum normalen Fernsehen zwei Bilder übertragen werden - eines für das rechte und eines für das linke Auge. Die Bilder sehen zwar inhaltlich gleich aus, sind aber aus etwas unterschiedlicher Perspektive aufgenommen. Den Stereoeffekt mit voller Farbwiedergabe erhält man z. B. durch die Polarisationstechnik. Dabei werden zwei Bildschirme mit in unterschiedlicher Richtung polarisierenden Folien versehen, und die beiden Strahlengänge mit einem halbdurchlässigen Spiegel deckungsgleich projiziert. Benutzt man nun eine Brille mit Folien entsprechender Polarisationsrichtung, so sieht jedes Auge nur das ihm zugeordnete Bild.

Digitale Kommunikations- und Verteilnetze bieten die Möglichkeit, Einzelbilder und Bildsequenzen "in Stereo" zu übertragen, zu verarbeiten und zu speichern. Dabei verdoppelt sich allerdings das Datenvolumen bzw. die zur Übertragung erforderliche Datenrate im Vergleich zu "Mono-Bildern": eine effektive und zur 2D-Übertragung kompatible Datenkompression ist daher von großem Interesse.

In Zukunft wird Stereo-TV nicht nur als Verteilfernsehen angeboten werden, sondern auch zur Übertragung spezieller Dienste wie z.B. in CAD und Medizin, als Abruf- oder Verteildienst genutzt werden.
In Bild 1 sind die Prinzipien einer Stereoaufnahme gezeigt: Die Aufnahme erfolgt mit zwei getrennten Kameras. Somit werden zwei Bilder, die zusammen ein Stereopaar ergeben, erzeugt. Ebenfalls erkennbar ist die Verschiebung der abgebildeten Objekte in horizontaler Richtung, der sogenannten Disparität.

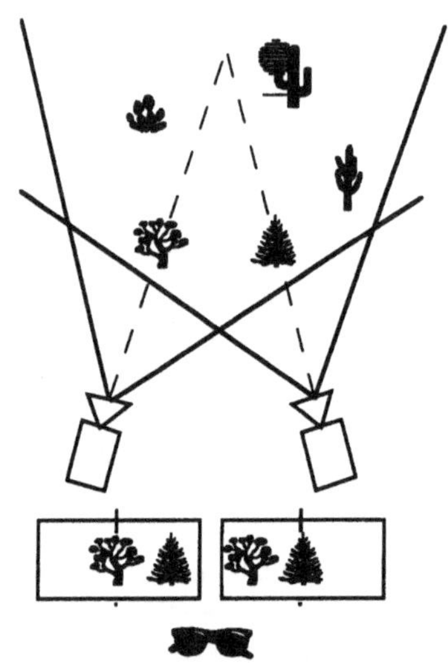

Bild 1: Stereoanordnung

Einer der zentralen Verarbeitungsschritte bei der Datenkompression von zweidimensionalen Bildsequenzen ist die Bewegungskompensation. Hierbei soll das nächste zu übertragende Bild aus dem vorhergehenden prädiziert werden. Mit Hilfe einer vektoriellen Verschiebung von Bildblöcken wird ein künstliches Bild erzeugt. Übertragen wird nur dessen Differenz zum Originalbild und das berechnete Bewegungsvektorfeld.
Ähnlich einer Bewegungskompensation in zeitlicher Richtung läßt sich diese örtliche Verschiebung ausnutzen, um das rechte Bild - ausgehend vom linken - zu schätzen. Übertragen werden dann ebenfalls der Schätzfehler und die berechneten Disparitätsvektoren. Mit Hilfe der Disparität kann auch die Tiefe der Szene, d.h. der Abstand zur Kamera, berechnet werden. Diese Tiefenberechnung ist mit ein Grund, daß Stereobildverarbeitung in vielen professionellen Bereichen, wie z.B. der Medizin, zur Anwendung kommt.

Stereoskopische Bilddatenkompression

Wie bereits erwähnt, werden stereoskopische Sequenzen mit zwei verschiedenen Kameras aufgenommen. Schwierigkeiten ergeben sich dabei zwangsweise, falls die beiden Kameras nicht ausreichend genau kalibriert wurden. Ein Hauptproblem ist vor allem eine unterschiedliche Helligkeit in den beiden Kanälen [Zie 91].

In [Fra 92] wird vorgeschlagen, den rechten Kanal *R* so anzupassen, daß er die gleichen Mittelwerte und die gleiche Varianz wie der linke Kanal *L* aufweist. Die notwendige lineare Transformation ist sehr einfach:

$$r' = a \cdot r + b$$

Die Parameter *a* und *b* können mit den folgenden Formel berechnet werden:

$$a = \sqrt{\frac{\mathrm{var(l)}}{\mathrm{var(r)}}} \quad \text{und}$$

$$b = E\{l\} - \sqrt{\frac{\mathrm{var(l)}}{\mathrm{var(r)}}} \bullet E\{r\} = E\{l\} - a \bullet E\{r\}$$

wobei *l* und *r* die Luminanzwerte des linken beziehungsweise des rechten Bildes sind, *var* die Varianz der Werte und *E* der Mittelwert.

Diese Anpassung des rechten Kanals kann als ein Vorverarbeitungsschritt gesehen werden, der in Zukunft bereits bei der Aufnahme mit professionellen Stereokameras durchgeführt wird. Im Rahmen des von der EG geförderten RACE[1] Projektes DISTIMA[2] wird u.a. eine solche Stereokamera entwickelt.

Ein weiteres Problem ergibt sich aus der Stereogeometrie. In der Theorie ist es möglich, die epipolaren Linien zu berechnen. Das Matching bei der Disparitätsschätzung wäre dann auf ein eindimensionales Problem reduziert, da nur noch entlang dieser Linie gematcht werden müßte. Dazu ist allerdings wiederum eine exakte Kamerakalibrierung notwendig. Die erforderliche Qualität ist aber im "täglichen Gebrauch" nicht zu gewährleisten. Immerhin ist es bei DISTIMA gelungen, den Suchbereich auf ± 3 Bildpunkte in vertikaler Richtung zu beschränken. Allerdings ist der horizontale Suchbereich auf Grund der erheblich größeren örtlichen Verschiebungen auf ± 63 Bildpunkte auszudehnen.
In [Vle 90] wird eine Möglichkeit beschrieben, einen künstlichen Mittelkanal zu erzeugen und die beiden Originalbilder mit Hilfe der Disparität aus diesem Mittelkanal beim Empfänger zu erzeugen. Der Nachteil eines solchen Verfahrens liegt in der Qualität des kompatibel übertragenen Kanals. Die übertragenen Bilder sind alle künstlich erzeugt und mit Fehlern behaftet.
Eine 2D-kompatible 3D-Übertragung ist dann mit einer ausreichenden Qualität möglich, wenn einer der beiden Kanäle aus dem anderen vorhergesagt wird. Auf diese Weise läßt sich ein Kanal kompatibel - nämlich durch Verwendung des 2D-Standards - codieren, während die Verarbeitung des zweiten, zusätzlichen Kanals keinen Beschränkungen unterliegt.
Ein Schritt in diese Richtung der 3D-Übertragung wurde in [Ten 91] vorgestellt. Hier wird das rechte Bild mit Hilfe einer Stereokompensation prädiziert und so die zu übertragende Datenmenge reduziert. Allerdings wird hier die zeitliche Information des rechten Kanals nicht ausgenutzt.
Eine qualitativ hochwertige, wenn auch deutlich aufwendigere Lösung, ist in [Zie 93] beschrieben: Eine Kombination aus Stereo- und Bewegungskompensation (Bild 2).

---

1 Research and Development on Advanced Communications Technologies in Europe
2 Digital Stereoscopic Imaging and Applications

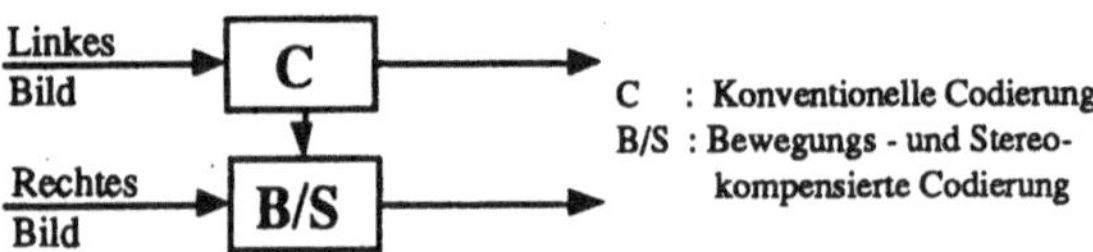

Bild 2: 3D-Codierschema

Hierbei wird im rechten Kanal sowohl die zeitliche (Bewegung) als auch die örtliche Information (Disparität) verwendet. Für jeden 16x16-Block wird auf Grund des quadratischen Fehlers des kompensierten Bildes entschieden, welche Prädiktion besser ist. Zusätzlich zu den Bewegungsvektoren bzw. den Disparitätsvektoren ist also auch noch die Art der gewählten Prädiktion zu übertragen.
Trotz dieser zusätzlichen Datenmenge wird der optische Eindruck deutlich verbessert.

Um eine Übertragung von stereoskopischen Bildsequenzen im Rahmen des bestehenden Standards zu gewährleisten, muß eine Gesamtdatenrate von 10 Mbit/s eingehalten werden, nun aber für zwei Videokanäle und einen Audiokanal. Für letzteren ist eine Übertragungsrate von 1 Mbit/s vorgesehen. Experimente mit 5.5 Mbit/s für den linken (kompatiblen) Videokanal und 3.5 Mbit/s für den rechten Videokanal zeigten, daß die erreichte Bildqualität für viele Anwendungen ausreichend ist.

Einbringen der Stereodaten in den MPEG-Datenstrom

In [Hor 93] wurde im Rahmen der bei MPEG[3] durchgeführten Arbeiten eine Möglichkeit vorgestellt, mit einer geringfügigen Änderung des Draft Standards Stereodaten in den MPEG-Datenstrom einzubringen. Hierbei wird die Ähnlichkeit der beiden Videosignale ausgenutzt, sowohl im "Simulcast Betrieb" (d.h. die zwei Signale werden unabhängig über unter Umständen verschiedene Leitungen gesendet), als auch im Fall von "Embedded Bitstreams" (d.h. die Daten des zweiten Kanals werden in den Datenstrom des ersten Kanals eingebettet). Diese Möglichkeiten bestehen sowohl für eine gleichzeitige Übertragung von TV und HDTV Signalen, als auch von stereoskopischen Bildsequenzen.
Hierzu wurde ein 3 bit Code "subsampling-ratio" (ssr) eingeführt, der den Subsampling-Faktor für das zweite Signal angibt. Im Fall ssr = 0 (Übertragung eines zweiten Signals mit der gleichen Auflösung wie das erste) kann dieses Prinzip für das rechte Signal eine Stereosequenz verwendet werden.
Im Fall des oben beschriebenen Verfahrens ist eine Übertragung als "Embedded Bitstreams" möglich (siehe Bild 3). Auf diese Weise kann der erste (= linke) Kanal als Prädiktion für den zweiten Kanal (= stereoskopische Erweiterung) benutzt werden. Sowohl die DCT[4]- und entropiecodierten Koeffizienten, als auch die benötigten Vektoren, können dann als "user-data" im MPEG-Datenstrom übertragen werden. Die generelle MPEG-Syntax wird daher nicht beeinflußt.

---

3 die Moving Picture Expert Group ist eine Gruppe von Fachleuten, die im Rahmen der International Standardization Organization (ISO) einen weltweiten Standard zur digitalen Bewegtbildübertragung entwickelt.

4 Discrete Cosine Transform

Diese Methode hat zur Folge, daß ein "normaler" MPEG-Decoder nach wie vor ein qualitativ hochwertiges Bild aus dem Datenstrom erzeugen kann (Kompatibilität), ein Stereoempfänger aber aus dem gleichen Datenstrom alle beiden Kanäle decodieren kann.

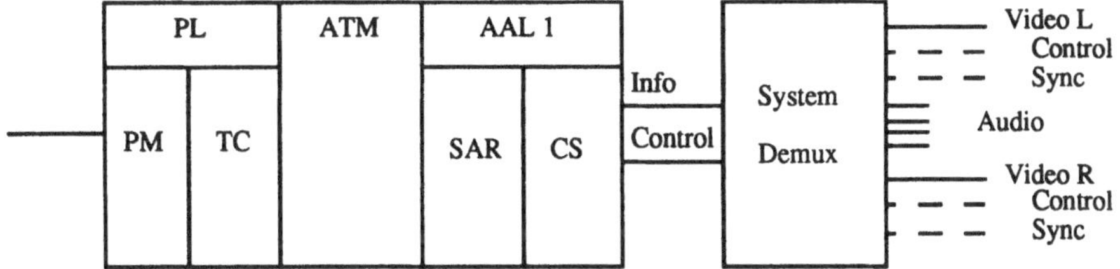

Bild 3: Übertragung von Stereoskopischen Bildsequenzen als "Embedded Bitstream"
PL: Physical layer, PM: Physical medium dependent layer,
ATM: Asynchronous Transfer Mode layer, AAL: ATM Adaptation Layer,
SAR: Segmentation And Reassambly, CS: Convergence Sublayer

Voraussetzung für diese Art der Übertragung ist ein Multiplexing der beiden Videokanäle und vorhandener Audiokanäle im System-Multiplex des Senders. Diese Möglichkeit ist im MPEG-Standard vorgesehen.

Anwendungen in der Medizin

Bildgebende medizinische Verfahren wie MR[5] oder CT[6] sind aus den Krankenhäusern nicht mehr wegzudenken. Diese Verfahren liefern dem Arzt Schnittbilder des Menschen, so daß es ihm möglich ist die dreidimensionalen Zusammenhänge besser zu erkennen. Durch die Anwendung von stereoskopischen Wiedergabeverfahren wird dieser Eindruck noch verstärkt. Denkbar ist auch die Übertragung im Rahmen von PACS[7], ebenfalls einem elektronischem Hilfsmittel, das in vielen Kliniken Einzug gehalten hat. Hiermit ist es möglich Bilder nicht nur zu speichern und bei Bedarf wieder abzurufen, sondern es erlaubt auch eine Konferenzschaltung mehrerer Spezialisten.
Eine sehr einfache Möglichkeit MR- oder CT-Bilder in das benötigte Datenformat (zwei aus verschiedenen Winkeln aufgenommene Bilder) zu wandeln, bieten bestehende Softwarepakete: Aus den Schnittbildern können verschiedene Projektionen eines 3D-Volumenmodells berechnet werden. Jeweils zwei dieser Projektionen werden nun wie ein Stereopaar behandelt, d.h. sie können wie beschrieben codiert und übertragen werden.
Im Rahmen der medizinischen Bildverarbeitung ist allerdings zu beachten, daß bezüglich Bilddatenkompressionsverfahren im Allgemeinen eine fehlerfreie Rekonstruktion gefordert wird. Somit ist das beschriebene Verfahren nicht zur Langzeitspeicherung geeignet, wohl aber zur Unterstützung einer Diskussion (z. B. per Videokonferenz) oder zu Ausbildungszwecken.

---

5 Magnet Resonanz
6 Computer Tomographie
7 Picture Archiving and Communications System

Literaturhinweise

[Fra 92] R. Franich
Balance Compensation for Stereoscopic Image Sequences
RACE DISTIMA, Document 45/TUD/WP3.2/DN/C/27.7.92/1

[Hor 93] R. ter Horst
Proposal for and Discussion on the Syntax to exploit the Correlation between similar Video Signals over a Pair of Video Codecs
ISO/IEC JTC1/SC 29/WG 11, Doc. MPEG 93/179, Rom, Italien, Januar 1993

[Ten 91] W. Tengler
Coding of Stereo-TV for Transmission in a 34 Mbit/s-Channel,
1991 Picture Coding Symposium, Tokyo, Japan, September 1991

[Vle 90] D. de Vleeschauwer
A Symmetric Scheme to Encode Stereo Images
2nd Workshop on Stereoscopic Television, Darmstadt, FRG, September 1990

[Zie 91] M. Ziegler, W. Tengler, P. Tabeling
Influence of Camera Calibration on the Coding of Stereo Sequences
1st International Festival of 3D-Images, Paris, Frankreich, September 1991

[Zie 93] M. Ziegler, F. Seytter
Coding of Stereoscopic Sequences using Disparity and Motion Estimation
1993 Picture Coding Symposium, Lausanne, Schweiz, März 1993

Autorenindex